Joint European Magnetic Symposia 2010

(JEMS 2010)

Journal of Physics: Conference Series Volume 303

Krakow, Poland
23-28 August 2010

ISBN: 978-1-61839-047-9
ISSN: 1742-6588

Printed from e-media with permission by:

Curran Associates, Inc.
57 Morehouse Lane
Red Hook, NY 12571

Some format issues inherent in the e-media version may also appear in this print version.

Printed by Curran Associates, Inc. (2011)

For permission requests, please contact the Institute of Physics
at the address below.

Institute of Physics
Dirac House, Temple Back
Bristol BS1 6BE UK

Phone: 44 1 17 929 7481
Fax: 44 1 17 920 0979

techtracking@iop.org

Additional copies of this publication are available from:

Curran Associates, Inc.
57 Morehouse Lane
Red Hook, NY 12571 USA
Phone: 845-758-0400
Fax: 845-758-2634
Email: curran@proceedings.com
Web: www.proceedings.com

TABLE OF CONTENTS

1. PLENARY, SEMI-PLENARY, TUTORIALS

012001 Surface Ferromagnetism in Non-magnetic and Dilute Magnetic Oxides 1
M Carmen Muñoz, Silvia Gallego, Nadiezhda Sanchez

012002 Crossover from Non-Fermi Liquid to Fermi Liquid Behavior in Heavy Fermion Systems 11
P Schlottmann

012003 Magnetism on a Mesoscopic Scale: Molecular Nanomagnets Bridging Quantum and Classical Physics 21
Nikolaos P Konstantinidis, Alexander Sundt, Joscha Nehrkorn, Anna Machens, Oliver Waldmann

2. MAGNETIZATION PROCESSES, SPIN EXCITATIONS AND ULTRAFAST DYNAMICS

012004 Magnon Excitations in Ultrathin Fe Layers: The Influence of the Dzyaloshinskii-moriya Interaction 39
Kh Zakeri, Y Zhang, J Prokop, T -H Chuang, W X Tang, J Kirschner

012005 Energy Thresholds in the Magnetic Vortex Core Reversal 46
Sebastian Gliga, Yaowen Liu, Riccardo Hertel

012006 Magnetization Processes in a Dot of Ferrimagnetic Garnet Near the Compensation Temperature 53
V Joch, P Molho, L Ranno

012007 Employing Magnonic Crystals to Dictate the Characteristics of Auto-oscillatory Spin-wave Systems 59
A D Karenowska, A V Chumak, A A Serga, J F Gregg, B Hillebrands

012008 Spin Wave Propagation in Ferromagnetic Wires with an Arbitrary Field Direction 65
Rémy Lassalle-Balier, Claude Fermon

3. HARD MAGNETIC MATERIALS AND MAGNETOCALORICS

012009 Structural and Magnetic Properties of NdFeB and NdFeB/Fe Films with Mo Addition 70
M Urse, M Grigoras, N Lupu, H Chiriac

4. MAGNETIC HYDRIDES

012010 Structural and Magnetic Transformations in $HoMn_2H_x$ Hydrides 76
A Budziak, M Balanda, H Figiel, L Kolwicz-Chodak, Z Tarnawski

012011 Onset of Magnetism Stimulated by H Sbsorption in UNiAl-UFeAl 83
S Maskova, L Havela, S Danis, K Kothapalli, F Nasreen, A V Andreev

5. INTERFACE OF MAGNETIC THIN FILMS

012012 Characterization of U-based Thin Films: The UFe_{2+x} Case 88
Nhu-T H Kim-Ngan, L Havela, A M Adamska, S Daniš, J Pešicka, J Macl, R Eloirdi, F Huber, T Gouder, A G Balogh

012013 Polarized Neutron Reflectivity Study of $NiFe_2O_4$ Films with Very Large Saturation Magnetization 96
F Rigato, X Martí, J Fontcuberta

012014 Effect of Magnetic Film Thickness on the Spatial Resolution of Magnetic Force Microscope Tips 102
Katsumasa Nagano, Kousuke Tobari, Mitsuru Ohtake, Masaaki Futamoto

012015 Epitaxial Growth of Permalloy Thin Films on MgO Single-Crystal Substrates 108
Mitsuru Ohtake, Takahiro Tanaka, Katsuki Matsubara, Fumiyoshi Kirino, Masaaki Futamoto

012016 Preparation of $Co(0001)_{hcp}$ and $(111)_{fcc}$ Films on Single-Crystal Oxide Substrates 114
Mitsuru Ohtake, Osamu Yabuhara, Yuri Nukaga, Masaaki Futamoto

6. MAGNONIC CRYSTALS

012017 One-dimensional Magnonic Circuits with Size-tunable Band Gaps and Selective Transmission 120
B Djafari-Rouhani, H Al-Wahsh, A Akjouj, L Dobrzynski

012018 Spin Information Transfer and Transport in Hybrid Spinmechatronic Structures 130
A D Karenowska, J F Gregg, A V Chumak, A A Serga, B Hillebrands

7. MAGNETISM OF METALS, ALLOYS AND INTERMETALLICS

012019 Magnetotransport in Tb_2Fe_{17} Single Crystals 135
Jolanta Stankiewicz, Konstantin P Skokov

012020 Anisotropy Study of Grain Oriented Steels with Magnetic Barkhausen Noise 141
M F De Campos, M A Campos, F J G Landgraf, L R Padovese

012021 Magnetic and Magneto-volume Properties of Fe_xM_{100-x} Solid Solutions with M = Pd and Pt 147
E Burzo, P Vlaic, D Eniu

012022 Neutron Diffraction Study of the Magnetic Order in $NdMn_2Ge_{1.6}Si_{0.4}$ 153
J L Wang, S J Campbell, J M Cadogan, A J Studer, R Zeng, S X Dou

012023 Magnetostriction and Transformation of Crystal Structure of Intermetallic Compound $NdCO_2$ 159
S A Nikitin, D Yu Karpenkov, I S Tereshina, Yu G Pastushenkov, T Palewski, J Cwik

012024 Magnetostriction in $(Tb_{0.45}Dy_{0.55})_{1-x}Er_xCo_2$ (x = 0.1, 0.2): High-field Investigation 163
I Tereshina, G Politova, E Tereshina, J Cwik, S Nikitin, O Chistyakov, A Karpenkov, D Karpenkov, T Palewski

012025 Magnetic Properties and Magnetocaloric Effect in $(Dy_{0.6}Er_{0.4})_{1-x}Gd_xCO_2$ Multicomponent Compounds 169
J Cwik, T Palewski, K Nenkov, J Lyubina, O Gutfleisch

012026 Neutron Powder Diffraction Determination of the Magnetic Structure of Nd_2Al 175
J M Cadogan, M Avdeev, P Kumar, K Suresh, D H Ryan

012027 Mössbauer and Calorimetric Studies of Spin Reorientation Processes in $Er_{2-x}Pr_xFe_{14}C$ 179
A T Pedziwiatr, K Krawiec, B F Bogacz, S Wróbel, D Kaczorowski, J Przewoznik

012028 Magnetic and Structural Properties of Nanocrystalline $PrCo_3$ 185
K Younsi, V Russier, L Bessais, J-C Crivello

012029 Influence of Microstructural Constituents on the Hysteresis Curves in 0.2%C and 0.45%C Steels 191
L F T Costa, F Girotto, R Baiotto, G Gerhardt, M F De Campos, F P Missell

012030 Magnetic Barkhausen Noise in Quenched Carburized Steels 197
M F De Campos, F A Franco, R Santos, F S Da Silva, S B Ribeiro, J F C Lins, L R Padovese

012031 The Change of Anisotropy in TbNi(Al,In) Compounds Studied by Low Temperature X-ray Diffraction 203
M Klicpera, P Javorský, S Daniš

012032 The Magnetostriction of the Intermetallic Compound $ErCO_2$ Near the Magnetic Phase Transition Paramagnetism-ferrimagnetism 209
S A Nikitin, D Yu Karpenkov, A Yu Karpenkov, N I Nizhankovskii, T Palewski, K P Skokov

8. MOLECULAR MAGNETISM

012033 Spin Entanglement in Supramolecular Systems 213
G Lorusso, F Troiani, V Bellini, A Ghirri, A Candini, S Carretta, P Santini, G Amoretti, W Wernsdorfer, G Timco, R E P Winpenny, M Affronte

012034 Studies of Critical Phenomena in Molecular Magnets by μSR Spectroscopy 223
T Wasiutynski, M Balanda, M Czapla, R Pelka, P M Zielinski, F L Pratt, T Korzeniak, R Podgajny, D Pinkowicz, B Sieklucka

012035 Magnetic Properties of Lanthanoid(III) Phthalocyaninato Triple-decker Complexes in an External Magnetic Field and Electronic Transport Properties for Molecular Spintronics 231
K Katoh, K Yamamoto, T Kajiwara, J Takeya, B K Breedlove, M Yamashita

012036 Single-chain Magnet Features in 1D $[MnR_4TPP][TCNE]$ Compounds 241
Maria Balanda, Zbigniew Tomkowicz, Wolfgang Haase, Michal Rams

012037 Octacyanoniobate(IV)-based Molecular Magnets Revealing 3D Long-range Order 255
R Pelka, D Pinkowicz, O Drath, M Balanda, M Rams, A Majcher, W Nitek, B Sieklucka

9. MAGNETOOPTICS OF NANOMAGNETS

012038 Intensity Magnetooptical Effect in Magnetoplasmonic Crystals 261
V I Belotelov, I A Akimov, M Pohl, A N Kalish, S Kasture, A S Vengurlekar, A V Gopal, V A Kotov, D Yakovlev, A K Zvezdin, M Bayer

012039 Modeling and Optimization of Nonreciprocal Transmission Through 2D Magnetophotonic Crystal 270
M Vanwolleghem, K Postava, L Halagacka, L Magdenko, P Beavillain, B Dagens

012040 Magnon-Photon Interactions for Quantum Key Distribution 276
P Kumar, A Prabhakar

012041 Ferromagnetic Nanoparticular Films Studied by Optical and Magneto-optical Ellipsometry 282
T Fördös, O Životský, K Postava, D Hrabovský, J Pištora, J Lancok, M Miglierini, M Klementová

012042 Magneto-optical and Optical Properties of Pt/Co GMR Films with Capping Layer 288
T Ishibashi, M Naganuma, S Tang, T Honma, T Komatsu, K Machida, K Furukawa, K Aoshima, N Funabashi, K Kuga, N Shimidzu

10. NANOMAGNETISM

012043 Influence of a Semiconducting Matrix on the Magnetic Behaviour of Iron Oxide Nanoparticles 293
P Granitzer, K Rumpf, M Venkatesan, L Cabrera, M P Morales, P Poelt, M Albu

012044 Self-assembled Strontium Ferrite on Au Nano-dot Arrays Using Rapid Thermal Annealing 299
A Morisako, X Liu, A Kaewrawang, A Ghasemi

012045 Magnetic and Structural Characterization of Nanosized $BaCo_xZn_{2-x}Fe_{16}O_{27}$ Hexaferrite in the Vicinity of Spin Reorientation Transition 303
A Pasko, F Mazaleyrat, M Lobue, V Loyau, V Basso, M Küpferling, C P Sasso, L Bessais

012046 Growth of Magnetic Cobalt/chromium Nano-arrays by Atom-optical Lithography 309
F Atoneche, D Malik, A Kirilyuk, A J Toonen, A F Van Etteger, Th Rasing

012047 Influence of Temperature and Mechanical Scratch on the Recorded Magnetization Stability of Longitudinal and Perpendicular Recording Media 315
Katsumasa Nagano, Kousuke Tobari, Masaaki Futamoto

012048 Interacting Low Dimensional Nanostructures Within a Porous Silicon Template 321
K Rumpf, P Granitzer, G Hilscher, P Poelt

012049 Estimate of the Anisotropy Field in Isotropic SmCo 2:17 Magnets with the Stoner-Wohlfarth CLC Model 327
M F De Campos, S A Romero, F J G Landgraf, F P Missell

012050 Discrete and Resonance States in Graphene Near the Dirac Point 333
M Inglot, V K Dugaev

012051 Magnetic Ground and Remanent States of Synthetic Metamagnets with Perpendicular Anisotropy 339
N S Kiselev, U K Robler, A N Bogdanov, O Hellwig

012052 Morphology, Magnetic and Resonance Properties of Fe/MgO Multilayers 345
A García-García, A Vovk, P Štrichovanec, J A Pardo, C Magén, V Golub, O Salyuk, P A Algarabel, M R Ibarra

012053 Preparation of hcp–Ni Epitaxial Films on Cr Underlayers 351
Jumpei Higuchi, Mitsuru Ohtake, Yoichi Sato, Masaaki Futamoto

11. OTHER TOPICS – BIOMAGNETISM, DOMAIN WALLS, INSTRUMENTATION

012054 Magnetocardiography with GMR-based Sensors 356
M Pannetier-Lecoeur, H Polovy, N Sergeeva-Chollet, G Cannies, C Fermon, L Parkkonen

012055 Low Field MRI with Magnetoresistive Mixed Sensors 361
N Sergeeva-Chollet, H Dyvorne, J Dabek, Q Herreros, H Polovy, G Le Goff, G Cannies, M Pannetier-Lecoeur, C Fermon

012056 Domain Wall Dynamics in a Single CrO_2 Grain 367
P Das, F Porrati, S Wirth, A Bajpai, Y Ohno, H Ohno, M Huth, J Müller

12. MAGNETIC PEROVSKITES AND MULTIFERROICS

012057 Parameters Controlling Magnetic Interactions in Perovskite Manganites 373
B Dabrowski, J Mais, S Kolesnik, O Chmaissem

012058 Substrate-induced Magnetic Anisotropy in $La_{0.7}sr_{0.3}MnO_3$ Epitaxial Thin Films Grown Onto (110) and (118) $SrTiO_3$ Substrates ... 379
P Perna, C Rodrigo, E Jiménez, N Mikuszeit, F J Teran, L Méchin, J Camarero, R Miranda

012059 Magnetic Tunnels Junctions for All-oxide Spin Valves Devices ... 385
A Solignac, R Guerrero, G Agnus, C Fermon, M Pannetier-Lecoeur, Ph Lecoeur

012060 Study of 2D Growth Stability of $SrRuO_3$ on $SrTiO_3$ ... 391
D Esteve, T Maroutian, V Pillard, G Agnus, Ph Lecoeur

012061 Finite-size Effects on the Magnetoelectric Response of Field-driven Ferroelectric/ferromagnetic Chains ... 395
Chenglong Jia, Alexander Sukhov, Paul P Horley, Jamal Berakdar

012062 Influence of Magnetic Heterogeneous State on Magnetocaloric Effect in $Sm_{0.55}Sr_{0.45}MnO_3$ ... 401
L I Koroleva, D M Zashichirinskii, A S Morozov

012063 Noise and Electric Field Characterization of Irradiated $SrTiO_3$ ... 405
R Guerrero, A Solignac, M Pannetier-Lecoeur, C Fermon, P Auban-Senzier, L Lemberger, C Pasquier, Y Apertet, Ph Lecoeur

012064 Experimental Evidence for Ising Spin-Glass Transition in the $YbCoGaO_4$ Single Crystal ... 411
R Szymczak, H A Dabkowska, A Dabkowski, G M Luke, P Aleshkevych, J Fink-Finowicki, H Szymczak

012065 Low Temperature Magnetic Behaviour of PZT-PFW Bulk Multiferroic Ceramics ... 416
Damir Pajic, Zvonko Jaglicic, Marko Jagodic, Robert Blinc, Janez Hole, Marija Kosec, Zvonko Trontelj

012066 Hopping Conductivity of $La_{1-x}Sr_xMN_{1-y}Fe_yO_3$ ... 421
V S Zakhvalinskii, R Laiho, A V Lashkul, K G Lisunov, E Lähderanta, Yu S Nekrasova, P A Petrenko

012067 Low-field Magnetic Properties of $La_{1-x}Sr_xMn_{1-y}Fe_yO_3$ ... 427
V S Zakhvalinskii, R Laiho, A V Lashkul, K G Lisunov, E Lähderanta, Yu S Nekrasova, P A Petrenko

012068 Andreev Reflection and Spin Polarization of $SrRuO_3$ Thin Films on $SrTiO_3$ (111) ... 433
M Foerster, S Piano, F Rigato, C J Mellor, J Fontcuberta

012069 Magnetic Properties of the Magnetoelectric Compound Cu_2OSeO_3: Magnetization and 77SE NMR Study ... 439
M Belesi, T Philippe, I Rousochatzakis, H C Wu, H Berger, S Granville, I V Shvets, J -Ph Ansermet

012070 Magnetocaloric Effect in Polycrystalline La Based Manganites Modified with Sr ... 445
A Kolano-Burian, R Szymczak, R Kolano, H Szymczak, A Burian, L Hawelek, P Zackiewicz, M Czepelak

012071 Structural and Magnetic Characteristics of Manganites in $La_{1-c+x}Sr_{c-x}Mn_{1-x}Ni_xO_3$ System ... 452
A M Smirnov, S Kh Estemirova, V K Karpasyuk

012072 Magnetic and Electrical Properties of $La_{1-c}Sr_cMn_{1-x}Ga_xO_3$ Manganites ... 457
A G Badelin, A A Shchepetkin, S Kh Estemirova, V V Sorokin, V K Karpasyuk

13. MAGNETIC SEMICONDUCTORS AND INSULATORS

012073 New Trends in the Discussion of CMR-manganites and Isostructural Ferrites ... 463
K Bärner, V Morsakov, I V Medvedeva

012074 Spontaneous Generation of Voltage in Single-crystal $Sm_{0.55}Sr_{0.45}MnO_3$ During Magneto-structural Phase Transition ... 469
D M Zashirinskii, L I Koroleva, A S Morozov, R Szymczak

012075 Spin Glass State in the Spinel Families and Stability Conditions ... 474
P Gusin, J Krok-Kowalski, J Warczewski, E Malicka

012076 Ab-Initio Calculation of the Basic Magnetic Properties of (Mn, Cr, V)-doped $MgSiN_2$ ... 480
J Rufinus

012077 Spin-driven Critical Fields in a Spinel Series Based on the Matrix $ZnCr_2Se_4$ ... 484
E Malicka, T Gron, A W Pacyna, A Gagor, T Mydlarz

14. MAGNETIC SHAPE MEMORY EFFECTS AND RELATED PHENOMENA

012078 Magneto-Mechanical Energy Conversion in Magnetic Shape Memory Alloys ... 491
Herbert Schmidt

012079 Ni-Mn-Ga Single Crystals with Very Low Twinning Stress ... 501
L Straka, H Hänninen, A Soroka, A Sozinov

012080 Influence of Grain Size and Training Temperature on Strain of Polycrystalline $Ni_{50}Mn_{29}Ga_{21}$ Samples ... 511
C Hürrich, S Roth, H Wendrock, M Pötschke, D Y Cong, B Rellinghaus, L Schultz

012081 Magnetic and Magnetoelastic Properties of Ni-Mn-Fa – Do They Need a Revision? ... 518
Oleg Heczko, Jaromír Kopecek, Dušan Majtás, Michal Landa

012082 Magnetic Refrigeration at Room Temperature – From Magnetocaloric Materials to a Prototype 526
L Theil Kuhn, N Pryds, C R H Bahl, A Smith
012083 Domain Model for the Magnetic Shape-memory Effect in Non-modulated Tetragonal Ni-Mn-Ga 536
A T Onisan, A N Bogdanov, U K Robler
012084 Monte Carlo Modeling of Exchange Bias Effect in $Ni_{50}Mn_{25+x}Sb_{25-x}$ Heusler Alloys 542
V Buchelnikov, V Sokolovskiy, I Taranenko, S Taskaev, P Entel

15. SOFT MAGNETIC MATERIALS

012085 Development of Magnetically Soft Microwires with GMI Effect 548
V Zhukova, M Ipatov, A Zhukov
012086 the Effect of Compaction Parameters and Dielectric Composition on Properties of Soft Magnetic Composites 556
Ling Xiao, Y H Sun, Lie Yu
012087 Magnetic and Thermomagnetic Studies of the Formation of the Rhometal Powders by High Energy Mechanical Milling 562
I Chicinas, O Isnard, H Chiriac, F Popa, V Pop, C V Prica, B V Neamtu, T F Marinca
012088 Efficient Dynamic Magnetization Measurement in a Py/Ru/Py Trilayer 568
A Bonneau-Brault, S Dubourg, V Dubuget, D Plessis, F Duverger
012089 Magnetic Properties of $Co_{1-x}Ni_{x/2}Sr_{x/2}Fe_2O_4$ (x = 0-1) Ferrite Nanoparticles Synthesized by Self-assembling Media of Reverse Micelles 574
A Ghasemi, A Kaewrawang, X Liu, A Morisako
012090 α-Fe Nanopowder Synthesised in Low-pressure Microwave Plasma and Studied by Mössbauer Spectroscopy 580
B David, N Pizúrová, O Schneeweiss, E Šantavá, O Jašek, V Kudrle
012091 The Influence of Assist Gas on Magnetic Properties of Electrotechnical Steel Sheets Cut with Laser 586
Dominika Gaworska-Koniarek, Bronislaw Szubzda, Wieslaw Wilczynski, Jerzy Drosik, Kazimierz Karas
012092 Subtractively Prepared Permalloy Nanowires for Spin-Torque Experiments 593
Gesche Nahrwold, Sedat Dogan, Lars Bocklage, Toru Matsuyama, Ulrich Merkt, Guido Meier, Kouta Kondou, Gen Yamada, Teruo Ono
012093 Microstructure and Magnetic Properties of Fe and Fe-alloy Thin Films Epitaxially Grown on MgO(100) Substrates 600
Katsuki Matsubara, Mitsuru Ohtake, Fumiyoshi Kirino, Masaaki Futamoto
012094 Magnetic Properties of FeCoC Thin Films Prepared by Various Sputtering Methods 606
V Edon, S Dubourg, J Vernières, J -F Bobo

16. SPINTRONICS

012095 Monte Carlo Study of Temperature and Bias dependence of Spin Transport in GaAs 613
Sheetal Kapoor, Gionni Marchetti, Fausto Rossi, Irene D'Amico
012096 Electric Detection of the Spin-seebeck Effect in Magnetic Insulator in the Presence of Interface Barrier 618
K Uchida, T Ota, Y Kajiwara, H Umezawa, H Kawai, E Saitoh
012097 Magnetoresistance in CuPc Based Organic Magnetic Tunnel Junctions 622
H Tokuc, K Oguz, F Burke, J M D Coey
012098 Annealing Effect on Low Frequency Noise in MgO-based Magnetic Tunnel Junctions 628
J F Feng, Z Diao, J M D Coey
012099 Annealing Effect on Perpendicular $[CoFe/Pd]_n$ Multilayers 634
J F Feng, J M D Coey
012100 Magnetic Tunnel Junctions Using Co/Pt Multilayered Free Layers with Perpendicular Magnetic Anisotropy 638
K Machida, K Furukawa, T Nakayama, N Funabashi, K Aoshima, K Kuga, H Kikuchi, T Ishibashi, N Shimidzu
012101 Simulation of Tunneling Magneto-resistance in Fe/MgO/Fe Junction 644
K Shiiki
012102 Change in Co/Al-oxide/Co Tunneling Junction under Constant Voltage Stress 649
K Shiiki, K Takeda, H Horikiri

012103 Temperature Dependence of Exchange Bias in Co/FeMn-structure Induced by Heating and Cooling in Magnetic Field ... 654
I O Dzhun, S A Dushenko, N G Chechenin, E A Konstantinova
012104 The Dependence of Magnetic Properties of Co/FeMn Bilayer Structure on the Magnitude of Magnetic Field Applied During the Layer Deposition ... 660
I O Dzhun, S A Dushenko, N G Chechenin, E A Konstantinova

17. STRONGLY CORRELATED ELECTRON SYSTEMS, MAGNETISM AND SUPERCONDUCTIVITY

012105 Chiral Skyrmionic Matter in Non-centrosymmetric Magnets ... 666
Ulrich K Roßler, Andrei A Leonov, Alexei N Bogdanov
012106 Easy Axis Switching in Magnetite ... 678
Z Kakol, G Król, W Tabis, T Kolodziej, A Wisniewski, H Stepankova, V Chlan, J Kusz, Z Tarnawski, A Kozlowski, J M Honig
012107 Non-collinear Ferromagnetic Luttinger Liquids ... 691
N Sedlmayr, S Eggert, J Sirker
012108 Exchange Interaction As the Source of Pairing in Correlated Electron Systems: A Brief Overview ... 697
Jozef Spalek
012109 Grüneisen Ratio in Kondo-lattice Compound CeRhSn ... 713
Andrzej Slebarski
012110 Entanglement and Position-space Information Entropy: Hubbard Model As an Approximation to Nanostructure Systems ... 717
J P Coe, V V França, I D'Amico
012111 Interorbital Pairs in Clean and Impure Superconductors ... 721
Anna Ciechan, Karol I Wysokinski
012112 Local Properties of an Inhomogeneous Two-component Correlated Superconductor ... 727
Jaromir Krzyszczak, Karol Izydor Wysokinski
012113 Quasiclassical Eilenberger Approach to the Vortex State in Pnictide Superconductors ... 733
P Belova, M Safonchik, K B Traito, E Lähderanta
012114 Coherence Length of Magnetic Field in the Mixed State of Type-II Superconductors ... 739
P Belova, M Safonchik, K B Traito, E Lähderanta
012115 Frustrated Magnetic Structure of Y-substituted CePdAl Studied by Powder Neutron Diffraction ... 745
Petr Cermák, Tommy Hofmann, Pavel Javorský
012116 Calculation of the Electronic Structure of the Mott Insulator NiO, $LaMnO_3$ and of GaN:Mn ... 749
R J Radwanski, Z Ropka
Author Index

PREFACE

The Joint European Magnetic Symposia JEMS 2010 took place in the complex *Auditorium Maximum* of the Jagiellonian University in Kraków, Poland, between 23–28 August 2010. It followed the series of the conferences in Grenoble (2001), Dresden (2004), San Sebastian (2006), and Dublin (2008). The next Symposia will be held in 2012 in Parma (Italy). The Symposia cover a broad range of aspects of magnetism and magnetic materials, as well as providing a forum for the magnetism community to discuss new concepts, properties, and developments in all branches of fundamental and applied magnetism.

The JEMS 2010 Symposia were organized by the Institute of Physics of Jagiellonian University, in cooperation with AGH University of Science and Technology (Kraków), Cracow University of Technology, Institute of Nuclear Physics of the Polish Academy of Sciences in Kraków, and the Silesian University in Katowice. I thank the Local Committee, and in particular Professor Krzysztof Tomala, for their hard work long before, during, and after the Conference. We dedicate this volume to Professor Henryk Szymczak from the Institute of Physics of the Polish Academy of Sciences for his long lasting service to the magnetism community and the organizational effort in bringing this Conference to our community. Thank you Henryk!

The Conference contained Plenary Sessions and 16 Symposia, which are listed below. Most of them had two chairpersons (also listed), one from abroad and one from Poland. I believe that a collective chairmanship of the Symposia is very helpful in both their organization, as well as in the reviewing process of the papers submitted to the Conference Proceedings. I would like to cordially thank all the persons listed below, who have contributed enormously to the success of our meeting.

The Proceedings comprises 116 invited and contributed papers. I thank the Co-editors for their continuing work long after the Conference.

Arrivederci in Parma!

Jozef Spałek
Chairman of JEMS 2010

Symposia

1. Plenary, Semi-plenary, Tutorials

2. Magnetization Processes Spin Excitations and Ultrafast Dynamics
Coordinator: Andrzej Maziewski (Bialystok)

3. Hard Magnetic Materials and Magnetocalorics
Coordinator: Henryk Figiel (Kraków)

4. Magnetic Hydrides
Coordinators: Ladislav Havela (Praha), Zbigniew Tarnawski (Kraków)

5. Interface of Magnetic Thin Films
Coordinators: Jürgen Fassbender (Dresden), N-T H Kim-Ngan (Kraków)

6. Magnonic Crystals
Coordinators: Bahram Djafari-Rouhani (Lille), Henryk Puszkarski (Poznan)

7. Magnetism of Metals, Alloys, and Intermetallics
Coordinator: Andrzej Szytula (Kraków)

8. Molecular Magnetism
Coordinators: Stephen Blundell (Oxford), Maria Balanda (Kraków)

9. Magnetooptics of Nanomagnets
Coordinators: Kamil Postava (Ostrava), Marek Kisielewski (Bialystok)

10. Nanomagnetism
Coordinators: Marek Przybylski (Halle), Jürgen Kirschner (Halle)

11. Other topics – Biomagnetism, Domain Walls, Instrumentation
Coordinator: Henryk Figiel (Kraków)

12. Magnetic Perovskites and Multiferroics
Coordinator: Henryk Szymczak (Warszawa)

13. Magnetic Semiconductors and Insulators
Coordinators: Klaus Baerner (Göttingen), Tadeusz Gron (Katowice)

14. Magnetic Shape Memory Effects and Related Phenomena
Coordinators: Oliver Gutfleisch (Dresden), Sebastian Fähler (Dresden)

15. Soft Magnetic Materials
Coordinators: Julian González (San Sebastian), Krzysztof Kulakowski (Kraków)

16. Spintronics
Coordinator: Maciej Sawicki (Warszawa)

17. Strongly Correlated Electron Systems, Magnetism and Superconductivity
Coordinator: Andrzej Slebarski (Katowice)

The next Joint European Magnetic Symposia, JEMS 2012, will be held in Parma, Italy, 9–14 September 2012.
www.jems2012.it
Co-Chairs:
Franca Albertini, Institute of Materials for Electronics and Magnetism (IMEM), CNR, Parma
Roberto De Renzi, Department of Physics, University of Parma

Chairman:
Jozef Spałek (Kraków)

Program Committee:
Stephen Blundell (Oxford), J Michael D Coey (Dublin), Dominique Givord (Grenoble), Dariusz Kaczorowski (Wrocław), Roman Micnas (Poznań), Marek Przybylski (Halle), Ludiwig Schultz (Dresden), Vladimir Sechovsky (Prague), Jozef Spałek (Kraków), Henryk Szymczak (Warszawa), Manuel Vázquez (Madrid)

Publication Committee:
Dariusz Kaczorowski, Robert Podsiadły, Jozef Spałek, Henryk Szymczak, Andrzej Szytuła

Local committee:
Maria Bałanda, Anna Majcher, Robert Podsiadły, Michał Rams, Andrzej Ślebarski, Krzysztof Tomala

Editors of the Proceedings:
Jozef Spałek, Krzysztof Tomala, Danuta Goc-Jagło, Robert Podsiadły, Michał Rams, Anna Majcher

Plenary, semi-plenary and tutorial speakers:
Ernst Bauer (Wien)
Stephen Blundell (Oxford)
J Michael D Coey (Dublin)
Russell P Cowburn (London)
Burkard Hillebrands (Kaiserslautern)
Claudine Lacroix (Grenoble)
Lluís Mañosa (Barcelona)
María del Carmen Muñoz (Madrid)
Bernard Raveau (Caen)
Pedro Schlottmann (Tallahassee)
Frank Steglich (Dresden)
Oliver Waldmann (Freiburg)

Invited speakers within symposia:

R Ahuja (Uppsala)	A Kirilyuk (Nijmegen)
M Albrecht (Vienna)	L Theil Kuhn (Roskilde)
K Bärner (Göttingen)	J Liu (Dresden)
U Bovensiepen (Duisburg)	G Lorusso (Modena)
V Buchelnikov (Chelyabinsk)	M M Maska (Katowice)

B Chevalier (Bordeaux)
O Chubykalo-Fesenko (Madrid)
A V Chumak (Kaiserslautern)
J M D Coey (Dublin)
B Dabrowski (DeKalb)
S Das (Aveiro)
A del Moral (Zaragoza)
V E Demidov (Muenster)
B Djafari-Rouhani (Lille)
H A Dürr (Menlo Park)
J Fassbender (Dresden)
J Fontcuberta (Barcelona)
V Garcia (Orsay)
J N Gonçalves (Aveiro)
M E Gruner (Duisburg)
G Gubbiotti (Perugia)
S Hamann (Bochum)
U Hannemann (Dresden)
L Havela (Prague)
O Heczko (Prague)
B Hernando (Oviedo)
O Isnard (Grenoble)
Z Kąkol (Kraków)
N-T H Kim-Ngan (Kraków)
Y Mukovskii (Moscow)
M Pannetier-Lecoeur (Saclay)
G Papavassiliou (Athens)
K R Pirota (Campinas)
P Przyslupski (Warszawa)
M Reiffers (Košice)
K Sandeman (London)
D Sander (Halle)
M Sawicki (Sendai/Warsaw)
J Schaefer (Würzburg)
H Schmidt (Wetzikon)
J Spałek (Kraków)
L Straka (Helsinki)
A Szewczyk (Warszawa)
Y Taguchi (Wako)
A Thiaville (Orsay)
R Varga (Košice)
P Vavassori (San Sebastian)
W Wulfhekel (Karlsruhe)
M Yamashita (Sendai)
R Zdyb (Lublin)
A Zhukov (San Sebastián)
A K Zvezdin (Moscow)

International Advisory Committee (2011):

Dominique Givord, President (Grenoble)
Manfred Albrecht (Chemnitz)
Agnés Barthélémy (Paris)
Roy Chantrell (York)
Russell Cowburn (London)
Tomasz Dietl (Warszawa)
Ludwig Schultz, Former President (Dresden)
Burkard Hillebrands (Kaiserslautern)
Andrei Kirilyuk (Nijmegen)
Ron Jansen (Tsukuba)
Nicoleta Lupu (Iasi)
Caroline A Ross (Cambridge, MA)

Claudia Felser (Mainz)
Josef Fidler (Wien)
Dino Fiorani (Roma)
Pietro Gambardella (Bellaterra)
Alberto Guimarães (Rio de Janeiro)
Stefano Sanvito (Dublin)
Vladimir Sechovsky (Praha)
Roberta Sessoli (Firenze)
Jozef Spałek (Kraków)

Surface ferromagnetism in non-magnetic and dilute magnetic oxides

M.Carmen Muñoz, Silvia Gallego, and Nadiezhda Sanchez
Instituto de Ciencia de Materiales de Madrid, CSIC, 28049 Madrid, Spain
E-mail: mcarmen@icmm.csic.es

Abstract. We explore hole-doping as an efficient route to develop magnetism in simple non-magnetic oxides. Based on ab-initio calculations of the prototypical material ZnO, we show that local spin polarization can be obtained even in the absence of magnetic dopants both in the bulk or at surfaces. However, the onset of long-range magnetic order additionally requires extended states, that can be achieved in surfaces without distorting the host lattice. We propose different possibilities which either profit the spontaneous polarization of Oxygen p states under defective charge conditions, or require the presence of adsorbates.

1. Introduction

The last decade has witnessed the emergence of the field of magnetism in non-magnetic materials [1]. Multiple nanostructures of either organic or inorganic systems have been reported to show magnetic properties in the absence of magnetic elements [2, 3, 4, 5, 6], and even the suggestion of magnetism as an universal feature at the nanoscale has been proposed [7]. Carbon-based structures also develop magnetic properties under different conditions [6, 8] and the possibility to introduce dilute concentrations of magnetic impurities in semiconductors to achieve relevant ferromagnetic responses has led to a new branch of research on the so-called dilute magnetic semiconductors (DMS) [9, 10, 11, 12]. Besides its evident fundamental interest, the investigation in this field is appealing for the development of new technological devices, which could exploit the advantages of a reduced scale, combine at the same time electric and magnetic fields, and ultimately lead to the development of spintronics.

One major drawback of most of these systems for the implementation of realistic devices is the low value of their critical temperature, usually well below room temperature. This is particulary true for DMS [13, 14]. In this sense, the prediction of Dietl at al. in 2000 that the critical temperature could be raised replacing traditional semiconductors by oxides or nitrides [15] launched a new and hot topic of research pursuing ferromagnetism in dilute magnetic oxides. Dilute magnetic oxides (DMO) are non-magnetic oxides doped with dilute amounts of magnetic transition metal impurities, so that the host lattice is essentially preserved and the main oxide properties can be retained. A large variety of systems has been considered, both concerning the oxide and the dopant, ZnO constituting a prototype [16]. Though initially magnetic elements such as Fe, Mn or Co were employed as impurities, soon appeared reports of magnetism in the absence of magnetic elements, such as non-magnetic transition metals (Cu, Ti, etc.) [17, 18], or even non-metals like N or C [19, 20]. After an intensive research effort, the initial excitement has opened way at the same time to some uncertainty about the ease to obtain

DMO showing intrinsic room temperature ferromagnetism (RT-FM) and to the identification of the basic mechanisms behind the development of magnetism in these systems [10, 11, 12]. Briefly, the appearance of local magnetic moments seems to be linked either to the introduction of magnetic impurities carrying their own spin polarization and slightly polarizing the adjacent O neighbours, or to the formation of holes at the top of the valence band of the oxide, which promote spontaneous magnetization. The holes can be originated by impurities, defects or boundaries [21, 22]. However, the magnetic interactions due to impurities (both magnetic and non-magnetic) or defects are short ranged and frequently antiferromagnetic [9], making difficult the achievement of long-range magnetic order and critical temperatures close to RT.

Several routes have been proposed to extend the range of the magnetic interactions. In the case of magnetic impurities, they imply codoping with other elements (for example, Li and Co in ZnO), introducing defects such as host cation or anion vacants, interstitials, etc., or increasing the concentration of the magnetic impurities [23, 24, 25]. This leads to a more difficult control of the final structure and to the alteration of the original host crystal characteristics. In fact, it has been shown that the formation of secondary phases or the occurence of spinodal decomposition are major issues that may alter the magnetic and conduction properties leading to a non-intrinsic origin of the measured ferromagnetic signal [26, 27, 28, 29]. More importantly, it has not been proved a general increase of the strength and range of the magnetic interactions. Also for magnetism in the absence of magnetic impurities only extended regions with a high concentration of impurities or defects may allow to obtain RT-FM, which again may influence the homogeneity and even the stability of the sample. In addition, the controlled incorporation of impurities and defects is arduous, and usually an inhomogeneous distribution occurs. To summarize, it remains an open question whether it would be possible to control the conditions to achieve uniform and reproducible RT-FM in bulk samples of DMO.

An alternative approach, which we proposed a few years ago, relies on the distinct properties of surfaces [22, 30]. The symmetry breaking imposed by the surface usually alters the position of the Fermi level, and for particular surfaces the system becomes metallic. Hence, the oxide can be either hole or electron doped at the surface and under specific conditions spontaneous magnetic moments develop. Furthermore, surface states, although confined to the topmost layers, are extended in two-dimensions unlike the localized impurity bulk states. Therefore, the subtle interplay between localization, necessary for the moment formation, and extended states leading to strong and long-range magnetic coupling can be reached at the surfaces. In fact, it is well-known that RT-FM is more easily achieved in thin films and nanoparticles of both DMO and undoped non-magnetic oxides. Further, there is experimental evidence that the FM signal seems to be linked to the near-surface regions, grain boundaries or nanostructure interfaces [31, 32] Moreover, magnetic impurities often tend to approach sample boundaries favouring magnetic order close to the frontiers.

In this article we will discuss the origin of magnetism and the possibility to achieve long range magnetic order at the surface of ionic oxides, based on our calculations on the prototypical material ZnO. First, we will show that the requisite to develop magnetic moments even in the bulk is to introduce hole-doping. Under these conditions, local magnetic states appear in the absence of magnetic elements. However, the magnetic interactions remain short-ranged. In order to achieve long-range magnetic order, extended magnetic states need to be formed. Second, we will show that surfaces appear as an unique scenario to achieve this goal.

2. Theoretical framework

The results presented here are based on ab-initio calculations within the density functional theory, using both the SIESTA [33] and VASP [34] computational packages. The first is based on basis sets formed by localized atomic orbitals. For the structures considered we have employed multiple-zeta polarized localized numerical atomic orbitals (AO). In the case of Zn we choose double-zeta (DZ) basis for the $4s$ and $3d$ AOs plus a single-zeta (SZ) $4p$ AO, whereas for O we employed DZ $2s$ and $2p$ AOs plus a SZ $2d$ AO. Besides undoped ZnO structures, we will also consider Cu doping and H adsorbates. Cu impurities have been described using the same type of basis set as for Zn, while Hydrogen is described by a DZ $1s$ AO. In VASP the electronic charge density is expressed in plane-wave basis sets [35]. We have used the projector augmented wave (PAW) method to describe the electron-ion interactions, and a plane-wave enery cutoff of 350 eV. More details about the conditions of the calculations can be found elsewhere [30, 36, 37].

Depending on the structure, to describe the exchange-correlation part we have used either the LSDA(local spin density approximation)+U with SIESTA [38] or HSE (Heyd-Scuseria-Ernzerhof) hybrid functionals with VASP [39], that introduce correlation and localization effects beyond LSDA and provide an improved description of the electronic band gap of ZnO. The empirical Coulomb U= 5.7 eV and exchange J= 1 eV parameters are chosen to model the orbital dependent Hubbard-like potential for the Zn $3d$ states [40]. The resulting parameters that characterize the wurtzite unit cell of ZnO are $a = 3.25Å$ and $c = 5.20Å$, and the band gap $E_g = 1.25$ eV, which can be compared to the experimental values of $a = 3.25Å$, $c = 5.20Å$, $E_g = 3.3$ eV [41, 42]. The corresponding shortest Zn-O distance between first neighbours is 1.98 Å. For HSE calculations we have employed a 25% of Hartree-Fock exchange and a value of 0.37 for the screening parameter, which provides $E_g = 3.55$ eV.

The structures are modelled as periodic slabs with the z-axis along [0001] and $(n \times m)$ unit cells in the xy plane, n and m ranging from 1 to 3 depending on the calculation. For surfaces, the thickness of the slab has been chosen in order to reproduce bulk features in the middle layers, typically between 15 and 17 ZnO planes. A vacuum region of at least 15 Å has been included, that guarantees the absence of interaction between both slab sides, and dipolar corrections have been applied for asymmetric slabs. The atomic positions have been relaxed until the forces on the atoms are always below 0.05 eV/Å and usually less than 0.03 eV/Å. In all cases we have ensured convergence in k-sampling, with reciprocal space integrations performed on Monkhorst-Pack supercells up to $12 \times 12 \times 1$ in particular cases.

3. Bulk (Cu,Zn)O

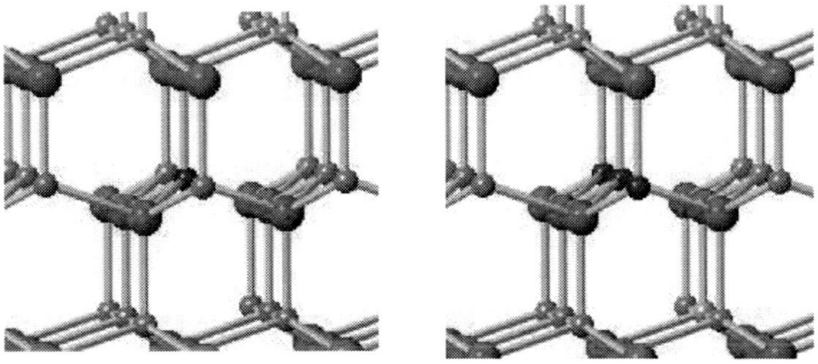

Figure 1. Scheme of the structure of (left) an isolated substitutional Cu impurity and (right) one-dimensional Cu aggregates contained in the [0001] plane in bulk ZnO. Oxygen, Zn and Cu atoms are represented respectively by large red balls and small gray and blue balls.

We have analyzed the substitution of Zn by Cu in bulk ZnO. First we have considered isolated Cu impurities (see left panel of figure 1) in a supercell of 32 atoms including eight [0001] atomic layers along the z-axis and a (2×2) unit cell in the xy plane, that represents a $Zn_{.94}$ $Cu_{.06}O$ structure. Under these conditions, adjacent Cu impurities are separated at least by an intermediate O-Zn-O sequence, which leads to a negligible interaction between them. The corresponding structure shows minor and asymmetric relaxations with respect to undoped ZnO, the Cu-O distance between first neighbours being 2.01 $\mathring{A}$ for the on top Oxygen along the [0001] axis and 1.95$\mathring{A}$ for the three Oxygens underneath. The analysis of the charge distribution as compared to bulk ZnO indicates that, similarly to Zn, Cu is in a +2 valence state, though the net charge transferred to the neighbouring Oxygen atoms is slightly reduced, which may be indicative of a reduced ionicity of the Cu-O bonds. Figure 2 provides the total density of states (DOS) of the supercell obtained using the HSE functional. The projected DOS on the Cu impurity and on one of its O neighbours are also shown. The Cu 3d states are spin splitted and appear mostly overlapping the O p valence band (VB), but additional impurity states fall in the bulk band gap. They correspond to the minority Cu t_{2g} states and the Fermi level crosses this impurity band. The hybridization of Cu and O is reflected in the weight on the O p states particularly of the t_{2g} Cu states. But the most important aspect is that, eventhough all elements in the structure are non-magnetic, the depletion of the Cu d electronic levels leads to a Cu magnetic moment of 0.65 μ_B, slightly polarizing the adjacent O neighbours with moments around 0.1 μ_B. The O moments are almost symmetrically distributed, with the spin density charge shared out among the p_x, p_y and p_z orbitals, as reflected in the spin density distributions at figure 3. Magnetic moments of Zn or more distant O atoms are negligible. The resulting total moment of the supercell is 1.00 μ_B, evidencing the local character of the magnetization. The magnetic interaction, measured as the difference between a ferro- and antiferro-magnetic configuration, is very small, less than 10 meV, and therefore short ranged.

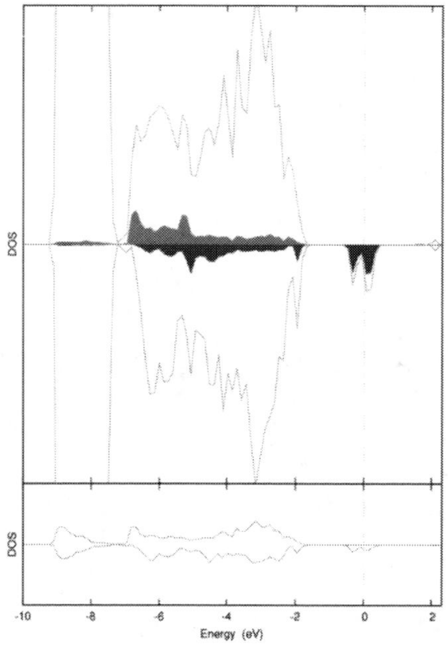

Figure 2. (Top) Spin polarized DOS of the modelled $Zn_{.94}Cu_{.06}O$ slab, where positive (negative) values correspond to majority (minority) spin projections. The energy zero is set at the Fermi level. The filled curve represents the Cu contribution. (Bottom) DOS projected on one of the O neighbours bonded to Cu. The scale at the y-axis has been enhanced 2.5 times with respect to the top panel.

Due to the short range of the interactions between impurities, and in order to explore the possibility of obtaining extended magnetic order, we have studied one-dimensional Cu structures in the [0001] plane with only one intermediate Oxygen atom between neighbouring Cu cations, as depicted in the right diagram of figure 1. Under these conditions, the local Cu magnetic moments increases to 0.85 μ_B, while the Oxygen polarization distributes as 0.20 μ_B for the Oxygen atoms bonded to 2 Cu impurities, and 0.07 μ_B for those with only one Cu bond. The resulting total magnetic moment at the supercell is 2.0 μ_B, thus equivalent to the induced magnetic moment of the isolated impurity. The strenght of the magnetic coupling, measured as the difference between a ferro- and antiferro-magnetic configuration, is around 20 meV. Based on simple mean-field models, this exchange interaction does not support the existence of RT-FM.

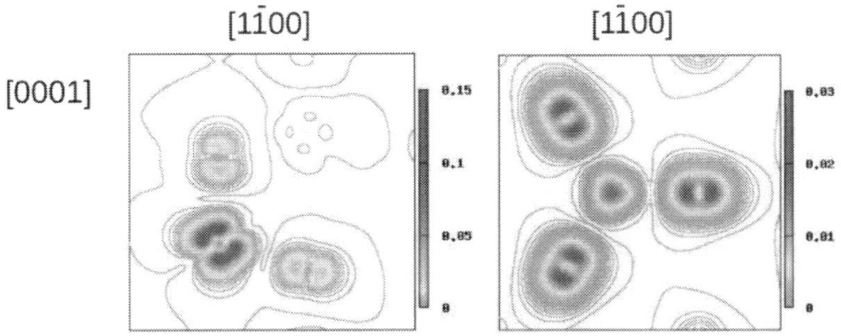

Figure 3. Spin density charge around a substitutional Cu impurity in bulk ZnO at two orthogonal sections.

To summarize, isolated Cu impurities in bulk ZnO are spin-polarized, the magnetic moments being associated with unpaired electrons of the Cu ions. The electronic states close to the VB edge are dominated by hybridization between O 2p and Cu 3d orbitals, which implies that the Cu-O bond is partially covalent instead of purely ionic. The Fermi level lies in the minority t_{2g} impurity band located in the ZnO band gap and then the magnetic coupling is short-ranged, and although ferromagnetic coupling is favoured for adjacent Cu impurities sharing an O neighbour, the corresponding Curie temperature is estimated to lie below RT.

4. Surfaces

Surfaces provide a suitable playground to manipulate and control magnetism in DMO. The high level of sophistication and development of surface science techniques allows the manipulation of structures at the atomic level much beyond the limits that can be controlled for bulk samples. Moreover, as we will show here, the intrinsic characteristics brought by surfaces (i.e., breakdown of symmetry, unsaturated bonds, etc.) offer an unique scenario where stable oxide magnetic states may develop, can be more easily controlled and, most importantly, become extended allowing the development of long-range magnetic order.

Here we investigate the polar ZnO (0001) surface. The [0001] is a preferential growth axis for ZnO, thus [0001] crystallographic terminations constitute the most investigated ZnO surfaces in spite of their polar character [43, 44]. A sketch of the possible terminations is provided in figure 4. Zn and O planes alternate along the [0001] direction and, due to the lack of inversion symmetry of the wurtzite structure along the c-axis, the polar (000$\bar{1}$) and (0001) ZnO surfaces are inequivalent. Two different terminations can be obtained, either Zn or O ended, hereafter

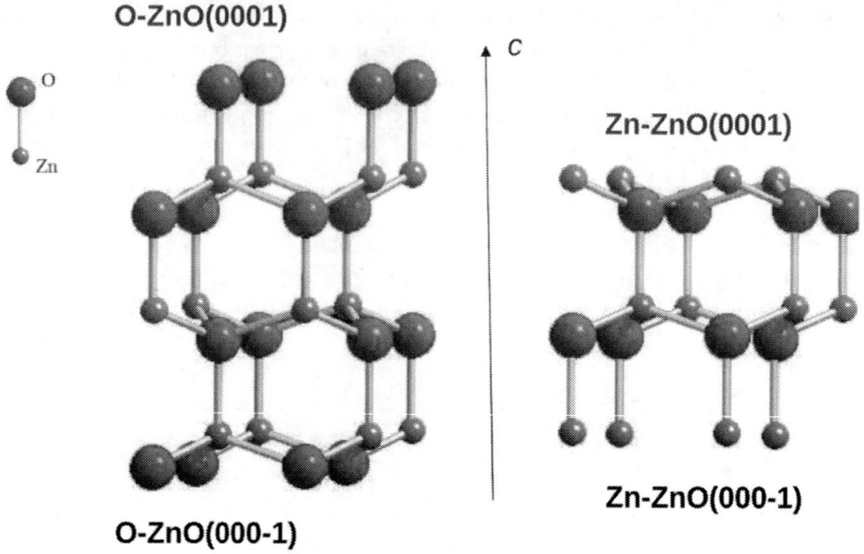

Figure 4. The four different possible surface terminations of wurtzite ZnO along the [0001] crystal axis.

labelled respectively as Zn-ZnO and O-ZnO. In principle, the most stable ones are those involving the minimum number of unsaturated bonds, therefore Zn-ZnO and O-ZnO are identified to correspond to the (0001) and $(000\bar{1})$ surfaces, respectively. The O-ZnO $(000\bar{1})$ surface is found to be the most stable from thermodynamic arguments, and also constitutes the most frequently observed termination of thin films and nanoparticles. Eventhough its polar character and high reactivity it usually does not show any reconstruction, and in general is well understood [44]. On the contrary, there is no consensus about the Zn-ZnO(0001) surface. Several studies evidence that its structure and chemical composition are highly dependent on growth mode, preparation conditions and environment [45, 46]. Diverse reconstructions and chemical compositions have been reported: it can show triangular islands and pits whose step edges are Oxygen terminated [47], highly ordered $(10\bar{1}0)$ nanofacet step arrays [48], or a (1×3) reconstruction which has been interpreted as arising from the removal of one third of the surface Zn atoms [49]. Also, partial Hydrogen adsorption, incomplete and even O-terminated surfaces have been observed [50]. In the following, we will concentrate in the (0001) surface.

4.1. H on Zn-ZnO(0001)

The interaction of H atoms with the polar Zn-ZnO(0001) surface is expected to be the weakest among the ZnO surfaces, since the binding energy of ZnH pairs should be considerably weaker than that of OH bonds. He-atom scattering measurements have shown that Hydrogen forms an ordered (1×1) overlayer on the surface at low Hydrogen coverages, and that larger exposures lead to a loss of lateral order and the random distribution of H adatoms [51].

The left panel of figure 5 shows the DOS of this surface, obtained after LSDA calculations. The introduction of correlation effects beyond the LSDA description provides a wider band gap, but does not alter the relevant electronic features described here [37]. The bare surface is metallic, due to the filling of surface states near the conduction band (CB) derived from the Zn unsaturated dangling bonds [45], and shows a significant contraction of about 60% of the

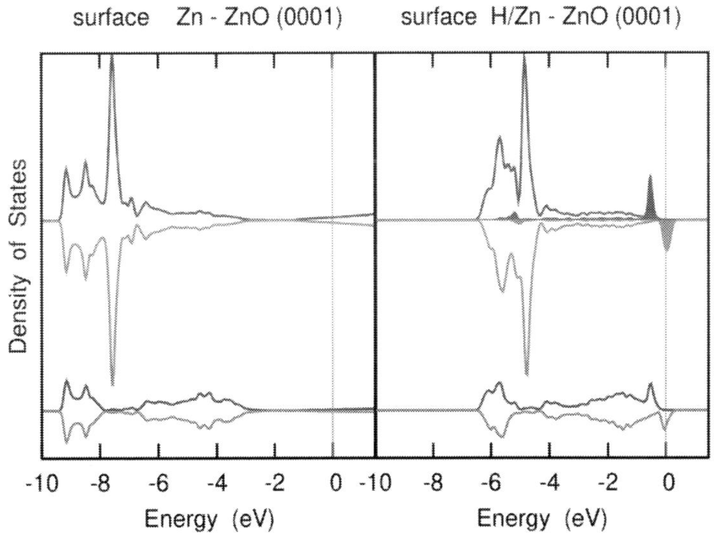

Figure 5. Spin-polarized and layer-resolved DOS corresponding to the two outermost surface layers of the Zn-ZnO(0001) surface either (left) bare or (right) covered by 1 H monolayer. The layers are ordered from top to bottom so that the DOS at the top corresponds to the surface Zn atoms. The H contribution on the right is represented by the filled curve superimposed to the surface layer DOS, and the scale has been enhanced 5 times for clarity. Energies are referred to the Fermi level.

first Zn-O interlayer distance. The adsorption of H serves to stabilize this surface avoiding the reconstruction.

As H is deposited, it occupies ontop positions forming strong H-Zn bonds, which create a shallow band at the VB edge. For intermediate coverages, the remaining (not bonded to H) Zn atoms strengthen then their interaction with the O layer underneath, a large corrugation appears in the Zn layer and up to half of a H monolayer coverage, the system becomes an insulator. When the H coverage exceeds half of a monolayer, the induced H band becomes spin splitted and robust ferromagnetism develops, with a difference of 46 meV between the ferro- and antiferro-magnetic configurations. In this situation, the H-Zn interaction promotes the formation of an extended magnetic band involving H, Zn and O states, as evidenced by the DOS corresponding to complete H coverage shown in the right panel of figure 5. The magnetic moment mostly resides in the H states, with a value of 0.29 μ_B. The outermost Zn and O atoms develop a weak polarization of 0.09 μ_B, reflecting the subsurface extension of the spin-splitted band and the involvement of H, Zn and O hybridized states.

The shift of the Fermi level from the CB at the clean surface to the energy gap at half monolayer of hydrogen coverage, and then to the VB edge upon the completion of a H monolayer can be viewed as a multipe metal-insulator transition, which is accompanied by a magnetic transition. The thermodynamic stability of H adsorption makes possible to tune reversibly this transition, controlling the H load at the Zn-ZnO(0001) surface.

4.2. Spontaneous magnetism at the O-ZnO(0001) surface

As mentioned above, the most usual and stable O-ended surface of ZnO is the $(000\bar{1})$, with one unsaturated bond per surface Oxygen atom. However, there are several reports revealing the presence of Oxygen atoms in the ZnO(0001) surface and the subsequent possibility to stabilize the O-ZnO(0001) termination in extended local areas. The presence of Oxygen is linked either to rough surfaces with Zn vacancies or incomplete O terminations [50], or even to smooth surfaces such as the (1×3) [48] or the triangular islands reconstruction [47], obtained under specific growth conditions. In addition, it has been shown both experimentally and theoretically that the stability of the O-ZnO(0001) termination is largely enhanced by the partial introduction of subsurface Co impurities [30, 50].

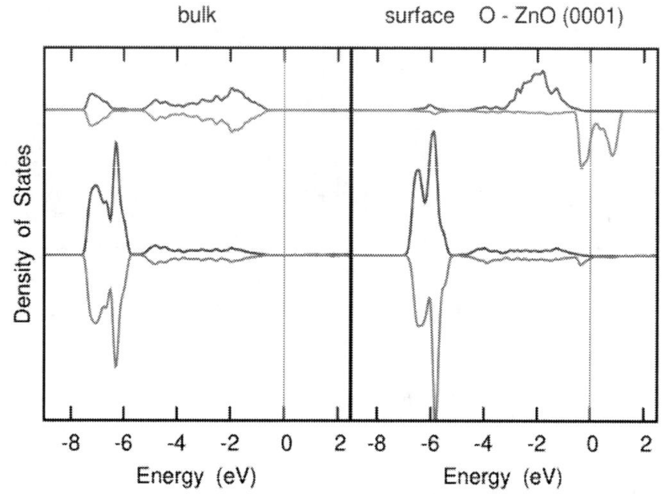

Figure 6. Same as figure 5 for the O-ended ZnO(0001) surface. For comparison, the layer-resolved bulk contribution of O (above) and Zn (below) is shown in the left panel.

Figure 6 shows the DOS at the outermost layers of the O-ZnO(0001) surface, as obtained from calculations based on LDA+U. The reduced O coordination causes a considerable decrease of the Mulliken charge of the surface O atoms, so that the Fermi level crosses the p states of the surface O atoms, which become spin splitted leading to a large magnetic moment of 1.5 μ_B. We would like to point out that analogous results were obtained even within the LDA approximation [30]. Since holes are in the VB, underestimation of the band gap does not affect their energy localization. The resulting surface is half-metallic, and FM order is favoured over antiferromagnetism by 160 meV. The large value of the magnetic energy proves that the exchange interactions are strong. In fact, a mean-field estimate of the critical temperature from the total energy difference between ferro- and antiferro-magnetic orders provides a value well above room temperature. Preliminary results obtained by Monte-Carlo simulations using self-interaction corrections confirm the high value of the Curie temperature [52].

The O magnetic moment in this surface is much larger than that induced in ZnO by hybridization with magnetic impurities. In fact, when subsurface Co impurities are introduced, two types of surface O atoms coexist: those bonded to Co showing a weak induced polarization, and the unbonded ones where the magnetic properties of the undoped surface pervive. Furthermore, the surface still shows robust ferromagnetic order, so that the combination of the intrinsic p surface magnetism and that induced by Co leads to a surface with improved stability showing ferromagnetism with high magnetization.

The origin of the O magnetic moments arising from the drastic reduction of coordination actually involves the introduction of a large number of p holes in the VB of the oxide. In this sense, it is similar to the cation vacancy induced moments found in bulk samples, but the surface offers the unique possibility to increase the number of holes without distorting the oxide lattice, and furthermore, to naturally extend the magnetic states in order to achieve a uniform ferromagnetic 2D sheet. Some peculiarities of this kind of surface magnetism are that it requires a critical number of $2p$ holes, and that it is associated to a highly anisotropic distribution of the magnetic charge [30]. For the (0001) O-surface of ZnO, holes mainly reside in the p_{xy} degenerated orbitals, since the crystal field determines where holes are localised.

This p magnetism is a general property of oxide surfaces, as we have proved in a previous systematic study of the surfaces of different highly ionic oxides [22]. Moreover, magnetic states associated to unsaturated p bonds have also been reported in other low dimensional systems as graphite and quasi-1D organic materials.

5. Conclusions

Spontaneous magnetization is predicted at the polar surfaces of ZnO, rooted in the intrinsic characteristics brought by the surface, namely symmetry breaking, unsaturated bonds or uncompensated charges. The physical mechanism underlying the spontaneous magnetization relies on the formation of holes at the top of the Oxygen derived p-like valence band of the oxide. In addition, surfaces states, although confined to the topmost layers, are extended in two-dimensions and thus the subtle interplay between localization and extended states required for the development of long-range magnetic order can be reached at the surface. As a consequence, high transition temperatures close to RT can be reached, with ferromagnetism associated to the topmost layers in the vicinity of the surface. The mechanism promoting the spontaneous magnetization of the surface is universal in ionic oxides, although the onset of magnetism requires a critical number of p-holes only achieved by specific surfaces or under particular environmental conditions.

In summary, we have explored the possibility that frontiers, and in particular surfaces, of ionic oxides may develop high temperature ferromagnetism. We propose that p surface magnetism can be responsible of the magnetic signals observed in a large variety of non-magnetic and dilute magnetic oxides.

6. Acknowledgments

Financial support from the Spanish Ministry of Science and Technology under grants MAT2006-05122 and MAT2009-14578-C03-03 is acknowledged.

References

[1] Awschalom D 2010 Physics **3** 53
[2] Nascimento R, de Oliveira A J A, Correa A A, Bulhoes L O S, Pereira E C, Souza V M and Walmsley L 2003 Phys. Rev. B **67** 144422
[3] de Paula F R, Walmsley L, Pereira E C, and Oliveira A J A 2008 J. Magn. Magn. Mater. **320** e193
[4] Gao D et al. 2009 J. Appl. Phys. **105** 113928
[5] Deng S et al. 2009 Appl. Phys. Lett. **93** 193111
[6] Yazyev O V 2010 Rep. Prog. Phys. **73** 056501
[7] Sundaresan A and Rao C N 2009 Sol. Sta. Comm. **149** 1197
[8] Makarova T and Palacio F 2006 *Carbon Based Magnetism: An Overview of the Magnetism of Metal Free Carbon-based Compounds and Materials* (Amsterdam: Elsevier)
[9] Lany S and Zunger A 2009 Phys. Rev. B **80** 085202
[10] Stoneham M 2010 J. Phys.: Condens. Matter **22** 074211
[11] Coey J M D, Stamenov P, Gunning R D, Venkatesan M and Paul K 2010 New J. Phys. **12** 053025
[12] Dietl T 2010 Nature Mater. **9** 965
[13] Ohno H, Shen A, Matsukura F, Oiwa A, Endo A, Katsumoto S and Iye Y 1996 Appl. Phys. Lett. **69** 363
[14] Wang M, Campion R P, Rushforth A W, Edmonds K W, Foxon C T and Gallagher B L 2008 Appl. Phys. Lett. **93** 132103
[15] Dietl T, Ohno H, Matsukura F, Cibert J and Ferrand D 2000 Science **287** 1019
[16] Sato K and Katayama-Yoshida H 2002 Semicond. Sci. Technol. **17** 367

[17] Coey J M D 2006 Curr. Opin. Solid State Mater. Sci. **10** 83
[18] Pan F, Song C, Liu X J, Yang Y C and Zeng F 2008 Mat. Sci. and Eng. R **62** 1, and references therein.
[19] Herng T S et al. 2009 Appl. Phys. Lett. **95** 012505
[20] Yu C-F et al. 2007 J. Phys. D: Appl. Phys. **40** 6497
[21] Das Pemmaraju C and Sanvito S 2005 Phys. Rev. Lett. **94** 217205
[22] Gallego S, Beltrán J I, Cerdá J and Muñoz M C 2005 J. Phys.: Condens. Matter **17** L451
[23] Sluiter MHF et al. 2005 Phys. Rev. Lett. **94** 187204
[24] Zhang X et al. 2009 Phys. Rev. B **80** 174427
[25] Wang Q et al. 2008 Phys. Rev. B **77** 205411
[26] Dietl T 2006 Nature Mater. **5** 673
[27] Dietl T, Andrearczyk T, Lipińska A, Kiecana M, Tay M and Wu Y 2007 Phys. Rev. B **76** 155312
[28] Iuçan D, Knut R, Sanyal B, Karis O, Eriksson O, Coleman V A, Westin G, Wikberg J M, and Svedlindh P 2008 Phys. Rev. B **78** 085319
[29] Iuçan D, Kabir M, Granäs O, Eriksson O and Sanyal B 2009 Phys. Rev. B **79** 125202
[30] Sanchez N, Gallego S and Muñoz M C 2008 Phys. Rev. Lett. **101** 067206
[31] Straumal B B, Mazilkin A A, Protasoval S G, Myatiev A A, Straumal P B, Schütz G, van Aken P A, Goering E, and Baretzky B 2009 Phys. Rev. B **79** 205206
[32] Khalid M, Setzer A, Ziese M, Esquinazi P, Spemann D, Pöppl A and Goering E 2010 Phys. Rev. B **81** 214414
[33] Soler J M, Artacho E, Gale J D, García A, Junquera J, Ordejón P and Sánchez-Portal D 2002 J. Phys.: Condens. Matter **14** 2745
[34] Kresse G and Hafner J 1993 Phys. Rev. B **47** 558(R); Kresse G and Furthmüller J 1996 Phys. Rev. B **54** 11169
[35] Kresse G and Joubert D 1999 Phys. Rev. B **59** 1758
[36] Beltrán J I, Gallego S, Cerdá J, Moya J S and Muñoz M C 2003 Phys. Rev. B **68** 075401
[37] Sanchez N, Gallego S, Cerdá J and Muñoz M C 2010 Phys. Rev. B **81** 115301
[38] Anisimov V, Aryasetiawan F, and Lichtenstein A 1997 J. Phys.: Condens. Matter **9** 767
[39] Heyd J, Scuseria G E and Ernzerhof M 2003 J. Chem. Phys. **118** 8207; Krukau A, Vydriv O, Izmaylov A and Scuseria G 2006 J. Chem. Phys. **125** 224106
[40] Khalid M et al 2009 Phys. Rev. B **80** 035331
[41] Kisi EH, Elcombe MM 1989 Acta Crystallographica Section C-Crystal Structure Communications **45** 1867
[42] Özgür Ü, Alivov YI, Liu C, Teke A, Reshchikov MA, Dogan S, Avrutin V, Cho S-J and Morkoc H 2005 J. Appl. Phys. **98** 041301
[43] Staemmler V, Fink K, Meyer B, Marx D, Kunat M, Girol S G, Burghaus U and Wöll C 2003 Phys. Rev. Lett. **90** 106102
[44] Wöll C 2007 Progress in Surf. Sci. **82** 55 , and references therein.
[45] Kresse G, Dulub O and Diebold U 2003 Phys. Rev. B **68** 245409
[46] Valtiner M, Borodin S and Grundmeier G 2008 Langmuir **24** 5350
[47] Dulub O, Diebold U and Kresse G 2003 Phys. Rev. Lett. **90** 016102
[48] Ostendorf F, Torbrugge S and Reichling M 2008 Phys. Rev. B **77** 041405(R)
[49] Torbrugge S, Ostendorf F and Reichling M 2009 J. Phys. Chem. C **113** 4909
[50] Matsui H and Tabata H 2007 Phys. Rev. B **75** 014438
[51] Becker T, Hovel S, Kunat M, Boas C, Burghaus U and Wöll C 2001 Surf. Sci. **486** L502; 2002 Surf. Sci. **511** 463
[52] Fischer G et al 2011 to be published

Crossover from non-Fermi liquid to Fermi liquid behavior in heavy fermion systems

P. Schlottmann
Department of Physics, Florida State University, Tallahassee, FL 32306, USA
E-mail: schlottmann@physics.fsu.edu

Abstract. Many heavy fermion systems display a crossover from Fermi liquid to non-Fermi liquid behavior due to a nearby quantum critical point (QCP). Some of the features are universal and do not depend on the nature of the QCP and the tuning mechanism. A simple one-band model of heavy electrons is considered to describe the crossover. The nesting of the spherical Fermi surfaces of an electron and a hole pocket separated by a nesting vector **Q** and the interaction between electrons gives rise to itinerant antiferromagnetism. The order can be gradually suppressed by mismatching the nesting and a QCP is obtained as T_N tends to zero. We review our results on the specific heat, the quasi-particle linewidth, the electrical resistivity, the amplitudes of de Haas-van Alphen oscillations and the dynamical spin susceptibility.

1. Introduction

Landau's Fermi liquid (FL) theory has been successful in describing the low energy properties of most normal metals. Numerous U, Ce and Yb based heavy fermion systems [1]-[3] display deviations from FL behavior, which manifest themselves as, e.g., a $\log(T)$-dependence in the specific heat over T, C/T, a singular behavior at low T of the magnetic susceptibility, χ, and a power-law dependence of the resistivity, ρ, with an exponent close to one. These deviations from FL are known as non-Fermi liquid (NFL) behavior. The breakdown of the FL can be tuned by alloying (chemical pressure), hydrostatic pressure or the magnetic field. In most cases the systems are close to the onset of antiferromagnetism (AF) and the NFL behavior is attributed to a quantum critical point (QCP) [4]-[14]. Some of the properties are quite universal and independent of the type of QCP. A model that is simple enough so that actual calculations can be performed can then provide insights even for more complex physical situations.

Recently we studied the pre-critical region of a heavy electron band with two parabolic pockets, one electron-like and the other hole-like, separated by a wave vector **Q** using (i) a field-theoretical multiplicative renormalization group (RG) approach [9] and (ii) the Wilsonian RG that eliminate the fast degrees of freedom close to an ultraviolet cutoff and rewrite the Hamiltonian in terms of renormalized slow variables [12]. The interaction is the remaining repulsion between heavy quasiparticles after the heavy particles have been formed in the sense of a Fermi liquid and is assumed to be weak. The interaction between the electrons induces itinerant AF or charge density waves (CDW) due to the nesting of the Fermi surfaces of the two pockets. For perfect nesting (electron-hole symmetry) an arbitrarily small interaction is sufficient for a ground state with long-range order. The degree of nesting is controlled by the mismatch parameter, $\delta = \frac{1}{2}|k_{F1} - k_{F2}|v_F$ [k_{F1} (k_{F2}) is the Fermi momentum of the electron (hole) pocket]. In this way the ordering temperature can be tuned to zero, leading to a QCP.

In this paper we summarize our main results. In the paramagnetic phase the effective mass, m^*, (specific heat over T, C/T) and the magnetic susceptibility increase logarithmically as T is lowered and diverge at the critical point signaling the breakdown of the FL [9, 12]. There is a crossover from the $-\ln(T)$ dependence of C/T to constant γ as T is lowered if the QCP is not perfectly tuned, in agreement with experiments on numerous systems. The quasi-particle linewidth shows a crossover from NFL ($\sim T$) to FL ($\sim T^2$) behavior with increasing nesting mismatch and decreasing temperature [15]. The electrical resistivity [16], the dynamical susceptibility [17] and the amplitudes of the de Haas-van Alphen oscillations [18] have also been studied.

The response function to superconductivity (of the FFLO type with one quasi-particle in each pocket) diverges as T_N is approached [19], but the dominating correlations are AF. NFL behavior, AF order and superconductivity in the neighborhood of a QCP have been observed in $CePd_2Si_2$ and $CeIn_3$ under pressure [20]. We have also investigated the renormalization of the electron-phonon coupling, the softening of the phonon with wave vector $\mathbf{Q}$ and the consequences of this softening on the thermal expansion [21].

2. Two-pocket model

A strong interaction between electrons gives rise to heavy fermion bands. In the spirit of the FL theory, there are weak remaining interactions between the heavy quasi-particles left after the heavy particles are formed. The heavy electron band is described by two pockets, one electron-like and the other one hole-like, separated by a wavevector $\mathbf{Q}$ [9, 12]

$$H_0 = \sum_{\mathbf{k}\sigma} \left[\epsilon_1(\mathbf{k})\, c^\dagger_{1\mathbf{k}\sigma} c_{1\mathbf{k}\sigma} + \epsilon_2(\mathbf{k})\, c^\dagger_{2\mathbf{k}\sigma} c_{2\mathbf{k}\sigma} \right] , \tag{1}$$

where $\mathbf{k}$ is measured from the center of each pocket, and assumed to be small compared to the nesting vector $\mathbf{Q}$. Here $\epsilon_1(\mathbf{k}) = v_F(k - k_{F1})$ and $\epsilon_2(\mathbf{k}) = v_F(k_{F2} - k)$, and for simplicity we assume that the Fermi velocity is the same for both pockets.

The weak remaining interactions between quasi-particles are given by [9, 12]

$$H_{12} = V \sum_{\mathbf{k}\mathbf{k}'\mathbf{q}\sigma\sigma'} c^\dagger_{1\mathbf{k}+\mathbf{q}\sigma} c_{1\mathbf{k}\sigma} c^\dagger_{2\mathbf{k}'-\mathbf{q}\sigma'} c_{2\mathbf{k}'\sigma'} + U \sum_{\mathbf{k}\mathbf{k}'\mathbf{q}\sigma\sigma'} c^\dagger_{1\mathbf{k}+\mathbf{q}\sigma} c^\dagger_{2\mathbf{k}'-\mathbf{q}\sigma'} c_{1\mathbf{k}\sigma'} c_{2\mathbf{k}'\sigma} , \tag{2}$$

where V and U represent the interaction strength for small ($|\mathbf{q}| \ll |\mathbf{Q}|$) and large (of the order of $\mathbf{Q}$) momentum transfer between the pockets, respectively. The limit of the Hubbard model is obtained by choosing $V = U$. Here interactions between electrons in the same pocket are neglected.

The leading order corrections to the vertex are the bubble diagrams with antiparallel propagator lines (zero-sound type), which are logarithmic in the external energy ω. Assuming that ω is small compared to the cutoff energy D, and that the density of states for electrons and holes is constant, ρ_F, we have

$$\tilde{V} = V/(1 - \rho_F V \xi) \quad , \quad 2\tilde{U} - \tilde{V} = (2U - V)/[1 + \rho_F(2U - V)\xi] , \tag{3}$$

where $\xi = \ln[D/(|\omega| + 2T + \delta)]$ [12]. A divergent vertex indicates strong coupling and signals an instability [9, 12].

Within the logarithmic approximation the linear response to a staggered magnetic field, $\chi_S(\mathbf{Q}, \omega)$, and to a CDW, $\chi_c(\mathbf{Q}, \omega)$, are given by [9]

$$\chi_S(\mathbf{Q}, \omega) = 2\xi\rho_F\tilde{V}/V \quad , \quad \chi_c(\mathbf{Q}, \omega) = 2\xi\rho_F(2\tilde{U} - \tilde{V})/(2U - V) , \tag{4}$$

which are closely related to the vertices in Eq. (3). Hence, if $V > 0$ a spin density wave is possible with a Néel temperature $T_N = \frac{1}{2}D\exp[-(\rho_F V)^{-1}] - \frac{1}{2}\delta$, and if $2U < V$ a CDW can be formed at $T_c = \frac{1}{2}D\exp\{-[\rho_F(V-2U)]^{-1}\} - \frac{1}{2}\delta$. The condition for a QCP is $T_N = 0$ or $T_c = 0$, and if $T_N < 0$ and $T_c < 0$ long range order has not developed. Thus, for sufficiently large Fermi surface mismatch the renormalization does not lead to an instability [12]. The QCP is an unstable fixed point and can only be reached by perfectly tuning the system [9].

In the disordered (paramagnetic) phase the γ-coefficient of the specific heat is given by the effective thermal mass [9, 12]

$$m^*(T)/m = \gamma/\gamma_0 = 1 + \frac{\xi\rho_F^2}{4}\left[3V\tilde{V} + (2U-V)(2\tilde{U}-\tilde{V})\right], \tag{5}$$

where γ_0 refers to the non-interacting system. Here we kept only the leading logarithmic contributions, and ξ is to be taken with $\omega = 0$.

The T-dependence of C/T as a function of $\ln(T)$ is shown in Fig. 1. Here $\delta_0 = 0.07$ corresponds approximately to the critical mismatch (the exact critical mismatch is a little smaller). For the perfectly tuned QCP, C/T increases logarithmically as T is lowered and diverges at the critical point signaling the break-down of the Fermi liquid [9, 12]. If $\delta > \delta_0$ there is a crossover from the logarithmic dependence (NFL) to a constant C/T (FL) as T is lowered [15]. The crossover temperature depends on the $\delta - \delta_0$ and is discussed below.

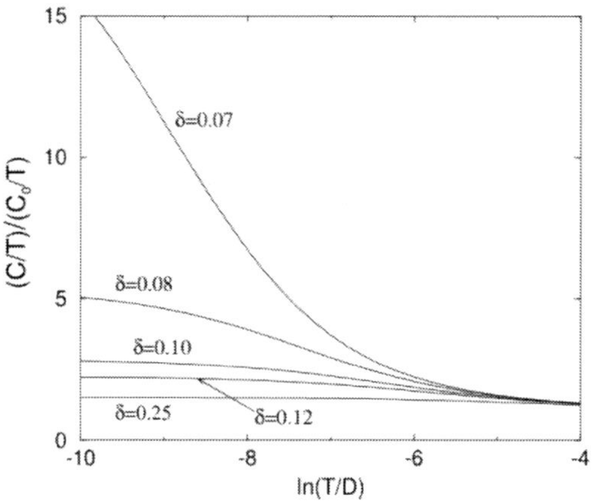

Figure 1. Enhancement of the thermal mass as a function of $\ln(T)$ for $V\rho_F = U\rho_F = 0.2$, $D = 10$, and several mismatch parameters δ. $\delta_0 \approx 0.07$ is approximately the critical mismatch. Note the crossover from NFL to FL for $\delta > \delta_0$ as T is lowered [15].

3. Quasi-particle linewidth

In an FL the damping of the quasi-particles is proportonal to T^2, while the nesting condition changes this behavior to a quasi-linear dependence in T. The linewidth Γ is calculated following a procedure outlined by Virosztek and Ruvalds [22] in the context of high-T_c superconductivity. In the disordered phase Γ is given by the imaginary part of the electron self-energy, which can be expressed as a convolution of a staggered susceptibility $\chi_S''(\omega/2T)$ with a fermion Green's function [15],

$$\Gamma_{NFL}(\omega, T) = \tfrac{1}{2}T\int dx\left[\coth(x) - \tanh\left(x - \frac{\omega}{2T}\right)\right]\chi_S''(x)\left[3|\tilde{V}|^2 + |2\tilde{U} - \tilde{V}|^2\right]\rho_F, \tag{6}$$

$$\chi_S''(\omega/2T) \approx \frac{\rho_F}{2}\sum_{s=\pm 1}\mathrm{Im}\psi\left(\frac{1}{2} + \frac{\Gamma_{NFL}}{2\pi T} + i\frac{\omega - 2s(\delta-\delta_0)}{4\pi T}\right), \tag{7}$$

where $\mathrm{Im}\psi$ is the imaginary part of the digamma function, ω is the external frequency, and δ_0 is the nesting mismatch corresponding to the QCP. The frequency in the vertices is $2T|x| + |\omega|/2$

and we use the analytic continuation of the vertex functions, i.e. $i\pi/2$ is added to ξ. The frequency of Γ_{NFL} in Imψ is $2T|x|$. The selfconsistent solution of Eqs. (6) and (7) yields the quasi-particle NFL linewidth as a function of ω and T [15].

There is also a FL contribution to the quasi-particle linewidth given by [15]

$$\Gamma_{FL}(\omega, T) = \frac{\pi}{8}[\omega^2 + (\pi T)^2][3V^2 + (2U - V^2)]\rho_F^3 \, , \tag{8}$$

which is added to F_{NFL} assuming that Matthiessen's rule is valid. The vertices in Γ_{FL} are not dressed, since this contribution does not arise from the nesting condition.

The ω and T dependence of the selfconsistent Γ_{NFL} can be understood from some limiting cases [15]. First, consider the perfectly tuned QCP, i.e. $\delta = \delta_0$, set $\omega = 0$ and neglect Γ_{NFL} in the digamma function, as well as the vertex renormalizations. The integral on the right-hand side of Eq. (6) is then independent of T and hence $\Gamma_{NFL} \propto T$, and not T^2 as for a FL. Similarly, as $T \to 0$, neglecting Γ_{NFL} in the digamma function and the vertex renormalizations, we obtain, for $\delta = \delta_0$, that the right-hand side of Eq. (6) is proportional to $|\omega|$, which again differs from the FL behavior ($\propto \omega^2$). The vertex renormalizations yield additional logarithmic corrections, so that to logarithmic order we have approximately

$$\Gamma_{NFL} \propto [3|\tilde{V}|^2 + |2\tilde{U} - \tilde{V}|^2]\rho_F^2 \max(|\omega|, T). \tag{9}$$

In the presence of an instability the vertex corrections strongly enhances Γ_{NFL}. The selfconsistent solution, as a consequence of the logarithmic corrections, yields a Γ_{NFL} that has a slightly sublinear T- and $|\omega|$-dependence [15].

Second, for $\delta \neq \delta_0$, neglecting again the selfconsistency and the vertex corrections, Γ_{NFL} is exponentially activated at low T and gradually crosses over to a linear T-dependence with increasing T. Hence, at low T the FL contribution (proportional to T^2) dominates, but at higher T there is NFL behavior (left panel of Fig. 2). Third, at $T = 0$, Γ_{NFL} vanishes identically for $|\omega| < 2(\delta - \delta_0)$ and is proportional to $|\omega| - 2(\delta - \delta_0)$ at larger frequencies (right panel of Fig. 2). Hence, again the FL contribution (proportional to ω^2) dominates at low energies [15]. As a consequence of this gap, the low energy behavior is FL like.

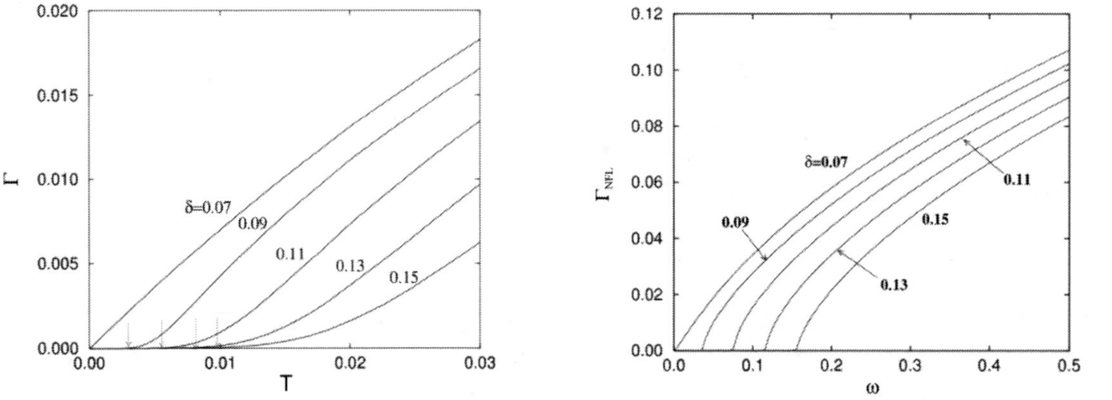

Figure 2. (left panel) Quasi-particle linewidth $\Gamma = \Gamma_{NFL} + \Gamma_{FL}$ as a function of T for $V\rho_F = U\rho_F = 0.2$, $D = 10$ and several values of δ. $\delta_0 = 0.07$ corresponds approximately to the tuned QCP. The arrows indicate the crossover from NFL to FL behavior with decreasing T. (right panel) Frequency dependence of Γ_{NFL} at $T = 0$ for the same parameters. The low energy gap is responsible of the FL behavior [15].

The selfconsistent solution of $\Gamma = \Gamma_{NFL} + \Gamma_{FL}$ for zero frequency is displayed in the left panel of Fig. 2. The crossover from FL to NFL is indicated by arrows, where the two contributions are equal. The crossover region is not a precisely defined quantity, especially at intermediate temperatures. In the specific heat, the crossover from the $-\ln(T)$ dependence of C/T to constant γ agrees with that of the linewidth. Both, $T_N = -(\delta - \delta_0)/2$ (separating the AF and paramagnetic phases) and the crossover temperature (shaded area) are shown in Fig. 3. Below the dashed curve the temperature dependence of Γ follows the FL T^2-behavior.

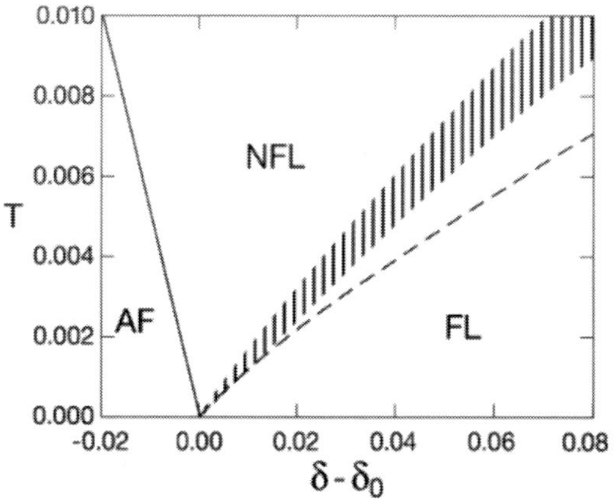

Figure 3. Phase diagram as a function of the nesting mismatch δ. The QCP corresponds to δ_0. The system orders AF for $\delta < \delta_0$ and T_N is indicated by the solid straight line separating the AF and NFL phases. The crossover region from FL to NFL behavior as obtained from the specific heat and the quasi-particle linewidth is shown as the shaded area [15]. Below the dashed curve the temperature dependence of Γ follows the FL T^2-behavior.

4. Resistivity

A proper definition of quasi-particles requires that their linewidth at low T is small compared to their energy. This is satisfied in a FL, where the width grows proportional to $(\omega/D)^2$, with D being the bandwidth. However, for the tuned QCP, the linewidth of the quasi-particles grows as fast as their energy, and the quasi-particles are not well-defined. The resistivity is then not necessarily proportional to Γ_{NFL}.

An appropriate approach to obtain the resistivity is the Kubo equation. The dynamical conductivity can be expressed in terms of a memory function $M(z)$ [23],

$$\sigma(z) = (i\omega_p^2/4\pi)/[z + M(z)] \ . \tag{10}$$

Here $\omega_p^2 = 4\pi e^2(N_1/m_1 + N_2/m_2)$ is the plasma frequency for the two-pocket model, and $N_1(N_2)$ and $m_1(m_2)$ are the particle (hole) density and mass, respectively. After a lengthy calculation [16] we obtained the imaginary part of the memory function. The full set of equations determining $M''(\omega)$ will not be presented here. In Fig. 4 we show the results for the resistivity $\rho_{NFL}(T) = 4\pi M''(0)/\omega_p^2$. Note that there is also a FL contribution to the resistivity, which strongly depends on impurity scattering [16].

The electrical resistivity is qualitatively similar to the quasi-particle linewidth and roughly proportional to $\Gamma_{NFL} + \Gamma_{FL}$ for zero frequency. $\rho(T)$ is slightly sublinear in T for the tuned QCP and for Fermi surface mismatch larger than the critical one, the resistivity displays a crossover from NFL ($\sim T$) to FL ($\sim T^2$) behavior with decreasing temperature, in agreement with experiments [16].

At low T the dynamical conductivity for the tuned QCP has strong deviations from the usual Lorentzian Drude behavior. This is the consequence of the NFL-dependence of $M''(\omega)$, which is roughly proportional to $|\omega|$, rather than to a constant (impurity scattering) as for a FL. Hence, instead of falling off as ω^{-2}, $\sigma(\omega)$ decreases roughly as ω^{-1} [16].

$CeRu_{0.48}Fe_{1.52}Ge_2$ is a system with QCP for which inelastic neutron scattering has revealed a linear T-dependence of the linewidth of the quasi-elastic peak [24]. At low T the quasi-elastic

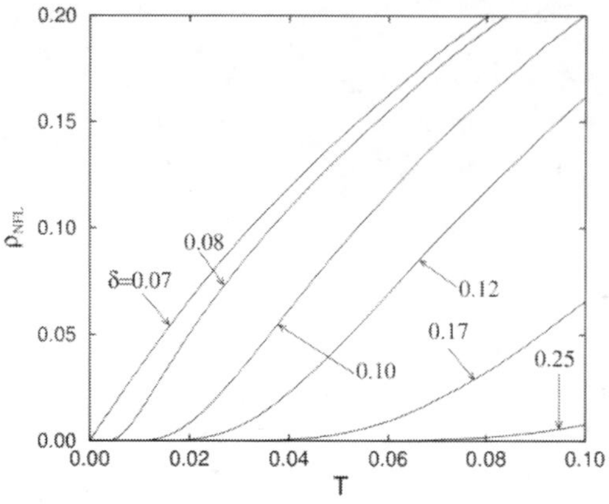

Figure 4. Selfconsistent resistivity in arbitrary units as a function of T for $V\rho_F = U\rho_F = 0.2$, $D = 10$, and several values of δ. $\delta_0 = 0.07$ corresponds approximately to the tuned QCP. ρ_{NFL} is qualitatively similar to Γ_{NFL} [16].

peak deviates from a Lorentzian (similarly to the conductivity in the present model) and when the Ru/Fe concentration is tuned away from quantum criticality, FL behavior and a Lorentzian peak are recovered.

5. Amplitudes of de Haas-van Alphen oscillations

Although the quasi-particles are not properly defined for the tuned QCP, there is a one-to-one correspondence between the excitations with those of a FL. Varying the Fermi surface mismatch parameter we can continuously interpolate between the states of the FL and the tuned QCP. Since all the states have free Fermi gas statistics, the extended Lifshitz-Kosevich equation can be applied [25].

There are two circular orbits corresponding to extremal cross-sectional areas of the Fermi surface of radii k_{F1} (electrons) and k_{F2} (holes), respectively, and hence two fundamental frequencies of oscillation. The amplitude of the oscillations modified by the quasi-particle linewidth is [18, 25]

$$A_r = 2\frac{2\pi^2 Tr}{\hbar\omega_c}\sum_{\xi_n>0}\exp\left[-\frac{2\pi r}{\hbar\omega_c}\xi_n\left(1+\int_{-D}^{D}\frac{d\omega}{\pi}\frac{\Gamma(\omega)}{\omega^2+\xi_n^2}\right)\right]. \tag{11}$$

Here $r = 1, 2, 3, \cdots$ labels the harmonics, $\omega_c = eB/m^*$, m^* is the effective heavy fermion mass before renormalization, and $\xi_n = \pi T(2n+1)$ for $n = 0, 1, 2, ...$ are the fermionic Matsubara frequencies. A_r is defined such that $A_r \to 1$ for $T \to 0$ if $\Gamma = 0$ (Lifshitz-Kosevich limit).

To simplify we assumed that both pockets have the same linewidth, $\Gamma(\omega) = \Gamma_{NFL}(\omega) + \Gamma_{FL}(\omega)$. For $\Gamma = 0$, i.e. no interactions, A_r reduces to $1/[2\sinh(2\pi^2 k_B Tr/\hbar\omega_c)]$. For interacting electrons A_r is always reduced with respect to the noninteracting system and the overall reduction of the amplitudes is largest close to the QCP [18]. Fig. 5 shows the logarithm of the oscillation amplitude for the first five harmonics as a function of T for the tuned QCP, $\delta_0 = 0.07$. If $\ln(A_r)$ vs. T is a straight line (except at very low T) the effective mass approximation (Lifshitz-Kosevich) is valid. Except for some curvature, this is approximately the case at higher T, but at low T there are deviations from the effective mass approximation. A very low Dingle temperature is necessary to observe the fundamental frequency.

The dHvA-oscillations are periodic as a function of B^{-1}, and are measured over a magnetic field interval. Hence, the amplitude of oscillation cannot be associated with a given B. The magnetic field frequently also acts as a tuning parameter for the QCP. The present discussion of the amplitudes is only meaningful if B, within the regime of measurement, does not affect the tuning of the QCP [18].

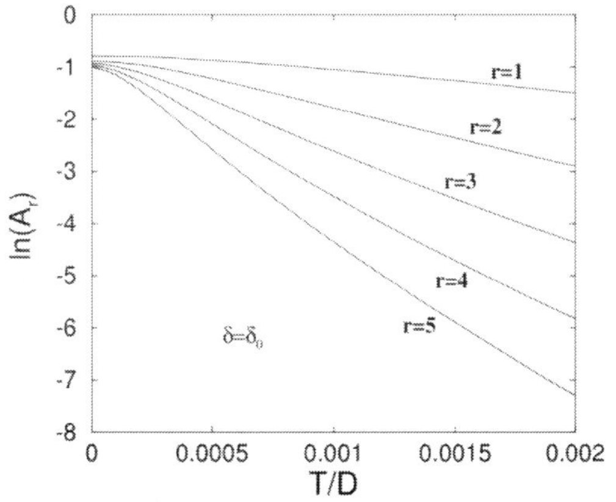

Figure 5. dHvA amplitudes for the first five harmonics r as a function of T for fixed B for the tuned QCP, $\delta_0 = 0.07$, and the same parameters as before. The Lifshitz-Kosevich equation with an effective mass would correspond to a straight line. For $D = 1000$ K the magnetic field is 40 T for $m^*/m = 20$, m being the free electron mass [18].

6. Dynamical spin susceptibility

The dynamical susceptibility is a function of the energy and momentum transfers, ω and $\mathbf{Q}+\mathbf{q}$. We first evaluate the response function for $\mathbf{q} = 0$ in the absence of interactions to get qualitative insight, and incorporate then the interactions by summing the ladder diagrams (leading order logarithms) and inserting the quasi-particle line-width (selfenergy).

For the non-interacting system the imaginary part of the dynamical susceptibility is given by [17]

$$\chi_0''(\mathbf{Q},\omega) = \frac{\pi\rho_F}{4}\left[\tanh\left(\frac{\omega+2\delta}{4\pi T}\right) + \tanh\left(\frac{\omega-2\delta}{4\pi T}\right)\right], \tag{12}$$

where for simplicity we used parabolic dispersions with $m_1 = m_2 = m$ and $v_{F1} = v_{F2} = v_F$. The nesting condition is $\epsilon_2(\mathbf{k}) = 2\delta - \epsilon_1(\mathbf{k})$ and $\rho_F = k_{F1}m/(2\pi^2)$ is the density of states. Eq. (7) is antisymmetric in ω, and at low T gapped for $-2\delta \le \omega \le 2\delta$. The gap is the consequence of the momentum and energy conservation, which cannot be satisfied simultaneously unless $|\omega| > 2\delta$. With increasing T the gap gradually closes.

We now incorporate the quasi-particle linewidth into the calculation. Eq. (7) is the bubble diagram with antiparallel propagator lines. These propagators are now broadened into Lorentzians. The momentum integrations can be carried out, yielding another Lorentzian with line-width given by the sum of the widths of the two original Lorentzians, i.e. $\Gamma_{LW} = \Gamma_{NFL}(\omega' + \omega/2, T) + \Gamma_{NFL}(\omega' - \omega/2, T)$. Here ω' is the energy integration variable for the bubble [22].

At low T the ω'-integration is limited to the interval $(-\omega/2, \omega/2)$. Hence, Γ_{LW} predominantly depends on ω and not on the integration variable ω' [15, 22]. The ω'-integration is then straightforward and we obtain approximately [17]

$$\chi_0''(\mathbf{Q},\omega) = \frac{\rho_F}{2}\mathrm{Im}\psi\left(\frac{1}{2} + \frac{\Gamma_{LW}}{4\pi T} + i\frac{\omega-2\delta}{4\pi T}\right) + \frac{\rho_F}{2}\mathrm{Im}\psi\left(\frac{1}{2} + \frac{\Gamma_{LW}}{4\pi T} + i\frac{\omega+2\delta}{4\pi T}\right) \tag{13}$$

with Im denoting imaginary part and $\Gamma_{LW} \sim 2\Gamma_{NFL}$. Note that expression (13) is different from $\chi_S''(\omega/2T)$ in Eq. (7).

We now choose opposite spins for the propagators, so that the correlation function involves a spin-flip (transversal susceptibility). The most important terms contributing to χ are the ladder diagrams in V, i.e. the random phase approximation (RPA) diagrams [17],

$$\chi(\mathbf{Q},\omega) = \chi_0(\mathbf{Q},\omega)/[1 - V\chi_0(\mathbf{Q},\omega)] \ . \tag{14}$$

Note that to leading order the U-interaction does not contribute to the transversal susceptibility (in RPA). The dissipative part of the transversal susceptibility is given by [17]

$$\chi''(\mathbf{Q},\omega) = \frac{\chi_0''(\mathbf{Q},\omega)}{[1 - V\chi_0'(\mathbf{Q},\omega)]^2 + V^2\chi_0''(\mathbf{Q},\omega)^2} . \tag{15}$$

The imaginary part for $\chi_0(\mathbf{Q},\omega)$ to be inserted in Eq. (15) is Eq. (13) and the expression for $\chi_0'(\mathbf{Q},\omega)$ is

$$\chi_0'(\mathbf{Q},\omega) = \ln(D/2\pi T) + c + \frac{\rho_F}{2}\sum_p \mathrm{Re}\ \psi(Z_p), \tag{16}$$

$$Z_p = \frac{1}{2} + \frac{\Gamma_{NFL} + \Gamma_{FL}}{2\pi T} + i\frac{\omega + 2p\delta}{4\pi T}, \tag{17}$$

where c is a constant, $p = \pm 1$ and Re denotes real part. Here Γ_{NFL} and Γ_{FL} are functions of ω and T. The additive constant c arises from the cut-off in the Cauchy transformation of χ_0'' and is in principle arbitrary. The constant is determined by the quantum criticality condition, i.e. $1 = V\chi_0'(\mathbf{Q},\omega = 0)$ for $T = 0$, and $\delta = \delta_0$. For the parameters used in this paper, we have $c = 0.038152$.

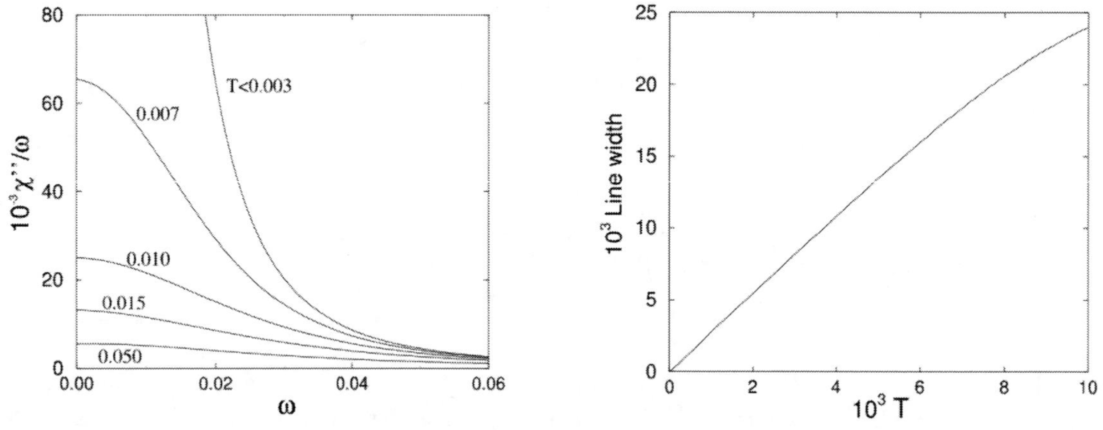

Figure 6. (left panel) Quasi-elastic peak in the imaginary part of the dynamical spin susceptibility divided by ω as a function of frequency for $\delta_0 = 0.07$ (critical mismatch) and the same parameters as before. (right panel) Linewidth of the central quasielastic peak as a function of T for critical mismatch [17].

Two features should be pointed out in $\chi''(\mathbf{Q},\omega)/\omega$ for the critical Fermi surface mismatch δ_0: (i) The quasi-elastic peak around $\omega = 0$ (Fig. 6) and (ii) the shoulder at $\omega = \pm 2\delta_0$ at low T (Fig. 7). The shoulder is reminiscent of the noninteracting case, Eq. (12), and is gradually smeared with increasing temperature. The central peak, on the other hand, arises from the interaction through the denominator in Eq. (14) and is displayed in the left panel of Fig. 6 as a function of ω for several temperatures. As $\omega \to 0$ the imaginary component of the susceptibility tends to zero and also the real part, $1 - V\chi'$, becomes small, thus giving rise to the peak. The maximum, defined as $\lim_{\omega\to 0}\chi''(\mathbf{Q},\omega)/\omega$, is proportional to T^{-3} at low T. The form of the resonance is approximately Lorentzian with the half-width of the peak at half of its height being proportional to T (right panel of Fig. 6).

For non-critical nesting mismatch the behavior of $\chi''(\mathbf{Q},\omega)/\omega$ is different, since the denominator in Eq. (14) does no longer vanish at $\omega = 0$. Hence, only for the critical mismatch δ_0 the peak is quasi-elastic, while for other δ the peak is inelastic. As seen in Fig. 7 the height of the peak is strongly reduced with increasing noncritical mismatch.

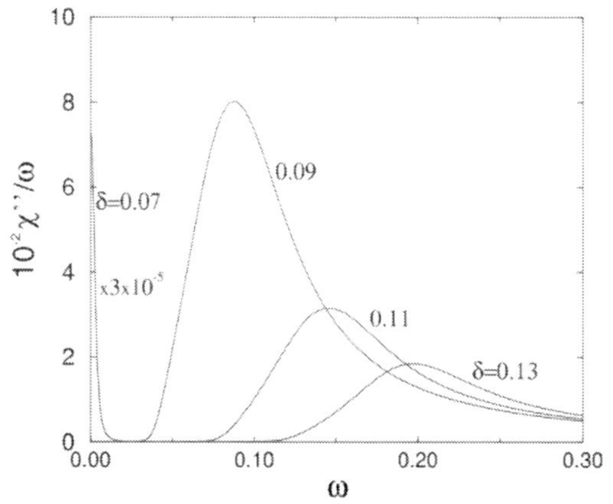

Figure 7. Imaginary part of the dynamical spin susceptibility divided by ω as a function of frequency for $T = 0.001$ and several values of δ. The parameters are the same as before. Note the strongly reduced scale for the central peak for $\delta = 0.07$ (critical mismatch) [17].

7. Conclusions

The nesting of a heavy electron Fermi surface can give rise to itinerant AF long-range order. The degree of nesting is controlled by a mismatch parameter. This way the ordering temperature can be tuned to zero, leading to a QCP. The QCP is an unstable fixed point and can only be reached by perfectly tuning the system. Otherwise, the RG flow will deviate to a phase with long-range order or the compound remains a heavy electron paramagnet. There is no characteristic energy scale associated with the QCP and a small perturbation to the system may give rise to a new physical situation at low energies. This fact is probably responsible for the *lack of universality* in NFL compounds close to the QCP.

We investigated the quasi-particle lifetime as the QCP is approached. Landau's FL theory predicts that for normal metals the linewidth is proportional to ω^2 and T^2. The QCP modifies this behavior to a linear $|\omega|$ and T dependence with logarithmic corrections. As a function of T and the Fermi surface mismatch there is a crossover from FL to NFL behavior. This crossover is also reflected in the specific heat over temperature and the electrical resistivity, leading to a phase diagram that is in qualitative agreement with experiments. For the tuned QCP the resistivity is approximately linear in T.

The dynamical spin susceptibility was selfconsistently calculated using the linewidth of the quasi-particles and the random phase approximation. For the tuned QCP a Lorentzian quasi-elastic peak with a width linear in T is obtained. For noncritical mismatch of the Fermi surface the peak becomes inelastic, i.e. it is shifted to $\omega \neq 0$. The dynamical susceptibility is not in agreement with the experimental observations (inelastic neutron scattering). The present model can also not account for the restructuring of the Fermi surface at δ_0 for $T \to 0$ [26]. A two-band model with one dispersionless band (Anderson lattice) is necessary to capture these features.

We calculated the amplitudes of the dHvA oscillations using a modified Lifshitz-Kosevich expression. As expected the amplitudes are strongly reduced and the oscillations are difficult to observe.

The results are valid in the disordered phase for weak and intermediate coupling. However, since the renormalization group does not allow a return to a weak-coupling fixed point once the system is strongly coupled, the present approach qualitatively describes the entire precritical regime.

Acknowledgments

The support by the U.S. Department of Energy under grant No. DE-FG02-98ER45707 is acknowledged.

References

[1] Stewart G R 2001 *Rev. Mod. Phys.* **73** 797
[2] von Löhneysen H 1994 *Physica* B **206 & 207** 101
[3] Maple M B *et al* 1994 *J. Low Temp. Phys.* **95** 225
[4] Andraka B and Tsvelik A M 1991 *Phys. Rev. Lett.* **67** 2886
[5] Tsvelik A M and Reizer M 1993 *Phys. Rev.* B **48** 4887
[6] Millis A J 1993 *Phys. Rev.* B **48** 7183
[7] Continentino M A 1993 *Phys. Rev.* B **47** 11587
[8] Moriya T and Takimoto T 1995 *J. Phys. Soc. Jpn.* **64** 960
[9] Schlottmann P 1999 *Phys. Rev.* B **59** 12379
[10] Si Q, Rabello S, Ingersent K, and Smith J L 2001 *Nature* (London) **413** 804
[11] Coleman P and Pèpin C 2002 *Physica* B **312-313** 383
[12] Schlottmann P 2003 *Phys. Rev.* B **68** 125105
[13] Custers J *et al* 2003 *Nature* **424** 524
[14] Gegenwart P, Si Q and Steglich F 2008 *Nature Physics* **4** 186
[15] Schlottmann P 2006 *Phys. Rev.* B **73** 085110
[16] Schlottmann P 2006 *Phys. Rev.* B **74** 235103
[17] Schlottmann P 2007 *Phys. Rev.* B **75** 205108
[18] Schlottmann P 2008 *Phys. Rev.* B **77** 195111
[19] Schlottmann P 2000 *J. Appl. Phys.* **87** 5140
[20] Mathur N D *et al* 1998 *Nature* (London) **394** 39
Walker I R *et al* 1997 *Physica* C **282** 303
[21] Schlottmann P 2004 *J. Appl. Phys.* **95** 7216
[22] Virosztek A and Ruvalds J 1990 *Phys. Rev.* B **42** 4064
Virosztek A and Ruvalds J 1999 *Phys. Rev.* B **59** 1324
[23] Götze W and Wölfle P 1972 *Phys. Rev.* B **6** 1226
[24] Montfrooij W *et al* 2003 *Phys. Rev. Lett.* **91** 087202
[25] Wasserman A and Springford M 1996 *Adv. in Phys.* **45** 471
[26] Paschen S *et al* 2004 *Nature* **432** 881

Magnetism on a Mesoscopic Scale: Molecular Nanomagnets Bridging Quantum and Classical Physics

Nikolaos P Konstantinidis, Alexander Sundt, Joscha Nehrkorn, Anna Machens and Oliver Waldmann

Physikalisches Institut, Universität Freiburg, 79104 Freiburg, Germany

E-mail: oliver.waldmann@physik.uni-freiburg.de

Abstract.
In recent years polynuclear transition metal molecules have been synthesized and proposed for example as magnetic storage units or qubits in quantum computers. They are known as molecular nanomagnets and belong in the class of mesoscopic systems, which are large enough to display many-body effects but small enough to be away from the finite-size scaling regime. It is a challenge for physicists to understand their magnetic properties, and for synthetic chemists to efficiently tailor them by assembling fundamental units. They are complementary to artificially engineered spin systems for surface deposition, as they support a wider variety of complex states in their low energy spectrum. Here a few characteristic examples of molecular nanomagnets showcasing unusual many-body effects are presented. Antiferromagnetic wheels and chains can be described in classical terms for small sizes and large spins to a great extent, even though their wavefunctions do not significantly overlap with semiclassical configurations. Hence, surprisingly, for them the transition from the classical to the quantum regime is blurred. A specific example is the Fe_{18} wheel, which displays quantum phase interference by allowing Néel vector tunneling in a magnetic field. Finally, the Co_5Cl single-molecule magnet is shown to have an unusual anisotropic response to a magnetic field.

1. Introduction

During the last fifteen years or so high-nuclearity molecules with magnetic exchange couplings between their transition metal ions have attracted enormous interest due to their novel magnetic properties. They are known as molecular nanomagnets (MNMs) and apart from their structural beauty, they present a challenge for physicists to interpret and understand their quantum magnetism. They have a relatively small size ranging from 2 to 80 metal centers, and belong in the class of mesoscopic quantum systems, which are large enough to show many-body effects, but small enough to be away from the finite-size scaling regime. With their small size they provide an ideal testing ground for studying basic questions of quantum mechanics in mesoscopic systems, such as which metal centers and topologies might be more efficient towards a desired magnetic property or the transition from the quantum to the classical regime when the ion spins are large. From the synthesis point of view they are a challenge for chemists to assemble molecules with a large number of magnetic subunits connected by bridging groups (ligands), which mediate the magnetic interactions between the metal centers. It is therefore necessary for the two disciplines

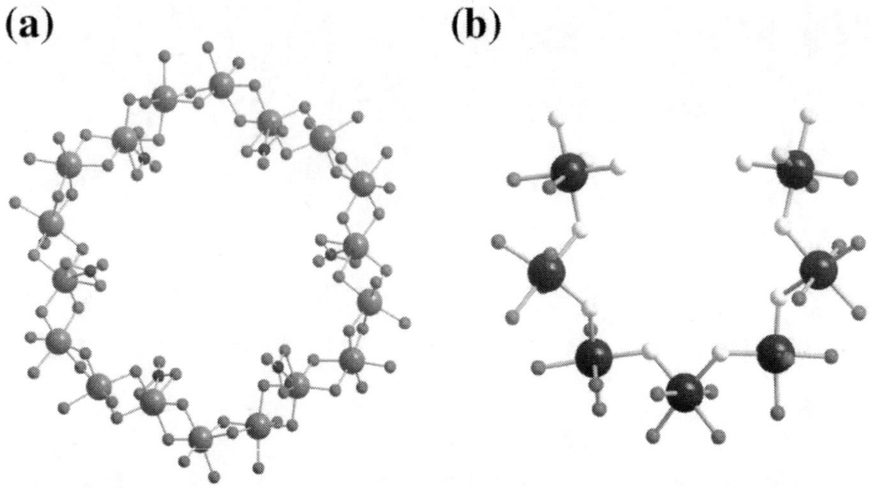

Figure 1. Molecular cores of **(a)** the AFM wheel Fe_{18}, and **(b)** the AFM chain Cr_7. Blue: nitrogen, red: oxygen, yellow: flour, green: iron, dark blue: chromium. Carbons and hydrogens are omitted for clarity.

to be in close collaboration to achieve MNMs with the desired properties. Importantly, intermolecular interactions for MNMs are very weak, consequently experiments probe directly the properties of individual molecules.

A very prominent representative of the MNM class are the so-called single-molecule magnets (SMMs) [1]. Their singular properties are slow relaxation and quantum tunneling of the magnetization (QTM), which make SMMs promising candidates for technical applications involving magnetic data storage [2, 3] and quantum computing on a molecular scale [4]. These appealing properties were observed only below a blocking temperature T_B less than ~ 4.5 K, which is indeed too low for applications. Thus the search for molecules showing SMM behavior at higher temperature is the subject of ongoing research. It is widely accepted that a high-spin ground state and large magnetic anisotropy are necessary for a SMM [5], with the latter being of higher importance to enhance in order to improve SMM behavior [6, 7].

Besides SMMs, the large class of MNMs are of special interest for basic sciences, as several striking magnetic quantum phenomena occur due to their mesoscopic size. For instance, in the Keplerate Fe_{30} strong magnetic frustration effects were observed [8, 9, 10, 11], in antiferromagnetic (AFM) molecular wheels rotations of the Néel vector and quantized AFM spin waves could be studied [12, 13, 14, 15, 16], and heteronuclear 'doped' AFM wheels have been in the focus for a number of phenomena such as coherent magnetic oscillations or thermal entanglement [17, 18, 19]. Two examples of molecules are shown in figure 1. Apart from the practical point of view, the understanding of the magnetism resulting from magnetic interactions between a dozen or so metal ions has often proved to be a challenge, as many-body quantum effects may play a crucial role in determining the properties [20]. Therefore, the MNMs can often support quite complex quantum states, and may show a drastically different behavior compared to extended systems (the finite-sized MNMs are practically 'zero-dimensional' systems).

The field of 'molecular nanomagnetism' appears hence as a very promising research area. This review focuses on three different topics that showcase the novel many-body effects and associated complex quantum many-body states that can be observed in MNMs despite their small size. Firstly, even AFM wheels and even and odd AFM chains are considered in a more general fashion. It has been found that for even wheels the physics can be understood by the concepts of Néel vector and spin waves that were characteristically put forward for infinite systems [12], and may be considered as classical in nature. These concepts can be carried over to the case of short AFM chains. However, even though the low-energy spectrum and operator expectation values can accurately be described by a semiclassical two-sublattice Hamiltonian, its overlap with the exact wavefunction is not approaching unity as would be expected. This motivated us to study the overlap of the exact wavefunctions with valence-bond states, which may be considered as quantum mechanical in nature. Surprisingly, both sets of wavefunctions, the 'classical' sublattice states and the 'quantum mechanical' valence-bond states work equally well in the small N and large S_i regime (where N is the number of ions in the wheel or chain, and S_i their spin lengths), revealing that for even AFM wheels and short AFM chain the line between classical and quantum physics is blurred.

Secondly, following the general discussion about even AFM wheels a particularly interesting example, the Fe_{18} wheel is considered, where quantum phase interference effects related to the dynamics of the Néel vector were observed [21, 22, 23]. Quantum interference was first found for the SMM Fe_8 [24], where it leads to zero tunneling between states in the same symmetry sector. In Fe_{18} it is related to quantum tunneling of the Néel vector, and this generalizes the standard QTM typically observed in a SMM. This appealing effect and the advantages and disadvantages of its descriptions in a semiclassical, a quantum mechanical and a two-sublattice model will be discussed.

The search for molecules with enhanced SMM behavior was recently extended to molecular clusters incorporating metal ions with unquenched orbital angular momentum such as Co^{2+}, as they can have large magnetic anisotropy [25]. Even though the obtained molecules show none or only weak SMM behavior, the description of the magnetism in them turned out to be an interesting challenge, leading to new insights. As the third example the molecule Co_5Cl will be discussed, which behaves as a SMM at temperatures less than 1.8 K. In apparent contradiction to the measured susceptibility, which would usually be interpreted as evidencing dominant AFM couplings, a ferromagnetic anisotropic interaction was found in the cluster. The subtle interplay of highly anisotropic exchange coupling tensors (**J** tensors) and gyromagnetic matrices (**g** matrices) is the reason for this apparent inconsistency, as explicated by negative gyromagnetic factors g that appear in the effective description of Co_5Cl.

In recent years not only the chemical route towards small quantum spin clusters has been advanced, but several beautiful examples of artificially engineered spin clusters also emerged [26, 27, 28]. Here, clusters of magnetic metal ions with desired topologies were fabricated directly on surfaces, and their magnetic properties were studied by means of scanning tunneling microscope (STM) techniques. A particularly relevant example is that of short chains of Mn ions with AFM next-neighbor Heisenberg interactions [26], which from the magnetic point of view are described by the same principles as the short AFM chains realized in some MNMs, such as the Cr_7 molecule shown in figure 1(b). Apparently in both fields, that of the MNMs and the artificially engineered spin clusters, one faces very similar scientific questions, yet the technological challenges in putting them forward in real-world applications are quite complementary. It is a second purpose of this review to point out these similarities as well as the complementariness, in a hope to inspire a cross fertilization of these two hitherto disjunct appearing research fields.

The paper is divided as follows: Section 2 discusses the cases of even AFM wheels and even and odd AFM chains and their semiclassical behavior. Section 3 deals with Néel vector tunneling originating in quantum interference in Fe_{18}. In section 4 the case of the SMM Co_5Cl is explained. Finally, section 5 presents the conclusions and a discussion of the connections between MNMs and artificially engineered clusters.

2. Quantum vs. Classical Description in the Many-Body Physics of Antiferromagnetic Wheels and Chains

The relatively small size of MNMs often enables their theoretical treatment with accuracy not easily attainable for extended systems. As a result, a classification of their properties relative to N, S_i and the type of interactions between the spins appears more promising compared to extended systems. However many-body effects, which can also appear for small clusters, merit closer investigation and often perplex such a classification. They also blur the limit between the classical and quantum behavior, as will be demonstrated here. The obvious distinction that exists for the two regimes in extended systems disappears for MNMs, which are zero dimensional and exhibit strong finite-size effects. Hence the states allowed by these small structures can be quite complex.

In AFM wheels the metal ions playing the role of magnetic centers form a close to perfect ring-like arrangement. Nearest-neighbor interactions are not frustrated if all $J_{ij} < 0$ and N is even [15, 29]. The Cr^{3+} and Fe^{3+} ions, which have relatively large spin $S_i = 3/2$ and $5/2$ respectively, are particulary useful. The decanuclear wheel $[Fe_{10}(OCH_3)_{20}(O_2CCH_2Cl)_{10}]$, most commonly known as Fe_{10}, is a prototype molecule for this class [30, 31], and there also exist hexanuclear, octanuclear, decanuclear and 18-membered ferric wheels [30, 32, 33, 34, 35, 36, 37, 38, 39, 40, 41, 42, 43, 44, 45]. Their magnetism is generally well described by the Hamiltonian

$$\hat{H} = -\sum_{i<j} J_{ij}\hat{\mathbf{S}}_i \cdot \hat{\mathbf{S}}_j + \sum_{i=1}^{N} D_i \hat{S}_{i,z}^2 + \mu_B \sum_{i=1}^{N} g_i \hat{\mathbf{S}}_i \cdot \mathbf{B}, \tag{1}$$

where three different effects are in competition, the isotropic Heisenberg exchange interactions between the magnetic centers with constants J_{ij}, the zero-field splitting (ZFS) term introducing magnetic anisotropy, and a Zeeman term that couples to an external magnetic field. In the case of the AFM wheels, an effective model describing the low-energy physics can be formed by assuming that the effect of quantum fluctuations is not so strong. Hence, grouping spins which are parallel in the classical limit into a single spin will capture the basic physics at least at the semiclassical level [29]. More generally, this situation applies for molecules with magnetic centers that are bipartite, and can be divided in two groups of spins ('sublattices') that classically point in opposite directions and all of a spin's nearest neighbors belong to the other sublattice. This results in two 'collective' spins S_A and S_B (with $S_A = S_B = NS_i/2$ for wheels) that effectively describe the low-energy physics as an interacting dimer via the Hamiltonian

$$\hat{H}_{AB} = -a_1 J \hat{\mathbf{S}}_A \cdot \hat{\mathbf{S}}_B + b_1 D(\hat{S}_{A,z}^2 + \hat{S}_{B,z}^2) + g\mu_B \hat{\mathbf{S}} \cdot \mathbf{B}, \tag{2}$$

where $a_1 = a_1^{sc}$ and $b_1 = b_1^{sc}$ in the semiclassical limit (see section 3 for wheels), but they can be modified as quantum fluctuations get stronger to provide a more accurate description [29]. Results generated with equation (2) compare very well with the predictions of equation (1) for even AFM wheels, and the semiclassical dimer Hamiltonian $\hat{H}_{AB}$ has proven to be a valuable tool for the calculation of the low-energy behavior for a large number of MNMs [32, 39, 46], see also the next section. It is noted here that this approach can be generalized for more than two sublattices, as is the case for the Fe_{30} Keplerate molecule [8, 9, 10, 11, 47, 48, 49].

The agreement of Hamiltonians (1) and (2) calls for a more systematic comparison of their properties. The ZFS term is typically weaker than the exchange couplings and may be neglected in a first approximation. In the special case where all $D_i = 0$ and uniform AFM interactions $J_{ij} \equiv J < 0$, Hamiltonian (1) simplifies to the AFM Heisenberg Model (AHM)

$$\hat{H} = -J \sum_{i<j} \hat{\mathbf{S}}_i \cdot \hat{\mathbf{S}}_j, \tag{3}$$

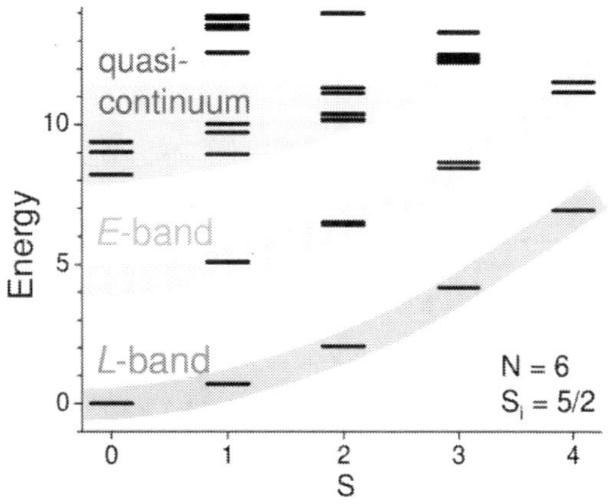

Figure 2. Energy spectrum of the $N = 6$, $S_i = 5/2$ AFM wheel. It is divided in the L band, the E band and the quasi-continuum.

written here in the absence of a magnetic field. The AHM, which is isotropic in spin space, has been the object of intense investigation over the years on different types of lattices as a prototype of strongly correlated electronic behavior in extended systems. In particular, it is closely associated with the investigation for the existence or not of long-range order in two dimensional lattices at zero temperature [50]. In the case of the square lattice, the AHM has been used to describe the behavior of the superconducting carriers above the critical temperature [51]. Existing evidence strongly points to a ground state where quantum fluctuations weakly renormalize the classical Néel state. The AHM has also been of great interest for frustrated lattices, where nearest-neighbor spins are not aligned in an antiparallel fashion at the classical level. Standard examples are the triangular and the Kagomé lattice [52]. These have been intensively studied with the hope of realizing the elusive spin-liquid ground state proposed by Anderson [53], which has pure quantum mechanical character and possesses no long-range order.

AFM wheels have been extensively studied within the framework of the AHM. Energy levels are characterized by S and M, the expectation values respectively of $\hat{\mathbf{S}}^2$ and $\hat{S}_z$, the projection of the total spin along the z axis. AFM wheels and chains provide a direct comparison with the infinite one-dimensional chain, their extended systems counterpart, by simply taking the limit $N \to \infty$. The infinite chain has been theoretically conjectured by Haldane to possess different physics depending on S_i being integer or half-integer [54, 55]. In the former case excitations are gapful, while in the latter gapless. In the infinite chain the distinction between the classical and quantum regimes is clear. Classical nearest neighbors are pointing in opposite directions, forming a state with perfect Néel long-range order in the staggered magnetization. In the full quantum limit $S_i = 1/2$ the correlations do not have long-range order, but rather decay exponentially as the ground state is in a critical phase (the ground-state energy can exactly be calculated with the Bethe ansatz [56]). This is in accordance with the Mermin-Wagner theorem [50], which states that quantum fluctuations are so strong in one dimension that they destroy any order for any temperature T.

In the case of AFM wheels, their finite size perplexes the problem. As their size is of the order of tenths of magnetic centers they are nowhere near the thermodynamic limit, but this size can be long enough to generate non-trivial many-body effects. The AFM interactions in equation (3) minimize the total spin S in the ground state, which is a singlet $S = 0$ for even AFM wheels. The lowest energies in the different S sectors follow Landé's interval rule $E(S) \propto S(S + 1)$ [30]. In fact, several parallel rotational bands were identified [14, 15, 57], see figure 2. The lowest

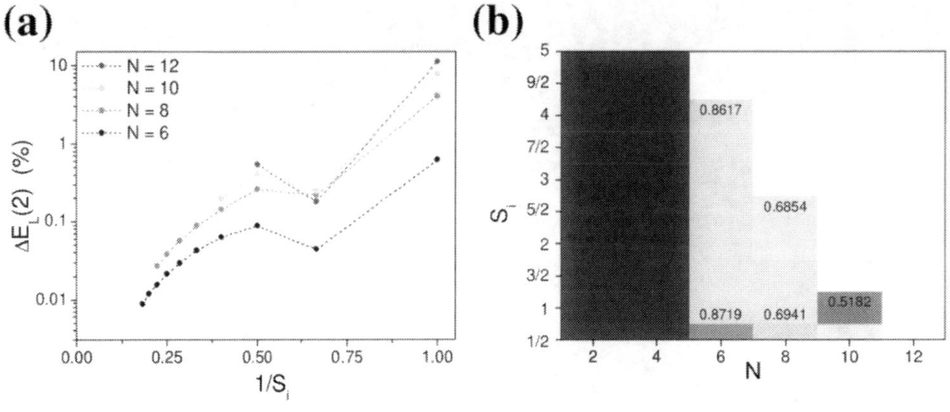

Figure 3. **(a)** Dependence of the quantity $\Delta E_L(2)$ discussed in the text as function of the inverse spin length for AFM wheels with $N = 6$, 8, 10, and 12. Lines are guides to the eye. **(b)** Overlap of the ground state wavefunction of the semiclassical model $\hat{H}_{AB}$ (4) with the corresponding wavefunction of the AHM (3). The overlap increases with a tendency to approach unity as N decreases and S_i increases. Nevertheless, except where equation (3) is exactly solvable, a significant deviation from unity remains even for S_i approaching the classical limit.

band is called L, while the next higher bands collectively form the E band, with the remaining states collected together in the so-called quasi-continuum, of little relevance for temperatures T less than $\sim |J|$ [14]. The two lowest bands have a simple physical interpretation. For an infinite antiferromagnet, arbitrary rotations in spin space are allowed but are extremely unlikely to be realized experimentally, as the size of the system implies a very large experimental time for such an observation. However for finite wheels these classical-type states are physically accessible and correspond to the L band. The E band is associated with the discretized versions of the familiar spin-wave excitations where a single spin is flipped from the ground state and allowed to propagate in the whole cluster [58]. The E band includes all the states accessible with significant probability from the L band by single-spin excitations, as observed for instance in inelastic neutron scattering experiments. Interestingly enough and unlike the infinite chain, the lowest energy excitations that determine the low-temperature magnetization such as the magnetization steps in an external field [15, 30, 38], relate to the L band and rotations of the Néel vector rather than the magnon type excitations.

The classical nature of the low-energy spectrum allows one to describe the low-energy physics of the L band with equation (2), written in the absence of magnetic anisotropy and external field,

$$\hat{H}_{AB} = -a_1 J \hat{\mathbf{S}}_A \cdot \hat{\mathbf{S}}_B. \tag{4}$$

This is in contrast with the extended AFM chain where quantum fluctuations are so strong that drive the system out of order. $\hat{H}_{AB}$ can be diagonalized and by construction its low-energy spectrum consists of an L and an E band and agrees with Hamiltonian (3) in first order in perturbation theory. It turns out that $\hat{H}_{AB}$ is a very good approximation for the AHM for small N and large S_i. This can be seen in figure 3(a), where the energy difference $\Delta E_L(S) = \frac{1}{2}\Delta_0 S(S+1) - E_{S,0}$ between the energy extrapolated from the two lowest states with $S \leq 1$ using the Landé rule (which determines Δ_0) and the lowest energy state with S is plotted for $S = 2$ [29]. This difference goes to zero as S_i increases towards its classical limit. The signature of the Haldane conjecture for larger spins with the oscillations as S_i alternates between integer and half-integer can also be seen [54, 55], however these oscillations also disappear as S_i

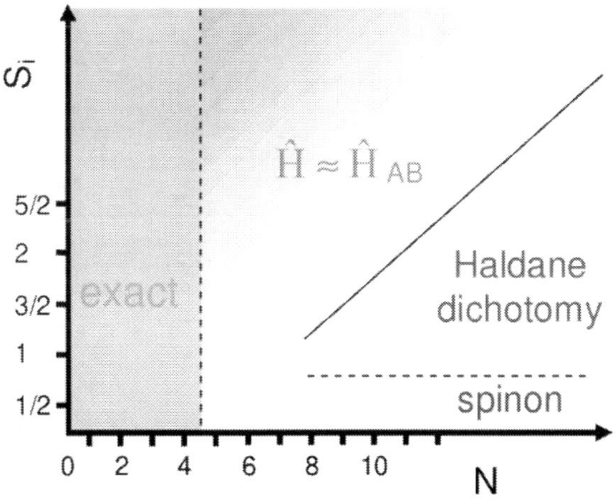

Figure 4. Properties of the AFM Heisenberg wheels as a function of their size N and the spin length S_i. The area in which the results of this work apply is blue shaded and marked by $\hat{H} \approx \hat{H}_{AB}$ and exact.

approaches the classical limit.

The behavior of even AFM wheels as function of N and S_i is summarized in figure 4 [29]. For large N, the system goes to the thermodynamic limit and the Haldane conjecture appears. For $S_i = 1/2$ the elementary excitations are spinons. As N is getting smaller the system behaves more and more classically, and for small N and large S_i $\hat{H}_{AB}$ provides an effective semiclassical description of the physics (for $N \leq 4$ Hamiltonian (3) can exactly be solved with the spin-coupling method [59]). This is the region where the correlation length of the Heisenberg wheels is much larger than the size of the system.

It appears quite natural that the semiclassical Hamiltonian (4) becomes more successful as the spins acquire a more classical character with increasing S_i. The reduction of N brings about a characterization of the states in terms of S, rather than the momentum k which is appropriate for large or infinite systems. In this regime the energies and the expectation values of operators are very well reproduced by equation (4) and the wavefunctions generated by it, however, the wavefunctions themselves, maybe surprisingly, do not approach the exact wavefunctions of equation (3), i.e. the overlap with the exact wavefunctions does not go to unity (figure 3(b)). It rather approaches a limiting value with decreasing N and increasing S_i, which is not very close to unity. Consequently the apparent semiclassical behavior of the AHM model (3) is not carried to a full extent, and there are still features of the system that call for a treatment where quantum fluctuations are more systematically taken into account.

Further insight into why the effective Hamiltonian (4) works to the extent it does comes from short AFM chains. In contrast to wheels, MNMs in the form of chains have open boundary conditions with free spins at their edges, and have been considerably less studied. Open boundary conditions combined with finite-size effects complicate the physics. The Cr_6 and Cr_7 horseshoes are AFM chains with an even and odd number of $S_i = 3/2$ magnetic centers [60]. The Cr_7 molecule is displayed in figure 1(b). Apart from their expected different ground-state spin of $S = 0$ and $S = 3/2$ respectively [61], the excited energy spectra are also intrinsically different. This difference is a more general feature of even and odd AFM chains [62]. In particular, for odd chains the Landé rule for the lowest states in each S sector is very weakly renormalized. In contrast, it is strongly renormalized for the even chains. It is the lowest states in the AFM region, where S is minimal, that deviate from the rule. This is due to a low-lying first excited state with

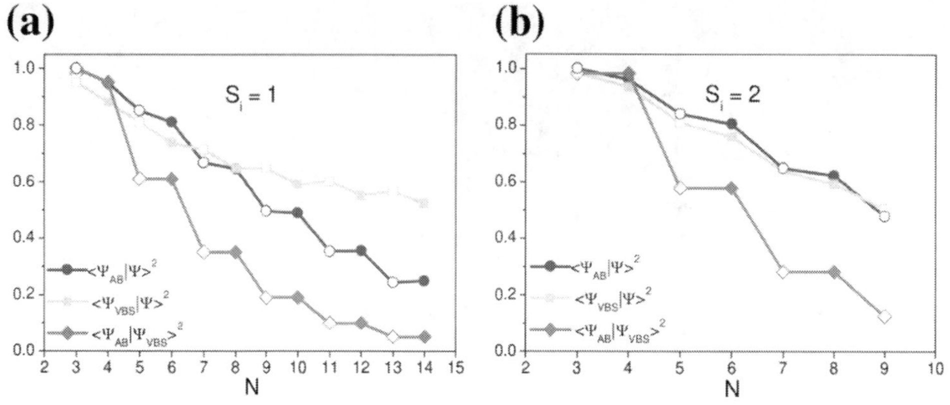

Figure 5. Overlap of approximating wavefunctions for the ground state of the AFM chain as functions of N for **(a)** $S_i = 1$ and **(b)** $S_i = 2$. Plotted are the overlap of the ground state of the AHM with the ground state of $\hat{H}_{AB}$, $|\langle\Psi_{AB}|\Psi\rangle|^2$ (blue), with the VBS state, $|\langle\Psi_{VBS}|\Psi\rangle|^2$ (green), and the overlap of the ground state of $\hat{H}_{AB}$ with the VBS, $|\langle\Psi_{AB}|\Psi_{VBS}\rangle|^2$ (red).

$S = 1$, which leads to a strong curvature in the band of the lowest S states. Consequently, the effective Hamiltonian $\hat{H}_{AB}$ approximates better the excited states for odd chains. Nevertheless, the quality of approximation of the ground state of equation (3) by $\hat{H}_{AB}$ is similar for chains as it is for even wheels. Another consequence of the open boundary conditions is that the E band is different for chains compared with even wheels, as the open boundary conditions split the running spin waves of the wheels to standing spin waves for the chains [63].

An alternative approach comes about naturally for AFM chains when considering the valence-bond-solid (VBS) model [64, 65]. A valence bond is a singlet coupling between two spins 1/2. If each spin S_i of a chain is considered to be made up of $2S_i$ spin-1/2 objects, then valence bonds can be formed between objects belonging to neighboring sites. The VBS model has been developed especially as a model for AFM chains and its ground state is the VBS state, which is named like this because its structure mimics the one of the lattice. If it has spin S, apart form the singlet bonds between nearest-neighbor spins there exist $2S$ unpaired spin-1/2 objects. If $S = 0$, the ground state is the same for AFM wheels. The VBS ground state can be generated by recursion or parent Hamiltonian techniques. Then it can be compared with the exact ground state of the AHM, by means of energy and overlap. The same can be done for the ground state of the semiclassical Hamiltonian $\hat{H}_{AB}$.

It turns out that both agree very well with the exact ground state for small N, which considering the apparently very different structure of the semiclassical and VBS wavefunctions is very surprising, and a main finding in this work. As N increases and S_i decreases the VBS state provides a better approximation for the energy and wavefunction. This is expected since when the AFM chains become larger than their correlation length the VBS model gives a better result, as it has exponentially decaying correlations in contrast to the infinite correlations of $\hat{H}_{AB}$. The overlaps of the ground states of the two models with the exact ground state of the AHM are shown in figure 5 for $S_i = 1$ and 2, and underpin the observation that for smaller N and larger S_i both approximate the ground state wavefunction equally well. It is also seen that their mutual overlap is also decreasing significantly with N, which points to the fact that even though the semiclassical and the VBS wavefunctions come from exactly opposite limits, the 'classical' and the 'quantum mechanical' respectively, they are almost equally successful in describing the physics of the short AFM chains. The even AFM wheels and the AFM chains

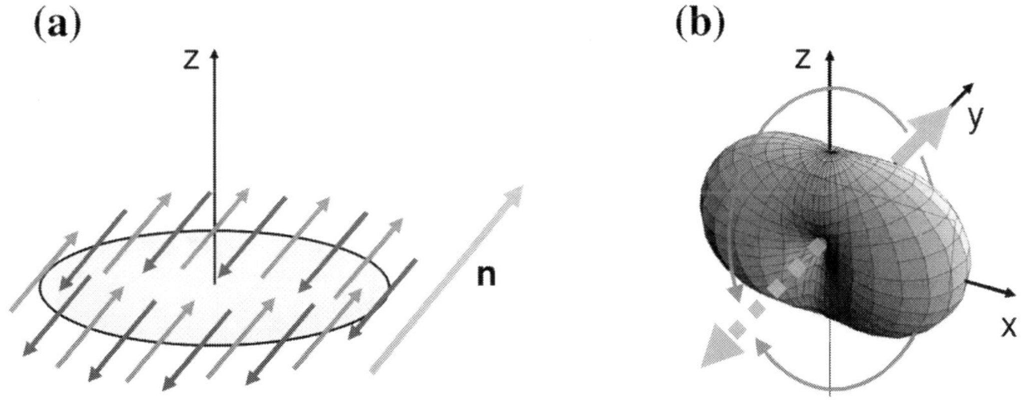

Figure 6. **(a)** Schematics of a classical ground state configuration for the AFM wheel Fe_{18} (red and blue arrows are the spins of the two sublattices) with the resulting Néel vector **n** (green arrow, length not scaled). **(b)** Shape of the potential $V(\mathbf{n})$ in the high field regime. The two tunneling paths of the Néel vector (green arrow) are indicated by the red arrows.

present therefore a class of MNMs where the distinction between classical and quantum behavior is not well-defined. Further investigations are needed for a full understanding of the complex finite-size many-body physics.

3. Quantum Phase Interference Effects in AFM Molecular Nanomagnets

Observing quantum phase interference effects (QPIEs), which may occur in the coherent tunneling between two classically degenerate magnetization directions, is difficult for mesoscopic particles. QPIE in a MNM was first observed in the molecule $[Fe_8O_2(OH)_{12}/tacn)_6]^{8+}$, known as Fe_8, which has an $S = 10$ ground state [24]. By applying a magnetic field along the anisotropy hard axis at very low temperatures oscillations in the tunneling splitting were observed. They correspond to QPIE due to different tunneling paths between the degenerate $M = \pm 10$ ground states. It took another 10 years to observe a similar phenomenon, this time in antiferromagnetically coupled MNMs with compensated magnetic moments. This is not only because of the experimental challenge to characterize a system with zero ground-state magnetization, but also due to chemical difficulties to synthesize an MNM with the 'right' physical properties [21, 22, 23]. The meaning of 'right' properties will be explained on a series of AFM wheels that seemed to be promising candidates to observe QPIE in the Néel vector tunneling (NVT). In NVT the Néel vector $\mathbf{n} = (\mathbf{M}_A - \mathbf{M}_B)/(2M_0)$, with sublattice magnetizations $\mathbf{M}_A$ and $\mathbf{M}_B$ of length M_0, has two tunneling paths through an anisotropy barrier (figure 6).

The phase of the ground-state wavefunction can generate interference which leads to oscillations of the tunnel splitting as function of an applied magnetic field. A generic spin Hamiltonian for AFM wheels is given by equation (1). For a magnetic field **B** parallel to the x axis and $D > 0$ the Néel-vector dynamics in a semiclassical theory (SCT) [21, 23] are governed by the effective semiclassical potential $V(\mathbf{n}) = N/|8J|(g\mu_B B)^2 n_x^2 + N S_i{}^2 D n_z^2$, which separates two minima at $\mathbf{n} = \pm\mathbf{y}$ by an energy barrier $\Delta U = N S_i \sqrt{2|D/J|}$. If this energy barrier is much larger than the attempt frequency $\hbar\omega_0 = S_i\sqrt{8D|J|}$, which relates to an exponentially small tunnel splitting Δ in the quantum-mechanical energy spectrum, NVT between the two ground states with opposite orientations of the Néel vector (or sublattice magnetization vectors) can occur. The phase of the ground-state wavefunction contains a $\pi/2$ term due to quantum fluctuations

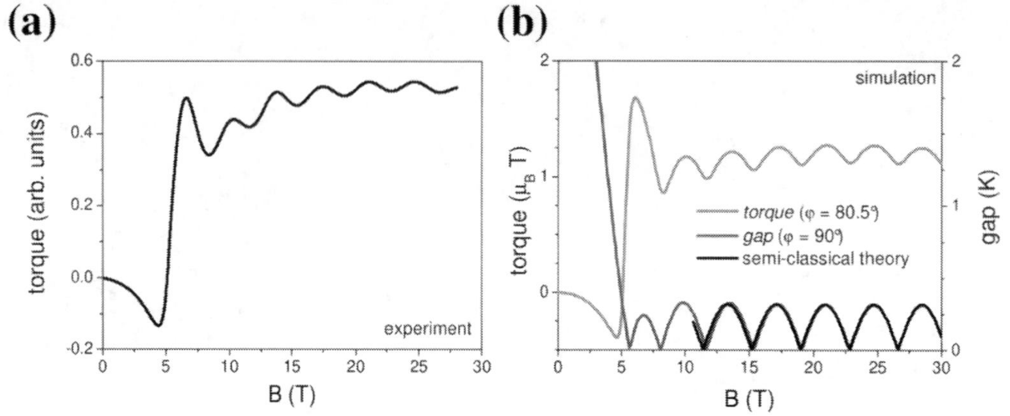

Figure 7. Magnetic torque as a function of B for (a) experiment, and (b) simulation ($\theta = 80.5°$ in red). Panel (b) also shows the gap between the two lowest quantum states vs B as simulated with equation (2) (blue curve), and as predicted by the SCT (black curve).

plus a topological term $\pi N g\mu_B B/|4J|$, which is proportional to the magnetic field [23]. Therefore one can repeatedly tune the phase through destructive and constructive interference by tuning the magnetic field. This results in tunneling splitting oscillations according to

$$\Delta(B) = \Delta_0 \left|\sin\left(\pi \frac{N g\mu_B B}{4|J|}\right)\right| \tag{5}$$

with the tunneling amplitude $\Delta_0 = 8\hbar\omega_0\sqrt{S_0/\hbar}\ exp(-S_0/\hbar)$ and the tunneling action $S_0/\hbar = NS_i\sqrt{2D|J|}$. In these quantities the condition for NVT reads $S_0/\hbar > 2$. It is important to point out that tunneling affects also the ground-state energy $\epsilon_0(B) = \epsilon(B) - \Delta(B)/2$. Hence, tunneling oscillations can be observed by static magnetization or torque measurements at zero temperature.

Since the Hilbert space of mesoscopic particles can be prohibitively large it is often necessary to use approximations to connect theory and experiment. One such approximation is to use the semiclassical two-sublattice Hamiltonian (2) (with $S_A = S_B = 45/2$ for Fe_{18}) [29]. This reduces the Hilbert space drastically so that exact numerical diagonalization may become possible. Another approximation is to use the SCT of the microscopic Hamiltonian (1) [23].

The QPIE in the NVT has been observed in the AFM wheel $[Fe_{18}(pd)_{12}-(pdH)_{12}(O_2CEt)_6(NO_3)_6](NO_3)_6$, or Fe_{18}, shown in figure 1(a) [43]. The experimentally measured magnetization curves for the magnetic field **B** parallel and perpendicular to the anisotropy axis z followed the predictions in [23], and thus provided strong indication for NVT in Fe_{18}. The measured torque (figure 7) for $\mathbf{B} \perp \mathbf{z}$ however, provided a clear-cut demonstration of the QPIE: after a first step it exhibits wiggles which directly correspond to the oscillations of the NVT splitting [66].

In the interpretation of the data and the simulation of the torque and magnetization curves in particular, it has to be realized that the two-sublattice Hamiltonian (2) and the SCT are approximations of equation (1), but at different levels. That is, they give quantitatively different results compared to each other and to equation (1) if one uses the same values for J and D in all three models. These values need to be 'adjusted' properly, which may be accomplished by using the appropriate values for a_1 and b_1 in equation (2). Then and only then do $\hat{H}_{AB}$ and SCT work with high accuracy.

Highly accurate results for $\hat{H}_{AB}$ are obtained by choosing a_1 and b_1 such that the lowest energies of equation (2) match the exact energies of equation (1), resulting in $a_1 = a_1^{qm} = 0.27$

and $b_1 = b_1^{qm} \approx 0.07$ for Fe_{18} (as derived from quantum Monte Carlo calculations [67]). If $\hat{H}_{AB}$ is derived in first-order perturbation theory the values $a_1 = a_1^{AB} = 4/N$ and $b_1 = b_1^{AB} = (2S_i - 1)/(NS_i - 1)$ are found [29]. The SCT provides $a_1 = a_1^{sc} = 4/N$ and $b_1 = b_1^{sc} = 2NS_i^2/[NS_i(NS_i + 2) - 3]$. Obviously these three results do not coincide with each other, and a careful analysis is needed to determine which set is to be used, since the experiment 'only' allows to determine the quantities $j = a_1 J$ and $d = b_1 D$. Hence different values for J and D, and thereby for $S_0/\hbar$, ΔU, $\hbar\omega_0$ and Δ_0, which are the key characteristics for NVT, may be derived and the question arises which set of parameters is the most appropriate one. For Fe_{18} the three resulting sets are collected in table 1.

Table 1. Comparison of the results of some key NVT quantities for Fe_{18}, as derived with the qm, AB, and sc values for a_1 and b_1.

	a_1	b_1	J (K)	D (K)	$S_0/\hbar$	ΔU (K)	$\hbar\omega_0$ (K)	Δ (K)
qm	0.27	≈ 0.07	-18.9	0.300	8.02	33.8	16.8	0.050
AB	0.2222	0.0909	-23.0	0.231	6.38	26.0	16.3	0.222
sc	0.2222	0.1065	-23.0	0.197	5.90	22.2	15.0	0.320

It turns out that the J and D values that have to be inserted in the microscopic Hamiltonian (1) should be derived from the values a_1^{qm} and b_1^{qm}. However, importantly, to use the formula of the SCT the 'corrected' J and D values should be inserted as derived with a_1^{sc} and b_1^{sc} in order to obtain quantitatively accurate results. The latter fact can be made plausible by looking for instance at the fields at which the magnetization/torque steps or zeros in the oscillation of the tunneling splitting respectively occur. In SCT these fields are controlled by the period $Ng\mu_B B/|4J|$ but actually appear with period $g\mu_B B/|j|$. Therefore, $J = j/a_1^{sc}$ should be inserted and not $J = j/a_1^{qm}$, and similarly $D = d/b_1^{sc}$ and not $D = d/b_1^{qm}$.

The next step is to consider the 'right' conditions for observing NVT, and in particular compare the situations for Fe_{18} and the two most promising previously studied AFM wheels, $[CsFe_8\{N(CH_2\text{-}CH_2O)_3\}_8]Cl$ or $CsFe_8$ [68], and $[Fe(OMe)_2(O_2CCH_2Cl)]_{10}$ or Fe_{10} [69]. As mentioned, the Néel-vector dynamics in the SCT are characterized by the following quantities:

$$\text{tunneling action:} \quad S_0/\hbar = NS_i\sqrt{2|D/J|}, \tag{6}$$

$$\text{attempt frequency:} \quad \hbar\omega_0 = S_i\sqrt{8|DJ|}, \tag{7}$$

$$\text{energy barrier height:} \quad \Delta U = NS_i^2|D|, \tag{8}$$

$$\text{tunneling amplitude:} \quad \Delta_0 = 8\hbar\omega_0\sqrt{\frac{S_0/\hbar}{2\pi}}\, exp(-S_0/\hbar). \tag{9}$$

Fe_{18} has a hard-axis single-ion anisotropy ($D > 0$), while Fe_{10} and $CsFe_8$ easy-axis type ($D < 0$). Furthermore, Fe_{18} has been investigated in the high-field regime, while the other two wheels in zero field (systems with a hard-axis anisotropy do not exhibit NVT in zero field). NVT is realized if the anisotropy barrier ΔU is larger than the ground state energy, which is $\hbar\omega_0/2$ for Fe_{18} and $\hbar\omega_0$ for Fe_{10} or $CsFe_8$. This criterion can be translated into $S_0/\hbar > 2$ and $S_0/\hbar > 4$, respectively. For Fe_{18} the tunneling splitting Δ shows an oscillation as function of field according to equation (5). For Fe_{10} and $CsFe_8$, to the best of our knowledge, Δ has not yet been calculated analytically, however numerical calculations indicate $\Delta \approx 2\Delta_0$. For a better comparison of the three wheels a constant c, with $c = 1$ for Fe_{18} and $c = 2$ for Fe_{10} and $CsFe_8$, is introduced such that the quantities to look at become $c\hbar\omega_0/2$, $\Delta = c\Delta_0$, and the NVT criterion $(S_0/\hbar)/c > 2$. With the experimentally determined values for j an d for each of the three AFM wheels, the NVT characteristics compiled in table 2 are obtained from the above relations (in the case of Fe_{10} one has to additionally take into account that it exhibits a significant E term and is far from uniaxial, its d value was hence determined as if it would be uniaxial).

Table 2. Comparison of the characteristic parameters for NVT in Fe_{18}, $CsFe_8$, and Fe_{10}. The last row sketches the energetic situation in the three AFM wheels. The solid line represents the potential barrier of height ΔU, the left and right bar the classical Néel vector states at energy $c\hbar\omega_0/2$, and the bars in the middle the tunneling-split quantum states at energies $c\hbar\omega_0/2 \pm c\Delta_0/2$.

	Fe_{18} [66]	$CsFe_8$ [68]	Fe_{10} [69]
experimental values (K)	$j = -5.1$ $d = 0.021$ Δ not directly measured	$j = -11.1$ $d = -0.104$ $\Delta = 5.92$	$j = -6.31$ $d = -0.0276$ $\Delta = 3.83$
a_1^{sc}	0.2222	0.5	0.4
b_1^{sc}	0.1065	0.2288	0.1860
J (K)	-23.0	-22.2	-15.8
D (K)	0.197	-0.454	-0.148
$S_0/\hbar$	5.90	4.03	3.42
$c\hbar\omega_0/2$ (K)	7.52	22.3	10.8
ΔU (K)	22.2	22.5	9.25
$\Delta = c\Delta_0$ (K)	0.320	5.08	4.18
$(S_0/\hbar)/c$	5.70	2.02	1.72

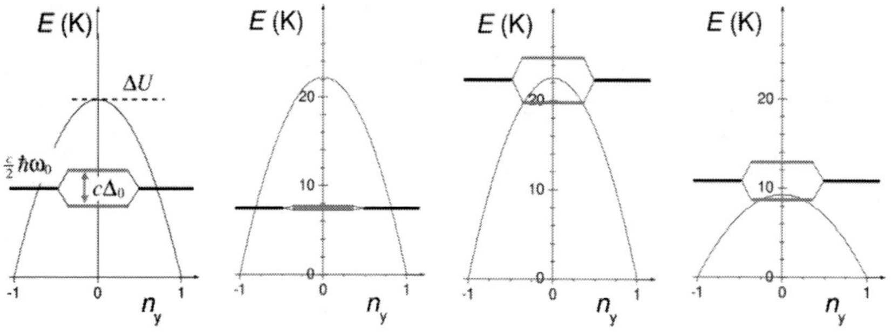

A further property that characterizes NVT is the 'localization' of the Néel vector. It can be measured e.g. by the matrix element $\langle 0|\hat{n}_y|1\rangle^2$ with the Néel-vector operator defined as $\hat{\mathbf{n}} = (\hat{\mathbf{S}}_A - \hat{\mathbf{S}}_B)/(S_A + S_B)$. The SCT yields the estimate $\langle 0|\hat{n}_y|1\rangle^2 \approx 1 - 1/(S_0/\hbar)$. For Fe_{18} this gives $\langle 0|\hat{n}_y|1\rangle^2 \approx 0.83$, i.e. the Néel vector is localized in Fe_{18} to 83%. For Fe_{10} and $CsFe_8$ the SCT yields the estimate $\langle 0|\hat{n}_z|1\rangle^2 \approx 1 - 2/(S_0/\hbar)$, or $\langle 0|\hat{n}_z|1\rangle^2 = 0.42$ and 0.50 respectively.

Apparently, while for Fe_{10} the NVT is at best only approximately fulfilled and similarly for $CsFe_8$, the AFM wheel Fe_{18} fully satisfies the criteria for it. This is probably most convincingly demonstrated by the energy diagrams in table 2.

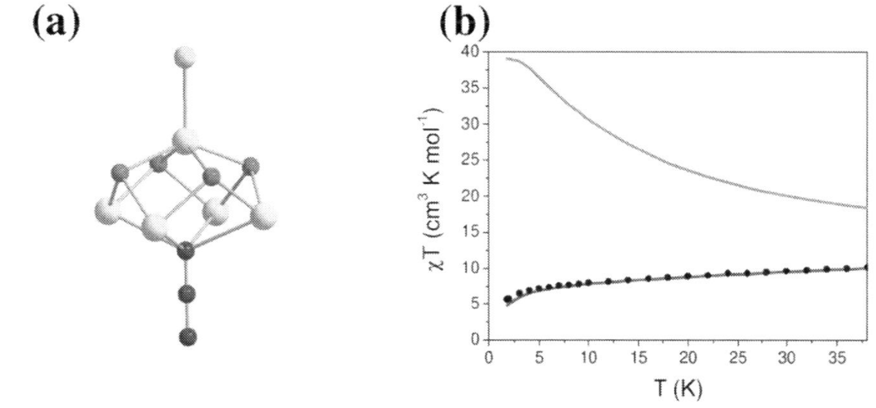

Figure 8. **(a)** Magnetic core of the molecule Co_5Cl. Turquoise: Co^{2+}, blue: nitrogen, red: oxygen, green: chloride. Carbon and hydrogen atoms are omitted for clarity. **(b)** Measured χT versus T curves for Co_5Cl (black dots) compared with the two simulations with a negative g_i factor as obtained in the best fit (blue line), and with the g_i factor set to be positive (red line). The other parameters were as found for the best fit.

4. Negative g-Factors and Apparent AFM Susceptibility for Ferromagnetically Coupled Co^{2+} Clusters

The studies of SMMs were mainly restricted in the past to molecules containing manganese or iron ions. Compared to them only a few SMMs have been reported in the literature which consist of high-spin Co^{2+} ions [70]. In principle Co^{2+} ions should be ideal for building SMMs, as they typically posses large magnetic anisotropy as a result of their unquenched orbital angular momentum. Due to this unquenched orbital angular momentum a spin-only approach for describing the magnetism, as in the cases of the previous sections, becomes invalid, and this complication is one of the major reasons why magnetic models were developed only for a few Co^{2+} clusters. Fortunately, the unquenched orbital angular momentum allows for a simplification in as much as that the low-temperature magnetism can be described by an effective spin-1/2 model. The ground state of a Co^{2+} ion is a Kramer's doublet and at temperatures below ca 40 K only this doublet is populated, consequently the dynamics of the ion may be described as that of an effective spin 1/2 [71]. The drawback is that this leads to highly anisotropic exchange coupling tensors **J** as well as **g** matrices, and hence to a large number of magnetic parameters even for the simplest clusters [72]. In many cases the anisotropy is thus restricted to uniaxial type [73, 74].

In our group the magnetic properties of three Co^{2+} clusters with similar core motif but surprisingly large differences in their magnetic properties, especially their SMM behavior, were studied [75]. The clusters contain five Co^{2+} ions arranged in a square-pyramidal fashion. $[Co_5^{2+}(\mu_4\text{-}N_3)(Cl)(t\text{bea})_2(\mu\text{-piv})_4(\text{pivH})_2$, or Co_5Cl in short, is shown in figure 8(a). In its magnetic susceptibility curve, plotted as χT versus T, Curie behavior was observed at the highest measured temperature of 250 K. With decreasing temperature χT decreases, which is usually taken as a clear signal of dominant AFM couplings in the cluster. The AC susceptibility measurements revealed an out-of-phase signal at the lowest temperatures, strongly evidencing SMM behavior of Co_5Cl. In view of the AFM type of behavior one would expect that the ground state of this system is weakly magnetic, and in particular that an energy barrier for spin reversal is small, disfavoring SMM behavior. After intense and careful efforts the appropriate model for Co_5Cl was established, and it describes the molecule as a cluster of five spin-1/2 centers. The model reproduces the magnetic data, i.e. the magnetization curves as well as the magnetic susceptibility below 40 K, excellently. Surprisingly it revealed dominant ferromagnetic couplings, and accordingly the simulated energy spectrum is characterized by a ferromagnetic ground state. Furthermore, it exhibits a sizeable energy barrier of ca 15 K, which explains the SMM behavior observed in the AC susceptibility measurements.

Ferromagnetic couplings and a ferromagnetic ground state are intuitively inconsistent with the apparent AFM curvature of χT. This contradiction is a result of the interplay of the highly anisotropic **J** tensors and **g** matrices, as they are typical for effective spin models. In fact, in the present magnetic model one component of the **g** matrix of the apical Co^{2+} ion was obtained as negative. Usually the possibility of negative g factors is not considered, with arguments that the g values enter only quadratically in measured observables such that the sign becomes irrelevant or that negative g values are unphysical. However, both arguments are incorrect and it was recently demonstrated that negative g factors can appear quite naturally in effective spin descriptions [72]. Indeed, as the χT curve was simulated for the Co_5Cl molecule with all entries $g_{i,xy}$ and $g_{i,z}$ in the local $\mathbf{g}_i$ matrices artificially set to positive, the typical increase at low temperatures in the χT curve was obtained as expected for dominant ferromagnetic couplings, see figure 8(b). The effect of the sign of g factors may be understood by analyzing the low-temperature value of χT. At these low temperatures only the ground state of the Co_5Cl cluster is populated, which because of the odd number of magnetic electrons in the molecule is again a Kramer's doublet, and may hence also be described by an effective spin-1/2, $\tilde{S} = 1/2$. This however should not be confused with the effective spin-1/2 introduced for each Co^{2+} center. The Zeeman splitting of this Kramer's doublet is described by the gyromagnetic factors

$$\tilde{g}_x = \tilde{g}_y = 0 \tag{10}$$

$$\tilde{g}_z = \sum_{i=1}^{5} g_{i,z} \tag{11}$$

and the Curie constant or χT constant can be calculated as

$$\chi T|_{T\to 0} = \frac{1}{8}\frac{\tilde{g}_x^2 + \tilde{g}_y^2 + \tilde{g}_z^2}{3}\tilde{S}(\tilde{S}+1) = \frac{1}{32}\tilde{g}_z^2 \tag{12}$$

Importantly, the $\tilde{g}_z$ value is given as as sum of the $g_{i,z}$ components of all Co^{2+} ions. Hence, if one or more of the $g_{i,z}$ factors have negative sign, then this tends to reduce the cluster g factor and thereby the χT constant. If it happens as in the case of Co_5Cl that this χT value is smaller than the χT value at high temperatures, then apparently the χT curve will show a drop towards lower temperatures reminiscent of an AFM coupling. In passing it is noted that the described effect may not only be produced by negative entries in the local $\mathbf{g}_i$ matrices. A negative g value is only the simplest representation of a different orientation of a $\mathbf{g}_i$ matrix with respect to the other $\mathbf{g}_i$ matrices. Indeed model calculations with anisotropic $\mathbf{g}_i$ matrices showed that this effect can be obtained even with $\mathbf{g}_i$ matrices with only positive entries, if they are rotated with respect to each other.

It is thus concluded that in clusters where ions may have different and anisotropic $\mathbf{g}_i$ matrices the usual interpretation of the curvature of χT versus T as to indicate ferromagnetic or aniferromagnetic interactions may be misleading, as in such cases an 'AFM' curvature can, opposite to common sense, be a result of the details of the individual $\mathbf{g}_i$ matrices and not the exchange couplings.

5. Conclusions

MNMs constitute an exciting class of relatively small systems where the combination of unusual magnetic properties and synthetic chemistry appears very promising for future applications.

Their comparatively small size places them away from the thermodynamic limit, while the relatively large spin of their centers is inbetween the classical and the full quantum limit, defining a mesoscopic scale. Consequently these molecules may support unusual complex quantum states. This is the main point demonstrated in this article. It may even happen that their properties do not have a predominant classical or quantum character, and rather belong to an intermediate regime. Hence, MNMs are obviously attractive objects for studying fundamental questions in basic sciences, but could also be potentially very useful in applications where classical or quantum behavior could be adjusted by the topology and spin length of the magnetic centers.

In this review a few examples of MNMs were considered, which showcase the unusual many-body character that comes about in this mesoscopic regime and borders between classical and quantum physics. Even AFM wheels as well as AFM chains were shown to behave classically in some aspects as N decreased and S_i increased, however their ground state is not exactly in the classical or quantum regime and the intuitively clear distinction between these two regimes becomes in fact blurred. The Fe_{18} AFM wheel has been shown to be one of the few magnetic examples where quantum phase interference effects occur, associated with tunneling of the Néel vector, a concept originating in semiclassical physics. Finally, a MNM incorporating high-spin Co^{2+} ions which behaves as a weak SMM was also considered. There the unquenched orbital angular momentum leads to highly anisotropic magnetic interactions and gyromagnetic matrices. The anisotropic **g** matrices lead to effects usually associated purely with the exchange coupling.

MNMs may also be important as magnetic units that can be potentially deposited on surfaces while preserving their magnetic properties. They are in general particularly fit for this purpose as the typical energy scale of their magnetic excitations is of the order of some Kelvins, which supersedes by orders of magnitude the tunneling gap of SMMs [5]. The latter is of the order of e.g. 10^{-7} K in the archetype SMM Mn_{12}-acetate [5], and therefore expected to be very easily influenced by environmental perturbations such as thermal vibrations or the different conditions found on surfaces as compared to in crystals. Indeed, observing magnetic hysteresis of a SMM deposited on a surface could only very recently be achieved [76, 77]. In contrast, the magnetism in MNMs is expected to be a lot more robust and to survive the change in environment on a surface.

A MNM will in general not exhibit a magnetic function obviously suitable for application as the magnetic hysteresis and/or QTM in the case of SMMs. This raises a serious question, and indeed innovative novel schemes need to be developed. A 'simple' example of a possible application for MNMs is their usage in magnetic refrigeration, where 'simple' refers to the fact that the 'complex quantum state' which is envisaged here is a ground state with a large spin and low magnetic anisotropy. By application of the magneto-caloric effect it is in principle possible to reach temperatures beyond the reach of liquid helium, by cooling on a molecular level [78]. However, more involved refrigeration schemes exploiting e.g. quantum spin frustration effects have also been suggested [79]. This points to the general idea of exploiting, yet to be identified, complex quantum states promoted by particular spin topologies for novel application schemes. This subject is currently much studied but still in its infancy. An example would be the chiral quantum spin states in spin triangles [80, 81, 82]. Obviously, MNMs may be the candidates of choice to realize experimentally such scenarios.

At this point it is interesting to note a potential link with another emerging research field. Recently, artificially engineered magnetic structures like AFM spin chains have been assembled on insulating layers, and their low-energy spectra have been measured with STM techniques with atomic scale precision [26]. In comparison with artificially formed atomic-scale magnetic structures, MNMs can be easily manipulated in size or spin length resulting in a wide variety of different topologies. In fact, hundreds of potentially interesting MNMs were already synthesized by chemists and are currently available for searching for complex quantum states, with potential for applications. In addition, the ground-state properties as well as the spin excitations can be

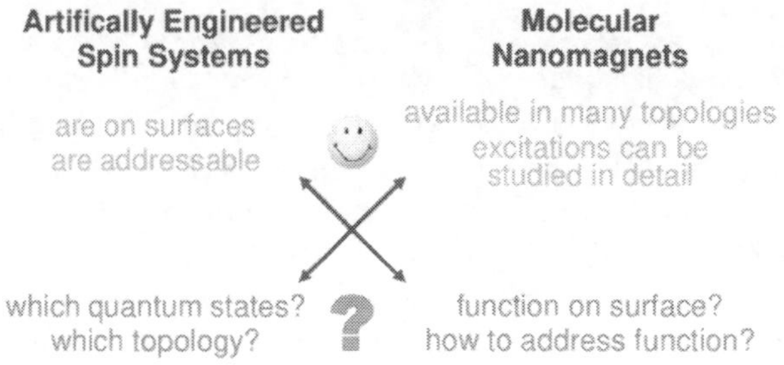

Figure 9. Comparison of achievements and opportunities as well as the challenges regarding application of artificially engineered atomic-scale spin systems and molecular nanomagnets, demonstrating the similarities as well as complementariness of these two research fields.

studied in any detail in such topologies thanks to the applicability of a full set of experimental techniques and bring about interesting effects, as were the ones discussed in this review. From this point of view MNMs have an advantage over artificial nanostructures. However, the latter have already been grafted on surfaces while maintaining their magnetic properties. Also, implicit in the use of the STM technique, the nanostructures and their magnetic function could be addressed individually, which is still a major obstacle in the case of MNMs.

In any case, both classes of systems are characterized by magnetic interactions of the order of few Kelvins or larger, thus they are quite robust when put on the surface against thermal vibrations and other types of external perturbations. MNMs and artificially engineered spin systems are complementary, as the former can be manipulated to many different topologies and assume complex many-body states, while the latter have already been shown to maintain their magnetic properties and their function to be addressable on the surface. Hence the advantages of each class can be an example for the other to overcome its problems, as summarized in pictorial fashion in figure 9.

References

[1] Gatteschi D and Sessoli R 2003 *Angew. Chem. Int. Ed.* **42** 268
[2] Cavallini M, Gomez-Segura J, Ruiz-Molina D, Massi M, Albonetti C, Rovina C, Veciana J and Biscarini F 2005 *Angew. Chem. Int. Ed.* **44** 888
[3] Affronte M 2009 *J. Mater. Chem.* **19** 1731
[4] Leuenberger M N and Loss D 2001 *Nature* **410** 789
[5] Gatteschi D, Sessoli R and Villain J 2006 *Molecular Nanomagnets* (Oxford: Oxford University Press)
[6] Waldmann O 2007 *Inorg. Chem.* **46** 10035
[7] Ruiz E, Cirera J, Cano J, Alvarez S, Loose C and Kortus J 2008 *Chem. Commun.* 52
[8] Schröder C, Nojiri H, Schnack J, Hage P, Luban M and Kögerler P 2005 *Phys. Rev. Lett.* **94** 017205
[9] Rousochatzakis I, Läuchli A M and Mila F 2008 *Phys. Rev. B* **77** 094420
[10] Schmidt R, Schnack J and Richter J 2005 *J. Magn. Magn. Mater.* **295** 164
[11] Kortz U, Müller A, Slageren J V, Schnack J, Dalal N S and Dressel M 2009 *Coord. Chem. Rev.* **253** 2315
[12] Anderson P 1952 *Phys. Rev.* **86** 694
[13] Bernu B, Lhuillier C and Pierre L 1992 *Phys. Rev. Lett.* **69** 2590
[14] Waldmann O, Guidi T, Carretta S, Mondelli C and Dearden A L 2003 *Phys. Rev. Lett.* **91** 237202
[15] Waldmann O 2002 *Phys. Rev. B* **65** 024424
[16] Dreiser J, Waldmann O, Dobe C, Carver G, Ochsenbein S T, Sieber A, Güdel H U, van Duijn J, Taylor J and Podlesnyak A 2010 *Phys. Rev. B* **81** 024408
[17] Carretta S, Santini P, Amoretti G, Guidi T, Copley J R D, Qiu Y, Caciuffo R, Timco G and Winpenny R E P 2007 *Phys. Rev. Lett.* **98** 167401
[18] Candini A, Lorusso G, Troiani F, Ghirri A, Carretta S, Santini P, Amoretti G, Muryn C, Tuna F, Timco G, McInnes E J L, Winpenny R E P, Wernsdorfer W and Affronte M 2010 *Phys. Rev. Lett.* **104** 037203
[19] Affronte M, Carretta S, Timco G A and Winpenny R E P 2007 *Chem. Commun.* 1789

[20] Laye R H, Larsen F K, Overgaard J, Muryn C A, McInnes E J L, Rentschler E, Sanchez V, Teat S J, Güdel H U, Waldmann O, Timco G A and Winpenny R E P 2005 *Chem. Commun.* 1125
[21] Barbara B and Chudnovsky E 1990 *Phys. Lett. A* **145** 205
[22] Tejada J, Garg A, Gider S, Awschalom D D, DiVincenzo D P and Loss D 1996 *Science* **272** 424
[23] Chiolero A and Loss D 1998 *Phys. Rev. Lett.* **80** 169
[24] Wernsdorfer W and Sessoli R 1999 *Science* **284** 133
[25] Palii A, Tsukerblatt B S, Coronado E, Clemente-Juan J M and Borrás-Almenar J J 2003 *Inorg. Chem* **42** 2455
[26] Hirjibehedin C F, Lutz C P and Heinrich A J 2006 *Science* **312** 1021
[27] Jamneala T, Madhavan V and Crommie M F 2001 *Phys. Rev. Lett.* **87** 256804
[28] Enders A, Skomski R and Honolka J 2010 *J. Phys.: Condens. Matter* **22** 433001
[29] Waldmann O 2002 *Europhys. Lett.* **60** 302
[30] Taft K L, Delfs C D, Papaefthymiou G C, Foner S, Gatteschi D and Lippard S J 1994 *J. Am. Chem. Soc.* **116** 823
[31] Gatteschi D, Caneschi A, Pardi L and Sessoli R 1994 *Science* **265** 1054
[32] Waldmann O 2005 *Coord. Chem. Rev.* **249** 2550
[33] Pilawa B, Desquiotz R, Kelemen M T, Weickenmeier M and Geisselmann A 1998 *J. Magn. Magn. Mat.* **177** 748
[34] Caneschi A, Cornia A, Fabretti A C, Foner S, Gatteschi D, Grandi R and Schenetti L 1996 *Chem. Eur. J* **11** 1379
[35] Low D M, Rajaraman G, Helliwell M, Timco G, van Slageren J, Sessoli R, Ochsenbein S T, Bircher R, Dobe C, Waldmann O, Güdel H U, Adams M A, Ruiz E, Alvarez S and McInnes E J L 2006 *Chem. Eur. J.* **12** 1385
[36] Winpenny R E P 2003 *Adv. Inorg. Chem.* **52** 1
[37] Dearden A L, Parsons S and Winpenny R E P 2001 *Angew. Chem. Int. Ed.* **40** 151
[38] Cornia A, Affronte M, Jansen A G M and G L Abbati D G 1999 *Angew. Chem. Int. Ed.* **40** 151
[39] Waldmann O, Schülein J, Koch R, Müller P, Bernt I, Saalfrank R W, Andres H P, Güdel H U and Allenspach P 1999 *Inorg. Chem.* **38** 5879
[40] Saalfrank R W, Bernt I, Uller E and Hampel F 1997 *Angew. Chem. Int. Ed.* **36** 2482
[41] van Slageren J, Sessoli R, Gatteschi D, Smith A A, Helliwell M, Winpenny R E P, Cornia A, Barra A L, Jansen A G M, Rentschler E and Timco G A 2002 *Chem. Eur. J.* **8** 277
[42] Yao H C, Wang J J, Ma Y S, Waldmann O, Du W X, Song Y, Li Y Z, Zheng L M, Decurtins S and Xin X Q 2006 *Chem. Commun.* **16** 1745
[43] King P, Stamatatos T C, Abboud K A and Christou G 2006 *Angew. Chem. Int. Ed. Engl.* **45** 7379
[44] Bärwinkel K, Hage P, Schmidt H J and Schnack J 2003 *Phys. Rev. B* **68** 054422
[45] Waldmann O, Koch R, Schromm S, Schülein J, Müller P, Bernt I, Saalfrank R W, Hampel F and Balthes E 2001 *Inorg. Chem.* **40** 2986
[46] Stamatatos T C, Christou A G, Mukherjee S, Poole K M, Lampropoulos C, Abboud K A, O'Brien T A and Christou G 2008 *Inorg. Chem.* **47** 9021–9034
[47] Schnack J, Luban M and Modler R 2001 *Europhys. Lett.* **56** 863
[48] Exler M and Schnack J 2003 *Phys. Rev. B* **67** 094440
[49] Waldmann O, Güdel H U, Kelly T L and Thompson L K 2006 *Inorg. Chem.* **45** 3295
[50] Mermin N D and Wagner H 1966 *Phys. Rev. Lett.* **17** 1133
[51] Manousakis E 1991 *Rev. Mod. Phys.* **63** 1
[52] Misguich G, Lhuillier C and Diep H T 2005 *Frustrated Spin Systems* (Singapore: World Scientific)
[53] Anderson P W 1987 *Science* **235** 4793
[54] Haldane F D M 1983 *Phys. Lett. A* **93** 464
[55] Haldane F D M 1983 *Phys. Rev. Lett.* **50** 1153
[56] Bethe H 1931 *Z. Phys.* **71** 205
[57] Schnack J and Luban M 2000 *Phys. Rev. B* **63** 014418
[58] Anderson P W 1958 *Phys. Rev.* **109** 1492
[59] Bencini A and Gatteschi D 1990 *Electron Paramagnetic Resonance of Exchange Coupled Clusters* (Berlin: Springer)
[60] Ochsenbein S T, Tuna F, Rancan M, Davies R, Myrun C, Waldmann O, Bircher R, Sieber A, Garver G, Mutka H, Fernandez-Alonso F, Podlesnyak A, Engelhardt L P, Timco G A, Güdel H U and Winpenny R E P 2008 *Chem. Eur. J* **14** 5144
[61] Lieb E and Mattis D 1962 *J. Math. Phys.* **3** 749
[62] Machens A, Waldmann O, Schneider I and Eggert S In preparation
[63] Ochsenbein S T, Waldmann O, Sieber A, Carver G, Bircher R, Güdel H U, Davies R S G, Timco G A,

Winpenny R E P, Mutka H and Fernandez-Alonso F 2007 *Europhys. Lett.* **79** 17003
[64] Affleck I, Kennedy T, Lieb E H and Tasaki H 1987 *Phys. Rev. Lett.* **59** 799
[65] Affleck I 1989 *J. Phys. Cond. Matt.* **1** 3047
[66] Waldmann O, Stamatatos T, Christou G, Güdel H U, Sheikin I and Mutka H 2009 *Phys. Rev. Lett.* **102** 157202
[67] Engelhardt L and Luban M 2006 *Phys. Rev. B* **73** 054430
[68] Waldmann O, Dobe C, Güdel H U and Mutka H 2006 *Phys. Rev. B* **74** 054429
[69] Santini P, Carretta S, Amoretti G, Guidi T, Caciuffo R, Caneschi A, Rovai D, Qiu Y and Copley J R D 2005 *Phys. Rev. B* **71** 184405
[70] Murrie M 2010 *Chem. Soc. Rev.* **39** 1986
[71] Palii A, Tsukerblatt B S, Clemente-Juan J M and Coronado E 2010 *Int. Rev. Phys. Chem.* **29** 135
[72] Chibotaru L F, Ceulemans A C and Bolvinm H 2008 *Phys. Rev. Lett.* **101** 033003
[73] Waldmann O, Hassmann J, Müller P, Hanan G S, Volkmer D, Schubert U S and Lehn J M 1997 *Phys. Rev. Lett.* **78** 3390
[74] Alley K G, Bircher R, Waldmann O, Ochsenbein S T, Güdel H U, Moubaraki B, Murray K S, Fernandez-Alonso F, Abrahams B F and Boskovic C 2006 *Inorg. Chem.* **45** 8950
[75] Klöwer F, Lan Y, Nehrkorn J, Waldmann O, Anson C E and Powell A K 2009 *Chem. Eur. J.* **15** 7413
[76] Mannini M, Pineider F, Danieli C, Totti F, Sorace L, Sainctavit P, Arrio M A, Otero E, Joly L, Cezar J C, Cornia A and Sessoli R 2010 *Nature* **468** 417
[77] Mannini M, Pineider F, Sainctavit P, Danieli C, Otero E, Sciancalepore C, Talarico A M, Arrio M A, Cornia A, Gatteschi D and Sessoli R 2009 *Nature Mater* **8** 194
[78] Nayak S, Evangelisti M, Powell A K and Reedijk J 2010 *Chem. Eur. J.* **16** 12865
[79] Schnack J, Schmidt R and Richter J 2007 *Phys. Rev. B* **76** 054413
[80] Trif M, Troiani F, Stepanenko D and Loss D 2008 *Phys. Rev. Lett.* **101** 217201
[81] Luzon J, Bernot K, Hewitt I J, Anson C E, Powell A K and Sessoli R 2008 *Phys. Rev. Lett.* **100** 247205
[82] Georgeot B and Mila F 2010 *Phys. Rev. Lett.* **104** 200502

Magnon excitations in ultrathin Fe layers: The influence of the Dzyaloshinskii-Moriya interaction

Kh. Zakeri [1], Y. Zhang[1], J. Prokop[1], T.-H. Chuang[1], W.X. Tang[1,2] and J. Kirschner[1]

[1] Max-Planck-Institut für Mikrostrukturphysik, Weinberg 2, 06120 Halle, Germany
[2] School of Physics, Monash University, Victoria 3800, Australia

E-mail: zakeri@mpi-halle.de

Abstract. High wave-vector magnon excitations in ferromagnetic Fe monolayer and double-layer grown on W(110) are investigated using spin-polarized electron energy loss spectroscopy. The magnon dispersion relation is obtained up to the Brillouin zone boundary. A direct comparison among different systems shows that the magnons in the Fe monolayer are extremely soft and are even softer than the acoustic surface mode of Fe(110).

By measuring the spectra in both energy loss and gain regions on a double-layer Fe film at room temperature and by reversing the sample magnetization, it is demonstrated that the magnon dispersion is asymmetric with respect to the sign of the wave-vector. The asymmetric dispersion relation is attributed to the degeneracy breaking of the magnons due to the presence of the Dzyaloshinskii–Moriya interaction.

1. Introduction

In a magnetically ordered solid the collective magnetic excitations are called magnons. Since in solids the properties are governed by collective phenomena, the investigation of the magnon spectrum would provide an access to the basic and core properties of a magnetic solid. This fact has made the study of spin dynamics in ferromagnets of fundamental interest [1, 2].

Magnon excitations are well established subjects in bulk magnets and have been intensively investigated both experimentally and theoretically over several decades [1, 2, 3, 4, 5, 6]. The experimental investigations are performed by means of various experimental techniques such as ferromagnetic resonance (FMR), Brillouin light scattering (BLS) and inelastic neutron scattering (INS). Only the later one allows a mapping of the magnon dispersion over the whole Brillouin zone in some cases, however, it cannot be employed to low dimensional systems because of its low sensitivity — due to the weak interaction of neutrons with matter—.
Since these kind of excitations in low dimensional ferromagnets are of great interest, a novel experimental technique was required to probe the magnons across the Brillouin zone. Now it is well-established that spin-polarized electron energy loss spectroscopy (SPEELS) is a powerful technique for investigating magnon excitations in ultrathin ferromagnets and ferromagnetic surfaces. The capability of SPEELS is well demonstrated by measuring the magnon dispersion relation of a few atomic layers of Co and Fe films [7, 8, 9].

In nanoscale magnetism, the investigation of the magnons with a wavelength of about (or less than) a nanometer is of great interest, since new physics will come into play in these systems. One particular example is that in low dimensional magnets the spin fluctuations can, in principle,

Published under licence by IOP Publishing Ltd

become so strong that ferromagnetic ordering will not take place at finite temperatures. In two dimensional systems this is known as Mermin–Wagner theorem, which will happen only in an ideal isotropic Heisenberg ferromagnet. In real ferromagnets the anisotropy fields contribute and thus real two dimenstional ferromagnetic objects do show ferromagnetic order at a finite temperature. However, the spin fluctuations remain the key point to explain most observed phenomena in low dimensional systems such as reduced Curie temperature (known as finite-size effect). Here, the fundamental questions are: if the magnons in real two-dimensional ferromagnets do exist, is it possible to see their signature of excitations in SPEELS spectra? How does the magnon dispersion relation look like in such a system?

In this contribution we present the results of magnon excitations in a prototype two dimensional magnetic system: one atomic layer of Fe grown on W(110). The choice of the Fe monolayer on W(110) is due to the fact that the system shows relatively good structural and morphological properties. The Fe monolayer possesses a high thermodynamic stability and no intermixing with the substrate. The magnon dispersion relation is measured along the $\bar{\Gamma}-\bar{H}$ of the Brilluin zone. A direct comparison among the different systems shows that the magnons in the Fe monolayer are extremely soft and are softer than the acoustic surface mode of Fe(110).

Since on the surface of an ultrathin film the inversion symmetry in broken, a large spin-orbit coupling can, in principle, lead to a large antisymmetric interaction —known as Dzyaloshinskii–Moriya (DM) interaction [10, 11]—. If the DM interaction is strong enough, it should change the static [12, 13, 14, 15, 16, 17] as well as dynamic [18] prosperities of the system. The consequence of the DM interaction on the spin dynamics and in particular on the magnon dispersion relation of low dimensional ferromagnets will be presented in this paper. By presenting the results of a two-atomic-layer-thick Fe film on W(110), we will show that, indeed, the DM interaction lifts the degeneracy of the magnons and causes an asymmetry in the magnon dispersion relation.

2. Experimental Details

The epitaxial Fe films with thicknesses of one and two atomic layers were deposited onto a clean W(110) single crystal at room temperature (RT) under ultrahigh vacuum (UHV) condition. Prior to film deposition the W(110) crystal was cleaned by cycles of low temperature flashing in oxygen atmosphere and a subsequent high temperature flash in UHV as it is well described in Ref. [19]. After preparation of the sample, it was annealed at about 500 K in order to improve the smoothness of the surface. Conventional UHV tools like low-energy electron diffraction and Auger electron spectroscopy were used to check the sample quality. A longitudinal magneto-optical Kerr effect measurement showed a well-ordered magnetic state of the films—a more detailed information about our sample preparation and their quality can be found elsewhere [20]—. The Fe films exhibit in-plane magnetization and a relatively large magnetic anisotropy with an effective easy axis along the $[\bar{1}10]$-direction. The SPEELS experiments were performed in the in-plane geometry, in which the polarization axis of the incoming and outgoing beam is parallel to the sample surface and thereby parallel or aniparallel to the magnetization of the film. The scattering plane was parallel to the (100)-plane.

A schematic illustration of the scattering geometry used in our experiments is shown in Fig. 1. The spin-polarized beam from a strained GaAs photo-cathode is focused on the sample surface. The intensity of the backscattered electrons is measured with respect to their energy for two different spin orientations of the incoming beam. The magnons can be excited only when the spin of incoming electrons is parallel to the sample magnetization (only by minority spins). The scattering process is commonly called as an inelastic electron scattering. In fact the scattering itself is elastic, but an apparent virtual energy loss is observed because the ejected electron stems from a lower energy level of the excited solid. This process is mediated by exchange interaction, i. e. Coulomb interaction. The wave-vector of the created magnons, $\mathbf{Q}_{\parallel}$, is provided by the wave-vector transfer of the electrons into the system, $\Delta k_{\parallel}$=k$_i \sin\theta$-k$_f \sin(\theta_0-\theta)$. Here θ (θ_0)

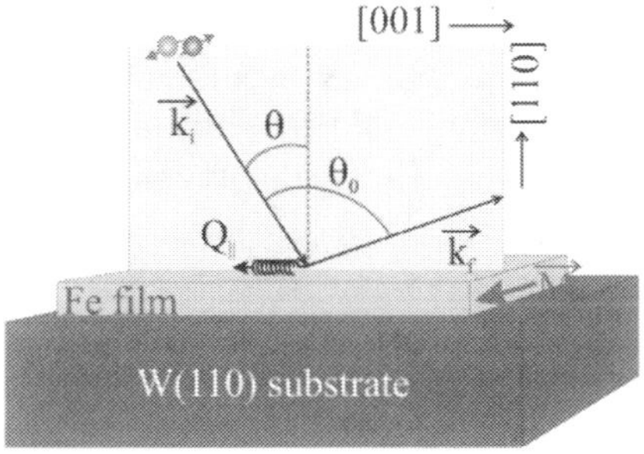

Figure 1. Schematic representation of the scattering geometry used in our SPEELS experiments.

denotes the angle between the incoming beam and the film normal (the outgoing beam), see Fig. 1.

Figure 2 shows examples of the intensity spectra measured at $\Delta k_{\parallel} = 0.6$ $\AA^{-1}$ on one-(a) and two- (b) atomic-layers of Fe films grown on W(110). The intensity of the backscattered electrons is plotted for the two possible spins of the incoming beam represented by $I_{\uparrow}$ and $I_{\downarrow}$ for spin up (↑) and down (↓), respectively. The spectra are dominated by the so-called quasi-elastic peak (large intensity at zero energy loss). Since only electrons with minority spin character are allowed to create magnons, the difference spectra ($I_{Diff} = \Delta I = I_{\downarrow} - I_{\uparrow}$) would provide all the necessary information. Therefore in the further data analysis we focus on the difference spectra. The magnon excitation appears as a peak in the energy loss region at a certain energy. The magnon dispersion relation can be measured by changing the wave-vector transfer. This is usually done by changing the angle of the incident beam with respect to the film normal via rotating the sample.

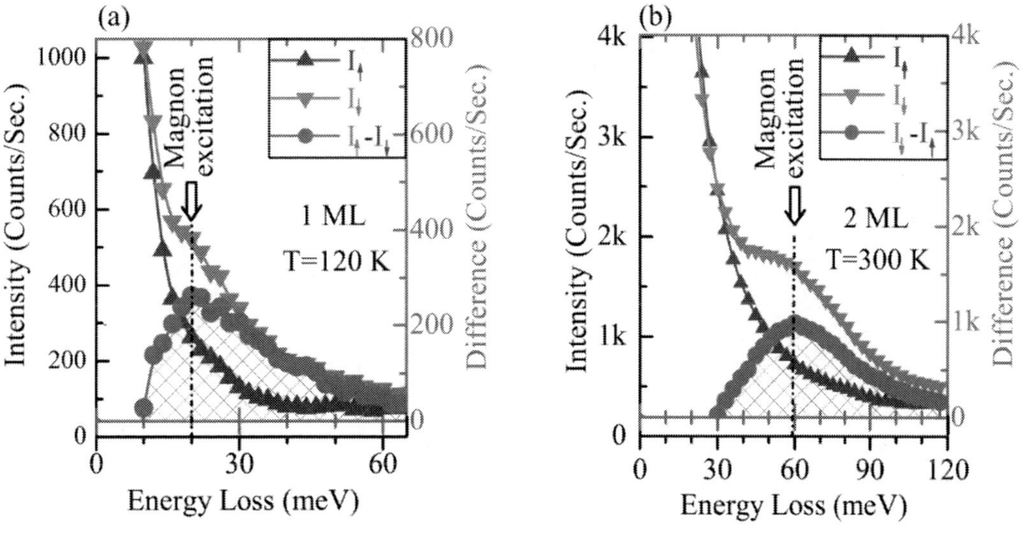

Figure 2. Spin-polarized electron energy loss spectra of one (a) and two (b) atomic layer(s) of Fe films on W(110) measured at a wave-vector transfer of $\Delta k_{\parallel} = 0.6$ $\AA$. All three spin up ($I_{\uparrow}$), spin down ($I_{\downarrow}$) and difference ($I_{\downarrow} - I_{\uparrow}$) spectra are shown. The spectra are recorded at 120 K (a) and 300 K (b). The energy of the incident beam was set to 3.8 eV in the case of monolayer and 4.0 eV for measuring the double-layer with an energy resolution of 11.2 meV and 13.9 meV, respectively.

3. Results and Discussion

3.1. Magnons in a monolayer of Fe

A direct comparison between the measured spectra of the Fe monolayer and double-layer, presented in Fig. 2, shows that for the monolayer Fe magnon excitations occur at much lower energies with respect to the ones in the Fe double-layer. In order to shed light onto the physics of the system, the magnon dispersion relation was measured over the whole Brillouin zone (see Fig. 3). Since the Curie temperature of the Fe monolayer is below RT (T_c^{ML}=223 K [21]), the spectra of the monolayer are measured at 120 K. For a better comparison the results of the surface mode of a thicker Fe film (with a thickness of 24 atomic layers) grown on the same substrate is also presented. As in the SPEELS experiments the electrons with very low energies are involved, therefore the technique is very surface sensitive. Hence measuring a thick film leads to a determination of the modes, which are mainly localized at the surface (acoustic surface modes). Figure 3 provides a direct comparison between the results of the monolayer Fe and the surface mode of Fe(110). It reveals that the magnon energies in the Fe monolayer are very small.

Taking a simple Heisenberg model with the Heisenberg spin Hamiltonian, $H = \frac{1}{2}J\sum_{i,j} S_i \cdot S_j - K_{eff}\sum_i(S_i \cdot \hat{n})^2$, one can derive the magnon dispersion relation. Here J is the exchange coupling between two spins, S, and K_{eff} is the effective magnetic anisotropy constant with an easy axis along $\hat{n}$ (the $[\bar{1}10]$-direction in this case). Fitting the experimental results with the calculated magnon dispersion relation within the Heisenberg model, $E = 4JS(1 - \cos\left(\frac{\Delta k_{\parallel} \cdot a_0}{2}\right) + 2K_{eff}S$; $a_0 = 3.165$ $\AA$, results in a very small exchange interaction (J=11±1 meV) and an effective magnetic anisotropy of $K_{eff} = 2.3 \pm 1.3$ meV [22]. These results are in reasonable agreement with the earlier theoretical and experimental works obtained using static magnetic measurements [18, 23, 24, 25, 26, 27].

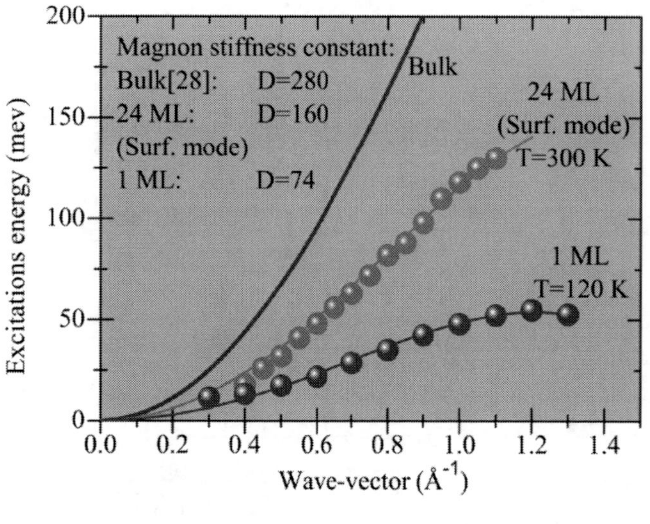

Figure 3. The magnon dispersion relation of one and twenty-four atomic layers of Fe films grown on W(110) measured along the $\bar{\Gamma} - \bar{H}$ direction of the Brillouin zone. The experiments are performed at 120 K and 300 K, respectively. The digitalized data of bulk Fe taken from [28] are also shown for a comparison. The values of magnon stiffness constants are obtained by a parabolic fit into the measured magnon dispersion and are given in meV/$\AA$.

Magnons in the Fe monolayer are even much softer than the acoustic surface mode of Fe(110). Fitting the experimental results, shown in Fig. 3, with the well-known parabolic dispersion ($E = DQ^2 + \beta Q^4$) leads to a determination of the magnon stiffness constant, D. We found that D for the monolayer system is by a factor of 2.2 smaller than the one obtained for the Fe(110) surface (D_{ML}=74 meV$\AA^2$ and $D_{surf.}$=160 meV$\AA^2$). It is only about one-fourth of the value measured for bulk Fe by INS [28]. A number of first principle calculations based on different approaches have predicted much larger energies for magnons in Fe monolayer

[23, 25, 29, 30, 31, 32, 33]. However, the theoretical results depend strongly on the parameters used for the calculations.

The magnon softening in the Fe monolayer may have different origins. The first one might be the temperature effect. Since the experimental results are obtained at 120 K, which is half of the Curie temperature of the system, this may cause the softening of the magnons. The seconded origin might be the strong hybridization with the substrate and change in the electronic structure of the film mediated by the beneath W(110) substrate. The third origin might be the influence of the DM interaction, which we will show that is very important for the Fe films grown on W(110).

Very recently, our experimental data are well reproduced by combining the first principle calculations with so-called atomistic spin dynamics simulations [34]. The authors could confirm that the temperature effect and the chemical relaxations, which influence the hybridization of the Fe film and W(110) lead to this magnon softening. According to their findings, although the DM interaction leads to an asymmetry in the magnon dispersion relation, it does not influence the overall softening too much. In the next section we will show how important is the DM interaction in the Fe/W(110) system and how does it influence the magnons.

3.2. The Dzyaloshinskii-Moriya interaction and magnons

Since the DM interaction is antisymmetric in its nature [10, 11], it would lift the degeneracy of magnons propagating along two opposite directions. In fact in the absence of the DM interaction the spin Hamiltonian is totally symmetric. By adding this interaction into the spin Hamiltonian the extended spin Hamiltonian reads as: $H = \frac{1}{2}J\sum_{i,j} S_i \cdot S_j - K_{eff}\sum_i (S_i \cdot \hat{n})^2 + \sum_{i,j} D_{ij} \cdot S_i \times S_j$. If one calculates the eigenvalues of this Hamiltonian, will find out that the cross product leads to a term, which is antisymmetric with respect to the sign of the wave-vector [23, 18]. In the other words, the DM interaction breaks the symmetry of the resulted dispersion relation, when changing Q to -Q.

In order to study the influence of the DM interaction on the magnons we have investigated the Fe double-layer in more detail [35]. The advantage of the Fe double-layer is that it is ferromagnetic at RT. This allows us to observe and analyze both magnon excitation and annihilation processes in energy loss and gain regions.

Figure 4(a) shows a set of difference SPEEL-spectra measured on a double-layer Fe film on W(110). The experiment is performed at wave-vector transfers of $\pm 0.6 \AA^{-1}$ and in both energy loss and gain regions. Please note that in a SPEELS experiment the annihilation processes occur in the energy gain region. In a real experiment, which is usually done at a finite temperature the thermally excited magnons do exist in the system. The probability for having such magnons is given by the Boltzmann distribution, which depends exponentially on the temperature.

In principle, electrons with majority character can annihilate the thermally excited magnons by gaining energy. This fact leads to a peak in the majority spin channel ($I_\uparrow$) and consequently a dip in the difference spectrum at a certain energy in the energy gain region. As this effect is very small a better representation of the data is plotting the asymmetry curve that is: $I_{Asy.} = \frac{\Delta I}{I_\downarrow + I_\uparrow}$.

Figure 4(b) shows the asymmetry curves of the data presented in Fig. 4(a). The magnon excitation and annihilation processes happening in energy loss and gain regions are clearly visible. Figure 4 reveals two facts: (i) The magnon excitation for the wave-vector transfer of +0.6 $\AA^{-1}$ takes place at a slightly higher energy with respect to the one of -0.6 $\AA^{-1}$. (ii) The magnon annihilation for the wave-vector transfer of +0.6 $\AA$ takes place at a lower energy with respect to the one at -0.6 $\AA^{-1}$. These facts clearly indicate that the degeneracy of the magnons is broken by the DM interaction because in the absence of the DM interaction the excitation and annihilation energies have to be exactly the same. Although these two facts clearly demonstrate the effect of the DM interaction on degeneracy breaking of the magnons, a measurement with a reversed magnetization would be another prove. Indeed the measurements performed with

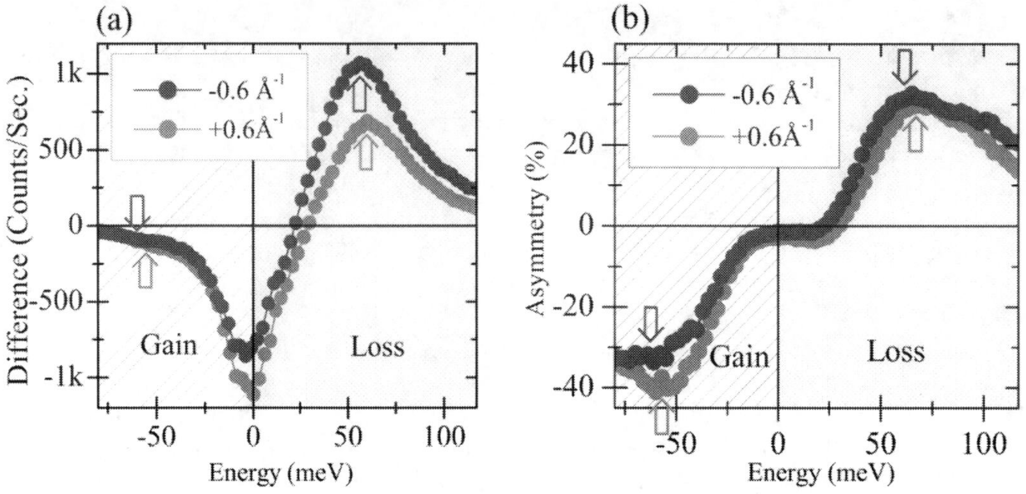

Figure 4. Difference (a) and asymmetry (b) energy loss and gain spectra measured on a two-atomic-layer thick Fe film on W(110). The sample magnetization was along $[\bar{1}10]$ direction. The arrows are showing the magnon excitation (energy loss) and annihilation peaks (energy gain).

magnetization pointing along the opposite direction confirmed our findings [35].

The first theoretical prediction of the effect of DM interaction on the magnon dispersion of Fe films on W(110) was done by Udvardi and Szunyogh [18]. They examined the Fe monolayer on W(110) and showed that the magnon dispersion in the presence of DM interaction is asymmetric. More recently, theoretical calculations performed by Costa *et al.*, based on multi band Hubbard model have also shown that in the Fe films grown on W(110) the spin-orbit coupling leads to the DM interaction and influences the magnon dispersion relation [36]. The results are in very good agreement with our experimental findings.

Using a Green's function technique Michael and Trimper could also confirm that the DM interaction leads to an asymmetry in the magnon dispersion [37]. Their method allows to consider the temperature effect on the asymmetry of magnon dispersion. According to their findings the temperature dependence of the asymmetry in the magnon dispersion is more important close to the transition temperature.

4. Conclusions

In conclusion, we have measured the magnon dispersion relation of one and two atomic layers of Fe on W(110) at 120 K and 300 K, respectively. It is shown that the magnons in one atomic layer of Fe are extremely soft in comparison to the Fe bulk and to the acoustic surface magnons of Fe(110). The magnon softening is attributed to the electronic structure of the Fe monolayer on W(110) and to the finite temperature effects.

By measuring an Fe double-layer on W(110) it is demonstrated that the antisymmetric DM interaction is very important in Fe films grown on W(110). The consequence of the DM interaction on the magnon dispersion relation of a low dimensional spin systems is measured. It is shown that the DM interaction lifts the degeneracy of the magnons leading to an asymmetric magnon dispersion relation. Our results are well confirmed by resent theoretical calculations.

5. Acknowledgments

The authors appreciate the fruitful discussions with D. L. Mills, L. M. Sandratskii, P. Buczek, L. Szunyogh, P. Weinberger, A. T. Costa and R. B. Muniz.

6. References

[1] Hillebrands B and Thiaville A 2006 Series: Topics in Applied Physics, Vol. 101 Volume package: Spin Dynamics in Confined Magnetic Structures III (Berlin: Springer, 2006).
[2] Kronmüller H and Parkin S 2007 Handbook of Magnetism and Advanced Magnetic Materials (New York: John Wiley & Sons, 2007).
[3] Frait Z 1959 *Czechoslovak Journal of Physics* **1** 750.
[4] Grünberg P A 1985 *Prog Surf. Sci.* **18** 1.
[5] Heinrich B 1994 Ultrathin magnetic structures, Ed Heinrich B and Bland J A C (Heidelberg: Springer 1994) Vol **2**, p 195.
[6] Fitzsimmons M R, Bader S D, Borchers J A, Felcher G P, Furdyna K, Hoffmann A, Kortright J B, Schuller I K, Schulthess T C, Sinha S K, Toney M F, Weller D and Wolf S 2004 *J. Magn. Magn. Mater* **271**, 103.
[7] Plihal M, Mills D L and Kirschner J 1999 *Phys. Rev. Lett.* **82** 2579.
[8] Vollmer R, Etzkorn M, Anil Kumar P S, Ibach H and Kirschner J 2003 *Phys. Rev. Lett.* **91** 147201.
[9] Tang W X, Zhang Y, Tudosa I, Prokop J, Etzkorn M and Kirschner J 2007 *Phys. Rev. Lett.* **99** 087202.
[10] Dzyaloshinskii I E 1957 *Sov. Phys. JETP* **5** 1259.
[11] Moriya T 1960 *Phys. Rev.* **120** 91.
[12] Bode M, Heide M, von Bergmann K, Ferriani P, Heinze S, Bihlmayer G, Kubetzka A, Pietzsch O, Blügel S and Wiesendanger R 2007 *Nature (London)* **447** 190.
[13] Vedmedenko E Y, Udvardi L, Weinberger P and Wiesendanger R 2007 *Phys. Rev. B* **75** 104431.
[14] Ferriani P, von Bergmann K, Vedmedenko E Y, Heinze S, Bode M, Heide M, Bihlmayer G, Blügel S and Wiesendanger R 2008 *Phys. Rev. Lett.* **101** 027201.
[15] Heide M, Bihlmayer G and Blügel S 2008 *Phys. Rev. B* **78** 140403(R)
[16] Heide M, Bihlmayer G and Blügel S 2009 *Physica B* **404** 2678.
[17] Meckler S, Mikuszeit N, Pressler A, Vedmedenko E Y, Pietzsch O and Wiesendanger R, 2009 *Phys. Rev. Lett.* **103** 157201.
[18] Udvardi L and Szunyogh L 2009 *Phys. Rev. Lett.* **102** 207204.
[19] Zakeri Kh, Peixoto T R F, Zhang Y, Prokop J and Kirschner J 2009 *Surf. Sci.* **604** L1.
[20] Zhang Y, Buczek P, Sandratskii L, Tang W X, Prokop J, Tudosa I, Peixoto T R F, Zakeri Kh and Kirschner J 2010 *Phys. Rev. B* **81** 094438.
[21] Przybylski M and Gradmann U 1987 *Phys. Rev. Lett.* **59** 1152.
[22] Prokop J, Tang W X, Zhang Y, Tudosa I, Peixoto, T R F, Zakeri Kh and Kirschner J 2009 *Phys. Rev. Lett.* **102** 177206.
[23] Udvardi L, Szunyogh L, Palotás K and Weinberger P 2003 *Phys. Rev. B* **68** 104436.
[24] Pini M G, Politi P and Stamps R L 2005 *Phys. Rev. B* **72** 014454.
[25] Costa A T, Muniz R B, Cao J X, Wu R Q and Mills D L 2008 *Phys. Rev. B* **78** 054439.
[26] Pratzer M, Elmers H J, Bode M, Pietzsch O, Kubetzka A and Wiesendanger R 2001 *Phys. Rev. Lett.* **87** 127201.
[27] Pratzer M and Elmers H J 2003 *Phys. Rev. B* **67** 094416.
[28] Mook H A and Nicklow R M 1973 *Phys. Rev. B* **7** 336.
[29] Muniz R B and Mills D L 2002 *Phys. Rev. B* **66** 174417.
[30] Muniz R B, Costa A T and Mills D L 2003 *J. Phys. Cond. Matter* **15** S495.
[31] Costa A T, Muniz R B and Mills D L 2003 *Phys. Rev. B* **68** 224435.
[32] Costa A T, Muniz R B and Mills D L 2006 *Phys. Rev. B* **74** 214403.
[33] Muniz R B, Costa A T and Mills D L 2008 *IEEE Trans. Mag.* **44** 1974.
[34] Bergman A, Taroni A, Bergqvist L, Hellsvik J, Hjorgvin B and Eriksson O 2010 *Phys. Rev. B* **81** 144416.
[35] Zakeri Kh, Zhang Y, Prokop, J, Chuang T- H, Sakr N, Tang W X and Kirschner J 2010 *Phys. Rev. Lett.* **104** 137203
[36] Costa A T, Muniz R B, Lounis S, Klautau A B and Mills D L 2010 *Phys. Rev. B* **82**, 014428.
[37] Michael Th and Trimper S 2010 *Phys. Rev. B* **82**, 052401.

Energy thresholds in the magnetic vortex core reversal

Sebastian Gliga[1,2,4], Yaowen Liu[3] and Riccardo Hertel[1]
[1]Institut für Festkörperforschung, IFF-9, Forschungszentrum Jülich, Jülich, 52428 Germany
[2]Center for Nanoscale Materials, Argonne National Laboratory, Argonne Illinois 60439 USA
[3]Physics Department, Tongji University, Shanghai, 200092 People's Republic of China

E-mail: sgliga@anl.gov

Abstract. Owing to its small size and high stability, the core of magnetic vortices has been proposed as a possible candidate for binary data storage. We have investigated the energetic aspect of the vortex core reversal in ferromagnetic thin-film elements by means of fully three-dimensional finite-element micromagnetic simulations. The simulations show that the reversal occurs once a well-defined threshold in the exchange energy density is reached. We establish the energetic origin of the core reversal by demonstrating that this threshold corresponds to the energy necessary for the production of a vortex-antivortex pair. The effect of the discretization size on the computed exchange energy of a magnetic singularity is discussed. Such a singularity is required for the core reversal process to unfold.

1. Introduction

The vortex is a naturally forming structure, which closes the flux in ferromagnetic thin-film elements. It is characterized by an in-plane curling of the magnetization around a narrow core only a few tens of nanometers in diameter where the magnetization is perpendicular to the film plane [1]. Owing to the very high stability of the core, its reversal is one of the most dramatic transformations of a vortex. One method for reversing the core consists in applying a static field of the order of a few hundreds of mT in the direction opposite to the core orientation [2,3]. Another method, which requires considerably weaker fields, consists in dynamically switching the core by means of in-plane AC excitations [4] or single in-plane field pulses [5,6,7]. In the same manner, the core reversal can be triggered using in-plane alternating spin-polarized AC currents [8] or pulses [9,10]. In all these cases, the reversal of the core orientation is mediated by the same series of complex processes: first, a transient vortex-antivortex pair is produced, followed by the annihilation of the original vortex with the newly formed antivortex [4,5,6]. The reversal unfolds over only a few tens of picoseconds and represents the fastest magnetization switching process known in micromagnetic theory [6].

The above results show that there are two distinct routes allowing to dynamically trigger the vortex core reversal using in-plane fields or spin-polarized currents. In the first case, the gyrotropic mode of the vortex is excited by applying an in-plane sinusoidal excitation tuned at the vortex resonant frequency [4,8]. This frequency is typically below one GHz in mesoscopic thin film elements [11], such that the switching time is of the order of a few to tens of nanoseconds. In contrast to this *resonant* switching, a second route, which consists in triggering the core reversal using *non-resonant* pulses,

4. Use of the Center for Nanoscale Materials was supported by the U.S. Department of Energy, Office of Science, Office of Basic Energy Sciences, under Contract No. DE-AC02-06CH11357.

allows to switch the core orientation within only a few hundreds of picoseconds. While the micromagnetic mechanism is identical in both cases, the switching times can therefore differ by up to an order of magnitude. To understand the origin of this difference, we searched for the fundamental physical parameter that triggers the core reversal. It has been suggested in Refs. [8,12] that the reversal is driven by a dynamic 'gyrofield' generated by the moving vortex. In this context, the core switch is predicted to occur for a critical velocity of the driven vortex [8,13] in analogy to the Walker breakdown of domain walls [14]. While this interpretation allows associating a measurable parameter to the core reversal, it does not explain the creation of a new vortex-antivortex pair, *i.e.* the micromagnetic mechanism of the reversal. In addition, at the critical velocity, the gyrofield was found to diverge [12]. We therefore adopted a different approach and studied the evolution of the vortex energy.

Considering the case of the current-driven core reversal, we found that the core orientation reverses once a well-defined critical value of the exchange energy is reached. Our simulations further show that under the influence of an in-plane spin-polarized current, the increase in exchange energy of the moving vortex is due to: 1) the distortion of the vortex in-plane structure, 2) the contraction of the moving core radius and 3) the formation of a new vortex-antivortex pair in the sample. The first two effects only account for less than 1% of the exchange energy increase at the moment of the core switch. The observed energy threshold is therefore due to the formation of a new pair, evidencing the energetic origin of the vortex core reversal.

2. Micromagnetic simulations

The results were obtained using our micromagnetic finite-element code [9] based on the Gilbert equation, including the adiabatic and non-adiabatic spin torque terms [15,16]:

$$\frac{d\boldsymbol{m}}{dt} = -\gamma(\boldsymbol{m}\times\boldsymbol{H}_{\mathrm{eff}}) + \alpha\left(\boldsymbol{m}\times\frac{d\boldsymbol{m}}{dt}\right) - (\boldsymbol{u}\cdot\nabla)\boldsymbol{m} + \beta\boldsymbol{m}\times[(\boldsymbol{u}\cdot\nabla)\boldsymbol{m}]$$

where $\boldsymbol{m}=\boldsymbol{M}/M_s$ is the local reduced magnetization, M_s the saturation magnetization, α the Gilbert damping constant, β the strength of the nonadiabatic torque term, and γ the gyromagnetic ratio. The vector $\boldsymbol{u}$ is in the direction of the electron flow and has an amplitude of $u=jPg\mu_B/(2eM_s)$, where j is the current density, P the degree of polarization, g the Landé factor, μ_B the Bohr magneton and e the electron charge. The effective field $\boldsymbol{H}_{\mathrm{eff}}$ is the negative variational derivative of the local energy density e with respect to the magnetization: $\mu_0\boldsymbol{H}_{\mathrm{eff}}=-\delta e/\delta\boldsymbol{M}$. The energy density takes into account contributions from the magnetostatic, exchange and anisotropy terms.

We performed simulations for a Permalloy (Py) disk of radius 100 nm and thickness 20 nm using the following parameters: A=13 pJ/m (exchange constant), $\mu_0 M_s$=1.0 T, α=0.01, K_s=0.10 mJ/m^2 [17] (surface anisotropy), β=0.02 [16] and P=0.7 [18]. The sample was discretized into 150,780 irregular tetrahedral elements, corresponding to an average cell size of (1.6 nm)^3. The magnetization is computed at the apexes of the elements and the vector field is linearly interpolated within the elements. The demagnetizing field is computed using a combined finite element-boundary element method [19]. The surface anisotropy energy is calculated by integrating over the triangular boundary elements with the same value of K_s assumed at the top and bottom surfaces. Time integration is performed using the Adams method with adaptive time intervals in the sub-picosecond range. A homogeneous current density distribution was assumed.

3. Energy thresholds for resonant and non-resonant switching

We investigated the time evolution of the vortex energy during the core switch for the resonant and non-resonant switching pathways. Our analysis shows that the core reversal is connected to a well-defined critical threshold value of exchange energy in both cases.

3.1. Resonant switching

We first consider the case when the vortex is driven by a resonant current $j(t) = J\sin(\omega_0 t)$, where J is the current amplitude and ω_0 the vortex resonant frequency, which is a function of the sample aspect ratio. For our sample, we determined numerically ω_0 = 776 MHz, in very good agreement with the analytical model in Ref. [11]. In this case, the minimum current density required to trigger the vortex core reversal is of $J = 4\times10^{11}$ A/m^2. The reversal occurs after approximately 14 ns. Increasing J leads to faster switching, of the order of a few nanoseconds as shown in figure 1.

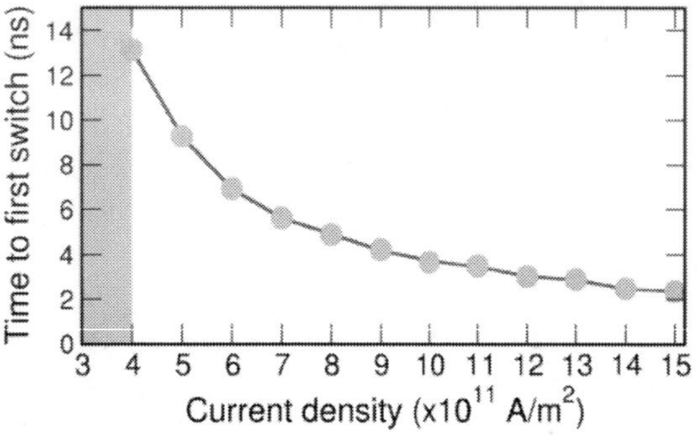

Figure 1. Vortex core switching time as a function of the resonant current amplitude. The switching time is defined from the moment an electric current is applied. No reversal occurs in the grey shaded region.

The time evolution of the average energy densities in the sample is shown in figure 2 for a current amplitude of $J = 1.2\times10^{12}$ A/m^2. As long as the current is applied, the core periodically reverses its polarization. Each switching event is clearly distinguishable by an increase in the total energy, followed by a sudden drop connected to the vortex-antivortex annihilation [20]. We found that the reversal is characterized by sharp peaks in the exchange energy. Considering in parallel the evolution of the magnetization, we have determined that these peaks correspond to the production of a new vortex-antivortex pair prior to each switching event. In contrast, the observed peaks in the magnetostatic energy density result from finite-size effects, *e.g.* the proximity of the excited vortex to the sample edge.

Following the reversal, the resulting spin waves [20,21] lead to magnetic ringing. Clearly, the switching events occur once specific total and exchange energy densities are reached, demonstrating the existence of well-defined switching thresholds. We also note that the surface anisotropy energy does not show important variations (maximum 2%) during the reversal process.

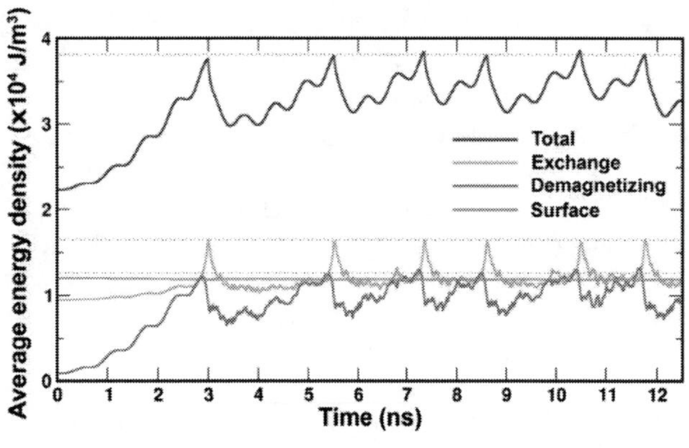

Figure 2. Evolution of the total and partial energy densities during the core reversal under the influence of an alternating current of maximal amplitude $J = 1.2\times10^{12}$ A/m^2. The dotted lines indicate the energy density thresholds (peak averages). The total energy is the sum of the exchange, demagnetizing and surface terms.

3.2. Non-resonant switching

The time evolution of the internal energy when the core switch is triggered by a non-resonant Gaussian pulse is shown in figure 3. Peaks in the exchange and total energy densities, followed by a sharp drop, equally accompany the reversal process. For the studied sample geometry, a minimum current of $J = 7.3\times10^{11}$ A/m^2 is required for the switch to occur.

Remarkably, we find that the exchange energy density threshold is the same within 2% for both the resonant and non-resonant reversal routes. We moreover found that it is independent of the applied current amplitude. From the data in sections 3.1 and 3.2, the average increase in exchange energy density at the moment of the reversal is $\Delta e_{\mathrm{thresh}} = (0.59\pm0.01)\times10^4$ J/m^3, where $\Delta e_{\mathrm{thresh}}$ is the difference between the static energy of the vortex and the energy at the moment of the switch.

Moreover, we find that the radius at which the core reversal occurs is the same for both the resonant and non-resonant pathways and is equally independent of the current strength. The threshold in magnetostatic energy density is also within 2% for both reversal pathways.

If energy is indeed the fundamental parameter determining the core reversal, the above threshold should not be affected by variations in the driving current parameters or in intrinsic properties of the sample, such as the damping constant. In the following, we study the influence of the current polarization and the damping constant α on the switching threshold.

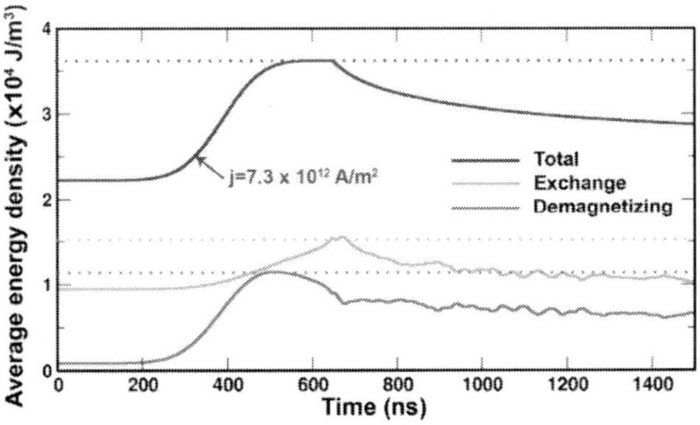

Figure 3. Evolution of the partial and total energy densities for the core switch triggered by a Gaussian current pulse of strength $J = 7.3\times10^{11}$ A/m^2 and width σ = 100 ps [9]. The dashed lines indicate the threshold energy densities.

4. Threshold invariance

4.1. Under variation of the current polarization

To verify the influence of the current polarization on the switching thresholds, the core was reversed by means of single in-plane field pulses as described in section 3.2. The width of the Gaussian pulses was kept at 100 ps and their strength varied for different values of the polarization P. Figure 4 shows that the value of the partial energy densities at the moment of the reversal are independent of the current polarization. Clearly, for the reversal to occur, the exchange energy in the sample must increase by $\Delta e_{\mathrm{thresh}}$ as found in section 3.2. In contrast, the inset of figure 4 shows the minimum switching current as a function of its polarization.

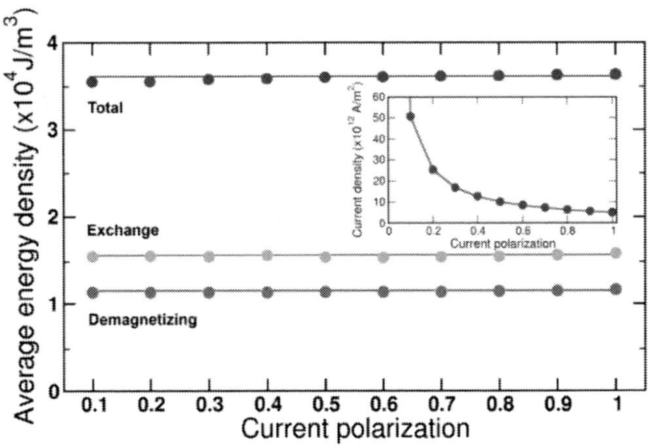

Figure 4. Average energy densities at the moment of the core switch as a function of the driving current polarization. The switch is triggered by non-resonant pulses as in Figure 3. The full lines indicate the average threshold value for each energy contribution. Inset: Current densities required to trigger the vortex core reversal as a function of the current polarization.

4.2. Under variation of the damping constant

The inset in figure 5 shows the behavior of the minimum current density required to reverse the core as a function of the damping coefficient α using a resonant excitation.

We show in Figure 5 that the switching energies are again invariant, with an average value within about 2% of the values found in figure 4. We note that following section 3, the threshold energies are independent of the excitation type, such that the results obtained for the energies in 4.1 and 4.2 hold for both resonant and non-resonant excitations.

A further analysis of the influence of other parameters such as the sample radius and thickness will be presented elsewhere.

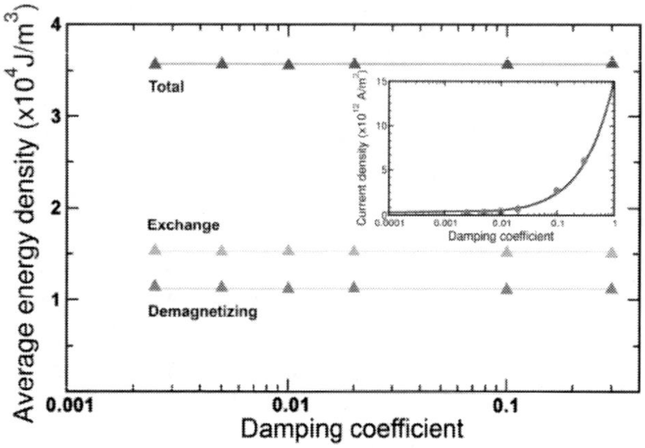

Figure 5. Average energy densities at the moment of the core switch as a function of the damping coefficient α. The switch is triggered by resonant excitations as in Figure 2. Inset: Evolution of the resonant current amplitude required to trigger the core reversal as a function of the damping coefficient α. The data (solid green circles) is fitted to a power function (blue line).

5. Vortex-antivortex pair energy

So far, we established the invariance of the switching energy thresholds with respect to extrinsic as well as intrinsic parameters of the vortex. In addition, we found that the exchange energy density Δe_{thresh} required for the core reversal to occur is connected the formation of a vortex-antivortex pair.

According to Refs. [22,23], the exchange energy of a pair is $E_{\text{pair}} = 8\pi Ah$, with A the exchange constant and h the sample thickness. In our sample, an increase in exchange energy by E_{pair} would correspond to an average exchange energy density increase of $\Delta e_{\text{pair}} = 1.04\times10^4$ J/m^3. In comparison, the threshold value Δe_{thresh} obtained in 3.2 from the simulations only represents about 60% of the energy of a vortex-antivortex pair in Py. As a further example, we have also investigated the energy evolution during the core switch in Iron (Fe: A= 21 pJ/m, $\mu_0 M_s$= 2.15 T). In this case, our simulations show that the switching threshold energy is of the order of only 40% of that required for pair production. We attribute these material-dependent discrepancies to the presence of a Bloch point in the sample. Indeed, the core reversal is mediated by the formation and propagation of a singularity [6,20]. In this case, the computed exchange energy is strongly dependent on the discretization cell size [3].

In order to determine the accurate value of the exchange energy density at the moment of the core reversal, we varied the distance between discretization nodes in our simulations and extrapolated the switching energy at zero cell size. The results are shown for both Py and Fe in figure 6 where the value of Δe_{thresh} is expressed in units of $8\pi Ah$. Fits to the data show an exponential behaviour of the switching energy as a function of the distance Δx between discretization nodes. At zero cell size we find that Δe_{thresh} indeed converges to $8\pi Ah$ for both materials.

6. Conclusions

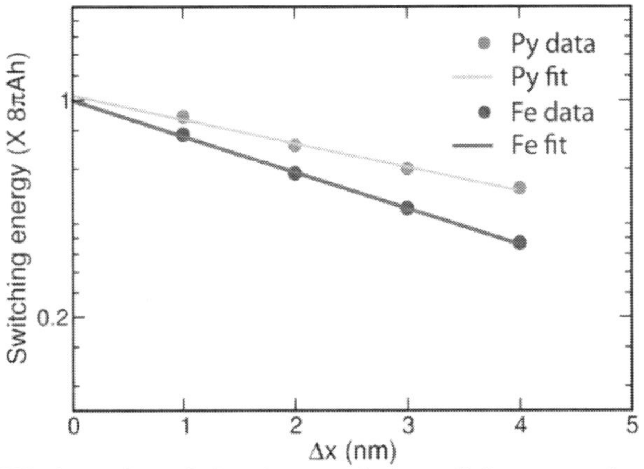

Figure 6. Evolution of the exchange energy threshold in Py and Fe as a function of the average distance Δx between neighboring discretization points. The best fit to the data is given by an exponential function. A distance of one nm between nodes approximately corresponds to an equivalent cubic cell of $(0.5\ \text{nm})^3$.

We investigated the time evolution of the magnetic energy during the vortex core reversal for the resonant and non-resonant switching pathways. The simulations demonstrate that in both cases a well-defined exchange energy barrier has to be overcome in order to trigger the vortex core reversal. This switching threshold does not depend on the type of excitation and is equally independent of the amplitude and polarization of the applied spin-polarized current as well as of the value of the damping constant.

Because the reversal process involves a magnetic singularity, in order to accurately determine the exchange energy at the moment of the reversal, we have extrapolated this energy for a vanishing distance between discretization nodes. We found that the value of the exchange threshold corresponds to the analytically predicted energy required for the formation of a new vortex-antivortex pair. This provides evidence for the energetic origin of the micromagnetic mechanism driving the core reversal. A link between the critical velocities found in Refs. [10,12,13] and the threshold energy remains to be established.

References

[1] Wachowiak A, Wiebe J, Bode M, Pietzsch O, Morgenstern M and Wiesendanger R 2002 *Science* **298** 577
[2] Kikuchi N, Okamoto S, Kitakami O, Shimada Y, Kim S G, Otani Y and Fukamichi K 2001 *J. Appl. Phys.* **90** 6548
[3] Thiaville G A, Garcia J M, Dittrich R, Miltat J, and Schrefl T 2003 *Phys. Rev. B* **67** 094410
[4] Van Waeyenberge B, Puzic A, Stoll H, Chou K W, Tyliszczak T, Hertel R, Fähnle M, Brückl H, Rott K, Reiss G, Neudecker I, Weiss D, Back C H and Schütz G 2006 *Nature* **444** 461
[5] Xiao Q F, Rudge J, Choi B C, Hong Y K and Donohoe G 2006 *Appl. Phys. Lett.* **89** 262507
[6] Hertel R, Gliga S, Fähnle M and Schneider C M 2007 *Phys. Rev. Lett.* **98** 117201
[7] Weigand M, Van Waeyenberge B, Vansteenkiste A, Curcic M, Sackmann V, Stoll H, Tyliszczak T, Kaznatcheev K, Bertwistle D, Woltersdorf G, Back C H and Schütz G 2009 *Phys. Rev. Lett.* **102** 077201
[8] Yamada K, Kasai S, Nakatani Y, Kobayashi K, Kohno H, Thiaville A and Ono T 2007 *Nature Mater.* **6**, 355
[9] Liu Y, Gliga S, Hertel R and Schneider C M 2007 *Appl. Phys. Lett.* **91** 112501
[10] Yamada K, Kasai S, Nakatani Y, Kobayashi K and Ono T 2008 *Appl. Phys. Lett.* **93** 152502
[11] Novosad V, Fradin F Y, Roy P E, Buchanan K S, Guslienko K Y and Bader S 2005 *Phys. Rev. B* **72** 024455
[12] Guslienko K, Lee K-S and Kim S-K 2008 *Phys. Rev. Lett.* **100**, 027203
[13] Lee K-S, Kim S-K, Yu Y-S, Choi Y-S, Guslienko K Y, Jung H and Fischer P 2008 *Phys. Rev. Lett.* **101** 267206
[14] Schryer N L and Walker L R 1974 *J. Appl. Phys.* **45** 5406
[15] Zhang S and Li Z 2004 *Phys. Rev. Lett.* **93** 127204
[16] Thiaville A, Nakatani Y, Miltat J and Suzuki Y 2005 *Europhys. Lett.* **69** 990
[17] Rantschler J O, Chen P J, Arrott A S, McMichael R D, Egelho W F and Maranville B B 2005 *J. Appl. Phys.* **97** 10J113

[18] Kasai S, Fischer P, Im M-Y, Yamada K, Nakatani Y, Kobayashi K, Kohno H and Ono T 2008 *Phys. Rev. Lett.* **101** 237203
[19] Fredkin D R and Koehler T R 1990 *IEEE Trans. Magn.* **26** 415
[20] Hertel R and Schneider C M 2006 *Phys. Rev. Lett.* **97** 177202
[21] Lee K-S, Kang B-W, Yu Y-S and Kim S-K 2005 *Appl. Phys. Lett.* **85** 1568
[22] Tretiakov O A and Tchernyshyov O *Phys. Rev. B* 2007 **75** 012408
[23] Belavin A A and Polyakov A M 1975 *JETP Letters* **22** 503

Journal of Physics: Conference Series **303** (2011) 012006 doi:10.1088/1742-6596/303/1/012006

Magnetization processes in a dot of ferrimagnetic garnet near the compensation temperature

V. Joch[1,2], P. Molho[1], L. Ranno[1]

Institut Néel, CNRS-UJF, BP 166, 38042 Grenoble cedex 9, France

Brno University of Technology, Brno, Czech Republic

pierre.molho@grenoble.cnrs.fr

Abstract. At the compensation temperature of ferrimagnetic materials, magnetization vanishes but domain wall energy does not. Magnetic domains and coercive field become then very large. We are interested in small objects of such materials, in order to study the effect of size, shape, edges... on the domain structure and on the magnetization processes. A magnetic garnet film was patterned by chemical etching, leading to dots 7 μm thick and 10 to 100 μm wide. Domain observations were performed close to compensation (213 K) under magnetic field up to 22 mT. In the case where the dots are connected by a thin continuous magnetic film (partial etching), magnetization reversal occurs via an intermediate spin-flop state, coexisting with perpendicular domains. In order to analyze this process, we performed a mean field computation to characterize the spin-flop state itself, by the amplitude and orientation of the magnetization of each sublattice.

1. Introduction

At the compensation temperature of ferrimagnetic materials, macroscopic magnetization vanishes but domain wall energy does not. Magnetic domains become then very large because dipolar energy vanishes, and also the coercive field because Zeeman energy vanishes. We are interested in observing small objects of such materials, in order to study the effect of size, shape, presence of edges, etc. on the domain structure itself, and on the magnetization processes. Temperature is then a parameter, which allows playing with the characteristic length of the system. We are studying the domain structure and hysteresis loops in a garnet film, using magneto-optical effect and image processing, to characterize the magnetization processes (nucleation, propagation, magnetization reversal).

The magnetic garnet sample, $(GdTmPrBi)_3(FeGa)_5O_{12}$, provided by CEA-LETI, is an epitaxial film grown on GGG substrate, about 7 μm thick. It has a strong uniaxial magnetic anisotropy perpendicular to the film, leading to up and down domains, separated by Bloch walls. At room temperature the usual maze domain structure is observed. This sample is ferrimagnetic and exhibits a compensation temperature of about 213 K, where the magnetization goes to zero. The resulting domain structure is characterized by very large domains (fig. 1).

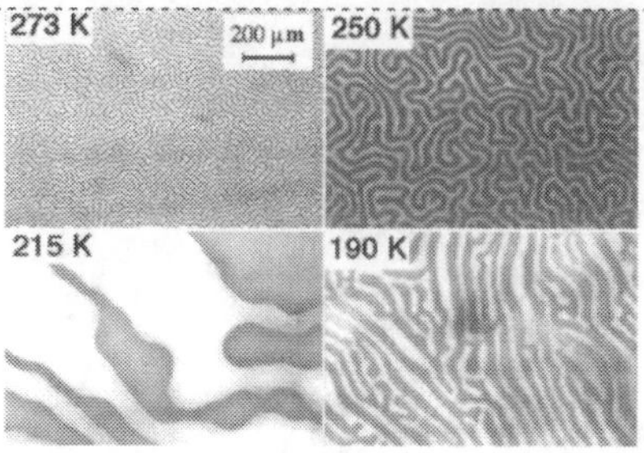

Figure 1. Domain structure as a function of temperature. Close to Tcomp (213 K) domains become very large.

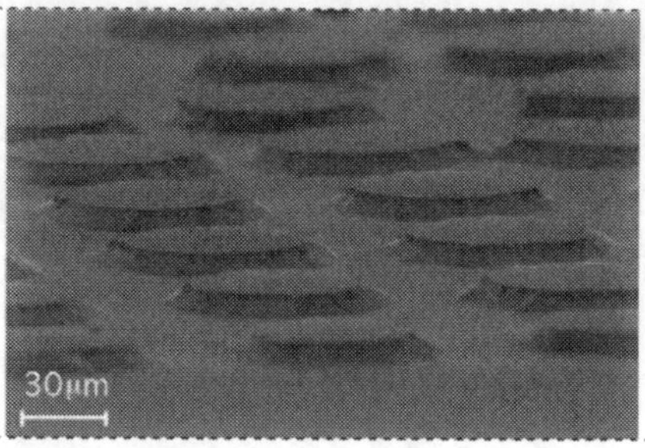

Figure 2. SEM image of the patterned film : lattice of dots 7 μm thick.

We optimized the patterning procedures (lithography, chemical etching…) to obtain lattices of ferrimagnetic dots, squares and disks, between 10 and 100 μm wide and 7 μm thick (fig. 2). The first step of the patterning process is the deposition of a 100 nm-thick layer of SiO_2 on the garnet film, patterned by UV lithography and Reactive-Ion Etching. After lifting off the photo resist, the magnetic film itself is patterned by chemical etching with hot phosphoric acid [1]. Partial etchings were also performed, leaving a thin magnetic layer between the dots.

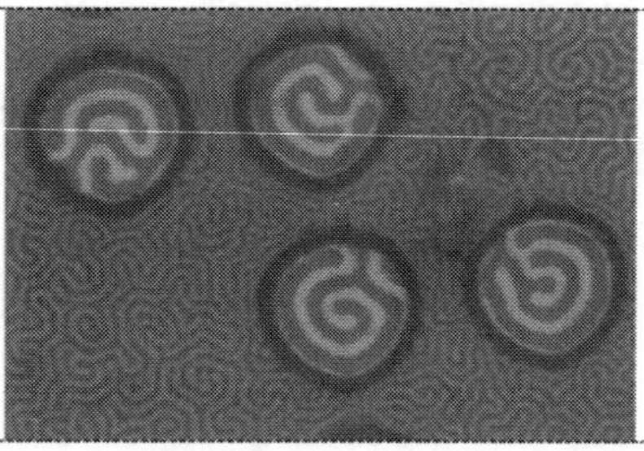

Figure 3. Domain structure at room temperature in a partially-etched sample. The domain size depends on thickness : 7 μm for the dots, 3 μm for the film (dot diameter : 50 μm)

The present study concerns a partially etched sample, especially four dots of diameter 50 μm, on top of a continuous film 3 μm thick. The domain structure in this thinner films appears like a 'background' for the domain structure in the dots (fig. 3). Beside the domain size reduction due to the film thickness, the first observations at 300 K seem to show no interaction between the dots, even through the 'background', but the shape and size of the dots seem to affect the domain morphology.

2. Experiment

Our goal is to analyze the magnetization processes close to Tcomp. Observations are realized by means of Faraday effect, using an optical polarizing microscope, under magnetic fields perpendicular to the film up to 22 mT, at temperatures close to Tcomp (213 K).

An example of a field cycle is shown in figure 4. It is not the magnetization as function of the field which is displayed, but the grey levels measured in the images, corresponding to the background and to the dots. The field cycle in term of grey level is quite complex, in particular due to optical effects (not magnetic) resulting from the patterning, so that intensity changes in the background affect the dot, and reciprocally. The interpretation of the field cycle in term of magnetization can be described the following way :

- The initial state, in zero field, is negative dots, negative background (100% remanence).
- When the field increases, there is nucleation of a reverse domain in the background, propagation of this domain (avoiding the dots), leading to a positive background, while the dots are still negative.

- Further increase of the field induces nucleation of a spin-flop state at the edge of the dots (ring), propagation in the dots, an intermediate state of coexistence between spin-flop and negative domain inside the dot, then collapse of this negative domain. The final observed state is then spin-flop in the dot, and positive background (fig. 5).
- Decreasing the field, the dots flip to positive domain state, so that in zero field the observed state is positive dots, positive background.
- Finally negative fields lead a similar evolution with a symmetrical situation.

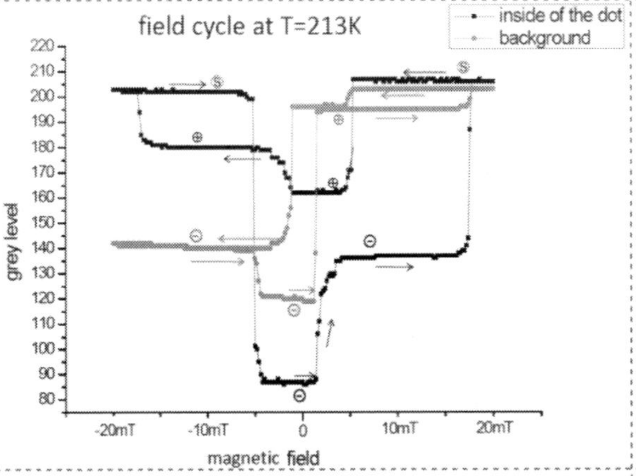

Figure 4. Optical field cycle, reflecting magnetization changes in the dots (black curve) and in the continuous background film (red curve).

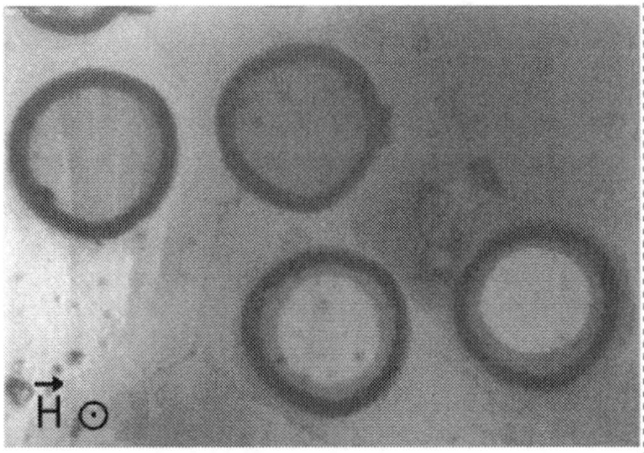

Figure 5. Example of a configuration during the field cycle (17 mT, increasing field). Bottom dots: coexistence between spin-flop state (darker ring) and normal domain opposite to the field. Top dots: the normal domain has collapsed.

The background is a continuous film, so nucleation of the reverse domain takes place on defects or edges, followed by propagation of the domain wall, leading to a low coercive field. In the dots, the nucleation probability is reduced, leading to a larger coercive field. The dots remain then negative in positive field, surrounded by a positive domain in the background, and with a domain wall in the edge of the dot. This situation is favorable for the nucleation of a spin-flop state when the temperature is close enough to Tcomp, which occurs here at about 20 mT.

The coexistence of negative domain and spin-flop within a dot is somehow analogous to a bubble domain opposite to the field, surrounded by domain in the field direction. Wall and Zeeman energies favor spin-flop. The energy barrier due to dipolar energy leads to a range of stability of such situation, before the collapse of the negative domain under a critical radius.

To go further in such description, it is necessary to better characterize the spin-flop state itself, the amplitude and orientation of the magnetization of each sublattice.

3. Model

The idea is a mean field approach, computing the magnetization of each sublattice as resulting from the local field (resulting from the magnetization of each sublattice).

The real formula for our sample is $(GdTmPrBi)_3(FeGa)_5O_{12}$ where a small amount of Bi was added to increase the magnetooptical effects, Pr to increase the anisotropy, and Tm to compensate for the lattice parameter mismatch with the substrate. We assume that concerning the magnetic properties, the magnetic elements in the rare-earth site can be described as being Gd, and the non magnetic as being Y, the resulting formula being $(YGd)_3(FeGa)_5O_{12}$.

The crystallographic cell (8 formula units) can be described as composed by 3 sublattices :
Fe-octa, containing 16 Fe^{3+}, surrounded by octahedron of oxygen, (referred as Fe1),
Fe-tetra, containing 24 Fe^{3+}, surrounded by tetrahedron of oxygen, (referred as Fe2),
and Gd, containing 24 Gd^{3+}, surrounded by dodecahedron of oxygen.

For each sublattice, the magnetization is defined by its amplitude at T= 0, a unit vector $\vec{m}$, the thermal variation being described by a Brillouin function :

$$\vec{M}(T) = M(0).\vec{m}(z).B_s(z)$$

$$B_s(z) = \frac{2S+1}{2S}\coth\left(\frac{2S+1}{2S}z\right) - \frac{1}{2S}\coth\left(\frac{z}{2S}\right) \quad \text{with } z = g\mu_B S.H_{eff}/kT$$

$$\vec{H}_{eff}^{Fe1} = \lambda_{11}\vec{M}_{Fe1} + \lambda_{12}\vec{M}_{Fe2} + \lambda_{13}\vec{M}_{Gd} + \vec{H}_0$$

$$\vec{H}_{eff}^{Fe2} = \lambda_{21}\vec{M}_{Fe1} + \lambda_{22}\vec{M}_{Fe2} + \lambda_{23}\vec{M}_{Gd} + \vec{H}_0$$

$$\vec{H}_{eff}^{Gd} = \lambda_{31}\vec{M}_{Fe1} + \lambda_{32}\vec{M}_{Fe2} + \lambda_{33}\vec{M}_{Gd} + \vec{H}_0$$

$\vec{H}_{eff}$ is the effective field on each sublattice. The exchange coupling constant λ_{ij} are taken from literature, based on study of $(Y)_3(FeGa)_5O_{12}$ [2] and $(Gd)_3(Fe)_5O_{12}$ [3]. $\vec{H}_0$ is the applied field.

Qualitatively, the stronger coupling is (Fe-tetra)-(Fe-octa) and is negative. The next stronger is (Fe-octa)-(Fe-octa), then (Fe-tetra)-(Fe-tetra). The next one is Gd-(Fe-tetra), also negative, and finally Gd-(Fe-octa). The coupling Gd-Gd is negligible) [2-4].

The values for exchange constant we have taken are the following :
λ_{11}= -721; λ_{12}= -1077; λ_{13}= -11; λ_{22}= -343; λ_{23}= -48; λ_{33}= -2.

Adjusting the composition to fit Tcomp (213 K) and Tc (450 K), taking into account the fact that 90% of Ga atoms substitute Fe-tetra [5], lead to the following formula :

$$(Y_{1.4}Gd_{1.6})(Fe_{1.94}Ga_{0.06})(Fe_{2.46}Ga_{0.54})O_{12}$$

The computed thermal variation of spontaneous magnetization for each sublattice is presented in figure 6. As expected it exhibits a compensation temperature close to the experimental value. What is remarkable is that the amplitude of Gd magnetization is very temperature dependent.

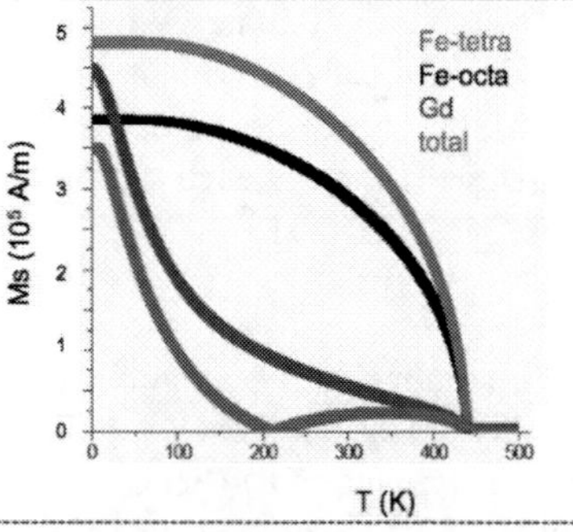

Figure 6. Computed spontaneous magnetizations (total and for each sublattice) as a function of T.

Under a constant magnetic field the thermal variation of the magnetization exhibits a spin-flop transition in a temperature range of about 20 K (fig. 7). The comparison with experiment is quite reasonable (fig. 8), apart from the field range which is very different. This can possibly be adjusted playing finely with the composition and the coupling constant values.

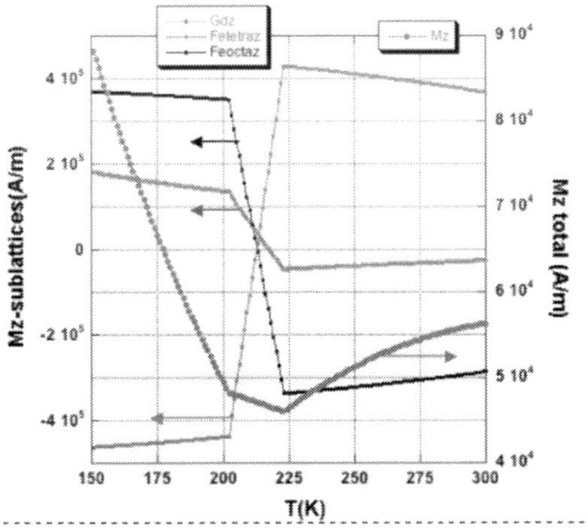

Figure 7. Model: thermal variation of Mz under constant field, for each sublattice and the total magnetization: spin-flop between 203 and 223 K, under 10 T.

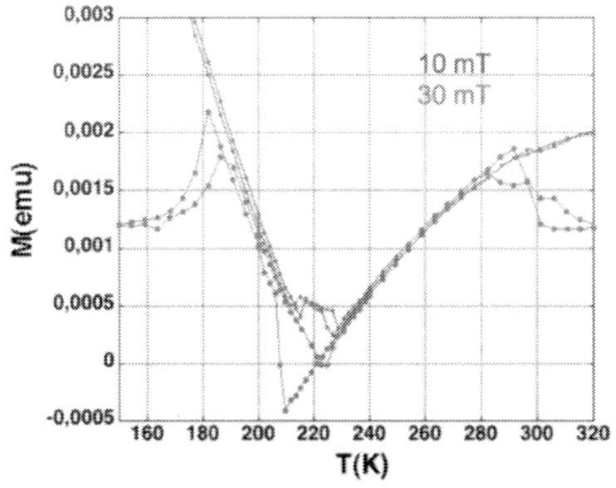

Figure 8. Experiment: thermal variation of Mz under constant field, for 10 and 30 mT. Spin-flop between 210 and 225 K under 30 mT. Far from Tcomp Mz increases, and 10 mT is a field too small to saturate the sample.

The main results from this computation is a description of the spin-flop state, as function of T, around Tcomp. The amplitude and orientation of the magnetization of each sublattice, as function of T, is shown in figure 9, and the angle between the sublattices in figure 10.

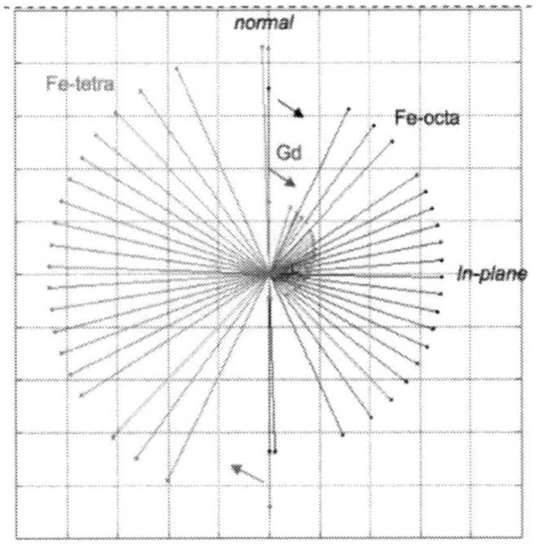

Figure 9. Rotation, with respect to the normal to the film, of the magnetization of each sublattice, as function of T, during spin-flop.

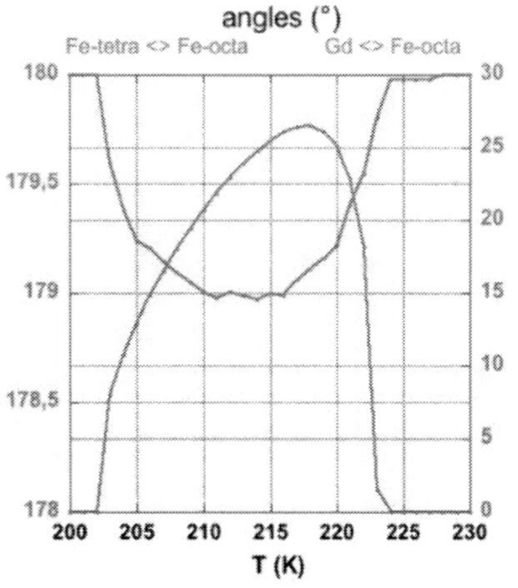

Figure 10. Angle between the two Fe sublattices (left), and between Gd and Fe-octa sublattices (right), as function of T, during spin-flop.

In the temperature range where spin-flop exists, the magnetization of Fe, in each sublattice, is not changing much, while the magnetization of Gd is decreasing a lot while rotating when T increases. Fe-tetra and Fe-octa remain almost antiparallel, and at the compensation they are almost lying in the plane. Gd and Fe-tetra, on the other hand, do not remain antiparallel. The maximum angle between Gd and Fe-tetra is 29°, and one can notice that this maximum deviation does not occur at Tcomp. This is due to the fact that the magnetization of Gd is in the direction of the applied field below Tcomp, but in the direction opposite to the field above Tcomp.

Gd sublattice is able to adjust its amplitude and orientation in order to increase Zeeman energy, which explain the existence of the spin-flop state. This state could occur in smaller field in a dot compared to the case of a continuous film, due to an easier nucleation mechanism.

4. Conclusion

Close to the compensation temperature, in this patterned system with "background", magnetization reversal occurs in the dot via an intermediate "spin-flop" state, which coexists with the normal domain opposite to the field. A mean field model, related to a film of our sample, lead to a description of the spin-flop state (amplitude and orientation of the magnetization of each sublattice with respect to the film), giving some elements in order to analyze quantitatively the magnetization processes in the dot.

Magnetization of the Gd sublattice is the variable allowing the increase of Zeeman energy. In our model the anisotropy is not yet included, but this will be done and should induce hysteresis. There is still a problem with the field range for spin-flop, which is very different in the model (10 T) and in the experiment (30 mT). We hope this could be solved tuning the composition and the exchange constant values.

The next step of our study will be the description of the domain wall between the spin-flop state and the normal domain (structure, width, energy). It will then be possible to analyze the equilibrium between these two domains, and the collapse of the normal domain when its radius get smaller than a critical value, by analogy with the problem of a magnetic bubble in opposite field.

References

[1] Wet etching of garnet films. Tolksdorf W, Bartels I, Dammann H, Pross E 1987 *J. Cryst. Growth*, **84** 323

[2] Molecular field coefficients of substituted yttrium iron garnets. Röschmann P, Hansen P 1981 *J. Appl. Phys* **52** 6257

[3] Canted spin phase in Gadolinium Iron Garnet. Bernasconi J, Kuse D 1971 *Phys. Rev. B* **3** 811

[4] Modern Magnetooptics and Magnetooptical Materials. Svezdin AK, Kotov VA 1997 *Studies in Condensed Matter Physics*, IOP Publishing Ltd

[5] Magnetic and magneto-optic properties of lead and Bi substituted YIG films. Hansen P, Witter, Tolksdorf W 1983 *Phys. Rev. B*, **27**

Employing magnonic crystals to dictate the characteristics of auto-oscillatory spin-wave systems

A D Karenowska[1], A V Chumak[2], A A Serga[2], J F Gregg[1] and B Hillebrands[2]

[1] Department of Physics, Clarendon Laboratory, University of Oxford, OX1 3PU Oxford, United Kingdom

[2] Fachbereich Physik and Forschungszentrum OPTIMAS, Technische Universität Kaiserslautern, 67663 Kaiserslautern, Germany

E-mail: `a.karenowska@physics.ox.ac.uk`

Abstract. Spin-wave active rings—positive-feedback systems incorporating spin-wave waveguides—provide important insight into fundamental magnetics, enable experimental investigations into nonlinear wave phenomena, and potentially find application in microwave electronics. Such rings break into spontaneous, monomode oscillation at a certain threshold value of feedback gain. In general, the wavenumber of this initially excited, threshold mode is impossible to predict precisely. Here we discuss how, by exploiting resonant spin-wave reflections from a magnonic crystal, an active ring system having a threshold mode with a well-defined and precisely predictable wavenumber may be realized. Our work suggests that study and development of active ring systems incorporating magnonic crystals may deliver useful insight into spin-wave transmission in structured magnetic films as well as devices with technological applicability.

1. Introduction

Magnetostatic spin waves (MSWs) typically have microwave (GHz) frequencies and wavelengths in the micron to millimetre range [1]. The study of MSW spectra has long been recognized for the important insight it provides into the underlying physics of static and dynamic magnetism. Moreover, because of the short wavelengths, low velocities, and rich dispersion characteristics of these excitations, MSW systems provide useful model environments for the study of general wave and quasi-particle phenomena (see for example [2–6] and references therein) and are widely acknowledged for their potential to provide a platform for spintronic information transfer and transport (see for example [7, 8]).

One material in particular, the ferrimagnet yttrium iron garnet $Y_3Fe_5O_{12}$ (YIG), dominates experimental studies into MSW dynamics. At around 0.04 kAm^{-1} (0.5 Oe) the ferrimagnetic resonance (FMR) linewidth of monocrystalline YIG is the narrowest of any known material. Magnon lifetimes of order hundreds of nanoseconds allow spin-wave propagation to be observed over centimetre distances, and large higher order terms in the garnet's dynamic magnetic susceptibility permit nonlinear propagation phenomena to be observed and studied at extremely low signal powers [6].

MSW propagation in YIG-based auto-exciting spin-wave oscillators or "active rings" has for some time been an area of interest within the field. Such systems comprise a thin-film YIG waveguide with spin-wave excitation and reception antennas connected together via a variable gain electrical positive-feedback path. If the feedback gain is sufficiently high, the ring supports noise-initiated, self-sustained propagating oscillations. As well as enabling investigations into the intrinsic dynamic properties of magnetic systems, YIG-based active rings also facilitate the experimental study of a range of exotic linear and nonlinear wave behaviours, for example soliton [3, 5, 9] and bullet [10] generation, fractal formation [11], and chaotic processes [12, 13], applicable to but not readily observable in other physical domains.

The majority of experimental studies of MSW dynamics in YIG are undertaken in thin-film waveguides under conditions of external, unidirectional magnetic bias. MSW propagation in such systems is strongly anisotropic and several distinct categories of waves, having markedly different characteristics, are identified [1, 6]. The waves are classified according to the relative orientation of the their propagation direction and the bias field. Backward volume magnetostatic spin waves (BVMSW), associated with an in-plane magnetic bias parallel to the spin-wave propagation direction, have latterly attracted particular interest in the context of active ring studies because of their unusual dispersion properties; the gradient of the BVMSW dispersion curve $\omega(k)$ is negative and thus the waves' group and phase velocities have opposite directions and short wavelengths are associated with low frequencies.

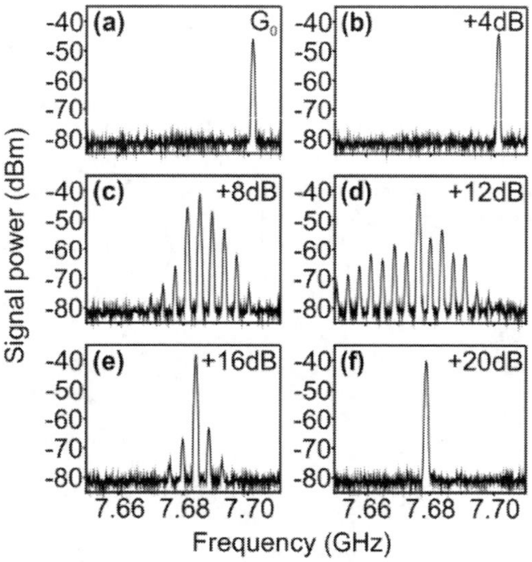

Figure 1. Excitation spectra of a simple BVMSW active ring as the feedback gain is varied from a starting value G_0 (a) to a value 20 dB above this (f). Excitation initiates at a threshold value of feedback gain. The frequency of the threshold mode lies in the lowest loss region of the spin-wave passband of the waveguide included in the loop but its precise wavenumber is almost impossible to predict. Monomode excitation persists ((a) and (b)) until a second threshold beyond which nonlinear splitting processes permit excitations at multiple k-values. The subsequent evolution of the spectrum is extremely complex and includes multi- and monomode regimes ((c) to (f)).

Excitation of a simple YIG-based BVMSW active ring initiates at a threshold value of feedback gain; the self-generation threshold. At this threshold, the ring signal is monochromatic, its wavenumber $k = k_t$ corresponds to the lowest loss phase-feasible mode, and its amplitude is regulated by the magnetic nonlinearity of the waveguide. If the gain is further increased, the presence of this threshold mode suppresses all others until a second, higher threshold beyond which nonlinear splitting processes permit the excitation of waves with multiple k-values and the system becomes multimoded. By way of example, Fig. 1 shows the excitation spectrum of such a system as the feedback gain is varied from a value G_0 in the monomode regime (a) to a value 20 dB above this level (f). The evolution of the spectral content of the ring oscillations in response to gain variations is too complex to be amenable to precise theoretical analysis, but certain signal categories can be identified and, depending on the magnetic bias conditions and the ring geometry, a range of interesting nonlinear behaviours can be observed, many of which, as alluded to above, have been subject to detailed study [3, 5, 9–13]. However, as well as these multimode, high-gain regimes, the monomode behaviour observed when the feeedback gain is at and just beyond the self-generation threshold is also very interesting. Not only does studying the competitive process of oscillation initiation offer physical insight into the magnetic properties of the waveguide in the loop, but, since the linewidth of the threshold mode is typically very narrow (of order kHz or tens of kHz), rings operating in this signal regime potentially find application in high-Q oscillators and other microwave devices.

The mode excited at the self-generation threshold k_t is the lowest loss mode for which the net phase shift around the ring is an integer multiple of 2π (phase feasibility). From a knowledge of the transmission characteristics of the spin-wave waveguide in a given ring subject to a given magnetic bias, it is possible to make a crude estimate of the frequency—and thus wavenumber—range within which this mode will fall. However, owing to a complex dependence on a wide range of intrinsic and extrinsic experimental parameters, its *precise* wavenumber is almost impossible to predict. This is disadvantageous from an applications perspective where, in many cases, a deterministic system is desirable.

A magnonic crystal is an artificially engineered spin-wave transmission structure with spatially periodic variations in its magnetic properties. The spin-wave analogue of an optical photonic or acoustic phononic crystal; the transmission characteristics of a magnonic crystal feature band gaps: frequency bands over which, due to resonant interactions with the artificial crystal lattice, the propagation of spin waves is strongly suppressed [14–21]. By analogy with X-ray or neutron diffraction in a natural crystalline solid, the wavenumbers corresponding to the nth BVMSW band gaps of a one dimensional magnonic crystal with lattice constant a are those which satisfy the Bragg condition at normal incidence:

$$k = n\frac{\pi}{a}, \qquad \text{where} \quad n = 1, 2, 3, 4, \ldots. \tag{1}$$

If spin waves with these k-values are incident on the crystal, they are strongly reflected.

In this paper, we describe how a magnonic crystal can be used to create a BVMSW active ring structure in which the wavenumber k_t of the threshold mode takes a single, well-defined value [22].

2. Methods

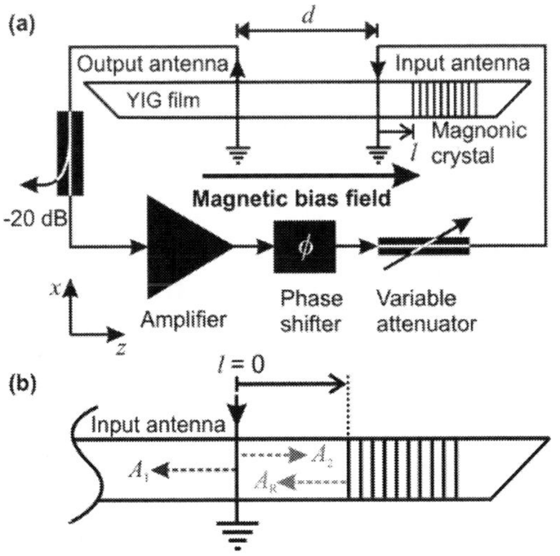

Figure 2. (a) Schematic diagram of the experimental BVMSW active ring system. Two microstrip antennae (width 50 μm), a distance d = 8 mm apart excite (input) and detect (output) propagating spin waves in a thin-film YIG waveguide. A magnonic crystal, etched into the surface of the film, is positioned a distance l from the input antenna. l can be varied with precision ±5 μm. A bias magnetic field of 160.4 kAm^{-1} (2016 Oe) is applied in-plane, parallel to the spin-wave propagation direction (the z-axis) (b) Expanded plan view of the region proximal to the input antenna.

The experimental active ring system is shown in Fig. 2. Microstrip spin-wave input (right) and output (left) antennae are in contact with the surface of a narrow YIG waveguide magnetically biased parallel to its long axis. The antennae are connected via an electrical feedback network comprising a phase shifter (phase shift ϕ), an amplifier, a variable attenuator and a directional coupler. The combination of the amplifier and the attenuator control the feedback gain whilst the coupler allows the ring signal to be sampled. The majority of the YIG film is of uniform thickness (5.5 μm) but, offset from its centre, is a short one-dimensional magnonic crystal [14–21] comprising ten grooves of width 30 μm, depth 300 nm and spacing 270 μm (lattice constant $a = 300$ μm) [20]. This crystal is positioned with its first groove a distance l from the input antenna. l can be varied by translating the film relative to the fixed antennae.

Figure 3 illustrates the BVMSW transmission characteristics of the homogeneous region of the film (solid, blue) and the magnonic crystal (dotted, red). The FMR frequency is approximately 7.710 GHz. Band gaps in the magnonic crystal characteristic corresponding to the first three Bragg orders (Eq. 1) are clearly visible. The low loss region of the spin-wave passband just below the FMR frequency (Band A) and the first band gap of the magnonic crystal (Band B) are the focus of the discussion which follows.

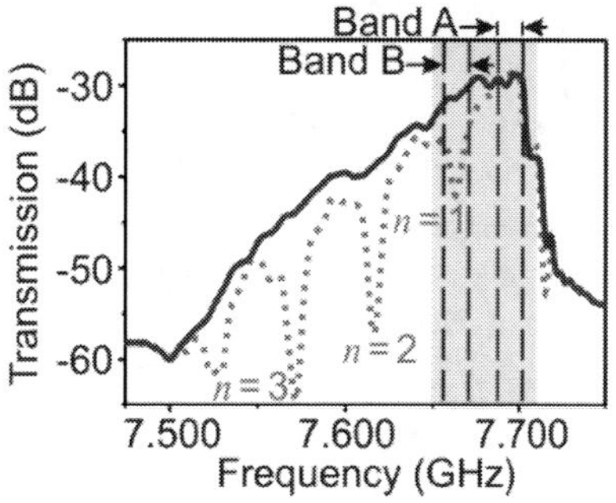

Figure 3. BVMSW transmission characteristics for the uniform (solid, blue) film and the magnonic crystal (dotted, red) used in the study. The nth order band gaps of the magnonic crystal are indicated. Band A identifies the lowest loss region of the spin-wave passband of the uniform film, Band B, the first band gap of the magnonic crystal.

Upon excitation, the input antenna drives propagating spin waves both toward the output antenna (blue, complex amplitude A_1 in Fig. 2(b)) and toward the magnonic crystal (red, complex amplitude A_2). If the spin-wave wavenumber satisfies Eq. 1 the wave incident on the magnonic crystal is reflected (green, complex amplitude A_{R}) and propagates back toward the input antenna.

Left and right-going waves have a relative phase of π [23] and it is also found experimentally that an effective phase shift of π (referenced to l) is associated with reflection from the magnonic crystal. Thus, the total phase accumulation of a reflected signal (satisfying Eq. 1) between transmission from the input antenna and arrival back at this point in the waveguide is $\phi_l = 2kl$. It follows that if it can be arranged that

$$\phi_l = 2\pi n\frac{l}{a} = 2\pi m, \qquad \text{where} \quad n = 1, 2, 3, 4, \ldots \quad \text{and} \quad m = 0, 1, 2, 3, \ldots, \tag{2}$$

the reflected signal will act to enhance the effective loop gain. If this enhancement is sufficient to increase the gain of a phase-feasible mode lying within a band gap of the magnonic crystal over and above the value associated with the lowest loss region of the passband of the uniform film (Band A, Fig. 3) then there is provided a mechanism by which this mode—the wavenumber of which is rigidly fixed by the geometry of the magnonic crystal—may be preferentially excited at the self-generation threshold.

The maximum possible increase in loop gain through reflection from the magnonic crystal is 3 dB. It is clear therefore from the data of Fig. 3, that for our particular experimental system, only spin-wave modes associated with the first band gap (Band B, Fig. 3) may be promoted to dominance via this mechanism, since the transmission loss at k-values corresponding to all higher-order gaps is at least 10 dB in excess of that associated with the lowest loss region of the passband (Band A). A theoretical model predicts an enhancement effect associated with $k = k_1 = \pi/a = 104.72$ rad/cm (Eq. 1) centred on 7.663 GHz, for the experimental ring, accessible for integer values of the ratio l/a (Eq. 2).

3. Results

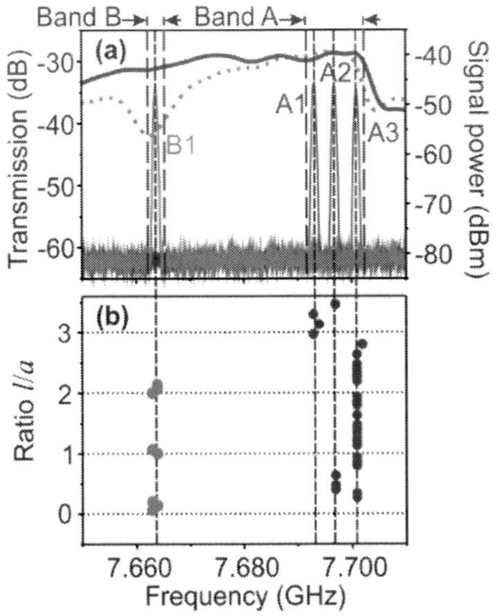

Figure 4. (a) The solid blue and dotted red curves (left ordinate axis) show the transmission characteristics of (respectively) the uniform region of the YIG film and the magnonic crystal over the frequency range 7.650–7.710 GHz (the grey shaded region of Fig. 3). The inset diagram is a composite of overlaid ring power spectra captured via the -20 dB directional coupler of Fig. 2(b). Each spectrum comprises a single peak corresponding to the threshold mode k_t. As indicated in (b), modes A1, A2 and A3 positioned within the low loss region of the spin-wave passband (Band A) correspond to general values of the ratio l/a (blue circles), whilst the mode B1 which lies within the first band gap of the magnonic crystal at $k = k_1$ (Band B) is observed at and only at integer values (red circles). Note that data has been compensated for temperature variations.

An expanded view of the transmission curves of Fig. 3 over the frequency range 7.650–7.710 GHz is shown in Fig. 4(a) (left ordinate axis). Overlaid, are experimental power spectra of threshold modes corresponding to a range of values of the ratio l/a between zero ($l = 0$) and 3 ($l = 900$ nm) (right ordinate axis). In order to collect these data, at each measured value of l, the feedback gain was increased via the variable attenuator (Fig. 2(a)) from a value below the self-generation threshold (i.e. from an inactive state of the ring) to a value 5 dB above it and the spectrum of the ring signal recorded. Each spectrum contains a single peak. (Note that the sharpness of the spectral peaks shown is limited by the scanning window of the spectrum analyzer, their true width is ≤ 25 kHz.) For general l/a, the wavenumber of the threshold mode k_t is determined solely by gain and phase feasibility conditions. As anticipated in the discussion above, the mode lies in the lowest loss region of the spin-wave passband (Band A, modes A1, A2 and A3), but a precise prediction of the value of k_t is not possible. By contrast, for integer l/a, the wavenumber selective magnonic crystal reflection mechanism enhances the effective ring gain of the mode $k = k_1$ associated with the first band gap of the magnonic crystal, forcing $k_t = k_1$. Figure 4(b) shows the l/a dependence of k_t for a fixed input phase ϕ at the phase shifter (tuned for feasibility of the k_1 mode). Each point corresponds to a separate experiment performed using an identical technique to that described in connection with the spectra of Fig. 4(a). The maximum integer value of l/a at which k_1 mode promotion is observed is 2. Beyond this value, the loss associated with the signal path of the reflected wave reduces the loop gain at k_1 below that for the lowest loss mode associated with the uniform region of the film.

In an ideal system, with fixed external phase shift ϕ we would expect to observe one of just two threshold modes at a given l: one in the low loss region of the spin-wave passband (Band A, Fig. 4(a)) associated with general k and l/a and a second at and only at integer l/a

with wavenumber k_1 (Band A). The fact that three separate modes are observed in Band A is an artifact of the experimental method; it is due to the fact that the thickness and thus the dispersion relationship $\omega(k)$ of the YIG film is not perfectly uniform along its length, providing a secondary mechanism for the phase accumulation around the ring to vary as a function of l independently of ϕ [22].

4. Conclusions

In summary, it has been demonstrated that the wavenumber selective reflectivity of a magnonic crystal can be exploited to create a BVMSW active ring system which will commence monomode auto-oscillation at precise wavenumber. This wavenumber corresponds the first band gap of the crystal and, as such, is readily determinable from its lattice constant. Such high-Q, wavenumber selective active ring arrangements may find practical application in microwave oscillators, spin-wave logic systems and sensor instrumentation. Moreover, this work suggests that by further pursuing their development and investigation we may obtain useful insight into the complex mechanism of spin-wave reflection from magnonic crystal structures.

Acknowledgments

Financial assistance from the DFG (grant numbers SE 1771/1-1 and GRK 792), and technical support from the Nano+Bio Centre, TU Kaiserslautern are gratefully acknowledged. ADK acknowledges the support of the IET Leslie H. Paddle Scholarship and Magdalen College, Oxford.

References

[1] Stancil D D 1993 *Theory of Magnetostatic Waves* (New York: Springer-Verlag)
[2] Schneider T, Serga A A, Chumak A V, Sandweg C W, Trudel S, Wolff S, Kostylev M P, Tiberkevich V S, Slavin A N and Hillebrands B 2010 *Phys. Rev. Lett.* **104** 197203
[3] Wu M and Patton C E 2007 *Phys. Rev. Lett.* **98** 047202
[4] Demokritov S O, Demidov V E, Dzyapko, O, Melkov, G A, Serga, A A, Hillebrands, B and Slavin A N 2006 *Nature* **443** 430
[5] Demokritov S O, Serga A A, Demidov V E, Hillebrands B, Kostylev M P and Kalinikos B A 2003 *Nature* **426**, 159
[6] Serga A A, Chumak A V and Hillebrands B 2010 *J. Phys. D* **43** 264002
[7] Kajiwara Y, Harii K, Takahashi S, Ohe J, Uchida K, Mizuguchi M, Umezawa H, Kawai H, Ando K, Takanashi K, Maekawa S and Saitoh E 2010 *Nature* **464** 262
[8] Schneider T, Serga A A, Leven B, Hillebrands B, Stamps R L and Kostylev M P 2008 *Appl. Phys. Lett.* **92** 022505
[9] Kalinikos B A, Kovshikov N G, and Patton C E 1998 *Phys. Rev. Lett.* **80** 4301
[10] Serga A A, Demokritov S O, Hillebrands B and Slavin A N 2004 *Phys. Rev. Lett.* **92** 117203
[11] Wu M, Kalinikos B A, Carr L D, and Patton C E 2006 *Phys. Rev. Lett.* **96**, 187202
[12] Wu M, Hagerstrom A M, Kondrashov A and Kalinikos B 2009 *Phys. Rev. Lett.* **102** 237203
[13] Hagerstrom A M, Tong W, Wu M, Kalinikos B A and Eykholt R 2009 *Phys. Rev. Lett.* **102** 207202
[14] Chumak A V, Serga A A, Wolff S, Hillebrands B and Kostylev M P 2009 *Appl. Phys. Lett.* **94** 172511
[15] Chumak A V, Serga A A, Wolff S, Hillebrands B and Kostylev M P 2009 *J. Appl. Phys.* **105** 083906
[16] Lee K S, Han D S and Kim S K 2009 *Phys. Rev. Lett.* **102** 127202
[17] Ustinov A B, Kalinikos B A, Demidov V E and Demokritov S O 2009 *Phys. Rev. B* **80** 052405
[18] Wu M, Hagerstrom A M, Eykholt R, Kondrashov A and Kalinikos B A 2009 *Phys. Rev. Lett.* **102** 237203
[19] Wang Z K, Zhang V L, Lim H S, Ng S C, Kuok M H, Jain S and Adeyeye A O 2009 *Appl. Phys. Lett.* **94** 083112
[20] Chumak A V, Serga A A, Hillebrands B and Kostylev M P 2008 *Appl. Phys. Lett.* **93** 022508
[21] Sykes C G, Adam J D and Collins J H 1976 *Appl. Phys. Lett.* **29** 388
[22] Karenowska A D, Chumak A C, Serga A A, Gregg J F and Hillebrands B 2010 *Appl. Phys. Lett.* **96** 082505
[23] Schneider T, Serga A A, Neumann T, Hillebrands B and Kostylev M P 2008 *Phys. Rev. B* **77** 214411

Spin wave propagation in ferromagnetic wires with an arbitrary field direction

Rémy Lassalle-Balier and Claude Fermon

DSM/IRAMIS/SPEC- URA2464, CEA Saclay, 91191 Gif sur Yvette Cedex France

Email : claude.fermon@cea.fr

Abstract. We report on the measurements of quantized spin waves spectra in thin ferromagnetic wires with an arbitrary field orientation obtained by using an original micro antenna configuration. The high efficiency of energy transfer from the micro antenna to the magnetic wires allows us fabricating devices using remanent magnetization configuration

1. Introduction

Spin wave propagation in ferromagnetic films has been used for a long time to produce RF devices. However, all these devices are made with garnet type films due to the extremely low spin wave absorption in these materials [1]. The development of devices using metallic films would open the way for low cost integrated RF devices like agile attenuators, circulators etc. Unfortunately, the spin wave attenuation in metallic films is large and practical devices should have tens of μm lengths. Several years ago, a technique based on micro-antennas has been introduced [2] which has allowed exciting and propagating spin wave in permalloy stripes with a reasonable efficiency when the magnetization is perpendicular to the stripes. When the magnetization is along the stripe, conventional geometry of micro-antennas cannot be used because the excitation field is along the magnetization. In ref [3], the authors have used a end of microantenna to create a transverse excitation field with a rather low efficiency. In our approach we have simply used microantennas at 45° from the stripe direction. This allows us exciting efficiently spin waves for any planar orientation of the magnetization. In this paper, after a small introduction on quantized spin waves, we describe the experimental system and we give the results obtained for different magnetization orientations and in particular at 45°.

2. Spin wave modes in confined stripes

Spin waves in confined stripes have been extensively studied by different experimental techniques like ferromagnetic resonance, Brillouin Light scattering and microantennas. A number of model have been developed when the stripe magnetization is homogeneous [4], [5], or in non saturated systems [6]. When a field is applied the frequency of the spin waves is increasing typically as $\sqrt{M_s(H+M_s)}$ where H is the applied field and M_s the magnetization of the film at saturation. The confinement induces a quantization of spin waves and for stripe width below 20μm different modes can be clearly separated. Figure 1 shows an example of resonance spectrum of two 5μm lines separated by 5μm measured with a microantenna.

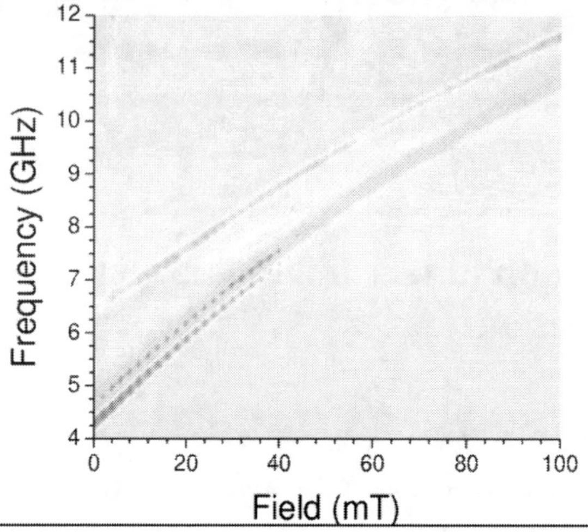

Figure 1. spin wave spectrum as function of field for a 5μm stripe of NiFe. The two first modes are clearly visible..

The dispersion of the spin waves as function of their wavevector (k) can be measured with BLS [7], [8] and show quantized k values for each mode.

3. Sample fabrication and characterization

Our samples are made of three layers deposited on glass substrate: Ferromagnetic system, Insulator layer, microantennas. The ferromagnetic material used in our samples is permalloy. First a full plane layer of 45nm deposited by RF sputtering at $3nm.min^{-1}$ rate is lifted off. Magneto-Optic Kerr effect measurements reveal a coercive field of 0.2mT and ferromagnetic resonance shows a saturation magnetization of 0.93T. The samples presented here have two Py lines of 6μm width separated by a 6μm space. The insulating layer is a full plane silicon nitride film. It is deposited on the pre-etched and patterned Py by sputtering at $65nm.min^{-1}$ rate. Finally, antennas are produced by lifting off a bilayer of titanium (9nm)/gold (90nm)

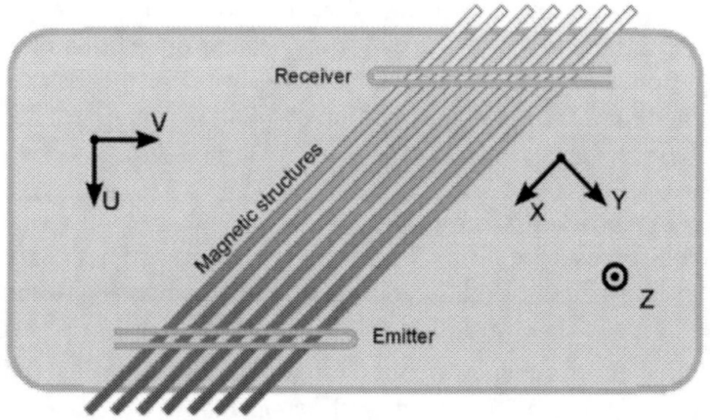

Figure 2. Sample structure. Permalloy network in light grey and antennas in orange.

Journal of Physics: Conference Series **303** (2011) 012008 doi:10.1088/1742-6596/303/1/012008

4. Microantenna properties and setup

The simplest microantenna is two parallel stripe lines shortened at the end. All the structure of a sample can be seen in top view in figure 2. On the electronic point of view, the antenna can be modeled by a resistance and inductance in series. The typical resistance of our antenna is 22 Ohms. These antennas are connected to picoprobes for signal injection and detection.

On a spin wave point of view, the microantenna is a field producer at a arbitrary frequency but a spatial shape determined by the dimensions of the antenna. Figure 3 shows the direction of the created magnetic field and the intensity of the excitation field as function of the k vector.

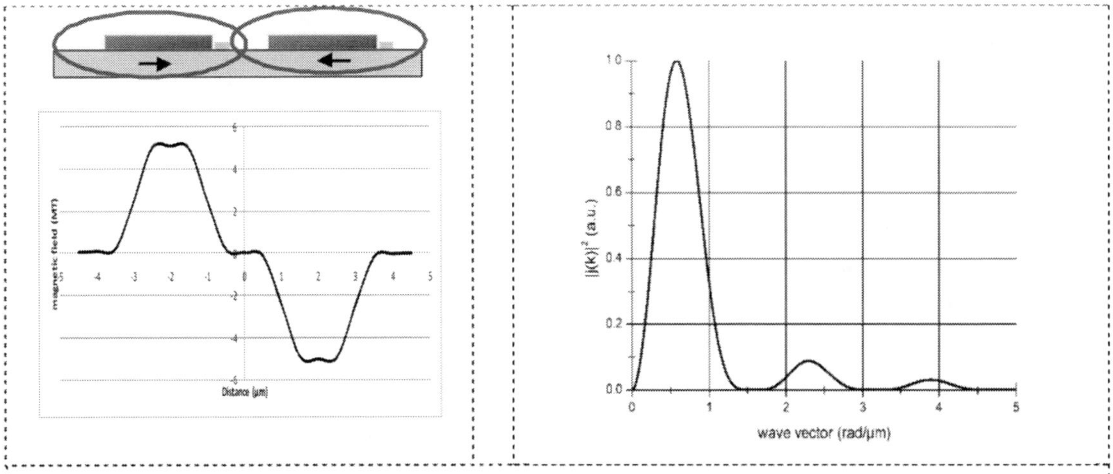

Figure 3. View of the created fields (left), intensity of the field in k space (right).

The effective efficiency of the antenna is hence dependant on the possible spatial waves which can be excited. The intensity of the field created by the antenna is rather large, typically 5mT for a 0dbm RF power. The detection antenna has exactly the same properties than the excitation one. Hence the selectivity in k space for the complete device is just proportional to the intensity given in fig 3.

P. Emtage has developed the theory of the coupling between a spin wave and a current. That theory can be applied to calculate the effective efficiency of the complete device owing the knowledge of the damping parameter of the spin wave. This calculation has been done for the case of MsFVW (Magnetostatic Forward Volume Waves) , MsSW (Magnetostatic Surface Volume Waves) and MsBVW (Magnetostatic Backward Volume Waves) [9] [10].

5. Experimental results

We present some results which show the good efficiency of the excitation and detection for three orientations of the magnetization of the stripe. The device is a pair of microantennas separated by 20µm and linked by only two 6µm wide stripes of permalloy. 0° is along the stripe (MsBVW). For 90° (MsSW), we see that the frequency is first decreasing as function of field. This is due to the rotation of the magnetization [11],[12]. At 20 mT the center of the stripe is perpendicular to the field and the frequency increases with the field. At 0° there is no rotation and the frequency is increasing with field as waited. For 45°, the rotation effect can be seen at very small fields. The ratio of intensities between 0° (or 90°) and 45° is 2, due to geometrical reasons.

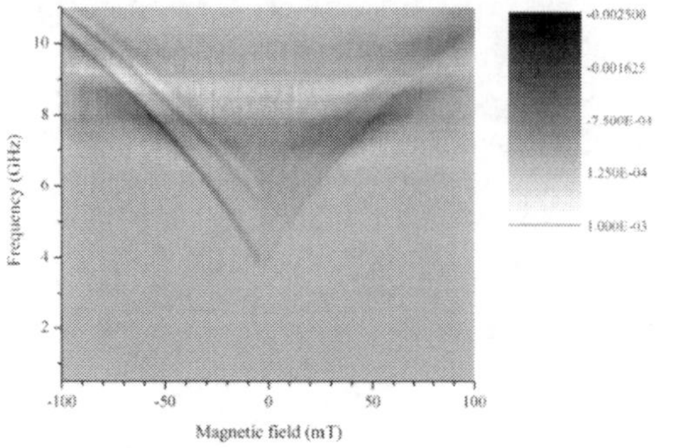

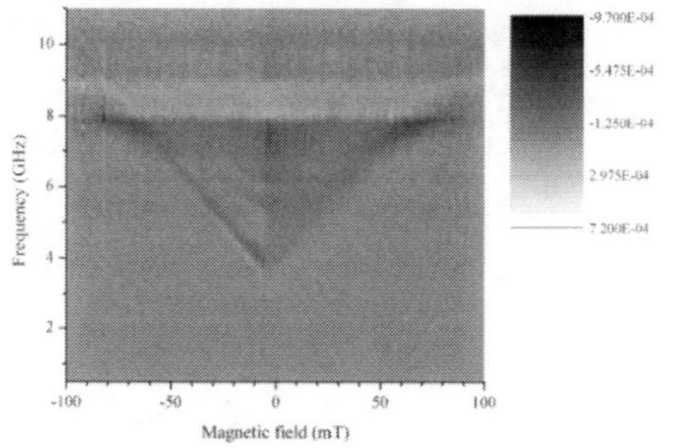

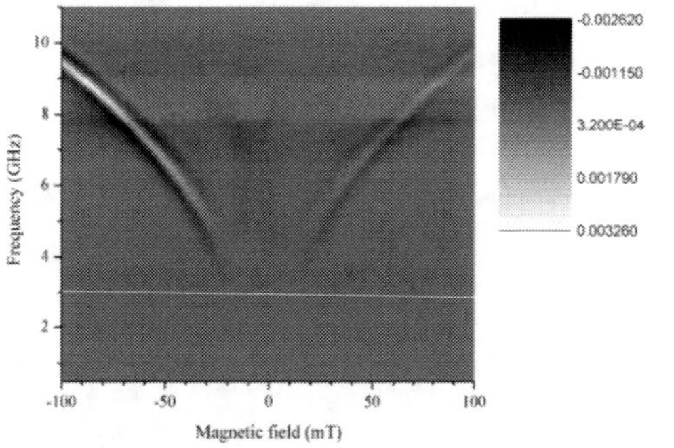

Figure 4. Transmission of the device as function of field and frequency for three orientations of the field(top left: MsBVW, top right: 45°, bottom: MsSW)

6. Conclusions

We have shown that spin waves can be reasonably excited and detected in micrometric metallic stripes even in their remanent state by 45° microantennas. This is the first step for the building of efficient integrated agile devices working without any external field. In particular that configuration can be used for the realization of micro-spin wave interferometers.

References

[1] . Conger, R.L. *Nonlinear damping losses in YIG* J. Appl. Phys., 1961, Vol. 32.

[2]. Bailleul, M., et al. *Spin waves propagation and confinement in conducting films at the micrometer scale.* s.l. : Europhys. Lett., Vol. 56. (2001)

[3] A. Kozhanov et al. *Dispersion and spin wave ``tunneling" in nanostructured magnetostatic spin waveguides.* J. Appl. Phys., Vol. 105, p. 07D311 (2009),

[4]. Guslienko, K.Y., Chantrell, R.W. and Slavin, A.N. *Dipolar localization of quantized spin-wave modes in thin rectangular magnetic elements.* 2, s.l. : Phys.Rev. B, Vol. 68 (2003).

[5]. Guslienko, K.Y., et al. *Effective dipolar boundary conditions for dynamic magnetization in thin magnetic stripes,* Phys. Rev. B, Vol. 66.(2002),

[6] Kalinikos, B.A. and Slavin, A.N *Theory of dipole-exchange spin wave spectrum for ferromagnetic films with mixed exchange boundary conditions..* s.l. : J. Phys. C: Solid State Phys., Vol. 19 (1986),.

[7] Demokritov, S.O., Hillebrands, B. and Slavin, A.N. *Brillouin light scattering studies of confined spin waves: linear and nonlinear confinement,* Phys. Rep., , Vol. 348. (2001)
[8] Jorzick, J., et al. *Brillouin light scattering from quantized spin waves in micron-size magnetic wires.* 22, s.l. : Phys. Rev. B, , Vol. 60 (1999).
[9] Emtage, P., *Interaction of magnetostatic waves with a current.* J. Appl. Phys., Vol. 49, pp. 4475-4484. (1978)
[10] Vlaminck, V. and Bailleul, M. *Spin-wave transduction at the submicrometer scale: Experiment and modeling.* s.l. : Phys. Rev. B, Vol. 81. (2010),
[11] Bailleul M, Olligs D, Fermon C *Propagating spin wave spectroscopy in a permalloy film: A quantitative analysis,* Appl. Phys. Lett., Vol83, (2003),
[12] Bailleul M., Olligs D, Fermon C, *Micromagnetic phase transitions and spin wave excitations in a ferromagnetic stripe* Phys. Rev. Lett, Vol91, (2003),

Joint European Magnetic Symposia – JEMS 2010 IOP Publishing
Journal of Physics: Conference Series **303** (2011) 012009 doi:10.1088/1742-6596/303/1/012009

Structural and magnetic properties of NdFeB and NdFeB/Fe films with Mo addition

M Urse, M Grigoras, N Lupu and H Chiriac
National Institute of R & D for Technical Physics, 47 Mangeron Blvd., 700050 Iasi, Romania

E-mail: urse@phys-iasi.ro

Abstract. The influence of the Mo addition on the microstructure and magnetic properties of Nd-Fe-B and Nd-Fe-B/Fe films was studied. The coercivity is a key parameter in the control of technical performances of Nd-Fe-B films. A small amount of about 1 at.% Mo can enhance the coercivity of Nd-Fe-B film by controlling the growth of soft and hard magnetic grains. A coercivity of 22.1 kOe, a remanence ratio, M_r/M_s, of 0.83 and a maximum energy product of 8 MGOe were obtained for Ta/[NdFeBMo(1at.%)(540nm)/Ta films annealed at 650°C for 20 minutes due to Mo precipitates formed at the $Nd_2Fe_{14}B$ phase boundaries which prevent the nucleation and expansion of the magnetic domains. Simultaneous use of Mo as addition and the stratification of Nd-Fe-B-Mo films using Fe as spacer layer are important tools for the improvement of the hard magnetic properties of Nd-Fe-B films. The Ta/[NdFeBMo(1at.%)(180nm)/Fe(1nm)]×3/Ta multilayer film annealed at 620°C exhibits an increase in the coercivity from 12.1 kOe to 22.8 kOe, in the remanence ratio from 0.77 to 0.80, and in the maximum energy product from 4.5 to 7.1 MGOe in comparison with Ta/Nd-Fe-B/Ta film. As compared to Ta/Nd-Fe-B/Ta film, the Ta/[NdFeBMo(1at.%)(180nm)/Fe(1nm)]×3/Ta film presents a decrease in the crystallization temperature of about 30°C.

1. Introduction

The thin-film Nd-Fe-B permanent magnets have been extensively investigated as promising candidates for magnetic microdevices actuation components [1]. Refinement of the Nd-Fe-B films microstructure to less than 40 nm for the hard grains is essential to obtain the high coercivity, H_c, and high-energy product, $(BH)_{max}$. A number of specific factors including sample composition, spacer layer material and annealing temperature can affect the structure of the thin films and its magnetic properties [2]. The addition of different metals with high and low melting points to the ternary Nd-Fe-B alloy is an effective method for microstructure improvement. For thin-film Nd-Fe-B permanent magnets a lower crystallization temperature is desired for applications which integrate such magnets with existing CMOS technologies. Many metallic elements with high melting point (such as Nb, Mo, Zr, V, Ti, etc) and low melting point (such as Cu, Al, Ga, In, Sn) can be added to the Nd-Fe-B system with the aim of enhancing the hard magnetic properties and decreasing the crystallization temperature [3]. Among them, Mo is known to act in the sense of preventing the grain growth and decreasing the crystallization temperature [4]. Supplementary, the Mo addition enhances the corrosion resistance [5] and leads to a more rectangular hysteresis loop.

In this paper, we report the influence of the Mo addition on the microstructure and magnetic properties of Nd-Fe-B and Nd-Fe-B/Fe films. For Ta/[Nd-Fe-B-Mo/Fe(t)]×n/Ta films, the thickness (t) of Fe spacer layers was varied from 1 to 3 nm. For all multilayered samples the total thickness of the Nd-Fe-B-Mo layers was maintained at about 540 nm. All the samples were sandwiched by two Ta layers with the thickness of 20 nm for the buffer layer and 40 nm for the capping layer.

2. Experimental details

Ta/Nd-Fe-B/Ta and Ta/[Nd-Fe-B-Mo/Fe]×n/Ta multilayer films have been deposited by r.f. sputtering and electron beam evaporation (i.e. Fe film) on silicium (111) substrates, at room temperature. The sputtering targets were mounted on separate guns and consisted of a disc of $Nd_{12}Fe_{82}B_6$ alloy with Nd and/or Mo chips on the surface and a disc of Ta for the buffer and capping layers. The thickness of the individual sputtered layers was measured 'ex-situ' by the KLA Tencor Alpha – Step IQ Profilometer and 'in-situ', during the deposition process, by a quartz-crystal deposition monitor (FTM 6 Film Thickness Monitor) having a resolution of about 0.1 nm.

The crystallographic structure was investigated using X-ray diffraction (XRD) analysis. A X-ray diffractometer (Diffractometer D8 Advance) with a monochromatized Cu-Kα radiation was used, in a Bragg-Brentano arrangement. The Warren-Averbach method [6] was used to estimate the crystalline grain sizes (with an error of ± 15%). The composition of the samples was investigated by a Scanning Electron Microscope (SEM) - JEOL JSM 6390 using Energy Dispersive Spectrometry (EDS) technique. The morphology of the samples was investigated by Transmission Electron Microscopy (TEM) technique using molibdenum 'microscope grids'.

The magnetic characteristics of the samples were determined by a vibrating sample magnetometer (Lake Shore VSM 7410) with a maximum magnetic field of about 31 kOe applied parallel with the film plane.

The as–deposited samples were annealed in vacuum atmosphere of 4 x 10^{-4} Pa, for different periods of time at temperatures between 500°C and 650°C, to obtain optimum hard magnetic properties.

3. Results and Discussion

Different compositions of Nd-Fe-B target were tested in view of obtaining the Ta/NdFeB(540nm)/Ta thin films with good hard magnetic properties. The composition of the Nd-Fe-B films was adjusted by changing the numbers of Nd chips on the surface of the $Nd_{12}Fe_{82}B_6$ target. The optimum hard magnetic properties of Nd-Fe-B thin films were obtained when the following configuration of sputtering target was used: disc of $Nd_{12}Fe_{82}B_6$ alloy with the diameter of 7.5 cm having on its surface Nd chips with the a total area of about 4.9 cm^2 and B chips with the a total area of about 5.3 cm^2. The composition of Ta(40nm)/Nd-Fe-B(540nm)/Ta(20nm) film obtained by EDS technique is as follows: Fe 69.63 at.%, Nd 15.02 at.%, Ta 9.43 at.%, B up to 100 at.%.

Different Mo contents were tested in order to obtained samples with good hard magnetic properties. Table 1 shows the evolution of the main magnetic characteristics of Ta(20nm)/Nd-Fe-B-Mo(540nm)/Ta(40nm) films annealed at two different temperatures as function of Mo content.

Table 1. The Mo content dependence of the magnetic characteristics of Ta/Nd-Fe-B-Mo(540nm)/Ta thin films annealed for 20 minutes at 620°C and 650°C.

Samples/ (all the thicknesses in nm)	Saturation magnetization M_s (emu/g)		Coercivity H_c (kOe)		Remanence ratio M_r/M_s		Energy product $(BH)_{max.}$ (MGOe)	
	620°C	650°C	620°C	650°C	620°C	650°C	620°C	650°C
NdFeB(540)	94.0	97.8	12.1	14.2	0.77	0.66	4.5	5.6
NdFeBMo(0.6 at.%)(540)	70.1	75.0	19.1	16.7	0.85	0.78	4.2	5.5
NdFeBMo(1 at.%)(540)	62.0	73.6	16.9	22.1	0.84	0.83	5.8	8.0
NdFeBMo(1.7 at.%)(540)	50.0	55.0	22.4	21.5	0.83	0.86	3.2	1.5

Good hard magnetic characteristics, especially for the energy product, $(BH)_{max.}$, are obtained for Ta/NdFeBMo(1at.%)/Ta samples annealed for 20 minutes at 650°C. It can be observed that by adding 1 at.% Mo, the coercivity, H_c, reaches a maximum value of about 22.1 kOe for samples annealed at 650°C, while a supplementary increase in the amount of Mo up to 1.7 at.% leads to a small decrease of coercivity down to 21.5 kOe. An increase in the maximum energy product value with increasing the Mo content from 0.6 at.% up to about 1 at.% followed by a decrease for a Mo content of 1.7at.% can be observed for both annealing temperatures.

In Figure 1 the typical magnetic hysteresis loops for Ta/$(NdFeB)_{1-x}Mo_x$(540nm)/Ta thin films annealed at temperatures of 650°C for 20 minutes are presented.

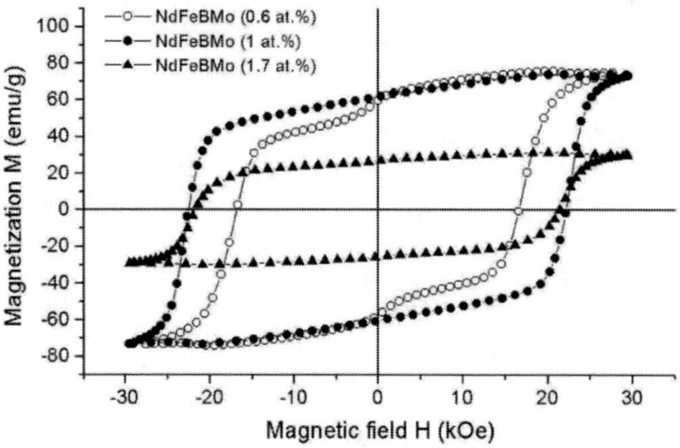

Figure 1. Typical magnetic hysteresis loops for Ta/[$(NdFeB)_{1-x}Mo_x$(540nm)/Ta thin films annealed for 20 minutes at 650°C.

It can be observed that in samples with a small Mo content (i.e. about 0.6 at.%), annealed at 650°C for 20 minutes, the soft and hard magnetic phases coexist. This fact is confirmed by two small shoulders in the II and IV quadrants. By increasing the Mo content over 0.6 at.%, the soft phase (i.e. Fe_3B, Fe_2B) amount is reduced and a good control on $Nd_2Fe_{14}B$ grain growth is obtained. For samples with a Mo content higher than 0.6 at.%, the saturation magnetization decreases and the coercivity increases due to the domain wall pinning effect as a result of the Mo-based precipitated fine grains at the $Nd_2Fe_{14}B$ boundaries. The optimum magnetic characteristics are obtained for Nd-Fe-B films with a Mo content of about 1 at. %.

The X-ray diffraction investigations indicate that Ta/NdFeB/Ta and Ta/[NdFeBMo/Ta thin films, in as–deposited state and after thermal treatments at temperatures below 500°C, have amorphous structure. For samples annealed at temperatures between 500°C and 560°C, the microstructure of the samples consists of a small number of Fe_2B or Fe_3B nanograins embedded in an amorphous matrix. Figure 2 shows the XRD patterns for Ta/NdFeB(540nm)/Ta film (curve a) and Ta/NdFeBMo(1at.%)(540)/Ta film (curve b), after annealing for 20 minutes at 650°C.

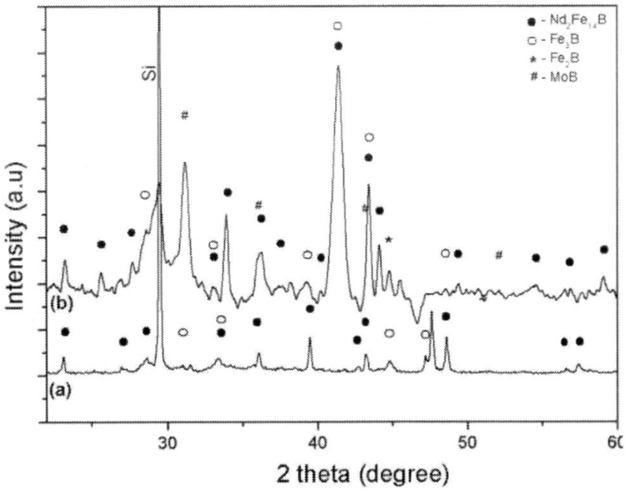

Figure 2. The X-ray diffraction patterns for Ta/NdFeB(540nm)/Ta film (curve a) and Ta/NdFeBMo (1at.%)(540)/Ta film (curve b), after annealing at 650°C for 20 minutes.

All the samples annealed at temperatures higher than 580°C exhibit a complex multiphase structure of tens of nanometers: Ta/[NdFeB(540nm)/Ta film annealed at 650°C, for 20 minutes (curve a), shows nanograins of $Nd_2Fe_{14}B$ hard magnetic phase and Fe_2B soft magnetic phase. The average grain size is in the range of 43-45 nm for the $Nd_2Fe_{14}B$ hard magnetic phase and of about 12 nm for the Fe_2B soft magnetic phase; Ta/NdFeBMo(1at.%)(540nm)/Ta multilayer film (curve b) presents a multiphase structure consisting of a mixture of nanograins of hard magnetic $Nd_2Fe_{14}B$, soft magnetic Fe_3B and non-magnetic MoB, phases. The average grain size for Ta/NdFeBMo(1at.%)(540nm)/Ta multilayer film is in the range of 30 -35 nm for the hard magnetic $Nd_2Fe_{14}B$ phase, of 13-14 nm for the soft magnetic Fe_3B phase and of 18-20 nm for the non-magnetic MoB phase.

TEM analyses were performed in order to understand the effect of the Mo addition on the physical and microstructural properties of the Nd-Fe-B. Figure 6 (a, b) shows the TEM micrographs of NdFeB(60nm) and NdFeBMo(1at.%)(60nm) films annealed at 650°C for 20 min.

The TEM micrograph of Nd-Fe-B film (figure 3a) reveals a grain structure with irregular shapes. The average grain size of the Nd-Fe-B film is of about 50 nm. The TEM micrograph of NdFeBMo(1at.%)(60nm) film (figure 3b) reveals a fine nanocrystalline structure with fine and almost spherical grains having an average size of about 35 nm.

From the analysis of the structural and morphological data it results that the Nd-Fe-B samples with a Mo content of about 1 at.% present a uniform distribution of grains with an average size of about 35 nm. As compared to Nd-Fe-B films, the NdFeBMo(≥1 at.%) films (see the Table 1) exhibit higher values of the coercive field due to the domain wall pinning effect within the intergranular region.

The coercivity is one of the key factors controlling the properties of nanocomposite materials. For multilayer [NdFeBMo/Fe]xn samples, the soft Fe grains act as nuclei for the formation of magnetic domains during the magnetization reversal. The dependence of the magnetic characteristics on the thickness (t) of Fe layers for Ta/[Nd-Fe-B-Mo(1at.%)/Fe(t)]×n/Ta films annealed at different temperatures between 600°C and 650°C, for 20 minutes is presented in Table 2.

Optimum hard magnetic characteristics, especially for maximum energy product, are obtained for Ta/[NdFeBMo(1at.%)(180nm)/Fe(1nm)]x3/Ta thin films annealed for 20 minutes at 620°C. From the analysis of the thermal treatment data presented in Table 2, it can be observed that the use of Fe film as spacer layer with a thickness of about 1nm takes down the optimal crystallization temperature of Ta/[NdFeBMo(1at.%)(180nm)/Fe(1nm)]x3/Ta films with about 30°C.

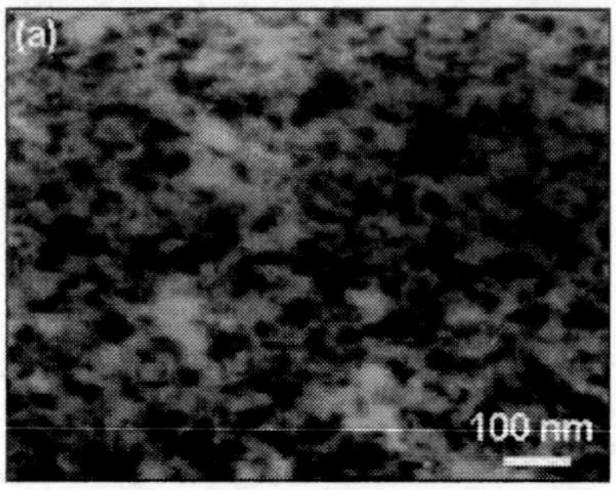

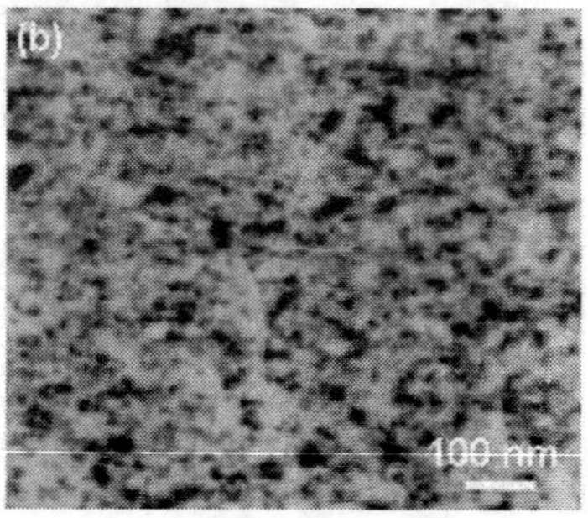

Figure 3. TEM micrographs of NdFeB(60nm) and NdFeBMo(1at.%)(60nm) films annealed at 650°C for 20 minutes.

Table 2. The dependence of the magnetic characteristics on the thickness of Fe layers for Ta/[Nd-Fe-B-Mo(1at.%)/Fe(t)]xn/Ta films annealed for 20 minutes at different temperatures between 600°C and 650°C.

Samples / (all the thicknesses in nm)	Coercivity H_c (kOe)			Remanence ratio M_r/M_s			Energy product $(BH)_{max.}$ (MGOe)		
	Annealing temperature (°C)			Annealing temperature (°C)			Annealing temperature (°C)		
	600	620	650	600	620	650	600	620	650
NdFeB Mo(540)	15.7	16.9	22.1	0.84	0.84	0.83	3.7	5.8	8.0
[NdFeBMo(180)/Fe(1)]x3	20.9	22.8	24.8	0.78	0.80	0.77	6.4	7.1	2.8
[NdFeBMo(180)/Fe(2)]x3	20.4	26.7	25.5	0.81	0.68	0.78	2.7	1.1	2.0
[NdFeBMo(180)/Fe(3)]x3	23.9	24.5	14.7	0.74	0.58	0.79	2.6	3.1	1.2

In Figure 4, the magnetic hysteresis loops for Ta/[NdFeBMo(1at.%)(540nm)/Ta single layer and Ta/[NdFeBMo(1at.%)(180nm)/Fe(t)]×3/Ta multilayer films with different thickness, t, of Fe spacer layer, annealed for 20 minutes at 620°C, are presented.

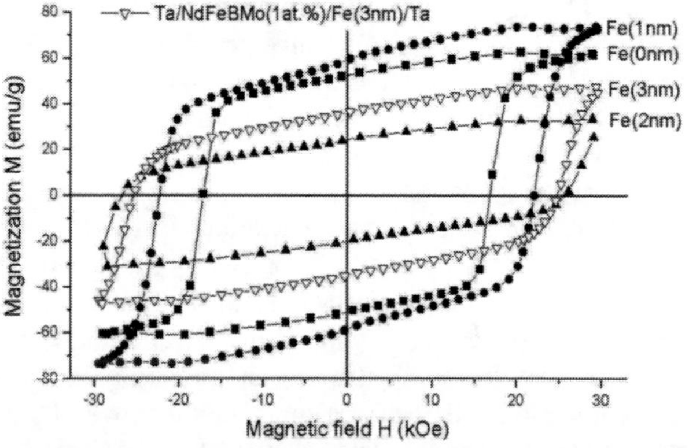

Figure 4. The magnetic hysteresis loops for Ta/[NdFeBMo(1at.%)(180nm)/Fe(t)]x3/Ta thin films with different thickness, t, of the Fe spacer layer, annealed for 20 minutes at 620°C.

The optimal magnetic properties are obtained for Ta/[NdFeBMo(1at.%)(180nm)/Fe(1nm)]×3/Ta film with a thickness of the Fe spacer layer of about 1 nm. It is clear that the Ta/[NdFeBMo(1at.%)(180nm)/Fe(1nm)]×3/Ta film with a Mo content of 1 at.% and multilayered with a Fe thinner soft magnetic layer may be available for the formation of nanostructure more favourable for exchange coupling between the soft and hard phases

4. Conclusions

The influence of the Mo addition on the structural and magnetic properties of Nd-Fe-B-Mo and [NdFeBMo/Fe]×n thin films has been investigated. The Mo addition refines the grain size of soft and hard magnetic phases. The optimum magnetic properties such as, coercivity H_c = 22.1 kOe, remanence ratio M_r/M_s = 0.83 and maximum energy product $(BH)_{max}$ = 8 MGOe are obtained for Ta/NdFeBMo (1at.%)(540nm)/Ta films annealed at 650°C for 20 minutes.

Simultaneous utilization of Mo as addition and Fe film as spacer layer is important for the enhancement of the hard magnetic characteristics of Nd-Fe-B films. As compared to Ta/NdFeB(540nm)/Ta single layer annealed at 620°C, the Ta/[NdFeBMo(1at.%)(180nm)/Fe(1nm)]×3/Ta multilayer film annealed at 620°C exhibits an increase in the coercivity from 12.1 kOe to 22.8 kOe, in the remanence ratio from 0.77 to 0.80, and in the maximum energy product from 4.5 to 7.1 MGOe. The Ta/[NdFeBMo(1at.%)(180nm)/Fe(1nm)]×3/Ta multilayer films exhibit a decrease in the crystallization temperature of about 30°C in comparison with Ta/NdFeBMo(1at.%)(540nm)/Ta thin film.

By analyzing the presented results it can be seen that the Ta/[NdFeBMo(1at.%)(180nm)/Fe(1nm)]×3/Ta multilayer films are interesting as actuation elements for magnetic microdevices.

References

[1] Okumoto T, Yamasawa K, Liu X, Matsumoto M and Morisako A 2005 *IEEE Trans. Magn.* **41** 3139.
[2] Jiang H and O'Shea M J 2000 *J. Magn. Magn. Mater.* **212** 59.
[3] Shandong L, Gu B X, Sen Y, Hong Bi, Yaodong D, Zongjun T, Guozhi X, Youwei D and Zuanru Y 2002 *J.Phys.D.: Appl.Phys.* **35** 732.
[4] Jiang H and O'Shea M J 2000 *J. Magn. Magn. Mater.* **212** 59.
[5] Hirosawa S, Tomizawa H, Mino S and Hamamura A 1990 *IEEE Trans. Magn.* **26** 1960.
[6] Klug H P and Alexander L E 1974 Diffraction Procedures for Polycrystalline and Amorphous Materials (Wiley, New York) p 665.

Structural and magnetic transformations in $HoMn_2H_x$ hydrides

A Budziak[1], M Bałanda[1], H Figiel[2], L Kolwicz-Chodak[2] and Z Tarnawski[2]

[1]The H. Niewodniczański Institute of Nuclear Physics PAN, ul. Radzikowskiego 152, 31-342 Kraków, Poland
[2]Faculty of Physics and Applied Computer Science, AGH University of Science and Technology, Al. Mickiewicza 30, 30-059 Kraków, Poland

E-mail: andrzej.budziak@ifj.edu.pl

Abstract. The $HoMn_2$ compound crystallizes in the C15 or C14 Laves phases depending on preparation. The effect of hydrogen absorption on structural, magnetic and thermodynamic properties of $HoMn_2H_x$ hydrides for both phases has been investigated by XRD, AC/DC magnetometry, and calorimetry in the temperature range of 80–300 K. In addition to general features revealed by RMn_2H_x compounds (*R*= rare earth or Yttrium), unusual behavior of these hydrides was found. In particular, a transformation from the cubic to the monoclinic structure was detected, which so far has not been reported for other cubic RMn_2H_x (*R* – rare earth or Yttrium) compounds. The structural transformations are reflected in the magnetic behavior. The change in ordering temperatures indicates a strong correlation between the magnetic interactions and the Mn-Mn distance modified by hydrogen absorption. Tentative magnetic and structural phase diagrams are proposed. The presented results are compared with the properties of other cubic and hexagonal RMn_2H_x hydrides.

1. Introduction

The hydrides of Laves phase type intermetallic compounds RMn_2 (*R* - rare earth or Yttrium) have been intensively studied due to their unusual physical properties. These compounds can easily absorb large amounts of hydrogen (about 4.5 H atoms per formula unit). In recent years, many investigations have proved that H atoms usually occupy tetrahedral A2B2 and AB3 sites. In most of the Laves phase hydrides H atoms occupy only A2B2 sites at low hydrogen concentration, whereas AB3 sites start to be filled if H-concentration is higher than $x \sim 3.0$ [1,2]. Significant increase of the volume of the unit cell due to hydrogen absorption causes a strong modification of structural and magnetic properties of those hydrides.

Systematic research of RMn_2 compounds shows the important role of Mn–Mn distance which determines the magnetic interactions in the RMn_2H_x systems and leads to complicated patterns of magnetic ordering. Above the critical value of Mn–Mn distance $d_c \approx 2.7$ Å [3] a localized magnetic moment at manganese sites is observed leading to a frustrated Mn sublattice [4]. The lattice parameter of the parent material $HoMn_2$ determined at room temperature is $a \approx 7.531$ Å implying that the Mn-Mn distance ($d_{Mn\text{-}Mn} = 2.66$ Å) is slightly smaller than the critical value d_c. Therefore upon the hydrogenation we would expect complex magnetic structures in this system.

The pure $HoMn_2$ compound crystallizes either in the cubic C15 or in the hexagonal C14 structure, depending on the heat treatment [5]. Structural and magnetic results presented here were obtained for the C15 structure (space group: *Fd-3m*). We tried to find the similarities to the other RMn_2 cubic structures, bearing in mind a strong affinity of both the C15 and C14 polytypes [6,7]. It is well known that those structures are closely related to each other, however, it is still unclear whether the difference in cubic or hexagonal packing affects the properties of hydrides.

2.Experimental details

The cubic C15 parent material $HoMn_2$ and its hydrides $HoMn_2H_x$ with hydrogen concentrations x = 0.55, 1.0, 1.65, 2.0, 2.5, 3.0, 3.5, and 4.3 and the hexagonal C14 parent material $HoMn_2$ and its hydrides with x = 1.0, 1.5, 2.0, 2.5, 3.0, 3.5, and 4.5 were prepared by the standard technique for sample preparation and hydrogenation process reported elsewhere [8,9].

The x-ray powder diffraction (XRD) measurements have been carried out on the Siemens D5000 diffractometer with Cu K_α radiation (λ = 1.54056 Å), and with continuous flow cryostat in the temperature range from 70 to 385 K. The data were collected during the heating-up of samples. At each step the temperature was stabilized (at least 40 min) in order to reach the thermal equilibrium. For identification of the XRD patterns the FullProf program [10] based on the Rietveld method [11] was used.

The AC/DC susceptibility at the temperature range of 4 ÷ 300 K were measured with the LakeShore VSM 7225 magnetometer. The magnetic measurements were performed during heating up.

In the paramagnetic state, the magnetic behavior was characterized by the Curie-Weiss law taking into account the correcting diamagnetic term χ_{dia} according to the equation $\chi(T) = \chi_{dia} + C/(T - \Theta)$.

3.Results

Because of many similarities between both series (C15 and C14) the results and their discussion is focused mainly on the C15 hydrides. Some selected results for the C14-type hydrides as well as the most important differences between the C15 and C14 series will be also presented.

*3.1.*The x-ray diffraction results

3.1.1. The $0 \leq x < 2$ range

For this hydrogen concentration range, the sample with x = 1.0 ($C15\_HoMn_2H_{1.0}$) reveals the most interesting behavior. The XRD-patterns for this hydride taken at different temperatures are presented in Figure 1.

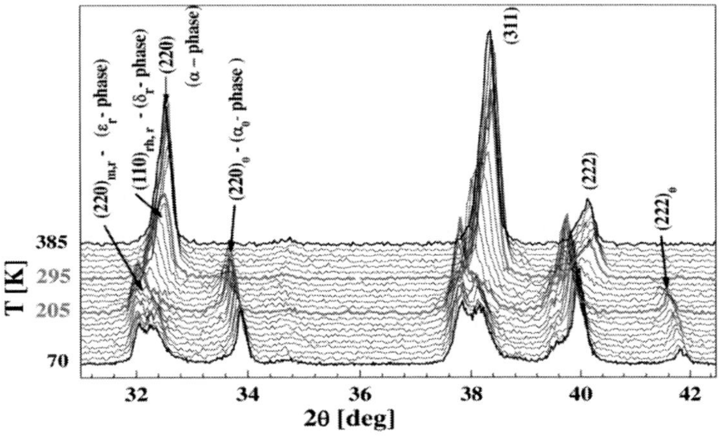

Figure 1. The temperature evolution of the (220), (311) and (222) diffraction lines for $HoMn_2H_{1.0}$ (more in text)

From 385K down to 295 K only one cubic structure was observed (the α phase, *Fd-3m* space group). Below 295 K we see a splitting of it into two structurally different phases, which is in contrast to other hydrides based on Gd,Tb or Er [9,12,13], where two structurally identical phases persisted down to the lowest temperatures. The lattice parameters of one of those phases are comparable with those of pure $HoMn_2$ compound (the α_0 phase). However, the other component transforms from the cubic structure to the rhombohedral one (the δ_r phase, *R-3m* space group) at 295 K and then to the monoclinic one (the ε_r phase, *C2/m* space group) at ~ 205 K. These structural transformations are fully reversible, i.e. heating the sample above the transition temperature leads to the reconstruction of the uniform cubic structure. These types of transformations were not observed for any RMn_2H_x (*R*: Gd,Tb,Er,Y,Dy,Sm) hydrides [9,12-16]. More detailed information of the structural transformations for x = 1.0 and other samples with $x \leq 1.65$ (C15) are presented in Figure 2a. For comparison Figure 2b shows results obtained for the C14-hydrides. For the sake of comparison, all the presented lattice parameters of monoclinic, rhombohedral, and hexagonal structures are converted to the pseudo-cubic system according to the relations: $a_m^* = [a_m\, b_m\, c_m\, (\sin\beta)]^{1/3}$, $a_{rh}^* = [2\,(a_{rh})^2 c_{rh} / (3)^{1/2}]^{1/3}$, $a_h^* = [(3)^{1/2} (a_h)^2 c_h]^{1/3}$.

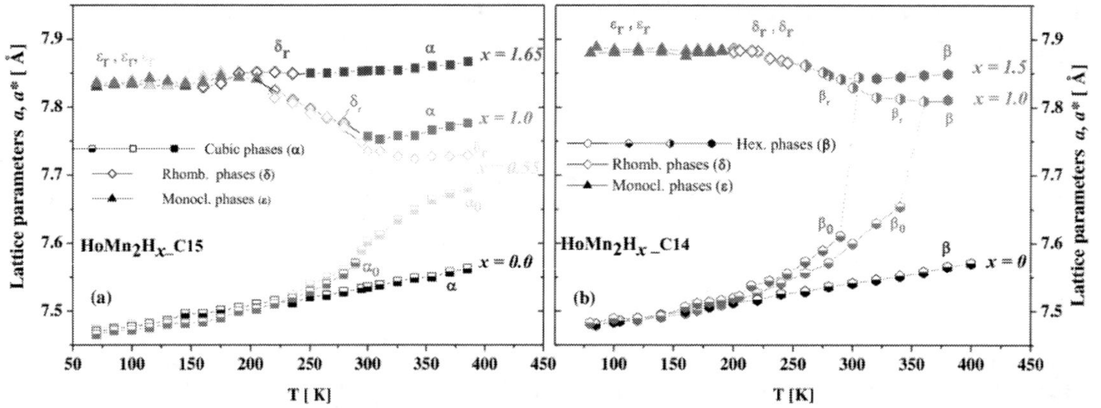

Figure 2. The lattice parameters of $HoMn_2H_x$ hydrides as a function of temperature: (a) C15, (b) C14.

For x = 0.55 (*C*15) two separate phases exist (Figure 2 a). The lattice parameters of one of them tend again towards those of the pure $HoMn_2$ system (the α_0 phase) at lower temperatures. For the other phase only one structural transformation at ~200 K is observed, from the monoclinic (ε_r, *sg: C2/m*) to the rhombohedral (δ_r, *sg: R-3m*) phase which holds to the highest temperatures. Comparing the lattice parameters of the pure $HoMn_2$ compound and its hydrides with x = 0.55 and 1.0 and their relative abundances, we could calculate the amount of hydrogen in the α_0 phase as ~ 0.04 H at./f.u. and in the ε_r phase as ~ 1.6 H at./f.u.

For hydride with x = 1.65 (*C*15) upon cooling, a sequence of single-phase transformations ($\alpha \rightarrow \delta \rightarrow \varepsilon$) is observed, similar to those detected for x = 0.55 and 1.0 (Figure 2a). The temperatures of these transformations are insignificantly shifted towards lower temperatures. The lattice parameters do not change monotonically but reveal characteristic jumps which correspond to structural transformations. The comparison of the lattice parameters for x = 1.65 with those of hydrogen rich phase for x = 0.55 and 1.0 (Figure 2a) below 205 K shows their mutual overlapping, which confirms the presence of the same monoclinic ε_r structure for all the hydrides. Such reversible transformations could be of martensitic type.

The spinodal decomposition is also observed for x = 1.0 (C14). More surprising is the structural transformation of the type $\beta \rightarrow \delta \rightarrow \varepsilon$, where β corresponds to the hexagonal structure (*sg: P6₃/mmc*). It means that independently of the initial structure (C15 or C14) the same structural transformations to the rhombohedral phases (δ) and then to the monoclinic structures (ε) in both the series are observed.

3.1.2. *The $2 \leq x \leq 4.3$ range*

The temperature dependences of the lattice parameters for $2 \leq x \leq 4.3$ (for C15 and C14) are presented in Figure 3 a and b.

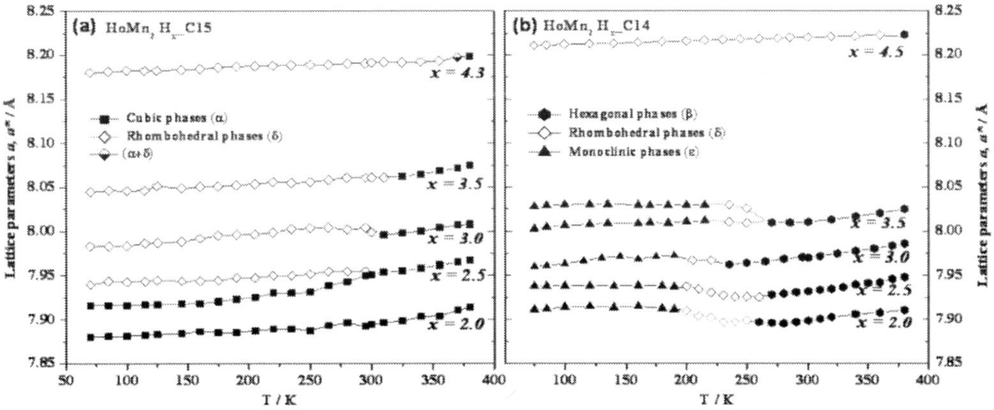

Figure 3. The lattice parameters of $HoMn_2H_x$ vs. temperature: (a) C15, (b) C14.

For this x range, for the *C*15 hydrides, a combination of two structures, the cubic one (α) and the rhombohedral one (δ) is observed. Lattice parameters of the δ phase were converted to the pseudo-cubic system. Only for the sample with x = 2.0, the cubic C15 structure persists in the whole temperature range without any traces of distortion. A small jump of the lattice parameters of the sample is visible at ~ 290 K, which will be shown to be related to a magnetic transition in what follows. The most complicated behavior is observed for the x = 2.5 hydride. From the lowest temperature up to ~ 305 K a splitting into two phases occurs again, which was not observed for any cubic RMn_2H_x hydrides with $2 < x < 3$.

The $HoMn_2H_{3.0}$ and $HoMn_2H_{3.5}$ hydrides show the same structural transitions but in different temperature ranges. For sample with x = 3.0 the rhombohedral unit cell exists up to 275 K and for x = 3.5 up to ~300 K. For the sample with hydrogen concentration x = 4.3 the single rhombohedral phase δ is observed from the lowest temperatures up to ~ 365 K (Figure 3a). Between ~ 365 K and ~380 K two phases (α+δ) are expected. Further increase of temperature leads to the appearance of the α phase. The process is reversible, i.e. decreasing temperature leads to a return to the δ structure.

The C14 hydrides reveal more complicated behavior than the C15 ones (Figure 3 b). The hydrides with x = 2.0, 2.5, 3.0 undergo the same type of structural transformations (β→δ→ε). The sample with x = 3.5 splits into two rhombohedral phases (δ) with different H content below ~ 270 K which then transform to monoclinic structures (ε). However, the $HoMn_2H_{4.5}$ maintains the rhombohedral structure almost for the whole temperature range. The structural transformation to the hexagonal phase is observed above ~ 385 K.

3.2. *The magnetic results*

The *M(T)* curves for different $HoMn_2H_x$ hydrides (C15 and C14) recorded at the external field of 200 Oe are presented in Figure 4 a and b.

The temperature dependence of *M/H* curves for the hydrides with x = 1.0 (both for C15 and C14) prove the two-phase decomposition. Beside the magnetic anomaly at T_N = 195 ± 5 K, which corresponds to a magnetic ordering in the hydrogen rich phase α_r, the second very clear maximum at $T \approx$ 20 K appears related to T_C of the pure $HoMn_2$ compound [17,18]. For the hydride with x =1.65 (C15) a magnetic transition was also observed. For x = 2.0 the estimated magnetic ordering temperatures are ~ 265 K and ~ 218 K (for C15 and C14, respectively), which is in agreement with a general tendency: increasing hydrogen content implies increasing $T_{N,C}$. The samples with x = 2.0 and 2.5 (both C15 and C14) reveal antiferromagnetic character of ordering. The hydrides (C15 and C14)

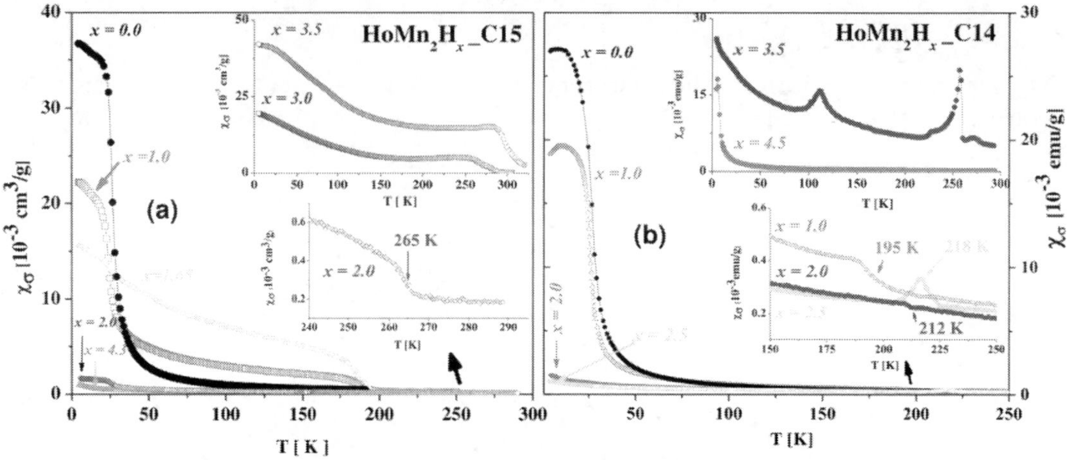

Figure 4 a and b. Magnetic susceptibility of the $HoMn_2H_x$ hydrides versus temperature in the external field of 200 Oe, for different hydrogen concentrations.

with x = 3.5 show the most complicated behavior, which can be also related to their complicated structural transformations. The samples with a maximal H content ($x \geq 4.3$, C15 and C14) reveal again a purely antiferromagnetic character and their magnetic ordering temperatures are expected to be higher than 370 K.

3.3. Comparison of XRD and magnetic results with Specific Heat (SH) measurements

The results and discussion of the SH measurements of the $HoMn_2H_x$ (C15) are reported in [19]. It is interesting to compare them with the XRD and magnetic results presented above.

Only for the $HoMn_2$ sample the phase transition was revealed by a single SH peak at T_N = 20K, which corresponds very well to the left maximum observed in the temperature dependence of the AC magnetic susceptibility measured along the crystal growth direction $[00\bar{1}]$ [17]. Almost all the SH anomalies observed for the $HoMn_2H_x$ have a peculiar double-peak (or even triple-peak) structure.

As hydrogenation involves additional degrees of freedom a contribution to the specific heat in the whole temperature range may be expected. The increase of hydrogen concentration causes a large increase of the magnetic ordering temperature (e.g. $RMn_2H_{1.0}$, T_N ~ 200 K), which was confirmed above by the magnetic and x-ray measurements. For example, the broad maxima of $C_{mol}(T)$ for the x = 1.0 and 1.65 hydrides between ~ 170 and 250 K [19] are located in the same temperature range where both the structural (monoclinic→rhombohedral→cubic) and magnetic transformations are observed (Figure 2 and 4). An additional peak at ~ 215 K is observed for the single-phase x = 2.0 hydride

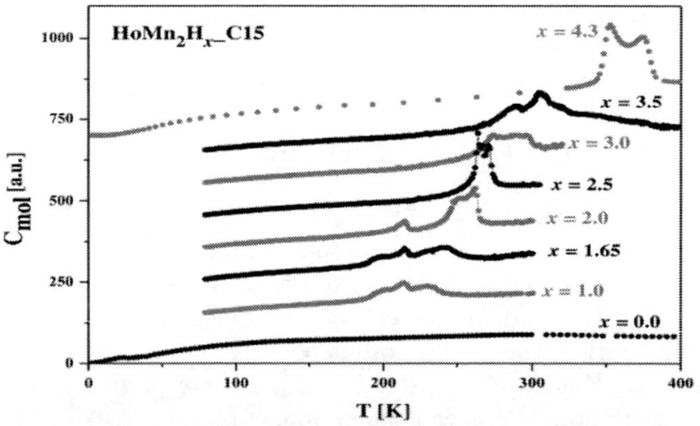

Figure 5. Specific heat of $HoMn_2H_x$ (C15) vs. temperature.

In our opinion, it may be related to hydrogen ordering. The more H atoms in a system the more energy (heat) should be invested to disorder them. That is why the SH values rise with increasing hydrogen concentration [19]. Also, for higher hydrogen concentrations, the transformations and characteristic temperatures (especially magnetic ones) are reflected in $C_{mol}(T)$ curves and their maxima. We expect magnetic ordering at ~ (370 - 375) K for the $HoMn_2H_{4.3}$ as well.

4. Conclusions

Both structural and magnetic properties of HoMn2Hx (C15 and C14) are strongly dependent on hydrogen concentration. On the basis of the x-ray measurements we propose structural phase diagrams shown in Figure 6 a and c. Generally, they can be divided into three areas: (1) the areas where parent phases exist (α and β), (2) the areas below x ~ (1.6-1.7) where the splitting into two phases of the same type ($\alpha/\beta + \varepsilon_r$) are observed, and (3) the areas with x > ~(1.9-2.0) where different structural transformations appear depending on the structure of the parent material.

The structural transformations are reflected in the magnetic phase diagrams (Figure 6 b and d). One can see that these diagrams are more similar to each other than the structural ones. The most important elements of both the diagrams are gray lines separating paramagnetic and ordered phases. A huge jump of the magnetic ordering temperature from ~ 24 K to ~ (200-400) K for the parent compound and its hydrides, respectively, is apparent.

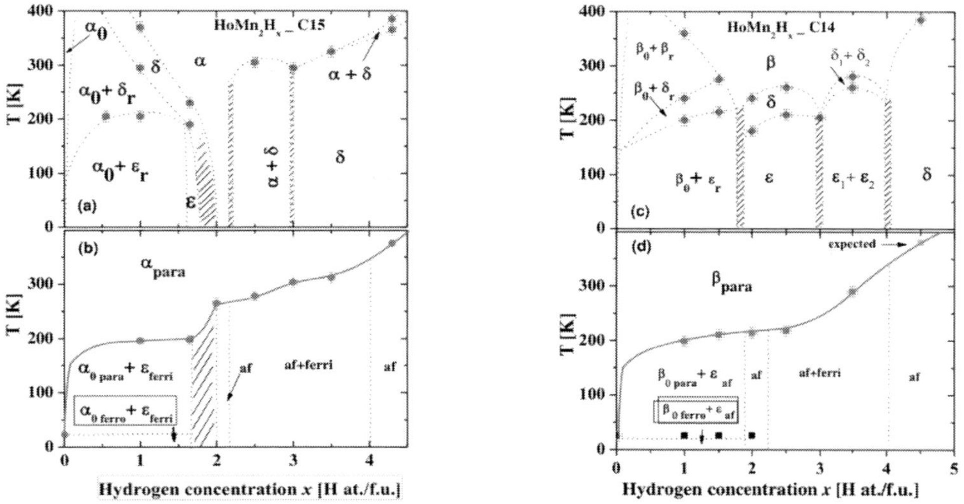

Figure 6. The proposed phase diagrams for the $HoMn_2H_x$ hydrides: (a,c) structural; (b,d) magnetic.

The steady increase of the magnetic ordering temperature with increasing hydrogen concentration was also observed for $(Gd,Y,Tb)Mn_2H_x$ [9,13,16]. Such a correlation indicates that the magnetic transitions are predominantly determined by the magnetic interactions within the Mn sublattice. Increase of the Mn-Mn distance, much more than the critical distance dc, leads to a better localization of the Mn magnetic moments and as a consequence to a substantial increase of the ordering temperatures in the hydrides.

The analysis of the complete set of data for these hydrides is in progress

References

[1] Somenkov A, Irodova AV 1984 *J. Less-Common Met.* **101** 481.
[2] Yvon K, Fisher P, in:) 1988 *Hydrogen in Intermetallic Compounds I* ed L Schlapbach (Berlin: Springer)
[3] Wada H, Nakamura H, Yoshimura Y,Shiga M, Nakamura Y 1987 *J. Magn. Magn. Mat.* **70** 134.
[4] Nunez Regueiro MD, Lacroix C, *J. Magn. Magn. Mater.* 140-144 (1995) 1753.
[5] Inoue K, Imai H 1997 *J. Phys. Soc. Japan* **66** 224.
[6] Berry RL, Raynor GV 1953 *Acta Cryst.* **6** 178.
[7] Johnston RL, Hoffmann R 1992 *Z. Anorg. Allg. Chem.* **616** 105.
[8] Żukrowski J, Strecker M, Wortmann G, Przewoźnik J, Krop K 1997 *J. Alloys Comp.* **261** 47.
[9] Żukrowski J, Figiel H, Budziak A, Zachariasz P, Fisher G, Dormann E 2002 *J. Magn. Magn. Mat.* **238** 129.
[10] Rodrigues-Carvajal J 1994 *FullProf ver.2.5, ILL* (unpublished).
[11] Rietveld MH 1969 *J. Appl. Crys.* **2** 65.
[12] Figiel H, Budziak A, Żukrowski J, Fisher G, Kelemen MT, Dormann E 2002 *J. Alloys Comp.* **335** 48.
[13] Figiel H, Budziak A, Zachariasz P, Żukrowski J, Fisher G, Dormann E 2004 *J. Alloys Comp.* **368** 260.
[14] Figiel H, Przewoźnik J, Paul-Boncour V, Lindbaum A, Gratz E, Latroche M 1998 *J. Alloys Comp.* **274** 29.
[15] Przewoźnik J, Żukrowski J, Freindl K, Japa E, Krop K 1999 *J. Alloys Comp.* **284** 31.
[16] Figiel H, Budziak A, Żukrowski J 1999 *Solid State Comm.* **111** 519–524.
[17] Talik E, Kulpa M, Winiarski A, Mydlarz T, Neumann M 2001 *J. Alloys Comp.* **316** 51–57.
[18] Hardman K, Rhyne JJ, Malik S, Wallace WE 1982 *J. Appl. Phys.* **53 (3)** 1944.
[19] Tarnawski Z, Kolwicz-Chodak L, Figiel H, Kim-Ngan NTH, Kozłowski A, Dawid T, Havelac L, Miliyanchuk K, Santava E 2007 *J. Alloys and Comp.* **446-447** 415-418.

Onset of magnetism stimulated by H absorption in UNiAl-UFeAl

S Maskova[1, a], L Havela[1], S Danis[1], K Kothapalli[2], F Nasreen[2], A V Andreev[3]

[1]Department of Condensed Matter Physics, Ke Karlovu 5, 12116 Prague 2, Czech Republic

[2]New Mexico State University, 88003-8001 Las Cruces, USA-NM

[3]Institute of Physics, Academy of Sciences, Na Slovance 2, 18221 Prague, Czech Republic

[a] e-mail: maskova@mag.mff.cuni.cz

Abstract. Weak itinerant antiferromagnetism of UNiAl is rapidly suppressed by Fe doping. Magnetic order reappears as weak ferromagnetism in $UNi_{0.75}Fe_{0.25}Al$. The onset of magnetism can be shifted by H absorption. $UNi_{0.85}Fe_{0.15}Al$ hydride is an antiferromagnet with T_N approx. 70 K. Hydrogenating $UNi_{0.70}Fe_{0.30}Al$ produces a ferromagnet ($T_C = 90$ K). Here we describe the compound with 20% of Fe and its hydride. The proximity to magnetism shows up as non-Fermi liquid behavior, detected in specific heat as upturn, followed down to 0.3 K, which is gradually suppressed in magnetic fields of several tesla. The hydride was found rather unstable, releasing H at ambient conditions, which made it unsuitable for the specific heat measurement. Magnetization study detected that the hydride shows up as ferromagnet $T_C \approx 98$ K.

1. Introduction

U(Ni,Fe)Al compounds crystallize in the ZrNiAl structure type (space group $P\bar{6}2m$) as many UTX compounds (T - late transition metal, X – *p*-metal). It is a hexagonal structure with two types of basal planes, one with U and T atoms, and the other with T and X atoms. The group exhibits interesting variations of magnetic properties. Weak itinerant antiferromagnetism of UNiAl is rapidly suppressed by Fe doping. Magnetic order reappears as weak ferromagnetism for 25% Fe (figure 1) [1]. Previous studies [2,3] showed that compounds with less than 30% of Fe can absorb hydrogen. The crystal structure preserves its hexagonal symmetry upon hydrogenation or is only slightly distorted. In all studied cases the hydrogen absorption leads to the enhancement of magnetic interactions. So far no findings were reported about the effect of hydrogenation of $UNi_{0.8}Fe_{0.2}Al$, which is close to the onset of magnetism.

Published under licence by IOP Publishing Ltd

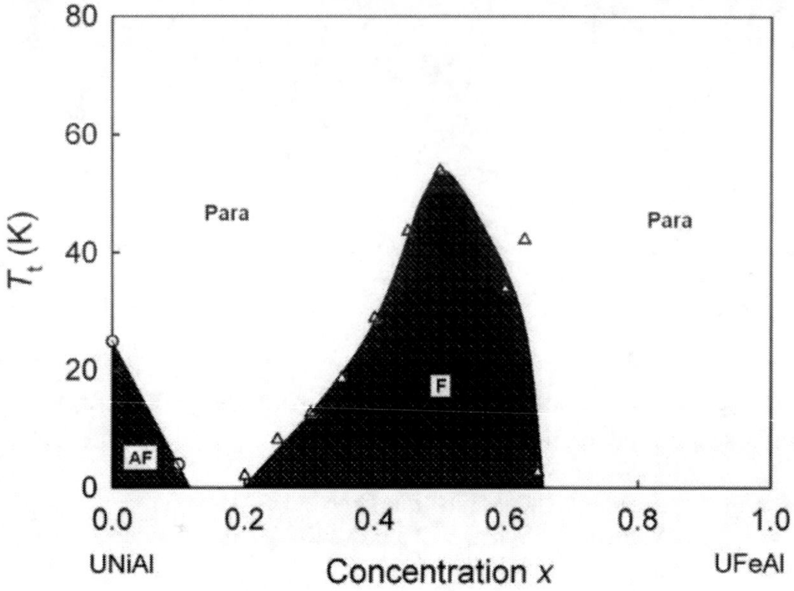

Figure 1. Magnetic phase diagram of U(Ni,Fe)Al: transition temperature T_t, circles mark the Néel temperature, triangles mark the Curie temperature.

2. Experiments

Intermetallic compound $UNi_{0.8}Fe_{0.2}Al$ was used as a starting material for the hydrogenation. The standard procedure of hydrogenation was performed. The sample was crushed into submillimeter particles, the surface of which was activated by annealing at $T = 473$ K for 2 h in a dynamic vacuum (10^{-6} mbar) in order to desorb surface contaminants. The reactor was subsequently filled with a pressurized hydrogen gas (120 bar), and then exposed to thermal cycling up to $T = 923$ K. The hydrogen absorption was registered by a drop of the pressure. The crystal structure was studied by X-ray diffraction (XRD). Quantum Design PPMS equipment was used for magnetic studies and specific heat measurements. Magnetic measurements of $UNi_{0.8}Fe_{0.2}Al$ and its hydride were performed in the temperature range 2-300 K and fields up to 9 T. The grains of the sample for magnetic measurement were fixed in random orientation by acetone-soluble glue. The specific heat was measured on the pellet, produced by pressing the powder, in the case of $UNi_{0.8}Fe_{0.2}Al$. The hydride was found rather unstable, releasing H at ambient conditions, which did not allow us to study specific heat, for which high vacuum conditions are needed.

3. Results and discussion

3.1 Crystal structure studies

The X-ray diffraction pattern of the hydrogenated sample shows that it consists of two phases, both hexagonal (the phase of the parent compound and the hydride phase, both having the same symmetry). It means that the hydride is unstable at ambient conditions and slowly decomposes. The hydrogenation leads to an anisotropic unit-cell expansion along the *a*-axis and compression along the *c*-direction what is the tendency revealed in all previously studied U(Ni,Fe)Al-hydrides. In the table 1 we compare the effect of hydrogen absorption on the crystal lattice of UNiAl [4] and $UNi_{0.8}Fe_{0.2}Al$. From the XRD we were not able to determine the amount of absorbed hydrogen. Also the instability of the hydride in time did not allow to perform controlled temperature-induced desorption in closed volume, what is usually used to determine the hydrogen concentration in the sample. But comparing the volume change upon hydrogenation with the volume change in UNiAl we can assume that the fully hydrogenated phase contains approx. 2 H atoms/f.u.

Table 1. Comparison of the crystal-structure parameters of UNiAl and $UNi_{0.8}Fe_{0.2}Al$ and their hydrides: lattice parameters *a* and *c*, unit cell volume *V* and volume change $\Delta V/V$ with respect to the parent compounds.

	UNiAl	$UNiAlH_{2.3}$	$UNi_{0.8}Fe_{0.2}Al$	$UNi_{0.8}Fe_{0.2}AlH_{\sim 2}$
a (Å)	6.733	7.181	6.736	7.215
c (Å)	4.035	3.988	4.003	3.858
V (Å^3)	158.413	178.103	157.321	173.940
$\Delta V/V$ (%)	-	12.4	-	10.6

3.2 Magnetic properties

The parent compound $UNi_{0.8}Fe_{0.2}Al$ does not order magnetically. Magnetic susceptibility increases monotonously with decreasing temperature and can be fitted by the modified Curie-Weiss (MCW) law down to ≈ 50 K with the parameters μ_{eff} = 1.9 μ_B/f.u., θ_P = -41.5 K and $\chi_0 = 7.4\cdot10^{-9}$ m^3 mol^{-1}. Below that temperature the susceptibility deviates from the MCW law towards higher values and becomes field dependent (figure 3). The hydrogenation dramatically changes magnetic properties. In contrast to the parent compound the hydride orders antiferromagnetically with T_C = 98 K. The temperature dependences of magnetization of the parent compound and of the hydrogenated sample are shown in the figure 2. We were not able to stabilize the hydride, hence we cannot see the pure hydride phase. We can see the ferromagnetic $UNi_{0.8}Fe_{0.2}Al$-H phase in the sample manifested as the anomaly at approx. 98 K corresponding to its T_C.

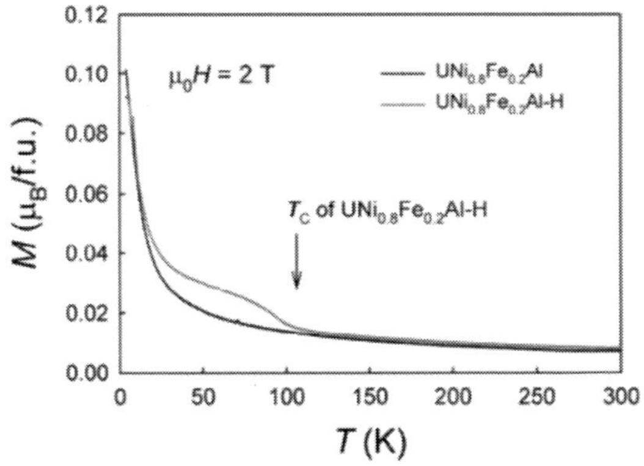

Figure 2. Temperature dependence of magnetization of $UNi_{0.8}Fe_{0.2}Al$ and its hydride.

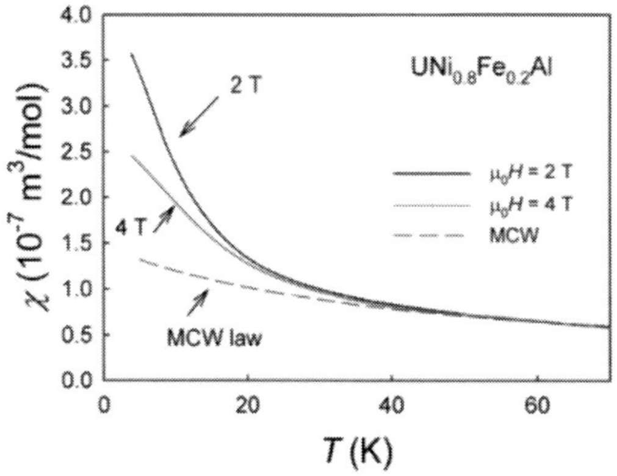

Figure 3. Low temperature part of the magnetic susceptibility measured in various fields for $UNi_{0.8}Fe_{0.2}Al$ compared to the modified Curie-Weiss behavior (MCW).

Comparing the impact of hydrogenation to the whole series of U(Ni,Fe)Al, we find that the hydrogenation leads to the enhancement of magnetic interactions in all cases. The transition temperature increases regardless of the type of magnetic order. The Néel temperature of UNiAl increases from 19 K to 95 K upon hydrogenation [4]. Also in the case of the ferromagnetically ordered example $UNi_{0.7}Fe_{0.3}Al$ the transition temperature increases from T_C = 15 K to T_C = 90 K for $UNi_{0.7}Fe_{0.3}AlH_{0.8}$ [2]. The magnetic properties can be also tuned from nonmagnetic to antiferromagnetic or ferromagnetic depending on the Fe concentration as for $UNi_{0.85}Fe_{0.15}Al$ and $UNi_{0.8}Fe_{0.2}Al$, respectively (table 2).

Table 2. Magnetic properties and ordering temperatures of U(Ni,Fe)Al compounds and their hydrides.

	UNiAl	$UNi_{0.85}Fe_{0.15}Al$	$UNi_{0.8}Fe_{0.2}Al$	$UNi_{0.7}Fe_{0.3}Al$
parent compound	AF, T_N = 19 K	no order	no order	F, T_C = 15 K
hydride	AF, T_N = 95 K	AF, T_N = 70 K	F, T_C = 98 K	F, T_C = 90 K
$\Delta V/V$ (%)	12.4	14.9	10.6	7.3
absorbed H (H at./f.u.)	2	2.4	~ 2	0.8

3.3 Specific heat measurement

The non-magnetic ground state of the parent compound $UNi_{0.8}Fe_{0.2}Al$ is corroborated by the results of the specific heat studies, which do not reveal any anomaly originating in magnetic order. Instead a pronounced upturn in the specific heat appears at low temperatures, which can be followed down to 0.3 K (figure 4). The upturn in specific heat is gradually suppressed in magnetic fields of several tesla. The non-Fermi liquid (NFL) behavior manifested by the upturn can be in our case better described by the paramagnon term $\delta T^2 \ln T$ than by $-\delta T^{1/2}$, predicted for weakly interacting spin fluctuations (figure 5). The extrapolated γ value reaches 274 mJ/mol·K^2. Such a high γ value is relatively common at the verge of magnetism. The observed type of NFL behavior contrasts for example with $U_2Co_2InH_{1.9}$, for which the other type of scaling, with $-\delta T^{1/2}$ term, was identified [5]. Because of the instability of the hydride in time we were not able to obtain the specific heat data for the hydride.

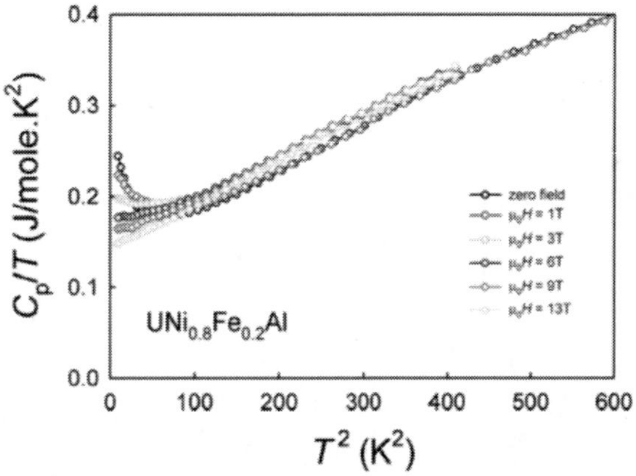

Figure 4. Temperature dependence of the specific heat (in the C/T vs. T^2 representation) of the parent compound $UNi_{0.8}Fe_{0.2}Al$.

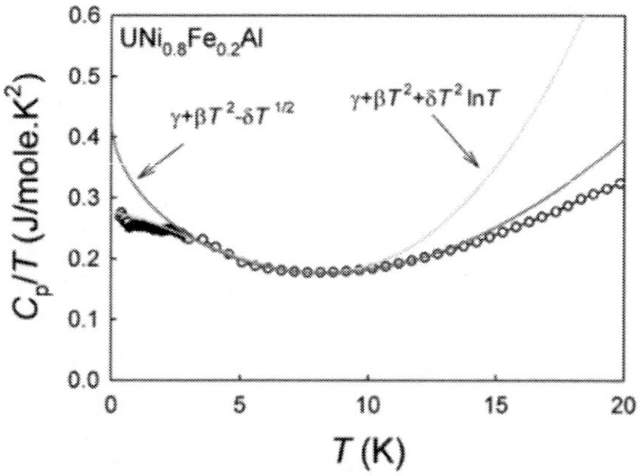

Figure 5. Low temperature part of the specific heat in the C/T vs. T representation of $UNi_{0.8}Fe_{0.2}Al$ measured in zero field. The green and red lines (in the color online version) represent two types of fits, as described in the text.

4. Conclusions

We found that $UNi_{0.8}Fe_{0.2}Al$ is very close to the onset of magnetism, which manifests in the NFL behavior with an enhanced γ value and $T^2 \ln T$ scaling of C/T. The volume increment of 10.6% due to the hydrogenation dramatically modifies magnetic properties, inducing ferromagnetism with T_C = 98 K. Such tendency is attributed to the 5*f* band narrowing.

Acknowledgments

This work was part of the research plan MSM 0021620834 financed by the Ministry of Education of the Czech Republic. It was supported by the Grant Agency of the Czech Republic under the grant No. P204/10/0330.

References

[1] Troc R, Tran V H, Vagizov F G and Drulis H 1995 *Phys. Rev.* B **51** 3003
[2] Raj P, Sathyamoorthy A, Shashikala K, Harish Kumar N, Venkateswara Rao C R and Malik S K 2000 *J. Alloys Compounds* **296** 20
[3] Raj P, Shashikala K, Sathyamoorthy A, Harish Kumar N, Venkateswara Rao C R and Malik S K 2002 *Physica* B **312-313** 885
[4] Bordallo H N, Nakotte H, Kolomiets A V, Christianson A, Havela L, Schultz A J, Drulis H and Iwasieczko W 2000 *Physica* B **276-278** 706
[5] Havela L, Miliyanchuk K, Pereira L C J and Santava E 2007 *Z. Naturforsch.* **62b** 977

Characterization of U-based thin films: the UFe_{2+x} case

Nhu-T H Kim-Ngan[1,2,a], L Havela[1], A M Adamska[1], S Daniš[1], J Pešička[1], J Macl[1], R Eloirdi[3], F Huber[3], T Gouder[3] and A G Balogh[4]
[1] Faculty of Mathematics and Physics, Charles University, 12116 Prague 2, The Czech Republic.
[2] Institute of Physics, Pedagogical University, 30-084 Kraków, Poland.
[3] European Commission, Joint Research Centre, Institute for Transuranium Elements, Postfach 2340, 76125 Karlsruhe, Germany.
[4] Institute of Materials Science, Technische Universität Darmstadt, 64287 Darmstadt, Germany.

[a] E-mail: tarnawsk@mag.mff.cuni.cz

Abstract. We have characterized UFe_{2+x} films prepared by sputter deposition onto fused silica (SiO_2) and Si(111) substrates with the film thickness ranging from 75 nm to 900 nm. The X-ray diffraction results showed an amorphous character of the deposited material. Some of the films showed in addition a pattern of highly textured cubic Laves phase. Rutherford Backscattering Spectroscopy with 2 MeV He^+ ions has been used to determine the composition, thickness and concentration depth profile of the films. A large ageing affect was observed within 1 month after that the films were exposed to air. Magnetic measurements revealed T_C increasing with relative Fe concentration and reaching approx. 450 K in $UFe_{3.0}$.

1. Introduction

Actinide elements with incomplete *f*-shell exhibit many interesting properties when alloyed with the transition metals or other elements, as e.g. strongly-correlated-electron behavior combined with superconductivity and magnetism. Increasing interest has been recently paid to actinide thin film systems with a great potential of tuning electronic and magnetic properties in comparison with the corresponding bulk materials.

UFe_2 is one of examples of combination of 3*d* and 5*f* magnetism in a compound with a relatively high Curie temperature (T_C = 162 K) and very small total U magnetic moments (μ_U = 0.01 μ_B/f.u.), as a consequence of the cancellation of spin and orbital moment. Iron magnetic moment is also reduced (μ_{Fe} = 0.60 μ_B/f.u.) due to a strong 5*f*-3*d* hybridization) [1]. An Fe-excess (i.e. off-stoichiometric UFe_{2+x} compounds with $x > 0$) is expected to increase the T_C value markedly. However, it is difficult to prepare the Laves phase with Fe excess. The quenching technique did not work. An attempt was made using a ball milling [2], which led to enhancement of T_C to 207 K in an amorphous UFe_{2+x} sample. We tried then the ultrafast cooling method (splat cooling) to incorporate more Fe into the Laves phase [3,4]. A nanocrystalline material was obtained up to the stoichiometry $UFe_{2.3}$, with the excessive Fe atoms entering the U sublattice. T_C was enhanced up to 230-240 K in this case. More Fe leads to the segregation of α-Fe. Recently, we continued the effort to embed more U into the UFe_{2+x} phase by sputter deposition of thin films (despite of amorphisation): the UFe_{2+x} thin film deposited on Si (111) substrate with $x \approx 2.0$ and with the film thickness of about 450 nm (sample I) reveals a ferromagnetic behavior with the T_C value of 240 K [5].

In the present work we concentrate to characterize various UFe_{2+x} films sputter deposited onto SiO_2 substrates. The thickness and composition of the layers especially the concentration depth of the films was determined by using Rutherford Backscattering Spectroscopy (RBS). We focus on analyzing the correlation between the film composition and the corresponding magnetic properties. In particular we have underlined the ageing effect in these films.

2. Thin-film synthesis and experimental techniques

UFe_{2+x} films were obtained by DC sputter deposition in a UHV chamber (base pressure better than 10^{-10} mbar) from U and Fe targets onto fused silica (SiO_2) and Si(111) substrates kept at the room temperature. The co-sputtering from the two U- and Fe targets allows the stoichiometry control. Sputtering time was in the range from 10 minutes to 3.5 hours. The stoichiometry of the films, before exposing the films to air, was estimated from in-situ X-ray photoelectron spectroscopy (XPS) and then confirmed by electron microprobe analysis (Table 1).

The X-ray diffraction (XRD) in the glancing angle geometry with an variable angle of incidence in the range of 3-15 degrees and in the Bragg-Brentano (B-B) geometry were used to characterize the film structure.

The RBS experiments were performed at the Institute of Nuclear Physics of the University Frankfurt/Main using 2 MeV He^+ ion beam with a 171° backscattering angle. Details of our RBS experiment conditions were reported elsewhere [6,7]. For the data evaluation the computer code SIMNRA [8] was used. It is more convenient to have the estimated layer thickness in nm and thus the simulated RBS areal density values (at./cm^2) were converted into the layer thickness value (nm) by using the bulk density of the stoichiometric UFe_2 (13.0 g/cm^3). However, the real layer thickness of different layers would be somewhat different, since the films do not have the same stoichiometry. Besides, as it turned out, the film composition is far from 1:2 ratio. The RBS measurements were performed about 1 week after the films were exposed to the atmosphere. In order to check the stability of the films, RBS has been performed again on selected films after 1, 6, 9, 12, and 18 months.

Magnetic properties were studied in the temperature range 2-500 K by means of a Vibrating Sample Magnetometer (VSM) in the Quantum Design PPMS System. For the high-temperature measurement ($T > 300$ K), the samples were fixed by the Zircar cement and tightly wrapped by the copper foil. The temperature dependence of magnetization was measured in the magnetic field of 0.1

Table 1. The composition (2+x) and thickness of the main U-Fe layer of the UFe_{2+x} amorphous films estimated from combined analysis of XPS, electron microprobe and RBS data.

Sample number	$x \approx$	$d \approx$ (nm)	Sample notation
I	2.0	450	$UFe_{4.0}$/Si(111)
II	2.0	310	$UFe_{4.0}/SiO_2$
III	6.0 (+ α-Fe)	370	$UFe_{6.0}$/Si(111)
IV	0.3	520	$UFe_{2.3}/SiO_2$
V	0.7	75	$UFe_{2.7}/SiO_2$
VI	0.7	900	$UFe_{2.7}/SiO_2$
VII	1.0	120	$UFe_{3.0}/SiO_2$

Tesla. Exposure of the samples to higher temperature $T > 500$ K led to a degradation of the films by the oxidation, which is revealed by an abrupt, irreversible increase of magnetization, this is why we suspect the segregation of α-Fe.

We notice here that it is very difficult to compare the in-situ XPS and other ex-situ data. Besides, the RBS data for instance can give the information about the element content in different sub-layers (depth profile), which is certainly different from other techniques estimated the average value for the entire film. For the sake of comparison, we concentrate on the main U-Fe layer. The composition of the films revealed in the film notation was determined as an average value of the Fe-content in the entire films from different methods (Table 1). Our results showed that x-value of the U-Fe layer varied between 0.3 and 2.0 (with the exception of the sample III indicating clearly the existence of the α-Fe phase) and the film-thickness is in the range from 75 nm to 900 nm. During the discussions of magnetic data when we regard only the U- and Fe content, we would skip the notation for the SiO_2 substrate, e.g. instead of $UFe_{3.0}/SiO_2$ film we label it as $UFe_{3.0}$ sample.

3. Results and discussion

3.1. Structure study

Figure 1 shows the XRD patterns obtained in the conventional B-B geometry for the UFe_{2+x} samples with $x = 0.3$ and 0.7 (i.e. $UFe_{2.3}/SiO_2$ (sample IV) and $UFe_{2.7}/SiO_2$ film (sample VI)), exhibiting a broad bubble between 35-40 deg., which is a fingerprint of an amorphous phase. The comparison with data obtained in the glancing angle geometry reveals that it is not only the amorphous substrate (in case of fused silica) giving rise to this feature, but that the deposited film must be largely amorphous. In some of the films we observed besides the amorphous feature also several sharp peaks, which can be attributed to the cubic Laves phases structure with the lattice parameter of $a = 7.15$ Å. The dominance of the (111) peak can be understood as strong texturing with the [111] direction perpendicular to the substrate. The same peaks are also revealed in the glancing-angle XRD patterns for $UFe_{2.7}/SiO_2$ film shown in figure 2. The diffraction lines of UO_2 were also detected. They partially overlap with the lines coming from the Laves phases. For all other remaining samples, XRD patterns in both geometries show no sharp diffraction peaks, indicating that the samples are purely amorphous. As an example, we showed in figure 2 the glancing-angle XRD patterns for amorphous $UFe_{3.0}/SiO_2$ film (sample VII).

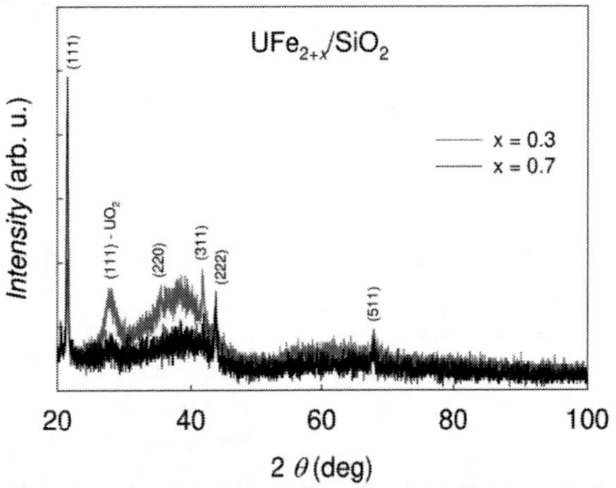

Figure 1. XRD patterns for UFe_{2+x}/SiO_2 films ($x = 0.3$ (sample IV), 0.7 (sample VI)) obtained in conventional Bragg-Brentano geometry using Cu-K_α radiation. The data were collected with 2θ-step size of 0.02° in a 2θ-range of 20°-100°.

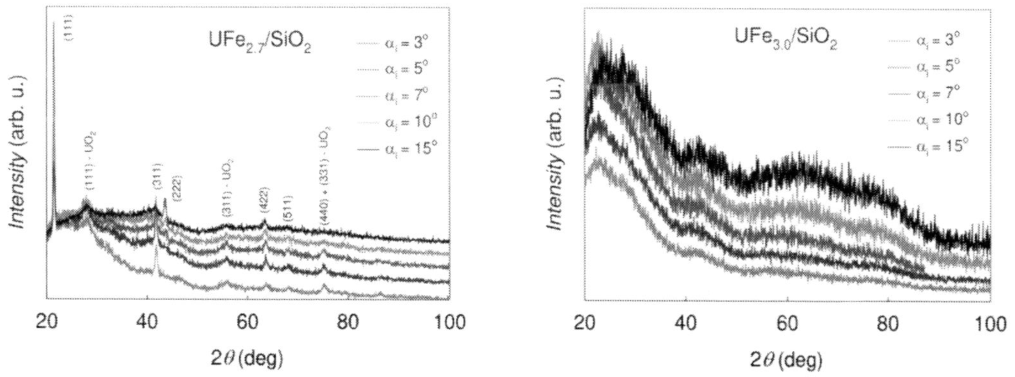

Figure 2. XRD patterns for $UFe_{2.7}/SiO_2$ (sample VI, left panel) and $UFe_{3.0}/SiO_2$ film (sample VII, right panel) obtained in the glancing angle XRD using Cu-K_α radiation with a variable angle of incidence increasing from 3° (bottom) to 15° (top).

3.2. RBS measurements

In figure 3, random RBS spectra of various films revealing some characteristic features are shown: the RBS spectrum with a sharp U- and Fe peaks completely separated from each other ($UFe_{3.0}/SiO_2$ film; sample VII), the spectrum with large but separated U- and Fe signals ($UFe_{4.0}/SiO_2$ film; sample II) (figure 3, left panel) and the spectrum with a partial overlap of the U- and Fe signals implying some fine structure of the combined large film-signal ($UFe_{4.0}/Si$ film; sample I, figure 3, right panel). The simulated spectra by SIMNRA are shown as solid lines in the same figures. The estimated layer thickness and composition are given in table 1. We notice here that one main difference between films deposited on Si(111) and SiO_2 substrate was that we observed only the backscattered signal from Si (Si-edge; Si(sub.) in the RBS spectra for films deposited on the Si(111) substrate, while both Si- and O-edge were observed for films deposited on SiO_2 ones (Si(sub.), O(sub.)). For the thin films with the thickness $d < 120$ nm, i.e. samples V and VII, a well-separated and sharp U- and Fe peak from the UFe_{2+x} film (U(film), Fe(film)) is always observed. The increase-slope of the peak at the right-hand side locates at 1850 keV and 1500 keV, i.e. the energy value related to the binary collision of the He ions with U- and Fe atoms in the front surface of the films, respectively. From the relative peak-intensity ratio, it was clear that the uranium-iron ratio is larger than 1:2. The Si- and O-edge from the SiO_2 substrate were revealed in the RBS spectra around respectively 1130 keV and 725 keV. Increasing the film thickness leads to a large increase of the peak widths. If the film is still not so thick, for instance for sample II with $d \approx 310$ nm (figure 3), the U- and Fe signal is still well separated and thus there exists a visible minimum between them. However, in this case, the O- and Si-edge moved down to a much lower energy. Increasing further the film thickness implied an overlap of the U- and Fe signal. For the film with the thickness 250 nm $< d <$ 500 nm, a partial overlap of the U- and Fe signals leads to a formation of a wide peak but with some features(step-rises) observed in the RBS spectrum (sample I and III). In figure 3, the step-rise at 1850 keV and the step-decrease around 1300 keV revealed the signals of the He ions backscattered from the U atoms from the front and rear surface of the films respectively (U(film-front), (U(film-rear)), whereas those at 1550 keV and 980 keV was related to the backscattered signals from Fe in the films respectively ((Fe(film-front), Fe(film-rear)). A total overlapping of U- and Fe signal resulted in a single but huge and very wide peak was observed for thicker films (sample IV with $d \approx 520$ nm and sample VI with $d \approx 900$ nm).

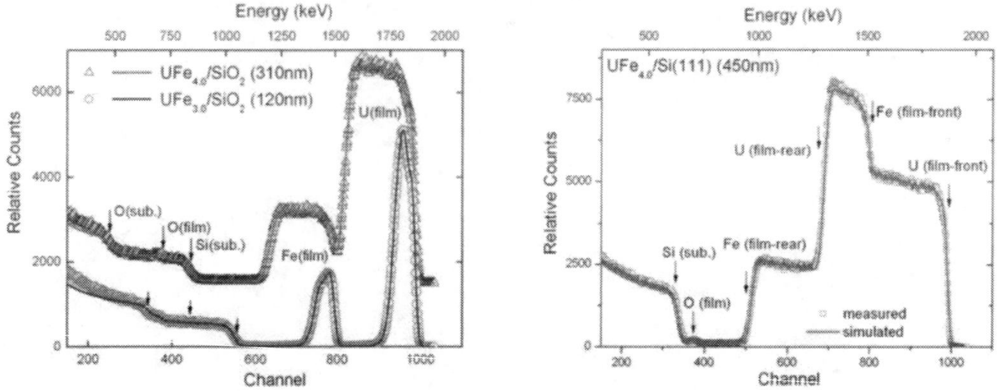

Figure 3. Random (markers) and simulated (lines) RBS spectra for selected UFe_{2+x} films: left) 120 nm-thick $UFe_{3.0}/SiO_2$ (sample VII) and 310 nm-thick $UFe_{4.0}/SiO_2$ (sample II), right) 450 nm-thick $UFe_{4.0}$/Si(111) (sample I). The exemplary spectra revealed different features of the U- and Fe backscattered signals: well-separated and sharp U- and Fe signals (U(film), Fe(film)) for thin films with $d < 120$ nm (as for sample V and VII), separated but wide U- and Fe signals for the film with $d \approx 310$ nm (sample II), a partial overlap of the U- and Fe signals implying some fine structures of the spectra such as steps in the large peak (sample I and III). In all films the existence of the surface oxidized layer was revealed by a visible oxygen signal from the film (O(film)) which was well separated from the Si- and O-edge from the SiO_2 substrate (Si (sub.), O(sub.)).

For all investigated films, a very pronounced oxygen peak from the film (O(film)) was revealed in the RBS spectra (the small peak around 720 keV), which was well separated from the oxygen signal from SiO_2 substrates. It is attributed to the presence of oxygen in the topmost surface layers of the films (i.e. oxidized surface). For a more detail analysis of the film composition we have constructed the depth profile. As example we show in figure 4 the depth profile for sample I. The surface oxidation leads to a formation of a mixed U-Fe-O layer with a thickness of 65 nm in which the O-content decreased from 40% (at the topmost layer) to zero (Figure 4). Probably it is UO_x mixed with the U-Fe phase. Beneath such the mixed layer, we obtained the pure U-Fe layer (oxygen–free layer) with the thickness $d \approx 450$ nm. At the interface (beneath the $UFe_{4.0}$ layer), an U-Fe-Si intermixing region (≈ 60 nm) could be detected. Probably it is USi_3 indicated also by XRD data. For the films deposited on the SiO_2 substrates, the RBS results reveal clearly the oxidation of the film surface as well as the Si diffusion and especially oxygen diffusion in the interface region. Besides, some oxidation of the main U-Fe layer for the films with thickness $d < 400$ nm (e.g. sample II, III, V and VII) was also observed. Only in the case of the film deposited on Si(111) (sample I), the composition of the main U-Fe layer determined from RBS with a thickness of about 450 nm is $U_{16}Fe_{84}$, in a good agreement with that from the in-situ XPS ($U_{14}Fe_{86}$). XPS also revealed 3 at. % Oxygen on the surface layer in this case. One cannot be conclusive about whether this effect can be related to the surface oxidation and subsequent diffusion. Increasing the deposition time and thus increase the film thickness, the pure U-Fe layers can be obtained. For example for sample IV, the main layer with a thickness of about 520 nm has the $U_{30}Fe_{70}$ composition (stoichiometry determined by RBS), while for sample VI, the main

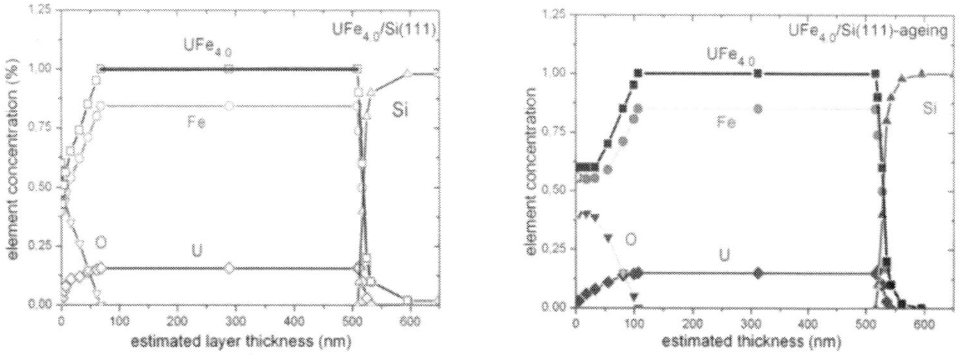

Figure 4. The concentration depth profile constructed from the RBS data for $UFe_{4.0}$/Si(111) film after one week (left) and after one month exposed to air (right) revealing a large ageing effect indicating by an increase of the thickness of the oxided surface layer from ≈ 65 nm to 110 nm.

layer with a thickness of about 900 nm has the $U_{25}Fe_{75}$ composition, i.e. in a good agreement with the XPS data ($UFe_{2.3}$ and $UFe_{2.7}$ respectively). However, we notice here that the RBS data revealed for both these films an oxidized surface, i.e. the mixed U-Fe-O layer with the thickness up to 100 nm. At the interface to the substrate, there exists a mixed U-Fe-SiO_x layer with the thickness in the range 200-300 nm. Certainly there is a difference in the U- and Fe content determined from different techniques. For instance the Fe-content determined from RBS for the main U-Fe layer is often higher than that from XPS. (A good reason for that is that this phase was with the highest temperature of magnetic ordering, besides α-Fe.) However, if taking into account the oxidized surface layer and the interface zone with a lower U-content, the average value for both U- and Fe content in the entire film certainly different to that of only the main U-Fe layer.

The RBS spectra performed on selected films after 1 month from the deposition have revealed a strong ageing effect of the films. A large increase of the oxygen signal in the film was observed within the first month. The depth profile analysis revealed clearly an increase of the thickness of the surface oxidized layer in e.g. $UFe_{4.0}$/Si(111) film (Figure 4), the thickness of the surface mixed U-Fe-O layer increased from 65 nm to 110 nm and the $UFe_{4.0}$ layer decreases from 450 nm to 410 nm. However, RBS measurements after 6, 9, 12 and 18 months indicated that no visible change in the RBS spectra was observed with further exposure to air (in comparison to that after one month). The results indicated that the ageing affect saturates relatively fast and the oxidation does not progress into the bulk. We may attribute it to the mechanical stability of the sputter deposited films, preventing the cracking. For a compact surface, any oxidation of uranium would stop the moment the surface is covered by UO_2, as the O molecule does not dissociate on UO_2 [9].

It is also interesting that a small amount of Fe (2%) and/or U (<2%) was found to diffuse into the Si and SiO_2 substrates in the investigated films, leading to a observed non-zero background between U- Fe peaks and Si-edge. This Fe can be possibly in the form of α-Fe, which is not easily detected by XRD due to the overlap with diffraction lines of USi_3.

3.3. Magnetic measurements

Although the studied films are heterogeneous objects, it is not difficult to identify the contribution of the U-Fe phase. They are all ferromagnetic at room temperature, unlike the Fe-rich splats, which have only a weak ferromagnetic signal at room temperature (room-*T*), clearly due to parasitic α-Fe [3,4], and the main increment corresponding to the bulk ferromagnetism is observed somewhat below 300 K. So as to recognize whether the observed room-*T* ferromagnetism is not due to a possible admixture of α-Fe, and to determine the actual T_C values, we had to undertake the magnetometry experiment using a furnace. The question was the upper temperature limit, which can be reached without desintegration of rather reactive deposited material in exchange gas, which contains besides He also traces of air. The experience showed that the magnetization measurement is reproducible up to approx. 450 K. Above this temperature, magnetization starts to increase with time even if temperature stays constant. Apparently this is the moment when excessive Fe forms a phase, while U forms either crystalline UFe_2 or more likely UO_2. Temperature dependence of magnetization shown in figure 5 for UFe_{2+x} films with $x = 0.3$, 0.7, 1.0 (UFe_2, $UFe_{2.7}$ and $UFe_{3.0}$ film) reveals that T_C values are enhanced and above the room temperature. $T_C = 400$ K was found for $UFe_{2.3}$, while it is further enhanced up to at least 450 K for $UFe_{3.0}$. For a comparison we show in figure 5 also the magnetization of the rather defected film V with a broad distribution of T_C values.

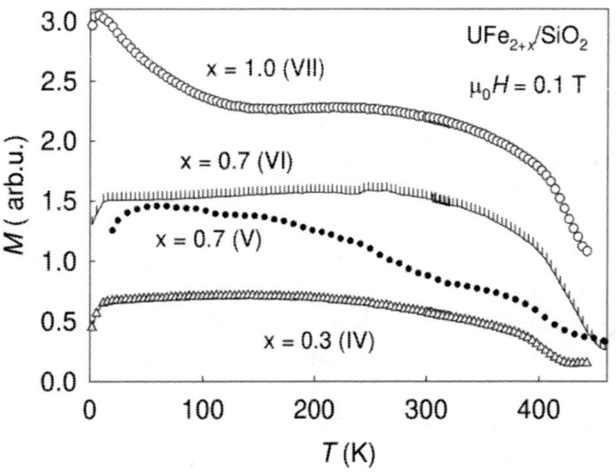

Figure 5. Temperature dependence of magnetization of sputter deposited UFe_{2+x} films on SiO_2 substrate with x ranging from 0.3 to 1.0. The very thin sample V is included to show that it exhibits a distribution of Curie temperatures. The absolute values of magnetization are irrelevant, as each sample was of different mass.

4. Summary

The obtained results on various UFe_{2+x} films indicate that films containing a U-Fe phase as a main ingredient could be successfully prepared by sputter deposition. Whether they are purely amorphous or show additional diffraction lines belonging to a Laves phase deposited with [111] direction perpendicular to the substrate, they reveal a pronounced enhancement of T_C values scaling with the excess of Fe over the UFe_2 stoichiometry and reaching approx. 450 K for the $UFe_{3.0}$, i.e. 1:3 ratio. The fact that ordering temperatures are not suppressed by the disorder well corresponds to the fact that Fe-Fe exchange interactions are responsible for the T_C value. The Fe magnetism is known to be quite insensitive to deviations from ideal crystal periodicity. The contribution of Fe moments remain to be determined by microscopic techniques.

Acknowledgments

The ion beam analysis have been performed within the scope of the German Academic Exchange Service (DAAD) project D/08/07729 (between Germany and Poland) and Polish MNiSW project Nr561/N-DAAD/2010/0. This work was also part of the research program MSM 0021620834. It was partially supported by the Grant Agency of the Czech Republic under the grant No. P204/10/0330 and by the Grant Agency of the Academy of Sciences under the grant No. IAA100100912.

References

[1] Lebech B, Wulff M, Lander G H, Rebizant J, Spirlet J C, and Delapalme A 1989 *J. Phys. Cond. Matter* **1** 10229
[2] Timko M, Szlafarek A, Zentko A, and Kováč J 1999 *J. Magn. Magn. Mater.* **655** 196
[3] Havela L, Miliyanchuk K, Pešička J, Gonçalves A P, Waerenborgh J C, Pereira L C J, Šantavá E, and Šebek J 2008 *Acta Phys. Polonica A* **113** 247
[4] Havela L, Miliyanchuk K, Gonçalves A P, Waerenborgh J C, Pereira L C J, Gaczynski P, Lopes E B, and Pešička J 2011 *Intermetallics* **19** 113
[5] Adamska A, Havela L, Daniš S, Pešička J, Macl J, Uhlířová K, Gouder T, Eloirdi R, Huber F, Kim-Ngan N-T H, and Balogh A G 2010 *Journal of Physics: Conference Series* **200** 012057
[6] Kim-Ngan N-T H, Balogh A G, Meyer J D, Brötz J, Hummelt S, Zajac M, Slezak T, and Korecki J 2008 *Surface Science* **602** 2358
[7] Kim-Ngan N-T H, Balogh A G, Havela L, and Gouder T 2010 *Nucl. Instrum. Methods B* **268** 1875
[8] SIMNRA; *Simulation Program* developed by Mayer M; http://www.rzg.mpg.de/~mam
[9] Gouder T, Seibert A, Havela L, and Rebizant J 2007 *Surface Science* **601** L77

Polarized neutron reflectivity study of $NiFe_2O_4$ films with very large saturation magnetization

F. Rigato, X. Martí, J. Fontcuberta

Institut de Ciència de Materials de Barcelona (ICMAB-CSIC), Campus UAB s/n, 08193 Bellaterra, Spain

E-mail: rigatofr@yahoo.it

Abstract. Epitaxial (001)-oriented $NiFe_2O_4$ films have been grown by RF magnetron sputtering on (001)$MgAl_2O_4$ substrates. In agreement with previous reports, magnetization loops of the thinner films reveal a saturation magnetization up to 3 times larger than the thicker films, which display values in agreement with the bulk. Here we discuss polarized neutron reflectivity experiments which are able to disregard that the magnetization overshoot is caused at the early stages of the growth.

1. Introduction

Spinel ferrites (AB_2O_4, B = Fe) are ferrimagnetic compounds intensively studied nowadays due to their potential use in spintronics. They have been studied both as insulating barriers [1] or as conducting magnetic electrodes [2] in magnetic tunnel junctions (MTJ), and spin filtering effects have been demonstrated for tunnelling barriers [1, 3, 4]. Combined in nanocomposites with ferroelectric perovskites, they also showed the possibility of controlling the room temperature magnetization by the application of electrical field, which exploitation is of major interest [5, 6]. From a more fundamental perspective, spinel thin films still raise one long standing question: why the magnetization of ultrathin films ($t < 10$ nm) can be up to 3 times larger than in the thicker films or the bulk samples [7, 8]. It turns out that most approaches to exploit spinel thin films in MTJ heterostructures will require films with thicknesses below 10 nm, which is just the thickness range where magnetization enhancing has been repeatedly observed. Understanding of this anomalous behaviour calls for more studies in this subject.

The magnetization of ferrites arises from the unbalance of the magnetic moments at antiferromagnetically coupled A (tetrahedral) and B (octahedral) sites of the spinel unit cell. In thin films, the presence of antiphase boundaries [9] seems to be predominant in reducing the saturation magnetization, and their relevance increases with the film thickness. On the other hand, as the magnetization is sensitive to the cation occupancies at A and B sites, anomalous magnetization, compared to equilibrium one, can be obtained both in bulk materials and thin films provided that cation distribution is somehow quenched in a form distinct from the equilibrium distribution. Indeed it has been proposed that this could be the origin of the exceedingly large magnetization reported in $NiFe_2O_4$ and $CoFe_2O_4$ thin films [8, 10, 11]. Cationic disorder could be produced by the out-of-equilibrium conditions associated with thin film growth and likely to be more important in thinner films as the growing time is typically shorter. Enhanced magnetization may also arise from the symmetry breaking at film surface or, most generally at any surface, as it is documented in antiferromagnetic systems [12]. In both cases, free surfaces may display magnetic properties distinct than deep into the film. Last, enhanced magnetization could also originate by some sort of chemical interdiffusion of film/substrate species leading to a further enhancement of the unbalancing between magnetization of A and B sublattices. Indeed it had been proposed, at least in the case of MgO substrates, that interdiffusion may occur [7].

We recall that, in order to be able to account for the observed dependence of the saturation magnetization with thickness in $NiFe_2O_4$ and $CoFe_2O_4$, surface effects are expected to be relevant as in thick films the enhanced magnetization is gradually washed out. In early works [10], we have analyzed thin films grown on several substrates, displaying different strain and surface roughness; we have shown that the same trend is always observed thus disregarding the epitaxial strain and surface roughness as major causes of the additional magnetic moment. Therefore it is of relevance to explore the magnetization profile of ferrite thin films aiming to elucidate the eventual presence of distinctive magnetic characteristics at surface and deep into the film.

Here we report on the use of polarized neutron reflectivity (PNR) [13], a technique particularly sensitive to the magnetization depth profile along the direction perpendicular to the sample surface, in contrast to the similar X-ray reflectivity (XRR). We present here the PNR study of (001)$NiFe_2O_4$ films grown on (001)$MgAl_2O_4$ substrate and we compare the collected patterns to several models where the excess of magnetic moment has been located at different parts of the sample. We conclude that the magnetic-rich layer cannot be located at the substrate/film interface thus disregarding the possibility of non-equilibrium A/B cationic distribution or interdifussion between the film and the substrate at the interface.

2. Experimental details

Thin films discussed here, with thickness t of 6 and 10 nm, were grown by radio frequency magnetron sputtering using a stoichiometric $NiFe_2O_4$ (NFO) target. $MgAl_2O_4$ (MAO) substrates, (001)-oriented, were placed at 6 cm from the magnetron. The growth was performed at a total pressure of 250 mTorr in a mixed atmosphere of argon and oxygen, in ratio 10:1. The substrate temperature during deposition was kept at 600 °C. The samples thickness was determined by regulating the deposition time, through a calibration of the growth rate (~0.28 nm/min) constructed by measuring with XRR the thickness of analogous films with $t > 15$ nm. PNR experiments were performed at CRISP instrument at ISIS spallation neutron source [14]. PNR curves simulations were performed using POLLY software [15].

Samples' structure was investigated by X-ray diffraction (XRD); magnetization data were collected using a superconducting quantum interference device (SQUID).

3. Results and discussion

XRD characterization revealed that the NFO films grown in this work are (001) textured and epitaxial on the MAO substrate. The reciprocal space map around the (206) NFO and MAO reflections for a typical sample (10 nm thick) is shown in Fig. 1a. The MAO(206) reflection is located just below the substrate maximum ($Q_x \sim 0.2460$ Å^{-1}) corresponding to a fully strained film and extends along a constant $Q_z \sim 0.7150$ Å^{-1} up to $Q_x \sim Q_z/3 \sim 0.2380$ Å^{-1} corresponding to a totally relaxed material. Although it may seem that strain may play a role in the first layers, in an early work [10] we have observed that fully relaxed thin films on STO substrates of similar thickness displayed the very same magnetization overshoot and hence we did not put the heading of this paper on an eventual strain-induced magnetization. Further details on identically prepared samples are available elsewhere [10]. Magnetization loops of two samples (6 and 10 nm thick) obtained at 10 K with the magnetic field applied along the [100] direction of the (001)MAO substrate are shown in Fig. 1b. The same linear diamagnetic contribution by the MAO substrate ($\chi_d = -1.09 \cdot 10^{-6}$ emu·Oe^{-1}·cm^{-3}), measured separately, has been removed in both cases from the rough data. The loops show a saturation magnetization well above the bulk expected value (signaled by a horizontal arrow), slightly larger for the thinnest sample in agreement with previous reports on similar ferrites [8, 10]. Importantly, the magnetization loops of the bare substrates do not show any measurable ferromagnetic contribution (see data in Inset of Fig 1b).

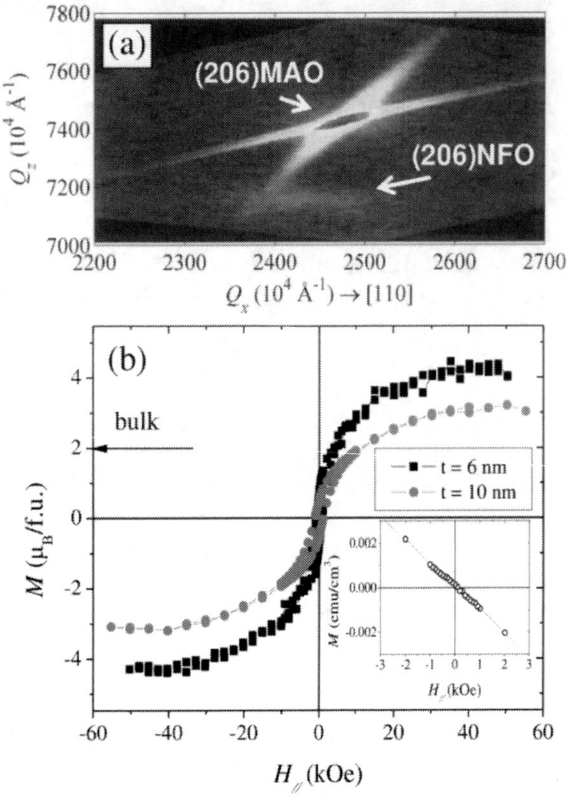

Figure 1. (a) Reciprocal space map around (206)MAO reflection of a 10 nm thick NFO//(001)MAO film. (b) Hysteresis loops obtained at 10 K with *H* parallel to the film surface, for 6 and 10 nm thick films; the inset shows the *M(H)* scan obtained for the bare substrate in the same conditions.

We now turn to the discussion on the preparation of the PNR experiment carried out in the mentioned samples. We first present PNR simulations corresponding to NFO samples, 10 nm thick, with three different distributions of magnetization along the depth of the film are considered. First, in Fig. 2(a) we shown two curves (R^+ and R^-) corresponding to the neutron reflectivity of a NFO film where an homogeneous distribution of magnetic moment (sketched in inset of Fig.2 (a)) and using polarized neutrons with magnetic moments pointing in opposite directions perpendicular to the beam axis and parallel to the sample surface. The splitting of both curves stems from the presence of net magnetic moment in the material [13]. Their difference, better evidenced in the so-called spin asymmetry $SA = (R^+ - R^-)/(R^+ + R^-)$, is shown also in inset of Fig. 2(a): the spin asymmetry is determined by the profile of the net magnetic moment along the sample depth, while the structural contributions are almost cancelled. The situations that we want to explore in detail in this manuscript does not correspond to the described homogeneous distribution of Fig. 2(a), but to a possible excess of magnetization confined either to the topmost region of thickness Δt_T (Top-rich model) or to a bottom layer of thickness Δt_B (Interface-rich model). Therefore we have also simulated the expected SA for these situations, considering several values of Δt_T and Δt_B in a range between 1 and 4 unit cells. Exemplary simulations of expected results for both models are presented in Fig. 2(b) (Δt_T = 2 u.c., Δt_B = 0 u.c.) and Fig. 2(c) (Δt_T = 0 u.c., Δt_B = 2 u.c.), respectively. Insets in Figs 2(b,c) sketch the profiles of magnetization in each case. Starting from a model of homogenous magnetization along the film thickness, the simulations were constructed by fixing two limiting conditions: (i) the total magnetic moment (m) of the film had to correspond to the value previously measured by SQUID at 3.5 kOe, the maximum field available in the PNR beamline, and 300 K, and (ii) the eventual regions with large magnetization are assumed to have a magnetization 4 times larger than the magnetization of the remaining part of the coating. This second condition arises because, within the picture of spinel inversion as source of the enhanced magnetization, the maximal magnetization of a normal $NiFe_2O_4$

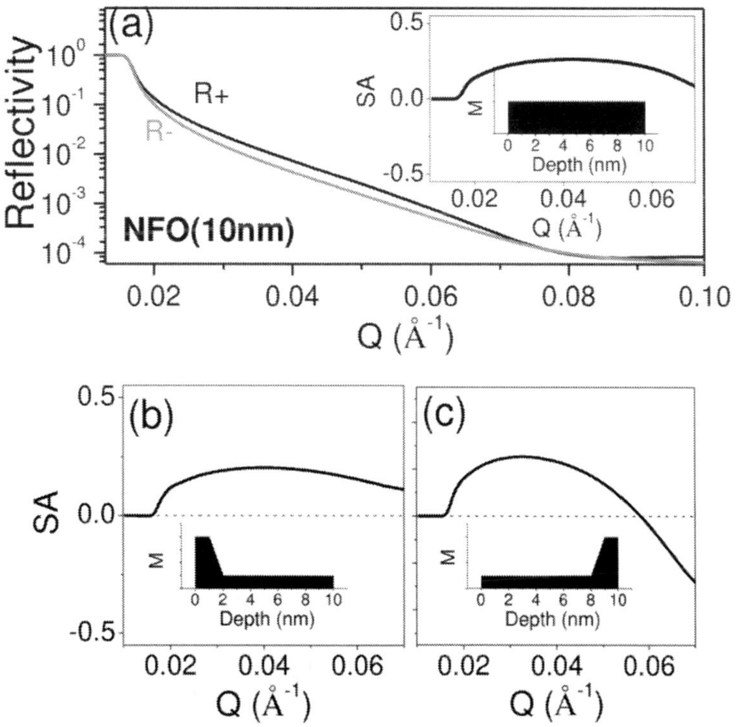

Figure 2. (a) Simulation of PNR curves for a 10 nm thick NFO//MAO film with homogeneous distribution of net magnetic moment along the profile. Spin asymmetry simulation is sketched for homogeneous (inset of (a)), interface-rich (b) and top-rich (c) magnetization models, together with their respective magnetization profiles.

spinel, having all Fe^{3+} ions at B sites, would be four times larger (8 μ_B) than that of a fully inverse structure with Fe^{3+} equally populating A and B sites (2 μ_B) per unit formula [10]. Resulting simulations are displayed in Fig. 2(b) and (c) respectively. In Fig 2(c) one first notice that a clear decay of SA parameter at larger Q values is expected, whereas the corresponding data in Fig. 2(b) displays a rather constant value even at larger Q. Therefore, an evident difference of SA should be observed experimentally in these two limiting cases. Notice that the SA curves cross the 0 value at very different Q's: while the top-excess curve (Δt_T = 2 u.c., Δt_B = 0 u.c.) barely crosses it in the simulated range, the interface-excess (Δt_T = 0 u.c., Δt_B = 2 u.c.) model clearly predicts a change of sign of SA. In short, in the first case the R^+ and R^- curves cross at much higher Q values than in the second.

Now we describe the PNR experimental results. As measurements were performed at relatively small magnetic field (H_0 = 3.5 kOe), in order to ensure the maximum magnetization possible of the NFO films and thus the largest possible contrast between R^+ and R^- curves, the films were ex-situ saturated in a magnetic field of 50 kOe applied parallel to the films surface and transferred to the sample holder in the neutron beamline. During the experiment, the maximum magnetic field available in the beamline was applied also in-plane and along the same direction used in the magnetic preparation of the samples. The PNR curves collected for the 10 nm and 6 nm thick samples are presented in Fig. 3. Data show that in both cases, the R^+ and R^- curves barely overlap in the analyzed

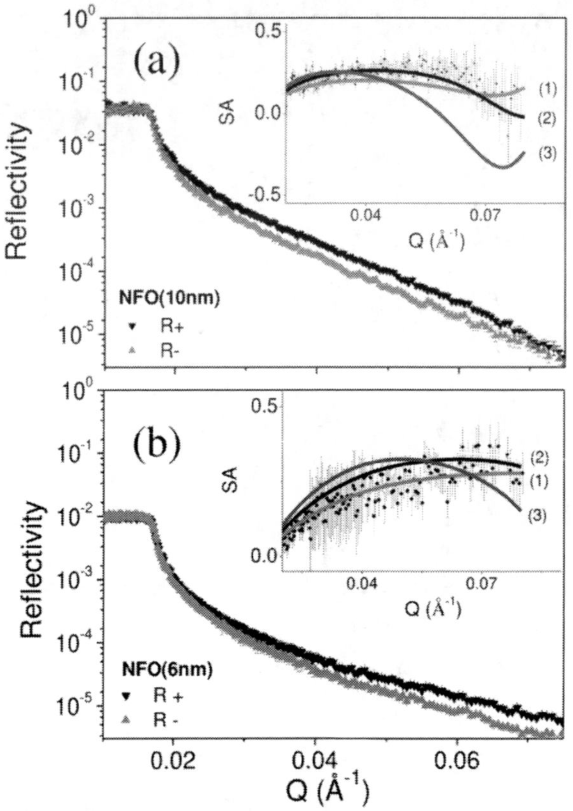

Figure 3. PNR curves for 10 nm (a) and 6 nm (b) thick NFO films; SA is presented in the respective insets, with simulations of the homogeneous (1), top (2) and interface rich (3) magnetization models.

Q-range and accordingly, the SA does not vanish. Indeed, as best seen in the insets of Fig. 3, the experimental SA for the t = 6 nm film clearly departs from zero whereas for the t = 10 nm SA only approaches zero gradually at large Q values. According the results of the modeling described above, that would indicate that the bottom-rich model cannot describe the data, as in this model SA should vanish within the explored Q range.

The solid lines through the SA data in Figs. 3(a,b) are the results of the best fits obtained assuming the distinct magnetization profiles describe above, that is: a homogenous (1), 2 u.c. Top-rich (2) or 2 u.c. Interface-rich model (3) respectively. While simulations for Δt_T or $\Delta t_B \leq 2$ u.c. lead to comparable results, both top and interface rich models considering thicker regions (not shown) result clearly inconsistent with the experimental data. Comparison of best fits with experimental data also points out that the Top-rich magnetization profile describes the data with much more accuracy. Detailed inspection of the homogeneous and Top rich models and the corresponding fits (lines (1) and (2)) in Figs. 3 shows that a clear distinction between these two situations cannot be achieved with the experimental sensitivity.

It thus follows that from the current set of data the possibility of a homogeneously distributed magnetic moment cannot be discriminated from the top-excess scenario. However, the PNR experiment presented here allows ruling out the possibility of an interface-excess scenario that could be produced by an eventual interdifussion of atoms at the interface or alternatively by a localized cationic only partial A/B (Ni/Fe) inversion.

Finally, we would like to point out that recently the origin of a giant magnetic moment in magnetite thin films has been also addressed by complementary magnetic techniques, including PNR. It was concluded from these studies that the excess magnetization is related to impurities (Fe) in the used MgO substrates [16]. Our results can not be interpreted in the same way as our substrates, which are $MgAl_2O_4$, do not reveal traces of any ferromagnetic impurity but a conventional diamagnetic behaviors as indicated above.

4. Conclusions

In summary, we have prepared (001)NFO//(001)MAO films to study the larger saturation magnetization reported for very thin films (t approximately below 10 nm). After a standard X-ray diffraction and magnetic characterization the films were analyzed using the PNR technique. Experiments pretended to discriminate among two different scenarios where the excess of magnetization is confined at the topmost surface or at the film/substrate interface. Data revealed that the excess of magnetization, if it is not homogenous, must be located in the topmost surface of the films thus ruling out the possibility of interface, substrate-related, effects.

5. Acknowledgements

Support by the Spanish Government [Projects MAT2008-06761-C03 and NANOSELECT CSD2007-00041] and Generalitat de Catalunya (2009 SGR 00376) are acknowledged.

References

[1] Ramos A V, Guittet M J, Moussy J B, Mattana R, Deranlot C, Petroff F and Gatel C 2007 *Appl. Phys. Lett.* **91** 12210

[2] Lüders U, Barthélémy A, Bibes M, Bouzehouane K, Fusil S, Jacquet E, Contour J P, Bobo J F, Fontcuberta J and Fert A. 2006 *Adv. Mater.* **18** 1733

[3] Chen Y F and Ziese M 2007 *Phys. Rev. B* **76** 014426

[4] Takahashi Y K., Kasai S, Furubayashi T, Mitani S, Inomata K, and Hono K 2010 *Appl. Phys. Lett.* **96** 072512

[5] Zheng H, Wang J, Lofland S E, Ma Z, Mohaddes-Ardabili L, Zhao T, Salamanca-Riba L, Shinde S R, Ogale S B, Bai F, Viehland D, Jia Y, Schlom D G, Wuttig M, Roytburd A and Ramesh R, 2004 *Science* **303**, 661

[6] Dix N, Muralidharan R, Guyonnet J, Warot-Fonrose B, Varela M, Paruch P, Sánchez F and Fontcuberta J 2009 *Appl. Phys. Lett.* **95** 062907

[7] Venzke S, van Dover R B, Phillips J M, Gyorgy E M, Siegrist T, Chen C H, Werder D, Fleming R M, Felder R J, Coleman E and Opila J 1996 *J. Mat. Res.* **11** 1187

[8] Lüders U, Bibes M, Bobo J F, Cantoni M, Bertacco R and Fontcuberta J, *Phys. Rev. B* 2005 **71** 134419

[9] Margulies D T, Parker F T, Spada F E, Goldman R S, Li J, Sinclair R and Berkowitz A E 1996 *Phys. Rev. B* **53** 149175

[10] Rigato F, Estradé S, Arbiol J, Peiró F, Lüders U, Martí X, Sánchez F and Fontcuberta J 2007 *Mater. Sci. Eng. B* **144** 43

[11] Rigato F, Geshev J, Skumryev V and Fontcuberta J 2009 *J. Appl. Phys.* **106** 113924

[12] Kodama R H, Makhlouf S H and Berkowitz A E 1997 *Phys. Rev. Lett.* **79** 1393

[13] Ott F. 2007 *C. R. Physique* **8** 763

[14] http://www.isis.stfc.ac.uk/instruments/crisp/

[15] Blundell S J and Bland J A C 1993 *J. Mag. Mag. Mater.* **121** 185

[16] Orna J, Algarabel P A, Morellón L, Pardo J A, de Teresa J M, López Antón R, Bartolomé F, García L M, Bartolomé J, Cezar J C and Wildes A 2010 *Phys. Rev. B* **81** 144420

Effect of Magnetic Film Thickness on the Spatial Resolution of Magnetic Force Microscope Tips

Katsumasa Nagano, Kousuke Tobari, Mitsuru Ohtake and Masaaki Futamoto

Faculty of Science and Engineering, Chuo University, 1-13-27 Kasuga, Bunkyo-ku, Tokyo 112-8551, Japan

E-mail: nagano@futamoto.elect.chuo-u.ac.jp

Abstract. Magnetic force microscope (MFM) tips were prepared by coating commercial atomic force microscope (AFM) tips of 5 nm radius with Co and CoCrPt magnetic thin films varying the thickness in a range of 10–80 nm. The structural and the magnetic properties of coated magnetic thin films were investigated by scanning electron microscopy, AFM, X-ray diffraction, and vibrating sample magnetometry. The tip radius and the film surface roughness increase with increasing the film thickness. With increasing the film thickness, the MFM signal sensitivity increases, whereas the resolution decreases due to increase of tip radius. The MFM observation resolutions of 10 nm and 23 nm are obtained with the tips coated with 20-nm-thick Co and 40-nm-thick CoCrPt films, respectively. The MFM resolution is influenced by both the tip radius and the magnetic moment of coated material.

1. Introduction

Magnetic force microscopy (MFM) [1] has been widely used to investigate the nano-scale magnetic domain structures of magnetic devices like high-density recording media [2,3]. The area density of hard disk drive (HDD) has been increasing through implementation of perpendicular magnetic recording. An MFM spatial resolution better than 10 nm is necessary to observe the recorded bits of future HDD around 1 Tb/in^2 area density. The tip is the key element in determining the resolution and the sensitivity of MFM observation. Commercially available MFM tips coated with CoCrPt films can offer resolutions of 20–40 nm. In order to improve the resolution, it is necessary to reduce the magnetic volume of tip interacting with a magnetic sample. Various methods have been tried to prepare MFM tips, such as magnetic material deposition on non-magnetic tips [4–7] or carbon-nano-tube tips [8–10], trimming the tips using focused ion beam [11–14], growing small volume of magnetic material on non-magnetic tips through electron-induced thermal decomposition of cobalt or iron carbonyl vapor flux [15,16], coating the tips with ordered alloy magnetic materials with high K_u values [14,17,18] or multilayers [11,19]. The magnetic material and the coating thickness give influence on the MFM signal sensitivity and resolution. However, there are few reports on the influences of the film thickness and the magnetic material on the MFM observation resolution. In the present study, MFM tips were prepared by coating atomic force microscopy (AFM) tips with Co and CoCrPt, and the effects of film thickness on the MFM spatial resolution were investigated for these magnetic materials.

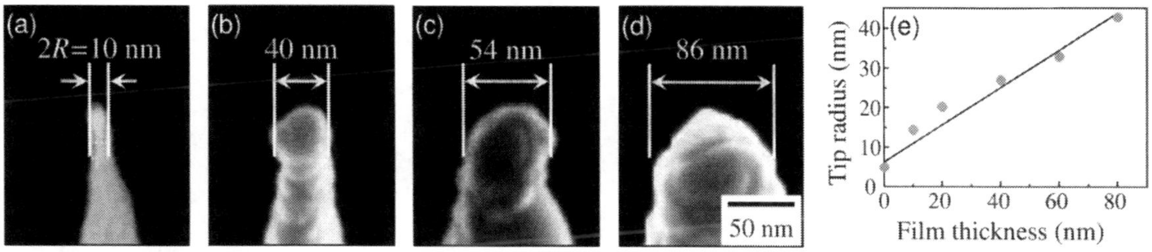

Figure 1. SEM images observed for (a) an AFM tip before coating and (b)–(d) tips coated with Co thin films. The film thicknesses are (b) 20 nm, (c) 40 nm, and (d) 80 nm. (e) Film thickness dependence on the tip radius estimated by SEM.

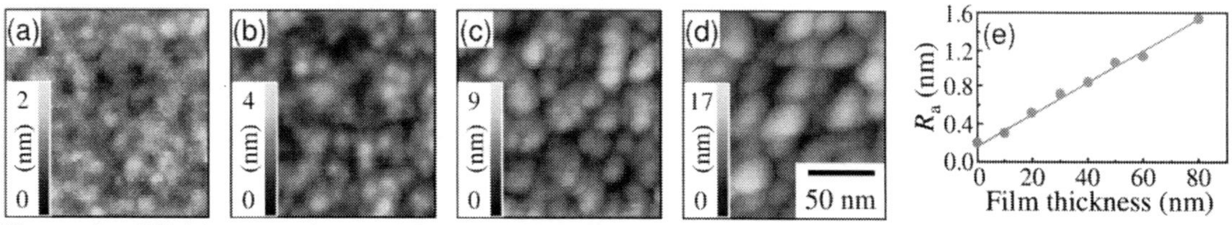

Figure 2. AFM images observed for (a) a Si substrate and (b)–(d) Co films deposited on Si substrates. The film thicknesses are (b) 20 nm, (c) 40 nm, and (d) 80 nm. (e) Film thickness dependence on the surface roughness estimated by AFM.

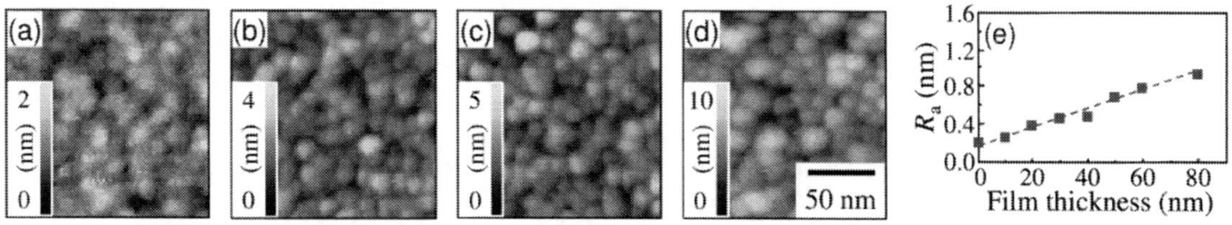

Figure 3. AFM images observed for (a) a Si substrate and (b)–(d) CoCrPt films deposited on Si substrates. The film thicknesses are (b) 20 nm, (c) 40 nm, and (d) 80 nm. (e) Film thickness dependence on the surface roughness estimated by AFM.

2. Experimental procedure

MFM tips were prepared by coating commercial AFM tips made of Si with Co and CoCrPt films at room temperature. The AFM tip radius was 5 nm. An rf magnetron sputtering system was used to coat the AFM tips with magnetic films. The rf powers for $Co_{75}Cr_{15}Pt_{10}$ (at. %) and Co targets were respectively fixed at 34 and 49 W, and the Ar gas pressure was kept constant at 0.7 Pa, where the deposition rate was 0.02 nm/s for both materials. The film thickness was varied between 10 and 80 nm. The CoCrPt film composition was determined to be Co –15 ± 0.5 at. % Cr –10 ± 2.5 at. % Pt by energy dispersive X-ray spectroscopy.

The tip shapes were observed by high-resolution scanning electron microscopy (SEM). In order to investigate the structural and magnetic properties of Co and CoCrPt films, Si substrates with natural oxide layers were employed as the substrates for magnetic thin film deposition. The surface roughness (R_a), the crystal structure, and the magnetic properties were studied by AFM, θ–2θ scan X-ray diffraction with Cu–$K\alpha$ radiation, and vibrating sample magnetometry, respectively.
Commercial perpendicular HDD disk media with the area densities of 163 and 528 Gb/in^2 were used as the MFM observation samples. Random data were recorded on the HDDs. MFM tips were magnetized along the tip axis so that the tip top to possess the south magnetic pole. Thus, the bright and the dark contrasts in an MFM image respectively correspond to the areas where repulsive and attractive forces are working between the tip and the sample.

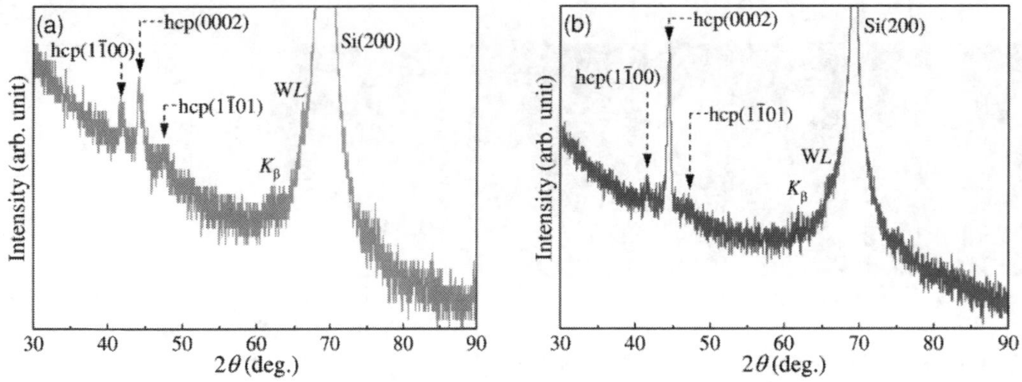

Figure 4. XRD spectra of (a) Co and (b) CoCrPt films of 80 nm thickness deposited on Si substrates. The intensity is shown in a logarithmic scale.

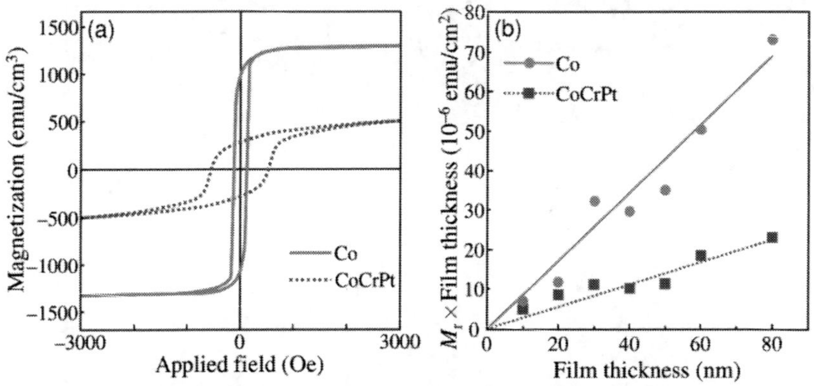

Figure 5. (a) Magnetization curves of 80-nm-thick Co and CoCrPt films deposited on Si substrates. (b) Film thickness dependence on the unit area (cm^2) remnant magnetization (M_r) of Co and CoCrPt films.

MFM observation was carried out at room temperature by using a scanning probe microscope (AFM/MFM) unit, SPI4000/SPA-300HV (SII Nano-Technology Inc.). The MFM chamber pressure during MFM observation was kept lower than 0.1 Pa vacuum. The quality factor (Q) value was controlled in a range between 4000 and 8000. The distance between the MFM tip and the sample was 5–10 nm. The scanning speed was 1.5–2.0 μm/s. The MFM resolution was carefully investigated by optimizing the observation condition.

3. Results and discussion

3.1. MFM tip profiles and the magnetic properties

Figure 1(a) shows the SEM image of AFM tip used for coating magnetic thin film. The tip radius of R = 5 nm is recognized. Figures 1(b)–(d) show the SEM images of tips coated with Co film with different thicknesses. Figure 1(e) shows the film thickness dependence on the tip radius. The tip radius increases linearly with increasing the film thickness. The tip profile suggests an involvement of island-like growth of Co film on the AFM tip made of Si. In order to investigate the effect of film thickness on the roughness, Co films were deposited on Si substrates by varying the thickness, and the surfaces were observed by AFM. Figures 2(a) and (b)–(d) respectively show the AFM images of a Si substrate and Co films with different thicknesses deposited on the Si substrate. The surface morphology of Co film is very similar to that of Co-coated tip for same thickness. Figure 2(e) shows the film thickness

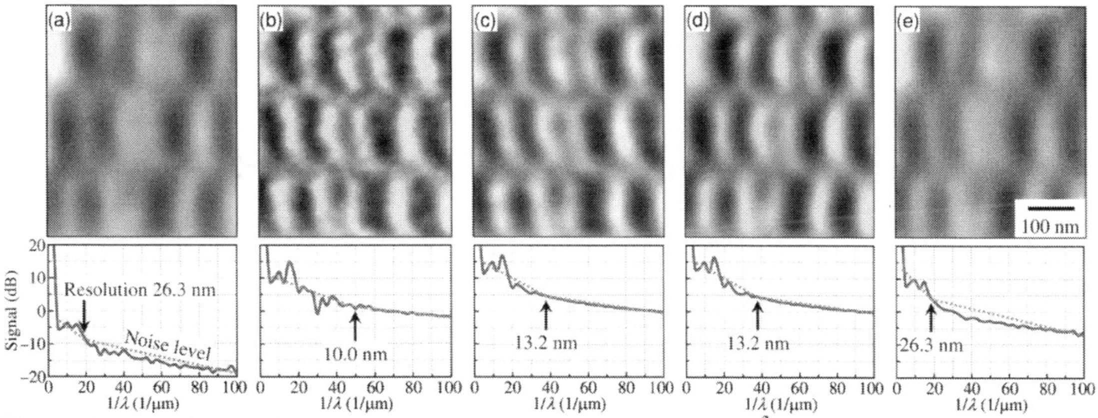

Figure 6. MFM images for a perpendicular HDD medium (163 Gb/in^2) observed by using tips coated with Co film with different thicknesses and the power spectra of MFM intensity profile measured along the track direction. The film thicknesses are (a) 10 nm, (b) 20 nm, (c) 30 nm, (d) 40 nm, and (e) 60 nm.

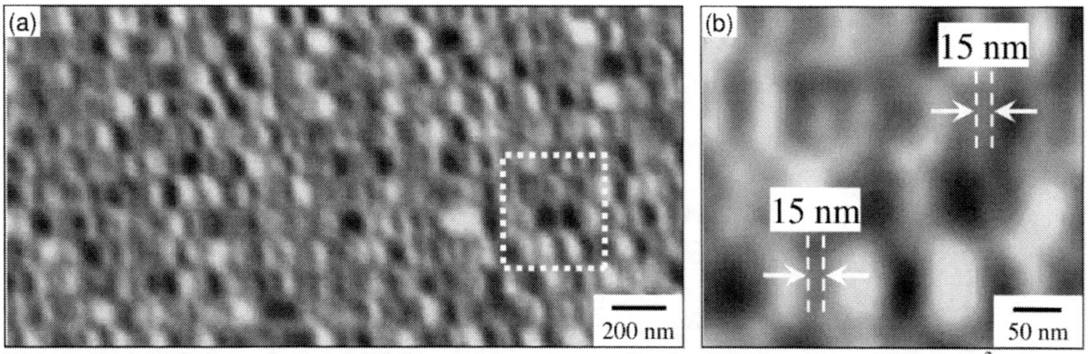

Figure 7. (a) High-resolution MFM image for a perpendicular HDD medium (528 Gb/in^2) observed by using a tip coated with 20-nm-thick Co film. (b) Enlarged view of the area surrounded by white square dotted lines in (a). The arrows in (b) show the recorded bits with shortest length of 15 nm.

dependence on the R_a value. The R_a increases linearly with increasing the film thickness. Similar tendency of R_a increase was also recognized for CoCrPt films, as shown in figure 3. These results show that the radius and the roughness of tip increase with increasing the film thickness. The magnetic volume of tip interacting with a magnetic sample increases with increasing the film thickness.

The crystal structure and the magnetic properties of Co and CoCrPt thin films were investigated by employing Si substrates with natural surface oxides, which are considered to be similar to the AFM tips made of Si. Figure 4 shows the XRD spectra of 80-nm-thick Co and CoCrPt films deposited on Si substrates. The XRD spectra show the hcp reflections of $(1\bar{1}00)$, (0002), and $(1\bar{1}01)$ in addition to Si(200) reflection. The crystal structures are determined to be hcp for the Co and the CoCrPt films. Figure 5(a) shows the magnetization curves of 80-nm-thick Co and CoCrPt films, where the magnetic field is applied along the in-plane direction. The saturation magnetization values of Co and CoCrPt films are similar to those of bulk hcp–Co crystal (1440 emu/cm^3) and bulk hcp–CoCrPt crystal (about 550 emu/cm^3), respectively. Figure 5(b) shows the film thickness dependence on the unit area remnant magnetization. For a same film thickness, the remnant magnetization value of Co film is three times larger than that of CoCrPt film. This result suggests that a Co-coated tip has a better MFM signal sensitivity than that for a CoCrPt-coated tip.

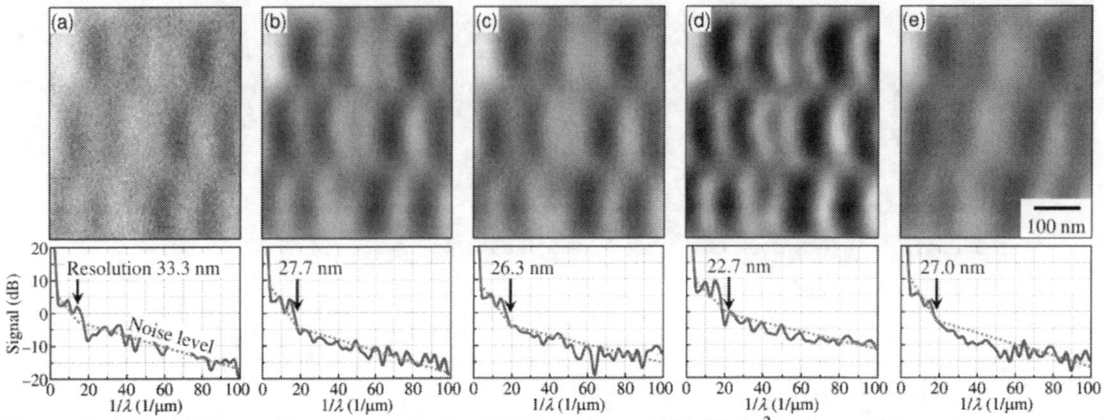

Figure 8. MFM images for a perpendicular HDD medium (163 Gb/in^2) observed by using tips coated with CoCrPt film with different thicknesses and the power spectra of MFM intensity profile measured along the track direction. The film thicknesses are (a) 10 nm, (b) 20 nm, (c) 30 nm, (d) 40 nm, and (e) 60 nm.

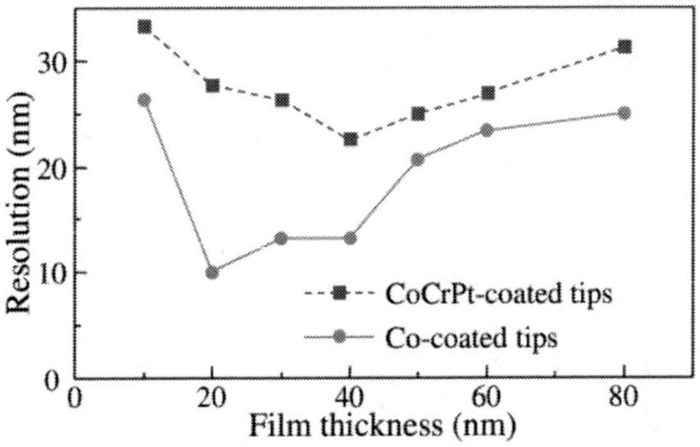

Figure 9. Film thickness dependence on the MFM observation resolution estimated for Co-coated and CoCrPt-coated tips.

3.2. Spatial resolution of MFM tips with different film thicknesses

Figure 6 shows a series of MFM image observed for a same area of perpendicular HDD medium (163 Gb/in^2) observed with Co-coated tips of different film thicknesses, and the power spectra of MFM intensity measured along the track direction. With a 10-nm-thick Co film, an unclear and low contrast image is observed, as shown in figure 6(a). However, with increasing the film thickness, MFM signal sensitivity increases possibly due to increase of magnetic volume reacting with the HDD medium sample. The MFM resolution is estimated as the half of wave length (λ), where the $1/\lambda$ value is estimated from the signal spectrum intersecting the noise line. The noise line is carefully determined by a wave-length cutoff value. A resolution of 10 nm is obtained from the intersection shown in figure 6(b), $1/\lambda$ = 50 (1/μm), for the tip coated with 20-nm-thick Co film. With further increasing the film thickness, the resolution decreases due to increase of tip radius and surface roughness. Figure 7 shows the high-resolution MFM images of a HDD medium (528 Gb/in^2) observed by using the tip coated with 20-nm-thick Co film. The shortest bit length of 15 nm is clearly observed. Similar tendencies of MFM signal sensitivity and resolution dependences on the film thickness are observed for the CoCrPt-coated tips, as shown in figure 8. A resolution of $\lambda/2$ = 23 nm is obtained with a tip coated with 40-nm-thick CoCrPt film.

Figure 9 shows the film thickness dependence on the resolution estimated for Co and CoCrPt tips. For a same film thickness, the resolution of Co tip is better than that of CoCrPt tip. The reason is considered to be related with the MFM signal sensitivity, where the Co-coated tip has three times larger magnetization compared with that of CoCrPt-coated tip for a same film thickness, as expected from the magnetic properties of coated materials shown in figure 5. With increasing the film thickness, the resolution difference between the Co and the CoCrPt coated tips decreases. This is possibly because that the tips have enough sensitivity to detect the MFM signal from the sample when the coating thickness is increased further. The tip coated with 20-nm-thick Co film shows the best resolution of 10 nm. The present study clearly indicates that the MFM resolution is apparently influenced by both the tip shape and the magnetic property of coated material.

4. Summary

MFM tips were prepared by coating commercial Si-AFM tips with Co and CoCrPt films. Influence of film thickness on the MFM resolution was investigated. With increasing the film thickness, the MFM signal sensitivity increases due to increase in magnetic volume of tip. Resolutions of 10 nm and 23 nm are achieved with the tips coated with 20-nm-thick Co and 40-nm-thick CoCrPt films. With increasing the film thickness further, the MFM resolution decreases due to increase of tip radius and surface roughness for both magnetic materials. Higher MFM signal sensitivity and higher resolution are recognized for the Co-coated tips. The MFM resolution is apparently influenced by both the tip shape and the magnetic property of coated magnetic material.

Acknowledgement

Authors thank Ms. Noriko Saidoh of RIMCOF for HR-SEM observation and Prof. Fumiyoshi Kirino of Tokyo National University of Fine Arts and Music for energy dispersive X-ray spectroscopy analysis. This work is partially supported by NEDO–Japan, MEXT–Japan (No. 22560302), and Chuo University Grant for Special Research. M. Ohtake is a Research Fellow of the Japan Society for the Promotion of Science.

References

[1] Martin Y, Wickramsinghe K, 1987 *Appl. Phys. lett.* **50** 1455
[2] Nagano K, Sasaki S, Futamoto M, 2010 *J. Phys. Conf. Ser.* **200** 102006
[3] Suzuki T, Zhang Z, Singh A, Yin J, Perumal A, Osawa H, 2005 *IEEE Trans. Magn.* **41** 555
[4] Boef A J D, 1990 *Appl. Phys. lett.* **56** 2045
[5] Rührig M, Porthun S, Lodder J C, 1994 *Rev. Sci. Instrum.* **65** 3224
[6] Skidmore G D, Dahlberg E D, 1997 *Appl. Phys. lett.* **71** 3293
[7] Yamaoka T, Watanabe K, Shirakawabe Y, Chinone K, 2003 *J. Magn. Soc. Jpn.* **27** 429
[8] Deng Z, Yenilmez E, Leu J, Hoffman J E, Straver E W J, Hongjie D, Moler K A, 2004 *Appl. Phys. Lett.* **85** 6263
[9] Akinaga H, Kuramochi H, Yasutake M, Uzumaki T, Tanaka A, 2005 *J. Magn. Soc. Jpn.* **29** 767
[10] Kuramochi H, Akinaga H, Semba Y, Kijima M, Uzumaki T, Yasutake M, Tanaka A, Yokoyama H, 2005 *Jpn. J. Appl. Phys.* **44** 2077
[11] Liu Z Y, Dan Y, Jinjun Q, Wu Y, 2002 *J. Appl. Phys.* **91** 8843
[12] Phillips G N, Siekman M, Abelmann L, Lodder J C, 2002 *Appl. Phys. Lett.* **81** 865
[13] Litvinov D, Khizrroev S, 2002 *Appl. Phys. lett.* **81** 1878
[14] Gao L, Yue L P, Yokota T, Skomski R, Liou H, Takahoshi H, Saito H, Ishio S, 2004 *IEEE Trans. Magn.* **40** 2194
[15] Utke I, Hoffmann P, Berger R, Scandella L, 2002 *Appl. Phys. lett.* **80** 4792
[16] Yoshida N, Yasutake M, Arie T, Akita S, Nakayama Y, 2002 *Jpn. J. Appl. Phys.* **41** 5013
[17] Saito H, Natsume T, Sunahara R, Rheem Y, Ishio S, 2005 *J. Magn. Soc. Jpn.* **29** 773
[18] Amos N, Lverenov A, Fernandez R, Ikkawi R, Litvinov D, Khizroev S, 2009 *J. Appl. Phys.* **105** 07D526
[19] Han G, Wu Y, Zheng Y, 2007 *Jpn. J. Appl. Phys.* **46** 4403

Epitaxial Growth of Permalloy Thin Films on MgO Single-Crystal Substrates

Mitsuru Ohtake[1], **Takahiro Tanaka**[1], **Katsuki Matsubara**[1], **Fumiyoshi Kirino**[2] **and Masaaki Futamoto**[1]

1 Faculty of Science and Engineering, Chuo University, 1-13-27 Kasuga, Bunkyo-ku, Tokyo 112-8551, Japan
2 Graduate School of Fine Arts, Tokyo National University of Fine Arts and Music, 12-8 Ueno-koen, Taito-ku, Tokyo 110-8714, Japan

E-mail: ohtake@futamoto.elect.chuo-u.ac.jp

Abstract. Permalloy (Py: Ni – 20 at. % Fe) thin films were prepared on MgO single-crystal substrates of (100), (110), and (111) orientations by molecular beam epitaxy. Py crystals consisting of fcc(100) and hcp(11$\bar{2}$0) orientations epitaxially nucleate on MgO(100) substrates. With increasing the substrate temperature, the volume ratio of fcc(100) to hcp(11$\bar{2}$0) crystal increases. The metastable hcp(11$\bar{2}$0) structure transforms into more stable fcc(110) structure with increasing the film thickness. Py(110)$_{fcc}$ single-crystal films are obtained on MgO(110) substrates, whereas Py films epitaxially grow on MgO(111) substrates with two types of fcc(111) variants whose orientations are rotated around the film normal by 180° each other. X-ray diffraction analysis indicates that the out-of-plane and the in-plane lattice spacings of these fcc–Py films agree within ±0.4% with the values of bulk fcc–Py crystal, suggesting that the strains in the films are very small. High-resolution transmission electron microscopy shows that periodical misfit dislocations are preferentially introduced in the films around the Py/MgO(100) and the Py/MgO(110) interfaces to reduce the lattice mismatches. The magnetic properties are considered to be reflecting the magnetocrystalline anisotropies of bulk fcc–Py and/or metastable hcp–Py crystals and the shape anisotropy caused by the surface undulations.

1. Introduction

Permalloy (Py: Ni – 20 at. % Fe) is a typical soft magnetic material with fcc structure. Py multilayer films combined with thin MgO layers are investigated for magnetic tunneling junction applications like tunneling magnetoresistance heads and magnetic random access memory devices [1]. For such applications, high-quality epitaxial films are strong candidates since the film uniformity and the magnetic anisotropy are well controlled. Epitaxial films are also useful to investigate the basic structural and magnetic properties [2]. However, the film structure and the magnetic properties vary depending on the substrate orientation and the substrate temperature [3]. In the present study, Py films are prepared on MgO single-crystal substrates with different orientations under similar experimental conditions. The effects of substrate orientation and substrate temperature on the film growth, the structure, and the magnetic properties are investigated.

2. Experimental procedure

Thin films were prepared on polished MgO substrates of (100), (110), and (111) orientations at temperatures between 100 and 500 °C by using a molecular beam epitaxy chamber under base pressures lower than 3×10^{-8} Pa. Substrates were heated at 500 °C for 1 h in the chamber to obtain clean surfaces. The surface structure was checked by reflection high energy electron diffraction

(RHEED). The RHEED patterns observed for the MgO substrates exhibited *Kikuchi* patterns [3]. Py films of 40 nm thickness were deposited by *e*-beam evaporation of an $Ni_{80}Fe_{20}$ (at. %) source. The Py film compositions were confirmed to be within Ni – 20 ± 2.5 at. % Fe by energy dispersive X-ray spectroscopy. The deposition rate was kept constant at 0.01 nm/s.

The surface structure during film deposition was studied by *in-situ* RHEED. The structural property was investigated by out-of-plane and in-plane X-ray diffraction with Cu–$K\alpha$ radiation (λ = 0.15418 nm). The surface morphology and the cross-sectional microstructure were observed by atomic force microscopy (AFM) and high-resolution transmission electron microscopy (HR-TEM), respectively. The magnetization curves were measured by using a vibrating sample magnetometer.

3. Results and discussion

Py epitaxial films were obtained on MgO(100) substrates. Figures 1(a)–(c) show the RHEED patterns observed during Py deposition on MgO(100) substrates heated at different temperatures. In early stages of Py film growth at temperatures between 100 and 500 °C, two kinds of RHEED patterns corresponding to fcc(100) and hcp(11$\bar{2}$0) textures respectively shown in the spot maps of figures 1(d) and (e) are overlapped, as shown in figures 1(a-1)–(c-1). The hcp RHEED pattern is analyzed to be an overlap of two hcp(11$\bar{2}$0) reflections, as shown by the spots, B and C, in figure 1(e). Py epitaxial films consisting of fcc(100) and hcp(11$\bar{2}$0) crystals are formed on the MgO(100) substrates, where the fcc structure is a stable phase, whereas the hcp structure is a metastable phase. The epitaxial orientation relationships are determined by RHEED as follows,

$$Py(100)[001]_{fcc} \parallel MgO(100)[001] \quad \text{(Type A)},$$
$$Py(11\bar{2}0)[0001]_{hcp} \parallel MgO(100)[001] \quad \text{(Type B)},$$
$$Py(11\bar{2}0)[1\bar{1}00]_{hcp} \parallel MgO(100)[001] \quad \text{(Type C)}.$$

The nucleation behavior is similar to the cases of Co [3] and CoNi [4] film growth on MgO(100). The hcp–Py crystal consists of two types of hcp(11$\bar{2}$0) variants whose orientations are rotated around the film normal by 90° each other. The lattice mismatch between the $Py(100)_{fcc}$ crystal and the substrate is as large as –15.9%. The mismatch between the $Py(11\bar{2}0)_{hcp}$ crystal and the substrate differs depending on the in-plane direction at the $Py(11\bar{2}0)_{hcp}$/MgO(100) interface, and the misfit values at the interface along the $Py[0001]_{hcp}$ and the $Py[1\bar{1}00]_{hcp}$ directions are –3.1% and +3.0%, respectively. These values

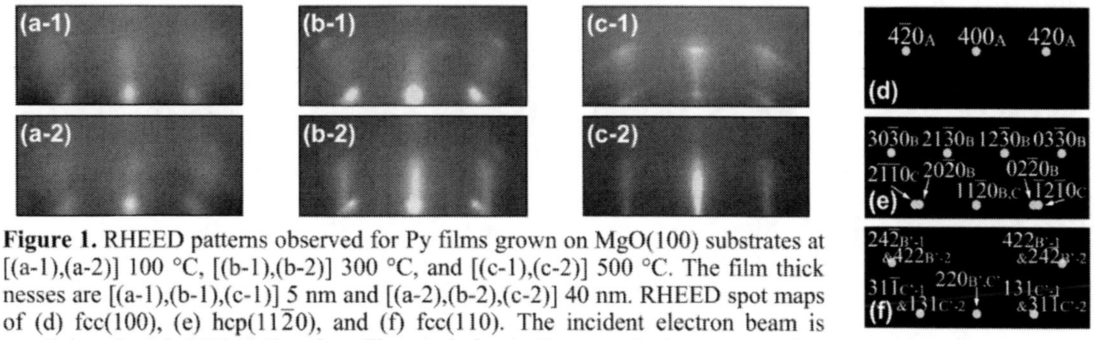

Figure 1. RHEED patterns observed for Py films grown on MgO(100) substrates at [(a-1),(a-2)] 100 °C, [(b-1),(b-2)] 300 °C, and [(c-1),(c-2)] 500 °C. The film thick nesses are [(a-1),(b-1),(c-1)] 5 nm and [(a-2),(b-2),(c-2)] 40 nm. RHEED spot maps of (d) fcc(100), (e) hcp(11$\bar{2}$0), and (f) fcc(110). The incident electron beam is parallel to the MgO[001] direction. The symbols, A–C, respectively correspond to the orientation relationships of Type A–C explained in the text.

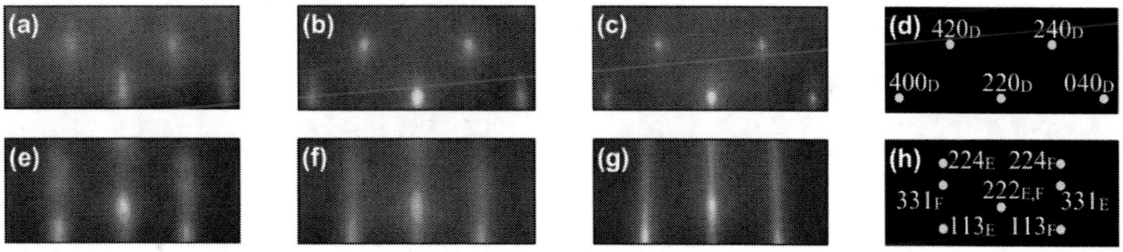

Figure 2. RHEED patterns observed for 5-nm-thick Py films grown on [(a),(b),(c)] MgO(110) and [(e),(f),(g)] MgO(111) substrates at [(a),(e)] 100 °C, [(b),(f)] 300 °C, and [(c),(g)] 500 °C. RHEED spot maps of (d) fcc(110) and (h) fcc(111). The incident electron beam is parallel to the MgO[001] or the MgO[$\bar{1}$10] direction. The symbols, D–F, respectively correspond to the orientation relationships of Type D–F explained in the text.

are calculated from the lattice constants of the bulk MgO crystal (a_{MgO} = 0.422 nm) [5], the bulk Py crystal ($a_{fcc\text{–}Py}$ = 3.55 nm) [6] and from those of the hcp–Py film prepared in the present study, which are estimated from the XRD analysis described later. The metastable hcp–Py crystal is considered to be stabilized through hetero-epitaxial growth. With increasing the film thickness, the peak intensities of $11\bar{2}0_{B,C}$, $20\bar{2}0_B$, $0\bar{2}20_B$, $30\bar{3}0_B$, $0\bar{3}30_B$, $2\bar{1}\bar{1}0_C$, and $\bar{1}2\bar{1}0_C$ hcp RHEED spots become stronger, whereas those of $21\bar{3}0_B$ and $12\bar{3}0_B$ spots weaken, as shown in figures 1(a-2) and (b-2). The RHEED data suggest that another pattern overlaps with the hcp(11$\bar{2}$0) pattern. The pattern of increasing spot intensity is analyzed to be an overlap of four fcc(110) reflections, as shown by the spots, B'–1, B'–2, C'–1, and C'–2, in the spot map of figure 1(f). The orientation relationships of

$$\mathrm{fcc}(110)[1\bar{1}1]_{\mathrm{Type\ B'-1}},\ \mathrm{fcc}(110)[\bar{1}1\bar{1}]_{\mathrm{Type\ B'-2}} \parallel \mathrm{hcp}(11\bar{2}0)[0001]_{\mathrm{Type\ B}} \quad \text{(Type B')}$$
$$\mathrm{fcc}(110)[1\bar{1}2]_{\mathrm{Type\ C'-1}},\ \mathrm{fcc}(110)[\bar{1}1\bar{2}]_{\mathrm{Type\ C'-2}} \parallel \mathrm{hcp}(11\bar{2}0)[1\bar{1}00]_{\mathrm{Type\ C}} \quad \text{(Type C')}$$

are determined by RHEED. In these configurations, the closed packed plane of hcp crystal is parallel to that of fcc crystal. The metastable hcp structure is apparently transforming into more stable fcc structure possibly by atomic displacement parallel to the hcp(0001) closed packed plane with increasing the film thickness, similar to the case of hcp–Py(11$\bar{2}$0) film grown on Au(100) [7]. For the Py films prepared at 100 and 300 °C, diffraction patterns of fcc(100) and hcp(11$\bar{2}$0)+fcc(110) are observed throughout the course of Py deposition, as shown in figures 1(a-2) and (b-2). When the film is prepared at 500 °C, with increasing the film thickness, the RHEED intensity from fcc(100) crystal increases, whereas that from hcp(11$\bar{2}$0)+fcc(110) crystal decreases. The result shows that the areal ratio of fcc(100) to hcp(11$\bar{2}$0)+fcc(110) crystal exposed to the film surface increases with increasing the film thickness. The pattern observed for the 40-nm-thick Py film shows only the fcc(100) reflection, as shown in figure 1(c-2). The hcp–Py phase is considered to be stable on the MgO(100) substrates at lower temperatures.

Py epitaxial films were also obtained on MgO(110) and MgO(111) substrates. Figures 2(a)–(c) and (d)–(f) show the RHEED patterns observed for 5-nm-thick Py films grown on MgO(110) and MgO(111) substrates at different temperatures, respectively. Clear RHEED patterns corresponding to fcc(110) and fcc(111) textures shown in the spot maps of figures 2(d) and (g) are observed from the beginning of Py depositions on MgO(110) and MgO(111) substrates, respectively, and they remain unchanged during film formation for all the films. The fcc(111) pattern is analyzed to be an overlap of two fcc(111) reflections, as shown by the spots, E and F, in figure 2(g). Py$(110)_{fcc}$ single-crystal films are formed on MgO(110) substrates, whereas Py$(111)_{fcc}$ bi-crystalline films epitaxially grow on MgO(111) substrates. The epitaxial orientation relationships of

$$\mathrm{Py}(110)[\bar{1}10]_{\mathrm{fcc}} \parallel \mathrm{MgO}(110)[\bar{1}10] \quad \text{(Type D)},$$
$$\mathrm{Py}(111)[\bar{1}10]_{\mathrm{fcc}} \parallel \mathrm{MgO}(111)[\bar{1}10] \quad \text{(Type E)},$$
$$\mathrm{Py}(111)[1\bar{1}0]_{\mathrm{fcc}} \parallel \mathrm{MgO}(111)[\bar{1}10] \quad \text{(Type F)},$$

are determined by RHEED. The Py$(111)_{fcc}$ film consists of two types of variants whose orientations are rotated around the film normal by 180° each other.

Figure 3 shows the out-of-plane and the in-plane XRD spectra of Py films grown on MgO substrates with different orientations at 300 °C. For the Py film grown on MgO(100) substrate, Py$(200)_{fcc}$ and Py$(11\bar{2}0)_{hcp}$+$(220)_{fcc}$ out-of-plane XRD reflections and Py$(002)_{fcc}$ and Py$(0002)_{hcp}$+$(111)_{fcc}$ in-plane reflections are clearly observed, as shown in figure 3(a) and (b). The

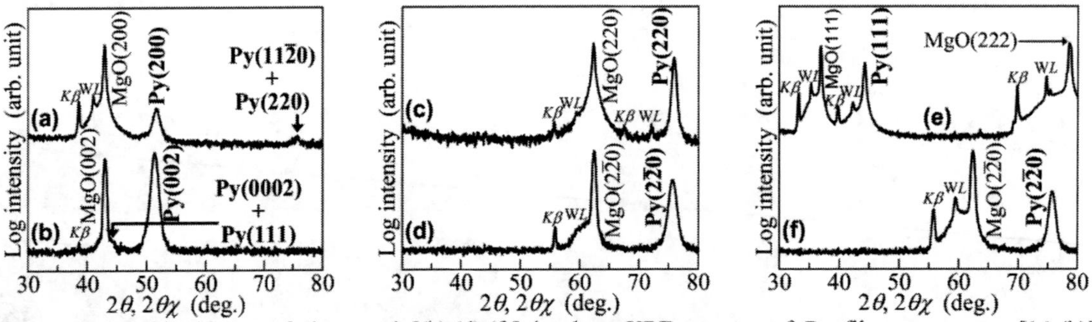

Figure 3. [(a),(c),(e)] Out-of-plane and [(b),(d),(f)] in-plane XRD spectra of Py films grown on [(a),(b)] MgO(100), [(c),(d)] MgO(110), and [(e),(f)] MgO(111) substrates at 300 °C. The scattering vector of in-plane XRD is parallel to (b) the MgO[001] or [(d),(f)] the MgO[1$\bar{1}$0] direction.

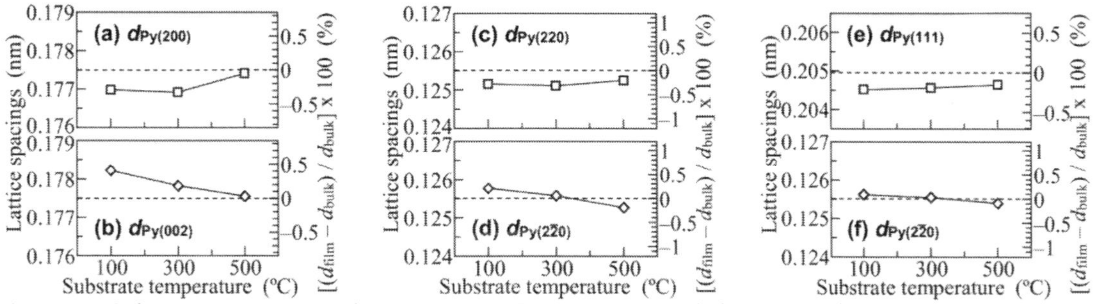

Figure 4. Substrate temperature dependence on [(a),(c),(e)] out-of-plane and [(b),(d),(f)] in-plane lattice spacings of Py films grown on [(a),(b)] MgO(100), [(c),(d)] MgO(110), and [(e),(f)] MgO(111) substrates. The lattice spacings are estimated from the XRD peaks, for example, shown in figure 3.

hcp(11$\bar{2}$0) and the fcc(220) out-of-plane and the hcp(0002)$_{hcp}$ and the fcc(1$\bar{1}\bar{1}$)$_{fcc}$ in-plane reflections are considered to be overlapped, as suggested by the RHEED study. The lattice constants of hcp–Py crystal are estimated from the XRD peaks to be $(a, c, c/a) = (2d_{hcp(11\bar{2}0)}, 2d_{hcp(0002)}, c/a)$ = (0.250 nm, 0.408 nm, 1.63). These values are almost in agreement with those of Py(11$\bar{2}$0) films grown on Au(100) [8] and Cr(100) [9]. For the Py films grown on MgO(110) and MgO(111) substrates, Py(220)$_{fcc}$ and Py(111)$_{fcc}$ out-of-plane and Py(2$\bar{2}$0)$_{fcc}$ and Py(2$\bar{2}$0)$_{fcc}$ in-plane XRD reflections are recognized, as shown in figures 3(c),(d) and (e),(f), respectively. Similar XRD patterns are observed for the Py films with the respective orientations prepared at 100 and 500 °C. Figure 4 shows the substrate temperature dependence on the lattice spacings of fcc–Py crystals estimated from the XRD peaks. The lattice spacings are in agreement within ±0.4% with those of bulk fcc–Py crystal. The XRD result shows that the strains in the films are very small, though fairly large mismatches of –15.9% exist between the fcc–Py films and the substrates. The film strain is further decreased by employing a higher substrate temperature. It is known that misfit dislocations are easily introduced in a film around the film/substrate interface to reduce the mismatch between immiscible elements when a weak binding force works between the deposited atoms and the substrate atoms. This type of epitaxial growth where misfit dislocations are introduced in the film around the interface is reported for the Au, Ag, Cr, Ni, FeCo/MgO systems [10–14]. The Py epitaxial film growth on MgO substrates are also considered to follow the growth mode.

Figures 5(a) and (d) show the cross-sectional HR-TEM images of Py(100)$_{fcc}$ and Py(110)$_{fcc}$ crystals grown on MgO(100) and MgO(110) substrates at 300 °C, respectively. Atomically sharp boundaries are recognized between the films and the substrates. Periodical misfit dislocations are preferentially introduced in the Py films around the Py/MgO interfaces, as shown by the ⊥ marks in figures 5(a) and (b). At the Py(100)$_{fcc}$/MgO(100) and the Py(110)$_{fcc}$/MgO(110) interfaces, the Py(200)$_{fcc}$ and the Py (020)$_{fcc}$ planes, respectively, match with the MgO(200) and the MgO(020) planes in integer ratios like 6:5, 7:6, etc. Figures 5(c) and (e) show the matching ratio distribution. When the matching distribution is considered, the effective lattice mismatches at the Py(100)$_{fcc}$/MgO(100) and the Py(110)$_{fcc}$/MgO(110) interfaces decrease from –15.9% to –0.5% and –0.3%, respectively. The small calculated misfit values thus coincide with the previous XRD results. The matching ratio distribution is considered to be delicately varied depending on the substrate temperature, since the film strain is further decreased with increasing the substrate temperature. The small strains in the films are apparently realized by the presence of such periodical misfit dislocations. On the contrary, dislocations are expected to be closed loops at the interface for the Py(111)$_{fcc}$/MgO(111) system, since fcc(111) is the slide plane. The in-plane lattice strains of Py films grown on MgO(111) substrates are very small, when compared with those of Py films grown on MgO(100) and (110) substrates at the respective substrate temperatures, as shown in figures 4(b), (d), and (f). The result supports that strain relaxation of Py(111)$_{fcc}$/MgO(111) system is faster.

Figure 6 shows the AFM images of Py films grown on MgO substrates of (100), (110), and (111) orientations. The film growth on all the substrates seems to follow the Volmer-Weber (island-growth mode). With increasing the substrate temperature, the R_a value increases. Py(100)$_{fcc}$ islands with {111} and {110} facets, Py(110)$_{fcc}$ islands with {111} and {100} facets, and Py(111)$_{fcc}$ islands with {001} and {110} facets are respectively recognized for the Py films formed on MgO substrates of (100), (110), and (111) orientations at 500 °C. The orientations of facets were estimated from the cross-

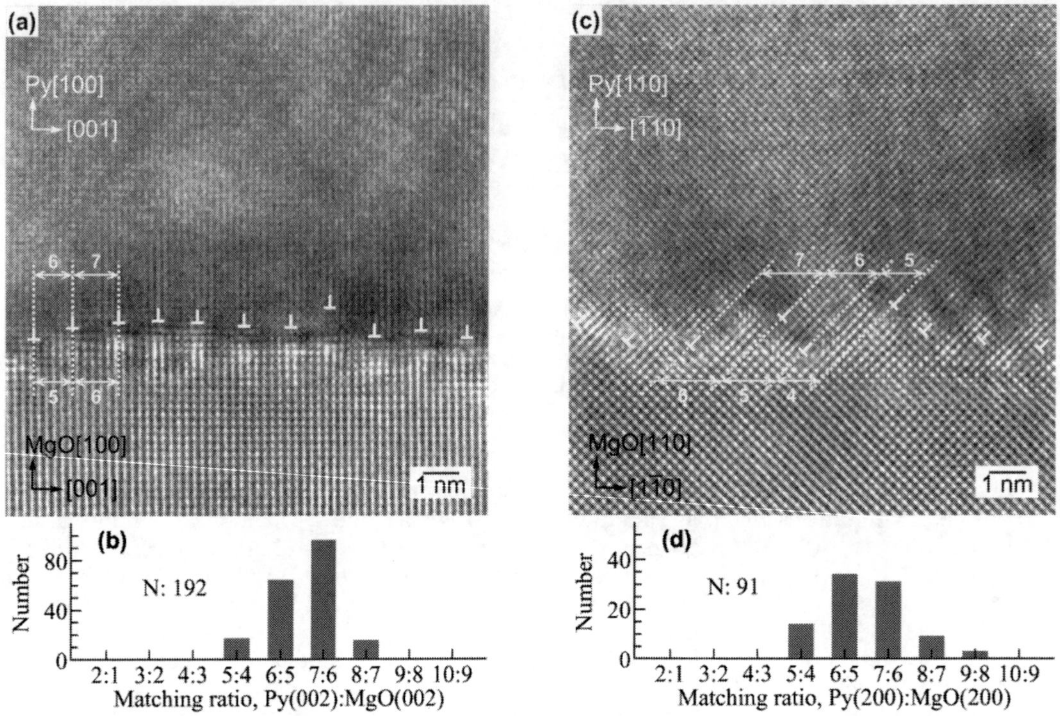

Figure 5. Cross-sectional HR-TEM images of (a) $Py(100)_{fcc}$ and (c) $Py(110)_{fcc}$ crystals grown on MgO(100) and MgO(110) substrates at 300 °C observed along (a) the MgO[001] and (d) the MgO[$1\bar{1}0$] directions, respectively. The ⊥ marks show the positions of misfit dislocations. Distributions of lattice matching ratio, (b) Py(200):MgO(200) and (d) Py(020):MgO(020).

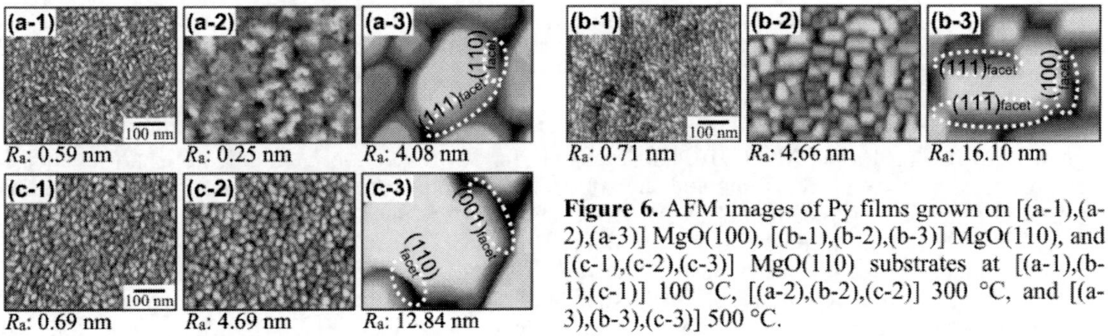

Figure 6. AFM images of Py films grown on [(a-1),(a-2),(a-3)] MgO(100), [(b-1),(b-2),(b-3)] MgO(110), and [(c-1),(c-2),(c-3)] MgO(110) substrates at [(a-1),(b-1),(c-1)] 100 °C, [(a-2),(b-2),(c-2)] 300 °C, and [(a-3),(b-3),(c-3)] 500 °C.

sectional profiles of AFM images (not shown here). Figure 7 shows the magnetization curves of Py films grown on MgO substrates of (100), (110), and (111) orientations. The in-plane magnetic anisotropy varied depending on the film orientations. For the Py/MgO(100) system, magnetic properties of $Py(100)_{fcc}$ and $Py(11\bar{2}0)_{hcp}+(110)_{fcc}$ are considered to be overlapped. Py films grown on MgO(100) substrates are easily magnetized when the magnetic field is applied along the MgO[011] direction, whereas the magnetization curves measured along the MgO[001] direction saturate at higher applied fields, as shown in figures 7(a-1)–(a-3). The magnetization behavior is similar to the cases of $Co(100)_{fcc}+(11\bar{2}0)_{hcp}$ [3] and $CoNi(100)_{fcc}+(11\bar{2}0)_{hcp}$ [4] films epitaxially grown on MgO(100) substrates. The Py films grown on MgO(110) substrates show an uniaxial anisotropy. The easy magnetization direction is observed along the $Py[1\bar{1}0]_{fcc}$ direction. The in-plane magnetization curves are almost isotropic in in-plane measurements for the Py films grown on MgO(111) substrates. The in-plane magnetization property is considered to be reflecting the magnetocrystalline anisotropies of bulk fcc–Py and/or metastable hcp–Py crystals. The film coercivity increases greatly for the films prepared

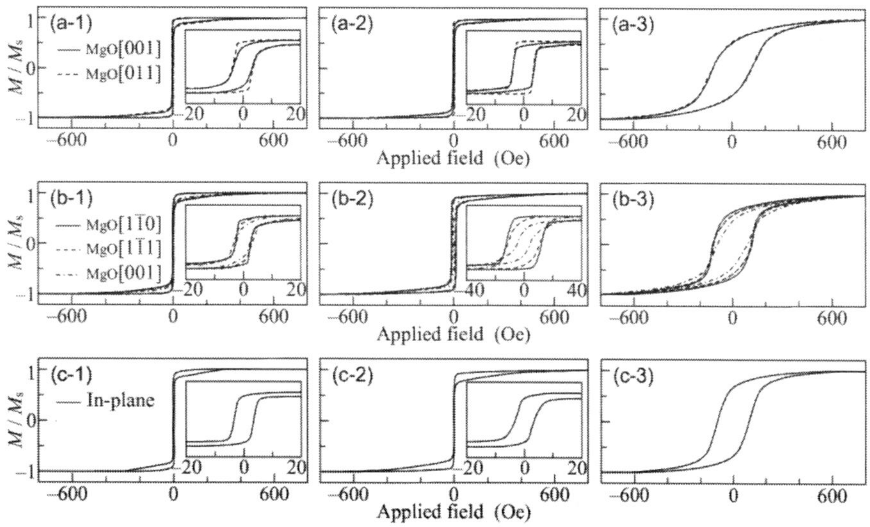

Figure 7. Magnetization curves of Py films grown on [(a-1),(a-2),(a-3)] MgO(100), [(b-1),(b-2),(b-3)] MgO(110), and [(c-1),(c-2),(c-3)] MgO(110) substrates at [(a-1),(b-1),(c-1)] 100 °C, [(a-2),(b-2),(c-2)] 300 °C, and [(a-3),(b-3),(c-3)] 500 °C.

at 500 °C, possibly due to suppression of domain wall motion by the crevasses in the films existing between large Py islands, as shown in figures 6(a-3), (b-3), and (c-3). The surface morphology influences the coercivities of Py films.

4. Conclusion

The epitaxial growth, the film structure, and the in-plane magnetization properties of Py films on MgO singe-crystal substrates of (100), (110), and (111) orientations were investigated. Py epitaxial films consisting of fcc(100) and hcp(11$\bar{2}$0)+fcc(110) crystals are formed on MgO(100) substrates. Py $(110)_{fcc}$ single-crystal and Py$(111)_{fcc}$ bi-crystalline films epitaxially grow on MgO(110) and MgO(111) substrates, respectively The lattice spacings of these Py films are in agreement within ±0.4% with the values of bulk Py crystal. The strains in the Py films formed on MgO substrates are very small, which is associated with periodical introduction of misfit dislocations. The in-plane magnetic properties are influenced by the magnetocrystalline anisotropies of bulk fcc–Py and/or metastable hcp–Py crystals and the shape anisotropy caused by the surface undulations.

Acknowledgements

A part of this work was supported by NEDO–Japan, MEXT–Japan, and STFJ (Science and Technology Foundation of Japan). M. Ohtake is a JSPS Research Fellow.

References

[1] Dave R W *et al.* 2006 *IEEE Trans. Magn.* **42** 1935
[2] Ohtake M, Futamoto M, Krino F, Fujita N and Inaba N 2008 *J. Appl. Phys.* **103** 07B522
[3] Nukaga Y, Ohtake M, Yabuhara O, Kirino F and Futamoto M 2010 *J. Magn. Soc. Jpn.* **34** 508
[4] Sato Y, Nishiyama T, Ohtake M, Kirino F and Futamoto M 2010 *IEEE Trans. Magn.* **46** 349
[5] Sasaki S, Fujino K and Takeuchi Y 1979 *Proc. Jpn. Acad.* **55** 43
[6] Person W B 1958 *A Handbook of Lattice Spacings and Structures of Metals and Alloys* (London, New York, Los Angeles, Paris: Pergamon Press)
[7] Sato Y *et al.* 2010 Structural Analysis of hcp–Ni and hcp–NiFe Thin Films Epitaxially Grown on Au(100) Single-Crystals *Abstr. ISAMMA 2010 (Sendai, Japan, 12–16 July 2010)* PF–03
[8] Ohtake M, Tanaka T, Kirino F and Futamoto M 2010 *J. Phys. Conf. Ser.* **200** 072072
[9] Sato Y, Ohtake M. Higuchi J, Nishiyama T, Futamoto M 2010 *Abstr. JEMS 2010* Abs. 250
[10] Hoei R H 1986 *Surf. Sci.* **169** 317
[11] Renaud G, Guénard P and Barbier A 1998 *Phys. Rev.* B **58** 7310
[12] Nakamura A, Koguchi M and Futamoto M 1995 *Jpn. J. Appl. Phys.* **34** 2307
[13] Lukaszew R A, Zhang Z, Stoica V and Clarke R 2003 *Appl. Surf. Sci.* **219** 74
[14] Ohtake M *et al.* 2010 *J. Magn. Magn. Mater.* **322** 1947

Preparation of Co(0001)$_{hcp}$ and (111)$_{fcc}$ Films on Single-Crystal Oxide Substrates

Mitsuru Ohtake, Osamu Yabuhara, Yuri Nukaga, and Masaaki Futamoto

Faculty of Science and Engineering, Chuo University, 1-13-27 Kasuga, Bunkyo-ku, Tokyo 112-8551, Japan

E-mail: ohtake@futamoto.elect.chuo-u.ac.jp

Abstract. Co thin film were prepared on oxide single-crystal substrates of $Al_2O_3(0001)_{D51}$, $MgO(111)_{B1}$, and $SrTiO_3(111)_{E21}$ by ultra high vacuum molecular beam epitaxy. Effects of substrate material and substrate temperate on the film growth and the crystallographic properties were investigated. Co epitaxial thin films of hcp(0001) and/or fcc(111) orientations are obtained on all the substrate. With increasing the substrate temperature, the volume ratio of hcp to fcc increases for the Co films grown on Al_2O_3 and MgO substrates, whereas the ratio decrease for the Co film grown on $SrTiO_3$ substrate. The in-plane lattice strain is larger than the out-of-plane strain due to accommodation of lattice mismatch between the film and the substrate

1. Introduction

Magnetic thin films with perpendicular magnetic anisotropy have been investigated for applications like magnetic recording media, magnetic random access memory devices, etc. Co has two stable crystallographic phases, hcp and fcc. Co-based alloy thin films with hcp structure and multilayer films combined with thin fcc–Co and noble metal (Pd, Pt, etc.) layers have been frequently used for such applications. These are mostly hcp(0001) or fcc(111) oriented polycrystalline films. The films generally include stacking faults existing parallel to the closed packed plane and even small volume of crystals with other orientations. The crystallographic quality gives strong influences on the magnetic properties like magnetic anisotropy [1], Gilbert's damping constant [2], etc. Well-defined epitaxial thin films are strong candidates for future magnetic applications [3, 4], since the film uniformity and the magnetic anisotropy are well controlled. Epitaxial films are also useful to investigate the basic structural and magnetic properties. Co(0001)$_{hcp}$ single-crystal films have been prepared by employing underlayers such as Au(111)$_{fcc}$ [3], Ti(0001)$_{hcp}$ [5], and Ru(0001)$_{hcp}$ [6], whereas Co(111)$_{fcc}$ thin films epitaxially grow on Cu(111) underlayers [7]. However, the epitaxial film structure delicately varies depending on the deposition condition like substrate material, substrate temperature, etc. Oxide single-crystal substrates generally offer stable surface structures. In the present study, Co thin films were prepared on three kinds of oxide single-crystal substrates under similar experimental conditions to investigate the effects of substrate material and substrate temperature on the film growth and the structure. The detailed film structures were investigated by *in-situ* reflection high energy electron diffraction (RHEED) and *ex-situ* X-ray diffraction (XRD).

2. Experimental procedure

Co thin films of 40 nm thickness were deposited on polished $Al_2O_3(0001)_{D51}$, $MgO(111)_{B1}$, and $SrTiO_3(111)_{E21}$ substrates at temperatures ranging between 100 and 500 °C by using a molecular beam epitaxy chamber with base pressures lower than 3×10^{-8} Pa. Substrates were heated at 500 °C for 1 hour in the ultra high vacuum chamber to obtain clean surfaces. The surface structure was checked by RHEED. The RHEED patterns observed for the substrates correspond to those from clean surfaces, as shown in figures 1(a), 3(a), and 5(a). Pure Co (99.99%) was evaporated by electron beam heating. The deposition rate was kept constant at 0.01 nm/s.

The surface structure during Co deposition was studied by *in-situ* RHEED. The film structure was investigated by out-of-plane ($2\theta/\omega$-scan), in-plane ($2\theta\chi/\phi$-scan), and pole figure XRD with Cu–$K\alpha$ radiation (λ=0.15418 nm). The cross-sectional microstructure was observed by transmission electron microscopy.

3. Results and discussion

Co epitaxial thin films with closed packed planes parallel to the substrate surfaces were obtained on $Al_2O_3(0001)_{D51}$ and $MgO(111)_{B1}$ substrates for the investigated substrate temperatures. On $SrTiO_3(111)_{E21}$ substrates, Co epitaxial thin films were formed at temperatures higher than 300 °C, whereas Co polycrystalline films grew at temperatures lower than 200 °C. Figures 1(c)–(f) show the RHEED patterns observed during Co deposition on $Al_2O_3(0001)_{D51}$ substrates heated at different temperatures. Clear RHEED patterns start to be observed from the beginning of Co deposition and they remain unchanged during 40-nm-thick Co film formation. With increasing the substrate temperature, the diffraction spots and streaks become sharper, which indicates that the strain in the film decreases further by employing high substrate temperatures. For the films prepared at 100 and 200 °C [figures 1(c) and (d)], the RHEED pattern consists of spots corresponding to fcc(111) texture [figure 1(g)] and streaks along the fcc[111] direction. The fcc(111) pattern consists of two reflections, as shown by the spots, A and B, in the RHEED spot maps of figure 1(g). The streaks indicate that the film has atomically flat terraces and a possibility of including hcp(0001) crystals. For the films prepared at temperatures higher than 300 °C [figures 1(e) and (f)], the RHEED spots are not recognized but only the streaks along the fcc[111] and/or the hcp[0001] direction are observed. The

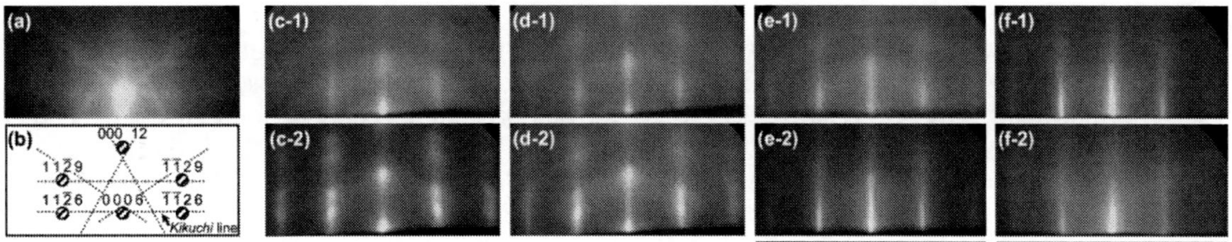

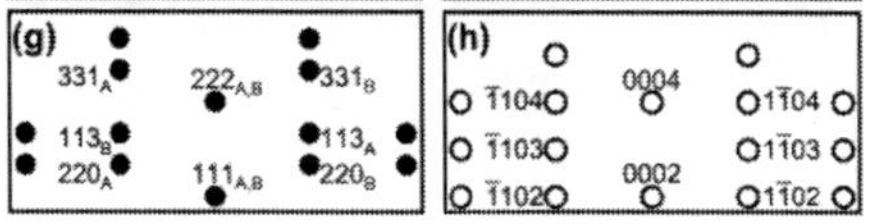

Figure 1. RHEED patterns observed for (a) an $Al_2O_3(0001)_{D51}$ substrate and (c)–(f) Co films grown on Al_2O_3 substrates at (c) 100 °C, (d) 200 °C, (e) 300 °C and (d) 500 °C. The film thicknesses are [(c–1),(d–1),(e–1),(f–1)] 2 nm and [(c–2),(d–2),(e–2),(f–2)] 40 nm. RHEED spot maps of (b) $D5_1(0001)$, (g) fcc(111), and (h) hcp(0001). The incident electron beam is parallel to the $Al_2O_3[1\bar{1}00]_{D51}$ direction.

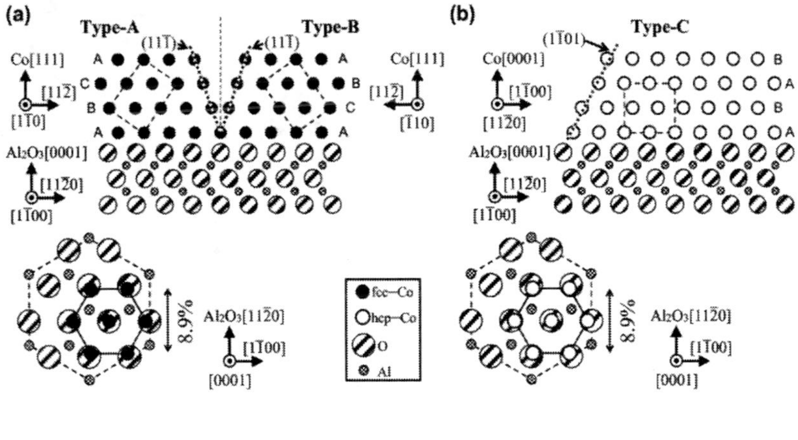

Figure 2. Epitaxial orientation relationships of (a) $Co(111)_{fcc}$ ‖ $Al_2O_3(0001)_{D51}$ and (b) $Co(0001)_{hcp}$ ‖ $Al_2O_3(0001)_{D51}$.

epitaxial orientation relationships of the fcc and the hcp crystals with respect to the substrate are respectively determined by RHEED to be Co(111)[1 $\bar{1}$ 0]$_{fcc\ (Type\ A)}$, (111)[1 $\bar{1}$ 0]$_{fcc\ (Type\ B)}$ || Al_2O_3(0001)[1$\bar{1}$00]$_{D51}$ and Co(0001)[11$\bar{2}$0]$_{hcp}$ || Al_2O_3(0001)[1$\bar{1}$00]$_{D51}$. The fcc–Co crystal has two kinds of epitaxial orientation relationships with respective to the substrate [figure 2(a)], whereas the hcp–Co crystal has one orientation relationship [figure 2(b)]. The orientations of two types of fcc(111) variants are rotated around the film normal by 180 degree each other. Al_2O_3 crystal has $D5_1$ crystal structure and the Al_2O_3(0001) consists of two alternate layers of Al and O. The atomic arrangements of Al and O layers are rotated around the film normal by 30 degrees each other. When the structure of Al_2O_3 is considered, there are two possibilities of Co crystal nucleation on $Al_2O_3(0001)_{D51}$. One possibility is that Co atoms of fcc(111) or hcp(0001) fit with the Al atoms of Al_2O_3(000 l_{odd}), where the lattice mismatch is calculated to be 5.6% ($|(2a_{fcc\text{–}Co}/2^{1/2} - a_{Al2O3}) / a_{Al2O3}| = 5.6\%$, $|(2a_{hcp\text{–}Co} - a_{Al2O3}) / a_{Al2O3}| = 5.6\%$). Another possibility is that Co atoms fit with the O atoms of Al_2O_3(000 l_{even}), where the mismatch increases up to 8.9% ($|(a_{fcc\text{–}Co}/2^{1/2} - a_{Al2O3}/3^{1/2}) / (a_{Al2O3}/^{1/2})| = 8.9\%$, $|(a_{hcp\text{–}Co} - a_{Al2O3}/3^{1/2}) / (a_{Al2O3}/^{1/2})| = 8.9\%$). The experimental result shows that the arrangement of Co atoms is similar to that of O atoms, as shown in figure 2. A theoretical simulation supports that 3*d* ferromagnetic transition metal film growth on Al_2O_3 substrate is strongly influenced by the O atoms located on the substrate surface [8]. The Co film growth on $Al_2O_3(0001)_{D51}$ is considered to be influenced by the atomic arrangement of O atoms of Al_2O_3. Furthermore, in the epitaxial orientation relationship, there exists a large mismatch of about 9% between the film and the substrate. However, clear RHEED patterns are observed throughout the course of Co deposition [figures 1(b)–(f)]. It is know that misfit dislocations are easily introduced in a film around the film/substrate to reduce the mismatch between immiscible elements when a weak binding force works between the deposited atoms and the substrate atoms. This type of epitaxial growth where misfit dislocations are introduced in the film around the interface is reported for the Cr/MgO [9] and the FeCo/MgO [10] systems. The epitaxial growth of Co films on $Al_2O_3(0001)_{D51}$ substrates is also considered to follow the growth mode. Figure 3 shows the high-resolution cross-sectional TEM image of a Co film grown on $Al_2O_3(0001)_{D51}$ substrate at 500 °C. A sharp boundary is recognized between the film and the substrate, where misfit dislocations are considered to be introduced so as to reduce the lattice mismatch existing at the Co/Al_2O_3 interface.

Similar tendency of film growth was recognized for the Co thin films formed on $MgO(111)_{B1}$ substrates. Figure 4 shows the RHEED patterns observed during Co deposition on $MgO(111)_{B1}$ substrates heated at different temperatures. Clear diffraction patterns corresponding to fcc(111) and/or hcp(0001) texture are observed throughout the course of Co deposition. The RHEED patterns observed for films prepared at 100 and 200 °C [figures 4(c) and (d)] consist of fcc(111) diffraction spots and the streaks along the fcc[111] direction, whereas those observed for films prepared at temperatures higher than 300 °C [figures 4(e) and (f)] consist of only the streaks. The epitaxial orientation relationships of Co(111)[1$\bar{1}$0]$_{fcc\ (Type\ A)}$, (111)[1$\bar{1}$0]$_{fcc\ (Type\ B)}$ || MgO(111)[1$\bar{1}$0]$_{B1}$ and Co(0001)[11$\bar{2}$0]$_{hcp}$ || MgO(111)[1$\bar{1}$0]$_{B1}$ are determined by RHEED. MgO crystal has *B*1 structure and the $MgO(111)_{B1}$ consists of two alternate layers of Mg and O. The atomic arrangement of O atoms located on the substrate surface is same as that of Mg atoms. The arrangement of Co atoms is also similar to those of O and Mg, as shown in figure 5. On MgO(111) substrates, the Co film growth is also considered to be influenced by the atomic arrangement of O atoms of MgO. Furthermore, there exists a fairly large mismatch of about 16% ($|(a_{fcc\text{–}Co}/2^{1/2} - a_{MgO}/2^{1/2}) / (a_{MgO}/2^{1/2})| = 15.9\%$, $|(a_{hcp\text{–}Co} - a_{MgO}/2^{1/2}) / (a_{MgO}/2^{1/2})| = 15.8\%$). Misfit dislocations are expected to be introduced in the films at the Co/MgO interfaces.

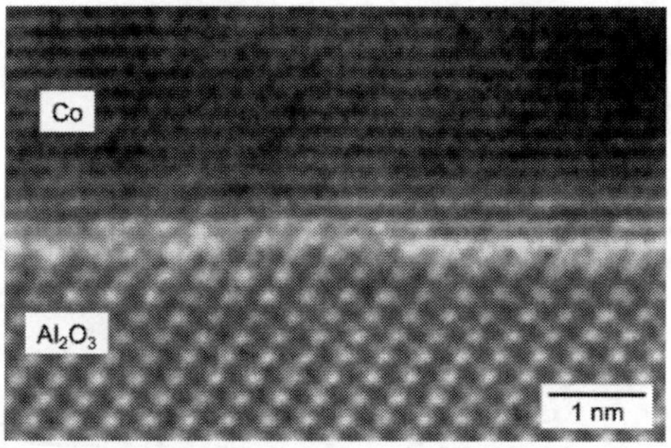

Figure 3. High-resolution cross-sectional TEM image of a Co film grown on $Al_2O_3(0001)_{D51}$ substrate at 500 °C.

Figures 6(c)–(f) shows the RHEED patterns observed during Co deposition on $SrTiO_3(111)_{E21}$ substrates heated at different temperatures. For the films prepared at 100 and 200 °C [figures 6(c) and

Journal of Physics: Conference Series **303** (2011) 012016 doi:10.1088/1742-6596/303/1/012016

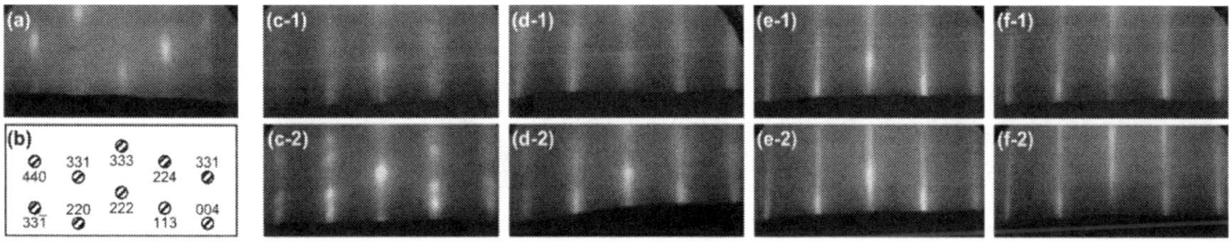

Figure 4. RHEED patterns observed for (a) an MgO(111)$_{B1}$ substrate and (c)–(f) Co films grown on MgO(111) substrates at (c) 100 °C, (d) 200 °C, (e) 300 °C and (d) 500 °C. The film thicknesses are [(c–1),(d–1),(e–1),(f–1)] 2 nm and [(c–2),(d–2),(e–2),(f–2)] 40 nm. (b) RHEED spot map of $B1$(111). The incident electron beam is parallel to the MgO[$1\bar{1}0$]$_{B1}$ direction.

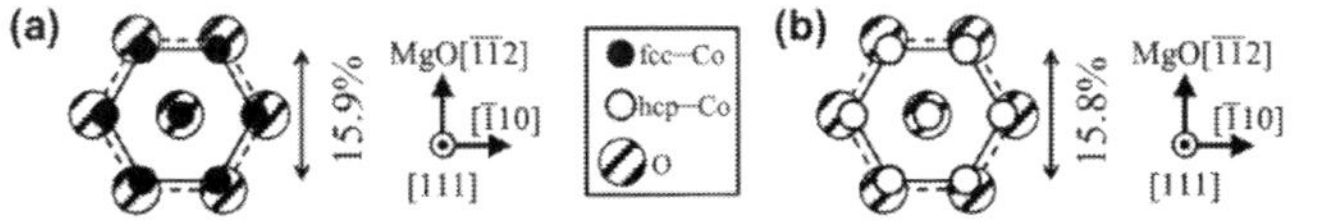

Figure 5. Epitaxial orientation relationships of (a) Co(111)$_{fcc}$ ‖ MgO(111)$_{B1}$ and (b) Co(0001)$_{hcp}$ ‖ MgO(111)$_{B1}$.

(d)], the ring RHEED patterns typical for polycrystalline structure are observed throughout the course of Co deposition. For the film prepared at 300 °C, a diffuse RHEED pattern is observed in an early stage of Co film growth [figure 6(e–1)]. With increasing the film thickness, an overlap of the ring pattern and the pattern corresponding to fcc(111) and/or hcp(0001) texture appears [figure 6(e–2)]. Co thin films grown on $SrTiO_3(111)_{E2_1}$ substrates are apparently strained. With increasing the substrate temperature [figure 6(f)], the diffraction pattern becomes sharper, which indicates that the strain in the film decreases by employing higher substrate temperatures. The epitaxial orientation relationships of Co(111)[$11\bar{2}$]$_{fcc\ (Type\ A)}$, Co(111)[$\bar{1}\bar{1}2$]$_{fcc\ (Type\ B)}$ ‖ $SrTiO_3$(111)[$1\bar{1}0$]$_{B1}$ and Co(0001)[$1\bar{1}00$]$_{hcp}$ ‖ $SrTiO_3$(111)[$1\bar{1}0$]$_{B1}$ are determined by RHEED. The lattice mismatch is calculated to be about 5% ($|[2a_{fcc\text{-}Co}/2^{1/2} - a_{SrTiO3}/(3/2)^{1/2}] / [a_{SrTiO3}/(3/2)^{1/2}]| = 4.8\%$, $|[2a_{hcp\text{-}Co} - a_{SrTiO3}/(3/2)^{1/2}] / [a_{SrTiO3}/(3/2)^{1/2}]| = 4.9\%$). $SrTiO_3$ has $E2_1$ structure and the $SrTiO_3(111)_{E2_1}$ consists of two alternate layers. One layer consists of Sr and O atoms. Another layer consists of only Ti atoms. The atomic arrangement of Co film is rotated around the film normal by 30 degrees with respective to that of Sr–O layer. The Co film growth mode on $SrTiO_3(111)_{E2_1}$ seems to be different with those on $Al_2O_3(0001)_{D5_1}$ and MgO(111)$_{B1}$.

In order to investigate the crystal structure of Co epitaxial films grown on oxide substrates, pole figure XRD analysis was performed. Figures 8(a)–(c) and (e)–(f) shows the pole figure XRD spectra of fcc–Co{$11\bar{1}$} and hcp–Co{$1\bar{1}01$} measured by making the optical axes of XRD system shown in figures 8(d) and (h), respectively. For the films prepared on $Al_2O_3(0001)_{D5_1}$ and MgO(111)$_{B1}$ substrates at 100 °C, six reflections, which originate from the two types of fcc–Co(111) variants (Types A and B), are observed with 60 degrees separation for the spectra of fcc{$11\bar{1}$} poles [figures

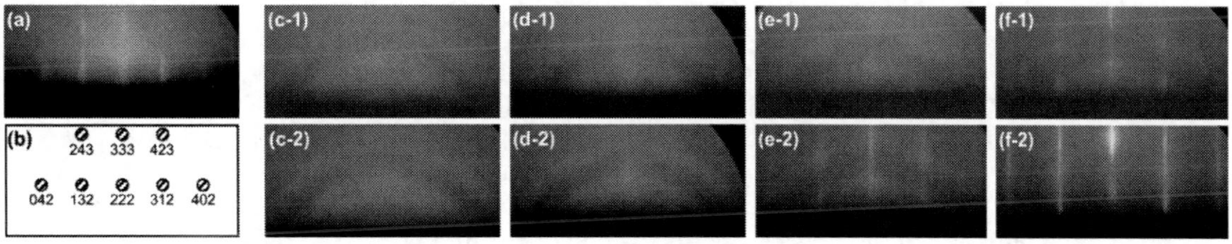

Figure 6. RHEED patterns observed for (a) an $SrTiO_3(111)_{E2_1}$ substrate and (c)–(f) Co films grown on $SrTiO_3$(111) substrates at (c) 100 °C, (d) 200 °C, (e) 300 °C and (d) 500 °C. The film thicknesses are [(c–1),(d–1),(e–1),(f–1)] 2 nm and [(c–2),(d–2),(e–2),(f–2)] 40 nm. (b) RHEED spot map of $E2_1$(111). The incident electron beam is parallel to the $SrTiO_3$[$11\bar{2}$]$_{E2_1}$ direction.

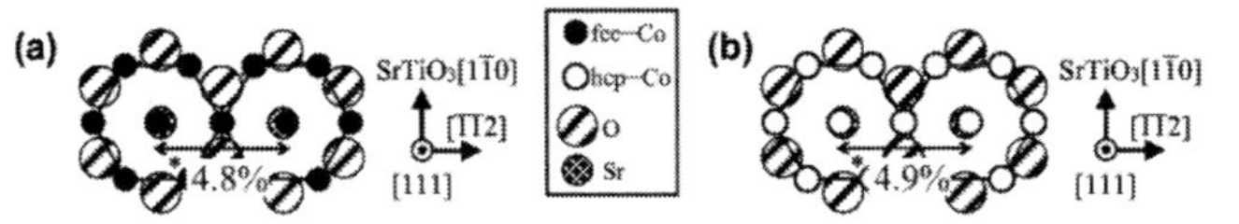

Figure 7. Epitaxial orientation relationships of (a) Co(111)$_{fcc}$ ‖ $SrTiO_3$(111) and (b) Co(0001)$_{hcp}$ ‖ $SrTiO_3$(111).

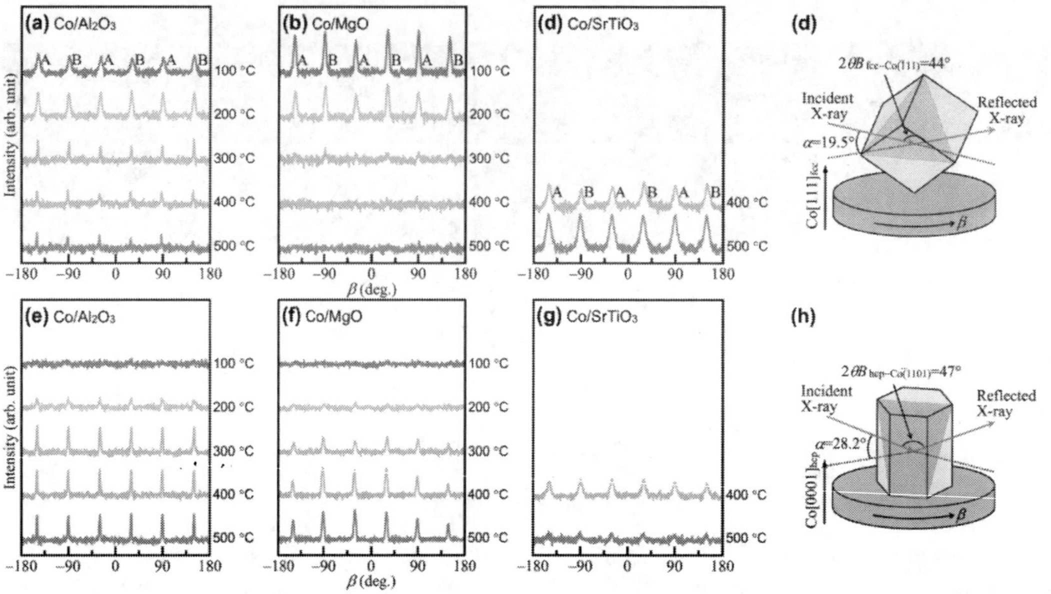

Figure 8. Pole figure XRD spectra of [(a),(b),(c)] Co$\{11\bar{1}\}_{fcc}$ and [(e),(f),(g)] Co$\{1\bar{1}01\}$ poles measured for Co films grown on [(a),(e)] $Al_2O_3(0001)_{D51}$, [(b),(f)] $MgO(111)_{B1}$, and [(c),(g)] $SrTiO_3(111)_{E21}$ substrates. The intensity is shown in a logarithmic scale. The symbols, A and B, respectively correspond to the epitaxial orientation relationship of Types A and B explained in the text. Schematic model showing the optical axes of pole figure XRD measurements for (d) Co$\{11\bar{1}\}_{fcc}$ and (h) Co$\{1\bar{1}01\}_{hcp}$ poles.

8(a) and (b)], whereas there are no reflections for the spectra of hcp$\{1\bar{1}01\}$ poles [figures 8(e) and (f)]. This result show that the films consist of pure fcc(111) crystals. When the substrate temperature exceeds 200 °C, six-fold symmetrical reflections appear in the spectra of hcp$\{1\bar{1}01\}$ poles [figures 8(a) and (b)]. The films consist primarily of fcc crystal and include very small amount of stacking faults (hcp crystals). With increasing the substrate temperature beyond 300 °C, the fcc-to-hcp volume ratio delicately varied depending on the substrate material. For the films prepared on Al_2O_3 substrates at temperatures higher than 300 °C, strong hcp$\{1\bar{1}01\}$ and weak fcc$\{11\bar{1}\}$ reflections are observed [figures 8(a) and (e)]. The films consist primarily of hcp crystals coexisting with small amount of fcc crystals. For the films prepared on MgO substrates at temperatures higher than 300 °C, only the hcp$\{1\bar{1}01\}$ reflections are observed, as shown in figures 8(b) and (f). The films consist of pure hcp crystals. On the other hand, for the Co films epitaxially grown on $SrTiO_3(111)_{E21}$ substrates at elevated temperatures of 400 and 500 °C, strong fcc$\{11\bar{1}\}$ and weak hcp$\{1\bar{1}01\}$ reflections are recognized [figures 8(c) and (g)], which indicates that the films consist primarily of fcc crystals. The volume ratio of hcp to fcc crystal is apparently influenced by the substrate material and the substrate temperature.

Figures 9(a) and (b) show the out-of-plane and the in-plane XRD spectra of Co films grown on Al_2O_3, MgO, and $SrTiO_3$ substrates at 600 °C. The $Co(111)_{fcc}$ and/or $(0002)_{hcp}$ out-of-plane and the $Co(2\bar{2}0)_{fcc}$ and/or $(11\bar{2}0)_{hcp}$ in-plane XRD reflections are clearly observed for all the substrates. Figure 10(a) shows the substrate temperature dependence on the lattice constant, c, estimated from the $Co(222)_{fcc}$ and/or $(0004)_{hcp}$ out-of-plane XRD peak diffraction angle. With increasing the substrate temperature, the c values of Co films grown on Al_2O_3 and MgO substrates decrease due to the increase in the volume ratio of hcp to fcc, whereas the c value of Co film grown on $SrTiO_3$ substrate increases due to the increase in the ratio of fcc to hcp. Figure 10(b) shows the substrate temperature dependence on the lattice constant, a, estimated from the $Co(2\bar{2}\bar{2})_{fcc}$ and/or $(11\bar{2}0)_{hcp}$ in-plane XRD peak diffraction angle. The a value is almost constant for all samples due to accommodation of lattice mismatch between the film and the substrate. Therefore, the $\Delta\theta\chi_{50}$ value [figure 10(c)] is larger than the $\Delta\theta_{50}$ value [figure 10(d)] for same substrate temperatures. With increasing the substrate temperature, $\Delta\theta_{50}$ and $\Delta\theta\chi_{50}$ decrease, indicating that the film strain decreases by employing higher substrate temperatures.

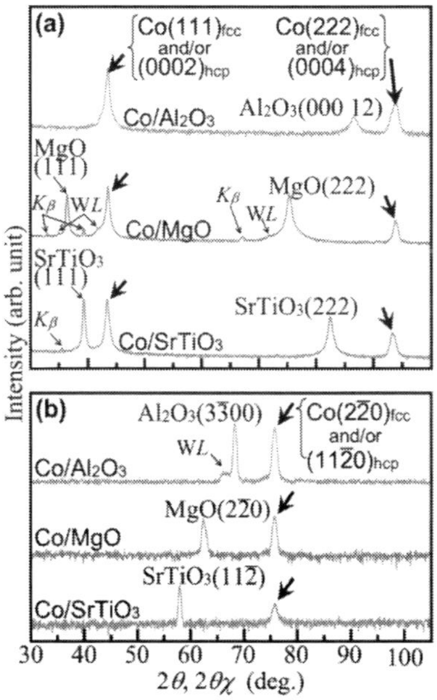

Figure 9. (a) Out-of-plane and (b) in-plane XRD spectra of Co films grown on $Al_2O_3(0001)_{D51}$, $MgO(111)_{B1}$, and $SrTiO_3(111)_{E21}$ substrates at 500 °C. The intensity is shown in a logarithmic scale. The arrows, $K\beta$ and WL, show the reflections caused by XRD with Cu–$K\beta$ and W–$L\alpha$ radiations.

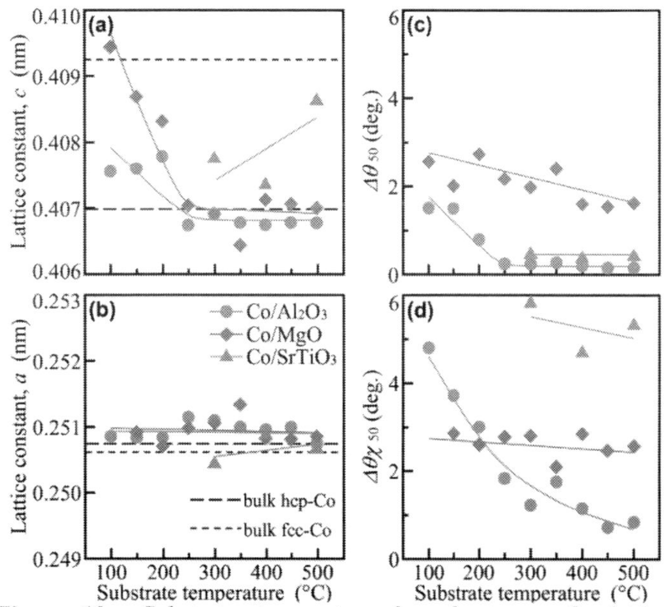

Figure 10. Substrates temperature dependences on the lattice constants, (a) c, (b) a. Substrate temperature dependence on the values of full width at half maximum of rocking curves observed for (c) the $Co(111)_{fcc}$ and/or $(0002)_{hcp}$ out-of-plane XRD peak ($\Delta\theta_{50}$) and (d) the $Co(2\bar{2}0)_{fcc}$ and/or $(11\bar{2}0)_{hcp}$ in-plane peak ($\Delta\theta\chi_{50}$).

4. Conclusion

Co epitaxial thin films were obtained on $Al_2O_3(0001)_{D51}$, $MgO(111)_{B1}$, and $SrTiO_3(111)_{E21}$ substrates. The film growth and the crystallographic properties were investigated by *in-situ* RHEED and *ex-situ* XRD. Co epitaxial films consist of fcc(111) and/or hcp(0001) crystals. With increasing the substrate temperature, the volume ratio of hcp to fcc increases for the Co films grown on Al_2O_3 and MgO substrates, whereas the ratio of fcc to hcp increases for the Co film grown on $SrTiO_3$ substrate. The lattice constant and the film strain are influenced by the substrate material and the substrate temperature.

Acknowledgment

A part of this work was supported by NEDO–Japan, MEXT–Japan and STFJ (Science and Technology Foundation of Japan). M. Ohtake is a Research Fellow of the Japan Society for the Promotion of Science.

References

[1] Hinata S, Yanagisawa R, Saito S and Takahashi M 2009 *J. Appl. Phys.* **105** 07B718
[2] Nukaga Y *et al.* 2009 *IEEE Trans. Magn.* **45** 2519
[3] Ohtake M, Futamoto M, Kirino F, Fujita N and Inaba N 2008 *J. Appl. Phys.* **103** 07B522
[4] Ohtake M, Nukaga Y, Kirino F and Futamoto M 2009 *J. Cryst. Growth* **311** 2251
[5] Gong H, Rao M, Laughlin D E and Lambeth D N 1999 *J. Appl. Phys.* **85** 4699
[6] Krishnan K M, Takeuchi T, Hirayama Y and Futamoto M 1994 *IEEE Trans. Magn.* **30** 5115
[7] Gronckel H A M de, Kopinga K and Jonge W J M de 1991 *Phys. Rev.* B **44** 9100
[8] Zhang W, Smith J R and Evans A G 2002 *Acta Mater.* **50** 3803
[9] Nakamura A, Koguchi M and Futamoto M 1995 *Jpn. J. Appl. Phys.* **34** 2307
[10] Ohtake M, Nukaga Y, Sato Y, Kirino F and Futamoto M 2009 *J. Appl. Phys.* **106** 123921

One-dimensional magnonic circuits with size-tunable band gaps and selective transmission

B. Djafari-Rouhani[1], **H. Al-Wahsh**[1,2], **A. Akjouj**[1], **L. Dobrzynski**[1]

[1]Institut d'Electronique, de Microélectronique et de Nanotechnologie (IEMN), UMR CNRS 8520, UFR de Physique, Université des Sciences et Technologies de Lille, 59655 Villeneuve d'Ascq Cédex, France.

[2] Faculty of Engineering, Benha University , 11241 Cairo, Egypt.

Abstract. We present a review of our theoretical calculations about magnon transport in quasi-one-dimensional (1D) magnonic circuits constituted by waveguides coupled to side resonators. Phenomena such as the existence of band gaps, rejective and selective transmissions and Fano resonances will be discussed as well as the applications of these structures in filtering and demultiplexing devices. The calculations are performed based on two types of models and in the frame of the Green's function method. First, the continuum long-wavelength Heisenberg model is studied in 1D monomode waveguide containing symmetric and asymmetric loops or coupled with grafted stubs. Then, we use the discrete dipole approximation in structures composed of a chain of nanometric magnetic clusters coupled to finite clusters on its vicinity. All such circuits exhibit a variety of interference effects in their transport properties which should have important consequences for designing integrated devices such as microwave filters.

Corresponding author: B. Djafari-Rouhani (bahram.djafari-rouhani@univ-lille1.fr)

1. Introduction

The propagation of waves in periodically structured materials is of fundamental interest in modern physics and technology. In particular, spin systems which have a regular distribution of scattering centers have been seen to possess a distinct and interesting array of magnonic properties, perhaps most strikingly frequency band gaps within which magnons cannot propagate through the structure-a so-called magnonic band gap [1, 2, 3, 4, 5, 6, 7, 8, 9]. The interest in these band gaps is related to a number of advantages where the magnonic crystals has in comparison with photonic crystals [1]. The wavelength of a spin wave and, hence, the properties of such crystals depend on the external magnetic field and can be controlled by this field. The wavelength of propagating spin waves for a wide class of ferromagnetic materials in the microwave range is on the order of tens or even hundreds of micrometers. The phase and group velocities of spin waves are also functions of the structure size and the applied external field and may vary over a wide range [1].

After a great deal of works devoted to the lamellar materials and superlattices, the first attempt to study the spin waves in 2D magnonic crystals was presented in [3]. Nowadays, the number of studies on this topic has surged and continues to grow at a fast pace [10].On the other hand, studies of lower dimensional systems such as 1D periodic layered media [2, 11, 12, 13] and periodic waveguide systems with different geometries [14, 15, 16, 17] are conducted as analogues of 2D and 3D systems and for applications in their own right. These structures are attractive since their production is more feasible and they require only simple analytical and numerical calculations. Spin waves are governed by different interactions dominating on different length scales, e.g. by the exchange and dipolar interactions dominating on nanoscopic and microscopic length scales, respectively. Spin waves propagation in micro- and nanopatterned magnonic waveguides have been discussed in some recent papers [18, 19, 20].

In this paper, we briefly review some of our theoretical calculations about magnon transport in two types of quasi-one-dimensional (1D) magnonic circuits constituted by waveguides coupled to side resonators, in analogy to the case of other excitations such as phonons, photons or electrons [21].In the first case, the continuum long-wavelength Heisenberg model is studied in 1D monomode waveguide containing symmetric and asymmetric loops or coupled with grafted stubs. Such structures can exhibit band gaps resulting both from their periodicity or from the existence of zeros of transmission associated with the resonant modes of each loop or grafted stub. By introducing a defect loop or a defect stub along the periodic waveguide, one can expect the existence of localized modes in the band gaps that can contribute to a selective filtering transmission. In this short paper, we emphasize (section 2) only some results[22, 23] about the so-called Fano resonances [24] which have been also reported in the electronic transport in mesoscopic systems using interferometric structures such as the Aharonov-Bohm systems[25, 26, 27, 28].

Then, in section 3, we use the discrete dipole approximation in structures composed of a chain of nanometric magnetic clusters coupled to finite clusters on its vicinity[29, 30, 31, 32]. Such circuits exhibit a variety of interference effects in their transport properties that may be useful in microwave filtering and multiplexing devices.

2. Magnon mono-mode circuits:Transmission gaps and Fano resonances

The simple magnonic filter with which we are dealing in this section is depicted in Fig.1. It consists of $N(N')$ dangling side branches (DSB), which play the role of resonators, grafted at two sites on an infinite 1D mono-mode waveguide. The length d_2 of the finite segment and the lengths d_1 and d_3 of the grafted resonators characterize the geometry of the structure. The media are assumed to be Heisenberg ferromagnets, which means that we are neglecting the dipole-dipole interactions as compared with the exchange contribution to the Hamiltonian. Moreover, we are dealing with long wavelength magnetic excitations and therefore take use of the continuum approximation of the Heisenberg model(see for details Refs. [14, 33]). The backbone and the grafted branches are assumed to be monomode waveguides for the propagation of magnons. With these ingredients, one can derive analytically the dispersion relation as well as the transmission and reflection coefficients through the waveguide structure. The transmission gaps have been established through an analysis of the transmission function (amplitude and phase) calculated within the framework of the Green's function method [22, 14]. We show that this simple structure can exhibit transmission gaps (their width depend on the number of dangling resonators) and Fano-like resonances. In particular, we show that the transmission amplitude through such a system can be written following the Fano-like shape around these resonances.

Although in our analytical calculation, the backbone and the DSB can be constituted by different magnetic materials, we shall assume, for the sake of simplicity, that all materials are constituted by the same ferromagnet. Indeed, the occurrence of magnonic gaps and Fano resonances in our structure does not require the use of two different materials, in contrast to the case of usual multilayered structures.

In Fig.2(a) we show the transmission amplitude versus the reduced frequency $\Omega_1 (= \omega d_1^2/D)$ for a simple structure consisting of one resonator grafted on an infinite guide (see Fig. 1(a)). Here

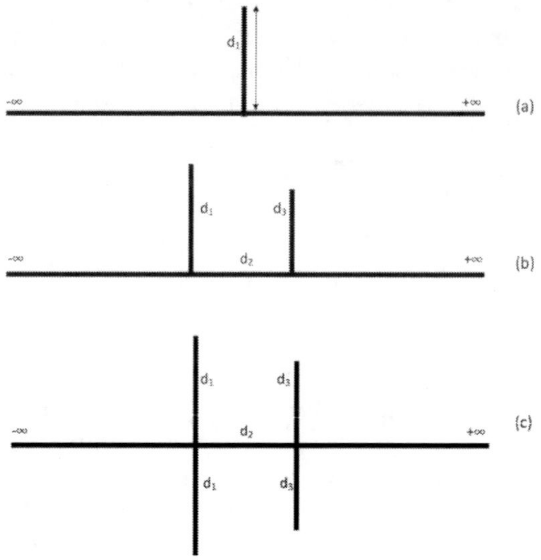

Figure 1. Schematic illustration of the one-dimensional waveguide with dangling resonators studied in the present section. The one-dimensional media constituting the infinite monomode guide and the finite segments are assumed to be of the same material. (a) An infinite line with one grafted segment of length d_1. (b) The same as in (a) but for an infinite line with two segments of lengths d_1 and d_3 grafted at two sites separated by a segment of length d_2. (c) The same as in (b) but for two side resonators grafted at two sites.

$D = (2Ja^2M)/(\gamma\hbar^2)$ where M, J, and γ stand, respectively, for the spontaneous magnetization, the exchange interaction between neighboring magnetic sites in the simple cubic lattice of lattice parameter a constituting the ferromagnetic medium, and the gyromagnetic ratio. One can notice that for this composite system there exist an infinite set of forbidden frequencies Ω_g corresponding to the eigenmodes of the grafted finite branch. This grafted branch behaves as a resonator and this simple composite system filters out these modes. This phenomenon is related to the resonances associated with the finite additional path offered to the spin-wave propagation. The variation in the phase versus the reduced frequency Ω_1 shown in Fig. 2(b) shows an abrupt change in π at the transmission zeros induced by the grafted resonator.

Figure 2(c) gives the transmission coefficient versus the reduced frequency $\Omega_2(=\omega d_2^2/D)$ in presence of two identical dangling resonators , namely, $N = N' = 1$ and $d_1 = d_3 = 0.5d_2$ (see the structure shown in Fig.1(b)). One can notice that the transmission coefficient presents well-defined dips induced by the grafted branches. These dips appears at $\Omega_2 = \tilde{H}+((l+0.5)\pi)^2$,where$\tilde{H} = \gamma H_0 d_2^2/D, l = 0, 1, 2, ...,$ and H_0 is the static external field. These dips transform into large transmission gap when the number of branches increase as it is illustrated in Fig. 2(e) for $N = N' = 2$. It is worth mentioning that because of the existence of two resonators, one can expect two phase drops of π (i.e., 2π) at the transmission zeros induced by the grafted branches. However, one can see in Figs. 2(d) and 2(f) that the phase presents only a phase drop of π. This is due to existence of a resonant state with zero width at these values of Ω_2 which induce a phase jump of π. These resonances collapse when $d_1 = d_3$ is taken exactly equal to $0.5d_2$. To enlarge these resonances, we have to take d_1 and d_3 slightly different from $0.5d_2$.

An example corresponding to the case where d_1 and d_3 is different from $0.5d_2$ is given in Fig.

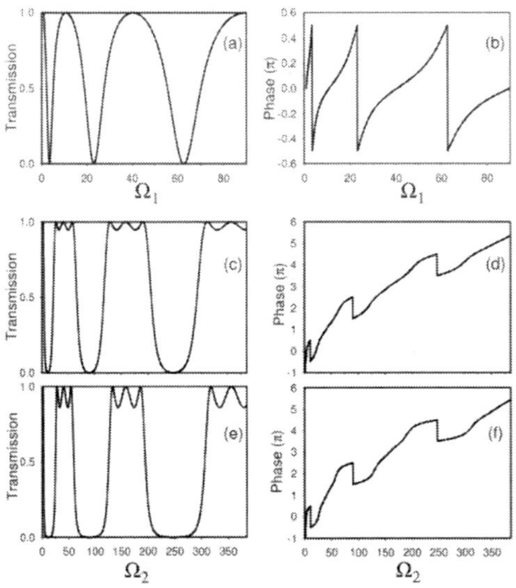

Figure 2. (a) Transmission coefficient vs. the reduced frequency Ω_1 for the structure depicted in Fig.1(a). The forbidden frequencies ω_g appears at $\omega_g = \gamma H_0 + D((l+0.5)\pi/d_1)^2$, or symbolically $\Omega_g = \tilde{H}' + ((l+0.5)\pi)^2$, where $\Omega_g = \omega_g d_1^2/D$ is a reduced frequency,H_0 is the static external field, l is a positive integer and $\tilde{H}' = \gamma H_0 d_1^2/D$. For convenience $\tilde{H}'$ is considered to be 1. (c) Transmission coefficient vs. the reduced frequency Ω_2 for the structure depicted in Fig.1(b) with $d_1 = d_3 = 0.5d_2$ and $N = N' = 1$. (e) The same as in (c) but for the structure depicted in Fig.1(c) ($N = N' = 2$). (b), (d) and (f) are respectively the same as (a), (c) and (e) but for the variation of the phase.

3(a) where $N = N' = 1$, $d_1 = 0.46d_2$, and $d_3 = 0.54d_2$ (with $d_1 + d_3 = d_2$). One can notice that the resonance is squeezed between two zeros (indicated by solid circles on the abscissa of Fig. 3(a)) induced by the dangling resonators as it is also illustrated in the plot describing the variation of the phase (Fig. 3(c)). The width of this resonance increases as far as d_1 and d_3 deviate from $0.5d_2$. It is worth mentioning that, in the particular case where the resonance as well as the two zeros induced by the resonators fall at the same position, the resonance collapses, the transmission coefficient vanishes and the phase drops by π (see also Figs. 2(d) and 2(f)).

The resonance in Fig. 3(a) shows the same characteristics as a Fano resonance but with two zeros of transmission around the resonance instead of one as it is usually the case [24, 25]. Indeed, one can obtain an approximate analytical expression for the transmission function in the vicinity of the resonance. A Taylor expansion around the resonance enables us to write the transmission coefficient T (following the Fano line shape[24, 25]) in the form

$$T = A\,\frac{(\varepsilon + q_1\Gamma)^2(\varepsilon - q_2\Gamma)^2}{\varepsilon^2 + \Gamma^2}, \tag{1}$$

where $A = (1-4\Delta^2/\pi^2)^2/4[N(N+1)+\Delta^2(2-N^2)]^2$ and Δ is the detuning of d_1 and d_3 from $0.5d_2$ (i.e., $\Delta = \pi(0.5 - d_1/d_2) = \pi(-0.5 + d_3/d_2)$). $\Gamma = 2\Delta^2/[N(N+1) + \Delta^2(2 - N^2)]$ characterizes the width of the resonance falling at $\varepsilon = 0$. $q_1 = [N(N+1) + \Delta^2(2 - N^2)]/\Delta(1 + 2\Delta/\pi)$ and $q_2 = [N(N+1)+\Delta^2(2-N^2)]/\Delta(1-2\Delta/\pi)$ are the coupling parameters, they give qualitatively the interference between the bound states and the propagating continuum states[24, 25, 26].

One can notice that when increasing Δ, Γ increases and $q_1(q_2)$ decreases. The results of the approximate expression (Eq. (1)) are shown in Fig. 3(b) by open circles. These results are in accordance with the exact ones (solid lines) and clearly show that the resonance is of Fano type

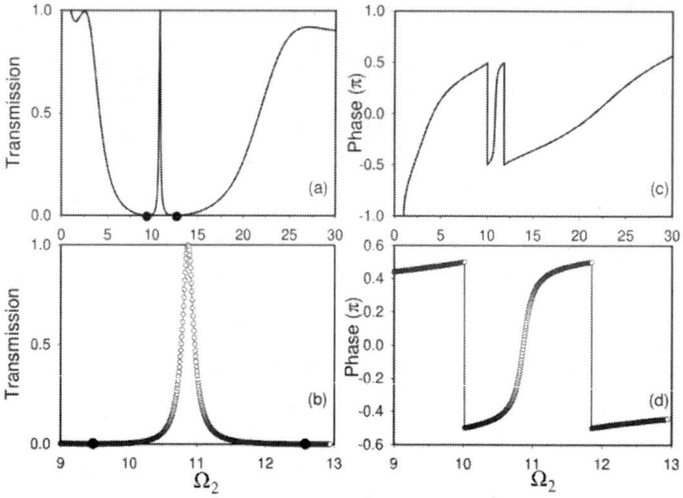

Figure 3. (a) The same as in Fig.2(c) but the lengths of the resonators are taken such that $d_1 = 0.46d_2$ and $d_3 = 0.54d_2$ and $N = N' = 1$. Solid circles on the abscissa axis indicate the positions of the transmission zeros induced by the dangling resonators on both sides of the resonance. (c) The same as in (a) but for the variation of the phase. (b) and (d) give the approximate results (open circles) around the resonance.

with $q_1 \simeq 14.85$ and $q_2 \simeq 17.43$ and width $2\Gamma \simeq 0.03$. The commonly studied Fano resonances are asymmetric because of the presence of only one transmission zero near the resonance (see below). In addition, in the electronic counterparts studies, a perturbation is often introduced to the system in order to create the resonance state [24, 25, 26, 27]. However, the above results shows that, *without introducing any perturbation in the structure, one can find a well-defined symmetric Fano resonance* with width 2Γ and coupling parameters q_1 and q_2 that can be adjusted by tailoring the lengths of the resonators (i.e., Δ). A Taylor expansion around the resonance enables us also to obtain an approximate expression for the phase [22]. The approximate result for the phase is plotted by open circles in Fig. 3(d) and clearly shows [in accordance with the exact results (solid line)] two abrupt phase changes in π at the transmission zeros around the resonance.

One can also create an asymmetric Fano resonance by adjusting the transmission zeros on only one side of the resonance, this can be obtained by considering a structure where the resonators are supposed to be identical with lengths slightly different from $0.5d_2$. This is shown in Fig.4(a) for $d_1 = d_3 = 0.46d_2$ and $N = N' = 1$. Indeed, an analytical Taylor expansion around the resonance enables us to write the following Fano line shape transmission coefficient

$$T \simeq \frac{B}{N^2}\frac{(\varepsilon - \varepsilon_R + q\Gamma)^4}{(\varepsilon - \varepsilon_R)^2 + \Gamma^2} \tag{2}$$

where $B = (1 + 2\Delta/\pi)^4/4(N + 1 + 2\Delta/\pi)^2$. $q = (N + 1 + 2\Delta/\pi)^2/\Delta(1 + 2\Delta/\pi)$ is the Fano parameter. $\Gamma = 2\Delta^2/N^2[1 + \frac{1}{N}(1 + 2\Delta/\pi)]^3$ and $\varepsilon_R = -2\Delta/(N + 1 + 2\Delta/\pi)$ characterize the width and the shift of the resonance respectively.

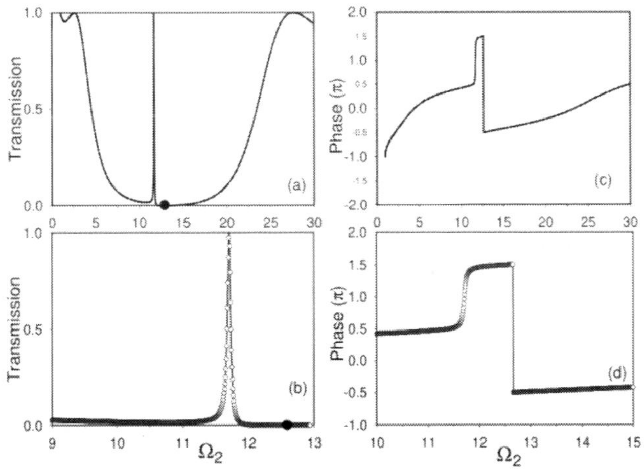

Figure 4. (a) The same as in Fig.3(a) but the resonators are taken to be of identical lengths $d_1 = d_3 = 0.46\, d_2$, (c) The same as in (a) but for the variation of the phase. (b) and (d) give the approximate results (open circles) around the resonance.

One can notice that the width of the resonance is small as compared to the preceding case, this is in accordance with the numerical results of Figs. 3(a) and 4(a). Also q increases when Δ decreases and tends to infinity when Δ vanishes. In this case the resonance falls at $\varepsilon_R = 0$ and, as expected, its width 2Γ reduces to zero (see Fig.2(c)). The results of the approximate expression (Eq.(2)) are sketched (open circles) in Fig.4(b) for $\Delta = \pi(d_1/d_2 - 0.5) = -0.04\pi$ (i.e., $d_1/d_2 = 0.46$) and $N = N' = 1$. These results are in accordance with the exact ones (solid lines) and clearly show that the resonance is of Fano type with $|q| \simeq 32$ and width $2\Gamma \simeq 0.0089$. Concerning the evolution of the phase of the spin waves in this structure, one can notice that (at $\Omega_2 = 12.66$, indicated by a filled circle on the abscissa of Fig.4(a)) the transmission zeros induced by the two identical resonators fall at the same frequency, therefore the phase (Figs. 4(c), 4(d)) shows a phase drop of 2π at these frequencies. Indeed, as the phase is defined modulo 2π, the 2π phase change can be observed if we take into account the absorption in the system [34, 35].

Finally, we would like to mention that Fano resonances and the transmission spectra can be significantly changed by using the pinning field at the end of resonators as a tuning parameter(see for details Ref.[23]). Indeed, the pinning field can vary over a wide range of values, depending on the nature of the ends of the resonators and the properties of the magnetic bearing ions[36].

3. Magnon in nanometric cluster chains

3.1. Coupling to side resonators

The object of this section is to investigate the magnon propagation in a nanometric magnetic cluster chain and focus on the effects of a few additional clusters near the chain. We take into account the dipole-dipole interactions between the nearest-neighbor cluster local magnetic moments. We show that an appropriate choice of the geometrical or magnetic parameters of the additional clusters can lead either to narrow peaks or to narrow dips in the transmission spectrum of magnons along the cluster chain. Such a device can be useful as a selecting or rejecting magnon filter. We consider such a linear chain of nanoparticles bearing a permanent magnetic moment and study the effect on the magnon transmission spectrum of three additional clusters coupled to the chain, as depicted in Fig.5(a). The effect of coupling the infinite wire to such a local resonator is to induce peaks and dips (or zeros) in the transmission coefficient. Our purpose is to discuss the possibility of narrow peaks or narrow dips in the magnon transmission by selecting appropriately

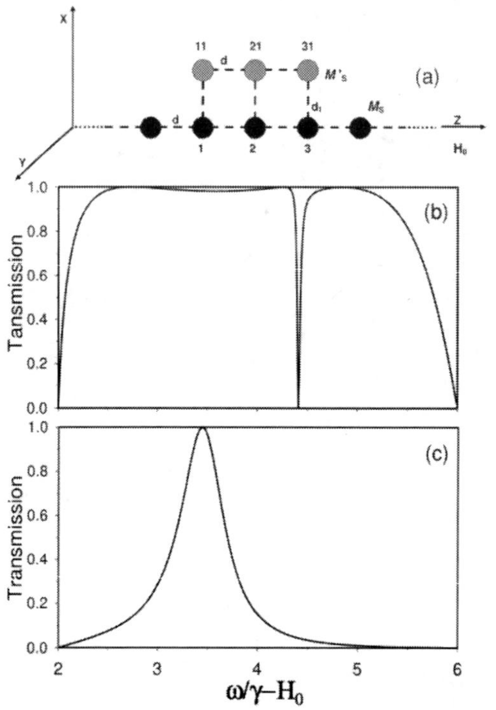

Figure 5. (a)Sketch of the geometry of the considered nanometric device. It consists of one cluster chain and three other clusters. The distances between these clusters are respectively d and d_1 as indicated on the figure. The static magnetic moments are M_s (for the chain clusters) and M_s' (for the three additional clusters) and the external magnetic field is H_0, along the z-axis). (b) Transmission through the structure of Fig. ?? as a function of ω/γ for $H_0 = 5$, $M_s = M_s' = 1, d = 1$ and $d_1 = 1.5$. (c)Transmission in function of ω/γ for $H_0 = 5$, $M_s = M_s' = 1, d = 1$ and $d_1 = 0.8$.

the geometrical or magnetic parameters of the problem. The mathematical derivation of the transmission coefficients, based on a Green's function method is given in Ref. [29].

Fig. 5(b) presents the intensity transmission coefficient T in function of ω/γ, for $d_1 = 1.5$ and $M_s' = 1$. Note the sharp zero of transmission a little above the middle of the bulk band. Such a narrow stop band could be useful to construct a rejecting signal device. When one decreases d_1 to 0.8, leaving all the other parameters as in Fig. 5(b), we obtain Fig. 5(c) which presents now an interesting transmission peak in between two zeros of transmission. This feature could be used as a transmission filter for a magnon signal.

It is interesting to underline that the sharp zeros of transmission are obtained when the resonator clusters are weakly coupled to the chain and can be related to the eigenfrequencies of the isolated three cluster resonator. The sharp transmission peaks appear when the resonator clusters are strongly coupled to the chain. In that case the transmission features can be

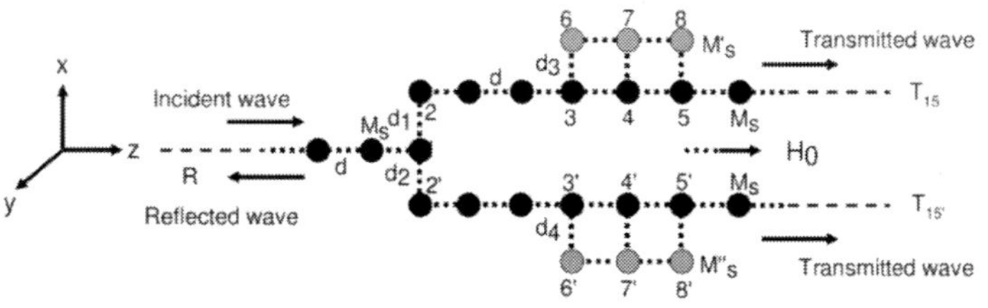

Figure 6. Sketch of the geometry of the considered nanometric device. It consists of three semi-infinite chains made out of a periodic sequence of equidistant nanometric magnetic clusters and three additional clusters attached to the upper (lower) output chain. The distances between these clusters are respectively d, d_1,d_2, d_3, and d_4 as indicated on the figure. The permanent static magnetic moments $\mathbf{M}_s$, $\mathbf{M'}_s$ ($\mathbf{M''}_s$) (for the three semi-infinite cluster chains, and for the additional clusters attached to the upper (lower) output chain respectively), and the external field $\mathbf{H}_0$ are assumed to be *along the z axis.*

interpreted as mainly due to the widening of the frequencies falling inside the bulk chain band of the eigenmodes of the ensemble of the six perturbed clusters.

Let us finally mention that the device presented in this section can be tuned in order to be a good magnon filter as well by transmission as by reflection. In order to achieve this the resonant structure must provide zeros and ones of transmission inside the bulk band. This was realized with the three adsorbed clusters device reported here. In order to get sharp transmission peaks one must tune the system parameters such that the transmission ones fall in between two zeros of transmission close one to the other.

3.2. Magnon multiplexers

A system exhibiting the spin wave transfer effect can be used as a multiplexer. A multiplexer is a system enabling to transfer one magnon frequency from one guide to another one. It is selective when only one magnon frequency is transferred, leaving the propagation of all the other incident magnon frequencies contained in a given energy window unperturbed. Such transfer processes are particularly important in wavelength multiplexing and in telecommunication routing devices.

In this section we describe a simple system, which under certain conditions, realizes the directional transfer of magnons with a good selectivity. The system is depicted in Fig.6. In order to illustrate the signal multiplexing results we present in Fig.7, the variations of the intensity transmission coefficients T_{15} and $T_{15'}$ versus ω/γ for the structure shown in Fig.6. Fig.7 shows together the direct transmission T_{15} from site 1 to site 5 (solid line) and $T_{15'}$ from site 1 to site $5'$ (dashed line). For the parameters $d = d_1 = 1, d_2 = 1.35, d_3 = 1.25, d_4 = 1.1$, $Ms = 1, M'_s = 4.3, M''_s = 3.6, H_0 = 5$, we obtain (Fig.7(a)) an interesting narrow dip (centered

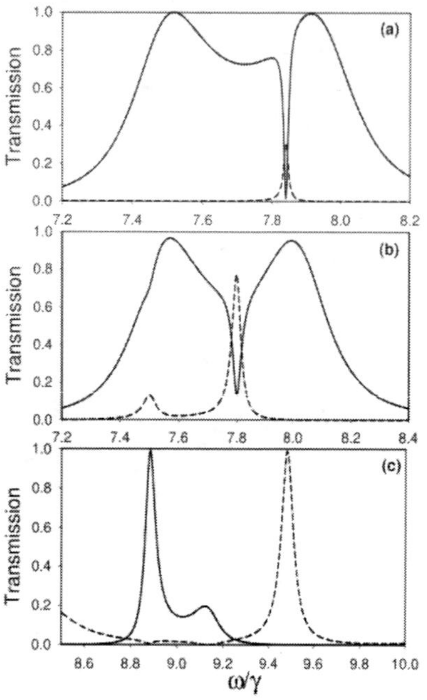

Figure 7. (a) Transmission in function of ω/γ for the structure shown in Fig. 6. The solid (dashed) line corresponds to the direct transmission T_{15} ($T_{15'}$) from site 1 to site 5 (from site 1 to site 5′). The parameters are $d = d_1 = 1, d_2 = 1.35, d_3 = 1.25, d_4 = 1.1$, $Ms = 1, M_s' = 4.3, M_s'' = 3.6, H_0 = 5$. (b) The same as in (a) but for the parameters $d_2 = 1.27, d_4 = 0.8$. The rest of the parameters have the same values as in (a). (c) The same as in (a) but for the parameters $d_2 = 1.2, d_3 = 1, d_4 = 1.2$, $M_s' = 4.35, M_s'' = 2.85$. The rest of the parameters have the same values as in (a).

at $\omega/\gamma \simeq 7.84$) squeezed between two transmission one in the upper output channel together with a peak with weak amplitude, at the same frequency, in the lower output channel. To increase the transmission amplitude in the lower channel we increased the coupling between the lower cluster chain and the resonators attached to it. This is done by shortening the distance d_4 (see Fig.6). In addition, a very slight decrease of the distance d_2 was also useful in enhancing the transmission probability. Fig.7(b) shows the ω/γ dependence of the transmission factor for the set of parameters $d_2 = 1.27, d_4 = 0.8$ (The rest of the parameters have the same values as in Fig.7(a)). In this plot one can notice that the transmission peak in the lower channel reaches almost 80 % of the input signal. One can also notice a slight shift in the frequency position of the resonance peak (dip). For other values of the parameters it is possible to obtain in the bulk transmission band two transmission peaks (one in each channel) where the amplitude in one channel goes to one, at a certain frequency, while the amplitude of transmission in the other channel goes to zero at the same frequency (and vice versa). This is effect, which could be useful in constructing a transmission filter for a magnon signal, is illustrated in Fig. 7(c) for $d_2 = 1.2, d_3 = 1, d_4 = 1.2$, $M_s' = 4.35, M_s'' = 2.85$ (The rest of the parameters have the same values as in Fig.7(a)). Before ending this paragraph it is worth mentioning that *a first step in the choice of the parameters is done on the basis of our previous calculation where a resonator (made of several clusters) is attached to a linear chain* [29, 30]. *Then, in a second step, the parameters have to be adjusted around these values in order to increase the transmission in one or both channels.*

Acknowledgments

H. Al-W. gratefully acknowledge the hospitality of the Université des Sciences et Technologies de Lille 1. The work of B. D.-R., A.A. and L.D was supported by Ministry of Higher Education and Research, Nord-Pas de Calais Regional Council and FEDER through the 'Contrat de Projets Etat Region (CPER) 2007-2013'

References

[1] S. L. Vysotskiĭ, S. A. Nikitov, and Yu. A. Filimonov, Journal of Experimental and Theoretical Physics, **101** (2005) 547.
[2] S. A. Nikitov, Ph. Tailhades, and C. S. Tsai, J. Magn. Magn. Mater. **236** (2001) 320.
[3] J. O. Vasseur, L. Dobrzynski, B. Djafari-Rouhani, and H. Puszkarski, Phys. Rev. B. **54** (1996) 1043.
[4] M. Krawczyk, J. C. Levy, D. Mercier, and H. Puszkarski, Phys. Lett. A. **282** (2001) 186.
[5] A. Figotin and I. Vitebsky, Phys. Rev. E. **63** (2001) 066609.
[6] D. S. Deng, X. F. Jin, and Ruibao Tao, Phys. Rev. B **66** (2002) 104435.
[7] L. Berger, Phys. Rev. B. **54** (1996) 9353.
[8] J.C. Slonczewski, J. Magn. Magn. Mater. **159** (1996) L1.
[9] G.A. Prinz, Phys. Today **48**(4)(1995) 58.
[10] V. V. Kruglyak, S. O. Demokritov and D. Grundler, J. Phys. D: Appl. Phys. **43** (2010) 264001.
[11] H. Puszkarski and M. Krawczyk, Phys. Lett. A **282** (2001) 106.
[12] H. Puszkarski, M. Krawczyk, J.-C. S. Levy, and D. Mercier, Acta Phys. Pol. A **100** (2001) 195.
[13] V. V. Kruglyak and A. N. Kuchko, Physica B **339** (2003) 130.
[14] H. Al Wahsh, A. Akjouj, B. Djafari-Rouhani, J.O. Vasseur, L. Dobrzynski, P.A. Deymier, Phys. Rev. B **59** (1999) 8709.
[15] A. Mir, H. Al Wahsh, A. Akjouj, B. Djafari-Rouhani, L. Dobrzynski, J.O. Vasseur, Phys. Rev. B **64** (2001) 224403.
[16] A. Mir, H. Al Wahsh, A. Akjouj, B. Djafari-Rouhani, L. Dobrzynski, and J. O. Vasseur, J. Phys.: Condens. Matter **14** (2002) 637.
[17] H. Al-Wahsh, Phys. Rev. B **69** (2004) 012405.
[18] Demidov V, Demokritov S, Rott K, Krzysteczko P and Reiss G 2008 Appl. Phys. Lett. **92** 232503.
[19] Topp J, Podbielski J, Heitmann D and Grundler D 2008 Phys. Rev. B **78** 024431.
[20] Topp J, Podbielski J, Heitmann D and Grundler D 2009 J. Appl. Phys. **105** 07D302.
[21] J. O. Vasseur, A. Akjouj, L. Dobrzynski, B. Djafari-Rouhani, and E. H. El Boudouti, Surf. Sci. Rep. **54**, 1 (2004).
[22] H. Al-Wahsh, E.H. El Boudouti, B. Djafari-Rouhani, A. Akjouj, T. Mrabti, L. Dobrzynski, Phys. Rev. B **78**, 075401 (2008)
[23] H. Al Wahsh, Eur. Phys. J. B **73**, 527 (2010)
[24] U. Fano, Phys. Rev. **124**, 1866 (1961).
[25] D. Goldhaber-Gordon, H. Shtrikman, D. Mahalu, D. Abush-Magder, U. Meirav, and M. A. Kastner, Nature (London) **391**, 156 (1988); J. Gores, D. Goldhaber-Gordon, S. Heemeyer, M. A. Kastner, H. Shtrikman, D. Mahalu, and U. Meirav, Phys. Rev. B **62**, 2188 (2000); Y. Ji, M. Heiblum, and H. Shtrikman, Phys. Rev. Lett. **88**, 076601 (2002).
[26] K. Kobayashi, H. Aikawa, S. Katsumoto, and Y. Iye, Phys. Rev. Lett. **88**, 256806 (2002); J. Kim, J.-R. Kim, Jeong-O. Lee, J. W. Park, H. M. So, N. Kim, K. Kang, K.-H. Yoo, and J.-J. Kim, ibid. **90**, 166403 (2003); B. Babic and C. Schönenberger, Phys. Rev. B **70**, 195408 (2004).
[27] Y. S. Joe, A. M. Satanin, and G. Klimeck, Phys. Rev. B **72**, 115310 (2005).
[28] M. L. Ladron de Guevara, F. Claro, and P. A. Orellana, Phys. Rev. B **67**, 195335 (2003); K. K. Voo and C. S. Chu, ibid. **72**, 165307 (2005); G. Hackenbroich and H. A. Weidenmuller, Phys. Rev. Lett. **76**, 110 (1996).
[29] H. Al-Wahsh, L. Dobrzynski, B. Djafari-Rouhani, G. Hernández-Cocoletzi, A. Akjouj, Surf. Sci. **600** (2006) 4883.
[30] H. Al-Wahsh, L. Dobrzynski, B. Djafari-Rouhani, A. Akjouj, Surf. Sci. **601** (2007) 4801.
[31] H. Al-Wahsh, B. Djafari-Rouhani, L. Dobrzynski, A. Akjouj, Surf. Sci. **602**, 1795 (2008)
[32] H. Al Wahsh, Eur. Phys. J. B **76**, 445 (2010)
[33] J. Barnas, Phys. Rev. B **45** (1992) 10 427.
[34] E. H. El Boudouti, N. Fettouhi, A. Akjouj, B. Djafari-Rouhani, A. Mir, J. Vasseur, L. Dobrzynski, and J. Zammouri, J. Appl. Phys. **95**, 1102 (2004)
[35] W. M. Robertson, J. Pappafotis, and P. Flannigan, Appl. Phys. Lett. **90**, 014102 (2007)
[36] R.C. Moul and M.G. Cottam, J. Phys. C **12** (1979) 5191.

Spin information transfer and transport in hybrid spinmechatronic structures

A D Karenowska[1], J F Gregg[1], A V Chumak[2], A A Serga[2] and B Hillebrands[2]

[1] Department of Physics, Clarendon Laboratory, University of Oxford, OX1 3PU Oxford, United Kingdom

[2] Fachbereich Physik and Forschungszentrum OPTIMAS, Technische Universität Kaiserslautern, 67663 Kaiserslautern, Germany

E-mail: a.karenowska@physics.ox.ac.uk

Abstract. Spin waves have long been recognized as potential signal carriers in spintronic devices. However, practical development of spin wave based information platforms is its infancy. To date, work in this area has focused on one-dimensional topologies based on purely magnetic thin-film transmission systems, typically exploiting interference phenomena to perform logical operations. In this paper, we describe an alternative approach in which spinmechatronic structures combining spin-wave transmission systems with magnetically loaded micro- and nano-mechanical elements provide spin-information processing functionality.

1. Introduction

Spinmechatronics alludes to the marriage of mechatronics—most particularly micro- and nano-mechanical systems (MEMS and NEMS)—with spin physics and spintronics. Although the term is relatively new, the idea of combining spin and mechanics is certainly not. Many spinmechatronic systems exist, among them a growing range of magnetic MEMS devices (see for example [1–3]) and scientific instruments such as the magnetic resonance force microscope [4–7]. However, as recent advances in fabrication and patterning techniques give life to a rapidly expanding range of micro- and nano-mechanical systems with novel physical properties and functionalities, it is timely to consider how as yet unexplored spin-mechanical synergies may open new routes to expanding our physical insight into magnetic systems of contemporary interest, and/or provide the basis for technologically relevant devices.

In this paper we seek to highlight fresh potential in this hybrid field of research through the description of a particular class of *spin-wave* spinmechatronic system which may find application in information processing.

2. The spin wave as an information carrier

Spin waves are eigenexcitations of the spin-lattice of a magnetic material; the quanta of which are known as magnons. The study of spin-wave systems offers unique fundamental insight into the physics of static and dynamic magnetism as well as a range of general wave and quasi-particle phenomena (see [8–12] and references therein) not readily accessible in other physical domains.

The spin-wave spectrum is traditionally divided into two parts; a region of short-wavelength, low-velocity waves with dynamics dominated by the magnetic exchange interaction, and a longer wavelength region of so-called magnetostatic waves which have higher velocities (typically of order $c/10^4$) and properties governed by the longer-range dipolar interaction [13]. Exchange dominated excitations have a dispersion relationship $\omega(k) \propto k^2$ as derived by the well-known treatment of a one-dimensional Heisenberg chain of spins (see for example [14]). Magnetostatic waves behave very differently and—in typical thin-film experimental systems under conditions of unidirectional magnetic bias—their dispersion depends strongly on the relative orientation of the bias field and their propagation direction [8].

Magnetostatic spin waves (MSWs)—which typically have frequencies in the gigahertz and wavelengths in the micron to millimetre range—are widely acknowledged as a prospective platform for spintronic information transfer and transport (see for example [15–17]); not just by virtue of their ability to function as direct carriers of spin angular momentum but because their unusual dynamic properties potentially conduce to sophisticated information processing capability.

The dispersion properties of linear MSWs can be tuned over a wide frequency and k-vector range using modest externally applied magnetic fields and, owing to their short wavelengths and low velocities in comparison with electromagnetic waves of the same frequency, they are well suited to the demands of micro- and nanoscale device development. Furthermore, in contrast with other physical domains including optics and conventional electronics where the feasibility of nonlinear signal processing techniques is hampered by the intrinsically weak nonlinearity of available materials, nonlinear dynamic phenomena including, for example, soliton and bullet formation, are accessible in certain MSW systems at extremely low power levels (see for example [12, 18, 19]). Artificial magnetic materials, in particular magnonic crystals [20–25]—the spin-wave equivalent of optical photonic crystals—represent a further potential route to information processing functionality, and the appeal of MSW systems is further heightened by their signal storage capability; information may be encoded in the spin-wave mode structure of magnetic elements, and a range of signal storage and restoration mechanisms based on parametric energy transfer between dynamically distinct magnon populations also exist (see [8, 26] and references therein).

3. A spinmechatronic vision for spin wave based information processing

To date, research in the area of magnetostatic spin wave based information platforms has focussed almost exclusively on the development of logical systems based on purely magnetic thin-film transmission systems operating in a linear signal regime. Conceptual gates typically function through the interference of MSWs in Mach-Zehender type structures, driven and detected inductively using microstrip antennas and controlled via current-induced magnetic fields [16, 27]. Whilst the device potential of such systems should not be understated, they are severely limited in the extent to which they capitalize on the rich spin-wave physics outlined above. Moreover, their operational robustness is restricted by the fact that they are low-Q systems in which the efficiency of excitation and detection of waves is intrinsically poor.

Resonant micro- and nano-mechanical elements with Q-factors several orders of magnitude in excess of those accessible in electrical or magnetic systems and operating frequencies in the kHz to MHz range are now readily available. Moreover, the technology to magnetically load or dope them, either via the deposition or application of hard or soft magnetic coatings, particles, or tips [28], is well developed. The exceptional frequency selectivity of these miniature mechanical devices is already exploited in a wide range of scanning probe microscopy techniques and—in an information processing context—strongly recommends them for use both as sensitive narrowband receivers and as a robust, low-power means to write information.

What we propose is a spintronic signalling platform in which information is encoded in the amplitudes and phases of propagating, amplitude modulated magnetostatic spin waves and written and read by resonant magno-mechanical elements. Hybridizing spin-wave physics and miniature mechatronics, such a system potentially combines operational robustness and low power requirements with sophisticated signal processing capability and integrated information storage.

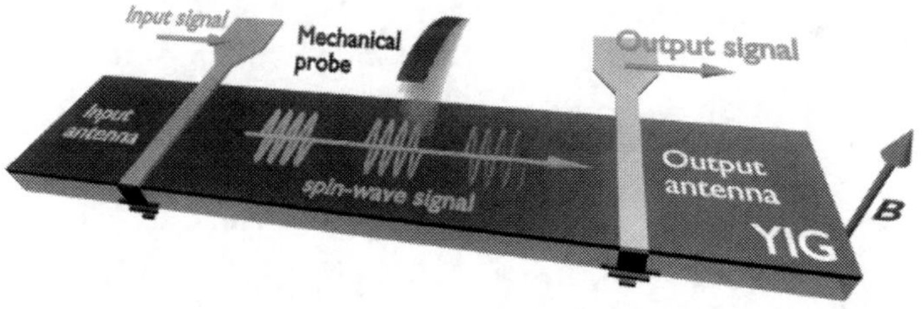

Figure 1. In a recent series of investigations we demonstrated the feasibility of mechanical coupling to propagating magnetostatic waves in a spin-wave transmission structure [29]. In the experiments, amplitude modulated spin waves, excited by a microstrip inductive antenna, were used to drive out-of-plane oscillations of a magnetic micro-mechanical probe (nickel, diameter 10 μm) suspended above the surface of an yttrium iron garnet thin-film spin-wave waveguide. The displacements of the probe were measured using a laser interferometric vibrometer. The same micro-mechanical device was found to be capable of measurably perturbing the amplitude and phase of waves transmitted a distance of 8 mm through the structure. The waveguide was biassed by an out-of-plane field so as to support the propagation of forward volume magnetostatic spin waves.

The success of our spinmechatronic scheme is predicated on the feasibility of two-way interactions between the stray magnetic fields from miniature magno-mechanical systems and *travelling* spin waves. The possibility of mechanical coupling to short wavelength, highly localized spin-wave excitations (generally exchange modes) is well established in the context of magnetic resonance force microscopy [4–7]. However, mechanical coupling to MSWs excited by spatially localized antennas in spin-wave transmission systems—the physics of which is subtly but importantly distinct—has until lately, remained largely unexplored.

Recently, we demonstrated the viability both of detecting amplitude modulated MSWs in a thin-film spin-wave transmission structure via the displacement of a magnetically coupled micro-mechanical probe, and of using the same magnetic probe to modify the phase and amplitude of waves transmitted through it [29]. In our proof-of-concept experiments, we employed an yttrium iron garnet (YIG) waveguide with a nickel probe (diameter 10 μm) suspended above its surface (Fig. 1). The waveguide was biased by a magnetic field applied parallel to the film thickness so as to support the propagation of forward volume magnetostatic spin waves (FVMSW) [13]. In our measurements of spin wave induced oscillations of the mechanical probe, on/off modulated travelling waves were excited in the YIG via an inductive microstrip spin-wave input antenna and displacements of the probe at the modulation frequency detected using a laser interferometric vibrometer (SIOS Meßtechnik GmbH, Nano Vibration Analyzer).

At present, the contemporary capabilities of miniature mechatronics restrict development of the spin-wave spinmechatronic concept to systems with functionality based on resonant interactions between mechanical elements and *amplitude modulated* spin waves driven inductively via microwave currents. However, looking ahead, although not yet widely available, mechanical systems with resonance frequencies well within range of GHz spin waves are now an emerging technology [30]. This invites us to speculate as to whether the spinmechatronic concept we describe could ultimately provide the foundation for a spintronic information platform in which phase-resolved, spin-wave read, write and *drive* operations are undertaken mechanically; entirely without the use of conventional currents.

4. Conclusions

In summary, we have described a proposal for a spinmechatronic signalling platform in which information encoded in the amplitudes and phases of propagating, amplitude modulated magnetostatic spin-waves is written and read by resonant magno-mechanical elements. The system potentially overcomes some of the key limitations of spin-wave information processing topologies proposed to date; affording a robustness and level of functionality surpassing that of purely magnetic, current-controlled schemes.

Proposing the hybrid spin-mechanical structures described in this paper, we seek not only to share a specific device concept, but to stimulate more general discussion around how, by exploring synergies between spintronics and other areas of physics not traditionally associated with the field, we may unearth both new physical insight and as yet unforeseen technological opportunity.

Acknowledgments

ADK is grateful for the support of the Institution of Engineering and Technology and Magdalen College, Oxford and all authors acknowledge the generosity of SIOS Meßtechnik GmbH (Ilmenau, Germany) and Armstrong Optical (Northampton, UK).

References

[1] Gibbs, M R J 2005 *J. Magn. Magn. Mater.* **290** 1298
[2] Gibbs, M R J 2004 *J. Phys. D: Appl. Phys.* **37** R237
[3] Niarchos, D 2003 *Sens. Actuators, A* **290** 12980
[4] Poggio M and Degen C L 2010 *Nanotechnology* **21** 342001
[5] Pigeau B, de Loubens G, Klein O, Riegler A, Lochner F, Schmidt G, Molenkamp L W, Tiberkevich V S and Slavin A N 2010 *Appl. Phys. Lett.* **96** 132506
[6] Klein O, de Loubens G, Naletov V V, Boust F, Guillet T, Hurdequint H, Leksikov A, Slavin A N, Tiberkevich V S and Vukadinovic N 2008 *Phys. Rev. B* **78** 144410
[7] Sidles J A 1991 *Appl. Phys. Lett.* **58** 24
[8] Serga, A A, Chumak, A V and B Hillebrands, B 2010 *J. Phys. D* **43** 264002
[9] Schneider, T, Serga, A A, Chumak, A V, Sandweg, C W, Trudel, S, Wolff, S, Kostylev, M P, Tiberkevich, V S, Slavin, A N and Hillebrands, B *Phys. Rev. Lett.* **104** 197203
[10] Wu, M and Patton, C E 2007 *Phys. Rev. Lett.* **98** 047202
[11] Demokritov, S O, Demidov, V E, Dzyapko, O, Melkov, G A, Serga, A A, Hillebrands, B and Slavin A N 2006 *Nature* **443** 430
[12] Demokritov, S O, Serga, A A, Demidov, V E, Hillebrands, B, Kostylev, M P and Kalinikos B A 2003 *Nature* **426**, 159
[13] Stancil, D D 1993 *Theory of Magnetostatic Waves* (New York: Springer-Verlag)
[14] Blundell, S 2003 *Magnetism in Condensed Matter* (Oxford, Oxford University Press) chapter 6 p122
[15] Kajiwara, Y, Harii, K, Takahashi, S, Ohe, J, Uchida, K, Mizuguchi, M, Umezawa, H, Kawai, H, Ando, K, Takanashi, K, Maekawa, S and Saitoh, E 2010 *Nature* **464**, 262
[16] Kostylev M P, Serga A A, Schneider T, Leven B and Hillebrands B 2005 *Appl. Phys. Lett.* **87** 153501
[17] Khitun A and Wang K L 2005 *Superlattices and Microstructures* **38** 184
[18] Kalinikos B A, Kovshikov N G, and Patton C E 1998 *Phys. Rev. Lett.* **80** 4301
[19] Serga A A, Demokritov S O, Hillebrands B and Slavin A N 2004 *Phys. Rev. Lett.* **92** 117203
[20] Chumak A V, Serga A A, Wolff S, Hillebrands B and Kostylev M P 2009 *Appl. Phys. Lett.* **94** 172511
[21] Lee K S, Han D S and Kim S K 2009 *Phys. Rev. Lett.* **102** 127202
[22] Ustinov A B, Kalinikos B A, Demidov V E and Demokritov S O 2009 *Phys. Rev. B* **80** 052405
[23] Wu M, Hagerstrom A M, Eykholt R, Kondrashov A and Kalinikos B A 2009 *Phys. Rev. Lett.* **102** 237203
[24] Wang Z K, Zhang V L, Lim H S, Ng S C, Kuok M H, Jain S and Adeyeye A O 2009 *Appl. Phys. Lett.* **94** 083112
[25] Chumak A V, Serga A A, Hillebrands B and Kostylev M P 2008 *Appl. Phys. Lett.* **93** 022508

[26] Serga A A, Chumak A V, André A, Melkov G A, Slavin A N, Demokritov S O and Hillebrands B 2007 *Phys. Rev. Lett.* **99** 227202
[27] Schneider, T, Serga, A A, Leven, B, Hillebrands, B, Stamps R L and Kostylev, M P 2008 *Appl. Phys. Lett.* **92** 022505
[28] Arnold D P and Wang N 2009 *JMEMS* **18** 1255
[29] Karenowska A D, Gregg J, Hillebrands B, Serga A A, Chumak A C and Vasyuchka V I *forthcoming*
[30] Huang X M H, Zorman C A, Mehregany M and Roukes M L 2003 *Nature* **421** 496

Journal of Physics: Conference Series **303** (2011) 012019 doi:10.1088/1742-6596/303/1/012019

Magnetotransport in Tb_2Fe_{17} single crystals

Jolanta Stankiewicz[1] **and Konstantin P. Skokov**[2]

[1] Instituto de Ciencia de Materiales de Aragón and Departamento de Física de la Materia Condensada, CSIC–Universidad de Zaragoza, 50009-Zaragoza, Spain

[2] Faculty of Physics, Tver State University, 170002 Tver, Russia

E-mail: jolanta@unizar.es

Abstract. We have performed measurements of the Hall effect, electrical resistivity, and magnetization on Tb_2Fe_{17} single crystals in the 5 to 300 K temperature range, and in magnetic fields of up to 5 T. We find that the anomalous Hall effect of this ferromagnet depends strongly on the magnetization direction relative to the crystal axes. The AHE resistivity, measured with an applied magnetic field **H** perpendicular to the c-axis, is very large and varies linearly with the longitudinal resistivity ρ, whereas the AHE resistivity for **H** along the hard magnetization direction is much smaller and increases as ρ^2. For the latter configuration, the electrical resistivity shows a sharp decrease at a field-induced first-order magnetization process (FOMP) which is observed in $H \sim 2.7$ T up to a temperature of 250 K.

1. Introduction

The Tb_2Fe_{17} intermetallic compound crystallizes in a Th_2Ni_{17}-type hexagonal structure [1] in which the Fe layers (perpendicular to the c-axis) are intercalated with dumb-bell Fe pairs along the c-axis. The Fe atoms occupy four nonequivalent crystallographic sites. However, only dumb-bell Fe atoms at $4f$ sites have a non-zero orbital magnetic moment. The Mössbauer effect experiments [2] show that the $3d$ bands coming from the $4f$ ions are distorted giving rise to a larger magnetic polarization and to a lower electron charge density. This agrees with results of band-structure calculations which find a highest orbital moment for Fe atoms at the dumbbell sites [3-4]. The Tb sublattice couples antiferromagnetically to the Fe sublattice in the Tb_2Fe_{17} alloy. A first-order magnetization process (FOMP) is observed at a magnetic field of about 2.5 T applied along the sixfold axis in this compound [5]. The FOMP does not affect the mutual orientation of the sublattices only their relative orientation to the crystallographic axes. Interestingly, the critical field for this transition is nearly independent of temperature up to ≈ 250 K even though the anisotropy constants vary with temperature as expected from existing models [6].

All these properties make Tb_2Fe_{17} compounds attractive for electron–transport studies which are very sensitive to the magnetic nature of the materials. In particular, intermetallic compounds with transition metals may show complex behavior as $3d$ itinerant magnetism and $4f$ localized magnetic moments compete in them. In addition, spin-orbit coupling, either intrinsic or extrinsic, of magnetic electrons gives rise to a magnetization-dependent Hall effect. This, in turn, depends crucially on scattering mechanisms [7, 8] and on the band structure [9]. However, no electronic transport studies have been reported for Tb_2Fe_{17}. Our aim is to fill this gap.

In this paper we report results of electrical resistivity and Hall effect measurements on Tb_2Fe_{17} single crystals in the temperature range of 5–300K and in magnetic fields of up to 6 T. The measurements have been performed for various magnetic field orientations with respect to the easy-magnetization axis. The anomalous Hall effect (AHE), both extrinsic and intrinsic, is very anisotropic in this ferromagnet. We attribute it to the band structure details and to a possible distortion of the $3d$ bands.

2. Experiment

Tb_2Fe_{17} single crystals were grown by inductive melting of a mixture of 99.9%–pure Tb and 99.99%–pure Fe in alundum crucibles under Ar atmosphere. The mixture was rapidly heated up to the melting point (T $\approx$ 1500 C), and subsequently cooled down at a rate of 40-60 C/min. Further heating on to approximately 1200 C, with an annealing time of approximately 20 hours, was performed in order to obtain large crystalline grains. The phase composition was checked by optical metallography. We carefully oriented the samples using x-ray back Laue diffraction. The magnetotransport measurements were performed on Hall bar shaped samples. The typical sample size was $0.5 \times 1 \times 5$ mm^3. We used a six probe method in order to check for samples' homogeneity both in resistivity and Hall coefficient. The Hall resistivity and magnetization were measured in a magnetic field of up to 6 T, in a temperature range from 5 to 300 K. Magnetization measurements were carried out in a superconducting quantum interference device (SQUID) magnetometer on the very same samples that were used in the magnetotransport studies. In this way, we expected to avoid sample–shape and domain dependent effects when comparing results of different experiments.

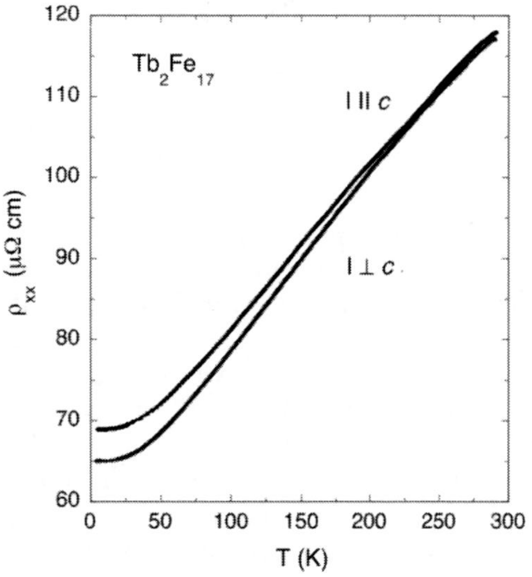

Figure 1. Longitudinal resistivity as a function of temperature for Tb_2Fe_{17} single crystals. The solid lines are guides to the eye.

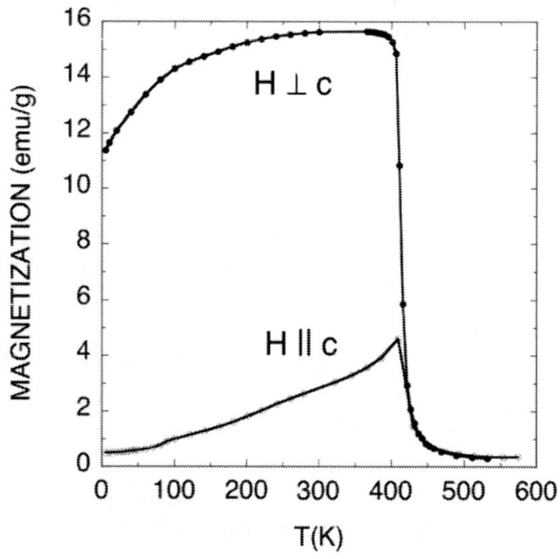

Figure 2. Magnetization measured in the field of 1 kOe as a function of temperature for the same Tb_2Fe_{17} single crystals. The solid lines are guides to the eye.

3. Results and Discussion

Figure 1 shows how the zero-field longitudinal resistivity $\rho_{xx}(T)$ depends on temperature T in Tb_2Fe_{17} single crystals. The resistivity, typical of metallic ferromagnet, is slightly anisotropic at low temperatures: it is approximately 6% lower for the current flowing in the easy plane than for the current along the c–axis. It increases with increasing temperature as spin–disorder and electron–phonon scattering become more important. The low-field magnetization variation with temperature and orientation is shown in Fig. 2. The magnetization M drops sharply at the Curie temperature of 411 K. How the electrical resistivity varies with magnetic field is shown in Figs. 3 and 4. Since an external magnetic field suppresses spin fluctuations, the electrical resistivity decreases when a magnetic field is applied. The observed negative magnetoresistance is very small (less than 3 per mil at $H = 5$ T) for magnetic fields on the easy plane. However, an appreciable jump is seen in $\rho_{xx}(H)$ at a field-induced first-order magnetization process (FOMP) for $\mathbf{H} \parallel c$–axis, as shown in Fig. 3. At 5 K, $\rho(H)$ drops approximately 3% at a field of ~ 2.8 T in this configuration. Such discontinuity is expected at the critical field of a first-order spin-reorientation transition, since the resistivity depends on the angle between $\mathbf{I}$ and the spontaneous magnetization $\mathbf{M}_s$ through spin–orbit coupling. As the temperature increases, the resistivity drop decreases. It becomes negligible for $T \gtrsim 250$ K. The critical field at which the abrupt resistivity change occurs is nearly independent of temperature. This behavior of the critical field is however incidental since the anisotropy constants vary with temperature as expected for intermetallic compounds [6].

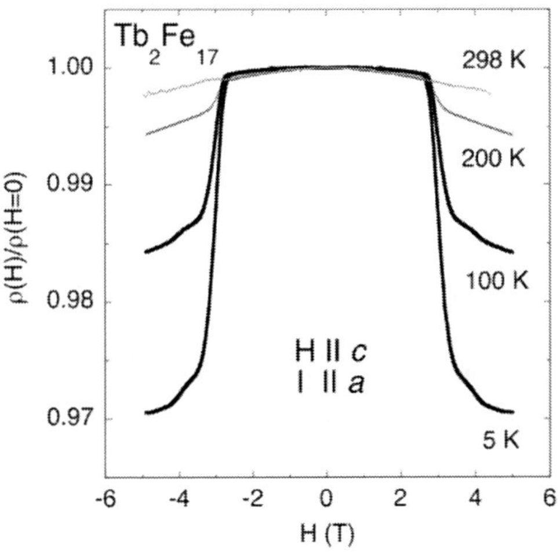

Figure 3. Transverse magnetoresistance in a Tb_2Fe_{17} single crystal for $\mathbf{H}$ along the c–axis and $\mathbf{I}$ along the a–axis.

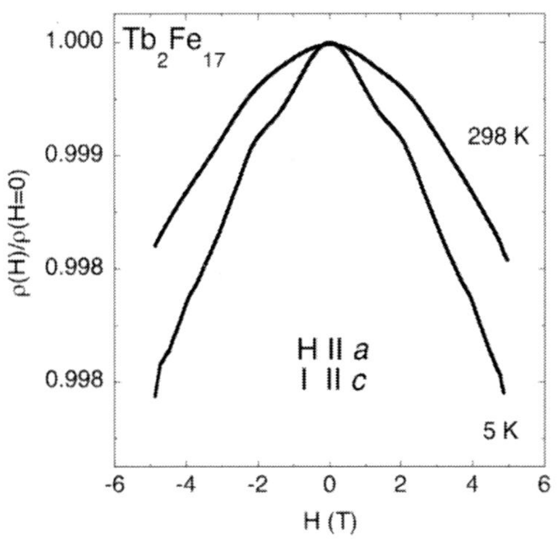

Figure 4. Transverse magnetoresistance in a Tb_2Fe_{17} single crystal for $\mathbf{H}$ along the a–axis and $\mathbf{I}$ along the c–axis.

We next discuss the Hall resistivity ρ_{xy} and magnetization data. Their variation with magnetic field are shown in Figs. 5 and 6 for T=5 and 300 K, respectively. Below technical saturation, ρ_{xy} follows the magnetization behavior of the sample. Above saturating field, ρ_{xy} remains almost constant since the negative, ordinary Hall resistivity is very small. The experimental values of ρ_{xy} and M_s are obtained by extrapolating the Hall resistivity and magnetization data from high magnetic fields back to $H = 0$. Since relatively small magnetic fields are sufficient to align magnetic moments, this procedure is quite straightforward. In ferromagnetic metals, the anomalous contribution to ρ_{xy} is usually proportional to the spontaneous magnetization because of spin-orbit interaction. Extrinsic mechanisms such as skew and side–jump scattering contribute terms linear [7] and quadratic [8] in the longitudinal resistivity, respectively. On the other hand, spin-orbit coupling of band electrons brings about intrinsic AHE which is independent of scattering and should give $\rho_{xy} \propto \rho_{xx}^2$ [10].

We plot the Hall resistivity vs the total longitudinal resistivity using our data in Fig. 7. The AHE resistivity varies quite linearly with ρ_{xx} for $\mathbf{M}$ lying in the ab plane. However, it

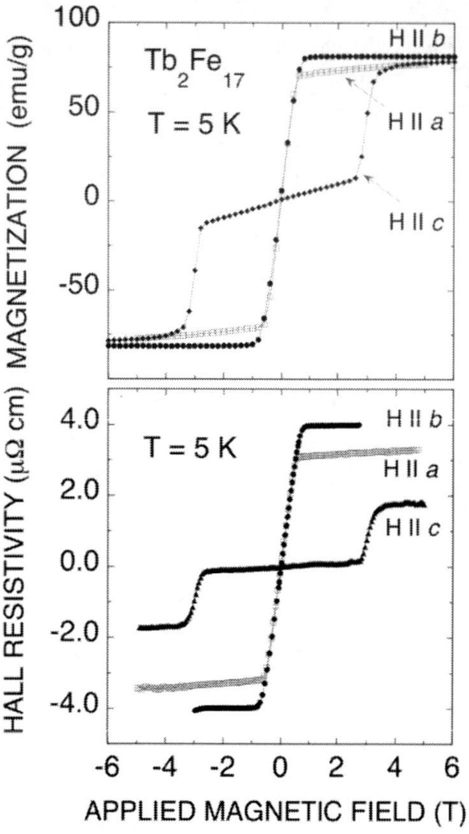

Figure 5. Magnetization and Hall resistivity as a function of magnetic field for Tb_2Fe_{17} single crystals at 5 K. The solid line is a guide to the eye.

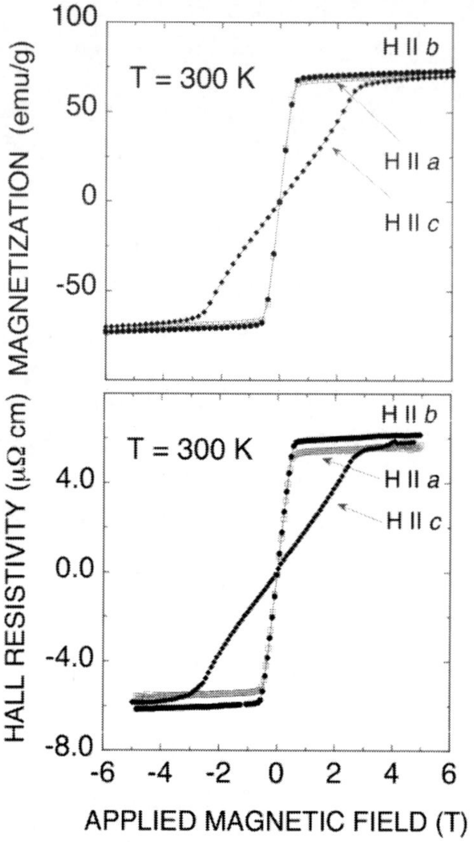

Figure 6. Magnetization and Hall resistivity as a function of magnetic field for Tb_2Fe_{17} single crystals at 300 K. The solid line is a guide to the eye.

varies as a ρ_{xx}^2 for **M** along the hard direction. In addition, the AHE depends strongly on orientation: it drops by a factor of ~ 3 as the magnetization is tilted from the easy *ab* plane to the *c*–axis at low temperatures. To interpret our data, consider the relation between the AHE and the longitudinal resistivity of the form: $\rho_{xy} = a(M_s)\rho_{xx} + b(M_s)\rho_{xx}^2$ [11-12]. The first term arises from skew scattering contribution that is usually linear in magnetization [7]. Assuming $a(M) \propto M$, and plotting $\rho_{xy}/M_s\rho_{xx}$ versus ρ_{xx} we can estimate a. The second term stands for the intrinsic contribution, related to Berry phase effects on conduction electrons. We consider side-jump scattering, which has identical ρ_{xx} dependence, negligible in Tb_2Fe_{17} on the basis of the observed orientation dependence of the AHE [13]. Such dependence is not expected for extrinsic side-jump mechanism which should lead to an isotropic Hall current [8]. The intrinsic Hall conductivity (AHC) is expressed as $\sigma_{xy}^a = \rho_{xy}/\rho_{xx}^2$, therefore $b(M) = \sigma_{xy}^a$. Since the AHC varies only weakly with temperature [12], a constant low–temperature value of σ_{xy}^a follows. Indeed, the plot of $\rho_{xy}/M_s\rho_{xx}$ vs ρ_{xx} is linear for $\mathbf{H}\| c$ configuration at low temperatures (see Fig. 8), hence $\sigma_{xy}^a \propto M$ in Tb_2Fe_{17}.

The variation of $[\rho_{xy}/\rho_{xx}][M_s^o/M_s(T)]$ with ρ_{xx} in Tb_2Fe_{17} single crystals is shown in Fig. 8. Here, $M_s^o = M_s(T=0)$. For **M** on the easy plane, this quantity is constant for $T \lesssim 250$ K. However, when **M** is lined up along the hard direction, $\rho_{xy}M_s^o/\rho_{xx}M_s(T)$ varies linearly with

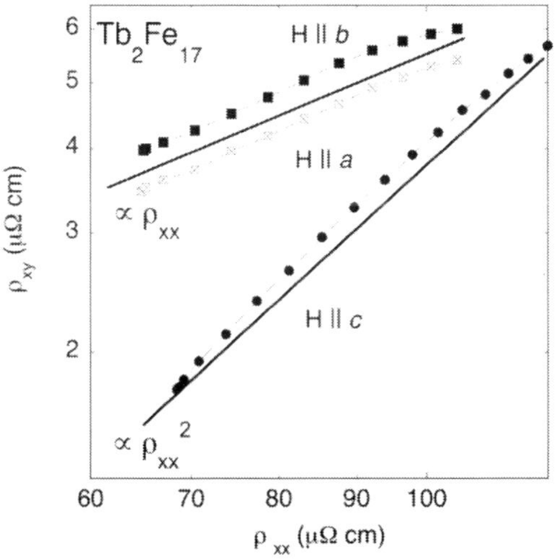

Figure 7. Hall resistivity of Tb_2Fe_{17} single crystals as a function of the longitudinal resistivity for various configurations. The solid lines show a linear and a quadratic in ρ_{xx} variation.

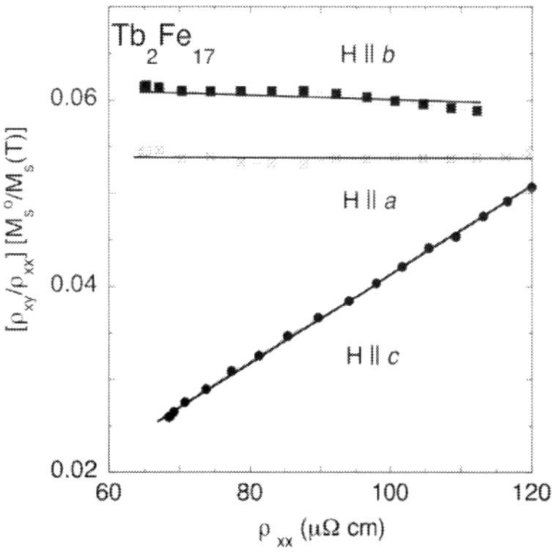

Figure 8. Plots of $[\rho_{xy}/\rho_{xx}][M_s^o/M_s(T)]$ as a function of the longitudinal resistivity for **H**⊥ c-axis and **H**|| c-axis in Tb_2Fe_{17} single crystals. The solid lines are linear fittings to the data points.

ρ_{xx}. This clearly suggests that large skew scattering entirely dominates the AHE resistivity for magnetic fields on the easy plane, but it has almost no effect otherwise. We find a value of approximately 0.055 for the skew–scattering coefficient a in Tb_2Fe_{17} when **M** lies on the easy plane. It is one order of magnitude larger than skew–scattering contributions to the AHE found for metallic ferromagnets [11]. This might be related to the distortion of d–bands corresponding to the dumb-bell Fe sites. However, a detailed knowledge of the impurity potential would be needed for a quantitative estimation of skew-scattering contributions. On the other hand, we find a value of 420 $\Omega^{-1}\mathrm{cm}^{-1}$ for σ_{xy}^a which is very close to theoretical estimations for the intrinsic Hall conductivity [14,15]. This agrees with the recent observations that the intrinsic mechanism is important for the AHE in moderately conductive ferromagnets [9]. In particular, variations of the total Berry curvature, coming from avoided crossings of nearly degenerate d–levels of Fe ions states near the Fermi energy, may resonantly enhance the AHC amplitude. Such scenario also explains the orientation dependence of the intrinsic AHC [13]. Band structure calculations for Y_2Fe_{17} [16], which should be not very different from those for Tb_2Fe_{17}, seem to give ground for the above conclusions. However, we cannot quite rule out that the large intrinsic AHC, observed in the system we study, follows from inter-band hopping between nearly degenerate d orbitals [15].

Acknowledgments

This work was supported in part by grant MAT2008/03074 from the Spanish Ministerio de Ciencia. Additional support from Diputación General de Aragón (DGA-IMANA) is also acknowledged.

References

[1] Givord D, Lemaire R, Moreau J M and Roudaut E 1972 *J. Less Common Met.* **29** 361
[2] Gubbens P C M, van Loef J J and Buschow K H J 1974 *J. Phys. (Paris) Colloq.* **35** C6-617
[3] Cochoorn R 1989 *Phys. Rev.* B **39**, 13072

[4] Sabiryanov R F and Jaswal S S 1998 *Phys. Rev.* B **57** 7767
[5] Verhoef R, Quang P H, Radwański R J, Marquina C and Franse J J M 1992 *J. Magn. Magn. Mater.* **104-107** 1473
[6] Kuz'min M D, Skourski Y, Skokov K P, Müller K -H and Gutfleisch O 2008 *Phys. Rev.* B **77** 132411
[7] Smit J 1955 *Physica (Amsterdam)* **21** 877
[8] Berger L 1970 *Phys. Rev.* B **2** 4559
[9] Nagaosa N, Sinova J, Onoda S, MacDonald A H and Ong N P 2010 *Rev. Mod. Phys.* **82** 1539
[10] Yao Y, Kleinman L, MacDonald A H, J. Sinova J, Jungwirth T, Wang D-S, Wang E and Niu Q 2004 *Phys. Rev. Lett.* **92** 037204
[11] Sales B C, Jin R and Mandrus D 2008 *Phys. Rev* B **77** 024409
[12] Zeng C, Yao Y, Niu Q and Weitering H 2006 *Phys. Rev. Lett.* **96** 037204
[13] Roman E, Mokrousov Y and Souza I 2009 *Phys. Rev. Lett.* **103** 097203
[14] Miyasato T, Abe N, Fujii T, Asamitsu A, Onoda S, Onose Y, Nagaosa N and Tokura Y 2007 *Phys. Rev. Lett* **99** 086602
[15] Kontani H, Tanaka T and Yamada K 2007 *Phys. Rev.* B **75** 184416
[16] Kulakowski K and del Moral A 1994 *Phys. Rev.* B **50** 234

Joint European Magnetic Symposia – JEMS 2010 IOP Publishing
Journal of Physics: Conference Series **303** (2011) 012020 doi:10.1088/1742-6596/303/1/012020

Anisotropy study of grain oriented steels with Magnetic Barkhausen Noise

M F de Campos[1], M A Campos[2], F J G Landgraf[2,3], L R Padovese[2]

[1]PUVR- Universidade Federal Fluminense, Av dos Trabalhadores 420, Vila Santa Cecília, Volta Redonda, RJ, 27255-125, Brazil

[2]Escola Politécnica, Universidade de São Paulo, São Paulo, SP, Brazil

[3]IPT, Instituto de Pesquisas Tecnológicas, São Paulo, SP, Brazil

E-mail: mcampos@metal.eeimvr.uff.br, mfdcampo@uol.com.br

Abstract. Grain oriented electrical steels present strong anisotropy, due to a {110} <001> texture (Goss), with [100] direction parallel to rolling direction (RD) and [110] direction parallel to transverse direction (TD). MBN (Magnetic Barkhausen Noise) were employed to measure magnetic properties in several angles towards RD using a 15° step. For 90° to the rolling direction (i.e., TD), the MBN signal changes, decreasing the MBN_{rms}. It is found a connection between initial permeability and MBN_{rms}. The lower initial permeability for the TD is related to a larger contribution of irreversible rotation in the hysteresis. The MBN procedure is non-destructive and provides rapid understanding of the anisotropy of the material, without the use of laborious methods like Epstein frame or toroidal coils.

1. Introduction

Magnetic Barkhausen Noise (MBN) is a very important tool for non-destructive characterization [1-3]. Many studies [1-5] have focused on establishing a relation between the microstructure of steels and the MBN, which can be affected by crystallographic orientation [6]. Grain oriented (GO) steels present only the (110) [001] texture component, named as Goss. This is confirmed in SEM-EBSD analysis, showing that the microstructure of the GO steels consists in large grains, with ~0.5 mm of diameter, having only Goss orientation as texture component [7]. In the same GO sheet, different crystalline orientations can be evaluated, as the easy axis [100], [110], and the hard axis [111]. Thus, crystallographic orientation can be isolated as the only possible variable. Although the MBN was reported almost 90 years ago, its origin and characteristics remain not plentifully understood [8].The MBN procedure is rapid and non-destructive and can provide understanding of the anisotropy of the material, without the use of laborious methods typically used for characterization of electrical steels like Epstein frame.

Experimental

Grain oriented steels sheets were kindly supplied by Acesita. The GO electrical steel has as physical characteristics: Silicon content 3.2%, thickness 0.27 mm and density d=7.65 g /cm^3. Hystereses of the

GO electrical steels were measured in Epstein frame under quasi-static conditions. The measurements followed the requirements of IEC 60404-2 standard, including the sinusoidal shape of the magnetic induction. MBN measurements were carried out with portable equipment, named "Barktech", developed at our laboratory in USP. Figure 1 shows a simplified sketch of this equipment. A sinusoidal wave current was applied to the MBN probe, with amplitude of 1 A and frequency of 10 Hz. The MBN signal was measured with a pick-up coil of high sensibility. The MBN sensor signal was amplified, band pass filtered, and sampled at a frequency of 400 kHz. The choice of 10 Hz as the alternating magnetic field frequency was due to the better signal to noise ratio. For this frequency (10 Hz), the effect of eddy currents is small (note also that the samples are thin and have high resistivity).

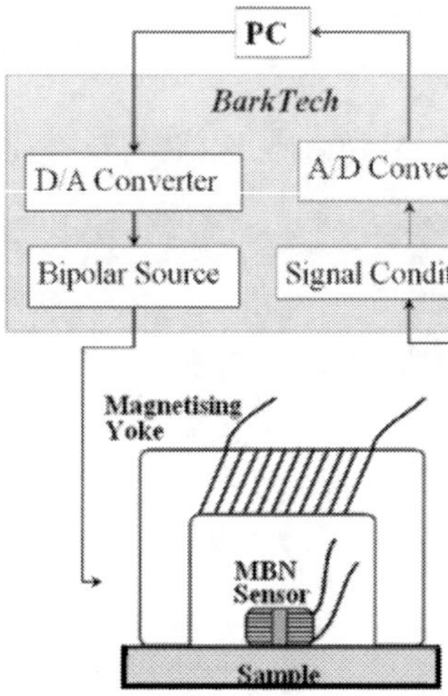

Figure 1. Experimental setup for the MBN measurement.

Results and Discussion

The anisotropy of magnetic properties is evident in the quasi-static hysteresis of Fig. 2. It is noteworthy the significant change of behavior about the permeability μ (Fig. 3). The permeability at 1.5 T (μ_{15}) is minimum at 45°, while the initial permeability μ_i continuously decreases with increasing angle, with a point of minimum at 90° (see Fig. 3).

A very simple analysis [9], focusing on the initial permeability μ_i (equation 1), provides insight about the competition between irreversible rotation and domain wall movement. Coercivity and initial permeability are closely related, and are inversely proportional, as seen at Fig. 3.

$$\mu_i \cong \chi_i = \left(\frac{\partial B}{\partial H}\right)_{H=0} \tag{1}$$

Note: $\mu=\chi+1$. χ is the magnetic susceptibility. B is magnetic induction. H is the applied magnetic field.

The Eq. 2 estimates the initial permeability supposing only occurrence of irreversible rotation, $\mu_{i(rotation)}$, without any domain wall (DW) displacement processes (in the deduction, it was supposed isotropic distribution of the magnetization of the domains) [9].

$$\mu_{i(rotation)} \cong \chi_{i(rotation)} = \frac{B_s^{\,2}}{3\mu_0 K_1} \tag{2}$$

Using the following values (~3%Si iron-silicon alloy): Induction of Saturation B_s = 2.03 T and First Order Magnetocrystalline Anisotropy Constant $K_1 = 3.6 \times 10^4$ J/m^2 (μ_0 is the permeability at the vacuum), it is obtained $\mu_{i(rotation)} \cong 30$. However, the measured initial permeability is much above this value (Fig. 3). Thus, the contribution of irreversible rotation seems to be less relevant than DW movement. It should be noted, anyway, that irreversible rotation dissipates more energy than domain wall movement and this should contribute to increase the coercivity for the 90° sample.

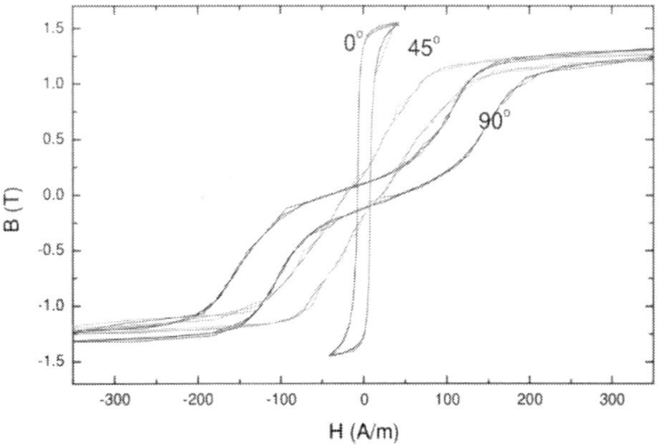

Figure 2. Hysteresis loops for the grain-oriented electrical steel measured for the directions 0°, 45° e 90° to RD. Measurements performed at quasi-static condition (5-10 mHz), Epstein frame. This is a detail of the complete measurements. Fields upto 5000 A/m were applied, until the Induction of 1.5 T could be attained.

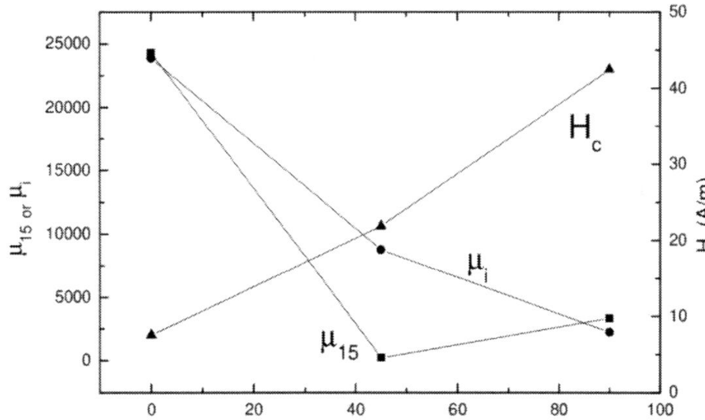

Figure 3. Permeability μ_{15}, initial permeability μ_i and coercivity Hc as function of the angle to RD for the Grain-oriented steel (with Goss (110) [001] texture)

The data of Fig. 3 indicate that, for the sample 90° , part of the quasi-static losses is dissipated by means of domain irreversible rotation, at low inductions. This is probably due to 90° irreversible rotation of the magnetization of domains. The domain wall structure of GO steels is very peculiar, with 180° domain walls parallel to the rolling direction [10,11], see Fig. 4. If a field is applied in the 90° direction (i.e., TD), which is transverse to RD, domain wall movement is not possible (Fig. 4). In this case, 90° irreversible rotation has to occur before any domain wall movement, and this implies in lower initial permeability for the 90° sample.

Without observing the domain structure of GO steels (Fig. 4), it would be impossible to explain the behaviour found in Figs. 2 and 3. The hard axis [111] is at arc cos (1/√3) or 54.74° from RD. The lower permeability is expected for the 45° sample, which is near the 54.74° angle. In fact, permeability at 1.5 T (μ_{15}) is lower for 45° sample, showing that dissipation by irreversible rotation is taking place in the higher inductions [12].

Figure 4. Domain structure of GO steels (schematic). The interface between the "stripes" represents 180° domain walls. The [100] axis is parallel to RD. Based on experimental pictures of refs. [10,11].

MBN signals are shown in Fig. 5. In the Figure 6, it is observed that the intensity of the MBN_{rms} (root mean square voltage of MBN signal) is lower for the 90° direction sample, and this indicates that irreversible rotation is probably contributing more significantly for the reversal of magnetization for this direction. MBN_{rms} is defined by the Eq. (3):

$$MBN_{rms} = \sqrt{\frac{\sum_{i}^{n}(V_i - V_m)^2}{n-1}} \quad (3)$$

Where Vi is the mean voltage, measured at a definite time; Vm the mean voltage of the entire signal and n the number of points of the MBN signal.

It can be seen from Fig.5 that the envelope of the MBN signal at 0° direction (RD) has larger amplitude than the envelopes from 45° and 90°. Barkhausen activity (avalanches) is more significant in this direction due to the large quantity of 180° Domains walls [6]. It is also observed that the MBN signals (Fig. 5) present a first and second maximum for RD, which was attributed by Hwang et al [13] to processes of DW nucleation and annihilation.

The MBN signal is mainly due to 180° domain wall moving [5,14]. Taking this into account, the MBN signals (frequency of 10 Hz) shown in Figure 5 should be compared with the initial permeability data (frequency of 5 mHz) shown in Figure 3. Steels with martensite (domain irreversible rotation is important reversal mechanism in this case, due to the nanocrystalline dimensions of the martensite) also present lower initial permeability, higher coercivity and lower RMS of MBN [15,16]. The same trend is observed here for the 90° direction of GO electrical steels, as it can be inferred from Figs. 3, 5 and 6.

Conclusions

When irreversible rotation of magnetization takes place, the MBN signal changes, decreasing the RMS of MBN (because the MBN is mainly due to domain wall motion, when irreversible rotation happens, the MBN activity is poor). It was found a connection between initial permeability and RMS of MBN. Lower initial permeability is related to a larger contribution of domain irreversible rotation in the magnetization reversal process. The MBN is a simple and rapid method to detect anisotropy in the steel sheets.

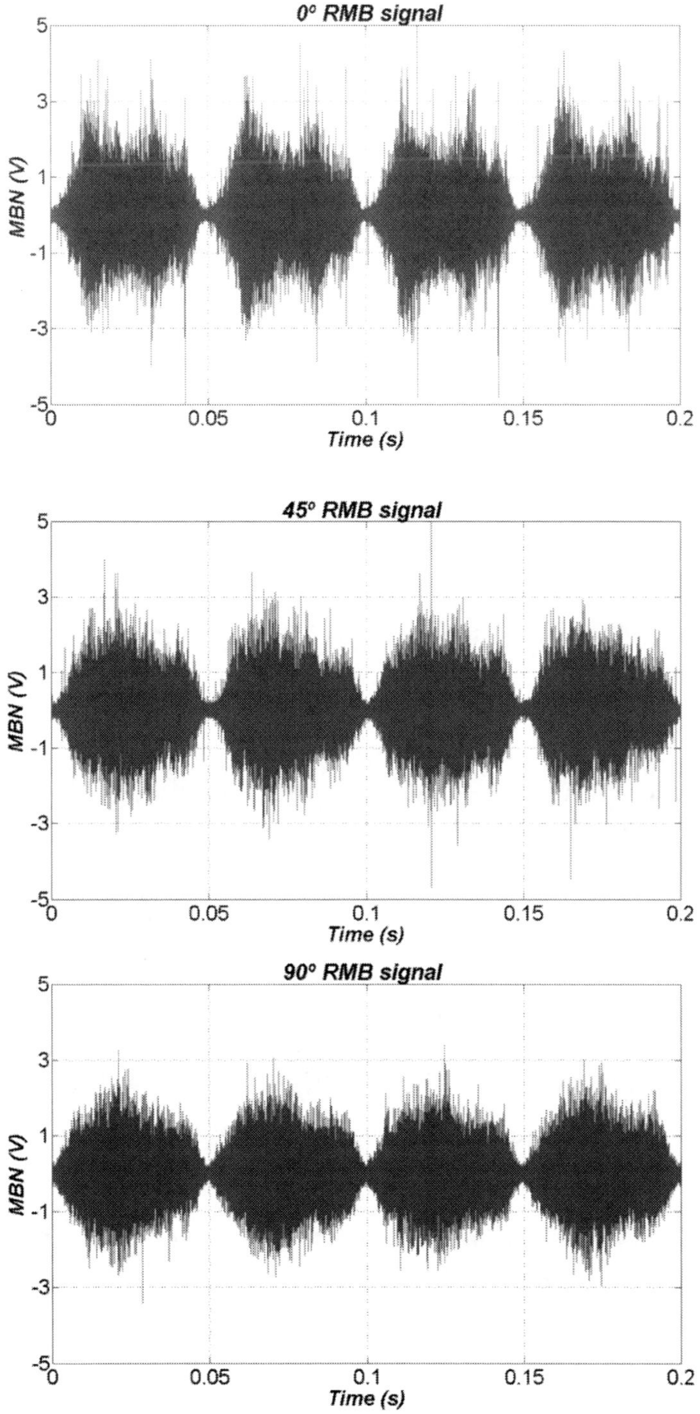

Figure 5. Magnetic Barkhausen Noise (MBN) signals for 0° (rolling direction), and for 45° and 90° from rolling direction. It can be noted that the signal is stronger for the rolling direction.

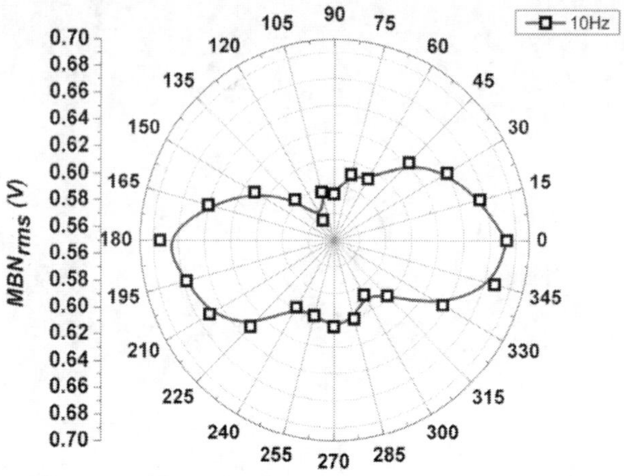

Figure 6 Polar plot of the RMS of MBN for the GO electrical steel. Angular step at 15° .

Acknowledgements

M F de Campos, FJG Landgraf and L R Padovese acknowledge financial support from CNPq. M F de Campos acknowledges CNPq contract 480891/2008-7. M. A. Campos acknowledge CAPES.

References

[1] Lo C C H, Jakubovics J P, Scruby C B 1997 *IEEE Trans. Magn.* **Mag-33** 4035
[2] Kikuchi H, Ara K, Kamada Y, Kobayashi S 2009 *IEEE Trans. Magn.* **Mag-45** 2744
[3] Hartmann K, Moses A J, Meydan T 2003 *J. Magn. Magn. Mater.* **254-255** 318.
[4] Anglada-Rivera J, Padovese L R, Capo-Sanchez J 2001 *J. Magn. Magn. Mater.* **231** 299
[5] Sablik M J, Augustyniak B, de Campos M F, Landgraf F J G 2008 *IEEE Trans. Magn.* **Mag-44** 3221.
[6] Krause T W, Szpunar J A, Atherton D L 2003 *IEEE Trans. Magn.* **Mag-39** 562
[7] Castro N A, de Campos M F, Landgraf F J G 2006 *J. Magn. Magn. Mater.* **304** e617
[8] Moses A J, Patel H V, Williams P I 2006 *J.of Electrical Eng.* **55** No 8/S 3.
[9] Chikazumi S 1964 *Physics of magnetism* John Wiley & Sons, New York.
[10] Ushigami Y, Mizokami M, Fujikura M, Kubota T, Fujii H, Murakami K 2003 *J. Magn. Magn. Mater.* **254–255** 307.
[11] Sato K, Ishida M, Hina E 1998 *Kawasaki Steel Technical Report* **39** 21.
[12] Landgraf F J G, de Campos M F, Leicht J 2008 *J. Magn. Magn. Mater.* **320** 2494.
[13] Hwang D G, Kim C G, Lee K H, Kim H C 1993 *J. Magn. Magn. Mater.* **125** 129.
[14] Alessandro B, Beatrice C, Bertotti G, Montorsi A 1990 *J. Appl. Phys.* **68** 2901.
[15] L F T Costa, F Girotto, R Baiotto, G Gerhardt, M F de Campos, F P Missell. Presented at the JEMS 2010 conference, Krakow, Poland. Journal of Physics Conference Series.
[16] M F de Campos, F A Franco, R Santos, F S da Silva, S B Ribeiro, J F C Lins, L R Padovese. Presented at the JEMS 2010 conference, Krakow, Poland. Journal of Physics Conference Series.

Magnetic and magneto-volume properties of Fe_xM_{100-x} solid solutions with M=Pd and Pt

E. Burzo[1)], P. Vlaic[2)] and D. Eniu[2)]

[1)] Faculty of Physics, Babes-Bolyai University, 400084 Cluj-Napoca, Romania
[2)] University of Medicine and Pharmacy "Iuliu Hatieganu", Physics and Biophysics Department Cluj-Napoca, Romania

E-mail: vlaic_pc@yahoo.com

Abstract: Band structure calculations were performed on Fe_xM_{100-x} solid solutions with M=Pd and Pt as well as on ordered phases present in the above systems. The equilibrium values of lattice parameters, magnetic moments, Curie temperatures, and spontaneous volume magnetostrictions were computed. A good agreement with experimental results has been shown. The magnetic properties, in particular Curie temperatures, were correlated with distances between iron atoms, as suggested by Néel-Slater curve.

1. Introduction

The Fe-Pd and Fe-Pt alloys systems have a rather complex phase diagrams [1]. Thus, in case of Fe-Pd one, at high temperatures, below solidus line, the presence of γ-Fe_xPd_{100-x} solid solutions, having fcc type structure, is shown in all the composition range. At room temperature, the α-Fe_xPd_{100-x} alloys, having bcc-type structure have been reported for $x \geq 70$ and fcc-type lattice for $x \leq 60$ [2]. In the compositions range $70 \leq x \leq 75$, four phases with different tetragonal distortions, following a Bain transformation path, from fcc to bcc-type lattice, have been reported [3]. When decreasing temperature, the alloys having compositions around x=50 and 25 at % Fe, exhibit a typical atomic disorder-order process. The disordered fcc-type alloys, with $40 \leq x \leq 50$, having low magnetic anisotropy, orders to tetragonal $L1_0$-type structure. The ordered $FePd_3$ phase crystallizes in $L1_2$-type structure having $Pm\bar{3}m$ space group. This phase can exist, at room temperature, in the composition range $14 \leq x \leq 38$ [1]. In the phase diagram of Fe-Pt system, the presence of three ordered compounds Fe_3Pt, FePt and $FePd_3$ was shown [1].

The magnetic properties of Fe_xPd_{100-x} alloys were experimentally studied [2,4-10]. In the composition range $20 \leq x \leq 70$, the resultant magnetizations decrease linearly with increasing Pd content [4]. The Curie temperatures, T_c, have a maximum at x=50. The T_c values of the atomic ordered alloys having $40 \leq x \leq 50$ at % Fe, decrease by $\cong$ 50 K, as compared to disordered phases [10]. Magnetic neutron scattering on Fe_xPd_{100-x} alloys with $10 \leq x \leq 15$ showed satellite diffuse peaks at $1 \pm \delta$, 0, 0.The wave vector of spin modulation varied with iron concentration. The oscillatory spin component homogeneously coexist with the ferromagnetic long-range order [9]. The effects of pressure on the Curie temperatures of γ-Fe_xPd_{100-x} with $66 \leq x \leq 72$ were also reported [8]. The T_c values decreased with pressure, while the changes of magnetizations, at 4.2 K, were small.

Band structure calculations were performed on some Pd-Fe alloys, particularly on those atomic ordered, having equatomic composition. There are only few studies on the electronic structure of Fe_xPd_{100-x} solid solutions [11,12]. The electronic structures and the magnetic properties of Fe_xPd_{100-x} alloys with $50 \leq x \leq 85$ were investigated in the framework of density functional theory using the full potential approximation [11]. The Korringa-Kohn-Rostoker Green's function method, with local density functional approximation, was used to obtain the

Published under licence by IOP Publishing Ltd

Journal of Physics: Conference Series **303** (2011) 012021 doi:10.1088/1742-6596/303/1/012021

effects of dilute substitutions of Pd in bcc Fe, on their magnetic properties.

Recently [13], we reported the band structure calculations performed on Fe-Pt alloys and compounds, where a review on previous experimental studies, including band structure calculations, were given. Consequently, in the following we focus mainly on the results obtained by band structure calculations on Fe-Pd system.

2. Computational method

The ground state electronic structures, total energies, magnetic properties and spontaneous volume magnetostriction, of α- and γ-Fe_xPd_{100-x} disordered alloys, as well as on FePd and $FePd_3$ ordered compounds, as in case of Fe-Pt system [13], have been studied by means of spin polarized and scalar relativistic tight-binding linear muffin tin orbital method (TB-LMTO), within atomic sphere approximation (ASA), together with coherent potential approximation (CPA) [14,15]. The LSDA was used for the exchange correlation potential of the electron gas, assuming the Vosko-Wilk-Nussair parameterization [16]. All band structure calculations have been performed by using a mesh of 24x24x24 k-points in the full Brillouin zone (BZ), resulting in 413 k-points in the irreductible wedge of BZ, that ensure the accuracy of the total energy better than 10^{-2} mRy. The magnetic disorder has been analysed within disorder local moment (DLM) formalism [17], by treating the binary alloys as pseudoternary $Fe^{\uparrow}_{x-c}Fe^{\downarrow}_{c}Pd_{100-x}$ ones, with (x-c) atoms having spin-up state and c atoms being in state with spin down. The DLM configuration with c=0 describes the ferromagnetic (FM) ground state, while that with c=x/2 has been considered as a paramagnetic (PM) state, which shows no resultant magnetic moment. Really, this state is antiferromagnetic (AF), in which the antiparallel alignment of magnetic moments are compensated, miming the PM state.

3. Results

The total energy calculations have been performed on γ-Fe_xPd_{100-x} with $0 \leq x \leq 80$ and α-Fe_xPd_{100-x} with $75 \leq x \leq 100$ alloys in both FM and NM states. The equilibrium values of the lattice parameters have been determined from a third-order polynomial interpolation of the total energy. The computed lattice parameters are in rather good agreement with experimental data [4,5,18], as shown in Fig.1a, for γ-Fe_xPd_{100-x} alloys. Although the same composition dependence is shown for α-Fe_xPd_{100-x} alloys, the experimental values are higher by ≅ 2 %. The difference between the two sets of data may be attributed to the LSDA approximation used in the present calculations and the well known underestimation of the equilibrium lattice constants in case of bulk iron alloys. Good agreement between the computed and experimental lattice parameters was shown in Fe_xPt_{100-x} alloys [13].

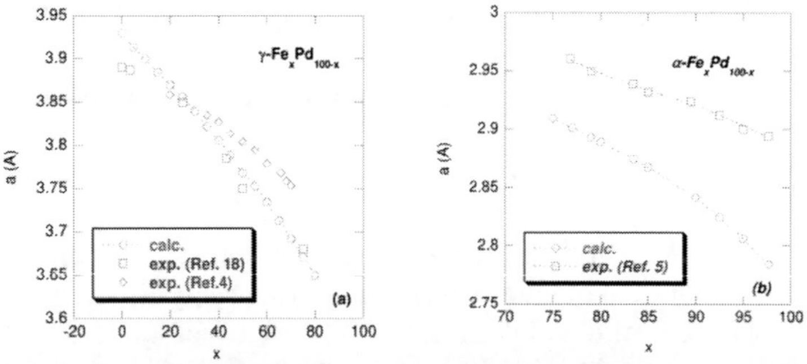

Fig.1 Computed and experimental [4,5,18] lattice constants determined in γ-Fe_xPd_{100-x} (a) and α-Fe_xPd_{100-x}(b) solid solutions.

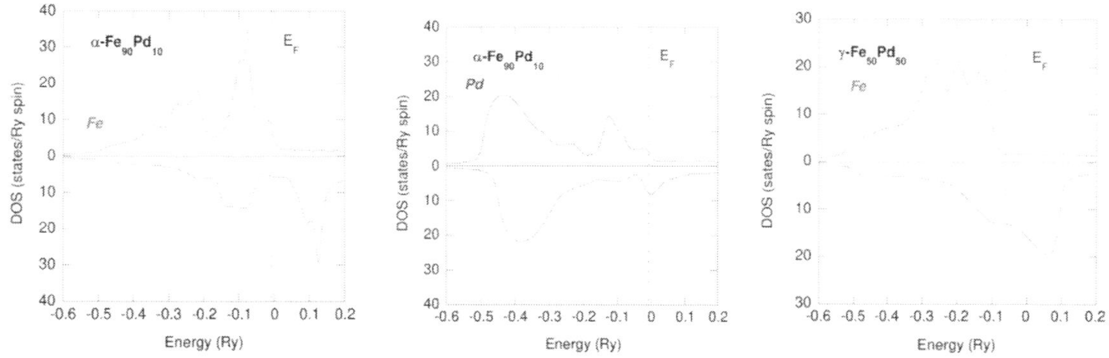

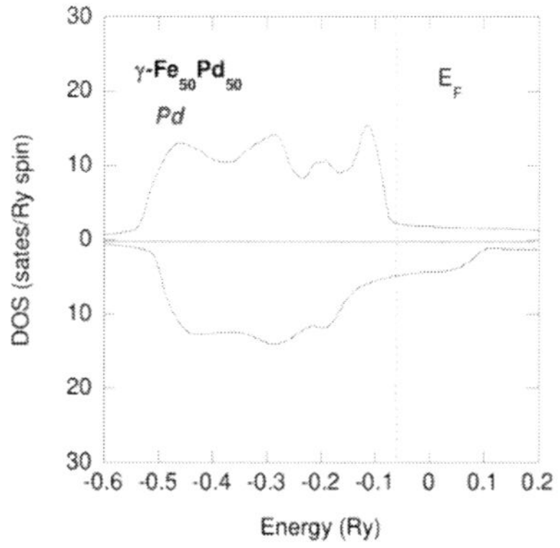

Fig.2 Densities of states for Fe and Pd in α-$Fe_{90}Pd_{10}$ and γ-$Fe_{50}Pd_{50}$ solid solutions

The atom resolved spin polarized density of states, DOS, for Fe and Pd in γ-$Fe_{50}Pd_{50}$ and α-$Fe_{90}Pd_{10}$ alloys are shown in Fig.2. The iron atoms in fcc solid solutions exhibit a strong ferromagnetism with almost completely filled majority spin sub-bands. A small contribution to DOS from majority spin sub-bands can be shown for iron in bcc type alloys. Similar behaviour was evidenced in Fe-Pt system. Polarizations of Pd4d and Pt5d bands, paralelly oriented to iron moments, are shown. These result from hybridization effects of Fe3d with Pd4d and Pt5d bands, respectively. The splitting of Pd4d and Pt5d bands can be correlated with their short range exchange interactions with the neighbouring Fe atoms. As a result, the Pd4d and Pt5d band polarizations increase with the iron content up to x=15 and then saturated to values $\cong 0.34\ \mu_B$/Pd atom and $0.30\mu_B$/Pt atom in fcc solid solutions. In α-Fe_xM_{100-x} alloys, the M=Pd and Pt saturated polarizations are somewhat higher, of $\cong 0.42\ \mu_B$/M atom. These values agree well with experimental data [19]. The iron magnetic moments increase as Pd or Pt content is higher-Fig.3. The computed composition dependences of the magnetizations in both Fe_xPd_{100-x} and Fe_xPt_{100-x} agree well with experimentally determined magnetic moments.

The exchange interactions in both Fe_xPd_{100-x} and Fe_xPt_{100-x} solid solutions were evaluated in the mean field approximation from the differences between the energies of the ferromagnetic and "paramagnetic" DLM states. From these values, Curie temperatures, T_c, were computed. In γ-Fe_xM_{100-x} solid solutions, with M=Pd,Pt these show a maximum located at x=60 mol %, while the experimentally determined values have the maxima at x=55 mol %-Fig.4. The same computational trend of the two sets of values are shown. The differences in the positions of the two maxima may be related with the single site approximation of CPA, used in the present calculations and also by neglecting the short-range order. A certain degree of chemical short-range order is generally present even in disordered alloys.

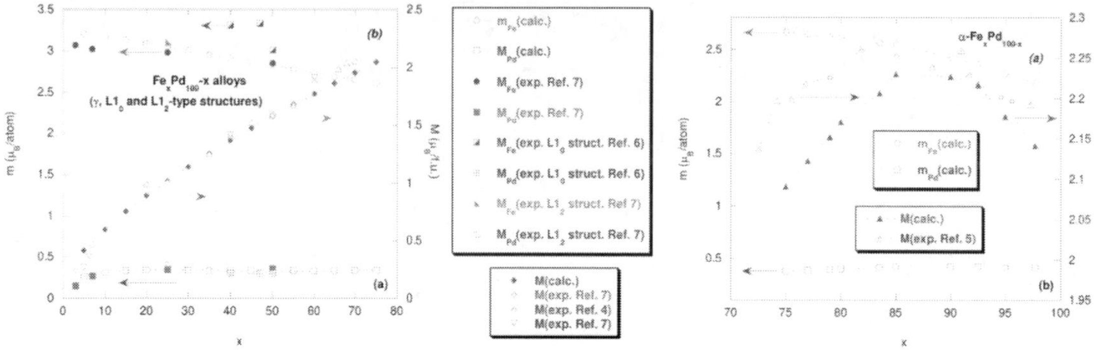

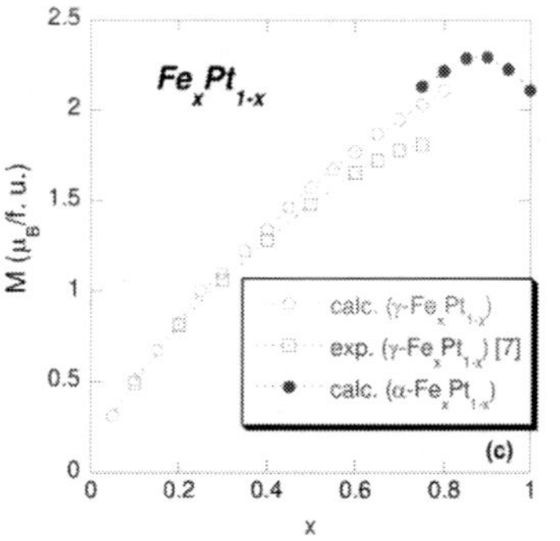

Fig. 3 Composition dependences of the iron and palladium moments as well as of resultant magnetizations in α-Fe_xPd_{100-x} (a), γ-Fe_xPd_{100-x}(b) and resultant magnetizations in Fe_xPt_{100-x} (c). The experimental data are also plotted [2,4-7].

The computed and experimentally determined Curie temperatures can be analysed starting from Néel–Slater curve [21]. When plotted T_c values, in Pd_xM_{100-x} alloys, as function of distances between iron atoms, d_{Fe}, the same type of dependence is shown as predicted by Néel for the evolution of exchange interactions. In a first order approximation, neglecting the contributions of Pd or Pt atoms to the exchange interactions, these will be proportional to T_c values. Thus, according to Néel [21], the Curie temperatures are expected to decrease from a maximum, situated at d_{Fe}=2.75 Å both by increasing or decreasing the distances between Fe atoms.

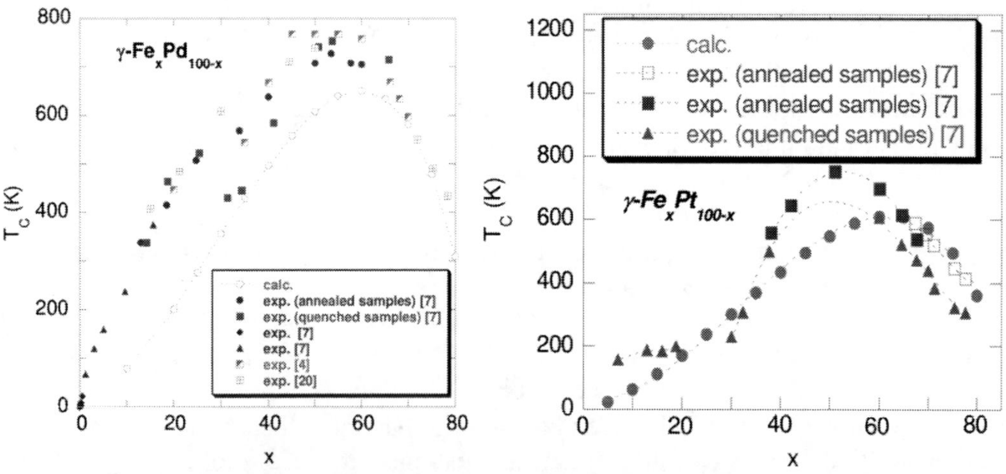

Fig.4 Composition dependences of the Curie temperatures in γ-Fe_xM_{100-x} M=Pd,Pt solid solutions. The experimental values are also plotted [4,7,20].

As the d_{Fe} values will be smaller than ≅ 2.50 Å, the presence of negative exchange interactions is expected. Such a behaviour has been observed in R_2Fe_{17} or $R_2Fe_{14}B$ compounds, where R is a rare earth or yttrium [22]. In the present systems the location of the maxima is at d_{Fe}=2.69 Å, little smaller value than predicted by Néel [21]. In this model, the diminution of Curie temperatures of Pd-Fe alloys as function of pressures, while keeping saturation magnetization at 4.2 K nearly constant, as experimentally observed [8], can be easily explained. In case of α-Fe_xPd_{100-x} with $10 \leq x \leq 15$, the distances between iron neighbouring atoms are situated between 2.52 and 2.539 Å, very close to d_{Fe} values where negative exchange interactions are expected to be present. Small local distortions of the lattices allow an antiferromagnetically coupling at some iron located in sites situated close to d_{Fe}=2.50 Å. The above, superposed on basic ferromagnetic ordering and in interaction with the ferromagnetically coupled atoms can change the antiferromagnetic coupling to an helicoidally arrangement, the modulation varying with iron content, the distances between iron atoms, respectively. Really such a behaviour was shown in Lu_2Fe_{17} compound [23].

The spontaneous volume magnetostrictions, ω, were determined from the relative volume difference between FM and DLM states [24]. This effect can be shown in $66 \leq x \leq 75$ composition range of Fe_xM_{100-x} alloys. In these alloys the paramagnetic DLM minimum energy with a smaller lattice constant (small magnetic moments), is situated slightly above the magnetic solution with a larger lattice constant. When increasing temperature, the local minimum with a small lattice constant can be populated by thermal fluctuations which counteract the usual thermal expansion due to lattice vibrations [25]. Néel [21] attributed the above anomaly to the magnetic energy variations, as function of the volume, the distances between magnetic atoms, respectively. Also showed that magnetostrictions at T_c, are proportional to the square of magnetizations. The computed ω values as function of the squares of the magnetic moments difference, between the ferromagnetic and DLM states, in Fe_xM_{100-x} alloys, are shown in Fig.5. Linear dependences are shown. The computed values are also close to those experimentally determined [6].

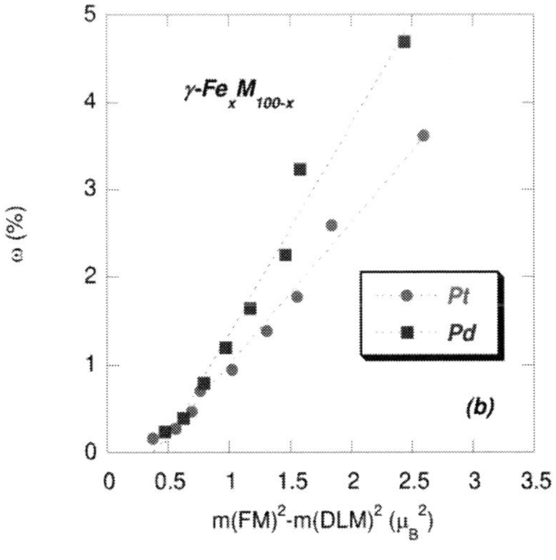

Fig. 5 Magnetostrictions of γ-Fe_xM_{100-x} solid solutions as function of the difference between the squares of the moments in ferromagnetic and paramagnetic DLM states.

The magnetic properties of Fe_3Pt, FePt and Fe_3Pt [13], as well as of FePd and Fe_3Pd ordered compounds were also computed. These described well the experimental values.

4. Conclusions

The total energy calculations on Fe_xM_{100-x} with M=Pd, Pt were performed. The determined equilibrium lattice constants describe rather well the experimental data. The composition dependences of the Fe magnetic moments and Pd4d and Pt5d band polarizations were determined. The computed magnetizations per formula unit agree well with experimental values. The composition dependences of the Curie temperatures were analysed according to Néel-Slater curve. In agreement with this curve, the effects of pressure on T_c's and magnetizations values were explained as well as the presence of an oscillatory spin component, which coexist with ferromagnetic long range order, in alloys with $10 \leq x \leq 15$. The volume magnetostrictions for Fe_xM_{100-x} with M=Pd and Pt scale linearly as function of the difference between the square of magnetizations in ferromagnetic and "paramagnetic" DLM states.

Realized under contract PNII APLIMAG 71-060/2007.

References

[1] Massalsky T B 1992 Binary Alloys Phase Diagrams, ASM International Materials Park, Ohio
[2] Yabe H, O'Handley RC and Kuji T 2007 J. Magn. Magn. Mat. 310 2500
[3] Sugiyama M, Oshima R and Fujita F E 1984 Trans. Jpn. Inst. Met. 25, 585
[4] Matsui M and Adachi K 1989 Physica B 161, 53
[5] Birsan M, Fulz B and Anthony L 1997 Phys. Rev. B 55 11502
[6] Lyubina J, Gutfleish O and Isnard O 2009 J.Appl. Phys. 105 07A717
[7] Adachi D et al 1992 Landolt Börnstein Handbuch, Springer Verlag, vol. 19a
[8] Matsushita M, Nishimura T, Endo S, Ishizuka M, Kindo K and Ono F 2002 J. Phys.: Cond. Matter 14, 10753
[9] Tsunoda Y and Abe R 1997 Physica B 237-238 458
[10] Wang L, Fan Z, Roy AG and Laughlin DE 2004 J. Appl. Phys. 95 7483
[11] Drittler B, Stefanou N, Blügel S, Zeller R and Dederichs PH 1989 Phys. Rev. B 40, 8203
[12] Opahle I, Koepernik K, Nitzsche U and Richter M 2009 Appl. Phys. Lett. 94 072508
[13] Vlaic P and Burzo E 2010 J. Opt. Adv. Mat. 12 1114; 2009 AIP Conf. Proc. 1203 3241
[14] Anderson OK and Jepsen O 1984 Phys. Rev. Letters 53 2571
[15] Turek I, Drchal V, Kudrnovský J, Šob M and Weinberger P 1997 Electronic Structures of Disordered Alloys, Surfaces and Interfaces, Kluwer Academic, Boston
[16] Vosko SH, Wilk L and Nusair M 1980 Can. J. Phys. 58 1200
[17] Pindor AJ, Stauton JB, Stocks GM and Winter H 1983 J.Phys. F: Metal Phys. 13 979
[18] Villars P and Calvert LD 1992 Pearsosn's Handbook of Crystallographic Data for Intermetallic Phases, 2nd ed, ASM International, Materials Park OH
[19] Collins MF and Low GC 1965 Proc. Phys. Soc. London 86 535
[20] Kussmann A and Jessen K 1962 J. Phys. Soc. Jpn. 17 136
[21] Néel L 1937 Ann. Phys. Phys. (Paris) 8, 237
[22] Burzo E 1998 Rep. Progr. Phys .61 1199
[23] Givord D and Lemaire R 1974 IEEE Trans. Magn. MAG-10 109
[24] Khmelevskyi B, Ruban AV, Kakehashi Y, Mohn P and Johansson B 2005 Phys. Rev. B 72 064510
[25]. Hayn R and Drchal V 1998 Phys. Rev. B 58, 4341

Journal of Physics: Conference Series **303** (2011) 012022 doi:10.1088/1742-6596/303/1/012022

Neutron diffraction study of the magnetic order in $NdMn_2Ge_{1.6}Si_{0.4}$

J L Wang[1,2], S J Campbell[1], J M Cadogan[3], A J Studer[2], R Zeng[4] and S X Dou[4]

[1]School of Physical, Environmental and Mathematical Sciences, University of New South Wales, Australian Defence Force Academy, Canberra ACT 2600, Australia
[2] Bragg Institute, ANSTO, PMB 1, Menai, NSW 2234, Australia
[3] Department of Physics and Astronomy, University of Manitoba, Winnipeg, MB, R3T 2N2, Canada
[4] ISEM, University of Wollongong, NSW 2522 Australia

E-mail: `cadogan@physics.umanitoba.ca`

Abstract. Here we report a detailed investigation of $NdMn_2Ge_{1.6}Si_{0.4}$; this forms part of our investigation of the magnetic order across the $NdMn_2Ge_{2-x}Si_x$ (x = 0–2.0) series by magnetometry, x-ray diffraction and neutron diffraction over the temperature range 6–465 K. On decreasing the temperature from 465 K, $NdMn_2Ge_{1.6}Si_{0.4}$ exhibits four magnetic transitions: (i) from paramagnetism to intralayer antiferromagnetism AFl at $T_N^{Intra} \sim 430$ K; (ii) AFl to canted ferromagnetism Fmc at $T_C^{Inter} \sim 330$ K; (iii) Fmc to conical magnetic ordering of the Mn sublattice Fmi at $T_{cc} \sim 178$ K and (iv) Fmi(Mn) to Fmi(Mn)+F(Nd) at $T_C^{Nd} \sim 72$ K.

1. Introduction

The tetragonal intermetallic compounds RT_2X_2 (R = rare-earth; T = transition metal and X = Si or Ge) have been studied extensively and present a wide range of magnetic behaviours. Of particular interest are those compounds with T = Mn which is the only T element to order magnetically, generally well above room temperature. The RT_2X_2 compounds form in the $ThCr_2Si_2$ structure (space group I4/mmm, #139) and the naturally layered nature of this crystal structure leads to an extreme sensitivity of the magnetic order to the interatomic distances. In particular, there is a critical intralayer Mn–Mn distance of ~2.87 Å above which the magnetic order is ferromagnetic (FM) and antiferromagnetic (AF) below. At least nine magnetic structures have been identified in the T = Mn compounds [1], encompassing collinear, canted, conical and incommensurate FM and AF order. Throughout this paper we will use the notation introduced by Venturini et al. [1] to describe our magnetic structures.

From a more practical viewpoint, the discovery of a giant magnetocaloric effect (MCE) near room temperature in the layered rare earth compound $Gd_5Si_2Ge_2$ [2] has led to increased efforts in the search for materials with a large MCE. $NdMn_2Ge_2$ and related compounds offer interesting prospects for enhanced magnetocaloric behaviour because the naturally layered crystal structure allows control of the intrinsic magnetism via inter- and intra-planar separations of the Mn atoms with the Nd atoms contributing to larger magnetization due to ferromagnetic coupling between the Mn and Nd sublattices below the Nd ordering temperature T_C^{Nd}.

We have carried out an extensive neutron powder diffraction study of the entire $NdMn_2(Ge, Si)_2$ system with the aim of determining the magnetic structures across the phase diagram from $NdMn_2Ge_2$ to $NdMn_2Si_2$. The two end-member compounds have previously been studied by Welter et al. [3, 4]. A tentative magnetic phase diagram for the $NdMn_2(Ge, Si)_2$ series has been proposed by Wang et al. [5, 6] on the basis of magnetometry measurements. In this paper, we present a detailed determination of the intrinsic magnetism of $NdMn_2Ge_{1.6}Si_{0.4}$. The magnetism of this Si-doped compound is found to be similar to that of $NdMn_2Ge_2$ [4].

2. Experimental Methods

The $NdMn_2Ge_{1.6}Si_{0.4}$ sample was prepared by arc melting stoichiometric amounts of the pure elements (Nd 99.9 wt.%; Mn, Si, Ge 99.99 wt.%). The sample was turned and remelted several times to ensure homogeneity. The alloyed button was then sealed under vacuum in a quartz tube, annealed for 1 week at 900 °C and quenched in water. Cu-Kα x-ray diffraction confirmed the majority phase to be the intended tetragonal $ThCr_2Si_2$–like $NdMn_2Ge_{1.6}Si_{0.4}$ phase. Magnetic characterization was carried out on a Quantum Design PPMS susceptometer/magnetometer operated down to 5 K.

Neutron diffraction experiments were carried out on the *Wombat* high-intensity powder diffractometer at the OPAL reactor in Sydney, Australia [7]. The neutron wavelength was 2.41726(2) Å. Temperatures down to 6 K were obtained using a closed-cycle refrigerator. All refinements of the x-ray and neutron diffraction patterns employed the FullProf/WinPlotr package [8, 9].

3. Results and Discussion

3.1. Basic characterisation and magnetometry

The tetragonal $ThCr_2Si_2$-type structure of $NdMn_2Ge_{1.6}Si_{0.4}$ has Nd atoms in the 2a site (0, 0, 0) while the Mn atoms occupy the 4d site (0, $\frac{1}{2}$, $\frac{1}{4}$) and the Si, Ge atoms occupy the 4e site (0, 0, z_e), with $z_e = 0.392$ at room temperature. The magnetization measurements (Figure 1) show that the temperature dependence of the magnetic behaviour of $NdMn_2Ge_{1.6}Si_{0.4}$ comprises five distinct regions:

- $T > T_N^{Intra} \sim 430$ K
- $T_C^{Inter} \sim 330$ K $< T < T_N^{Intra} \sim 430$ K
- $T_{cc} \sim 178$ K $< T < T_C^{Inter} \sim 330$ K
- $T_C^{Nd} \sim 72$ K $< T < T_{cc} \sim 178$ K
- $T < T_C^{Nd} \sim 72$ K

Magnetization measurements at 10 K in fields up to 5 T yield a saturation magnetization of 70 $JT^{-1}kg^{-1}$, indicating substantial FM order. To determine the magnetic order in $NdMn_2Ge_{1.6}Si_{0.4}$ we have carried out neutron powder diffraction experiments in each of these temperature regions.

4. Neutron powder diffraction

4.1. $T > 430$ K

Figure 2 shows examples of the refined neutron powder diffraction patterns of $NdMn_2Ge_{1.6}Si_{0.4}$ obtained over the temperature range 6 K to 465 K. An expanded view of the neutron diffraction pattern of $NdMn_2Ge_{1.6}Si_{0.4}$ at 6 K from $2\theta = 20°$ to $2\theta = 60°$ is shown in Figure 3. In Table 1

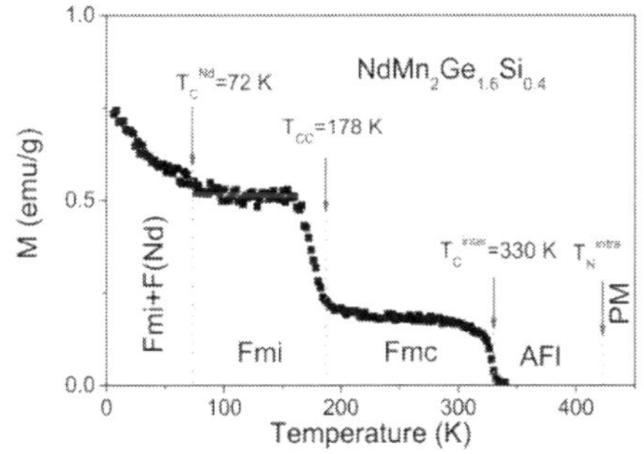

Figure 1. Temperature dependence of the magnetization of $NdMn_2Ge_{1.6}Si_{0.4}$ obtained in a field of 10 Oe. $T_N^{Intra} \sim 430$ K is derived from the temperature dependence of the intensity of the (101) neutron diffraction peak. The easy magnetization is along the c-axis for $T > T_{cc}$, while the direction of easy magnetization turns into the ab-plane with $T < T_{cc}$.

we present a summary of the crystallographic and magnetic parameters determined from our refinements and in Figure 4 we show the temperature dependence of three key diffraction peaks (002), (101) and (112). As discussed below, these reflections allow us to follow the thermal evolution of the intrinsic magnetic behaviour of $NdMn_2Ge_{1.6}Si_{0.4}$. The $NdMn_2Ge_{1.6}Si_{0.4}$ phase is paramagnetic at 465 K with the refined diffraction pattern purely nuclear, as expected.

4.2. 330 K < T < 430 K

Upon cooling, the magnetic contributions to the $NdMn_2Ge_{1.6}Si_{0.4}$ diffraction patterns arising from the AF ordering of the Mn sublattice appear at $T_N^{Intra} \sim 430$ K. In figure 2 we show the refinement to the neutron diffraction pattern obtained at 405 K. At 405 K the magnetic contributions occur only at nuclear peak positions ($\mathbf{k} = [0\ 0\ 0]$). The effect is most clearly seen in the increase in the (101) peak at $2\theta = 38°$. We find no evidence for additional magnetic-only peaks from the $NdMn_2Ge_{1.6}Si_{0.4}$ phase at 405 K. The Mn moments order in the tetragonal basal plane and form an AFl magnetic structure.

4.3. 178 K < T < 330 K

The magnetization measurements (Figure 1) show a sudden increase at 330 K corresponding to a rearrangement of the Mn order, resulting in a ferromagnetic component. In the neutron diffraction patterns, this rearrangement manifests as an increase in the (112) peak at $2\theta = 57°$, together with the continuing increase in the (101) intensity. The presence of both (101) and (112) magnetic intensities shows that the Mn order comprises both AF and FM contributions. The 4d special position of the Mn moments is effectively a 'C' symmetry position so the Mn nuclear/FM structure factor for the (101) peak is zero and non-zero for (112). Our refinement of the patterns in the temperature range 178 K < T < 330 K indicates that the Mn sublattice magnetic order is Fmc, i.e. a canted ferromagnet with the FM order along the tetragonal c-axis. The pattern at 300 K is shown in Figure 2 as a typical example of the behaviour in this region.

4.4. 72 K < T < 178 K

The magnetization measurements show a second dramatic increase at 178 K; this corresponds to a further rearrangement of the Mn order, resulting in a stronger FM component. In the neutron

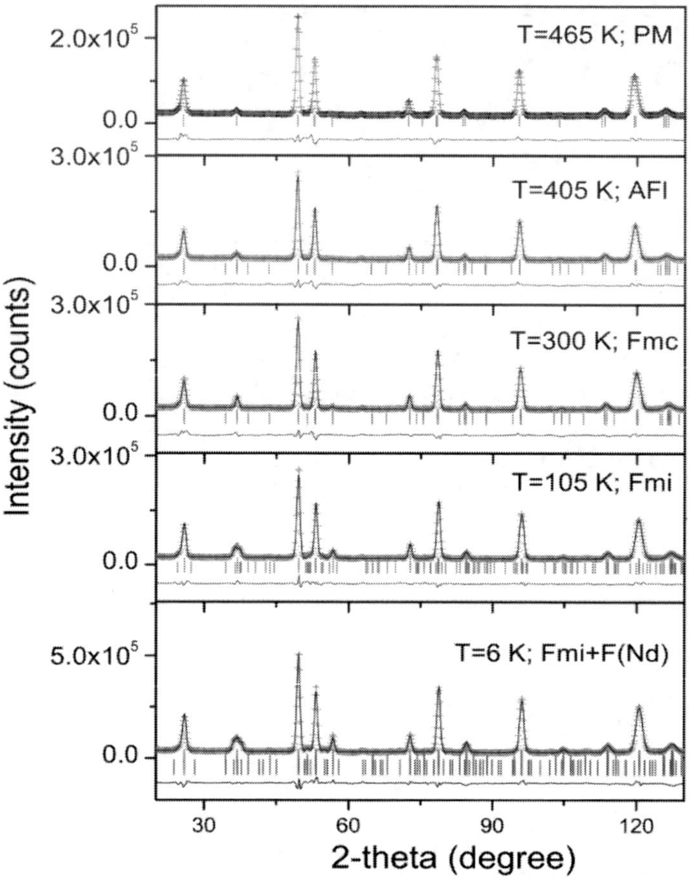

Figure 2. Neutron diffraction patterns of $NdMn_2Ge_{1.6}Si_{0.4}$ at the temperatures indicated (neutron wavelength $\lambda = 2.41726(2)$Å). Refinements to the magnetic structures indicated are given by the full lines and peak markers in the usual way.

diffraction patterns this rearrangement manifests as a sudden increase in the intensities of the (101) and (002) peaks (see Figure 4). The increase in the nuclear-allowed (002) peak intensity shows that the FM component of the Mn order has reoriented from the c-axis to the basal plane. Our refinement of the patterns in the temperature range 72 K $<$ T $<$ 178 K indicates that the Mn sublattice magnetic order is Fmi, i.e. a conical ferromagnet with the FM order perpendicular to the tetragonal c-axis. The incommensurate propagation vector is [0 0 0.111(1)] at 105 K.

4.5. T < 72 K

Below 72 K, the Nd sublattice orders magnetically in the tetragonal basal plane while the Mn order remains Fmi, albeit with a larger propagation vector component [0 0 0.159(1)] at 6 K. The effect of the FM ordering of the Nd sublattice is seen in the magnetization curve as a pronounced upturn in magnetization (see Figure 1). The change in the propagation vector describing the

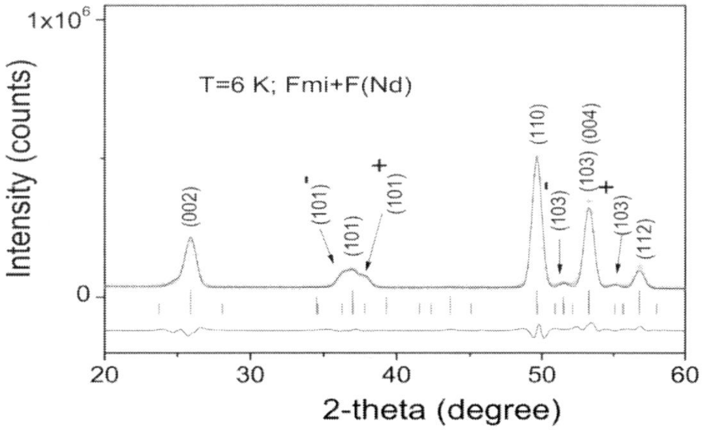

Figure 3. Expanded view of the neutron diffraction pattern of $NdMn_2Ge_{1.6}Si_{0.4}$ at 6 K.

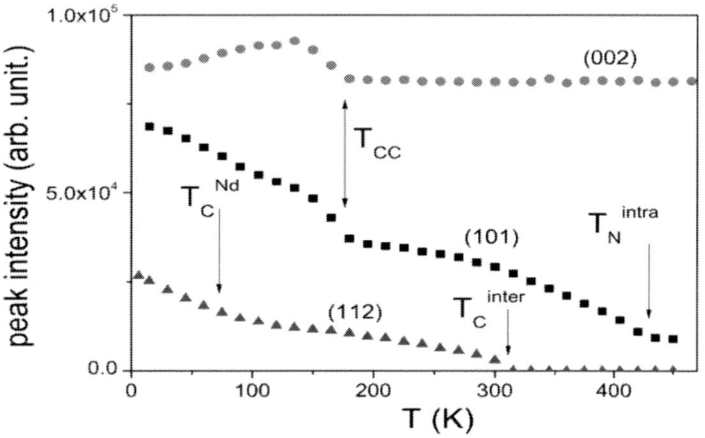

Figure 4. Temperature dependence of the (101), (112) and (002) neutron diffraction peak intensities of $NdMn_2Ge_{1.6}Si_{0.4}$.

ordering of the Mn sublattice is clearly seen in the broadening of the (101) peak as the **k**-vector satellites progressively separate from the central peak. The Nd and Mn magnetic moments at 6 K are 2.8(2) μ_B and 3.7(1) μ_B, respectively.

5. Conclusions

We have determined the various magnetic structures of $NdMn_2Ge_{1.6}Si_{0.4}$ by neutron powder diffraction over the temperature range 6–465 K. The magnetic ordering temperature is $T_N^{Intra} \sim 430$ K, involving AF order of the Mn sublattice. Upon cooling, the Mn order reorients at $T_C^{Inter} \sim 330$ K and $T_{cc} \sim 178$ K. At $T_C^{Nd} \sim 72$ K, the Nd sublattice orders ferromagnetically.

Table 1. Structural and magnetic parameters derived from Rietveld refinements of the neutron diffraction patterns for $NdMn_2Ge_{1.6}Si_{0.4}$. The errors are shown for T = 6 K as a typical example.

T (K)	6 K	15 K	105 K	300 K	405 K	465 K
Magnetic state	Fmi+F(Nd)	Fmi+F(Nd)	Fmi	Fmc	AFl	PM
a (Å)	4.071(1)	4.071	4.072	4.078	4.084	4.085
c (Å)	10.798(2)	10.798	10.796	10.829	10.863	10.879
z_e	0.384(1)	0.383	0.385	0.392	0.395	0.397
μ_{ab} (μ_B)	2.95(2)	2.94	2.65	1.42	0.98	–
μ_c (μ_B)	2.27(2)	2.29	2.31	0.54	–	–
Canting angle	38(2)	38	41	62	90	–
μ_{total} (μ_B)	3.7(1)	3.7	3.5	1.5	1.0	–
k_z	0.159(3)	0.157	0.111	–	–	–
μ_{Nd} (μ_B)	2.8(2)	2.8	–	–	–	–
R_{wp}	5.29	5.56	6.08	6.59	7.99	9.22
R_p	3.84	4.13	4.18	4.61	5.80	6.82

The behaviour of $NdMn_2Ge_{1.6}Si_{0.4}$ is similar to that of $NdMn_2Ge_2$ [4]. Replacement of Ge with 20% of Si leads to a reduction in the transition temperatures and the k_z propagation vector component compared with the behaviour shown by $NdMn_2Ge_2$.

Acknowledgments

Financial support for this work was provided by the Australian Research Council and the Australian Institute for Nuclear Science and Engineering. This work is supported by a joint agreement between ANSTO and UNSW. Finally, JMC acknowledges support from the Canada Research Chairs programme.

References

[1] Venturini G, Welter R, Ressouche E and Malaman B 1995 *J. Magn. Magn. Mater.* **150** 197.
[2] Gschneidner Jr. K A, Pecharsky V K, Pecharsky A O, Ivtchenko V V and Levin E M 2000 *J. Alloys and Compounds* **303–4** 214.
[3] Welter R, Venturini G, Fruchart D and Malaman B 1993 *J. Alloys and Compounds* **191** 263.
[4] Welter R, Venturini G, Ressouche E and Malaman B 1995 *J. Alloys and Compounds* **218** 204.
[5] Wang Y G, Yang F, Chen C, Tang N and Wang Q 1997 *J. Alloys and Compounds* **257** 19.
[6] Wang Y G, Yang F, Chen C, Tang N and Wang Q 1997 *J. Appl. Phys.* **81** 7909.
[7] Studer A J, Hagen M E and Noakes T J 2006 *Physica B* **385–6** 1013.
[8] Rodríguez-Carvajal J 1993 *Physica* B **192** 55.
[9] Roisnel T and Rodríguez-Carvajal J 2001 *Mater. Sci. Forum* **378–81** 118.

Magnetostriction and transformation of crystal structure of intermetallic compound $NdCo_2$

S A Nikitin[1], D Yu Karpenkov[2], I S Tereshina[3], Yu G Pastushenkov[2], T Palewski[4], J Cwik[4]

[1] Department of Physics, Lomonosov Moscow State University, Leninskie Gory, 19992, Moscow, Russia

[2] Physicotechnical Department, Tver State University,Sadovy lane,35,170002,Tver, Russia

[3] Baikov Institute of Metallurgy and Material Sciences RAS, 119991 Moscow, Russia

[4] International Laboratory of High Magnetic Fields and Low Temperatures, Gajowicka 95, 53- 421, Wroclaw,Poland

E-mail: Karpenkov_d_y@mail.ru

Abstract. Ferromagnetic compound $NdCo_2$ has complex magnetic transformation, which are accompanied by changes of the crystal lattice. The purpose of this paper was the investigation of influence of these transformations on the field and temperature dependence of magnetostriction. It was found that near the Curie temperature Tc = 98K, where the transition from cubic to tetragonal phase is observed, there is a maximum in the temperature dependence of the transverse magnetostriction $\lambda_\perp(T)$ in the field B=4T. In this field $\lambda_\perp$ reaches high values of $\lambda_\perp \sim 200 \cdot 10^{-6}$ at temperatures higher than T_c. This shows that in this field, the tetragonal distortion and band magnetism of 3d-subsystem remain even at T>Tc. At T<42K $\lambda_\perp$ decreases in field B=0.5 T due to phase transition from tetragonal phase to orthorhombic. The curve $\lambda_\perp$(H) at T = 10 K anomalies are observed. They are associated with the fact that under the action of the external field there are consistent transitions from orthorhombic phase to tetragonal and then to cubic. Thus, it was found that the compound $NdCo_2$ crystal structure is reconstructed with magnetic field.

1. Introduction

These In the past 15 years the attention of many researchers [1-4] was attracted by magnetically ordered materials in which there is coexistence of magnetic and structural phase transitions. They usually have gigantic values of the magnetocaloric effect, magnetostriction and magnetoresistance [2-3]. Among these materials there are RCo2 compounds, in which there is a sharp dependence the magnetic phase transitions from the lattice parameters [5-9].

Transformation of the crystal and magnetic structure as a result of magnetic phase transitions in $NdCo_2$ on polycrystalline samples was studied in [1], with methods of X-ray and neutron diffraction.

Temperature dependence of the magnetostriction constants λ_{100} for RCo_2 was calculated by measuring the lattice constant in [1,8,10] and it was shown that λ_{100} for $NdCo_2$ reaches very high values of $\lambda_{100}=-3.2 \cdot 10^{-3}$. However there is no data about the temperature dependence of linear and volume magnetostriction induced by magnetic field in $NdCo_2$. That hinders the development of physical models, taking into account the relationship between structural transitions and the strong

magnetoelastic interactions. In ferrimagnetic $NdCo_2$ compound complex magnetic transformations occur, which is accompanied by the transformation of crystal structure [1, 10].
The purpose of this investigation was to study the influence of transformations of the crystal lattice on the field and temperature dependence of magnetostriction and to determine magnetoelastic contributions of magnetostrictive deformation in $NdCo_2$.
In this work, longitudinal $\lambda_{\|}$ and transverse $\lambda\perp$ magnetostriction induced by the magnetic field up to 10T in the temperature range 4.2 - 135K in $NdCo_2$ were measured. Measurements were carried out with resistance strain gauges.

2. Experimental details

As cast $NdCo_2$ samples were prepared with method of high-frequency induction melting in an atmosphere of argon. The samples have plate shape and the linear sizes 4x5x2 mm. The crystalline structure and phase composition of the samples were checked by X-ray diffraction. The research of microstructure of as cast samples was performed by a method of an optical microscopy (magnification in 1000). The measurements of both longitudinal $\lambda_{\|}$ and transverse $\lambda\perp$ magnetostriction induced by the magnetic field up to 10T in the temperature range 4.2 - 130K in $NdCo_2$ were carried out with resistance strain gauges. The Vishay Micromeasurements SR-4 strain gauges were pasted on a surface of the sample.

3. Results and discussion

Figures 1,2 and 3 shows the temperature dependence of $\lambda_{\|}$, $\lambda\perp$ and ω of $NdCo_2$ in magnetic fields up to B = 4-10T. As shown in figures the value of ω reaches high values of $\sim 600\cdot 10^{-6}$ near the Curie temperature Tc = 98K, where the transition from the paramagnetic to the ferromagnetic state is observed. It is accompanied by a transition from cubic to tetragonal phase according to [1].

a) b)

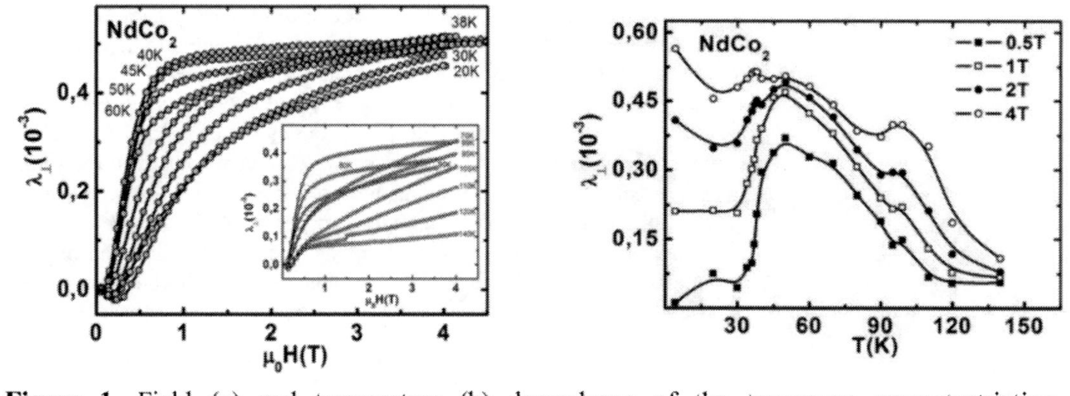

Figure 1. Field (a) and temperature (b) dependence of the transverse magnetostriction of polycrystalline samples $NdCo_2$

It should be noted the presence of high values of magnetostriction $\lambda_{\|}$, $\lambda\perp$ and ω in the paramagnetic region at T>Tc. Thus at temperatures 20K above Tc (Fig. 1) magnetostriction ω reaches $200\cdot 10^{-6}$, that can be explained with the influence of magnetic field on the Co magnetic sublattice.
While further decreasing of temperature below Tc and the achievement of a phase transition from tetragonal to orthorhombic phase there is an even higher positive maximum $\omega=750\cdot 10^{-6}$ at T_{SR}=42K (Fig.3).
Isotherms of the magnetostriction $\lambda_{\|}(H)$ and $\lambda\perp(H)$ (Figures 1a and 2a) measured in the temperature range T_{SR}-Tc, show a change of magnetostriction with increasing field from zero up to B = 1T. This contribution to the $\lambda_{\|}$, $\lambda\perp$ dues to the orientation of the magnetostrictive deformations in magnetic

Journal of Physics: Conference Series **303** (2011) 012023 doi:10.1088/1742-6596/303/1/012023

field. As a result the tetragonal crystal structure transforms into a cubic, it was established with X-ray and neutron diffraction studies [1].

a) b)

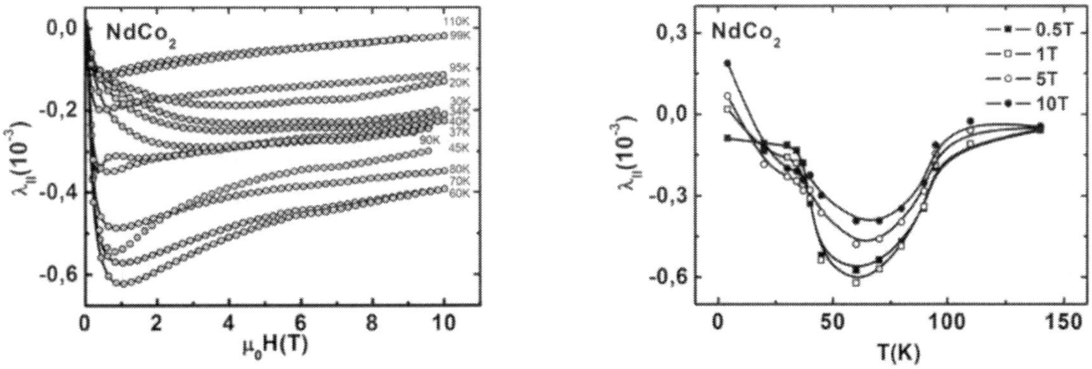

Figure 2. Field (a) and temperature (b) dependence of the longitudinal magnetostriction of polycrystalline samples $NdCo_2$

The temperature range below T_{SR} = 42K attract attention. At $T<T_{SR}$ transverse magnetostriction decreases sharply in a field B=0,5T (Fig. 1b), due to phase transitions from tetragonal phase to orthorhombic. With increasing of the field jump $\lambda_\perp$ at T_{SR} smooth, and in the field of 4T curve $\lambda_\perp(T)$ reveals only a small anomaly at $T=T_{SR}$. This can be explained by the fact that the orthorhombic structure is suppressed by the field and converted to cubic.

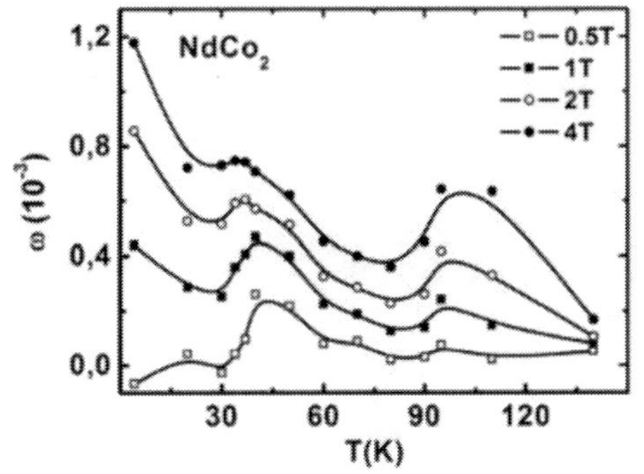

Figure 3. Temperature dependence of the volume magnetostriction in polycrystalline samples $NdCo_2$.

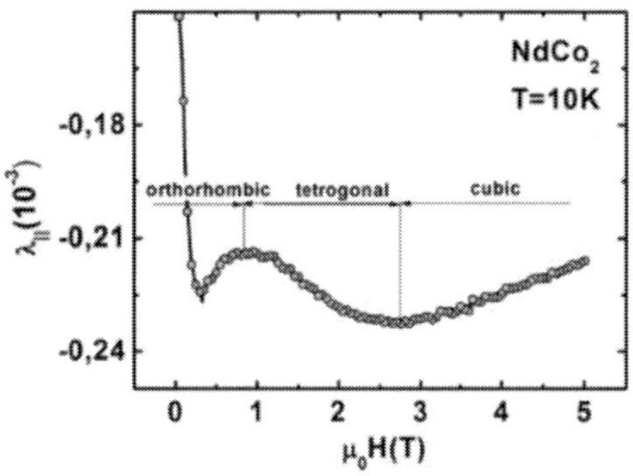

Figure 4. The field dependence of $\lambda_{||}$ measured at T=10K in polycrystalline samples $NdCo_2$.

The details of this transformation are observed on the curve of the field dependence of $\lambda_{||}$ measured at 10K(fig.4). Based on the data obtained with the field dependence of neutron reflections [1], it can be argued that the maximum in the curve of the field dependence $\lambda_{||}(H)$ at B = 0.8T corresponds to the transition from orthorhombic phase to tetragonal, and the minimum at curve $\lambda_{||}(H)$ at B = 2.8T corresponds to the transition from tetragonal to cubic phase. Hence, our experimental data of temperature and field dependence of magnetostriction $NdCo_2$ and neutron diffraction data [1] indicate the unusual for RCo_2 compounds phenomenon - the restructuring of the crystal structure under the influence of a magnetic field.

In figure 5 measurements of thermal expansion of polycrystalline sample $NdCo_2$ are presented. At T<98K anomalies of the thermal expansion are observed, caused by imposing of spontaneous volume magnetostriction on phonon part of thermal expansion. In magnetic field the curve of temperature expansion transforme because of the appearance of additional magnetostriction deformations. It is necessary to note, anomaly of thermal expansion reaches high values $\Delta l/l=1000\cdot10^{-6}$.

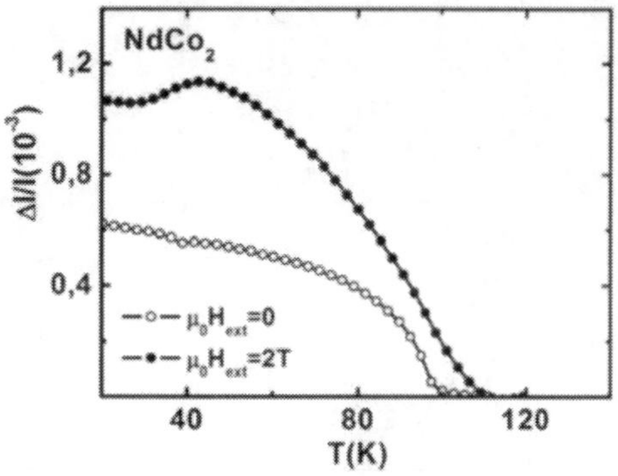

Figure 5. Thermal expansion of polycrystalline samples $NdCo_2$.

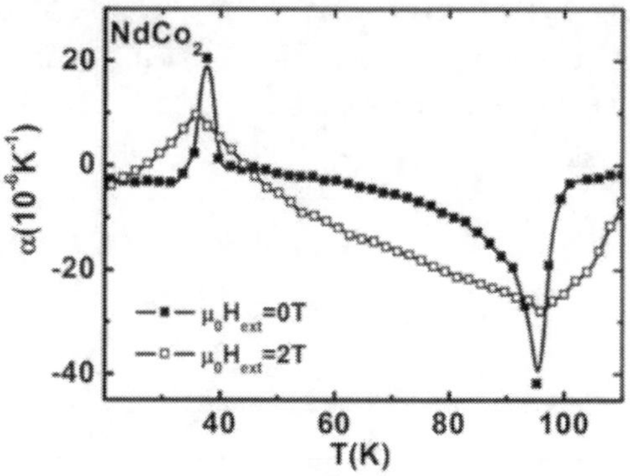

Figure 6. Temperature dependence of the coefficient of thermal expansion of polycrystalline samples $NdCo_2$.

On the curve of temperature dependence coefficient of thermal expansion α(T) (fig. 6) at B=0T the extrema are observed, connected with magneto-structural phase transitions at T=42K and T=98K. In the field B=2T these extrema are smoothed because of arisen additional magnetostriction deformations.

4. Conclusion

$NdCo_2$ is a ferromagnet, which combines the band magnetism of the Co sublattice and localized magnetism of the Nd sublattice.

Below the Curie temperature Tc=98K there are giant values of linear and volume magnetostriction, which sharply increases with decreasing temperature.

The temperature and field dependence of linear and volume magnetostriction at T<Tc can be attributed to the transformation of the crystal lattice from cubic to tetragonal phase during the magnetic ordering. At T<Tc the tetragonal phase is transformed into a cubic under the influence of an external magnetic field. It is manifested with the huge values of linear and volume magnetostriction.

While the increasing of the external field at $T<T_{SR}=42K$ temperature and field dependence of magnetostriction show transitions between orthorhombic - tetragonal - cubic phases. Obtained data confirmed that in the $NdCo_2$ compound magnetic and structural phase transitions co-exist.

References

[1] Ouyang Z W, Wang F W, Huang Q, Liu W F, Xiao Y G, Lynn J W, Liang J K, and Rao G H, 2005 *PhysRevB.***71** 064405

[2] Pecharsky V K and Gschneidner K A, 2001 *Jr.Adv. Mater.* Weinheim, Ger. **13**, 683 .

[3] Gratz E and Markosyan A S, 2001 *J. Phys.: Condens. Matter* **13**, R385.

[4] Baranov N, Gratz E, Nowotny H, and Steiner W, 1983 *J. Magn. Magn. Mater.* **37**, 206.

[5] Duc N H and Brommer P E, in Handbook of Magnetic Materials, 1999 edited by Buschow K H *J. North-Holland*, Amsterdam, , Vol. **12**, p. 259.

[6] Khmelevskyi S and Mohn P, 2000 *J. Phys.: Condens. Matter* **12**, 9453.

[7] Gratz E and Markosyan A S, 2001 *J. Phys.: Condens. Matter* **13**, R385.

[8] Gratz E, Lindbaum A, Markosyan A S, Mueller H, and Sokolou A Y, 1994 *J. Phys.: Condens. Matter* **6**, 6699.

[9] Gratz E, 1983 *Solid State Commun.* **48**, 825.

[10] Xiao Y G and others, 2006 *J.Alloys and compounds* **420** 29-33

Acknowledgments

This work is supported by the grant RFBR 10-02-00721-a

Magnetostriction in $(Tb_{0.45}Dy_{0.55})_{1-x}Er_xCo_2$ ($x = 0.1, 0.2$): high-field investigation

I Tereshina[1,2], G Politova[1,2], E Tereshina[2,3], J Cwik[2], S Nikitin[4,2], O Chistyakov[1], A Karpenkov[5,2], D Karpenkov[5,2] and T Palewski[2]

[1] Baikov Institute of Metallurgy and Material Sciences RAS, 119991 Moscow, Russia
[2] International Laboratory of High Magnetic Fields and Low Temperatures, 53-421 Wroclaw, Poland
[3] Institute of Physics ASCR, 18221 Prague, Czech Republic
[4] Faculty of Physics, Lomonosov Moscow State University, 119991 Moscow, Russia
[5] Physicotechnical Department, Tver State University, 170002 Tver, Russia

E-mail: teresh@imet.ac.ru

Abstract. Intermetallic compounds $(Tb_{0.45}Dy_{0.55})_{1-x}Er_xCo_2$ ($x = 0.1; 0.2$) were synthesized using high purity rare-earth metals. Multicomponent compositions with compensated magnetic anisotropy (MA) were chosen with respect to theoretical predictions based upon the single-ion MA model. Magnetostriction was measured in magnetic fields up to 10 T within the temperature range of 1.5–200 K by means of strain gauges. In the vicinity of magnetic phase transitions (160–170 K), giant volume magnetostriction ($\sim 2\times10^{-3}$ in 10 T) provided by the band magnetism of a Co sublattice was observed in the compounds studied.

1. Introduction

The rare earth (R) RCo_2 compounds have been a subject of extensive theoretical and experimental study in past decades owning to their relatively simple crystallographic (cubic $MgCu_2$-type Laves phase (C15, space group $Fd\bar{3}m$)) and magnetic structures [1,2]. The magnetic structure of RCo_2 can be considered as consisting of two magnetic sublattices: one is formed by localized moments of rare earth atoms and therefore, the model of localized electrons is applicable for its description; and the second is the Co sublattice analyzed within the band electron model (the magnetism of the latter is strongly influenced by the R sublattice in RCo_2). Although YCo_2 and $LuCo_2$ are nonmagnetic, they show features characteristic of exchange-enhanced paramagnetism. The magnetic state of the Co-sublattice can be stabilized by applying magnetic fields higher than 70 T or by a molecular field arising from the magnetic R-sublattice.

As a rule the coupling between the R and Co magnetic moments is antiparallel in compounds with heavy rare earths. At the temperatures of magnetic ordering, $GdCo_2$ and $TbCo_2$ pass through the second-order magnetic phase transitions from ferri- to the paramagnetic state whereas the first-order transition takes place for $DyCo_2$, $HoCo_2$ and $ErCo_2$ [3]. Large magnetoelastic effects inducing the crystal structure distortions in the magnetically ordered state were observed and studied in RCo_2 previously [4-9] with relatively little attention paid to the investigation of mechanisms of the large

[1] I S Tereshina

Published under licence by IOP Publishing Ltd

spontaneous and field-induced magnetostrition formation. Furthemore, a rather large anisotropy field (H_a ~ 10 T) in RCo_2 with non-zero orbital moment of R makes it difficult to interpret correctly the results obtained in weak magnetic fields.

In the present work, the first magnetostriction study of the multicomponent $(Tb_{0.45}Dy_{0.55})_{1-x}Er_xCo_2$ (x = 0.1; 0.2) compounds with compensated (in the rare-earth sublattice) magnetic anisotropy (MA) was performed at temperatures 1.5–200 K and in the fields up to 10 T. The compositions of the alloys were chosen with respect to theoretical predictions based upon the single-ion model of magnetic anisotropy [10,11].

2. Experimental details

The compounds $(Tb_{0.45}Dy_{0.55})_{1-x}Er_xCo_2$ (x = 0.1 and 0.2) were obtained in an arc furnace using a non-consumable tungsten electrode on a water-cooled copper bottom plate in a helium atmosphere with the high-purity metals as starting components. The synthesized alloys were characterized by metallographic, chemical, X-ray fluorescence and thermomagnetic methods, and the phase composition was checked by the X-ray diffraction analysis. The details of the sample preparation and performed quality control can be found elsewhere [12].

The study of the field dependencies of longitudinal and transverse magnetostriction at different temperatures was conducted by the strain-gauge method on the plate-shaped samples of 4×5×2 mm^3. For the measurement, the strain gauges (Vishay Micromeasurements SR-4) were glued to the sample surface. The magnetic field was created by a superconducting magnet (Oxford).

3. Results and discussion

The study of the crystal field effect on localized 4f rare-earth orbitals plays an important role in the interpretation of magnetic properties of both the RFe_2 and RCo_2 compounds [13, 14]. The R-sublattice magnetic anisotropy constant in R-(Co,Fe) with a combination of several rare-earth ions is a sum of the single-ion MA constants of R-ions multiplied by their concentration. The total lth-order constant can be calculated from the equation [10]

$$K_l = \sum X_n K^0_{nl} \hat{I}_{l+(1/2)} \{L^{-1}_m\}, \tag{1}$$

where K^0_{nl} is the lth-order single-ion anisotropy constant at 0 K, X_n is the R ion concentration, $\hat{I}_{l+(1/2)}$ is the ratio of the hyperbolic Bessel function of the $l+1/2$ order to that of the ½-order, $L^{-1}_m(m)$ is the inverse Langevin function of reduced magnetization $m = I_s(T)/I_s(0)$, where $I_s(0)$ and $I_s(T)$ are the spontaneous magnetizations at 0 K and at a temperature T, respectively. Moreover,

$$m(T,H) = L(x) = \hat{I}_{3/2}(x) = \coth x - 1/x. \tag{2}$$

In the low-temperature region the following equation holds [16]

$$\hat{I}_{l+(1/2)}\{L^{-1}_m\} = m^{l(l+1)/2} \tag{3}$$

and the contribution of the 3d transition-metal sublattice to the total magnetic anisotropy can be omitted [15].

In order to compensate the magnetic anisotropy within the rare-earth sublattice, the compositions with R = Tb, Dy and Er were chosen due to different signs of the single-ion MA constants of the R ions [14, 17] and the multicomponent $(Tb_{0.45}Dy_{0.55})_{1-x}Er_xCo_2$ (x = 0.1 and 0.2) alloys were prepared. The performed complex analysis confirmed the single-phase $MgCu_2$ structure with a given composition for the samples under study [12]. Both the thermomagnetic investigation and the analysis carried out in terms of Landau theory showed that compounds demonstrate the first order magnetic transitions close to the Curie temperatures T_C [12]. Worth noting is that the field and temperature dependencies of magnetization show a rather weak hysteresis.

Let us consider the field dependencies of longitudinal $\lambda_\parallel$ and transversal $\lambda_\perp$ magnetostriction as well as the volume ($\omega = \lambda_\parallel + 2\lambda_\perp$) and anisotropic ($\lambda_a = 2/3(\lambda_\parallel - \lambda_\perp)$) magnetostriction for $(Tb_{0.45}Dy_{0.55})_{0.8}Er_{0.2}Co_2$. As seen from Fig. 1, the longitudinal magnetostriction reaches 1.5×10^{-3} at T = 4.2 K in the magnetic field of 10 T. Such a high value of magnetostriction (for reference in YCo_2 and

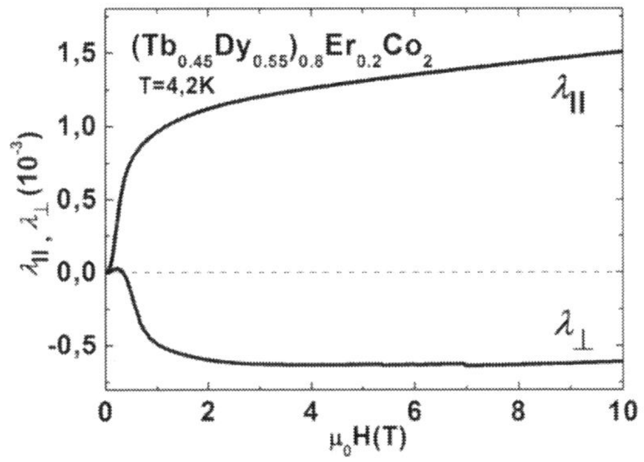

Figure 1. Field dependencies of longitudinal $\lambda_{\parallel}$ and transverse $\lambda_{\perp}$ magnetostriction of $(Tb_{0.45}Dy_{0.55})_{0.8}Er_{0.2}Co_2$ at $T = 4.2$ K.

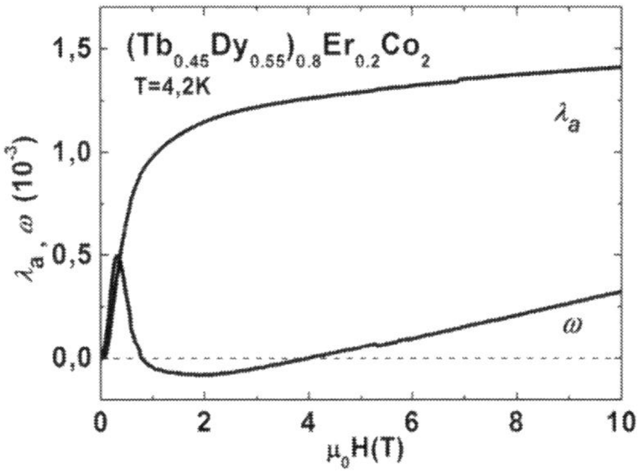

Figure 2. Field dependencies of anisotropic λ_a and volume ω magnetostriction of $(Tb_{0.45}Dy_{0.55})_{0.8}Er_{0.2}Co_2$ at $T = 4.2$ K.

$LuCo_2$, $\lambda_{\parallel}$ equals to 2×10^{-6} [18] and to -80×10^{-6} in $GdCo_2$ [19] in the 10 T field) confirms considerable contribution of the rare-earth sublattice to the total magnetostriction and single-ion nature of the phenomenon for $(Tb_{0.45}Dy_{0.55})_{0.8}Er_{0.2}Co_2$. However with the increased field, the value of longitudinal magnetostriction continues to grow, i.e. a large magnetostriction of paraprocess is observed at 4.2 K. In case of a totally localized magnetic moment at 0 K, the saturation magnetization should be field-independent with the magnetostriction of paraprocess equal to zero. Therefore the experimental fact reveals the band nature of magnetism of the compound studied. A strong magnetic field is capable of changing the splitting of the spin-up and spin-down bands resulting in the Co magnetic moments change and as such the appearance of the magnetostriction of paraprocess at low temperatures.

The transverse magnetostriction of $(Tb_{0.45}Dy_{0.55})_{0.8}Er_{0.2}Co_2$ shows a rather saturated behavior (Fig. 1) with a negative $\lambda_{\perp}$ value of -0.6×10^{-3} in a field of 10 T. (For comparison, in compounds with non-magnetic rare earths as Y and Lu, transversal magnetostriction is positive and reaches the value of 2×10^{-6} [18], while for the Gd-containing alloy, $\lambda_{\perp}$ is 10×10^{-6} [19] in the 10 T field).

As for the anisotropic and volume magnetostriction, the large value of λ_a is observed in $(Tb_{0.45}Dy_{0.55})_{0.8}Er_{0.2}Co_2$ at $T = 4.2$ K (see Fig. 2) whereas the volume magnetostriction has a much lower absolute value and demonstrates complex behavior. So that in the low-field region where the processes of magnetization vector rotation and the domain wall shifts dominate, ω increases and then drops down to zero. With the further magnetic field rise, the volume magnetostriction increases. Similar behavior is observed for the $(Tb_{0.45}Dy_{0.55})_{0.9}Er_{0.1}Co_2$ compound. Therefore at low temperatures, the volume magnetoelastic effect provided by the Co subsystem in these compounds is rather small and on the contrary, the anisotropic effect caused by the rare-earth subsystem is much larger.

Figures 3 and 4 show the temperature dependencies of longitudinal, transverse and volume magnetostriction for $(Tb_{0.45}Dy_{0.55})_{0.9}Er_{0.1}Co_2$ and $(Tb_{0.45}Dy_{0.55})_{0.8}Er_{0.2}Co_2$, respectively. It can be seen that all of the presented curves demonstrate a pronounced maximum at the Curie temperature. The giant values of volume magnetostriction of 2×10^{-3} and 1.6×10^{-3} in the 10 T field were observed for $(Tb_{0.45}Dy_{0.55})_{0.9}Er_{0.1}Co_2$ at $T_C = 168$ K and $(Tb_{0.45}Dy_{0.55})_{0.8}Er_{0.2}Co_2$ at $T_C = 162$ K, respectively. Such values of volume magnetostriction are provided by positive values of both the longitudinal and transverse magnetostriction in the vicinity of the transition.

As a rule, the absolute values of longitudinal and transverse magnetostriction depend on the structural state, i.e. on the single-, polycrystalline or directional structure of the samples under study. In the case of the $(Tb_{0.45}Dy_{0.55})_{0.9}Er_{0.1}Co_2$ compound, we suppose that the sample had the directional structure whereas $(Tb_{0.45}Dy_{0.55})_{0.8}Er_{0.2}Co_2$ was polycrystalline. However, the structural state does not influence

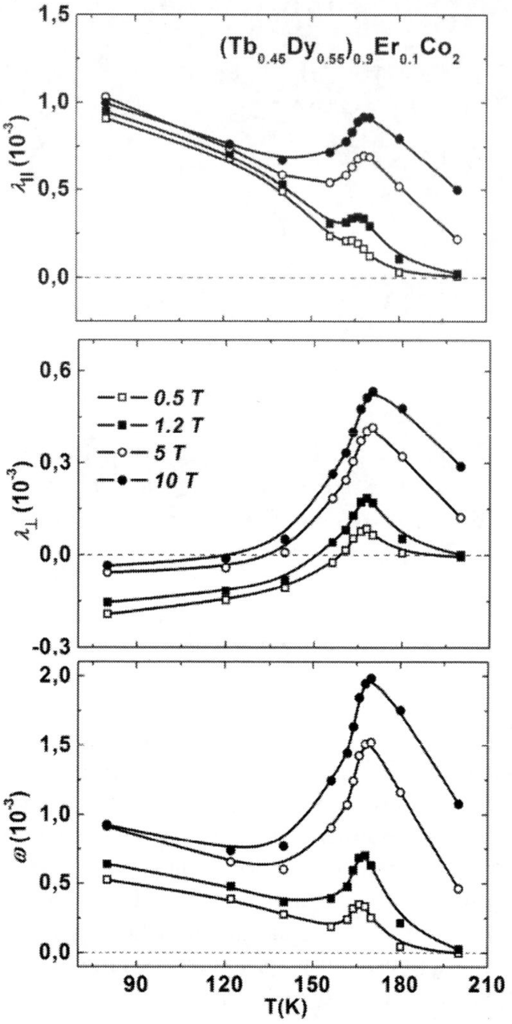

Figure 3. Temperature dependencies of longitudinal $\lambda_{\parallel}$, transverse $\lambda_{\perp}$ and volume ω magnetostriction of $(Tb_{0.45}Dy_{0.55})_{0.9}Er_{0.1}Co_2$ in various fields.

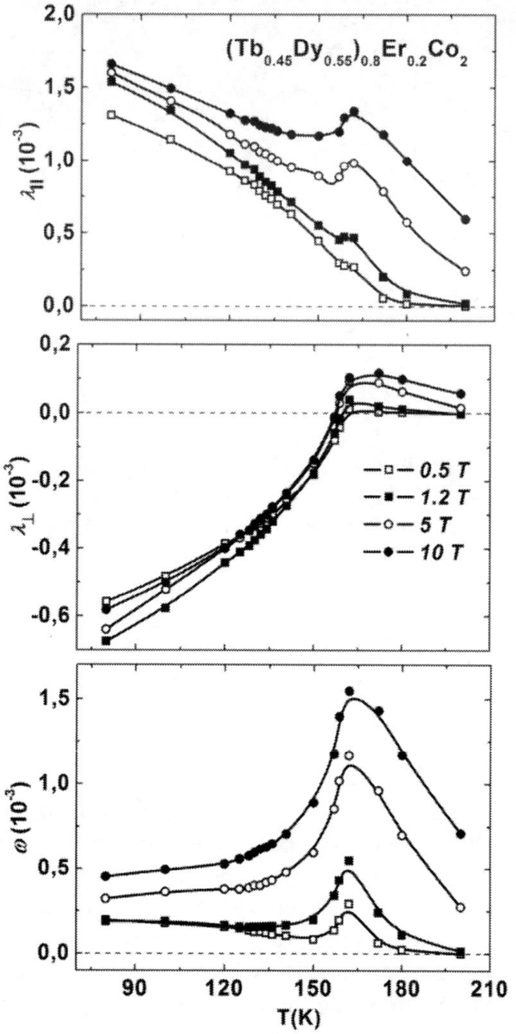

Figure 4. Temperature dependencies of longitudinal $\lambda_{\parallel}$, transverse $\lambda_{\perp}$ and volume ω magnetostriction of $(Tb_{0.45}Dy_{0.55})_{0.8}Er_{0.2}Co_2$ in various fields.

the value of volume magnetostriction and therefore the unit cell volume of both samples increases significantly with the magnetic field increase. From our point of view the observed giant magnetostriction effect is due to the larger role of the Co-sublattice 3*d* band magnetism despite the disordering of the rare-earth sublattice at the temperatures close to T_C. At the Curie temperature and above (in the paramagnetic region), an additional (band) magnetic moment of 3*d*-electrons is induced by the magnetic field and the "band" magnetostriction can be observed. This fact is supported by the analysis of the field dependencies of the volume and anisotropic magnetostriction in the vicinity of the Curie temperature. As seen in Fig. 5, the field dependencies of λ_a and ω for $(Tb_{0.45}Dy_{0.55})_{0.8}Er_{0.2}Co_2$ at T = 164 K, the values of volume magnetostriction exceed those of the anisotropic by a factor of two whereas the opposite was observed at the lowest temperatures (see Fig. 2). Below T_C, magnetostriction demonstrates a strong anisotropic character, i.e. the sign of the transverse magnetostriction changes

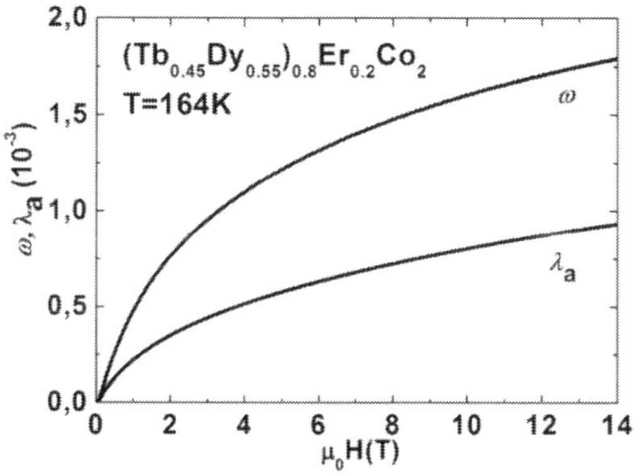

Figure 5. Field dependencies of anisotropic λ_a and volume ω magnetostriction of $(Tb_{0.45}Dy_{0.55})_{0.8}Er_{0.2}Co_2$ at $T = 164$ K.

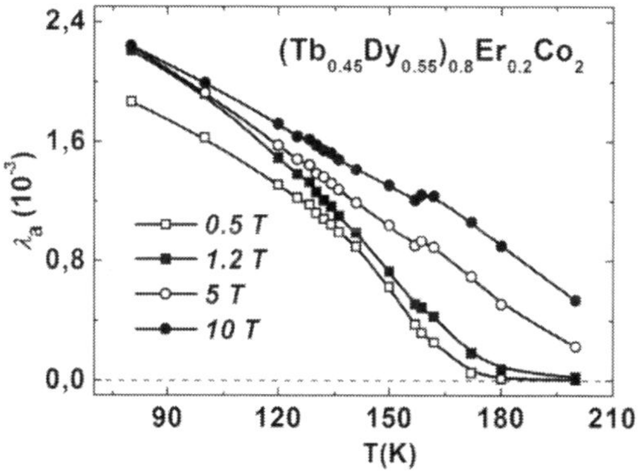

Figure 6. Temperature dependencies of anisotropic magnetostriction of $(Tb_{0.45}Dy_{0.55})_{0.8}Er_{0.2}Co_2$.

from positive to negative one whereas the longitudinal magnetostriction remains positive (see Figs. 3 and 4).

The temperature dependencies of the anisotropic magnetostriction of $(Tb_{0.45}Dy_{0.55})_{0.8}Er_{0.2}Co_2$ are shown in Fig. 6. With the temperature decrease, anisotropic magnetostriction provided by the interaction of anisotropic 4*f*-charge cloud of the rare-earth ions of Tb, Dy, Er with the crystal field again reaches giant values [15]. Therefore we can conclude that in high magnetic fields, two competing magnetostriction mechanisms are to be realized in the studied Co-containing compounds. In the example of $(Tb_{0.45}Dy_{0.55})_{1-x}Er_xCo_2$ ($x = 0.1$ and 0.2) alloys with compensated (in the rare-earth sublattice) magnetic anisotropy, we have revealed those mechanisms by analyzing the field and temperature dependencies of longitudinal, transverse, volume and anisotropic magnetostriction. The single-ion mechanism provided by the rare-earth sublattice is shown to prevail in the low-temperature range. Conversely, at temperatures close to the Curie temperature, the giant volume magnetostriction appears mainly due to the band magnetism of 3*d*-electrons of the Co sublattice.

The work is supported by RFBR, pr. No. 10-03-00848.

References

[1] Duc N C and Brommer P E Formation of 3d-moments and spin fluctuations in some rare-earth-cobalt compounds, in: *Handbook of Magnetic Materials* K.H.J. Buschow (Ed.) 1999 vol. 12 (Amsterdam, Elsevier Science)
[2] Gratz E and Markosyan A S 2001 *J. Phys.:Condens. Matter* **13** R385
[3] Khmelevski S and Mohn P 2000 *J. Phys.:Condens. Matter* **12** 9453
[4] Levitin R Z and Markosyan A S 1988 *Usp. Fiz. Nauk* **155** 623 (Engl. Transl. 1988 *Sov. Phys.-Usp.* **31** 730).
[5] Andreev A V Thermal expansion anomalies and spontaneous magnetostriction in rare-earth intermetallics with cobalt and iron, in: *Handbook of Magnetic Materials* K.H.J. Buschow (Ed.) 1995 vol. 8 (Amsterdam, Elsevier Science).
[6] Minakata R, Shiga M, Nakamura Y 1976 *J. Phys. Soc. Japan* **41** 1435
[7] Markosyan A S 1981 *Solid State Physics* **23** 1656
[8] Gratz E 1983 *Solid State Comm.* **48** 825
[9] Xiao Y G, Huang Q, Ouyang Z W, Lynn J W, Liang J K, Rao G H 2006 *J. Alloys Compd.* **420** 29
[10] Koon N C, Williams C M, Das B N 1991 *J. Magn. Magn. Matter.* **100** 173

[11] Nikitin S A Magnetic properties of rare-earth metals and their alloys 1989 (Moscow, Moscow State University, Russia)
[12] Tereshina I, Politova G, Tereshina E, Nikitin S, Burkhanov G, Chistyakov O, Karpenkov A 2010 *J. Phys.: Conf. Series* (to be pulished)
[13] Atzmony U., Dublon G. 1977 *Physica* **86-88B** 167
[14] Dariel M.P., Atzmony U. 1973 *Int. J. Magnetizm* **4** 213.
[15] Belov K P Magnetostriction Phenomena and their Technical Application 1987 (Moscow, Nauka, Russia).
[16] Klmker H., Dariel M.P., Rosen M. 1980 *J. Phys. Chem. Solids* **41** 215
[17] Callen E and Callen H B 1965 *Phys. Rev.* **139** 455
[18] Markosyan A.S., Snegirev V.V. 1985 *Phys. Met. Metallogr.* **59** 99.
[19] Levitin R. Z., Markosyan A.S., Snegirev V.V. 1982 *JETP Lett.* **36** 445.

Magnetic properties and magnetocaloric effect in $(Dy_{0.6}Er_{0.4})_{1-x}Gd_xCo_2$ multicomponent compounds.

J. Ćwik[1], T. Palewski[1], K. Nenkov[1,2], J. Lyubina[3], O. Gutfleisch[2]

[1] International Laboratory of High Magnetic Fields and Low Temperatures, Gajowicka 95, 53-421 Wroclaw

[2] IFW Dresden, Institute of Metallic Materials, Postfach 270016, 01171 Dresden, Germany

[3] Department of Materials, Imperial College London, Exhibition Road, London,

SW7 2AZ, United Kingdom

E-mail: cwikjac@alpha.ml.pan.wroc.pl

Abstract. We report magnetic and magnetocaloric properties of the polycrystalline series of $(Dy_{0.6}Er_{0.4})_{1-x}Gd_xCo_2$ (x = 0.2, 0.3 and 0.4) intermetallic compounds. These samples were synthesized using high purity rare earth metals. X-ray diffraction patterns taken at room temperature revealed that all the $(Dy_{0.6}Er_{0.4})_{1-x}Gd_xCo_2$ compounds have the C15 cubic Laves phase superstructure. Magnetization measurements were carried out using a vibration sample magnetometer with a step motor in applied fields up to 14 T using a Bitter-type magnet. Heat capacity measurements have been performed in temperature range 2-300 K. The Debye temperatures, phonon and conduction electron contributions as well as the magnetic part of heat capacity were estimated. The magnetocaloric effect has been estimated both in terms of isothermal magnetic entropy change and adiabatic temperature change for selected solid solutions in magnetic fields up to 3 T. The raising amount of the Gd in samples on magnetic and magnetocaloric properties will be discussed.

1. Introduction

In recent years, the intermetallic compounds RM_2 (R = rare earth, M = transition metal) belonging to the Laves-phase family with a large magnetocaloric effect (MCE) were the subject of extensive investigations because of their intrinsic magnetic and transport properties, which make them suitable candidates for magnetic refrigeration technology [1,2]. Attractive materials for magnetic refrigeration applications are compounds, which display first order or combined magnetic phase transitions [3]. In this context, RCo_2 intermetallic compounds are one of the promising candidates. It has been found that in the case of RCo_2 with heavy rare earth elements (R = Ho, Dy and Er) the ferromagnetic - paramagnetic phase transitions are of first order and these compounds exhibit a large magnetic entropy change [4,5].

Numerous investigations of the MCE on RCo_2 compounds with substitutions either in rare - earth sublattice $R_{1-x}R'_xCo_2$ or in Co - sublattice $RCo_{2-x}T_x$ (where T = Si, Ga, Ge, Al, Fe, Ni, Mn) have been performed [6]. However, no information on investigation of multicomponent compounds

$(R_xR'_yR''_z)Co_2$ (where x + y + z = 1) is available. From a large number of intermetallic compounds of RM_2-type, $DyCo_2$ and $ErCo_2$ were selected and their Curie temperature is 140 and 35 K, respectively [7]. The $GdCo_2$ compound is a ferromagnet with T_C = 400 K and exhibits a second order phase transition [7]. It should be noted that only in the case two first compounds their MCE properties were investigated. According to Tishin et al. [6] for $DyCo_2$ the maximum ΔT_{ad} at 138 K reaches 9.6 K in 10 T applied magnetic field. Giguere et al. [8] reported the maximum ΔT_{ad} = 14.4 K for $ErCo_2$ at 33 K in 14 T.

Here, we investigate a $Dy_{1-x}Er_xCo_2$ solid solution with x = 0.4. The influence of the partial replacement of $Dy_{0.6}Er_{0.4}Co_2$ by Gd in $(Dy_{0.6}Er_{0.4})_{1-x}Gd_xCo_2$ solid solutions on the phase composition, magnetic properties and MCE is reported.

2. Experimental details

Polycrystalline samples of $(Dy_{0.6}Er_{0.4})_{1-x}Gd_xCo_2$ (x = 0.2, 0.3 and 0.4) solid solutions were prepared by arc-melting on a water-cooled copper crucible under high purity argon atmosphere. The starting materials are of at least 99.9 % purity. The alloys were remelted four times to ensure good homogeneity. The mass losses after the melting were less than 1 wt. %. The buttons obtained were wrapped in a tantalum foil, sealed in an evacuated quartz ampoule, and annealed at 850°C for two weeks.

The structure of the samples was characterized using X-ray powder diffraction with CoK_α radiation. Magnetic measurements were carried out using a superconducting quantum interference device (SQUID) magnetometer (Quantum Design) and a vibrating sample magnetometer with a step motor in applied fields up to maximum 14 T using a Bitter-type magnet. The heat capacity was measured using Quantum Design PPMS 14 Heat Capacity System in a temperature range of 2-295 K without the magnetic field and in a magnetic field of 1 and 1.5 T.

3. Results and discussion

The X-ray powder diffraction data revealed that the polycrystalline $(Dy_{0.6}Er_{0.4})_{1-x}Gd_xCo_2$ solid solutions are single phase and have the cubic C15 superstructure (space group F-43m). The lattice parameters obtained by Rietveld refinement are 1.441 nm for $(Dy_{0.6}Er_{0.4})_{0.8}Gd_{0.2}Co_2$, 1.445 nm for $(Dy_{0.6}Er_{0.4})_{0.7}Gd_{0.3}Co_2$, and 1.451 nm for $(Dy_{0.6}Er_{0.4})_{0.6}Gd_{0.4}Co_2$. The substitution of Gd for $Dy_{0.6}Er_{0.4}Co_2$ by Gd in $(Dy_{0.6}Er_{0.4})_{1-x}Gd_xCo_2$ solid solutions leads to the increase of the lattice parameter; this dependence is largely linear.

In all the samples the ferromagnetic order has been observed. The magnetization curves measured at 4.2 K for these samples with various gadolinium contents are presented in Fig. 1. Saturation is not reached even in the maximum applied magnetic field of 14 T. Values of the saturation magnetic moment were determined from the magnetization corresponding to a magnetic field of 14 T by extrapolation to $\mu_oH = 0$. The μ_S values are well below 7.8 μ_B/f.u, which is expected for an assembly of free Dy, Gd and Er atoms. In the case of $(Dy_{0.6}Er_{0.4})_{0.8}Gd_{0.2}Co_2$ μ_S reaches 5.9 μ_B/f.u in 14 T and it decreases with increasing gadolinium content. The relatively low saturation values are apparently connected to the presence of antiferromagnetic components, which naturally would account for the low saturation moments [9].

The temperature dependence of the magnetic susceptibility χ_g measured in a static magnetic field of 0.03 T is shown as inset in Fig. 1. All the solid solutions under study have two magnetic phase transitions. The first one is observed in the low temperature region (T_1), below 20 K and the second one is observed in the high temperature region (T_2), below 270 K. The substitution of Gd for $Dy_{0.6}Er_{0.4}Co_2$ in the $(Dy_{0.6}Er_{0.4})_{1-x}Gd_xCo_2$ solid solutions leads to an increase of the magnetic ordering temperature. The Curie temperature of $(Dy_{0.6}Er_{0.4})_{0.8}Gd_{0.2}Co_2$ is equal T_1 ~ 13 K, T_2 ~ 179 K; in the case of $(Dy_{0.6}Er_{0.4})_{0.6}Gd_{0.4}Co_2$ the magnetic ordering is observed around 17 K and 264 K.

Magnetization isotherms of the $(Dy_{0.6}Er_{0.4})_{1-x}Gd_xCo_2$ (x = 0.2, 0.3 and 0.4) solid solutions were measured at selected temperatures near T_1 and T_2, and the maximum applied field of 3 T.

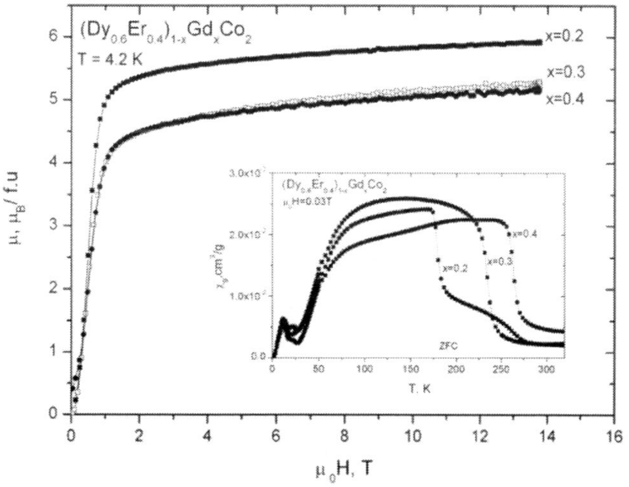

Figure 1. Magnetization curves measured at 4.2 K for the $(Dy_{0.6}Er_{0.4})_{1-x}Gd_xCo_2$ (x = 0.2, 0.3 and 0.4) solid solutions. The insets show temperature dependence of the magnetic susceptibility for x = 0.2, 0.3 and 0.4 solid solutions measured in applied magnetic field of 0.03 T.

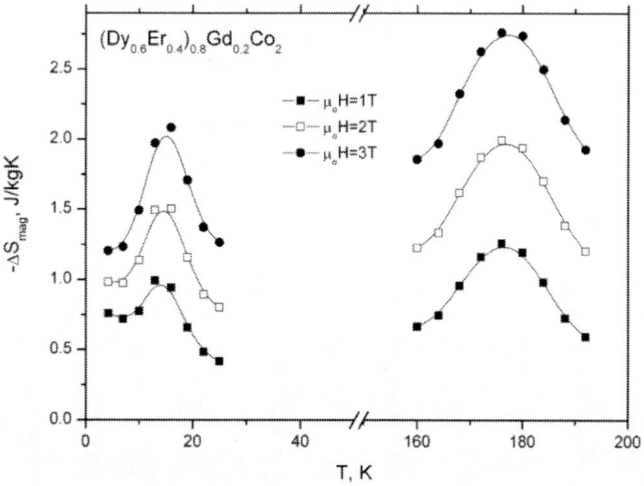

Figure 2. Temperature dependencies of $-\Delta S_{mag}$ for $(Dy_{0.6}Er_{0.4})_{0.8}Gd_{0.2}Co_2$ solid solutions for $\Delta\mu_0 H$ = 1, 2 and 3 T.

The magnetic entropy change ΔS_{mag} was calculated based on the results of magnetization isotherms using the integrated Maxwell relation [10]:

$$\Delta S_{mag}(T,H) = \int_{H_i}^{H_j} \left(\frac{\partial M(T,H)}{\partial T} \right)_H dH$$

$$\approx \frac{1}{2}\sum_i \left[\left(\frac{\Delta M}{\Delta T}\right)_{i+1} - \left(\frac{\Delta M}{\Delta T}\right)_i\right]\Delta H$$

Fig. 2 shows the variation of ΔS_{mag} for x = 0.2 as a function of temperature close to T_1 and T_2, for $\Delta\mu_0 H$ = 1, 2 and 3 T. Furthermore, both the peak height and the peak width are dependent on the magnetic field and they grow with an increasing magnetic field. As can be seen from the Fig. 2, the maximum entropy change for $(Dy_{0.6}Er_{0.4})_{0.8}Gd_{0.2}Co_2$ reaches 2.1 $Jkg^{-1}K^{-1}$ near the transition temperature T_1=13 K and 2.7 $Jkg^{-1}K^{-1}$ near T_2 = 179 K. In the case of sample with x = 0.3 maximum ΔS_{mag} reaches 1.2 $Jkg^{-1}K^{-1}$(T_1 = 15 K) and 1.5 $Jkg^{-1}K^{-1}$ (T_2 = 234 K). For $(Dy_{0.6}Er_{0.4})_{0.6}Gd_{0.4}Co_2$ maximum ΔS_{mag} reaches 1.1 $Jkg^{-1}K^{-1}$ (T_1 = 17 K) and 1.4 $Jkg^{-1}K^{-1}$ (T_2 = 264 K).

The temperature dependence of the heat capacity of all solid solutions confirm that the sample have two magnetic transitions. As an example, the temperature dependence of the heat capacity $C_{tot}(T)$, the sum of electron and phonon $C_{el+ph}(T)$ and magnetic contributions of solid solution $(Dy_{0.6}Er_{0.4})_{0.8}Gd_{0.2}Co_2$ are shown in Fig. 3. The sharp λ-like maximum on the heat capacity curves at T_1 ~ 10 K and T_2 ~ 170 K corresponds to the magnetic ordering temperatures. Heat capacity measurements in magnetic fields (insets in Fig.3 near T_2) show that the magnetic field causes the broadening and displacement of the λ-kind maximum. In order to estimate the magnetic contribution, a theoretical calculation of the Debye function was made [11]. The best fit for wide temperature range of 2-300 K could be obtained by fixing the parameters γ = 65 $mJ/molK^2$ for all the measured samples, while the Debye temperature is fluctuating around Θ_D = 235 K.

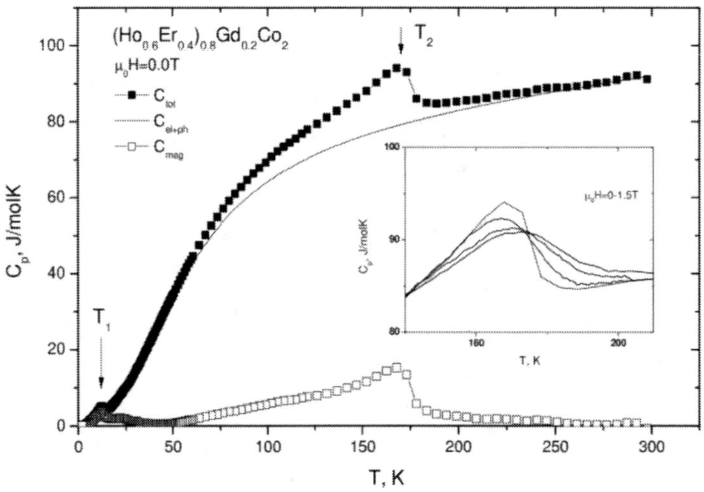

Figure 3. Total heat capacity $C_{tot}(T)$ of $(Dy_{0.6}Er_{0.4})_{0.8}Gd_{0.2}Co_2$ measured in zero magnetic field. The insets show a total heat capacity near T_2 for a magnetic field change 0-1.5 T.

Another important parameter for MCE material is the adiabatic temperature change ΔT_{ad}. The ΔT_{ad} can be estimated using the formula suggested by P.J. von Ranke et al.[2]:

$$\Delta T_{ad}(T)_{\Delta H} \cong \left[T(S)_{H=i} - T(S)_{H=0}\right]$$

As an example the temperature dependence $\Delta T_{ad}(T)$, which was calculated using this technique for selected sample near its ordering temperature (T_2) under 1 and 1.5 T magnetic field is shown in Fig.4. In the case of 1 T magnetic field, the maximum magnetocaloric effect ΔT_{ad} reaches ~5 K at 183 K. In the case of 1.5 T magnetic field, the maximum magnetocaloric effect ΔT_{ad} ~8 K. It should be noted that ΔT_{ad} values obtained in $(Dy_{0.6}Er_{0.4})_{0.8}Gd_{0.2}Co_2$ are comparable and even exceed those observed in materials with a first-order transition [12]. The maximum MCE near ordering temperature (T_1) under

1.5 T magnetic field is equal 1.5 K. As expected, the magnetic field leads to an increase of ΔT_{ad}. The adiabatic temperature changes existing over a wide temperature range, what is essential for a practical magnetic refrigerant. The sharp peaks in the ΔT_{ad} curves are related to the paramagnetic-ferromagnetic phase transition at the Curie point, where the magnetization change is maximal.

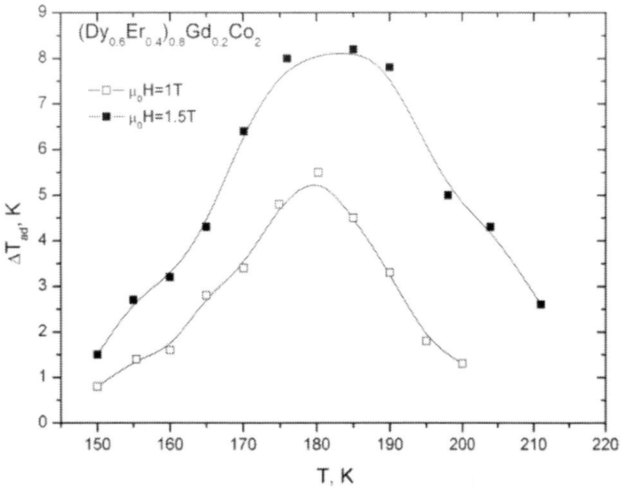

Figure 4. Temperature dependence of ΔT_{ad} of $(Dy_{0.6}Er_{0.4})_{0.8}Gd_{0.2}Co_2$ for a magnetic field change of 1 and 1.5 T

4. Conclusion

The effect the partial replacement of $Dy_{0.6}Er_{0.4}Co_2$ by Gd on the structure and physical properties in $(Dy_{0.6}Er_{0.4})_{1-x}Gd_xCo_2$ solid solutions has been studied. It was confirmed that the main phase in all solid solutions is cubic C15 superstructure. The substitution of Gd in $Dy_{0.6}Er_{0.4}Co_2$ results in the increase in the ordering temperatures, T_1 and T_2, as observed both in the magnetization and heat capacity measurements. The T_1 temperature increases from 13 K to 17 K, and T_2 increases from 179 K to 264 K. The maximum MCE estimated in terms of isothermal magnetic entropy change reaches 2.8 $Jkg^{-1}K^{-1}$ (magnetic field of 3 T) for x = 0.2. The large MCE of 8 K in a magnetic field of 1.5 T in $(Dy_{0.6}Er_{0.4})_{0.8}Gd_{0.2}Co_2$ is highly attractive for magnetic cooling applications.

References

[1] N.K. Singh, K.G. Suresh, A.K. Nigam, S.K. Malik, A.A. Coelho, S. Gama, J.Magn. Magn. Mater.317 (2007) 68.
[2] P.J. von Ranke, D.F. Grangeia, A. Caldas, N.A. de Oliveira, J. Appl. Phys. 93 (2003) 4055
[3] M. Foldeaki, A. Giguere, R. Chahine, T.K. Bose, Adv. Cryogenic Eng. 43 (1998) 1533.
[4] N.H. Duc, D.T. Kim Anh, J.Magn. Magn. Mater.873-875 (2002) 242
[5] D.H. Wang, S.L. Tang, H.D. Liu, W.L. Gao, Y.W. Du, Intermetallics, 10 (2002) 819
[6] A.M. Tishin, Y.I. Spichkin, The Magnetocaloric Effect and its Applications, IOP Publishing 2003.
[7] H.R. Kirchmayr, E. Burzo, in: H.P.J. Wijn (Ed.), Landolt Börnstein, New Series III/19d2, Berlin (1990).
[8] A. Giguere, M. Foldeaki, W. Schnelle, E. Gmelin; J. Phys. Condens. Mater. 11 (1999) 6969.
[9] H. Oesterreicher, J. Stanley, R. Pitts, Phys. Stat. Sol. (a) 12 (1972) K65.

[10] V.K. Pecharsky, K.A. Gschneidner Jr., J.Appl. Phys. 86 (1999) 565.
[11] M. Bouvier, P. Lethuillier, D. Schmitt; Phys. Rev. B 43, (1991) 13137
[12] J. Lyubina, R. Schäfer, N. Martin, L. Schultz, O. Gutfleisch, Adv. Mater., in press (http://dx.doi.org/10.1002/adma.201000177).

Neutron powder diffraction determination of the magnetic structure of Nd_2Al

J M Cadogan[1], M Avdeev[2], P Kumar[3], K Suresh[3] and D H Ryan[4]

[1] Department of Physics and Astronomy, University of Manitoba, Winnipeg, Manitoba, R3T 2N2, Canada
[2] Bragg Institute, ANSTO, PMB 1, Menai, NSW 2234, Australia
[3] Department of Physics, I I T Bombay, Mumbai 400076, India
[4] Department of Physics, McGill University, Montreal, Québec, H3A 2T8, Canada

E-mail: cadogan@physics.umanitoba.ca

Abstract. We have determined the magnetic structure of Nd_2Al by neutron powder diffraction. This orthorhombic intermetallic compound orders ferromagnetically below 36 K with the Nd moments aligned along the b-axis. Even at 1.7 K, the larger of the two Nd moments is only 2.3(2) μ_B, about 70% of the 'free-ion' value of 3.27 μ_B. This reduction is a consequence of the substantial crystal-field effects at the Nd^{3+} sites.

1. Introduction

The R_2Al intermetallic compounds (R = rare earth) crystallize in the orthorhombic Co_2Si-type structure (space group *Pnma*, #62) in which the R atoms occupy two crystallographic sites (both 4c) and the Al occupies a third 4c site. In 1978, Sill and Biggers [1] showed that Nd_2Al is a ferromagnet with a Curie temperature of 36 K and they quoted a "fully-stretched" Nd magnetic moment of 3.27 μ_B. More recent work [2, 3, 4] suggested that the Nd magnetic moment at low temperatures is actually quite strongly reduced from the free-ion value of 3.27 μ_B and it was proposed that this reduction in moment may be due to either strong crystal-field quenching or antiferromagnetic components in the magnetic order, either intrinsic or as clusters. In this paper we present neutron powder diffraction measurements we recently made on Nd_2Al. In particular, we confirm that Nd_2Al is ferromagnetic with a strongly quenched Nd moment. No antiferromagnetic components in the magnetic order were observed.

2. Experimental Methods

The Nd_2Al sample was prepared by arc melting stoichiometric amounts of the pure elements (Nd 99.9 wt.%, Al 99.99 wt.%). The sample was turned and remelted several times in order to ensure homogeneity. The alloyed button was then sealed under vacuum in a quartz tube, annealed for 3 weeks at 700 °C and quenched in water. Cu-K_α x-ray powder diffraction and EDAX analysis confirmed the majority phase to be the intended orthorhombic Nd_2Al phase. It proved impossible to prepare a single-phase sample. Refinement of the x-ray diffraction pattern using the GSAS/EXPGUI package [5, 6] showed the presence of impurities of about 5 wt% each of NdAl (orthorhombic *Pbcm* [7]) and Nd_3Al (cubic $Pm\overline{3}m$ [8]), with a trace of unreacted Al (cubic $Fm\overline{3}m$) also present. Basic magnetic characterization was carried out on a Quantum Design PPMS susceptometer/magnetometer operated down to 1.8 K.

Neutron diffraction experiments were carried out on the *Echidna* high-resolution powder diffractometer at the OPAL reactor in Sydney, Australia [9]. The neutron wavelength was 2.44160(2) Å, calibrated against a standard Al_2O_3 sample (NIST SRM676). All refinements of the neutron diffraction patterns employed the GSAS/EXPGUI package [5, 6]. The neutron diffraction data were corrected for absorption effects.

3. Results and Discussion

In figure 1 we show the refined neutron diffraction pattern of Nd_2Al obtained at 106 K, at which temperature Nd_2Al is paramagnetic and the neutron diffraction pattern exhibits only nuclear scattering.

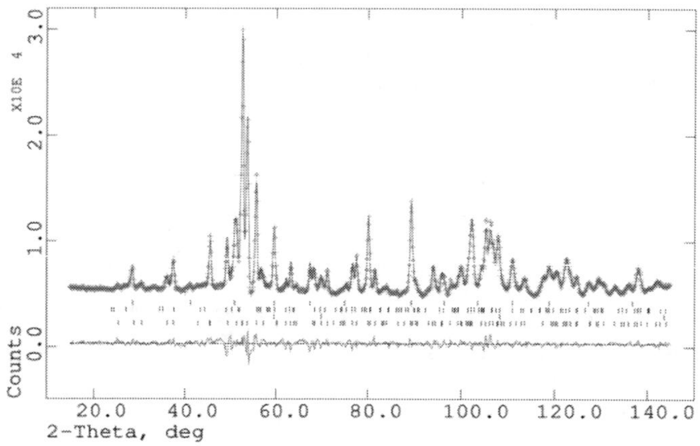

Figure 1. Neutron diffraction pattern of Nd_2Al obtained at 106 K ($\lambda =$ 2.44160(2) Å). The Bragg markers (bottom to top) represent Nd_2Al, Al, NdAl and Nd_3Al.

The refined lattice parameters at 106 K are a = 6.6825(4) Å, b = 5.2342(3) Å and c = 9.7321(7) Å. In table 1 we give the refined atomic position parameters of Nd_2Al, deduced from the refinement of the 106 K neutron powder diffraction pattern. The conventional refinement R-factors (%) are R(p) = 3.4 and $R(F^2)$ = 3.0.

Table 1. Crystallographic data for Nd_2Al obtained by refinement of the 106 K neutron powder diffraction pattern.

Atom	Site	x	y	z
Nd	4c	0.0184(5)	$\frac{1}{4}$	0.7040(3)
Nd	4c	0.1929(5)	$\frac{1}{4}$	0.0724(4)
Al	4c	0.2045(12)	$\frac{1}{4}$	0.4026(6)

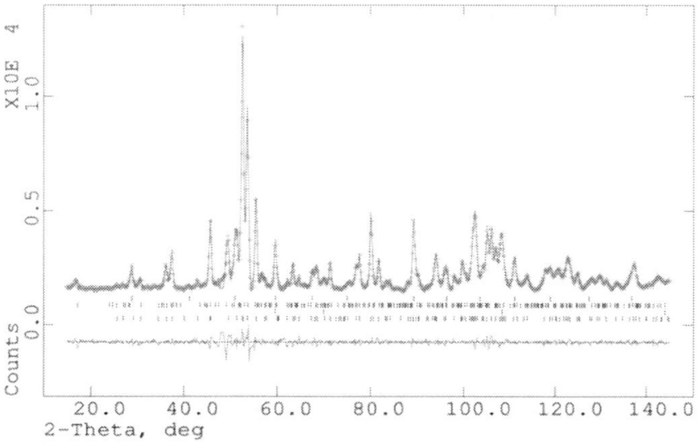

Figure 2. Neutron diffraction pattern of Nd_2Al obtained at 1.7 K (λ = 2.44160(2) Å). The Bragg markers (bottom to top) represent Nd_2Al, Al, NdAl and Nd_3Al.

In figure 2 we show the refinement to the neutron diffraction pattern obtained at 1.7 K. The magnetic contributions to the 1.7 K Nd_2Al diffraction pattern occur only at the nuclear peak positions and correspond to the propagation vector $\mathbf{k}$ = [0 0 0]. We find no evidence for additional *magnetic-only* peaks from the Nd_2Al phase. In order to consider all possible magnetic structures allowed for Nd_2Al, we carried out Representational Analysis for the Nd site using the SARA*h* program [10]. The decomposition of the magnetic representation comprises eight one-dimensional representations:

$$\Gamma^{4c}_{Mag} = 1\Gamma^{(1)}_1 + 2\Gamma^{(1)}_2 + 2\Gamma^{(1)}_3 + 1\Gamma^{(1)}_4 + 1\Gamma^{(1)}_5 + 2\Gamma^{(1)}_6 + 2\Gamma^{(1)}_7 + 1\Gamma^{(1)}_8 \quad (1)$$

and the basis vectors of these irreducible representations are given in table 2.

Table 2. Representational Analysis for the Nd(4c) site in Nd_2Al with a propagation vector [0 0 0]. The respective atomic positions are $(x,\ y,\ z)$, $(\frac{1}{2}+x,\ y,\ \frac{1}{2}-z)$, $(-x,\ -y,\ -z)$ and $(\frac{1}{2}-x,\ -y,\ z+\frac{1}{2})$.

Representation	Ordering Mode	First component	Second component
Γ_1	G_Y	+ − + −	0
Γ_2	C_X A_Z	+ + − −	+ − − +
Γ_3	F_X G_Z	+ + + +	+ − + −
Γ_4	A_Y	+ − − +	0
Γ_5	F_Y	+ + + +	0
Γ_6	A_X C_Z	+ − − +	+ + − −
Γ_7	G_X F_Z	+ − + −	+ + + +
Γ_8	C_Y	+ + − −	0

We can immediately rule out the purely antiferromagnetic representations Γ_1, Γ_2, Γ_4, Γ_6 and Γ_8 because magnetometry measurements make it clear that Nd_2Al is at least "predominantly ferromagnetic" [4]. This leaves three possible magnetic structures, namely Γ_3, Γ_5 and Γ_7. The only allowed ordering directions for the Nd(4c) magnetic sublattices with $\mathbf{k} = [0\ 0\ 0]$ are either along the b-axis (Γ_5) or in the ac-plane (Γ_3 and Γ_7). The best refinement to the measured diffraction pattern is with the Nd(4c) sites ordered ferromagnetically in the F_Y mode, along the crystal b-axis, corresponding to the Γ_5 representation. Of particular note are the Nd^{3+} magnetic moments, 1.2(2) μ_B and 2.3(2) μ_B, both of whch are substantially smaller than the 'free-ion' value of 3.27 μ_B for the Nd^{3+} ion.

Our refinements show that there are no antiferromagnetic components associated with the magnetic order of the Nd sublattices, either as an intrinsic canting or as clusters. The reduction in the Nd magnetic moments in Nd_2Al is therefore most likely the result of crystal-field quenching. It is known that the crystal-field acting on the R^{3+} sites in the orthorhombic R_2Al compounds is quite large and this, coupled with the fact that the magnetic exchange interaction is relatively weak (T_C = 36 K for Nd_2Al), leads to the observed quenching. In a ^{169}Tm Mössbauer study of Tm_2Al, for example, one of us showed that the strong effect of the crystal-field leads to unusually slow electronic relaxation of the Tm^{3+} ion [11].

4. Conclusions

We have determined the magnetic structure of Nd_2Al by neutron powder diffraction. The magnetic ordering temperature is 36(2) K. At 1.7 K, the magnetic order of the Nd(4c) sublattices is ferromagnetic along the orthorhombic b-axis. Significant crystal-field quenching of the Nd^{3+} magnetic moments is present.

Acknowledgments

JMC and DHR are grateful to the staff of the Bragg Institute, OPAL reactor, Australian Nuclear Science and Technology Organisation, Sydney for their technical and scientific support provided during the course of these neutron diffraction measurements. In particular, we wish to thank Dr R. Robinson, Director of the Bragg Institute, for giving us the opportunity to carry out these experiments which were originally scheduled for the NRU reactor at Chalk River, Ontario.

JMC acknowledges support from the Canada Research Chairs programme and the Natural Sciences and Engineering Research Council of Canada.

References

[1] Sill L R and Biggers R R 1978 *J. Appl. Phys.* **49** 1500–2.
[2] Li X G, Chiba A, Aoki K and Masumoto T 1997 *Mat. Sci. Eng.* **A222** 197–200.
[3] Li X 1999 *J. Alloys and Compounds* **288** 76–8.
[4] Kumar P, Suresh K G and Nigam A K 2008 *J. Phys. D: Appl. Phys.* **41** 105007 (9pp).
[5] Larson A C and von Dreele R B 2004 *Los Alamos National Laboratory* LAUR 86–748 (unpublished).
[6] Toby B H 2001 *J. Appl. Cryst.* **34** 210–21.
[7] Bécle C, Lemaire R and Paccard D 1970 *J. Appl. Phys.* **41** 855–863.
[8] Fukuhara T, Yamagata R, Li L, Nishimura K and Maezawa K 2009 *J. Phys. Soc. Japan* **78** 034723 (4 pp).
[9] Liss K-D, Hunter B A, Hagen M E, Noakes T J and Kennedy S J 2006 *Physica B* **385-6** 1010-2.
[10] Wills A S 2000 *Physica B* **276–8** 680–1 (program available from www.chem.ucl.ac.uk/people/wills/).
[11] Bowden G J, Cadogan J M, Day R K and Dunlop J B 1988 *Hyp. Int.* **39** 359-367.

Mössbauer and calorimetric studies of spin reorientation processes in $Er_{2-x}Pr_xFe_{14}C$.

A. T. Pędziwiatr[1], K. Krawiec[1], B. F. Bogacz[1], S. Wróbel[1], D. Kaczorowski[2], J. Przewoźnik[3]

[1]M. Smoluchowski Institute of Physics, Jagiellonian University, Reymonta 4, 30-059 Kraków, Poland

[2]W. Trzebiatowski Institute for Low Temperature and Structure Research, Polish Academy of Sciences, Okólna 2, 50-950 Wrocław, Poland

[3]AGH University of Science and Technology, Al. Mickiewicza 30, 30-059 Kraków, Poland

antoni.pedziwiatr@uj.edu.pl

Abstract. The $Er_{2-x}Pr_xFe_{14}C$ (x = 0.25, 0.5) polycrystalline compounds have been synthesized and investigated by means of X-ray diffraction, ^{57}Fe Mössbauer spectroscopy and Differential Scanning Calorimetry (DSC). The influence of carbon on the spin reorientation phenomena with increasing praseodymium content was of special interest in these compounds. Investigations of spin reorientation phenomena were performed by means of ^{57}Fe Mössbauer spectroscopy with a narrow step temperature scanning in the neighborhood of the spin reorientation temperature. Obtained Mössbauer spectra were analyzed using a procedure of simultaneous fitting and the transmission integral approach. Consistent description of Mössbauer spectra were obtained. Temperature dependencies of hyperfine interaction parameters and subspectra contributions were derived from fits and the transition temperatures were determined for all the compounds studied. DSC studies revealed weak thermal effect (heat absorption) during the transition process. Heat absorption was detected only for the compound with the lowest content of Pr (x = 0.25). The results obtained with different methods were analyzed and data were compared with those for $Er_{2-x}Pr_xFe_{14}B$ series. The spin arrangement diagram was constructed.

1. Introduction

Compared with $R_2Fe_{14}B$ – where R stands for rare-earth, carbides $R_2Fe_{14}C$ are known to be less stable and having somewhat lower Curie temperature. However, higher anisotropy field of carbides as well as fact that carbon doesn't have negative influence on health and environment during processing, made $R_2Fe_{14}C$ compounds worthy of investigations. We have made an attempt to synthesize 2-14-1 phase, substituting heavy rare earth Er with light one – Pr.

$Er_{2-x}Pr_xFe_{14}C$ (x = 0.25, 0.5) compound belongs to $Nd_2Fe_{14}B$ structure type, with a tetragonal crystal lattice of the $P4_2/mnm$ space group with 68 atoms in elementary cell, where Fe atoms occupy six different positions denoted: $16k_1$, $16k_2$, $8j_1$, $8j_2$, 4e, 4c; rare earth atoms locate themselves at 4f and 4g positions; and carbon atom at 4g position [1]. Planar anisotropy of rare earth sublattices has stronger temperature dependences than axial anisotropy of iron sublattice. Due to the weak coupling between sublattices in comparison with coupling within the particular sublattice and a competition between planar and axial tendency in rare earth and Fe sublattice [2, 3], the direction of easy magnetization changes from planar (in basal plane) to axial (along the *c*-axis) with increasing temperature. This phenomenon, called spin reorientation phenomenon, was studied previously for different borides and for few carbide compounds prepared by different techniques [4-11]. For $R_2Fe_{14}B$ compound spin reorientation phenomenon was confirmed by neutron diffraction investigations on single crystals [12]. Similar change of magnetic structure with increasing temperature (axis/plane) takes place also in $Er_{2-x}Fe_{14+2x}Si_3$ compounds [13] with $P6_3/mmc$ structure type due to different exchange interactions in various sublattices.

The spin reorientation temperature, T_{SR}, determines the onset of axial anisotropy region and it is sensitive to the Pr content in the sample.

The main goal of this work was to synthesize $Er_{2-x}Pr_xFe_{14}C$ carbides and study the influence of carbon on the spin reorientation phenomena in substitutions R = Pr – using ^{57}Fe Mössbauer spectroscopy and Differential Scanning Calorimetry methods. Additional tasks were: to establish spin structure phase diagram, to investigate influence of spin reorientation process on hyperfine interaction parameters and to compare this process with the reorientation process in borides of the same composition.

2. Experimental methods

The $Er_{2-x}Pr_xFe_{14}C$ alloys were prepared by arc melting stoichiometric proportions of the pure starting materials under protective argon atmosphere followed by annealing at 900 °C for two weeks and then rapid cooling in water to room temperature.

X-ray diffraction analysis was performed at room temperature on randomly oriented powdered samples with the use of Cu-radiation. Data were analyzed by using Rietveld method implemented within the Fullprof program. Apart from the main phase ($R_2Fe_{14}C$), X-ray data exhibited a small amount of natural iron impurity in the samples (up to 5%). Additionally, small amount (up to 2%) of other rare earth containing phase was detected. Iron impurity patterns were consequently subtracted from the experimental spectra by the numerical procedure.

For Mössbauer investigations, samples were prepared in the form of thin layer of powdered material. There is a random distribution of orientations of magnetic moments for polycrystalline material, so – in a thin absorber approximation – it would give 3:2:1 ratio for line intensities in a single Zeeman pattern.

The Mössbauer spectra were recorded in the temperature range 80–330 K, with 5 K step in the vicinity of reorientation temperature T_{SR}, using a ^{57}Co (Rh) source and a computer driven constant acceleration mode spectrometer. The velocity scale was calibrated with a high purity iron foil. Isomer shift was established with respect to the center of the room temperature iron Mössbauer spectrum.

DSC (Differential Scanning Calorimetry) measurements were performed on the calorimeter Pyris 1. Measurement procedure for all compounds was repeated several times in the temperature range 100-373 K, with different scanning rates (100 K/min – 10 K/min) for heating and cooling cycles.

3. Results and discussion

3.1. Mössbauer spectroscopy method.

Elementary cell for the $R_2Fe_{14}C$ compounds is smaller than in the case of borides of the same composition, due to smaller size of carbon atoms. In case of larger Pr ion replacing Er ion on its crystallographic position, lattice parameters increase in both carbides and borides (Fig. 1, Table 1).

A large number of spectra in the temperature region 80-340 K was obtained by performing ^{57}Fe Mössbauer measurements with a narrow step temperature scanning in the neighborhood of the spin reorientation temperature. Similar like in our previous studies [14,15], obtained spectra were analysed by fitting six Zeeman subspectra according to six iron occupations of the crystallographic sublattices, using "exponential" approximation [16] of the transmission integral, which takes into account the influence of sample thickness on the ratio of line intensities in Mössbauer spectrum. According to the assumption that inside the temperature region of reorientation some part of magnetic moments already switch to axial arrangement, and some part still remains in (a-b) plane, each Zeeman sextet splits into two subspectra in the transition region, which gives twelve Zeeman sextets in the whole spectrum. These sextets, called "low" and the "high temperature" subspectra, are related to planar and axial magnetic arrangements, respectively.

Each subspectrum was characterized by the three hyperfine interaction parameters: isomer shift – IS, quadrupole splitting – QS, define as $[(v_6 - v_5) - (v_2 - v_1)]/4$, where v_i are velocities corresponding to Mössbauer line positions, and hyperfine magnetic field – B. One common set of three line widths was used for all Zeeman subspectra. A procedure of simultaneous fitting of several spectra with interconnected parameters was applied in order to get a consistent description of spectra throughout the series.

Exemplary spectra are presented in Fig. 2. The spectra at the top and the bottom of the figure are related to the planar and axial spin arrangements, respectively. They were described with six Zeeman sextets with relative intensities taken as 4:4:2:2:1:1, according to iron occupations of the crystallographic sublattices. The intermediate spectra described with twelve Zeeman sextets, including "low" and "high temperature" components, are related to the region of changing magnetic structure. The separation of the sixth line of $8j_2$ "high temperature" sextet from other lines in a spectrum allows to establish precisely its contributions to the spectrum and to estimate the stage of reorientation process. During the transition, "low" and "high temperature" Zeeman sextets exchange gradually (between themselves) their contributions C_l, C_h to the total spectrum, $(C_l + C_h = 1)$. Spin reorientation temperature, $T_{SR,}$ is taken as intersection points of C_l and C_h curves (Fig. 3).

A common linear temperature dependence of IS caused by second order Doppler shift effect was assumed for all sextets (Fig.4), whereas for quadrupole splitting QS and hyperfine magnetic field B there is a shift in the values of temperature dependencies between "low" and "high temperature" Zeeman sextets (Fig.5-6). As quadrupole splitting QS is related to the angle between the easy axis of magnetization and the electric field gradient direction [17], its temperature dependence is very weak (Fig.5). For all iron sites the values of QS parameters increase during the transition process (plane-axis) and the largest values of QS before and after reorientation belong to $8j_2$ sublattice. The values of hyperfine field B have temperature dependences of square polynomial character, and they decrease with the increase in temperature (Fig.6). During the transition process parameters B increase their values only for the $8j_2$ and 4e iron sites. The largest value of B after reorientation belongs to $8j_2$ sublattice.

3.2. Differential Scanning Calorimetry (DSC) method.

Considering thermal effects, which may accompany the process of change in magnetic structure [14], a number of DSC measurements were performed in the temperature range 100-373 K, with different scanning rates (100 K/min – 10 K/min) for heating and cooling cycles of measurements. With increasing scanning rate up to 100K/min for the sample with the lowest content of Pr x=0.25, we observed a peak on the endothermic curve, corresponding to the heat absorption during the reorientation process (Fig.7). Hence, reorientation temperature for this compound was estimated as 283 K, which is 20 degrees higher than the result obtained from Mössbauer measurements, however there is a relatively large error in determining ($\Delta T = \pm$ 10 K) arising from broad peak shape. The area under the DSC peak, defined as the transformation enthalpy $\Delta H = 0.302 \pm 0.05$ J/g, is close to the value, obtained for the $Er_2Fe_{14}C$ compound [14]. Measurements performed on other compounds with different scanning rates didn't give any significant results, similar as in the case of boron compounds [15]. Apparently, substitution Er with even small amount of Pr change the character of reorientation process reducing strongly calorimetric effects, making them undetectable.

3.3. Conclusions.

The reorientation temperatures obtained for the series with different methods are listed in Table 1. Fig.8 shows the partial magnetic phase diagram for the studied $Er_{2-x}Pr_xFe_{14}C$ series. Spin reorientation temperature separates axial magnetic arrangement area, dominating in high temperatures, from planar arrangement, taking place in low temperature. Pr substitution influences strongly the reorientation temperature, shifting it to the lower values and making the reorientation region broader.

In the case of sample $Er_{2-x}Pr_xFe_{14}C$ with x = 0.5, the obtained value of T_{SR} is higher than in the case of boride of the same composition, contrary to expectations.

References

[1] Herbst J F, Croat J J, Pinkerton F E and Yelon W B 1984 Phys. Rev. B 29 4176–8.
[2] Givord D, Li H S and Perrier de la B^ athie R 1984 Solid State Commun. 51 857–60.
[3] Hirosawa S, Matsuura Y, Yamamoto H, Fujimura S, Sagawa M and Yamauchi H 1986 J. Appl. Phys. 59 873–9.
[4] J. Herbst, Review of Modern Physics, Vol. 63, No. 4 (1991) p.819.
[5] E. Burzo, Rep. Prog. Phys. 61 (1998), 1099.
[6] K.H.J. Buschow, in: E.P. Wohlfarth, K.H.J. Buschow (Eds.), Ferromagnetic Materials, vol.4, Amsterdam, 1988.
[7] K.H.J. Buschow, Rep. Prog. Phys. 54 (1991), 1123.
[8] M.R. Ibarra, Z. Arnold, P.A. Algarabel, L. Morellon, J. Kamarad, J. Phys., Condens. Matter 4 (1992), 9721.
[9] A.T. Pędziwiatr, W.E. Wallace, J. Less-Common Met. 126 (1986), 41.
[10] C. Piqué, R. Burriel, J. Bartolomé, J. Magn. Magn. Mater 154 (1996), 71.
[11] M. Takahashi, M. Tobise, Y. Sugawara, IEEE Transl. J. on Magnetics in Japan, vol.7, No.1 (1992), 51-56.
[12] P. Wolfers, M. Bacmann, D. Fruchart, J. Alloys Comp. 317-318 (2001), 39-43.
[13] J. Żukrowski, A. Błachowski, K. Ruebenbauer, J. Przewoźnik, D. Sitko, N.-T. H. Kim-Ngan, Z. Tarnawski, and A.V. Andreev, J. Appl. Phys. 103 (2008), 123910.
[14] A.T. Pędziwiatr, B.F. Bogacz, K. Krawiec, S. Wróbel, J. Alloys Comp 490 (2010), p.11-14.
[15] A.T. Pędziwiatr, A. Wojciechowska and B.F. Bogacz, J. Phys.: Condens. Matter. 18 (2006), 8891-8898.
[16] B.F. Bogacz, Mol. Phys. Rep. 30 (2000), 15.
[17] P. Gütlich, R. Link, A. Trautwein, Mössbauer Spectroscopy and Transition Metal Chemistry, Springer-Verlag, Berlin, Heidelberg, New York, 1978.

Tables and figures.

x, Pr	*a,(Å)*	*c,(Å)*	T_{SRM} *(K)*	T_{SRC} *(K)*	T_{SRM} *(K), boron compounds*
0	8,740	11,788	316	313	324
0.25	8,749	11,813	263	283	269
0.5	8,755	11,838	224	-	167

Table 1.
Values of the lattice parameters and spin reorientation temperatures for $Er_{2-x}Pr_xFe_{14}C$ compounds: T_{SRM} – determined from Mössbauer measurements, T_{SRM} – from calorimetric studies. Values for compound with x = 0 are taken from [8], values for boron compounds are taken from [11]. T_{SRM} error is ± 2 K.

Journal of Physics: Conference Series **303** (2011) 012027 doi:10.1088/1742-6596/303/1/012027

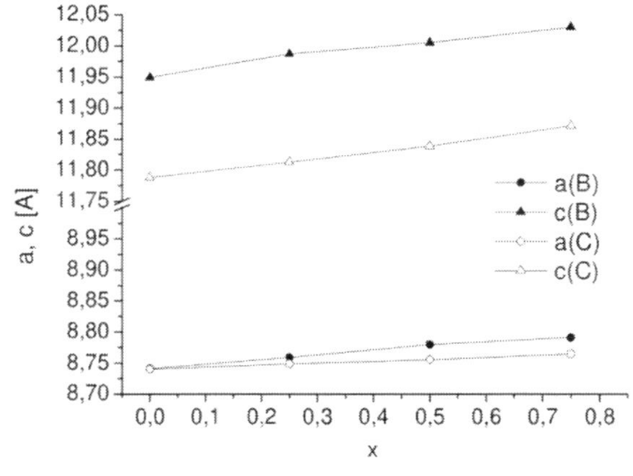

Fig. 1. The composition dependencies of lattice parameters for $Er_{2-x}Pr_xFe_{14}C$ and $Er_{2-x}Pr_xFe_{14}B$ compounds. a(C) and c(C) are lattice parameters for the carbides and a(B) and c(B) – for borides, taken from [11].

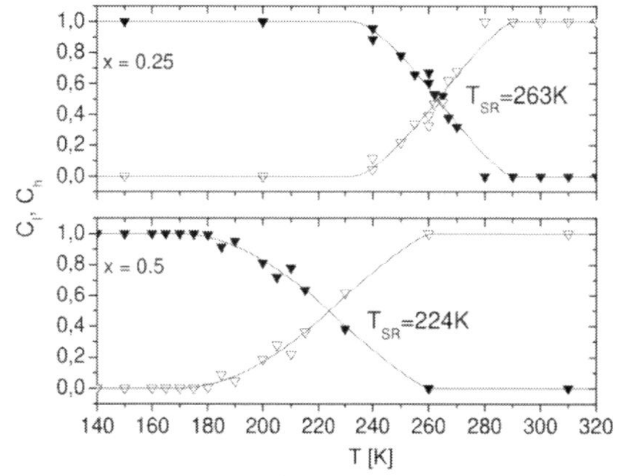

Fig. 3. The temperature dependences of subspectra contributions for C_l – "low temperature" (solid triangle) and C_h – "high temperature" (open triangle) Zeeman sextets for the samples x = 0.25, 0.5.

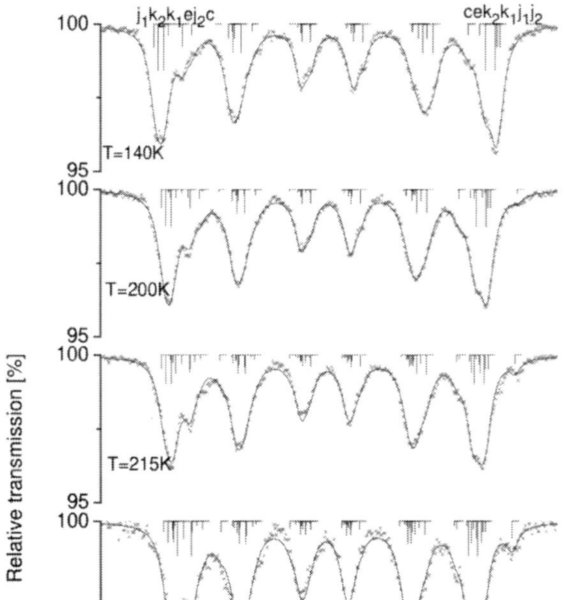

Fig.2. The selected experimental ^{57}Fe Mossbauer spectra of $Er_{2-x}Pr_xFe_{14}C$, x = 0.5 compound. The solid lines are fits to the data. The stick diagrams show the line positions and their relative intensities.

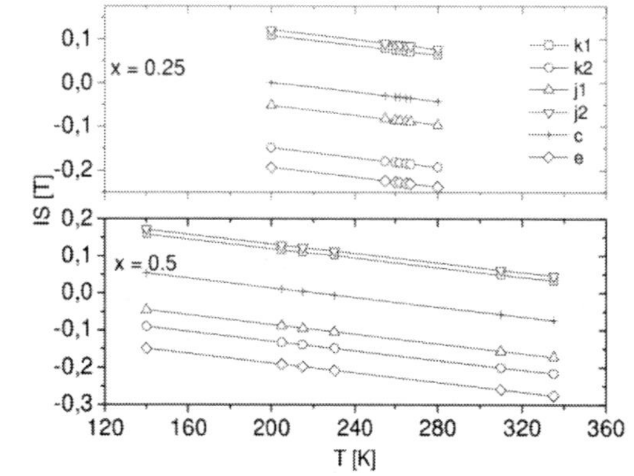

Fig. 4. The temperature dependences of the isomer shift IS for different crystal sites of $Er_{2-x}Pr_xFe_{14}C$.

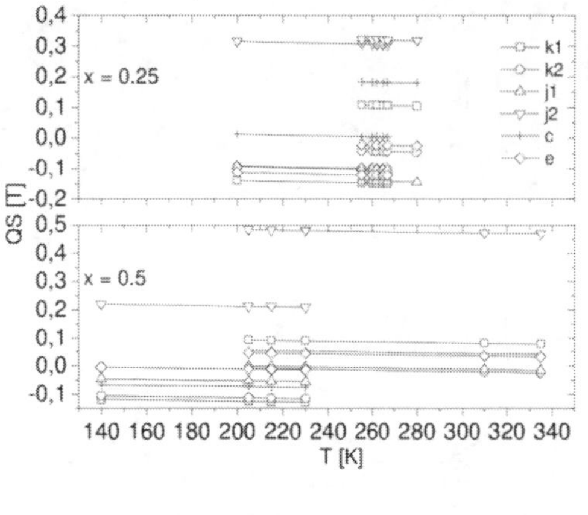

Fig.5. The temperature dependences of the quadrupole splitting QS for different crystal sites of $Er_{2-x}Pr_xFe_{14}C$.

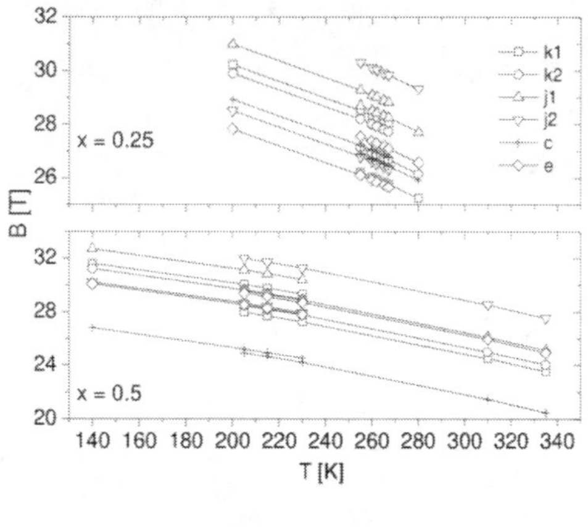

Fig.6. The temperature dependences of the hyperfine field B for different crystal sites of $Er_{2-x}Pr_xFe_{14}C$

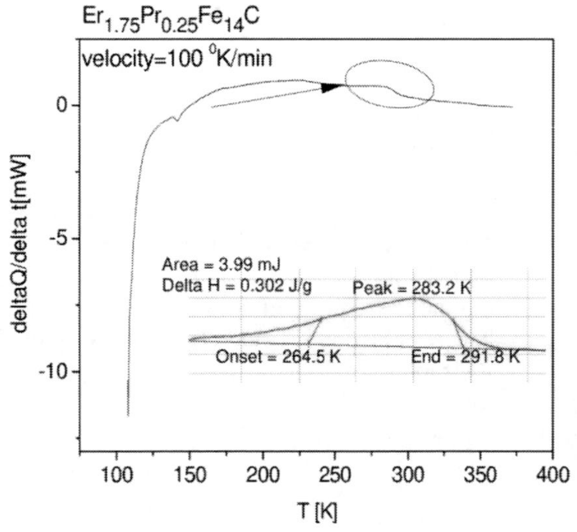

Fig. 7. Endothermic curve for the $Er_{1.17}Pr_{0.25}Fe_{14}C$ compound. Peak corresponds to the heat absorption of sample during the transition process. Measurements performed on heating with the rate of 100 K/min.

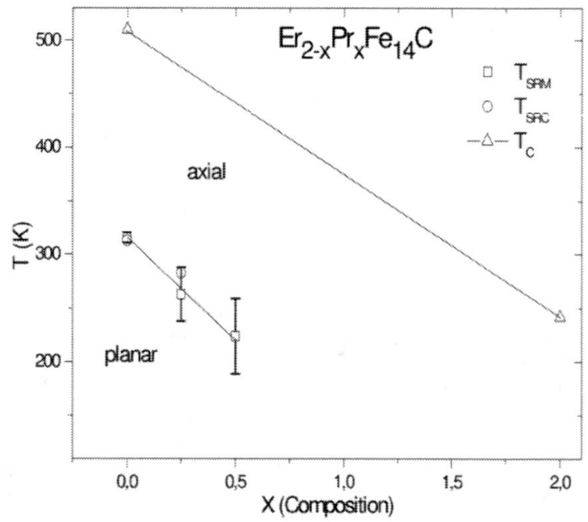

Fig. 8. Partial spin structure phase diagram for the $Er_{2-x}Pr_x$ $Fe_{14}C$ compounds. T_C — Curie temperature, T_{SRM}, T_{SRC} —spin reorientation temperatures determined from Mössbauer and calorimetric measurements, respectively. Solid vertical lines (not error bars) mark area of the coexistence of axial and planar arrangements. The value of Curie temperature for $Pr_2Fe_{14}C$ compound is taken from [11].

Magnetic and structural properties of nanocrystalline $PrCo_3$.

K Younsi[1], V Russier[1], L Bessais[1], J-C Crivello[1]
[1] ICMPE, UMR 7182 CNRS-Université Paris Est, 2-8 Rue Henri Dunant F-94320 Thiais, France.

E-mail: `russier@icmpe.cnrs.fr`

Abstract.
The structure and magnetic properties of nanocrystalline $PrCo_3$ obtained from high energy milling technique are investigated by X-ray diffraction, Curie temperature determination and magnetic properties measurements are reported. The as-milled samples have been annealed in a temperature range of 1023 K to 1273 K for 30 mn to optimize the extrinsic properties. The Curie temperature is 349 K and coercive fields of 55 kOe at 10 K and 12 kOe at 293 K are obtained on the samples annealed at 1023 K. A simulation of the magnetic properties in the framework of micromagnetism has been performed in order to investigate the influence of the nanoscale structure. A composite model with hard crystallites embedded in an amorphous matrix, corresponding to the as-milled material, leads to satisfying agreement with the experimental magnetization curve. [K. Younsi, V. Russier and L. Bessais, J. Appl. Phys. **107**, 083916 (2010)]. The microscopic scale will also be considered from DFT based calculations of the electronic structure of $R\mathrm{Co}_x$ compounds, where R = (Y, Pr) and x = 2,3 and 5.

1. Introduction

The permanent magnetic materials with high performance require appropriate intrinsic hard magnetic properties *i.e* a high uniaxial magnetocrystalline anisotropy field, a high Curie temperature and a high saturation magnetization. Among nanocrystalline magnets based on rare earth and transition metal compounds the $R\mathrm{Co}_3$ compounds (R as a rare earth) have been the subject of many studies during the last decades. These ones crystallize in rhombohedral $PuNi_3$-type (space group: $R\bar{3}m$). The unit cell contains two non-equivalent crystallographic sites for R ions, $3a$ and $6c$, and three sites for Co: $3b$, $6c$ and $18h$. $R\mathrm{Co}_3$ alloys present excellent magnets properties such as large magnetocrystalline anisotropy and important saturation magnetization. These properties are resulting from to combination of the $3d$ itinerant magnetism of the Co sublattice and $4f$ localized magnetism of the sublattice.

In this study, we report the elaboration of nanocrystalline $PrCo_3$ by high-energy milling, and their structural and magnetic properties in amorphous state and after recrystallization. Hard magnetic properties of the as milled $PrCo_3$ have been improved by a controlled nanocrystallization, which optimize the grain size in order to obtain a high coercivity.

In addition to the experiments, results from simulation and microscopic calculations are presented. We have proposed a simple model for the system after annealing based on a lattice of crystallites embedded in an amorphous matrix. The simplest realization of such a model is considered here, in order to bring out at the qualitative level the microstructure effect on the

Published under licence by IOP Publishing Ltd

magnetization curve. This model is treated in the framework of micromagnetism, *i.e.* through a continuous medium type of approach, and the calculations are performed by using the MAGPAR [1] code based on the finite elements method. We also present the results of *ab initio* band structure calculations and magnetic properties based on the spin polarized local approximation of the density functional theory (DFT) of RCo_x, where R=(Y,Pr) and x=2,3 and 5 intermetallic compounds. We will focus on a quantitative analyse of the localized $4f$ states and itinerant $3d$ magnetism.

2. Experimental studies

A polycrystalline $PrCo_3$ was prepared by melting high-purity starting elements ($> 99.9\%$) in an induction furnace under an atmosphere of high purity Ar. The ingot was remelted five times to ensure homogeneity. These ingots were used as pre-alloys to manufacture samples by high-energy milling. Technique and conditions of milling has been described in [2]. After milling, the powder mixtures was wrapped in tantalum foil and sealed in silica tubes under a vacuum of 10^{-6} Torr, then annealed for 30 min at temperature between 1023 K and 1323 K followed by quenching in water. X-ray diffraction (XRD) patterns of as-cast, as-milled and annealed powders were carried out with Cu-Kα radiation source on a Bruker diffractometer. The unit-cell parameters were measured with Si as the standard (a= 5.4308 Å) leading to a unit-cell parameter accuracy of $\pm 1\times 10^{-3}$ Å. The data treatment was made by Rietveld refinement with FULLPROF computer code in the assumption of ThompsonCox-Hastings line profile alloying multiple-phase refinement. This refinement gives the weight percentage of each of the coexisting phases, the line broadening leads to the autocoherent domain size owing to Scherrer formula. The magnetic ordering temperatures T_C were obtained on a differential sample magnetometer MANICS in a field of 1 kOe on powder, in vacuum-sealed silica tube. Curie temperature was determinate by extra-polating the linear part of the MT curve and extending the baseline: intersection point corresponds to the value of T_C. The $M - H$ curves (Magnetic hysteresis measurement) were plotted at T= 10 and 293 K with a Physical Properties Measurement System (PPMS) Quantum Design equipment and a maximum applied field of 90 kOe on sample in epoxy resin.

2.1. Structure analysis

Before milling, the ingot is analyzed by X-ray diffraction (XRD). The Fig. 1 presents the analysis result of XRD pattern of $PrCo_3$ ingot using the Rietveld analysis. The $PrCo_3$ ingot is nearly single phase with the rhombohedral $PuNi_3$-type structure (space group: $R\bar{3}m$) and is an appropriate precursor for using as pre-alloys to manufacture samples by high-energy milling.

In order to check the crystalline structure of $PrCo_3$ powders, the XRD measurement of samples annealed at distinct temperature have been performed. The XRD patterns of both as-quenched and annealed $PrCo_3$ powder are shown in Fig.2. It can be seen that as-quenched powder is essentially quasiamorphous. Such quasiamorphous matrix suggests the presence of various types of small crystallites. The diffraction peaks become more pronounced with increasing annealing temperature. The annealed samples at 773 and 1273 K for 30 min reveal the presence of $PrCo_3$ phases crystallized.

However, when the annealed is at temperature between 773 and 1023 K, we observe in the low angles a large line around 32°. The large peak widths points to a small grain size. When the powders are annealed at temperature higher than 1023 K, the XRD patterns show the presence of $PrCo_3$ phases crystallized. We have observed a small amount (around 0.5 wt%) of Pr_2O_3 in all samples annealed after milling. The mean diffraction crystallite size of the powder after annealing at 1023 and 1273 K for 30 min obtained by fitting the XRD diagram with Rietveld method, ranges between 31 and 45 nm respectively, this implies that synthesized materials are noncrystalline.

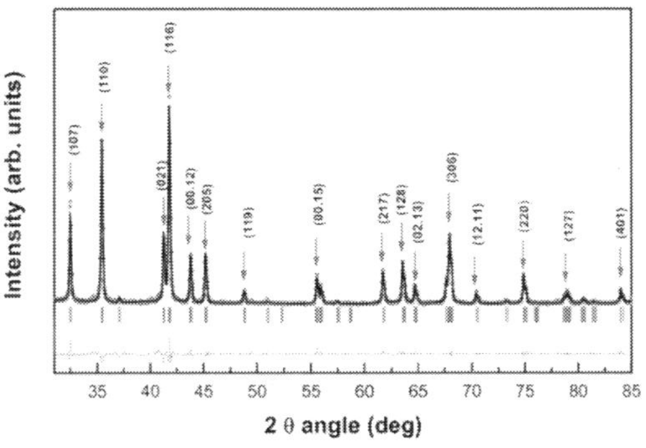

Figure 1. Rietveld analysis for $PrCo_3$ ingot.

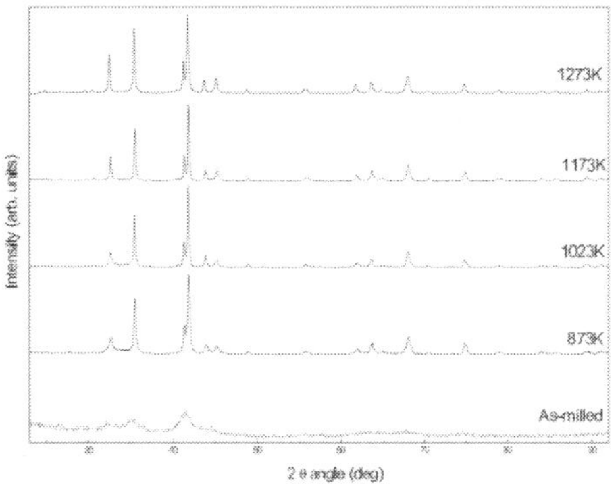

Figure 2. XRD pattern of the $PrCo_3$ as-milled and annealed samples at 773 and 1273 K during 30 min.

The Curie temperature is determined by the exchange-coupling interaction between the nearest-neighbor Co-Co atoms, which is the origin of ferromagnetism. This interaction depends markedly on the interatomic distance. The Curie temperature of $PrCo_3$ compound is 349 K, [2] this result is in agreement with the data given by Lemaire [3].

The low Curie temperature in $PrCo_3$ compounds is due to the short Co-Co interatomic distances at the sites $6c - 18h$ and at $18h - 18h$ [2], where the Co atoms couple antiparallel. In this structure, the $6c - 18h$ and at $18h - 18h$ interactions are strongly negative.

Isothermal magnetization curves $M - H$ obtained at 293 K is represented in Fig. 3 for $PrCo_3$. This figure shows that at 293 K the saturation is not reached, which gives evidence for a magnetocrystalline anisotropy among the highest known in those kinds of systems. The pattern obtained on sample oriented under an external magnetic field, applied perpendicularly to the plane of the sample, shows only $(00l)$ Bragg peaks, which amount to suggest that the anisotropy strengthens the peaks of diffraction of type $(00l)$ and at room temperature the easy magnetization direction is parallel to the c-axis. A similar result is obtained for $YCo_{3-x}Fe_x$ by [4] obtained by oriented powder under extern field.

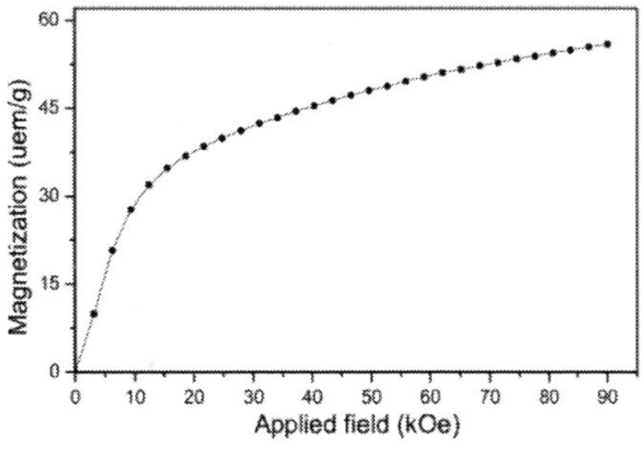

Figure 3. Magnetization curves of $PrCo_3$ obtained after high-energy milling and annealed at 750°C for 30 min, measured at 293 K.

2.2. Extrinsic magnetic properties

The best magnetic properties were reached after a treatment at 1023 K for 30 min. Fig. 4 show as an example the hysteresis loops measured at 10 and 293 K. This sample exhibits the highest coercivities we have obtained, for $T = 10\,\mathrm{K}$, $H_c = 55\,\mathrm{kOe}$, $M_R = 56\,\mathrm{emu/g}$ and $M_R/M_{max} = 0.75$. For T= 293 K, $H_c = 12\,\mathrm{kOe}$ and $M_R = 44\,\mathrm{emu/g}$ and $M_R/M_{max} = 0.60$ inherent to nanocrystalline state. The highest coercivity of 55 kOe at 10 K confirms the large magnetocrystalline anisotropy.

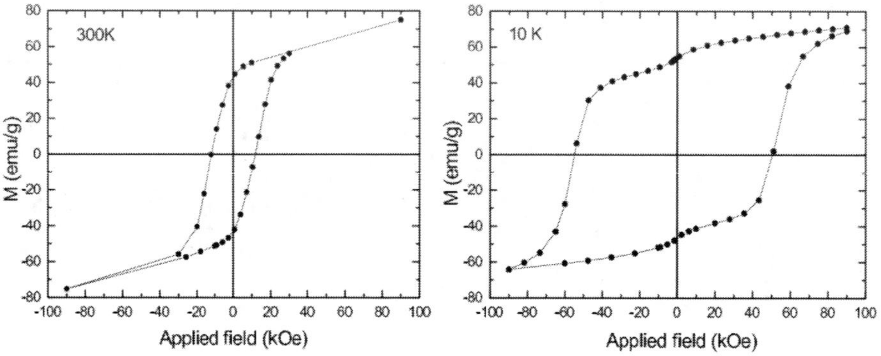

Figure 4. Hysteresis loop of $PrCo_3$ milling before and after annealing at 1023 K for 30 min, measured at 293 and 10 K.

3. Computational studies

3.1. Micromagnetism simulation

In order to get a picture of the link between the structure at the nanometer scale and the extrinsic magnetic properties we have performed a micromagnetic simulation on a model for the system. The picture we have in mind is that the crystallization takes place locally on dispersed seeds in the amorphous matrix representing the as-milled sample. Hence we consider an assembly of crystallites presenting a uniaxial anisotropy comparable to what is expected in hard magnetic R-TM compounds. The saturation magnetization, J_s of the crystallites is that of the bulk $PrCo_3$, namely $J_s = 0.723\ T$. We consider the simplest realization of the model, by using identical crystallites, located on the nodes of a simple cubic lattice. The crystallites are either spherical or cubic; however, the later case is likely to be closer to the experimental sample. We have chosen somewhat arbitrarily a volumic fraction occupied by the crystallites of $\phi = 0.73$ (in the case of cubic crystallites) and the crystallite size corresponds to what has been estimated experimentally, namely, a cubic edge $a = 35\ nm$. The amorphous character of the matrix is translated by a zero valued anisotropy constant, while its saturation magnetization ($J_s^{(mat)}$) is a free parameter of the model. The value we get for $J_s^{(mat)}$, namely c.a. $J_s/5$ seems reasonable in our picture of an amorphous matrix. The important point is that we clearly get the signature of a two uncoupled phases behavior for $M(H)$; the amplitude of the kink at zero field being determined by ϕ and $J_s^{(mat)}$.

3.2. DFT calculations

3.2.1. Methodology Band structure spin polarized calculations were carried out by using the DFT scheme implemented in the projector-augmented wave (PAW) method, performed with the VASP package [5, 6]. In the present study, the generalized gradient approximation (GGA) is used for the exchange and correlation energy functionals with the Perdew-Burke-Ernzerhof

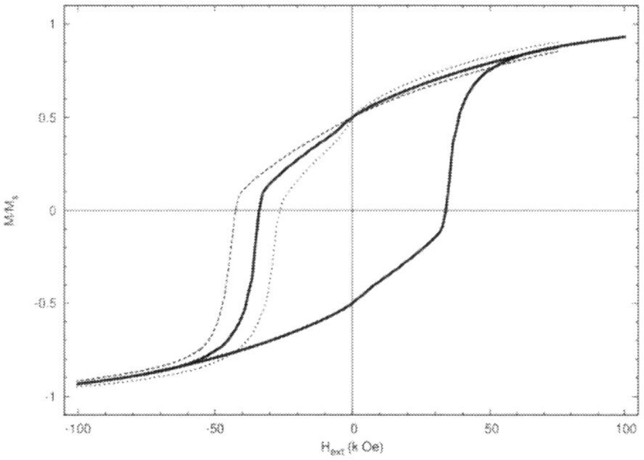

Figure 5. Calculated de-magnetization curve in terms of the external field for the systems of cubic inclusions with $\varphi = 0.73$. Randomly distributed axes and matrix saturation magnetization $J_s = 0.05$ T, dashed line; $J_s = 0.125$ T, solid line and $J_s = 0.25$ T, dotted line.

(PBE) [7] functional. A plane wave basis set with a cutoff energy of 600 eV was used in all calculations converged within 0.1 eV in total energy. Brillouin zone sampling using a mesh of 0.5 k points for each primitive cell of $R\mathrm{Co}_x$ compounds, described respectively in $Fd\bar{3}m$ space group for x=2, $R\bar{3}m$ for x=3 and $P6_3/mmm$ for x=5 compounds. Structural optimization were carried out under the condition that residual forces should be smaller than 0.1 eV.Å^{-1}.

3.2.2. Electronic structures Figure 6 shows the density of states (DOS) for the YCo_x and PrCo_x (x=2, 3 and 5) in the two directions of spin. For both series, the majority occupied states are dominated by the 3d states of Co, with a small electronic charge transfert of R to Co, which is more important from R=Y to Co than from Pr to Co (calculated by Bader's theory). The increase of x leads to a distinct shift of the Co-3d states in spin majority and minority DOS as expected in the rigid-band Stoner theory. Whereas $R\mathrm{Co}_2$ shows the paramagnetic state as the most stable in energy, the ferromagnetic behavior is confirmed for x=3 and x=5 compounds, with an increase of the angular magnetic moment from 1.45 to 7.34 and 3.59 to 4.70 μ_B f.u.$^{-1}$ for R=Y and Pr respectively. The magnetic anisotropy has been confirmed for $R\mathrm{Co}_3$ compounds, since non-collinear spin calculation along the z axis has been found ~0.2 eV f.u.$^{-1}$ more stable than calculation along x axis direction. Since the DOS at Fermi level is high enough for x=3 and 5, the itinerant magnetism properties of theses compounds can be explained by intra-atomic $3d-3d$ exchange. In contrast to YCo_x, PrCo_x compounds present tiny and high density structures above the Fermi level, corresponding to localized empty 4f states of Pr. It is clear that the Co-3d states are modified compared to YCo_x, due to the hybridization with the Pr-4f states. As expected in $3d-5d$ hybridization compounds, the energy between the bonding and anti-bonding sub-bands becomes different for the two-direction of spins [8]. The consequence is a larger occupation of minority 5d states as compared to majority 5d of the rare earth, that means a ferrimagnetic coupling between 3d and 5d states occurs and increases with the increasing of x composition.

4. Conclusion and outlook

The PrCo_3 intermetallic compound crystallizes in rhombohedral PuNi_3-type structure and presents a very high uniaxial magnetocrystalline anisotropy. The coercivity of this compound,

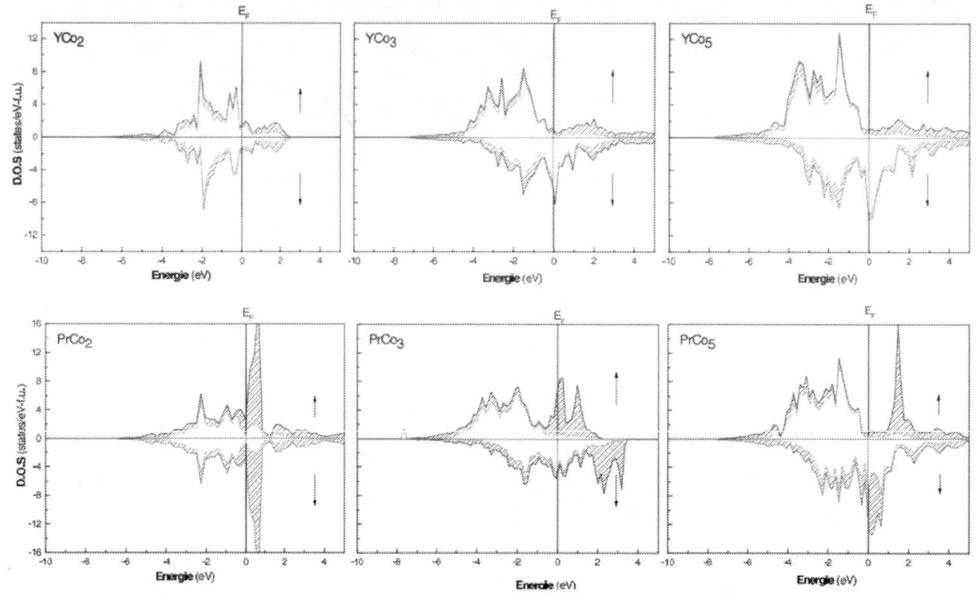

Figure 6. Total density of states (shaded region) of $R\mathrm{Co}_x$ (R=Y,Pr ; x=2, 3 and 5). The white area is corresponding to the only $3d$-states of Co. The origin of the energy scale is located at the Fermi energy E_F.

prepared by high energy milling followed by annealing (1023 K, 30 mn), takes a very high value, up to 55 kOe at 10 K. A micromagnetic simulation and atomic scale *ab-initio* calculations have been performed to understand on the one hand the way in which the electronic structure and especially the interplay between $3d$ and $4f$ electronic states, leads to the observed anisotropy and magnetic moment, and on the other hand the relation between the nanoscale structure and the extrinsic magnetic properties. DFT calculations will be improved with spin-orbit coupling considerations, and with Hubbard corrections in order to better describe cohesive properties of materials with strong electronic correlations.

References

[1] W. Scholz, T. Schrefl, D. Suess, R. Dittrich, H. Forster, and V. Tsiantos. *Comput. Mater. Sci.*, 28:366, 2003.
[2] K. Younsi, V. Russier, and L. Bessais. *J. Appl. Phys.*, 107:083916, 2010.
[3] R. Lemaire. *Cobalt*, 33:201, 1966.
[4] M. I. Bartashevich, Tsuneaki. Goto, and Katsuhiko. Koui. *Physica B*, 292(1-2):9, October 2000.
[5] G. Kresse and J. Furthmüller. Efficient iterative schemes for ab initio total-energy calculations using a plane-wave basis set. *Phys. Rev. B*, 54(16):11169–11186, Oct 1996.
[6] G. Kresse and D. Joubert. From ultrasoft pseudopotentials to the projector augmented-wave method. *Phys. Rev. B*, 59(3):1758, 1999.
[7] John P. Perdew, Kieron Burke, and Matthias Ernzerhof. Generalized gradient approximation made simple [phys. rev. lett. 77, 3865 (1996)]. *Phys. Rev. Lett.*, 78(7):1396, Feb 1997.
[8] Manuel Richter. Band structure theory of magnetism in 3d-4f compounds. *J. Phys. D: Appl. Phys.*, 31(9):1017, 1998.

Joint European Magnetic Symposia – JEMS 2010 IOP Publishing
Journal of Physics: Conference Series **303** (2011) 012029 doi:10.1088/1742-6596/303/1/012029

Influence of microstructural constituents on the hysteresis curves in 0.2%C and 0.45%C steels

L F T Costa[1,2], F Girotto[3], R Baiotto[3], G Gerhardt[3], M F de Campos[1], F P Missell[3]

[1]PUVR- Universidade Federal Fluminense, Av dos Trabalhadores 420, Vila Santa Cecilia, Volta Redonda, RJ, 27255-125, Brazil

[2]Eletrobras, Av. Rio Branco 53, 20th floor, Rio de Janeiro, RJ, Brazil

[3]Centro de Ciências Exatas e Tecnologia, Universidade de Caxias do Sul, Caxias do Sul, RS, 95070-560 Brazil

E-mail: fmissell@yahoo.com, mcampos@metal.eeimvr.uff.br

Abstract. Steels with 0.2% and 0.45%C were submitted to different types of heat treatment, leading to different microstructures, with different amounts of the microstructural constituents: ferrite, pearlite and martensite. Measurements of magnetic hysteresis and Barkhausen noise were performed at different frequencies. A relationship was found between coercive force and the volume fraction of the microstructural constituents. The influence of the microstructural constituents on the shape of hysteresis curve is also discussed. The presence of martensite leads to more magnetization reversal by domain rotation, reducing the relevance of domain wall movement as a reversal mechanism. When domain wall movement predominates, permeability increases and coercivity decreases. As a consequence, when the martensite volume fraction increases, coercivity also increases. The results help clarify the relation between microstructure of steels and Barkhausen noise measurements in non-destructive testing.

1. Introduction

Non-destructive evaluation (NDE) is becoming increasingly important in modern industry, contributing both to quality control on the production line and in monitoring equipment during operation. In ferromagnetic materials – carbon steels in general – the microstructural characteristics play a decisive role in determining structure sensitive magnetic properties, for example, magnetic hysteresis curves or Barkhausen noise. For this reason, a better understanding of the relation between microstructure and magnetic properties is always useful. Some time ago, Néel [1] had argued that it is possible to interpret the effects of iron carbide precipitates on the magnetic properties in terms of the pinning of domain walls by the inclusions. Dijkstra and Wert [2] assumed that magnetostatic energy is stored around second phase particles (carbides and inclusions) and that magnetostatic energy can be decreased when domain walls intersect the particles, causing the walls to be pinned by the inclusions. More recently, Jiles [3] presented a comprehensive discussion of the influence of carbon content on the properties of the hysteresis loop for several carbon steels with up to 1wt% C. That work concluded that domain wall motion was mainly influenced by the presence of iron carbide particles, either in lamellae or spheroids, while grain boundaries were of lesser importance. The models of [1,2] were recently applied to the phenomenon of magnetic aging [4], a decline of the magnetic properties due to the growth of carbide particles.

Since the measurement of Barkhausen noise is an important NDE technique, the present work was undertaken as an initial step to investigate the effect of microstructure on Barkhausen noise and magnetic properties for a couple of important steels: steel containing 0.45wt% C can be hardened by heat treatments or other means, while steel with 0.2 wt%C is a common structural material. The effect of microstructure on Barkhausen noise peak height, position, and frequency spectrum has been considered previously [5-8].

2. Experiment

Two steels were studied in this work, nominal AISI 1020 (actually C: 0.26; Si: 0.19; Mn: 0.59; P: 0.02; S: 0.02 in weight %) and AISI 1045 (actually C: 0.48; Si: 0.24; Mn: 0.80; P: 0.02; S: 0.03 in weight %). Samples were prepared in the form of toroids to facilitate magnetic measurements. All toroids were austenitized at 910°C (±10°C). After austenitization they were placed in another furnace for 20 minutes at temperatures of 700°C, 650°C, 600°C, 550°C, 500°C and then cooled in air. Normalization consists of complete austenitization followed by air cooling. Heat treatments are listed in the table below.

Table 1 Heat treatments

Number	Heat Treatment
1	Normalization
2	Water quench
3	Oil quench
4	Cool from 700°C
5	Cool from 650°C
6	Cool from 600°C
7	Cool from 550°C
8	Cool from 500°C

Metalographic preparation was with Trident and Buehler Microcloth with diamond suspensions of 15, 9, 6, 3, 1, ¼ μm, followed by attack with Nital and examination with a Zeiss AX10 optical microscope and/or a Zeiss EVO 40 scanning electron microscope. After heat treatment, toroidal samples were wound with a primary and secondary for magnetic measurements. A hysteresis loop tracer from Globalmag Ltda. generated a sinusoidal current (f = 0.05 – 400 Hz) which drove the primary. The secondary voltage was integrated to furnish the magnetic induction. The coercive field is always taken to correspond to B = 0. For Barkhausen noise measurements an additional, localized secondary was wound on each toroid. Noise measurements were made on the toroids with a home made system, using a sinusoidal magnetic field varying at 1 Hz. The signal was filtered (5–100 kHz) and amplified, after which noise data were collected (16 bits, 250 000 samples per second) and stored for later analysis. The values presented for the rms Barkhausen noise are the result of averaging over 10 pulses.

3. Magnetic measurements

Hysteresis curves for samples quenched in oil are shown in Fig. 1 for a measuring frequency of 0.5 Hz. Quenching results in the introduction of martensite into the microstructure which has the effect of increasing the coercive field substantially in regard to those samples which were slowly cooled. In Fig.

Journal of Physics: Conference Series **303** (2011) 012029 doi:10.1088/1742-6596/303/1/012029

2, H_c is shown for two frequencies (0.5 Hz and 1 Hz) for all of the heat treatments. Eddy current effects were very strong and led to large increases in the coercive field. The eddy current effects were larger than the small difference between the coercive fields of 1020 and 1045 steels which were slowly cooled from 500-700°C. Previously, for lamellar pearlite samples, Jiles [3] had encountered H_c = 10.5 Oe for the 1045 steel, but only H_c = 6.5 Oe for the 1020 steel. Of course, our steels and those of Jiles are not exactly the same, since both the residual elements and the grain size are different. The amount of Mn and Si should influence the number and size of pinning sites which might lead to increased domain wall pinning and increased H_c, for example.

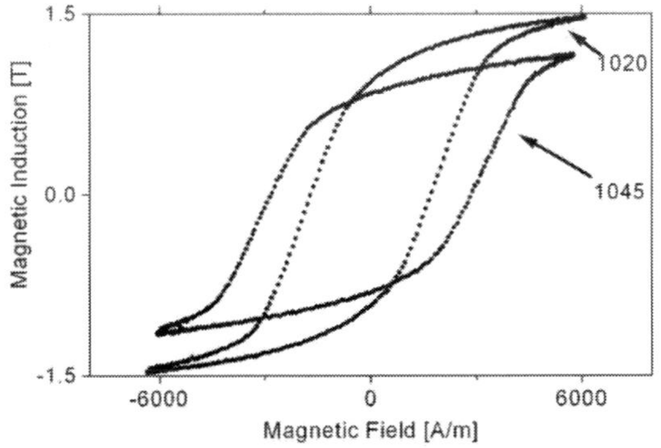

Fig. 1 Hysteresis curves for 0.5 Hz.

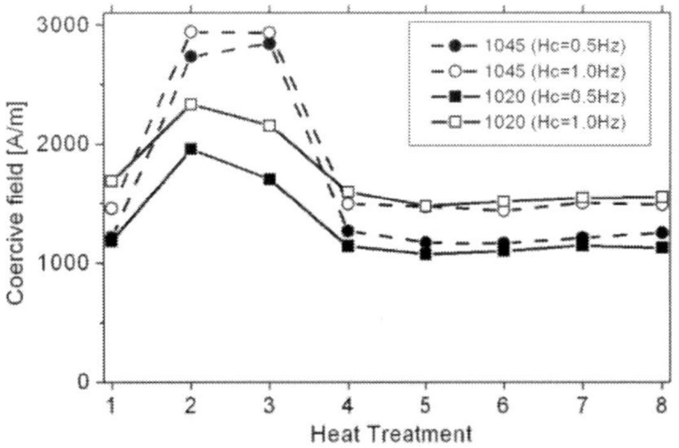

Fig. 2 H_c vs. heat treatment.

Eddy current effects lead to a strong dependence of the coercive field upon frequency and eventually result in the distortion of the hysteresis curves. This frequency dependence of H_c below 20 Hz can be seen in Fig. 3 below.

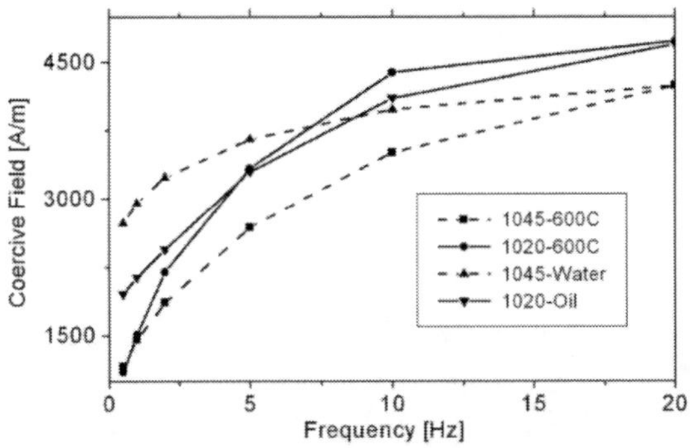

Fig. 3 Frequency dependence of H_c.

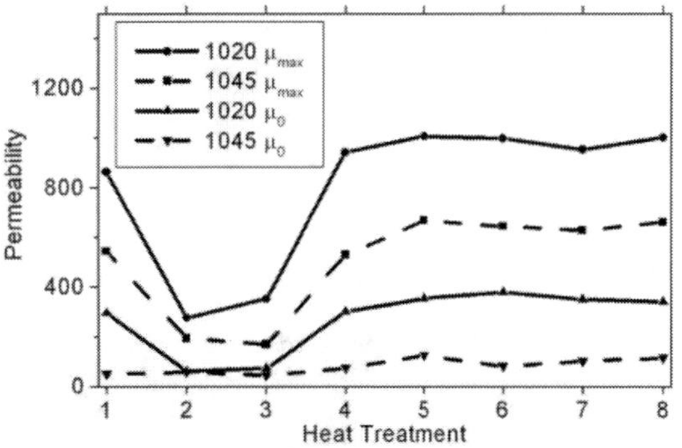

Fig. 4 Permeability vs. HT.

Initial permeability μ_o and maximum permeability μ_{max} are shown above in Fig. 4 for the 1020 and 1045 steels. The quenched samples (2 and 3) have lower permeability values indicating that wall movement is becoming more difficult. Since martensite consists of nanocrystalline, lens-shaped grains of α-Fe with carbon inducing stress in the lattice and with a high dislocation density [9], domain wall movement in such nanocrystalline grains is difficult, and domain rotation becomes relevant as an inversion mechanism. This leads to low permeability and high coercivity.

In Fig. 5 below, the rms Barkhausen voltage associated with the two steels is shown. Since the rms Barkhausen noise is related to domain wall movement, this explains the clear relationship between Figs. 4 and 5. The martensite is much harder magnetically than the ferrite-pearlite and its contribution to the Barkhausen noise is smaller than that of the softer material. The 1045 steel has more martensite (Compare Figs. 9 and 10 below.) and therefore shows a reduced Barkhausen noise voltage for samples 2 and 3. A large (magnetic field) separation of the Barkhausen contributions in two-phase steels has been observed previously [7]. Analyzing this two-peaked structure, Kleber *et al.*

[7] found a good correlation between the martensite volume fraction and the Barkhausen noise amplitude. Gür and Çam [8], studying SAE 4140 steel, found a reduced rms Barkhausen contribution from martensite, but much larger contributions from the ferrite-pearlite. Furthermore, the peak fields associated with the several constituents were not widely spaced.

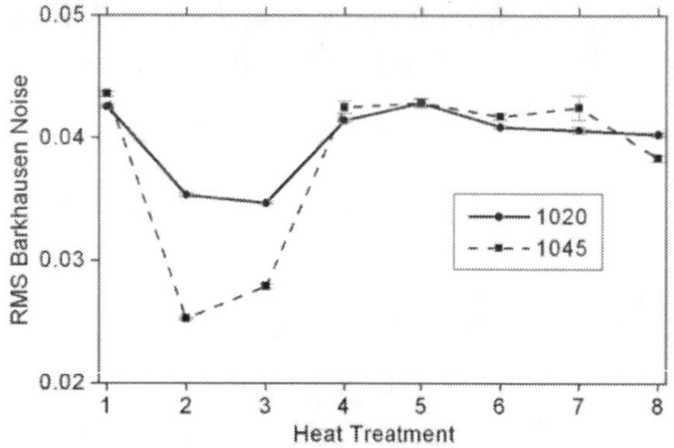

Fig. 5 RMS Barkhausen noise voltage

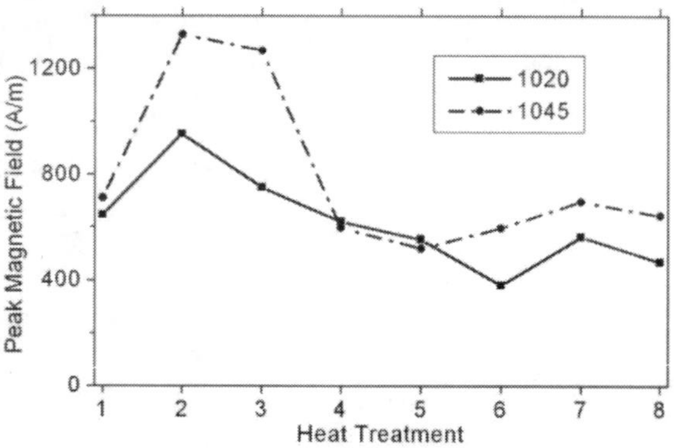

Fig. 6 Peak magnetic fields for HT

In Fig. 6 we see that the peak fields for the 1020 steel after treatments 2 and 3 are larger than those for the other treatments. But this difference is more striking in the case of the 1045 steel which contains more martensite in its microstructure. See Figs. 9 and 10 below. Recently, a detailed discussion of the factors which influence the peak fields has been reported by Kleber et al. [10] in their study of commercial dual-phase steels. Their results show [10] a linear decrease of H_p with increasing volume fraction of ferrite. This result might be consistent with the larger values of H_p that we observe for the quenched samples (heat treatments 2 and 3) since, for these samples, we find much less ferrite and more martensite. The authors of Ref. 10 attributed the shift in peak position to the distribution of magnetic flux in the two phases and presented a model for the phenomenon which seemed to be consistent with their measurements. The microstructures in the present samples, however, are more complex than those found in [10]. We also see in Fig. 6 that the peak fields are lower than the corresponding coercive fields of Fig. 2 above. This seems to indicate that there is significant domain wall movement before the magnetic field reaches H_c.

Microstructure

Optical micrographs of the normalized samples are shown in Figs. 7 and 8 below.

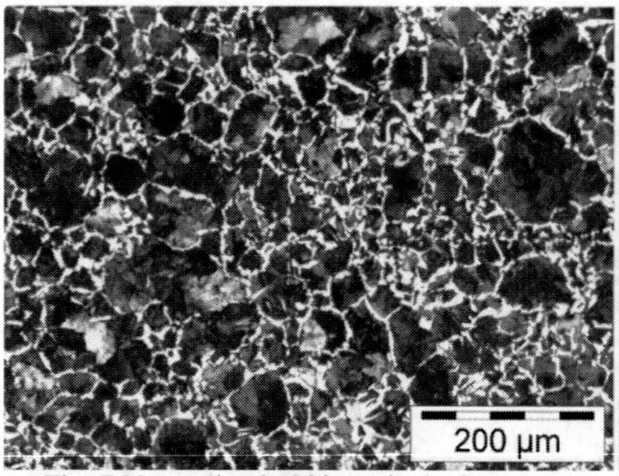

Fig. 7 Normalized 1020, optical microscopy

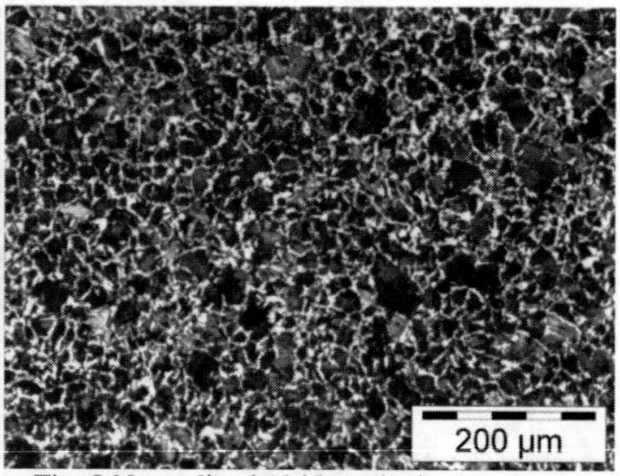

Fig. 8 Normalized 1045, optical microscopy

Normalized samples do not contain martensite, only pearlite and ferrite. However, normalized hypoeutectoid steels generally show less pro-eutectoid ferrite and a finer pearlite than their corresponding numbers which have been austenitized and treated isothermally (samples 4 – 8). Thus, from the microstructure, one might have expected H_c to be slightly greater for the normalized samples (number 1) than for those steels which were treated isothermally. However, the H_c values found seem to be very similar to those of materials treated in the range 500-700°C.

Microstructures of the oil-quenched samples are shown in Figs. 9 and 10, as obtained by optical microscopy. The samples with a higher C content always present more martensite when quenched and more pearlite for other heat treatments. Nital leaves the ferrite white in the micrographs, the darker phase being the martensite. The martensite can also be identified by the needle-shaped grains, a consequence of their lenticular shape in three dimensions.

Fig. 9 Sample 1020 oil-quenched

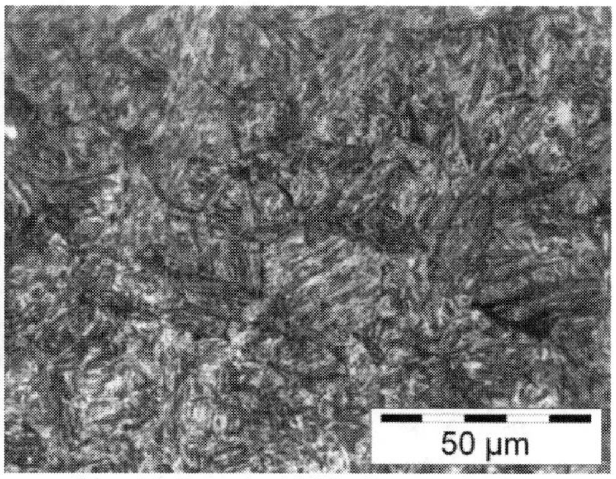

Fig. 10 Sample 1045 oil-quenched

In the samples which were slowly cooled from 500-700°C, we expect to find ferrite and pearlite. Typical microstrucutres from materials cooled from 700°C are found in Figs. 11 and 12 below. The white regions correspond to ferrite and the darker regions with a lamellar structure are pearlite.

Fig. 11 Sample 1020 cooled from 700°C

Fig. 12 Sample 1045 cooled from 700°C

Typical microstructures from material cooled from 500°C are found in Figs. 13 and 14 below. The spacing of the lamellae is slightly smaller than the spacing in the material cooled from 700°C. The interlamellar distance can be calculated from the Zener equation [11,12]. Thus it might be expected that a smaller value of H_c would be encountered in the material cooled from 700°C. A reduction in H_c with increasing interlamellar distance has been reported by Byeon and Kuwn [13]. However, the effect must be small and we did not observe a significant difference in H_c between heat treatments 4 and 8.

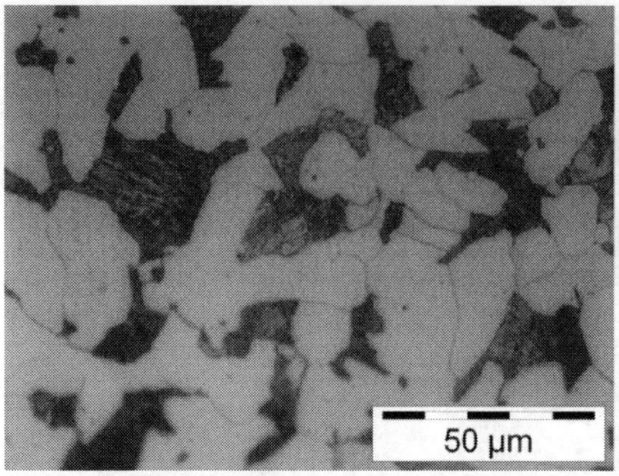

Fig. 13 Sample 1020 cooled from 500°C

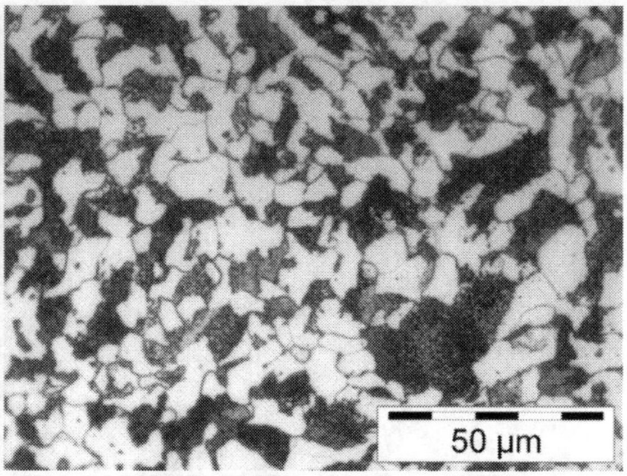

Fig. 14 Sample 1045 cooled from 500°C

4. Conclusion

A correlation between microstructure and magnetic properties has been observed which confirms an ordering of magnetic hardness: martensite > pearlite > ferrite. Samples having greater amounts of C have more martensite upon quenching and more pearlite in other heat treatments. Thus the 1045 steel had a larger H_c than the 1020 steel for all of the heat treatments considered.

Martensite is much harder magnetically than the ferrite-pearlite and its contribution to the Barkhausen noise is reduced and displaced in magnetic field. In two-phase steels, the contributions to the Barkhausen noise from martensite and ferrite are well separated in magnetic field and have been used to determine the relative proportions of the two phases and the carbon content of the martensite [7]. In the three constituent steels considered here, the problem is more complex, in part because the contributions of the individual constituents are not so easily separated.

Acknowledgments

M F de Campos, F P Missell, and R Baiotto acknowledge financial support from CNPq. M F de Campos acknowledges CNPq contract 480891/2008-7. F S B Girotto acknowledges a fellowship from FAPERGS and L F de T Costa acknowledges a graduate fellowship from CAPES. L F de T Costa acknowledges the use of the metallographic laboratory of CEPEL.

References

[1] Néel L 1949 *Physica* **15** 225
[2] Dijkstra L J, Wert C 1950 *Phys. Rev.***79** 979
[3] Jiles D C 1988 *J. Phys. D: Appl. Phys.* **21** 1186
[4] de Campos M F, Emura M, Landgraf F J G 2006 *J. Magn. Magn. Mater.* **304** e593
[5] Clapham L, Jagadish C, Atherton D L 1991 *Acta metall. mater.* **39** 1555
[6] Anglada-Rivera J, Padovese L R, Capo-Sanchez J 2001 *J. Magn. Magn. Mater.* **231** 299
[7] Kleber X, Hug A, Merlin J, Soler M 2004 *ISIJ International* **44** 1033
[8] Gür C H, Çam I 2006 *Proc. ECNDT 2006* Mo.2.2.4
[9] Hosford W F 2005 *Physical Metallurgy* Taylor & Francis
[10] Kleber X, Hug-Amalric A, Merlin J 2008 *Metal. Mater. Trans.* **39A** 1308
[11] Hillert M 1986 *J. Appl. Phys.***60** 1868
[12] Caballero F G, Capdevila C, Garcia de Andres C 2000 *Scripta Materialia* **42** 537
[13] Byeon J W, Kwun S I 2004 *Phys. Stat. Sol. (b)* **241** 1697

Magnetic Barkhausen Noise in quenched carburized steels

M F de Campos[1], F A Franco[2], R Santos[1], F S da Silva[1], S B Ribeiro[2], J F C Lins[1], L R Padovese[2]

[1]PUVR- Universidade Federal Fluminense, Av dos Trabalhadores 420, Vila Santa Cecilia, Volta Redonda, RJ, 27255-125, Brazil

[2]Escola Politécnica, Universidade de São Paulo, São Paulo, SP, Brazil

E-mail: mcampos@metal.eeimvr.uff.br, mfdcampo@uol.com.br

Abstract. Steels with different carbon content, 0.11%C and 0.48%C were submitted to a heat treatment for carburization in the surface. The samples were analyzed after several types of heat treatment, including quenching for producing martensite. The Magnetic Barkhausen Noise (MBN) is directly related to the microstructure. Samples with lower carbon content, have ferrite, a constituent where domain walls can move freely and present higher amplitude in the envelope of MBN. It is also found that the MBN peaks are quite distinct for the samples with martensite, which have lower permeability, and the results suggest that domain rotation contributes as mechanism for reversal of magnetization in martensite. The results also indicate that MBN is very suitable for monitoring the carburizing heat treatment.

1. Introduction

Magnetic Barkhausen Noise is a very important tool for non-destructive characterization [1-4]. Most of the MBN signal is attributed to 180° domain wall movement [1-4]. In steels, microstructural defects like grain boundaries, inclusions and dislocations promote both mechanical and magnetic hardening, increasing the area of hysteresis curve and reducing permeability (especially the initial permeability).
This happens because the same defects which "pin" dislocations also may "pin" domain walls. There is strong relationship between microstructure and magnetic properties and this can be used for non-destructive evaluation. However, an important remark should be made: Dislocations are linear defects (i.e., 1 dimension=1D) with ~5 Angstroms of thickness. Domain walls, sometimes considered as planar 2D, are in fact 3D (volume), and have ~1000 Angstrons of thickness (in steels). Thus, atoms in solid solution can pin dislocations, but they cannot pin domain walls. On the other hand, atoms in solid solution can generate stress in the lattice, and this has effect on the hysteresis curve and on the MBN signal [5].
Quenching of steels promotes martensite and other fine microstructural constituents, like bainite. In such fine microstructures domain wall movement is more difficult, and domain rotation can be important as magnetization reversal mechanism. The following Mechanical Hardness relationship holds: Martensite > Pearlite > Ferrite. Martensite microstructure is very hard, due to: (i) Large number of interfaces (ii) High density of dislocations (iii) Carbon in solid solution generating stress in the lattice [6].

The carburizing process is an important industrial process, where a high carbon content layer is produced in the surface, promoting high resistance against abrasion in the surface, after a subsequent quenching. The MBN can be used to non-destructive evaluation of the carburizing process, and formation of that above mentioned high-resistance high-carbon layer, as it will be discussed thoroughly in this study.

Experimental

The nominal chemical composition of the studied steels is presented in Table I. The heat treatments were made in 2 steps, with the 1st step having the objective of austenitizing the steel (austenitizing is the name of the heat treatment in the gama-iron (face centered cubic) field of the Fe-C phase diagram).
1st step: (850°C for 2 hours)
2nd step : carburization at 1050°C during 8 hours (in a CO-rich atmosphere)
Group 1: Carburized + quenching in oil. Step 1 + Step 2+ quenching in oil
Group 2: Only Carburized. Step 1 + Step 2, plus cooling in air.
Group 3: Normalized*. Only step 1, followed by a normalization heat treatment
Group 4: As received.
*) When an annealed sample is removed from the furnace and allowed to cool in air, this is called a *normalizing* heat treatment

Table I Chemical composition for the samples (AISI Steels) (wt%)

Alloy	C	Mn	Si	P	S	Cr	Ni	Mo	V	Cu	Al	Ti	Nb	Sn
1048NH	0.47	0.83	0.33	0.010	0.026	0.16	0.11	0.029	0.007	0.05	0.013	0.0009	0.014	0.008
1011E	0.10	0.45	0.27	0.012	0.004	0.22	0.09	0.029	0.006	0.07	---0	0.0010	0.014	0.009

MBN measurements were carried out with a portable equipment, named "Barktech", developed at our laboratory in USP. Figure 1 shows a simplified sketch of this equipment. The MBN sensor signal was amplified, band pass filtered. Each measurement condition was repeated 10 times.

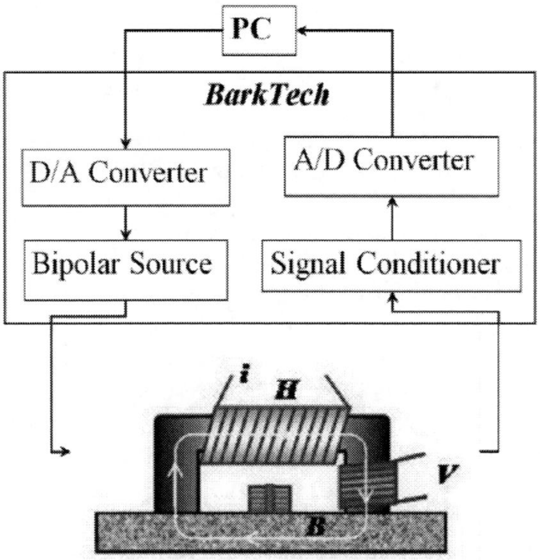

Figure 1. Experimental setup showing the U-shaped yoke and sample. The magnetic circuit is also indicated in the figure. It should be noted there are three different coils in the system.

Journal of Physics: Conference Series **303** (2011) 012030 doi:10.1088/1742-6596/303/1/012030

In the Yoke of Figure 1, an alternate current *I* circulates in a coil, producing a field *H*, and this current has controlled amplitude, shape and frequency. This current produces a magnetization cycle in the studied material. There is also a very sensitive reading coil between the Yoke poles, to detect the magnetic pulses caused by the abrupt movement of the domain walls (The Magnetic Barkhausen Noise). A third coil is placed in the Yoke (see Fig. 1), monitoring the magnetic induction and measuring the induced Voltage V. The envelope of MBN is calculated using the MBN signal. A filter is used to eliminate very high frequencies of the envelope. The upper cut-off frequency was 200 Hz. The signals were acquired in the following range of frequencies: 1Hz - 200 Hz.
Microstructures were obtained in SEM Zeiss EVO MA 10, 25 kV. X-Ray Diffraction spectra were measured in equipment Shimadzu 6000, with Co K-alpha radiation.

Results and Discussion

The Microstructures of the samples are shown in Fig 2 and Fig. 3. For both samples, 1011 and 1048, the microstructures of groups 3 and 4 are very similar, indicating that the as-received samples probably received the normalizing heat-treatment, which is typical in the steel industry. The microstructure of group 1 is martensitic for both samples 1011 and1048. The microstructures of samples group 2 is pearlitic, for samples 1001 and 1048. The microstructures of samples 1011, groups 3 and 4 are ferrite (i.e., alpha-iron) with some islands of pearlite, and it should be noted that in ferrite the free movement of domain walls is likely. Nevertheless, the microstructure of samples 1048 is basically pearlite with some islands of ferrite, and the moving of domain walls is more difficult. X-ray diffraction (XRD) has revealed large peak broadening for the samples of group 1(carburized and quenched), which have martensite. This is due to the high density of dislocations and high internal stress in the martensite.

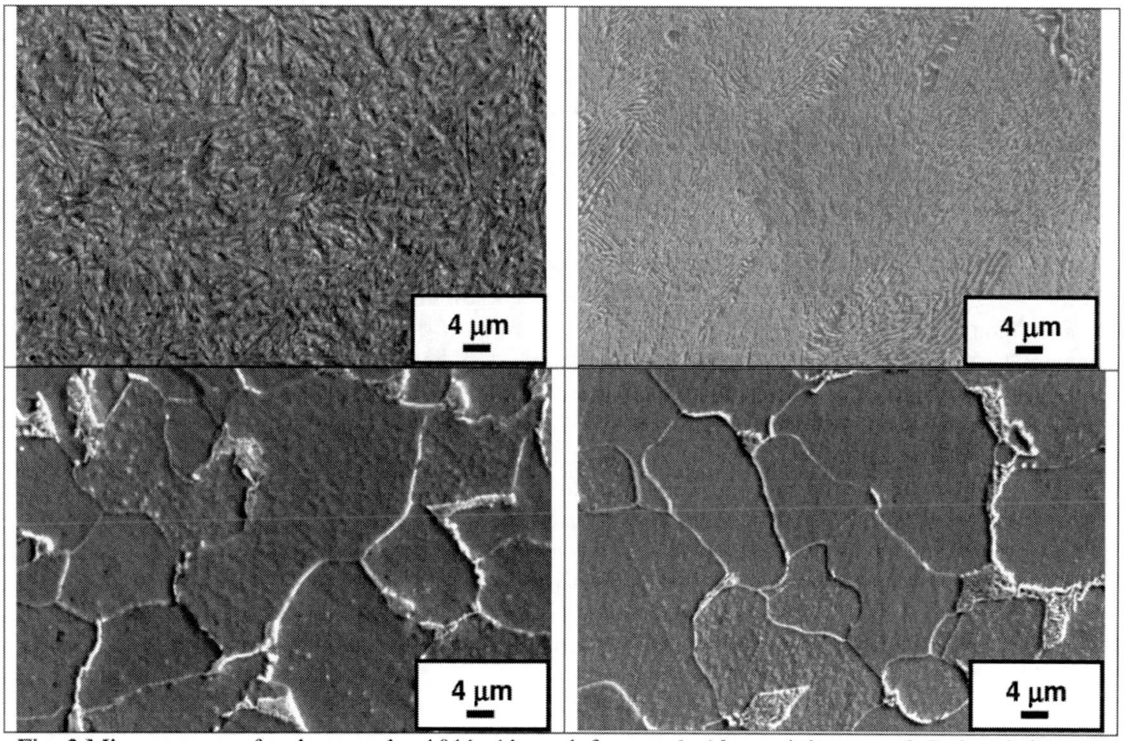

Fig. 2 Microstructure for the samples 1011. Above left group1. Above right group 2. Below left group 3. Below left group 4. Images obtained near the carburized surface for groups 1 and 2.

Journal of Physics: Conference Series **303** (2011) 012030 doi:10.1088/1742-6596/303/1/012030

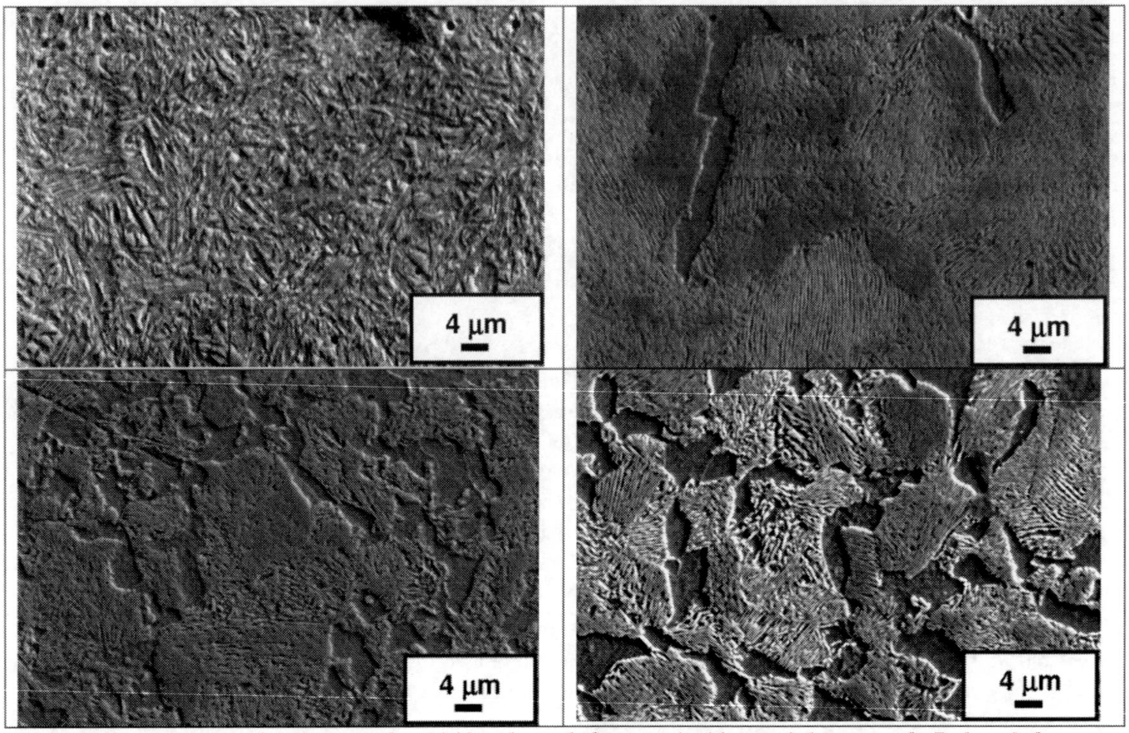

Fig. 3 Microstructure for the samples 1048. Above left group1. Above right group 2. Below left group 3. Below left group 4. Images obtained near the carburized surface for groups 1 and 2.

The MBN envelope (see Fig. 4) is quite distinct for the quenched samples (group 1), which have martensite, and this is in agreement with several other studies [5,7,8]. Martensite consists in lens shaped nanocrystalline grains of alpha-iron supersatured in carbon [6], and in this fine constituent the domain wall movement is difficult and domain rotation processes may contribute more significantly. Besides, carbon in solid solution introduces stress in the body-centered cubic (bcc) lattice of iron and affects magnetic properties. When martensite is present, it reduces the permeability (see Figure 5). Thus, Figures 4 and 5 confirm each other. The hysteresis (see Fig 5) obtained from the signals of the coils in the yoke is complementary for the MBN data, and indicate the same trend observed with the MBN: samples with lower permeability also have lower RMS of MBN. It has been reported that changing carbon content and pearlite volume fraction, the envelope of MBN is affected [9,10]. The data of Figure 4 also shows that the applied current may have effect on the envelope. For example the envelope of martensitic samples (group 1) present double peak for high current, and only one peak for the lower current, see Fig. 4. The following values (910.7 A/m and 535.7 A/m) of field intensity were chosen because they allowed signal strong enough for the calculation of the envelopes, with also good repeatability. The excitation frequency for the hysteresis (10 Hz, see Fig. 5) was chosen because it was also used for the MBN measurements. The frequency of 10 Hz allows rapid measurements, with good penetration of the applied field. However, when increasing the excitation frequency, the area of the hysteresis increases, due to eddy currents. Thus, eddy currents also affect the shape of the hysteresis, but in this case this effect is small, because the frequency is not very high.

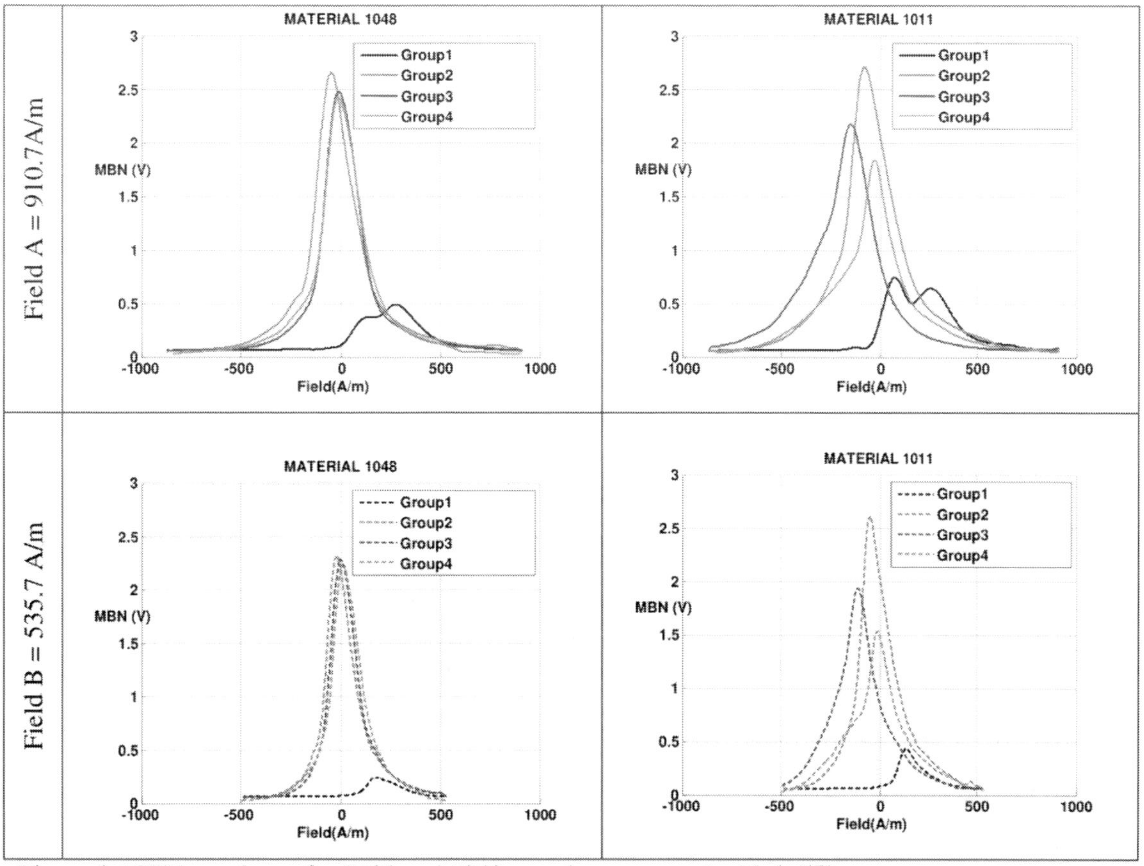

Figure 4 .MBN envelopes for 1001 and 1048 samples, measured with 2 different currents

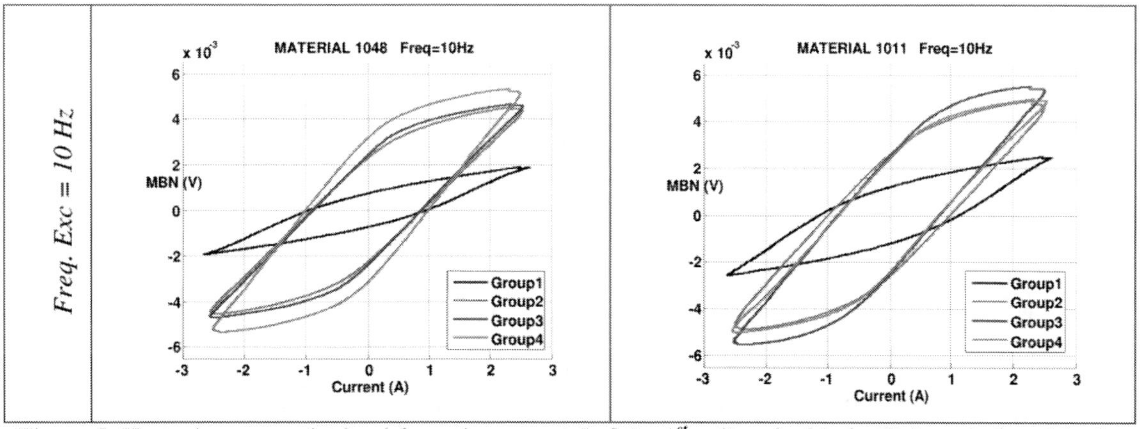

Figure 5. Hystesis curves obtained from the current *I* of the 1st coil and with the Voltage induced in the 3rd coil (see Fig. 1). These hysteresis curves are qualitative, but provide an interesting comparison between the samples, since the samples have the same geometry. This analysis indicates that the martensitic samples (group 1) have lower permeability.

In the Figure 5, hysteresis curves obtained from the magnetic circuit comprising the U-shaped yoke and sample surface are presented (see Fig. 1). A hypothetical *H* field could be obtained from the data of figure 5 using the expression: $H=n\,I/L$, where n is the number of turns in the excitation coil, I is the applied current and L is the length of the magnetic circuit. A hypothetical magnetic induction B could also be estimated using data from Fig. 5 with the expression:

$$B = \frac{1}{nA}\int\left(\frac{d\phi}{dt}\right)dt \therefore [T] \qquad (1)$$

where: n is the number of turns in the third coil (see Fig. 1); A is the transverse area of the yoke [m^2]; and $\frac{d\phi}{dt}$ is the variation of magnetic flux, or the Voltage measured in the third coil (see Fig. 1).

It should be also noted that the analysis presented in Figure 5 only makes sense if the magnetic permeability of the samples is lower than the permeability of the material of the yoke.

Conclusions

It is suggested the use of the BxH curve obtained from coils of the yoke as complementary information, besides the MBN. The employed analysis (MBN and Hysteresis with excitation coil) requests samples with identical geometry, because of the magnetic circuit. The qualitative hysteresis obtained from the yoke indicate that the martensitic samples have lower permeability. It is also found that the MBN envelopes are quite distinct for the martensitic samples when compared to the ferritic/pearlitic samples. The results allow establishing a relationship between the MBN and the microstructure, as function of the carbon content, and indicate that MBN is very suitable for monitoring the carburizing heat treatment.

Acknowledgements

M F de Campos and L R Padovese acknowledge financial support from CNPq. M F de Campos acknowledges CNPq contract 480891/2008-7. The authors would like to thank the LMME / UFF for technical support during electron microscopy work.

References

[1] Ranjan R, Jiles D C, Rastogi P K 1987 *IEEE Trans. Magn.* **Mag-23** 1869.
[2] Alessandro B, Beatrice C, Bertotti G, Montorsi A 1990 *J. Appl. Phys.* **68** 2901.
[3] Sablik M J, Augustyniak B, de Campos M F, Landgraf F J G 2008 *IEEE Trans. Magn.* **Mag-44** 3221.
[4] Pérez-Benitez J A, Capó-Sánchez J, Padovese L R 2008 *J. Appl. Phys.* **103**, 043910.
[5] Blaow M, Evans J T, Shaw, BA 2005 *Acta Materialia* **53** 279
[6] Hosford W F 2005 *Physical Metallurgy* Taylor & Francis.
[7] Kaplan M, Gür, C H, Erdogan M 2007 *J Nondestruct. Eval.* **26** 79.
[8] Kleber X, Hug A, Merlin J, Soler M 2004 *ISIJ International* **44** 1033
[9] Ng D H L, Cheng K, Cho K S, Ren Z Y, Ma X Y, Chan S L I 2001 *IEEE Trans. Magn.* **Mag-37** 2734
[10] Lo C C H, Scruby C B 1999 *J. Appl. Phys.* **85** 5193.

submitted to J.Phys.: Conference Series

The change of anisotropy in TbNi(Al,In) compounds studied by low temperature x-ray diffraction

M. Klicpera, P. Javorský, S. Daniš

Charles University, Faculty of Mathematics and Physics, Department of Condensed Matter Physics, Ke Karlovu 5, 121 16 Prague 2, The Czech Republic.

E-mail: mi.klicpera@seznam.cz

Abstract. We have investigated the transition between the easy-axis and easy-plane type of magnetocrystalline anisotropy in the TbNi(Al,In) series. These compounds crystallize in the hexagonal ZrNiAl-type structure and order magnetically at low temperatures. TbNiAl is an antiferromagnet below 47 K with the moments aligned along the c-axis, whereas TbNiIn orders magnetically below 70 K with moments perpendicular to the c-axis. The presented study is based mainly on the low temperature x-ray powder diffraction performed in presence of external magnetic field. We have revealed that the transition between both types of anisotropy occurs in compounds with In concentration around 30%.

1. Introduction

The ternary RTX compounds (R = rare earth, T = transition d -metal, X = p -metal) crystallizing in the hexagonal ZrNiAl-type structure (space group P$\bar{6}$2m; see figure 1) form a large group of compounds showing interesting magnetic properties. The type of magnetic order is very sensitive to the number of conduction electrons. On the other hand, the magnetocrystalline anisotropy seems to be strongly influenced by the ratio of lattice parameters. This feature is most noticeable in the Tb-based compounds showing the strongest anisotropy. The substitution of Al by In in the $TbNiAl_{1-x}In_x$ series is isoelectronic, but involves a large change of lattice parameters. Therefore, the changes of magnetic properties are primarily driven by this structural change.

TbNiAl orders antiferromagnetically below T_N = 47 K with a propagation vector of (1/2 0 1/2) [1, 2]. The Tb magnetic moments are aligned along the hexagonal c-axis, but one third of moments remains reduced to almost zero due to a geometrical frustration. The frustrated moments start to propagate along different equivalent direction, e.g. with propagation vector (-1/2 1/2 1/2), below further magnetic phase transition at T_1 = 23 K and finally all the Tb moments reach the same value at 2 K. The metamagnetic transition to a ferromagnetic state occurs when applying external magnetic field of $\simeq$ 0.4 T along the c-axis, whereas the magnetic structure is almost insensitive to magnetic fields applied perpendicular to the c-axis [2]. TbNiIn orders magnetically around T_{ord} = 70 K and the AC-susceptibility indicated additional magnetic phase transitions at T_1 = 59 K and T_2 = 29 K [3]. As inferred from neutron diffraction, the Tb moments lie within the basal plane and form a non-collinear structure with (0 0 0) propagation below T_{ord} and additional weaker component develops below T_2 [4, 5].

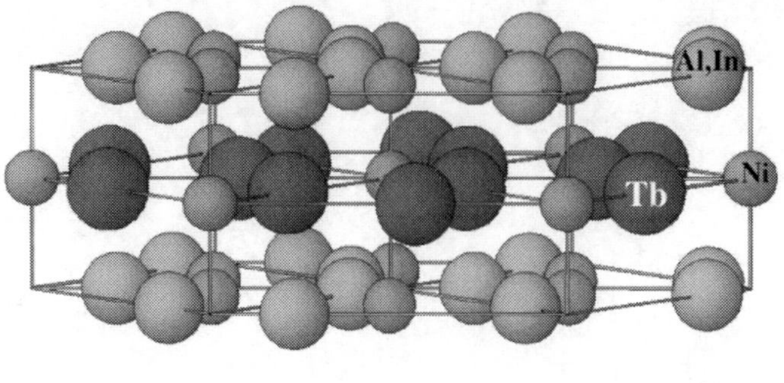

Figure 1. ZrNiAl-type structure (space group P$\bar{6}$2m). Green balls present Tb-atoms, blue balls Ni-atoms and orange balls Al- or In-atoms.

2. Experimental

Polycrystalline $TbNiAl_{1-x}In_x$ samples were prepared by arc-melting of pure elements (3N for Tb, 4N5 for Ni, 5N5 for Al and 6N for In) in a mono-arc furnace under the protection of an argon atmosphere. A higher evaporation of In (approximately two or three orders of magnitude) was considered before weighing individual elements. The X-ray diffraction patterns obtained using Cu K_α radiation showed single-phase samples with the ZrNiAl-type structure.

The low-temperature X-ray diffraction measurements were performed on a Bragg-Brentano diffractometer (Siemens D-500) equipped with helium gas flow cryostat (Oxford Instruments CF1108T) on powder samples fixed by acetone on the Si plate. The diffraction patterns were recorded for 2θ range of 20 - 140 degrees with the step of 0.05 degree by using position-sensitive detector (Braun). Filtered radiation of the cobalt tube (Co K_α) was employed. The temperature of the samples was stabilized within 0.1 K by the ITC-503 temperature controller operating with the RhFe resistor. The diffraction patterns were taken for various temperatures: all samples at 300, 100 and 20 K, selected samples also at 200 and 5 K. Measurements were performed first at the room temperature in zero magnetic field. Consequently, permanent magnet made of sintered Sm_2(Co, Fe, Cu, Zr)$_{17}$ alloy (Vacomax 225HR, Vakuumschmelze) was placed under the sample holder. It produced a magnetic field of approximately 1 T perpendicular to the sample surface. The room-temperature pattern was then taken again and revealed identical results as the zero-field measurement. The temperature was then gradually lowered to the selected temperatures and the diffraction patterns were taken always at stabilized temperature. The refinement of the structure parameters from the diffraction patterns was done using the Rietveld analysis employing the Fullprof program [6].

3. Results and discussion

Diffraction pattern of pure TbNiAl measured at 20 K (i.e. below T_N in the presence of the magnetic field reveals a strongly textured powder (see figures 2 and 3). Observed increase in intensity of the (0 0 l) reflections indicated macroscopic ordering of crystallites with the c-axis parallel to the direction of the external magnetic field. This is in agreement with the results of neutron diffraction and magnetization showing ferromagnetic order with moments aligned along the c-axis in 1 T [2]. Diffraction patterns of In poor compounds ($x < 0.3$) show similar behavior as pure TbNiAl, but the increase of intensity on the (0 0 l) peaks becomes less prominent. We can observe also the strongest peaks other than (0 0 l) (see figure 4). The intensity of (001) and (002) reflections are still clearly enhanced in the sample with 30% of In, but the preferred orientation is much weaker compared to TbNiAl (see figures 3 and 4). Very small difference between the 300 and 20 K patterns is observed for 40% of In and the texture seems to disappear completely for $x > 0.4$ as can be seen from figure 4. This is in agreement with the complex non-collinear magnetic structure reported for TbNiIn [4, 5].

Figure 2. The textured powder on $TbNiAl_{0.9}In_{0.1}$ sample after measurement and heating to the room temperature.

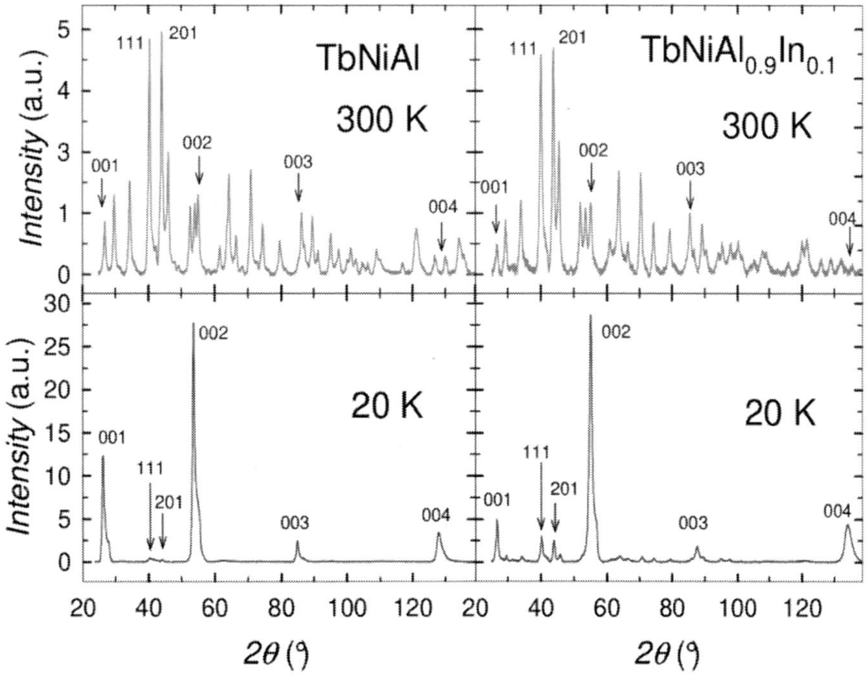

Figure 3. X-ray diffraction patterns at temperatures above and below magnetic phase transition. Strongest peaks are described by indices (h k l).

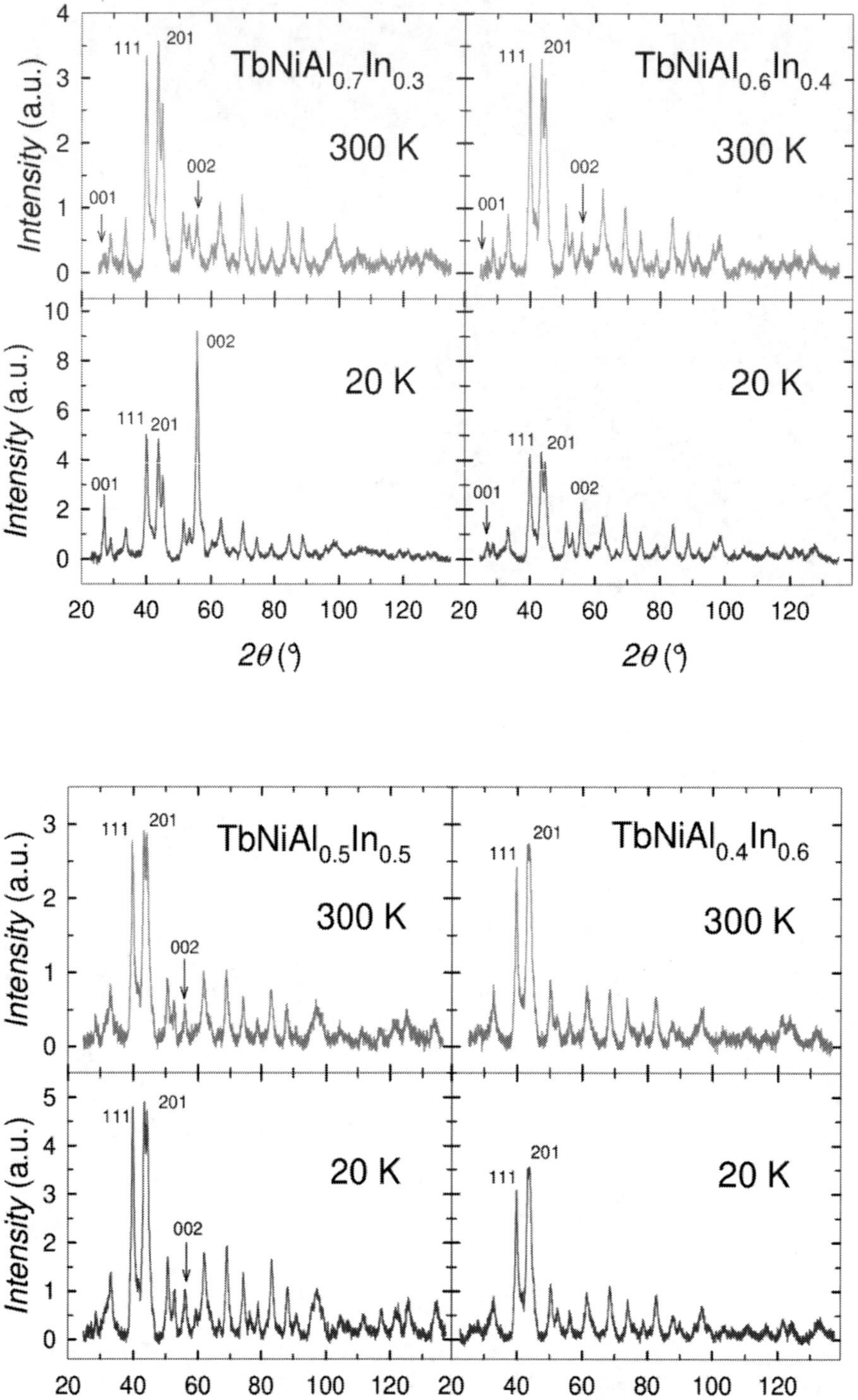

Figure 4. X-ray diffraction patterns at temperatures above and below magnetic phase transition. Strongest peaks are described by indices ($h\ k\ l$).

The concentration dependence of lattice parameters is approximately linear as can be seen from table 1 and figure 5. The lattice parameter a increases with increasing the In content, whereas the parameter c decreases. This observation can be well understood as the Al(In) atoms lie only within one of the two layers (see figure 1) and the replacement of Al by larger In causes expansion in this layer. The large difference between pure TbNiAl and $x = 0.1$ sample is due to the fact that TbNiAl undergoes structural change (change of the lattice parameters only, not the crystallographic structure itself) around 100 K [7]. Our study revealed that no such change occurs for the rest of the series as can be seen from figure 5 for the $x = 0.1$ compound. This change, in fact, could not be expected here because the c/a ratio is far from the critical region [7].

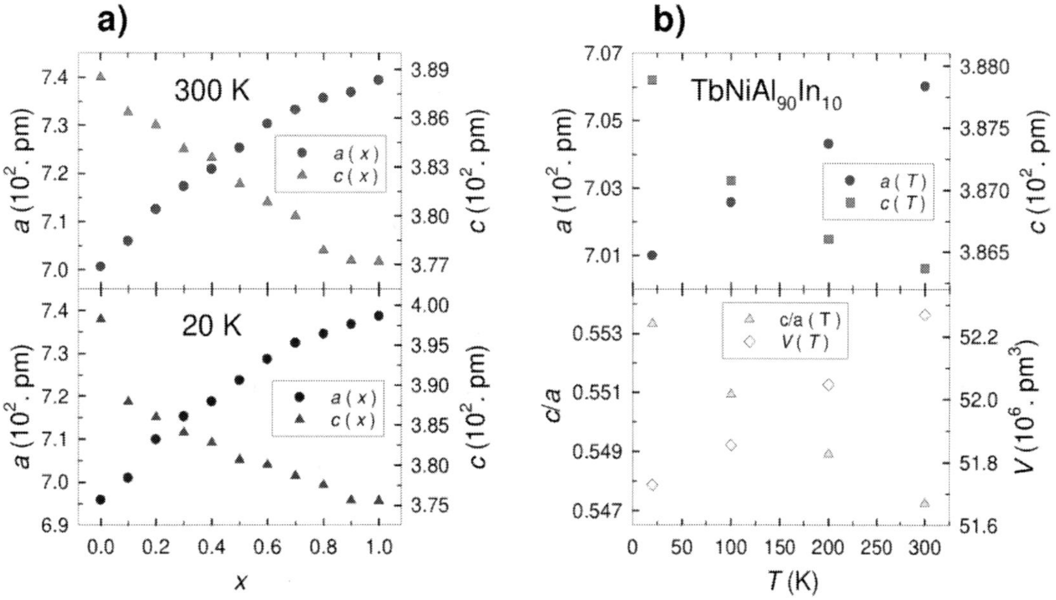

Figure 5. **a)** - Lattice parameters of $TbNiAl_xIn_{1-x}$ compounds at temperatures above and below the magnetic phase transition. **b)** - Temperature dependence of lattice parameters of $TbNiAl_{0.9}In_{0.1}$.

The temperature development of the lattice parameters reflects also the change of the magnetocrystalline anisotropy. When comparing the data above and below the temperature of magnetic ordering (300 and 20 K, respectively), the c/a ratio increases for compounds with x up to 0.2 and stays nearly unchanged for the rest of the series (see table 1 and figure 5). This is probably related to the change of magnetocrystalline anisotropy around $x = 0.3$ as indicated by the observed change of the preferred orientation in magnetic field. More detailed temperature dependence of the lattice parameters should be performed to make an unambiguous statement about the relation between structural and magnetic parameters (see reference [8] for comparison).

Table 1. Lattice parameters of $TbNiAl_{1-x}In_x$ at temperature above and below the magnetic phase transition.

	300 K			20 K		
x	a (pm)	c (pm)	c/a	a (pm)	c (pm)	c/a
0.0	700.5(3)	388.5(2)	0.5545(3)	696(24)	398(17)	0.572(23)
0.1	706.0(9)	386.4(5)	0.5472(8)	701(6)	388(4)	0.553(6)
0.2	713(1)	385.6(6)	0.541(1)	710(1)	386(1)	0.544(1)
0.3	717(1)	384.1(6)	0.535(1)	715(4)	384(2)	0.537(3)
0.4	721(2)	383.6(7)	0.532(1)	719(3)	383(2)	0.533(3)
0.5	725(2)	382.0(8)	0.527(2)	724(2)	381(1)	0.526(1)
0.6	730(2)	381(1)	0.522(2)	729(2)	380(1)	0.522(2)
0.7	733(3)	380(1)	0.518(2)	733(3)	379(1)	0.517(2)
0.8	736(3)	378(1)	0.514(2)	735(3)	378(1)	0.514(2)
0.9	737(3)	377(1)	0.512(2)	737(3)	376(1)	0.510(2)
1.0	739(3)	377(1)	0.510(2)	739(3)	376(2)	0.508(2)

4. Conclusions

The X-ray diffraction patterns of the $TbNiAl_{1-x}In_x$ series collected in presence of magnetic field below the respective magnetic ordering temperatures clearly indicated transition from the uniaxial anisotropy of TbniAl to a more complex non-collinear magnetic structures of the In rich compounds. The anisotropy change occurs for compounds with In concentration around 30%.

Acknowledgments

The work was supported by the Grant Agency of the Czech Republic under the Grant no. 202/08/0711, the work of M. K. was also supported by the grant SVV-2010-261303. This work is a part of the research program MSM 0021620834 financed by the Ministry of Education of the Czech Republic.

References

[1] Ehlers G and Maletta H 1996 *Z. Phys. B* **101** 317.
[2] Javorský P, Burlet P, Sechovský V, Andrejev A V, Brown J, Svoboda P, 1997 *J. Magn. Magn. Mater.* **166** 133-140.
[3] Tyvanchuk Yu B, Kalyczak Ya M, Gondek L, Rams M, Szytuła, Tomkowicz Z, 2004 *J. Magn. Magn. Mater.* **277** 368.
[4] Gondek L, Szytula A, Baran S, Hernandez-Velasco J, 2004 *J. Magn. Magn. Mater.* **272 276** e443-e444.
[5] Ehlers G, 1996 *Phd. Thesis.*
[6] RodriguezCarvajal J, 1993 *Physica B* **192** 55.
[7] Prchal J, Javorský P, Rusz J, de Boer F, Diviš M, Kitazawa H, Dönni A, Daniš S, Sechovský V, 2008 *Phys. Rev. B* **77** 134106 and references therein.
[8] Daniš S, Javorský P and Rafaja D, 2002 *J. Alloys and Compounds* **345** 10-15.

Journal of Physics: Conference Series **303** (2011) 012032 doi:10.1088/1742-6596/303/1/012032

The magnetostriction of the intermetallic compound $ErCo_2$ near the magnetic phase transition paramagnetism-ferrimagnetism

S A Nikitin[1], D Yu Karpenkov[2], A Yu Karpenkov[2], N I Nizhankovskii[3], T Palewski[3], K P Skokov[2]

[1]Department of Physics, Lomonosov Moscow State University, Leninskie Gory, 119992, Moscow, Russia
[2]Physicotechnical Department, Tver State University,Sadovy lane,35,170002,Tver, Russia
[3]International Laboratory of High Magnetic Fields and Low Temperatures, Gajowicka 95, 53- 421, Wroclaw, Poland

E-mail: Karpenkov_d_y@mail.ru

Abstract. The purpose of this work was the investigation of magnetostriction near magnetic phase transition paramagnetism-ferrimagnetism in ferrimagnetic intermetallic compound $ErCo_2$. It is known that 3d-electrons are delocalized and form a 3d-zone in it, whereas 4f-electrons localized and create an effective exchange field which acts on the Co sublattice. It leads to magnetic ordering of the last. We have found there is a sharp jump in the curve of thermal expansion near T = 37K, which indicates the presence of first order phase transition. Volume magnetostriction ω reach huge values (3.10^{-3} in the field μ_0H = 4T near T_C). At temperatures $T > T_C$ $\lambda_{\parallel}$ and $\lambda_{\perp}$ have small values in the absence of an external magnetic field. However, in fields $H>H_{crit}$ $\lambda_{\parallel}$ and $\lambda_{\perp}$ rapidly increase and get the values that have the same order of magnitude as at $T \approx T_C$. This indicates that the external magnetic field can restore the magnetism of 3d-subsystem at $T>T_C$.

1. Introduction

Magnetism of intermetallic RCo_2 compounds is the object of intense research due to the unique combination of itinerant magnetism of the cobalt 3d-sublattice and localized 4f-sublattice magnetism of rare-earth ions [1-5]. RCo_2 compounds with nonmagnetic rare-earth elements (Y, Lu) are exchange-enhanced paramagnetics, which do not have the spontaneous moment in the cobalt sublattice. However, the rare-earth sublattice of the ions having magnetic moments induces a moment of the cobalt sublattice due to the R-Co exchange interaction. The effective exchange field H_{R-Co} is defined by the spin and total angular momentum of the RE ion and electron density of states at the Fermi level. The magnitude of the magnetic moment of the itinerant Co sublattice strongly depends on the magnitude of H_{R-Co}. If H_{R-Co} is large, the Co sublattice has a stable zone magnetic moment ($GdCo_2$, $TbCo_2$). Reduction of the H_{R-Co} leads to situation when the magnetic moment M_{Co} disappears above the Curie temperature ($ErCo_2$, $HoCo_2$). But it can occur near Tc due to the external magnetic field, which is added to the H_{R-Co}. In this case phase transitions of the first order and restoration of itinerant magnetism of the Co sublattice above Tc are observed due to the influence of an external magnetic field. To find out the physical nature of this transition it is important to determine the energy contributions to the thermodynamic potential near Tc and to determine the influence of lattice parameters on order of magnetic phase transitions [6].

The purpose of this investigation was to study the magnetoelastic characteristics (longitudinal, transverse and volume magnetostriction, thermal expansion) near the magnetic phase transition (paramagnetism-ferrimagnetism) of ferrimagnetic intermetallic compound $ErCo_2$.

$ErCo_2$ compound has a cubic crystal lattice Laves' phase type, and it is the two-sublattice magnet. It is known that there 3d-electrons are delocalized and form a 3d-zone, whereas 4f-electrons, on the contrary, are localized and create an effective exchange field which acts on the Co sublattice, that leads to the magnetic ordering of the last[1].

Taking into account the abovementioned, in the cubic intermetallic compounds RCo_2 (unlike RFe_2, where the single-ion magnetostriction (MS) dominates) two mechanisms of MS are observed. In addition to the single-ion (anisotropic) MS there is a volume (isotropic) magnetostriction [7, 8]. This phenomenon was studied in detail by us during measurements of longitudinal and transverse magnetostriction $ErCo_2$ compound in fields up to 100 kOe in the temperature range 4.2-120K.

2. Experimental details

As cast $ErCo_2$ samples were prepared with method of high-frequency induction melting in an atmosphere of argon. The samples have plate shape and the linear sizes 4x5x2 mm. The crystalline structure and phase composition of the samples were checked by X-ray diffraction. The measurements of both longitudinal $\lambda_{||}$ and transverse $\lambda_{\perp}$ magnetostriction induced by the magnetic field up to 10T in the temperature range 4.2 - 120K in $ErCo_2$ were carried out with resistance strain gauges. The Vishay Micromeasurements SR-4 strain gauges were pasted on a surface of the sample.

3. Result and discussion

Fig. 1 shows the field (a) and temperature (b) dependence of the longitudinal magnetostriction of $ErCo_2$ in fields up to B = 10T. It is evident that $\lambda_{||}<0$ and increases in absolute value below Tc = 38K, where the transition from the paramagnetic state to a ferromagnetic occur [3]. The presence of the critical magnetic field Hc at T> Tc attracts attention, at which $\lambda_{||}$ increases dramatically. Hc appears on the isotherms of the transverse magnetostriction $\lambda_{\perp}$ (Fig. 2a) at T>Tc more clearly. There are huge positive values of $\lambda_{\perp}$, reaching values of $\sim1800\cdot10^{-6}$ near Tc = 38K. The increase of temperature at T>Tc decreases the jump on the isotherms $\lambda_{\perp}(H)$ little, since the value of Hc increases with increasing of temperature. The temperature dependence of Hc is shown in Fig. 2(b). $\lambda_{\perp}$ (T) has a huge maximum of $\sim1900\cdot10^{-6}$ near Tc.

a) b)

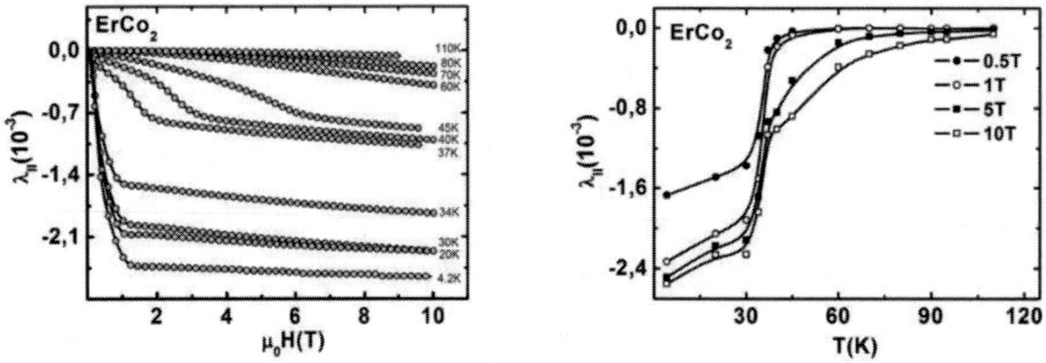

Figure 1. Field (a) and temperature (b) dependence of the longitudinal magnetostriction of polycrystalline samples $ErCo_2$

Using the obtained data of $\lambda_{\parallel}$ and $\lambda_{\perp}$ we received temperature dependence of $H_{crit}(T)$ of metamagnetic transition (fig.3). It is seen that H_{crit} linearly increases in area of temperatures above the Tc and reaches values from small, at a point of transition at Tc=33K, to H_{crit}=5T at T=44K. The results of investigations show, the energy barrier, that separate paramagnetic and ferrimagnetic phases, increases with temperature increasing.

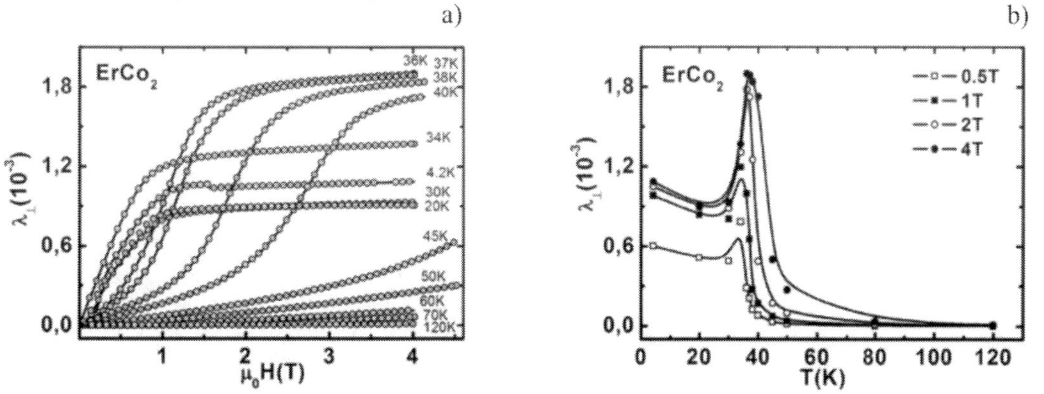

Figure 2. Field (a) and temperature (b) dependence of the transverse magnetostriction of polycrystalline samples $ErCo_2$

Hence, $\lambda_{\parallel}$ and the $\lambda_{\perp}$ reveal two magnetostrictive contributions. The first - when T <Tc is characterized by strong anisotropy $\lambda_{\parallel}<0$ and $\lambda_{\perp}> 0$. This contribution is caused by the single-ion anisotropic magnetostriction, which is due to the magnetocrystalline interaction, especially its magnetoelastic part. The second contribution is caused by the band magnetostriction, which is induced by the magnetic field near and above Tc at H>Hc. Significantly, in $ErCo_2$ volume magnetostriction is extremely high near Tc have a maximum $\omega \sim 2700 \cdot 10^{-6}$(fig.4). The measured curve of thermal expansion shows that giant magnetostriction values exist in the sample even in the absence of a magnetic field (fig.5). The influence of magnetic field on band structure of the compound leads to additional volume effect.

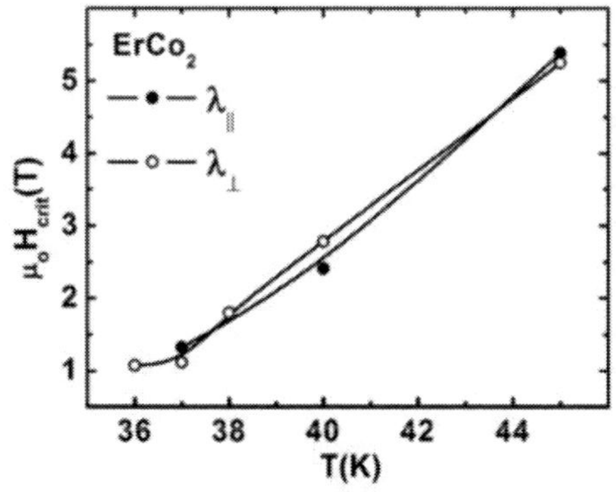

Figure 3. Temperature dependence of critical field of polycrystalline samples $ErCo_2$

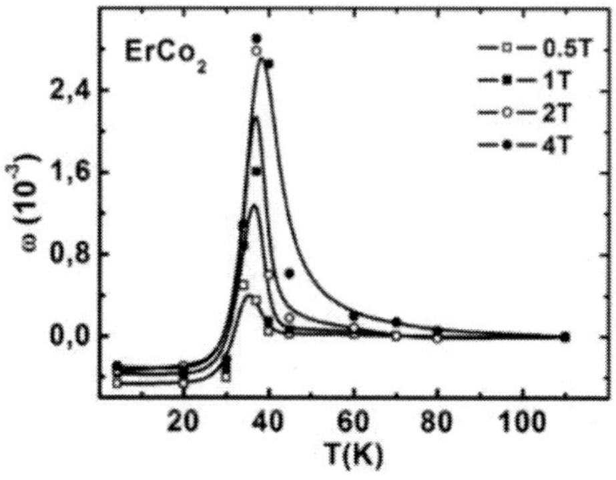

Figure 4. Temperature dependence of the volume magnetostriction of polycrystalline samples $ErCo_2$

It is seen that in $ErCo_2$ at T<Tc there is a huge spontaneous magnetostriction and invar anomaly of the coefficient of thermal expansion (fig.6). Because of the additional to usual thermal expansion volume expansion, caused by spontaneous band magnetostriction, the coefficient of thermal expansion reveals anomaly - sharp negative peak on a curve $\alpha(T)$ (fig.6). Since the magnetic field displaces

temperature of metamagnetic transition to the higher temperature region, the negative maximum on curve B=2T is displaced to the right.

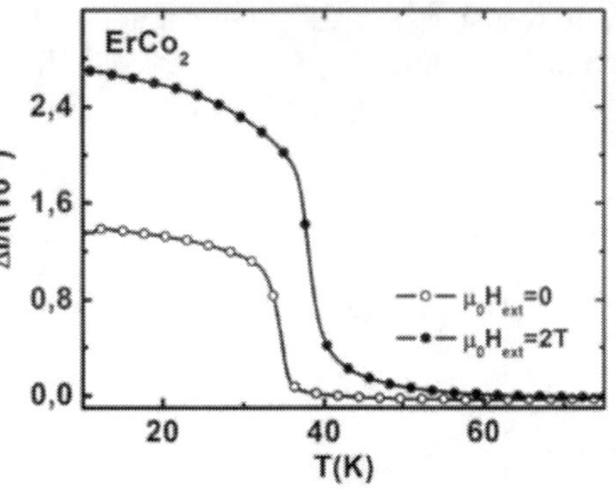

Figure 5. Thermal expansion of polycrystalline samples $ErCo_2$

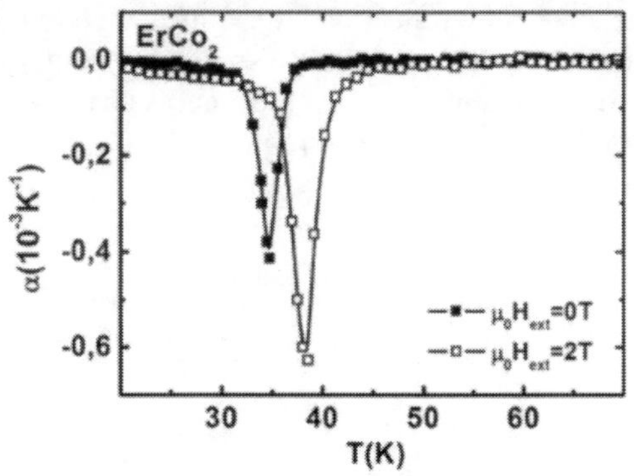

Figure 4. Temperature dependence of the coefficient of thermal expansion of polycrystalline samples $ErCo_2$

4. Conclusion

In intermetallic $ErCo_2$ compounds magnetostriction has two contributions: 1) the anisotropic single-ion magnetostriction is caused by the magnetoelastic energy due to magnetocrystalline interaction, 2) the band magnetostriction, which is due to the influence of the magnetostriction deformations on band structure of 3d-electrons.

The band magnetostriction has a giant value of more than 10^{-3}. It is restored at T>Tc in the fields above critical as a result of the metamagnetic transition, at which external magnetic and exchange fields induced the band magnetism of 3d-sublattice. Band magnetostriction caused by the dependence of energy 3d-band from the interatomic distances.

References

[1] Levitin R Z, Markosyan A S 1988 *Sov.Phys.-Usp.* **31** 730
[2] Gratz E, Markosyan A S 2001 *J. Phys.: Condens. Matter* **13** R385–R413.
[3] Tureka, Ruszb J, Diviš M 2005 *Journal of Magnetism and Magnetic Materials* 357-363.
[4] Belov K P 1987 *Magnetostriction phenomena and their technical applications* (Moscow) p 160
[5] Duct N H, Hient T D, Brommer P E and J J M Franse 1988 *J.Phys F: Met. Phys.* **18** 275-294
[6] Khmelevski S and Mohn P 2000 *J. Phys.:Condens. Matter* **12** 9453
[7] Pourarian F 1978 *Phys.Lett 67A* **5,6** 407-409
[8] Levitin R Z, Markosyan A S and Snegirev V V 1982 *JETP* **36** (10) 367-369

Acknowledgments

This work is supported by the grant RFBR 10-02-00721-a.

Journal of Physics: Conference Series **303** (2011) 012033 doi:10.1088/1742-6596/303/1/012033

Spin entanglement in supramolecular systems.

G. Lorusso[1,2], **F. Troiani**[1], **V. Bellini**[1], **A. Ghirri**[1] **A. Candini**[1], **S. Carretta**[1,3], **P. Santini**[1,3], **G. Amoretti**[1,3], **W. Wernsdorfer**[4], **G. Timco**[5], **R.E.P. Winpenny**[5], **M. Affronte**[1,2]

[1]CNR-Institute of nanoSciences S3, via G. Campi 213/a, 41100 Modena, (Italy).
[2]Dipartimento di Fisica, Università di Modena e Reggio Emilia, via G. Campi 213/a, 41100 Modena, Italy
[3]Dipartimento di Fisica, Università di Parma and Unità CNISM di Parma, via Usberti 7/a, 43100 Parma, Italy.
[4]Institut L. Néel CNRS, BP166, 25 Av des Martyrs, 38042 Grenoble, France.
[5]School of Chemistry, University of Manchester, Oxford Road, Manchester M139PL, United Kingdom.

E-mail: giulia.lorusso@unimore.it

Abstract. Entanglement is an intriguing property of quantum systems that is essential for the implementation of quantum information processing. We review recent results on the establishment of magnetic coupling and spin entanglement between molecular clusters of Cr_7Ni antiferromagnetic rings. We firstly present the magnetic features of the molecular building blocks and the strategy to link them without altering their properties. A systematic study of the magnetic interaction in $(Cr_7Ni)_2$ dimers through different heteroaromatic linkers is then reported. Finally, for the dimer with strongest coupling, the entanglement is directly investigated both theoretically and experimentally.

1. Introduction

Entanglement is a peculiar property of quantum mechanics. Given a composite system, entanglement is established when the states of the parts are so highly correlated each other that it is impossible to describe one quantum component without the knowledge of the rest of the system [1, 2]. Consequently, having two entangled qubits, if we carry out an experiment and we measure one of them, we have information about the state of the other qubit. This property is a great resource for fields like quantum cryptography, teleportation, information and computation [3].

The interest for the exploitation in quantum information science, together with the appeal of quantum entanglement itself, has stimulated intensive research activity for the controlled generation of entanglement between nanoscaled objects. For photons [4] or cold atoms, as well as for a few solid state systems [5], entanglement is largely investigated, both theoretically and experimentally trying to unveil both its qualitative and quantitative aspects [1, 2].

For what concern the entanglement in magnetic systems there is an extensive theoretical literature [6, 7, 8, 9, 10, 11] but very few experiments [12, 13] and the latter are referred to collections of spin systems. Thus, the problem of measuring entanglement in finite spin systems is still very open. Molecular spin clusters represent a suitable test bed in this context. In fact, they represent complex but finite systems whose structural and physical features can be tailored

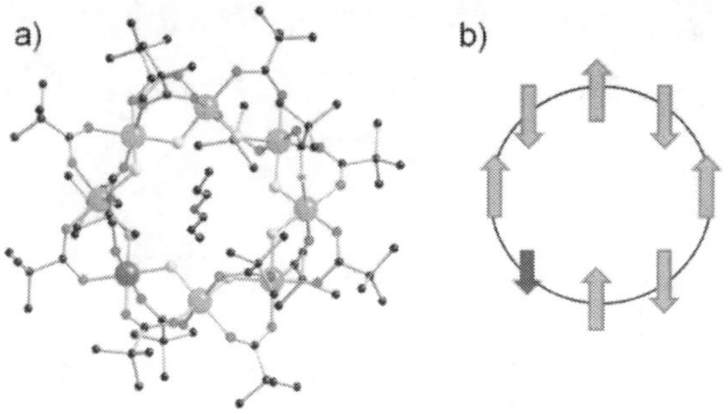

Figure 1. a) Crystallographic structure of Cr_7Ni obtained from X-ray diffraction; b) sketch of the classical AF spin arrangement in the molecules.

at synthetic level and whose properties can be predicted by microscopic, albeit demanding, models. Recent achievements in supramolecular chemistry, experiments and modeling appear extremely encouraging in this field [14, 15, 16].

Here we review some results recently obtained, enlightening the steps needed to establish and measure entanglement between spins in supramolecular structures. In the following, Cr_7Ni molecular spin clusters are firstly introduced showing that they are good building blocks with well defined state. Afterwards, the problem of establishing magnetic coupling between the units is discussed analyzing how the propagation of spin information at the supra-molecular (nanometer) scale works in four *purple*-Cr_7Ni dimers that differs each other only for the heteroaromatic linker between the rings. Finally, for a particular molecular dimer the entanglement is experimentally measured and theoretically quantified.

2. Molecular spin clusters and Cr_7Ni.

Molecular spin clusters are ideal test bed for the study of nanoscale magnetic phenomena. These systems are molecules of few nanometers, constituted by a magnetic nucleus sorrounded by organic chemical groups. The magnetic core is typically constituted by transition metals, rare-earth ions or even organic radicals, while the organic bridges comprise light elements like carbon, oxygen, hydrogen and nitrogen. The deep level of control achieved at synthetic level allows to produce collections of identical non-interacting molecules, embedded in crystals and oriented along specific crystallographic directions. Although the final product of synthesis appears as bulk crystals or powders, i.e. macroscopic samples, we can assume these materials as statistical ensembles of quantum systems. Thus, solid state experimental techniques, like the measurements of thermodynamic properties, reflect mainly ensemble properties of the single molecule. Moreover, the experimental conditions needed to reveal the quantum behavior are easily achievable in laboratory: molecular spectra are well resolved at liquid helium temperatures, while level crossings can be observed at few teslas fields. All these reasons contributed to the success of molecular magnets in the last 20 years.

Structural and physical features of molecular magnets can be tailored at synthetic level and a huge variety of molecules can be produced. One possible classification is based on their different magnetic properties: single molecules magnets (SMM), photomagnetic and spin-cross-over derivatives, ferromagnetic or antiferromagnetic (AF) chains and rings.

SMM is a wide class of molecules with high spin. Among them, Mn_{12}-acetate has spin $S = 10$ and high anisotropy barrier [17, 18, 19] and it is one of the first ones that has been extensively studied and probably also the most popular. Together with Fe_8, for Mn_{12}-ac key experimental results were obtained like: tunneling of magnetization [20, 21, 22, 23, 24], the observation of quantum phase interference (Berry phase) [25] and coherent quantum oscillations of spin between two energy minima [26].

In this review paper the attention is focused on another interesting subclass of molecules, that of AF spin rings. These, differently to the SMM, are low spin systems and with lower dipolar interaction, that are advantages, as discussed later, for applications in quantum information processing. The first AF rings synthesized were homometallic and based on trivalent metals. As examples, there are rings based on Fe^{+3} ($s = 5/2$), like Fe_6 [27, 28], F_{10} [29, 28], Fe_{12} [30] and Fe_{18} [31]; those on Cr^{+3} ($s = 3/2$), like Cr_8[32, 33] and Cr_{10} [34]; and finally those based on V^{+3}, like V_8 and V_{10} [35]. These even-membered AF rings have a fully compensated total spin and singlet ground state at low temperature. More recently an extra spin was introduced, in order to break the compensation and control the total spin of the AF wheels. This strategy was used for the Cr-based cyclic systems to synthesize heterometallic molecular wheels in which one Cr ion is substituted with a divalent transition metal M[36]. If $M = Ni^{+2}$ with $s = 1$, the spin balance gives an effective total $S = 1/2$ to the molecule; two examples are Cr_9Ni[37], obtained from Cr_{10}, and Cr_7Ni, obtained from Cr_8.

The structure of Cr_7Ni is depicted in Fig. 1; the seven Cr^{+3} ($s = 3/2$) ions and the extra Ni^{+2} ($s = 1$) are disposed in a symmetric octagonal planar structure, with the spins antiferromagnetically coupled. This molecule, synthesized in 2003 [36], has been deeply investigated by specific heat and torque [38], nuclear magnetic resonance [32], neutron scattering [39] and micro-SQUID magnetization at very low temperature [40] measurements. The experimental results were fitted by spin-Hamiltonian calculations [41] founding that Cr_7Ni is an effective two-level system with $S = 1/2$ ground state well separated from $S = 3/2$ first excited state. Cr_7Ni has been then proposed as a molecular qubit, suitable candidate for Quantum Information Processing (QIP) [38, 42]. This proposal has been corroborated by very promising theoretical [43] and experimental [44] studies of coherence. The time of coherence measured for the molecule of Cr_7Ni results indeed about 3 μs at 2 K, at least two order of magnitude greater than the time needed for spin manipulation. Finally, successful grafting of isolated molecules on surfaces was achieved, with only minor changes in the magnetic features from the bulk-material [45, 46, 47, 48].

3. Linking the rings

To have entanglement between molecular magnets, we firstly need to establish inter-molecular magnetic interaction in a controlled way. The easiest strategy is by chemical linkers. Examples of supramolecular dimers of single-molecule magnets (SMM) were already reported in the early 1990s by Christou group [50], but the magnetic coupling of the linked Mn_4 cages through C–H$\cdots$Cl hydrogen bonds were found much later in 2002 [51]. In this case, the possibility to change the type of linker is not simple, using diketonates [52]. Aromatic groups could be suitable linkers to realize supramolecular systems with tailored and controlled intermolecular interactions. One critical point in their use at synthetic level, is the risk of obtaining polymeric networks with long-range magnetic order; in the case of Mn_4 linked through dicyanamide $N(CN)_2$ the SMM were collected into a 3-dimensional polymer[53]. The crucial factor to avoid this problem is having molecules with small dipolar interaction. In this respect, the Cr_7Ni rings offer an unique opportunity.

The first attempt to bind together Cr_7Ni rings was realized in 2005 [49]; the link was organo-metallic and it was attached to the Cr_7Ni wheels via hydrogen-bonds between F-bridges of the cage and the terminal H–N groups of the link (see Fig. 2(a)). Unfortunately, this type of connection is not magnetically efficient [37], indeed it was shown that no significant exchange of spin can be achieved through this link. Progresses were achieved linking the rings via the external carboxylates at the Ni edge, where appreciable spin density resides. Examples are the dimer structures represented in Figs. 2(b) and (c) where the magnetic coupling between the

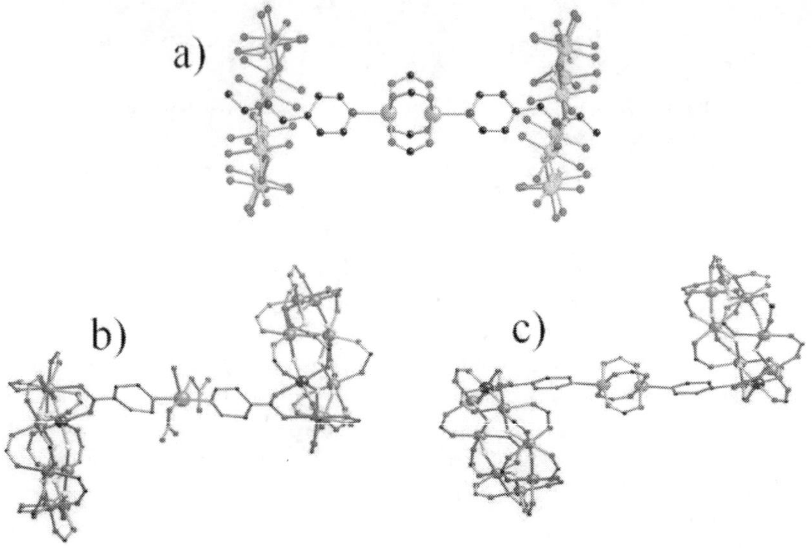

Figure 2. Crystallographic structures of the first Cr_7Ni dimers synthesized. In a) the organo-metallic link is connected at the centers of the wheels via hydrogen-bonds (from [49]); in b) and c) the link is still organo-metallic but connected via the carboxylate bridge at the Ni edge (from [14]). Moreover, in c) the magnetic coupling can be permanently coupled switched on and off through Cu_2 dimer (from [14]).

rings is permanently established by one or two copper ions in the link [14].

In 2008 a new derivative of Cr_7Ni, hereafter called *purple*-Cr_7Ni, was synthesized [55]. In this new derivative, with respect to the original *green*-Cr_7Ni ring [36], five bridging fluorides are substituted by alkoxides. The replacement of fluoride with alkoxide within the coordination sphere of the majority of Cr^{+3} ions causes a change in color of the crystals from dark green to purple, from this the names. This also affects the Cr –Cr exchange coupling and consequently magnetic levels of each ring [15]. Most importantly, thank to the presence of a terminal ligand, *purple*-Cr_7Ni offers an ideal opportunity to link the rings anchoring directly onto the Ni sites organic linkers with or without metal ions (see Fig. 3(a)). Finally, exploiting faster chemical reaction at Ni site, it is possible to systematically change the aromatic linker, thus tuning the degree of conjugation.

In Figs. 3(b)-(e) four representative linkers of *purple*-$(Cr_7Ni)_2$ dimers are depicted, respectively: L== 4,4'-*bipyridyl* (*bipy*, in short), $trans - 1,2 -$ *bipyridylethene* (*bipyet*), *bipyridyltetrazine* (*bipytz*) and *bidimethylpyrazolyl* (*bipz*). The structures of the individual rings in the molecular dimers such synthesized do not vary between the derivatives, and in each case the nitrogen from a di-imine is coordinated to nickel, giving the nickel a 2F, 3O, 1N coordination sphere. Chemical synthesis and structure information of these molecules are reported in Refs. [55, 54].

Low temperature specific heat for the four derivatives, taken in presence of an external 1 T magnetic field, is reported in Fig. 4. Data, relative to microcrystalline samples, are plotted as a function of temperature and normalized to the gas constant R. For comparison, specific heat multiplied by two of the single *purple*-Cr_7Ni ring that constitutes each dimer (black triangles) and spin-Hamiltonian calculations [15] of its magnetic contribution (dashed line) are reported. The curves of the four compounds superimpose each other almost perfectly, reproducing the

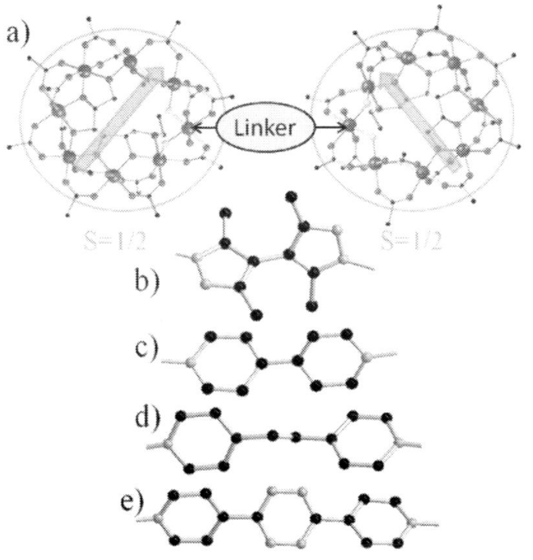

Figure 3. a) Sketch of the *purple*-$(Cr_7Ni)_2$ dimer structure, the linker is anchored directly to the Ni ions. b)-e) four heteroaromatic groups used to link the molecules with through-space-distance (d) between the Ni sites that progressively increases: b) $bis-dimethylpyrazole$ ($bipz$ in short) $d = 9.95$Å; c) $4,4'-bipyridyl$ ($bipy$) $d = 11.22$Å; d) $trans-1,2-bipyridylethene$ ($bipyet$) $d = 13.43$Å and e) $bipyridyltetrazine$ ($bipytz$) $d = 15.17$Å [54]. Colors: purple, Cr; green, Ni; red, O; black, C; yellow, F; light blue, N.

behavior of the single ring at moderate field (1 T). This demonstrates that the pattern of the lowest lying states of the single *purple*-Cr_7Ni rings is preserved also in the supramolecular systems. In particular the presence of well defined Schottky anomaly below 1 K shows that energy split between $S = 1/2$ and $S = 3/2$ states is conserved in spite of inter-molecular interaction that should be in energy scale below 1 K.

Evidence of the coupling between the rings can emerge from the specific heat measurements in zero magnetic field. In the inset of Fig. 4 we present the comparison between the single ring and $(Cr_7Ni)_2 - bipy$ dimer. A Schottky anomaly around 4K is visible for both the derivatives, this is due to the energy gap between the ground and excited states. Conversely, at low temperature the shape of the curves is different. In the case of the single ring (black circles) the specific heat goes to zero, as expected for a Kramers doublet, while for $(Cr_7Ni)_2 - bipy$ (red circles) there is an upturn below 0.5 K due to the energy level splitting in the ground state multiplet.

To explicitly reveal the presence of inter-rings interaction, micro-SQUID magnetization measurements at 40 mK have been performed. In Fig. 5 the resulting curves are plotted for all the derivatives; data are normalized to the saturation value M_S (in correspondence of about $H = 1$ T) and measured with a sweeping rate $dH/dt = 0.28$T/s. The magnetization curves referred to *bipy* and *bipyet* dimers present inflection points that can be related to the energy gap between $S = 0$ and $S = 1$ states. As depicted in the inset indeed, if $S = 0$ state is the ground state in zero-field, it should be possible to see a spin-state crossing by increasing the field, with one sub-level of $S = 1$ state falling in energy as field increases. As evident from magnetization curves, the level crossings occur at $H = \pm 0.05$ T for *bipy* and at $H = \pm 0.04$ T for *bipyet*. For the other two compounds (*bipytz* and *bipz*), the magnetic coupling is too small (must be weaker than 100 mK), and no inflection points are visible in the curves. These results are fully consistent

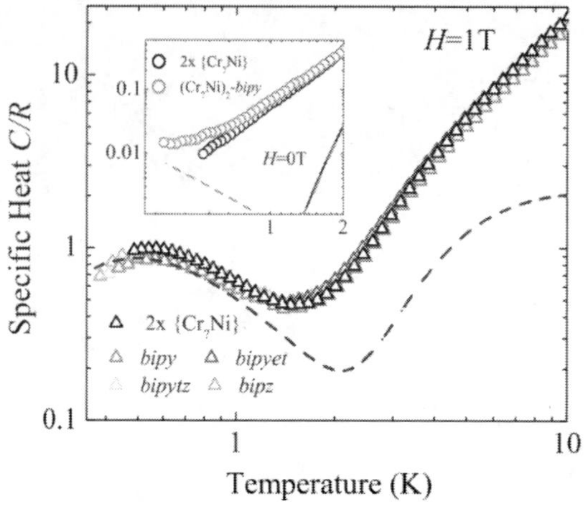

Figure 4. Specific heat at $H = 1$T for the four *purple*-$(Cr_7Ni)_2$ dimers discussed in the text: bipy (red), bipyet (blue), bipytz (green) and bipz (orange). The curves are compared to the signal for the single *purple*-Cr_7Ni ring [15] multiplied by two (black). Data are normalized to the gas constant R. (Dashed line) simulation of magnetic contribution of the single ring calculated using the eigenvalues of the microscopic spin-Hamiltonian and multiplied by two. (Inset) specific heat at zero field for single *purple*-Cr_7Ni ring multiplied by two (black circles) and $(Cr_7Ni)_2-$*bipy* dimer (red circles). Theoretical magnetic (lines) contributions are reported. The common bump at higher temperatures is the Schottky anomaly due to the main energy splitting between the ground doublet and the first excited state, while the Schottky at lower temperature, present only for the dimer, is the energy splitting arising from the inter-ring coupling. Measurements were performed on thin pellets ($\sim$2 mg) of polycrystalline sample by a ^{3}He cryostat and by using the two-tau method.

with EPR data [54] and spin-Hamiltonian calculations [15, 54]. Finally, Density Functional Theory calculations explain how the magnetic coupling is established through the conjugated linkers [54].

In summary, both experimental and theoretical evidences indicate that, among the supramolecular dimers investigated, the strongest coupling between the rings is achieved through *bipy* linker. Thus, this is the ideal system in which explore also the presence of entanglement.

4. Entanglement between molecular spins

Experimental detection and quantification of entanglement is a demanding task [56]. Formally a state is entangled if its wavefunction cannot be factorized. This definition is not helpful, because, in general, telling whether a state of n parties is separable or not is a non trivial mathematical problem. In theory, to reconstruct experimentally the overall density matrix it is necessary to measure the correlations between the observables of all the parties. This approach can be used if a small number of particles are involved, as in the case of photons [4]. Conversely, for magnetic bulk systems, i.e. ensemble of the order of Avogadro's number of spins, one may not refer to individual properties of every constituent of a solid sample or correlation between them. Fortunately, an alternative and much simpler experimental approach exists to certify if a certain state is entangled directly. This tools are the so-called *entanglement witnesses*. An entanglement witness (EW) is a Hermitian operator which is able to detect entanglement in a state. The basic idea is that the expectation value of the EW for the state ρ under consideration exceeds certain

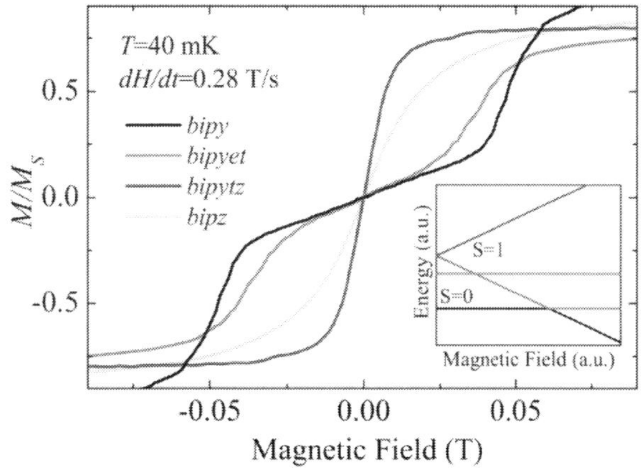

Figure 5. Micro-SQUID magnetization for the four *purple*-$(Cr_7Ni)_2$ dimers discussed in the text. The curves are normalized to the saturation values M_S. The experiments were performed at 40 mK and with a magnetic field sweeping rate $dH/dt = 0.28\mathrm{T/s}$. To enlighten the transitions, we have zoomed in a region around zero field and only data with decreasing field are shown, the complete loops are reported in Ref. [54]. (Inset) schematic representation of the magnetic field dependence for $S = 0$ and $S = 1$ spin levels resulting from two interacting $S = 1/2$ and taking into account axial anisotropy.

bounds only when ρ is entangled [1, 2]. The limits of EW approach are essentially two: it gives only a sufficient condition for entanglement and it is poor in quantitative information. For the cases in which it is possible to know the eigenvalues of the Hamiltonian a quantification of the entanglement can be done theoretically by the calculation of the concurrence ($\mathcal{C}$)[1, 2, 56]. By definition[2] the concurrence is a monotonous function of the entanglement of formation and it can assume values comprised between zero and one, which represent the extreme situations of a not entangled or maximally entangled state, respectively.

Recently, the magnetic susceptibility, insofar as sum over all microscopic spin correlation functions, has been proposed as entanglement witness by Wiesniak et al. [57]. Using the complementary principle for the components of the susceptibility χ_x, χ_y, χ_z, measured along three perpendicular direction, it can then be demonstrated [57] that a magnetic system is entangled if:

$$EW = \frac{1}{3}\left(\frac{\chi_x}{g_x^2} + \frac{\chi_y}{g_y^2} + \frac{\chi_z}{g_z^2}\right) - \frac{N_A \mu_B^2 S}{k_B T} \leq 0 \qquad (1)$$

where g_x, g_y and g_z are the components of gyromagnetic factors, N_A is Avogadro's number, μ_B the Bohr magneton, k_B is the Boltzmann constant and finally S is the spin of the systems taken into account. As visible from Eq. 1, the entanglement witness depends on the temperature T, meaning that the same system can be entangled for certain temperature ranges and not entangled in others. This is a direct consequence that entanglement is a property of the state of the system. Moreover, this enlightens the distinction between entanglement and simple interaction that, being related to the Hamiltonian of the system, is instead independent from the temperature.

In the specific case of $(Cr_7Ni)_2$ – *bipy* dimer, we have estimated the entanglement witness described in Eq. 1 considering: $\chi_z = \chi_\perp$, $\chi_x = \chi_y = \chi_\parallel$, $g_z = g_\perp$ and $g_x = g_y = g_\parallel$. $\chi_\perp$ is the component perpendicular to the largest surface of the crystal; while $\chi_\parallel$ refers to the directions within the crystal plane. Rotation of magnetic field within this plane does not evidence changes in the magnetic response. The susceptibility components were obtained as dM/dH by derivative

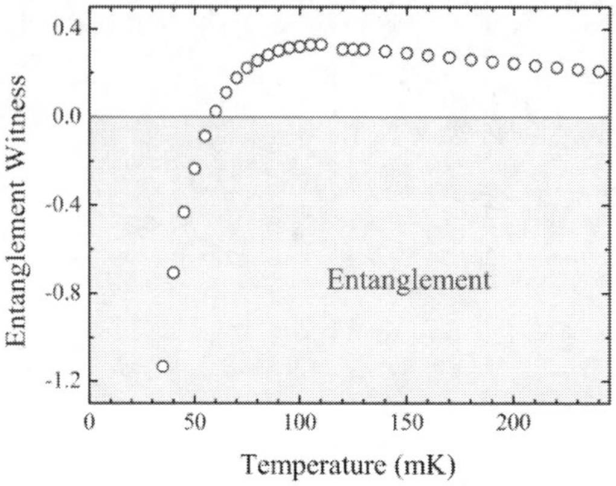

Figure 6. Temperature dependence of the entanglement witness (EW) for $(Cr_7Ni)_2$-*bipy*. The EW is extimated from susceptibility measurements, as explained in the text.

of micro-SQUID magnetization on single crystals. Measurements were performed with very small magnetic field and for different temperatures [15]. The result is plotted in Fig. 6: for $T \lesssim 50$ mK $EW < 0$, thus providing a direct experimental evidence of entanglement between the two Cr_7Ni rings at this temperature range. This result is in agreement with concurrence calculations obtained using an effective two-qubit picture (see Ref. [15]). Furthermore, the concurrence gives quantitative estimation of entanglement founding the dependence of the entanglement also on the magnetic field. At vanishing temperatures, a sharp transition is found from maximum entanglement to not entangled state for magnetic fields that exceeds the value of crossing field between $S = 0$ and $S = 1$ states.

It should be noted that the expression (1) is valid in the condition $[\mathcal{H}, \mathcal{H}_1] = 0$, where $\mathcal{H}$ is the Hamiltonian of the system, and $\mathcal{H}_1$ the part of the Hamiltonian that describes the interaction with magnetic field [57]. In the present case, the coupling between the effective spins $\mathbf{S}_\alpha$ includes an anisotropic contribution ($D_{AB} \neq 0$). However, for small enough values of D_{AB}/J_{AB}, with J_{AB} exchange constant between the clusters, the difference between the susceptibility and the entanglement witness is negligible with respect to the threshold, as the case presented here (see Ref. [16] for the complete calculations).

5. Conclusions

In summary, Cr_7Ni appears to be a molecular two-level system very promising for qubit encoding. The magnetic coupling between the molecular rings in the $(Cr_7Ni)_2$-dimers were experimentally characterized in detail by heat capacity, micro-SQUID magnetization and EPR. Finally, for the dimer with the strongest coupling, $(Cr_7Ni)_2 - $ *bipy*, spin entanglement between the rings is experimentally demonstrated below 50 mK using magnetic susceptibility as entanglement witness, in agreement with concurrence calculations.

Acknowledgments

This work has been carried out supported by FP7-ICT FET MolSpinQIP project, Contract No. 211284 and by the PRIN project of the Italian Ministry of Research and by the EPSRC (UK).

References

[1] R. Horodecki, P. Horodecki, M. Horodecki, and K. Horodecki. *Rev. Mod. Phys.*, 81:865, 2009.
[2] L. Amico, R. Fazio, A. Osterloh, and V. Vedral. *Rev. Mod. Phys.*, 80(2):517–576, 2008.

[3] M. A. Nielsen and I. L. Chuang. volume 10. Cambridge: Cambridge University Press, 2000.
[4] D. Bouwmeester et al. *Phys. Rev. Lett.*, 82(7):1345–1349, 1999.
[5] J. R. Petta et al. *Science*, 309:2180, 2005.
[6] T. Roscilde, P. Verrucchi, A. Fubini, S. Haas, and V. Tognetti. *Phys. Rev. Lett.*, 93(16):167203, 2004.
[7] M. C. Arnesen, S. Bose, and V. Vedral. *Phys. Rev. Lett.*, 87(1):017901, 2001.
[8] A. Osterloh, L. Amico, G. Falci, and R. Fazio. *Nature*, 416:608, 2002.
[9] T.J. Osborne and M.A. Nielsen. *Phys. Rev. A*, 66(3):032110, 2002.
[10] G. Vidal, J. I. Latorre, E. Rico, and A. Kitaev. *Phys. Rev. Lett.*, 90(22):227902, 2003.
[11] F. Verstraete, M. Popp, and J. I. Cirac. *Phys. Rev. Lett.*, 92(2):027901, Jan 2004.
[12] S. Ghosh, T. F. Rosenbaum, G. Aeppli, and S. N. Coppersmith. *Nature*, 425:48, 2003.
[13] A. M. Souza et al. *Phys. Rev. B*, 77(10):104402, 2008.
[14] G. Timco et al. *Nature Nanotec.*, 4(3):173–178, 2009.
[15] A. Candini, G. Lorusso, F. Troiani, A. Ghirri, et al. *Phys. Rev. Lett.*, 104(3):037203, 2010.
[16] F. Troiani, V. Bellini, A. Candini, G. Lorusso, and M. Affronte. *Nanotechnology*, 21:274009, 2010.
[17] T. Lis. *Acta Crystall. B*, 36:2042, 1980.
[18] A. Caneschi et al. *J. Am. Chem. Soc.*, 113:5873–5874, 1991.
[19] J. R. Friedman, M. P. Sarachik, J. Tejada, and R. Ziolo. *Phys. Rev. Lett.*, 76(20):3830–3833, 1996.
[20] R. Sessoli, D. Gatteschi, A. Caneschi, and M. A. Novak. *Nature (London)*, 365:141, 1993.
[21] L. Thomas, F. Lionti, R. Ballou, D. Gatteschi, R. Sessoli, and B. Barbara. *Nature*, 383:145, 1996.
[22] L. Thomas and B. Barbara. *J. Low Temp. Phys.*, 113:1055–1060, 1998.
[23] D. Gatteschi and R. Sessoli. *Angew. Chem. Int. Ed.*, page 268, 2003.
[24] C. Sangregorio et al. *Phys. Rev. Lett.*, 78(24):4645–4648, 1997.
[25] W. Wersdorfer and R. Sessoli. *Science*, 284:133, 1999.
[26] F. Luis, F. L. Mettes, J. Tejada, D. Gatteschi, and L. J. de Jongh. *Phys. Rev. Lett.*, 85(20):4377–4380, 2000.
[27] G. L. Abbati et al. *Inorg. Chem.*, 36:6443, 1997.
[28] M. Affronte, J. C. Lasjaunias, A. Cornia, and A. Caneschi. *Phys. Rev. B*, 60(2):1161–1166, 1999.
[29] K. L. Taft et al. *Journal of the American Chemical Society*, 116(3):823–832, 1994.
[30] G. L. Abbati, A. Caneschi, A. Cornia, A. C. Fabretti, and D. Gatteschi. *Inorg. Chim. Acta*, 297:291, 1999.
[31] S. P. Watton et al. *Angew. Chem. Int. Ed.*, 36:2774, 1997.
[32] E. Micotti et al. *Phys. Rev. Lett.*, 97(26):267204, 2006.
[33] S. Carretta et al. *Phys. Rev. B*, 67(9):094405, 2003.
[34] D.M. Low et al. *Chem.-Eur. J.*, 12:1385, 2005.
[35] R. H. Laye et al. *Chem. -Eur. J.*, 9:6215, 2003.
[36] F. K. Larsen et al. *Angew. Chem. Int. Ed.*, 42:101, 2003.
[37] M. Affronte, F. Troiani, A. Ghirri, S. Carretta, et al. *Dalton Trans.*, pages 2810–2817, 2006.
[38] F. Troiani, A. Ghirri, M. Affronte, S. Carretta, et al. *Phys. Rev. Lett.*, 94(20):207208, 2005.
[39] R. Caciuffo, T. Guidi, G. Amoretti, S. Carretta, E. Liviotti, P. Santini, C. Mondelli, G. Timco, C. A. Muryn, and R. E. P. Winpenny. *Phys. Rev. B*, 71(17):174407, 2005.
[40] W. Wernsdorfer, D. Mailly, G. A. Timco, and R. E. P. Winpenny. *Phys. Rev. B*, 72(6):060409, 2005.
[41] S. Carretta, P. Santini, G. Amoretti, M. Affronte, et al. *Phys. Rev. B*, 72(6):060403, 2005.
[42] M. Affronte, F. Troiani, A. Ghirri, A. Candini, V. Evangelisti, M. Corradini, S. Carretta, P. Santini, G. Amoretti, F. Tuna, G. Timco, and R. E. P. Winpenny. Single molecule magnets for quantum computation. *J. Phys. D: Appl. Phys.*, 40:2999, 2007.
[43] Arzhang Ardavan et al. *Phys. Rev. Lett.*, 98(5):057201, 2007.
[44] F. Troiani, V. Bellini, and M. Affronte. *Phys. Rev. B*, 77(5):054428, 2008.
[45] V. Corradini, A. Ghirri, U. del Pennino, R. Biagi, et al. *Dalton Trans.*, 39:4928, 2010.
[46] A. Ghirri, V. Corradini, C. Cervetti, A. Candini, et al. *Adv. Func. Mat.*, 2010.
[47] V. Corradini, F. Moro, R. Biagi, V. De Renzi, et al. *Phys. Rev. B*, 79(14):144419, 2009.
[48] V. Corradini, R. Biagi, U. del Pennino, V. De Renzi, et al. *Inorg. Chem.*, 46:4937, 2007.
[49] M. Affronte, I. Casson, M. Evangelisti, A. Candini, et al. *Angew. Chem. Int. Ed.*, 44:6496, 2005.
[50] D. N. Hendrickson, G. Christou, E.A. Schmitt, E. Libby, et al. *J. Am. Chem. Soc*, 114:2455, 1992.
[51] Wolfgang Wernsdorfer, Nria Aliaga-Alcalde, David N. Hendrickson, and George Christou. Exchange-biased quantum tunnelling in a supramolecular dimer of single-molecule magnets. *Nature*, 416:406, 2002.
[52] E. C. Sanudo, T. Cauchy, E. Ruiz, R. H. Laye, et al. *Inorg. Chem.*, 46:9045, 2007.
[53] H. Miyasaka, K. Nakata, K.-I. Sugiura, M. Yamashita, and R. Clerac. *Angew. Chem.*, 116:725, 2004.
[54] T. Faust, V. Bellini, G. A. Timco, G. Lorusso, A. Candini, S. Carretta, F. Tuna, G. Whitehead, E.J.L. McInnes, R.G. Pritchard, S.J. Teat, W. Wernsdorfer, M. Affronte, and R.E.P. Winpenny. *submitted*, 2010.
[55] G. A. Timco, E. J. L. McInnes, R. G. Pritchard, F. Tuna, and R. E. P. Winpenny. *Angew. Chem. Int. Ed.*,

47(50):9681–9684, 2008.
[56] F. G. S. L. Brandão. *Phys. Rev. A*, 72:022310, 2005.
[57] M. Wieśniak, V. Vedral, and Č. Brukner. *N. J. Phys.*, page 258, 2005.

Studies of critical phenomena in molecular magnets by μSR spectroscopy

T Wasiutyński[1], M Bałanda[1], M Czapla[1], R Pełka[1], P M Zieliński[1], F L Pratt[2], T Korzeniak[3], R Podgajny[3], D Pinkowicz[3] and B Sieklucka[3]

[1]Institute of Nuclear Physics PAN, Radzikowskiego 152, 31-342 Kraków, Poland
[2] ISIS Facility, Rutherford Appleton Laboratory, Chilton, Oxfordshire OX11 0QX, UK
[3] Faculty of Chemistry, Jagiellonian University, Kraków, Poland

Abstract. The rapidly developing field of molecular magnetism supplies a multitude of novel compounds of unprecedented properties and structure. Molecular magnets predominantly belong to the class of compounds involving well localized magnetic moments. This feature together with the fact that the nature and symmetry of magnetic interactions is encrypted in the critical behaviour makes them a perfect testing ground of the existing theoretical spin models. It is demonstrated that the experimental technique of the μSR spectroscopy is perfectly suited to study magnetic fluctuations and spin dynamics in the neighbourhood of a phase transition. This unique method can even dispense with the complementary measurements of the AC susceptibility or heat capacity to supply a complete set of the static and dynamic critical exponents. It can thus be used to pinpoint the universality class of the material of interest.

1. Introduction

Novel magnetic materials based on molecular building blocks offer unprecedented properties which are due to the interplay of unique architectures of molecular networks and magnetic anisotropy induced by different coordination patterns. Magnetic interactions mediated by molecular bridges may be of different nature and symmetry, which is inscribed in the critical bahaviour near the transition point to the ordered state. Due to the fact that molecular magnets display low density of magnetic moments and weak couplings between them even a low magnetic field may destroy the ground state. The experimental technique of μSR (muon spin rotation/relaxation) [1] represents a unique method for studying magnetic properties of materials in zero applied magnetic field. The present paper reports the results of a study of the critical behaviour in the family of octacyanometallates $M(CN)_8$ [2] which is of special interest in the field of molecular magnetism.

2. Experimental

2.1. Muon implantation

Our experiments were performed on the positive muon beam at ISIS facility with the spectrometers MuSR and Argus. Pions produced in the collisions of sufficiently high-energy protons with a graphite or other light element target decay quickly to muons and neutrinos ($\tau_\pi = 26$ ns). Muons are then implanted in a sample. The key point of the method is that muons are 100 % spin polarized: μ^+ has spin 1/2 aligned antiparallel to its momentum. This high level of polarization is preserved during the beam transport and, as the spin state is insensitive to electrostatic interactions, also during the thermalization process in a sample. The muon itself is an unstable particle with the lifetime of $\tau_\mu \approx 2.2\,\mu$s and decays to positrons

Published under licence by IOP Publishing Ltd

$(\mu^+ \to e^+ + \nu_e + \overline{\nu}_\mu)$ emitted preferentially in the spin direction of the parent muon with the probability distribution $p(\theta) = 1 + a\cos\theta$, where θ is the angle between the instantaneous muon spin orientation and the positron momentum, and $a \approx 0.3$. The asymmetry function

$$A(t) = \frac{N_B(t) - N_F(t)}{N_B(t) + N_F(t)}, \tag{1}$$

where $N_B(t)$ and $N_F(t)$ denote the numbers of decay positrons detected by the backward and forward counters, respectively, gives us information about the time evolution of the muon spin. It is governed by the interaction of the muon magnetic moment with the dipolar magnetic field due to electrons and nuclei of the species of interest. This provides a unique opportunity to study the distribution of the local magnetic field static on the microsecond scale at the stopping sites of the muons without the need to apply external magnetic field. Moreover, the in-field experiments provide further insights into spin correlations and spin dynamics in the system. Applying magnetic field so that it is parallel to the initial muon spin polarization (longitudinal mode) does not affect the spin of the muon but modifies the local field in the sample. Contrary to that, a small field applied perpendicular to the initial muon spin polarisation (transverse mode) does not affect spins in the sample but may affect the muon spins which precess in it in phase. If the field varies from site to site, the individual precessions dephase and the initial precession signal is damped. In addition, there is always relaxation of the muon spin due to the hyperfine interactions with the nuclear spins of a sample.

2.2. *Relaxation functions*

The muon asymmetry function (1) is proportional to the average muon-spin polarization $G_z(t)$. The muon-spin precesses around the local magnetic field $\vec{B}$ with angular frequency $\gamma_\mu|\vec{B}|$, where $\gamma_\mu = 2\pi \times 135.5$ MHz T^{-1} is the muon gyromagnetic ratio. If the local magnetic field at the muon site is at an angle θ to the initial muon-spin-direction, the muon-spin will precess around the end of a cone of semi-angle θ about the magnetic field. The normalized decay positron asymmetry is then given by $G_z(t) = \cos^2\theta + \sin^2\theta\cos(\gamma_\mu|\vec{B}|t)$. If the direction of the local magnetic field is entirely random, as for polycrystalline samples, the averaging over all possible directions of field $\vec{B}$ yields $G_z(t) = \frac{1}{3} + \frac{2}{3}\cos(\gamma_\mu|\vec{B}|t)$. Next, if the strength of the local field $|\vec{B}|$ is itself a random variable distributed around zero according to the Gaussian distribution of width Δ/γ_μ, the corresponding averaging gives $G_z(t) = \frac{1}{3} + \frac{2}{3}(1 - \Delta^2t^2)\exp(-\frac{1}{2}\Delta^2t^2))$, a result first obtained by Kubo and Toyabe and known as the Kubo-Toyabe function $f_{\mathrm{KT}}(t)$. In the studies reported in this paper the experimental asymmetry function (1) was fitted to the following formula

$$G_z(t) = A_0[\frac{2}{3}e^{-\lambda_1 t}\cos(\gamma_\mu Bt + \phi) + \frac{1}{3}\exp^{-\lambda_2 t}] + A_1 f_{\mathrm{KT}}(t) \tag{2}$$

where A_0, A_1, λ_1, λ_2, B, ϕ are to be determined, and the last contribution is taken into account only in the paramagnetic regime above the transition temperature. The output data of our μSR experiments were analysed with the WIMDA software [3].

2.3. *Critical behaviour*

The phase transition to a magnetically ordered state is clearly signalled by the onset of spontaneous oscillations in the time dependence of the asymmetry function as exemplified by figure 1 This fact enables us to study the critical behaviour. Different spatial dimensions of a spin network as well as different dimensions of the spin space (due to exchange anisotropy) lead to a variety of critical patterns displayed by molecular magnets. By selecting appropriately the experimental set-up one can uncover diverse aspects of the critical behaviour:

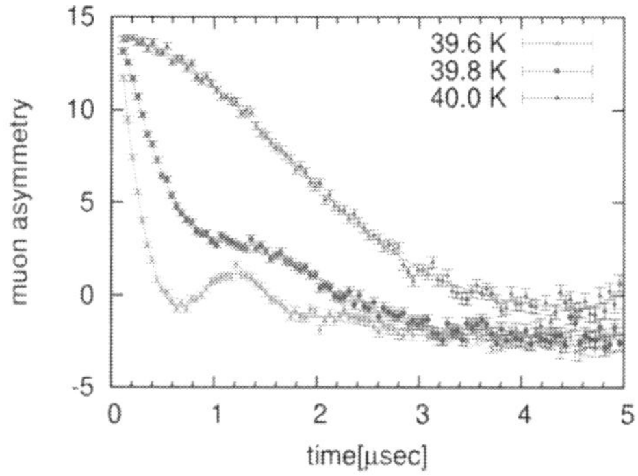

Figure 1. Zero field scan across the phase transition for $Cu_6[W(CN)_8]_4$.

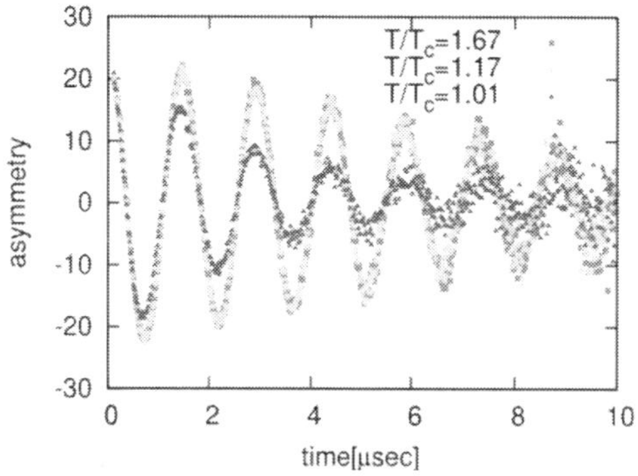

Figure 2. Transverse field scan at the vicinity of the phase transition for $Cu_6[W(CN)_8]_4$.

- zero field mode revealing spontaneous oscillations of the muon-spin polarization provides a direct insight into the thermal dependence of the order parameter $M(T) \sim (1 - T/T_c)^\beta$, the behaviour of the tail of the relaxation function observed at long times permits the measurement of spin dynamics $\tau \sim (T/T_c - 1)^{-w}$;
- transverse field mode: the frequency shift due to magnetic fluctuations is related to the spin-spin correlation function $(f - f_0)/f_0 \sim \chi \sim (T/T_c - 1)^{-\gamma}$;
- longitudinal field mode: the internal field distribution, metamagnetic behaviour, decoupling from the hyperfine fields.

3. Results

3.1. $Mn_2Nb(CN)_8$

$[Mn^{II}(H_2O)_2(pydz)][Mn^{II}(H_2O)_2][Nb^{IV}(CN)_8]\cdot 2H_2O$ (pydz=pyridazine) crystallizes in the monoclinic system with space group $P2_1/c$, and with unit cell parameters a = 10.605 Å, b = 15.575 Å, c = 14.386 Å, and $\beta = 108.3°$. The compound represents a three dimensional (3D) network of cyanidobridged framework composed of corrugated square-grid motifs parallel to the **bc** crystallographic plane cross-linked at the Nb centers by the ladder motifs running along the **a** crystallographic axis. Figure 3 shows the projection of the crystal structure onto the **ac** crystallographic plane. Each Nb ion is coordinated by eight cyanide ligands, of which seven link it to the Mn ions. The eighth terminal NC^- group together with pyridazine ligand and crystallization water is involved in the network of hydrogen bonds ensuring the stabilization of the crystal structure. There are two distinct Mn sites; sites of the first type build the ladder motifs and are bridged to three Nb ions, whereas those of the second type are coordinated to four Nb ions and are incorporated in the square-grid motifs. The Mn^{II} ions carry spin 5/2, whereas the Nb^{IV} ions carry spin 1/2. The temperature dependence of the AC susceptibility, see figure 4, shows an unusually sharp anomaly at $T_c \approx 42$ K revealing the transition to an ordered phase. Large values of the AC signal persisting below the transition temperature indicate that the compound can be classified as a soft magnet. The analysis of the DC susceptibility (not shown) in the paramagnetic regime implies antiferromagnetic coupling between the Nb ions and the Mn ions. This is consistent with the saturation value of the isothermal magnetization amounting to 9.5 μ_B, which is slightly higher than the value expected for a perfect antiparallel alignment of the magnetic moments suggesting the presence of weak non-collinearity. The scaling analysis of χ_{ac} performed using the classical scaling relation $\chi = c(T/T_c - 1)^{-\gamma}$ yielded $\gamma = 1.18$, the value placed between that for the mean-field model (1.0) and the 3D Ising model ($\approx$ 1.24). The μSR experiment performed in zero magnetic field reveal the onset of spontaneous oscillation of the asymmetry function below the transition temperature, see figure 5. A single precession frequency was observed in the measured asymmetry spectra pointing to a

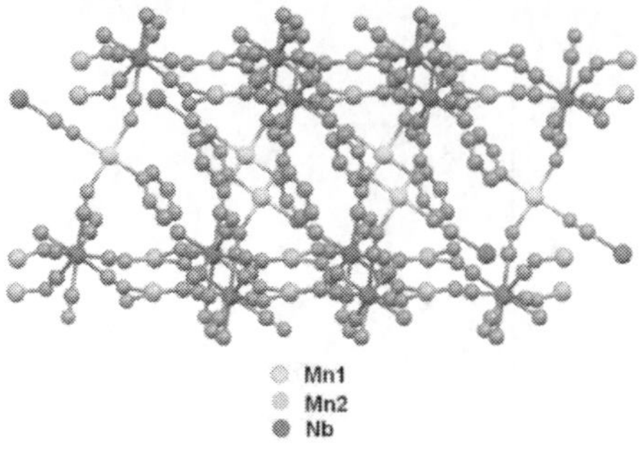

Figure 3. Projection of the crystal structure of $Mn_2Nb(CN)_8$ onto the **ac** crystallographic plane.

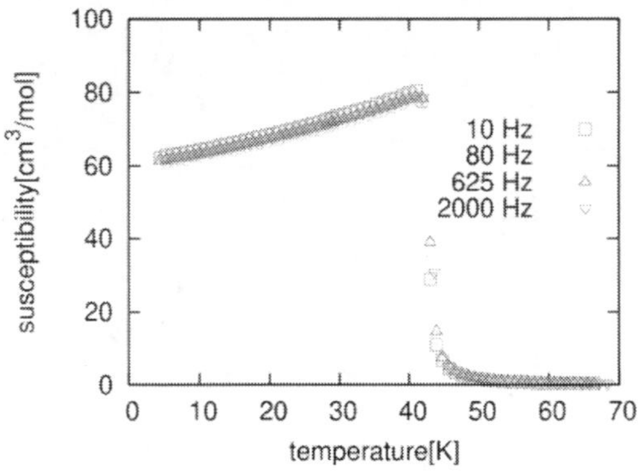

Figure 4. AC susceptibility of $Mn_2Nb(CN)_8$ for four different frequencies.

magnetically unique stopping site in this material. Most probably it is located near the terminal CN^- ligands which due to their uncompensated negative charge are good candidats to capture the incident muons. The temperature dependence of the local field inferred from the μSR experiment was fitted to the phenomenological form $B(T) = B(0)[1 - T/T_c]^\beta$ yielding $\beta = 0.53(2)$, see figure 6. This finding corroborates the mean-field character of the transition. The transverse field mode of the μSR experiment in the field of 50 G permitted the estimation of the γ critical exponent. The value of $\gamma = 1.1(1)$ agrees with that found from the ac susceptibility analysis. The temperature dependence of the muon longitudinal relaxation rate λ_L, originating from the muon-lattice interaction, revealed a sharp peak (not shown) at ~ 42 K. In the fast fluctuation limit it is directly proportional to the diverging spin-spin correlation time $\tau \propto |T - T_c|^{-w}$ corresponding to the slowing down of critical fluctuations. The scaling analysis of λ_L gave the critical exponent $w = 0.12(1)$. The exponent w is related to the dynamic exponent z and two static exponents by $z = d(2\beta + w)/(2\beta + \gamma)$, where $d = 3$ is the dimension of the spin network. Using this relation and the values of the exponents β, γ and w one obtains $z = 1.6(4)$. This estimate is close to the value of $d/2 = 1.5$ expected for dynamical model G of isotropic antiferromagnet [4]. In conclusion, it was found that the $Mn_2Nb(CN)_8$ compound is a soft ferrimagnet displaying the critical features of the mean-field model.

3.2. $Cu_4[W(CN)_8]_4$

The crystal structure of the compound $(tetrenH_5)_{0.8}Cu_4^{II}[W^V(CN)_8]_4 \cdot 7.2H_2O$ was solved by the single-crystal x-ray diffraction and is depicted in figure 7 (left). It crystallizes in the orthorhombic system (space group $Cmc2_1$) with unit cell parameters a=7.3707 Å, b=31.725 Å, c=7.017 Å. It consists of anionic double layers formed by copper and tungsten ions distributed in an alternating fashion and linked through the cyanide bridges. Within the double-layer each Cu ion is surrounded by five NC bridges connecting it to the neighbouring W ions. The coordination sphere of each W ion comprises eight cyanido ligands. Five of them mediate the linkage to the neighbouring Cu ions, and the three remaining ones stand out of the double-layers and are presumably involved in the hydrogen-bond network. The space between the double-layers is filled with water molecules and the $tetrenH_5$ molecules to maintain the electrostatic charge balance. Due to the the sizeable interbilayer distance of 10 Å the magnetic interaction between the double-layers is significantly smaller than the intrabilayer coupling. The real component of the AC susceptibility shows a sharp peak at $T_c \approx 33$ K marking the transition to an ordered state, see figure 8. The single crystal study of $Cu_4[W(CN)_8]_4$ [5] revealed a strong easy-plane anisotropy and a weak metamagnetism with spin-flop field on the order of 50 Oe. The classical scaling analysis of the AC signal

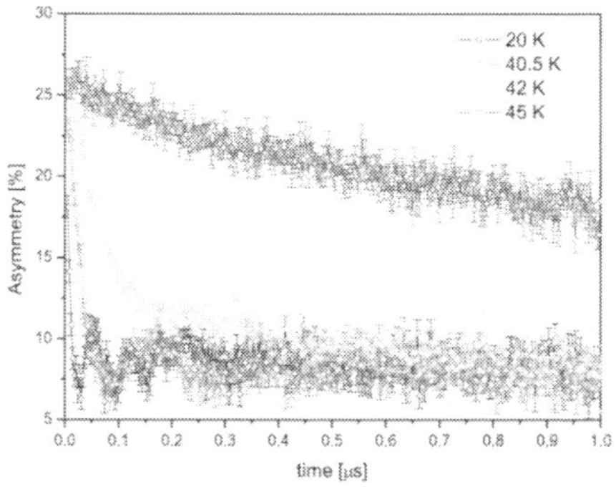

Figure 5. The time dependence of the muon asymmetry function in $Mn_2Nb(CN)_8$ measured in the vicinity of the phase transition.

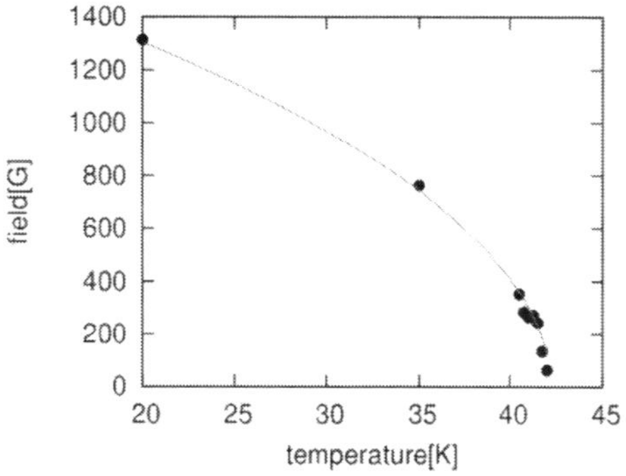

Figure 6. Local field at the muon stopping site for $Mn_2Nb(CN)_8$ deduced from the oscillatory part of muon asymmetry.

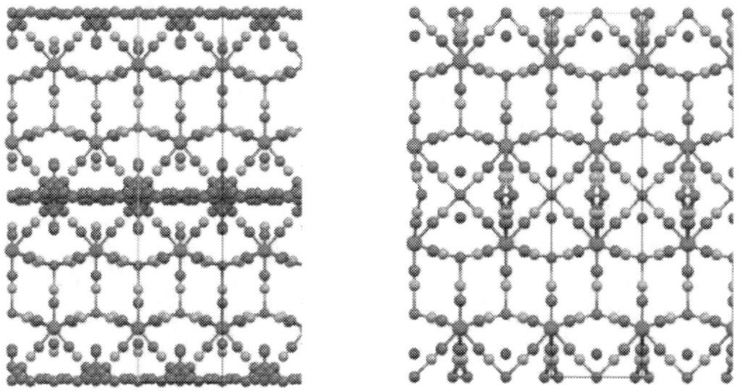

Figure 7. Left: The structure of the quasi-2D magnet $Cu_4(tetrenH_5)[W(CN)_8]_4 \cdot 7H_2O$ with the $tetrenH_5$ molecules in between the double-layers. Right: The structure of the 3D magnet $Cu_6[W(CN)_8]_4$.

yielded a large value of the critical exponent $\gamma = 1.63(1)$. It is higher than the values expected for the standard universality classes ($\gamma \sim 1.24$ for the 3D Ising model, 1.32 for the 3D XY model, and 1.38 for the 3D Heisenberg model). However, it is placed consistently between $\gamma_\perp(\approx 0.67)$ and $\gamma_\parallel(\approx 2.2)$ found in the single crystal study for the direction perpendicular and parallel to the double-layers, respectively. That coincides with the Monte Carlo results for systems with a strong easy-plane anisotropy, thus with a reduced dimensionality of the spin space [5]. The onset of an ordered state is also apparent in the temperature dependence of the transverse relaxation rate λ_1 inferred from the μSR experiment. It can be seen from figure 9 that λ_1 diverges at the transition point. The time evolution of the muon asymmetry function in the vicinity of the phase transition is shown in figure 10. The measured signals in the whole temperature range below the transition were satisfactorily reproduced if two oscillating amplitudes in Eq. (2) were introduced, which indicates that there are two distinct muon stopping sites. The temperature dependence of the two local fields corresponding to those stopping sites are shown in figure 11. The scaling analysis indicated that both of them have the same temperature dependence yielding a single value of the critical exponent $\beta = 0.23(2)$. The zero-field μSR experiment carried out for the analog of $Cu_4[W(CN)_8]_4$, where the W atoms were replaced by the Mo atoms, revealed a similar value of

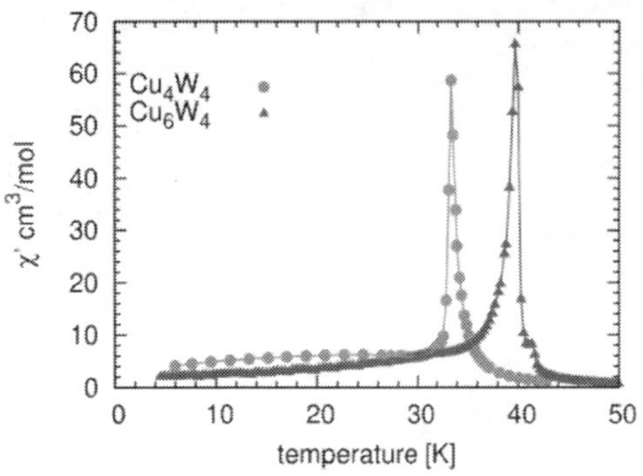

Figure 8. Comparison of AC susceptibility of $Cu_4[W(CN)_8]_4$ and $Cu_6[W(CN)_8]_4$.

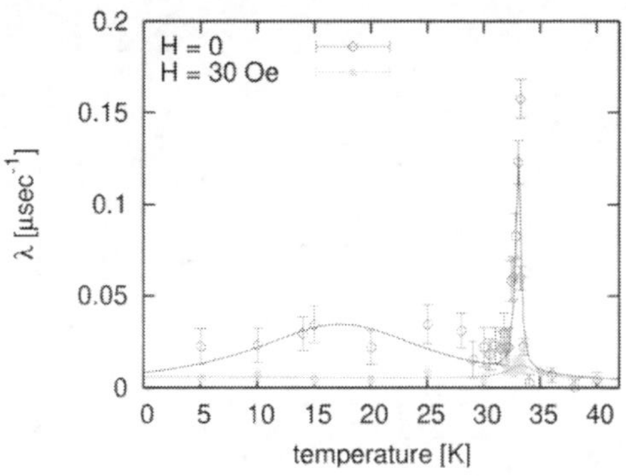

Figure 9. The temperature dependence of the relaxation rate λ_1 detected for $Cu_4[W(CN)_8]_4$.

the exponent $\beta = 0.27$. These values imply that the compounds can be classified to the group of 2D systems with XY exchange anisotropy. The anomalous part of the specific heat shown in figure 12 has a symmetric shape, which is typical for the transition involving reduced dimensionality of the spin network. The scaling analysis can be further extended by combining the results of the μSR spectroscopy with those obtained by a complementary calorimetric measurement. It was demonstrated [6] that the excess entropy and the square of the order parameter satisfy a combined scaling law involving the critical exponents κ and κ'. Their values determined for $Cu_4[W(CN)_8]_4$ pointed consistently to the XY anisotropy of the exchange interaction. The compound can thus be classified as a quasi-2D magnet with the XY exchange anisotropy. To our knowledge it is the first example of a molecular magnet displaying the signatures of the Berezinskii-Kosterlitz-Thouless topological transition [5].

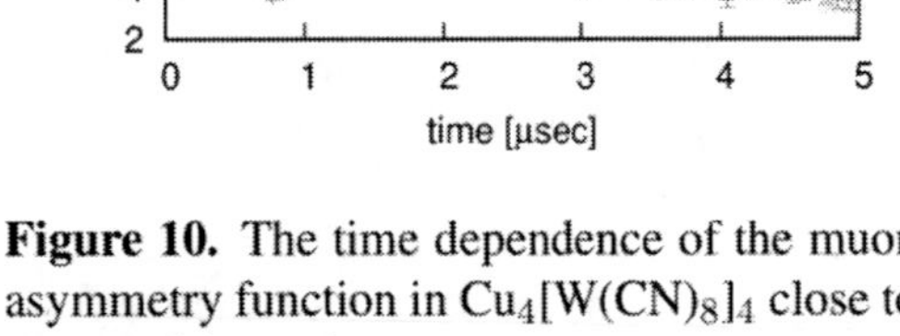

Figure 10. The time dependence of the muon asymmetry function in $Cu_4[W(CN)_8]_4$ close to the transition point.

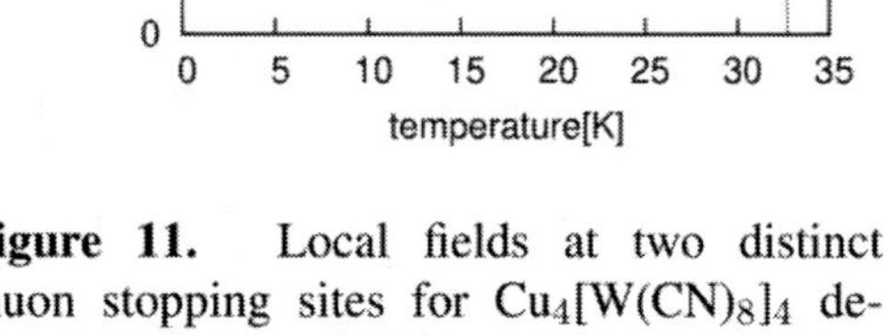

Figure 11. Local fields at two distinct muon stopping sites for $Cu_4[W(CN)_8]_4$ deduced from the oscillatory part of the muon asymmetry function.

3.3. $Cu_6[W(CN)_8]_4$

The molecular magnet $Cu_6^{II}[W^V(CN)_8]_4{\cdot}4H_2O$ was obtained by withdrawing the tetren molecule from the synthesis path of the previous compound $Cu_4[W(CN)_8]_4$ with the view to enhance the magnetic coupling between the structural double-layers [6]. In consequence additional Cu ions were located in the interbilayer spaces and the magnetic transition temperature was raised up to 40 K. The compound crystallizes in the tetragonal system with space group I4/mmm and unit cell parameters $a = b = 7.2678$ Å, $c = 28.2373$ Å. Figure 7 (right) shows the projection of the crystal structure onto the **bc** crystallographic plane. Each W ion is coordinated by eight CN ligands providing a linkage to the Cu ions. Four Cu ions of the formula unit are involved with the W ions in the formation of the double-layers. The remaining two Cu ions are located between the double layers providing direct exchange pathways between the double-layers. Both ions W^V and Cu^{II} carry spin 1/2. Temperature dependence of the AC susceptibility shows a rather sharp peak at $T_c \approx 40$ K, see figure 8, revealing the transition to an ordered phase. There is an apparent difference in the high-temperature slope of the AC signal between this compound and $Cu_4[W(CN)_8]_4$, which suggests distinct critical behaviour related to the change in the dimensionality of the spin network. Classical scaling analysis of the AC signal yielded the critical exponent $\gamma = 1.05(2)$. It is close to the value predicted in the mean-field model and consistent with the value of 1.063(8) found from the transverse field scan obtained in the μSR experiment, see figure 2. On the other hand, the zero-field asymmetry spectra, see figure 1, processed for the temperature dependence of the local field (figure 13) yielded $\beta = 0.373(8)$ which agrees with the value expected for the 3D Heisenberg model.

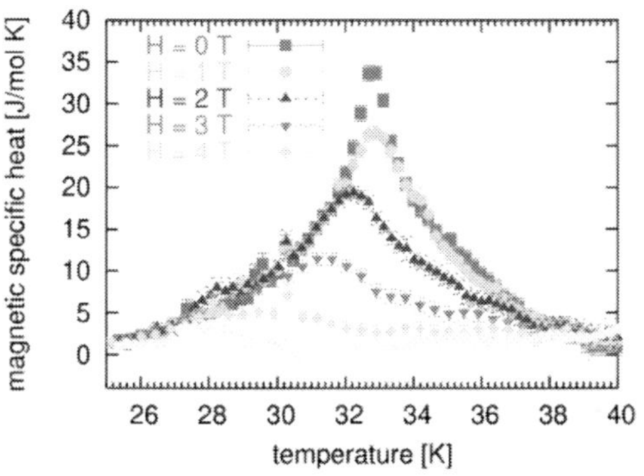

Figure 12. Anomalous part of the specific heat of $Cu_4[W(CN)_8]_4$.

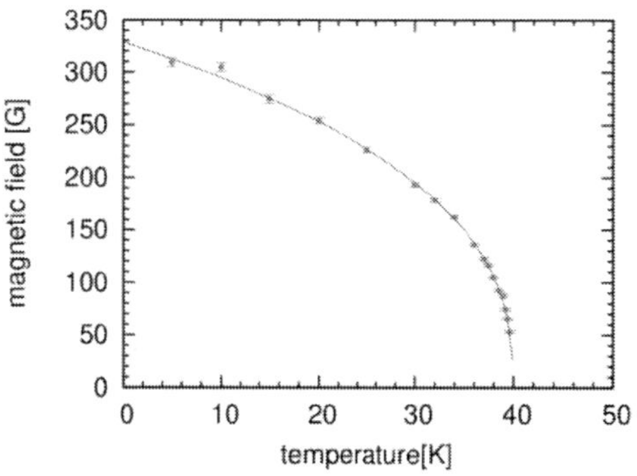

Figure 13. Local field in $Cu_6[W(CN)_8]_4$ seen at muon stopping site.

4. Conclusions

It was demonstrated that the μSR spectroscopy represents a unique experimental technique perfectly suited to the investigations of the critical behaviour of molecule based magnets. What is most important in this technique is that one can dispense with subjecting a sample to external magnetic field to get a signal. If only the critical region is sufficiently broad, μSR provides a matchless means to extract a complete set of static and dynamic critical exponents. Firstly, implanted muons in magnetically ordered phase precess in the local magnetic field with frequency proportional to the magnitude of that field, which permits the determination of the critical exponent β. Secondly, the frequency shift in a small transverse field detected in the paramagnetic regime yields the exponent γ. Finally, the longitudinal muon spin relaxation is a probe sensitive to the spin correlation time giving the dynamical exponent w. The remaining static and dynamic exponents can be deduced from the scaling or hyperscaling relations, such as $\alpha + 2\beta + \gamma = 2$ (the Rushbrook equality), $\nu = (2\beta + \gamma)/d$, $\eta = 2 - \gamma/\nu$, and $w = \nu(z + 2 - d - \eta)$, where d is the dimension of the spin network. Table 1 summarizes the critical exponents determined experimentally for four representatives of molecular magnets mentioned in this work. For comparison, the static and dynamic exponents for five model systems have also been quoted. It can be seen that Mn_2Nb can be classified as the mean-field magnet, and Cu_4W_4 and Cu_4Mo_4 are consistent with the 2D XY model. For Cu_6W_4 a discrepancy is observed, as according to the β exponent this material corresponds to the 3D Heisenberg model, while the γ exponent is close to the mean-field value. Interestingly, for both molecular magnets with the 3D spin network reduced values of the γ exponent were observed. Further studies of molecule based magnets are bound to shed some light on this conjecture.

Table 1. Experimental critical exponents compared to different models. The values of the dynamical exponent w for the model systems are quoted from [1].

model	β	w	γ
Ising d=2	1/8	3/2	7/4
XY d=2	(0.24)	0.309	1.80 (2.17 for XXY)
Ising d=3	0.326	0.717(FM)	1.239
		0.596(AF)	
Heisenberg d=3	0.369	1.026(FM)	1.36
		0.328(AF)	
mean field	0.5	0	1
Mn_2Nb	0.53(2)	0.12(1)	1.15(5)
Cu_6W_4	0.373(8)		1.05(2)
Cu_4W_4	0.23(2)	1.35(5)	1.63(1)
Cu_4Mo_4	0.27(2)	0.18(2)	1.495(2)

Acknowledgments
This work has been partially supported by the Polish Ministry of Science and Higher Education within Research Projects 0087/B/H03/2008/34 and 1535/B/H03/2009/36.

References

[1] Pratt F L, Lancaster T, Baker P, Blundell S, Kaneko W, Ohba M, Kitagawa S, Ohira-Kawamura S and Takagi S 2009 *Physica* B **404** 585–589
[2] Przychodzeń P, Korzeniak T, Podgajny R and Sieklucka B 2006 *Coord. Chem. Rev.* **250** 2234–2260
[3] Pratt F L 2000 *Physica* B **289-290** 710–714
[4] Hohenberg P C and Halperin B I 1977 *Rev. Mod. Phys.* **49** 435
[5] Bałanda M, Pełka R, Wasiutyński T, Rams M, Nakazawa Y, Miyazaki Y, Sorai M, Podgajny R, Korzeniak T and Sieklucka B 2008 *Phys. Rev.* B **78** 174409
[6] Czapla M *et al* 2010 *Phys. Rev.* B **82** 094446

Magnetic properties of lanthanoid(III) phthalocyaninato triple-decker complexes in an external magnetic field and electronic transport properties for molecular spintronics

K Katoh[1], K Yamamoto[1], T Kajiwara[2], J Takeya[3], B K Breedlove[1] and M Yamashita[1]

[1] Department of Chemistry, Graduate School of Science, Tohoku University, Aramaki-Aza-Aoba, Aoba-Ku, Sendai 980-8578, Japan

[2] Department of Chemistry, Faculty of Science, Nara Women's University, Nishi-Machi, Kita-Uoya, Nara 630-8506, Japan

[3] The Institute of Scientific and Industrial Research, Osaka University, Mihogaoka, Ibaragi 567-0047, Japan

e-mail: yamashita@agnus.chem.tohoku.ac.jp

Abstract. Lanthanoid phthalocyaninato complexes of $Dy_2(obPc)_3$ (**1**) and $Tb_2(obPc)_3$ (**2**) are shown to be a single-molecule magnet. The relationships among the magnetic relaxation properties of **1** in a direct current (dc) magnetic field in comparison to **2** and the electronic properties of a cast film are discussed.

1. Introduction

Scientists have envisioned using a single spin as a 'bit' of information to prepare high-density storage and quantum-computing devices[1]. Quantum tunneling of the magnetization (QTM) in systems with double-well potentials[2], which is a characteristic property of single-molecule magnets (SMMs), underpins this idea[3]. On the basis of the properties of lanthanoid-phthalocyaninato ($LnPc_2$) SMMs[4], we believe that $TbPc_2$ can be used as a 'bit' of information in high-density storage technology by taking advantage of the single up-spin/down-spin property, which is equivalent to 2^1. For triple- and quadruple-decker type SMMs, this is equivalent to 2^2 and 2^3, respectively, in relation to the number of spins.

We have recently reported the characteristics of $LnPc_2$ and LnPc (Ln = Tb and Dy) deposited on an Au(111) surface in an ultrahigh vacuum (UHV) using a dry process technique[5]. Both $LnPc_2$ and LnPc are present on the Au(111) surface on the basis of height profiles and dI/dV mapping obtained by using scanning tunneling microscopy (STM) and spectroscopy (STS). A Kondo peak, which is due to coupling between magnetic impurities, including Tb^{3+} ions, and conduction electrons from the STS, is only observed at the center of TbPc at a Kondo temperature (T_K) of ~250 K. More recently, we have observed a Kondo peak for $TbPc_2$ on an Au(111) surface[6]. We believe that the relation between T_K and the blocking temperature (T_B) must be studied further. In addition, the properties of SMMs and the Kondo effect can be modulated with an external magnetic field[7].

Although Ln-Pc double-decker complexes have been studied in detail, little is known about Ln(III)-Pc triple-decker complexes Ln_2Pc_3. The first lanthanoid phthalocyaninato triple-decker, Ln_2Pc_3 (Ln^{3+} = Y, Nd, Gd, Er, and Lu) was reported in 1986[8]. However it is not well characterized. Ishikawa et al. have reported a series of studies on the magnetic properties of the Ln(III)-Pc triple-decker complex [(Pc)Ln(Pc)Ln(obPc)] (obPc = dianion of 2,3,9,10,16,17,23,24-octabutoxy- phthalocyanine), which were the first reports on the dynamic magnetism of a coupled $4f^8$ system[9]. The effects of the *f-f* interactions on the temperature and frequency dependence of the ac magnetic susceptibility were discussed by comparing experimental data for isostructural bis-Tb and mono-Tb complexes. In addition, they have reported that Tb(III)-Pc triple-decker SMMs with asymmetric Tb^{3+} sites exhibit two $\chi_M''T$ peaks at 20 and 27 K. However, they have not reported its crystal structure. Therefore, the relationship between the molecular structure and the magnetic relaxation mechanisms in the Tb(III)-Pc triple-decker SMMs is not clear. Very recently, we have reported a new magnetic relaxation phenomenon for an Ising dimer of the Tb-phthalocyaninato triple-decker SMMs $Tb_2(obPc)_3$ (**2**)[10]. In Argand plots for **2**, the magnetic relaxation splits from a one-component system into a two-component system (temperature-independent and temperature-dependent regimes) in a dc magnetic field.

Chen et al. have recently reported OFETs based on an amphiphilic heteroleptic tris(phthalocyaninato) rare earth triple-decker complex with high carrier mobilities of 0.24–0.60 $cm^2\ V^{-1}\ s^{-1}$ and an on/off current ratio of 10^5[11]. It is important to examine the relationship between the SMMs and electron transport properties of Ln(III)-Pc triple-decker complexes as well as Ln(III)-Pc double-decker complexes[5]. Here we present the results of studies on the SMMs properties of $Dy_2(obPc)_3$ (**1**). The relationships among the magnetic relaxation properties in a direct current (dc) magnetic field in comparison to **2** and the electronic properties of a cast film are discussed.

2. Sample and technique

$Ln_2(obPc)_3$ was synthesized in one step starting from $Ln(acac)_3{\cdot}4H_2O$ and H_2obPc by using a published procedure[8]. $Dy(acac)_3{\cdot}4H_2O$ (57 mg, 0.13 mmol) was added to a refluxing 1-octanol suspension (2.5 mL) of H_2obPc (55 mg, 0.05 mmol). After 4 h of refluxing, the reaction mixture was added to methanol (100 mL). The crude product was purified by using column chromatography on silica gel with $CHCl_3$ as the eluent. The collected deep blue fraction was evaporated to dryness to obtain a dark blue solid. Column chromatography (C-200 silica gel, Wako and Sephadex G-10, Pharmacia Biotech) was used to remove the remaining impurities. Dark blue fine needle crystals were obtained from chloroform/ethanol (30 mg, 8.15×10^{-6} mol, 16%). ESI-MS: *m/z* (%): 3593 (30) [M^+]; elemental analysis calcd (%) for $C_{192}H_{240}N_{24}O_{24}Dy_2$: C 64.18, H 6.71, N 9.36; found: C 64.0, H 6.56, N 9.20.

Single crystals of **1** were prepared by the method described above. A single crystal was mounted on a loop rod coated with Paratone-N (HAMPTON RESEARCH). The crystal dimensions were 0.31 × 0.14 × 0.09 mm^3 for **1**. Data collection was performed on a Rigaku CCD diffractometer (Saturn70) with graphite monochromated Mo Kα radiation (λ = 0.71069 Å) at a temperature of −180 ± 1 °C. An empirical absorption correction based on azimuthal scans of several reflections was applied. The data were corrected for Lorentz and polarization effects. All non-hydrogen atoms were refined anisotropically using a least-squares method, and hydrogen atoms were fixed at calculated positions and refined using a riding model. SHELXTL[12] was used for structure refinement, and the structure was expanded using Fourier techniques[13]. The weighting scheme was based on counting statistics. Full-matrix least-squares refinements on F_2 based on unique reflections were employed, where the unweighted and weighted agreement factors of $R = \Sigma|\,|F_o| - |F_c|\,|/\Sigma|F_o|$ ($I > 2.00\sigma(I)$), and $wR = [\Sigma w(F_o^2 - F_c^2)^2/\Sigma w(F_o^2)^2]^{1/2}$ were used. Crystal data for **1**·2EtOH (CCDC-807110): $C_{196}H_{250}N_{24}O_{26}Dy_2$, Mr = 3683.20, triclinic, *P*–1, a = 1.32329(15), b = 1.7835(2), c = 2.0575(3) nm, α = 107.421(2), β = 90.530(2), γ = 100.852(2)°, V = 4.5391(10) nm^3, T = 93(1) K, Z = 1, ρ_{calcd} = 1.347 Mg m^{-3}, F_{000} = 1934, λ = 0.7107 Å, $\mu(Mo_{K\alpha})$ = 0.893 mm^{-1}, 28306 measured reflections, 19941 unique which were used in all calculations. $R_1 = 0.0506$ ($I > 2\sigma(I)$) and $wR_2 = 0.1396$ (all data) with GOF = 1.087. Powder X-ray diffraction (XRD) patterns of **1** and **2** were collected at room temperature using a Rigaku X-ray diffractometer (AFC-7R/LW) operated at 50 kV and 300 mA. Powder samples in a capillary tube were scanned in the diffraction angle range of 3°–60° (2θ) in steps of 0.02° at 2s/step.

UV-Vis-NIR spectra for $CHCl_3$ solutions in a quartz cell with a pathlength of 1 cm and for cast films on a quartz plate were acquired on a SHIMADZU UV-3100PC spectrometer at 298 K.

Magnetic susceptibility measurements were performed on Quantum Design SQUID magnetometers MPMS-5S and MPMS-XL. The dc measurements were performed in the temperature range of 1.8–300 K and dc fields of –7–7 T. Ac measurements were performed in the frequency range of 1–1500 Hz with an ac field amplitude of 3 Oe in the presence of a dc field (zero, 0.1, 0.2, 0.3, 0.5, and 1.0 T). Measurements were performed on a randomly oriented powder sample of **1**, which was suspended in *n*-eicosane, heated to 330 K, and then cooled to 300 K for solidification. All data were corrected for sample holder, *n*-eicosane, and diamagnetic contributions from the molecules by using Pascal's constants.

Cyclic voltammetry (CV) was carried out on a Bioanalytical Systems CV-50W. Solutions of **1** (10^{-4} mol dm^{-3}) in dry CH_2Cl_2 with *n*-Bu_4PF_6 (0.1 mol dm^{-3}) as the supporting electrolyte were used with glassy-carbon working, Pt wire counter, and Ag/AgCl reference electrodes. A scan rate of 100 mV s^{-1} was used. In the case of the cast film of **1**, films were deposited on a glassy-carbon working electrode. The CVs were acquired in an aqueous solution of $KClO_4$ (0.1 mol dm^{-3}), which was used as the supporting electrolyte. A Pt wire was used as the counter electrode, and an SCE was used as the reference electrode. A scan rate of 100 mV s^{-1} was used.

Field-effect transistors (FETs) were fabricated from **2** to determine the charge transport properties the assembly. Drop-casting from heptane solutions on doped-Si/SiO_2 substrates was used to prepare the FET, which was used a doped-Si layer as a gate-electrode. Source and drain electrodes were formed from the top of the semiconductor layer affording a top-contact geometry is adopted. The effective width and length of the channel were 150 μm and 20 μm, respectively. The FET characteristics were measured using a Keithley 4200 semiconductor parameter analyzer.

3. Result and discussion

Although Takahashi et al. have reported the Dy-Pc triple-decker complex **1**, they have not reported its crystal structure and magnetic properties[8]. We used a Pc ligand with 2,3,9,10,16,17,23,24-octabutoxy substituents because it should have a higher solubility, thus making crystallization easier.

Complex **1**, shown in Fig. 1, crystallized with ethanol in the crystal lattice in the triclinic space group *P*–1 and was isomorphous with **2**. Crystal-packing diagrams of **1** are shown in Fig. 2.

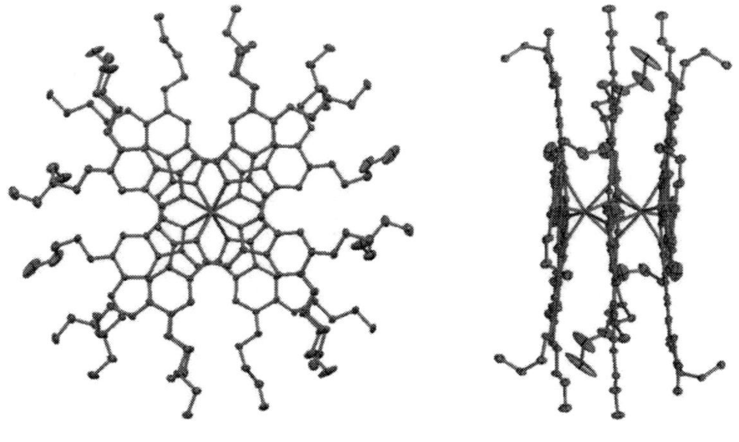

Figure 1. View of the molecular structure of $Dy_2(obPc)_3$ (**1**) with ellipsoids at 50% probability.

Complex **1** has two Dy^{3+} ions sandwiched between three Pc ligands with eight isoindole-nitrogen donor atoms (N_{iso}) and a center of symmetry. The center of the square formed by the four pyrrolic nitrogens of Pc form a crystallographically imposed inversion center, making the two Ln^{3+} ions and outer-Pcs equivalent. The crystal structure of **1** is similar to that of **2**, and both outer Pc ligands of **1** are equally distorted from planarity, adopting a biconcave shape[10]. The Dy^{3+} ions are unevenly spaced between the two Pc ligands, being displaced 0.258–0.262 nm (**2**: 0.258–0.263 nm) from the mean plane of the four N_{iso} of the inner Pc ligands and 0.234–0.237 nm (**2**: 0.235–0.237 nm) from the mean plane of the four N_{iso} of the outer ones. These bond lengths are different from those in $DyPc_2$ (0.239–0.242 nm)[5]. The intramolecular Dy–Dy distance is 0.350 nm (Tb-Tb distance is 0.352 nm in **2**). The twist angle between the outer rings and the center one was determined to be 32°, causing a pseudo 4-fold axis. The intermolecular Dy–Dy distance along the *a*, *b*, and *c* axes were determined to be 1.104, 1.588, and 1.861 nm, respectively. Each molecule of **1** is rather well separated from neighboring molecules due to the *n*-butoxy chains. The size and height of **1** were estimated to be ~2.4 and ~0.7 nm same as **2** ($DyPc_2$: ~1.6 and ~0.4 nm[5]) on the basis of the distance between two hydrogen atoms of the *n*-butoxy chains at both ends of the molecules and between the oxygen atoms of the line splitting the two Pc rings, respectively. Powder XRD of **1** and **2** (Fig. 3) are similar to simulated patterns based on X-ray single crystallographic data of **1** at room temperature[10].

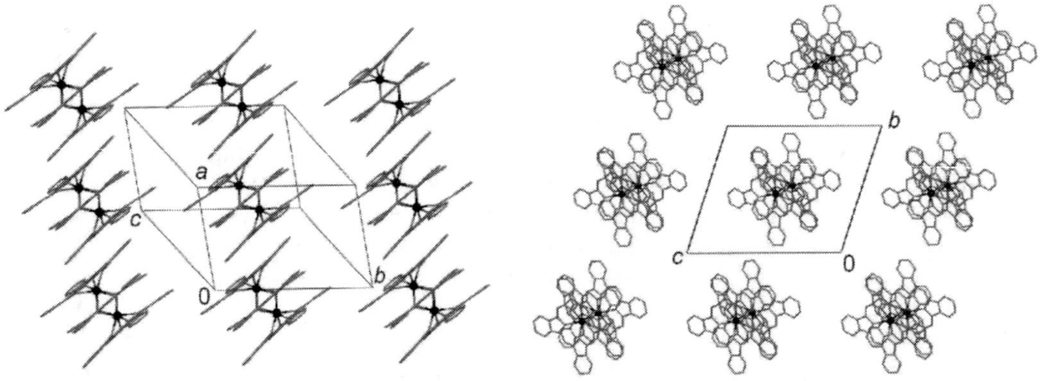

Figure 2. Packing diagrams of **1**. Solvent molecules located between the *n*-butoxy chains and the *n*-butoxy chains have been omitted for clarity. Each column of **1** is rather well separated from neighboring columns by the *n*-butoxy chains.

$\chi_M T$ versus T plots for **1** increased with a decrease in T and reached a maximum of ~37 cm^3 K mol^{-1} at 1.8 K, which suggests the existence of a ferromagnetic interaction between the Dy^{3+} ions (Fig. 4a). The magnetic dipolar term, not the exchange term, is dominant. **2** exhibits the same behavior[10].

The M-H curve for **1** as well as **2** at 1.8 K showed no magnetic hysteresis (Fig. 4b). At this temperature the spins are not frożen. In the case of **2**, below 1.5 K, hysteresis loops were observed by using a micro-SQUID. However, there was a sharp drop at near zero-field leaving practically no remanent magnetization because the relaxation time, which was strongly temperature dependent down to ~0.4 K and field sweep-rate dependent even at 0.04 K, was not sufficiently long[10]. This behavior is typical for an SMMs with a crossover temperature of ~0.4 K.

Fig. 5 shows ac susceptibility measurements on a powder sample of **1**. A maximum $\chi_M''T$, which was dependent on the applied ac field, indicating that **1** is an SMMs, appeared at 4.5 K at an ac frequency (*f*) of 1488 Hz, whereas that for **2** was observed at 24 K[10]. It is clear that the magnetic relaxation process is related to the sublevel structures of the ground-state multiplets of the Ln^{3+}

complexes[4]. In addition, $\chi_M''T$ did not disappear below 3 K in a zero field, indicating that the magnetic moment is not frozen (Fig. 5). In other words, as in the case of **2**, a different relaxation process becomes dominant in the low-temperature region.

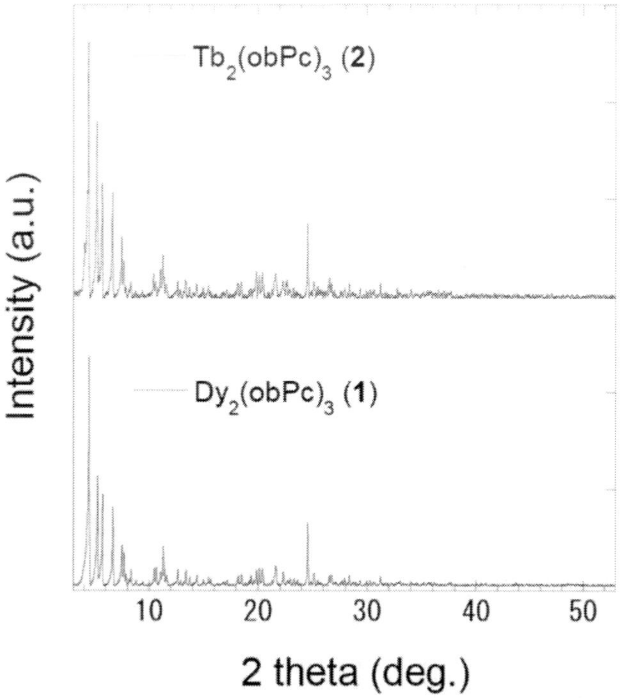

Figure 3. Powder XRD patterns of **1** and **2** at 298 K.

Dependence of $\chi_M''T$ on the ac field frequency was clearly observed in the low-temperature region in several dc magnetic fields. At f = 1 Hz, in a 0.2 T dc field, only one peak at 2.5 K was observed, whereas, at f = 1488 Hz, two $\chi_M''T$ peaks were observed at 4.5 and ~2.5 K, which correspond to two different magnetic relaxation processes. The $\chi_M''T$ peak for **1** at 4.5 K (f = 1488 Hz) was independent of the dc field.

On the other hand, the shape of the peak at ~2.5 K (f = 1, 100, and 1488 Hz) drastically changed, and the peak shifted to a higher temperature with an increase in the strength of the dc magnetic field. In Dy^{3+} SMMs, the lowest substates are $J_z = \pm 13/2$[14], and in Dy-Pc as well as Tb-Pc triple-decker complexes, a Zeeman effect involving the ground-state energy levels plays a role in the observed changes[10]. Investigations to clarify the mechanism of these observations, including determination of the multiplet substructures of **1**, are in progress.

Since **1** and **2** are soluble in most organic solvents, except for alcohols, we could prepare cast films for potential application in devices. They were prepared on quartz plates directly by using a hand-casting method. UV-vis-NIR spectra of a solution and cast films of **1** are shown in Fig. 6. The electronic spectrum of **1** is similar to that of **2**[10]. The spectra of the cast films are similar to that of the solution; however, the bands are broader. In addition, the Q band region is red-shifted. In the solution (cast film) spectra, two intense Q bands were observed at 643 nm (666 nm) with a shoulder peak at 547 nm (561 nm) and 760 nm (824 nm). These bands drastically changed, and a broad band in the near-IR region (600–1200 nm) appeared in the spectra of the cast film. On the basis of Kasha's theory and previously published results, the red shift suggests that **1** forms *J*-aggregates in the cast films[15].

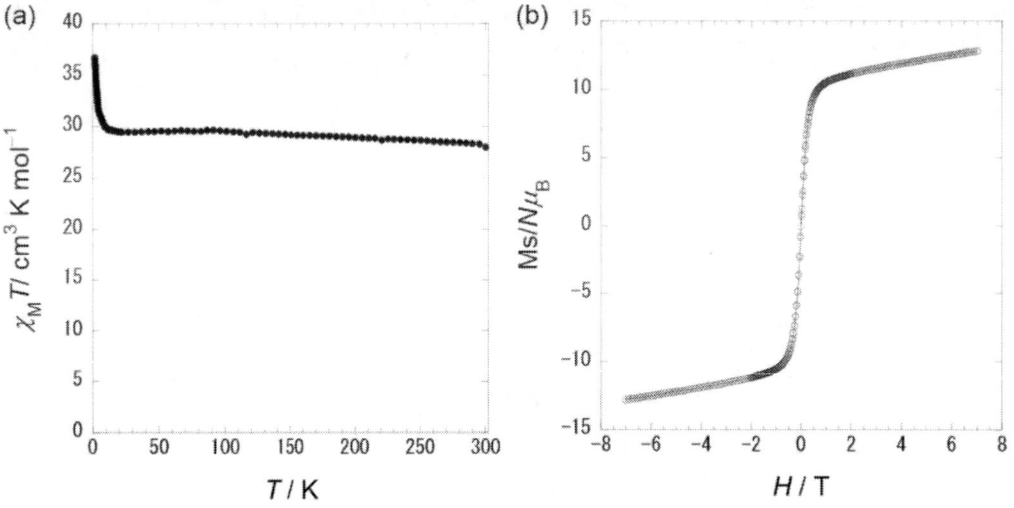

Figure 4. dc magnetic measurements on a powder sample of **1**. (a) $\chi_M T$ versus T plots. In the room temperature region, the $\chi_M T$ value as asymptotically approached that of two free Dy^{3+} ions (14.18 cm^3 K mol^{-1} × 2). (b) Field dependence of the magnetization at 1.8 K.

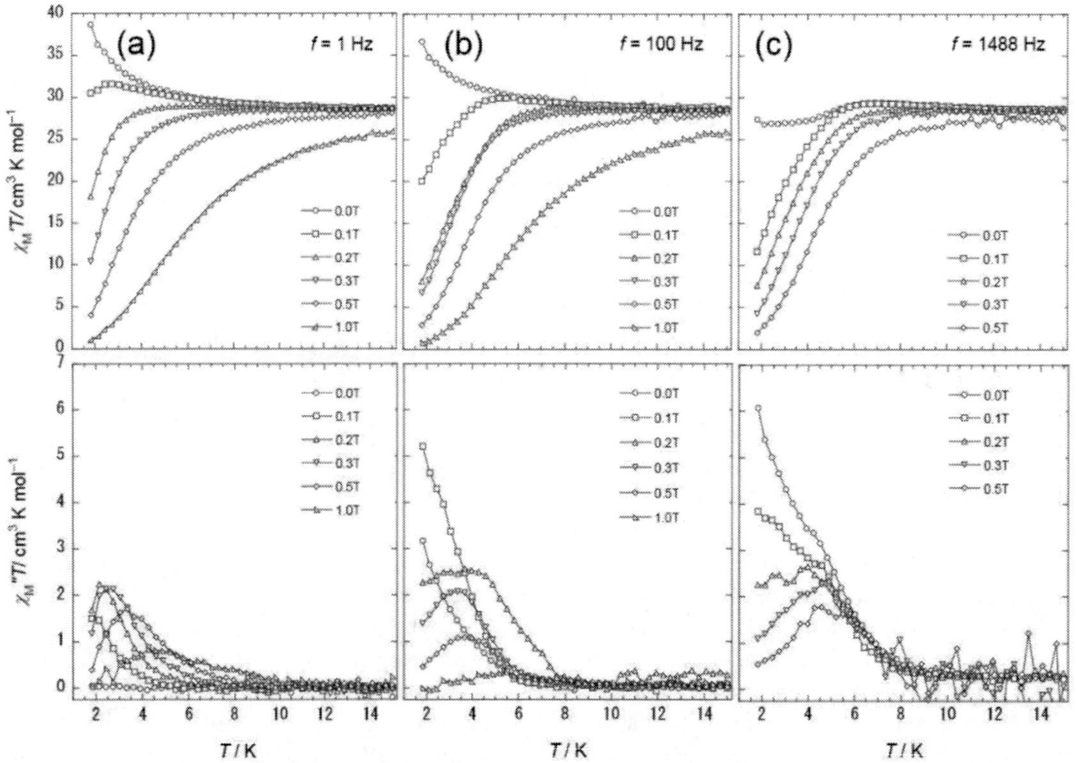

Figure 5. $\chi_M'T$ and $\chi_M''T$ versus temperature (T) for **1** in dc fields of 0, 0.1, 0.2, 0.3, 0.7, and 1.0 T measured in a 3 Oe ac magnetic field with frequencies (f) of (a) 1, (b) 100, and (c) 1488 Hz.

The spectra clearly show that there is intermolecular π–π stacking (edge-to-edge interaction) in the cast films, and the results agree with those from X-ray crystal analysis, in which **1** is arranged in columns along the *a* axis with π–π stacking (0.343 nm, Fig. 2).

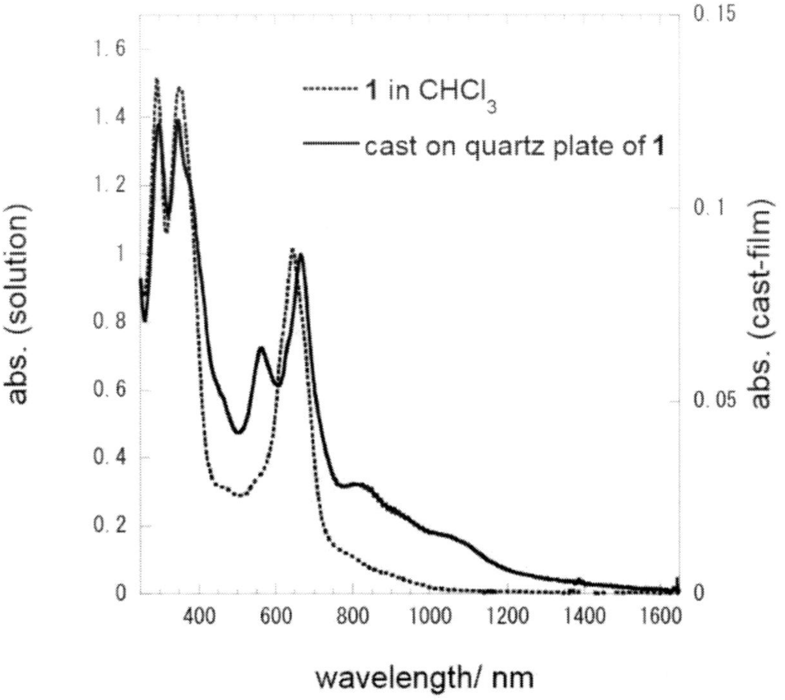

Figure 6. Electronic absorption spectra of **1** in chloroform (10^{-5} mol dm^{-3}) and cast film at 298 K. The cast film of **1** was prepared on a quartz plate from a dichloromethane solution (10^{-4} mol dm^{-3}) by evaporation of the solvent.

Fig. 7a shows a cyclic voltammogram (CV) of a cast film of **1** in a 0.1 mol dm^{-3} aqueous solution of $KClO_4$. The cast film was prepared on a glassy-carbon working electrode (geometric area 0.07 cm^2) directly by using a hand-casting method. The CVs of **1** and **2** film showed only two irreversible oxidation waves Ox1 ($E_{1/2}{}^1$ = 294 mV) and Ox2 ($E_{1/2}{}^2$ = 476 mV). This is clear evidence that only the Pc ligands in the triple-decker structure undergo redox reactions. The CVs did not change even after more than 10 cycles. On the other hand, the CVs of **1** in dichloromethane showed four reversible waves (Ox1[$E_{1/2}{}^1$ = 111 mV], Ox2[$E_{1/2}{}^1$ = 369 mV], Ox3[$E_{1/2}{}^1$ = 855 mV], Ox4[$E_{1/2}{}^1$ = 1217 mV]), as shown in Fig. 7b. The oxidation processes in solution are similar to those previously reported[8]. The differences between the CVs of the cast film and the solution of **1** show that *J*-aggregates are present. In order to obtain stabler films, we are investigating other CV conditions. However, the electrochemical stability of the cast films of **1** and **2** in these studies are satisfactory for preliminary studies on their use in field-effect transistor (FET) devices.

Fig. 8 shows an optical image of the device. Fiber-like **2** was coated on the SiO_2 gate dielectric layer. A plot of the transfer characteristics, i.e., drain current (I_D) vs. gate voltage (V_G), of the device is shown in Fig. 9. The drain voltage V_D was fixed at –150 V. I_D increased when V_G was increased in the negative direction, indicating that the device shows *p*-type FET behavior. In other words, *p*-type charges (holes) accumulate in the semiconductor layer, and the charges are mobile in the fiber-like assembly of **2**. The mobility (μ) of the semiconductor was estimated using the standard formula for

FETs to be 2×10^{-5} $cm^2\ V^{-1}\ s^{-1}$ and 1×10^{-5} $cm^2\ V^{-1}\ s^{-1}$ when $V_G > 100$ V and $V_G < 100$ V, respectively. The gate leakage current, which was measured simultaneously, was negligibly small. We note that the current is induced only at the bottom surface of the semiconductor film by the electric field. Since the rest of the semiconductor layer has pronounced conductivity, only 10% of the total current is modulated.

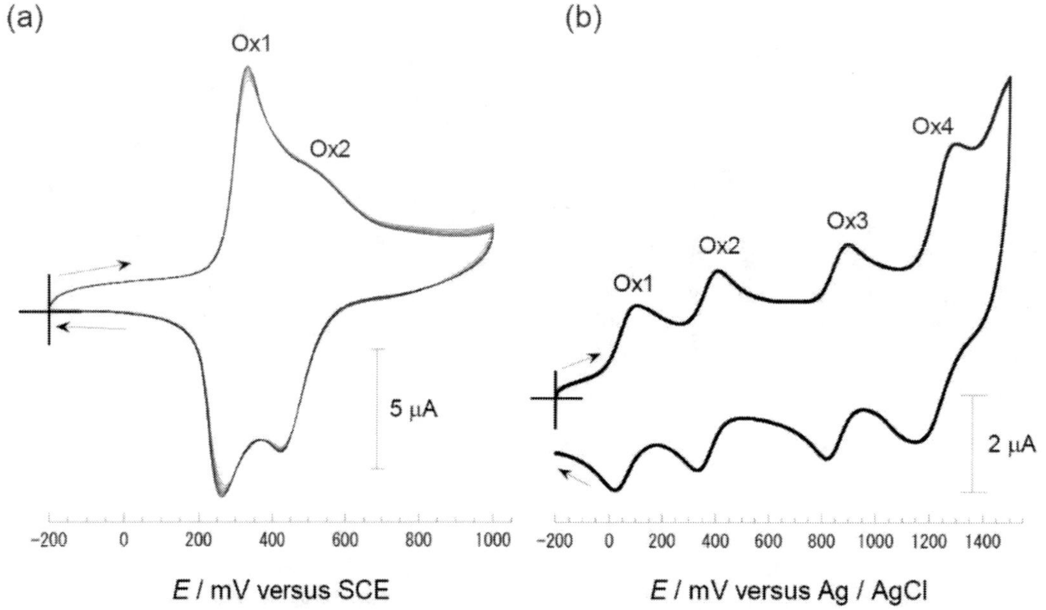

Figure 7. Cyclic voltammogram of (a) a cast film of **1** on a glassy-carbon working electrode versus SCE in 0.1 mol dm^{-3} aqueous solution of $KClO_4$. Start at –200 mV, v = 100 mV s^{-1}, 298 K. (b) 10^{-4} mol dm^{-3} dichloromethane solution of **1** using glassy-carbon working electrode versus Ag/AgCl in 0.1 mol dm^{-3} of $TBAPF_6$, Start at –200 mV, v = 100 mV s^{-1}, 298 K.

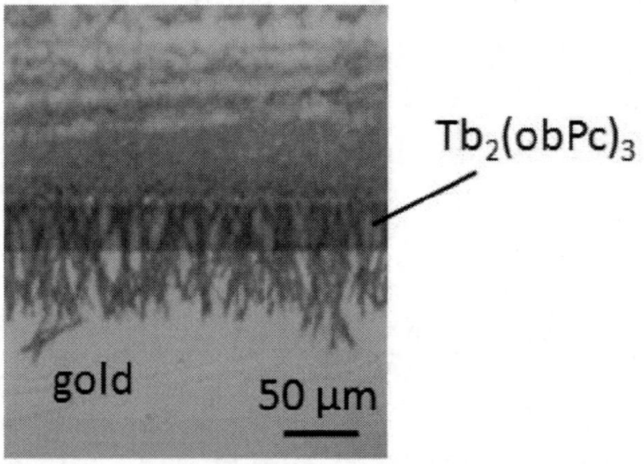

Figure 8. Optical image of field-effect transistors of **2**.

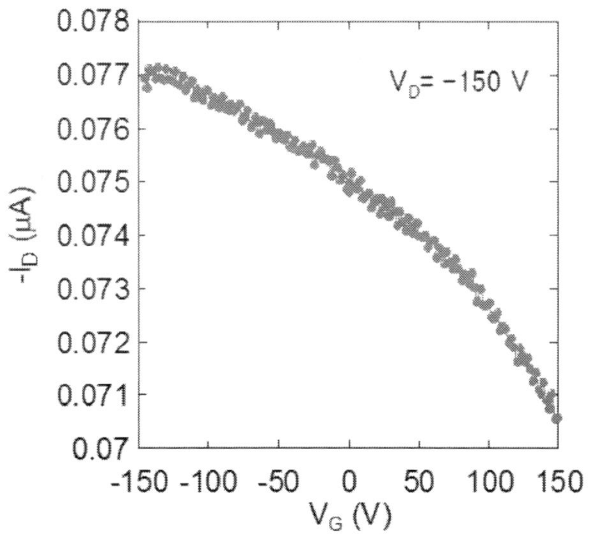

Figure 9. Transfer characteristics of field-effect transistors of **2**.

4. Conclusions

Complexes **1** and **2** showed two magnetic relaxation processes in the low temperature region in the presence of a dc magnetic field. Our results suggest that the SMMs/magnetic properties of **1** and **2** significantly change in a dc magnetic field. These relaxation mechanisms are related to the energy gap of the ground-state and QTM[10]. In order to elucidate fully the magnetic relaxation mechanism, further experimental and theoretical studies are required.

Furthermore, our work clearly shows that the cast films of **1** and **2** can be studied for use in devices, such as molecular electronic sensors, electrochromic displays, field-effect transistor devices etc. For preparing devices, the relationship between the SMMs and electron transport properties of Ln_2Pc_3 as well as $LnPc_2$ must be clarified. The carrier mobility increased by reducing the lattice thermal vibrations of the atoms at low temperature. It might be possible to switch the magnetism of OFETs containing Ln(III)-Pc multiple-decker SMMs between SMMs ↔ paramagnetic by changing the gate voltage[16] and the magnetoresistance between the SMMs as a localized spins and the conduction electrons at low temperature[17]. We believe that these studies will lead to an understanding of the effect of the magnetic field on the magnetic relaxation mechanism and electronic properties. Further studies involving molecular spintronics are now in progress.

Acknowledgments

This work was financially supported by a Grant-in-Aid for Scientific Research (S) (Grant No. 20225003) from the Ministry of Education, Culture, Sports, Science, and Technology, Japan.

References

[1] Aviram A and Ratner M A 1974 *Chem Phys lett.* **29** 277
[2] Gatteschi D, Sessoli R and Villain J 2007 *Molecular Nanomagnets*; Oxford Univ. Press, New York
[3] Bogani L and Wernsdorfer W 2008 *Nature materials* **7** 179
[4] Ishikawa N, Sugita M, Ishikawa T, Koshihara S and Kaizu Y 2003 *J. Am. Chem. Soc.* **125** 8694
[5] Katoh K, Yoshida Y, Yamashita M, Miyasaka H, Breedlove B K, Kajiwara T, Takaishi S, Ishikawa N, Isshiki H, Zhang Y F, Komeda T, Yamagishi M and Takeya 2009 J, *J. Am. Chem. Soc.* **131** 9967

[6] Komeda T, Isshiki H, Zhang Y F, Lorente N, Katoh K, Breedlove B K and Yamashita M private communication
[7] Romeike C, Wegewijs M R, Hofstetter W and Schoeller H 2006 *Phys. Rev. Lett.* **97** 206601-1
[8] Takahashi K, Itoh M, Tomita Y, Nojima K, Kasuga K and Isa K 1993 *Chem. Lett.* 1915
[9] Ishikawa N, Iino T and Kaizu Y 2002 *J. Am. Chem. Soc.* **124** 11440
[10] Katoh K, Kajiwara T, Nakano M, Nakazawa Y, Wernsdorfer W, Ishikawa N, Breedlove B K and Yamashita M *Chem. Eur. J.* in press
[11] Chen Y, Su W, Bai M, Jiang J, Li X, Liu Y, Wang L and Wang S 2005 *J. Am. Chem. Soc.* **127** 15700
[12] Sheldrick G M 1994 *SHELXTL, program for Crystal Structure Determination*; Siemens Analytical X-ray Instruments Icn.: Madison, WI
[13] Beurskens P T, Admiraal G, Beurskens G, Bosman W P, de Gelder R, Israel R and Smits J M M 1994 DIRDIF94
[14] Ishikawa N, Mizuno Y, Takamatsu S, Ishikawa T and Koshihara S 2008 *Inorg. Chem.* **47** 10217
[15] Kasha M, Rawls H R and EL-Bayomi M A 1965 *Pure Appl. Chem.* **11** 371
[16] Chiba D, Yamanouchi M, Matsukura F and Ohno H 2003 *Science* **301** 943
[17] Hiraga H, Miyasaka H, Nakata K, Kajiwara T, Takaishi T, Oshima Y, Nojiri H and Yamashita M 2007 *Inorg. Chem.* **46** 9661

Joint European Magnetic Symposia – JEMS 2010 IOP Publishing
Journal of Physics: Conference Series **303** (2011) 012036 doi:10.1088/1742-6596/303/1/012036

Single-chain magnet features in 1D [MnR_4TPP][TCNE] compounds

Maria Bałanda[1], Zbigniew Tomkowicz[2], Wolfgang Haase[3] and Michał Rams[2]

[1] Institute of Nuclear Physics PAN, Radzikowskiego 152, 31-342 Kraków, Poland
[2] Institute of Physics, Jagiellonian University, Reymonta 4, 30-059 Kraków, Poland
[3] Institute of Physical Chemistry, Darmstadt University of Technology, 64287 Darmstadt, Germany

E-mail: Maria.Balanda@ifj.edu.pl

Abstract. Molecular chains of antiferrimagnetically coupled Mn^{III}-ion (S = 2) and TCNE (tetracyanoethylene) radical moments (s = ½) show different behaviour depending on group R substituted to TPP (tetraphenylporphyrin) and on the substitution site. The compound with R = F in *Ortho* position is a Single-Chain Magnet (SCM) with blocking temperature T_b = 6.6K, while that with R = F in *Meta* position shows both blocking (T_b = 5.4 K) and magnetic ordering transition (T_c = 10 K). For bulky groups R = OC_nH_{2n+1}, the magnetically ordered phase is observed ($T_c \approx 22$ K), which does not however prevent slow relaxation at T <8 K. Magnetic hysteresis with coercive field H_c of 2 T at 2.3 K is like that of SCM. The frequency dependent AC susceptibility in the superimposed DC field reveals common features of all systems. The energy of intrachain ferromagnetic coupling between effective spin units 3/2, relevant at low temperatures, is determined for all compounds and the interchain dipolar coupling is estimated. It is concluded that slow relaxation is inherent for all quasi one-dimensional compounds and for the magnetically ordered ones shows up in the high enough magnetic field.

1. Introduction

Discovery of slow relaxation and magnetic bistability in anisotropic high spin molecules [1] entered upon the research on molecular nanomagnets. As an analogue to single-molecule magnets (SMM), 1D single-chain magnets (SCM) [2] also show the unique feature of slow magnetic relaxation and magnetic hysteresis without long range order. Slow relaxation of magnetization occurs when the energy barrier E_a in the Arrhenius law $\tau = \tau_0 \exp(E_a / k_B T)$ for the thermally activated process is large. The temperature at which relaxation is so slow that no magnetization change can be observed in the laboratory scale, is called the blocking temperature T_b and reads $T_b = E_a / (k_B \ln(\tau_m/\tau))$, where τ_m is a characteristic time of the method. The search for SCMs was motivated with the increase of the blocking temperature, in view of potential application of such materials as molecular memory units.

Relaxation of molecular nanomagnets may be compared to that of conventional single-domain particles, which are superparamagnetic at high temperatures, while below T_b show hysteresis and remanence. The activation energy E_a for single-domain particles is proportional to the anisotropy of the material and the volume of the particle. For SMMs E_a is the well known product $-DS^2$, where D is the easy-axis anisotropy, and S is the total spin of the molecular cluster. For SCMs, apart from spin and anisotropy, the activation energy is basically proportional to the intrachain exchange integral J, as it hampers the spin reversal in the ferromagnetically coupled spin array [1,2,3]. The interchain interactions should be negligible to prevent the transition to 3D magnetic ordering. It was also found [4] that the preexponential factor τ_0 representing the flipping rate of an isolated spin, scales with the size L of the system, which is a new important property.

The subject of the paper are the compounds from the family of 1-dimensional charge-transfer chains based on Mn-porphyrin. The [Mn^{III}(TPP)][TCNE]*solvent (TPP-tetraphenylporphyrin, TCNE-tetracyanoethylene) was first reported by Miller et al. [5]. Magnetic moments of Mn^{III} cations of spin S = 2 and delocalized moments of $TCNE^-$ radicals s = 1/2 are antiparallely coupled by strong kinetic superexchange J_{intra}, usually equal to about 100 K [5,6,7]. Magnetic properties of the substance may be tuned using various chemical modifications, e.g. substituting various functional groups R to the phenyl rings at the periphery of the porphyrin disc. Figure 1 shows the structure of the [MnTPP][TCNE] molecular chain and possible *Para*-, *Meta*- or *Ortho*-position substitution sites. One-dimensional ferrimagnetic behaviour at high temperatures may be evidenced by the fit of magnetic susceptibility χ with the Seiden's model [8]. The model deals with an infinite chain of classical spins coupled by the isotropic exchange with alternating quantum spin $s = ½$ according to the Hamiltonian

$$H = -J\sum_{i=1}^{N-1}(S_i + S_{i+1})s_i \tag{1}$$

Seiden's expression for the χT product as a function of T, allows to determine the intrachain exchange coupling $J = J_{intra}$ and the Landé g-factor. At low temperatures a number of compounds of this type are magnetically ordered what means that besides the interaction along the chain, J_{intra}, and the single ion anisotropy D, the interchain coupling J_{inter}, mainly of dipolar origin, is present. A series of [MnR_4TPP][TCNE] samples using different functional groups R substituted at *Para*-position was synthesized and investigated [6, 9, 10,11]. Most of them showed transition to the long range order at temperatures about ≈ 20 K. The long alcoxy OC_nH_{2n+1} functional groups were substituted to the phenyl rings with the aim to investigate the role of dipolar forces in setting the magnetic order. It appeared that even in the case of 30 Å distant chains the compound could be magnetically ordered and T_c did not change much with interchain range [11,12]. Magnetic transition detected with AC susceptibility χ_{AC} was very sharp (critical exponent γ close to that of the 3D Ising model), however its strong dependence on the frequency of the oscillating field was astonishing. Applying weak constant field pushed dependence on frequency to lower temperatures, where irreversibility of magnetization was observed [10,12]. The other strange thing was the absence of heat capacity C_p singularity at T_c [13] and a large magnetization hysteresis at lowest temperatures. It was also interesting to test the dependence of the intrachain coupling and the Mn(III) anisotropy on the position into which group R is substituted. To this end a strongly electronegative fluorine ion was chosen. A striking difference in magnetic behaviour, dependent on the substitution site, was observed by us for compounds modified with R = F. Figure 2 presents real part of χ_{AC} for three modifications of [MnF_4TPP][TCNE]. The *Para*-F compound with fluorine in *Para*-position, is a cluster-glass system [10], [MnF_4TPP][TCNE] with fluorine in *Ortho*-position (hereafter abbreviated as Ortho-F compound) is a Single Chain Magnet [7]. Meta-F compound with fluorine in *Meta*-position is an intermedium system showing magnetic phase transition and simultaneously SCM-like slow relaxation at low temperature, which can be observed with or without magnetic field.

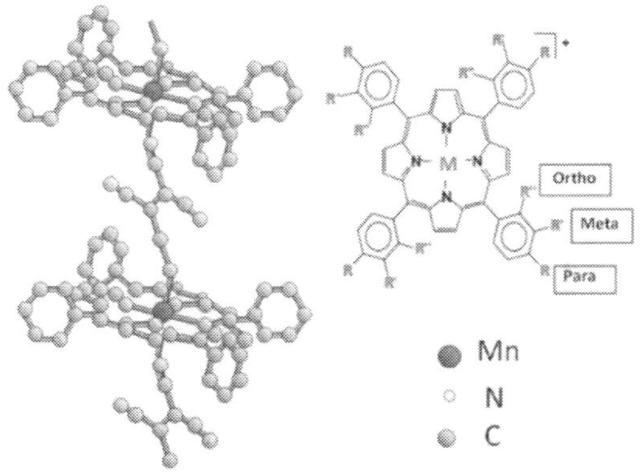

Figure 1. Structure of the [MnTPP][TCNE] molecular chain and possible *Para-*, *Meta-* or *Ortho*-position substitution sites.

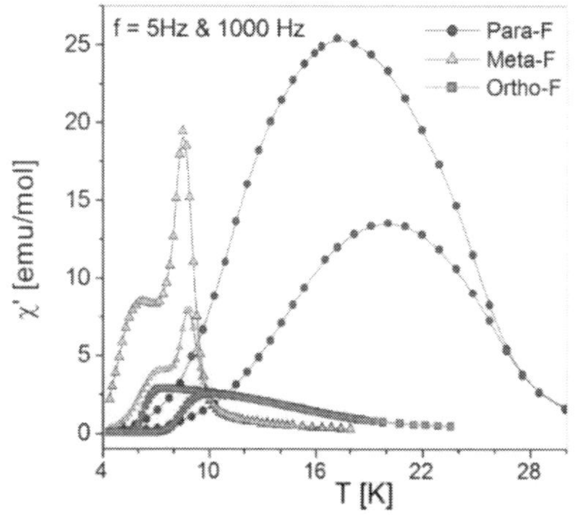

Figure 2. Real component of AC susceptibility measured with frequency 5 Hz and 1000 Hz for three phases of $[MnF_4TPP][TCNE]$: fluorine substituted into *Para-*, *Meta-* and *Ortho-*position.

In light of the data obtained thus far, we are going to discuss magnetic relaxation in the Mn-porphyrin chains and to reveal the Single Chain Magnet features in compounds showing magnetic transition to the ordered phase. All measurements were carried out at the Lake Shore 7225 AC susceptometer / DC magnetometer. We present three compounds of different relaxation and/or magnetic order type, as directly reflected in the dynamic susceptibility result given in Fig. 3. The first system is the Ortho-F Single Chain Magnet (figure 3a); the second one, with fluorine in *Meta*-position (Meta-F) shows both magnetic transition and blocking of relaxation (figure 3b). The third one, $[Mn(OC_nH_{2n+1})_4TPP][TCNE]$, represented with compound with $OC_{12}H_{25}$ group in *Para*-position (hereafter abbreviated as $OC_{12}H_{25}$), has a sharp magnetic transition (figure 3c) but reveals slow relaxation of the SCM type at low temperatures in the presence of superimposed DC magnetic field. The common feature of all compounds is the 1D ferrimagnetic character compatible with the Seiden isotropic model [8] in the temperature range ~50 - 300 K. As will be shown, at low temperatures (~ 20 - 70 K) these systems may be described by the anisotropic Heisenberg model [14], which predicts the ferromagnetic Ising-like behaviour in the limit of zero field. Therefore, at low temperatures (e.g. 20 - 50 K), the chains may be described as composed of effective units of spin S = 2 - ½ = 3/2, which interact ferromagnetically. Exchange integral J_{eff} of such ferromagnetic (FM) coupling may be determined from the formula

$$\chi = \frac{2Ng^2\mu_B^2}{3J_{eff}}\left(\frac{J_{eff}S}{k_BT}\right)^2 \tag{2}$$

derived in Ref.[14]. Values of J_{eff} determined from the low-temperature susceptibility will help us to explain the value of activation energy E_a observed in relaxation measurements.

It is known that parallel magnetic susceptibility for an anisotropic Heisenberg or Ising-like 1D system measured in the limit of zero field is proportional to ξ / T, where ξ is the correlation length. This implies the relation

$$\chi T \approx C_{eff} \exp\left(\frac{\Delta_\xi}{k_BT}\right) \tag{3}$$

where Δ_ξ is the energy necessary to create a domain wall in the chain and C_{eff} is the Curie constant of the effective spin unit [15]. In the paper we determine Δ_ξ for several $[Mn^{III}R_4(TPP)][TCNE]$

compounds and use its values for discussion of competition between magnetic order and SCM behaviour (see Table 1.).

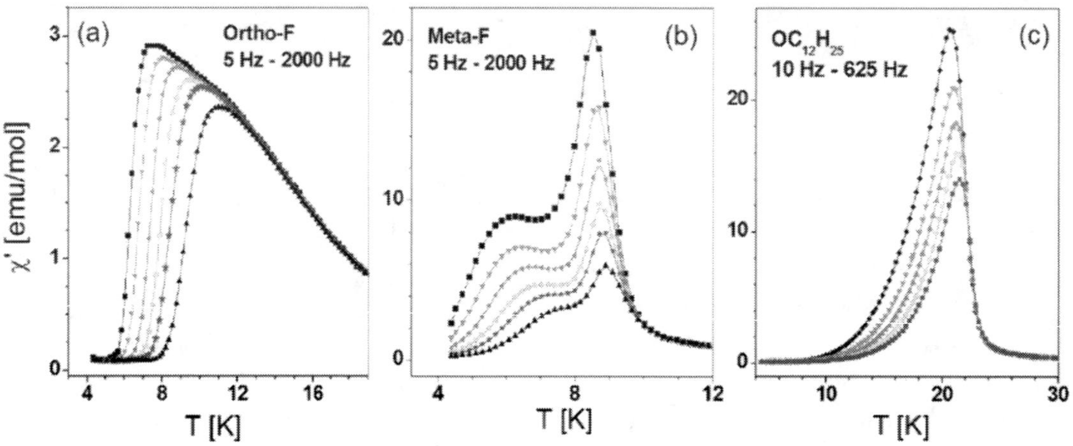

Figure 3. Comparison of frequency dependent AC susceptibility (real component χ') for (a) Ortho-F Single Chain Magnet, (b) Meta-F showing magnetic transition and blocking of relaxation and (c) magnetically ordered compound with $OC_{12}H_{25}$ group in *Para*-position. For higher frequency maximum of the χ' peak shifts to higher temperature.

2. Main characteristics of Ortho-F SCM

Structural and morphological characterization of Ortho-F compound together with the magnetic properties studied by AC and DC techniques have been already reported in [7]. The intrachain antiferromagnetic exchange integral derived with the Seiden model is J_{intra} = −217 K. The ferromagnetic intrachain coupling of effective spin units 3/2 determined from χ' in the temperature region 22 - 68 K using Eq. 2 is J_{eff} = 19.5 K. The energy Δ_ξ for domain wall creation determined from Eq. 3 in the range 38 - 5 K is equal to 22.5 K (see Table 1) [16]. χ_{AC} of this compound shows strong dependence on frequency in accord with slow time decay of thermoremanence. Slow relaxation evokes the hysteresis loop with the coercive field H_c equal to 20 kOe at 2 K and decreasing linearly with temperature.

As may be seen from figure 3a and Ref.[7], upon lowering temperature, χ' reaches maximum and abruptly falls down, while the imaginary component χ" [7] for increased frequencies moves as a whole to higher temperatures. Value of the relative variation of the temperature T_p of the χ" peak per decade of frequency, $X = (\Delta T_p/T_p)/\Delta(\log_{10} f)$, is equal to 0.128. This is much bigger than that of spin glasses and points to the blocking phenomenon. Absence of higher harmonics of the χ_{AC} anomaly means that there is no spontaneous magnetic moment in the sample and that no phase transition occurs. The χ" vs χ' Cole-Cole plots for T = 6 K up to T = 9 K given in figure 4 deviate insignificantly from perfect semicircles and form arcs of size $(1 - \alpha)\pi$. Parameter α represents distribution of relaxation times. In Ortho-F it does not change with temperature and is equal to 0.120±0.002, which means that in good approximation Ortho-F has a single relaxation time. This result is a fingerprint of the Single Chain Magnet behaviour.

From the frequency shift of the χ" maxima one obtains a temperature dependence of the relaxation time τ and then, using the fit to the Arrhenius law, the activation energy E_a. Based on the data of the 6 K÷9.5 K range, one gets E_a = (122 ± 2) K, where $\tau_0 = 1*10^{-9}$ s. The field dependence of the activation energy deduced from the χ"(T,f) data obtained under static magnetic field is shown in log scale in figure 5. As seen, for H_{DC} less than ≈ 1.5 kOe, E_a does not change but for higher field gradually decreases. Simultaneously, τ_0, which is the characteristic flipping time for a magnetically isolated spin unit, increases (see figure 5, inset on the left). The ratio of τ_0 at a given field to that at zero field, $\tau_0(H)/\tau_0(0)$, shown in figure 5 (inset on the right) is a quadratic function of H given: $\tau_0(H)/\tau_0(0) = 1+(H/H_0)^2$, with $H_0 = (2.1 \pm 0.1)$ kOe. This result is consistent with that reported for Mn/Fe and Mn/Ni chains [17].

In the temperature range 6 K÷9.5 K the finite-size effects [4,17] are relevant for Ortho-F compound (Figure 7 in Ref.[7]). As follows from the numerical study of the finite chain presented in [18], the highest field (10 kOe) applied by us stays in the low field limit, as $\tanh(\mu H/k_B T) \leq 0.3$, where μ ($= 3\ \mu_B$) is the magnetic moment of one [MnF$_4$TPP][TCNE] unit. Therefore, the single-time approximation should be valid. Despite the increase of τ_0 in the presence of the superimposed magnetic field, the decrease of E_a results in shortening of the relaxation time, e.g. at T = 5 K from $\tau = 49$ s to $\tau = 0.17$, and at T = 8 K from $\tau = 0.005$ s down to $\tau = 0.0004$ s for H = 0 and H = 10 kOe respectively. As reported in [7], the E_a vs H dependence determined experimentally for Ortho-F, could be explained as associated with solitary waves, i.e. with creation and propagation of domain walls in this one-dimensional anisotropic Heisenberg system. It is known that the value of activation energy E_a at zero field is a function of the exchange constant J between the spin units and of the single-ion anisotropy D. According to [3], for the infinite chain (L >> ξ) $E_a=(4J+|D|)S^2$, whereas for L << ξ one has $E_a = (2J+|D|)S^2$, because the energy cost to flip the spin in the domain is approximately twice as much as to flip the spin at the end of the chain. In order to estimate these two energy barrier limits we take $J = J_{eff} = 19.5$ K, obtained above, and the easy axis anisotropy parameter D = -3.3 K determined by means of HFEPR spectroscopy [19] for MnTPPCl complex. We believe that this D value is adequate here because, as concluded from *ab initio* studies for manganese porphyrin complexes [20], TCNE ligands hardly affect magnetic anisotropy of the manganese porphyrin complex, and they are important only to the ferrimagnetic alignment of the spins. Using the values above, we get $E_a = 183$ K and 95 K for two cases respectively. In view of this, the experimentally determined $E_a = (122 \pm 2)$ K attests to L << ξ case relevant for our SCM.

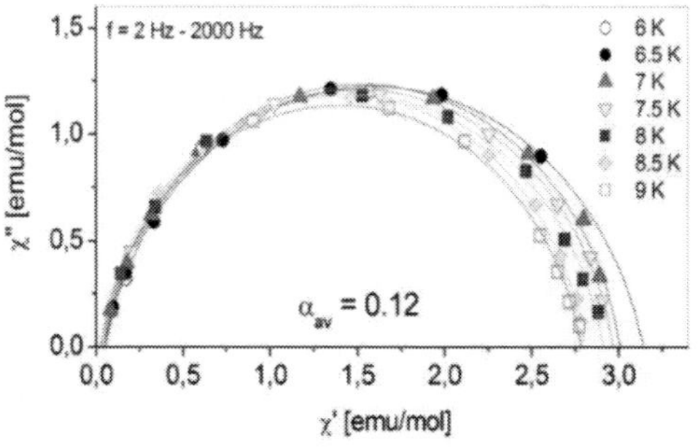

Figure 4. Cole-Cole diagrams for Ortho-F. The solid curves represent the least-squares fit to a generalized Debye model with α = 0.12.

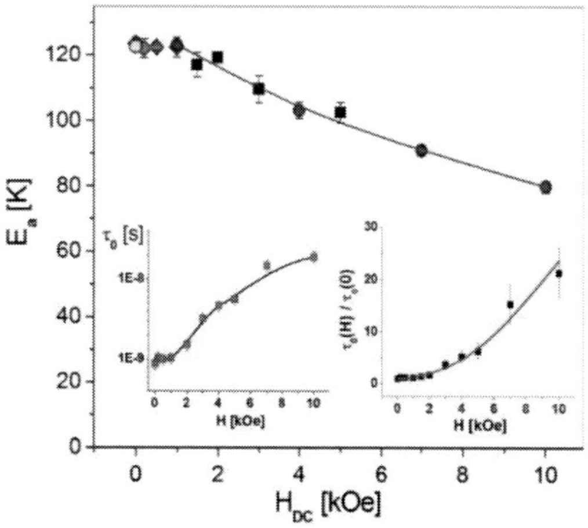

Figure 5. Effect of applied magnetic field on activation energy for Ortho-F; the solid line is a guide. Left inset: field dependence of time τ_0; solid line is a guide. Right inset: field dependence of $\tau_0(H)/\tau_0(0)$, solid line is the fit to a quadratic dependence (see text).

3. Blocking and magnetic order in Meta-F compound

Synthesis and crystallographic data of [MnF_4TPP][TCNE] with fluorine in *Meta*-position have been described in [21]. Meta-F compound crystallizes in the monoclinic system, while that of the Ortho-F compound is the triclinic one [22,7]. Magnetic measurements were performed for two types of the Meta-F sample: the ground crystalline sample (C) and the powder one (P). Both C and P samples show the phase transition to a 3D ordered magnetic state, that is easily suppressed by DC magnetic field. The intrachain antiferromagnetic exchange coupling between Mn^{III} ion and TCNE radical, determined from the fit of the Seiden's expression to the experimental χT data for $50 \leq T \leq 300$ K, is $J_{intra} = -100$ K for both samples. The low temperature fit of χ' according to Eq. 3 gives ferromagnetic coupling of spin 3/2 effective units $J_{eff} = 28$ K for sample C and only $J_{eff} = 21$ K for sample P.

AC susceptibility of the two samples is different. Figure 6 presents data measured with frequency 10 Hz and 1000 Hz for sample C and sample P. As may be seen, the temperature dependence of χ' and χ'' for powder sample is not typical and consists of two anomalies, one at $T_c = 10$ K (χ'' onset), reflecting transition to the ordered state, and the other related to blocking of relaxation, with $T_b = 5.4$ K (χ'' maximum at f = 10 Hz). The different character of two anomalies was checked by the measurement of susceptibility nonlinear components: the second and third harmonic components accompanied only the peak at T_c, therefore blocking is not a phase transition in the thermodynamic sense. Parameter X of frequency shift for phase transition is 0.015, while the one for blocking is much larger and equals 0.058, similarly to that of cluster [23] or fractal spin-glasses [24]. We would like to mention that the complex behaviour of the Meta-F specimen, relying on parallel SCM- and ordered magnet features, is like that recently observed for [Mn(3,5-Cl_2saltmen)Ni(pao)$_2$-(phen)]PF_6 [25] and cobalt-radical coordination magnet [26]. On the other hand, unlike the sample P, the C sample shows one anomaly (Figure 6), whose double nature comes forward only in the applied DC field [21]. Below, we present results obtained for Meta-F powder sample. Samples of all other compounds discussed in this paper were also powders.

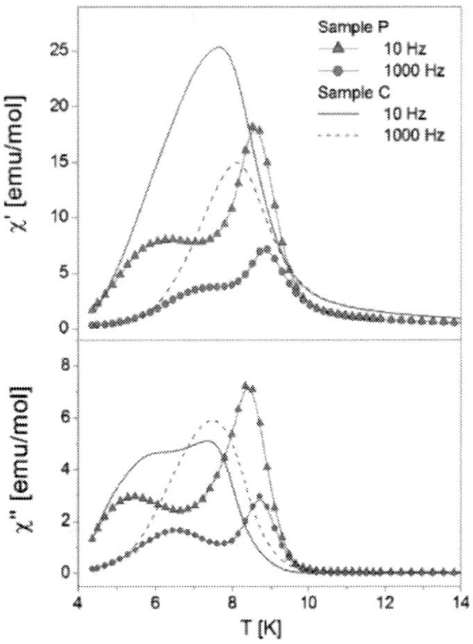

Figure 6. AC susceptibility measured with frequency 10 Hz and 1000 Hz for sample P and sample C of the Meta-F compound.

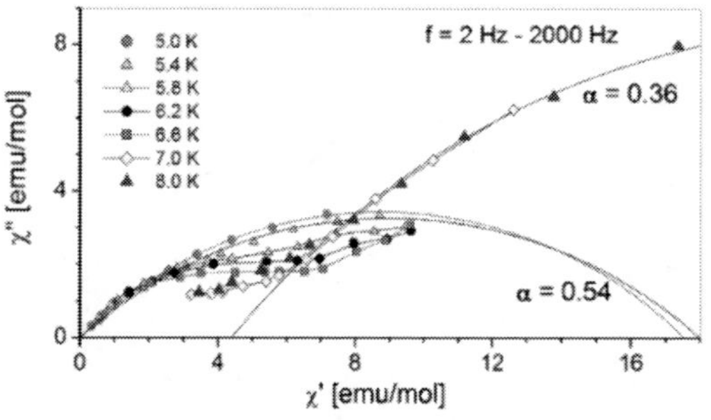

Figure 7. Cole-Cole diagram for Meta-F sample P with two distributions of relaxation times: α = 0.54 for blocking, α = 0.36 at phase transition.

The Cole-Cole diagram obtained from measurements with f = 2 Hz - 2000 Hz in the temperature range 5 K - 8 K is presented in Fig. 7. The two processes are visible with distribution of relaxation times $\alpha = 0.54$ for blocking and $\alpha = 0.36$ for phase transition. The flat segments of curves for 5.8 K, 6.2 K and 6.6 K, where χ" does not change, are certainly responsible for simultaneous reversal of all the spins in short segments of the chains. Such relaxation mechanism was observed and theoretically explained for CoPhOMe SCM, highly doped with diamagnetic impurities [4]. Therefore, the magnetically correlated segments of chains in our powder Meta-F sample are short.

Figure 8 shows the influence of the external field on the AC susceptibility. It can be seen that at H = 0.8 kOe the anomaly related to phase transition is completely suppressed, while that coming from blocking is less field sensitive. As the magnitude of the external magnetic field is increased, the χ" maximum shits to lower temperatures, i.e. T_b decreases. Activation energy determined from the Arrhenius plot is $E_a = 139(3)$ K where $\tau_0 = 5*10^{-13}$ s. Despite the similar value of E_a at zero field to that of Ortho-F, the dependence of activation energy on applied field looks quite different (Figure 9). Already at H ≈ 200 Oe E_a drops down to the half value and then stays constant at the value $E_a = 70$ K, while the preexponential factor increases up to $\tau_0 \approx 5*10^{-9}$ s and then does not change. This clear effect undoubtedly reflects crossover between L >> ξ and L << ξ case. Magnetic field, which increases the correlation length in chains of parallel (to field) oriented spins, at the same time will reorient the other spins in adjacent chains and will divide magnetic domains into shorter segments. Hence, the interchain dipolar coupling gets attenuated and phase transition disappears.

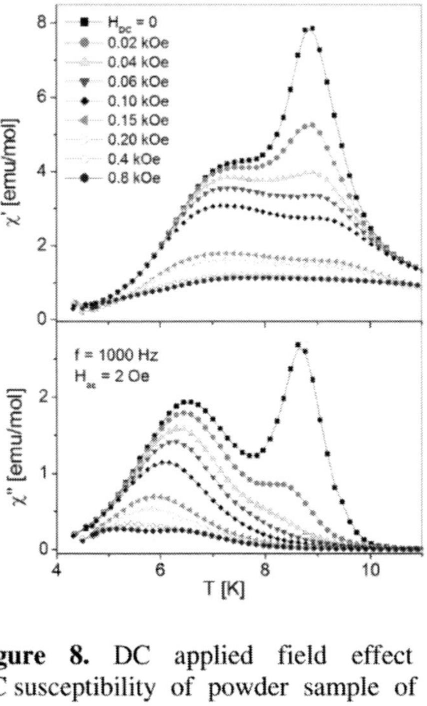

Figure 8. DC applied field effect on AC susceptibility of powder sample of the Meta-F compound.

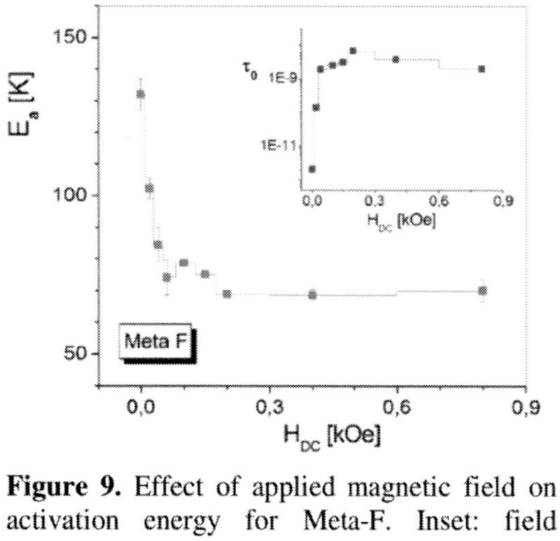

Figure 9. Effect of applied magnetic field on activation energy for Meta-F. Inset: field dependence of time τ_0. Solid lines are guides.

4. Magnetic phase transition and slow relaxation in the OC_nH_{n+1}-substituted compound

The present paragraph will deal with the properties of $[Mn(OC_nH_{2n+1})_4TPP][TCNE]$ (n = 10, 12, 14), which are somehow unique due to the very large (up to 30 Å) interchain spacing [9,11]. Despite the structural isolation of the chains the compounds show transition to the magnetically ordered state with $T_c = 21.7$ K, 22.0 K and 20.5 K respectively [11,12]. Interestingly, no heat capacity anomaly due to the onset of the magnetic order was detected [13], like it also occurred for other members of the family of the Mn-porphyrin chains [27]. It was therefore concluded that the most part of the magnetic entropy is retained above the phase transition temperature in the form of the dominant short-range order. At low temperature, the divergence of the in-chain correlation length promotes correlated spin blocks which interact from chain to chain through dipolar forces. As shown in [28], the rate of increase of the correlation length is the essential parameter in order to reach the observed transition temperatures. For the T_c values above an exponential divergence of the 1D correlation length is required, which implies the existence of single-ion anisotropy, whereas the power-law divergence for 1D Heisenberg coupled spins yields transition temperatures one order of magnitude smaller than observed.

Some results obtained for the samples with n = 10 and 14 have been already described in [12]. Here we report investigations for n = 12, as a representative for alcoxy-substituted modifications, because it shows the highest AC susceptibility at T_c, equal to 22 emu/mol for f = 40 Hz (see Fig. 3c), compared to 8 emu/mol and 9 emu/mol for n = 10 and 14 [12]. The one-dimensional character of these ordered molecular magnets is markedly visible in measurements performed in magnetic field. Figure 10 shows χ' and χ" of $OC_{12}H_{25}$ measured at two frequencies of the oscillating field for several values of the external static field. When applied field increases, the AC peak reveals its double nature: the anomaly at T_c gradually disappears and the frequency dependent part is shifted to lower temperatures. As given in figure 11, at H = 5 kOe AC susceptibility is of the SCM-type, yet with the broad distribution of relaxation times ($\alpha_{av} = 0.22$, not shown). Activation energy for this process is $E_a = 52$ K, with $\tau_0 = 3.5*10^{-10}$ s. We would like to notice the different in-field behaviour of $OC_{12}H_{25}$ (Figure 10) as compared to that of Meta-F (Figure 8). For the latter, the low temperature χ' and χ" components responsible for blocking were visible already in zero field and when the field increased, the χ' and χ" components were getting weaker. For $OC_{12}H_{25}$ in turn, the χ' and χ" components responsible for blocking developed with the field, which means that applied field attenuates the interchain coupling and brings in the magnetic isolation of the chains.

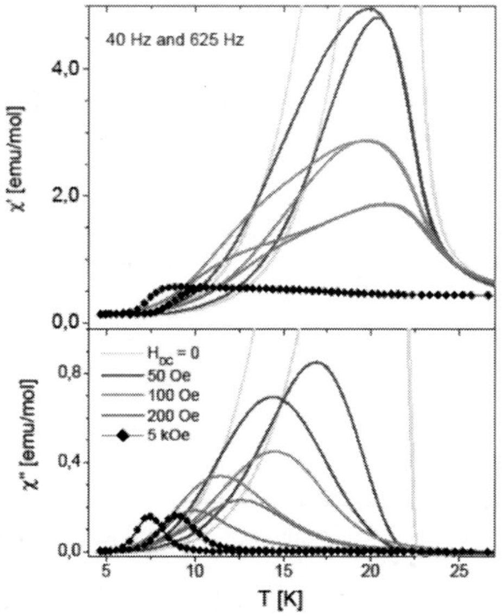

Figure 10. Influence of applied magnetic field up to 5 kOe on χ' and χ" of $OC_{12}H_{25}$ measured for f = 40 Hz and 625 Hz.

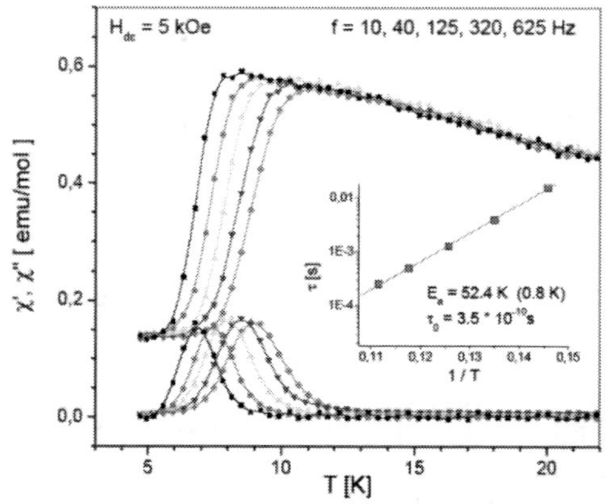

Figure 11. AC susceptibility for $OC_{12}H_{25}$ at H_{DC}= 5 kOe, measured for f = 10, 40, 125, 320 and 625 Hz. Inset: Arrhenius fit.

The other striking feature of $OC_{12}H_{25}$ is a large value of the coercivity field H_c at T= 2.3 K and its quick decrease with temperature. Figure 12 shows hysteresis curves for $OC_{12}H_{25}$ measured with the magnetic sweep rate ≈ 60 s per point (1 kOe/min) at several temperatures. The curve for Ortho-F SCM recorded at T = 2 K with the rate ≈ 100 s per point (0.6 kOe/min) is shown for comparison [7]. Hysteresis of $OC_{12}H_{25}$ at T = 2.3 K is very similar. The drop in H_c with temperature is however not the same (see Figure 13). For Ortho-F a fast linear drop is observed, which, at T approaching zero, leads to $H_c(T=0) \approx 40$ kOe. The exponential $H_c(T) = H_0\exp(-0.67*T)$ dependence for $OC_{12}H_{25}$ shown in figure 13 suggests much higher value of H_0, equal to ~ 80 kOe. Upon heating the sample, H_c markedly decreases and at T = 16 K is equal to 20 Oe.

One should notice, that due to metamagnetism and spin flop transition displayed by most of compounds from the family of Mn-porphrins, the hysteresis loop opens only for the field higher than the critical one H_{sf}. The value of H_{sf} at 2.4 K, 3.5 K and 4.3 K is equal to 28 kOe, 14.5 kOe and 5 kOe respectively. Figure 14 shows M(H) dependence during the spin flop transition at 4.3 K together with the time evolution of magnetization M(t) measured at H = 4 kOe. The time dependence of magnetization could be fitted with the stretched exponential function $M(t) = A\text{-}M*\exp[-(t/\tau_{av})^{\beta}]$, with β=0.74 and mean relaxation time $\tau_{av}=4*10^3$s. It was also checked that τ_{av} decreased under illumination with light [29].

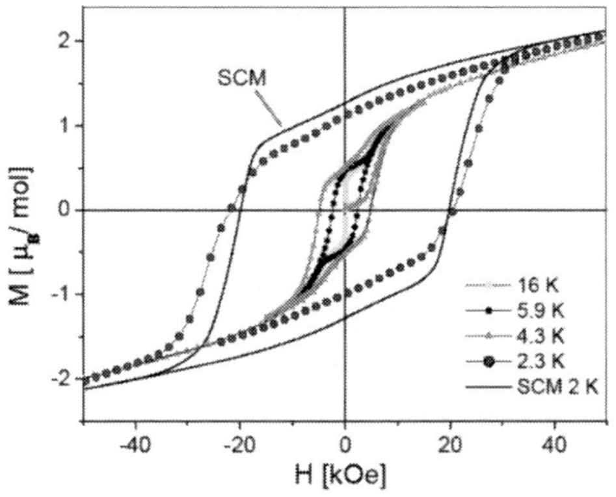

Figure 12. Hysteresis curves at several temperatures for $OC_{12}H_{25}$. The curve for Ortho-F SCM is shown for comparison. See text for magnetic field sweep rate.

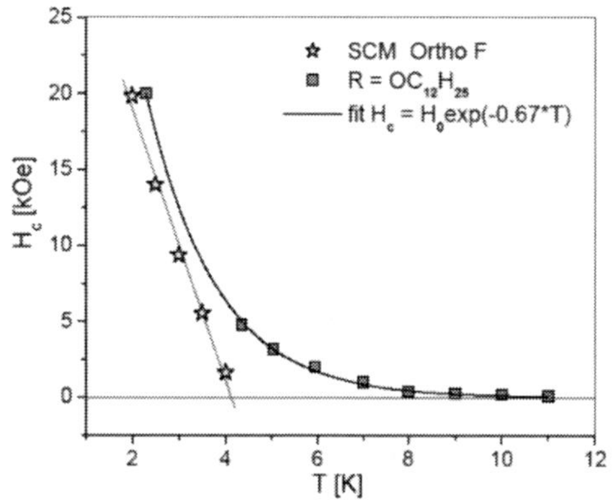

Figure 13. Temperature dependence of the coercivity field for Ortho-F (linear fit) and for $OC_{12}H_{25}$ (exponential fit).

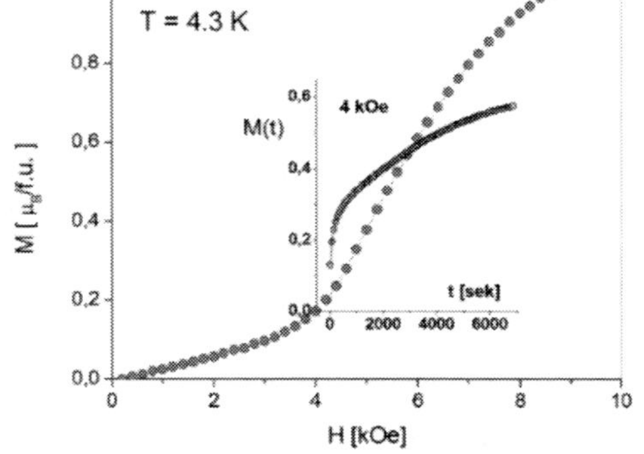

Figure 14. M(H) dependence at the spin flop transition at 4.3 K. Inset: Time evolution of magnetization M(t) measured at H = 4 kOe fitted to the stretched exponential function.

5. Discussion

All materials presented above are well approximated by quasi one-dimensional ferrimagnetic Heisenberg chains with small anisotropy. In this group, depending on type of the substituent R and on the substitution site, one finds Single Chain Magnet, magnetic cluster glass, as well as magnetically ordered phases, which show slow relaxation at low temperatures. In case of magnetically ordered compounds, long range correlations along the chains are present above T_c and phase transition is triggered due to the weak interchain dipolar forces acting between the correlated chain segments. This is the reason why such transitions were not detected by calorimetry despite the big singularity in magnetic susceptibility coming from formation of magnetic domains. As shown above, transition at T_c undergoes attenuation by the applied DC field (Figure 8 and Figure 10). Two types of magnetically ordered phases were mentioned: the one showing the blocking anomaly below T_c without applied field (Meta-F P) and the other, in which slow relaxation and blocking could be revealed only in non-zero DC field (OC_nH_{2n+1}). The first case (Meta-F P) concerns the sample consisting in some part of magnetically isolated chains, as number of defects prevents formation of long correlated segments, which are necessary for creation the relatively strong interchain dipolar forces. In the second case, the respectively strong DC field is needed to bring in the magnetic isolation of the chains.

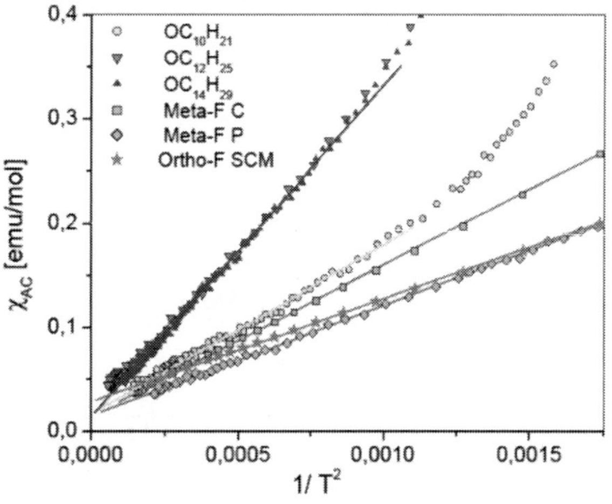

Figure 15. The χ' vs T^{-2} plots for compounds under study; the tilt angle of the stright line fits is proportional to the effective ferromagnetic intrachain coupling J_{eff} between the net spin units 3/2.

In order to discuss magnetic relaxation in the family of [MnR$_4$TPP][TCNE] compounds and the competition between the tendency to ordering vs the tendency to blocking we have determined the coupling constants and other crucial values. Figure 15 shows the χ' vs T^{-2} plots from which the intrachain effective FM coupling J_{eff} between effective spin units 3/2 was obtained according to Eq.2. The results and temperature ranges in which linear fits were carried out are given in Table 1. The dipolar interaction J_{dip} can be estimated from the ordering temperatures using the relationship developed for anisotropic exchange interaction [30]

$$k_B T_c = 4S(S+1)\sqrt{J_{eff} J_{dip}} \tag{4}$$

where S = 3/2 is the net spin. The values obtained are dozens of mK (see Table 1.). For the assessment of J_{dip} in Ortho-F Single Chain Magnet, instead of T_c in Eq.4, the blocking temperature T_b = 6.6 K, determined from the maximum of the χ" maximum for f = 10 Hz, was used.

The one-dimensional character of the compounds under study is illustrated in Fig. 16 with the ln(χ_{mol}T) vs 1/T dependences. According to Eq.3, for anisotropic Heisenberg or Ising-like 1D systems in the low temperature region this dependence should be linear with the tilt angle equal to the energy Δ_ξ necessary to create a domain wall in the chain [15]. Figure 16 shows such a behaviour down to the lowest temperatures for Ortho-F SCM and for Meta-F, while for the other compounds of the family the linear ln(χ'T) vs T^{-1} dependence is observed at temperatures much higher than T_c. The only exception is the $OC_{12}H_{25}$ system, for which the $\ln(\chi_{mol}T)$ vs $1/T$ dependence was not linear in a suitably long temperature range. This exception is in accord with the overall appearance of this magnetically ordered compound. On the other hand, the presented above experimental evidence of the in-field slow relaxation in $OC_{12}H_{25}$ point to the common mechanism of magnetic coupling and relaxation in $[Mn^{III}R_4(TPP)][TCNE]$ molecular compounds.

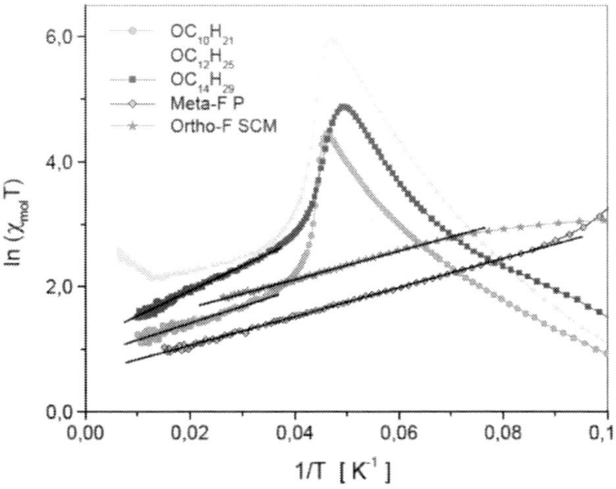

Figure 16. The $\ln(\chi'T)$ vs T^{-1} dependences for Ortho-F SCM and for other compounds of the family; the energy Δ_ξ needed to create a domain wall in the chain is determined from the Eq.3.

In order to find the characteristic, common for SCM and magnetically ordered compounds, we plot Δ_ξ values as a function of intrachain effective FM coupling J_{eff} as shown in figure 17. The Δ_ξ vs J_{eff} dependence obtained is linear and reads $\Delta_\xi = 14\text{ K} + 0.416*J_{eff}$. The lowest Δ_ξ and J_{eff} are observed for Ortho-F Single Chain Magnet. While the intersection point (14 K) may represent the $|D|S^2$ term, the proportionality factor (0.416) is much less than $2S^2$ value expected for domain creation in a ferromagnetic Ising chain. Such small value could be expected for quasi one-dimensional ferrimagnetic Heisenberg systems of small anisotropy which posses wide domain walls.

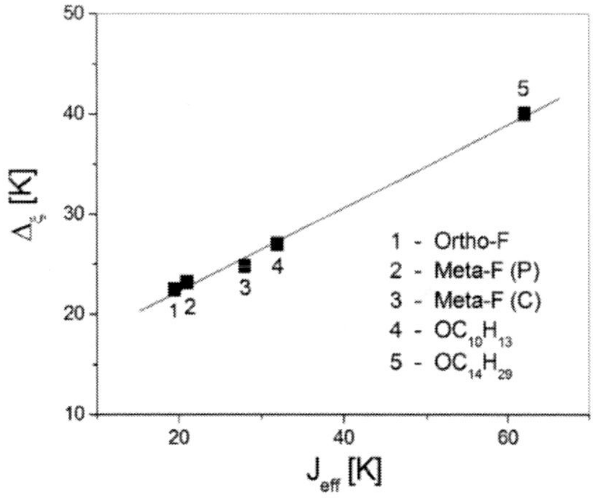

Figure 17. Linear dependence of Δ_ξ vs J_{eff} obtained for the compounds of the $[MnR_4TPP][TCNE]$ family.

Table 1. Characteristic temperatures, magnetic interactions and activation energies for compounds under study: T_c - ordering temperature; T_b - blocking temperature, T_f - freezing temperature, J_{intra} - intrachain AFM coupling between Mn^{III} ion and TCNE radical, J_{eff} - low temperature intrachain effective FM coupling between effective spin units 3/2 (Eq.2), J_{dip}- interchain dipolar coupling (Eq.4), E_a - activation energy of Arrhenius relaxation, τ_0 - prefactor in the Arrhenius law, Δ_ξ - the energy of creation a domain wall in the chain (Eq.3).

Compound	Type	J_{intra}	J_{eff}	J_{dip}	E_a, τ_0	Δ_ξ [K]
Ortho-F	T_b=6.6 K	217 K [7]	19.5 K fit 14-60 K	≤0.01 K	123 K $1*10^{-9}$s	22.5 K
Meta-F (P)	T_c=10 K T_b=5.4 K	100 K	21 K fit 22-68 K	~0.02 K	132 K $2*10^{-12}$s	23.2 K
Meta-F (C)	T_c=10 K	100 K [21]	28 K fit 22-68 K	~0.02 K		24.8 K
Para-F	T_f=26 K	236 K [6]			336 K $1.8*10^{-12}$s	
$OC_{10}H_{21}$	T_c=22.5 K	117 K [6]	32 K fit 30-100 K	~0.07 K		27 K
$OC_{12}H_{25}$	T_c=22 K	178 K [6]	63 K fit 36-106 K	~0.04 K	52 K[a] $3.5*10^{-10}$s	
$OC_{14}H_{29}$	T_c=21.2 K	148 K [6]	62 K fit 34-100 K	~0.03 K		40 K

6. Conclusion

On the basis of the presented results we conclude that slow relaxation is an inherent feature for all compounds of the [MnR_4TPP][TCNE] family. For the magnetically ordered compounds it is apparent in the AC susceptibility as a frequency dependent bump at temperatures below T_c or may show up in the high enough magnetic field. The superimposed DC magnetic field destroys the antiferromagnetic coupling between the chains and brings in magnetic isolation of the chains. The preexponential factor τ_0 in the Arrhenius law increases in the applied field, however, as the energy barrier E_a decreases, the in-field relaxation accelerates. As observed for Meta-F P and Meta-F C samples, the quality of the sample relying on the undefected structure is crucial for setting magnetic order. For the compounds of the [MnR_4TPP][TCNE] family, including Ortho-F SCM, the linear dependence of the energy Δ_ξ on the intrachain effective ferromagnetic coupling J_{eff} is observed. A strong dependence of the AC susceptibility on frequency, observed in a number of ordered quasi one-dimensional compounds, previously interpreted in the frames of spin (spin-cluster) glass, originates from SCM features. However, the true Single Chain Magnet behaviour is a unique property which rules out any magnetic transition to the collective phase.

Acknowledgements
The authors acknowledge receiving the samples under investigation from Dr. K. Falk and Dr. K. Griesar (Institute of Physical Chemistry, Darmstadt University of Technology) as well as from Dr. S. Najak (working temporally in Darmstadt TU, on leave from Bhabha Atomic Research Centre, Trombay, India). M.B. acknowledges partial support from the Polish Ministry of Science and Higher Education within the Research Project No. 0087/B/H03/2008/34. Thanks are also due to Dr. A. Pacyna (INP, Kraków) for his kind assistance in the lab.

References

[1] Sessoli R, Gatteschi D, Caneschi A and Novak M A 1993 *Nature* **365** 141
Thomas L, Lionti F, Ballou R, Gatteschi D, Sessoli R and Barbara B 1996 *Nature* **383** 145
Sangregorio C, Ohm T, Paulsen C, Sessoli R and Gatteschi D 1997 *Phys. Rev. Lett.* **78** 4645
[2] Caneschi A, Gatteschi D, Lalioti N, Sangregorio C, Sessoli R, Venturi G, Vindigni A, Rettori A, Pini M G and Novak M A 2001 *Angew. Chem. Int. Ed.* **40** 1760
Caneschi A, Gatteschi D, Lalioti N, Sangregorio C, Sessoli R, Venturi G, Vindigni A, Rettori A, Pini M G and Novak M A 2002 *Europhys. Lett.* **58** 771
Clérac R, Miyasaka H, Yamashita M and Coulon C 2002 *J. Am. Chem. Soc.* **124** 12837
Hao-Ling Sun, Zhe-Ming Wang and Song Gao 2010 *Coord. Chem. Rev.* **254** 1081
[3] Miyasaka H, Julve M, Yamashita M and Clérac R 2009 *Inorganic Chemistry* **48** 3420
[4] Bogani L, Caneschi A, Fedi M, Gatteschi D, Massi M, Novak M A, Pini M A, Rettori A, Sessoli R and Vindigni A 2004 *Phys. Rev. Lett.* **92** 207204
[5] Wynn C M, Girtu M A,. Sugiura K-I, Brandon E J, Manson J L, Miller Joel S and Epstein A J 1997 *Synth. Metals* **85** 1695
Girtu M A, Wynn C M, Sugiura K-I, Miller Joel S and Epstein A J 1997 *Synth. Metals* **85** 1703
Wynn C M, Girtu M A, Brinckehoff W B, Sugiura K-I, Miller J S and Epstein A J 1997 *Chem. Mater.* **9** 2156
[6] K. Falk, Ph.D. Thesis, TU Darmstadt 2002 http://elib.tu-darmstadt.de/diss/000220/
[7] Bałanda M, Rams M, Nayak S, Tomkowicz Z, Haase W, Tomala K and Yakhmi J 2007 *Phys. Rev.* B **74** 224421
[8] Seiden J 1983 *J. Phys.-Lett.* **44** L-947
[9] Griesar K, Athanassopoulou M A, Soto Bustamante E A, Tomkowicz Z, Zaleski A J and Haase W 1997 *Adv. Mater* **9** 45
Griesar K, Athanassopoulou M A, Tomkowicz Z and Bałanda M 1997 *Mol. Cryst. Liq. Cryst.* **306** 57
Falk K, Werner R, Tomkowicz Z, Bałanda M and Haase W 1999 *J. Magn. Magn. Mater.* **196-197** 564
[10] Bałanda M, Falk K, Griesar K, Tomkowicz Z and Haase W 1999 *J. Magn. Magn. Mater.* **205** 14
Bałanda M, Falk K, Griesar K, Tomkowicz Z, Wasiutyński T and Haase W 1999 *Mol. Cryst. Liq. Cryst.* **335** 845
[11] Falk K, Bałanda M, Tomkowicz Z, Mascarenhas F, Schilling J, Klavins P and Haase W 2001 *Polyhedron* **20** 1521
[12] Tomkowicz Z, Bałanda M, Falk K and Haase W 2003 *„Relaxation Phenomena–Liquid Crystals, Magnetic Cystems, Polymers, High-Tc Superconductors, Metallic Glasses"* ed W Haase and S Wróbel (Berlin Heidelberg: Springer) p 569
[13] Bhattacharjee A, Falk K, Haase W and Sorai M 2005 *J. Phys. Chem. Sol.* **66** 147
[14] Suzuki F, Shibata N and Ishii C 1994 *J. Phys. Soc. Jpn.* **63** 1539
[15] Loveluck M, Lovesey S W and Aubry S 1975 *J. Phys. C* **8** 3841
Nakamura K and Sasada T 1978 *J. Phys.* C **11** 331

[16] Eq.3 used by us for estimating Δ_ξ in all compounds under study does not take into account the interchain interaction and the finite length of the chains; Δ_ξ value for Ortho-F reported in [7] determined in the more precise approach is bigger
[17] Coulon C, Clérac R, Lecren L, Wernsdorfer W and Miyasaka H 2004 *Phys. Rev.* B **69** 132408
[18] Coulon C, Clérac R, Wernsdorfer W, Colin T, Saitoh A, Motokawa N and Miyasaka H 2007 *Phys. Rev* B. **76** 214422
[19] Krzystek J, Telser J, Pardi L A, Goldberg D P, Hoffman B M and Brunel L C 1999 *Inorg. Chem.* **38**, 6121
[20] Takeda R, Koizumi K, Shoji M, Nitta H, Yamanaka S, Okumura M and Yamaguchi K 2007 *Polyhedron* **26** 2309
[21] Tomkowicz Z et al. in preparation
[22] Brandon E J, Rittenberg D K, Arif A M and Miller J S 1998 *Inorg. Chem.* **37** 3376
[23] Novak M 2004 *J. Magn. Magn. Mater.* **272-276** E707
[24] Etzkorn S J, Hibbs W, Miller J S and Epstein A J 2004 *Phys. Rev.* B **70** 134419
[25] Miyasaka H, Takayama K, Saitoh A, Furukava S, Yamashita M and Clérac R 2010 *Chem. Eur. J.* **16** 3656
[26] Oshii N, Okamura Y, Chiba S, Nogami T and Ishida T 2008 *J. Am. Chem. Soc.* **130** 24
[27] Nakazawa Y, Hoffman W, Miller J S and Sorai M 2005 *Sol. St. Comm.* **135** 71
[28] Ostrovsky S, Haase W, Drillon M and Panissod P 2001 *Phys. Rev.* B **64**, 134418
[29] Zieliński P, Bałanda M, Pełka R and Wasiutyński T 2005 *J. Phys.: Conf. Ser.* **21** 237
[30] Richards P M 1974 *Phys. Rev.* B **10** 4687

Octacyanoniobate(IV)-based molecular magnets revealing 3D long-range order

R Pełka[1], D Pinkowicz[2], O Drath[2], M Bałanda[1], M Rams[3], A Majcher[3], W Nitek[2] and B Sieklucka[2]

[1] Institute of Physics PAN, Radzikowskiego 152, 31-342, Kraków, Poland
[2] Faculty of Chemistry, Jagiellonian University, Ingardena 3, 30-060 Kraków, Poland
[3] Institute of Physics, Jagiellonian University, Reymonta 4, 30-059 Kraków, Poland

robert.pelka@ifj.edu.pl

Abstract. Isostructural series of chemical formula $\{[M^{II}(pirazol)_4]_2[Nb^{IV}(CN)_8]\cdot 4H_2O\}_n$ (M^{II}=Mn (**1**), Fe (**2**), Co (**3**), Ni (**4**)) has been obtained by the self-assembly technique. Its unique crystallographic structure consists in the formation of a 3D extended network of magnetic centers braced by geometrically identical cyanido bridges. Magnetic measurements reveal the transitions to the 3D order at temperatures 23.7, 8.3, 5.9, 13.4 K for **1**, **2**, **3**, and **4**, respectively. The character of order is demonstrated to be ferrimagnetic for **1** and **2** and ferromagnetic for **3** and **4**. The mean-field approach is used to determine the corresponding exchange coupling constants. The observed interactions are discussed within the magnetic orbital model.

1. Introduction

The field of molecular magnetism has been attracting attention of physicist as a potential source of materials to be used in memory storage devices, magnetic sensors or switches, etc., ever since the first successful synthesis of two magnetic molecular compounds in 1986 [1]. Until present a vast number of molecule-based magnets displaying a striking variety of structural topologies and technologically relevant properties have been reported [2]. Among them a prominent position is taken by those based on cyanido-bridges linking the transition metal ions which are known to display a wide spectrum of structures and functionalities [3]. This rich family of compounds has recently been joined by four 3D isostructural compounds $\{[M^{II}(pyrazole)_4]_2[Nb^{IV}(CN)_8]\cdot 4H_2O\}_n$, where M^{II} = Mn, Fe, Co, Ni, obtained by self assembly of $[Nb^{IV}(CN)_8]^{4-}$ with different 3*d*-metal centers in aqueous solution and excess of pyrazole [4]. All four coordination polymers crystallize in the same I $4_1/a$ space group, see [4] for crystallographic details. Due to specific symmetry of the crystal lattice the M^{II}-NC-Nb^{IV} linkages providing the main exchange coupling pathways between metal centers are geometrically equivalent, which implies a single exchange coupling constant J_{MNb}. The present contribution shows two simple approaches based on the mean-field theory employed to estimate the values of J_{MNb} for **1-4**. To this end three magnetic properties have been measured using 7227 Lakeshore AC Susceptometer/DC Magnetometer (**1**, **3**, **4**) or S600X SQUID Magnetometer (**2**): dc susceptibility, isothermal magnetization and thermal dependence of magnetization. The magnetic susceptibility data were corrected for

Published under licence by IOP Publishing Ltd

the diamagnetic contribution using Pascal constants [5]. Isothermal magnetization curves in the field up to 56/50 kOe were measured at T=4.3 K for **1**, **3**, and **4** and at T=2 K for **2**.

2. Theoretical background

The first approach consists in analyzing the thermal dependence of magnetization for **1**, **2**, **3**, and **4**. The calculations follow the lines reported in [6]. The mean field model consists of two sublattices, one sublattice for the Nb^{IV} centers and the other formed by the M^{II} ions. Next, only the superexchange coupling between the nearest-neighbor sites is taken into account. The thermally averaged values of the sublattice spins are given by the following set of coupled equations

$$\langle S_{\rm M}\rangle = S_{\rm M0}B_{S_{\rm M0}}\left(\frac{g_{\rm M}\mu_B S_{\rm M0}H_0}{k_BT}+\frac{Z_{\rm MNb}J_{\rm MNb}S_{\rm M0}}{k_BT}\langle S_{\rm Nb}\rangle\right) \tag{1}$$

$$\langle S_{\rm Nb}\rangle = S_{\rm Nb0}B_{S_{\rm Nb0}}\left(\frac{g_{\rm Nb}\mu_B S_{\rm Nb0}H_0}{k_BT}+\frac{Z_{\rm NbM}J_{\rm MNb}S_{\rm Nb0}}{k_BT}\langle S_{\rm M}\rangle\right) \tag{2}$$

where B_S is the Brillouin function, $S_{\rm M0}$ and $S_{\rm Nb0}$ are the values of $\langle S_{\rm M}\rangle$ and $\langle S_{\rm Nb}\rangle$ at $T=0$ K, k_B is the Boltzmann constant, μ_B is the Bohr magneton, H_0 is the external magnetic field, Z_{ij} are the numbers of the nearest-neighbor j-site ions surrounding an i-site ion, and $g_{\rm M}$ and $g_{\rm Nb}$ are the spectroscopic factors. The total molar magnetization is $M_{\rm total}=N_A\mu_B[2g_{\rm M}\langle S_{\rm M}\rangle+g_{\rm Nb}\langle S_{\rm Nb}\rangle]$, where the means $\langle S_{\rm M}\rangle$ and $\langle S_{\rm Nb}\rangle$ differ in sign for antiferromagnetic coupling between the M^{II} and Nb^{IV} centers. The numbers of the nearest neighbors are $Z_{\rm MNb}$=4, $Z_{\rm NbM}$=2, and we assume $S_{\rm Nb0}$=1/2, $g_{\rm Nb}$=2.0. The spin quantum number $S_{\rm M0}$ of the M^{II} center and the corresponding spectroscopic factor $g_{\rm M}$ are case-dependent. The crucial point in the first approach is that the mean-field curve is chosen so that the corresponding peak of the derivative dM/dT coincides with that obtained on the basis of the measured magnetization values. The error of $J_{\rm MNb}$ thus obtained is estimated assuming the uncertainty of 0.5 K in the peak position defining the transition temperature. The calculated mean-field curves display generally higher values of magnetization at low temperatures. This is due to the fact that the calculated magnetizations are essentially the saturated values, because the molecular-field theory does not account for the complex ordering process involving domain formation. Figures 2, 4, 6, and 8 show thermal dependence of magnetization for **1**, **2**, **3**, and **4**, respectively, measured in the external magnetic field H_0=0.5 kOe. The solid lines show the corresponding mean field curves. Inserts of these figures show the derivative dM/dT used to pinpoint the corresponding exchange coupling constants $J_{\rm MNb}$.

The second approach is based on the mean-field prediction for the magnetic susceptibility at high temperatures which reads

$$\chi = \frac{(2C_{S_{\rm M0}}+C_{S_{\rm Nb0}})T+4\lambda_{\rm MNb}C_{S_{\rm M0}}C_{S_{\rm Nb0}}}{T^2-2\lambda_{\rm MNb}^2C_{S_{\rm M0}}C_{S_{\rm Nb0}}} \tag{3}$$

where $C_S=N_Ag^2\mu_B^2S(S+1)/(3k_B)$ denotes the Curie constant for the corresponding spins, and $\lambda_{\rm MNb}$ is the molecular field constant. This high-temperature law is a counterpart of the Curie-Weiss law for unequal occupation numbers of the constituent magnetic sublattices. The positive (negative) sign of the molecular field constant corresponds to (anti)ferromagnetic character of the coupling. The exchange coupling constants are calculated from the relation $\lambda_{\rm MNb}=2J_{\rm MNb}/(N_Ag_{\rm av}^2\mu_B^2)$.

3. Results

3.1. $\{[Mn^{II}(pirazol)_4]_2[Nb^{IV}(CN)_8]\cdot 4H_2O\}_n$

Figure 1 shows the temperature variation of the inverse dc susceptibility for **1**. The fit to the mean-field prediction given by equation (3) in the range of 240-300 K yielded $\lambda_{\rm MnNb}$=-28(4) mol cm^{-3} and $g_{\rm Mn}$=1.99(1). The sign of the molecular-field constant indicates that the manganese and niobium sublattices are coupled antiferromagnetically. This is consistent with the magnetization detected at T=4.3 K (insert of figure 1) saturating at the value of 8.95 ($\approx 2g_{\rm Mn}S_{\rm Mn0}$-$g_{\rm Nb}S_{\rm Nb0}$=9) μ_B. The calculated mean-field $M(T)$ curve (figure 2, solid line) assuming the antiferromagnetic interaction between Mn^{II} and Nb^{IV} sublattices yielded J_{MnNb} = -6.8(2) cm^{-1}.

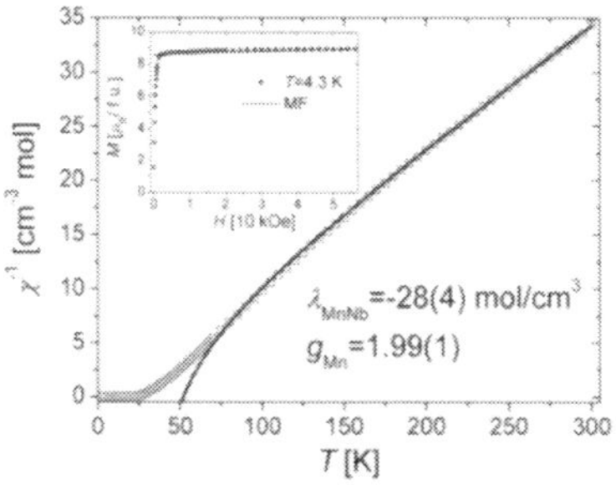

Figure 1. χ^{-1} vs. T for **1**. Solid line - fit to equation (3). Insert: M vs. H at T= 4.3 K.

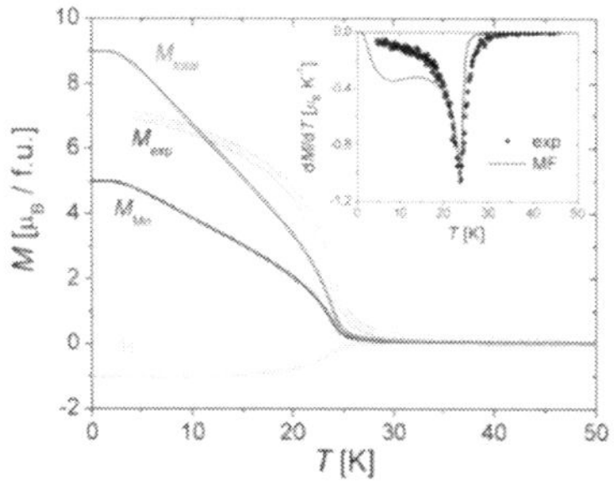

Figure 2. M vs. T obtained in H_{DC} = 0.5 kOe for **1**. Solid lines – the mean-field curves. Insert: dM/dT vs. T used to determine J_{MnNb}.

3.2. {[Fe^{II}(pirazol)$_4$]$_2$[$Nb^{IV}(CN)_8$] ·4H_2O}$_n$

Figure 3 shows the inverse of the dc susceptibility for **2**. The fit to equation (3) in the temperature range 150-300 K yielded λ_{FeNb}=+16.2(3) mol/cm^3 and g_{Fe}=2.163(2) [7]. The isothermal magnetization measured at T=2 K (insert of figure 3) shows a rapid increase followed by a steady linear grow for higher fields. In the field of 50 kOe the magnetization attains the value of 7 μ_B, which is substantially lower than that expected for the ferromagnetic coupling between the Fe^{II} and Nb^{IV} ions ($2g_{Fe}S_{Fe0}+g_{Nb}S_{Nb0}$=9.6), and comparable to that for the antiferromagnetic coupling ($2g_{Fe}S_{Fe0}-g_{Nb}S_{Nb0}$=7.6). The deviation of the measured magnetization values from those obtained within the mean-field model as well as the shape of the magnetization curve imply the presence of local anisotropy of the Fe^{II} ions and/or noncollinearity of the constituent magnetic moments. The calculated mean-field $M(T)$ curve (figure 4, solid line) assuming the antiferromagnetic interaction between Fe^{II} and Nb^{IV} sublattices yielded J_{FeNb} = -3.1(2) cm^{-1}.

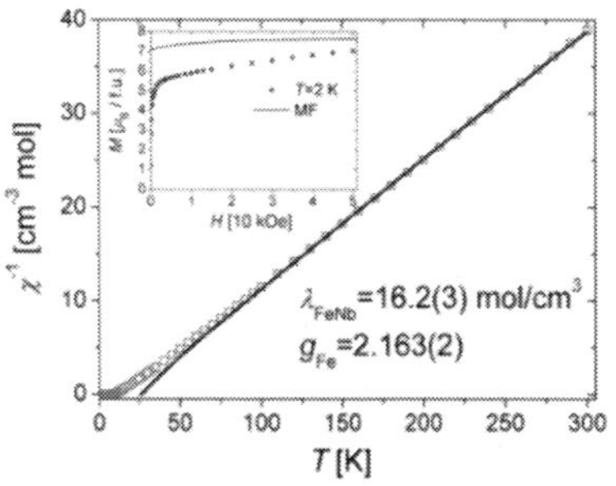

Figure 3. χ^{-1} vs. T for **2**. Solid line - fit to equation (3). Insert: M vs. H at T= 2 K.

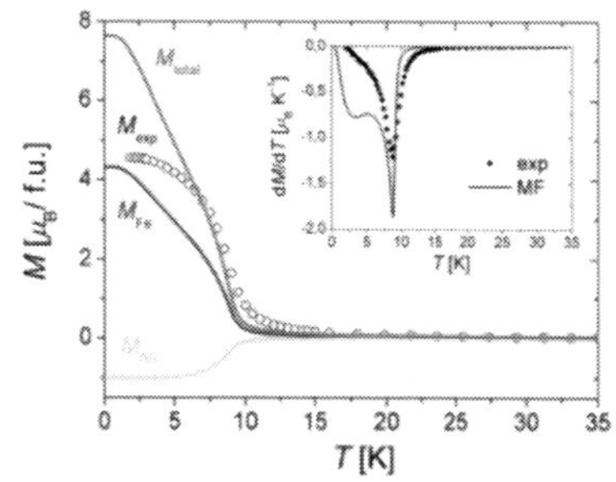

Figure 4. M vs. T obtained in H_{DC} = 0.5 kOe for **2**. Solid lines – the mean-field curves. Insert: dM/dT vs. T used to determine J_{FeNb}.

3.3. {[Co^{II}(pirazol)$_4$]$_2$[$Nb^{IV}(CN)_8$] ·4H_2O}$_n$

Isothermal magnetization measured at T=4.3 K for **3** (insert of figure 5) attains the value of 5.55 μ_B in the field of 50 kOe , which indicates that the Co^{II} ion ($3d^7$) must be in the high spin state $(t_{2g})^5(e_g)^2$. In such a case the orbital contribution to the magnetic moment cannot be neglected [8] and a special treatment is necessary. It can be shown (see [4] for more detail) that in octahedral ligand field the ground state of the Co^{II} ion may be described by an effective spin Σ_{Co}=1/2 with Landé factor g_{Co}=13/3≈4.33. Moreover, due to the orbital contribution the exchange coupling should be rescaled by factor 5/3 [9]. Figure 5 shows the thermal variation of the inverse susceptibility for **3**. Fit to equation (3) in the temperature range 150-300 K implies λ_{CoNb}=+15(1) mol/cm^3 and g_{Co}=4.55(5), which indicates the ferromagnetic coupling between the Co^{II} and Nb^{IV} sublattices. This is consistent with the value of magnetization attained in 50 kOe (5.55 μ_B) as the expected value is $2g_{Co}\Sigma_{Co}+g_{Nb}S_{Nb}$=5.33 μ_B. The calculated mean-field $M(T)$ curve (figure 6, solid line) assuming the ferromagnetic interaction between Co^{II} and Nb^{IV} sublattices yielded J_{CoNb} = +3.5(3) cm^{-1}.

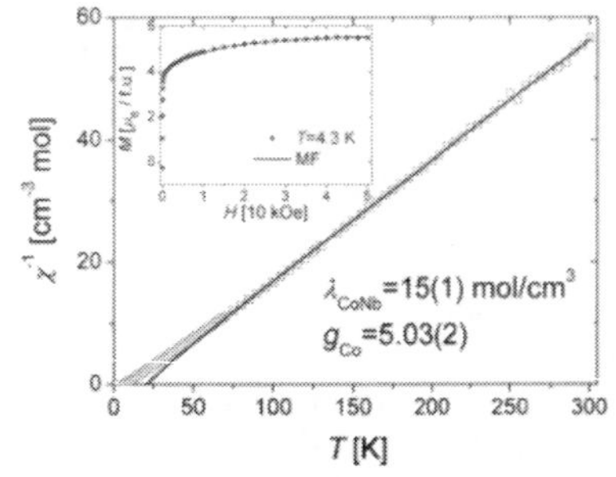

Figure 5. χ^{-1} vs. T for **3**. Solid line - fit to equation (3). Insert: M vs. H at T= 4.3 K.

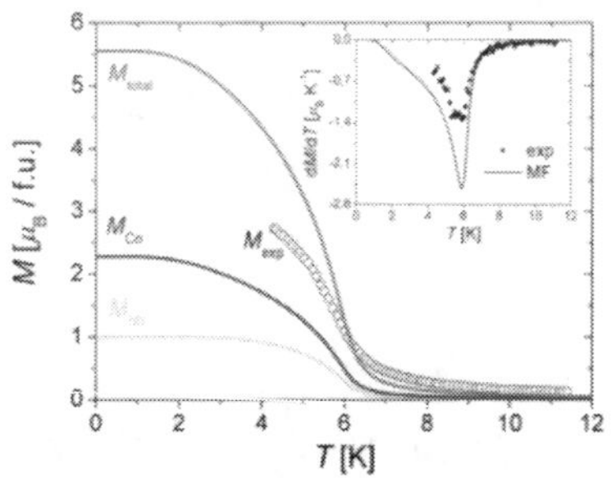

Figure 6. M vs. T obtained in H_{DC} = 0.5 kOe for **3**. Solid lines – the mean-field curves. Insert: dM/dT vs. T used to determine J_{CoNb}.

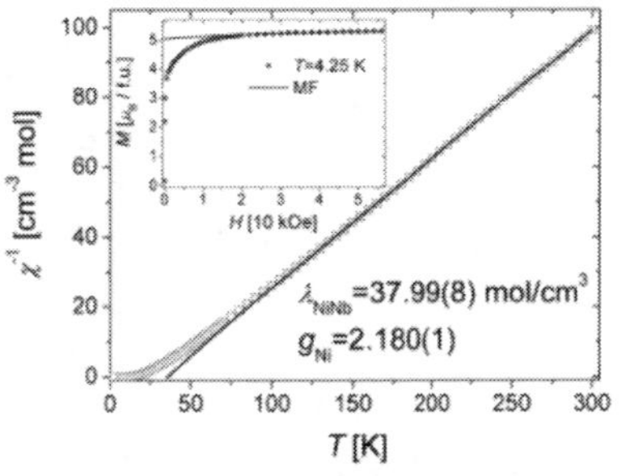

Figure 7. χ^{-1} vs. T for **4**. Solid line - fit to equation (3). Insert: M vs. H at T= 4.25 K.

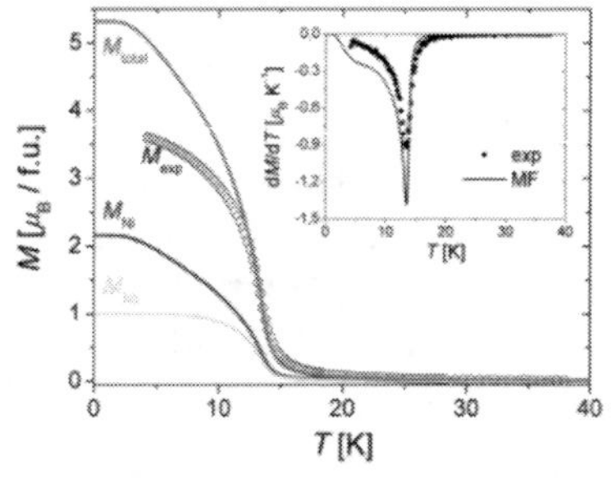

Figure 8. M vs. T obtained in H_{DC} = 0.5 kOe for **4**. Solid lines – the mean-field curves. Insert: dM/dT vs. T used to determine J_{NiNb}.

3.4. {[Ni^{II}(pirazol)$_4$]$_2$[Nb^{IV}(CN)$_8$] ·4H$_2$O}$_n$

For the Ni^{II} ion (3d^8) placed in the octahedral ligand field the orbital degeneracy of the corresponding ground term ^3F (L=3, S=1) is partially lifted and the term splits into two triplets Γ_4 ($^3T_{1g}$) and Γ_5 ($^3T_{2g}$), and a singlet Γ_2 ($^3A_{2g}$) [9], of which the orbital singlet is the lowest lying level. Hence the quantum state of the Ni^{II} is described by spin quantum number S_{Ni0}=1. Due to the spin-orbit interaction, which couples the ground orbital singlet with the triplet Γ_5, one can expect the enhancement of the mean spectroscopic factor for the Ni^{II} g_{Ni}>2.0. Fitting equation (3) in the temperature range 150-300 K (see figure 7) yielded λ_{NiNb}=+37.99(8) mol/cm^3, g_{Ni}=2.180(1), which implies the ferromagnetic coupling between the Ni^{II} and Nb^{IV} sublattices. Insert of figure 7 shows the relatively rapid saturation of magnetization detected at T=4.25 K. The magnetization in H=56 kOe attains the value of 5.31 μ_B, which is consistent with the expected value of $2g_{Ni}S_{Ni0}+g_{Nb}S_{Nb0}$=5.36 μ_B. The calculated mean-field $M(T)$ curve (figure 8, solid line) assuming the ferromagnetic interaction between Ni^{II} and Nb^{IV} sublattices yielded J_{NiNb} = +8.1(3) cm^{-1}.

4. Discussion

Table 1 recapitulates the results obtained for compounds **1-4**. Two last columns list the exchange coupling constants J_{MNb} obtained from the adjustment of the dM/dT peak and on the basis of the molecular-field constant λ_{MNb}. It can be seen that the values obtained by the latter approach are consistently higher than those given by the former one. These differences may originate from the fact that the mean-field model overestimates the transition temperature and hence the lower values for J_{MNb} obtained on the basis of the dM/dT peak position. For **2** the positive sign of λ_{FeNb} is inconsistent with antiferromagnetic character of exchange coupling implied strongly by low magnetization values, which suggests that the fits to equation (3) may not always be reliable. Therefore in further discussion we will refer to the exchange couplings determined by the former approach.

Table 1. Parameters of the mean-field calculation for **1-4**.

No.	M^{II}	Coupling assumed	S_{M0}	T_c [K]	g_M	λ_{MNb} [mol cm^{-3}]	J_{MNb} [cm^{-1}] dM/dT peak	J_{MNb} [cm^{-1}] from λ_{MNb}
1	Mn	AF	5/2	23.7(5)	1.99(1)	-28(4)	-6.8(2)	-14(2)
2	Fe	AF	2	8.3(5)	2.163(2)	+16.2(3)	-3.1(2)	+9.1(2)
3	Co	F	1/2	5.9(5)	4.55(5)	+15(1)	+3.5(3)	+11.8(8)
4	Ni	F	1	13.4(5)	2.180(1)	+37.99(8)	+8.1(3)	+21.6(1)

The values of exchange couplings found for **1**, **2**, **3**, and **4** indicate the presence of antiferromagnetic pathways within the Nb^{IV}-CN-M^{II} unit. They are expected to arise due to a "through bond" interaction, i.e. the overlap between the magnetic orbital centered on the $4d_{x2-y2}$ orbital of niobium (the lowest state for the DD coordination geometry) and some of the 3d orbitals of the metal ion. The monotonic increase of exchange constants $J_{MnNb}<J_{FeNb}<0<J_{CoNb}<J_{NiNb}$ revealed consistently by both approaches points out that the ferromagnetic terms contributing to the total exchange constant arising from the orthogonality of the $4d_{x2-y2}$ niobium orbital with some of the M^{II}-centered orbitals may not be negligible, despite the fact that the Nb^{IV} ion and the M^{II} ion is located at the sizable distance of about 5.6 Å. There is an evident correlation between the decreasing number of singly occupied 3d orbitals on the M^{II} ion (5, 4, 3, and 2 for **1**, **2**, **3** and **4** respectively) and the increasing trend in the exchange coupling constant. For the DD geometry, the $4d_{x2-y2}$ orbital of the Nb^{IV} ion is σ nonbonding because all the ligand atoms are located in the nodal planes of this orbital [10]. On the other hand, the lowest d level substantially mixes with the π system of the cyanido ligands. It was shown [10] that the overlap between the d_{x2-y2} orbital and the π orbital of the ligand orthogonal to the nodal plane in which the ligand is located is more efficient for the ligands on the vertexes with θ=69.46° than those on the vertexes with θ=36.85°. As for all the compounds the bridging ligands are located on the former vertexes, one can expect a relatively stronger exchange coupling. Among the magnetic orbitals (singly occupied molecular orbitals) centered on the partially occupied orbitals of the M^{II} ion, there are two of the t_{2g} type (d_{xy}, d_{xz}) which can be expected to be effectively delocalized on the nitrogen and the carbon of the cyanido group (π symmetry). Even for the remaining t_{2g} orbital (d_{yz}) a weak overlap with the orbitals centered on the bridging cyanido group can be expected due to its bending. At the same time, the magnetic orbital centered on the $4d_{x2-y2}$ orbital of Nb^{IV} is also partially delocalized on the carbon and nitrogen (with the same π symmetry) of the cyanido group. For two electrons, we can apply the Kahn's model ($J=2k+4\beta S$) which predicts two competing contributions to the resulting exchange coupling. The first contribution corresponding to the exchange integral k (>0) is of ferromagnetic character, whereas the second contribution, proportional to the resonance integral β (<0) and the overlap integral S is antiferromagnetic. This model provides an argument for a sizable antiferromagnetic interaction for the d_{xy}, d_{xz} orbitals ($2k<<|4\beta S|$) and a weak antiferromagnetic interaction for the remaining d_{yz} orbital ($2k<|4\beta S|$). Let us denote the corresponding averaged contribution to the total exchange coupling by J_{AF}. The d_{z2} and d_{x2-y2} orbitals (e_g orbitals) of the M^{II} ion have zero overlap integrals and each results in an averaged ferromagnetic contribution J_F to the total exchange coupling. The total exchange integral in the Nb^{IV}-CN-M^{II} unit can be expressed by the well known Kahn formula [5]:

$$J = \frac{1}{n_A n_B} \sum_{\mu=1}^{n_A} \sum_{\nu=1}^{n_B} J_{\mu\nu} \qquad (4)$$

In equation (4) the A site corresponds to the Nb^{IV} ion and the B site to the M^{II} ion. Because there is only one electron on the $4d_{x2-y2}$ orbital of the Nb^{IV} center, we have always $n_{Nb} = 1$. The situation is case-dependent for the M^{II} center. Let us start with the analysis for **1** where M = Mn. The electronic configuration of Mn^{II} is $(t_{2g})^3(e_g)^2$, so there are five singly-occupied orbitals ($n_{Mn} = 5$). Hence for J_{MnNb} we obtain $J_{MnNb} = 1/5(3J_{AF} + 2J_F)$. This expression alone is insufficient to estimate both contributions from the subsequent pathways. We apply the same formula for compound **2** with M = Fe and the electronic configuration $(t_{2g})^4(e_g)^2$, which implies $n_{Fe} = 4$. Hence, the expression for the exchange coupling constant has the following form $J_{FeNb} = 1/4(2J_{AF} + 2J_F)$. Using J_{MnNb} and J_{FeNb} obtained from respective molecular field models one finds $J_{AF} = -21.6$ cm^{-1}, $J_F = +15.4$ cm^{-1}. Using these values one can calculate analogously the exchange couplings for the two remaining compounds. The electronic configuration of the Co^{II} ion in **3** is $(t_{2g})^5(e_g)^2$ and in this case $n_{Co} = 3$, and consequently $J_{CoNb} = 1/3(J_{AF} + 2J_F)=+3.1$ cm^{-1}. For **4** with M = Ni the corresponding electronic configuration is $(t_{2g})^6(e_g)^2$. The number of unpaired electrons is again reduced by 1, hence $n_{Ni} = 2$. The exchange coupling constant is given by $J_{NiNb} = 1/2(2J_F)=J_F=+15.4$ cm^{-1}. These values compare well with those of $J_{CoNb}=+3.5$ cm^{-1} and $J_{NiNb}=+8.1$ cm^{-1} estimated within the corresponding mean-field models.

5. Conclusions

Two approaches based on the mean-field approximation were taken to get insight into the exchange couplings in four isostructural compounds $\{[M^{II}(\text{pyrazole})_4]_2[Nb^{IV}(CN)_8]\cdot 4H_2O\}_n$, where M^{II} = Mn, Fe, Co, Ni. The observed couplings were discussed using the simple-counting model (4) proposed by Kahn. This interpretation allows to predict the strength of exchange coupling for not yet reported Cr^{II} and V^{II} analogues of the $\{[M^{II}(\text{pyrazole})_4]_2[Nb^{IV}(CN)_8]\cdot 4H_2O\}_n$ series: $J_{CrNb} = 1/4(3J_{AF} + J_F) \approx -12$ cm^{-1} and $J_{VNb} = 1/3(3J_{AF}) \approx -21$ cm^{-1}.

Acknowledgments

This work has been partially supported by the Polish Ministry of Science and Higher Education within Research Projects 0087/B/H03/2008/34 and 1535/B/H03/2009/36.

References

[1] Miller J S, Calabrese J C, Epstein A J, Bigelow R B, Zang J H and Reiff W M 1986 *J. Chem. Soc.: Chem. Commun.* 1026–28; Pei Y, Verdaguer M, Kahn O, Sletten J and Renard J P 1986 *J. Am. Chem. Soc.* **108** 7428–30
[2] Yakhmi J V 2009 *Bull. Mater. Sci.* **32** 217–25; Férey G 2008 *Chem. Soc. Rev.* **37** 191–214
[3] Sieklucka B, Podgajny R, Przychodzeń P and Korzeniak T 2003 *Coord. Chem. Rev.* **249** 2203–21; Przychodzeń P, Korzeniak T, Podgajny R and Sieklucka B 2006 *Coord. Chem. Rev.* **250** 2234–60
[4] Pinkowicz D, Pełka R, Drath O, Nitek W, Bałanda M, Majcher A M, Poneti G and Sieklucka B 2010 *Inorg. Chem.* **49** 7565–76
[5] Kahn O 1993 *Molecular Magnetism* (New York: VCH Publishers, Inc.)
[6] Ohkoshi S I, Iyoda T, Fujishima A and Hashimoto K 1997 *Phys. Rev.* B **56** 11642–652
[7] Arai M, Kosaka W, Matsuda T and Ohkoshi S I 2008 *Angew. Chem. Int. Ed.* **47** 6885–87
[8] Kettle S F A 1996 *Physical Inorganic Chemistry. A Coordination Chemistry Approach* (Oxford: Oxford University Press)
[9] Abragam A and Bleaney B 1970 *Electron Paramagnetic Resonance of Transition Ions* (Oxford: Clarendon Press)
[10] Burdett J K, Hoffmann R and Fay R C 1978 *Inorg. Chem.* **17** 2553–68

Intensity magnetooptical effect in magnetoplasmonic crystals

V I Belotelov[1,2], **I A Akimov**[3,4], **M Pohl**[3], **A N Kalish**[1,2], **S Kasture**[5], **A S Vengurlekar**[5], **A V Gopal**[5], **V A Kotov**[6], **D Yakovlev**[3,4], **A K Zvezdin**[2] and **M Bayer**[3]

[1] M V Lomonosov Moscow State University, 119992 Moscow, Russia
[2] A M Prokhorov General Physics Institute, Russian Academy of Sciences, 119991 Moscow, Russia
[3] TU Dortmund University, D-44221 Dortmund, Germany
[4] A F Ioffe Physical-Technical Institute, Russian Academy of Sciences, 194021 St. Petersburg, Russia
[5] Tata Institute of Fundamental Research, Mumbai, 400005, India
[6] V A Kotelnikov Institute of Radio Engineering and Electronics, Russian Academy of Sciences, 125009 Moscow, Russia

Email: Belotelov@physics.msu.ru

Abstract. Significant increase of the intensity magnetooptical effect (transversal Kerr effect) is observed in transmission for a magnetoplasmonic crystal consisting of the perforated noble-metal film attached to a smooth magnetic dielectric layer. It is largely due to the magnetic field induced shift of the Fano resonances in transmission associated with the surface plasmon polaritons excited at the metal/magnetic-dielectric interface. It is shown that the quasi-guided modes of the magnetic layer also lead to the enhancement of the intensity magnetooptical effect. The considered magnetoplasmonic structures are of great importance for applications in telecommunication and molecular sensing as they also drastically enhance other magnetooptical effects.

1. Introduction

The magnetic field influence on light was discovered by M. Faraday in the mid of the 19th century [1]. Nowadays, magnetooptical effects are considered for nanophotonics applications requiring manipulation of light at gigahertz frequencies [2,3]. A prospective example is the intensity magnetooptical effect - transverse magnetooptical Kerr effect (TMOKE) providing an intensity change of light reflected from a ferromagnetic metal if an applied external magnetic field is reversed [4]. TMOKE allows one to investigate magnetic material properties and can be utilized in magnetooptical data storage [3,5]. For its observation the magnetic field has to be perpendicular to the plane of light incidence. The TMOKE is then characterized by the relative change of reflected light intensity $I(\mathbf{M})$ when a medium's magnetization $\mathbf{M}$ is reversed:

$$\delta = \left(I(\mathbf{M}) - I(-\mathbf{M})\right)/I(0). \quad (1)$$

This change originates from the magnetic field induced change of boundary conditions at the magnetic layer surface, and attains maximum values for oblique incidence of p-polarized light. Usually, for smooth ferromagnetic metals δ is on the order of 10^{-3} limiting its applicability [3,6]. The

TMOKE's counterpart in transmission may also occur, for which a necessary condition is a difference between the opposite magnetic-film boundaries. However, it is difficult to observe because of the small transmission through ferromagnetic metals in addition to its small magnitude [7].

In this paper we investigate both experimentally and theoretically a magnetic structure in which transmission is not vanishing and the TMOKE effect attains significantly enhanced values. It is achieved by using a special metal-dielectric structure – a magnetoplasmonic crystal. It consists of two thin films: a smooth ferromagnetic dielectric and a noble metal periodically perforated with subwavelength slits (figure 1). Surface plasmon polaritons (SPP) can be efficiently excited in this material. So the material combines gyrotropic, plasmonic and nanostructuring features. At the same time its constituents possess large value of the magnetooptical parameter (specific Faraday rotation) due to a ferromagnetic dielectric (bismuth rare-earth iron garnet) and on the other hand they have record small optical losses for the long wavelengths visible and near IR ranges.

Alliance between magnetooptics and plasmonics has been investigated since 70s of the previous century [8-25]. Early papers on the interplay between SPP and magnetooptics addressed SPPs propagating along the smooth surface of a ferromagnetic film [8-9] or along a smooth semiconductor surface in a transverse external magnetic field [10]. In that case, the magnetic field modifies the SPP wave vector but leaves its transverse magnetic polarization (TM) unchanged. An SPP assisted increase of the TMOKE at a ferromagnetic metal surface was reported [11-14] as well as an enhancement of the TMOKE in bimetallic systems consisting of a noble and a ferromagnetic metal [15-17]. Nanostructured plasmonic systems were investigated in [18-23], where a resonance increase of the TMOKE was reported for one-dimensional Co and Fe gratings [18,19], for noble metals in a high external magnetic field [20], and for noble-metal/ferromagnetic-metal multilayers with periodic perforations or protrusions [21,22]. The main disadvantage of most of the aforementioned approaches is that the optical losses associated with the presence of a ferromagnetic metal are relatively high. If the ferromagnetic metals were avoided as in case of pure semiconductor or noble metal systems, huge external magnetic fields exceeding several Tesla would be necessary to make the TMOKE at least comparable with the effect in ferromagnets. In this respect the case of the periodic structures with low loss magnetic dielectric and a noble metal studied in this work seem much more advantageous. This conclusion is supported by the previous theoretical studies [24,25].

In this paper we demonstrate experimentally the SPP assisted significant enhancement of the TMOKE in the magnetoplasmonic crystals and investigate the influence of the quasi-guiding modes of the magnetic film on the TMOKE.

2. Eigenmodes of the magnetoplasmonic crystal

Optical properties of the considered magnetoplasmonic crystal are strongly dependent on the eigenmodes excited in the system by the incident light. These are SPPs propagating along the metal/air and metal/magnetic-dielectric interfaces, cavity modes in the metal grating slits and guiding modes of the magnetic film. Let us consider the influence of these modes on the magnetooptical response and on the TMOKE effect, in particular. At this we characterize the magnetic medium by the permittivity tensor $\hat{\varepsilon}^{(m)}$ having the following non-zero components: $\varepsilon_{11} = \varepsilon_{22} = \varepsilon_{33} = \varepsilon_m$, $\varepsilon_{13} = \mathrm{i}g$, $\varepsilon_{31} = -\mathrm{i}g$, where g is the value of medium's gyration [3]. At this we assume that the magnetization of the layer is along y-axis (figure 1). The typical values for g are about 0.001, so further we neglect effects that are nonlinear on magnetization.

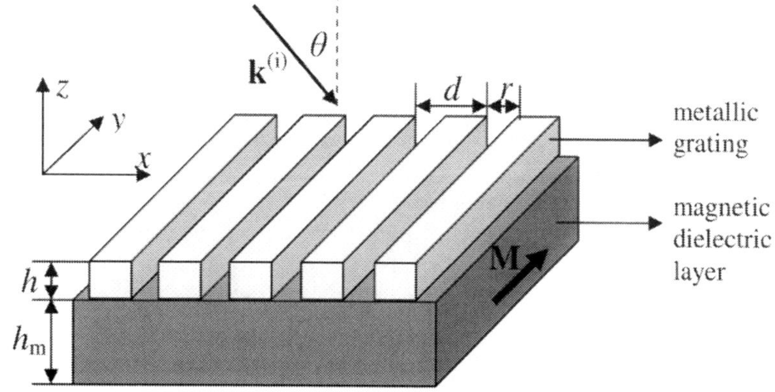

Figure 1. Magnetoplasmonic crystal. $\mathbf{k}^{(i)}$ is the wave vector of incident light.

Let us start from the SPPs at the horizontal interfaces of the metallic grating. It is well known that excitation of the SPPs in the metal-dielectric structures causes so called Wood anomalies [26]. In the case of sufficiently high grating (h>50nm) their dispersion is well approximated by

$$\kappa = k_0 \sqrt{\varepsilon_{\mathrm{me}} / (1 + \varepsilon_{\mathrm{me}})}, \tag{2}$$

where k_0 is the vacuum wave number and $\varepsilon_{\mathrm{me}}$ is the dielectric constant of the metal. There is no gyrotropic input here so one can expect rather weak influence of the magnetization on the metal/air SPP and consequently there are no any pronounced TMOKE features related to these SPPs excitation.

In this respect, the SPPs at the metal/magnetic-dielectric interface and guiding modes in the magnetic dielectric are more promising since they are localized in the vicinity of the gyrotropic part of the magnetoplasmonic crystal. Considering the magnetic layer placed between the metal without perforation and a substrate the following equations for their dispersion relations can be obtained from Maxwell's equations with the corresponding boundary conditions:

$$\frac{\gamma_{\mathrm{m}}}{\varepsilon_{\mathrm{m}}}\left(\frac{\gamma_{\mathrm{me}}}{\varepsilon_{\mathrm{me}}} + \frac{\gamma_{\mathrm{s}}}{\varepsilon_{\mathrm{s}}}\right) + \left[\frac{\gamma_{\mathrm{me}}}{\varepsilon_{\mathrm{me}}}\frac{\gamma_{\mathrm{s}}}{\varepsilon_{\mathrm{s}}} + \left(\frac{\gamma_{\mathrm{m}}}{\varepsilon_{\mathrm{m}}}\right)^2 + \frac{g\kappa}{\varepsilon_{\mathrm{m}}^2}\left(\frac{\gamma_{\mathrm{me}}}{\varepsilon_{\mathrm{me}}} - \frac{\gamma_{\mathrm{s}}}{\varepsilon_{\mathrm{s}}}\right)\right]\tanh(\gamma_{\mathrm{m}} h_{\mathrm{m}}) = 0 \tag{3a}$$

for the SPPs and

$$\frac{k_{\mathrm{zm}}}{\varepsilon_{\mathrm{m}}}\left(\frac{\gamma_{\mathrm{me}}}{\varepsilon_{\mathrm{me}}} + \frac{\gamma_{\mathrm{s}}}{\varepsilon_{\mathrm{s}}}\right) + \left[\frac{\gamma_{\mathrm{me}}}{\varepsilon_{\mathrm{me}}}\frac{\gamma_{\mathrm{s}}}{\varepsilon_{\mathrm{s}}} - \left(\frac{k_{\mathrm{zm}}}{\varepsilon_{\mathrm{m}}}\right)^2 - \frac{g\kappa}{\varepsilon_{\mathrm{m}}^2}\left(\frac{\gamma_{\mathrm{me}}}{\varepsilon_{\mathrm{me}}} - \frac{\gamma_{\mathrm{s}}}{\varepsilon_{\mathrm{s}}}\right)\right]\tan(k_{\mathrm{zm}} h_{\mathrm{m}}) = 0 \tag{3b}$$

for guiding modes, where h_{m} is the magnetic layer height, $\gamma_i = \left(\kappa^2 - \varepsilon_i k_0^2\right)^{1/2}$, $i = \mathrm{m, me, s}$, $k_{\mathrm{zm}} = \left(\varepsilon_{\mathrm{m}} k_0^2 - \kappa^2\right)^{1/2}$, $k_0 = \omega c^{-1}$, index 's' refers to the substrate. It follows from (3) that the propagation constant κ of both types of eigenmodes is dependent on magnetization.

If the magnetic layer is rather thin the SPPs dispersion law is close to one for the interface between the metal and the substrate: $\kappa = k_0 \left[\varepsilon_{\mathrm{s}} \varepsilon_{\mathrm{me}} \left(\varepsilon_{\mathrm{s}} + \varepsilon_{\mathrm{me}}\right)^{-1}\right]^{1/2}$, so it is almost magnetization independent. With the increase of h_{m} the influence of magnetization on the SPP dispersion gets more pronounced according to (3b) and also the optical spectra acquire additional features related to the excitation of guiding modes that are also magnetic dependent as seen from (3a). If h_{m} exceeds the value $\left[-\left(\varepsilon_{me}+\varepsilon_{m}\right)\right]^{1/2}\left(k_0\varepsilon_m\right)^{-1}$ that is about hundred nanometers the SPPs dispersion becomes independent of the layer height and the dispersion law (3a) takes the following form:

$$\kappa = k_0\left[\varepsilon_m\varepsilon_{me}\left(\varepsilon_m+\varepsilon_{me}\right)^{-1}\right]^{1/2}\left[1+g\left(-\varepsilon_m\varepsilon_{me}\right)^{-1/2}\left(1-\varepsilon_m^2/\varepsilon_{me}^2\right)^{-1}\right], \tag{4}$$

The essential point here is the effect of magnetooptical nonreciprocity. It implies that the dependence of the SPP and guiding modes propagation constants on magnetization is odd and hence for fixed magnetization the two waves propagating in opposite direction have different phase velocities.

If an eigenmode is excited by an incident light it leads to appearance of the resonant features in optical reflection and transmission spectra. For that the projection of the incident wave number on the sample surface $k_x^{(i)} = k_0\sin\theta$, where θ is the incident angle, should be equal to the eigenmode's propagation constant κ. As the latter is now dependent on magnetization the resonant wavelengths are shifted with respect to the non-magnetic case. Moreover, according to magnetooptical nonreciprocity the shifts are of opposite signs for opposite directions of magnetization. That leads to emergence of the TMOKE described by (1). Its value is dependent not only on the shift value but also on the shape of the resonance curve. It can be shown that the TMOKE value is proportional to the derivative of reflection/transmission on wavelength, so the spectral shape of the TMOKE resonance has negative minimum followed by positive maximum or visa versa. The TMOKE value can be rather high in comparison with that in case of a smooth magnetic film.

For the case of the plasmonic crystal depicted in figure 1 the excitation condition for eigenmodes is:

$$\kappa = k_0\sin\theta + mG, \tag{5}$$

where $G = 2\pi d^{-1}$ is the reciprocal lattice vector, d is the period of the plasmonic crystal and m is an integer. In this case the eigenmodes are leaky and they also contribute to the transmitted wave in the far field. That is why, it is more correct to call the guiding waves the quasi-guiding ones. The correspondent resonant features in transmission and reflection spectra have the typical shape consisting of minimums followed by maximums or visa versa. The shapes of this kind are called Fano resonances [27]. They appear in cases when a resonant process is accompanied by a non-resonant one. In our case the former is the eigenmode excitation via one of the diffraction orders and the latter is the wave propagating through the structure.

The dispersion laws for the quasi-guiding modes and the SPPs are influenced by periodicity and the slits in the metal grating. However, if the slits are rather narrow being compared to the eigenmode wavelength the empty lattice approximation can be applied, according to which κ can be estimated from (2), (3) or (4). Near the centre and the edges of the Brillouin zone ($\kappa = mG/2$) this approximation is not applicable as the wave propagation is strongly influenced by periodicity. The excitation of the quasi-guiding modes and the SPPs at metal/magnetic interface is accompanied by resonant enhancement of TMOKE as described above. Due to the periodicity of the structure nonreciprocal propagation is prohibited at the centre and the edges of the Brillouin zone, so TMOKE near these points of the Brillouin zone vanishes.

Besides there are eigenmodes of another kind that are Fabry-Perot cavity modes. Since they are localized mainly inside the slits they are almost magnetization independent and don't significantly contribute to TMOKE.

To approve the above qualitative analysis we need to conduct a comprehensive numerical modelling of the electromagnetic properties of the structure. It is accomplished on the basis of the rigorous coupled waves analysis (RCWA) extended to the instance of gyrotropic materials [28-29]. At this, special factorization rules are used for the improvement of the algorithm's convergence.

3. Magnetoplasmonic sample and experimental set up

Magnetic part of the magnetoplasmonic structure is a 2.5 micron thick bismuth-substituted rare-earth iron garnet film of composition $Bi_{0.4}(YGdSmCa)_{2.6}(FeGeSi)_5O_{12}$ grown by liquid phase epitaxy from Bi_2O_3 : PbO : B_2O_3 melt on the $Gd_3Ga_5O_{12}$ substrate with orientation (111). The film possesses uniaxial magnetic anisotropy in the direction perpendicular to the film plane. The specific Faraday rotation is 0.46 deg/micron at wavelength 633 nm. The magnetoplasmonic sample of structure shown in figure 1 was fabricated by the thermal deposition of the gold layer on the bismuth-substituted rare-earth iron garnet film and subsequent electron beam lithography combined with the reactive ion etching in Ar plasma. The sample was characterized by AFM and SEM imaging. The grating parameters obtained are gold layer height h is 150 nm, period d is 595 nm and the air groove width r is 110 nm.

For magnetooptical measurements we used a halogen lamp as a source of white light. The collimated light was focused at the sample into a spot with diameter of about 300-500 μm and aperture angle below 1°. In order to perform the measurements at different angles of light incidence the sample was mounted on rotation stage. Zero order transmission signal was spectrally detected with a single monochromator (linear dispersion 6.28 nm/mm) equipped with charged coupled device detector. The overall spectral resolution was below 0.3 nm. The polarization direction of the transmitted light was selected with a polarizer in the detection path. The magnetic fields up to 4 kOe were applied in transverse geometry using water cooled electromagnet.

4. Extraordinary TMOKE related to SPPs

Results of the experimental observation of TMOKE in zero order transmitted light are shown in figure 2 for different angles of light incidence.

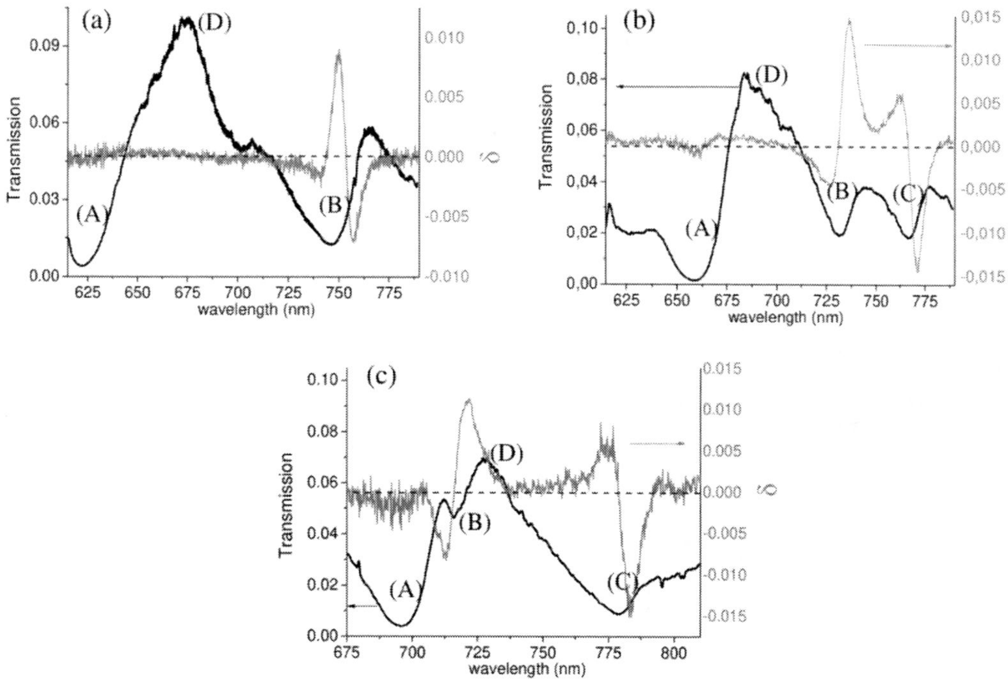

Figure 2. Experimentally measured transmission (thick black curves) and TMOKE (thin red curves) for three different incidence angles: (a) θ =0.8°; (b) θ =5°; (c) θ =10°. Light is p-polarized and is incident perpendicularly to the slits.

Since height of the magnetic film in the investigated sample is rather large (h_m=2.5 μm) influence of the guided modes is not noticeable and main features of the spectra are related to SPPs and cavity modes. Indeed, it follows from (2) and (4) that the Fano resonance (A) is due to the SPP of the zero order at the air/gold interface, while the Fano resonances (B) and (C) are related to the SPPs of the second order at the gold/magnetic-film interface. The prominent transmission peak (D) is attributed to the collective Fabry-Perot cavity mode inside the slits.

To ensure that sample magnetization is nearly in-plane relatively large external magnetic field of 2000 Oe was applied. It can be seen that away from resonances δ is very small. Actually, δ cannot be measured experimentally which means that it does not exceed 10^{-3}. On the contrary, at the resonances δ reaches $1.5\cdot10^{-2}$ demonstrating an increase by at least one order of magnitude. Electromagnetic modeling for the nonresonant case gives $\delta \sim 10^{-4}$. If compared to the case of uncovered bare iron garnet film the enhancement factor is even larger, namely it is up to 10^3. So we can announce that a giant TMOKE in transmission is observed. The effect can be made larger if a magnetic film with higher specific Faraday rotation is used [30].

It is apparent that no noticeable TMOKE increase is observed for other resonant regions. It goes exactly in line with the discussion above and highlights the TMOKE's sensitivity to excitation of different eigenmodes. For normal incidence the TMOKE is zero because of the full symmetry of the incident light with respect to the structure. As soon as the incidence is not normal the symmetry is broken and two SPP propagating into opposite directions are excited at slightly different frequencies and the TMOKE appears. The value of δ is almost 10^{-2} even if the incidence angle is as small as 0.8° (figure 2(a)). At the transition from normal to slightly oblique incidence transmission spectrum does not change noticeably while TMOKE spectrum does. It demonstrates TMOKE sensitivity to slight changes of the illumination conditions. If the incidence angle is larger (e.g. it is 5°, figure 2(b)) two eigenfrequencies of the SPP propagating in opposite directions differs significantly and give birth to two Fano resonances in transmission. The TMOKE accompanies both resonances, but δ has opposite signs at them reflecting the fact that these resonances are due to the SPP propagating in opposite directions. So the sign of the TMOKE at the magnetoplasmonic crystals is also sensitive to the direction of the SPP propagation. Further increase of the incidence angle shifts the plasmonic resonances further apart, but the value of the TMOKE remains almost steady (figure 2(c)).

5. Transmission and TMOKE versus height of the magnetic film

The results of the numerical simulations showing the transmission and TMOKE dependence on the magnetic layer height are presented in figure 3.

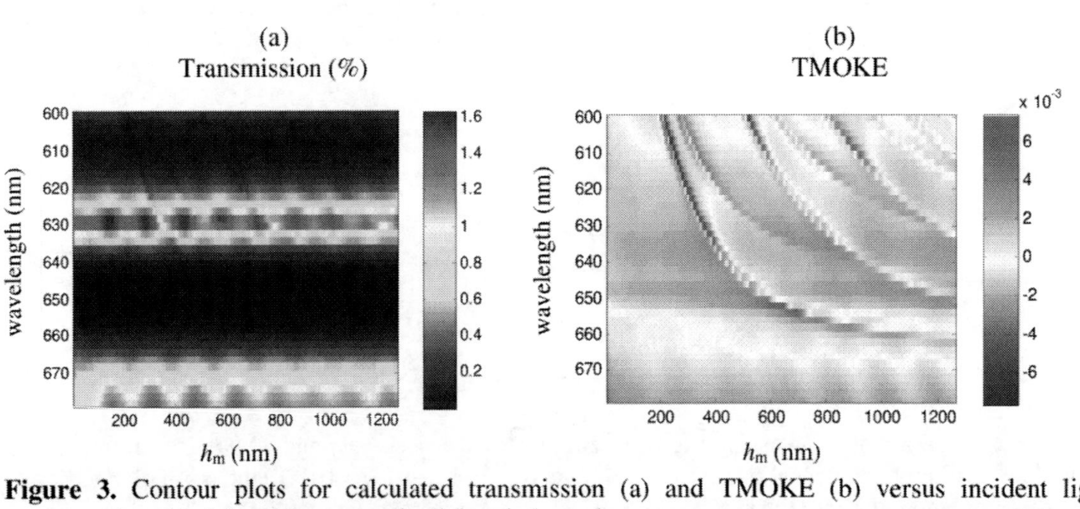

Figure 3. Contour plots for calculated transmission (a) and TMOKE (b) versus incident light wavelength and height of the magnetic dielectric layer h_m.

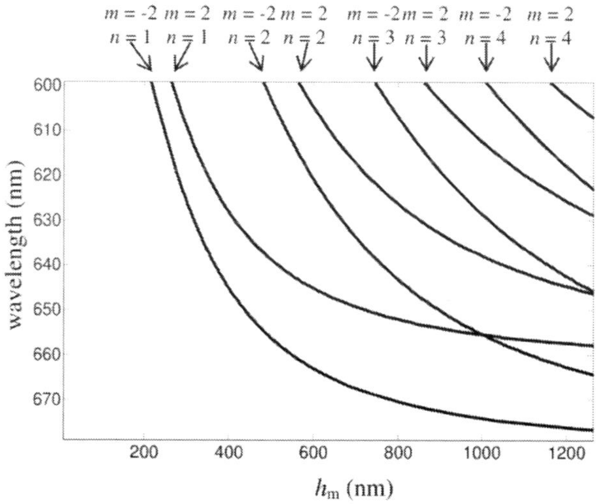

Figure 4. The dispersion curves for guiding modes of magnetic layer calculated from (3b) and (5) for $m = \pm 2$.

The principal features on the TMOKE spectrum (figure 3(b)) are series of resonances that are dependent on the magnetic layer height and they are accompanied by transmission dips. These are the contribution of the magnetic layer quasi-guiding modes excited by ±2nd diffraction orders. Their excitation conditions estimated from (3b) and (5) are shown in figure 4 and there is rather nice correspondence with transmission and TMOKE resonances. The difference can be explained by the only approximate applicability of the empty lattice approximation.

According to (3b) and (5) each quasi-guiding mode in the plasmonic crystal is characterized by two integers m and n, the former denoting the diffraction order via which the mode is excited and the latter coming from solution of (3b) for fixed m. Within the spectral range shown in figures 3 and 4 the quasi-guiding modes only with m=±2 are excited. It should be noted that opposite signs of m correspond to opposite directions of modes propagation and hence according to the discussion above the signs of TMOKE are opposite for $m = 2$ and $m = -2$, that is clearly seen in figure 3(b). For the case of normal incidence the difference in dispersion between modes with the same values of n but opposite values of m vanishes and so does TMOKE since this situation corresponds to the quasi-momentum κ=0 from the first Brillouin zone (see Section 2).

As we have already discussed in Section 4 the other features in transmission spectrum (figure 3(a)) don't demonstrate strong magnetooptical response. One of them is increase in transmission at wavelengths about 680 nm that is independent of the magnetic layer thickness. It is Fabry-Perot resonance inside the slits of metallic grating. The others are transmission dips at wavelengths about 610 and 650 nm. They are related to the SPPs excitation at the metal/air interface and therefore they don't contribute much to TMOKE.

It should be noted that for rather large values of magnetic layer height the excitation of modes is less effective. Moreover, if the structure is illuminated with light with small coherence length the modes are not excited at all, so in our experiments they were not observed.

6. Conclusion

We demonstrate experimentally that the magnetooptical effect can be drastically increased via excitation of the SPPs in the magnetoplasmonic crystals – periodic structures consisting of metal and magnetic dielectric perforated layers. The increase of the TMOKE in transmission is measured to be up to tree orders of magnitude if compared with a smooth magnetic dielectric film. The main reason for this enhancement is the magnetization induced shift of the SPP associated Fano resonances in transmission spectra. This shift makes more significant input in the TMOKE signal for larger gradients of the Fano resonances. However, noticeable influence of the SPPs on the TMOKE is observed only for the SPP propagating along the metal/magnetic-dielectric interface. Plasmonic excitations at the other interface hardly contribute into the TMOKE enhancement because they are not affected much by the gyrotropic properties of the underlying magnetic layer.

The TMOKE increase takes also place if quasi-guided TM modes of the magnetic layer are excited. Similarly to the SPPs their resonance frequency is linearly dependent on the magnetic layer transverse magnetization.

So TMOKE in magnetoplasmonic crystals is sensitive to the type of the eigenmodes of the structure. Moreover, it changes sign for SPPs or quasi-guiding modes traveling in opposite directions. Thus, TMOKE can become an important tool for complete characterization of plasmonic nanostructures.

Even though here we paid here only attention to TMOKE, our previous study [31] allows us to claim that the proposed class of magnetoplasmonic heterostructures can be used to enhance also other magnetooptical effects such as the Faraday effect or the longitudinal Kerr effect. With these parameters the structures are very promising for applications in ultra high sensitivity devices and optical data processing.

Acknowledgements
The work is supported by Russian Foundation for Basic Research and "Dynasty" foundation.

References

[1] Faraday M 2005 *Experimental Researches in Electricity* vol 3 (New York: Adamant Media Corp.)
[2] Prasad P N 2004 *Nanophotonics* (Hoboken, NJ: J. Wiley & Sons)
[3] Zvezdin A and Kotov V 1997 *Modern Magnetooptics and Magnetooptical Materials* (Bristol: IOP Publ.)
[4] Kerr J 1877 On the rotation of the plane of the polarization by reflection from the pole of a magnet *Philosophical Magazine Series 5* **3** 321–43
[5] Mansuripur M 1998 *The Physical Principles of Magneto-optical Recording* (Cambridge: Cambridge University Press).
[6] Krinchik G S and Artem'ev V A 1968 Magneto-optical properties of Ni, Co and Fe in ultraviolet visible and infrared parts of spectrum *J. Exper. Theor. Phys.* **26** 1080–85
[7] Druzhinin A V, Lobov I D, Mayevskiy V M and Bolotin G 1983 Transverse magnetooptical Kerr effect in transmission *The Physics of Metals and Metallography* **56** 58-65
[8] Ferguson P E, Stafsudd O M and Wallis R F 1977 Surface magnetoplasma waves in nickel *Physica* B & C **86–88**, 1403–05
[9] Burke J J, Stegeman G I and Tamir T 1986 Surface-polariton-like waves guided by thin, lossy metal films *Phys. Rev.* B **33**, 5186–201
[10] Aers G C and Boardman A D 1978 The theory of semiconductor magnetoplasmon-polariton surface modes: Voigt geometry *J. Phys. C: Solid State Phys.* **11**, 945–59
[11] Hickernell R K and Sarid D 1987 Long-range surface magnetoplasmons in thin nickel films *Opt. Lett.* **12** 570–2
[12] Olney R D and Romagnoli R J 1987 Optical effects of surface plasma waves with damping in metallic thin films *Appl. Opt.* **26** 2279–82
[13] Newman D M, Wears M L and Matelon R J 2004 Plasmon transport phenomena on a continuous ferromagnetic surface *Europhys. Lett.* **68** 692–8
[14] Bonod N, Reinisch R, Popov E. and Neviere M 2004 Optimization of surface-plasmon-enhanced magneto-optical effects *J. Opt. Soc. Am.* B **21** 791–7
[15] Gonzalez-Diaz J B, Garcia-Martin A, Armelles G, Garcia-Martin J M, Clavero C, Cebollada A, Lukaszew R A, Skuza J R, Kumah D P and Clarke R 2007 Surface magneto-plasmon nonreciprocity effects in noble-metal/ferromagnetic heterostructures *Phys. Rev.* B **76** 153402

[16] Vila E F, Sueiro X M B, Gonzalez-Diaz J B, Garcia-Martin A, Garcia-Martin J M, Navarro A C, Reig G A, Rodriguez D M and Sandoval E M 2008 Surface plasmon resonance effects in the magneto-optical activity of Ag–Co–Ag trilayers *IEEE Trans. Magn.* **44** 3303–6
[17] Temnov V, Armelles G, Woggon U, Guzatov D, Cebollada A, Garcia-Martin A, Garcia-Martin J M, Thomay T, Leitenstorfer A and Bratschitsch R 2010 Active magnetoplasmonics in hybrid metal/ferromagnet/metal microinterferometers *Nature Photon.* **4** 107
[18] Newman D M, Wears M L, Matelon R J and Hooper I R 2008 Magneto-optic behavior in the presence of surface plasmons *J. Phys.: Condens. Matter* **20** 345230
[19] Wurtz G A, Hendren W, Pollard R, Atkinson R, Le Guyader L, Kirilyuk A, Rasing Th, Smolyaninov I I and Zayats A V 2008 Controlling optical transmission through magneto-plasmonic crystals with an external magnetic field *New J. Phys.* **10** 105012
[20] Strelniker Y M and Bergman D J 2008 Transmittance and transparency of subwavelength perforated conducting films in the presence of a magnetic field *Phys. Rev.* B **77** 205113
[21] Clavero C, Yang K, Skuza J R and Lukaszew R A 2010 Magnetic-field modulation of surface plasmon polaritons on gratings *Opt. Lett.* **35** 1557–9
[22] Armelles G, Cebollada A, Garcia-Martin A, Garcia-Martin J M, Gonzalez M U, Gonzalez-Diaz J B, Ferreiro-Vila E and Torrado J F 2009 Magnetoplasmonic nanostructures: systems supporting both plasmonic and magnetic properties *J. Opt. A: Pure Appl. Opt.* **11** 114023
[23] Buchin E Y, Vaganova E I, Naumov V V, Paporkov V A and Prokaznikov A V 2008 Enhancement of the transversal magnetooptical Kerr effect in nanoperforated cobalt films *Technical Phys. Lett.* **35** 589–93
[24] Belotelov V I, Bykov D A, Doskolovich L L, Kalish A N and Zvezdin A K 2009 Extraordinary transmission and giant magneto-optical transverse Kerr effect in plasmonic nanostructured films *J. Opt. Soc. Am.* B **26** 1594–8
[25] Khanikaev A B, Mousavi S H, Shvets G and Kivshar Y S 2010 One-way extraordinary optical transmission and nonreciprocal spoof plasmons *Phys. Rev. Lett.* **105** 126804
[26] Wood R W 1935 Anomalous diffraction gratings *Phys. Rev.* **48** 928–36
[27] Luk'yanchuk B, Zheludev N I, Maier S A, Halas N J, Nordlander P, Giessen H and Chong C T 2010 The Fano resonance in plasmonic nanostructures and metamaterials *Nature Mater.* **9** 707–15
[28] Moharam M G, Pommet D A, Grann E B and Gaylord T K 1995 Stable implementation of the rigorous coupled-wave analysis for surface-relief gratings: enhanced transmittance matrix approach *J. Opt. Soc. Am.* A **12** 1077–86
[29] Li L 2003 Fourier modal method for crossed anisotropic gratings with arbitrary permittivity and permeability tensors *J. Opt. A: Pure Appl. Opt.* **5** 345–55
[30] Vasiliev M, Nur-E-Alam M, Kotov V A, Alameh K, Belotelov V I, Burkov V I and Zvezdin A K 2009 RF magnetron sputtered $(BiDy)_3(FeGa)_5O_{12}$:Bi_2O_3 composite garnet-oxide materials possessing record magneto-optic quality in the visible spectral region *Opt. Expr.* **17** 19519–35
[31] Belotelov V I, Doskolovich L L, Zvezdin A K 2007 Extraordinary magnetooptical effects and transmission through the metal-dielectric plasmonic systems *Phys. Rev. Lett.* **98** 77401

Modeling and optimization of nonreciprocal transmission through 2D magnetophotonic crystal

M. Vanwolleghem[1], K. Postava,[2] L. Halagačka[1,2], L. Magdenko[1], P. Beavillain[1], B. Dagens[1]

[1]Institut d'Electronique Fondamentale, UMR CNRS 8622, Universite Paris-Sud, Orsay, France
[2]Department of Physics, Technical University of Ostrava, 708 33 Ostrava, Czech Republic

E-mail: `mathias.vanwolleghem@u-psud.fr, kamil.postava@vsb.cz`

Abstract. A combination of unique magneto-optic (MO) non-reciprocity and photonic band gap in periodic structures is promising for efficient enhancement of optical isolation and integrated isolator applications [M. Vanwolleghem et al., Phys. Rev. B 80 (2009) 121102(R)]. In this paper we model and optimize a novel magneto-photonic crystal structure consisting air holes in transparent magneto-optic material in transverse geometry (Bismuth iron garnet ($\varepsilon_{xx} = 6.25$ and $\varepsilon_{yz} = 0.1\,i$) at wavelength $\lambda = 1300$ nm). Such a system with reduced symmetry shows an unidirectional bandgap. The model is based on plane wave Fourier expansion of the field inside the periodic system using RCWA. While in the forward direction the structure transmit the light in the backward direction it shows a band gap and transmission is almost forbidden.

1. Introduction

The need for an integrated and miniaturised version of an optical isolator is making itself increasingly felt. The drive towards ever higher degrees of all optical on-chip integration is often hindered by the absence of an element that induces a one-way sense in the long optical path of the signal in such an integrated circuit. Without such an isolating element, long path interferences can lead to large changes somewhere in the circuit due to small amplitude oscillations at a remote distance. A competitive integrated optical isolator is still not available today despite many years of intense research effort by several groups worldwide [1]. Among other reasons this is mainly due to the limited magneto-optical strength of the common magnetic material in the infrared part of the spectrum, leading to large integrated devices with important sensitivity to fabrication errors. Over the past decade increasing attention has therefore been devoted to artificial enhancement of MO effects using photonic crystal layouts [2]. One very interesting concept that appears by combining magneto-optics and photonic crystal (PhC) effects is the possibility to obtain one-way band gaps. We have recently theoretically proven the existence of such unidirectional regimes for TE modes in 2D integrated uniformly transversely magnetized magneto-optic photonic crystals [3]. In this communication we present a numerical optimization study of one flexible type of such a 2D photonic crystal.

2. Structure of photonic crystal

In Ref. [4] it is proven that the necessary condition for band structure asymmetry (of which local one way band gaps are an extreme manifestation) is the absence of any symmetry operation

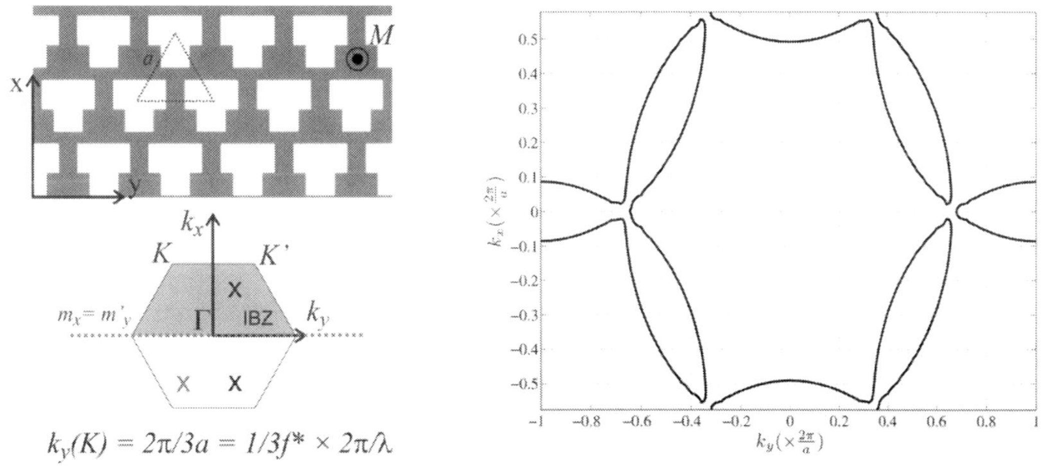

Figure 1. (left panel) Schematic representation of the cm' hexagonal crystal and the resulting symmetries in the Brillouin zone. The black crosses indicate equivalent points in the BZ. The red cross shows the time reversed point. (right panel) The isofrequency contour at a reduced frequency of $f^* = \frac{a}{\lambda} = 0.4545$ of the uniformly magnetized MO PhC of the left panel (material parameters and geometrical details as below). A one-way forbidden gap around the K-points is clearly observed. Incident plane waves with $k_y = k_y(K)$ would be totally reflected at this frequency. Those with the opposite k_y would be transmitted.

inside the magnetic space group describing the MOPhC that transforms an arbitrary reciprocal space vector $\overrightarrow{k}$ into its opposite. We have derived in [3] the 5 plane magnetic Shubnikov groups (out of 80 possible ones [5]) that fulfill this condition under uniform magnetization and derived the recipe to identify the location of the high nonreciprocity regions in the Brillouin zone. By appropriately reducing the symmetry of the motif of the crystal lattice, one can in this way artificially create strong band structure asymmetry close to "crystallographic" directions of the PhC. Fig. 1 shows a possible layout that corresponds to the cm' plane group. This precise motif symmetry with its purely rectangular features is perfectly adapted for RCWA modelling. Other symmetries (e.g. $p3m'1$) might lead to stronger effects. Its Brillouin zone along with its irreducible part is given in Fig. 1. It can be seen that due the axial character of the (uniform) magnetization and the reduced symmetry, the only symmetry present, besides the unity operator, is m'_y ($\equiv m_x$), or thus the combination of time reversal and the mirror operation perpendicular to the y-axis. As a result the irreducible part of the Brillouin zone of this hexagonal Bravais lattice is extended to half of the hexagon. For this reason the K ans K' are not equivalent anymore. The only remaining reciprocal direction is $\Gamma - M$, i.e. the k_x axis.

The right panel of Fig. 1 shows the impact of the symmetry reduction on the resulting isofrequency contours in the Brillouin zone. These have been calculated using a standard plane wave expansion technique taking into account the anisotropic nonreciprocal character of the permittivity [6]. The material and geometrical details of the crystal are detailed in a following paragraph. We have plotted a single isofrequency contour (reduced frequency $f^* = \frac{a}{\lambda} = 0.4545$) of the second photonic band at a frequency halfway those of the now non-degenerate K and K' points. As expected from the symmetry arguments elaborated in the left panel of the Figure, a local interior gap without inversion symmetry in the reciprocal space opens up. A TE plane wave at this frequency and incident under the appropriate angle ($k_y = \frac{2\pi}{3a} = \frac{2\pi}{\lambda}\frac{1}{3f^*}$) on this crystal

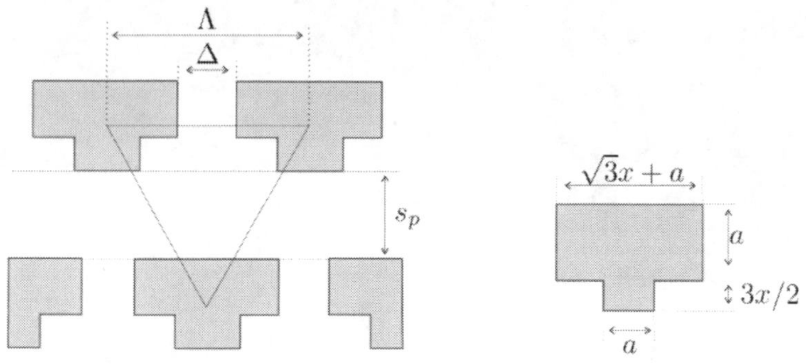

Figure 2. Scheme of photonics crystal structure.

cut along a $\Gamma - K$ direction will be totally reflected in one sense while partially transmitted in the other. This makes for an extremely compact one-way mirror that can be used as for instance a novel integrated isolator.

In the following paragraph we will numerically validate these qualitative band structure arguments using a RCWA method.

Figure 2 shows a crossection of the photonic crystal structure and its parametrization. We consider two-dimensional structure consisting of magnetic garnet with the refractive index of $n = 2,25$ for the wavelength $\lambda = 1.3$ μm with T-shaped air holes. The centers of the holes constitute a hexagonal lattice with the period Λ. Δ denotes the spacing between T-shaped elements, s_p is the thickness of separation layer between gratings layers, and a and x are the scale and the shape parameters of the element, respectively. In our notation one period of photonics crystal consists of two layers of the air holes in magnetic garnet shifted by $\Lambda/2$ and separated by the thickness s_p. The filling factor, defined as the surface fraction of the air holes, can be obtained from the geometrical parameters $f = (1 + \sqrt{3}/2 + 2/\sqrt{3})\, a^2/\Lambda^2$.

We expect that the magnetic garnet is homogeneously magnetized in x direction (perpendicular to the 2D crossection plane) and the permittivity tensor of the magnetic garnet is in the form

$$\hat{\varepsilon} = \begin{bmatrix} n^2 & 0 & 0 \\ 0 & n^2 & i\,\varepsilon_1 \\ 0 & -i\,\varepsilon_1 & n^2 \end{bmatrix}, \qquad (1)$$

where $\varepsilon_1 = 0.1$ denotes the off-diagonal magneto-optic parameter. The periodic structure is surrounded by nonmagnetic garnet (in 1 $\varepsilon_1 = 0$) in our model.

3. Modeling of photonic-crystal structure

Transmission through the finite magneto-photonic crystal was modeled using Rigorous Coupled Wave Analysis (RCWA). Optical field in the anisotropic structures containing a periodic grating can be on the basis of Bloch theorem expressed using a Fourier expansion into plane waves [7, 8, 9]. The Maxwell's equations are converted into the eigenvalue problem [7, 10, 11, 12]. Improvement of the convergence speed can be achieved by appropriate application of boundary conditions also for the permittivity tensor elements representations. This improvement for the lamellar gratings is denoted as the Li factorization rules [13, 14] or Fourier factorization [7, 8]. We found that the Fourier expansion can be truncated without loss of required precision for $\mathcal{N} = 30$ Fourier harmonics.

In the following we consider the structure with the period of $\Lambda = 590.85$ nm ($f^* = \Lambda/\lambda = 0.4545$), the angle of incidence $\varphi = 17.94$ ($N_y = n \sin\varphi = 0.693$), and the filling factor of the air holes $f = 0.4$.

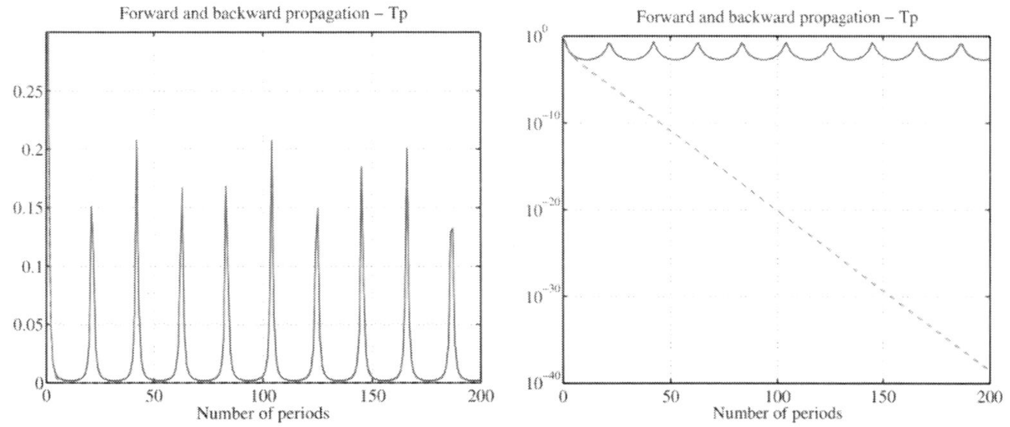

Figure 3. Transmission of the magneto-photonic crystal as a function of number of periods. Solid (blue) and dashed (red) lines show the transmission of p-polarized light in forward and backward direction, respectively. Right subplot shows transmission in logarithmic scale. The period $\Lambda = 590.85$ nm, the wavelength $\lambda = 1.3$ μm, and the angle of incidence of $\varphi = 17.94°$ (corresponding $N_y = n \sin\varphi$) were used in the calculation.

Figure 3 (left panel) shows the forward (solid blue line) and backward (dashed red line) transmission as a function of the number of periods in the photonic crystal corresponding to increased thickness of the stack. Forward transmission shows oscillatory behavior similar to Fabry-Perot resonances. However, in the backward direction, the transmission rapidly decreases. Right subplot shows linear decrease of the transmission in the logarithmic scale, which corresponds to exponential decay of optical field and photonic bandgap behavior. The first maximum of the transmission in forward direction is around 22 periods.

Figure 4 shows dependence of the forward and backward transmission for 22 periods of air holes as a function of N_y. The backward transmission is blocked in the interval from about $N_y = 0.67$ to 0.72 corresponding to photonic bandgap. However the structure is transparent in the forward direction and transmission of 30% is obtained.

Figure 5 shows calculated field (the x-component of the magnetic field H_x) inside the structure. Left subplot shows the complete structure of 22 periods in forward and backward regime. Note that the color scale is logarithmic to visualize strong decrease of the field in backward direction. In the forward direction (structure is illuminated from the top) the optical field is amplified in the structure and the structure is partially transparent (orange color on the bottom). However, in backward direction (illumination from the bottom) the field is exponentially decreasing and the photonic structure block the transmission (blue color on the top). Right subplot shows a detail of the field distribution in the central area of the structure. It is clearly visible that the field is concentrated in between air holes.

4. Conclusions

Nonreciprocal photonic crystal mirror was designed and modeled. Due to unique behavior of transverse magneto-optic effect, the photonic bandgap was obtained only in backward direction, while in the forward direction the device is transparent.

Journal of Physics: Conference Series **303** (2011) 012039 doi:10.1088/1742-6596/303/1/012039

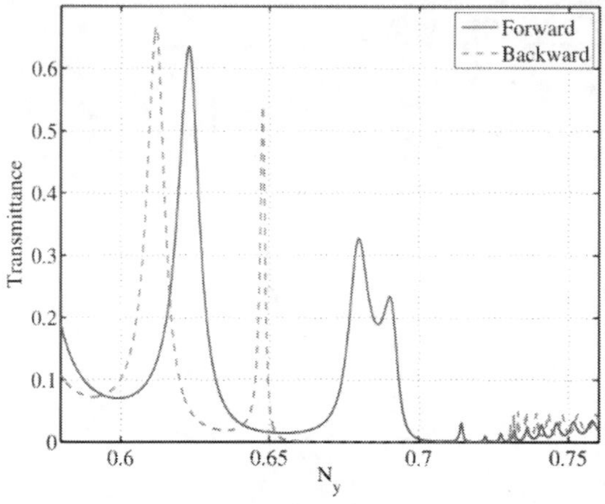

Figure 4. Transmittance of the *p*-polarized light through the photonic crystal structure as a function of $N_y = n \sin\varphi$ for 22 periods. Solid (blue) and dashed (red) lines show the transmission in forward and backward direction, respectively. In the bandgap ranging from about $N_y = 0.67$ to 0.72 the transmission in backward direction is almost blocked, while the forward transmission reach more than 30%.

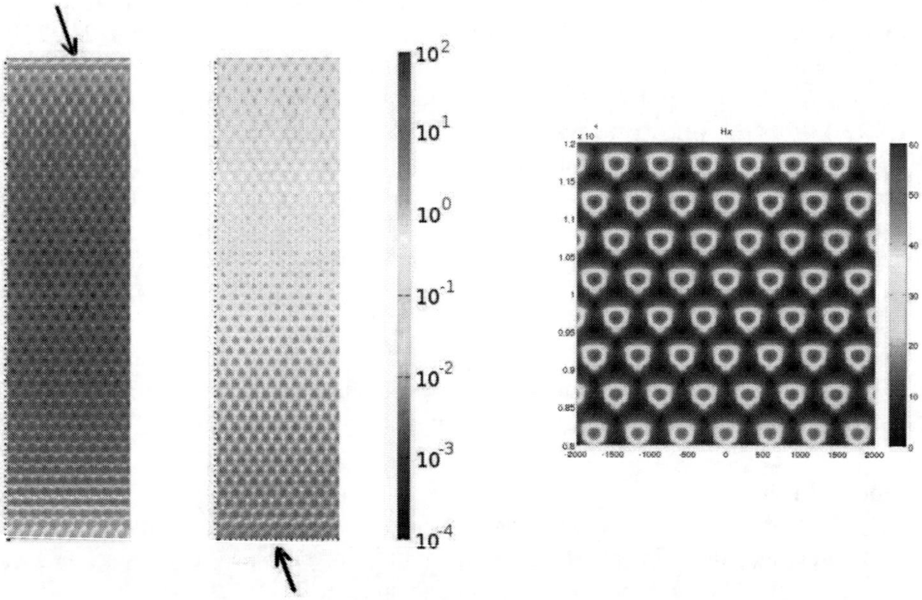

Figure 5. Field in the photonic-crystal structure

Acknowledgments

Partial support from the projects KAN 400100653 (Grant Agency of Academy of Sciences of the Czech Republic), MSM6198910016 (Ministry of Education of the Czech Republic), and CZ.1.05/2.1.00/01.0040 (RMTVC), and SP/2011132 is acknowledged.

References

[1] Dötsch H, Bahlmann N, Zhuromskyy O, Hammer M, Wilkens L, Gerhardt R, Hertel P and Popkov A F 2005 *J. Opt. Soc. Am. B* **22** 240–253
[2] Inoue M, Fujikawa R, Baryshev A, Khanikaev A, Lim P B, Uchida H, Aktsipetrov O, Fedyanin A, Murzina T and Granovsky A 2006 *J. Phys. D* **39** R151–R161
[3] Vanwolleghem M, Checoury X, Smigaj W, Gralak B, Magdenko L, Postava K, Dagens B, Beauvillain P and Lourtioz J M 2009 *Phys. Rev. B* **80** 121102(R)–1–4
[4] Figotin A and Vitebsky I 2001 *Phys. Rev. E* **63** 066609
[5] Litvin D B 2008 *Acta Crystallographica Section A: Foundations of Crystallography* **64** 419–424 ISSN urn:issn:0108-7673 URL `http://dx.doi.org/10.1107/S010876730800768X`
[6] Johnson S and Joannopoulos J 2001 *Optics Express* **8** 173 ISSN 1094-4087 URL `http://www.opticsinfobase.org/abstract.cfm?URI=oe-8-3-173`
[7] Nevière M and Popov E 2002 *Light Propagation in periodic media: Differential theory and design* (Marcel Dekker)
[8] Watanabe K, Petit R and Nevière M 2002 *J. Opt. Soc. Am. A* **19** 325–334
[9] Li L 2003 *J. Opt. A: Pure Appl. Opt.* **5** 345–355
[10] Moharam M G, Pommet D A and Grann E B 1995 *J. Opt. Soc. Am. A* **12** 1077–1086
[11] Mori S, Mukai K, Yamakita J and Rokushima K 1990 *J. Opt. Soc. Am. A* **7** 1661–1665
[12] Š Višňovský 2006 *Optics in magnetic multilayers and nanostructures* (CRC Press, Taylor & Francis)
[13] Li L 1996 *J. Opt. Soc. Am. A* **13** 1870–1876
[14] Li L 1998 *J. Mod. Opt.* **45** 1313–1334

Magnon-Photon interactions for Quantum Key Distribution

P. Kumar[1] **and A. Prabhakar**[2]

[1]Department of Electrical Engineering, I.I.T. Kanpur, Kanpur-208016, India
[2]Department of Electrical Engineering, I.I.T. Madras, Chennai-600036, India

E-mail: [1]pradeepk@iitk.ac.in [2]anilpr@iitm.ac.in

Abstract. We propose an implementation of quantum key distribution (QKD) protocol using magnon-photon interactions in dielectric waveguides. Starting from the Hamiltonian of spin wave–optical interactions in a symmetric dielectric waveguide, we derive the single-photon transformations that allow us to define two non-orthogonal basis sets, which are isomorphic to polarization basis of single-photon. Implementation of the BB84 and B92 protocols is then straightforward. The principal advantage of our scheme is the higher quantum bit error rate (QBER) (37.5%) compared to polarization-coded scheme (25%) for simple intercept/resend attack. This considerably relaxes the specifications of optical components. Finally, we consider the effect of low conversion efficiency on QBER.

1. Introduction

Spin waves (SWs) are excitations in magnetic materials. SW–optical interactions in waveguides, specifically in heterostructures made of magnetic material such as YIG, is used in applications such as optical isolators, mode converters, filters, frequency shifters etc., and has been widely studied [1–6]. More recently, there has been a lot of work on spin wave propagation in metallic thin films. However, the magnon-photon interactions in metallic films are typically based on Brillouin scattering (BLS). In this paper, we describe the use of collinear propagating optical and spin waves to implement BB84 and B92 quantum key distribution (QKD) protocols. We also estimate the quantum bit error rate (QBER) of the proposed scheme, demonstrating how a similar estimate can be obtained in other magnon-photon interactions.

In the BB84 QKD protocol [7], key bits, $\{0,1\}$ are encoded as one of the four non-orthogonal states–$\{|0\rangle, |1\rangle, |+\rangle, |-\rangle\}$, grouped into two basis sets, $\mathcal{B}_1 = \{|0\rangle, |1\rangle\}$ and $\mathcal{B}_2 = \{|+\rangle, |-\rangle\}$ such that,

$$\langle 0|1\rangle = \langle +|-\rangle = 0, \quad \text{and} \quad \langle i|j\rangle \neq 0, \; i = \{0,1\}, \; j = \{+,-\}.$$

In the B92 protocol [8], key bits are encoded using only two non-orthogonal states, say $\{|0\rangle, |+\rangle\}$. SW–optical interactions provide a convenient way to generate the non-orthogonal states by coupling the TE and TM optical modes. The TE$\rightleftharpoons$TM mode coupling has been studied both theoretically and experimentally in the classical regime [6, 9]. Recently, we extended the treatment to the interaction of SWs with quantized optical TE and TM modes [10] and described magnon-photon interactions as:

$$|\omega_0 - \Omega\rangle_{\rm TM} \to \frac{1}{\sqrt{2}}\left(|\omega_0\rangle_{\rm TE} - |\omega_0 - \Omega\rangle_{\rm TM}\right) \equiv |-\rangle$$
$$|\omega_0\rangle_{\rm TE} \to \frac{1}{\sqrt{2}}\left(|\omega_0\rangle_{\rm TE} + |\omega_0 - \Omega\rangle_{\rm TM}\right) \equiv |+\rangle\,,$$

where Ω is the angular frequency of SWs (rad/m). These states are formally isomorphic to the states of polarization of a single-photon and can therefore be used to implement both BB84 and B92 QKD protocols. The principal advantages of our proposed scheme are:

- improved key rate since SW–Optical interaction occurs at over 25GHz,
- single-sideband scheme without use of optical filtering at the transmitter as opposed to the schemes that use electro-optic modulators, and
- a QBER of 37.5% for intercept/resend attack, which relaxes specifications on the optical components.

These advantages are offset to somewhat from a low TE⇌TM mode conversion efficiency which introduces an "intrinsic" QBER.

2. SW–Optical interactions in a waveguide

The optical modes of the symmetric waveguide can be quantized starting with Maxwell equations and following a canonical procedure leading to the Hamiltonian,

$$H_0 = \sum_m \hbar\omega_m \left(a_{xm}^\dagger a_{xm} + a_{ym}^\dagger a_{ym}\right), \tag{1}$$

where the first and second terms represent transverse magnetic (TM) and transverse electric (TE) modes respectively [11]. In the collinear geometry, spin waves (SWs) propagating parallel to the optical modes lead to TE⇌TM mode conversion [6, 12]. When the SW has the form, $\mathbf{M}(t) = M_S\hat{\mathbf{x}} + m_y\hat{\mathbf{y}} + m_z\hat{\mathbf{z}}$, where the small signal components are time varying, and M_S is the saturation magnetization, the SW–optical interaction can be modeled by the effective permittivity tensor,

$$\bar{\epsilon} = \epsilon_0 \begin{pmatrix} \epsilon_r & -jfm_z & jfm_y \\ jfm_z & \epsilon_r & jfM_S \\ -jfm_y & jfM_S & \epsilon_r \end{pmatrix},$$

where we have assumed that the medium is isotropic in the absence of SW excitation and $m_y, m_z \ll M_s$. Assuming that TE ⇌TM mode conversion is due only to the SW–optical interactions, the interaction Hamiltonian is given by,

$$H_{\rm int} = \int_V dv \frac{1}{2}\epsilon_0(-jfm_z E_x^* E_y),$$

where $m_z(t) = m_{0z}(x)\exp(j(\mathbf{k}_{\rm SW}\cdot\mathbf{r} - \Omega t))$ is the SW and $\mathbf{k}_{\rm SW}$ and Ω are the SW vector and angular frequency respectively. Replacing the classical fields by the respective quantized mode operators and carrying out the integration, we obtain,

$$H_{\rm int} = j\hbar \sum_{m,n}\left(\gamma_{mn} a_{ym}^\dagger a_{xn} - \gamma_{mn}^* a_{xm}^\dagger a_{yn}\right), \tag{2}$$

where γ_{mn} is the coupling coefficient between TE_m and TM_n modes. In writing (2), we have assumed energy and momentum conservation relations,

$$\omega_{\rm TE}^{(m)} - \omega_{\rm TM}^{(n)} \mp \Omega = 0, \text{ and } \left(\mathbf{k}_{\rm TE}^{(m)} - \mathbf{k}_{\rm TM}^{(n)} \mp \mathbf{k}_{\rm SW}\right)\cdot\hat{\mathbf{z}} = 0.$$

Under the *tight-binding approximation* [12], we can write (2) as

$$H_{\text{int}} = j\hbar \left(\gamma \left| x \right\rangle \left\langle y \right| - \gamma^{*} \left| y \right\rangle \left\langle x \right| \right), \tag{3}$$

where $|x\rangle$ and $|y\rangle$ denote the TM and TE modes respectively and we have dropped the subscript m for notational simplicity. A general state vector,

$$|\psi(t)\rangle = A(t)\,|x\rangle + B(t)\,|y\rangle\,, \quad |A(t)|^2 + |B(t)|^2 = 1 \;\; \forall t,$$

evolves according to the Schrödinger's equation:

$$j\hbar \frac{\partial}{\partial t} |\psi(t)\rangle = H_{\text{int}} |\psi(t)\rangle\,. \tag{4}$$

Using (3), the general solution of the coefficients, $A(t)$ and $B(t)$ are

$$A(t) = C\cos(\kappa' t) + D\sin(\kappa' t), \tag{5a}$$

$$B(t) = e^{-j\phi}\left[C\sin(\kappa' t) - D\cos(\kappa' t)\right], \tag{5b}$$

where $\gamma = \kappa' e^{j\phi}$, and C and D are constants which depend on the initial conditions and we have absorbed $j\hbar$ into the definition of γ. Taking $\phi = 0$, the TE and TM mode conversions are indicated in Table. 1.

Initial state ($t = 0$)	Final state ($t = T$)
TE, $\|\omega_0\rangle$	$\cos(\kappa' T)\,\|\omega_0\rangle - \sin(\kappa' T)\,\|\omega_0 - \Omega\rangle$
TM, $\|\omega_0 - \Omega\rangle$	$\sin(\kappa' T)\,\|\omega_0\rangle + \cos(\kappa' T)\,\|\omega_0 - \Omega\rangle$

Table 1. TE and TM mode conversions.

3. Implementing BB84 and B92 protocols

Adjusting the values of coupling coefficient so that $\kappa' T = \pi/4$, we obtain the transformations:

$$|\omega_0 - \Omega\rangle_{\text{TM}} \rightarrow \frac{1}{\sqrt{2}}\left(|\omega_0\rangle_{\text{TE}} - |\omega_0 - \Omega\rangle_{\text{TM}}\right) \equiv |-\rangle \tag{6a}$$

$$|\omega_0\rangle_{\text{TE}} \rightarrow \frac{1}{\sqrt{2}}\left(|\omega_0\rangle_{\text{TE}} + |\omega_0 - \Omega\rangle_{\text{TM}}\right) \equiv |+\rangle\,, \tag{6b}$$

where we have explicitly identified the TE and TM modes. We define two non-orthogonal bases, $\mathcal{B}_1 = \{|\omega_0\rangle\,, |\omega_0 - \Omega\rangle\}$ and $\mathcal{B}_2 = \{|+\rangle\,, |-\rangle\}$ which transform a second time after interacting with the SWs to

$$|+\rangle \rightarrow |\omega_0\rangle\,, \;\text{ and }\; |-\rangle \rightarrow -|\omega_0 - \Omega\rangle\,. \tag{7}$$

This is equivalent to measuring the kets $|+\rangle$ and $|-\rangle$ in the "diagonal" basis $\mathcal{B}_2$. Detecting without remodulating $|\omega_0\rangle$ and $|\omega_0 - \Omega\rangle$ is equivalent to measuring them in the basis $\mathcal{B}_1$. Thus, the states in bases $\mathcal{B}_1$ and $\mathcal{B}_2$ are isomorphic to the state of polarization of a photon and are suited to implement both BB84 and B92 protocols.

Fig. 1 shows the transmitter and receiver structures to implement BB84 protocol. We first fix the encoding scheme. In basis $\mathcal{B}_1$ and $\mathcal{B}_2$, bit 0 is encoded by $|\omega_0\rangle$ and $|+\rangle$, and bit 1 is encoded by $|\omega_0 - \Omega\rangle$ and $|-\rangle$, respectively.

- Alice picks a basis $\mathcal{B}_1$ or $\mathcal{B}_2$ at random with equal probability. She does so by choosing between the two lasers emitting photons at frequencies ω_0 and $\omega_0 - \Omega$, polarized TE and TM respectively as shown in Fig. 1.
- Alice transmits the key bit 0 or 1 in the chosen basis. She does so by choosing to modulate or not the photons. Modulating the photons generates the states in basis $\mathcal{B}_2$ as given in Table. 1. This is shown in Fig. 1 where a RNG drives the VCO of frequency Ω rad/s. The setup is similar to the classical TM⇌TE conversion as described in the literature [6, 12].
- Alice transmits the state generated in the above step over an optical fiber.
- Bob uses a filter, say Fabry-Perot fibre grating, to separate the photons into two channels ω_0 and $\omega_0 - \Omega$.
- Bob corrects for polarization fluctuations in each channel using a polarization compensation unit (PCU), re-combines them and recovers the state transmitted by Alice.
- Bob modulates the state only half the time, but at random. He is thus choosing between basis $\mathcal{B}_1$ and $\mathcal{B}_2$ at random.
- Bob separates the photons, again using a filter, and feeds it to two detectors D_0 and D_1.
- Bob communicates to Alice whether he modulated the photons, or not, as bits 1 and 0. However, he does not communicate the results of his detection.
- Alice and Bob retain the time slots in which the basis match to form the sifted bits.
- Alice and Bob perform privacy amplification and error correction on the sifted bits to arrive at the final key.

Alice and Bob retain only 4 out of the 8 possible cases, i.e., the link has a 50% efficiency. These four cases are summarized in Table. 2. Note however that the QBER is estimated not on the full system, but after sifting, i.e., on these 4 cases. To implement B92 protocol, Alice and Bob encode the bits as two non-orthogonal states, say $|\omega_0\rangle$ and $|+\rangle$.

Table 2. Possible cases in the SW–optical interaction based BB84 protocol.

Alice's basis	$\mathcal{B}_2$	$\mathcal{B}_2$	$\mathcal{B}_1$	$\mathcal{B}_1$
Alice's bit	0	1	0	1
Txd state	$\|+\rangle$	$\|-\rangle$	$\|\omega_0\rangle$	$\|\omega_0 - \Omega\rangle$
Bob modulation?	Y	Y	N	N
Bob's state	$\|\omega_0\rangle$	$-\|\omega_0 - \Omega\rangle$	$\|\omega_0\rangle$	$\|\omega_0 - \Omega\rangle$
$\langle\omega_0\rangle$	1	0	1	0
$\langle\omega_0 - \Omega\rangle$	0	1	0	1
Sifted bits	0	1	0	1

4. QBER for intercept/resend attack

In a conventional intercept/resend attack, Eve intercepts a subset of the quantum bits transmitted by Alice, measures them in her chosen basis, and transmits to Bob a fabricated pulse in the same state as her result. In doing so, she introduces errors in the sifted bits, characterized by quantum bit error rate (QBER). It can be shown that Eve introduces a QBER of 25% in polarization coded scheme, if her basis is the same as Alice. Assuming that Eve uses the same basis as Alice and Bob, we can determine the QBER of our scheme subject to intercept/resend attack. We note the following:

- Alice is transmitting in basis $\mathcal{B}_1$. If Eve does not modulate the incoming photon, she does not change the state. However, she must still subject the intercepted photons to detectors

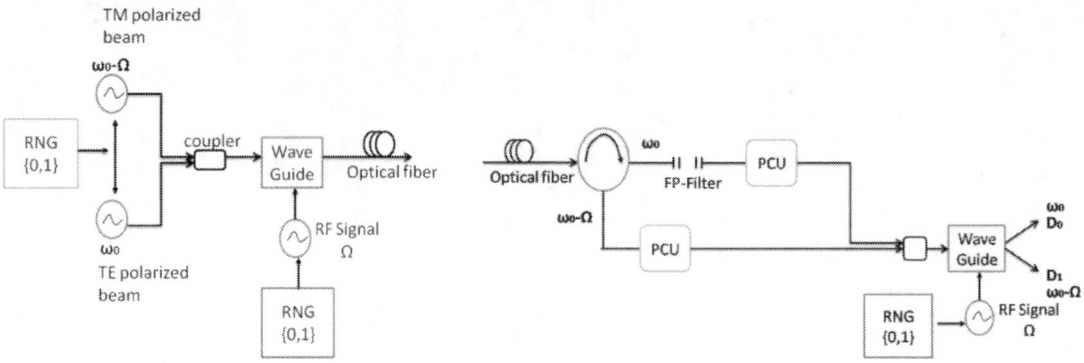

Figure 1. Transmitter and receiver to implement BB84 protocol. RNG=random number generator, FP=Fabry-Perot, D_0 and D_1 are single-photon detectors [10]. (©IEEE.)

D_0 and D_1 in order to distinguish between $|\omega_0\rangle$ and $|\omega_0 - \Omega\rangle$. For the intercepted pulse, Eve transmits a photon in the same state that she detects.

- Alice is transmitting in basis $\mathcal{B}_2$. If Eve does not modulate the incoming photon, she collapses the state into one of the sidebands ω_0 and $\omega_0 - \Omega$. She then transmits in the same state that she detects. On the other hand, if she modulates the incoming photon, she again collapses the state into one of the sidebands, depending on the state Alice has transmitted.
- Thus, in both cases, Eve effectively transmits in basis $\mathcal{B}_1$. This is completely different from the polarization-coded setup in which Eve sends fabricated pulse in both the bases.

In Fig. 2, we show the results of Eve's intercept/resend attack. We have not indicated the outputs when Alice and Bob's bases differ, since they do not contribute to the sifted key. Similar results are obtained if Alice transmits the remaining kets: $|\omega_0\rangle$ and $|+\rangle$. We see that Eve's intercept/resend attack introduces a QBER of $\simeq 37.5\%$ in the sifted bit string. This is different from the QBER of polarization coded scheme with intercept/resend attack. The difference is due to the fact that Eve introduces errors even if she chooses the same basis as Alice(See Fig. 2). Further, we can show that any other strategy for Eve will result in higher error probability than 37.5%. For example, if Eve sends Bob a modulated photon whenever she modulates the photon from Alice, she introduces error of at least 50%. Thus, the intercept/resend strategy described in Fig. 2 represents the minimum error probability introduced by Eve as long as she uses the same basis as Alice and Bob.

It is instructive to consider the effect of low conversion efficiency on the BEP of the scheme. Due to low conversion efficiency, Alice will not be able to prepare states in $\mathcal{B}_2$ by modulating the states in $\mathcal{B}_1$. We take the worst case scenario in which Alice cannot transform the states $|\omega_0 - \Omega\rangle$ and $|\omega_0\rangle$ into the states $|+\rangle$ and $|-\rangle$ respectively. The "intrinsic" QBER, when Alice sends $|\omega_0 - \Omega\rangle$ instead of the ideal state $|+\rangle$ can be seen to be $\simeq 33.33\%$ from Table. 3. We note that ideally, Bob would transform the states $|\omega_0 - \Omega\rangle$ and $|\omega_0\rangle$ into $|+\rangle$ and $|-\rangle$. However, due to low conversion efficiency the state remains unchanged.

Summary

Spin wave excitations in magnetic thin films are a well documented phenomena. While optical modulation using spin waves was also researched, devices that leveraged the phenomena could not compete with electro-optic modulators. The unique feature of magnon-photon interaction is its single-sideband nature at GHz frequencies. This is the primary requirement in FC-QKD systems and we have quantified the secure nature of the protocol, against an Intercept-Resend

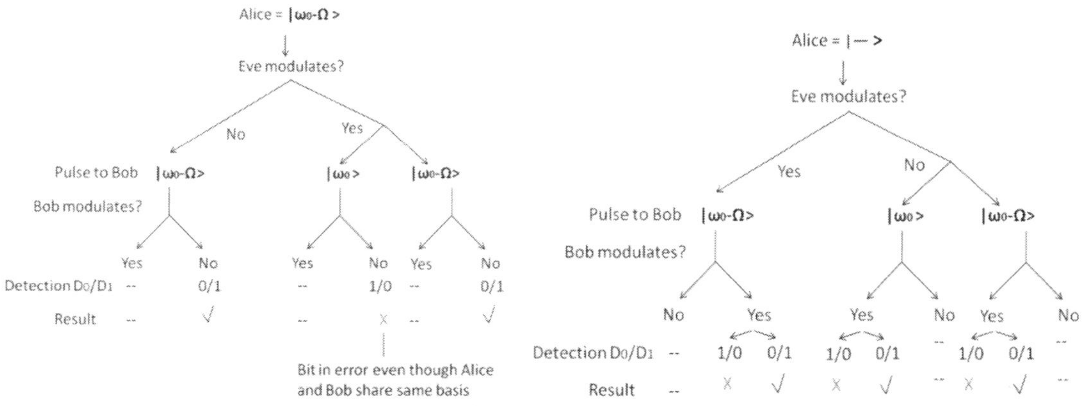

Figure 2. Intercept/Resend attack by Eve.

Table 3. Intrinsic error in sifted key due to low conversion efficiency. Alice and Bob retain slots only when Bob modulates the photon.

Alice $= \|\omega_0 - \Omega\rangle$			
Eve modulates?	N	Y	
Pulse to Bob	$\|\omega_0 - \Omega\rangle$	$\|\omega_0 - \Omega\rangle$	$\|\omega_0\rangle$
Bob modulates?	N Y	Y N	Y N
Detection $\omega_0/\omega_0 - \Omega$	– 0/1	0/1 –	1/0 –
Result	– ×	× –	√ –

attack. We hope this encourages further development of magnon-photon interaction systems in transverse and suface spin wave geometries, in both insulating and metallic thin films.

Acknowledgments

The authors acknowledge financial support from the Department of Science and Technology, Govt. of India under the India-EU collaborative project "DYNAMAG" (grant number INT/EC/CMS(24/233552))

References

[1] Fisher A D Lee J N Gaynor E S and Tveten A B 1982 *Appl. Phys Lett.* **41** 779
[2] Tsai C S Young D Chen W Adkins L Lee C C and Glass H 1985 *Appl. Phys. Lett.* **47** 651
[3] Tamada H Kaneko M and Okamoto T 1988 *J. Appl. Phys.* **64** 554
[4] Bilaniuk N Stancil D D Talisa S H 1990 *J. of Applied Phys.* **67** 1 508-10
[5] Stancil D D 1991 *IEEE J. of Quant. Electron.* **27** 1 61-70
[6] Prabhakar A and Stancil D D 1996 *IEEE Trans., on Magnetics* **32** 3 191823
[7] Bennett C H and Brassard G 1984 *Proc., of IEEE Int., Conf., on Computers Systems and Signal Processing* Bangalore 175-9
[8] Bennett C H 1992 *PRL.* **68** 3121-24
[9] Butler J C 2007 *Optical Engineering* **46** 127-203
[10] Kumar P and Prabhakar A 2010 *IEEE J. of Quant. Electron.* **46** 11 1542-48
[11] Żakowicz, Władysław and Janowicz, Maciej 1995 *Phys. Rev. A* **52** 2 1640-50
[12] Stancil D D and Prabhakar A 2009 *Spin Waves: Theory and Applications* (Springer) chapter 8

Ferromagnetic nanoparticular films studied by optical and magneto-optical ellipsometry

T. Fördös,[1] O. Životský,[1] K. Postava,[1] D. Hrabovský,[1] J. Pištora,[1] J. Lančok,[2] M. Miglierini,[3] M. Klementová[3]
[1]Department of Physics, Technical University of Ostrava, 708 33 Ostrava, Czech Republic
[2]Institute of Physics, AS CR, Na Slovance 2, 182 21 Prague, Czech Republic
[3]Institute of Inorganic Chemistry of the AS CR, 250 68 Husinec-Řež 1001, Czech Republic
E-mail: kamil.postava@vsb.cz, lancok@fzu.cz

Abstract. Fe and Co nanoparticles embedded in SiO_2 matrix were prepared using dual radiofrequency magnetron sputtering. Films were characterized by high resolution transmission electron microscopy (HRTEM). Magnetic properties were measured using longitudinal magneto-optical Kerr effect. Magneto-optical hysteresis loop measurements show differences between Fe particular film with strong in-plane uniaxial anisotropy and Co one, which is almost isotropic in the plane of the film. Nanoparticular film thicknesses and volume fractions of particles were obtained using spectroscopic ellipsometer UVISEL. Ellipsometric data were fitted to the model based on the Bruggeman effective medium approximation.

1. Introduction

Ferromagnetic metal particles in dielectric matrix possess interesting magnetic and optical properties [1, 2]. Self-organized single crystal metal particles can be produced using three-dimensional (3D) epitaxial growth [3], which is promising technique for preparation of single-domain particular nanostructure, magnetic metamaterials, and magnetophotonic crystals. On the other hand optical and magneto-optical ellipsometry are very sensitive nondestructive methods suitable for study of magnetic anisotropy and magnetization reversal properties of nanostructured systems [4, 5]. Magneto-optic vector magnetometry enables measurement of the magnetization components in the structure during magnetic reversal.

In this paper, we describe preparation and characterization of layers consisting of the ferromagnetic Co and Fe single crystal particles embedded in SiO_2 matrix. Magnetic properties of the particular layers are studied using magneto-optical ellipsometry and optical properties of the layers are measured using spectroscopic ellipsometry. Section 2 deals with brief description of the sample structure, preparation technique, and structure of obtained particles. Magneto-optic setup and ellipsometer are described. Main results are summarized in Sec. 3.

2. Experimental

2.1. Samples preparation

Nanoparticular Co and Fe films were prepared by using dual radiofrequency (RF) magnetron sputtering with alternating targets. The vacuum chamber was evacuated down to 6×10^{-4} Pa and filled with Ar of pressure 1 Pa. Figure 1 shows structure of the samples. First, the SiO_2 film of

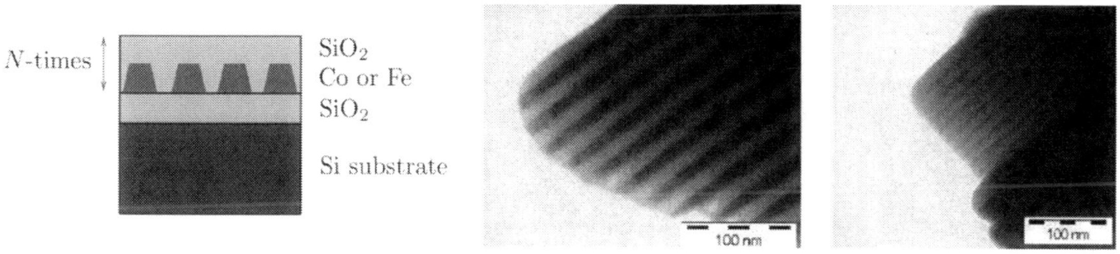

Figure 1. Left subplot shows schematically the layer structure of the samples. Observations of Fe multilayer structure – 10 nm of SiO_2 (light layers) sandwiched with 10 nm (left TEM image) and 3 nm (right TEM image) layers of Fe nanoparticles embedded in SiO_2 matrix.

nominal thickness of 10 nm was deposited on the single-crystal Si substrate at room temperature. Particles of ferromagnetic metal was formed on the top of the SiO_2 film. The system of particles were covered by SiO_2, which filled interparticle volume, capped the nanoparticular layer, and smoothed the top surface. We have prepared systems with one ($N = 1$) and ten ($N = 10$) nanoparticular layers (see Fig. 1). Nominal effective thickness of the ferromagnetic nanoparticular layers are 3, 5, and 10 nm. The growth rates were 2.22 nm/min, 3.76 nm/min, and 1.48 nm/min for Fe, Co, and SiO_2, respectively. The HRTEM observations were made by using JEOL JEM 3010 (300 kV, LaB_6 cathode and 1.7 Å point resolution) with an energy dispersive X-ray detector. As one can see in Figure 2, Fe particles are joined to each other, while Co particles are mostly completely surrounded by host SiO_2. Average size of the Co particles and Fe clusters determined from HRTEM images is 8.4 nm and 12.6 nm, respectively, for the 3 nm thick layers.

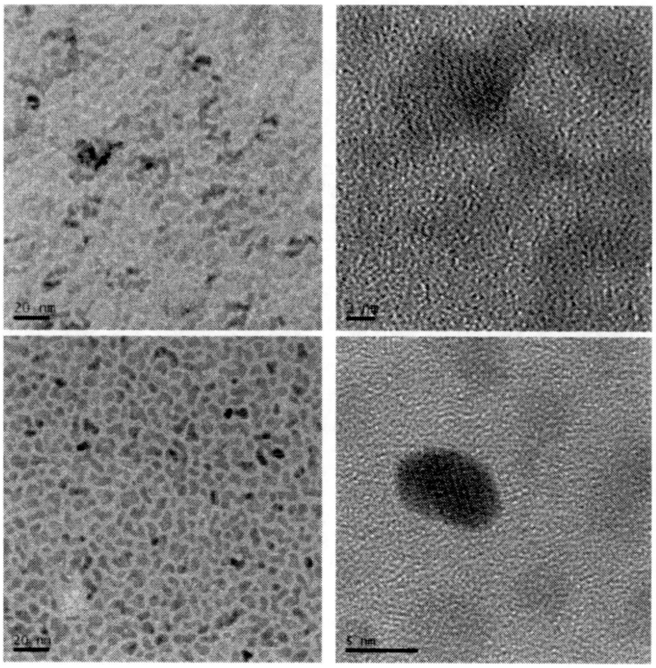

Figure 2. Observations of 3 nm layers of Fe nanoparticles (in the above pictures) and Co nanoparticles (in the below pictures) embedded in SiO_2 matrix.

2.2. Magneto-optical ellipsometry

Magneto-optical ellipsometry is a very sensitive measurement technique based on a polarization change of light reflecting by magnetic sample [4]. We can describe magneto-optical response of a sample by using the Jones reflection matrix

$$R = \begin{bmatrix} r_{ss} & r_{ps} \\ r_{sp} & r_{pp} \end{bmatrix} \tag{1}$$

where r_{ij} is the amplitude reflection coefficient. Now, two magneto-optical angles can be defined for incident s- and p-polarized light using approximation of small angles

$$\frac{r_{sp}}{r_{ss}} = \frac{\tan\,\theta_s + i\,\tan\,\epsilon_s}{1 - i\,\tan\,\theta_s\,\tan\,\epsilon_s} \approx \theta_s + i\,\epsilon_s \tag{2}$$

$$\frac{r_{ps}}{r_{pp}} = \frac{\tan\,\theta_p + i\,\tan\,\epsilon_p}{1 - i\,\tan\,\theta_p\,\tan\,\epsilon_p} \approx \theta_p + i\,\epsilon_p \tag{3}$$

In this paper we use the s-polarized incident light, thus the reflected light is polarized elliptically with the ellipticity ϵ_s and the azimuth rotation θ_s.

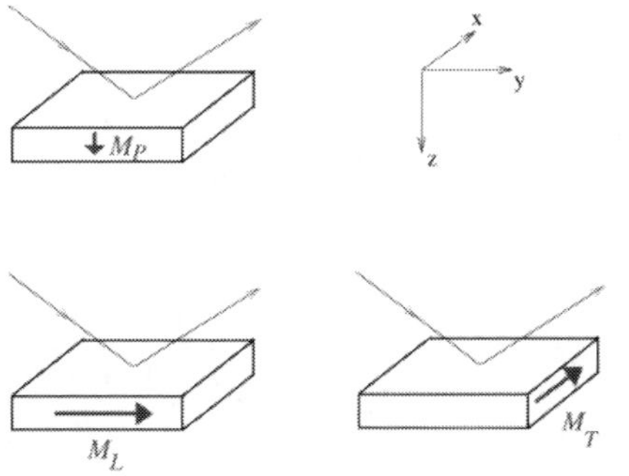

Figure 3. Basic MO configurations – polar, longitudinal and transverse component of magnetization vector.

Figure 3 shows three components of the magnetization vector. However, their contributions (frequently polar and longitudinal) can be mixed together. Therefore it is very important to make good separation of different contributions. One of the ideas is separating by using 180° sample and magnetic field rotation. This idea is based on fact, that the polar contribution is symmetric, while longitudinal is antisymmetric [6]. Thus, separation can be done by addition and substraction of measured data.

Our experimental set-up consists of semiconductor laser working at wavelength of $\lambda = 670$ nm, polarizer, quarter-wave plate as a retarder, Wollaston prism, and two photodiodes. The method is so called differential intensity method. We have measured longitudinal and transversal hysteresis loops for incident s-polarized light as a function of the external in-plane magnetic field H. The angle of incidence was set to $\varphi = 60°$ and the maximum external magnetic field of $H_{max} = 1$ kOe was applied.

2.3. Optical ellipsometry

Ellipsometric characterization is based on the change of the incident light polarization by a sample. Classical ellipsometry enables to measure the amplitude and the phase of the complex reflection coefficients ratio

$$\frac{r_{pp}}{r_{ss}} = \tan\,\psi\,\mathrm{e}^{i\,\Delta} \tag{4}$$

where r_{pp} and r_{ss} are the Fresnel reflection coefficients for p- and s-polarized light and ψ, Δ are the ellipsometric angles.

We use the ellipsometer UVISEL produced by Jobin-Yvon Horiba company, which is based on phase modulation technology and covers the spectral range from 190 nm to 2100 nm. Our ellipsometric data were measured at the angle of incidence of 70° and were fitted to the model based on the Maxwell-Garnett and Bruggeman effective medium approximation describing nanoparticular films.

3. Results

Longitudinal and transverse hysteresis loops of single layer (N=1) of Co and Fe particles were measured for different sample azimuth angles. Figure 4 shows the hysteresis loops for easy and hard axis, and the polar plot of remanence effect. Clear two-fold in-plane uniaxial anisotropy confirms that coherent rotation of the magnetization dominates.

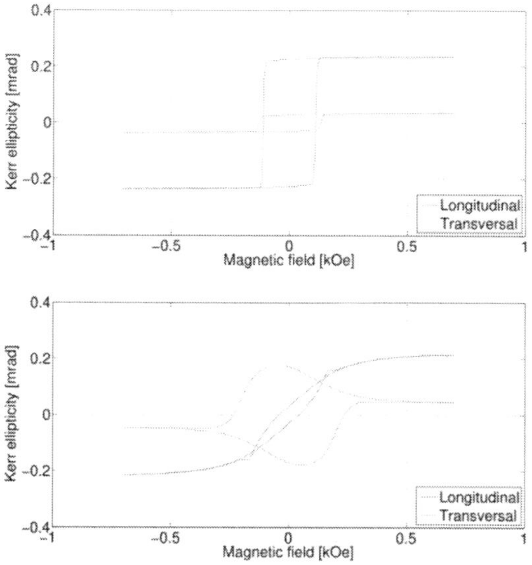

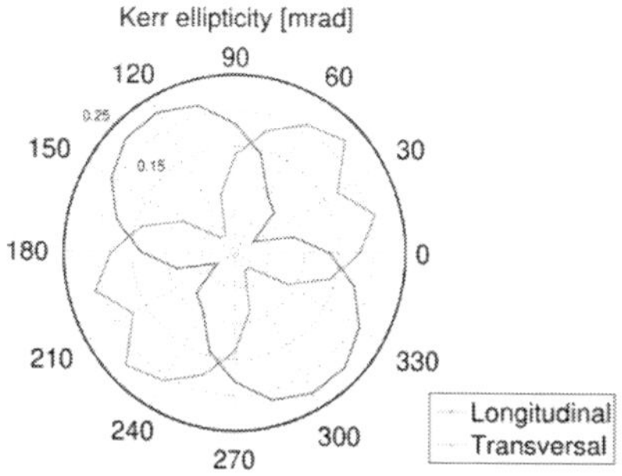

Figure 4. Hysteresis loops for magnetic field along in-plane easy (ϕ = 120°) and hard axis (ϕ = 30°) of Fe 3 nm thick monolayer in longitudinal and transversal directions. Right figure shows corresponding polar plot.

In contrast to the Fe sample, the Co nanoparticular film is almost magnetically isotropic, which is evident from the polar remanence plot shown in Fig. 5. Also the shape of the longitudinal hysteresis loop is almost independent on the in-plane magnetic field direction. The origin of this magnetic isotropy is associated with the random distribution of the magnetic properties of nanoparticles, which is also confirmed by rounded shape of the longitudinal hysteresis loop (see Fig. 5).

We have studied multilayer samples (N = 10) using spectroscopic ellipsometry to increase sensitivity to optical properties of ferromagnetic particular layers. Complex refractive indexes of the Co and Fe particles embedded in SiO_2 matrix were modeled using the Maxwell-Garnett and Bruggeman effective medium approximations. The Maxwell-Garnett approach describes a system of very diluted collections of small particles completely surrounded by host material [7]. Let us consider the mixture where spherical inclusions, much smaller than the optical wavelength, with permittivity ϵ_i occupy random positions in the environment of permittivity ϵ_e. Then the Maxwell-Garnett mixing formula can be written as

$$\frac{\epsilon_{eff} - \epsilon_e}{\epsilon_{eff} + 2\epsilon_e} = f\frac{\epsilon_i - \epsilon_e}{\epsilon_i + 2\epsilon_e} \qquad (5)$$

where ϵ_{eff} is the effective permittivity and f denotes the volume fraction of the inclusions in the mixtures. The Bruggeman approach describes a composite of aggregated phases or random-

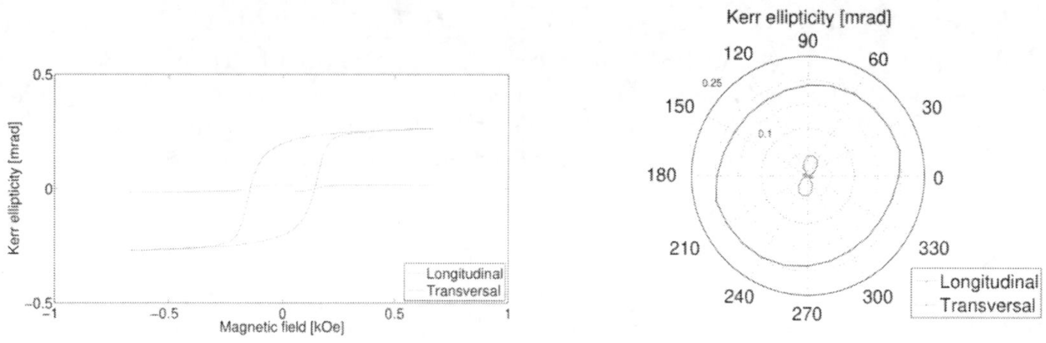

Figure 5. In-plane hysteresis loops ($\phi = 30°$) of 3 nm thick Co monolayer in longitudinal and transversal directions. Right figure shows corresponding polar plot.

mixture structures. The essence is the absolute equality between the phases in the mixture. The Bruggeman formula can be written as

$$(1-f)\frac{\epsilon_e - \epsilon_{eff}}{\epsilon_e + 2\epsilon_{eff}} + f\frac{\epsilon_i - \epsilon_{eff}}{\epsilon_i + 2\epsilon_{eff}} = 0 \qquad (6)$$

We have applied a simple model for multilayer sample ($N = 10$) consisting of the 3 nm thick Fe films by using basic optical constants of Fe and SiO_2 [8, 9]. As can be seen in Figure 6, the Bruggeman model is more appropriate for Fe particles. The reason is associated with the fact, that most of the nanoparticles are joined to each other. The obtained thicknesses are: $t_{SiO_2} = 11.5$ nm, $t_{Fe} = 7.3$ nm and the volume fraction of Fe is $f_{Fe} = 0.36$. On the other hand, ellipsometric data of Co particular film can be sufficiently well fitted also using the Maxwell-Garnett model, which indicates that the particles are more separated.

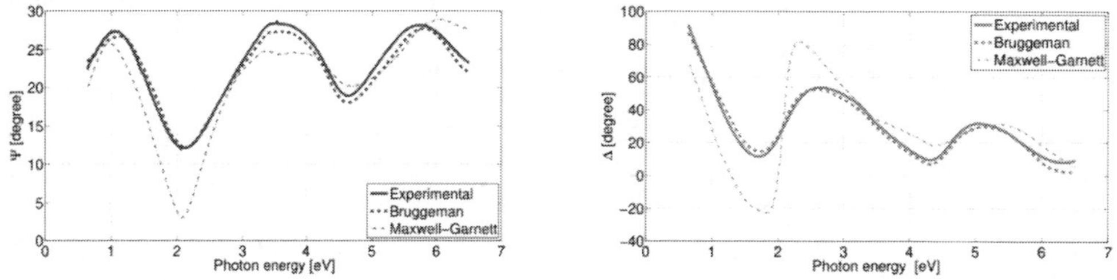

Figure 6. Ellipsometric angles ψ and Δ of 3 nm thick Fe multilayers. Comparison between the Maxwell-Garnett model and the Bruggeman model shows that the later fits the data better.

4. Conclusions

Ferromagnetic nanoparticles of Co and Fe embedded in SiO_2 matrix were prepared using the RF magnetron sputtering with alternating targets. Magnetic nanoproperties were studied using the magneto-optic vector magnetometry. In the case of Co nanoparticles smooth hysteresis curve was observed almost independent on magnetic field direction, which indicate randomly oriented magnetic moments of the particles. Strong in-plane uniaxial magnetic anisotropy was observed in the case of the Fe nanoparticles. Origin of the anisotropy is going to be a subject of further study. Also applicability of the Bruggenmam effective medium approximation to fit the ellipsometric data indicates particle coalescence.

Acknowledgements

Partial support from the projects KAN 400100653 (Grant Agency of Academy of Sciences of the Czech Republic), MSM6198910016 (Ministry of Education of the Czech Republic), and CZ.1.05/2.1.00/01.0040 (RMTVC), and SP/2010150 is acknowledged.

References

[1] Terris B D and Thomson T 2005 *J. Phys. D: Appl. Phys.* **38** R199–R222
[2] Pawlak D A 2008 *Scientia Plena* **4** 014801
[3] Zhou H, Kumar D, Kvit A, Tiwari A and Narayan J 2003 *J. Appl. Phys.* **94** 4841–4846
[4] Postava K, Hrabovský D, Hamrlová J, Pištora J, Wawro A, Baczewski L T, Sveklo I and Maziewski A 2011 *Thin Solid Films* (to be published)
[5] Postava K, Hrabovský D, Životský O, Pištora J, Dix N, Muralidharan R, Caicedo J M, Sánchez F and Fontcuberta J 2009 *J. Appl. Phys.* **105** 07C124 ISSN 0021-8979
[6] Ding H F, Pütter S, Oepen H P and Kirschner J 2001 *Phys. Rev. B* **63** 134425–1–134425–7
[7] Sihvola A 2000 *Electromagnetic Mixing Formulas and Applications* IEE Publication Series (Institution of Engineering and Technology (IET))
[8] Palik E D (ed) 1991 *Handbook of Optical Constants of Solids I, II, III* (Academic Press)
[9] Johnson P B and Christy R W 1974 *Phys. Rev. B* **9** 5056–5070

Magneto-optical and optical properties of Pt/Co GMR films with capping layer

T. Ishibashi[1], M. Naganuma[1], S. Tang[1], T. Honma[1], T. Komatsu[1], K. Machida[2], K. Furukawa[3], K. Aoshima[2], N. Funabashi[2], K. Kuga[2], N. Shimidzu[2]

[1]Nagaoka University of Technology, 1603-1 Kamitomioka, Nagaoka, Niigata 940-2188, Japan
[2]Science & Technology Research Lab., Japan Broadcasting Corporation, 1-10-11 Kinuta Setagaya, Tokyo 157-8510, Japan
[3]Tokyo Denki Univ., 2-2 Kanda-Nishiki-cho, Chiyoda-ku, Tokyo 101-8457, Japan

E-mail: t_bashi@mst.nagaokaut.ac.jp

Abstract. As a light modulating device used in high resolution magneto-optical spatial light modulator (MO-SLM), giant magneto resistance (GMR) devises driven by spin-injection is promising. In this paper, we report on optical and magneto-optical (MO) properties of GMR films, Si/SiO_2/bottom electrode/TbFeCo (9 nm)/Cu (4 nm)/(Pt/Co (6 nm))/X (X = Ta, Au, Pt (3 nm)). Kerr rotation spectra of those films with Ta, Au, Pt capping layer ranging between 260 - 700 nm have features similar to those of Pt/Co multilayers reported previously. However, peak positions and rotation angles are slightly different; peak positions of Ta, Au, and Pt are 4.1, 4.3 and 3.8 eV, respectively, and maximum Kerr rotations are 0.35, 0.25 and 0.23°, respectively. Optical constants measured by ellipsometry revealed that those values were changed due to the capping layers with a thickness of 3 nm. These results indicate that the non-magnetic layers should be mentioned as well as the magnetic layers to design MO properties of the light modulating devices for MO-SLM.

1. Introduction

Magneto-optical spatial light modulator (MO-SLM) is a promising device for applications of holography, such as 3 dimensional (3D) holographic display, because its pixel size can be reduce to less than 1 $_m$m which is required for practical use of 3D holographic display [1, 2, 3]. In order to realize the MO-SLM, a giant magneto resistance (GMR) devise using an Magneto-optical (MO) material as a free layer driven by spin-injection has been proposed as a light modulating device because of its high speed operation, submicron-size, as well as MO effect. However, MO and optical properties of the GMR devices has not been investigated, because it has a thickness thin enough for a visible light to pass through and it has a complicated structure including a bottom electrode, pinned layer, nonmagnetic layer, free layer, capping layer and top electrode. To understand and control the MO and optical properties, contribution from each layers should be understand. In this paper, we focus on the MO and optical properties of GMR films that has complicated structures of Si/SiO_2/bottom electrode/TbFeCo (9 nm)/Cu (4 nm)/(Pt/Co (6 nm))/(capping layers of Ta, Au and Pt), because capping layer made of normal metals deposited on the GMR devices could change its properties while it is indispensable to protect the GMR devices from a deterioration suffered by device fabrication processes.

2. Experiments

A Pt/Co GMR structure, Si/SiO_2/bottom electrode/TbFeCo (9 nm)/Cu (4 nm)/(Pt/Co (6 nm)), with perpendicular magnetic anisotropy was deposited on a Si/SiO_2 substrate by DC magnetron sputtering with a base pressure of less than 2×10^{-6} Pa. The film was annealed in vacuum at 190°C for an hour in the absence of a magnetic field. Finally, capping layers of Ta, Au and Pt with a thickness of 3 nm were deposited on top of the films. A schematic drawing of the structure is shown in Fig. 1(a).

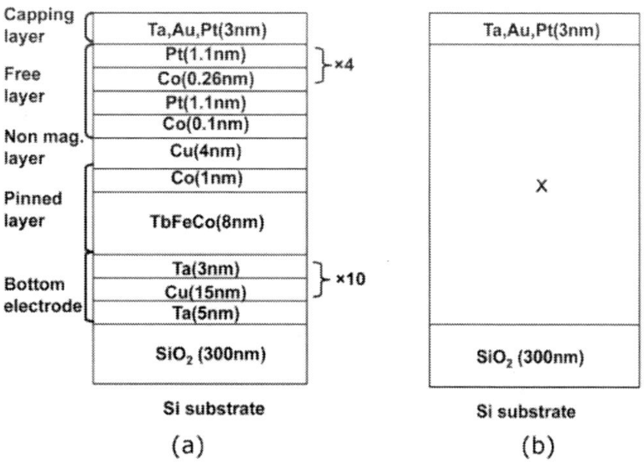

Figure 1. (a) a schematic drawing of the GMR structures, and (b) a simplified structure consisting of Si/SiO_2/X/Caping layer for an analysis.

Kerr rotation spectra were measured by an MO spectrometer using the polarization modulation technique with a wavelength between 260 - 900 nm. To measure Kerr rotation of free layers, MO spectra were obtained as an average of two spectra measured with magnetic fields of ± 2 kOe that is higher enough than the coercivity of the free layers (~ 1 kOe) and lower than that of pinned layers (> 5 kOe).

Optical constants were measured by using ellipsometer (HORIBA JOBIN YVON: MM-16) with a wavelength between 430 - 700 nm. Optical properties were analyzed by using optical simulation by using software (HORIBA: Delta Psi2). In this experiment, to simplify the optical analysis, the complicated structure was assumed to be 4 layer consisting of Si/SiO_2/X/Caping layer, where X includes bottom electrode/TbFeCo (9 nm)/Cu (4 nm)/(Pt/Co (6 nm)) as shown in Fig. 1(b). Optical constants of Au, Ta and Pt used in the optical analysis were measured for each films deposited on Si substrate by the ellipsometer, and those of Si and SiO_2 were obtained from a reference [8].

3. Results and discussion

Figure 2 shows MO spectra of free layers with different capping layers, Ta, Au, and Pt. A feature that has peak around 4 eV is similar to those of Pt/Co multilayered reported previously [4, 5]. This result indicates that those Kerr rotation spectra are mostly showing a characteristic of Pt/Co multilayer used as free layers. However, it was found that values of Kerr rotation and peak positions seem to depend on material of the capping layers. The peak positions of those spectra are 4.1, 4.3 and 3.8 eV and the values of maximum Kerr rotation are 0.35, 0.25 and 0.23° for samples with Ta, Au and Pt capping layer, respectively.

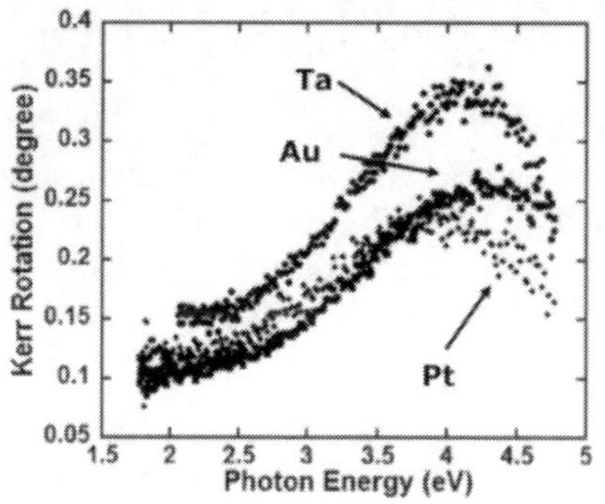

Figure 2. MO spectra of the GMR films with different capping layers, Ta, Au, and Pt.

Is known that MO effect, such as Kerr effect and Faraday effect, is evaluated using a real and an imaginary part of the off-diagonal element of the optical conductivities, σ'_{xy}, σ''_{xy}, and the refractive index n and the extinction coefficient κ [5]. It means that a value of MO effect could be changed not only by the off-diagonal element of the optical conductivities, but also by optical constants, n and κ. In Fig. 3, refractive indices and extinction coefficients of those samples measured by means of ellipsometry are shown. It is found that they have different values and structures in those spectra as well as Kerr rotation spectra shown in Fig. 2. In order to find a contribution from the capping layers, optical constants of each layers should be analyzed. However, we had to simplify them as 4 layers consisting of Si/SiO_2/X/Caping layer as shown in Fig. 1(b), since the samples used in this study have complicated structures.

Therefore, we obtained optical constant of the effective layer X by analyzing the optical constants of the GMR structure with Pt and Au as shown in Figure 4. Although the fitting was well done for both samples with Au and Pt, they were quite different for each other. We consider that it could happen, since it is known that the multilayer of metal have a peculiar electronic structure which can not be explained only by assuming a stack of metals as observed in Pd/Co and Fe/Au multilayers [6, 7]. Therefore, we consider that the differences in Fig. 4 may be due to the change in electronic structures. For Ta, we could not obtained good result in the optical analysis. We consider that the reason is due to oxidation of Ta. If Ta were oxidized, it could contribute to large Kerr rotation as observed in Fig. 2.

The optical constants of the GMR structures were reconstructed by the simulation using the optical constants of Si, SiO_2, X obtained from the GMR structure with Pt capping layer, and the caping layer (Ta, Pt and Au) as shown in Fig. 5. It is found that the optical constant of the sample with Pt capping layer is reproduced properly. In addition, the optical constants for the sample with Au capping layer are rather consistent with the experimental data shown in Fig. 3. In paticular, the refractive index of those two samples are well agree with those of experimental data. This result indicate that a capping layer with a thickness of 3 nm could make a large contribution for the optical constants. On the other hand, the optical constants for Ta did not agree with the experimental data. It must be due to the oxidization of Ta as mentioned above. In the case of Au capping layer, we could not obtain good results from the simulation, it suggests that the electronic structure of Pt/Co multilayer underneath the Au capping layer was affected resulting in the sift of peak of the Kerr spectrum.

These results suggest that capping layers contribute the change in the optical constants even if it has a thickness of several nm and it also could change the MO properties. In order to understand the effect, further study for all the layers in the GMR structures and an MO simulation are needed.

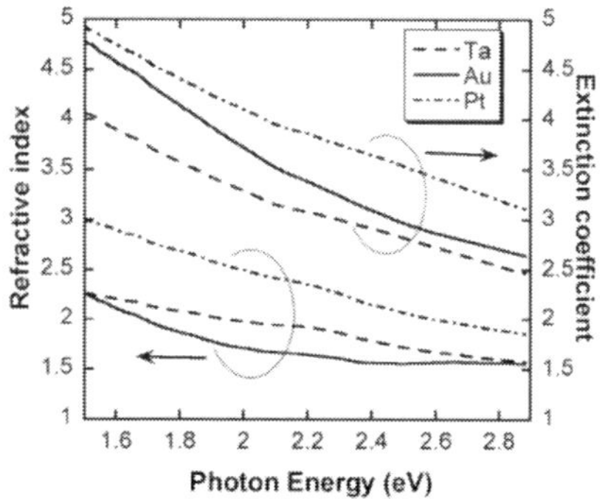

Figure 3. Refractive indices and extinction coefficients of the GMR films with different capping layers, Ta, Au, and Pt, measured by an ellipsometry.

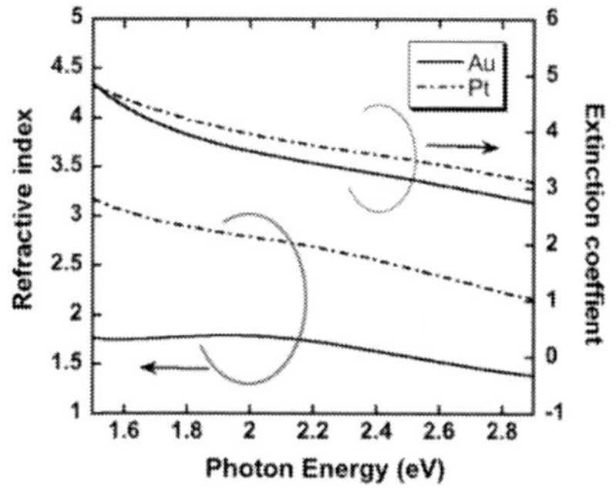

Figure 4. Refractive index and extinction coefficient of the effective layer X obtained by analyzing the optical constant of the GMR structure with Pt and Au capping layer.

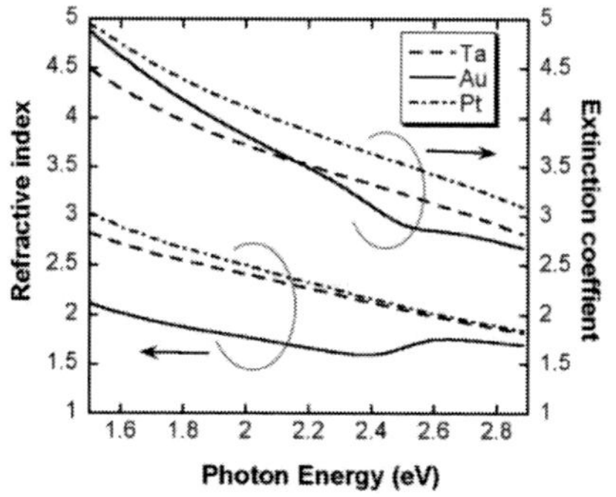

Figure 5. Refractive indices and extinction coefficients of the GMR films with different capping layers, Ta, Au, and Pt, reconstructed by a simulation.

4. Summary

MO and optical properties of the GMR structures with different capping layers with a thickness of 3 nm were studied. It was found that MO spectra and optical constants had different values and structures. Simulation was revealed that the capping layers could change the optical constants of the GMR structures. We consider that the change in the optical constants is one of the reason of the change in the MO spectra.

Acknowledgments

This research is supported by the National Institute of Information and Communications Technology (NICT).

References

[1] K.Aoshima, N. Funabashi, K. Machida, Y. Miyamoto, N. Kawamura, K. Kuga, N. Shimidzu, F. Sato, T. Kimura, Y.Ohtani, 2007 *Appl. Phys. Lett.* **91** p.052507.
[2] K. Machida, N. Funabashi, K.Aoshima, Y. Miyamoto, N. Kawamura, K. Kuga, N. Shimidzu, 2008 *J. Appl. Phys.* **103** p.07A713.
[3] K.Aoshima, N. Funabashi, K. Machida, Y. Miyamoto, N. Kawamura, K. Kuga, N. Shimidzu, Y. Ohtani, F. Sato 2009 *IEEE trans. magn.* **44** p.2491.
[4] W. B. Zeper, F. J. A. M. Greidanus, P. F. Carcia, C. R. Fincher, 1989 *J. Appl. Phys.* **65** p.4971.
[5] K. Sato, H. Hongu, H. Ikekame, J. Watanabe, K. Tsuzukiyama, Y. Togami, M. Fujisawa, T. Fukazawa, 1992 *Jpn. J. Appl. Phys.* **31** p.3603.
[6] Y. Tosaka, H. Ikekame, K. Urago, S. Kurosawa, K. Sato, S. C. Shin, 1994 *J. Magn. Soc. Jpn.* **18** p.389.
[7] K. Sato, E. Takeda, M. Akita, M. Yamaguchi, K. Takanashi, S. Mitani, H. Fujimori, Y. Suzuki, 1998 *J. Magn. Magn. Mater.* **18** p.177.
[8] G.E.Jellison, Jr., 1992 *Optical Materials* **1**, p.41.

Influence of a semiconducting matrix on the magnetic behaviour of iron oxide nanoparticles

P. Granitzer[1], K. Rumpf[1], M. Venkatesan[2], L. Cabrera[3], M.P. Morales[3], P. Poelt[4], M. Albu[4]

[1] Institute of Physics, Karl Franzens University Graz, Universitaetsplatz 5, A.-8010 Graz, Austria

[2] Institute of Physics, Trinity College, College Green, Dublin 2, Ireland

[3] Institute de Ciencia de Materiales de Madrid, CSIC, Cantoblanco, 28049 Madrid, Spain

[4] Institute for Electron Microscopy, University of Technology Graz, Steyrergasse 17, 8010 Graz, Austria

petra.granitzer@uni-graz.at

Abstract. To achieve a magnetic nanocomposite, iron oxide nanoparticles are prepared by high temperature decomposition of iron organic precursor and then they are infiltrated into oriented pores of a porous silicon (PS) matrix. The PS is fabricated by anodization of a n^+ silicon wafer in aqueous HF solution. The magnetic behavior of this composite is determined by the morphology of the matrix as well as by the distribution of the particles within the individual pores. The composite offers FM/SPM switching due to dipolar coupling between the particles, whereas the transition temperature can be tuned in a broad range by the coating of the particles and the concentration of the particle solution. The anisotropy between two magnetization directions (magnetic field parallel and perpendicular to the pores) can be modified by varying the morphology of the PS matrix.

1. Introduction

The fabrication of nanostructured materials is important in many of today's research areas not only because of the downscaling and miniaturization of devices but exceptionally also due to the modified and arising completely new physical properties compared to the bulk materials. Magnetic nanostructures such as nanowires, nanoparticles or nanodots are promising candidates for applications in magnetic data storage [1], electronic devices [2] as well as in biomedicine [3]. Often there is a high interest on metallic nanoparticles and their large magnetization but they are not stable in air and oxidize easily which leads to a loss of their magnetization [4]. Furthermore magnetic oxides, such as iron oxides are under extensive investigation, e.g. for magnetic recording [5], optical devices [6] and gas sensors [7]. They appear in various structures and hydration states and also their size and shape can be modified by the preparation conditions [8]. One of the most investigated iron oxides is Fe_3O_4

due to its inverse spinel structure, its ferromagnetic behaviour and its low toxicity which renders it a promising candidate for various biomedical applications as drug delivery or imaging. A hexane solution can be used to stabilize Fe_3O_4-particles and the treatment by oleic acid is widely spread [9]. In general quite monodisperse particles can be fabricated which is an important requirement because the physical properties strongly depend on the size. Particles in a size range between 5 nm and 20 nm are superparamagnetic but due to dipolar coupling a ferromagnetic behaviour can be observed. Furthermore the magnetic properties of magnetite particles are related with their surrounding, e.g. a non-magnetic material which is used as substrate or matrix. In this work magnetite nanoparticles of 5 and 8 nm in size are mixed with a hexane solution in the presence of oleic acid-and infiltrated into a mesoporous silicon matrix to achieve a nanomaterial with specific magnetic properties which can be tuned by the morphology of the matrix as well as by the density distribution of the infiltrated particles.

2. Experiments

The formation of the porous silicon matrix is performed by anodization of an n^+-silicon wafer in 10 wt% aqueous hydrofluoric acid solution [10]. All samples are produced under room temperature and the same electrolyte concentration and the pores exhibit a length of about 30 µm, grown in (100)-direction. The average pore-diameter of the investigated samples is 50 nm and 90 nm, respectively and the mean distance between the pores is in the range of 50 nm and 35 nm. In the fabricated diameter range the growth of side-pores in (111)-direction can not be suppressed completely but it is assured that the pores are separated from each other. The magnetite nanoparticles are produced by high temperature decomposition in the presence of an organic precursor following the procedure described in [11]. To prevent oxidation, the particles are mixed with hexane and oleic acid. The average size of the Fe_3O_4-particles is 5 nm (figure 1a) and 8 nm (figure 1b), respectively, whereas the size of the particles is varied by changing the organic solvent. An increase of the boiling temperature of the solvent results in an increase of the particle size.

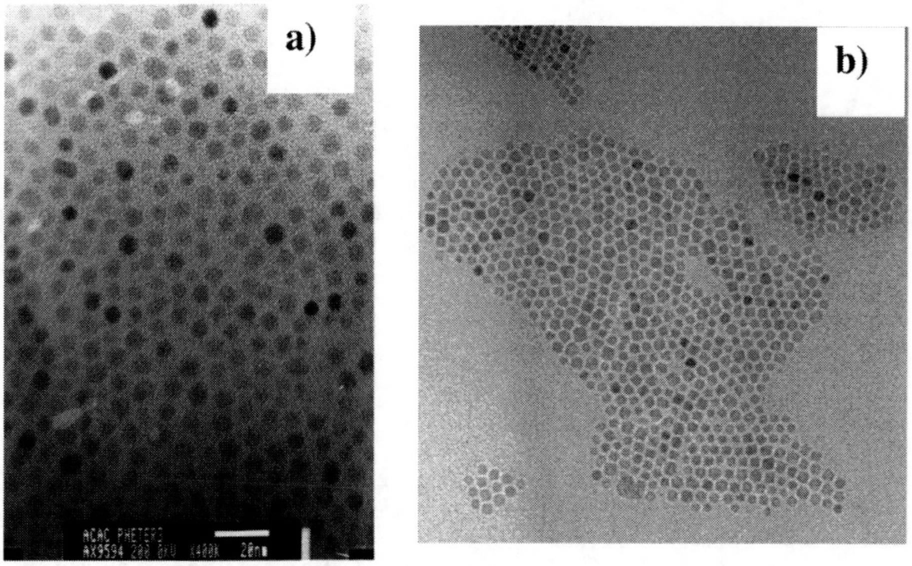

Figure 1: Transmission electron microscopy images of magnetite nanoparticles covered with oleic acid a) particle-size 5 nm b) particle-size 8 nm.

A nanocomposite material is produced by infiltration of the Fe_3O_4-nanoparticles into the pores of the porous silicon matrix-and this procedure is carried out at room temperature for a time of about 30

minutes. To gain information about the nature of the particles infiltrated into the porous silicon matrix XRD-diffractograms were recorded by a Philips 1710 diffractometer using CuKα radiation.
Evidence of the presence of iron oxide within the pores has been investigated by EDX-spectroscopy showing the distribution of the elements.

The magnetic behaviour of the fabricated nanocomposite has been investigated by SQUID-magnetometry in a temperature range between 4.2 K and 250 K whereas the transition between superparamagnetic and ferromagnetic behaviour has been figured out. Information about the anisotropy of the system has been elaborated by the magnetization versus magnetic field measured in two directions, namely with the external field applied normal and parallel to the sample surface.

3. Discussion

The utilized particle-sizes of 5 nm and 8 nm show a quite monodisperse distribution and XRD-diffractogramms of the infiltrated particles correspond to an inverse spinel structure with a lattice parameter of 8.38 Å which correlates very well with the published magnetite pattern (JCPDS 19-629).The presence of the iron oxide particles inside the pores has been investigated by EDX-spectroscopy, whereas a line-scan at a cross-sectional region of the sample gives the intensities of the elements (Fe, O) along the pores which is shown in figure 2. One can see that the oxygen as well as the iron is slightly decreasing towards the pore-tips but nevertheless the pores are filled with magnetite till down to the tips.

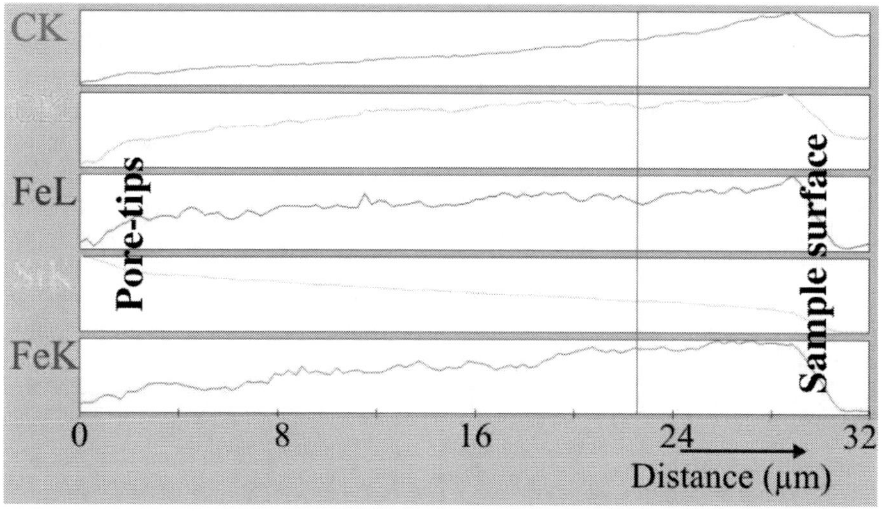

Figure 2: Line scan across a porous layer from the pore-tips to the pore-mouth (sample surface) filled with magnetite particles, showing the intensity of the various elements (carbon, oxide, iron and silicon).

Individual particles could be resolved by scanning electron microscopy (SEM) along the whole pore-length, which verifies a more or less homogeneous filling of the entire porous layer (figure 3).
For the greater particles infiltrated into the matrix a size of about 8 nm has been identified which is in good agreement with the transmission electron microscopy (TEM) images of the pure magnetite particles (see figure 1a, b).
In the case of the smaller used particles the individual magnetite particles of 5 nm in size could not be resolved by SEM. TEM investigations of such porous silicon/magnetite samples is not an adequate method because the particles are adhered weakly to the pore-wall and thus they are released during the preparation of the sample by focused ion beam (FIB).

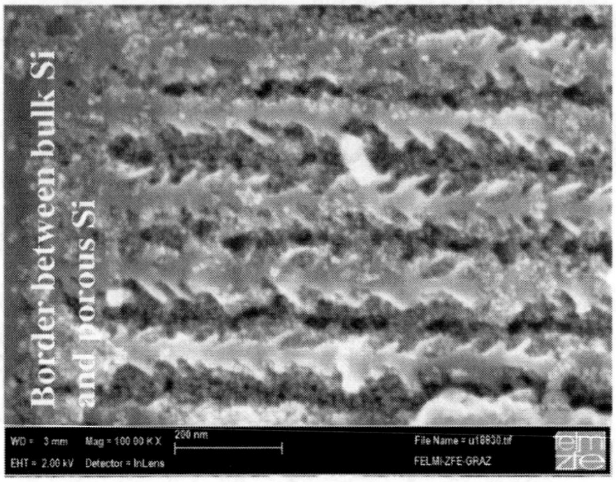

Figure 3: Scanning electron micrograph of a cross-section of porous silicon with infiltrated magnetite nanoparticles of 8 nm in size.

The magnetic behaviour of such nanocomposites consisting of porous silicon matrices with infiltrated magnetite particles of 8 nm in size and of 5 nm in size has been investigated by magnetization measurements performed in a broad temperature range between 4.2 K and 250 K. The temperature dependent magnetization gives information about the blocking temperature T_B which indicates the transition between superparamagnetic and ferromagnetic behaviour. This transition can be influenced not only by the size [12] of the particles but also by changing their interaction in using different concentration of the Fe_3O_4 solution which is infiltrated into the pores. Magnetic field dependent measurements have been performed in two directions of magnetization, with an applied field perpendicular and parallel to the sample surface. The magnetic anisotropy varies with the morphology of the porous silicon template as well as with the particle size.

Considering a matrix with an average pore-diameter of 90 nm and a mean distance between the pores of 35 nm the magnetization shows only a small anisotropy between the two magnetization directions. In the case of an average pore-diameter of 50 nm and a pore-distance of about 50 nm there is a clear anisotropy between easy axis (magnetic field along the pores) and hard axis (magnetic field normal to the pores) magnetization (table1).

Table 1: Magnetic anisotropy between magnetic field applied parallel (II) or perpendicular ($\perp$) to the surface. The coercivities H_C vary with the morphology of the matrix as well as with the particle size.

Particle-size (nm)	**Pore-diameter (nm)**	**Pore-distance (nm)**	$H_{C,II}$ **(Oe)**	$H_{C,\perp}$ **(Oe)**
8	90	35	230	250
8	50	50	275	370
5	50	50	60	60

This anisotropy is caused by dipolar coupling between the magnetite particles which arises differently in the two magnetization directions because the distance between the particles within individual pores is in minimum twice the thickness of the coating which is about 6 nm, whereas the minimum distance

between particles of adjacent pores is the thickness of the pore-walls which is in average 50 nm. Therefore the coupling between particles within the pores is stronger than between particles of neighbouring pores. In the case of larger pore-diameters and smaller pore-distances this behaviour becomes weaker which results in a diminished anisotropy of the magnetization between the two directions. Infiltration of magnetite nanoparticles of 5 nm in size shows an isotropic behaviour in both direction of magnetization. Furthermore the blocking temperature (T_B) which indicates the transition between superparamagnetic and ferromagnetic behaviour is around 10 K (figure 4). T_B is significantly decreased compared to the one of specimens with infiltrated particles of 8 nm. In the latter case T_B is around 135 K which indicated dipolar coupling between the particles.
Both, the isotropic behaviour as well as the low blocking temperature reveals that the 5 nm particles are not interacting. The dipolar coupling of nanoparticles is strongly correlated to the size of the particles and decreases linearly with the size. The oleic acid coating is for both particle sizes around 3 nm, whereas in the case of the 5 nm particles this separation has a stronger influence due to a weaker dipolar coupling of the smaller particles.

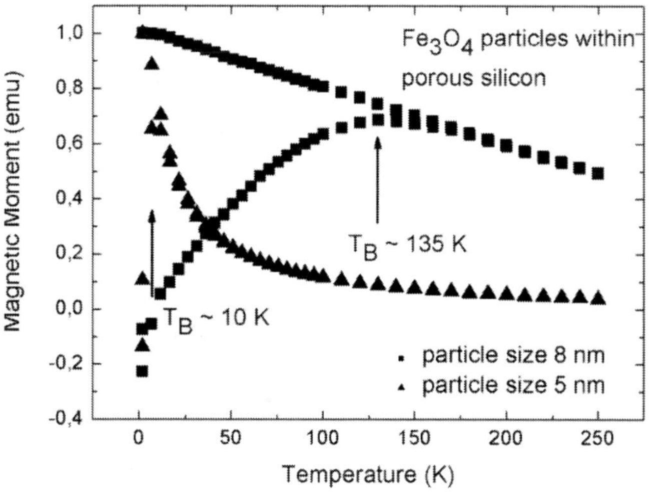

Figure 4: ZFC/FC measurements of 8 nm and 5 nm particles infiltrated within porous silicon. T_B is shifted towards lower temperatures for smaller particles indicating less coupling between them.

4. Conclusions

In this work the magnetic behaviour of porous silicon matrices with infiltrated Fe_3O_4-nanoparticles is presented. Magnetite particles of 5 nm and 8 nm in size which are quite monodisperse are incorporated into the oriented pores of a porous silicon matrix resulting in a composite material exhibiting magnetic anisotropy between applying a magnetic field parallel or perpendicular to the sample surface. The anisotropy depends on the structure of the porous silicon matrix (e.g. pore-diameter, pore-distance) and the dipolar coupling between the particles, respectively. In the case of densely packed particles within the pores the coupling is stronger within the pores than between adjacent pores resulting in various coercivities. The minimum distance between particles within one pore is determined by the thickness of the oleic acid coating whereas the distance between adjacent pores is assigned to the thickness of the pore-walls. If the relation between particle-size and distance between them becomes small enough by reducing the particle size (same coating) the interaction and the anisotropy, respectively is no longer observed.

Acknowledgements

This work is supported by the Austrian science Fund FWF under project P21155. Furthermore the authors thank M. Dienstleder from the Institute for Electron Microscopy for focused ion beam preparation of the samples.

References

[1] M. Tanase, J.G. Zhu, C. Liu, N. Shukla, T.J. Klemmer, D. Weller, D.E. Laughlin, Metallurgical and Materials Transactions A 38, 798, 2007.
[2] S.M. Wang, C.W. Leung, P.K.L. Chan, Appl. Phys. Lett. 97, 023511, 2010.
[3] A.G. Roca, R. Costo, A.F. Rebolledo, S. Veintemillas-Verdager, P. Tartaj, T. Gonzalez Carreno, M.P. Morales, C.J. Serna, J. Phys. D: Appl. Phys. 42, 224002, 2009.
[4] S.P. Gubin, Magnetic Nanoparticles, Wiley-VCH, Weinheim 2009.
[5] L.C. Veranda, G.F. Goya, M.P. Morales, R.F.C. Marques, R.H.M. Godoi, M. Jafelicci Jr., C.R. Serna, IEEE Transact. Magnetics 38, 1907, 2002.
[6] U.T. Lam, R. Mammucari, K. Suzuki, N.R. Foster, Ind. Eng. Chem. Res. 47, 599, 2008.
[7] M. Nidhin, R. Indumathy, K.J. Sreeram, B.U. Nair, Bulletin of Materials Science 31, 93, 2008.
[8] S.P. Gubin (Ed.), Magnetic Nanoparticles, Wiley-VCH, Weinheim, 2009
[9] L. Zhang, R. He, H.-Ch. Gu, Appl. Surf. Sci. 253, 2611, 2006.
[10] P. Granitzer, K. Rumpf, P. Poelt, M. Albu, B. Chernev, Phys. Stat. Sol. (C) 6, 2222, 2009.
[11] A.G. Roca, R. Costo, A.F. Rebolledo, S. Veintemillas-Verdaguer, P. Tartaj, T. Gonzalez-Carreno, M.P. Morales, C.J. Serna, J. Phys. D: Appl. Phys. 42, 224002 (2009).
[12] J.R. Correa, E. Bordallo, D. Canetti, V. Leon, L.C. Otero-Diaz, C. Negro, A. Gomez, R. Saez-Puche, Materials Reasearch Bulletin 45, 946, 2010.

Self-assembled strontium ferrite on Au nano-dot arrays using rapid thermal annealing

A Morisako[1], X Liu[1,*], A Kaewrawang[1] and A Ghasemi[1]
[1]Spin Device Technology Center, Faculty of Engineering, Shinshu University, 4-17-1 Wakasato, Nagano City, Nagano 380-8553, Japan.

*E-mail: liu@cs.shinshu-u.ac.jp

Abstract. The self-assembled SrM dot array on Au nano-dots has been prepared by using DC magnetron sputtering system without any substrate heating. Then, flash annealing was carried out to crystallize the SrM film. *C*-axis oriented SrM perpendicular to the films can be observed only at annealing temperature of 800 °C for annealing time of 7 min. The coercivity and remanent squareness ratio in perpendicular direction are 2.8 kOe and 0.72, respectively. The reversal magnetization of SrM on patterned Au nano-dots is closer the Stoner-Wohlfarth model compared to continuous films. The peak of ΔM of SrM film deposited on patterned Au nano-dots is lower compared to SrM deposited on continued Au films.

1. Introduction

To extend data storage capabilities beyond superparamagnetic limit of conventional granular media, patterned media [1,2], in which each bit is stored in a single domain islands, have been considered as one of prominent candidates for high density magnetic recording media. Several varieties of patterned films fabrication e.g. lithography [3], irradiation through a mask [4], milling with focused ion beam [5] and imprint technology [6] were reported. However, these four techniques are unsuitable for large scale, low throughput and expensive method. Another method for preparation of patterned film is self-assembling method. In this work, self-assembly prepared by sputtering deposited method is used because it has the potential for making very fine-scale structures over large areas in a rapid process and no chemical used in process.

The Co-alloys, $L1_0$ phases, rare-earth transition metals and ferrites are the candidate magnetic materials for high density recording. In this work, we choose hexagonal strontium ferrite ($SrFe_{12}O_{19}$ or SrM in short) because it has fairly large magnetocrystalline anisotropy, excellent chemical stability, magnetic moment aligned along the *c*-axis and minimum stable grain size about 8.8 nm which is computed using the media stability criterion of $KV > 60k_BT$. Due to these characteristics, SrM is an attractive material for high density magnetic recording application.

In the previous works, self-assembled strontium ferrite (SrM) on random distribution of Au islands structures that are prepared by sputtering method at high substrate temperature was studied [7,8]. In this study, an attempt to prepare the patterned SrM thin films on Au nano-dots with an approximately uniform distribution has been carried out by rapid thermal annealing (RTA) process.

2. Experiments

The SrM film was deposited on Au underlayer using a magnetron sputtering apparatus. The Au film is expected to be used as the underlayer for promotion of c-axis orientation of hexagonal ferrite layer.

The misfit ratio between (111) plane of fcc Au and c-plane of hcp SrM is 2.1%. The self-assembled Au nano-dots are prepared by chemical process. Au dot arrays were deposited onto the substrate by sputtering at an elevated substrate temperature. The average Au dot size is about 15 nm in diameter and uniform distribution. The thickness of SrM layer was kept at 35 nm. A sintered ferrite disk with stiochiometric composition of M-type, i.e. Sr:Fe = 1:12, was used as the target for SrM film. After evacuating the chamber to a pressure below 2×10^{-6} Torr, the mixture gas of Ar and O_2 with gas pressure of 1.98 and 0.02 mTorr, respectively, was used as sputtering gas. The samples were prepared without any substrate heating. An infrared image furnace was used for flash annealing. The as-deposited SrM films were put in the closed chamber of the annealing furnace, and then annealing was carried out within at temperature of 800 °C for annealing time of 3-10 min in atmospheric pressure without applying any vacuum. The crystallographic properties of the films were characterized by an x-ray diffractometer (XRD) with Cu K_α radiation. The magnetic properties were measured by vibrating sample magnetometer (VSM) with maximum field strength of 24 kOe. The surface morphologies of the films were observed by using scanning electron microscope (SEM) and atomic force microscope (AFM). The magnetic domain structures of the films were evaluated by using magnetic force microscope (MFM).

3. Results and Discussion

In Fig. 1, the SrM thin films with and without Au underlayer deposited on Corning glass can be crystallized for annealing time, *t*, over 5 and 10 min, respectively. It means that the underlayer can be used to enhance the crystallization degree of SrM. For SrM/Au/SiO_2/Si thin films, the crystallization of SrM/Au deposited on SiO_2/Si is lower than that for Corning glass. It is related to the nature of the SiO_2/Si substrate which can absorbs the heat energy but it penetrates the Corning glass substrate. So, the temperature of Au islands for SiO_2/Si substrate is higher than that for glass substrate. From these results, an attempt to prepare the isolated SrM patterned films has been carried out by using Au underlayer [7,8].

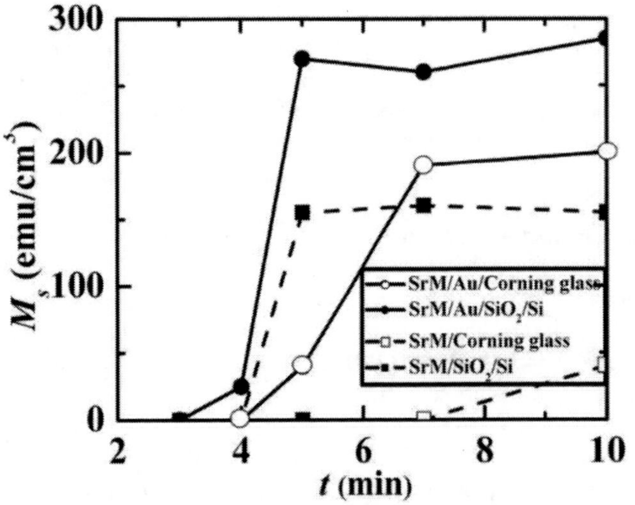

Figure 1. Saturation magnetization, M_s, of SrM thin films at T_a of 800 °C for various annealing times, *t*.

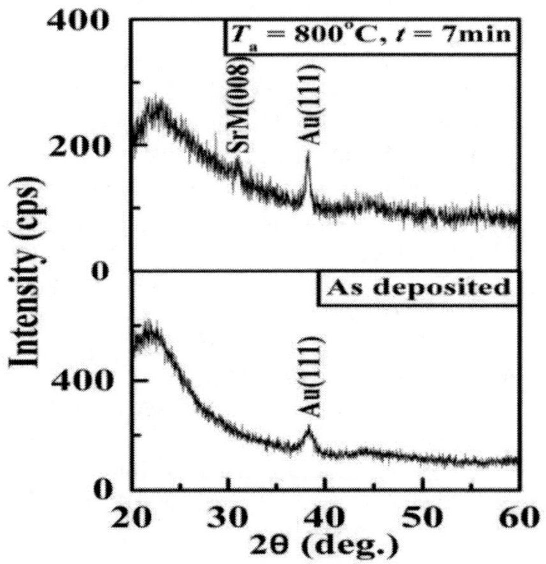

Figure 2. XRD patterns of SrM/Au/Pyrex glass at as deposited state and annealed sample at T_a of 800 °C for 7 min.

Figure 2 shows XRD diagrams for SrM films prepared on Au nano-dots at as-deposited and annealed state. The preferred orientation of (111) *fcc* lattice of Au underlayer can be observed for all of the films. As-deposited SrM/Au/Pyrex glass thin films are completely amorphous with the saturation magnetization, M_s, value of zero and no diffraction line from SrM planes was observed. Upon annealing at T_a of 800 °C for 7 min, the Au grains may act as nucleation sites for the formation of SrM crystallites in the SrM/Au/Pyrex glass thin films. The presence of diffraction lines from SrM (00*l*) planes can be observed in Fig. 2. These results indicate that the annealing process can be used to promote the crystallized SrM thin films and the (111) orientation of Au underlayer promotes *c*-axis oriented SrM.

Magnetic properties are also in a good agreement with the XRD patterns. As shown in Fig. 3, the coercivity in perpendicular direction, $H_{c\perp}$, is higher than that for in-plane direction, $H_{c//}$. It means that all of the films exhibit perpendicular anisotropy. The perpendicular coercivity, $H_{c\perp}$, and perpendicular squareness ratio, $S_\perp$, are 2.8 kOe and 0.72, respectively. The saturation magnetization, M_s, is about 90 emu/cm^3. This value of M_s is lower than that of the theoretical value of standard SrM film (360 emu/cm^3). It might be due to partially crystallization of SrM thin films.

Figure 4 shows the angular dependence of $H_c/H_c(0^\circ)$ for SrM thin films prepared on Au nano-dots compared to the continuous SrM thin films that are SrM/Au/SiO_2/Si, SrM/Au/Corning glass and SrM/SiO_2/Si. In comparison, the results of Stoner – Wohlfarth (S – W) and domain wall motion (DWM) modes are included. Apparently, the angular dependence does not agree with those of S – W and DWM modes. The magnetization reversal process is accompanied by the combination of DWM and S – W rotation. However, the magnetization reversal process of SrM thin films prepared on Au nano-dots is nearly the S – W model than the others.

The interactions between magnetic particles of the thin films have been evaluated with the ΔM method [9]. Following this method, the presence and strength of interactions is described as this equation: $\Delta M(H) = \dfrac{I_d(H)}{I_r(\infty)} - \left[1 - 2\dfrac{I_r(H)}{I_r(\infty)}\right]$ where $I_d(H)$ is the dc demagnetization remanence (DCD), $I_r(H)$ is the isothermal remanent magnetization (IRM) and $I_r(\infty)$ is the isothermal remanent magnetization at saturation. The finite negative and positive values of ΔM indicate the magnetostatic and exchange coupling, respectively.

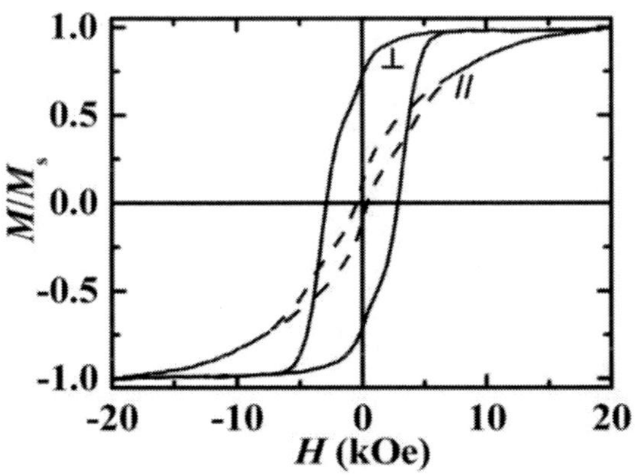

Figure 3. Hysteresis loop of SrM/Au/Pyrex glass annealed at 800 °C for 7 min.

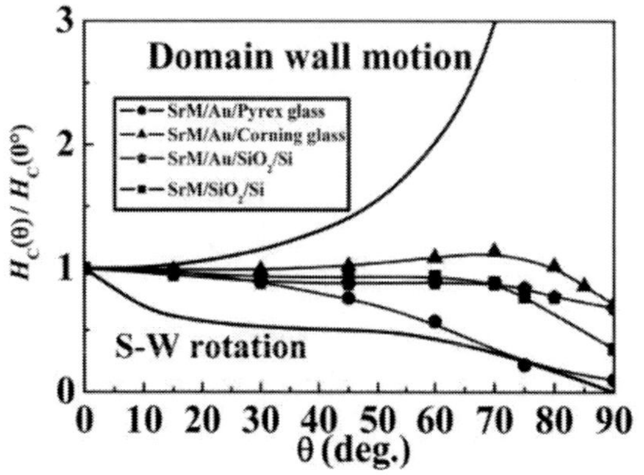

Figure 4. The Angular dependence of coercivity of the SrM/Au and SrM thin films on various substrates.

As shown in Fig. 5, the ΔM(H) deviations of the SrM thin films prepared on Au nano-dots compared to the continuous SrM thin films that are SrM/Au/SiO_2/Si, SrM/Au/Corning glass and SrM/SiO_2/Si are reported. It indicates magnetostatic interactions in the films. The peak of ΔM of

SrM/Au/Pyrex glass thin films is lower compared to SrM/Au/SiO_2/Si, SrM/Au/Corning glass and SrM/SiO_2/Si. The weaker interaction might be contributed from non-magnetic phase existed in the space between the magnetic particles and hence the intergranular interactions are reduced.

Figure 6 shows surface of the SrM thin films prepared on Au nano-dots and its magnetic domain structure. The average size of domain structure is 200 – 500 nm larger than Au dot size because of reformation of Au underlayer to be island structure when annealing the sample.

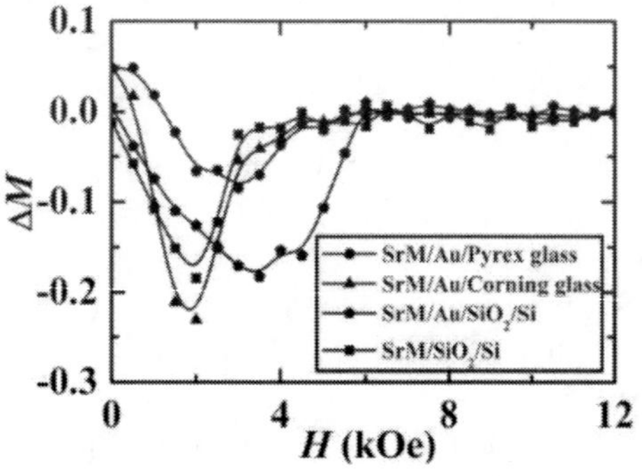

Figure 5. The ΔM deviation of the SrM/Au thin films on various substrates.

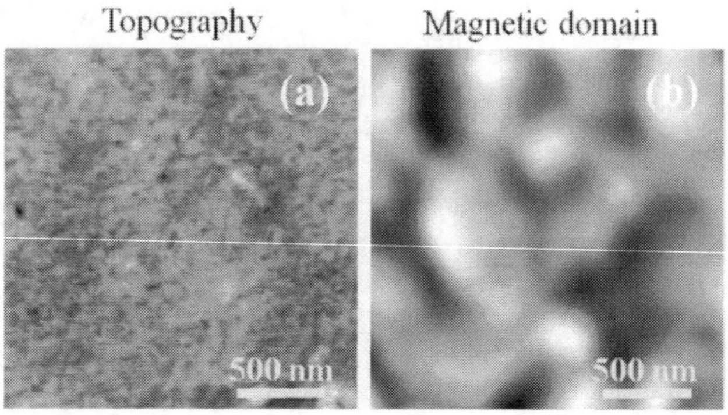

Figure 6. (a) Surface morphology and (b) magnetic domain structure of SrM/Au/Pyrex glass annealed at 800 °C for 7 min.

4. Conclusions

The *c*-axis oriented SrM thin films can be prepared on patterned Au nano-dots with an approximately uniform distribution by rapid thermal annealing process. The magnetization reversal process of SrM on patterned Au thin film is accompanied by the combination of domain wall motion and S – W rotation. However, it is closer the S – W model compared to continuous films. The magnetic interaction of SrM grains on Au nano-dots is reduced in compare with SrM films deposited on continuous Au thin films.

Acknowledgments

This work was supported in part by Grants-in-Aid for Scientific Research (B) and Grant-in-Aid for JSPS Fellows from Japan Society for Promotion of Science (JSPS).

References

[1] Chou S Y 1997 *Proc. IEEE* **85** 652
[2] White R L, New R M H and Pease R F W 1997 *IEEE Trans. Magn.* **33** 990
[3] Castano F J, Hao Y, Haratani S, Ross C A, Vogrli B, Walsh M and Smith H I 2001 *IEEE Trans. Magn.* **37** 2073
[4] Chappert C, Bernas H, Ferre J, Kottler V, Jamet J P, Chen Y, Cambril E, Devolder T, Rousseaux F, Mathet V and Launois H 1998 Science **280** 1919
[5] Rettner C T, Best M E and Terris B D 2001 *IEEE Trans. Magn.* **37** 1649
[6] Heidari B, Maximov I and Montelius L 2000 *J. Vac. Sci. Technol.* B **18** 3557
[7] Kaewrawang A, Ghasemi A, Liu X and Morisako A 2010 *J. Magn. Magn. Mater.* **322** 2043
[8] Kaewrawang A, Ghasemi A, Liu X and Morisako A 2010 *J. Alloys Comp.* **492** 44
[9] Kelly P E, Grady K O, Mayo P I and Chantrell R W 1989 *IEEE Trans. Magn.* **25** 3881

Magnetic and structural characterization of nanosized $BaCo_xZn_{2-x}Fe_{16}O_{27}$ hexaferrite in the vicinity of spin reorientation transition

A Pasko[1], F Mazaleyrat[1], M LoBue[1], V Loyau[1], V Basso[2], M Küpferling[2], C P Sasso[2] and L Bessais[3]

[1] SATIE, ENS Cachan, CNRS, UniverSud, 61 av President Wilson, F-94235 Cachan, France
[2] INRIM, Strada delle Cacce 91, I-10135 Torino, Italy
[3] ICMPE, CNRS, Université Paris 12, 2-8 rue Henri Dunant, F-94320 Thiais, France

E-mail: pasko@satie.ens-cachan.fr

Abstract. Numerous applications of hexagonal ferrites are related to their easy axis or easy plane magnetocrystalline anisotropy configurations. Certain W-type ferrites undergo spin reorientation transitions (SRT) between different anisotropy states on magnetic field or temperature variation. The transition point can be tuned by modifying the chemical composition, which suggests a potential application of hexaferrites in room temperature magnetic refrigeration. Here we present the results of structural and magnetic characterization of $BaCo_xZn_{2-x}Fe_{16}O_{27}$ ($0.7 \leq x \leq 2$) doped barium ferrites. Fine powders were prepared using a sol-gel citrate precursor method. Crystal structures and particle size distributions were examined by X-ray diffraction and transmission electron microscopy. The optimal synthesis temperature ensuring complete formation of single W-phase with limited grain growth has been determined. Spin reorientation transitions were revealed by thermomagnetic analysis and AC susceptibility measurements.

1. Introduction

Hexaferrites have been the subject of intensive studies due to an appealing combination of good magnetic properties and low cost. This large family of oxides with hexagonal crystal structure contains ferrimagnetic compounds with easy axis of magnetization (*e.g.* M-type ferrites) and easy plane of magnetization (*e.g.* Y-type ferrites). Hence, hexaferrites have been widely adopted in two distinct fields: permanent magnets and microwave technology components [1]. On the other hand, W-type ferrites $BaM_2Fe_{16}O_{27}$ (M = Mg, Mn, Fe, Co, Ni, Cu, Zn) can undergo spin reorientation transitions (SRT) between different anisotropy configurations (easy plane ↔ easy cone ↔ easy axis) induced by change of temperature or applied magnetic field [2–9]. The transition temperatures can be tuned by modifying the chemical composition (substitution of bivalent metal M). Moreover, some SRT are expected to be of the first order, which suggests a potential application of W-type ferrites in room temperature magnetic refrigeration [10].

Conventional ceramic method of hexaferrite synthesis is efficient, but requires elevated temperatures for solid state reaction to occur between premixed powders. Alternative production routes (aerosol pyrolysis, chemical co-precipitation, glass crystallization, hydrothermal synthesis, *etc.*)

are intended to improve mixing of initial components down to atomic level and thereby to facilitate diffusion. In particular, a sol-gel technique enables to obtain sufficiently homogeneous precursors for low-temperature synthesis of nanosized simple (spinel) ferrites. However, hexagonal ferrites with complex layered crystal structures still require relatively high temperatures to form because of thermodynamic stability conditions. Trying to avoid rapid growth of grains, we have used this soft-chemistry approach for production of W-type ferrites. An important goal was to determine the heat treatment regimes ensuring complete transformation of a precursor into the smallest particles of single W-phase. In this paper the results of structural and magnetic characterization of $BaCo_xZn_{2-x}Fe_{16}O_{27}$ powders are presented. The chemical compositions were chosen so that SRT occur near room temperature ($x = 0.7, 0.75, 0.8$) or much higher ($x = 2$).

2. Experimental

Hexaferrite powders have been prepared by a sol-gel citrate precursor method [11–13]. High purity iron(III) nitrate, barium hydroxide, cobalt(II) oxide, zinc oxide and citric acid were used as starting materials with the molar ratio of citrate to metal ions 2:1. Iron(III) nitrate was dissolved in deionized water and quantitatively precipitated with excess of ammonia solution as iron(III) hydroxide. The precipitate was filtered and washed with water until neutrality. Then the obtained iron(III) hydroxide was dissolved in a vigorously stirred citric acid solution at 60–70 °C. Barium hydroxide and other metal sources were added according to stoichiometry. At this stage 3 samples of each composition were separated: (A) pH value of the solution was adjusted to 6 for better chelation of cations [13]; (B) no modification was done [12]; (C) ethylene glycol was added to increase viscosity by polycondensation reaction [11]. Water was slowly evaporated at 80–90 °C with continuing stirring until a highly viscous residue is formed. The gel was dried at 150–170 °C and heat treated for 2 h at 450 °C for total elimination of organic matter. Finally, the inorganic precursor with homogeneous cationic distribution was calcined at temperatures up to 1300 °C for 2 h with heating/cooling rate 200 K/h to synthesize a hexaferrite phase. Details of the procedure will be discussed elsewhere.

Crystal structures were examined by PANalytical X'Pert Pro X-ray diffractometer (XRD) in Co-K_α radiation with X'Celerator detector for rapid data acquisition. Magnetization curves were recorded on Lake Shore vibrating sample magnetometer (VSM). Direct observations of powder particles were carried out by FEI Tecnai G^2 F20 transmission electron microscope (TEM) operating at 200 keV. Thermogravimetric analyzer (TGA) PerkinElmer Pyris 6 equipped with a permanent magnet was used for thermomagnetic measurements above room temperature. AC magnetic susceptibility at low temperatures was studied in Quantum Design PPMS. Rietveld analysis of XRD spectra was performed using MAUD software [14].

3. Results and discussion

Characterization had a twofold purpose: to find a correlation between the calcination temperature and the final product properties (phase composition, mean grain size, coercivity); to confirm the existence of SRT at expected temperatures in fine hexaferrite powders.

3.1. Structural characterization and phase analysis

Figure 1 represents XRD patterns from $BaCo_2Fe_{16}O_{27}$ powder synthesized at different temperatures. W-ferrite becomes the major phase at 1300 °C, while lower calcination temperatures lead to a mixture of W-ferrite, M-ferrite (both are hexagonal) and S-ferrite (with spinel structure) in the final product. Traces of α-Fe_2O_3 are sensitive to the method of preparation and dissolve completely with increase of calcination temperature; however, other secondary phases may appear together with W-ferrite. The results of quantitative phase analysis based on Rietveld method of XRD full profile fitting are shown in figure 2. At 900 °C only M-ferrite can be formed, with excess of cobalt giving also S-ferrite. Increase of temperature creates favorable conditions for W-ferrite synthesis, directly from the precursor components (nanocrystalline or amorphous oxides) or through the intermediate reaction

$$BaFe_{12}O_{19} + 2\,CoFe_2O_4 \rightarrow BaCo_2Fe_{16}O_{27}$$

The proportion between 3 ferrite phases depends on calcination temperature and production route (as the precursor composition is affected, for instance, by pH value of the solution). However, at 1300 °C M-ferrite and S-ferrite almost completely disappear.

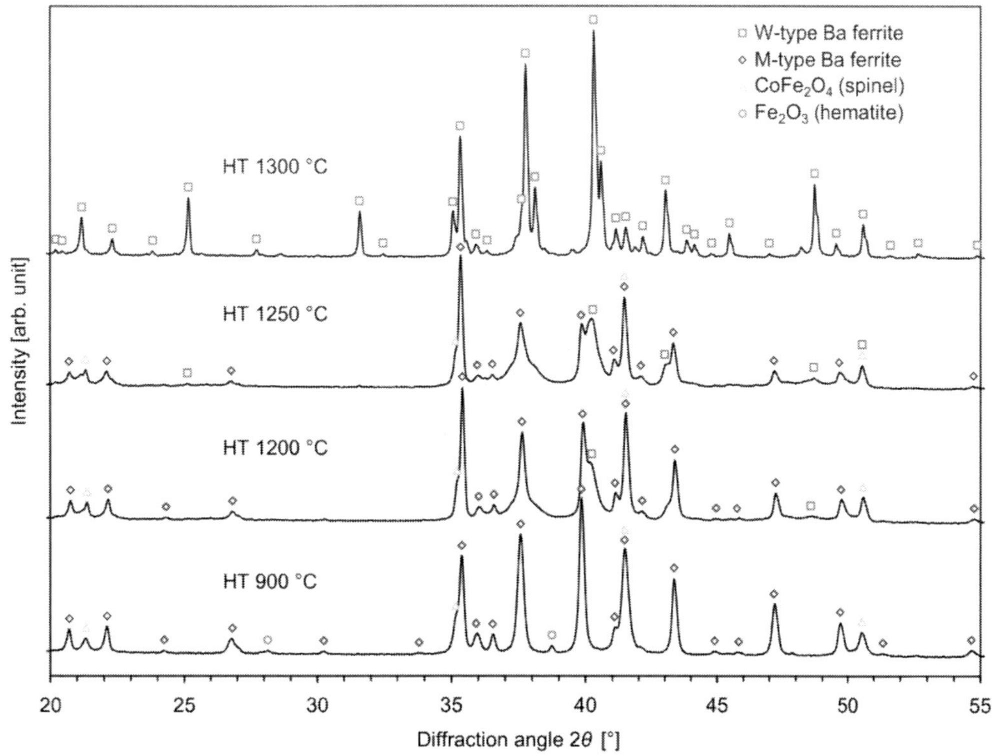

Figure 1. XRD patterns of $BaCo_2Fe_{16}O_{27}$ powder A taken after different heat treatments.

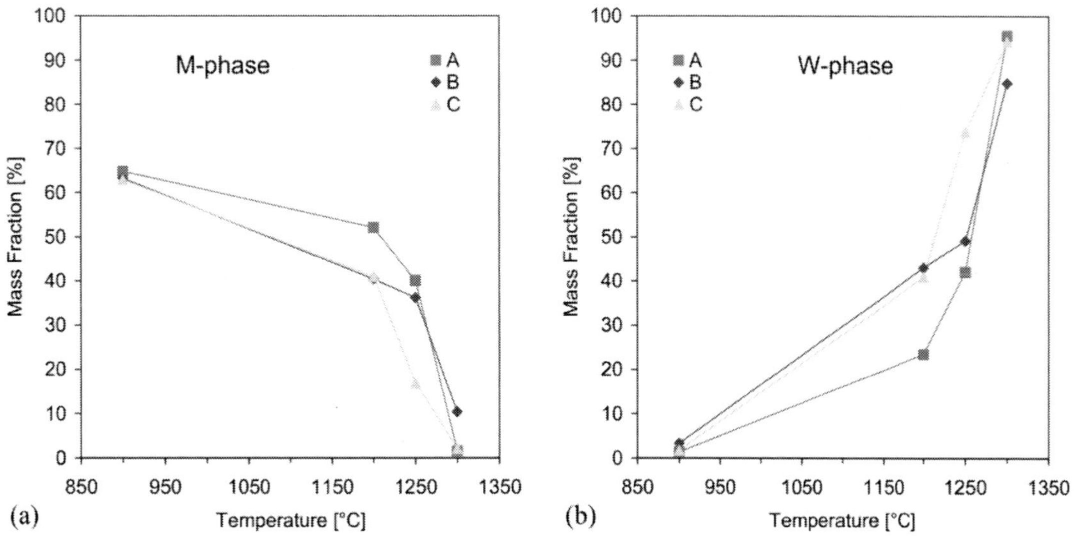

Figure 2. Phase composition of $BaCo_2Fe_{16}O_{27}$ powders A, B, C (from Rietveld fitting of XRD spectra) *vs.* calcination temperature: (a) M-ferrite fraction; (b) W-ferrite fraction.

3.2. Morphology and average size of particles

The size of diffracting crystallites in W-ferrite powders is derived from XRD data by analysis of line broadening using Rietveld refinement as shown in figure 3(a). It gives a minimum estimate for average particle size (a particle can contain several crystallites). The mixture of M-ferrite and S-ferrite formed at 900 °C have a mean size of crystallites ~ 50 nm. With increase of calcination temperature the calculated values slightly decrease, this can reflect contribution to line broadening caused by generation of defects in the transforming mixture of phases. At 1300 °C, when W-ferrite phase rapidly grows and becomes dominant, the apparent size of crystallites significantly increases. In addition, XRD spectra of the powders synthesized at this temperature exhibit a specific texture characteristic for plate-like particles pressed in a holder.

Magnetic measurements give an illustration to these phase transformations. Figure 3(b) shows that after calcination at 900 °C coercivity is high due to the presence of M-ferrite. When W-ferrite forms, the coercivity decreases and reaches ~ 15 mT at 1300 °C, a typical for this phase value.

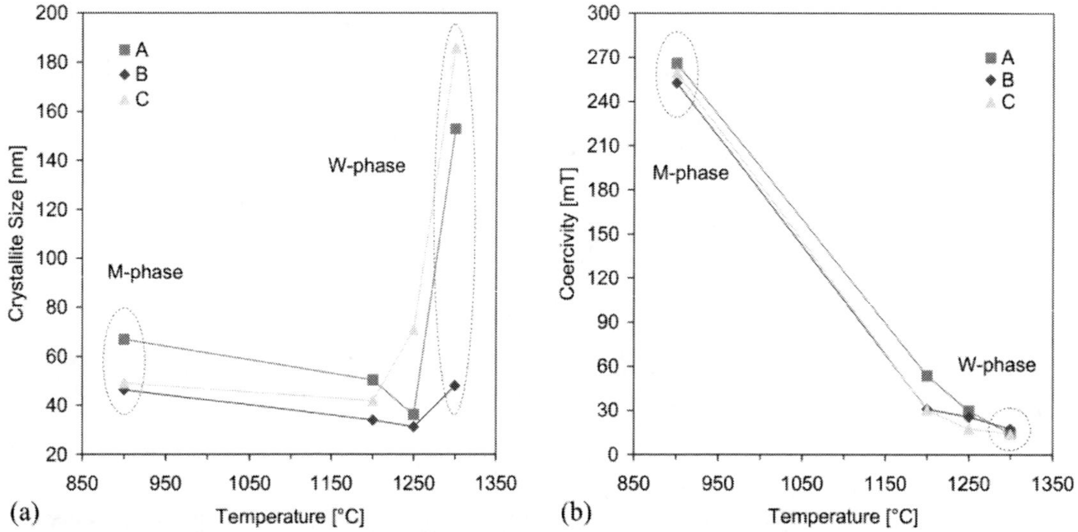

Figure 3. $BaCo_2Fe_{16}O_{27}$ powders A, B, C calcined at different temperatures: (a) mean size of crystallites (from XRD line broadening); (b) coercivity (from VSM measurements).

Direct observations of W-ferrite particles were performed by TEM. Powders calcined at 900 °C consist of aggregates of clean and almost round particles not exceeding ~ 100 nm, an example is shown in figure 4(a). The shape of crystallites reflects their hexagonal crystal structure. Powders calcined at temperatures high enough to form W-ferrite exhibit absolutely different morphology as shown in figure 4(b). The particles have a wide size distribution from 50 nm to 2000 nm and are mostly plate-like. This morphology is observed not only for 1300 °C when W-ferrite is almost single phase, but also for lower temperatures when W-ferrite appears in the precursor. The average particle size of $BaCo_xZn_{2-x}Fe_{16}O_{27}$ ($x = 0.75$) powder estimated by laser light scattering analysis is ~ 700 nm, which agrees with TEM observations.

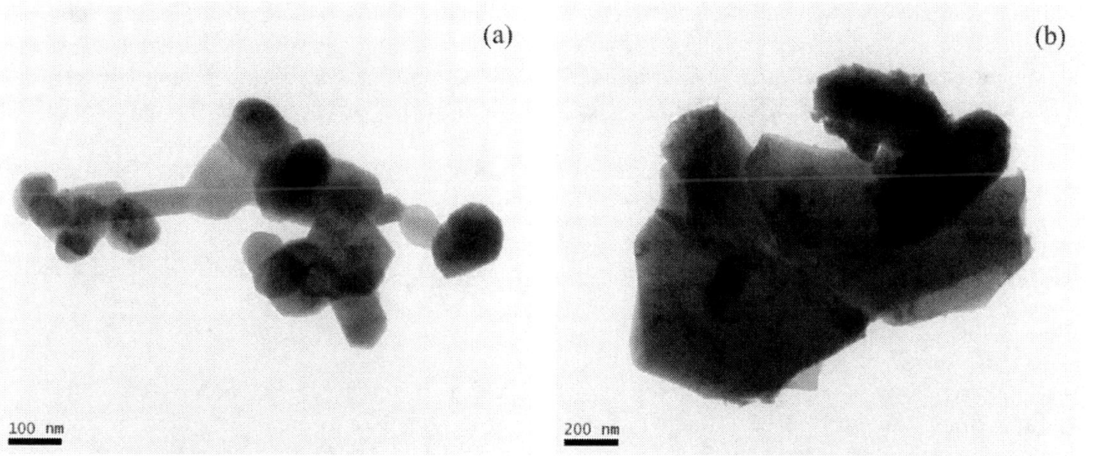

Figure 4. Typical TEM images of powder particles: (a) $BaCo_2Fe_{16}O_{27}$ precursor after calcination at 900 °C (mixture of ferrites); (b) W-type $BaCo_xZn_{2-x}Fe_{16}O_{27}$ ($x = 0.75$) ferrite synthesized at 1275 °C.

3.3. *Spin reorientation transitions*

Thermomagnetic measurements were performed on loose powders in a weak (1–10 mT) magnetic field with cycling from room temperature up to 600 °C. An example of magnetization as a function of temperature for W-type $BaCo_xZn_{2-x}Fe_{16}O_{27}$ ferrite with $x = 0.8$ is shown in figure 5(a). Heating and cooling segments do not match (typical hysteretic behaviour), and the first heating differs from others because of particular magnetization history. Magnetic susceptibility is higher in the easy plane state than in the easy axis one, therefore both spin reorientation (at ~ 55 °C) and ferrimagnetic-paramagnetic (at ~ 430 °C) transitions are well resolved. Small amplitude of the anomalies observed above W-ferrite Curie point confirms that fractions of magnetic secondary phases are negligible.

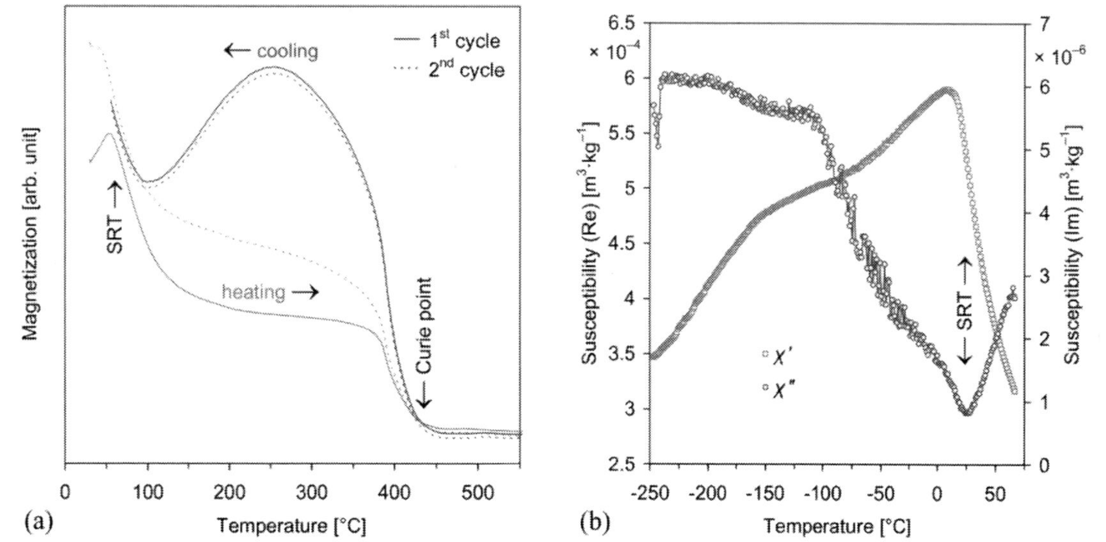

Figure 5. Magnetic transitions in W-type $BaCo_xZn_{2-x}Fe_{16}O_{27}$ ferrites: (a) magnetization measurements for $x = 0.8$ (higher temperatures); (b) AC susceptibility measurements for $x = 0.7$ (lower temperatures).

On the first heating, a sharp drop of magnetization related to SRT is preceded by a distinct peak. The spin reorientation temperature of polycrystalline powder is also characterized by inflection point of magnetization dependence [15]. Another reference can be the point where subsequent cooling and heating curves converge (clearly seen in figure 5(a) on the left of magnetization hump). Unlike [6], the latent heat of SRT has not been reliably detected by differential scanning calorimetry (DSC), while Curie point is well visible.

AC magnetic susceptibility as a function of temperature for W-type $BaCo_xZn_{2-x}Fe_{16}O_{27}$ ferrite with $x = 0.7$ is shown in Fig. 5(b). The real part of susceptibility has a maximum at ~ 6 °C followed by a sharp drop attributed to SRT. Its imaginary part has a minimum at ~ 27 °C which can also be taken as the spin reorientation temperature. Moreover, the minimum of imaginary part turns out to approach the inflection point of the real part.

4. Conclusions

Single-phase W-type ferrite submicron powders with different magnetocrystalline anisotropy configurations are synthesized through a sol-gel citrate precursor route. The effect of calcination temperature and other production parameters on the phase composition, average particle size and magnetic properties of the powders is established. Chemical compositions of substituted hexaferrites undergoing spin reorientation transitions near room temperature are determined.

Acknowledgements

This work is supported by EC Seventh Framework Programme funding (project SSEEC, contract FP7-NMP-214864). The authors are grateful to P. Audebert (ENS Cachan) for assistance in sol-gel technique and G. Wang (ICMPE) for TEM observations.

References

[1] Smit J and Wijn H P J 1959 *Ferrites* (Eindhoven: Philips Technical Library)
[2] Asti G, Bolzoni F, Licci F and Canali M 1978 *IEEE Trans. Magn.* **14** 883–5
[3] Albanese G, Calabrese E, Deriu A and Licci F 1986 *Hyperfine Interact.* **28** 487–9
[4] Paoluzi A, Licci F, Moze O, Turilli G, Deriu A, Albanese G and Calabrese E 1988 *J. Appl. Phys.* **63** 5074–80
[5] Samaras D, Collomb A, Hadjivasiliou S, Achilleos C, Tsoukalas J, Pannetier J and Rodriguez J 1989 *J. Magn. Magn. Mater.* **79** 193–201
[6] Naiden E P, Maltsev V I and Ryabtsev G I 1990 *Phys. Status Solidi* A **120** 209–20
[7] Naiden E P and Ryabtsev G I 1990 *Russ. Phys. J.* **33** 318–21
[8] Sürig C, Hempel K A, Müller R and Görnert P 1995 *J. Magn. Magn. Mater.* **150** 270–6
[9] Turilli G and Asti G 1996 *J. Magn. Magn. Mater.* **157–158** 371–2
[10] Naiden E P and Zhilyakov S M 1997 *Russ. Phys. J.* **40** 869–74
[11] Licci F and Besagni T 1984 *IEEE Trans. Magn.* **20** 1639–41
[12] Srivastava A, Singh P and Gupta M P 1987 *J. Mater. Sci.* **22** 1489–94
[13] Zhang H, Liu Z, Yao X, Zhang L and Wu M 2003 *J. Sol-Gel Sci. Technol.* **27** 277–85
[14] Lutterotti L 2000 *Acta Crystallogr.* A **56** s54
[15] Boltich E B, Pedziwiatr A T and Wallace W E 1987 *J. Magn. Magn. Mater.* **66** 317–22

Growth of magnetic cobalt/chromium nano-arrays by atom-optical lithography

F. Atoneche, D. Malik, A. Kirilyuk, A. J. Toonen, A. F. van Etteger, and Th. Rasing

Radboud University Nijmegen, Institute for Molecules and Materials, Heyendaalseweg 135, 6525 AJ, Nijmegen, The Netherlands

E-mail: `f.atoneche@science.ru.nl`

Abstract. Arrays of magnetic cobalt/chromium (Co-Cr) nanolines are grown by depositing an atomic beam of Co-Cr alloy through a laser standing wave (SW) at $\lambda/2 = 212.8$ nm onto a substrate. During deposition, only the chromium atoms are resonantly affected by the optical potential created by the SW, causing a periodic modulation of the chromium concentration and consequently of the magnetic properties. Magnetic force microscopy and magneto-optical Kerr effect studies reveal a patterned magnetic structure on the substrate surface.

1. Introduction

The fabrication and characterization of micro- and nano-structured magnetic dots and arrays is an area of intense research [1, 2, 3] in recent years. Their main interest stems not only from their importance as model systems for the study of magnetic interactions, including coupling and ordering phenomena [3], but also from their potential for the development of future storage devices. Different techniques (most of them involving two-steps or more) like electro-deposition, the very slow electron beam lithography process, and anisotropic etching of substrates, have been implored for the fabrication of these arrays. Despite all these efforts, a one-step process for the fabrication of magnetic arrays remained a challenge.

Here we present such a one-step technique of fabricating magnetic nanostructures via atom-optical modulation of the concentration of a binary alloy, demonstrating a new method of nanoscale modulated magnetic material composition. The advantage of this technique is that it offers inherent parallelism and ultrahigh vacuum compatibility, with unrivaled periodicity [4].

The method is based on the interaction of a thermal atomic beam with a slightly detuned resonant laser standing wave (SW) resulting in a direct optical manipulation of atoms as described in [5]. By using a combination of laser beams, various structures can be created, such as nanowires and nanodots with 2, 3, 4 or even 5-fold symmetry [6]. Iron nonowires grown in this way however showed no periodic magnetic behavior [5], presumably because of a large background layer.

In this study, we utilize the simultaneous deposition of chromium and cobalt atoms through a spatially varying optical SW potential (for Cr) to create a concentration modulated alloy structure on a silicon substrate. During deposition, selective manipulation of chromium is achieved by using a laser wavelength (of 425.55 nm) that is near resonant with the ^{52}Cr transition $^7S_3 \rightarrow {}^7P_4^0$ while Co is uniformly deposited. Because the magnetic properties of

Published under licence by IOP Publishing Ltd

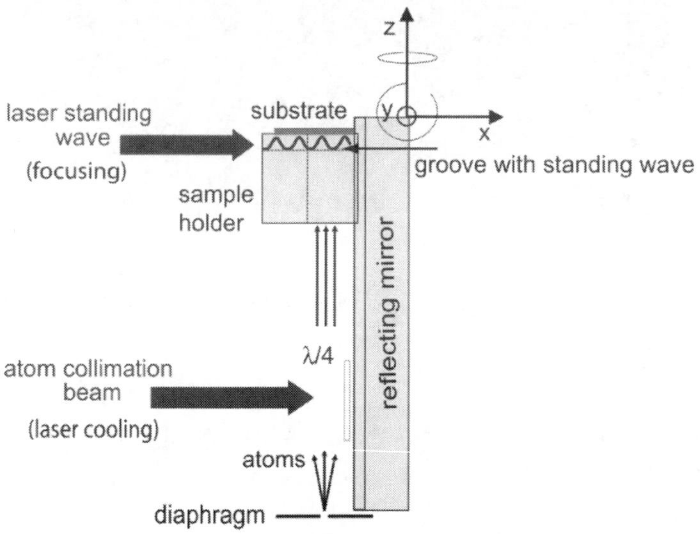

Figure 1. (color online) *In vacuo* optics showing substrate and positions of atom collimation beam and laser standing wave. The required degrees of freedom of the manipulator in x, y and z are also shown.

Co-Cr strongly depend on the Cr concentration, and by carefully selecting the concentration in the starting alloy material, this results in a uniform array of nanolines revealed by atomic force microscopy studies, with periodically varying magnetic properties observable in magnetic force microscopy. The period is 212.8 nm and the full width at half maximum (FWHM) is 90 nm. MOKE measurements reveal a uniaxial anisotropy parallel to the lines and a vanishing saturation magnetization perpendicular to the lines.

2. Experiment

For the experiment, blue continuous wave (CW) laser light at 425.55 nm was obtained by frequency doubling the output of a titanium-doped sapphire ring laser in an external enhancement cavity fitted with a lithium triborate (LBO) crystal [7]. For the deposition, we used a beryllium oxide crucible holding a Co-Cr alloy of composition Co-86/Cr-14 wt percent and particle size of $\leq$ 150 μm. The alloy composition was carefully chosen so that, in the alloy form, the mixture would still be magnetic [8]. We also note that evaporation of the alloy constituents at the same temperature should also be possible. In order to evaporate the alloy for deposition, a high temperature effusion cell holding the crucible with the alloy was heated up to a temperature of 2030 K. This assembly, below called the oven, produced a thermal beam of neutral chromium and cobalt atoms. The base pressure in the deposition chamber is routinely below 10^{-8} mbar and around 10^{-7} mbar during deposition. The thermal beam was collimated by means of transverse laser cooling, after it had gone through a defining diaphragm of 0.7 mm diameter. The power in the cooling beam was 55 mW. Its frequency was tuned by about 10 MHz below the Cr resonance, allowing for the atoms to only lose energy and momentum by scattering of the laser light [9]. The atomic beam could be transversally collimated to a divergence angle of 0.6 $\pm$0.1 mrad FWHM. It was then deposited on a silicon substrate through a one-dimensional SW potential. An *in vacuo* optical alignment unit shown in Fig. 1, could be rotated around the y and z axes and included translations in the x-, y- and z-directions. The mirror was made of a zerodur substrate coated with aluminum. The sample rested on the top part of a sample holder cut from a prism. Two grooves were cut into the holder in the x- and z directions for the laser SW and atom beam respectively.

The SW laser frequency was tuned approximately 200 MHz above the ^{52}Cr transition $^7S_3 \rightarrow ^7P_4^0$ at 425.55 nm using an acousto-optic modulator. The positive detuning causes the atoms to be focused to the intensity minima (nodes) of the SW [10]. The power in the SW was 30 mW focused to $W_0 = 200\ \mu$m. The optical unit (Figure 1) was then moved into position so that the vertical atomic beam was directed into the groove and through the SW. The substrate was then moved down to the center of the standing wave. We note here that only the chromium atoms are affected by the light. The deposition time was 30 minutes.

3. Results and Discussion

We studied the fabricated samples *ex vacuo* by atomic force microscopy (AFM), magnetic force microscopy (MFM) and the magneto-optical Kerr effect (MOKE). Figure 2(a) shows an AFM picture of an array of periodic nanolines with a height above background (film formed from atoms not focused by the SW) of 3 nm. Note that the background film thickness was determined by ellipsometry to be 10 $\pm$ 0.04 nm. This gives a height to background ratio of $\approx 1:3$. The periodicity is 212.8 nm which is that of the interference pattern created by the standing wave laser beam at 425.55 nm. Notice an obstacle (probably dirt particle) on the surface of the sample. Figures 2(b,d) show a zoomed-in AFM and an MFM picture of the obstacle, respectively. While no topographic contrast is visible through this round obstacle (Figure 2(b)), the MFM image, of the same area as in the AFM picture, taken at a lift height of 30 nm, shows a clear magnetic contrast. (Figure 2(d)).

In the phase detection scheme of MFM, changes $\Delta\phi$ in the phase shift of the cantilever's phase oscillation relative to a piezo drive are measured. If a force F acts on the cantilever in the opposite direction to the vibration direction z, then the changes in the phase shift can be expressed as [11]:

$$\Delta\phi \approx -Q/K(\partial F/\partial z), \tag{1}$$

where K is the cantilever spring constant and Q is the quality factor of the vibrating system. For an attractive force, the force gradient, $\partial F/\partial z > 0$. This leads to a negative phase shift, $\Delta\phi$. The negative phase shift means that the phase ϕ is reduced, corresponding to a darker contrast. For a repulsive force, $\partial F/\partial z < 0$, leading to an increase in the phase which corresponds to a bright contrast.

The MFM measurements shown in Figures 2(c) and (d) fit the description of standard MFM behavior above, as the dark contrast regions are magnetic and correspond to regions where an attractive interaction exists between the sample and the tip. In our sample, the ridges have higher chromium concentration due to the focused chromium atoms. A higher Cr concentration leads to non magnetic ridges. On the other hand, the valleys have a higher concentration of cobalt since most of the chromium atoms are directed away from these regions by the laser standing wave pattern. The result is a magnetic contrast between the magnetic valleys and non-magnetic ridges, revealed by MFM imaging.

In order to further characterize the magnetic properties of the nanolines, static longitudinal MOKE measurements were performed using a HeNe laser polarized perpendicular to the plane of incidence. The inset of Figure 3(a) shows a schematic picture of the experimental layout. The laser beam was focused to a spot size of 40 microns. This allowed for a good spatial resolution to obtain a signal related to the structured region of the sample only. Figure 3(a) shows MOKE measurements at different positions, 1, 2 and 3, across the sample for ϕ= 0o . No magnetic signal was observed at position 1, where the Co-Cr film is homogeneous. This was further confirmed by the non saturation of the magnetization in this region even at fields as high as 0.2 kG. This means that during the deposition, a higher percentage of Cr than contained in the original material, might have been evaporated. This can be explained by a higher Cr vapour pressure

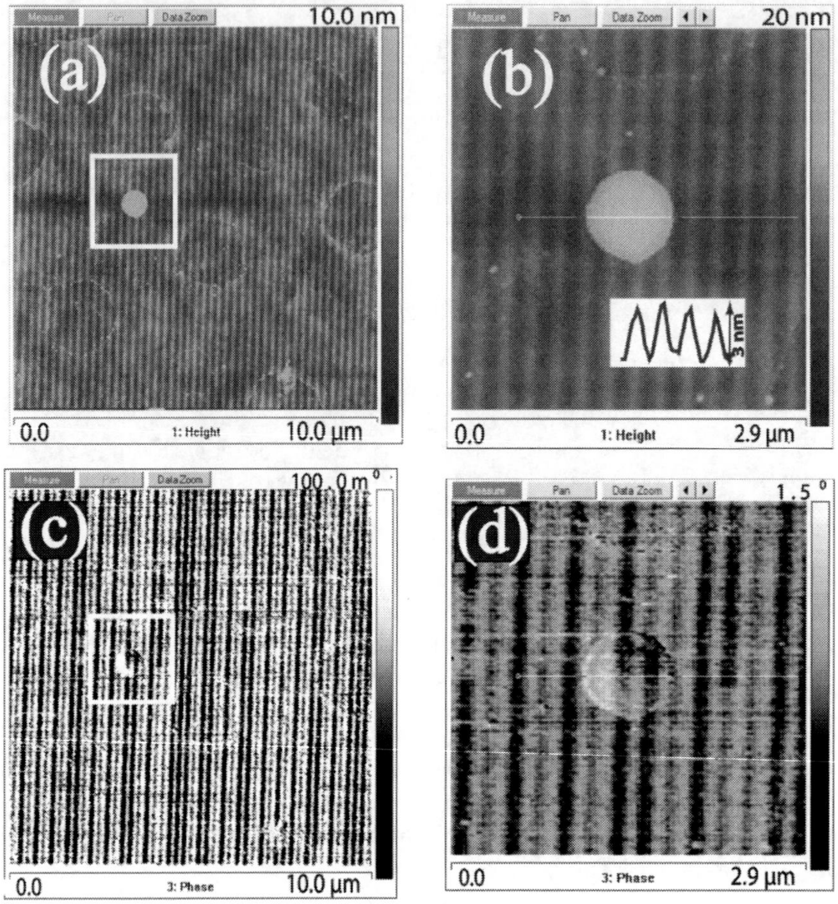

Figure 2. (color online)Laser-focused Co-Cr nanolines a) Patterned Co-Cr alloy lines as observed by AFM. b) zoom in showing area enclosed by rectangle in (a). c) MFM image. d) shows enclosed area in (c) with magnetic contrast visible under round spot in middle. Black areas are magnetic and white areas are non-magnetic.

at the evaporation temperature [12]. At position 2, on the edge of the modulated area, a weak magnetic contrast can be observed which further increases towards the center of the nanolines (position 3).

The sample could also be rotated around an axis perpendicular to its surface, leading to a variation of the angle ϕ of the applied field with respect to the direction of the nanolines.

Figure 3(b) shows MOKE hysteresis loops measured for different orientations ϕ of the applied magnetic field, relative to the nanolines. The maximum Kerr rotation is observed when $\phi = 0°$. In this case the hysteresis loop is almost square-like. This behavior is characteristic of a specimen magnetized to saturation along an easy axis. For this case, the saturating field is about 0.2 G. Increasing the angle ϕ leads to a smaller Kerr rotation which disappears completely for $\phi = 90°$. The inset of Figure 3(b) shows a plot of the Kerr angle θ_k at magnetization saturation, normalised by the maximum Kerr angle θ_m, at parallel alignment (ϕ=0°) as a function of the angle ϕ. The normalized Kerr rotation at magnetization saturation decreases with increasing ϕ. The magnetization saturation and coercivity fields are extremely low compared to typical large saturation fields and coercivities of thin film Co-Cr alloys of comparable thickness [14]. Note

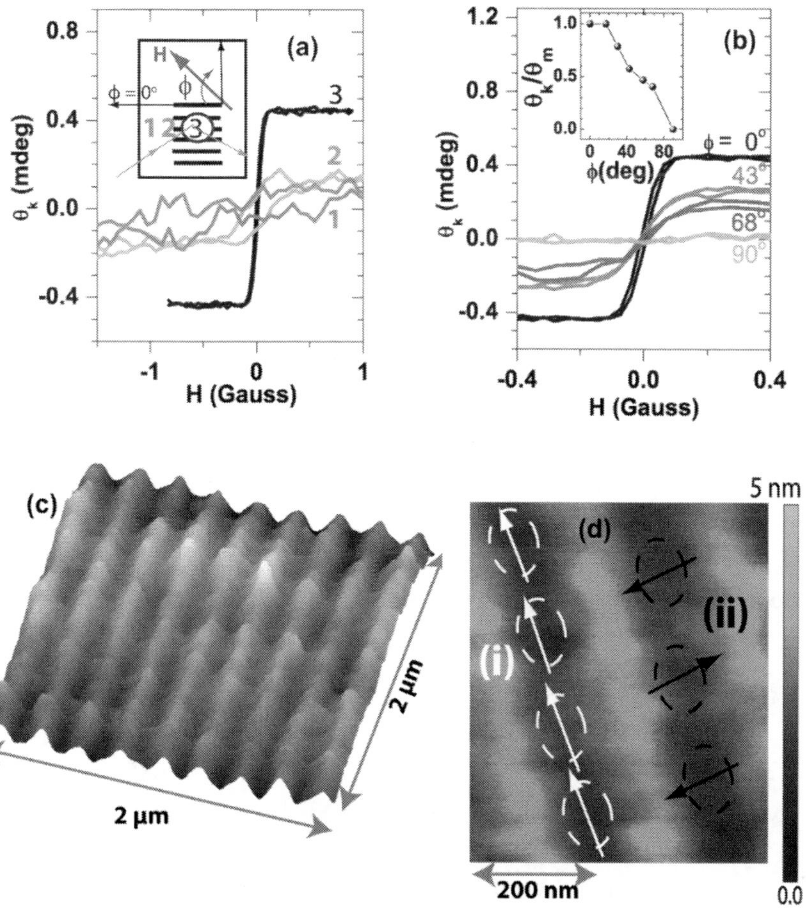

Figure 3. (color online) Inset of a) schematic picture of the sample for MOKE measurements. a) Hysteresis loops across the sample for ϕ=0°. b) Hysteresis loops measured for different angles ϕ. Inset of b) Normalized plot of Kerr angle θ_k, at saturation as a function of ϕ. c) High resolution 3D AFM image showing disconnected "droplets". d) AFM image showing disconnected magnetic "droplets" and alignment scenario in parallel field (i), and in perpendicular field (ii).

that the earth's magnetic field was compensated for during deposition and ex-vacuo magneto optical studies. We observe not just an increase of the saturation field, as is expected when going from the easy- to the hard magnetization axis. In addition, the saturation magnetization vanishes, which was also verified by increasing the applied magnetic field range up to 0.4 kG. Therefore, the patterned area exhibits an anisotropy parallel to the nanolines. This effect is nontrivial and cannot be explained by treating the nanolines as being fully continuous. We can explain this behavior by assuming that our nanolines consist of small magnetic "droplets" that are magnetostatically coupled to each other. A strong argument for the presence of such a non-continuous structure is given by the MFM picture: Moreover, if the nanolines were continuous, no contrast in the MFM images would be observed at all because stray fields "felt" by the MFM tip would only be present at the ends of the nanowires. Figure 3(c,d) appears to reveal these magnetic "droplets". The coupling is ferromagnetic if the magnetization is along the "droplets" chain. Rotating the magnetization by 90° changes the coupling to antiferromagnetic, destroying the net magnetization of the chain. Note that given the subtle balance of Co-Cr concentration ratio, the formation of such magnetic droplets is quite probable.

The results of magnetic characterization presented above only provide a qualitative understanding of the magnetic properties of the concentration modulated patterned nanolines on the substrate. Moreover, although the composition of the alloy is carefully chosen before evaporation, that of the material deposited on the substrate is largely unknown because of the uneven evaporation of Cr and Co due to the difference in their vapor pressures at the evaporation temperature, and the vacuum conditions during evaporation. Therefore a further quantitative analysis of the observed magnetic behavior cannot be made without further studies, such as compositional analysis.

4. Conclusions

In conclusion, we fabricated Co-Cr magnetic nanostructures by modulating the concentration of chromium in a Co-Cr alloy via atom-optics at the nanoscale, thus demonstrating the proof-of-principle of a new method for nanoscale-modulated magnetic material composition. AFM investigations show a global periodic pattern of lines with a period of 212.8 nm and a height above the background of 3 nm while detailed MFM and MOKE studies reveal an alternating pattern of magnetic and nonmagnetic nanolines. Concentration modulation of an alloy constituent via atom-optics for nanofabrication is therefore a valid technique for the production of periodic magnetic nanostructures. By extending this method to 2-dimensional patterns [15], magnetic nanostructures with a variety of symmetries can be produced.

Acknowledgments

This work was supported by the Dutch organization for Fundamental Research on Matter (FOM), the NanoNed consortium and NWO.

References

[1] Mallet J, Yu-Zhang K, Chien C, Eagleton T S, and Searson P C 2004 Appl. Phys. Lett. **84**, 3900
[2] Choi J, Oh S J, Ju H, and Cheon J 2005 Nano Letts. **5**, 2179
[3] Wang Z K, Kuok M H, Ng S C, Lockwood D J, Cottam M G, Nielsch K, Wehrspohn R B, and Gsele U 2002 Phys. Rev. Lett. **89**, 027201
[4] McClelland J J, Scholten R E, Palm E C, and Celotta R J 1993 Science **262**, 877
[5] Myskiewicz G, Hohfeld J, Toonen A J, van Etteger A F, Shklyarevskii O I, Meerts W L, and Rasing Th 2004 Appl. Phys. Lett. **85**, 3842
[6] Jurdik E, Myszkiewicz G, Hohlfeld J, Tsukamoto A, Toonen A J, van Etteger A F, Gerritsen J, Hermsen J, Goldbach-Aschemann S, Meerts W L, van Kempen H, and Rasing Th (2004) Phys. Rev. B **69**, 201102,
[7] Jurdik E, Hohlfeld J, van Etteger A F, Toonen A J, Meerts W L, van Kempen H, and Rasing Th 2002 J. Opt. Soc. Am. B **19**, 1660
[8] Maeda Y, Asahi M and Seki M 1986 Jpn. J. Appl. Phys. **25**, L668
[9] Hansch T W and Schawlow A L 1975 Optics Comm. **13**, 68
[10] McClelland J J and Sheinfein M R 1991 J. Opt. Soc. Am. B **8**, 502
[11] Hartmann U 1999 Annu. Rev. Mater. Sci. **29**, 53
[12] Lide D R (ed) 2003 CRC Handbook of Chemistry and Physics, 84th Edition, CRC Press, Boca Raton, Florida, Section 6, Fluid Properties; Vapor Pressure.
[13] Theis-Brohl K, Schmitte T, Leiner V, Zabel H, Rott K, Brukl H and McCord J 2003 Phys. Rev. B **67**, 184415
[14] Yamamoto S, Wada H, Kurisu H and Matsuura M 2001 J. Magn. Magn. Mater. **235**, 133
[15] Bradley C C, Anderson W R, McClelland J J, and Celotta R J 1999 Appl. Surf. Science **141**, 210

Influence of Temperature and Mechanical Scratch on the Recorded Magnetization Stability of Longitudinal and Perpendicular Recording Media

Katsumasa Nagano, Kousuke Tobari and Masaaki Futamoto

Faculty of Science and Engineering, Chuo University, 1-13-27 Kasuga, Bunkyo-ku, Tokyo 112-8551, Japan

E-mail: nagano@futamoto.elect.chuo-u.ac.jp

Abstract. Stability of recorded magnetization of hard disk drive (HDD) is influenced by external environments, such as temperature and magnetic field. Small scratches are frequently formed on HDD medium surface upon contacts with the magnetic head. The influences of temperature and mechanical scratch on the magnetization structure stability are investigated for longitudinal and perpendicular recording media by using a magnetic force microscope. PMR media remained almost unchanged up to about 300 °C for the area with no scratches, whereas the areas near and under mechanical scratches started to change around 250 °C. The magnetization structure of LMR media started to change at about 100 degrees lower temperature under mechanical scratches when compared with no scratch areas. A quantitative analysis of magnetization structure variation is carried out by measuring the recorded magnetization strength difference estimated from the MFM images observed for a same sample area before and after exposing the sample to different temperatures.

1. Introduction

The area density of hard disk drive (HDD) has been increasing through implementation of perpendicular magnetic recording. As the area density increases, stability of recorded magnetization is becoming an important research topic. Various studies have been carried out by using computer simulation [1,2], spin-stands [3,4] and magnetic force microscopes (MFMs) [5–7]. MFM is useful to directly observe the detailed variation of recorded magnetization structure. The stability of recorded magnetization will be influenced by several factors, such as temperature, magnetic field, stress, impact, etc. Increasing magnetic recording density requires lower flying height of head slider. A modern HDD generally requires head-disk spacing of less than 10 nm. With such a narrow spacing, small scratches are frequently formed on the medium surface upon contacts with the magnetic head. The influence of mechanical scratch on the magnetization of recording media has been studied for longitudinal [8,9] and perpendicular media [10–12]. In the present study, the influences of temperature and mechanical scratch on the recorded magnetization structure were investigated for longitudinal (LMR) and perpendicular (PMR) media under similar experimental conditions using a magnetic force microscope equipped with a heating stage. The magnetization stability was compared between PMR and LMR recording media.

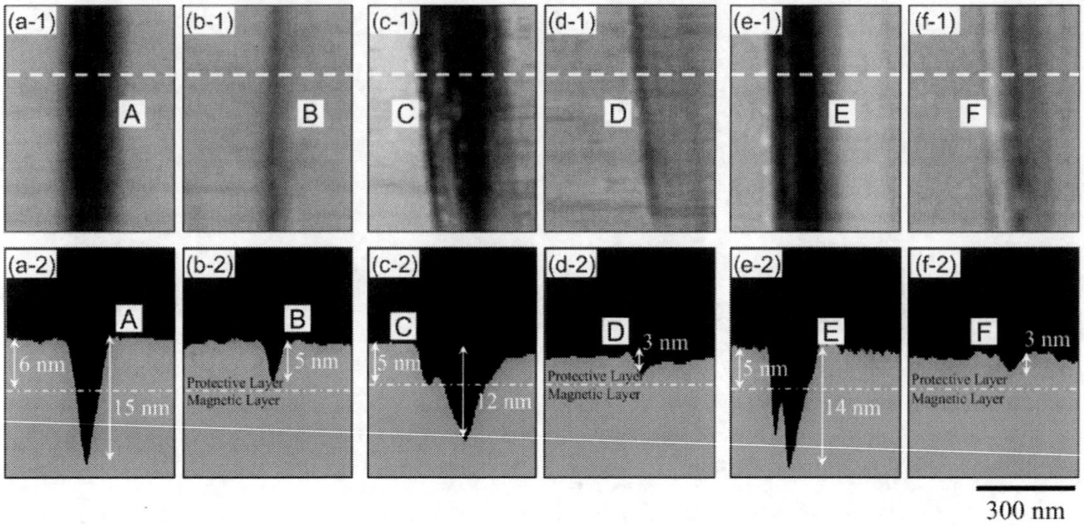

Figure 1. AFM images of (a-1, b-2) LMR medium (120 Gb/in^2), (c-1, d-1) PMR medium (163 Gb/in^2), and (e-1, f-1) PMR medium (528 Gb/in^2) with two mechanical scratches. Cross-sectional surface profiles measured along the respective dotted lines are shown in the lower figures (a-2, b-2, c-2, d-2, e-2, f-2).

2. Experimental procedure

A commercial LMR HDD disk with 120 Gb/in^2 area density, a PMR HDD disk with 120 Gb/in^2 area density, and a PMR HDD disk with 528 Gb/in^2 were used as the samples. The HDD disks were cut into small pieces to observe the magnetization structure under an MFM. Before cutting the disks, random data were recorded on the HDDs. Mechanical scratches were formed on the small piece sample with a diamond-coated cantilever using an atomic force microscope (AFM). The cantilever was moved along a line while applying an adequate force to form a mechanical scratch. The depth was controlled by adjusting the force. The sample surface was observed by using the AFM. The protective layer thickness of HDD disk sample was estimated by employing a reflectivity measurement using X-rays. The samples were then set on a heating stage in an MFM chamber which could be evacuated down to 10^{-4} Pa vacuum. The sample temperature was controlled in a range between room temperature (RT) and 300 °C. Same sample areas were observed with the MFM repeatedly after cooling the sample to RT.

3. Results and discussion

3.1. Formation of mechanical scratches

Figure 1 shows AFM images of LMR and PMR HDD disk samples with mechanical scratches. The bottom figures show the cross-sectional profiles of sample surface measured along the dotted lines shown in the upper figures. The depths are 15 nm and 5 nm for the two mechanical scratches (A and B) of the LMR medium. The protective layer thickness is estimated to be 6 nm by X-ray analysis. Figure 1(c-1), (d-1) and Figure 1(e-1), (f-1) respectively indicate the AFM images of PMR HDD disk samples (120 Gb/in^2 and 528 Gb/in^2) with mechanical scratches. The depths are 12 nm, 3 nm, 14 nm, and 3 nm for the mechanical scratches of C, D, E, and F, as respectively shown in Figs. 1(c-2), (d-2), (e-2), (f-2). For both PMR samples, the protective layer thickness is estimated to be 5 nm. A deeper scratch is penetrating the protective layer and is causing some plastic deformation in the recording layer, while a shallow scratch is within the protective layer thickness.

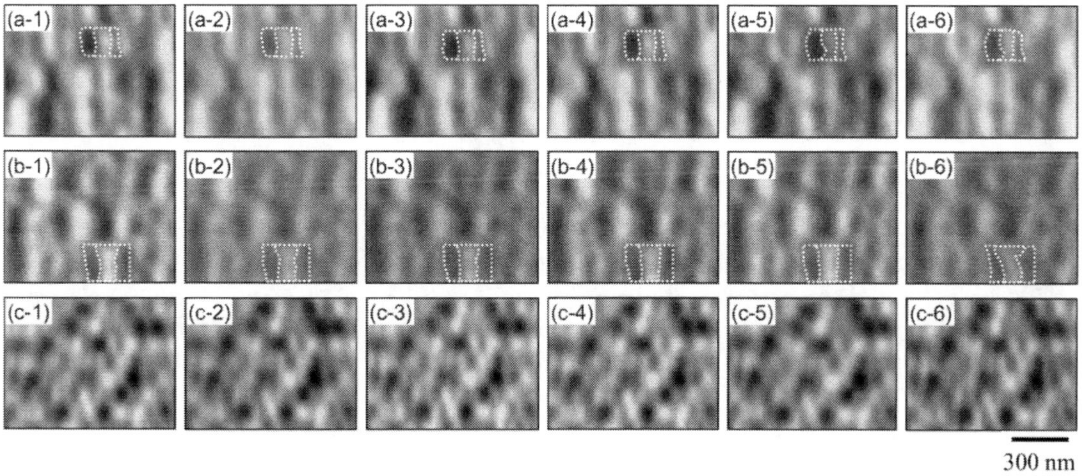

Figure 2. MFM images of (a) LMR medium (120 Gb/in^2), (b) PMR medium (120 Gb/in^2), and (c) PMR medium (528 Gb/in^2) with no mechanical scratches. The samples were heated at various temperatures for about 40 min. and cooled down to RT for MFM observation. The temperatures are (1) RT, (2) 100 °C, (3) 150 °C, (4) 200 °C, (5) 250 °C, and (6) 300 °C. The white dotted lines are added to indicate clearly the variations of magnetization structure for the samples of LMR (120 Gb/in^2) and PMR (120 Gb/in^2) media.

3.2. Influence of temperature on the recorded magnetization

Figure 2(a-1)–(a-6) shows a series of MFM image observed for a same position area after exposing the LMR sample at elevated temperatures. The sample temperature was increased in a stepwise up to 300 °C. The duration of heating at respective temperature was kept nearly constant at 40 minutes. The MFM images were observed at RT by cooling from high temperatures to prevent sample drift at elevated temperatures. In the MFM images, the bright contrast corresponds to the area where a repulsive force is working between the MFM cantilever and the medium, while the dark contrast corresponds to the region where an attractive force is working. The magnetization structure of LMR sample is kept unchanged up to 200 °C. Magnetization structure variation is recognizable when the sample was heated at 250 °C as shown in Figure 2(a-5). It is notable, for example in the areas shown with dotted white lines, that the magnetization transition starts to shift parallel to the track direction. The magnetization transition profile is apparently different between the images shown in Figure 2(a-4) and (a-5).

Figures 2(b-1)–(b-6) and (c-1)–(c-6) respectively indicate the magnetization structure variations of PMR samples (120 Gb/in^2 and 528 Gb/in^2). The low area density PMR sample is notable that the recorded bit shape starts to change when the sample was heated at 300 °C as shown in Figure 2(b-6), whereas the high area density PMR sample shows no variation.

3.3. Influence of temperature and mechanical scratch on the recorded magnetization

Figure 3 shows an AFM and a series of MFM images observed for a same area of LMR sample with mechanical scratches when the sample temperature was increased. Recorded magnetization structure is visible for the area under the shallow mechanical scratch of B, whereas it is apparently deformed under the deep scratch of A as shown in Figure 3(a-1). The recording layer under the deep scratch A is considered to have damaged where the depth is 15 nm and exceeds the carbon protective layer thickness. The AFM image (Figure 3(a)) shows the topological information of the disk surface while the MFM image (Figure 3(a-1)) shows the magnetic information. It is notable that the recorded bit shape near the shallow mechanical scratch starts to change at around 150 °C. The result suggests that

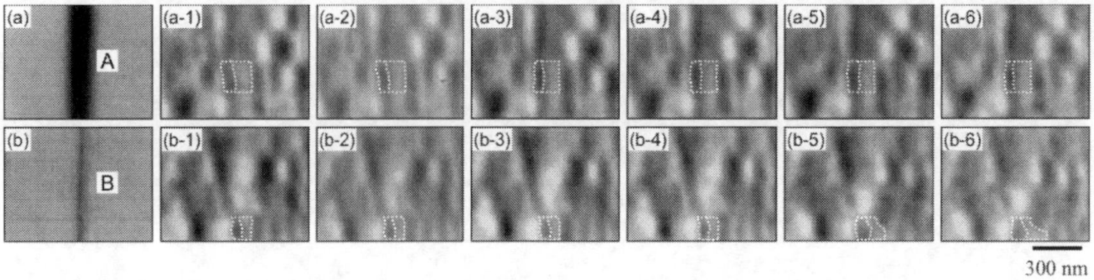

Figure 3. AFM and MFM images of LMR sample with mechanical scratches. (a) and (b) are the AFM images of the sample shown in Figure 1 for the areas of A and B, respectively. (1)-(6) are the MFM images of the areas corresponding respectively to the AFM images of (a) top row, (b) bottom row, observed after exposing the sample to (1) RT, (2) 100 °C, (3) 150 °C, (4) 200 °C, (5) 250 °C, and (6) 300 °C. The white dotted lines are added to indicate clearly the variations of magnetization structure.

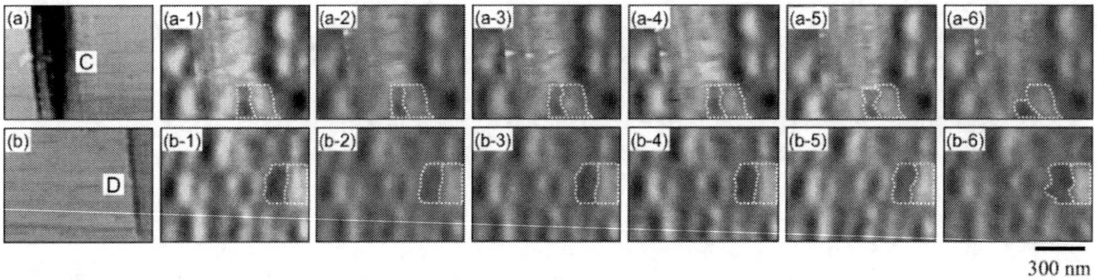

Figure 4. AFM and MFM images of PMR sample (120 Gb/in^2) with mechanical scratches. (a) and (b) are the AFM images of the sample shown in Figure 1 for the areas of C and D, respectively. (1)-(6) are the MFM images of the areas corresponding respectively to the AFM images of (a) top row, (b) bottom row, observed after exposing the sample to (1) RT, (2) 100 °C, (3) 150 °C, (4) 200 °C, (5) 250 °C, and (6) 300 °C. The white dotted lines are added to indicate clearly the variations of magnetization structure.

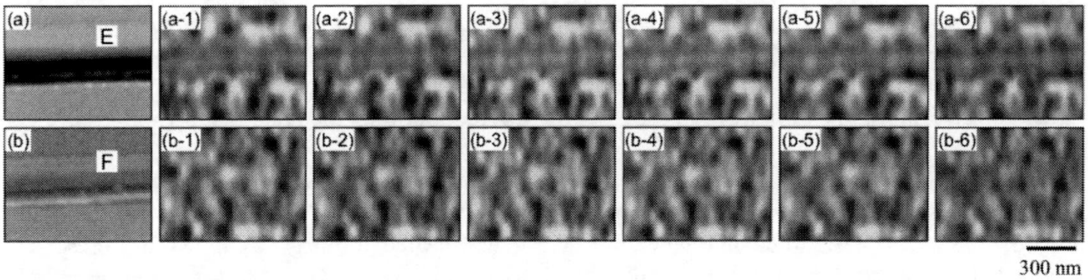

Figure 5. AFM and MFM images of PMR sample (528 Gb/in^2) with mechanical scratches. (a) and (b) are the AFM images of the sample shown in Figure 1 for the areas of E and F, respectively. (1)-(6) are the MFM images of the areas corresponding respectively to the AFM images of (a) top row, (b) bottom row, observed after exposing the sample to (1) RT, (2) 100 °C, (3) 150 °C, (4) 200 °C, (5) 250 °C, and (6) 300 °C.

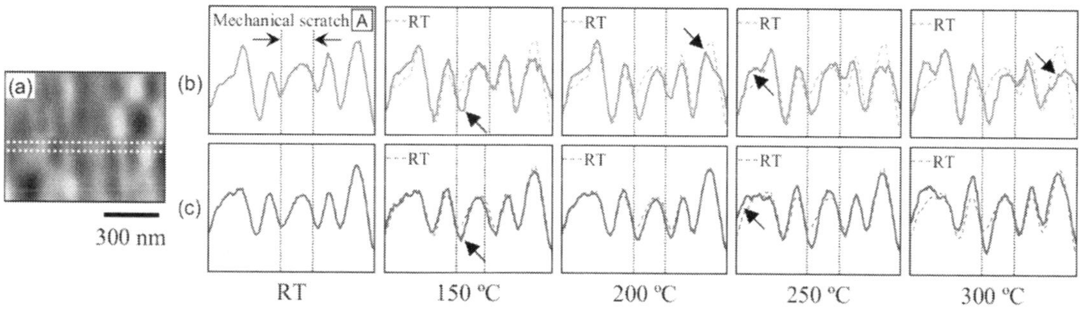

Figure 6. (a) MFM image of LMR sample with mechanical scratches and (b), (c) MFM intensity line profiles. MFM line profiles were measured after exposing the sample to RT, 150 °C, 200 °C, 250 °C, and 300 °C: (b) along upper track edge and (c) along track center. The arrows show the positions where bit shifts are observed.

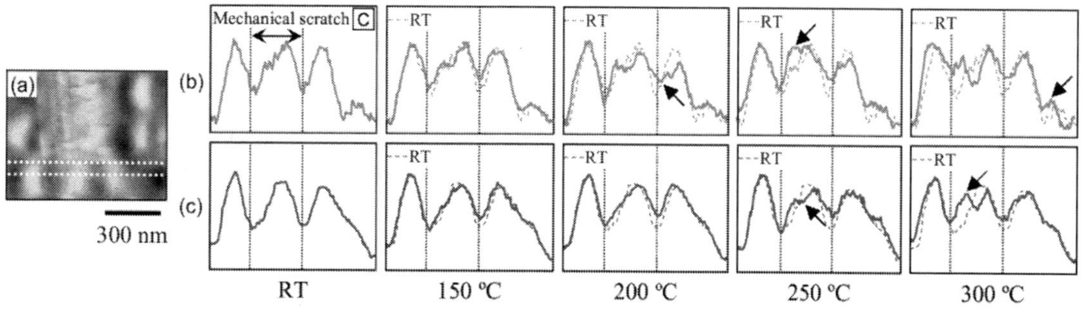

Figure 7. (a) MFM image of PMR sample (163 Gb/in^2) with mechanical scratches and (b), (c) MFM intensity line profiles. MFM line profiles were measured after exposing the sample to RT, 150 °C, 200 °C, 250 °C, and 300 °C: (b) along upper track edge and (c) along track center. The arrows show the positions where bit shifts are observed.

the mechanical stress caused by the scratch destabilizes the recorded magnetization.

Figure 4 shows an AFM image and the magnetization structure variation of PMR sample (120 Gb/in^2) with mechanical scratches when the sample temperature was increased. The recorded magnetization structure is deformed under the scratch of C similar to the case of LMR medium sample (Figure 4(a-1)). From the series of MFM images shown in Figure 4, it is clear that the recorded magnetization structure starts to change noticeably near and under the mechanical scratches when the sample was heated at 250 °C. The temperature above which the recorded magnetization structure starts to change is lowered near the deep scratch and around the shallow scratch when compared with other normal areas for both types of media. The result suggests that a stress distribution around the scratch delicately influences the demagnetization behavior of recorded magnetization.

Figure 5 indicates an AFM image and the magnetization structure variation of PMR sample (528 Gb/in^2) with mechanical scratches when the sample temperature was controlled in a range between room temperature and 300 °C. There is no variation in the area around the deep scratch E and the shallow scratch F of perpendicular magnetization for the medium with increased area density. The result indicates that the recorded magnetization of perpendicular medium tends to be stabilized with in creasing the recording density.

3.4. Analysis of magnetization structure variation

The recorded magnetization variation is estimated from the MFM contrast intensity measured along the track centers. Examples are shown in Figure 6 and Figure 7. The analysis lines along the track center and the along upper track edge are shown as the white dotted lines in the respective MFM images. It is possible to estimate the bit shift caused by magnetic domain wall movement on heating. Bit shifts are apparently enhanced by the presence of mechanical scratches. When the LMR sample is exposed to 150 °C, the bit shift starts to be observed in the area near the mechanical scratch as indicated by an arrow. In the area with no mechanical scratch, the MFM intensity peak positions tend to move in the direction parallel to the recording track when the sample is exposed to 200 °C. When the magnetization variations are compared between near and under a scratch, it is notable that MFM contrast intensity variation tends to be observed more preferentially around the scratch fringes. The magnetization structure changes easily near the track edge rather than at the track center. Such influences are noted up to 45 nm distance from the scratch fringe as shown in Figure 7(b). For both media, the MFM contrast intensity became reduced and degraded at 300 °C.

4. Summary

Influences of temperature and mechanical scratch on the recorded magnetization structure of LMR and PMR media have been investigated using an MFM equipped with a heating stage. The magnetization structure starts to change at lower temperature in the area near and under the mechanical scratch when compared with a normal area with no scratch for both the LMR and PMR media with similar area density (120 Gb/in^2). With increasing the area density of PMR medium, the magnetization structure is stabilized due to a decrease of demagnetization and shows no variation up to 300 °C. The recording layer under a mechanical scratch where the depth exceeds the carbon protective layer tends to lose the recorded magnetization structure, while that of under a shallow mechanical scratch less than the protective layer thickness tends to keep the recorded magnetization structure. The recorded magnetization stability decreases preferentially around the fringes of mechanical scratch.

Acknowledgement

A part of this work was supported by NEDO and MEXT (No. 22560302), Japan, and Chuo University Grant for Special Research.

References

[1] Igarashi M, Hara M, Suzuki Y, Nakamura A, Sugita Y, 2003 *IEEE Trans. Magn.* **39** 2303
[2] Wang J W, Yuan Z M, Liu B, Ding J, 2000 *IEEE Trans. Magn.* **36** 2686
[3] Valcu B, Brockie R, Nan-Hsiung Y, Richter H.J, 2007 *IEEE Trans. Magn.* **43** 2115
[4] Abarra E N, Sato H, Suzuki M, Okamoto I, 2000 *IEEE Trans. Magn.* **36** 2450
[5] Yasui Y, Shimomai K, Futamoto M, 2007 *J. Magn. Soc. Jpn.* **31** 312
[6] Takahoshi H, Saito H, Ishio S, Yamaoka T, Okamoto I, 2002 *J. Magn. Soc. Jpn.* **26** 284
[7] Yoshimura T, Oshika S, Futamoto M, 2009 *J. Magn. Soc. Jpn.* **33**, 199
[8] Liew T, Siong W W, Shiau K C, Chwee T L, 2000 *Tribol. Int.* **33** 611
[9] Suk M, Jen D, 1998 *IEEE Trans. Magn.* **34** 1711
[10] Xu J, Furukawa M, Nakamura A, Honda M, 2009 *IEEE Trans. Magn.* **45** 893
[11] Nagano K, Sasaki S, Futamoto M, 2010 *J. Phys. Conf. Ser.* **200** 102006
[12] Nagano K, Sasaki S, Futamoto M, 2010 *J. Magn. Soc. Jpn.* **34** 185

Interacting low dimensional nanostructures within a porous silicon template

K. Rumpf[1], P. Granitzer[1], G. Hilscher[2], P. Poelt[3]

[1] Institute of Physics, Karl Franzens University Graz, Universitaetsplatz 5, A-8010 Graz, Austria

[2] Institute of Solid State Physics, Vienna University of Technology, Wiedner Hauptstr. 8, 1040 Vienna, Austria

[3] Institute for Electron Microscopy, University of Technology Graz, Steyrergasse 17, A-8010 Graz, Austria

klemens.rumpf@uni-graz.at

Abstract. Ni nanostructures of different geometries as spheres, ellipsoids or wires, as well as with various spatial distributions are deposited electrochemically within the pores of porous silicon (PS). The hybrid material offers magnetic properties which depend on the morphology of the composite and the interactions between the metal precipitates, respectively. So the magnetic behavior can be correlated with the process parameters. On the one hand the morphology of the PS template determines the coupling of the metal structures located within adjacent pores and on the other hand the spatial distribution and geometry of the deposits influences the interaction of nanostructures within one pore. Thus also the anisotropy of the nanocomposite can be modified not only by the shape of the ferromagnetic nanostructures but also by their spatial density within the oriented pores of the porous matrix.

1. Introduction

The fabrication of materials having their characteristic length scale in the nanometer regime are of great interest due to their specific properties. Top-down or bottom-up processes in using lithographic or self-organizing methods are widely spread to achieve such nanosized objects whereas self-organization is of special interest especially to obtain large area samples of nanostructures and nanoparticles on surfaces and in 3-dim arrays. A lot of works report on metal or oxide particles dispersed in a host material as another metal [1] or a nonmagnetic matrix [2]. Polymers are often utilized as matrix to investigate the properties of the obtained composite as well as the interaction between the magnetic particles [3] within the investigated system. Furthermore there are reports on the formation of mixed metal complexes leading to a phase-homogeneous product at low temperatures which is obtained by dispersing magnetic particles within silica [4]. Silicon oxide is also used as host for the fabrication of a superparamagnetic composite material with particle-size dependent properties [5]. The presented work reports on a magnetic/semiconducting nanocomposite material which is achieved by electrodeposition of ferromagnetic metal nanostructures into the pores of the silicon matrix whereas the self-organized pores are grown straight and normal to the silicon surface. The shape and size of the deposits as well as their spatial distribution within the pores can be controlled by the electrochemical process parameters and thus samples with desired magnetic properties can be achieved. Semiconductor and ferromagnetic materials combined in one system join the electronic properties of the semiconductor and the ferromagnetic behaviour of the Ni nanostructures.

2. Experiments

A wet etching process of a highly n-doped silicon wafer under certain conditions leads to porous silicon in the mesoporous regime which is acting as template material. The formation of straight grown pores with an average pore-diameter of 55 nm is achieved by anodization in an 8 wt% aqueous hydrofluoric acid solution in applying a constant current density of 80 mA/cm^2. The pores are oriented perpendicular to the surface and clearly separated from each other. The pore-arrangement is self-assembled but nevertheless offers a four-fold symmetry of domains in the micrometer regime under certain conditions. Such a quasi regular morphology is obtained by applying a constant current density between 75 mA/cm^2 and 100 mA/cm^2 [6]. For higher current densities the critical current density J_{PS} is reached and instead of pore formation electropolishing at the porous silicon/silicon border takes place [7], whereas in the case of lower current densities the pore-diameter becomes smaller and less regular. In the case of smaller pore-diameters (< 40 nm) the distance between the pores is enhanced and exceeds twice the thickness of the space charge region and thus nucleation of pores in the remaining silicon walls is possible.
Within the fabricated highly oriented pores a ferromagnetic metal e.g. Ni is electrochemically deposited resulting in a nanocomposite material consisting of a silicon matrix and precipitated Ni nanostructures. The deposition is performed in a pulsed way under cathodic conditions. As electrolyte an adequate metal salt solution is used whereas in the case of Ni-deposition the so called Watts electrolyte ($NiCl_2$, $NiSO_4$) with boric acid as buffer is utilized. A variation of the pulse frequencies and the current density leads to modifications of the geometry of the metal deposits [8]. The shape of the metal deposits can be controlled by varying the pulse-frequency of the applied current density between 0.025 Hz and 0.2 Hz which results in a modification from spheres to wires of an aspect ratio of 100 (ratio of length to diameter). The spatial distribution of the precipitates can be adjusted by varying the current density between 5 mA/cm^2 and 25 mA/cm^2 whereas the distance between precipitated particles within the pores is modified by the time between the pulses. In the case of densely packed particles (mean distance about 10 nm) the time between the pulses is 5 s and for loosely packed particles (mean distance about 200 nm) it is 20 s. This opens a broad avenue to achieve desired magnetic properties of the specimens but requires accurate control of the deposition parameters. The temperature of the electrolyte during deposition is also an important factor but it has been kept constant at room temperature during the presented experiments.
Magnetic measurements are carried out by SQUID-magnetometry (SQUID magnetometer Cryogenics) up to a magnetic field of 6 T and in a temperature range between 4 and 250 K. The structure and composition of the nanocomposite is investigated by scanning electron microscopy (SEM) and transmission electron microscopy (TEM). Especially the shape of the metal deposits and their average size is gained from many successively recorded back scattered electron (BSE) images over a broad cross-sectional region of the samples and using image processing.

3. Discussion

In order to investigate the correlation between magnetic characteristics and structure of the nanocomposite we deposit different metal nanostructures within a silicon matrix. The porous silicon structure offers pores which are arranged in a quasi regular way depending on the anodization conditions and the deposited metal structures are distributed randomly within these pores. The morphology of the porous silicon matrices can be varied in pore-diameter, pore-distance and pore-length as discussed in the experimental section. To assure a clear separation of the channels the distance between the pores must not exceed twice the thickness of the space charge region of the substrate which is around 10 nm for highly doped silicon. In this investigated morphology regime always a dendritic growth of pores is observed which means that small side-pores (length smaller than the pore-radius) grow in (111)-direction additionally to the main-pores [9]. Pulsed electrodeposition is employed to deposit Ni structures within mesoporous silicon whereas this technique allows to vary the pulse frequency of the applied current and the time between the pulses. The precipitated Ni structures within the pores are also self-assembled which results in a randomly arranged but more or less homogeneously over the entire porous matrix distributed metal nanostructure array. The cross-sectional image of a specimen gained by scanning electron microscopy in figure 1 shows a homogeneous distribution of sphere-like Ni structures within the pores whereas the distance between the particles is small enough for dipolar coupling.

Temperature dependent magnetization measurements show that the coercivity (H_C) of porous silicon matrices with precipitated Ni-particles (figure 2a) is always higher than H_C of samples with deposited elongated Ni-structures (figure 2b) due to the higher demagnetizing field of the latter ones which exhibit an average elongation of 1 µm. H_C varies in the case of Ni-particles between 580 Oe (T = 4 K) and 410 Oe (T = 250 K), if the magnetic field is applied perpendicular to the sample surface meaning parallel to the pores. In the case of elongated structures H_C decreases to values between 330 Oe (T = 4 K) and 200 Oe (T = 250 K). If the magnetic field is applied parallel to the sample surface a further decrease of the coercivity is always observed (figure 2 a, b) which is mainly caused by the shape-anisotropy of the precipitates in case of elongated metal-structures. In the case of Ni particles the observed magnetic anisotropy is also a quasi shape effect because sufficient densely packed Ni particles interact dipolar within one pore (average distance about 10 nm) but the coupling of particles between adjacent pores (distance between the pores around 50 nm) is neglectable which leads to quasi elongated magnetic structures within the oriented pores.

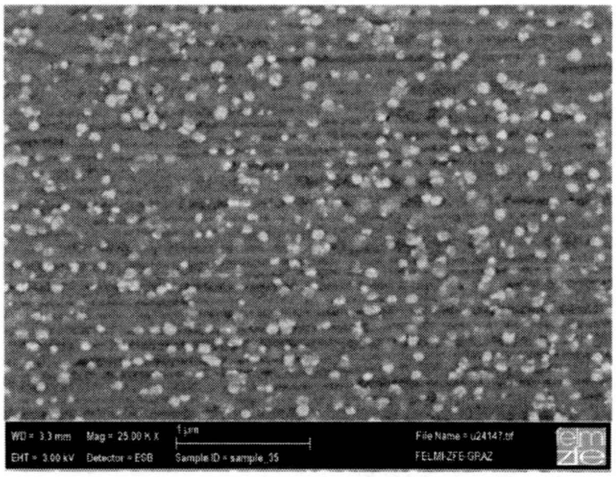

Figure 1: Ni particles with a diameter of about 55 nm (equal to the pore-diameter) and an average length of 70 nm (maximum 100 nm) within the pores of a porous silicon matrix. The SEM-image (back scattered electrons) shows a cross-sectional region of the sample.

Journal of Physics: Conference Series **303** (2011) 012048 doi:10.1088/1742-6596/303/1/012048

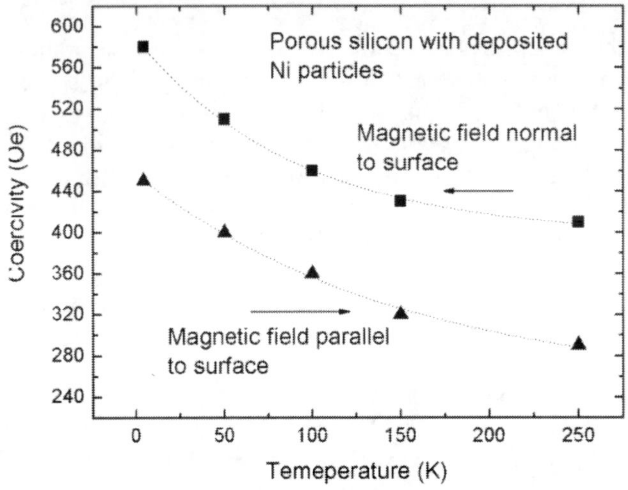

Figure 2 a: Decrease of the coercivity H_C with increasing temperature measured on porous silicon with embedded Ni-particles. The curves show an exponential behaviour which is up to now an experimental result. The measurements are performed with a magnetic Field applied perpendicular to the surface (squares) as well as parallel to the surface (triangles). The magnetic field has been varied between ± 1Tesla.

The magnetic anisotropy can be modified by varying the geometry of the precipitated metal structures or by changing the spatial distribution of the deposits within the pores. Densely distributed particles meaning a distance between them in the range of their size or smaller interact magnetically within the pores and thus show a higher anisotropy between the two magnetization directions than loosely packed particles.

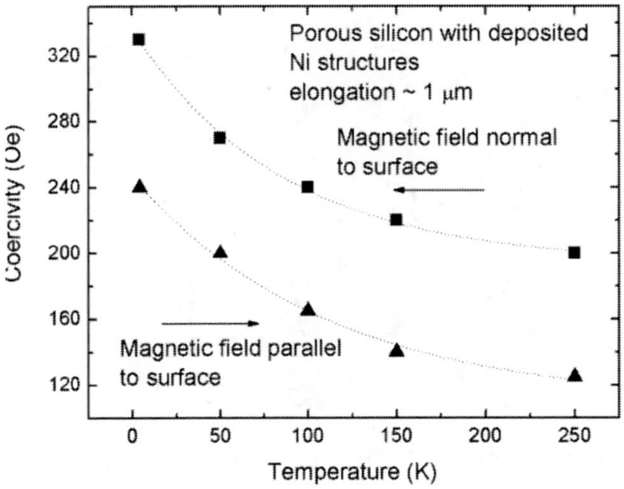

Figure 2 b: Temperature dependency of porous silicon with incorporated elongated Ni structures. The experiments show that H_C decreases with increasing temperature in an exponential way for both magnetization directions, magnetic field perpendicular to the surface (squares) and parallel to the surface (triangles).

The deposited particles corresponding to figure 2a show a value of the coercivity H_C for a magnetic field applied parallel to the surface which is about 23% less than for a field applied normal to the surface. In the case of the elongated Ni structures corresponding to figure 2b the anisotropy is in the same range (~ 27%). The elongated structures are packed less densely than in the case of particles but the anisotropy is caused by the shape. The packing density of the Ni structures is obtained by image processing using BSE images of the samples. Estimations of elongated Ni-structures with an aspect ratio of about 20 (ratio of length to diameter) are in good agreement with the literature [10]. If the particles interact within the pores they form a magnetic quasi chain also resulting in a quasi shape anisotropy. Particles (average diameter 55 nm) deposited within the pores with a distance between them much greater than their size (by adequate control of the deposition parameters as mentioned above) offer a weak dipolar coupling and therefore no significant magnetic anisotropy between the two magnetization directions can be observed (figure 3).

Considering the anisotropy between easy axis (magnetic field perpendicular to the surface) and hard axis (magnetic field parallel to the surface) magnetization the differences of the coercivities at each temperature of the whole temperature range (between 4 K and 250 K) are in the same range for any morphology which can be seen in figure 2a and figure 2b. Phenomenologically it is found from the experimental data that both curves follow an exponential law.

The presented nanocomposite system offers ferromagnetic properties whereas the magnetic characteristics as coercivity, remanence and magnetic anisotropy can be tuned by the fabrication parameters (e.g. the pulse-frequency of the current, current density and electrolyte concentration).

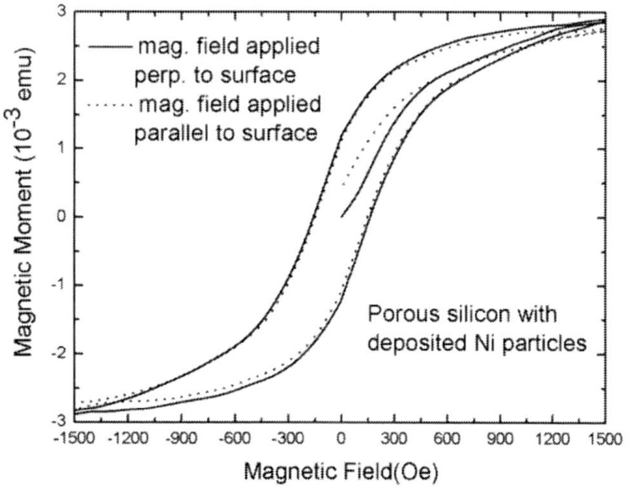

Figure 3: Hystereses curves of porous silicon with precipitated Ni particles (average diameter 55 nm) whereas the distance between them is much greater than their size offering weak interactions (interparticle distance > 200 nm). Measurements are carried out at T = 250 K in two directions of magnetization exhibiting coercivities which are nearly equal (Magnetic field perpendicular to the surface: $H_{c\perp}$ = 160 Oe; magnetic field parallel to the surface: $H_{c\parallel}$ = 150 Oe).

4. Conclusions

The fabrication of a nanocomposite system consisting of a semiconducting matrix and embedded ferromagnetic nanostructures has been presented. The ferromagnetic properties of the system can be controlled by the electrochemical process parameters. On the one hand various geometries of the metal structures can be precipitated leading to different magnetic characteristics as coercivity, remanence and shape anisotropy. On the other hand the density of the spatial distribution of the deposits within the pores can be modified which influences the magnetic interactions between the particles. In the case of densely packed particles within the pores dipolar coupling between them occurs and results in quasi magnetic chains which offer a much larger magnetic anisotropy than non-interacting particles. The investigated nanocomposite offers the possibility to vary its magnetic properties in a broad range and therefore it is adequate for magnetic applications e. g. sensors, whereas it is due to the silicon based host-material proper for the integration in existing process technology.

Acknowledgements

This work is supported by the Austrian Science Fund under project P 21155.

References

[1] P. Nordblad, J. Phys. D: Appl. Phys. 41, 134011, 2008.
[2] B. Bräuer, Y. Vaynzof, W. Zhao, A. Kahn, W. Li, D.R.T. Zahn, C. de J. Fernandez, C. Sangregorio, G. Salvan, J. Phys. Chem. B 113, 4565, 2009.
[3] N. Guskos, J. Typek, T. Bodziony, Z. Roslaniec, U. Narkiewicz, M. Kwiatkowska, M. Maryniak, Rev. Adv. Mater. Sci. 12, 133, 2006.
[4]] M.M. Milanova, M.G. Arnaudov, M.M. Getsova, D.S. Todorovsky, J. Alloys Compd. 264, 95, 1998.
[5] S. M. Montemayor, L.A. Garcia-Cerda, J.R. Torres-Lubian, O.S. Rodriguez-Fernandez, Mater. Res. Bull. 43, 1112, 2008.
[6] P. Granitzer, K. Rumpf, P. Poelt, M. Albu, B. Chernev, Phys. Stat. Sol. (C) 6, 2222, 2009.
[7] V. Lehmann, J. Electrochem. Soc. 140, 2836, 1993.
[8] K. Rumpf, P. Granitzer, P. Pölt, ECS Transactions 25, 157, 2010.
[9] V. Lehmann, S. Rönnebeck, J. Electrochem. Soc. 146, 2968, 1999.
[10] R. Skomski, J. Phys.: Cond. Matter. 15, R841, 2003.

Estimate of the anisotropy field in isotropic SmCo 2:17 magnets with the Stoner-Wohlfarth CLC model

M F de Campos[1], S A Romero[2], F J G Landgraf[3], F P Missell[4]

[1]PUVR- Universidade Federal Fluminense, Av dos Trabalhadores 420, Vila Santa Cecilia, Volta Redonda, RJ, 27255-125, Brazil

[2]Instituto de Física, Universidade de São Paulo, São Paulo, SP, Brazil

[3]Escola Politécnica, Universidade de São Paulo, São Paulo, SP, Brazil

[4]Centro de Ciências Exatas e Tecnologia, Universidade de Caxias do Sul, Caxias do Sul, RS, 95070-560 Brazil

E-mail: fmissell@yahoo.com, mcampos@metal.eeimvr.uff.br

Abstract. The Callen-Liu-Cullen (CLC) modification of the Stoner-Wohlfarth model was found able to describe properly the hysteresis curves of isotropic Sm(CoFeCuZr)z magnets. The SW-CLC model uses three parameters, and all of them have physical meaning. One of the parameters is related to the saturation magnetization, another to the anisotropy field, and another is 1/d, which evaluates the interaction between grains or particles. The model was applied for several magnets, indicating an anisotropy field of 6-7 T, which is compatible with other methods for anisotropy field determination. The model also gives insight into the abnormal temperature dependence of the coercivity found in SmCo 2:17 magnets. For compositions with a low z, the parameter 1/d is significant. These compositions with a low z are those showing the most abnormal coercivity behavior with temperature.

1. Introduction

It is expected that randomly distributed, non-interacting, single domain particles of a high uniaxial anisotropy phase, embedded in a paramagnetic matrix, might exhibit a behavior similar to that predicted by Stoner and Wohlfarth (SW) [1-5]. In real magnets, we expect interactions between grains. The SW model was modified by Callen, Liu, and Cullen (CLC) [6] in order to treat amorphous ferromagnetic alloys. The SW result can be presented in terms of a reduced magnetization $m = M/M_s$ as a function of a reduced magnetic field $h = H/H_a$. The experimental variables M and H are scaled by the saturation magnetization M_s and the anisotropy field H_a. CLC [6] added mean-field interactions to the SW model. Hadjipanayis and Gong [7] applied the SW-CLC in melt spun NdFeBAlSi with grain size below the single domain particle size. McCallum [8,9] employed the mean-field calculation of CLC to model the first quadrant demagnetization curve of several NdFeB magnets, thereby obtaining values for M_s and H_a as well as determining the mean-field interaction parameter 1/d . This treatment leads to a straight forward mean-field equation:

$$h_{CLC} = h_{SW} + (1/d)\, m \qquad (1)$$

This equation says that the reduced field h_{CLC} for the CLC model [6-9] is obtained for any magnetization state m by correcting the SW reduced field h_{SW} with a term proportional to m. In this equation, 1/d is the mean-field interaction coefficient, where d is the ratio of anisotropy to exchange energy. Equation (1) can be applied to fit experimental curves and was used by McCallum to obtain M_s and H_a for two NdFeB magnets [8]. To fit experimental curves with the CLC model, one must vary M_s and H_a, the scale factors, as well as the mean-field interaction parameter (1/d). Figure 1 shows the effect of varying 1/d, while maintaining M_s and H_a constant.

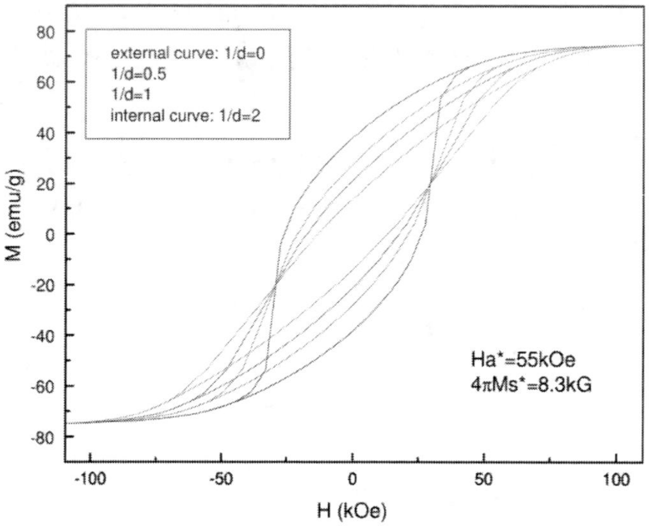

Figure 1. Hysteresis curves calculated from the SW-CLC model showing the effect of variations in the interaction parameter 1/d. For the example shown, the anisotropy field was taken to be H_a = 55 kOe.

The SW model with the CLC modifications was recently used to study some compositions of SmCo 2:17 magnets. The hysteresis curves were well-described by the SW-CLC model and this indicated the presence of interactions [10]. The model also gave a very reasonable estimate of the anisotropy field. The reported anisotropy field at room temperature is around 6-7 T [11,12]. The present study again focuses on Sm(CoFeCuZr)z magnets and we report the results for an alloy with higher z than the previous one [10]. This alloy exhibits almost perfect SW behavior, indicating coherent rotation as a reversal mechanism. But this is only possible if the grains are below the single domain particle size. In the case of the 2:17 phase, the single domain particle size is around 0.6 μm [12]. Previous TEM studies have indicated nanograin size around 100 nm [13,14] in these magnets.

Experimental

One set of samples was prepared from pure elements in an arc-furnace. Those samples have the chemical composition $Sm(Co_{bal}Fe_{0.2}Cu_{0.1}Zr_x)_8$ (bal=balance) with x= 0.02 and, 0.04. Those samples were encapsulated in quartz tubes under an Argon atmosphere and underwent a solubilization heat treatment at 1175°C for 4h. The samples were quenched in water, reheated up to 820°C, remaining at this temperature for 7h, and were then slow cooled (-1°C/min) to 400°C, remaining for an additional 3 h at this temperature. Another set of samples was produced from a commercial alloy powder with composition $Sm_{0.104}Co_{0.60}Fe_{0.195}Cu_{0.072}Zr_{0.027}$ and received the same heat treatment.

The samples for magnetic measurement were parallelepipeds with dimensions 5 x 1 x 1 mm , for minimizing demagnetizing field effects. Magnetic measurements were performed at room temperature in a 9 T superconducting coil, using an EG&G Princeton Applied Research model 4500 vibrating sample magnetometer (VSM).

Results and Discussion

In Figure 2, we present the experimental room-temperature hysteresis curve (dashed line) for the isotropic magnet obtained from the commercial alloy. We note that the curve obtained from the SW-CLC model (solid line) is in good agreement with the experimental curve when we use the values of the scale factors given in the insert. The saturation magnetization $4\pi M_s$ = 10.5 kG and the anisotropy field H_a = 69 kOe, are in good agreement with experimental determinations of these quantities. We note also the slight curvature of the initial magnetization curve. In the SW model, the initial magnetization curve is the average between 1st and 4th quadrant. This was already predicted by Stoner and Wohlfarth in 1948 [1]. Thus the initial magnetization curves serves as additional confirmation of coherent rotation as the coercivity mechanism. The same was observed in our previous study [10], where the Callen-Liu-Cullen modification was employed [6].

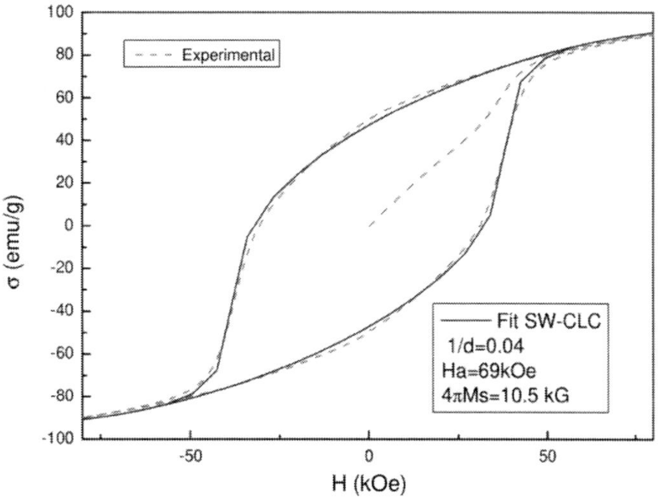

Figure 2. Experimental room-temperature hysteresis curve (dashed curve) fitted to SW-CLC model (solid curve). The small value of the interaction parameter (1/d) indicates nearly non-interacting Stoner-Wohlfarh behavior.

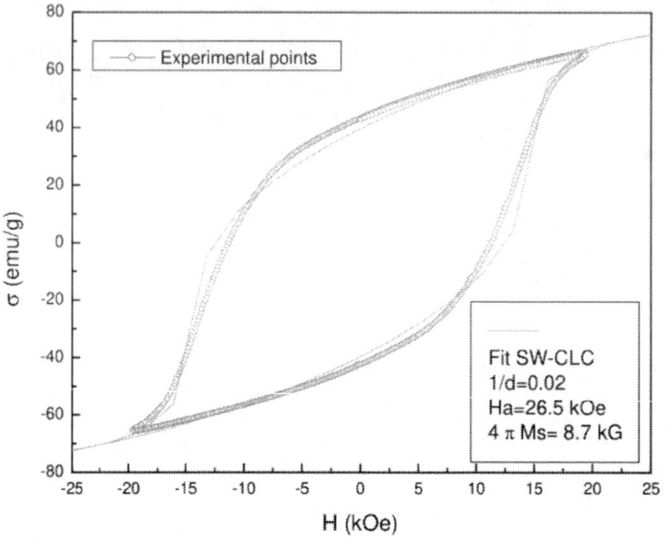

Figure 3. Experimental hysteresis curve compared to the prediction of the SW-CLC model for the sample of Figure 1, now measured at 300ºC. The sample was initially magnetized by a pulsed field of 7T.

There is an additional difficulty in high temperature measurements, because the samples are small (5x1x1 mm). The samples were well protected for this high temperature measurement. However, when samples are measured at high temperatures, a Sm-depleted zone forms near surface. Sm has high vapor pressure [15] and vaporizes easily and also oxidizes easily. This Sm-depleted layer has lower coercivity and may explain the slight disagreement between experiment and the model, in Figure 3.
The single domain particle size decreases when temperature increases. However, the behavior displayed in Figure 3 shows that D_c at 300ºC is above the nanocrystalline particle size of the samples, which should be in the range 100-200 nm [13]. The critical size for a single domain is given by Eq. (2), and depends on the domain wall energy given by Eq. (3). All these parameters are function of the temperature.

$$D_c = 9/2\,\pi\,(\gamma / M_s^2) \qquad (2)$$

$$\gamma = 4\,(A\,K_1)^{1/2} \qquad (3)$$

In these equations γ is the domain wall energy, M_s is the saturation magnetization, K_1 is the first order anisotropy constant, A is the exchange constant. Hysteresis curves for the magnets of Table I at room temperature are shown in Figure 4. The fits obtained using the model coincide almost exactly with the experimental curves.

Table I. Chemical compositions and SW-CLC parameters for the samples studied

Sample	z	1/d	H_a (kOe)	$4\pi M_s$ (kG)	Measurement temperature
$Sm_{0.104}Co_{0.60}Fe_{0.195}Cu_{0.072}Zr_{0.027}$.	8.6	0.04	69.0	10.5	300 K
$Sm_{0.104}Co_{0.60}Fe_{0.195}Cu_{0.072}Zr_{0.027}$.	8.6	0.02	26.5	8.7	573 K
$Sm(Co_{bal}Fe_{0.2}Cu_{0.1}Zr_{0.02})_8$	8	0.85	59.5	8.8	300 K
$Sm(Co_{bal}Fe_{0.2}Cu_{0.1}Zr_{0.04})_8$	8	1.05	57.0	8.3	300 K

Note: Zirconium replaces samarium in phases like 1:3, 5:19 or 2:7. Thus zirconium frees samarium, which leads to an increase of the $Sm(Co,Cu)_5$ volume fraction.

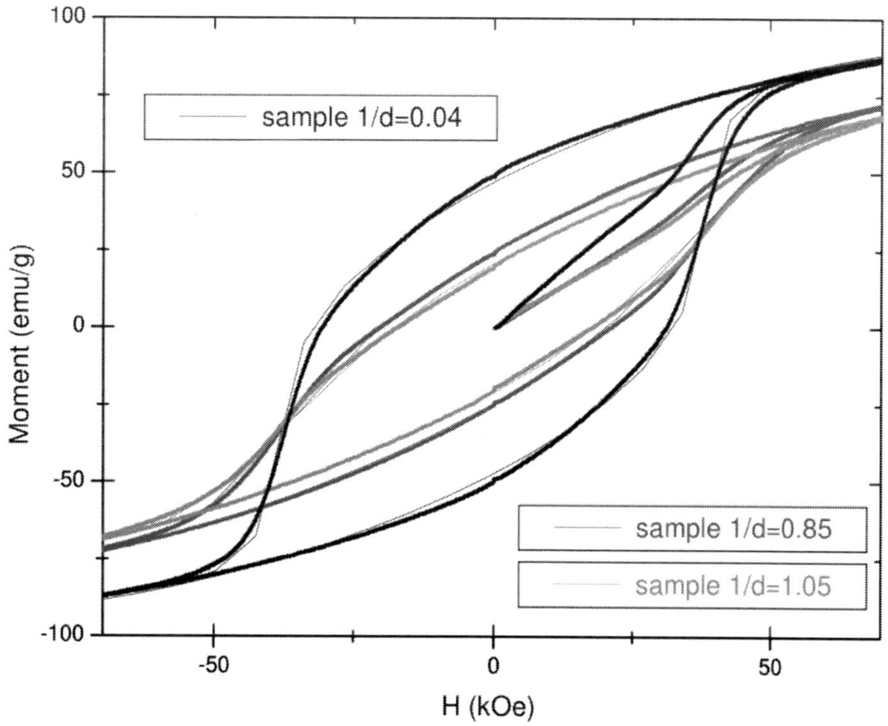

Figure 4. A comparison of some measured hysteresis curves at 300K where the SW-CLC model was applied. These magnets exhibited different values of 1/d (see Table I). Heavy curves are experimental. The fine solid lines represent the calculated model curves.

It is possible that the higher values of 1/d are associated with abnormal coercivity behavior. The abnormal coercivity behavior takes place when the $Sm(Co,Cu)_5$ phase is not paramagnetic (or when there is not perfect isolation of the 2:17 grains). If z is small in Sm(CoCuFeZr)z, this increases the volume fraction of 1:5 phase. Increasing the volume fraction of the 1:5 phase, the Co content in 1:5 also increases, and the 1:5 is no longer paramagnetic, leading to abnormal coercivity behavior. This was experimentally verified in a previous study [13].

It is worthy of note that the model can be applied for non-isotropic magnets, since the texture is taken into account. Thus oriented magnets can also be modeled. The isotropic Stoner-Wohlfarth model is the average of the hysteresis curves of all orientations in the magnet. An oriented magnet can be modeled, for example, considering most of grains in a small angle between the applied field and the easy axis, let us say 0-15 degrees from the easy axis, as previously determined with SEM-EBSD for $SmCo_5$ magnets [16]. As already reported, SEM-EBSD can also be used to study crystallographical texture in 2:17 magnets [17,18].

Conclusions

As the values found with the SW-CLC are very close to experimental determinations using other methods, for example, the singular point detection method, this model can be considered another method for estimating the anisotropy field in 2:17 nanocrystalline magnets.

Our results further indicate that coherent rotation is the coercivity mechanism in these materials.

Acknowledgements

MF de Campos, FP Missell, and F J G Landgraf acknowledge CNPq for financial support. We also thank RN Faria, who kindly supplied the commercial Sm-Co-Fe-Zr-Cu alloy.

References

[1] Stoner E C and Wohlfarth E P 1948 *Phil. Trans. Royal Soc. London* **A240** 599
[2] Lipert K, Kazmierczak J, Pelech I, Narkiewicz U, Slawska-Waniewska A, and Lachowicz H K 2007 *Materials Science-Poland* **25** 399
[3] Kumar G, Kerschi P, Rößler U K, Nenkov K, Müller K-H, Schultz L, and Eckert J 2006 *J. Appl. Phys.* **99** 083904
[4] Girt Er, Krishnan K M, Thomas G, Girt E, and Altounian Z 2001 *J. Magn. Magn. Mater.* **231** 219
[5] Pinkerton F E 1988 *J. Appl. Phys.* **63** 5427
[6] Callen E, Liu Y J and Cullen J R 1977 *Phys. Rev. B* **16** 263
[7] Hadjipanayis G C, Gong W 1988 *J. Appl. Phys.* **64** 5559
[8] McCallum R W 2005 *J. Magn. Magn. Mater.* **292** 135
[9] McCallum R W 2006 *J .Magn. Magn. Mater.* **299** 472
[10] Romero S A, de Campos M F, Rechenberg H R, Missell F P 2008 *J. Magn. Magn. Mater.* **320** e73
[11] Tellez-Blanco J C, Kou X C, Grossinger R, Estevez-Rams E, Fidler Jand Ma B M 1997 *J. Appl. Phys.* **82** 3928
[12] Perkins R S, Gaiffi S, Menth A 1975 *IEEE Trans. Magn.* **Mag-11** 1431
[13] Corte-Real M M, de Campos M F, Zhang Y, Hadjipanayis G C and Liu J F 2002 *Phys. Status Solidi (a)* **193** 302
[14] Liu J F, Ding Y, Hadjipanayis G C 1999 *J.Appl.Phys.* **85** 1670
[15] Herrick C C 1964 *J. Less Common Met.* **7** 330
[16] de Campos M F, Yonamine T, Fukuhara M, Machado R, Romero S A, Landgraf F J G, Rodrigues D, Missell F P 2007 *J. Appl. Phys.* **101** 09K516
[17] Yonamine T, Fukuhara M, Machado R, Missell F P 2008 *J. Magn. Magn. Mater.* **320**, e77
[18] Romero S A, de Campos M F, Landgraf F J G, Yonamine T, Fukuhara M, Neiva A C 2010: Proc. 21th Int. Workshop on Rare Earth Magnets and Their Applications (Bled, Slovenia) Eds S Kobe and P J McGuinness, pp. 124-126.

Discrete and resonance states in graphene near the Dirac point

M Inglot[1] **and V K Dugaev**[1,2]

[1] Department of Physics, Rzeszów University of Technology, Al. Powstańców Warszawy 6, 35-959 Rzeszów, Poland

[2] Department of Physics and CFIF, Instituto Superior Técnico, Technical University of Lisbon, Av. Rovisco Pais, 1049-001 Lisbon, Portugal

E-mail: ming@prz.edu.pl

Abstract. We consider the problem of impurity states in the vicinity of Dirac point in graphene taking into account the spin-orbit (SO) interaction. Two main contributions to the SO interaction, namely the internal and Rashba, are considered. Even though the internal SO interaction is relatively small, its effect is crucial because a very small perturbation potential can create both discrete and resonance impurity states located near the gap. In the case of strong Rashba SO interaction like in the case of graphene on a substrate, the resonance impurity states can be created producing some anomalies in transport properties.

1. Introduction

Electric and magnetic properties of graphene are mostly determined by the electronic states in the vicinity of Dirac points [1, 2, 3, 4, 5]. If the spin-orbit (SO) interaction is neglected [7, 8, 9, 10], the energy spectrum in these points is degenerate, and the elementary excitations belong to the linear spectrum, which can be described by the two-dimensional relativistic Dirac model [6]. In reality, the SO interaction is not zero and, correspondingly, there is an energy gap [11]. However, it was found that in case of internal SO interaction the magnitude of SO gap is very small, and correspondingly it can be neglected [12, 13, 14, 15]. In this paper we demonstrate that even though the energy gap is really very small, the presence of SO interaction of any origin leads to formation of additional low-energy impurity states, which are absent if the SO interaction is neglected. Besides, we find that the impurity creating very small perturbation like, for example, some neutral adatoms or molecules, can be responsible for these states.

In the case of SO interaction related to the Rashba term, the magnitude of SO is not necessarily small, and the energy spectrum at low energies is completely different from the Dirac spectrum. In this case one can also study the problem of impurity states, which can be effectively controlled by electric gate, as it is known that the electric field of the gate modifies the magnitude of Rashba SO coupling.

2. Nonmagnetic impurity: the case of internal spin-orbit interaction

We start from considering the internal SO interaction in graphene, using the following Hamiltonian [11]

$$\hat{H}_0 = v\boldsymbol{\tau}\cdot\mathbf{k} + \tau_z\sigma_z\Delta. \tag{1}$$

Here the first term is the relativistic two-dimensional Dirac Hamiltonian for the K point in the Brillouin zone of graphene, and the second term is due to the internal SO interaction. Hamiltonian (1) is 4×4 matrix, in which the Pauli matrices $\boldsymbol{\tau}$ and $\boldsymbol{\sigma}$ are acting in the sublattice and spin spaces, respectively. The constant v is the velocity parameter and Δ is the constant of SO interaction. The energy spectrum consists of two bands, $\varepsilon = \pm\varepsilon_k$, and $\varepsilon_k \equiv (v^2k^2 + \Delta^2)^{1/2}$.

Let us consider now the impurity states created by a single nonmagnetic impurity which is described by a perturbation in the basis of eigenfunctions of the Hamiltonian without SO interaction (first term in Eq. (1)). For definiteness we assume that the impurity atom is located at the site of sublattice A. We also assume that the perturbation is strongly localized within the site, so that the matrix elements of the perturbation with the eigenfunctions of basis do not depend on electron momentum k for the states in a vicinity of the single Dirac point, where the Hamiltonian (1) can be used. Thus, the matrix of perturbation in our model is

$$\hat{V}_{\mathbf{k}\mathbf{k}'} = \frac{1}{2}\left(1 + \tau_z\right) V_0 \equiv \hat{V}_0. \tag{2}$$

It should be noted that this model can be justified for relatively short range potential of neutral impurities. As we do not consider Umklapp scattering, then in this model, the matrix elements of perturbation with large transition of momentum are assumed to be strongly suppressed, which restricts the consideration of impurities of not too small potential range.

We use the T-matrix method [16], which accounts for the scattering in all orders of the impurity perturbation. The general equation for the T-matrix has the following form

$$\hat{T}_{\mathbf{k}\mathbf{k}'}(\varepsilon) = \hat{V}_{\mathbf{k}\mathbf{k}'} + \sum_{\mathbf{k}_1} \hat{V}_{\mathbf{k}\mathbf{k}_1} \hat{G}_{\mathbf{k}_1}(\varepsilon)\, \hat{T}_{\mathbf{k}_1\mathbf{k}'}(\varepsilon), \tag{3}$$

where $\hat{G}_{\mathbf{k}}(\varepsilon)$ is the electron Green's function corresponding to Hamiltonian (1),

$$\hat{G}_{\mathbf{k}}(\varepsilon) = \left(\varepsilon - \hat{H}_0\right)^{-1} = \frac{\varepsilon + \tau_z\sigma_z\Delta + v\boldsymbol{\tau}\cdot\mathbf{k}}{\varepsilon^2 - \varepsilon_k^2}, \tag{4}$$

and we assume that the energy parameter ε has a small imaginary part, $\varepsilon \to \varepsilon + i\delta$, where $\delta = 1/2\tau$ and τ is the momentum relaxation time. It corresponds to consideration of the retarded Green's function.

Using (3) and (4) we calculate the T-matrix as $\hat{T}(\varepsilon) = [1 - \hat{V}_0\hat{F}(\varepsilon)]^{-1}\hat{V}_0$ where the function $\hat{F}(\varepsilon) = \sum_{\mathbf{k}} \hat{G}_{\mathbf{k}}(\varepsilon)$. This function is diagonal in spin space, and we can find for the spin up state the real and imaginary parts (here we take the limit of $\delta \to 0$)

$$\mathrm{Re}\,\hat{F}_\uparrow(\varepsilon) \simeq -\frac{\varepsilon + \tau_z\Delta}{4\pi v^2}\,\ln\frac{v^2k_m^2}{|\Delta^2 - \varepsilon^2|}, \tag{5}$$

$$\mathrm{Im}\,\hat{F}_\uparrow(\varepsilon) \simeq \frac{\varepsilon + \tau_z\Delta}{4v^2}\,\theta(\varepsilon^2 - \Delta^2), \tag{6}$$

where $\theta(x)$ is the Heaviside step function. The location of impurity level is determined by the poles of T-matrix, and we can find for the spin up and spin down levels the following equations

$$\varepsilon_{\uparrow,\downarrow} = \mp\Delta - 4\pi v^2 \left(V_0 \ln\frac{v^2k_m^2}{\left|\Delta^2 - \varepsilon_{\uparrow,\downarrow}^2\right|}\right)^{-1}. \tag{7}$$

If the imaginary part of $\hat{F}(\varepsilon)$ for corresponding energy is nonzero, then we obtain the resonance state, and the width of resonance is proportinal to $\mathrm{Im}\,\hat{F}(\varepsilon)$.

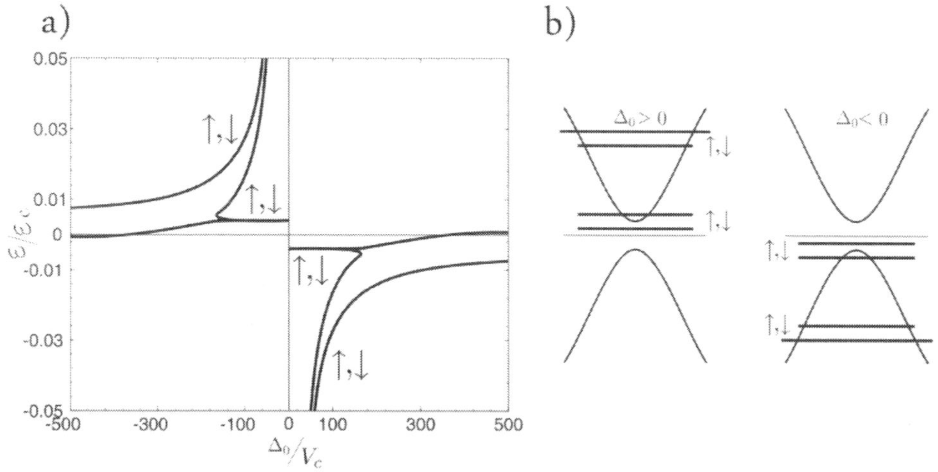

Figure 1. The location of impurity levels as a function the impurity strength parameter V_0 (a) and the schematic presentation of the impurity levels (b) in the case of non-magnetic impurity and internal SO interaction in graphene. In numerical calculations we use the units: $\varepsilon_c = \hbar^2/m_e a_0^2$ and $V_c = \varepsilon_c a_0^2$, where m_e is the electron mass and a_0 is the lattice constant.

There is another nonequivalent Dirac point K' in graphene, which can be also described by Hamiltonian (1) with the substitutions $k_x \to -k_x$ and $\Delta \to -\Delta$. We performed the analogous calculations for the K' point, and the results of numerical calculations for both Dirac points are presented in Fig. 1. As we see, in the case of very small impurity potential V_0, there are impurity states located near the edge of the SO gap. In the absence of SO interaction, only very strong impurity potential leads to creation of the resonance state in the vicinity of Dirac point.

It should be noted that the spin degeneration in this figures is due to the levels related to different Dirac points. The most important is presence of impurity states in the vicinity of the SO gap. These states appear even for very weak impurity potential.

3. Nonmagnetic impurity: Rashba spin-orbit interaction

The Hamiltonian of low-energy excitations near the Dirac point, which takes into account the Rashba SO innteraction has following form [11]

$$\hat{H}_0 = v\boldsymbol{\tau}\cdot\mathbf{k} + \lambda_R\,(\tau_x\sigma_y - \tau_y\sigma_x), \tag{8}$$

where λ_R is the Rashba coupling constant. This type of SO interaction appear for the graphene on substrate, and the magnitude of the constant λ_R depends strongly on the material of substrate. The energy spectrum corresponding to Hamiltonian (8) has four branches, and there is no gap in the energy spectrum at $k \to 0$, $E_{1,2}(k) = \lambda_R \pm \sqrt{\lambda_R^2 + v^2k^2}$ and $E_{3,4}(k) = -\lambda_R \pm \sqrt{\lambda_R^2 + v^2k^2}$.

We do not present the details of calculations in this case because they are quite similar to the ones presented before. The perturbation is assumed in the same form of Eq. (2) corresponding to short-range neutral impurities. The results of calculations of the location of impurity levels as a function of V_0 for the Dirac points K and K' are presented in Figs. 2 and 3, respectively. Here 1 and 2 label different impurity states, each of them being certain superposition of the spin up and down states belonging to different sublattices (i.e., superposition of spins and pseudospins).

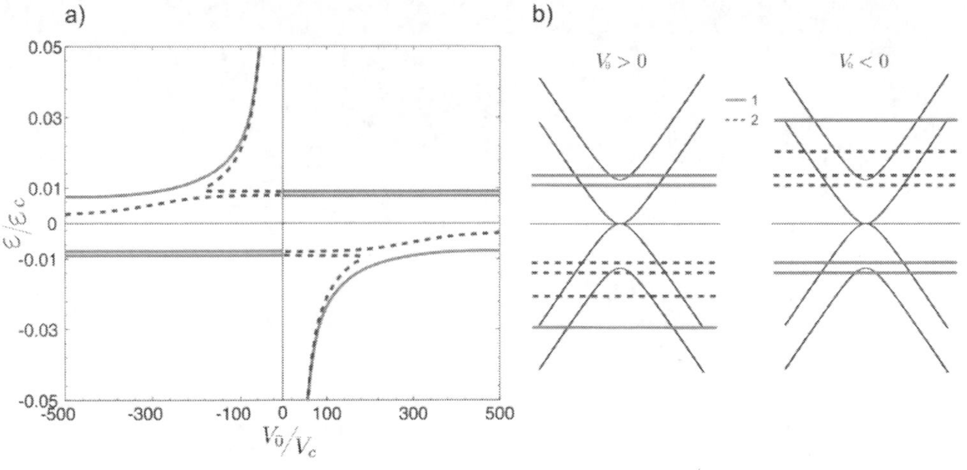

Figure 2. Energy levels for $\lambda_R = 0.004$ (in units of ε_c) at the K - point

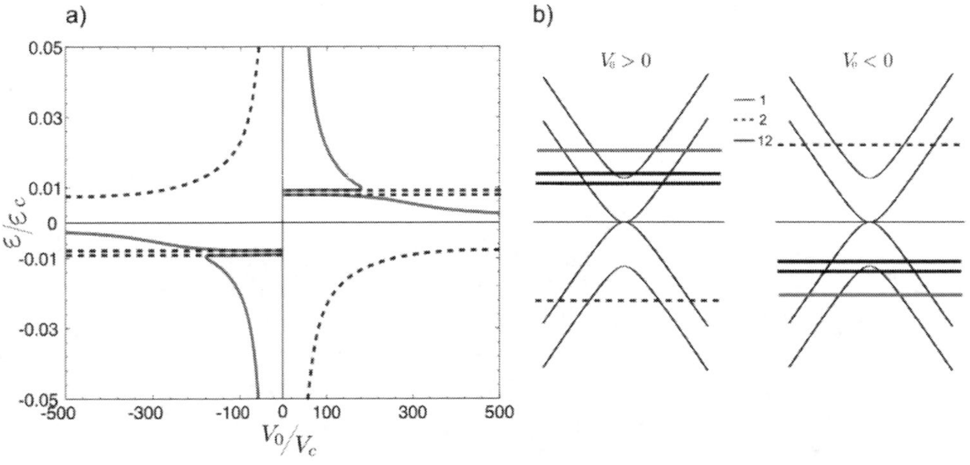

Figure 3. Energy level for $\lambda_R = 0.004$ at the K' - point

As we see, the asymmetry with respect to reverse of sign of the perturbation, $V_0 \to -V_0$, is absent in each of the figures 2 and 3, but it is completely restored if we take into account both Dirac points. The location of the low-energy impurity states is determined by the magnitude of Rashba SO interaction.

4. Nonmagnetic impurity with strong SO interaction

Using the same formalism, one can consider the impurity, which locally enhances the internal SO interaction. Physically, it can be the case of heavy neutral atoms like, for example, gold or platinum. In this case the corresponding perturbation for the K point has the form

$$\hat{V}_\Delta = \frac{\Delta_0}{2}\left(1 + \tau_z\right)\sigma_z, \tag{9}$$

where Δ_0 is the constant of local SO perturbation. The perturbation for the K' points differs from (9) by the sign of Δ_0. It should be noted that both the spin and sublattice index are the good quantum numbers in this case. Using again the T-matrix method and performing the calculations similar to presented above we come to the following equations determining the spin polarized states belonging to the Dirac points K and K'

$$\varepsilon^{K}_{\uparrow,\downarrow} = \mp\Delta \mp 4\pi v^2 \left(\Delta_0 \ln \frac{v^2 k_m^2}{\left|\Delta^2 - \varepsilon^2_{\uparrow,\downarrow}\right|}\right)^{-1}, \tag{10}$$

$$\varepsilon^{K'}_{\uparrow,\downarrow} = \pm\Delta \pm 4\pi v^2 \left(\Delta_0 \ln \frac{v^2 k_m^2}{\left|\Delta^2 - \varepsilon^2_{\uparrow,\downarrow}\right|}\right)^{-1}. \tag{11}$$

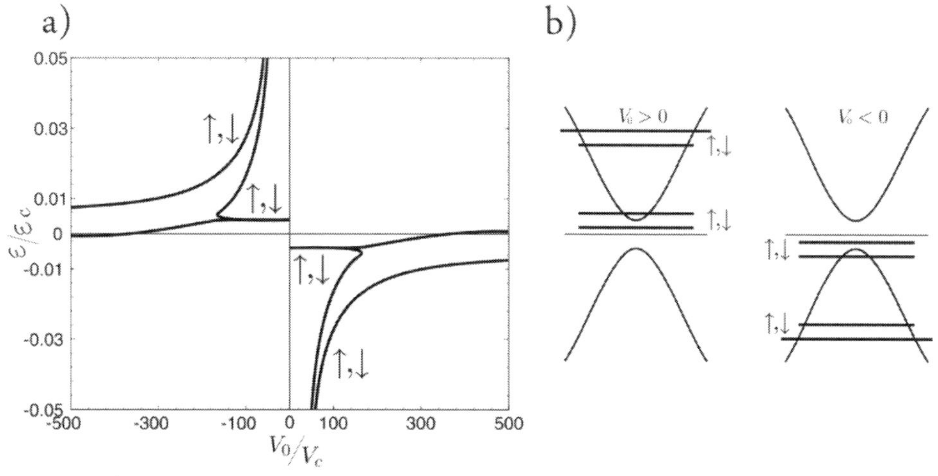

Figure 4. The same dependences as in Fig. 1 in the case of spin-orbit interaction on impurity.

The results are presented in Fig. 4 as a function of Δ_0. Here the constant Δ refers to the internal SO interaction like in the case considered in Sec. 2. The important point is that each of the states for a given Dirac point is spin polarized, but the spin degeneration is restored for both Dirac points.

5. Discussion

Spin-orbit interaction in graphene can be responsible for creation of a number of additional impurity states near the Dirac point. As we see from the results of calculations, in many cases they are resonance states in the continuum, which can affect substantially the electronic and magnetic properties of graphene. In this paper we considered the case of nonmagnetic impurity. Correspondingly, there is no net magnetization on the site. However, spin degeneracy is related to the presence of two nonequivalent Dirac points. It means that for a single Dirac point the impurity can be magnetically polarized. It can be used in the "valley electronics" [17] exploring the difference between the properties related to separated valleys. As we see, in the case under consideration, the valley electronics has exactly the same meaning as spin electronics because the localized electron spin is related to the certain Dirac point.

Acknowledgments

This work was supported by the Polish Ministry of Science and Higher Education as a research project in years 2007 – 2010.

Journal of Physics: Conference Series **303** (2011) 012050 doi:10.1088/1742-6596/303/1/012050

References

[1] Novoselov K S, Geim A K, Morozov S V, Jiang D, Zhang Y, Dubonos S V, Grigorieva I V and Firsov A A 2004 *Science* **306** 666–669
[2] Novoselov K S, Geim A K, Morozov S V, Jiang D, Katsnelson M I, Grigorieva I V, Dubonos S V and Firsov A A 2005 *Nature* **438** 197–200
[3] Geim A K and Novoselov K S 2007 *Nat Mater* **6** 183–191
[4] Katsnelson M I 2007 *Materials Today* **10** 20–27 ISSN 1369-7021
[5] Schedin F, Geim A K, Morozov S V, Hill E W, Blake P, Katsnelson M I and Novoselov K S 2007 *Nat Mater* **6** 652–655
[6] Castro Neto A H, Guinea F, Peres N M R, Novoselov K S and Geim A K 2009 *Rev. Mod. Phys.* **81** 109–162
[7] Pereira V M, Guinea F, Lopes dos Santos J M B, Peres N M R and Castro Neto A H 2006 *Phys. Rev. Lett.* **96** 036801
[8] Peres N M R, Guinea F and Castro Neto A H 2006 *Phys. Rev. B* **73** 125411
[9] Hu B Y K, Hwang E H and Das Sarma S 2008 *Phys. Rev. B* **78** 165411
[10] Pereira V M, Lopes dos Santos J M B and Castro Neto A H 2008 *Phys. Rev. B* **77** 115109
[11] Kane C L and Mele E J 2005 *Phys. Rev. Lett.* **95** 226801
[12] Min H, Hill J E, Sinitsyn N A, Sahu B R, Kleinman L and MacDonald A H 2006 *Phys. Rev. B* **74** 165310
[13] Huertas-Hernando D, Guinea F and Brataas A 2006 *Phys. Rev. B* **74** 155426
[14] Yao Y, Ye F, Qi X L, Zhang S C and Fang Z 2007 *Phys. Rev. B* **75** 041401
[15] Gmitra M, Konschuh S, Ertler C, Ambrosch-Draxl C and Fabian J 2009 *Phys. Rev. B* **80** 235431
[16] Ziman J M 1969 *Elements of Advanced Quantum Theory* (Cambridge Univ. Press, Cambridge)
[17] Nakanishi T, Koshino M and Ando T 2010 *Phys. Rev. B* **82** 125428

Joint European Magnetic Symposia – JEMS 2010 IOP Publishing
Journal of Physics: Conference Series **303** (2011) 012051 doi:10.1088/1742-6596/303/1/012051

Magnetic ground and remanent states of synthetic metamagnets with perpendicular anisotropy

N. S. Kiselev[1], **U. K. Rößler**[1], **A. N. Bogdanov**[1] **and O. Hellwig**[2]

[1] IFW Dresden, Postfach 270116, D-01171 Dresden, Germany

[2] San Jose Research Center, Hitachi Global Storage Technologies, San Jose, CA 95135, USA

E-mail: m.kyselov@ifw-dresden.de

Abstract.
In this work, we summarize our theoretical results within a phenomenological micromagnetic approach for magnetic ground state and nonequilibrium states as topological magnetic defects in multilayers with strong perpendicular anisotropy and antiferromagnetic (AF) interlayer exchange coupling (IEC), e.g. [Co/Pt(Pd)]/Ru(Ir, NiO). We give detailed analysis of our model together with the most representative results which elucidate common features of such systems. We discuss phase diagrams of the magnetic ground state, and compare solutions of our model with experimental data. A model to assess the stability of so-called *tiger tail* patterns is presented. It is found that these modulated topological defect cannot be stabilized by an interplay between magnetostatic and IEC energies only. It is argued that *tiger tail* patterns arise as nuclei of ferro-stripe structure in AF domain walls and that they are stabilized by domain wall pinning.

1. Introduction

AF coupled superlattices with strong perpendicular anisotropy, as [Co/Pt(Pd)]/Ru (Ir) or [Co/Pt]/NiO [1–4], are considered as promising candidates for nonvolatile magnetic recording media, spin electronics devices, high-density storage technologies, and other applications [5]. Owing to the strong competition between antiferromagnetic interlayer exchange (IEC) and magnetostatic coupling these nanoscale superlattices are characterized by a large variety of multidomain states, unusual magnetization processes [1–3,8], and other specific effects [3,4,6,7] which have no counterpart in common layered systems with perpendicular magnetization [9]. Theoretical description of magnetic states and reversal magnetization process, however, can be achieved within the known phenomenological framework of micromagnetism and magnetic domain theory [9]. Here we give an overview of our model used to describe magnetic ground states in AF coupled multilayers.

2. The ground magnetic state

In ferromagnetic exchange decoupled or ferromagnetically coupled multilayers, e.g., pure Co/Pt or Co/Pd multilayers the magnetic ground state is a multidomain state [13]. For the theoretical description of this state a model of magnetic *stripe domains* is usually used. First this model has been formulated for a single ferromagnetic plate with perpendicular anisotropy by Kittel [10]. Later, it has been developed for ultra thin ferromagnetic single films by Malek and Kambersky [11] and independently by Kooy and Enz [12]. This model is based on a few assumptions: (i) The

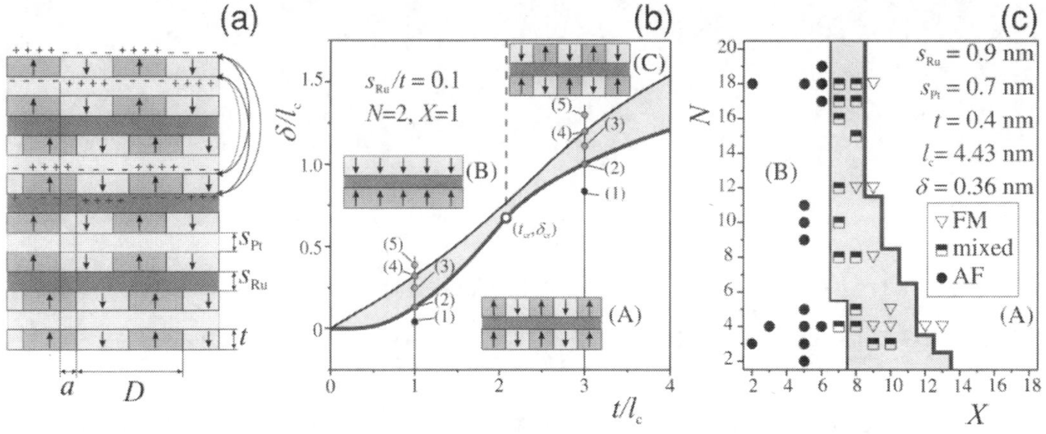

Figure 1. (a) Schematic representation of regular stripe domains with period D in a perpendicularly magnetized $\{[Co/Pt]_{X-1}/Co/Ru\}_N$ multilayer with $N = 4$, $X = 2$. AF coupling favor a lateral shift a of domain walls in adjacent [Co/Pt] stacks. The arrows show four pairs of interacting charged surfaces, for details see the text. (b) Magnetic phase diagram of the ground state for AF coupled bilayers in terms of reduced magnetic layer thickness t/l_c and interlayer exchange δ/l_c ($\delta = J_{ex}/(2\pi M_s)$). Thick blue line indicates the first-order transitions between ferromagnetic state (shifted ferro-stripes (A)) and antiferromagnetic states (single domain state (B) and AF-stripe state (C)). At the vertical dashed line $t_{cr} = 2.08\,l_c$ the AF-stripes continuously transform into the single domain state. These lines meet in a critical point $t = t_{cr}$, $\delta = \delta_{cr} = 0.639\,l_c$, hollow circle. The stability region of the shifted ferro-stripes (A) is limited by the lability line (thin black). Shaded area correspond to the *mixed* state where FM stripe and AF states can coexist. The energy profiles corresponding to enumerated points (1)-(5) along the sections of the phase diagram at $t/l_c = 1$ and 3 are shown in Fig. 2 (a)-(b). (c) Magnetic phase diagram of the ground state of multilayers depending on number of magnetic layers in each ferromagnetic stack X and number of stacks N. All notations are the same as in figure (b). Symbols correspond to experimental data for $\{[Co/Pt]_{X-1}/Co/Ru\}_N$, Ref. [3].

material is uniaxial with the direction of the magnetization normal to the surface of the plate. (ii) The plate has the uniform thickness t and its lateral dimensions in the x and y directions are infinite. (iii) The anisotropy K_u is strong ($Q = K_u/(2\pi M_s^2) \gg 1$), so that no closure domains occur. (iv) The domain structure consists of straight periodical domains with the period D and the magnetization as well as the 180° *uncharged* Bloch walls are perpendicular to the surface.

The stripe domain model has been extended for the case of multilayers with alternate ferromagnetic and nonmagnetic layers by Draaisma and de Jonge [13]. In these multilayers the dipolar coupling between magnetic layers can be considered as result of interactions between magnetostatic "charges" on the surface of each magnetic layer. As a result, the magnetization of the layers are strictly parallel across the multilayer and their direction alternates laterally in a *ferro-stripe* configuration. In this model it is assumed that the domain walls in the layers sit exactly on top of each other [13]. For the case of exchange decoupled and ferromagnetically coupled multilayers this assumption is reasonable because there are no intrinsic forces which favor a distortion of the domain wall configuration. However, for the case of AF IEC which favors antiparallel alignment of magnetic moments at the Co-Ru-Co interfaces this assumption becomes invalid in general. In Ref. [14] we have introduced a model for the magnetic ground state in such multilayers. It considers that AF coupling causes a lateral shift of the domain walls in the adjacent ferromagnetic stacks. As is seen from Fig. 1 (a), within the region of the lateral shift the magnetic moments at Co-Ru-Co interfaces are ordered antiparallel. This gives an energy gain due to AF coupling. The energy contribution from IEC is then proportional to the width of this area a. Exchange energy density per one magnetic layer is then:

$$e_{\text{ex}} = \frac{(N-1)J_{\text{ex}}}{2\pi M_s^2}\left(1 - 4\frac{|a|}{D}\right) \quad (1)$$

where J_{ex} is the interlayer exchange constant. The magnetostatic energy density per magnetic layer of ferro-stripe domains with lateral shift is:

$$e_{\text{d}} = \frac{4D}{\pi^3}\sum_{\substack{n=1\\ \text{odd}}}^{\infty}\left[f_n(t) + \sum_{k=1}^{NX-1}\sum_{j=k+1}^{NX}\left\{\cos\left(\frac{2\pi na}{D}\Gamma_{kj}\right)\frac{f_n(L_{kj}+t)+f_n(L_{kj}-t)-2f_n(L_{kj})}{NX}\right\}\right], \quad (2)$$

with $f_n(t) = [1-\exp(-2\pi ntD)]\,n^{-3}$, $L_{kj} = \sum_{i=k}^{j-1}[t+s_i]$, $s_i = \begin{cases} s_{\text{Ru}}, & i = mX, m = 1, ..., N, \\ s_{\text{Pt}}, & \text{otherwise,}\end{cases}$ and

$$\Gamma_{kj} = \begin{cases} 1, & 2mX+1 \le k \le (2m+1)X,\ (2m+1)X+1 \le j \le (2m+2)X,\ m = 0,1,2,3..., \\ 0, & \text{otherwise.}\end{cases}$$

In Eq. (2) the summation is over all pairs of magnetic layers. The expression under the double summation sign is the magnetostatic interaction energy between two Co layers separated by distance L_{kj} between the centers of the layers. It equals the sum of interactions between the four charged surfaces of the layers pairs. Figure 1 (a), illustrates one such pair. Four double-ended arrows correspond to four coupled surfaces. As AF coupling is provided by Co-Ru-Co interfaces, the lateral shift equals zero for the pairs of layers which belong to the same stack or for two layers belonging to different stacks, if they have both even or odd number in the multilayer (see Fig. 1 a). This assumption is reflected by the factor Γ_{kj}. Thereby, the total energy density per surface unit of the AF coupled multilayer is

$$e_{\text{tot}} = e_{\text{ex}} + e_{\text{d}} + NX4l_{\text{c}}t/D, \quad (3)$$

where e_{ex} and e_{d} are defined by Eqs. (1) and (2), respectively, $l_{\text{c}} = \gamma_{\text{w}}/(4\pi M_s^2)$ is the *characteristic length*, γ_{w} is the domain wall energy density.

The simplest case of a multilayer with $N = 2$ and $X = 1$ is representative for multilayers with arbitrary X and even N [14,16,17]. The results of minimization of the total energy with respect to both parameters a and D are reflected in the phase diagram Fig. 1 (b), and in Figs. 2. The main features of the solutions are the following: (i) Ferro-stripes are the ground state of an FM coupled ($J_{\text{ex}} < 0$) and exchange decoupled ($J_{\text{ex}} = 0$) multilayer up to $t \to 0$, see Fig. 1 (b). (ii) For $J_{\text{ex}} > 0$ ($\delta/l_{\text{c}} > 0$, AF coupling) a ferro-stripe state with a small shift remains the ground state up to some critical value of δ and then transforms to the state with AF arrangement by a first order transition marked by thick blue line in Fig. 1 (b), compare energy profiles (2) in Fig. 2. (iii) In wide range of the thickness t and exchange δ ferro-stripes can exist as metastable states. Here, solutions for the ferro-stripes correspond to local energy minima of the system that are separated by a barrier from the global minimum corresponding to the homogeneous ($D \to \infty$) AF state (see energy profiles (3) in Fig. 2). (iv) At the lability line (thin black line in Fig. 1 (b)) the barrier vanishes and the solution for the ferro-stripe state becomes unstable with respect to the lateral shift a, curves (4) and (5) in Fig. 2. (v) For thick ferromagnetic layers t above the transition line the ground state of the system is a stripe configuration with antiparallel arrangement, see inset (C) in Fig. 1 (b). These *AF-stripes* are *always* stable with respect to

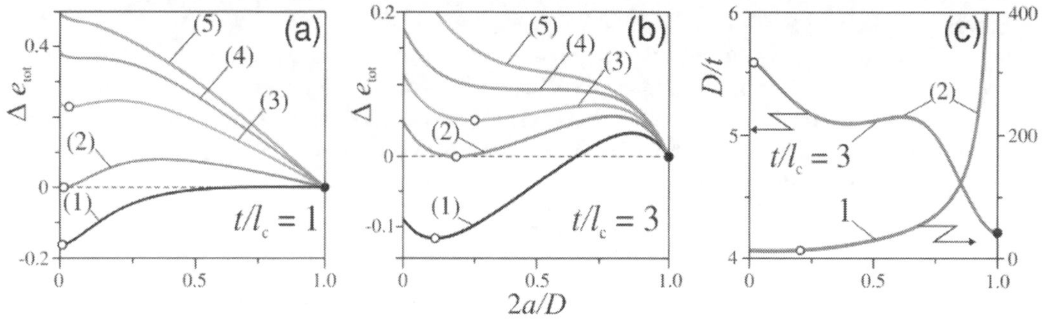

Figure 2. (a) and (b) are the energy profiles with respect to reduced parameter $2a/D$ measured from the AF state ($a = D/2$). The numbers of the profiles correspond to the points along the sections of the phase diagram shown in Fig. 1 (b) for the fixed values $t/l_c = 1$ (a) and 3 (b). For each point of the profiles we find an equilibrium value of D defined by the condition $\partial e_{tot}/\partial D = 0$, $\partial^2 e_{tot}/\partial D^2 > 0$. Stable or metastable shifted ferro-stripes correspond to the potential wells marked by hollow circles in the energy profiles (1)-(5). Solid circles correspond to the potential well for the AF single domain state ($D \to \infty$) in (a), and the *unshifted* AF stripes in (b). Figure (c) illustrates a strong variation of the equilibrium value of D with respect to the lateral shift a at the transition line between FM and AF which correspond to the energy profiles marked by (2) in figures (a) and (b).

a lateral shift, see potential well for $a = D/2$ in Fig. 2 (b). On the contrary, the solutions for the ferro-stripe phase has a finite shift, see inset (A) in Fig. 1 (b), and Figs. 2 (a), (b). These results are supported by the analytical calculation and simple physical arguments [14]. (vi) With decreasing magnetic layer thickness t the period of AF-stripes increases and at some critical thickness, see dashed vertical line in Fig. 1 (b), AF-stripes continuously transform into the AF homogeneous state ($D_{AF}(t_{cr}) \to \infty$). Thus, AF-stripes form equilibrium phases *only* for thicker layers, $t > t_{cr}$, and for sufficiently strong AF coupling $\delta > \delta_{cr}$.

In Fig. 1 (c) the ground state phase diagram for $[Co/Pt]_{X-1}/Co/Ru_N$ multilayers depending on number of Co layer per each stack X and number of stacks N is shown together with experimental data for a corresponding multilayer system studied in Refs. [2, 3]. The ground state of the system small X on the left of the lability line (black thin line) is the homogeneous AF state. For large X on the right of the transition line (blue thick line) the ferro-stripes are the ground state. Theoretically calculated lability and transition lines define the region for a *mixed state*, where metastable ferro-stripes and homogeneous AF state can coexist (shaded area). This phase diagram results from direct minimization of Eq. (3) with respect to both free parameters a and D for the fixed material parameters indicated in the figure. We have estimated the value of $l_c = 4.43$ nm, as a best fit of the experimentally measured dependence of stripe domain period versus number of cobalt layers X in a pure $[Co/Pt]_X$ multilayer [3, 15]. For $\delta = 0.363$ nm which fits well experimental data, and $M_s = 1420$ emu/cm^3, typical value for a bulk cobalt, we estimated J_{ex}=0.46 erg/cm^2 which is in a good agreement with 0.45 erg/cm^2 found in the experiment [3]. Therefore, for [Co/Pt]/Ru multilayers $\delta/l_c = 0.08$ which is about one order of magnitude smaller than the critical value $\delta_{cr}/l_c = 0.639$, see phase diagram Fig. 1 (b). For $[Co/Pt]_3/NiO/[Co/Pt]_3$ with the largest values of $J_{ex} = 0.044$ erg/cm^2 [7] the value of δ is even about one order of magnitude smaller than in [Co/Pt]/Ru systems. Thus, in [Co/Pt]/Ru and [Co/Pt]/NiO multilayer systems the ground state of the system is either ferro-stripe state or homogeneous AF state, area (A) and (B) respectively in Figs. 1 (b) and (c). This is in agreement with the experimental observations [2, 3].

3. The mystery of "tiger tails"

In Refs. [17,18] we have shown that within homogeneous AF state different types of topological defects can appear. The main types of these defects include ferrobands, sharp domain wall, and metamagnetic band. The last one is formed in the system during magnetization reversal processes and essentially is the trapped metamagnetic domain which can exist as a metastable state even at zero field, for details see Refs. [17–19]. The isolated sharp domain wall and the ferroband exist in remanent states as planar defects separating antiferromagnetically arranged regions. The ferroband, Fig. 3 (a), with finite width w and sharp domain wall ($w = 0$) are separated by a finite energy barrier. The ferroband arises in AF coupled multilayers due to the interplay between magnetodipole and interlayer exchange interactions [17,18].

There is a lot of evidence for an instability of ferromagnetic band defects into a modulated "tiger tail" pattern, which is a periodical multidomain state with ferromagnetic arrangement through the whole multilayer, see inset in Fig. 3 (c). To investigate the stability of the homogeneous ferroband with respect to a transformation into tiger tail patterns we consider the model schematically shown in Fig. 3 (b), for $N = 2$, $X = 1$. Magnetization distribution in each ferromagnetic layer is considered as two semi-infinite plates with opposite magnetization, separated by the bar-shape region of width w_{tt} and infinite length in y-direction. We assume that these bars are composed of regular up-down domains with period D_{tt} and infinitely thin domain walls perpendicular to the edges of the bar. The total energy of the tiger tail patterns e_{tt} then consists of dipolar interaction energy between semi-infinite planes, self magnetostatic energy of two bars split into domains, dipolar interactions between these bars, interlayer exchange coupling energy which is proportional to w_{tt} and domain wall energy along bent profiles as shown in the inset in Fig. 3 (c). Magnetostatic energy interactions in this problem can be expressed in term of magnetic charges and functions G_{20} and F_{200} introduced by Hubert and Schäfer for infinite band-shaped and rectangular-shaped elements, respectively [9]. Note, because of the symmetry of the problem, the interaction energy between bars and semi-infinite plates equals zero. Minimization of the total energy of the tiger tail e_{tt} with respect to period of domains D_{tt} and width of band w_{tt} gives the solutions for equilibrium states. The rigorous mathematical treatment of this problem will be reported elsewhere [20]. Here, we highlight only as a representative result a solution for $N = 2$, $X = 1$ and corresponding material and geometrical parameters in zero magnetic field, Fig. 3 (c). The main features of the solution for the tiger tail model are as follows: (i) For $0 < t < t_{\mathrm{b}}$ the global minimum of the system is a sharp domain wall ($w_{\mathrm{tt}} \equiv 0$) which theoretically exists up to $t \to \infty$. For $0 < t < t_{\mathrm{a}}$ a sharp domain wall is the *sole* solution of the system. (ii) For $t_{\mathrm{a}} < t < t_{\mathrm{c}}$ there are stable solutions for a *homogeneous* ferroband with both finite w_{tt} and $D_{\mathrm{tt}} \to \infty$. In Refs. [17,18] we derived analytical equations for the *lower limit* for ferroband existence t_{a}. In the interval $t_{\mathrm{b}} < t < t_{\mathrm{tr}}$ up to the transition line (vertical blue line in Fig. 3 (c)) ferrobands correspond to the global minimum of the system. (iii) Above $t > t_{\mathrm{tr}}$ the global minimum of the system corresponds to the regular ferro-stripe state with finite $D_{\mathrm{tt}} = D_{\mathrm{FS}}$ and $w_{\mathrm{tt}} \to \infty$. (iv) At $t = t_{\mathrm{c}}$ the energy barrier between metastable ferroband (doted line) and ferro-stripe state disappears. t_{c} is the *upper limit* for the ferroband existence. For $t > t_{\mathrm{c}}$ the global minimum of the system is strictly speaking the *shifted* ferro-stripe, see Sec. 1. (v) A equilibrium solutions for the tiger tail pattern with finite w_{tt} and D_{tt} (inset in Fig. 3 (c)) which satisfy the conditions for a global or a local energy minimum of the system *do not exist.* (vi) However, on the assumption that width of the defect band w_{tt} is a *fixed* geometrical parameter, and D_{tt} is the only free parameter of the system, the solutions for tiger tail pattern with finite D_{tt} *do exist* in a wide range of thickness t, dashed area in Fig. 3 (c).

Tiger tail patterns, therefore, appear in the region of the phase diagram where the homogeneous AF state becomes instable towards the ferro-stripe state. Along a topological ferroband defect an instability can create a modulated domain configuration, which can be achieved by a local meandering of the domain walls in the magnetic layer, while a widening

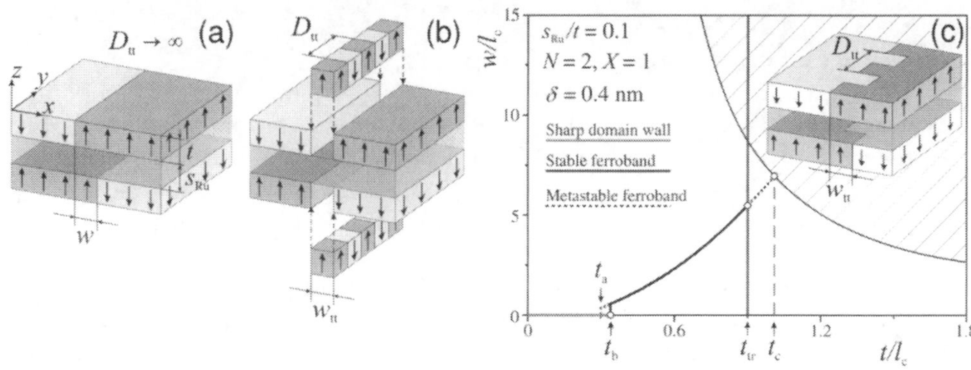

Figure 3. (a) Model of ferrobands in AF coupled multilayers. (b) Constitutents for the tiger tail pattern. (c) Exemplary result for existence range and width w of topological defects in dependence layer thickness t, see text for details. Vertical blue line tr phase transition between ferro-stripe and homogeneous AF state. In dashed area solutions for tiger tail patterns with finite modulation period D_{tt} exist for *fixed* widths w_{tt}.

of the band is prevented by domain wall pinning. This suggests that tiger tail patterns are modes of an inhomogeneous nucleation of ferro-stripes in 1-dimensional topological defects of the antiferromagnetic state. In AF multilayer systems, as $\{[Co/Pt]_{X-1}/Co/Ru\}_N$, topological defects with *freely mobile* or with *pinned* domain walls have been observed simultaneously [3]. Likely, crystallite grain boundaries act as pinning centers in these materials [3].

In conclusion, we have demonstrated that, according to our theoretical model and experimental observation, the ground state of AF coupled multilayers is either AF homogeneous or a ferromagnetic multidomain state, see Fig. 1 (c). Within the AF state there are different topological defects. Contrary to sharp domain walls and ferrobands, the tiger tail pattern cannot be stabilized by an interplay between magnetostatic and IEC energies only, but it can be stabilized by domain wall pinning.

References

[1] S. Hamada, K. Himi, T. Okuno, and K. Takanashi, J. Magn. Magn. Mater. **240**, 539 (2002).
[2] O. Hellwig, T.L. Kirk, J.B. Kortright, A.Berger, E.E. Fullerton, Nature Mater. **2**, 112 (2003).
[3] O. Hellwig, A. Berger, J. B. Kortright, E. E. Fullerton, J. Magn. Magn. Mater. **319**, 13 (2007).
[4] Z. Y. Liu, N. Li, F. Zhang, et al. Appl. Phys. Lett. **93**, 032502 (2008).
[5] E. E. Fullerton, D. T. Margulies, N. Supper, et al. IEEE Trans. Magn. **39**, 639 (2003).
[6] Olav Hellwig, Andreas Berger, and Eric E. Fullerton Phys. Rev. Lett. **91**, 197203 (2003).
[7] A. Baruth, D. J. Keavney, J. D. Burton, et al. Phys. Rev. B **74**, 054419 (2006).
[8] Z. Y. Liu, J. L. He, and Y. J. Tian, G. H. Yu and Y. Jiang Appl. Phys. Lett. **88**, 022504 (2006).
[9] A. Hubert, R. Schäfer, *Magnetic Domains* (Springer-Verlag, Berlin, 1998).
[10] C. Kittel, Rev. Mod. Phys. **21**, 541-583 (1949); C. Kittel,J. K. Galt, Solid State Phys. **3**, 437-565 (1956).
[11] Z. Malek, V. Kambersky, Czech. J. Phys. **8**, 416 (1958).
[12] C. Kooy, U. Enz, Philips Res. Repts. **15**, 7 (1960).
[13] H. J. G. Draaisma, W. J. M. de Jonge, J. Appl. Phys. **62**, 3318 (1987).
[14] N. S. Kiselev, I. E. Dragunov, U. K. Rößler, A. N. Bogdanov, Appl. Phys. Lett. **91**, 132597 (2007).
[15] N.S. Kiselev, I.E. Dragunov, U.K. Rößler, and A. N. Bogdanov Pis'ma Zh. Tekh. Fiz. 33 (23), 87 (2007) [Tech. Phys. Lett. 33, 1028 (2007)].
[16] S. V. Tarabtseva, N. S. Kiselev, Pis'ma Zh. Tekh. Fiz. **36** (6), 87 (2009) [Tech. Phys. Lett. **35**, 289 (2009)].
[17] N. S. Kiselev, U. K. Rößler, A. N. Bogdanov, O. Hellwig, Appl. Phys. Lett. **93**, 162502 (2008).
[18] N. S. Kiselev, U. K. Rößler, A. N. Bogdanov, O. Hellwig, J. Magn. Magn. Mater. **322**, 1340-1342 (2010).
[19] N. S. Kiselev, C. Bran, U. Wolff, et al. Phys. Rev. B **81**, 054409 (2010).
[20] N. S. Kiselev, U. K. Rößler, A. N. Bogdanov, O. Hellwig, to be published.

Morphology, magnetic and resonance properties of Fe/MgO multilayers

A García-García[1], A Vovk[2, 3], P Štrichovanec[4], J A Pardo[4,5], C Magén[4], V Golub[3], O Salyuk[3], P A Algarabel[1], M R Ibarra[1,4]

[1]Instituto de Ciencia de Materiales de Aragón, Universidad de Zaragoza-CSIC 50009 Zaragoza, Spain

[2]Centro de Fisica da Matéria Condensada Unversidade de Lisboa, Campo Grande Ed. C8, 1749-016, Lisboa, Portugal

[3]Institute of Magnetism NAS of Ukraine, 36-b Vernnadsky blvd., 03142, Kyiv, Ukraine

[4]Instituto de Nanociencia de Aragón, Universidad de Zaragoza 50018 Zaragoza, Spain

[5]Departamento de Ciencia y Tecnología de Materiales y Fluidos, Universidad de Zaragoza, 50018-Zaragoza, Spain

E-mail: ayvovk@fc.ul.pt

Abstract. Magnetic, resonance and transport properties of Fe(t nm)/MgO(3.0 nm) multilayers prepared by pulsed laser deposition were investigated. Comparison of the data allows conclusions on Fe layers morphology. For t<0.61 nm typical features of granular cermet films in dielectric regime are observed, i.e. high electrical resistance, isotropic magnetoresistance and strong temperature dependence of magnetization. For higher t coalescence of Fe granules occurs and metallic percolation cluster is formed at t~0.81 nm. This is manifested by rapid decrease of films resistance and formation of multipeak ferromagnetic resonance spectra. For t>1.25 nm a continuous coverage of MgO by Fe takes place. However, the morphology of Fe layers is rough. This causes the appearance of magnetostatic resonance modes analogous to those observed for continuous films deposited on embossed surfaces.

1. Introduction

Magnetic metal-insulator nanostructures and thin films were intensively studied in past decades. They show outstanding magneto-transport and magneto-optical properties which makes them promising candidates for next generation magnetic recording media, low magnetic field sensors and other spintronics applications. Granular films and multilayers of various magnetic metals (Fe, Co, Ni) and alloys (CoFe, FeNi) embedded in different insulator matrixes (SiO_2, Al_2O_3, TiO_2, ZrO_2) have been studied previously [1-8]. Multilayers, especially discontinuous metal-insulator multilayers (DMIMs), prepared by sequential deposition are of particular interest [4, 5]. DMIMs show relatively high tunnelling magnetoresistance (TMR) and enhanced low field magnetic sensitivity since two dimensional (2D) arrangements of magnetic particles favours dipolar interactions within the layer [4]. The structural state of metallic phase in DMIMs varies with increasing nominal thickness of metallic layer (t) from an ensemble of metallic granules incorporated in insulator matrix to a continuous metallic films separated by an insulator. This causes dramatic changes in physical properties. Recently enhanced TMR values have been reported in Fe/MgO/Fe epitaxial magnetic tunnel junctions (MTJ) [9]. This effect was associated with the crucial role of band symmetry and coherent tunnelling. On the other hand, experimental data on DMIMs of Fe/MgO are generally lacking. Only a limited number of experiments on Fe/MgO/Fe trilayers and DMIMs have been carried out recently [10-13]. Isotropic TMR values of ~-3.3% at room temperature and ~-10% at 30K in magnetic field H=18 kOe were reported [13]. However details of Fe layer morphology and microstructure are still not clear.

Transmission electron microscopy (TEM) and X-ray diffraction are basic techniques to investigate microstructure of thin films. However, both techniques have drawbacks. X-ray reflectivity (XRR) proves that investigated Fe/MgO multilayers have periodic structure [11]. Bilayer thickness derived from the fitting of the experimental XRR profiles using LEPTOS software (BRUKER AXS) in assumption of strictly periodic structure was slightly lower than expected from the deposition rates measured for continuous films. The difference between nominal and fitted Fe/MgO bilayer thickness was about 5% for $t>0.81$ nm. It was supposed that the difference is caused by surface roughness. On the other hand, XRR fitting is hardly reliable in discontinuous regime (for small t values) due to non-continuous Fe/MgO interfaces. Cross-section TEM confirms that the structure of Fe layers changes from an ensemble of small granules barely distinguishable from the MgO matrix for $t=0.40$ nm, to continuous Fe layers for $t=1.25$ nm. The polycrystalline state of the MgO layers was also evident [11]. But cross-section TEM provides information only from small area of the sample. Also, the TEM specimen thickness (generally a few tens of nanometers) is much larger than typical Fe grain size (2-3 nm). As a result superimposition of the grains two dimensional projections within specimen thickness gives the impression of a continuous Fe layer. Thus, to determine the evolution of the morphology of metallic layers and establish correlations between structure and physical properties of DMIM one should use complementary experimental methods. Here we report the results of indirect structure investigations of Fe/MgO DMIMs using magnetic, transport and resonance techniques.

2. Experiment

Granular multilayers MgO(3 nm)/[Fe(t nm)/MgO(3 nm)]$_N$ (0.4 nm$<t<$1.50 nm, N is the number of bilayers) were deposited on glass substrates by pulsed laser deposition (PLD). The details of films preparation, structural, transport and magnetic characterizations were reported earlier [11-13]. Ferromagnetic resonance (FMR) measurements were carried out at room temperature at 9.45 GHz using a Bruker ELEXSYS-E500 spectrometer equipped with a goniometer.

3. Results and Discussion

Comparison of magnetic and transport properties is made in figure 1. The values of magnetization (M) measured at 5K and in H=50 kOe practically do not change for $t>0.81$. This indicates presence of continuous ferromagnetic layers in the films [11]. However M values are lower comparing to bulk Fe. The reduction was attributed to formation of mixed Fe-O layer on Fe/MgO interface. Such an intermixture seems to be typical for Fe based multilayers [7, 8]. Electrical resistivity of the films (ρ) shows sharp increase as t drops below 0.81 nm. This is a characteristic of metallic layer structure transition from continuous to discontinuous state [4, 10, 12]. The values of ρ for $t>0.81$ nm are higher comparing to bulk Fe due to effects associated with dimensionality confinements [14, 15] but correspond well to earlier published data on thin continuous Fe films [10, 16-18]. Thus, both magnetometry and transport measurements pointed to the transition from continuous to discontinuous state of Fe layers at $t\sim0.81$ nm.

Zero field cooled (ZFC) and field cooled (FC) magnetic susceptibility and magnetization measurements at different temperatures allow estimation of an average granule size (D_{avr}) for films with discontinuous Fe layers [11]. Samples with $t\leq0.61$ nm demonstrate superparamagnetic properties. The bifurcation temperature (T_b) above which ZFC and FC curves superimpose strongly depends on t and increases from 25K to 100 K when t varies from 0.40 nm to 0.61 nm. This indicates increase of D_{avr}. For $T>T_b$ estimations using Curie-Weiss law give $D_{avr}\sim2$ and $D_{avr}\sim2.8$ nm for $t=0.4$ nm and $t=0.61$ nm, respectively. The same results were obtained by fitting magnetization loops using weighted sum of Langevin functions. In the latter case a broadening of the particle size distribution was also found [11]. This effect is associated with coalescence and aggregation of metallic particles in cermet systems approaching percolation threshold.

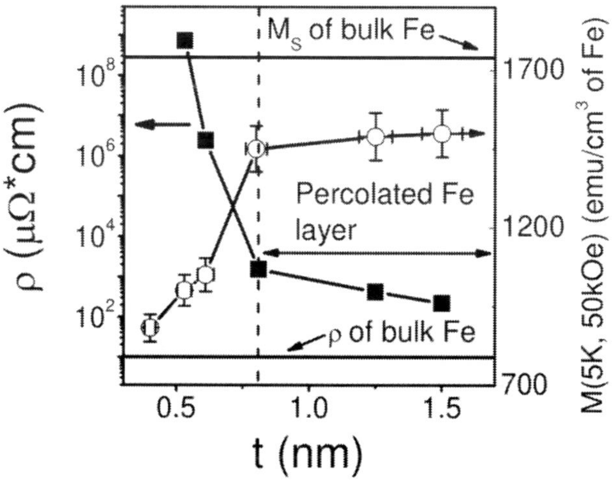

Figure 1. Dependencies of electrical resistivity (ρ) (solid squares) and magnetization (M) (open circles) vs. Fe layer nominal thickness (t) for Fe (t nm)/MgO (3 nm) DMIMs. Measurements of ρ were carried out at room temperature. The error bars at M vs. t plot are associated with 5% accuracy thickness determination.

Below t=0.61 nm electron tunneling govern the conductance and isotropic TMR effect was observed. Average intergranular distance (s) for films with discontinuous Fe layers was estimated in Ref. 13 from temperature dependence of conductance using approach of Abeles et al. [1]. The values of s~2.1 nm and s~1.1 nm were evaluated for films with t=0.53 nm and t=0.61 nm, respectively. The derived intergranular distance is smaller than the thickness of MgO layer (3 nm). This result supports the assumption that coalescence of Fe granules and tunneling between them takes place mainly within the same layer.

For the films with $t \geq 0.81$ nm, T_b is above room temperature and ferromagnetic order is present in the whole temperature range. Anisotropic magnetoresistance (AMR) typical of continuous ferromagnetic films was also found [12]. For $t \geq 1.25$ nm predominantly metallic conductance was observed. In the range of 0.81 nm $\leq t \leq$ 1.25 nm a crossover from the 2D disordered conductance to metallic regime takes place [12, 18]. Judging magnetic and transport properties the films within this t range could be described as 2D percolated structures for which conditions of continuous coverage are not fulfilled.

To provide additional information about the evolution of the morphology of Fe layers with t FMR technique was used. Out-of-plane angular dependences of the resonance fields as well as resonance linewidths were studied. The results are presented in figure 2. The analysis of the FMR data for such films is based on the simple Kittel formula [19]

$$\omega^2 = \gamma^2 \left[H + (N_x - N_z)M_s\right]\left[H + (N_y - N_z)M_s\right] \quad (1)$$

which is analogous to the one for continuous films. Here ω is the FMR frequency, γ is the gyromagnetic ratio, M_s is the saturation magnetization, and N_i are the components of demagnetizing tensor in the coordinate frame with principal z-axis along the external field H. However, the internal dipolar fields in nanostructured materials (especially for DMIMs) do not always fit the classical concept of demagnetizing tensor [20]. It has been shown earlier [21] that the construction of demagnetizing tensor for such systems is very similar to that for a three-dimensional granular sample:

$$\hat{N}_{eff} = f\hat{N}_s + (1 - f)\hat{N}_g \quad (2)$$

where $\hat{N}_s$ and $\hat{N}_g$ are the tensors for sample and granule respectively, and f the volume filling factor. Although the main results in [20] were obtained for square grids of granules the approach presented in that work allows qualitative analysis of our systems.

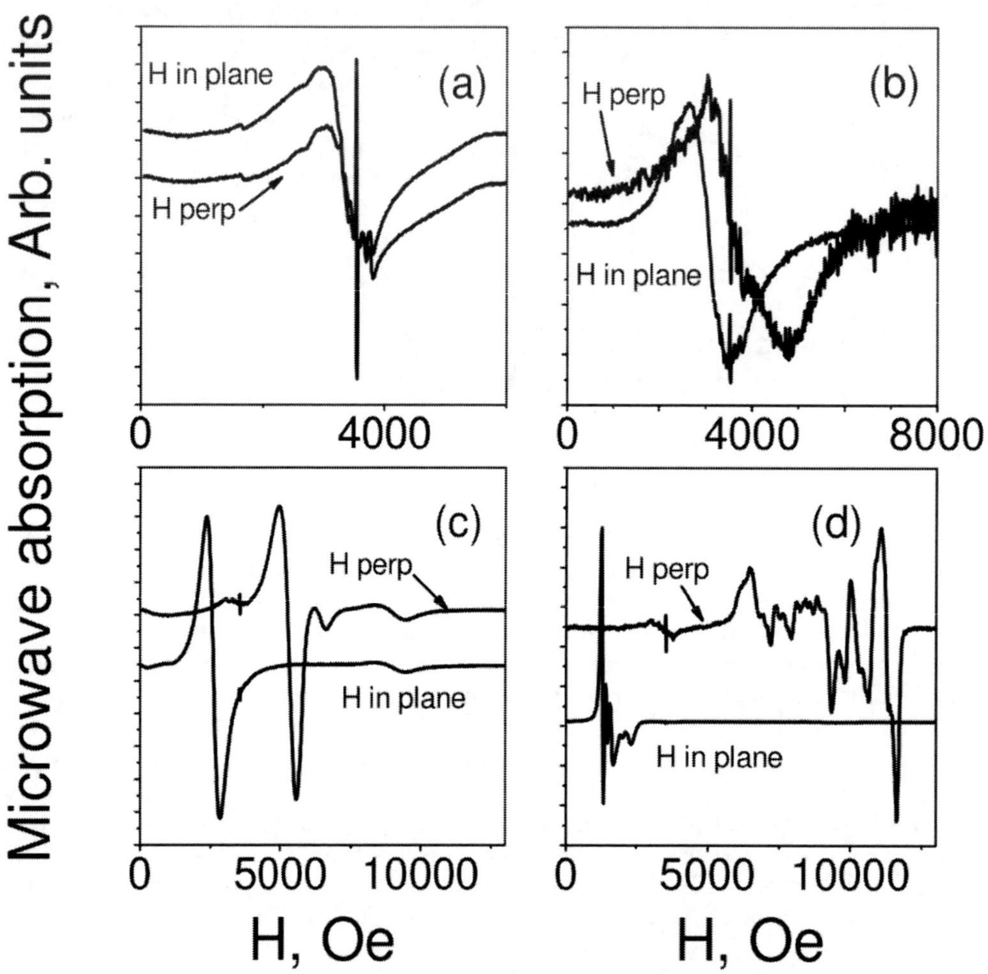

Figure 2. FMR spectra of Fe (t nm)/MgO (3 nm) DMIMs. (a) t=0.40 nm; (b) t=0.61 nm; (c) t=0.90 nm; (d) t=1.50 nm. Sharp resonance line at H~3500 Oe corresponds to sample holder.

For the sample with t=0.40 nm the resonance line is similar to granular films with weak magnetostatic interaction between granules. There is practically no difference between the resonance fields for perpendicular and in-plane configuration (Fig. 2a). The increase of t up to 0.61 nm (Fig. 2b) enlarges the splitting between perpendicular and in plane resonance fields. This should be ascribed to the boost of magnetostatic interaction between granules. It is worth noting that the resonance linewidth increases twice comparing to the sample with t=0.40 nm (1700 Oe and 780 Oe for t=0.61 nm and t=0.40 nm, respectively). It is clearly seen from Eq. 2 that the influence of individual granule contribution to the resonance line broadening decreases with decreasing distances between granules (increase of the volume filling factor). This mechanism should cause reduction of the resonance linewidth with the increase of the filling factor for the granules of the same shape. At the same time the decrease of the distances between granules in disordered system leads to the increase in dispersion of magnetostatic fields on granules coming from their local environment, which is not taking into account in [20]. Also one should consider that the increase of granules concentration in layers leads to their coalescence. As a result, the latter reasons associated with the increase of dispersion of demagnetizing factors due to the particle shape distribution favor the broadening of resonance linewidth.

The dramatic decrease of the resonance linewidth (down to 450 Oe) is observed for the film with t=0.9 nm (Fig. 2c). This manifests an aggregation of granules in the layers and formation of extended magnetic clusters. This result is in good agreement with magnetic and transport measurements that demonstrate formation of percolation cluster of Fe in the films.

Presence of continuous ferromagnetic layer is confirmed by FMR for t>1.25 nm. The resonance peaks are narrow (~70 Oe in the in-plane configuration) and their linewidth are close to those of continuous thick iron films. However, FMR spectra even for the film with the thickest Fe layer (t=1.50 nm, Fig. 2d) are not identical to those of thick films. First of all, the effective magnetization (M_{eff}) of the film derived from the angular dependences of resonance field is more than twice smaller (~650 emu/cm^3) comparing to one of continuous iron layer (1740 emu/cm^3). The decrease of M_{eff} is usually ascribed to formation of perpendicular anisotropy ($H_{A\perp}$) due to the layer surfaces and magnetoelastic contribution. We can not also exclude the reduction of effective demagnetizing factor of Fe layers as a result of the surface roughness, which leads to decrease of perpendicular to the film plane demagnetizing factor $N_{\perp}$ and increase of in-plane ones $N_{\parallel}$. Both these contributions are symmetrically entering into the resonance equations and can not be separated from FMR experiments; only the value of $4\pi M_{eff} = (N_{\perp} - N_{\parallel})M_s - H_{A\perp}$ can be derived. The film roughness is confirmed by multipeak FMR spectra which are analogous to those observed for continuous films deposited on embossed surfaces [22]. However these mechanisms can hardly explain such a strong decrease of M_{eff} (see for instance [23]). On the other hand the results of magnetometry measurements show a significant reduction in saturation magnetization comparing to bulk iron. Considering the fact that intermixed Fe-O layer likely exists on the Fe/MgO interface we should suppose the formation of a surface layer with low magnetization. The exchange interaction between disordered surface and internal iron atoms leads to formation of magnetic layer with reduced saturation magnetization [24]. However it seems the main mechanisms of M_{eff} reduction are the surface anisotropy and magnetoelastic contribution. Taking into consideration plausible values of effective magnetization reduction due to surface roughness and partial oxidation of metal layers the perpendicular anisotropy field can be evaluated as 9-10 kOe.

4. Conclusions

The combination of magnetometry, FMR and transport measurements allows to follow the evolution of the morphology of iron layers in Fe(t nm)/MgO(3 nm) with t. The increase of t from 0.40 nm to 1.50 nm leads to a modification of the films from DMIMs with practically non-interacting metallic granules separated by oxide (t=0.40 nm) through percolated structures (t~0.81 nm) to multilayers with continuous metallic layers (t>1.25 nm). The comparison of FMR and magnetometry data allows to reveal a formation of an oxidized shell with weak magnetization on Fe/MgO interface that cause the reduction of saturation magnetization of Fe layers. The multipeak FMR spectra of continuous films for t up to 1.50 nm give evidence that the surfaces of the Fe layers are very rough. It has been shown a formation of strong perpendicular to the films plane anisotropy due to the layers surfaces and magnetostriction. However additional investigations are needed to separate these contributions.

Acknowledgements

Financial support by Spanish Ministry of Science (through project MAT2008-06567-C02) and by the Aragón Regional Government (through projects E26, PI059/08 and ARAID foundation) is acknowledged. Work of A. Vovk was supported by FCT "Ciencia 2008" program. Work of V. Golub and O. Salyuk was partially supported by NASU through the Grant X-15-247.

References

[1] Abeles B, Sheng P, Coutts M D and Arie Y 1975 *Adv. Phys.* **24** 401
[2] Niklasson G A and Granqvist C G 1984 *J. Appl. Phys.* **55** 3382
[3] Fujimori H, Ohnum S, Kobayashi N and Matsumoto T 2006 *J. Magn. Magn. Mater.* **304** 32
[4] Kakazei G N, Pogorelov Yu G, Lopes A M L, Sousa J B, Cardoso S, Freitas P P, Pereira de Azevedo M M and Snoeck E 2001 *J. Appl. Phys.* **90** 4044
[5] Dieny B, Sankar S, McCartney M R, Smith D J, Bayle-Guillemaud P and Berkowitz A E 1998 *J. Magn. Magn. Mater.* **185** 283
[6] Vovk A, Wang J Q, He J, Zhou W, Pogoriliy A, Shypil' O, Kravets A and Khan H 2002 *J. Appl. Phys.* **91** 10017
[7] Dempsey N M, Ranno L, Givord D, Gonzalo J, Serna R, Fei G T, Petford-Long A K, Doole R C and Hole D E 2001 J. Appl. Phys. **90** 6268
[8] Auric P, Micha J S, Proux O, Giacomoni L and Regnard J R 2000 *J. Magn. Magn. Mater.* **217** 175
[9] Parkin S S P, Kaiser C, Panchula A, Rice P M, Hughes B, Samant M and Yang S H 2004 *Nature Materials* **3** 862
[10] Arita M, Wakasugi K, Ohta K, Hamada K, Takahashi Y and Choi J B 2008 *Microelectronic Engineering* **85** 2445
[11] García-García A, Vovk A, Pardo J A, Štrichovanec P, Magén C, Snoeck E, Algarabel P A, De Teresa J M, Morellón L and Ibarra M R 2009 *J. Appl. Phys.* **105** 063909
[12] García-García A, Vovk A, Štrichovanec P, Pardo J A, Magén C, Algarabel P A, De Teresa J M, Morellón L and Ibarra M R 2010 *J. Phys.: Condens. Matter.* **22** 056003
[13] García-García A, Vovk A, Pardo J A, Štrichovanec P, Algarabel P A, Magén C, De Teresa J M, Morellón L and Ibarra M R 2010 *J. Appl. Phys.* **107** 033704
[14] Fuchs K 1938 *Proc. Camb. Phil. Soc.* **34** 100
[15] Sondheimer E H 1952 *Adv. Phys.* **1** 1
[16] Schad R, Belien P, Verbanck G, Moshchalkov V V and Bruynseraede Y 1998 *J. Phys.: Condens. Matter* **10**, 6643
[17] Sangiao S, Morellon L, Simon G, De Teresa J M, Pardo J A, Arbiol J and Ibarra M R 2009 *Phys. Rev.* B **79** 014431
[18] Liu C, Park Y and Bader S D 1991 *J. Magn. Magn. Mater.* **111** L225
[19] Kittel C 1996 Introduction to Solid State Physics (Wiley, New York)
[20] Kakazei G N, Pogorelov Yu G, Costa M D, Golub V O, Sousa J B, Freitas P P, Cardoso S and Wigen P E 2005 J. Appl. Phys. **97** 10A723
[21] Dubowik J 1996 *Phys. Rev. B* **54** 1088
[22] Vovk A, Malkinski L, Golub V, Whittenburg S, O'Connor C, Jung J S and Min S H 2005 *J. Appl. Phys.* **97** 10J506
[23] Schlömann E 1970 *J. Appl. Phys.* **41** 1617
[24] Guslienko K Y, Lesnik N A, Mitsek A I and Vozniuk B P 1991 *J. Appl. Phys.* **69** 5316

Journal of Physics: Conference Series **303** (2011) 012053 doi:10.1088/1742-6596/303/1/012053

Preparation of hcp–Ni epitaxial films on Cr underlayers

Jumpei Higuchi, Mitsuru Ohtake, Yoichi Sato, and Masaaki Futamoto

Faculty of Science and Engineering, Chuo University, 1-13-27 Kasuga, Bunkyo-ku, Tokyo 112-8551, Japan

E-mail: higuchi@futamoto.elect.chuo-u.ac.jp

Abstract. Ni epitaxial films are prepared by ultra-high vacuum rf magnetron sputtering on Cr underlayers of (100) and (211) planes grown hetero-epitaxially on MgO substrates of (100) and (110) orientations, respectively. The film growth behavior and the crystallographic properties are studied by *in-situ* reflection high energy electron diffraction and pole figure X-ray diffraction. Metastable hcp–Ni($11\bar{2}0$) and hcp–Ni($1\bar{1}00$) crystals respectively nucleate on Cr(100) and Cr(211) underlayers, where the hcp crystals are stabilized through hetero-epitaxial growth. With increasing the film thickness, the hcp structure in the Ni films starts to transform into more stable fcc structure by atomic displacement parallel to the hcp(0001) closed packed plane. The resulting films consist of hcp and fcc crystals.

1. Introduction

hcp–Ni is metastable and does not appear in the bulk phase diagram. A theoretical calculation shows that the magnetic anisotropy of bulk hcp–Ni crystal is similar to that of bulk hcp–Co crystal [1]. Preparation of hcp–Ni crystal has been reported in the forms of nano-particles and polycrystalline thin films [2]–[4]. However, the hcp crystals easily transform into more stable fcc structure during sample preparation processes, and these samples generally include large volume of fcc phase. Recently, epitaxial film growth has made it possible to stabilize metastable structures. Epitaxial films are useful to investigate the basic structural properties, since the film uniformity is well controlled. Metastable hcp–Ni and hcp–permalloy epitaxial films have been prepared by employing *B*1–MgO(100) [5,6] and fcc–Au(100) [7]–[10] underlayers. However, parts of these epitaxial hcp crystal also transformed into fcc structure. X-ray pole figure and high-resolution transmission electron microscopy analyses [10] show that the crystallographic orientation relationships between the hcp and the transformed fcc crystals are hcp($11\bar{2}0$)[$1\bar{1}00$] || fcc(110)[$1\bar{1}\bar{1}$] and fcc(110)[$1\bar{1}1$]. bcc–Cr is frequently used as the underlayer to prepare hcp–Co films with the *c*–axis parallel to the substrate surface. In the present study, Ni films were prepared on Cr underlayers of (100) and (211) hetero-epitaxially grown on single-crystal substrates. The effects of underlayer orientation on the crystallographic properties were investigated.

2. Experimental procedure

Thin films were prepared on MgO single-crystal substrates of (100) and (110) by using an rf magnetron sputtering system of which base pressures were lower than 4×10^{-7} Pa. The rf powers for Ni and Cr targets were respectively fixed at 50 and 40 W, and the Ar gas pressure was kept constant at 0.7 Pa, where the deposition rate was 0.02 nm/s for both materials. Substrates were heated at 600 °C for 1 hour in the UHV chamber to obtain clean surfaces. The surface structure was checked by reflection high energy electron diffraction (RHEED). Cr underlayers of (100) and (211) were prepared by hetero-epitaxial growth on MgO substrates of (100) and (110), respectively. The sample structure was Ni(40 nm)/ Cr(10 nm)/MgO. By depositing Cr material on the substrates, high-quality epitaxial

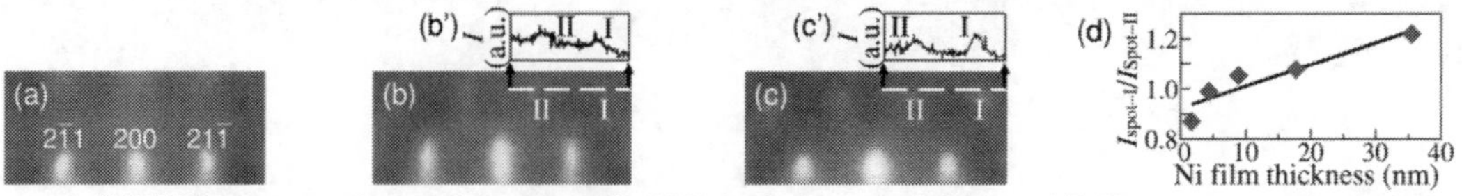

Figure 1. RHEED patterns observed for (a) an Cr(100) underlayer and (b,c) an Ni film grown on the underlayer. The Ni film thicknesses are (b) 2 nm and (c) 40 nm. The incident electron beam is parallel to the MgO[001] direction. (b',c') RHEED intensity profiles along the dotted lines in (b,c), respectively. (d) Film thickness dependence on the RHEED spot peak intensity ratio of Spot–I to Spot–II shown in (a)–(c).

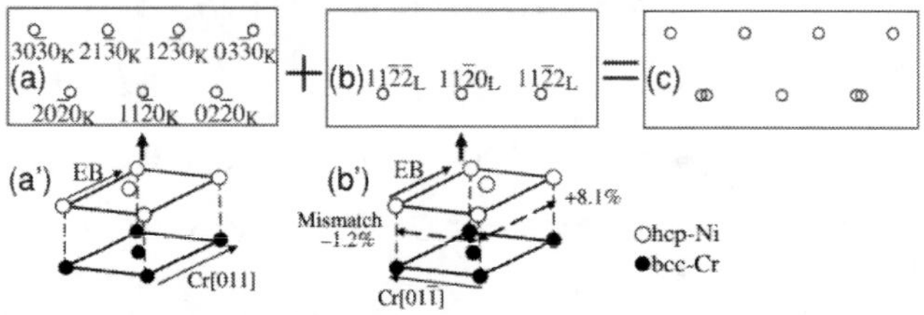

Figure 2. (a)–(c) RHEED spot maps and (a',b') epitaxial orientation relationships between the Ni(11$\bar{2}$0) crystal and the Cr(100) underlayer. The incident electron beam is parallel to (a) the Ni[0001]$_{\text{Type K}}$, (b) the Ni[1$\bar{1}$00]$_{\text{Type L}}$. (c) A spot map overlapped with (a,b). The lattice mismatches are calculated from the lattice constants of the bulk Cr crystal and the hcp–Ni(11$\bar{2}$0) film prepared in the present study.

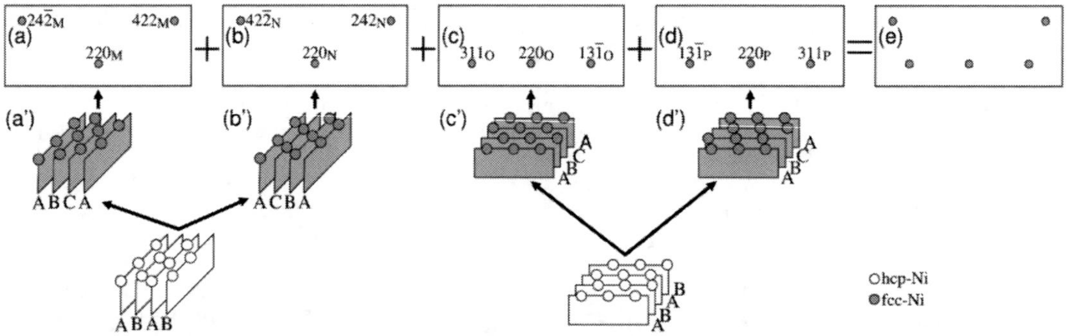

Figure 3. (a)–(e) RHEED spot maps and (a')–(d') crystallographic orientation relationships between the hcp–Ni(11$\bar{2}$0) crystal and the fcc–Ni(110) crystal transformed from the hcp crystal. The incident electron beam is parallel to (a) the Ni[1$\bar{1}\bar{1}$]$_{\text{Type M}}$, (b) the Ni[1$\bar{1}$1]$_{\text{Type N}}$, (c) the Ni[$\bar{1}$1$\bar{2}$]$_{\text{Type O}}$, or (d) the Ni[1$\bar{1}\bar{2}$]$_{\text{Type P}}$. (e) A spot map overlapped with (a)–(d).

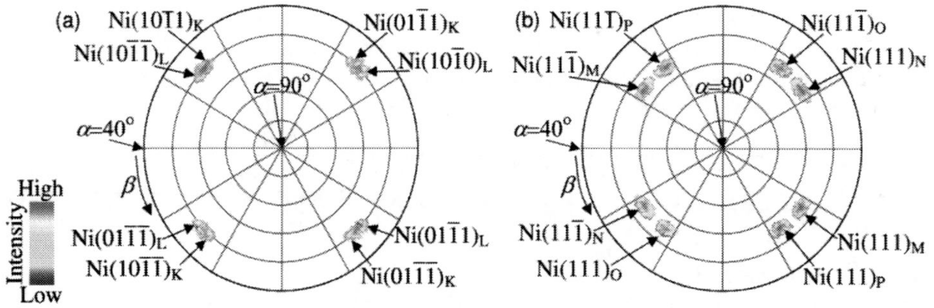

Figure 4. Pole figure XRD profiles measured by making the diffraction angle, $2\theta B$, fixed at (a) 47.6° and (b) 44.4°, respectively. The intensity is shown in a logarithmic scale.

Cr underlayers of (100) and (211) orientations were prepared. A substrate temperature of 300 °C was used for the underlayer formation. The epitaxial orientation relationships were determined by RHEED observations (figure 1(a) and 5(a)) as follows,

$$\mathrm{Cr}(100)[110] \parallel \mathrm{MgO}(100)[100],$$
$$\mathrm{Cr}(211)[\bar{1}11],\ \mathrm{Cr}(211)[1\bar{1}\bar{1}] \parallel \mathrm{MgO}(110)[\bar{1}10].$$

The Cr(211) underlayer consists of two variants whose orientations are rotated around the film normal by 180° each other, whereas the Cr(100) underlayer is a single-crystal. 40-nm-thick Ni films were deposited on the underlayers at room temperature. The surface structure during Ni rf–sputter deposition process was studied by RHEED. The structural properties were investigated by out-of-plane, in-plane, and pole figure X-ray diffraction (XRD) with Cu–Kα radiation (λ=0.15418 nm).

3. Results and discussion

An Ni epitaxial film was obtained on a Cr(100) underlayer. Figures 1(b) and (c) show the RHEED patterns observed during rf–sputter deposition process of Ni on the Cr(100) underlayer. A clear diffraction pattern consisting of two types of hcp(11$\bar{2}$0) reflections shown in the RHEED spot maps of figure 2 is observed in an early stage of Ni film growth, as shown in figure 1(b). An hcp–Ni(11$\bar{2}$0) epitaxial film is obtained on the Cr underlayer. There are two possibilities for hcp–Ni(11$\bar{2}$0) plane nucleation on the Cr(100) underlayer in terms of atomic arrangement matching between the two materials. In this case, the c-axes of two hcp–Ni(11$\bar{2}$0) crystals are in plane and perpendicular each other. The epitaxial orientation relationships between the hcp–Ni crystals and the underlayer are determined by RHEED observation as follows,

$$\mathrm{Ni}(11\bar{2}0)[0001] \parallel \mathrm{Cr}(100)[011]\ \text{(Type K)},$$
$$\mathrm{Ni}(11\bar{2}0)[1\bar{1}00] \parallel \mathrm{Cr}(100)[0\bar{1}1]\ \text{(Type L)}.$$

With increasing the film thickness, the peak intensity of $03\bar{3}0_{\mathrm{Type\ K}}$ (Spot–I) becomes stronger, whereas that of $12\bar{3}0_{\mathrm{Type\ K}}$ (Spot–II) weakens, as shown in figure 1(c). The RHEED data suggest a possibility that parts of the hcp crystal have transformed into fcc crystal. The reason is described below by considering the atomic displacement parallel to the hcp (0001) closed packed plane. When the orientation relationships between the hcp crystal and the transformed fcc crystal of Ni films formed on fcc–Au(100) underlayers [10] are considered, the fcc–Ni crystal is expected to be formed on the Cr(100) underlayer in the following orientation relationships (figures 3(a')–(d')),

$$\mathrm{Ni}(110)[1\bar{1}\bar{1}] \parallel \mathrm{Ni}(11\bar{2}0)[0001]\ \text{(Type M)},$$
$$\mathrm{Ni}(110)[1\bar{1}1] \parallel \mathrm{Ni}(11\bar{2}0)[0001]\ \text{(Type N)},$$
$$\mathrm{Ni}(110)[\bar{1}1\bar{2}] \parallel \mathrm{Ni}(11\bar{2}0)[1\bar{1}00]\ \text{(Type O)},$$
$$\mathrm{Ni}(110)[1\bar{1}\bar{2}] \parallel \mathrm{Ni}(11\bar{2}0)[1\bar{1}00]\ \text{(Type R)}.$$

These orientation relationships are confirmed by pole figure X-ray profiles, as shown in figure 4. The fcc RHEED pattern consisting of four types of (110) reflections shown in the spot maps of figure 3 overlaps with the hcp pattern, where the Spot–I consists of the hcp spots, $03\bar{3}0_{\mathrm{Type\ K}}$ and the fcc spots, $422_{\mathrm{Type\ M}}$ and $242_{\mathrm{Type\ N}}$, while the Spot–II corresponds to only the hcp spots, $12\bar{3}0_{\mathrm{Type\ K}}$. Figure 1(d) shows the film thickness dependence on the RHEED spot peak intensity ratio of Spot–I to Spot–II estimated from the RHEED intensity profiles. The increase in the spot peak intensity ratio, $I_{\mathrm{Spot-I}}/I_{\mathrm{Spot-II}}$, thus indicates that the volume ratio of fcc to hcp phase exposed to the film surface increases with increasing the film thickness. In the pole figures shown in figure 4, the XRD peak intensities from the hcp and fcc structures are very similar. The result may indicate that the fcc volume in Ni film is comparable to that of hcp at 40 nm thickness.

In the present study, clear sharp RHEED patterns are observed in early stages of Ni film growth, though there is a fairly large mismatch of 8.1 % at the Ni(11$\bar{2}$0)/Cr(100) interface. (Figure 2(b')). The mismatch seems to be reduced by introduction of misfit dislocations similar to the case of Ni(11$\bar{2}$0) film growth on Au(100) layer [8].

An Ni epitaxial film was also obtained on a Cr(211) underlayer. Figures 5(b) and (c) show the RHEED patterns observed during deposition process of Ni on a Cr(211) underlayer. In an early stage of Ni film growth, the RHEED pattern of hcp(1$\bar{1}$00) shown in the spot maps of figures 6(a) is observed on the Cr underlayer. The epitaxial orientation relationships are determined as follows,

$$\mathrm{Ni}(1\bar{1}00)[0001] \parallel \mathrm{Cr}(211)[\bar{1}11],\ \mathrm{Cr}(211)[1\bar{1}\bar{1}].\ \text{(Type Q)},$$

An hcp–Ni(1$\bar{1}$00) single-crystal film is formed on the Cr(211) underlayer with variant structure.

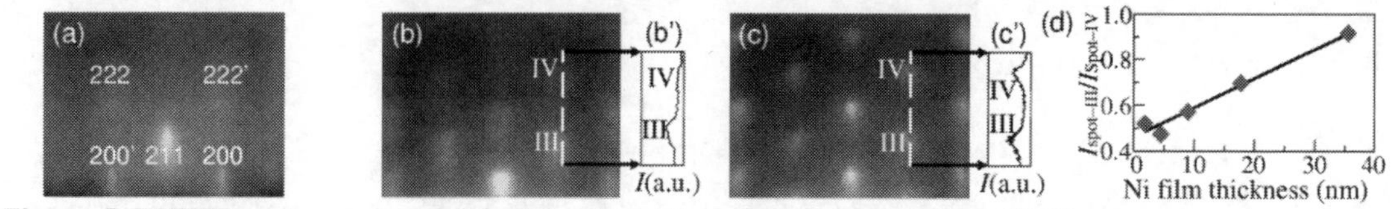

Figure 5. RHEED patterns observed for (a) a Cr(211) underlayer and (b,c) an Ni film grown on the underlayer. The Ni film thicknesses are (b) 2 nm and (c) 40 nm. The incident electron beam is parallel to the MgO$[\bar{1}10]$ direction. (b',c') RHEED intensity profiles along the dotted lines in (b,c), respectively. (d) Film thickness dependence on the RHEED spot peak intensity ratio of Spot–III to Spot–IV shown in (a)–(c).

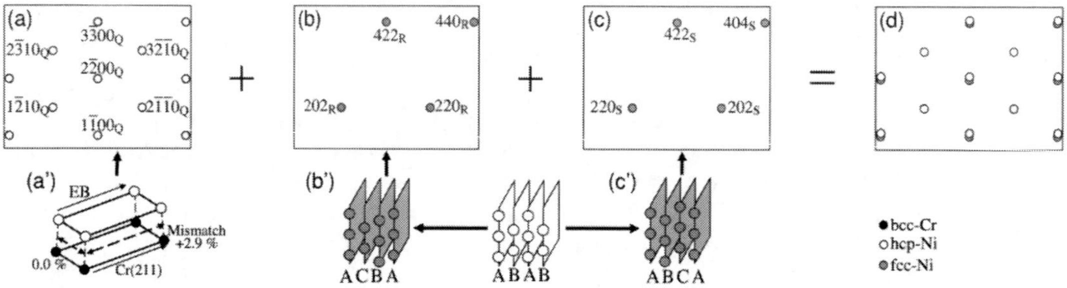

Figure 6. (a)–(c) RHEED spot maps and (a')–(c') crystallographic orientation relationships between the hcp–Ni$(1\bar{1}00)$ crystal and the Cr(211) underlayer and between the hcp–Ni$(1\bar{1}00)$ crystal and the transformed fcc–Ni(211) crystal. The incident electron beam is parallel to (a) the Ni$[0001]_{\mathrm{Type\ Q}}$, (b) the Ni$[\bar{1}11]_{\mathrm{Type\ R}}$ and (c) the Ni$[1\bar{1}\bar{1}]_{\mathrm{Type\ S}}$. (d) A spot map overlapped with (a)–(c). The lattice mismatches are calculated from the lattice constants of the bulk Cr crystal and the hcp–Ni$(1\bar{1}00)$ film prepared in the present study.

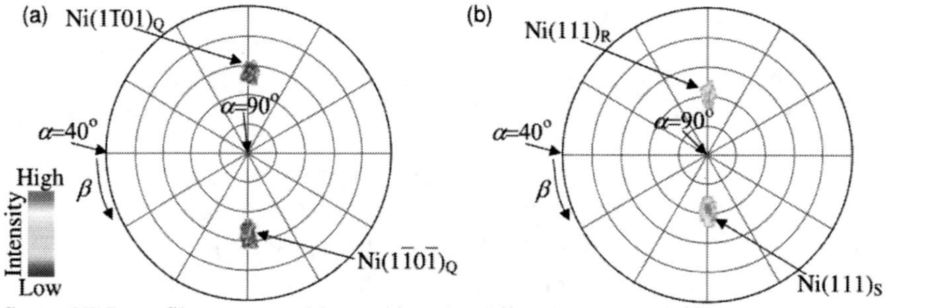

Figure 7. Pole figure XRD profiles measured by making the diffraction angle, $2\theta B$, fixed at (a) 47.6° and (b) 44.4°, respectively. The intensity is shown in a logarithmic scale.

This is because that the surface atomic arrangement of Cr is the same for the two Cr variants and the atomic arrangement of hcp–Ni$(1\bar{1}00)$ matches with the Cr layer under the Type Q orientation relationship. Figure 5(d) shows the film thickness dependence on the RHEED spot peak intensity ratio of $2\bar{1}\bar{1}1_{\mathrm{Type\ Q}}$ (Spot–III) to $3\bar{2}\bar{1}0_{\mathrm{Type\ Q}}$ (Spot–IV). With increasing the film thickness, the spot peak intensity ratio, $I_{\mathrm{Spot-III}}/I_{\mathrm{Spot-IV}}$, increases, suggesting that parts of the hcp crystal have also transformed into fcc crystal, similar to the case of Ni film growth on Cr(100) underlayer. The orientation relationships between the hcp$(1\bar{1}00)$ and the transformed fcc crystals formed on the Cr(211) underlayer are determined from RHEED as follows,

$$\mathrm{Ni}(211)[\bar{1}11] \parallel \mathrm{Ni}(1\bar{1}00)[0001] \text{ (Type R)},$$
$$\mathrm{Ni}(211)[1\bar{1}\bar{1}] \parallel \mathrm{Ni}(1\bar{1}00)[0001] \text{ (Type S)}.$$

These orientation relationships are confirmed by X-ray pole figure analysis, as shown in figure 7. The Spot–III consists of the $2\bar{1}\bar{1}0_{\mathrm{Type\ Q}}$, the $220_{\mathrm{Type\ R}}$, and the $202_{\mathrm{Type\ S}}$, while the Spot–IV corresponds to only the $3\bar{2}\bar{1}0_{\mathrm{Type\ Q}}$. With increasing the film thickness, the volume ratio of fcc(211) to hcp$(1\bar{1}00)$ increases. The XRD pole figure peak intensity data shown in figure 7 suggest that the fcc volume is larger than that of hcp structure at the Ni layer thickness of 40 nm.

Figures 8(a)–(c) and 9(a)–(c) show the XRD spectra of Ni films grown on Cr underlayers of (100) and

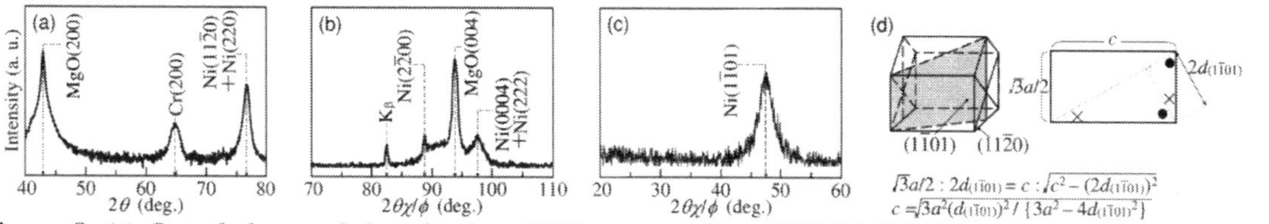

Figure 8. (a) Out-of-plane and (b,c) in-plane XRD spectra of an Ni/Cr/MgO(100) specimen. The scattering vector of in-plane XRD is parallel to (b) the MgO[002] and (c) the Ni[1$\bar{1}$01] directions. The intensity is shown in a logarithmic scale. (e) Schematic model of relationship between Ni(11$\bar{2}$0) and Ni(1$\bar{1}$01).

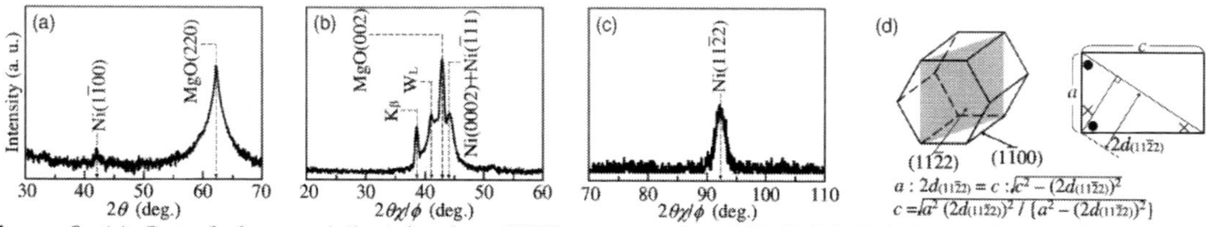

Figure 9. (a) Out-of-plane and (b,c) in-plane XRD spectra of an Ni/ Cr/MgO(110) specimen. The scattering vector of in-plane XRD is parallel to (b) the MgO[002] and (c) the Ni[11$\bar{2}$2] directions. The intensity is shown in a logarithmic scale. (d) Schematic model of relationship between Ni(1$\bar{1}$00) and Ni(11$\bar{2}$2).

(211) orientations, respectively. XRD reflections corresponding to the hcp–Ni and the transformed fcc–Ni crystals are clearly recognized. The lattice constants, *a*, of the hcp–Ni crystals formed on Cr underlayers of (100) and (211) orientations are determined from the Ni(2$\bar{2}$00) XRD peaks to be $4d_{(2\bar{2}00)}/(3)^{1/2}$ = 0.254 nm and 0.249 nm, respectively. By considering the structures shown in figures 8(d) and 9(d), the lattice constants, *c*, of the hcp–Ni crystals formed on Cr underlayers of (100) and (211) orientations are calculated to be 0.402 nm and 0.419 nm, respectively. The lattice constants are delicately influenced by the mismatches between the hcp–Ni crystals and the Cr underlayers.

4. Conclusion

The film growth behavior and the crystallographic properties of Ni epitaxial films grown on Cr underlayers of (100) and (211) orientations were investigated by in-situ RHEED and XRD. hcp–Ni epitaxial crystals of (11$\bar{2}$0) and (1$\bar{1}$00) parallel to the substrate surface are respectively obtained on Cr underlayers of (100) and (211) orientations. With increasing the film thickness, the hcp–Ni crystal transforms into more stable fcc structure by atomic displacement parallel to the hcp (0001) closed packed plane. The lattice constants of hcp–Ni crystals formed on Cr underlayers of (100) and (211) orientations are determined.

Acknowledgements

A part of this work was supported by NEDO–Japan and STFJ (Science and Technology Foundation of Japan). M. Ohtake is a Research Fellow of the Japan Society for the Promotion of Science.

References

[1] Papaconstantopoulos D A, Fry J L and Brener N E 1989 *Phys. Rev.* B **39** 2526
[2] Hemenger P and Weik H 1965 *Acta Crystallogr.* **19** 690
[3] González V R *et al.* 2007 *J. Solid State. Chem.* **180** 22
[4] Carturan G, Cocco G, Enzo S, Ganzerla R, and Lenarda M 1988 *Mater. Lett.* **7** 47.
[5] Tian W *et al.* 2005 *Appl. Phys. Lett.* **86** 131915
[6] Tanaka T, Ohtake M, Kirino F and Futamoto M 2009 *IEEE Trans. Magn.* **45** 2515
[7] Ohtake M, Tanaka T, Kirino F and Futamoto M 2010 *J. Phys. Conf. Ser.* **200** 072072
[8] Ohtake M, Tanaka T, Kirino F and Futamoto M 2010 *J. Appl. Phys.* **107** 09E310.
[9] Ohtake M, Tanaka T, Kirino F and Futamoto M 2010 *J. Magn. Soc. Jpn.* **34** 267.
[10] Sato Y *et al.* 2010 *Abstr. ISAMMA2010* PF–03

Magnetocardiography with GMR-based sensors

M. Pannetier-Lecoeur*, H. Polovy, N. Sergeeva-Chollet, G. Cannies, C. Fermon,
DSM/IRAMIS/SPEC- CNRS URA 2464, CEA Saclay, 91191 Gif sur Yvette Cedex France

L. Parkkonen

Elekta Oy, 00530 Helsinki, Finland

*correponding author: myriam.lecoeur@cea.fr

Abstract. We have developed a sensor based on the Giant Magneto-Resistive (GMR) effect, associated to a superconducting flux-to-field transformer, which allows femtotesla field detection in a wide range of frequencies [1]. Such sensors are good candidates for measurements of biomagnetic signals generated by the brain or the heart. Here we present Magneto-Cardiographic (MCG) recordings over the chest of healthy volunteers at 4K and 77K with these sensors in a magnetically shielded room. The sensitivity is now limited by the 1/f noise of the GMR element which appears below few kHz. We present a technique based on supercurrent switching which reduces this low frequency noise.

1. Introduction

Biomagnetic signals are generated either by ferromagnetic particles in the body either by the electrical activity of cells. In particular, the heart electrical activity generates magnetic signal with amplitude in the picotesla range (1pT=10^{-12}T) at the chest surface. Electrical activity of bunches of neurons in the brain generates signals even weaker, typically of tens or hundreds of femtotesla (1fT=10^{-15}T). These signals can be measured non-invasively and without electrodes by extremely sensitive magnetometers. Most of the MCG recordings so far have been acquired with Superconducting Interference Devices (SQUIDs), mainly low-Tc SQUIDS for brain imaging.

New proposals for alternative detection have been proposed [1], [2]. Here we present first detection of biomagnetic signals using a spin-electronics based sensor.

In the first part of this paper, we give an example of MCG recordings made with a mixed sensor and in the second part we present a supercurrent modulation technique to repel the 1/f noise outs from the low frequency measurement range.

2. Experimental set-up

2.1. Mixed sensor

A mixed sensor associates one or several magnetoresistive elements (Fig. 1), here a GMR spin valve, whose sensitivity at room temperature is of the order of 0.5nT/√Hz, with an efficient flux-to-field transformer realized with a superconducting loop containing a constriction at the position of the GMR element. When a field is applied perpendicularly to the loop in its superconducting state, a supercurrent is generated to prevent the entrance of the flux. The current density in the loop constriction becomes high enough to be detected by the spin valve element. Sensitivity in the femtotesla range can be reached in the thermal noise [3]. The typical GMR stack composition is Ta(2)/NiFe(3)/CoFe(1)/Cu(2)/CoFe(2.5)/MnIr(6)/Ta(5), thicknesses given in nm, and the superconducting loop is either made of Nb or YBCO. Limitation to performances of this device at low frequencies is due to the small magnetic volume of the GMR element, which leads to a 1/f noise contribution which is dominant below typically few hundred Hertz. Nevertheless, a 1cm² surface sensor exhibit field equivalent noise of about 1pT/√(Hz) at 1Hz [1], which is sensitive enough to record biomagnetic signals as those produced by the heart electrical activity. A mixed sensor resistance varies linearly on a field range of about 1μT which is far larger than the detected signals.

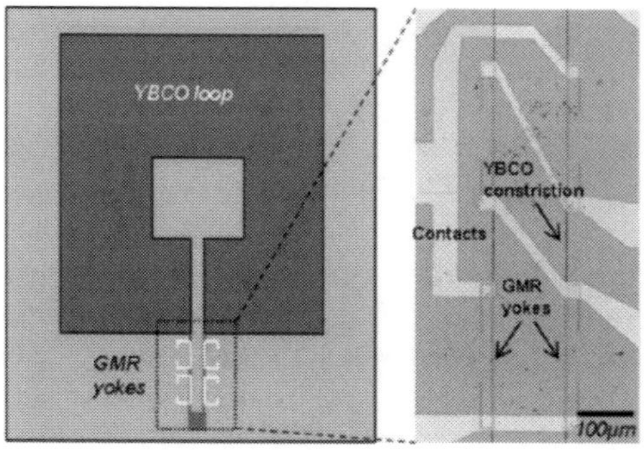

Figure 1. Schematic (left) and micrograph (right) picture of a hybrid sensor with YBCO superconducting loop with two constrictions in parallel and four GMR sensing elements mounted in Wheatstone bridge configuration. The length of each GMR element is 250 μm, its width 5μm.

2.2. Experimental setup

To realize the MCG recording, the sensor is cooled in a home-made fiberglass dewar allowing a distance of 2.5cm between the sensor and the outer surface of the cryostat. The sensor is in a magnetometer configuration and the experiment is carried out in a magnetically shielded room to reduce the contribution of the environmental noise. The volunteer is laying under the measurement system, the sensor being located over his/her chest. The measurement system can be translated along the x-y plane. Electrocardiographic (ECG) recordings are performed at the same time with three electrodes, located on the volunteer's wrists and shoulder to avoid magnetic contamination from the electrodes on the MCG.

The output voltage of the GMR elements is amplified by a first stage low noise INA 103 with a gain of 500, followed by a SR560 with an adjustable gain, and filtering between 0.3Hz and 1 kHz. The acquisition is realized on one channel of a commercial Elekta 306- channel MEG system. Real time signals are acquired. The signal is also averaged during 1mn

In the present experiment, the acquisition has been done with a Nb based sensor of 15x15mm the field equivalent noise level of the sensor was of 2pT/√Hz at 1Hz and 100fT/√Hz above 300Hz.

3. MCG recordings

MCG recordings have been successfully realized (see Figure 2) on a healthy male volunteer. In these recordings, the T-wave and the QRS-complex of the cardiac cycle were clearly visible with few averages.

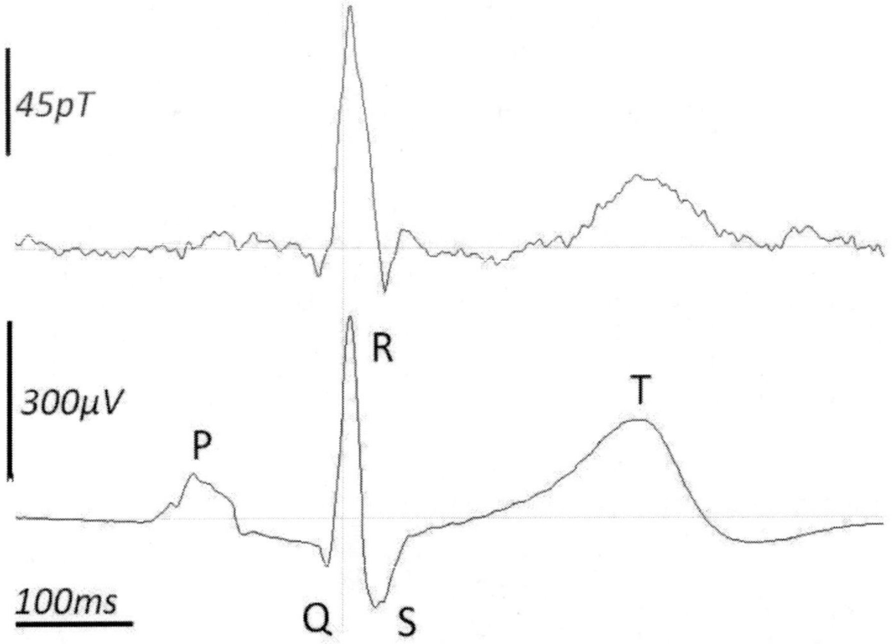

Figure 2: MCG (top) and ECG (bottom) signals averaged during 30s. QRS complex and T-wave are clearly visible. The differences in shape are expected since the ECG records the sum of the electrical signal while the MCG depends on the sensor position with respect to the electrical source in the heart.

4. 1/f noise reduction

In order to improve the sensitivity of the sensor for biomagnetic signal measurements, 1/f noise reduction techniques have been developed. In the case of non linear sensors, like SQUIDs, modulation techniques can be successfully used by an external modulator [4]. In our device, as the response is linear, another strategy should be applied to allow modulation of the signal in the kHz range, where the sensitivity is given by the thermal noise level. The technique proposed here is based on a modulation of the supercurrent passing through two constrictions in parallel of the loop by a local Joule heating (Fig.3).

Without heating, the supercurrent is split equally in the two constrictions, inducing the same resistance change on both GMRs. When one branch is heated, all the supercurrent is running in the other branch and the GMR experiences a change due to this current, whereas the other GMR experiences only the environmental noise. By switching the heating to the other branch, the effect is reversed from the first GMR to the other. This modulation can be achieved in the kHz range, and the relevant signal can be extracted from the demodulated signal.

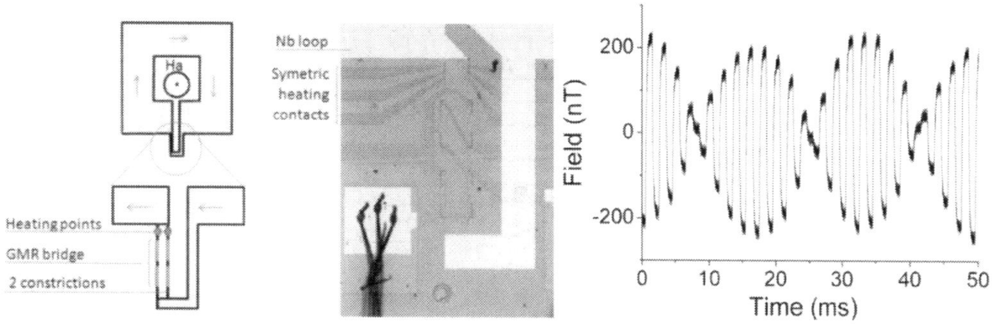

Figure 3: Parallel mixed –sensor designed for Joule effect heating modulation. Left: schematic of an antiparallel mixed sensor. In response to the applied field H_a, a supercurrent runs into the superconducting loop (arrows). Field measured in nT by four GMRs mounted in bridge configuration on a parallel mixed sensor that toggles at 2kHz, while a 200nT sinusoidal magnetic field at 30 Hz is applied perpendicularly to the loop.

If one considers the field-equivalent noise power spectrum S_B with and without toggling (Figure 4), it appears that the low frequency noise is shifted towards the 2kHz modulation peak, and that the signal-to-noise ratio of a 1nT test signal at 30Hz is enhanced by a factor of 8. During the modulation, a low pass filter effect is also observed, due to the loss in flux during the switching.

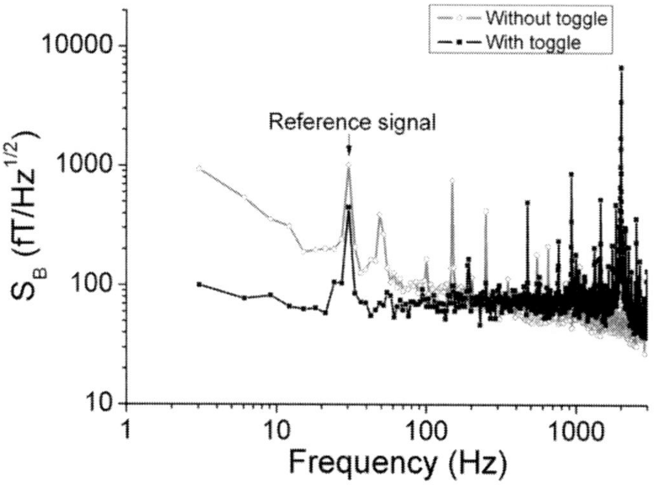

Figure 4: Detection limit of a parallel mixed sensor in bridge configuration with (plain squares) and without (open dots) toggle at 2 kHz. The signal with toggle has been demodulated and exhibits a noise level 8 times lower than without toggle. Hence the 1/noise of the GMR sensor appears around 2kHz.

This high pass filtering effect limits at present the use of this technique for very low frequency applications. It is due to the quantity of flux which enters the loop when a toggling is done. This flux is roughly proportional to the length of the constriction divided by the perimeter of the sensor. For our present sensor, each toggle leads to a 5% loss of supercurrent. It can be demonstrated that this effect is formally equivalent to a first order high pass filtering of the signal with a cut-off frequency equal to 5% of the toggle frequency [5]. We are presently working on new designs to improve this effect by reducing the length of the GMR elements.

5. Conclusion

MCG recordings have been successfully acquired with GMR superconducting mixed sensors. This opens the possibility of developing that technique with rather cheap, nitrogen cooled sensors. The limitation given by the intrinsic 1/f noise of GMR elements can be partially overcome with a toggle technique based of supercurrent switching.

[a]This work has been supported by European contracts Biomagsens NMP4-CT-2005-017210 and MEGMRI – FP7-Health-200859

References

[1] M. Pannetier, C. Fermon, G. Goff, J. Simola, and E. Kerr, Science **304**, 1648 (2004).
[2] I. K. Kominis, T. W. Kornack, J. C. Allred, M. V. Romalis, Nature 422, 596 (2003).
[3] N. Sergeeva-Chollet et al, JEMS 2010 proceedings.
[4] O. Doessel, B. David, M. Fuchs, W. Kullmann, KM. Luedecke, IEEE Trans. Magn. **27**, 2797(1991).
[5] H. Polovy, PhD thesis, and *"1/f noise cancellation in mixed sensors for femtoTesla detection at very low frequencies."* H. Polovy, M. Pannetier-Lecoeur, C. Fermon, submitted.

Low field MRI with magnetoresistive mixed sensors

N Sergeeva-Chollet[1*], H Dyvorne[1], J Dabek[2], Q Herreros[1], H Polovy[1], G Le Goff[1], G Cannies[1], M Pannetier-Lecoeur[1] and C Fermon[1]

[1] DSM/IRAMIS/SPEC-CNRS URA 2464 CEA Saclay, 91191 Gif-sur-Yvette, Cedex, France

[2] Aalto University/BECS Department, P.O. Box 122000, FI-00076 Aalto, Finland

E-mail: natalia.sergeeva-chollet@cea.fr

Abstract. Low-field magnetic resonance imaging (MRI) requires very sensitive magnetometers for detection of weak magnetic signals in femtoTesla range. Magnetoresistive mixed sensors could be used as the sensitive magnetometers [1]. These sensors are the combination of a Giant Magnetoresistive (GMR) field sensor and a superconducting loop used as a flux-to-field transformer. A MRI set-up based on mixed sensor has been mounted. This system operates in static field up to 10mT without shielded room. We will present the LF-MRI setup with the implementation of mixed sensors and the first MRI images obtained at low field without prepolarization.

1. Introduction

Low-field magnetic resonance imaging is a very promising imaging method for biological tissue study. This technique is less expensive and operates in lower frequencies than conventional clinical MRI systems. Nevertheless, as the working frequency is lower the magnetic signals from tissues are very weak requiring femtotesla range very sensitive magnetometers to detect them. Superconducting Quantum Interference Devices (SQUID) [2] are mostly used as detection devices for low-field MRI applications. New devices have been proposed such as atomic magnetometers [3] and mixed sensors [1].

In this paper we present mixed sensors which combine a field GMR sensor and a superconducting flux-to-field transformer. This sensor is sensitive in the femtotesla range and is broadband. This gives the possibility to detect the low-field Nuclear Magnetic Resonance (NMR) signals and to perform MRI imaging.

2. Mixed sensors

Thin film technology is applied to fabricate mixed sensors. This device is a combination of the GMR element and of a superconducting loop which acts as a flux-to-field transformer. Mixed sensors could be fabricated either with a low T_c superconducting loop (Nb) or with a high T_c loop (YBaCuO-YBCO) combined with GMR element. First one operates at liquid helium temperature (4K) whereas the second one needs to be cooled to liquid nitrogen temperature (77K). GMR element is a spin valve which consists of a hard magnetic layer which magnetization orientation is fixed, a copper spacer and a free magnetic layer whose magnetization orientation could be changed under small magnetic fields. This GMR element has a form of yoke to reduce the magnetic noise (figure1).

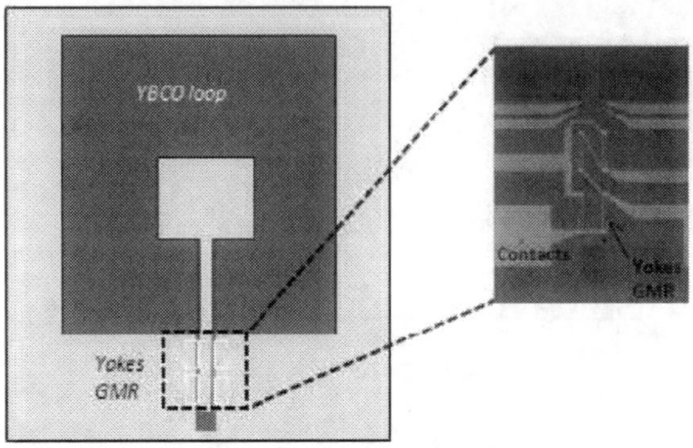

Figure 1. Schematic (left) and micrograph (right) picture of a hybrid sensor with YBCO superconducting loop with two constrictions in parallel and GMR sensing elements. The length of each GMR element is 250 μm.

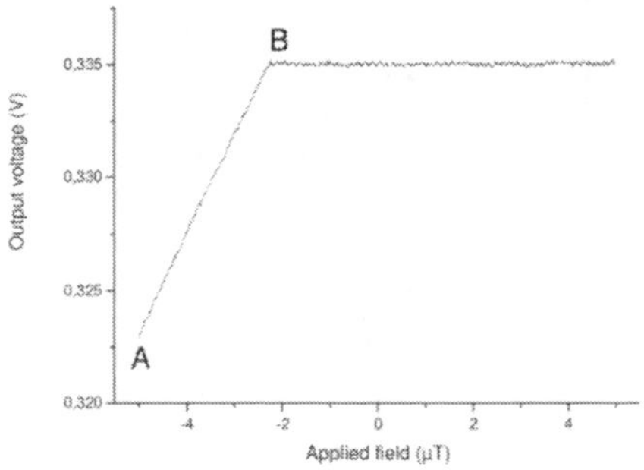

Figure 2. Typical response of a mixed sensor to a perpendicular applied field. From A to B, the output voltage increases as supercurrent in the constriction increases. At point B, the supercurrent reaches its critical value and the local field does not increase anymore.

The typical response of a mixed sensor to magnetic field applied perpendicular to the superconducting loop is shown in figure 2. The applied field generates the supercurrent to prevent the entrance of the flux. As the loop has a constriction, supercurrent flowing in the constriction creates locally high field lines which could be detected by GMR sensors located on top or under the constriction.

Mixed sensor performance could be estimated from its gain and its noise characteristics. Sensor gain is the ratio between the slope of magnetoresistance when the superconducting loop is acting (magnetic field is applied perpendicular to the loop) and the slope of GMR elements (magnetic field is applied on the GMR element sensitive axis with a non acting superconducting loop).

The detectivity of a mixed sensor (figure 3) gives information about the signal-to-noise ratio as a function of frequency. It is clearly seen from the detectivity characteristics that the 1/f noise is dominant below kHz range, leading to a loss of sensitivity at low frequency. This contribution to noise is given by the GMR element of small volume. To reduce this contribution, a modulation technique is proposed in [4] which consists of the local heating of the constriction. At higher frequencies the thermal noise is dominant, which is very important for MRI measurements.

3. Experimental setup

The experimental setup [5] shown in figure 4 consists on double pair of coils generating main static magnetic field of 8mT (corresponding to 340 kHz). The RF pulses are applied by a coil situated around the sample perpendicularly to the main field. The mixed sensor is placed on a liquid nitrogen dewar on top of room temperature sample. The distance between the sensor and the sample is about 30 mm. The experimental setup is an open system without shielding.

Low noise preamplifier is used for the signal amplification. Acquisitions are done with homemade NMR spectrometer. First the ^{1}H NMR signal detection with mixed sensor has been performed to verify the robustness of the sensor.

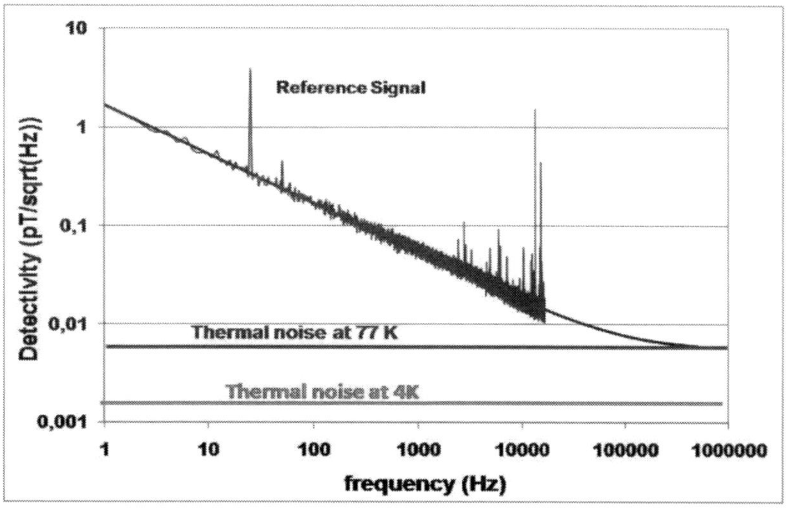

Figure 3. Field equivalent noise as a function of frequency for a YBCO-based mixed sensor. Indication of thermal noise is given with straight line.

The mixed sensor used for NMR measurements is high-T_c sensor with a gain of 716 and critical temperature of T_c=80K. The sensitivity of the sensor is about 20 fT/sqrt(Hz).

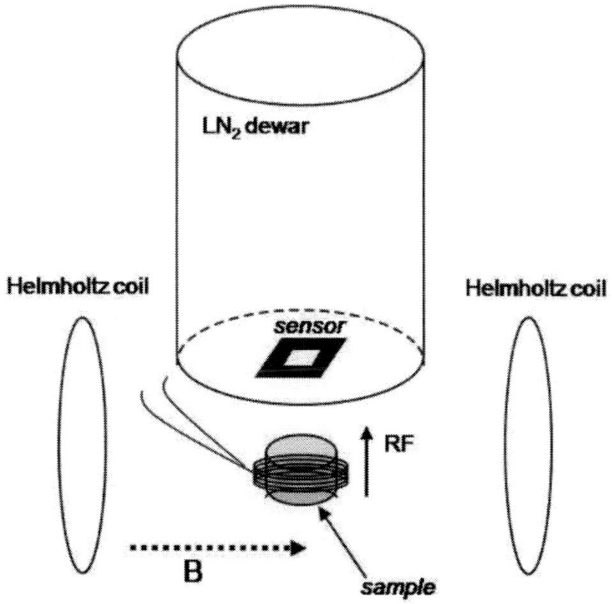

Figure 4. Schematics of experimental setup. A main field (B) is generated by 2 pairs of coils. An RF pulse is applied at corresponding frequency by coil wounded around the sample. The signal is detected by the mixed sensor placed into liquid nitrogen cooled dewar.

Journal of Physics: Conference Series **303** (2011) 012055 doi:10.1088/1742-6596/303/1/012055

4. Results

4.1. NMR measurements

Mixed sensor is used for the ^{1}H NMR signal detection. The NMR resonance line for a single acquisition is given in figure 5. The linewidth is 3Hz which corresponds to 10ppm of homogeneity. To give a scale, this corresponds only to 70nT of inhomogeneity. If we perform a 100 averaging of the signal in time, the linewidth is doubled. This effect is due to the 50Hz and magnetic fluctuations present in the laboratory which creates a low frequency random field of about 100nT.

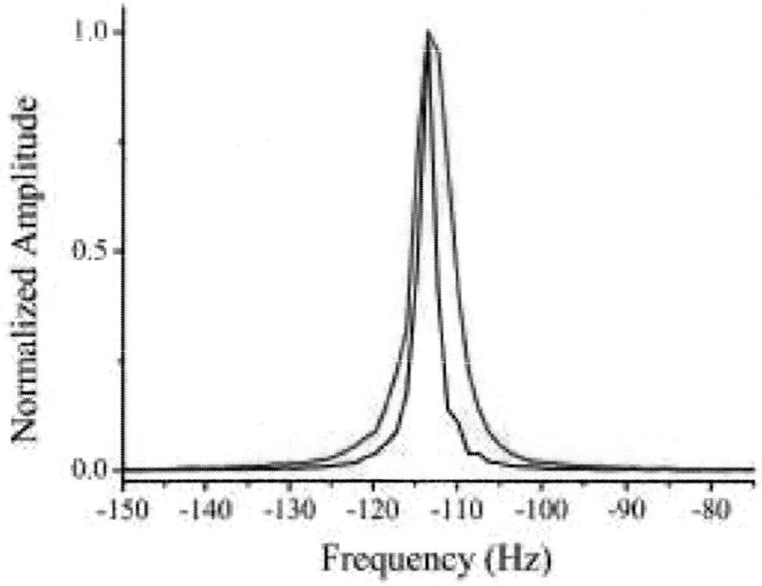

Figure 5. ^{1}H NMR signal measured on doped water at 340kHz with a mixed sensor. The black curve is a single acquisition on 5ml of water

4.2. MRI measurements

For MRI experiments, we have added to the setup a set of three gradients made of a Helmholtz inverted pair for the z gradient and two X and Y figure-of-eight square coils. The linearity of the gradients is better than 1% on the detection volume. First a classical MRI with tuned detection coil has been performed to optimize imaging sequence. This detection coil is placed around the sample perpendicular to the RF coil. The first gradient is applied to perform the slice selection, the second one for the phase encoding and the third gradient is applied during the acquisition. Once the slice is selected, the covering of 2D reciprocal space is performed and then 2D Fast Fourier transform is applied to reconstruct the image. The size of coils is 8cm and the homogeneity zone is $6x6x6cm^3$.

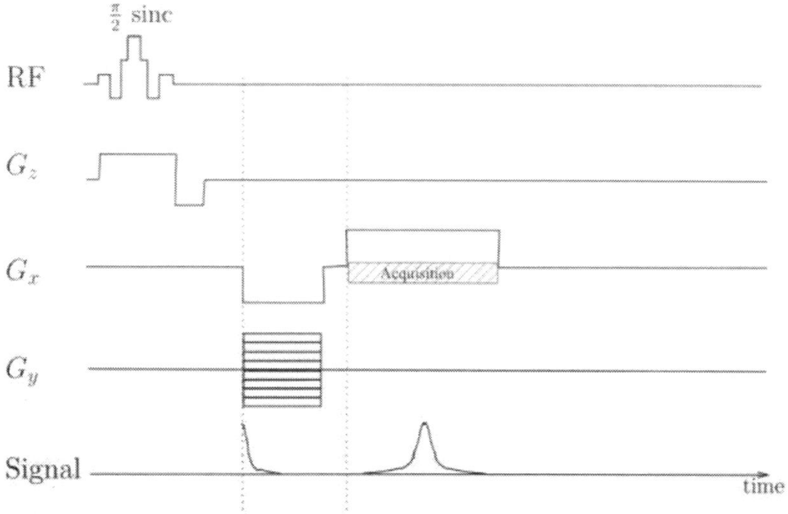

Figure 6. 3D imaging gradient echo-sequence. G_z – slice selection, G_y – phase encoding, G_z- reading gradient.

The MRI image of a phantom is performed using a gradient echo sequence as shown in figure 6. The selective excitation pulse has a bandwidth of 160 Hz. The phantom is a plastic holder with holes of different diameter (from 7mm to 2 mm) filled with doped water ($NiCl_2$) with different doping concentrations (figure 7) varying from 0 to 80 nM. The five 4mm holes (2nd line on the left) have been filled with concentrations varying from 5nM to 20nM. The brightest holes are filled with 80nM $NiCl_2$ concentration. We used a fast T2 (20ms),T1 (200ms) acquisition scheme so the rapid relaxation times correspond to brighter holes. Exact quantization of the link between brightness and precise relaxation times will be published elsewhere [6].

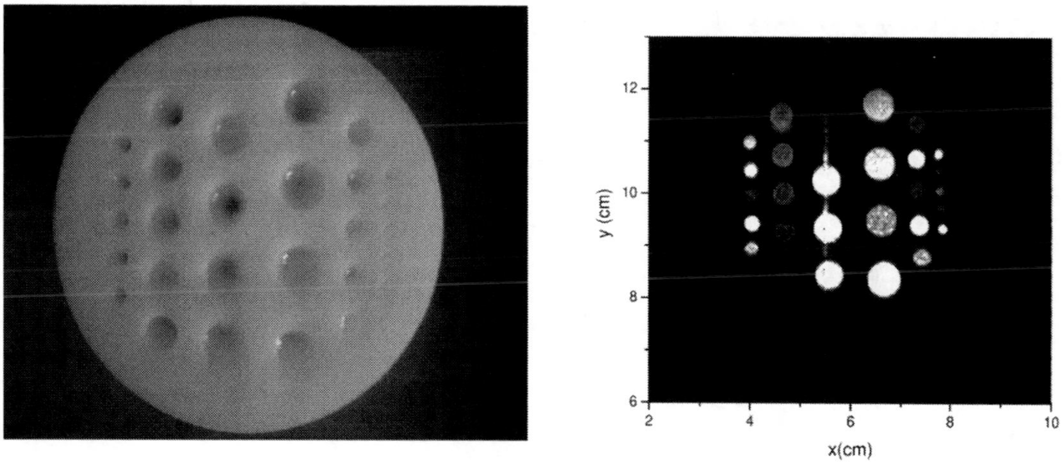

Figure 7. Photo (left) and MRI image (right) of phantom. The size of the phantom is 5cm with holes varying from 2 to 7 mm.

From Figure 7, one can see the good spatial resolution obtained $1mm^3$. 128 k_y values along y (phase encoding) have been used. This corresponds to a pitch of 0.4mm. 20 averages on each k value have been done. The slice thickness, determined by the Gz gradient, is about 2.5mm.

5. Conclusions

Mixed sensors composed of superconductor with a magnetoresistive element are very promising devices for ultra-low signal detection. Main advantages of these sensors are their robustness and wideband operation. These properties allow us performing NMR and MRI sequences without prepolarization as these sensors could operate in mT range static magnetic field and do not require a magnetic shielding room. The effective surface of the sensor and the distance to sample play an important role in MRI performance. It could be enhanced by using of flux transformer with a large pick-up loop [7]. Mixed sensors are also good candidates for MEG; however the 1/f noise cancellation technique should be applied. The combination of simultaneous MEG and MRI acquisition are targeted using mixed sensor for simultaneous anatomic and functional information of human brains.

Acknowledgements
This work is supported by the European project MEGMRI (HEALTH-F5-2008-200859).

References

[1] Pannetier M, Fermon C, Legoff G, Simola J, Kerr E 2004 Femtotesla Magnetic field measurement with magnetoresistive sensors *Science* **304** 1648-1650
[2] Clarke J, Braginskii A I 2004 The SQUID handbook Wiley-vch
[3] Kominikos I K, Kornack T W, Allred J C, Romalis M V 2003 A subfemtotesla multichannel atomic magnetometer *Nature* **422** 596
[4] Pannetier-Lecoeur M, Fermon C, Dyvorne H, Jacquinot J F, Polovy H, Walliang A L 2010 Magnetoresistive-superconducting mixed sensors for biomagnetic applications *J. Magn. Magn. Mater.* **322** 9-12 1647-1650
[5] Dyvorne H, Fermon C, Pannetier-Lecoeur M, Polovy H, Walliang A-L 2009 NMR with superconducting-GMR mixed sensors *IEEE Trans.Appl.Supercond.* **19** 3 819-822
[6] Herreros Q, Fermon C, Pannetier-Lecoeur M T1and T2 concentration dependence at low frequency. in preparation
[7] Dyvorne H, Scola J, Fermon C, Jacquinot J F, Pannetier-Lecoeur M 2008 Flux transformers made on commercial high critical temperature superconducting wires *Rev.Sci.Instrum.* **79** 025107-025111

Domain wall dynamics in a single CrO_2 grain

P Das[1,2], F Porrati[1], S Wirth[2], A Bajpai[3], Y Ohno[4], H Ohno[4], M Huth[1] and J Müller[1]

[1]Institute of Physics, Goethe University, Max von Laue Str. 1, 60438 Frankfurt (M), Germany
[2] Max Planck Institute of Chemical Physics of Solids, Nöthnitzer Str. 40, 01187 Dresden, Germany
[3] Institute of Solid State Research, IFW-Dresden, 01171 Dresden, Germany
[4] Laboratory for Nanoelectronics and Spintronics, Research Institute of Electrical Communication, Tohoku University, Sendai 980-8577, Japan

E-mail: `das@physik.uni-frankfurt.de`

Abstract. Recently we have reported on the magnetization dynamics of a single CrO_2 grain studied by micro Hall magnetometry (P. Das et al., *Appl. Phys. Lett.* **97** 042507, 2010). For the external magnetic field applied along the grain's easy magnetization direction, the magnetization reversal takes place through a series of Barkhausen jumps. Supported by micromagnetic simulations, the ground state of the grain was found to correspond to a flux closure configuration with a single cross-tie domain wall. Here, we report an analysis of the Barkhausen jumps, which were observed in the hysteresis loops for the external field applied along both the easy and hard magnetization directions. We find that the magnetization reversal takes place through only a view configuration paths in the free-energy landscape, pointing to a high purity of the sample. The distinctly different statistics of the Barkhausen jumps for the two field directions is discussed.

1. Introduction

In the realm of magnetism, the concept of domains and domain walls (DWs) are at the core of the understanding of the spontaneous magnetization of a ferromagnetic material. The domains, which are formed as a result of minimization of the total free energy, undergo structural changes under the influence of an external field. In real materials, the structural or compositional imperfections lead to several minima in the free energy landscape. These imperfections act as pinning centers for the DWs. An understanding of the dynamics of DWs in the presence of pinning centers is critical for the understanding of parameters like coercivity and remanence, which is essential for improving materials for practical applications. However, this is often difficult to realize in real samples consisting of several domains and DWs, where one measures an average magnetic signal and the details of the configuration path followed by the various DWs (due to local effects of pinning centers) are averaged out. Therefore, in order to gain detailed and quantitative information on the magnetization processes, as e.g. the pinning center distribution in a sample, it is essential to investigate samples where a single DW is present.

In literature, several studies on the dynamics of a single DW have been reported where, in most cases, a single wall is artificially created [1, 2, 3]. In the present study, we explore the possibilities of quantitative investigation of the dynamics of a single DW in an individual CrO_2 micro-grain of dimensions of approximately $5 \times 1.2 \times 0.7\,\mu m^3$. The DW in the present case is not artificially created (see below). CrO_2 has long been known as a material used for information storage in magnetic tapes. However, this half-metallic ferromagnet with a T_C above room temperature ($\sim 395\,K$) attracted a renewed interest because of its very high spin polarization ($\sim 100\%$), a property which is interesting for spintronics applications [4]. Our magnetic studies of a single CrO_2 grain indicate that the ground state of the grain corresponds to a closure domain pattern with a single cross-tie DW [5]. We observe that upon application of an external field $\vec{H}_{ext}$ along the long axis of the grain, which is the easy magnetization axis (EMD), the magnetization reverses due to the motion of the cross-tie DW. In this paper, we show that when the external field is applied along the hard magnetization direction (HMD) of the grain, the reversal involves the motion of the 90° DWs at the end caps. Accordingly, the histograms of the magnitudes of Barkhausen jumps strongly differ for the different field directions.

2. Experimental

High purity CrO_2 samples for the present studies were synthesized in a two step process. As a first step, intermediate precursor oxide is prepared by heating CrO_3 flakes or granules and then pellets formed from the precursor undergo a further heat treatment at appropriate temperature and at ambient pressure. The method is successfully used to tune the parameters like grain size and the grain boundary density. The grains thus prepared are of rod like shape. The details of the synthesis is described in Ref. [6, 7].

In order to study the magnetization dynamics in a single grain, we employed a micro-Hall magnetometer based on two-dimensional electron gas (2DEG) that forms at the interface of high-mobility material GaAs/AlGaAs heterostructures [8]. Hall crosses with active areas of $5 \times 5\,\mu m^2$ were prepared using standard photolithography followed by wet etching technique. A single grain of CrO_2 obtained by crushing a pellet was then picked and placed between two Hall crosses (1 and 2, see Fig 1) aiming to simultaneously measure the Hall signals due to the stray fields emanating from the two grain ends. The 2DEG is sensitive to the perpendicular component of the average stray field $\langle B_z \rangle$. The measured Hall voltage is converted to $\langle B_z \rangle$ using the Hall coefficient of the device measured at the corresponding temperature.

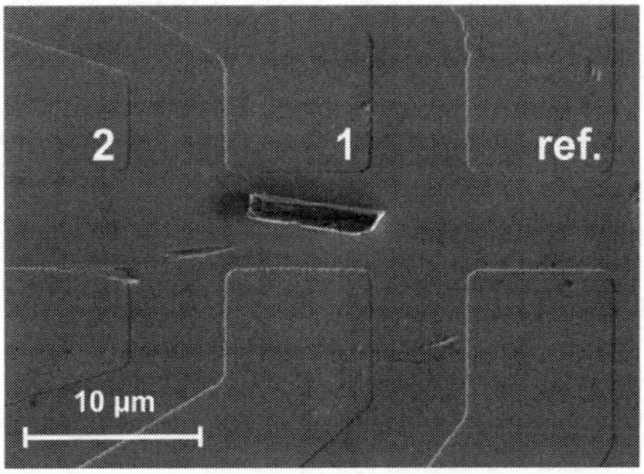

Figure 1. SEM image of a CrO_2 grain placed on a Hall device. Note that the position of the grain is slightly misaligned ($\sim 7°$) with respect to the two Hall crosses. A reference cross is used for the background correction (see text).

3. Results and discussions

In order to obtain quantitative information on the magnetization dynamics of the grain, the Hall voltages at both grain ends were simultaneously measured while sweeping the external magnetic field $\vec{H}_{ext}$ applied perpendicular and parallel to the long axis of the grain. As mentioned above, the EMD of the grain lies along the long axis, which is the [001] direction. The changes in the stray field were determined from the measured Hall signals after removing the background (due to a small misalignment of the device with respect to $\vec{H}_{\text{ext}}$), which was measured at an empty reference cross. Figure 2 shows the hysteresis loops measured in $\langle B_z \rangle$ versus H_{ext} with $\vec{H}_{\text{ext}}$ applied along a HMD. The two loops correspond to the signals measured simultaneously at the two ends of the grain, denoted by 1 and 2, see Fig. 1. For these measurements, a slow continuous field sweep was applied such that a complete hysteresis loop was measured in $\sim$ 22 hrs. Such a slow sweep rate allows to track and measure the DW motion in details. The magnetization saturates at the anisotropy field $\mu_0 H_{\text{a}} \sim 250\,\text{mT}$. Using this, we obtain an effective uniaxial anisotropy constant of $K_{\text{eff}} \sim 7.3 \times 10^4\ \text{J/m}^3$ from which a uniaxial magnetocrystalline anisotropy constant $K_1 \sim 1.3 \times 10^4\,\text{J/m}^3$ follows (see ref. [5] for details). From Fig. 2, we make the following observations: First, the magnetic signal at grain end 1

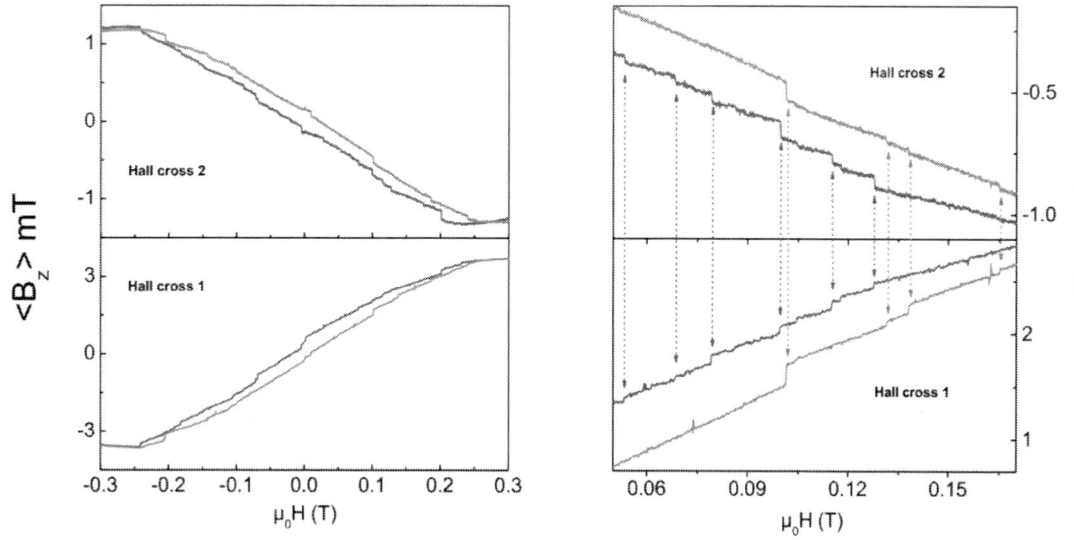

Figure 2. (left) Hysteresis loops measured at $T = 5\,\text{K}$ with $\vec{H}_{\text{ext}}$ applied along one of the HMD (field in the plane of the Hall device, cf. Fig. 1). Zoomed-in figure (right) show the sharp correlated Barkhausen jumps at two grain ends.

(i.e. measured at Hall cross 1) is larger and opposite in sign to that measured at Hall cross 2. The larger signal at Hall cross 1 is due to the asymmetry in the placement of the grain with respect to the crosses (see Fig. 1). Secondly, the magnetization reversal is accompanied by a series of sharp Barkhausen jumps. Interestingly, we find that the jumps at both grain ends occur simultaneously as is evident from Fig. 2. Correlated jumps were also observed, when the external field applied was applied along the EMD [5]. These simultaneous jumps in magnetic signal from both ends of the grain are consistent with a single DW that is being pinned/depinned by the force exerted by the field sweep. In order to find out about the variety of paths in configuration space of the free-energy landscape, a series of 50 successive hysteresis loops, shown in Fig. 3, has been measured. The set of data demonstrates that the magnetization reversal takes place along only a few $(2-3)$ distinct, stochastically chosen paths corresponding to different configurations of the energy landscape. These measurements have been carried out in a gradiometry setup where the Hall voltage at the cross carrying the sample is compensated by applying a current of same magnitude but opposite sign through an empty reference cross. This results in a differential signal ΔV_H solely due to the sample's magnetization [9, 10]. The data in Fig. 3 are shown without background correction. The data suggest the motion of a single DW,

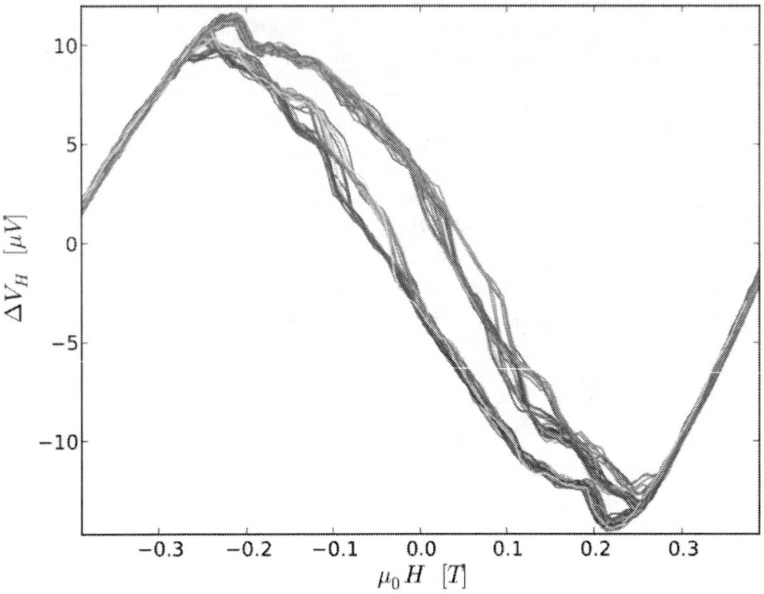

Figure 3. 30 out of 50 successively measured hysteresis loops (gradiommetry setup, see text) with the applied field along one of the hard directions. The data are shown without background correction.

since for multiple walls the magnetization reversal wouldn't be expected to follow only a few distinct paths. This finding was confirmed by micromagnetic simulations, which revealed two different field regimes of the grain's magnetization. The ground state corresponds to a closure domain configuration (labeled as configuration A) with a single cross-tie DW parallel to the long axis of the grain. However, there exists a metastable state (configuration B) with three domains, which is slightly higher in energy. Application of a small field along the direction opposite to that of the main domain brings the system back to the lowest energy state. We refer to Ref. [5] for the details on the domain configurations. As mentioned above, we observed Barkhausen jumps simultaneously at both grain ends for fields applied both along the EMD and HMD. For $\vec{H}_{\rm ext} \parallel$ EMD, the magnetization reversal takes place through DW motion in a direction perpendicular to the direction of the applied field. For perpendicular fields smaller than the anisotropy field $H_{\rm a}$ most likely the metastable three-domain configuration B is formed. At $\mu_0\vec{H}_{\rm ext} \approx \pm 200\,{\rm mT}$ we observe downward jumps (see Figs. 2 and 3) in the magnetic signal where the single cross-tie DW with closure domain configuration A is formed. We relate this transition to the observed transition from configuration A to B (see Fig. 3 in Ref. [5]) at $\mu_0\vec{H}_{\rm ext} \approx \pm 20\,{\rm mT}$ for $\vec{H}_{\rm ext}$ along the EMD.

In the present case where $\vec{H}_{\rm ext} \parallel$ HMD, the observation that the stray field changes simultaneously at both grain ends indicates that the motion of the 90° wall at one end influences that at the other end of the grain via the connecting cross-tie DW. Such a behavior would be expected in case of a rigid connecting wall. In order to understand this in more detail, we analyze the Barkhausen jumps observed for $\vec{H}_{\rm ext}$ along both the EMD and HMD. In Fig. 4, we show the histograms of the magnitude of the magnetization changes (i.e. size of the Barkhausen jumps) observed for both field directions at $T = 5\,{\rm K}$. The behavior, which reflects the distribution of pinning centers is qualitatively different for the two field directions. For $\vec{H}_{\rm ext} \parallel$ EMD the step heights can be directly converted to the distance (Δx) of DW motion between two pinning centers using $\Delta\phi = 2M_s d\Delta x$, where $\phi \propto \langle B_z \rangle$ is the flux emanating from the grain of thickness d, and M_s is the saturation magnetization, see Ref. [5] for details. For $\vec{H}_{\rm ext} \parallel$ EMD the data show an asymmetric distribution, whereas for $\vec{H}_{\rm ext} \perp$ EMD the distribution of the magnitude of Barkhausen jumps is more symmetric with mean values of the step heights of $\sim 107\,\mu$T in the former case and $\sim 65\,\mu$T in the latter case, respectively. Jumps with much larger magnitudes correspond to the switching between domain configurations A and B. From the easy-axis measurements we roughly estimated a density of pinning centers of $1 - 10\,\mu\text{m}^{-3}$ implying a high purity of the investigated grain, in agreement with the small number of configuration paths as seen from Fig. 3. We find that about 1/3 of the magnetization changes through DW pinning/depinning and about 2/3 through free motion of DWs and magnetization rotation. Different heights observed for the different field directions at the same temperature are in agreement with the assumption that the direction of DW motion is different for $\vec{H}_{\rm ext}$ along the EMD and HMD. For the latter case, the observed Barkhausen jumps involve the motion of the 90° DW being affected by pinning centers.

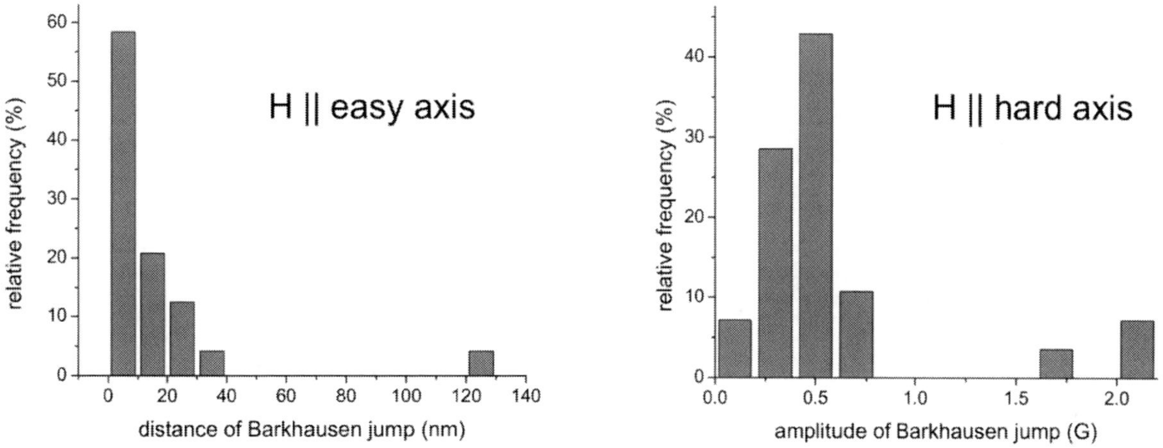

Figure 4. Histograms of Barkhausen jump heights observed for $\mu_0\vec{H}_{\rm ext}$ along both EMD and HMD at $T = 5$ K. The amplitude of Barkhausen jumps is directly proportional to the spatial distance traversed by the domain wall (for $\vec{H}_{\rm ext} \parallel$ EMD, see text and Ref.[5]). The relative frequency in the ordinate refers to the no. of jumps of certain size with respect to the total number of jumps. The smallest jump which was observed at $T = 5$ K is ~ 0.8 nm.

Conclusions

We have studied the DW dynamics in a single CrO_2 grain using micro-Hall magnetometry. The motion of the DW was tracked and quantitative information on the pinning-center distribution was obtained. We find that the magnetization reversal takes place through only a view configuration paths in the free-energy landscape, pointing to a high purity of the sample. The observed Barkhausen jumps for the two field directions exhibit at distinctly different statistical behavior, since for the field along the hard axis the magnetization reversal involves the motion of the 90° DWs (forming closure domains). The data suggest that the cross-tie DW along the long axis of the grain is a rigid wall.

References

[1] Atkinson D, Allwood D A, Xiong G, Cooke M D, Colm C F and Cowburn R P 2003 *Naturematerials* **2** 85.

[2] Bedau D, Kläui M, Hua M T, Kryzk S, Rüdiger U, Faini Gand Vila L 2008 *Phys. Rev. Lett.* **101** 256602 (2008).
[3] Hayashi M, Thomas L, Rettner C, Moriya R, Jjiang X and Parkin S S P 2006 *Phys. Rev. Lett.* **97** 207205.
[4] Coey J M D and Venkatesan M 2002 *91* 8345.
[5] Das P, Porrati F, Wirth S, Bajpai A, Ohono Y, Ohono H and Müller J 2010 *Appl. Phys. Lett.* **97** 042507.
[6] Bajpai A and Nigam A K 2005 *Appl. Phys. Lett.* **87** 222502.
[7] Bajpai A and Nigam A K , US Patent 7276226.
[8] Kent A D, von Molnár S, Gider S and Awschalom D D 1994 *Journ. Appl. Phys.* **76** 6656.
[9] Wirth S, Anane A and von Molnár S 2001 *Phys. Rev. B* **63** 012402.
[10] Li Y, Xiong P, von Molnár S, Ohno Y and Ohno H 2005 *Phys. Rev. B* **71** 214425.

Parameters controlling magnetic interactions in perovskite manganites

B Dabrowski, J Mais, S Kolesnik and O Chmaissem

Department of Physics, Northern Illinois University, DeKalb, IL, USA

E-mail: dabrowski@anl.gov

Abstract. Magnetic superexchange interactions in single-valent perovskites $(R,A)MO_3$ (R,A = Rare and Alkaline Earth's, M = transition metal) are known to dependent on the M-O-M bond angle θ and the interatomic distances <M-O> through the band-width relation $W \sim \cos\theta/\langle M\text{-}O\rangle^n$ (n ~ 3.5). It was observed previously that $T_N \sim W^2 \sim \langle\cos^2\theta\rangle$ for RMO_3 with M = Cr^{3+} (t^3), Fe^{3+} (t^3e^2), and Ni^{3+} (t^6e^1) when <M-O> are constant. We have observed similar dependence for Mn^{3+} (t^3e^1). When the <M-O> bonds were stretched beyond their equilibrium values in mixed-valent $La_{0.5}Sr_{0.5-x}Ba_xMn^{3.5+}O_3$, a strong suppression of magnetic interactions was observed. In addition, size disorder on the A-site suppressed T_N for $Sr_{1-x}Ca_xMn^{4+}O_3$ because of a local variance of bond angles Δθ. By removing the effects of stretched <M-O> bonds and size disorder we have increased Curie temperatures of the layer ordered manganites $R_{1-y}Ba_yMnO_3$. Frequently interesting compounds with θ ~ 180 deg are not possible to obtain due to the limited range of sizes and charges of the A and R ions. Recently, in an attempt to increase the bond angle θ in compounds, which are not possible to obtain through simple A-site substitutions, we have studied the double substituted R- and Mn-site perovskites. Properties of these compounds depend mostly on dilution of the magnetic Mn-O lattice in addition to the previously discussed parameters.

1. Introduction

Perovskites ABO_3 have been studied comprehensively for years because of intriguing magnetic, electronic and other properties resulting from competing charge, exchange, and phonon interactions [1]. The properties can be tuned over a wide range through the choice of the sizes and charges of the large size 12-coordinated A cations, which control the degree of structural distortions (bandwidth, W) and the formal valence (band filling, x) of the smaller size 6-coordinated B ions, here the B-site ions of interest are Mn. The relatively rigid MnO_6 octahedra sharing corners and forming Mn–O–Mn bond-angles θ are mostly responsible for electronic and magnetic properties. Based on systematic study of the relationship between structural and magnetic properties of perovskite manganites we have developed phenomenological correlation between magnetic transition temperature and structural parameters for compounds with fixed valence of Mn: $T_M \sim W^2 \sim \{\cos\theta/\langle Mn\text{-}O\rangle^n\}^2 - ps^2$ (n ~ 3.5), where <Mn-O> is bond length and s^2 is variance of the A-site sizes causing local variance of bond angles Δθ. Understanding this correlation enabled the preparation of improved materials and predicting new exciting compounds. However, many of them were not possible to obtain through simple A-site substitutions due to the limited range of sizes and charges of available ions. To synthesize interesting compounds with $\theta \sim 180^o$ we have studied the double substituted R- and Mn-site perovskites and found that magnetic properties of these compounds depend mostly on dilution of the magnetic Mn-O lattice in addition to the previously discussed parameters.

2. Magnetic transitions and structural features for single-valent systems

Our description of magnetism and spin-transport in manganites is based on concepts of local sites of individual magnetic ions and crystal-field theory, which grasp most of essential physics. This approach is justified by the inability of current band structure calculations to capture all aspects of the strong coupling among electrons, spins, and phonons as well as to describe strong correlations of the 3d electrons in distorted crystal fields [2]. The description uses internal structural parameters of the Mn-O-Mn bond angles and the A-O and Mn-O bond lengths that are obtained from neutron diffraction measurements. These structural measurements are combined with detailed magnetic measurements that are performed over a wide temperature range to provide accurate composition-structure-property phase diagrams.

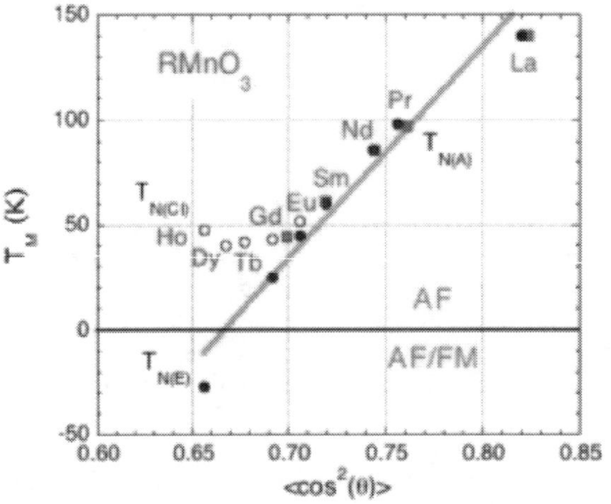

Figure 1. Magnetic transition temperatures versus $<\cos^2\theta>$ for $RMn^{3+}O_3$ perovskites.

The range of the Mn oxidation states considered here spans from Mn^{3+} to Mn^{4+}. In the cubic crystal field, which is characteristic for perovskites, these oxidation states of Mn display well-defined ionic ground states of t^3e^1 and t^3e^0 for Mn^{3+} and Mn^{4+}, respectively. These single-valent states can be achieved by using rare earth's (R = La, Pr, Nd, …) and alkaline earth's (Ca, Sr, and Ba) on the A-sites. The filling of atomic electronic d-orbitals, t_{2g} and e_g, with electrons that is consistent with the Hund's and crystal-field splitting rules has been described in several textbooks. Because the spin-orbit coupling is quenched for perovskite manganites [2], the antiferromagnetic (AF) and ferromagnetic (FM) superexchange describe very well the coupling between magnetic moments for the single-valent states. The Goodenough-Kanamori rules of superexchange interactions predict the sign of the interaction for straight Mn-O-Mn bond angles [3]. When the Mn-O-Mn bonds are not straight, the interactions can be described as a superposition of 180 and 90 deg bonds [4]. The correlation of antiferromagnetic transition temperatures to the Mn–O–Mn bond-angle θ, $T_N \approx T_{Nmax} - T_1(1 - <\cos^2\theta>)$, has been clearly established for single valent RBO_3 systems with B = Fe^{3+} (t^3e^2) and Cr^{3+} (t^3) when there is only one kind of Rare Earth on the A–site and the <B-O> bonds lengths are constant [4,5]. We have recently shown that similar relationship between the Mn–O–Mn bond-angle and T_N holds for the $RMnO_3$ perovskites with a single valent Mn^{3+} (t^3e^1) while for the solid solution system $Sr_{1-x}Ca_xMnO_3$ with Mn^{4+} (t^3), the structural disorder on the A-site introduces a measurable

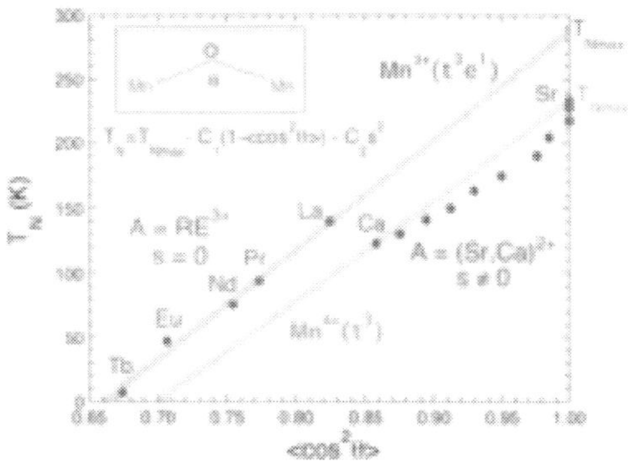

Figure 2. Dependence of T_N on $<cos^2\theta>$ for the solid solution system $Sr_{1-x}Ca_xMn^{4+}O_3$.

deviation from this relationship [6,7]. Figure 1 shows magnetic transition temperatures versus $<cos^2\theta>$ for $RMnO_3$ perovskites. The complex AF A-type magnetic structure of these compounds is a result of the Jahn-Teller distortion and the orbital order present below temperatures, T_{JT} [8]. Figure 2 shows that the linear dependence of T_N on $<cos^2\theta>$ observed for $RMn^{3+}O_3$ is less regular for the solid solution system $Sr_{1-x}Ca_xMn^{4+}O_3$. The size difference between interatomic distances of the Sr-O and Ca-O introduces local variance of the bond angles $\Delta\theta$ that suppresses T_N by ~ 20 K at x = 0.5 where the variance of sizes is the largest [9].

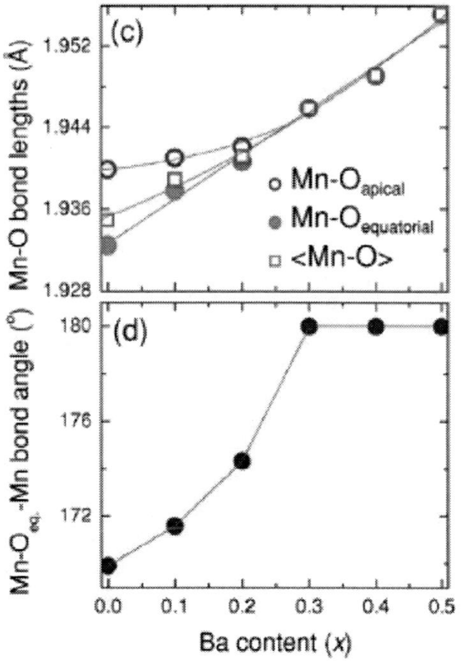

Figure 3. Neutron powder diffraction measured bond lengths [$Mn^{3.5+}$-O] and bond angles θ as a function of Ba content for the $La^{3+}{}_{0.5}Sr^{2+}{}_{0.5-x}Ba^{2+}{}_xMn^{3.5}O_3$ system at room temperature.

3. Magnetic transitions and structural features for mixed-valent systems

Decreasing Curie temperatures have been observed for several mixed-valent manganites as a function of increased substitution of the large-size Ba, and interpreted in terms of increasing variance of the A-site sizes s^2. Recently, by using highly accurate interatomic distances we have shown that the explanation of T_C behavior only in terms of bond angle and size variance is not sufficient, especially in the regime where structural distortions from the cubic structure are very small [10].

Figure 3 shows measured bond lengths [$Mn^{3.5+}$-O] and bond angles θ as a function of Ba content for the $La^{3+}{}_{0.5}Sr^{2+}{}_{0.5-x}Ba^{2+}{}_xMn^{3.5}O_3$ system at room temperature. Compositions x = 0 – 0.2 and 0.3 – 0.5 are tetragonal (slightly distorted) and cubic, respectively. Bond lengths [$Mn^{3.5+}$-O] exhibit small increase already in non-cubic phase and more rapid increase imposed by requirement of matching bond lengths [A-O] and $\sqrt{2}$[$Mn^{3.5+}$-O] in cubic phase. Figure 4 shows T_C's and the variance of sizes s^2 as a function of the bond length [A-O] for several $(La_{0.5}A_{0.5})MnO_3$ compounds (A=Ca, Sr, Ba). We observe that T_C decreases for both tetragonal and cubic phases despite increasing bond angle and decreasing variance of sizes. Thus, the explanation of the observed T_C reduction in terms of increasing ionic size variance cannot be employed. Additional cause is required when the crystal structure is cubic or nearly cubic. Such a cause can arise from presence of the [Mn-O] bonds that are stretched beyond their equilibrium values. Fraction of such bonds would be present because the equilibrium $\sqrt{2}$[Mn-O] bonds are smaller than the longest [A-O] bonds (i.e., the [Ba-O] bonds) even if the average structure is non-cubic, i.e., when the average $\theta \neq 180$ deg. As the formula for the band-width W shows, elongation of these bonds would cause very strong and non-linear decrease of $T_C \approx [Mn\text{-}O]^{-7}$ that is not modeled by the variance of sizes. As the materials become cubic considerable internal strains develop that stretch the [Mn^{3+x+y}-O] bonds, thus, lowering T_c by narrowing the bandwidth W.

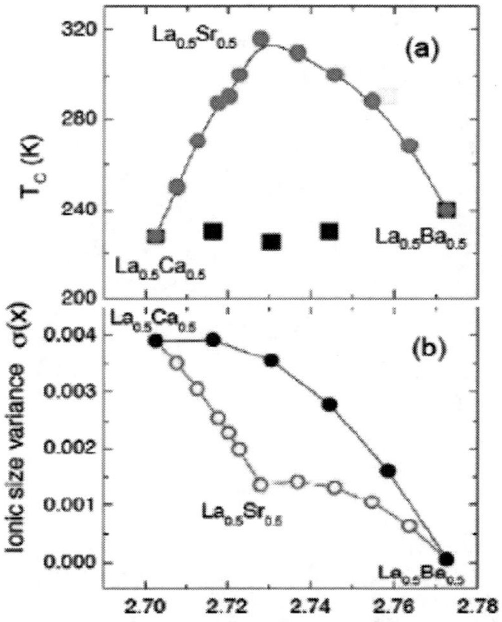

Figure 4. T_C's and the variance of sizes s^2 as a function of the bond length [A-O] for several $(La_{0.5}A_{0.5})MnO_3$ compounds.

We have used derived formula for W to increase T_c of $La_{1-x}Ba_xMnO_3$ compounds that can be obtained in two forms: randomly mixed perovskite and layered-ordered $LaBaMn_2O_{6-\delta}$ phase [7,9].

Comparison of layered-ordered and randomly substituted compounds with the same charge doping x revealed much greater effects of structural disorder on the ferromagnetic transition temperature for mixed valent perovskite manganites than was observed for the T_N of single-valent $Sr_{1-x}Ca_xMnO_3$ system [9,11]. Because the measured variance of size is small s^2 = 0.009Å at x=0.5, the effects of structural disorder are anticipated to be small. Nonetheless, we have observed a large, ~ 80 K, difference of ferromagnetic T_C between the two structural forms. We attribute the different properties of the randomly substituted and layered-ordered phases to the local A-site charge disordering that disturb the coherent electronic states of the Mn-O bonds. The layer-ordered phase with composition x = 0.42 shows the second highest T_C of 365 K ever observed for the perovskite manganites. Similar effects of local structural/charge order no doubt have profound effects on the properties of other complex oxides and can be used to optimize their useful functionalities.

4. Magnetic transitions and structural features of double substituted perovskites

Frequently interesting compounds with θ ~ 180 deg are not possible to obtain due to the limited range of sizes and charges of the available A and R ions. Recently, in an attempt to increase the bond angle θ in $RMnO_3$ compounds, which show significantly distorted orthorhombic structure even for the largest R = La, we have studied the double substituted R- and Mn-site perovskites. Figure 5 shows an example of the measured T_N for isoelectronically substituted on the Mn site perovskites by B^{3+} = Al, Ga, Sc, and In ions with increasing ionic size. Despite achieved increase of θ that should increase Neel temperature a rapid decrease of T_N is observed. Similar effects were observed for a double substituted $La_{1-3x}Ba_{3x}Mn_{1-x}Mo_xO_3$, however due to much large increase of $<\cos^2\theta>$ for small x, an increase of T_N and transition to ferromagnetic phase have been observed for the first time with Mn^{3+}. Our results show that the effects of bond angle $<\cos^2\theta>$ and variance of sizes, s^2 on T_N are clearly observable; however the most dominant effect is dilution of the magnetic network by the non-magnetic impurities [12].

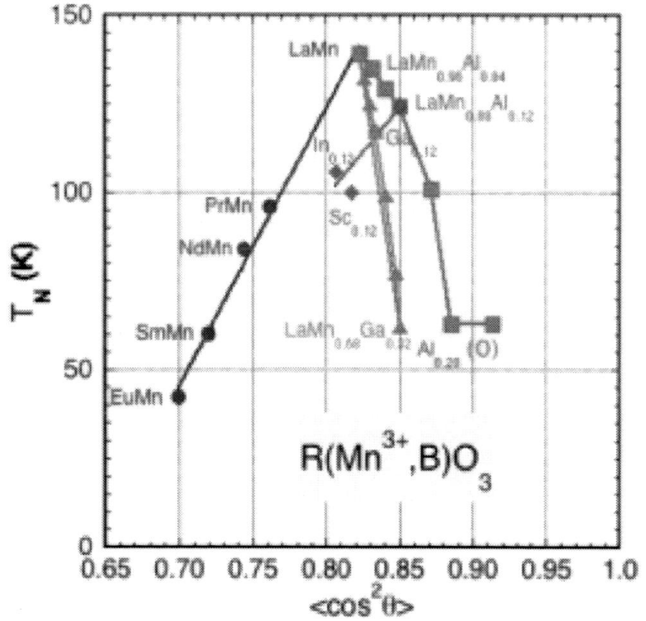

Figure 5. T_N for isoelectronically substituted on the Mn site perovskites with B^{3+} = Al, Ga, Sc, and In ions.

5. Conclusions

Bond angles Mn-O-Mn and bond lengths [A-O] have been used to describe T_N of the single-valent manganites through the relation to the band-width. Maximum T_N's were found for straight bonds [$\cos^2(\theta)\sim 1$] and minimal variance of sizes ($s^2\sim 0$). Study of the correlations between chemical, structural, and physical properties for mixed-valent manganites show that T_C and T_N depend strongly on the bond angle, charge doping, structural and charge disorder, and bond lengths through the relation $T_M \sim W^2 \sim \{\cos\theta/\langle\text{Mn-O}\rangle^n\}^2 - ps^2$. Properties of double substituted R- and Mn-site perovskites depend mostly on dilution of the magnetic Mn-O lattice in addition to the previously discussed parameters. Understanding of similar effects of local structural/charge order can be used to optimize magnetic, electronic, thermoelectric, and other properties of several perovskites.

Acknowledgment

The work was supported by the NSF-DMR-0706610.

References

[1] Tokura Y and Nagaosa N 2000 *Science* **288**, 462
[2] Dionne GF 2003 in: *Magnetic Interactions and Spin Transport*, Eds.: Chtchelkanova A Wolf S and Idzerda Y (Kluwer Academic), pp. 1-130
[3] Goodenough JB1963 *Magnetism and Chemical Bond*, Interscience Publishers, John Wiley, New York.
[4] Boekema C, Van Der Woude F and Sawatzky GA 1972 *Int. J. Magnetism* **3**, 341
[5] Goodenough JB and Longo JM 1970 Crystallographic and magnetic properties of perovskites and perovskite-related compounds, in: *Landolt-Bornstein Tabellen*; New Series III/4a, edited by Hellwege K (Springer, Berlin) p.231.
[6] Dabrowski B, Chmaissem O, Mais J, Kolesnik S, Jorgensen JD and Short S 2003 *J. Solid State Chem.* **170**, 154
[7] Chmaissem O, Dabrowski B, Kolesnik S, Mais J, Brown DE, Kruk R, Prior P, Pyles B and Jorgensen JD 2001 *Phys. Rev. B* **64**, 134412
[8] Goodenough JB 1955 *Phys. Rev.* **100**, 564
[9] Dabrowski B, Chmaissem O, Mais J, Kolesnik S, Jorgensen JD and Short S 2002 *Mat. Res. Soc. Symp. Proc.* **718**, 169
[10] Chmaissem O, Dabrowski B, Kolesnik S, Mais J, Maxwell T, Suescun L and. Jorgensen JD 2006 *Phys. Rev. B* **74**, 144415
[11] Chmaissem O, Dabrowski B, Kolesnik S, Mais J, Jorgensen JD and Short S 2003 *Phys. Rev. B* **67**, 094431
[12] Mais J, Dabrowski B, Kolesnik S, and Chmaissem O unpublished

Substrate-induced magnetic anisotropy in $La_{0.7}Sr_{0.3}MnO_3$ epitaxial thin films grown onto (110) and $(1\bar{1}8)$ $SrTiO_3$ substrates

P Perna[1], C Rodrigo[1,2], E. Jiménez[1,2], N Mikuszeit[1,2], F J Teran[1], L Méchin[3], J Camarero[1,2] and R Miranda[1,2]

[1] Instituto Madrileño de Estudios Avanzados en Nanociencia, IMDEA-Nanociencia, Campus Universidad Autónoma de Madrid, 28049 Madrid, Spain

[2] Departamento de Fisica de la Materia Condensada and Instituto "Nicolás Cabrera", Universidad Autónoma de Madrid, 28049 Madrid, Spain

[3] GREYC (UMR6072) CNRS-ENSICAEN and Université de Caen Basse-Normandie, Bd. de Marèchal Juin, 14050 Caen, France

E-mail: paolo.perna@imdea.org

Abstract. We show a detailed magneto-optical Kerr study at room temperature of well characterized epitaxial $La_{0.7}Sr_{0.3}MnO_3$ (LSMO) thin films grown onto (110) and $(1\bar{1}8)$ $SrTiO_3$ substrates. The films present a well-defined uniaxial (two-fold) magnetic anisotropy ascribed to substrate-induced anisotropy. In particular, the in-plane uniaxial anisotropy in the(110)-oriented LSMO films originates from the existence of elongated in-plane [001]-oriented structures. Similar elongated structures, parallel to the [110] crystallographic direction, are found for LSMO films grown on $(1\bar{1}8)$ STO surfaces. In all films, such a uniaxial magnetic anisotropy is characterized by an easy axis lying along the elongated structures. Furthermore, the vectorial-resolved hysteresis loops as a function of the in-plane applied field direction are interpreted in terms of rotation and propagation and nucleation of magnetic domains processes. Our results demonstrate the tailoring of magnetic anisotropy by exploiting the substrate-induced anisotropy in epitaxial thin films.

1. Introduction

In epitaxial magnetic thin films the cubic crystal symmetry can be broken by atomic steps on the surface or by anisotropic lattice relaxation, inducing an additional in-plane uniaxial magnetic anisotropy contribution. Such a symmetry breaking has been demonstrated in epitaxial magnetic thin films with cubic symmetry in half metallic manganese-based oxides [1, 2], as well as in metal systems [3] and in diluted semiconductors [4]. The competition between the biaxial (four-fold) and the additional uniaxial (two-fold) anisotropy results in a magnetic reorientation, which depends on many parameters, such as substrate step density [3], thickness [1], angle of deposition [5], and even the temperature range [1]. Thus, the symmetry breaking in magnetic systems provides an additional contributions to the magnetic anisotropy, which could alter both magnetization easy and hard axis and reversal processes [6]. In order to control and tailor the magnetic properties of materials, it is possible to resort to the fabrication of artificial nanostructures [7], thin films [8, 9, 10] and superlattices [11].

In this work, we resort to the interesting properties of the magneto-resistive mixed-valence manganese oxides, $La_{0.7}Sr_{0.3}MnO_3$ (LSMO) that shows both a Curie temperature above 300 K and an almost 100% spin polarization. We have engineered the growth of epitaxial films in order to obtain purpose-designed magnetic anisotropy exploiting to anisotropy induced by substrates with different crystallographic orientation, which can be employed for the fabrication of spintronics devices, operating at room temperature (RT), such as read-heads magnetic hard disks and non-volatile magnetic memories [12]. In this system, the substrate can induce tensile or compressive strain to the film depending on the film-substrate lattice mismatch, determining in-plane or out-of-plane easy magnetization directions, respectively [13]. In particular, the strain in LSMO thin films deposited on $SrTiO_3$ (STO)(001) is in-plane tensile and an in-plane biaxial magnetic anisotropy is generally observed, with the easy in-plane direction along ⟨110⟩, and the hard in-plane direction along ⟨100⟩ [13, 14]. In the particular case of LSMO grown onto STO(110), the substrate induces a strain that is anisotropic in-plane, i.e. the two in-plane directions of strain are inequivalent. This causes an in-plane uniaxial magnetic anisotropy with the easy axis (e.a.) of magnetization along the [001] crystallographic direction [9]. Another possibility to induce in-plane magnetic anisotropy is to create artificially periodic stepped surfaces by exploiting vicinal substrates [2]. These substrates are intentionally misoriented to a (near) low index surface, therefore determining surface step edges. In such a way, the high symmetry of the low index surface is broken, and an additional uniaxial anisotropy is expected [15]. Matthews *et al.* have reported an in-plane uniaxial magnetic anisotropy at RT in 25 nm and 7 nm thick LSMO films deposited on very low miscut STO substrates (0.13° and 0.24°), which vanishes at low temperatures [1]. Uniaxial magnetic anisotropy with easy axis along the step edges has been found at 80 K for a 12.6 nm thick LSMO film deposited on a vicinal STO(001) substrate with a 10° miscut off the [001] plane toward the [010] crystallographic direction [16].

2. Structural and morphological characterizations

70 nm thick LSMO thin films were deposited by pulsed laser deposition (PLD) from a stoichiometric target onto commercially available STO(110) and vicinal STO($1\bar{1}8$) substrates. In the latter case the vicinal angle was 10° from the [001] surface towards the [$1\bar{1}0$] crystallographic direction, thus inducing step edges along the [110] direction [2]. The optimization of the growth conditions was performed on standard STO (001) substrates [11]. The laser fluence was $1-2$ J cm^{-2}, the target-to-substrate distance was 50 mm, the oxygen pressure was 0.35 mbar and the substrate temperature was 720 °C.

The crystal structure was investigated by means of X-ray Diffraction (XRD). Standard θ-2θ scans were routinely performed in order to determine the out-of-plane lattice parameters. The XRD $\theta - 2\theta$-scans indicate that the LSMO films were epitaxially grown on the substrates. In particular, the LSMO films grown onto STO(110) present the (110) axis coincident with the (110) axis of the substrate, as well as the LSMO film grown onto vicinal STO(001) presents the (001) axis parallel to the (001) axis of the substrate [2]. In the case of vicinal LSMO films, the offset angle was checked to be equal to the substrate vicinal angle within ±0.05°. The out-of-plane and the in-plane lattice parameters were determined by XRD measurements around symmetric and asymmetric crystallographic peaks. In the case of the LSMO/STO(110), the two inequivalent in-plane directions of strain induced by the STO are the [001] and [$1\bar{1}0$], which determine two different in-plane strain tensor componentes, $\epsilon_{[001]} = 0.80\%$ and $\epsilon_{[1\bar{1}0]} = 0.30\%$. The measured lattice parameters and the strain tensor component along the [100] ($\epsilon_{[100]}$) in-plane crystallographic direction for the LSMO grown onto vicinal (001)-oriented STO substrates present the two in-plane lattice parameters of the LSMO cell equally tensile strained ($\epsilon_{[100]} = \epsilon_{[010]} = 0.80\%$) by the substrate. The crystalline quality of the films was checked by measuring the Full-Width-Half-Maximum (FWHM) of the rocking curves (ω-scan), which are always found below 0.15°, and the in-plane crystal plane alignement (ϕ-scan).

The morphology of the samples were investigated at room temperature by means of atomic force (AFM) and scanning tunnel microscopies (STM), using a Nanoscope microscope [2]. The average roughness (RMS) of the samples was found always in the range of few unit cells (u.c.). In general, the morphology of the LSMO films replicates that of the substrates for all samples investigated. In the case of the LSMO/STO(110), the particular morphology of the substrate surface induces film structures elongated along the in-plane [001] crystallographic direction, corresponding to the direction of the higher in-plane tensile strain value (see inset in Fig. 1). LSMO films deposited onto 10° vicinal STO(001) substrate present similar elongated structures on the top surface oriented along the $[1\bar{1}0]$ crystallographic direction (see inset in Fig. 2).

A detailed characterization of the films is reported in Ref. [2]. It is worth noting that in all the investigated samples, the Curie temperature was always found above RT.

3. Magneto-optical Kerr characterization

The magnetic anisotropy of the films and the angular dependence of the magnetization reversal were investigated at RT by high-resolution vectorial-Kerr magnetometry measurements. In our setup the combination of *p*-polarized incident light in Kerr experiments and the simultaneous detection of the two orthogonal components of the reflected light allow the simultaneous determination of the components of the in-plane magnetization, parallel ($M_{\parallel}$) and perpendicular ($M_{\perp}$) to the magnetic field direction [17].

The samples were mounted in a stepper-motorized eucentric goniometer head that keeps the reflection plane fixed for the whole set of experiments. In magnetooptical measurements this is important to be able to compare the values of the magnetization components measured at different rotation angles and between different samples. The study of the magnetization reversal processes as well as magnetic anisotropy of the films was performed by measuring simultaneously the in-plane vectorial-resolved hysteresis loops, i.e., $M_{\parallel}(H,\theta)$ and $M_{\perp}(H,\theta)$, as a function of the sample in-plane angular rotation angle (θ), keeping fixed the external magnetic field direction. The whole angular range was probed every 4.5°, with 0.5° angular resolution.

3.1. LSMO grown onto STO(110)

Representative in-plane vectorial-resolved Kerr hysteresis loops of 70 nm thick LSMO film grown onto STO(110) acquired at selected angles θ are shown in Fig. 1. The angle $\theta = 0°$ is taken when the external field is aligned parallel to the [001] in-plane crystal direction, i.e. lying along the elongated structures visible on the film surface.

Both magnetization components, i.e., parallel ($M_{\parallel}$) and perpendicular ($M_{\perp}$) to the external magnetic field, show either sharp irreversible transitions and/or smoother fully reversible transitions, in dependence of θ. Taking into account the extended character of the film, the irreversible transitions correspond to nucleation and further propagation of magnetic domains and the reversible ones correspond to magnetization rotation processes [2]. For $\theta = 0°$ the parallel component presents a perfect squared shape hysteresis loop (central top graph of Fig. 1). $M_{\parallel}$ does not change from the saturation (M_S) to the remanence ($M_{\parallel,R}$), i.e., $M_{\parallel,R}/M_S \approx 1$, and there is only a sharp irreversible jump at the coercive field $\mu_0 H_C = 1.50$ mT, in which the magnetization reverses. In turn the perpendicular component is negligible in the whole field loop, i.e., $M_{\perp}(H) \approx 0$. Both are expected behaviors of a magnetization e.a. direction, in which the magnetization reversal takes place via nucleation and further propagation of magnetic domains oriented parallel to the field direction [2].

For $\theta \neq 0°$, clear $M_{\perp}(H)$ loops with both reversible and irreversible transitions are found, in correspondence to the $M_{\parallel}(H)$ loops, as shown by the left and right graphs of Fig. 1. In particular, for $\theta \pm 18°$ the irreversible switching field of the perpendicular component is $\mu_0 H_S(\pm 18°) = 1.55$ mT, identical to $\mu_0 H_C(\pm 18°)$. In addition, the $M_{\perp}(H)$ loops acquired at opposite angles present similar shape but different sign. The latter arises from the sensitivity

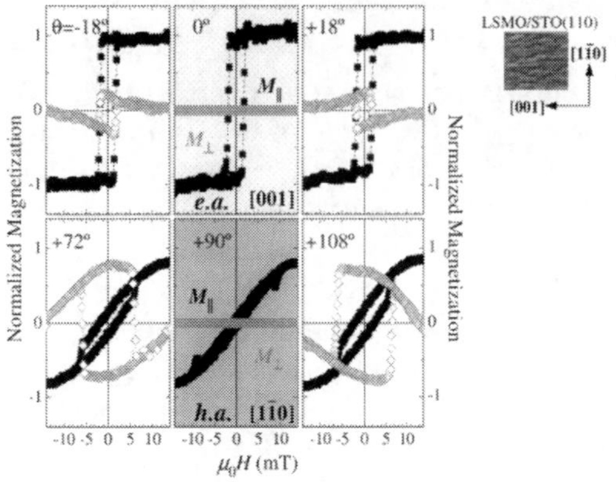

Figure 1. Magnetization reversal study of a 70 nm thick LSMO film grown onto a (110)-oriented STO substrate around the e.a. (left panel) and h.a. (right panel) directions. The corresponding applied field angles θ are indicated in the graphs. The experimental $M_{\parallel}(H)$ and $M_{\perp}(H)$ loops are given by filled black and open red circles, respectively. The inset shows the STM (300×300 nm^2) image of the LSMO surface.

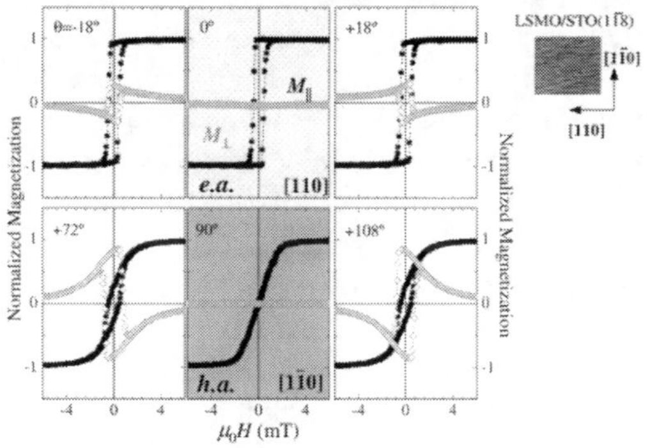

Figure 2. Magnetization reversal study of the 70 nm thick LSMO film grown onto vicinal 10° STO(001) substrate around the e.a. (left panel) and h.a. (right panel) directions. The corresponding applied field angles θ are indicated in the graphs. The experimental $M_{\parallel}(H)$ and $M_{\perp}(H)$ loops are given by filled black and open red symbols, respectively. The inset shows the STM (300×300 nm^2) image of the LSMO surface.

of $M_{\perp}(H)$ to the anisotropy direction [18]. Therefore, around the e.a. direction, the reversible transitions correspond to a reversal by magnetization rotation whereas the irreversible ones correspond to nucleation a propagation of magnetic domains not oriented parallel to the field direction but to the e.a. direction which, in the present case, is aligned parallel to the elongated structures.

3.2. LSMO grown onto STO(1$\bar{1}$8)

Well-defined uniaxial anisotropy is also found in LSMO/STO(1$\bar{1}$8) systems. Fig. 2 shows representative Kerr hysteresis loops of a 70 nm thick film grown onto vicinal STO(001) surface, with a 10° miscut off the [001] plane toward the [1$\bar{1}$0] crystallographic direction, at selected angles θ between the magnetic field and the surface steps direction. $\theta = 0°$ is hence taken when the external field is aligned parallel to the [110] in-plane crystal direction. As before, the characteristic axes are located precisely at the change of sign of the $M_{\perp}(H)$ loops. Hence, the e.a. direction is along the direction of the steps (i.e., [110]) and the h.a. is perpendicular to it (i.e., [1$\bar{1}$0]).

The evolution of the vectorial-resolved hysteresis loops is analogous to the flat LSMO(110) film. Hence, the relevant mechanism for the magnetization reversal close to the e.a. direction is the nucleation and propagation of magnetic domains oriented parallel to the elongated structures, i.e., parallel to the steps, whereas reversal by rotation processes are the relevant mechanism close to the h.a. direction, i.e., perpendicular to the steps. This picture has been strongly supported recently in real space by means of angular dependence Kerr microscopy measurements [19].

4. Summary

For both the investigated systems the angular dependence of the normalized remanence values of both magnetization components, i.e., $M_{\|,R}/M_S$ and $M_{\perp,R}/M_S$ shows a pronounced oscillation of both magnetization components with periodicity of 180°, the parallel component follows a $|\cos\theta|$ law, the perpendicular component changes the sign when a characteristic direction, i.e., e.a. and h.a. directions, is crossed (Fig. 3(a),(b)). The polar-plot of $M_{\|,R}/M_S$ shown in Fig. 3 shows the characteristic "two-lobe" behavior, originated from a two-fold symmetry. All these features confirm the uniaxial magnetic anisotropy behavior of the film, where the anisotropy axis is aligned parallel with the direction of the elongated grains, i.e. parallel to the in-plane [001] crystallographic direction. Both experimental hysteresis loops and remanence magnetization have been properly reproduced in the whole angular range with the Stoner-Wolfhart (SW) model depicted in Ref. [2, 6] (Fig. 3(a),(b)). In order to reproduce satisfactorily the experimental data, in such a simple coherent rotation SW model, only the uniaxial anisotropy term (K_u), arising from the two-fold film morphology, has been taken into account, neglecting completely the biaxial term (K_b), arising from the four-fold crystal symmetry.

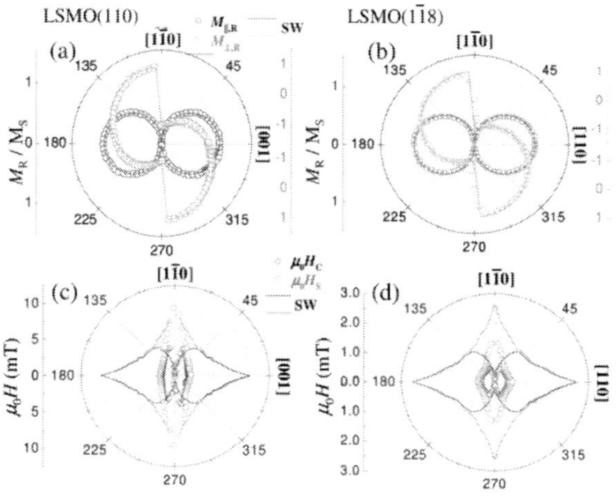

Figure 3. (Color online) Top graphs show the polar plot representations of the angular evolution of the in-plane normalized remanence of magnetization ($M_{\|,R}/M_S$ and $M_{\perp,R}/M_S$) of 70 nm thick LSMO films grown onto (a) STO(110) and (b) vicinal STO(001) substrates. Bottom graphs show polar plot representation of both coercive (H_C) and switching (H_S) fields of 70 nm thick LSMO films grown onto (c) STO(110) and (d) vicinal STO(001) substrates. The symbols are the corresponding experimental values obtained from the vectorial-resolved hysteresis loops. The continuous lines represent the expected behavior from the SW model described in the text, which only considers the uniaxial anisotropic term derived from the experimental data.

It is worth noting that such a theoretical model, which assumes a single particle behavior, takes into account only coherent magnetization reversal by switching (irreversible process) and/or rotation (reversible), where it fails where other irreversible magnetization reversal processes, like nucleation and propagation of magnetic domains, dominate. The latter can be energetically more favorable in systems in which defects—structural or morphological—play the main role to activate irreversible magnetic domain nucleation processes at magnetic fields that are significantly lower than expected by the SW model. This defects can act as pinning centers for the created magnetic domain walls and the reversal further continues via domain wall propagation. Hence, in our case, the simulated loops of both coercive (H_C) and switching (H_S) fields, extracted numerically by the SW model depicted in Ref. [2], overestimates the coercive field near the e.a., where irreversible transitions for the magnetization reversal due to the presence of pinning centers dominate (Fig. 3(c),(d)). The angular evolution of the remanence magnetization (Fig. 3(a),(b)) as well as both coercive and switching fields (Fig. 3(c),(d)), presents 180° periodicity. The expected behavior from the numerical simulations are also superimposed to the experimental data in order to verify the good qualitative agreement of model and experiment. Again the rotation model reproduces satisfactorily both $M_{\parallel,R}$ and $M_{\perp,R}$ in the whole angular range whereas it fails to reproduce $\mu_0 H_C$ and $\mu_0 H_S$ close to e.a. directions. In contrary, close to the h.a. direction, where reversible processes are the relevant mechanism during reversal, the rotation model reproduces satisfactorily the coercive and switching fields behavior.

In conclusion, we have investigated the magnetic anisotropy of LSMO thin films by high-resolution vectorial Kerr magnetometry. The dominance of the uniaxial anisotropy over the biaxial anisotropy is achieved in LSMO thin films grown onto STO(110) and STO($1\bar{1}8$) susbtrates. In both systems, the surface-induced anisotropy determines a well defined uniaxial magnetic anisotropy, with the e.a. lying along the direction of surface step edges. We have also shown the angular dependence of the magnetization reversal processes from the detailed analysis of the vectorial-resolved Kerr loops. Nucleation and further propagation of magnetic domains and rotation processes are the relevant mechanism during magnetization reversal around the e.a. and h.a. directions, respectively. The angular dependence of both switching and coercive fields for θ near the e.a. and h.a. directions can be understood in the framework of the pinning and SW models, respectively. We demonstrated the ability to control and tailor the magnetic properties of LSMO thin films which is an important task for the design of novel devices based on thin film technology.

ACKNOWLEDGEMENTS
This work was supported in part by the Spanish MICINN through Project No. CSD2007-00010 and by the Comunidad de Madrid through Project No. S2009/MAT-1726. P.P. thanks the European Science Foundation (ESF) through the activity entitled 'Thin Films for Novel Oxide Devices' (http://www.ims.tnw.utwente.nl/thiox/) for partial financial support through exchange grants.

5. References
[1] M. Mathews, *et al.*, Appl. Phys. Lett. **87**, 242507 (2005).
[2] P. Perna, *et al.*, "Tailoring magnetic anisotropy in epitaxial half metallic $La_{0.7}Sr_{0.3}MnO_3$ thin films", *submitted to* (2010).
[3] A. Berger, *et al.*, Phys. Rev. Lett. **68**, 839 (1992).
[4] U. Welp, *et al.*, Phys. Rev. Lett. **90**, 167206 (2003).
[5] S. van Dijken, *et al.*, Phys. Rev. B **63**, 104431 (2001).
[6] D. Ecija,*et al.*, Phys. Rev. B **77**, 024426 (2008); *ibid.* J. Magn. Magn. Mater. **316**, 321 (2007).
[7] J. J. de Miguel and R. Miranda, J. Phys.: Condens. Matter **14**, R1063 (2002).
[8] M. Radovic, *et al.*, Appl. Phys. Lett. **94**, 022901 (2009)
[9] A. Ruotolo, *et al.*, Appl. Phys. Lett. **88**, 252504 (2006).
[10] A. Ruotolo, *et al.*, Appl. Phys. Lett. **91**, 132502 (2007).
[11] P. Perna, *et al.*, J. Phys.: Condens. Matter **21**, 306005 (2009).
[12] J.-H. Park, *et al.*, Nature **392**, 794 (1998).
[13] F. Tsui, *et al.*, Appl. Phys. Lett. **76**, 17 2421 (2000).
[14] M. Ziese, *et al.*, J. Mag. Mag. Mat. **246**, 327-334 (2002).
[15] R.A. Hyman, *et al.*, Phys. Rev. B **58**(14), 9276 (1998).
[16] Z.-H. Wang, *et al.*, Appl. Phys. Lett. **82**(21), 3731 (2003).
[17] J. Camarero, *et al.*, Phys. Rev. Lett. **95**, 057204 (2005).
[18] E. Jiménez, *et al.*, Phys. Rev. B **80**, 014415 (2009).
[19] P. Perna, *et al.*, "Imaging the magnetization reversal of step-induced uniaxial magnetic anisotropy in vicinal epitaxial $La_{0.7}Sr_{0.3}MnO_3$ films", *submitted to* (2010)

Journal of Physics: Conference Series **303** (2011) 012059 doi:10.1088/1742-6596/303/1/012059

Magnetic tunnels junctions for all-oxide spin valves devices

A Solignac[1], R Guerrero[1], G Agnus[2], C Fermon[1], M Pannetier-Lecoeur[1,3], Ph Lecoeur[2]
[1] DSM/IRAMIS/SPEC- CNRS URA 2464, CEA Saclay, 91191 Gif sur Yvette Cedex, France
[2] Institut d'Electronique Fondamentale, Univ Paris-Sud, CNRS UMR8622, F-91405 Orsay, France

E-mail: myriam.pannetier@cea.fr

Abstract. We have developed $La_{0.7}Sr_{0.3}MnO_3$ (LSMO) based spin valve tunnel junctions where the hard layer is composed of Ru doped $La_{0.7}Sr_{0.3}MnO_3$ (LSMRO). These perovskite oxide layers have been grown on $SrTiO_3$ by pulsed laser deposition. X ray diffraction, transmission electron microscopy and atomic force microscope measurements show the crystalline quality of the $La_{0.7}Sr_{0.3}MnO_3$ and $La_{0.7}Sr_{0.3}MnO_3$ doped Ruthenium thin films. Magnetization, resistance and noise characterisation of these layers are presented. The doping of Ruthenium increases the coercitive field of the $La_{0.7}Sr_{0.3}MnO_3$ but degrades the metallic properties of the layer and leads to a noise increase of one order of magnitude. We have studied the magnetoresistance and magneto-transport properties of LSMO /STO/ LSMRO–based junctions which show the importance of the interfaces quality.

Introduction

Perovskite oxides have a variety of interesting properties going from high temperature superconductivity to insulating behavior passing through magnetism, half-metallicity, ferrorelectricity. Among these materials, YBaCuO (YBCO) is a cuprate exhibiting superconductivity below 92K. The manganite $La_{1-x}Sr_xMnO_3$ with a doping level of x=0.3 (LSMO) is metallic and ferromagnetic under 340K. A high spin polarization in transport is also observed [1] and as a consequence LSMO-based tunnel junctions present a very high magnetoresistance [2], [3]. Both YBCO and LSMO present a perovskite crystalline structure and a close lattice parameter in the (a, b) plane, and are, therefore, good candidates for integrated devices. In particular, integrated epitaxial system can be designed with these oxides to develop an all oxide mixed sensor [4].

A mixed sensor is a high sensitive magnetic field sensor which uses a flux-to-field transformer to increase up to a factor of 1000 the sensitivity of a magnetoresistive sensor. The flux-to-field transformer is a superconducting ring with a small constriction which, combined to conventional metallic Giant Magneto-resistance (GMR) structure, exhibits sensitivity of few $fT/\sqrt{Hz}$ in the thermal noise [5]. The gain is given by the geometry of the loop [6]. The aim is to develop a new mixed sensor which combines the performances of a high-Tc (YBCO) superconducting loop, with the high magnetoresistance of $LaSrMnO_3$-based tunnel junction, which would lead to a subfemtotesla magnetometer operating at 77K. A typical all oxide mixed sensor would contain

[3] To whom any correspondence should be addressed.

LSMO/$SrTiO_3$/LSMO tunnel junction separated from the superconducting loop by a thicker $SrTiO_3$ (STO) insulating layer (figure 1).

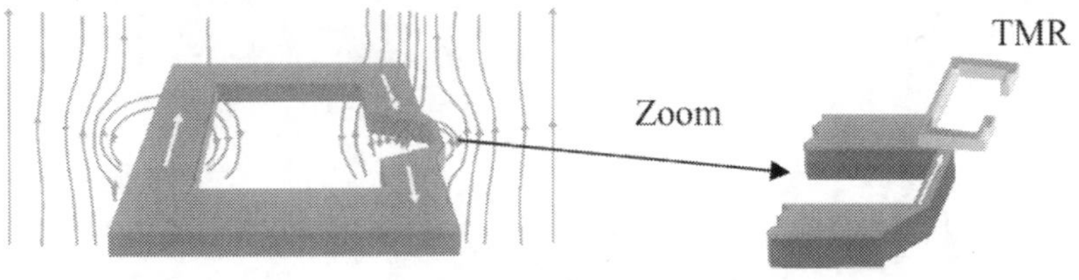

Figure 1. All oxide mixed sensor principle: a YBCO superconducting loop with a constriction is associated to a LSMO-based TMR sensor. The field is enhanced around the ring constriction and then is measured by the TMR.

A tunnel magnetoresistive sensor is generally composed of two magnetic layers with different coercitive fields separated by an insulating layer in order to stabilize the antiparallel state. This structure is called a spin valve. The LSMO coercitive field can be increased by doping it with Ruthenium (Ru) (LSMRO) [7], [8] and thus a LSMRO layer can be used to create a spin valve structure [9]. In this paper, we present results on LSMO/STO/LSMRO tunnel junctions. LSMO and LSMRO films crystalline, magnetic, transport and noise properties are first described. Then, we present results on complete patterned junctions.

1. Growth

The oxide layers are grown by pulsed laser deposition (PLD) with a KrF laser (wavelength of 248nm). LSMO layers are deposited on STO vicinal substrate (100) under 120 mTorr of pure oxygen atmosphere at 600°C. In situ annealing is done at 300Torr and 600°C after the deposition. Laser energy of 200mJ allows a maximum fluence of 3J/cm² on the target. The energy is adjusted using an attenuator to obtain an average deposition flux of 0.13ML/s. An homogenizer is used to ensure a top flat profile of the laser beam on the target, and avoids formation of droplets on the surface. The thickness of the films deposited is between 25 and 44 nm. The lattice mismatch between the LSMO (a_{bulk} (bulk lattice parameter) = 3.875Å) and the STO (a_{bulk}=3.905Å) is approximately 0.8%. The lattice parameter of the LSMRO depends of Ru doping [7], [10], but it is higher than the LSMO lattice parameter.

2. Single films characterizations

2.1. Crystalline quality of the thin LSMO films

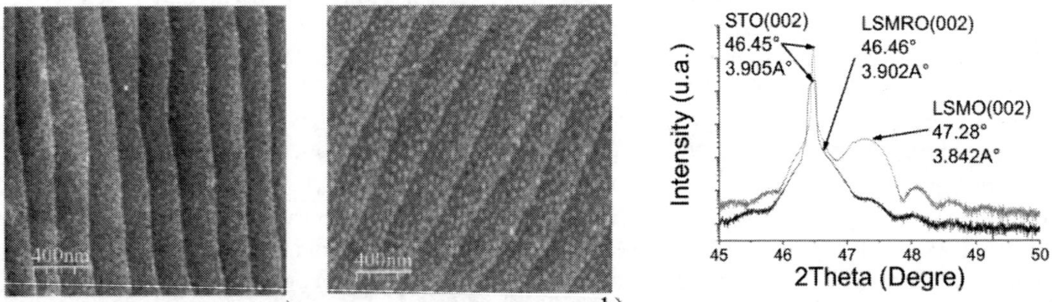

a) b) c)

Figure 2. a) AFM (atomic force microscope) image of LSMO layer after deposition on a vicinal STO film. b) AFM image of a LSMRO layer grown in the same conditions. c) X-Ray diffraction measurement of the same layers

Journal of Physics: Conference Series **303** (2011) 012059 doi:10.1088/1742-6596/303/1/012059

Figure 2 a) and b) show that the vicinal steps are transmetted from the substrate to the layers and correspond to one LSMO or LSMRO monolayer (~ 0.4 nm). The roughness of the layers is less than one atomic layer and is 0.1 nm RMS for the LSMO and 0.2 nm RMS for the LSMRO. The finite size effect, visible from the Kiessing oscillations around the LSMO diffraction peak (47.28°) and the LSMRO diffraction peak (46.46°) in figure 2 c), is characteristic of a good cristalline structure and of the absence of extended defects in the layers.

2.2. Magnetic and transport properties of LSMO and LSMRO

Yoke shape structures have been processed on the different layers (figure 3 c)), using Ar-plasma etching neutralized under O_2 and photolithography. Hundred nanometers of Pd were deposited by sputtering for metallic contacts. The resistance of the contacts is less than 1kΩ. The dimensions of the structure are 5µm x 400µm. It has been shown that the yoke structure stabilizes the magnetization along the long arm of the structure for metallic GMR [11].

Resistance measurements were performed using four probes configuration. We used a Keithley 6221 current source to bias our samples; the voltage was detected using an INA103-based DC preamplifier and an acquisition card. Magnetization versus field measurements were done with SQUID magnetometry.

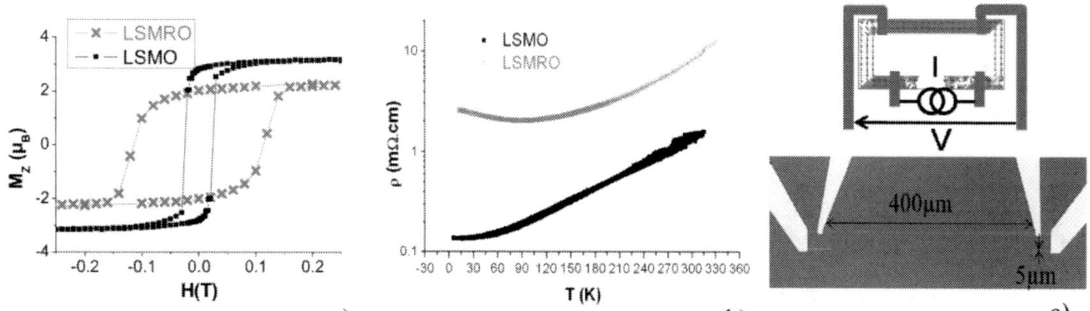

a) b) c)

Figure 3. a) Magnetic moment per unit cell (M_Z) versus field at 5K. b) Resistivity versus temperature of a LSMO (square points) and a LSMRO (cross points) thin films deposited on STO substrates. c) Schematic and picture of the yoke structure which had been defined on LSMO and LSMRO layers, and allowed measuring the resistivity in four points.

We observed, in figure 3 a) an enhancement of the coercitive field by a factor of five in the LSMRO film, due to Ru doping, and also a decrease of the magnetic moment. Indeed, the Ru atoms, in the LSMRO, are in substitution on the Mn sites, and give the chemical formula $La^{3+}_{0.7}Sr^{2+}_{0.3}(Mn_{1-y}Ru_y)^{3+}_{0.7}(Mn_{1-y}Ru_y)^{4+}_{0.3}O^{2-}_3$ for low doping level (y<0,5). In our case the doping level is 0.05. LSMRO magnetic properties could be explained in terms of charge transfer and antiferromagnetic exchange coupling between Mn^{3+}/Mn^{4+} and Ru^{3+}/Ru^{4+} centers [7,10].

The resistivity of the LSMO varies from 1.4 mΩ.cm at 300K down to 0.14 mΩ.cm at 4 K. An increase of resistance is observed for the LSMRO film around 90K, probably due to polaron hopping, and charge career localization [10]. So doping by Ru degrades the transport properties of the layer, despite the good crystalline features given by X-Ray analysis and surface roughness.

2.3. Noise measurements

For sensor applications, noise is as important as sensitivity. Noise is also another way to probe transport properties. Johnson noise due to thermal excitations and shot noise due to current fluctuations are independent of the frequency f. Another noise source is the 1/f noise. It is due to resistance fluctuations in small volumes and hence is proportional to the current and inversely proportional to the square root of the number of carriers. Noise measurements are done with an AC coupled low noise amplifier and a battery fed current source. From Figure 4 a), it appears that 1/f noise is dominant on the yoke structure at frequency lower than 1kHz. Johnson noise level can be measured from the noise power spectrum at zero bias current.

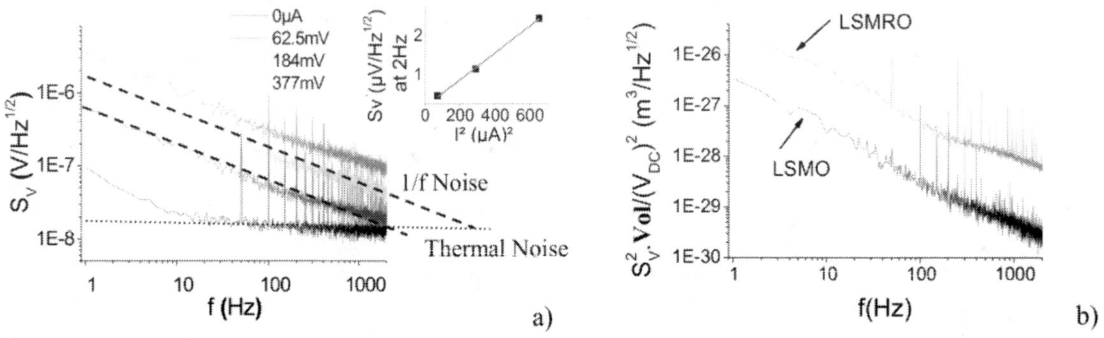

Figure 4. a) Voltage noise spectral density in a LSMO yoke (thickness 28nm) at 110K. Johnson noise is present with zero current. 1/f noise appears when a current is applied and increases linearly with the current square (inset). b) Voltage noise spectral density normalized by the voltage and the volume in LSMO (thickness 28nm) and LSMRO (thickness 44nm) yokes at 110K.

To compare noise from different samples, the α parameter from the empirical Hooge formula (equation (1)) is used.

$$\alpha = \frac{S_V^2(f).f^{\alpha}.Vol}{V_{DC}^2} = \frac{\gamma_H}{n_C} \quad (1)$$

The noise spectral density, $S_V(f)$, depends on the sample voltage V_{DC} , the sample volume Vol, the density of careers n_c and the Hooge constant γ_H.
For our LSMO samples, α is 3.10^{-27} m^3 at 150K; this noise is higher than the best results presents in the literature. For instance, A. Palanisami et al. achieved 2.10^{-32} m^3 [12] and L. Mechin et al. 8.10^{-31} m^3 [13] but in the average of the broad range of noise value [14] published (from 10^{-22} m^3 to 10^{-32} m^3). The noise level measured in LSMRO yoke is one order of magnitude bigger than in the LSMO yoke (figure 4 b)). The origin of the degradation of the conductivity might also explain the increase of noise in the LSMRO. Polaron hopping and charge careers localization could be responsible of this larger low frequency noise.

3. Magnetic tunnel junctions results

LSMRO/STO/LSMO junctions on STO substrate, with thicknesses of 30nm/4nm/30nm, have been produced by PLD with the same conditions than before. These junctions were processed with UV lithography to design square junctions of 8μm x 8μm. The shape of the overall structure is shown in figure 5 a). The junctions' resistance were measured in four points. The typical RA product (where R is the tunnel resistance and A is the area of the junctions) measured are in the GΩ.μm² range. For practical application, this product should be reduced, by decreasing the thickness of the STO barrier. We estimate the barrier height to 0.1 eV by fitting the IV curves measured at 300K with the Brinkmann model [15]. This value is smaller than the barrier height reported in the literature, around 0.5eV [16], and could indicate the presence of impurities in our barrier, probably oxygen vacancies. TMR versus field, voltage bias and temperature, is shown in figure 5.

Journal of Physics: Conference Series **303** (2011) 012059 doi:10.1088/1742-6596/303/1/012059

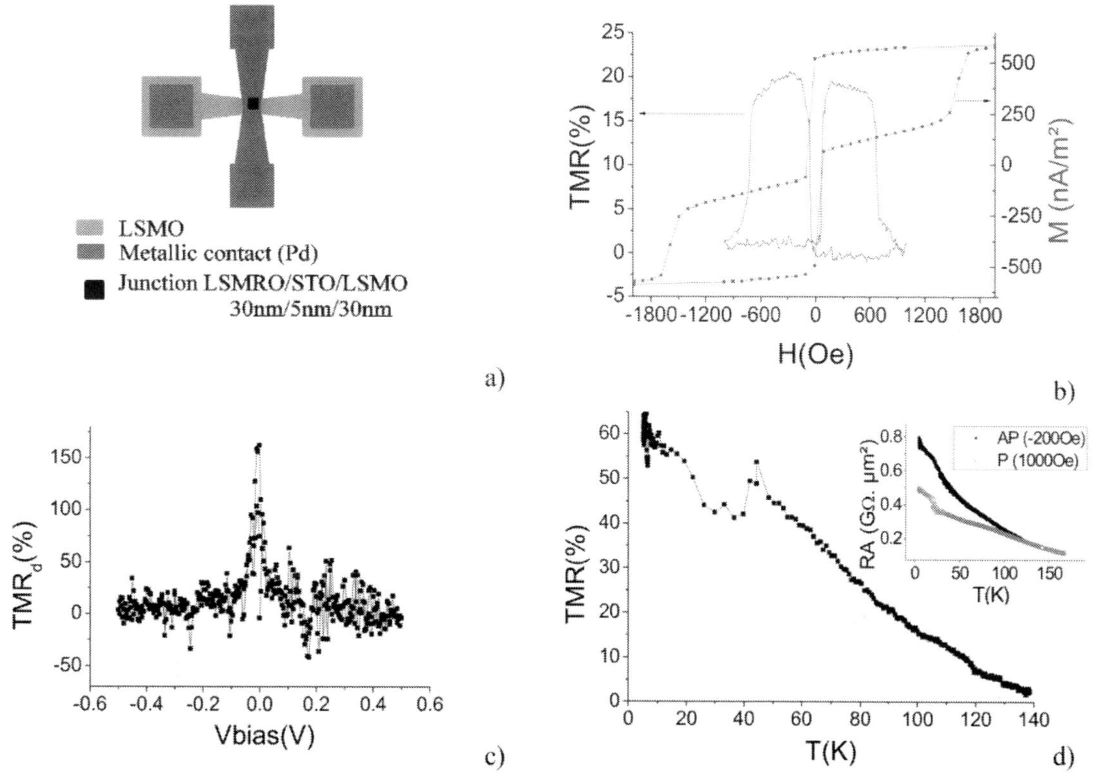

Figure 5: a) Schematic of the junction realized by UV photolithogaphy. The junction is a square of 8µmx8µm. b) TMR at 5K and the magnetization versus field of the same sample before processing at 5K. The LSMRO layer is the hard layer but the coercitive field is reduced by the process. a) TMR as function of the voltage bias applied. b) TMR and RA product in the parallel and the antiparallel state (inset) of the junction versus temperature. A 50 mV constant voltage bias is applied, and the current is measured.

4. Discussion

LSMO/STO/LSMRO junction has a spin valve structure behavior before processing in the full layer, but also after patterning in a 8x8µm² junction. We observe a decrease of the coercitive field (H_c) due to the processing. Ishii et al. [9] observed in LSMO/LAO/LSMRO junctions an H_c of the full layer smaller than the processed junctions, and they did not find correlation between H_c and the geometric shape anisotropy. We could explain the decrease of H_c by a non homogeneity of Ru doping or a migration of the Ru during the junction processing. Nevertheless, the LSMO doped by Ru totally stabilizes the anti-parallel state.

A peak of the TMR is observed at zero bias up to 160%, as well as a sharp decrease of the TMR with the voltage bias. At 100mV, no more TMR is observed. This behavior is usually explained by the interaction of electron with magnons in the tunneling process [17], i.e. inelastic tunneling, but also by the presence of metallic particles, magnetic impurities or multi-step tunneling stage in the barrier or at the interface [18]. In our opinion, the presence of Ru at the interface between STO and the LSMRO could be at the origin of the quick decrease of the TMR with voltage observed in these junctions.

We observed an increase of the resistance versus temperature for most of the junctions measured (Figure 5 d) insert) which is a typical behavior for junctions [19].

The TMR decreases when the temperature is raised and disappears above 140K. J. S. Moodera [18] proposed that the two principal mechanisms that could explain the temperature dependence of the TMR: spin-flip scattering of tunneling electrons from impurities in the barrier or a reduction of the magnetic moment in the ferromagnet due to excitation of magnons. Nevertheless, in LSMO, it is known that a loss of the magnetization appears at the interfaces [20]. This could explain the TMR temperature variation in a more relevant way.

Conclusion

We have grown LSMO and LSMO doped Ru thin films and we have studied their crystalline, magnetic and transport properties. The Ru doping hardens the LSMO layer but degrades the transport and noise properties in the studied films. We have processed and measured tunnel junctions based on LSMO/STO/LSMRO. Firstly the LSMRO layer allows creating a spin valve structure. But it seems that the Ru plays an important role at the interface of the junction and may be at the origin the TMR peak at zero bias and the rapidly decrease of the TMR with voltage bias and temperature. To avoid these effects a simple solution could be to add a LSMO spacer (~2nm) between the STO and the LSMRO. This spacer could avoid the effect of the Ru present at the interface while keeping the enhancement of the coercitive field.

Acknowledgments

We acknowledge the French funding RTRA-Triangle de la Physique, project 2009-086T, ATTOS, and ANR-09-NIRF009 for partial support.

[1] Nadgorny B, Mazin I I,Osofsky M, Soulen R J, Broussard P, Stroud R M, Singh D J, Harris V G, Arsenov A and Mukovskii Ya 2001, *Phys. Rev. B* **63** 184433
[2] Viret M, Drouet M, Nassar J, Contour J P, Fermon C and Fert A 1997 *Europhysics Letters* **39** 545
[3] Bowen M, Bibes M, Barthélémy A, Contour J P, Anane A, Lemaître Y and Fert A 2003 *Appl. Phys. Lett.* **82** 233
[4] Pannetier M, Fermon C, Le Goff G, Simola J and Kerr E 2004 *Science* **304** 1648
[5] Pannetier M, Fermon C, Polovy H, Dyvorne H, Sergeeva-Chollet N and Paul J 2009 *IEEE SENSORS 2009 Conference* 1856
[6] Pannetier M, Fermon C,Vedrine P, Welling M S and Wijngaarden R J 2006 *Sensors and Actuators A* **129** 146
[7] Sundar Manoharan S, Ju H L and Krishnan K M 1998 *J. Appl. Phys.* **83** 7183
[8] Yamada H, Kawasaki M and Tokura Y 2005 *Appl. Phys. Lett.* **86** 192505
[9] Ishii Y, Yamada H, Sato H, Akoh H, Kawasaki M and Tokura Y 2005 *Appl. Phys. Lett.* **87** 022509
[10] Wang L M, Lai J H, Wu J I, Kuo Y K and Chang C L 2007 *J. Appl. Phys.* **102** 023915
[11] Pannetier M, Fermon C, Le Goff G, Simola J, Kerr E and Coey J M D 2005 J. *Magn. Magn. Material* **290** 1158-60
[12] Palanisami A, Merithew R D, Weissman M B, Warusawithana M P, Hess F M, and Eckstein J N 2002 *Phys. Rev. B* **66** 092407
[13] Méchin L, Routoure J M, Mercone S, Yang F, Flament S, and Chakalov R A 2008 *J. Appl. Phys.* **103**, 083709
[14] Gaucher F 2008 PhD
[15] Brinkman W F, Dynes R C and Rowell J M 1970 *J. Appl. Phys.* **41** 1915
[16] Vera Marún I J, Postma F M, Lodder J C and Jansen R 2007 *Phys. Rev. B* **76** 064426
[17] Bowen M, Maurice J L, Barthélémy A, Bibes M, Imhoff D, Bellini V, Bertacco R, Wortmann D, Seneor P, Jacquet E, Vaurès A, Humbert J, Contour J P, Colliex C, Blügel S and Dederichs P H 2007 *J. Phys.: Condens. Matter* **19** 315208
[18] Moodera J S and Mathon G 1999 *J. Magn. Magn. Material* **200** 248
[19] Rüdiger U, Calarco R, May U, Samm K, Hauch J, Kittur H, Sperlich M and Güntherodt G 2001 *J. Appl. Lett.* 89, 7573
[20] Ott F, Viret M, Borges R, Lyonnet R, Jacquet E, Fermon C and Contour J P 2000 *J. Magn. Magn. Material* **211** 200

Journal of Physics: Conference Series **303** (2011) 012060 doi:10.1088/1742-6596/303/1/012060

Study of 2D growth stability of $SrRuO_3$ on $SrTiO_3$

D. Esteve[1], T. Maroutian[1], V. Pillard[1], G. Agnus[1], Ph. Lecoeur[1]

[1] Institut d'Electronique Fondamentale, Univ Paris-Sud, CNRS UMR 8622, F-91405 Orsay Cedex, France

thomas.maroutian@u-psud.fr

Abstract. In order to control the surface roughness and chemical termination of an oxide film, a promising route is to use a vicinal substrate and to control the atomic step positions during growth in the so-called step flow mode. The stability of this mode is studied in the case of $SrRuO_3$ heteroepitaxy on $SrTiO_3$, focusing on the influence of deposition flux and substrate mean terrace width. Transitions from unstable step bunching to stable step flow and from step flow to island nucleation are observed and discussed in terms of the mean step velocity. The step bunching is traced back to a strain-based driving force in competition with a stabilizing asymmetry for incorporation of adatoms at steps. A thin film obtained after stable step flow gives a greater finite size effect than a step-bunched one, as seen by X-ray diffraction

1. Introduction

The growth of perovskite oxides using pulsed Laser deposition (PLD) allows for the control of the surface roughness down to the atomic scale, which is a great asset in particular in the field of oxide-based electronics where interfaces play a huge part in device properties. The interface structure, in particular its sharpness and the presence of oxygen vacancies, will determine for example the magnetic coupling between layers [1] or the formation of a two-dimensional electron gas [2]. A first step to achieve such control of the interface between heteroepitaxial oxide layers is to use a substrate with homogeneous surface termination such as a TiO_2-terminated $SrTiO_3$ (001) substrate (STO) [3]. Then, a two-dimensional (2D) growth will often be preferred to have a low level of roughness. The PLD technique has proven the ability to build fully strained $SrRuO_3$ (SRO) films with the target stoechiometry on single-terminated STO substrates [4].

During PLD, each nanosecond-duration laser pulse brings a quantity N_p of atoms to the surface, which is adsorbed as single adatoms or form small unstable clusters. These species then diffuse to stable incorporation sites, either at pre-existing step edges or through nucleation of stable islands thus forming a new layer. This transient diffusion stage is typically over before the next pulse. When all impinging atoms are able to reach the step edges, the growth proceeds in the so-called step flow mode. At variance with this 2D scenario, island nucleation on the terraces can lead to tri-dimensional (3D) growth if the nucleation of a new layer occurs before the completion of the previous one. Step flow on a vicinal surface thus appears as an optimal growth mode to control surface morphology. However, in the case of heteroepitaxy it has been shown that step flow growth can be unstable with respect to step bunching [5]. The occurence of this instability is not yet well understood in particular regarding the transition towards stable step flow upon increasing the deposition flux. In this paper, we present a study of the stability of step flow growth using SRO on vicinal TiO_2-terminated STO substrate as model system, for which step bunching has already been observed [6]. For a fixed temperature, we have mapped SRO growth as a function of deposition flux and STO mean terrace width.

2. Experimental details

Substrates of STO were prepared according to an established procedure [3,4] to ensure a TiO_2 chemical termination. The substrates were systematically characterized with atomic force microscopy (AFM) in contact mode prior to their introduction into the growth chamber, in particular to measure their mean terrace width (L). Thin films of SRO were grown on these substrates by PLD using a KrF excimer laser (λ = 248 nm) at a growth temperature of 610°C under 120 mTorr of pure oxygen atmosphere. Laser energy was set to 200 mJ and the fluence was further refined with an attenuator to get 3 J/cm² on the target. A homogenizer is used to give the laser beam a flat-top profile to avoid droplet formation on the sample surface. To ensure a good control of both film thickness and mean deposition flux, an *in situ* optical reflectometry technique is used [7]. Using X-ray diffraction (XRD) measurements, all SRO films were found fully strained on STO with unit cell parameter perpendicular to the surface close to 0.395 nm. The deposition flux F is taken as $F = pN_p$ with p the laser repetition rate and N_p the (fixed) amount of SRO deposited per pulse. A repetition rate between 0.6 Hz and 10 Hz thus gives a flux ranging from 3.3×10^{-2} ML/s to 5.5×10^{-1} ML/s.

3. Results and discussion

Figure 1 shows 4 x 4 µm² AFM topographs of typical morphologies observed after growth of 9 nm-thick SRO films in the step bunching instability region (a), after stable step flow (b), and with island nucleation (c). The initial difference of terrace width between samples (a) and (b) is rather small, but the deposition flux was increased by about a decade, from 3.6×10^{-2} ML/s for the bunched morphology to 3.8×10^{-1} ML/s for the step flow sample. This huge increase of the flux stabilizes the structure, promoting an equal width for all terraces. For the same mean step velocity $FL \sim 40$ nm/s but with larger terrace width, islands are formed which can give rise to 3D growth upon further deposition.

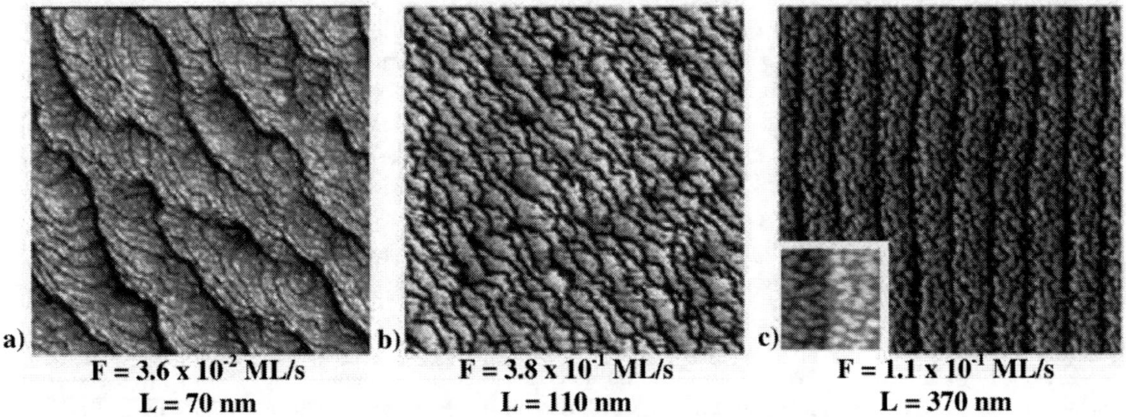

Figure 1: 4 x 4 µm² AFM topographs of 9 nm-thick SRO films on STO after (a) unstable step flow giving step-bunched surface, (b) stable step flow and (c) island formation. Deposition flux and substrate terrace width are given below each image. Inset of (c) is a close-up on a terrace.

In the case of heteroepitaxy, the misfit parameter between the substrate and the film (~ 0.6% for SRO/STO) leads to an increase of elastic energy in the structure. There are different ways to release this energy, for example through dislocations or by gathering the steps together. Tersoff *et al.* [5] have expressed the elastic energy between steps as $E = \alpha_1 \mathrm{Ln}(L) + \alpha_2/2L^2$, where α_1 is linked to the stress

Journal of Physics: Conference Series **303** (2011) 012060 doi:10.1088/1742-6596/303/1/012060

discontinuity at step edges, L is the terrace width, and α_2 is a dipolar interaction. The first term expresses an attractive interaction between step edges, whereas the second one points a repulsive force. These two interactions together lead to an optimal distance between steps with characteristic length $L_0 = (\alpha_2 / \alpha_1)^{1/2}$. Typically steps with $L > L_0$ will benefit from moving closer together.

On most surfaces diffusing adatoms have a higher probability to attach to ascending step edges due to an excess energy barrier for diffusion over a step, the so-called Ehrlich-Schwöbel (ES) effect. In a kinetic picture, the larger the terrace width and deposition flux, the faster this incorporation asymmetry at steps will damp terrace width fluctuations. Increasing the flux or more generally the mean step velocity is thus a stabilizing factor with respect to step bunching. Combined with the aforementioned elastic driving force, step bunching is thus expected for small deposition flux and small terrace widths with a transition to stable step flow upon increasing these parameters, as seen experimentally (**figure 1**). At higher temperature the instability should occur over a wider (F,L) range due to a decreasing ES effect with temperature. As this yields an increased adatom diffusion length, larger terrace widths can be used to obtain stable step flow without going into the island nucleation regime.

Upon closer scrutiny of the morphology shown in **Figure 1 (c)**, all islands are found to be of one monolayer height (see inset). The adatom diffusion length is estimated of the order of 100 nm at our growth temperature, thus island nucleation is found on 370 nm-wide terraces Since our films are 9 nm thick, it is not possible at this stage to predict whether 2D growth will continue through nucleation and coalescence or if 3D growth will set in with formation of multi-layer islands.

We have elaborated two SRO films, one in the step bunching regime (60 nm thick) and the other in the stable step flow mode (50 nm thick). **Figure 2** shows the XRD spectra of these two samples. The width difference of the SRO main peak comes from the thickness variation between the two thin films. To fit our data, we used a grating function corresponding to the Fourrier transform of the grating mesh along the perpendicular axis, which takes into account the strained SRO parameter (0.395 nm) and the finite film thickness. This model gives a very good fit to the data for both films. The main difference between these two XRD patterns is the finite size effect which is very well defined for the step flow sample and barely visible for the step-bunched one.

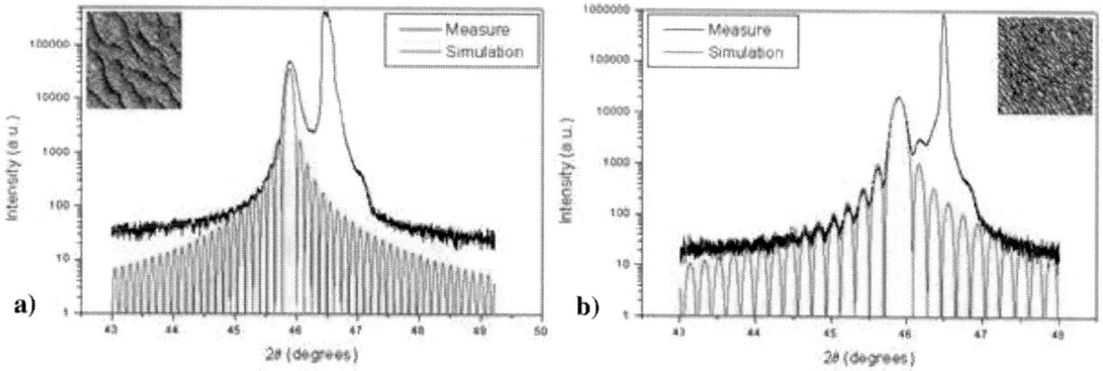

Figure 2: XRD spectra of the (002) peaks of SRO (left peak) and STO (right peak) for (a) the 60 nm-thick step-bunched film and (b) the 50 nm-thick step flow sample.

The attenuation of the finite size effect in **Figure 2 (a)** can be related to the Root Mean Square roughness of the sample, which is higher in the case of a step bunched morphology (~ 0.4 nm) than for a step flowed sample (~ 0,2 nm). Such roughness and associated strain field can explain variations of the unit cell parameter in the near-surface region which would give a reduced size effect. XRD here allows to evidence a link between growth-induced surface morphology and film properties.

4. Conclusion

In this paper we have shown that upon increasing the deposition flux and/or the substrate mean terrace width the 2D growth of SRO on STO goes from unstable step bunching to stable step flow and finally to island nucleation and coalescence. We have evidenced an influence of surface morphology and thus of growth mode on the finite size effect as measured by XRD. This work opens the way to further studies to address the temperature dependence and the late stage of growth.

5. Acknowledgments

This work was partly supported by the French Agence Nationale pour la Recherche (ANR) under "MINOS" project.

References

[1] X. Ke, M. S. Rzchowski, L. J. Belenky, C. B. Eom, App. Phys. Lett. 84, 5458 (2004)
[2] H. P. R. Frederikse, W. R. Thurber, W. R. Hosler, Phys. Rev., vol. 134, number 2A (1964)
[3] M. Kawasaki, K. Takahashi, T. Maeda, R. Tsuchiya, M. Shinohara, O. Ishiyama, T. Yonezawa, M. Yoshimoto, H. Koinuma, Science 266, 1540 (1994)
[4] G. Rijnders, D. H. A. Blank, J. Choi, C. B. Eom, Appl. Phys. Lett. 84, 505 (2004)
[5] J. Tersoff, Y. H. Phang, Z. Y. Zhang, and M. G. Lagally, Phys. Rev. Lett. 75, 2730 (1995)
[6] M. Yoon, H. N. Lee, W. Hong, H. M. Christen, Z. Zhang, Z. Suo, Phys. Rev. Lett. 99, 055503 (2007)
[7] D. Estève, K. Postava, P. Gogol, G. Niu, B. Vilquin, P. Lecoeur, Phys. Stat. Sol. B 247, 1956 (2010)

Finite-size effects on the magnetoelectric response of field-driven ferroelectric/ferromagnetic chains

Chenglong Jia[1], Alexander Sukhov[1,2], Paul P. Horley[3] and Jamal Berakdar[1]

[1]Institut für Physik, Martin-Luther Universität Halle-Wittenberg, 06120 Halle, Germany
[2]Max Planck Institute of Microstructure Physics, Weinberg 2, 06120 Halle, Germany
[3]Centro de Investigatión en Materiales Avanzados, S.C. (CIMAV), Chihuahua/Monterrey, 31109 Chihuahua, Mexico

E-mail: Jamal.Berakdar@physik.uni-halle.de

Abstract. We study theoretically the coupled multiferroic dynamics of a finite one-dimensional ferroelectric/ferromagnet chain driven by harmonic magnetic and electric fields as a function of the chain length. We consider the case of a linear magnetoelectric coupling that results from the spin-polarized screening charge at the interface. We performed Monte-Carlo simulations and calculations based on the coupled Landau-Lifshitz-Gilbert and Landau-Khalatnikov equations showing that the net magnetization and the total polarization of thin heterostructures, i.e. with up to ten ferroelectric and ferromagnetic sites counted from the interface, can be completely reversed by external electric and magnetic fields, respectively. However, for larger systems merely a limited magnetoelectrical control is achievable.

1. Introduction

Multiferroics, i.e. materials in which ferroelectric (FE) and ferromagnetic (FM) orders coexist, have attracted considerable attention in recent years [1]. The strong magnetoelectric (ME) coupling in these materials renders possible manipulation by external magnetic and electric fields[2, 3, 4, 5, 6]. This brings about entirely new concepts in the design of next generation devices. Currently studied multiferroics are of two kinds: single-phase systems and two-phase systems. Single phase multiferroics, such as $RMnO_3$ (e.g. R=Tb, Dy) [3, 7] have successfully been exploited to control the magnetic order by electrical means and vice versa. A number of interesting phenomena such as an electrically controllable persistent spin current [8] and a modified Datta-Das spin-field-effect transistor [9] associated with the two dimensional electron gas formed at the interface of multiferroic oxides was reported. However, the small electric polarization and the low transition temperature in single-phase multiferroics[7] are adverse for applications. On the other hand, two-phase multiferroics [1, 10], including artificially synthesized ferroelectric and ferromagnetic materials, are very promising as a room-temperature ME system in which large ferroic orders coexist.

Very recently, the dynamic response to applied external fields in a multiferroic chain with a linear ME coupling induced by electrostatic screening at the ferroelectic/ferromagnet interface was theoretically investigated [11]. It was demonstrated that for material parameters corresponding to $BaTiO_3$/Fe the total polarization and the net magnetization are controllable by external magnetic and electric fields, respectively. In this work we study the size-dependence of the multiferroic dynamics of the electric polarization and of the magnetization in several heterostructures. Calculations are performed within the framework of the coupled Landau-Khalatnikov [12] and zero Kelvin Landau-Lifshitz-Gilbert equations [13]. We find that the polarization (magnetization) are magnetically (electrically) switchable for chains with a small size.

2. Theoretical formulation

As sketched in Fig.1, we consider a two-phase multiferroic chain consisting of an artificially grown ferroelectric-ferromagnetic heterostructure such as $BaTiO_3$/Fe. This system is similar to that in which the magnetoelectric effect was recently predicted theoretically[11, 14, 15] and realized experimentally[16]. The total energy of the FE/FM-system in a general case falls into three parts [11]

$$F_\Sigma = F_{\rm FE} + F_{\rm FM} + E_{\rm c}, \tag{1}$$

The ferroelectric energy contribution reads

$$F_{\rm FE} = a^3 P_{\rm S} \sum_{j=1}^{N_{\rm FE}} \left(\frac{\alpha_{\rm FE} P_{\rm S}}{2} \mathbf{p}_j^2 + \frac{\beta_{\rm FE} P_{\rm S}^3}{4} \mathbf{p}_j^4 + \frac{\kappa_{\rm FE} P_{\rm S}}{2} (\mathbf{p}_{j+1} - \mathbf{p}_j)^2 - \mathbf{p}_j \cdot \mathbf{E}(t) \right), \tag{2}$$

and the ferromagnetic energy part is

$$F_{\rm FM} = \sum_{i=1}^{N_{\rm FM}} \left(-J\, \mathbf{S}_i \cdot \mathbf{S}_{i+1} - D\,(S_i^z)^2 - \boldsymbol{\mu}_i\, \mathbf{S}_i \cdot \mathbf{B}(t) \right), \tag{3}$$

where $\mathbf{E}(t)$ and $\mathbf{B}(t)$ are the time-dependent applied external electric and magnetic fields, respectively. Various parameters and the values of reduced $\mathbf{p}_j$ and $\mathbf{S}_i$ entering eqs. (2) and (3) are discussed in details in Ref. [11]. In general there are several sources for the ME coupling. Here the ME coupling arises due to the accumulation of spin-polarized electrons at the FE/FM interface when the FE subsystem is polarized[17]. Since the electric field does not penetrate into the bulk of FM metals and the induced electric charges decay exponentially away from the interface, the ME effect is confined to a depth on the order of atomic dimensions from the surface. Here, we consider an ideal FM metal, the exchange interaction between the induced surface magnetization and the FM part is thus limited to the first site only. The linear ME-coupling energy is given as

$$E_{\rm c} = \lambda \mathbf{p}_1 \cdot \mathbf{S}_1. \tag{4}$$

with λ being the coupling strength.

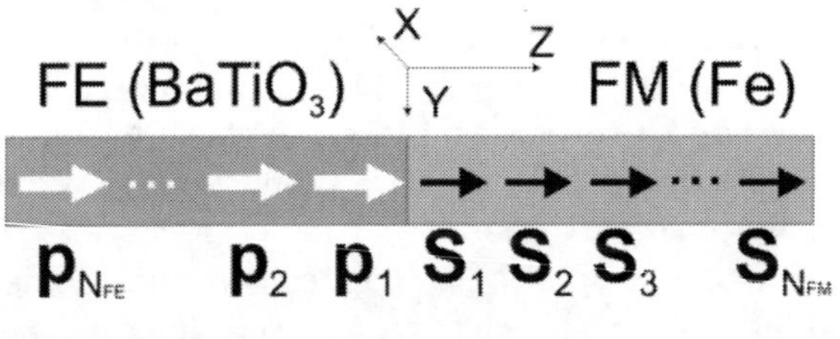

Figure 1. Multiferroic chain consisting of N_{FE} FE polarization sites and N_{FM} localized FM moments.

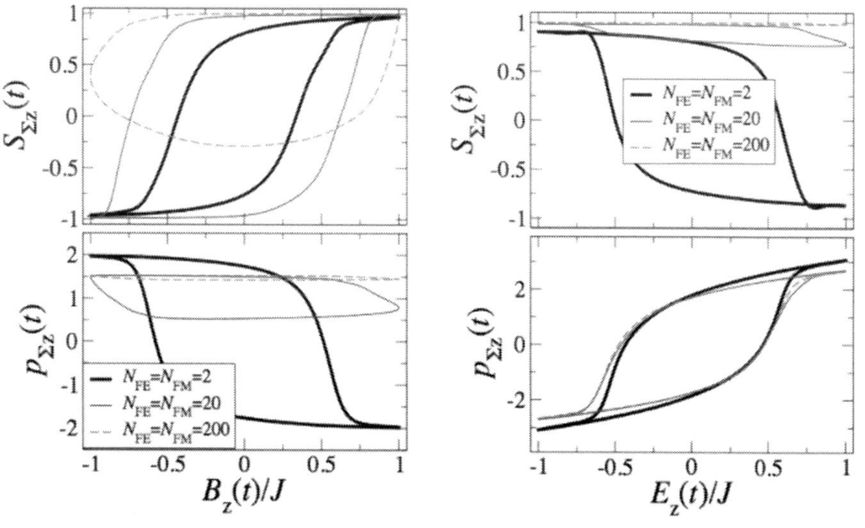

Figure 2. The hysteresis loops of the reduced total polarization/magnetization of different ME chains in the presence of a harmonic external magnetic field $B_z(t) = B_{0z}\cos\omega t$ (*left*) or electric field $E_z(t) = E_{0z}\cos\omega t$ (*right*). The parameters are chosen such, that $\lambda = J$, $\omega = 3.61\times10^{11}\ s^{-1}$, $\alpha_{FM} = 0.5$. Amplitudes of external fields expressed in the energy units are $B_{0z} = J$ and $E_{0z} = 0$ (*left*) and $B_{0z} = 0$ and $E_{0z} = J$ (*right*).

The magnetization dynamics proceeds according to the zero Kelvin Landau-Lifshitz-Gilbert[13] (LLG) equation of motion

$$\frac{d\mathbf{S}_i}{dt} = -\frac{\gamma}{1+\alpha^2{}_{\rm FM}}\left[\mathbf{S}_i \times \mathbf{H}_i^{\rm FM}(t)\right] - \frac{\alpha_{\rm FM}\gamma}{1+\alpha^2{}_{\rm FM}}\left[\mathbf{S}_i \times \left[\mathbf{S}_i \times \mathbf{H}_i^{\rm FM}(t)\right]\right], \tag{5}$$

where $\gamma = 1.76\cdot10^{11}\ (T\ s)^{-1}$ is a gyromagnetic ratio, $\alpha_{\rm FM}$ a Gilbert damping parameter. The total effective field acting on $\mathbf{S}_i$ is defined as $\mathbf{H}_i^{\rm FM}(t) = -\frac{1}{\mu_i}\frac{\delta F_\Sigma}{\delta \mathbf{S}_i}$.
The polarization dynamics is governed by the Landau-Khalatnikov (LKh) equation as it was considered in the literature [12] for ferroelectric materials

$$\gamma_\nu P_{\rm S}\frac{d\mathbf{p}_j}{dt} = \mathbf{H}_j^{\rm FE} = -\frac{1}{P_{\rm S}}\frac{\delta F_\Sigma}{\delta \mathbf{p}_j}, \tag{6}$$

where γ_ν is the viscosity constant. Numerically both eqs. (5) and (6) are solved in reduced units, renormalizing the energy (1) over double anisotropy strength $2D$. Then, the dimensionless time in both equations is $\tau = \omega_{\rm A}\,t = \gamma\,B_{\rm A}\,t = \gamma\,2D\,t/\mu_{\rm S}$ and the reduced effective fields are $\mathbf{h}_i^{\rm FM}(\tau) = \mathbf{H}_i^{\rm FM}(\tau)/B_{\rm A}$, $\mathbf{h}_j^{\rm FE} = \mathbf{H}_j^{\rm FE}/(\gamma\gamma_\nu P_{\rm S} B_{\rm A})$.

3. Results and discussions

The multiferroic response to external fields is investigated in two cases: (I) FE/FM chains with different lengths, $N_{FE} = N_{FM} = 2$, 20, and 200; (II) the first n_i ($n_i = 1$, 2 and 20) polarization sites and magnetic moments around the interface of the long chain consisting of $N_{FE} = N_{FM} = 200$ sites in total. Temperature is set to zero. This ensures the FM-subsystem remains away from its superparamagnetic limit and the FE-subsystem to be far from its phase transition temperature at nearly $T = 10$ K for the whole time scale of simulations.

Aiming at fast switching devices we apply harmonic fast oscillating magnetic and electric fields, as in Ref. [11]. We stress, however, that the switching by a magnetic field is not due to the resonance absorption of the magnetic field, as described in Refs. [18]. The strong magnetic field induces changes of the potential barrier compared with the height of it.

In the case (I) with different sizes of heterostructures, the system is described via the reduced net magnetization $\mathbf{S}_\Sigma(t)$ and reduced total polarization $\mathbf{p}_\Sigma(t)$

$$\mathbf{S}_\Sigma(t) = \frac{1}{N_{\rm FM}} \sum_{i=1}^{N_{\rm FM}} \mathbf{S}_i(t), \quad \mathbf{p}_\Sigma(t) = \frac{1}{N_{\rm FE}} \sum_{j=1}^{N_{\rm FE}} \mathbf{p}_j(t). \tag{7}$$

Fig. 2 shows the ME response of the total reduced polarization/magnetization to harmonic magnetic field $B_z(t) = B_{0z}\cos\omega t$ and electric field $E_z(t) = E_{0z}\cos\omega t$, respectively. Clearly, the FE polarization (FM magnetization) is completely switched by the external electric (magnetic) field. Additionally, the hysteresis loops indirectly driven by the applied field due to the ME

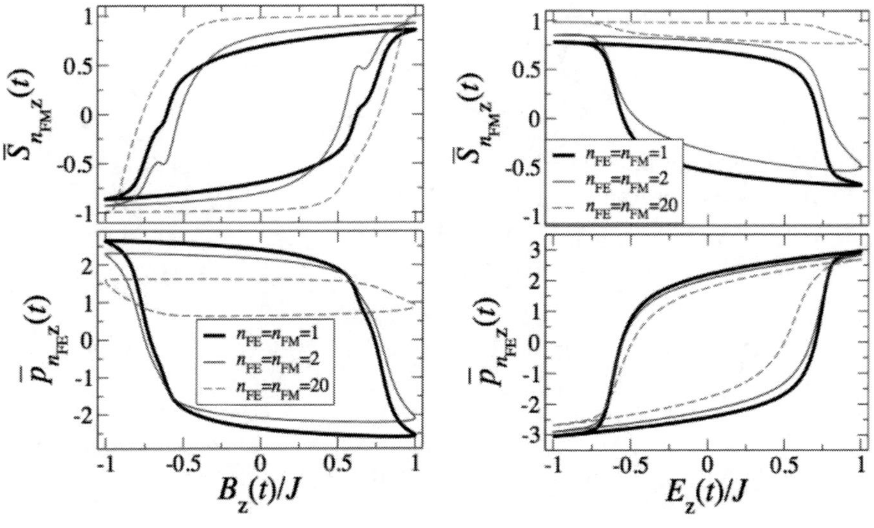

Figure 3. The hysteresis loops of the renormalized polarization/magnetization of the first n_i (n_i =1, 2 or 20) FE and FM sites as a function of a harmonic external magnetic field $B_z(t)$ (*left*) or electric field $E_z(t)$ (*right*) in the ME chain consisting of 200 polarization sites and 200 magnetic moments. The parameters are chosen to be the same as in Fig. 2.

coupling are presented. For the long ME chain, such as $N_{FE} = N_{FM} = 200$, the total FE polarization (magnetization) can not be reversed by the magnetic (electric) field since the active area of the ME effect is limited to the FE/FM interface only. A way around this problem is a multilayer stacking of alternating FE and FM structures coupled at their interfaces.

In order to investigate whether only the several first polarization (spin) sites in the long ME chain follow the magnetic (electric) field, the first n_i (n_i= 1, 2 and 20) sites are extracted from the $N_{FE} = N_{FM} = 200$ chain. For convenience, we define the following renormalized parameters

$$\bar{\mathbf{S}}_{n_{FM}}(t) = \frac{1}{n_{FM}}\sum_{i=1}^{n_{FM}} \mathbf{S}_i(t), \quad \bar{\mathbf{p}}_{n_{FE}}(t) = \frac{1}{n_{FE}}\sum_{j=1}^{n_{FE}} \mathbf{p}_j(t). \tag{8}$$

The dynamical behavior is shown in Fig. 3. Although the total FE polarization (FM magnetization) driven indirectly by the external magnetic (electric) field, is not completely switched, it is indeed observed that the several first sites react strongly to the applied fields through the interface ME coupling.

4. Conclusion

In summary, the dynamical multiferroicity in a one-dimensional FE/FM heterostructure is investigated by means of the coupled LLG and LKh method as a function of the length of the heterostructure. These results are supported by our kinetic Monte-Carlo simulations, not shown here for brevity. We find that the FE polarization and the magnetization are switchable for thin MF-chains (e.g. $N_{FE} = N_{FM} = 1, 2$ (Fig. 2) and up to 10) by applied magnetic and electric fields, respectively. The complete MF-switching mediated by the multiferroic coupling is not achievable for larger MF-chains (e.g. for $N_{FE} = N_{FM} = 20$, as inferred from Fig. 2). Finally, we mention that there are several additional effects that may affect the multiferroic dynamics. In Ref. [11], the effect of the depolarizing fields and the effect of the induced electric (magnetic) fields are discussed in some details. These effects are quite small for the $BaTiO_3$/Fe chain and for the chosen range of frequencies. The temperature- and the frequency-dependence of the dynamical response deserves further considerations and is currently under study.

References

[1] W. Eerenstein, N. D. Mathur, and J. F. Scott, Nature(London) **442**, 759 (2006); L.W. Martin, S. P. Crane, Y-H Chu, M. B. Holcomb, M. Gajek, M. Huijben, C-H Yang, N. Balke and R. Ramesh, J. Phys.: Condens. Matter **20** 434220 (2008).
[2] T. Lottermoser, T. Lonkai, U. Amann, D. Hohlwein, J. Ihringer, and. M. Fiebig, Nature (London) **430**, 541 (2004).
[3] T. Kimura, T. Goto, H. Shintani, K. Ishizaka, T. Arima, and Y. Tokura, Nature (London) **426**, 55 (2003).
[4] P. Borisov, A. Hochstrat, X. Chen, W. Kleemann, and C. Binek, Phys. Rev. Lett. **94**, 117203 (2005).
[5] M. Weisheit, S. Fähler, A. Marty. Y. Souche, C. Poinsignon, and D. Givord, Science **315**, 349 (2007).
[6] E. Y. Tsymbal and H. Kohlstedt, Science **313**, 181 (2006).
[7] Y. Tokura and S. Seki, Adv. Mater. **21**, 1(2009).
[8] C. Jia and J. Berakdar, Phys. Rev. B **80**, 014432 (2009).
[9] C. Jia and J. Berakdar, Appl. Phys. Lett. **95**, 012105 (2009); Proc. Spie **7603**, 76030O (2010).
[10] C.-G. Duan, S. S. Jaswal, and E. Y. Tsymbal, Phys. Rev. Lett. **97**, 047201 (2006); J. M. Rondinelli, M. Stengel, and N. A. Spaldin, Nat. Nanotechnol. **3**, 46 (2008); M. Fechner, I. V. Maznichenko, S. Ostanin, A. Ernst, J. Henk, P. Bruno, and I. Mertig, Phys. Rev. B **78**, 212406 (2008); M. K. Niranjan, J. P. Velev, C.-G. Duan, S. S. Jaswal, and E. Y. Tsymbal, Phys. Rev. B **78**, 104405 (2008).
[11] A. Sukhov, C.L. Jia, P. P. Horley and J. Berakdar, J. Phys.: Condens. Matter **22** 352201 (2010).
[12] D. Ricinschi, C. Harnagea, C. Papusoi, L. Mitoseriu, V. Tura and M. Okuyama, J. Phys.: Condens. Matter **10**, 477 (1998).
[13] L. Landau and E. Lifshitz, Phys. Z. Sowjetunion **8**, 153 (1935); T.L. Gilbert Phys. Rev. **100**, 1243 (1955). Phys. Rev. Lett. **102**, 057204 (2009); Phys. Rev. B **79**, 134433 (2009).

[14] C.-G. Duan, S. S. Jaswal, and E. Y. Tsymbal, Phys. Rev. Lett. **97**, 047201 (2006).

[15] M. Fechner, I. V. Maznichenko, S. Ostanin, A. Ernst, J. Henk, P. Bruno, and I. Mertig, Phys. Rev. B **78**, 212406 (2008).

[16] S. Sahoo, S. Polisetty, C.-G. Duan, S. S. Jaswal, E. Y. Tsymbal, and C. Binek, Phys. Rev. B **76**, 092108 (2007).

[17] T. Cai, S. Ju, J. Lee, N. Sai, A. A. Demkov, Q. Niu, Z. Li, J. Shi, and E. Wang, Phys. Rev. B **80**, 140415(R) (2009).

[18] C. Thirion, W. Wernsdorfer and D. Mailly, Nat. Mater. **2**, 524 (2003); Z. Z. Sun and X. R. Wang, Phys. Rev. B **74**, 132401 (2006); A. Sukhov, and J. Berakdar, J. Phys.: Condens. Matter **20**, 125226 (2008),

Joint European Magnetic Symposia – JEMS 2010 IOP Publishing
Journal of Physics: Conference Series **303** (2011) 012062 doi:10.1088/1742-6596/303/1/012062

Influence of magnetic heterogeneous state on magnetocaloric effect in $Sm_{0.55}Sr_{0.45}MnO_3$

L I Koroleva[1], D M Zashichirinskii, A S Morozov

M.V. Lomonosov Moscow State University, Vorobyevy Gory, 119992 Moscow, Russia

E-mail[1]: koroleva@phys.msu.ru

Abstract. Magnetocaloric effect (ΔT-effect) was studied by straight method in two samples of $Sm_{0.55}Sr_{0.45}MnO_3$ manganite: ceramic sample and single crystal, annealed in air atmosphere. The temperature dependence of ΔT-effect $\Delta T(T)$ of both of these samples has maximum at T_{max} equal to 143.3 K for single crystal sample and 143 K for ceramic sample. In these maxima ΔT-values are 0.8 K for the former and 0.4 K for second samples. In addition the $\Delta T(T)$ curve of single-crystal sample has minimum at T_{min} = 120 K and ΔT-value in minimum equal to - 0.1 K. The maximum value of ΔT-effect increases with H up to the maximum field of measurement 14.2 kOe. At this takes place the rate of this increasing is more at H > 8 kOe than at H < 8 kOe. The above listed peculiarities of ΔT-effect are explained of presence in samples of ferromagnetic, antiferromagnetic A-type and antiferromagnetic CE-type clusters.

Introduction

The temperature change induced in magnetic material during an adiabatic magnetic field change is the magnetocaloric effect (MCE). MCE provides the physical basis of magnetic refrigeration technology. Magnetic cooling has a potential to reduce global energy consumption, minimize the need for ozone depleting and green house chemicals, and may soon become an alternative to the vapor-compression approach. In our work we studied MCE in $Sm_{0.55}Sr_{0.45}MnO_3$ manganite.

Experimental Details

MCE was studied on two $Sm_{0.55}Sr_{0.45}MnO_3$ samples. The ceramic sample was prepared by standard ceramic techniques and single crystal was grown by the floating-zone method. Single crystal was annealed in air atmosphere. Using X-ray diffraction, the samples were found to be pure single-phase perovskite with the orthorhombic Pnma structure. Two cube-shaped samples were cut from ceramic and single-crystal sample. The sides of single-crystal sample were parallel to *c*-axis and *ab*-plane. MCE was measured by straight method. The magnetic field H for quasistatic measurement was ranged from 0 to 14.2 kOe and was created by an electromagnet. Due to relatively large magnetic induction, the time of a field sweep from 0 to 14.2 kOe was t > 4 s.

The equilibrium temperature T of the sample was measured using copper-constantan thermocouple before and after the sweep. The thermocouple joint was put inside the sample. The MCE was determined as the difference between these two equilibrium temperatures. The MCE-values of single-crystal sample were measured when the field was applied along *c*-axis and *ab*-plane. They are the same. It is known that the value of MCE is connected with the magnetization M and the specific heat c the in the following way:

$$\Delta T = -\int_0^{H_{max}} \frac{T}{c_H}\left(\frac{dM}{dT}\right)_H dH \qquad (1)$$

The large value of the specific heat of manganites (relative to gadolinium) results in reduction of ΔT, that is a source of a large error. So it is necessary to minimize the heat losses. It was reached by high vacuum (not less than 10-5 mm Hg), heat screen around the sample and the use of super-thin detector wires. The experimental setup was tested with the gadolinium sample with 99.99% purity.

Results and Discussion

Figures show the temperature dependence of MCE $\Delta T(T)$ in magnetic field 14.2 kOe of the single-crystal (figure 1) and ceramic (figure 2) samples. As we can see from figure 1 the $\Delta T(T)$ curves has maximum at T_{max} = 143.3 K and minimum at T_{min} = 120 K with ΔT = 0.8 K in maximum and ΔT = -0.1 K in minimum. The maximum on the $\Delta T(T)$ curve is observed too in the ceramic sample at T_{max} = 143 K. In this maximum MCE value is 0.41 K. In this time the negative MCE is absent on the $\Delta T(T)$ curve of ceramic sample.

The large value of the specific heat of manganites (relative to gadolinium) results in reduction of ΔT, that is a source of a large error. So it is necessary to minimize the heat losses. It was reached by high vacuum (not less than 10^{-5} mm Hg), heat screen around the sample and the use of super-thin detector wires. The experimental setup was tested with the gadolinium sample with 99.99% purity.

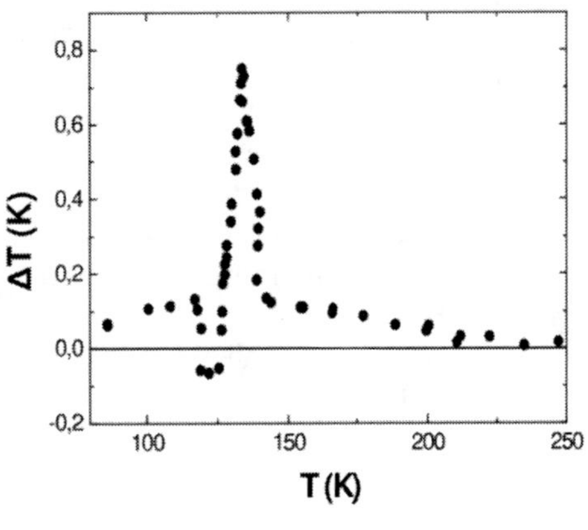

Figure 1. The temperature dependence of magnetocaloric effect of $Sm_{0.55}Sr_{0.45}MnO_3$ single crystal, annealed in air atmosphere, in magnetic field equal to 14.2 kOe.

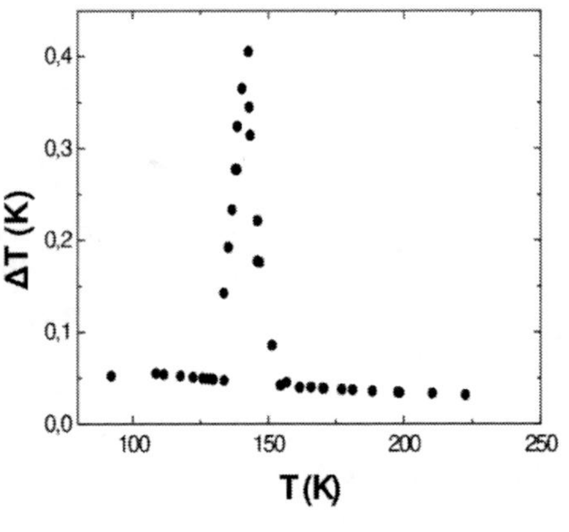

Figure 2. The temperature dependence of magnetocaloric effect of $Sm_{0.55}Sr_{0.45}MnO_3$ ceramic in magnetic field equal to 14.2 kOe.

Figure 3 shows the dependence of ΔT versus magnetic field H at T = 143 K of single-crystal sample. Similar dependence of ΔT versus H at T_{max} is observed in ceramic sample too.

Recent neutron and electron diffraction studies of $Sm_{0.5+x}Sr_{0.5-x}MnO_3$ ($0 \leq x \leq 0.1$) [1, 2] have shown that a magnetic inhomogeneous state consisting of ferromagnetic (FM) clusters with Curie point T_C, A-type antiferromagnetic (AFM) clusters with temperature Neel T_{NA} and CE-type AFM clusters with temperature Neel T_{NCE} is present in it. Moreover the next relation occurs: $T_C \leq T_{NA} < T_{NCE}$. It is shows [3] that these compounds exhibit a jamplike increase in the magnetization in a certain interval of field strength $H_{C1} < H < H_{C2}$ at $T > T_C$ in which the sample transfers to FM state. It is found the above mentioned interval 8 < H < 14.2 kOe disposes in region $H_{C1} < H < H_{C2}$, and $H_{C2} >$ 14.2 kOe.

The T_C-value, determined by us from the DC magnetization measurements in magnetic field of 100 Oe, is 132 K. The study of paramagnetic susceptibility shows that it obeys the Curie-Weiss law at temperatures far above T_C with the paramagnetic Curie points θ = 228 K (single crystal) and 234 K (ceramic) respectively. This is evidence that in region from T_C to θ the abovementioned AFM clusters present.

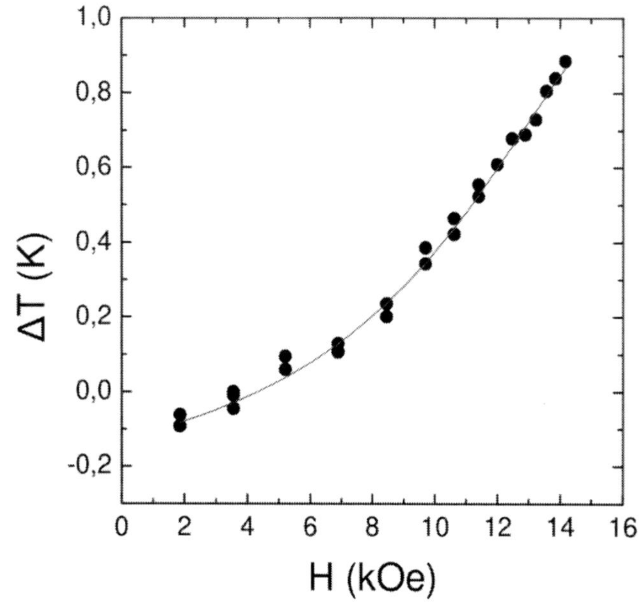

Figure 3. The magnetic field dependence of magnetocaloric effect of $Sm_{0.55}Sr_{0.45}MnO_3$ single crystal, annealed in air atmosphere, at temperature equal to 143 K.

As it shown in ration (1), which describes MCE in the paraprocess region, MCE is positive at destruction of FM ordering and negative at destruction of AFM ordering. Obviously the maximum on the $\Delta T(T)$ curve is connected with the sharp decrease of the magnetization of FM clusters in the T_C - region whereas the destruction of AFM ordering in clusters with the AFM ordering A-type causes the little minimum start. It is possible that latter destruction is continued in the region of the maximum on the $\Delta T(T)$ curve decreasing it. It is apparent that the contribution from AFM clusters A-type decreases the total MCE of $Sm_{0.55}Sr_{0.45}MnO_3$; this decreasing is smaller in single crystal than in ceramic sample. This clearly demonstrates that technology of preparation of samples may be modified the relation between FM and AFM portions of sample. As suggested in figure 3 the maximum value of MCE increases with H up to the maximum field of measurement 14.2 kOe. At this takes place the rate of this increasing is more at H > 8 kOe than at H < 8 kOe. In reference [3] shows that the identical compound exhibits a jamplike increase in the magnetization in a certain interval of field strength $H_{C1} < H < H_{C2}$ at $T > T_C$ in which the clusters with AFM order of CE-type transfers to FM state. Figure 3 shows that the above mentioned interval $8 < H < 14.2$ kOe disposes in region $H_{C1} < H < H_{C2}$, and $H_{C2} > 14.2$ kOe. By this is meant that the maximum value of MCE can be increases with H at 8 kOe $\leq H \leq H_{C2}$ at the sacrifice of FM phase formed from the AFM CE-type phase.

Conclusion

In $Sm_{0.55}Sr_{0.45}MnO_3$ manganite the positive magnetocaloric effect is found which maximum is reached: 0.8 K in H = 14.2 kOe and at the temperature T_{max}. As this takes place T_S –value is a trifle over T_C. This positive MCE represents the sum of MCE fractional quantities over all the FM, A-type AFM and CE-type AFM clusters.

References

[1] Runov V V, Chernyshev D Yu, Kurbakov A I, Runova M K, Trunov V A and Okorokov A I 2000 *J. Exp. and Theor. Phys.* **118** 1174
[2] Martin C, Maignan A, Hervieu M and Raveau B 1999 *Phys. Rev.* B **60** 12191
[3] Abramovich A I, Koroleva L I and Michurin A V 2002 *J. Exp. and Theor. Phys.* **95** 917

Noise and electric field characterization of irradiated $SrTiO_3$

R. Guerrero[1,3 a], A. Solignac[1], M. Pannetier-Lecoeur[1], C. Fermon[1], P. Auban-Senzier[2], L. Lemberger[2], C. Pasquier[2], Y. Apertet[3], Ph. Lecoeur[3]

[1] DSM/IRAMIS/SPEC- CNRS URA 2464, CEA Saclay, 91191 Gif sur Yvette Cedex France

[2] Laboratoire de Physique des Solides, Université Paris-Sud, CNRS, 91405 Orsay, France

[3] Institut d'Electronique Fondamentale, Université Paris-Sud, CNRS, 91405 Orsay, France

[a] Corresponding author: Ruben.guerrero@cea.fr

Abstract. SrTiO3 (STO) is a band insulator with a perovskite crystallographic structure, widely used as substrate for the deposition of other perovskite materials, for instance manganites or cuprates. STO surface becomes conductive under Ar^+ ion irradiation, exhibiting similar properties to STO/$LaAlO_3$ interface or bulk reduced STO. We report the transport and noise properties as function of temperature and electric field of Ar^+ irradiated STO exhibiting at its surface a high mobility electron gas. This system shows metallic properties with low charge densities allowing mobility measurements. The low frequency 1/f noise presents an unexpected linear dependence with the mobility which cannot be described by the usual Hooge model. The effect of the electric field at a given temperature on the mobility, magnetoresistance and low frequency noise is also presented.

1. Introduction

Oxide based electronics is a new emergent field that shows a great variety of new applications, ranging from magnetism to high temperature superconductivity. Among them $SrTiO_3$ (STO) is one of the most interesting systems. It is a band insulator with a perovskite crystallographic structure, which has been widely used as substrate for other perovskites, for example manganites or high critical temperature superconductors (cuprates). Recently, field effect transistors have been developed based on this material [1-4]. Due to the possibility of being metallic, while it is transparent, it has been proposed as an interesting component for optoelectronic devices [5]. Besides that, the possibility of controlling the growing processes of STO and its surface has enabled the development of a high mobility electron gas in the interface of STO and $LaAlO_3$ (LAO) [6]. Even if the origin of this phenomenon is unclear [7-9], the properties of the interface are very interesting and promising for new potential applications.

In this work we present results on transport on Ar^+ irradiated STO. Under Ar^+ irradiation a nanometric layer of the material becomes conductive [5,10,11] exhibiting similar properties than STO/LAO interfaces [6] and bulk reduced STO [12-15]. Although those materials have been extensively studied, the irradiated STO properties have been poorly studied. Here we present in a first part general characterization using DC transport noise measurements. Secondly we study the effect of an electric field applied through a backgate in the transport characteristics. Finally we investigate the magnetotransport in the studied samples.

2. Sample preparation and experimental techniques.

Samples were prepared using commercial STO substrates from Crystec, with dimensions 10mmx10mmx0.5mm and oriented along the (100) direction. On top of the substrate we defined several lines of different widths: 10, 20, 50 and 200μm using standard photo-lithography. After that, we irradiated the substrate during 15 minutes in order to define lines of conductive STO. The irradiation was done using an Ar plasma accelerated by 0.7kV, inducing a current of 8mA incident on the surface. In order to protect the samples from a long term re-oxygenation we covered the irradiated surface with a 25nm thick layer of Si_3N_4. This procedure maintains the properties of the samples during months when they are stocked in ambient conditions. Finally we deposited contacts on the top of the structure using a sputtering machine. Six contacts were deposited: 2 for current, 2 for longitudinal voltage and other two for Hall effect measurements. The contacts were composed of Ta(10nm)/Cu(150 nm)/Ta(10 nm) . All the samples were contacted to a sample holder through Al wire bonding in order to perform their characterization.

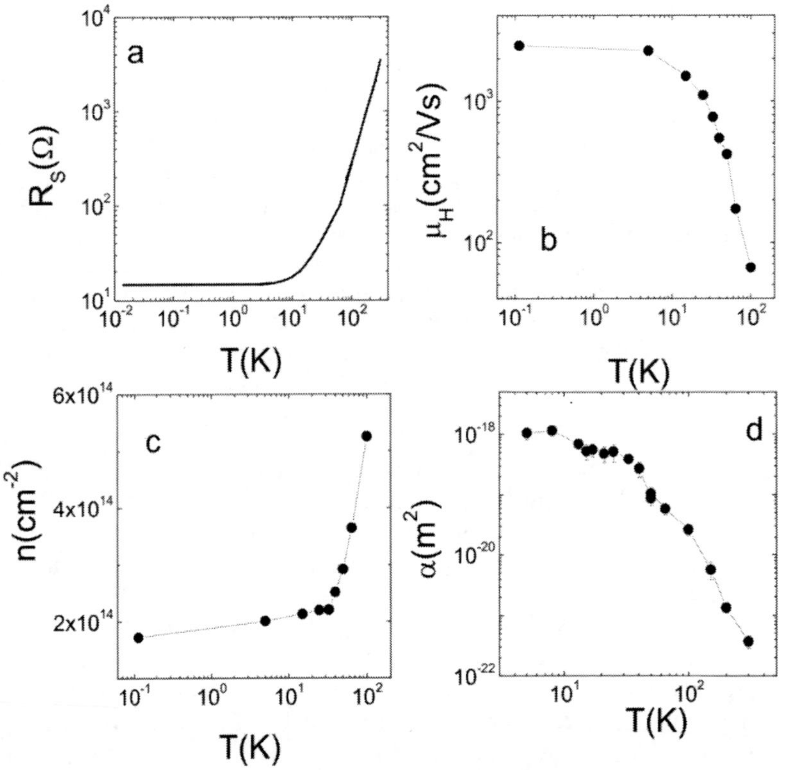

Figure 1 : (a): Dependence on temperature of the sheet resistance of one of the studied samples; there is no indication of any superconducting transition down to 30mK. b): Hall mobility of the irradiated film, the metallic like temperature dependence shown in a) is due to the increased mobility. The number of carriers (c) is diminished at low temperatures. This behavior is related to the thermally activated character of the generation of carriers. (d): Normalized noise as a function of the temperature.

Transport measurements were carried out in a He cryostat. A temperature controller allows us stabilizing the temperature between 200K and 4K. Resistance measurements were realized using the four probes method. We use a Keithley 6221 current source to bias our samples, the voltage was detected using a DC amplifier and an acquisition card. Hall effect measurements were done in a magnetic field up to 1T using the same current source and a differential amplifier. In order to perform the noise measurement we used an AC coupled low noise amplifier and a battery fed current source. We also performed high field measurement measurements (up to 8T) in a dilution cryostat with a base temperature of 30mK. In this case a lock-amplifier was used to measure the resistance of the sample and the magnetic field was measured using a calibrated Hall probe.

3. Experimental results

3.1. Transport measurements and noise

In the patterned samples we have performed resistance measurements as function of the temperature. The dependence in all the studied samples was metallic-like, this behavior being attributed to the increase of the mobility when the sample temperature was diminished, instead of due to the increment of the number of carriers. This data are displayed in Fig. 1, panels a,b and c.

Reduced bulk STO exhibit a superconducting transition at temperatures below 0.5K. In this material the critical temperature (T_C) depends on the density of carrier. T_C is maximal at about 10^{-20} $cm^{-3}$and it extend about one order of magnitude above and below that value. Due to the pairing mechanism T_C diminishes for carrier densities one order of magnitude above and below the maximum one [16]. In our samples we have estimated the bulk carrier density in about $10^{-20}cm^{-3}$. To obtain such a data we used the irradiation parameters to estimate the thickness of our conducting layer, which yields a thickness of about 10nm. Therefore we would expect to observe a superconducting transition at about 0.5K. Nevertheless, such a transition is not observed down to 30mK. In our opinion the disorder induced during the irradiation process might destroy the superconductivity. We cannot rule out the possibility that the extension of the conductive region is in the μm range as it is proposed in ref. [17]. In this case the volume density of carriers would be smaller than the minimum density required to observe a superconducting transition.

The noise in the studied samples is mainly of 1/f type. We have not found any generation-recombination noise, which usually exhibits a Lorentzian like dependence on the frequency. In order to characterize the noise in each temperature we use the parameter α, defined as:

$$\alpha = \frac{S_V(f)A}{V^2}$$

Where V is the voltage in the sample and A is the corresponding area. This parameter normalizes the 1/f noise and allows us comparing the noise at different temperatures. Interestingly the noise increases when the temperature decreases and is directly correlated with the mobility. A full discussion about this behavior has been published in [18].

3.2. Effect of the electric field

In order to test further the low temperature regime we apply an electric field in the samples. We do that by placing a counter electrode in our substrates that allow us applying a voltage up to 350V. In figure 2 we present the dependence of the resistance (a), the mobility (b) and the surface density of carriers (inset) as a function of the backgate voltage ($V_{backgate}$).

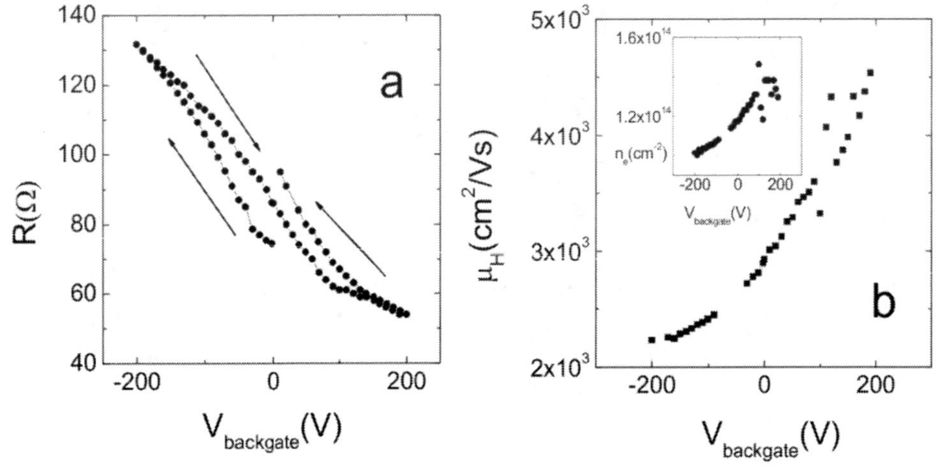

Figure 2: Dependence on the backgate voltage of the resistance (a) and the mobility (b). The resistance graph shows several consecutive sweeps, the arrows show the direction of the sweep. The non reproducibility of the resistance could be explained by diffusion and pinning of oxygen vacancies, thus making the effect of applied electric field non reversible. Graph b and the inset shows the Hall mobility and surface density of carriers obtained in one sweep.

Firstly we can observe that the resistance of the conducting layer does not return back to the 0V value after several sweeps. This effect has been already reported in STO [2] and it is attributed to the movement of oxygen vacancies in the substrate/conducting layer system, the movement changes locally the electric field felt by the conduction electrons, thus changing the resistance in a different way each sweep of the electric field. In the figure2a it is clear that the resistance of the samples decreases as $V_{backgate}$ increases. This change in resistance is due to the increase in the mobility, Fig. 2b, whereas the number of carriers is slightly modified, inset in Fig 2b.

3.3. Magnetoresistance

At low temperatures the samples exhibit a positive magnetoresistance (MR) when the magnetic field is applied perpendicular to the substrate. An example of the observed perpendicular MR is shown in Fig. 3a. At low magnetic field (H<1T) the dependence of MR with the magnetic field is proportional to μ^2H^2, which is the usual behavior in metals when the product $\omega_c\tau<<1$ [19], where ω_c is the cyclotron frequency and τ is the relaxation time. This has been tested using the Kohler law, which is followed down to 5K. At high magnetic fields the behavior of the magnetoresistance depends on the material and the crystalline orientation. In our case the magnetoresistance seems to saturate into a linear regime. Typically this effect happens when the number of electrons (holes) is much bigger than the number of holes (electrons) [19].

In the studied samples we can also notice that we do not observe Shubnikov-de Haas (SdH) oscillations, like it has been been reported in bulk reduced and Nb doped STO [20], or in STO/LAO systems [21] and high mobility STO [22]. Our samples do not exhibit such a behavior down to temperatures of 100mK as can be seen on Fig. 3a. This oscillations in the resistance as function of the magnetic field are due to the splitting of the Landau levels, given by $\delta E=\hbar\omega_c$. In order to be observable at reasonably low magnetic fields the temperature should be small and the product $\omega_c\tau>>1$, those conditions disable the smearing of the different levels due to the thermal energy and the disorder, respectively. In our case seems that at 8T the smearing of the levels due to the collisions is

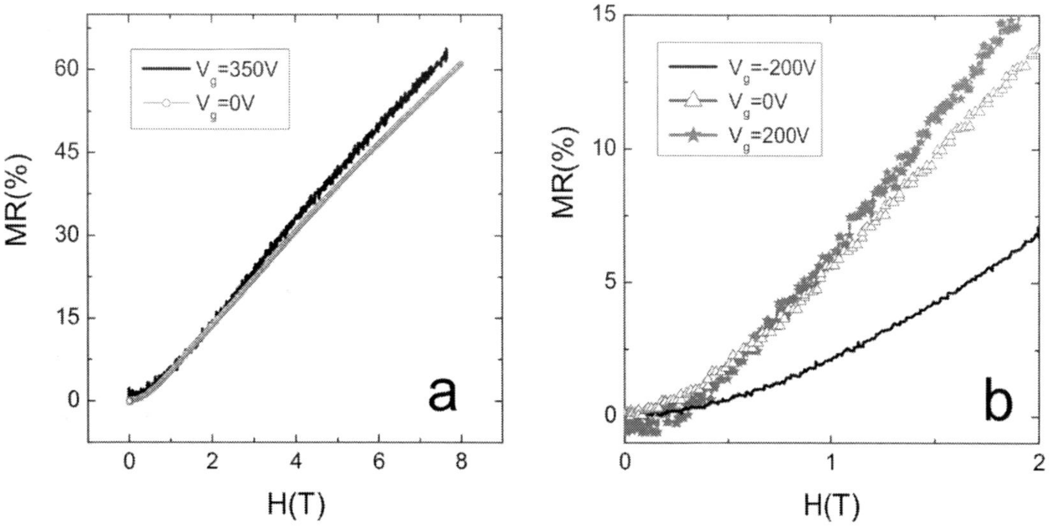

Figure 3: Magnetoresistance of a STO irradiated sample in perpendicular magnetic field at 100mK. At low field (b) the parabolic behavior is evident while at H>1T (a) the MR trends to a linear behavior. In order to increase the mobility we applied an electric field by using the backgate. As we increase the mobility the magnetoresistance increases, however at high field the SdH oscillations still are not visible.

big enough to attenuate the oscillations. Considering that the mobility in sample shown is about 2500cm^2/Vs, at 8T the factor $\omega_c\tau\approx1.8$ which is not enough to observe the SdH oscillations. An applied magnetic field up to 25T would be then necessary to see the expected behavior. The increase of the mobility by applying 350V in the backgate is still not enough to observe such phenomenon.

4. Conclusions

In conclusion we have studied in detail the transport in irradiated STO. Changes in the resistance are improved by the enhancement of the mobility rather than the generation of carriers, both as function of the temperature and the use of a backgate. The magnetotransport measurements display a behaviour that is proportional to μ^2H^2 at low field, whereas at high field it trends to a linear regime. SdH oscillations have not been observed in fields up to 8T, probably due to the relatively low mobility.

References

[1] Newns D M, Misewich, Tsuei C C, Gupta A. Scott B A, and Schrott A, 1998 *App. Phys. Lett.* **73**, 780.
[2] Bellingeri E, Pellegrino L, Marre D, Pallecchi I, and Siri A S, 2003 *Journal of Applied Physics* **94**, 5976.
[3] Pallecchi I, Grassano G, Marre D, Pellegrino L., Putti M. and Siri A S, 2001 *App. Phys. Lett.* **78**, 2244.
[4] Ueno K, Inoue I H. Akoh H, Kawasaki M, Tokura Y and Takagid H, 2003 *App. Phys. Lett.*, **83**, 1755.
[5] Reagor D W and Butko V Y, 2005 *Nat. Mat.* **4**, 593.
[6] Ohtomo A and Hwang H Y, 2004 *Nature* **427**, 423.
[7] Kalabukhov A, Gunnarsson R, Börjesson J, Olsson E, Claeson T and Winkler D, 2007 *Phys. Rev. B*, **75**, 121404R.
[8] Siemons W, Koster G, Yamamoto H, Harrison W A, Lucovsky G, Geballe T H, Blank D H A and Beasly M R, 2007 *Phys. Rev. Lett.* **98**, 196802.
[9] Siemons W, Koster G, Yamamoto H. Geballe T H, Blank D H A, Beasley M R, 2007 Phys Rev. *B*, **76**, 155111.
[10] Schultz M and Klein L, 2007 *App. Phys. Lett.* **91**, 151104.
[11] Kan D, Terashima T, Kanda R, Masuno A, Tanaka K, Chu S, Kan K, Ishizumi A, Kanemitsu Y. Shimakawa Y and Takano M., 2005 *Nat. Mat.*, **4**, 816.
[12] Frederikse H P R. Thurber W R and Hosler W R, 1964 *Phys. Rev.* **134**, A442.
[13] Frederikse H P R and Hosler W R, 1967 *Phys. Rev.* **161**, 822.
[14] Tufte O N and Chapman P W, 1967 *Phys. Rev.* **155**, 796.
[15] Lee C, Yahia J and Brebner L, 1971 *Phys. Rev. B*, **3**, 2525.
[16] Schooley J F, Hosler W R, Ambler E and Becker J H, 1965 *Phys.* Rev. Lett. **14**, 305. (1965).
[17] Herranz G, Copie O, Gentils A, Tafra E, Basletić M, Fortuna F, Bouzehouane K, Fusil S, Jacquet É, Carrétéro C, Bibes M, Hamzić A, and Barthélémy A, 2010 *Journal of App. Phys.* **107**, 103704.
[18] Guerrero R, Solignac A, Pannetier-Lecoeur M, Apertet Y, Lecoeur P and Fermon C, 2010 *Phys. Rev. B*, **81**, 035102.
[19] Kittel C. 1963 *Quantum Theory of Solids*, (New York: John Wiley and sons), ch 12.
[20] Frederikse H P R. Hosler W R. Thurber W R, Babiskin J and Siebenmann P G, 1967 *Phys. Rev.* **158**, 775.
[21] Herranz G, Basletic M. Bibes M, Carretero C, Tafra E. Jacquet E, Bouzehouane K., Deranlot C, Hamzic A, Broto J M, Barthelemy A and Fert A, 2007 Phys. Rev. Lett. **98,** 216803.
[22] Herranz G, Basletic M. Bibes M., Ranchal R, Hamzic A, Tafra E, Bouzehouane K, Jacquet E, Contour J P, Barthélémy A, and Fert A, 2006 Phys. Rev. B **73**, 064403.

Joint European Magnetic Symposia – JEMS 2010 IOP Publishing
Journal of Physics: Conference Series **303** (2011) 012064 doi:10.1088/1742-6596/303/1/012064

Experimental Evidence for Ising Spin-Glass Transition in the $YbCoGaO_4$ Single Crystal

R Szymczak[1], H A Dabkowska[2], A Dabkowski[2], G M Luke[2], P Aleshkevych[1], J Fink-Finowicki[1], H Szymczak[1,3]

[1] Institute of Physics, Polish Academy of Sciences, 02668 Warsaw, Poland

[2] The Department of Physics and Astronomy, McMaster University, Hamilton, Ontario, Canada

E-mail: szymh@ifpan.edu.pl

Abstract. Magnetic measurements (ac and dc susceptibilities and magnetization) on $YbCoGaO_4$ single crystal as a function of frequency and temperature are presented. The studied system behaves like a three-dimensional Ising-like spin glass. A cusp is seen only in the longitudinal susceptibility, whereas a paramagnetic behavior is observed in the transverse susceptibility. Typical spin glass features are observed in the frequency and temperature dependence of the ac susceptibility. Dynamical scaling is performed and critical exponents are deduced and compared to those obtained for Ising and Heisenberg systems. The origin of the Ising anisotropy of Yb^{3+} and Co^{2+}ions is briefly discussed.

1. Introduction

In the present article, we report the magnetic properties of a single crystal of $YbCoGaO_4$ layered cobaltite. Its crystal structure is described by space group R-3m [1,2]. It consists of double layers of Co/GaO_5 triangular bipyramids and triangular layers of YbO_6 octahedra. In the $YbCoGaO_4$ structure Co^{2+} and Ga^{3+} cations are distributed randomly on the sites of the bilayers and consequently this crystal is both spin and charge frustrated. Moreover, this crystal is unusual in that it has two magnetic sublattices, both of which have triangular geometries. All these reasons are expected to lead to unconventional ground states such as a spin liquid. The main purpose of this paper is attempt to clarify this problem.

2. Experimental

The single crystal of $YbCoGaO_4$ was grown by floating zone image furnace techniques [1.2]. The crystal structure was investigated at room temperature by the powder X-ray diffraction method. It was found that samples investigated are rhombohedral (space group R-3m). The site symmetry of all the atoms is 3m. Magnetic measurements (dc and ac susceptibilities and magnetization) were performed using a commercial superconducting quantum interference device squid magnetometer MPMS-7 XL (Quantum Design). Figure 1 shows the temperature dependence of the zero-field-cooled (ZFC) and

[3] To whom any correspondence should be addressed : Henryk Szymczak, Institute of Physics, Polish Academy of Sciences, Al. Lotnikow 32/46, 02-668 Warsaw, Poland ; e-mail: szymh@ifpan.edu.pl

Journal of Physics: Conference Series **303** (2011) 012064 doi:10.1088/1742-6596/303/1/012064

field-cooled (FC) magnetization of $YbCoGaO_4$ crystal recorded in magnetic field H=5 kOe applied along the *c* axis and perpendicularly to this axis. As is seen in this figure the longitudinal (with field applied along the *c* axis) susceptibility shows a typical spin-glass-like cusp, whereas no cusp is observed in the transverse direction. Even stronger cusp is observed in lower magnetic fields (see figure 2) at temperature T_0=20.3 K. Above T_0 magnetic susceptibility χ is well described by the Curie–Weiss formula $\chi=C/(T+\theta)$ with the paramagnetic Curie–Weiss temperatures θ equal to: θ_C=33 K for H || *c* and $\theta_{\perp C}$=56 K for H ⊥ *c*. Despite high value of a Curie-Weiss temperature, no magnetic ordering is observed down to 2.4 K and a broad peak in the susceptibility is currently attributed to a spin glass transition.

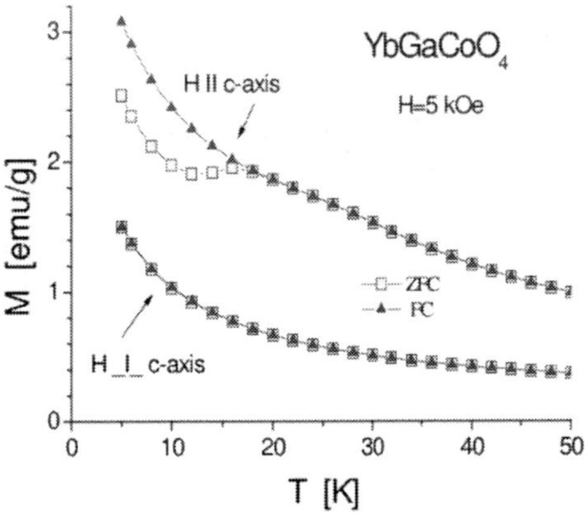

Figure 1. Temperature dependence of the ZFC (squares) and FC (triangles) magnetization recorded in H=5 kOe applied along the *c* axis and perpendicularly to this axis

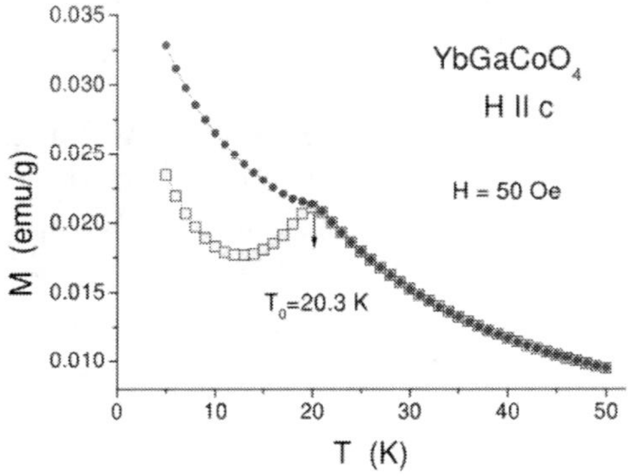

Figure 2. Temperature dependence of the ZFC magnetization recorded in H=50 Oe applied along the *c* axis

The observed difference between θ_C and $\theta_{\perp C}$ arises due to the strong magnetocrystalline anisotropy of easy axis type in both sublattices. From the Curie-Weiss law one can determine magnetic moments p_{eff} of the studied system per formula unit: p_{eff}=6.6 μ_B/f.u. for H || *c* and p_{eff}=4.28μ_B/f.u. for H ⊥ *c*. The strong anisotropy of the paramagnetic Curie–Weiss temperatures and magnetic moments indicates on the Ising-like character of the system. The results of p_{eff} measurements on polycrystalline magnets of the same structure: $LuCoGaO_4$,$YbCuGaO_4$ and $LuCuGaO_4$ [3] suggest that both Yb^{3+} and Co^{2+} are strongly anisotropic (probably Ising-like) ions in $YbCoGaO_4$ lattice. The mechanism responsible for this strong anisotropy is related to the crystal field effects. Because of strong trigonal crystal field the ground state of Yb^{3+} is doublet with extremely anisotropic g-tensor in full analogy to the behavior of Tb^{3+} in $TbBaCo_4O_7$ [4]. For Co^{2+} ions it is sufficient to conclude that their ground state in $YbCoGaO_4$ lattice is orbital triplet $^4T_1$, and consequently their magnetic properties are determined mainly by orbital moment contribution to this state.

To further characterize the system near T_0 the real (χ') and imaginary (χ'') components of ac susceptibility were measured using SQUID magnetometer with a small ac field (h) superimposed on a static field (H). Both magnetic fields were applied along the *c* axis of the crystal. The susceptibility was measured in the frequency interval 1 Hz $\leq\omega/2\pi\leq$ 1000 Hz. Figure 3 presents the temperature dependences of the real part of the ac susceptibility with zero bias dc field. The different curves correspond to different frequencies $\omega/2\pi$ of the ac field. For fixed frequency $\omega/2\pi$ a sharp cusp is observed at $T_0(\omega)$. The temperature dependence of the imaginary part of ac susceptibility presents figure 4.

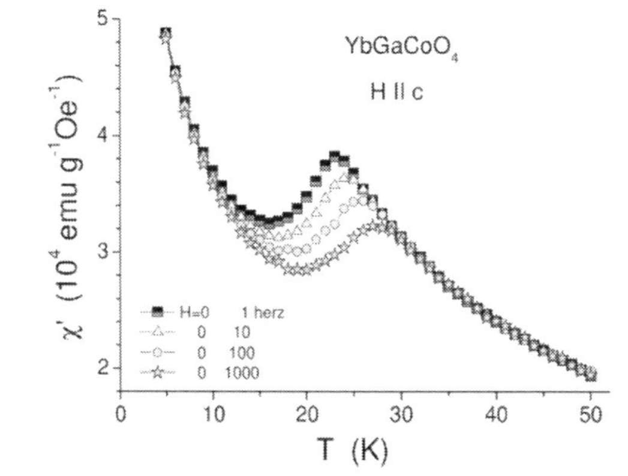

Figure 3. The real part, χ', of the ac susceptibility vs temperature at different frequencies

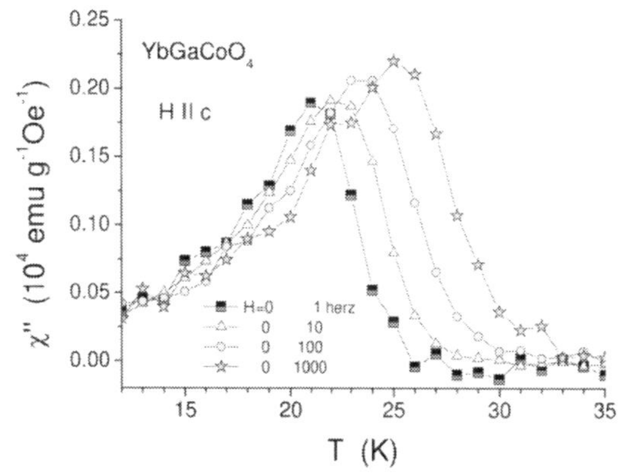

Figure 4. The imaginary part, χ'', of the ac susceptibility vs temperature at different frequencies

The positions of the cusps in χ' correspond to the maximum slope in χ'' and shift to higher temperature with increasing frequency, while the magnitude of χ' decreases for higher frequency. Such behavior is a signature of spin-glass behavior. A quantitative measure of the observed frequency shift is introduced by Mydosh [7] parameter $\Delta T_0/T_0 \ln\omega$. This parameter varies in the range 0.001÷0.02 for spin glasses and is considerably higher (of the order of 0.3) for superparamagnets. In the case of $YbCoGaO_4$ this parameter is of about 0.02. This value confirms the spin-glass character of the studied crystal.

The most relevant support of this conclusion may be achieved by analyzing the frequency, field and temperature dependences of the magnetic susceptibility $\chi(\omega,T)$. The fundamental question is whether the observed transition in T_0 temperature is a true phase transition or a gradual freezing of the magnetic moment.

According to [5,6] the following dynamic scaling equation should be valid for spin glass systems:

$$\chi'' T / \omega^{\beta/z\nu} = f(\varepsilon / \omega^{1/z\nu}) \tag{1}$$

where $\varepsilon = T/T_0 - 1$, β is a critical exponent, $z\nu$ is a dynamic exponent and $f(x)$ is a scaling function. Using a fitting parameters, all $\chi''(\omega)$ vs T curves should collapse onto the same curve. Figure 5 shows such scaling plot, performed with the parameters $z\nu=18\pm1$ and $\beta=1.00\pm0.01$. Interesting relation is obtained for $\varepsilon=0$. In this case the relation for the maximum, χ''_{max}, of each curve in figure 4 is given:

$$\chi''_{max} T(\chi''_{max}) \propto \omega^{\beta/z\nu} \tag{2}$$

Using the relation (2) the parameter $\beta/z\nu$ was determined to be $\beta/z\nu=0.055\pm0.02$. This value is very comparable to that obtained using relation (1).

The critical exponents for the three-dimensional Ising-like spin glasses is expected [8] to be $z\nu$=10-12 and β=0.54 while for Heisenberg-like spin glasses these values are $z\nu$=6-8 and β=0.9. It means that the critical exponents of $YbCoGaO_4$ essentially lie between those obtained for the Heisenberg and Ising spin glasses. The possible reason for the deviation from Ising model may be related to the two-sublattices structure of $YbCoGaO_4$ not taken into account in theoretical considerations.

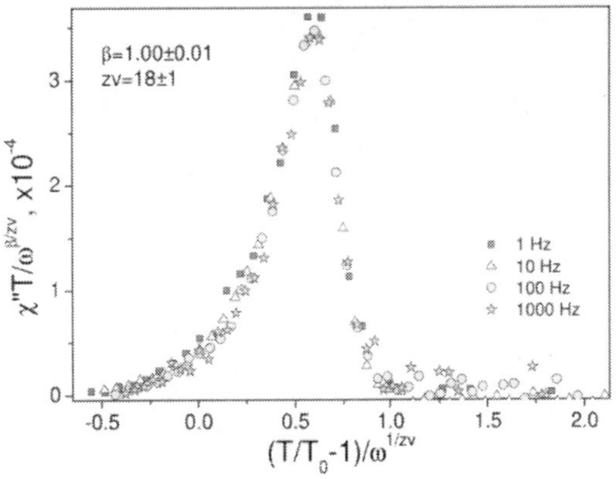

Figure 5. Scaling of $\chi''(\omega,T)$ performed according to the relation (1) with $z\nu$=18±1 and β=1.00±0.01

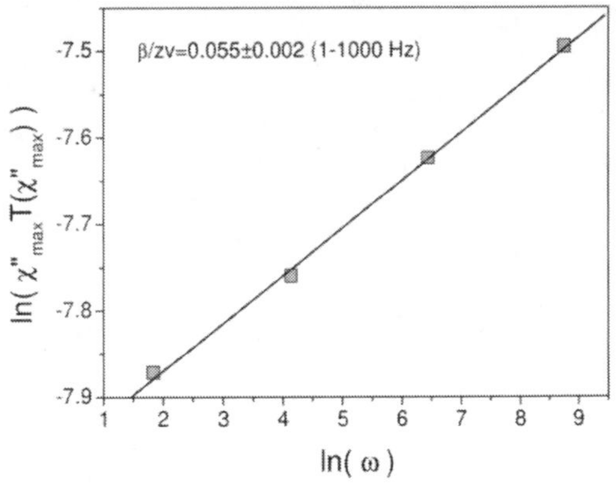

Figure 6. Determination of $\beta/z\nu$ parameter from $\chi''_{\max}T(\chi''_{\max})$ vs $\omega^{\beta/z\nu}$ using dependence (2)

3. Discussion and conclusion

It was shown that $YbCoGaO_4$ single crystal exhibits a strong uniaxial anisotropy observed in its dc and ac magnetic susceptibilities. A sharp cusp is seen in the longitudinal susceptibility measured with the field directed along the *c* axis, whereas a paramagnetic behavior is observed in transverse direction. Transverse susceptibility is considerably weaker than the longitudinal one because of Ising-like anisotropy. Considering all experimental data it was concluded that $YbCoGaO_4$ is Ising-like spin glass. The spin glass nature of this crystal is due to frustration arising from random distribution of Co^{2+} and Ga^{3+} ions . Yb^{3+} subsystem exhibits also geometrical frustration. Ising anisotropy of both kinds of ions is due to crystal fields. The scaling analysis performed on $YbCoGaO_4$ single crystal and for magnetic field applied along the c axis proves that the relation (1) for spin-glass systems describes properly the frequency and temperature dependences of $\chi''(\omega,T)$.

The results obtained are in agreement with theoretical models based on mean-field calculations developed for uniaxial crystals[9, 10]. These models predict spin-glass ordering for the longitudinal spin components and not for the transverse ones under the condition that the uniaxial magnetic anisotropy is large enough with respect to the exchange interactions. In the case of $YbCoGaO_4$ crystals two such subsystems exist but only one cusp is observed in longitudinal susceptibility. It suggests strong Yb-Co exchange coupling inside the studied system.

Acknowledgments
This paper was partially supported by the Ministry of Science and Higher Education (Poland) through grant No. N202 125135

References

[1] Dabkowska H A, Dabkowski A, Luke G M and Gaulin B D 2002 *J. Cryst. Growth* **234** 411
[2] Dabkowska H A, Gaulin B D, Dabkowski A, Luke G M and Britten J F 2002 *Acta Cryst.* **E58** i1
[3] Cava R J, Ramirez A P, Huang Q and Krajewski J J 1998 *J. Sol. State Chem.* **140** 337
[4] Szymczak R, Bludov A N, Gnatchenko S L, Barilo S N, A. Jezierski A and Szymczak H 2010 *Acta Phys. Pol. A* **118** 299
[5] Continentino M and Malozemoff A P 1986 *Phys. Rev.* B **34** 471
[6] Geschwind S, Huse D A and Devlin G E 1990 *Phys. Rev.* B **41** 4854
[7] Mydosh J A 1993 *Spin glasses:an experimental introduction*, (London:Taylor & Francis)
[8] Mathieu R, Asamitsu A, Kaneko Y, He J P and Tokura Y 2005 *Phys. Rev.* B **72** 014436
[9] Cragg D M and Sherrington D 1982 *Phys.Rev.Lett.* **49** 1120

[10] Yeshurun Y and Sompolinsky H 1985 *Phys. Rev.* B **31** 3191

Low temperature magnetic behaviour of PZT-PFW bulk multiferroic ceramics

Damir Pajić[1,2], Zvonko Jagličić[1,3], Marko Jagodič[1,4], Robert Blinc[5], Janez Holc[5], Marija Kosec[5] and Zvonko Trontelj[1]

[1] Institute of Mathematics, Physics and Mechanics, Jadranska 19, SI-1000 Ljubljana, Slovenia
[2] Department of Physics, Faculty of Science, University of Zagreb, Bijenička c. 32, HR-10000 Zagreb, Croatia
[3] Faculty of Civil and Geodetic Engineering, University of Ljubljana, Jamova c. 2, SI-1000 Ljubljana, Slovenia
[4] EN-FIST Centre of Excellence, Dunajska 156, SI-1000 Ljubljana, Slovenia
[5] Jožef Stefan Institute, Jamova c. 39, SI-1000 Ljubljana, Slovenia

E-mail: dpajic@phy.hr , zvonko.trontelj@fmf.uni-lj.si

Abstract.
Combining the ferroelectric and magnetic components into the solid solution, it is expected to obtain the magneto-electric multiferroic material. Novel bulk ceramic compounds xPZT+$(1-x)$PFW, where PZT = $Pb(Zr_{0.575}Ti_{0.425})O_3$ and PFW = $Pb(Fe_{2/3}W_{1/3})O_3$, for x = 0.15, 0.20, 0.25, 0.50, 0.70, 0.80 were synthesized and their magnetic investigation was performed, including the measurement of zero-field cooled (ZFC) and field-cooled (FC) magnetization curves, magnetic hysteresis loops and AC susceptibility.

In the low temperature region (below 20 K) splitting between the ZFC and FC curves together with the frequency dependent peaks in AC susceptibility point to the glassy freezing/blocking of magnetic moments and magnetic relaxor-like behaviour.

1. Introduction

Multiferroics are of great interest in solid state physics [1] because of coexistence of the magnetic and ferroelectric order and their coupling [2, 3]. One of the most known classes of multiferroics are the perovskite oxides [4], where the spin-spiral order is of crucial importance for the multiferroicity. Combining the ferroelectric and magnetic materials into single compound/structure, one hopes to achieve the coupling between the two orders and formation of the magnetoelectric multiferroic.

Therefore, we prepared and investigated the PZT-PFW compounds, where PZT = $Pb(Zr_{0.575}Ti_{0.425})O_3$ and PFW = $Pb(Fe_{2/3}W_{1/3})O_3$. PZT is known as a tetragonal ferroelectric below 500-750 K depending on the $PbTiO_3$ concentration [5], whereas PFW is known as a weak ferromagnet below 20 K and antiferromagnet below 350 K [6].

The magneto-electric coupling was already observed in the PZT/PFW layered nanostructure [7]. The aim of present research was to elucidate some open questions in the interpretation of magnetic properties in this class of compounds, which are smearing over ferrimagnetism, weak (anti)ferromagnetism, spin glass, short range order, frustration, etc. We performed magnetic studies on ceramic PZT-PFW solid solutions with different concentrations, with emphasis on the low temperature magnetic behaviour.

2. Experimental

Novel bulk solid solution ceramics between the PZT and PFW were synthesized using the same method as for the similar 0.65 $Pb(Mg_{1/3}Nb_{2/3})O_3$–0.35 $PbTiO_3$ system [8]. The compounds xPZT+$(1-x)$PFW (x = 0.15, 0.20, 0.25, 0.50, 0.70 and 0.80) were prepared by the solid-state mechano-chemical synthesis, including the high energy milling of the metallic oxides for 4 hours in isopropanol and sintering at 850 °C for 2 h in oxygen atmosphere. The single phase polycrystalline perovskite structure was obtained, as confirmed by the X-ray diffraction [9].

Measurements of magnetization were mostly performed at low temperature using the QD MPMS-XL-5 SQUID magnetometer. The temperature dependence of magnetization $M(T)$ was measured in applied magnetic field H after zero field cooling (ZFC) and after field cooling (FC). The magnetic hysteresis loops $M(H)$ were measured with maximum field of 5T. AC magnetic susceptibility $\chi(T)$ was measured in a broad frequency range (0.1 Hz – 1500 Hz) with the excitation field of 6.5 Oe.

Other measurements are still in progress, including the heat capacity, electrical (magneto)resistivity and dielectric susceptibility.

3. Results

The temperature dependence of magnetization $M(T)$ measured in field of 100 Oe for all compounds is presented in Fig.1. Splitting between the ZFC and FC $M(T)$ curves points to the blocking/freezing of magnetic moments. Similar behaviour of the FC curves was reported in another solid solution $PbFe_{2/3}W_{1/3}O_3$–$PbTiO_3$ ceramic [6]. Appearance of the magnetic hysteresis, presented in Fig.2, below the splitting temperature is in accordance with such kind of irreversibility.

AC in phase susceptibilities $\chi'(T)$ of xPZT+$(1-x)$PFW systems exhibit the peaks in the low temperature region, depending on x as shown in Fig.3. The shift of the peaks toward higher

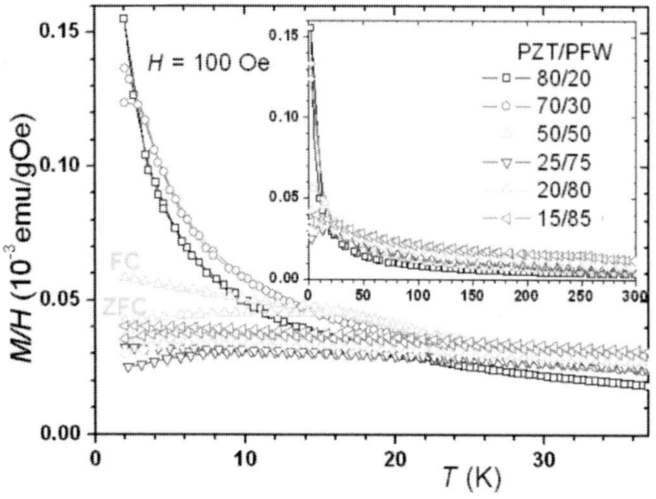

Figure 1. The temperature dependence of PZT-PFW magnetization $M(T)$ in magnetic field of 100 Oe after the zero-field cooling (ZFC) and field cooling (FC). Inset: The same data extended up to room temperature.

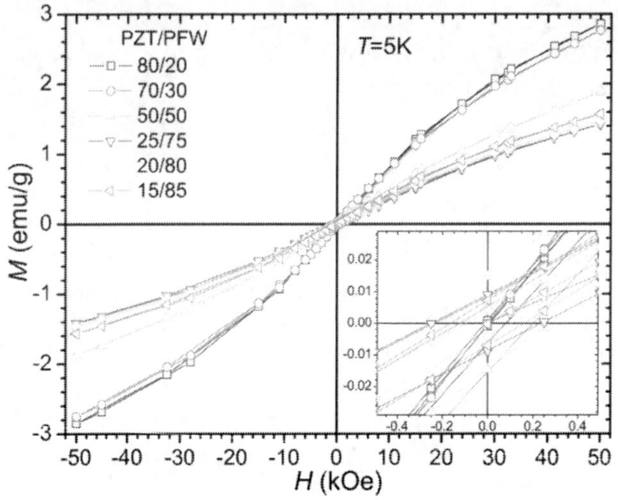

Figure 2. Magnetic hysteresis $M(H)$ loops of PZT-PFW at 5 K temperature. Inset: Enlarged region around the origin.

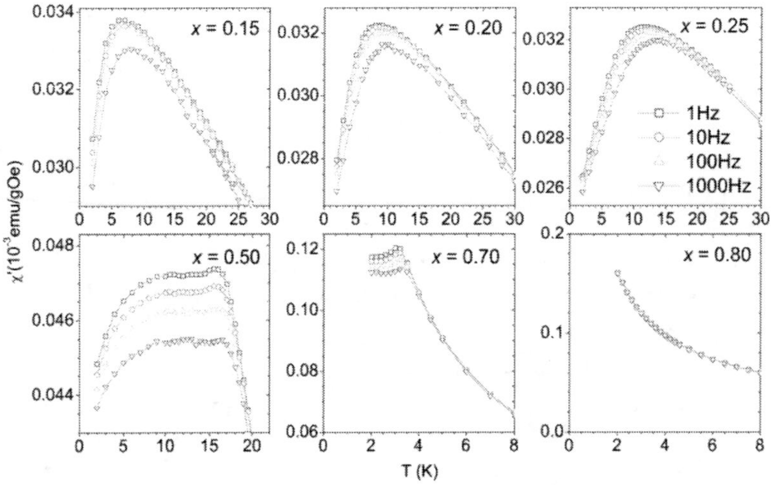

Figure 3. Low temperature in phase AC susceptibilities $\chi'(T)$ of xPZT+$(1-x)$PFW at different frequencies f.

temperature with increase of the frequency is clearly observable. The out of phase components $\chi''(T)$ for corresponding x (not shown here) have similar shapes, and their maxima also move to higher temperatures with increase of the frequency. Peaks of $\chi''(T)$ are at temperatures slightly below the peaks of $\chi'(T)$, while the values of χ'' are considerably smaller than the values of corresponding χ'. The $\chi'(T)$ at higher temperatures follow the same profile as the FC curve for corresponding x.

4. Discussion

The dispersion of the AC susceptibility peaks with frequency is characteristic of several different magnetic systems, and is a sign of the thermal activation of magnetic moments, or unfreezing of the short range ordered magnetic regions, as well as the spin glass disorder unfreezing. Certainly, the long range magnetic order can be excluded, because it does not exhibit such shift at so low frequencies [10]. The dispersion of peaks is ordinarily accompanied by the observed ZFC-FC splitting.

In our case, the relative shift of the temperature of peak, T_m, over decade of the frequency f, expressed as $\Delta T_m/(T_m \Delta \log f)$, amounts 0.068, 0.051, 0.042, 0.017 and 0,023, for $x = 0.15$, 0.20, 0.25, 0.50 and 0.70, respectively. These values are one order of magnitude bigger than in most of spin glasses, but also fit well within the values for some rare earth compounds and those with non metallic ingredients, both accepted as the spin glasses [10]. However, the obtained values are somewhat smaller than in the pure superparamagnets, excluding thus the thermal activation of independent single domain magnetic entities.

The frequency dependence of the peak temperature $T_m(f)$ is described very well with the Vogel-Fulcher type of activation $f = f_0 \exp(E_a/(T_m - T_f))$, where E_a is the activation energy, T_f the freezing temperature, and f_0 the microscopic reversal attempt frequency [10]. The agreement between this model and the experimental data is presented in Fig.4. From the least square fitting procedure it is obtained E_a = 53 K, 98 K, 132 K, 58 K and 22 K, for $x = 0.15$, 0.20, 0.25, 0.50 and 0.70, respectively. Parameter f_0 stays around $10^{10} - 10^{11}$ Hz order of magnitude, and T_f is consistently lower than the ZFC-FC splitting temperature of corresponding compound. The same model was used for the interpretation of dielectric (and magnetic) susceptibility for slightly different system where polar (and magnetic) nano-regions were found to be responsible for the relaxor behaviour [11, 12]. Therefore, the observed magnetic behaviour could be understood within the activation of magnetic moments of short range ordered regions that possibly interact. Similar magnetic fluctuations of the short range ordered regions were found in other compounds and experiments, as is the electron spin resonance in $Pb(Fe_{1/2}Nb_{1/2})O_3$ [13]. Additional analysis of $T_m(f)$ using the Arrhenius activation law gave the unphysical values for the parameters, especially f_0, that supports the exclusion of pure superparamagnetic blocking.

Furthermore, from Fig.1 it can be seen that higher magnetization is attained for smaller content of magnetic PFW component at low temperature, as confirmed by the $M(H)$ loops in Fig.2. Oppositely appears at room temperature, as can be observed in the inset of Fig.1. Based on the above AC analysis, this behaviour could be explained with development/coexistence of the superparamagnetic regions and weak/canted antiferromagnetic order. Smaller concentration

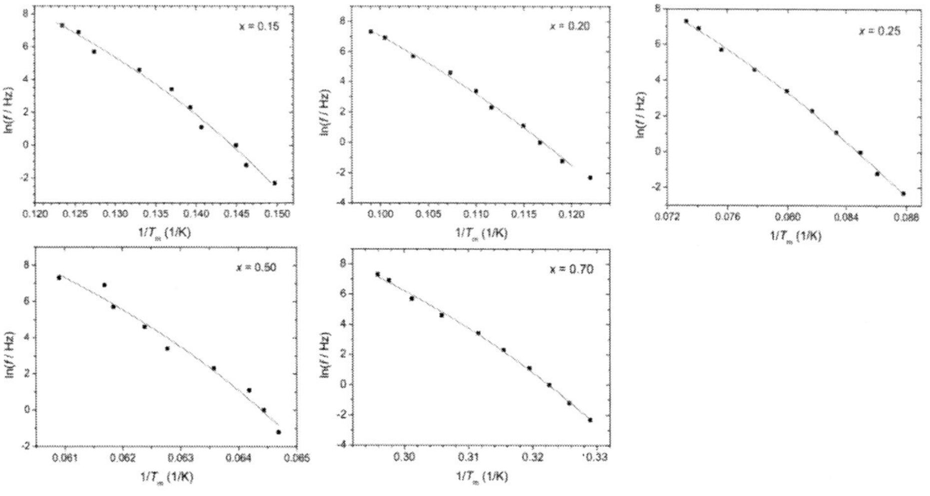

Figure 4. Vogel-Fulcher type plots of xPZT+$(1-x)$PFW. Lines are the fitting curves.

x of PZT gives more place for the long range magnetic ordering of PFW, which is basically antiferromagnetic, and therefore reducing the total magnetization. Higher concentration of PZT enhances the localized ordering of PFW into finite magnetic regions, having the magnetic moments that contribute appreciably to the total magnetization.

5. Conclusion

ZFC-FC DC magnetization curves and AC susceptibility indicate the magnetic moment blocking/freezing in xPZT+$(1-x)$PFW in low temperature region, that could also be the reason for magnetic hysteresis. Frequency dispersion of the AC susceptibility peaks is in favour of formation of the short range magnetically ordered regions responsible for the relaxor-like magnetic behaviour.

Vogel-Fulcher approach to the activation of magnetic moments describes very well the shift of AC susceptibility peaks with frequency, for all PZT-PFW, supporting thus the unfreezing of the locally ordered magnetic moments.

The magnetization reached at the low temperature for different PZT/PFW concentration ratio indicates the competition between the formation of magnetic regions and long range weak ferromagnetic ordering. This is evident from the magnetic hysteresis loops, as well as from ZFC-FC magnetization curves.

Presented results are of importance for design of the novel multiferroic materials in form of bulk solid solution ceramics, in particular with the relaxor-like behaviour [14].

Acknowledgments

This research was supported by the Slovene Research Agency (Program P2-0348).

The work of D.P. was financially supported by the Slovene Human Resources and Scholarship Fund within the Ad Futura Programme and the National Foundation for Science, Higher Education and Technological Development of the Republic of Croatia within the Brain Gain Programme.

References

[1] Fiebig M 2005 Revival of the magnetoelectric effect *J. Phys. D: Appl. Phys.* **38** R123

[2] Kreiseli J and Kenzelmann M 2009 Multiferroics - the challenge of coupling magnetism and ferroelectricity *Europhys. News* **40** 17

[3] Khomskii D 2009 Classifying multiferroics: Mechanisms and effects *Physics* **2** 20

[4] Catalan G and Scott J F 2009 Physics and applications of bismuth ferrite *Adv. Mater.* **21** 2463

[5] Jaffe B, Roth R S and Marzullo S R 1954 Piezoelectric properties of lead zirconate - lead titanate solid-solution ceramics *J. Appl. Phys.* **25** 809

[6] Mitoseriu L, Marré D, Siri A S and Nanni P 2003 Magnetic properties of $PbFe_{2/3}W_{1/3}O_3$–$PbTiO_3$ solid solution *Appl. Phys. Lett.* **83** 5509

[7] Kumar A, Katiyar R S, Premnath R M, Rinaldi C and Scott J F 2009 Strain-induced artificial multiferroicity in $Pb(Zr_{0.53}Ti_{0.47})O_3/Pb(Fe_{0.66}W_{0.33})O_3$ layered nanostructure at ambient temperature *J. Mater. Sci.* **44** 5113

[8] Kuščer D, Holc J and Kosec M 2007 Formation of 0.65 $Pb(Mg_{1/3}Nb_{2/3})O_3$–0.35 $PbTiO_3$ using a high-energy milling process *J. Am. Ceram. Soc.* **90** 29

[9] Details of the preparation and structural investigation will be published elsewhere.

[10] Mydosh J A 1993 *Spin glasses: an experimental introduction* (London, Washington: Taylor & Francis) pp 64-76

[11] Levstik A, Filipič C, Bobnar V, Holc J, Drnovšek S, Trontelj Z and Jagličić Z 2008 0.3$Pb(Fe_{1/2}Nb_{1/2})O_3$–0.7$Pb(Mg_{1/2}W_{1/2})O_3$: A magnetic and electric relaxor, *J. Appl. Phys.* **104** 054113

[12] Kumar A, Murari N M and Katiyar R S 2007 Diffused phase transition and relaxor behavior in $Pb(Fe_{2/3}W_{1/3})O_3$ thin films *Appl. Phys. Lett.* **90** 162903

[13] Blinc R, Cevc P, Zorko A, Holc J, Kosec M, Trontelj Z, Pirnat J, Dalal N, Ramachandran V and Krzystek J 2007 Electron paramagnetic resonance of magnetoelectric $Pb(Fe_{1/2}Nb_{1/2})O_3$ *J. Appl. Phys.* **101** 033901

[14] Bokov A A and Ye Z-G 2006 Recent progress in relaxor ferroelectrics with perovskite structure *J. Mater. Sci.* **41** 31

Hopping conductivity of $La_{1-x}Sr_xMn_{1-y}Fe_yO_3$

V S Zakhvalinskii [1,2], R Laiho [3], A V Lashkul [1], K G Lisunov [1,4], E Lähderanta [1], Yu S Nekrasova [2] and P A Petrenko [4]

[1] Department of Mathematics and Physics, Lappeenranta University of Technology, PO Box 20, FIN-53852 Lappeenranta, Finland
[2] Department of Physics, Belgorod State University, RUS-308015 Belgorod, Russia
[3] Wihuri Physical Laboratory, University of Turku, FIN-20014 Turku, Finland
[4] Institute of Applied Physics ASM, Academiei Str. 5, MD-2028 Kishinev, Moldova

E-mail: erkki.lahderanta@lut.fi

Abstract. The resistivity, ρ, is investigated in ceramic $La_{1-x}Sr_xMn_{1-y}Fe_yO_3$ samples with $x = 0.3$ and $y = 0.03 - 0.25$ at temperatures $T \sim 5 - 310$ K in magnetic fields B up to 8 T. The metallic conductivity at $y = 0.03$ and the variable-range hopping (VRH) conductivity between $y = 0.15 - 0.25$ are observed. The Mott VRH conductivity takes place in all investigated samples with $y \geq 0.15$ in different temperature intervals. In addition, at $y = 0.25$ the interval of the Shklovskii-Efros VRH conductivity, characteristic of the Coulomb gap in the density of the localized states (DOS), is observed. Analysis of the VRH conductivity yields the values of the microscopic parameters, including the DOS features and the localization radii of charge carriers.

1. Introduction

The mixed-valence ($Mn^{3+/4+}$) hole-doped manganite perovskites attract a stable interest due to remarkable electronic and magnetic properties, including the colossal magnetoresistance (CMR) effect [1]. Such materials are characterized by coexistence of various microscopic phases, determined by interplay between ordering of the spin, the charge and the orbital degrees of freedom and the phase separation effect [2]. This leads to correlations of the macroscopic properties of the manganite perovskites, such as the resistivity, ρ, and the magnetization, M, including a similar behavior and close values of the ferromagnetic (FM) transition temperature, T_C, and the metal-insulator transition (MIT) temperature, T_{MI} [1]. At $T < T_{MI}$ the conductivity of manganites is metallic and above T_{MI} it is realized by hopping of small polarons [1]. The nearest-neighbor hopping (NNH) conductivity mechanism is observed usually above the room temperature, whereas with lowering T the microscopic disorder becomes important leading to the variable-range hopping (VRH) conduction [1, 3 – 6].

$La_{1-x}Sr_xMn_{1-y}Fe_yO_3$ (LSMFO) belongs to the family of the manganite perovskite compounds and is characterized by strong damping of the FM properties by doping with Fe [7 – 11, 12]. The latter leads to substantial decrease of T_C with y [12], which takes place along with the decay of T_{MI} stimulating the onset of the non-metallic character of ρ (T) at higher temperatures.

In this paper are presented investigations of the hopping conductivity of LSMFO. Mechanisms of the hopping charge transfer are identified and information about microscopic parameters of charge carries is obtained.

2. Results and discussion

Ceramic LSMFO samples with $x = 0.3$ and $y = 0.03$ (# 03), 0.15 (# 15), 0.20 (# 20) and # 25 (# 25), investigated in this work, were prepared and characterized in a same way as described in [12]. Measurements of the resistivity were made in the temperature interval of $T \sim 5 - 310$ K with the conventional four-probe technique in transversal magnetic field (B) configuration for $B = 0 - 8$ T.

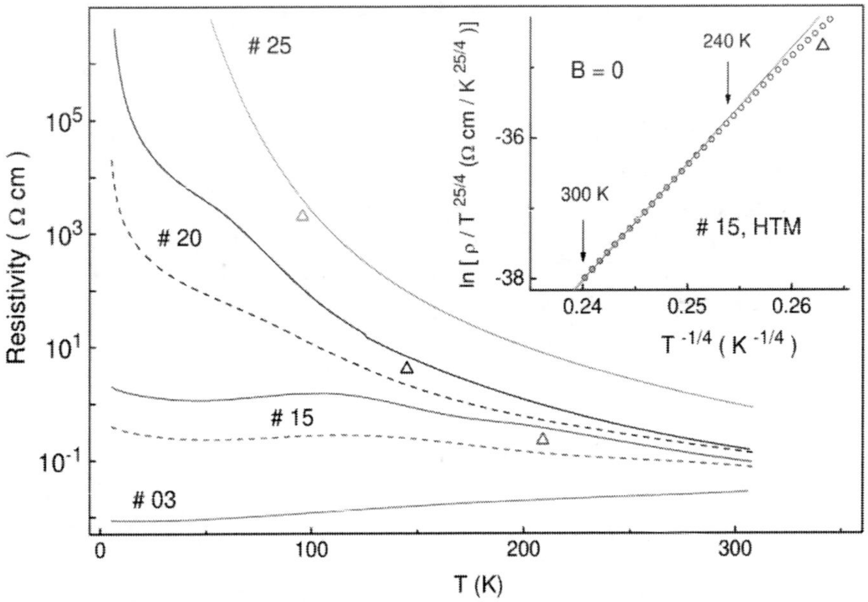

Figure 1. Temperature dependence of the resistivity in the investigated samples at $B = 0$ (solid lines) and $B = 8$ T (dashed lines). Inset: the plot of ln $(\rho / T^{25/4})$ versus $T^{-1/4}$ at $B = 0$ for the interval HTM of # 15. The straight line is a linear fit. Open triangles mark T_C.

As follows from figure 1 ρ (T) depends strongly on y. Metallic conductivity in # 03 is changed eventually to activated one in # 25. The behavior of ρ (T) in # 15 and # 20 is more complicated, exhibiting two extremes divided by the interval of the metallic behavior (downturn) in # 15, and two inflections of ρ (T) in # 20 within similar intervals ΔT below ~ 100 K. It can be seen also the weak inflection of ρ (T) in # 15 and # 20 near T_C, taken from [12] and shown by open triangles in figure 1. Large drop of the resistivity of LSMFO in the magnetic field, characteristic of CMR in manganites [1], is increased with increasing y from 0.15 to 0.20. In addition, the CMR effect within the interval ΔT, defined above, is stronger than at temperatures around T_C, which is enhanced in # 20, as well.

The behavior of the resistivity in figure 1 is attributable to the hopping conduction mechanism in # 20 and # 25 above T_C and below it, and in # 15 at $T > T_C$. In the domain of the hopping charge transfer ρ (T) is given by the universal expression,

$$\rho (T) = AT^m \exp [(T_0 / T)^p], \qquad (1)$$

where A is the prefactor constant and T_0 is the characteristic temperature [13]. The exponents m and p in equation (1) are interrelated and depend on the type of the hopping conduction. The NNH conductivity is characterized by $m = p = 1$ [13]. The VRH conductivity takes place, when the hopping of charge carriers between the nearest sites becomes energetically unfavorable due to a microscopic disorder. The VRH conductivity is given by the values of $p = 1/4$ (the Mott mechanism) and $p = 1/2$ (the Shklovskii-Efros or SE mechanism). The Mott VRH conductivity sets in when the long-range Coulomb interactions between the carriers can be neglected. In the opposite case these interactions lead to a soft Coulomb gap with width Δ in the density of localized states (DOS) around the Fermi level, μ, and to the SE VRH conductivity [13]. At any VRH conduction regime of manganite perovskites the relation $m = 8 - p\,(7 + q)$ has been predicted, where $q = 4$ or 0, depending on whether or not a fluctuating short-range potential is important for the disorder [3].

In equation (1) one has $T_0 = T_{0M}$ and $T_0 = T_{SE}$ for the Mott and the SE VRH conduction mechanisms, respectively, satisfying the expressions

$$T_{0M} = \beta_M / [k\, g\,(\mu)\, a^3] \text{ and } T_{SE} \approx [\, T_\delta^{1/2} + (T_\delta + T_{0SE})^{1/2}]^2, \tag{2}$$

respectively [3, 13, 14]. Here $T_\delta = \delta_V{}^2/(4k^2T_{VSE})$, δ_V is width of the rigid gap, existing inside Δ due to polaron effects in manganites [3], T_{VSE} is the onset temperature of the SE VRH conductivity, $T_{0SE} = \beta_{SE}\, e^2 / (k\, \kappa\, a)$, $\beta_M = 21$ and $\beta_{SE} = 2.8$ are constants [13], $g\,(\mu)$ is the DOS value at μ in the Mott VRH model, a is the localization radius and κ is the dielectric permittivity [13]. The prefactor constant in equation (1) is given for the manganite perovskites by the equation

$$A = C_0\, \alpha^{11}\, T_0^{\,p\,(7+q)}, \tag{3}$$

where $T_0 = T_{0M}$ and T_{SE} for $p = 1/4$ and $1/2$, respectively, and C_0 is a constant [3]. The DOS outside the Coulomb gap, g_0, and Δ satisfy the expressions [3, 13]

$$g_0 = (3/\pi)\,(\kappa^3 / e^6)\,(\,\Delta - \delta_V)^2 \text{ and } \Delta \approx k\,(T_{VSE}\, T_{SE})^{1/2}. \tag{4}$$

Identification of the hopping conduction mechanisms can be done by linearization of $\rho\,(T)$ in coordinates $\ln\,(\rho / T^{\,m})$ versus T^{-p}, taking into account equation (1), the relation between m and p and possible values of m and p as discussed above. In # 15 the best linearization of $\rho\,(T)$ above T_C is achieved for $p = 1/4$ and $m = 25/4$ (inset to figure 1), giving evidence for the Mott VRH conductivity at high-temperature interval HTM ($T > T_C$) with $T_{0M} = 7.43\times10^8$ K and $q = 0$. As follows from the top panel of figure 2, the Mott VRH conductivity with $p = 1/4$ and $m = 25/4$ ($q = 0$) is observed in # 20, as well. This regime takes place in # 20 within three different temperature intervals HTM ($T > T_C$), ITM (T around T_C) and LTM ($T < T_C$), given by different values of $T_{0M} = 9.43\times10^8$ K, 8.16×10^8 K and 3.19×10^7 K, respectively. In # 25 two intervals of the Mott VRH conductivity are observed, namely at high (interval HTM between $\approx$ 220 K and 170 K) and low (interval LTM with the onset near $T_C \approx 95$ K) temperatures, as follows from the bottom panel of figure 2. However, the intervals of the Mott VRH conductivity in # 25 are separated by the region of the SE VRH conductivity (interval SE in figure 2 between $T = T_{VSE} \approx 150$ K and T near T_C). The values of $m = 25/4$ in the intervals HTM and LTM, as well as $m = 9/2$ in the interval SE yield $q = 0$ in # 25, too. The value of $q = 0$ gives evidence for absence of the fluctuating short-range potential in LSMFO in the whole interval of $y = 0.15 - 0.25$. It is worth mentioning quite another situation ($q = 4$) in $La_{0.7}Ca_{0.3}Mn_{1-y}Fe_yO_3$, indicating influence of the short-range potential on the microscopic disorder already at $y \geq 0.03$ [3].

The values of T_{0M}, found above, yield the localization radius a of the charge carriers in the intervals of the Mott VRH regime. If $g\,(\mu)$ is known, evaluation of a can be done with the first of equations (2). The DOS can be found with the expression $g\,(\mu) \approx N_s\, c\,(1 - c) / (2\,W)$ [1], where $N_s \approx N_0\,(1 - y)$, $N_0 = 1.42\times10^{22}$ cm^{-3} is the concentration of Mn sites, $c \approx c_0 - y$ is the relative concentration of Mn^{4+} involved in the hopping, $c_0 \approx 0.31$ and $W \approx 2.6$ eV is width of the electron band of LSMFO [12]. This gives $g\,(\mu) \approx 3.1$, 2.1 and 1.2 (in units of 10^{20} eV^{-1} cm^{-3}) for # 15, # 20 and # 25, respectively.

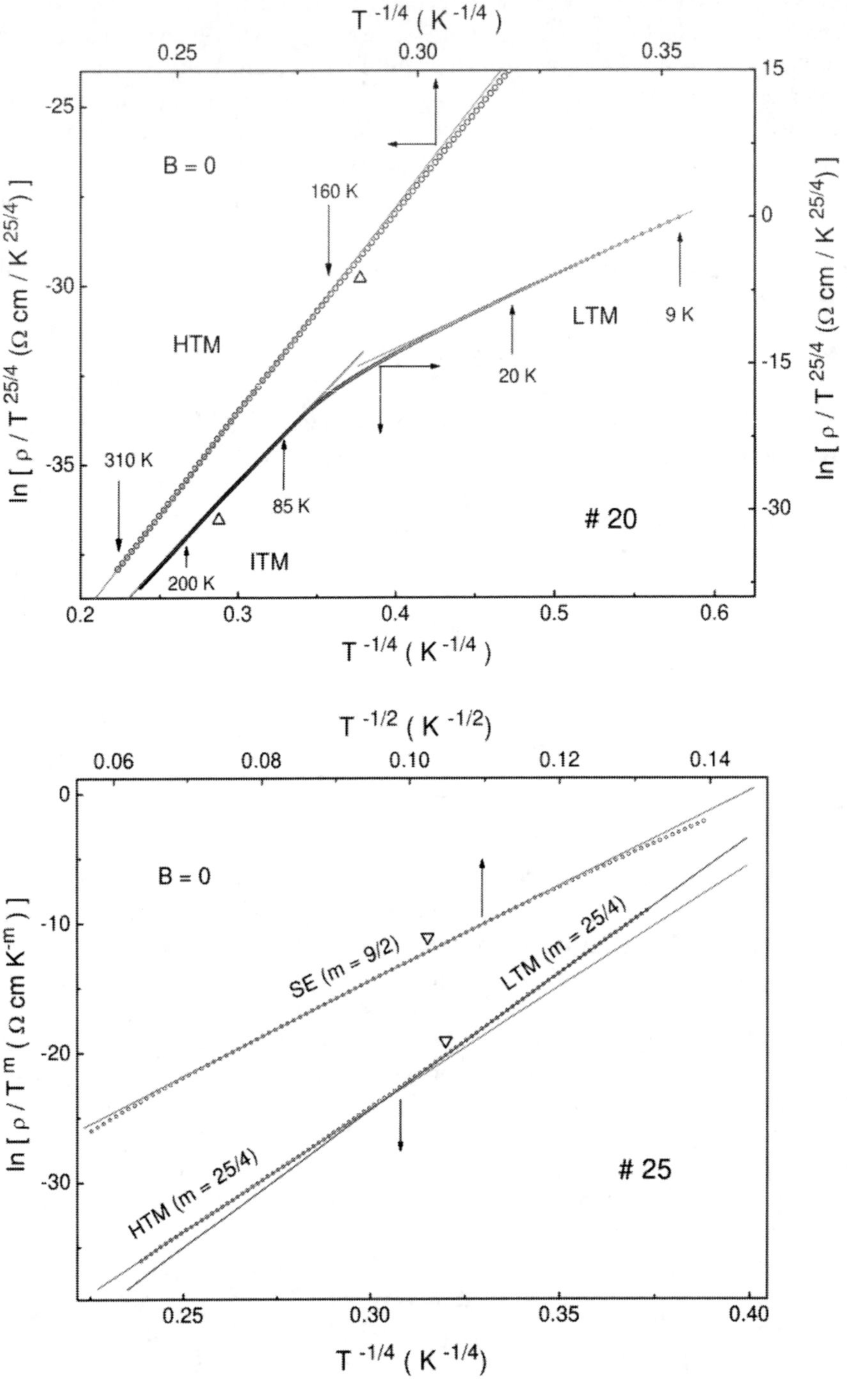

Figure 2. The plots of ln (ρ / $T^{25/4}$) versus $T^{-1/4}$ at B = 0 for the intervals HTM, ITM and LTM of # 20 (top panel) and the plots of ln (ρ / T^{m}) versus $T^{-1/2}$ and $T^{-1/4}$ for # 25 (bottom panel). The lines are linear fits. T_C is given by open triangles.

Eventually, the following values of the localization radius are found: $a \approx 1.0$ Å in the interval HTM of # 15, $a \approx 1.1$, 1.1 and 3.3 Å in the intervals HTM, ITM and LTM of # 20, respectively, and $a \approx 1.2$ and 1.0 Å in the intervals HTM and LTM of # 25, respectively.

In the interval SE of # 25, corresponding to the SE VRH conductivity (the upper plot in the bottom panel of figure 2), one obtains the value of $\Delta \approx 0.30$ eV with the second of equations (4) and the data of T_{SE} and T_{VSE} above. Then the value of $\kappa \approx 3.6$ is evaluated with the relation $\Delta \approx U$, where $U \approx e^2 / (\kappa R)$ is the mean energy of the Coulomb interactions and $R \approx 2[4\pi N_0(c_0 - y)/3]^{-1/3}$ is the mean distance between the charge carriers or Mn^{4+} involved in the hopping. Finally, $\delta_V \approx 0.07$ eV is found with the first of equations (4), whereas $a \approx 2.0$ Å in the interval SE of # 25 Å is obtained with the second of equations (2).

The value of $\Delta \approx 0.3$ eV in # 25 is smaller than $\Delta \approx 0.4 - 0.5$ eV in $La_{1-x}Ca_xMn_{1-y}Fe_yO_3$ [3], $La_{1-x}Ba_xMnO_3$ [4] and $LaMnO_{3+\gamma}$ [5, 6]. This is attributable to the smaller concentration of the holes or Mn^{4+}, involved in the hopping, $c \approx 0.061$ in # 25 versus $c \approx 0.20 - 0.30$ [3, 4 – 6] in the aforementioned compounds, leading to weaker Coulomb interactions of charge carriers. The value of $\kappa \approx 3.6$ in LSMFO is close to those of $\kappa \approx 3.0 - 3.7$ [3, 4 – 6], determined in other manganites from analysis of the hopping conductivity (the reason to smallness of κ in manganites is discussed in [4]). The value of $\delta_V \approx 0.07$ eV in # 25 is smaller, but quite comparable with the width of the rigid gap, $\delta_V \approx 0.12 - 0.28$ eV, whereas a is typical of the localization radius, $a \approx 1.2 - 2.9$ Å, of small polarons, found in the manganite perovskites mentioned above [3, 4 – 6].

An appreciable difference between the values of a in the intervals HTM at $T > T_C$ (1.1 Å) and LTM at $T < T_C$ (3.3 Å) of # 20 is attributable to the narrow FM transition and well-defined FM properties of # 20 below T_C [12]. At this point, the FM transition in # 25 is considerably expanded and FM properties of # 25 are damped [12]. This correlates well with close values of $a \approx 1.2$ and 1.0 Å in the HTM and LTM intervals of # 25, respectively. One can also see that the value of $a \approx 2.0$ Å in the SE interval of # 25 above T_C is enhanced with respect to $a \approx 1.0 - 1.2$ Å in # 15 and # 20 at $T > T_C$. Such enhancement suggests proximity to the metal-insulator transition, induced by doping with Fe (y-MIT). Indeed, the SE interval of # 25 can be treated as a result of transformation of the interval of the metallic behavior of ρ (T) in # 15 via the intermediate interval between two inflections of ρ (T) in # 20 (these intervals are defined as ΔT above). Therefore, a critical behavior of $a = a^* (1 - c/c_{cr})^{-\nu}$ alike that near the Anderson transition [14, 15], is expected inside ΔT. Here $c_{cr} \approx c_0 - y_{cr}$, y_{cr} is the value at y-MIT, $\nu \approx 1$ and a^* is the localization radius far from y-MIT. Taking into account the universal Mott criterion, $N_{cr}^{1/3} a^* \approx 0.25$ [14, 15], where $N_{cr} = N_0 (1 - y_{cr})$, the equation $a(y) \approx 0.25 [N_0 (1 - y_{cr})]^{-1/3} [(c_0 - y_{cr})/(y - y_{cr})]$ can be obtained. Then, putting $a (0.25) \approx 2.0$ Å, one finds with this equation $y_{cr} \approx 0.18$, lying between $y = 0.15$ in # 15 and $y = 0.20$ in # 20. This means proximity of # 15 to y-MIT from the metallic side and closeness of # 20 to y-MIT from the insulating side. Such relations are in a complete agreement with metallic behavior of ρ (T) in # 15 and activated behavior, accompanied with two inflections, of ρ (T) in # 20 in the interval ΔT. In addition, the value of $a^* \approx 0.25 [(1 - y_{cr}) N_0]^{-1/3} \approx 1.1$ Å is quite close to $a \approx 1.0 - 1.2$ Å, characterizing the temperature intervals $T > T_C$ outside ΔT, where the y-MIT does not take place.

The magnetic field dependence of the microscopic parameters a and g (μ) for # 15 and # 20 can be obtained with the linearization of ρ (T) in non-zero field, similar to that in the inset to figure 1 and in the top panel of figure 2 at $B = 0$, and using equations (2) and (3). The general tendency is increase of a and decrease of g (μ), when B is increased. Decrease of DOS with B can be expected due to the increase of the effective electron bandwidth. This is the broader the smaller is the angle between the spins in the Mn^{3+} – Mn^{4+} pairs [1], decreasing by alignment of these spins in the applied magnetic field. The increase of W with B stimulates increase of a (B), as well, making the hopping of electrons between $Mn^{3+,4+}$ sites easier, when the localized magnetic moments of the neighboring ions are aligned better [1]. In addition, different dependences of a and g (μ) on B in various temperature intervals are observed, indicating different sensitivity of the Mn ion spins to the magnetic field.

3. Conclusions

We have investigated resistivity of $La_{1-x}Sr_xMn_{1-y}Fe_yO_3$ with $x = 0.3$ and $y = 0.03 - 0.25$, paying major attention to the mechanisms of the hopping charge transfer. The transformation of the metallic conduction at $y = 0.03$ into the activated conductivity at $y = 0.25$, accompanied with strong overall increase of the resistivity with increasing y, is observed between $T \sim 5 - 310$ K. The Mott and the Shklovskii-Efros variable-range hopping conductivity regimes have been identified. The values of the microscopic parameters such as the localization radius, the density of the localized states and the width of the soft Coulomb gap and of the rigid gap have been obtained and analyzed.

References

[1] Coey J M D, Viret M and von Molnar S 1999 *Adv. Phys.* **48** 167

[2] Dagotto E 2002 *Nanoscale Phase Separation and Colossal Magnetoresistance* (Berlin: Springer)

[3] Laiho R, Lisunov K G, Lähderanta E, Petrenko P A, Salminen J, Shakhov M A, Safontchik M O, Stamov V S, Shubnikov M V and Zakhvalinskii V S 2002 *J. Phys.: Condens. Matter* **14** 80439.

[4] Laiho R, Lisunov K G, Lähderanta E, Shakhov M A, Stamov V N, Zakhvalinskii V S, Kozhevnikov V L, Leonidov I A, Mitberg E B and Patrakeev M V 2005 *J. Phys.: Condens. Matter* **17** 3429

[5] Laiho R, Lisunov K G, Lähderanta E, Stamov V N, Zakhvalinskii V S, Colomban Ph, Petrenko P A and Stepanov Yu P 2005 *J. Phys.: Condens. Matter* **17** 105

[6] Laiho R, Lisunov K G, Lähderanta E, Shubnikov M L, Stepanov Yu P, Petrenko P A, Khokhulin A and Zakhvalinskii V S 2006 *J. Phys.: Condens. Matter* **18** 10291

[7] Tiwari A and Rajeev K P 1999 *J. Appl. Phys.* **86** 5175

[8] Huang Q, Li Z W and Ong C K 2001 *J. Appl. Phys.* **89** 7410

[9] Chashka K B, Fisher B, Genossar J, Patlagan L and Reisner G M 2001 *Phys. Rev.* B **63** 064403

[10] Yanchevsjii O Z, V'yunov O I and Belous A G 2006 *Low Temp. Phys.* **32** 134

[11] Pashchenko V P, Shemyakov A A, Pashchenko A V, Prokopenko V K, Revenko Yu F, Turchenko V A and Varyukhin V N 2007 *Low Temp. Phys.* **33** 663

[12] Zakhvalinskii V S, Laiho R, Lashkul A V, Lisunov K G, Lähderanta E, Nekrasova Yu S and Petrenko P A 2010 Low-field magnetic properties of $La_{1-x}Sr_xMn_{1-y}Fe_yO_3$ *J. Phys.: Conference Series* (Submitted to Proceedings of JEMS 2010)

[13] Shklovskii B I and Efros A L 1984 *Electronic Properties of Doped Semiconductors* (Berlin: Springer)

[14] Mott N F and Davies EA 1979 *Electron Processes in Non-Crystalline Materials* (Oxford: Clarendon)

[15] Mott N F 1990 *Metal-Insulator Transitions* (London: Taylor and Francis)

Low-field magnetic properties of $La_{1-x}Sr_xMn_{1-y}Fe_yO_3$

V S Zakhvalinskii [1,2], **R Laiho** [3], **A V Lashkul** [1], **K G Lisunov** [1,4], **E Lähderanta** [1], **Yu S Nekrasova** [2] **and P A Petrenko** [4]

[1] Department of Mathematics and Physics, Lappeenranta University of Technology, PO Box 20, FIN-53851 Lappeenranta, Finland
[2] Department of Physics, Belgorod State University, RUS-308015 Belgorod, Russia
[3] Wihuri Physical Laboratory, University of Turku, FIN-20014 Turku, Finland
[4] Institute of Applied Physics ASM, Academiei Str. 5, MD-2028 Kishinev, Moldova

E-mail: erkki.lahderanta@lut.fi

Abstract. Low-field (B = 10 G – 1 kG) magnetic properties are investigated in ceramic samples of $La_{1-x}Sr_xMn_{1-y}Fe_yO_3$ (LSMFO) with x = 0.3 and y = 0.15, 0.20 and 0.25. A substantial decrease of the ferromagnetic (FM) Curie temperature, T_C, with increasing y is observed, which is connected to breaking of the FM double-exchange interactions by doping with Fe. Strong magnetic irreversibility at B = 10 G gives evidence for a frustrated magnetic state of LSMFO. The asymptotic Curie-Weiss behaviour of the magnetic susceptibility, χ (T), observed well above T_C, yields the values of the effective Bohr magneton per magnetic ion, exceeding considerably those of single ions. Critical behaviour of $\chi (T) \sim (T/T_C - 1)^{-\gamma}$ is influenced at y = 0.15 and 0.20 by a percolative ($\gamma \approx 1.8$) and non-percolative or Heisenberg ($\gamma \approx 1.4$) spin systems. At y = 0.25 the percolative contribution to χ (T) is not observed. The low-field magnetic properties above can be explained by the phase separation effect or generation of nanosize FM particles in the paramagnetic host matrix of LSMFO.

1. Introduction

$La_{1-x}Sr_xMn_{1-y}Fe_yO_3$, briefly LSMFO, belongs to a family of mixed-valence ($Mn^{3+/4+}$) manganite perovskites, exhibiting the colossal magnetoresistance effect (CMR) [1]. The mixed valence of Mn is realized by the hole doping, including substitution of a divalent element for La^{3+} or formation of cation vacancies. Rich magnetic phase diagram and interesting transport properties of such compounds are connected basically to competition between the ferromagnetic (FM) Mn^{3+} – Mn^{4+} double-exchange (DE) and the antiferromagnetic (AF) Mn^{3+} – Mn^{3+} superexchange interactions [1]. In addition, the properties of manganites and related CMR compounds are determined by interplay between orderings of the spin, the charge and the orbital degrees of freedom, as well as by a phase separation or generation of the nanosize hole-rich FM particles in the paramagnetic (PM) or AF host matrix [1, 2].

Doping with Fe influences strongly the magnetic and the transport properties of manganite perovskites, including LSMFO [3, 4]. In $La_{1-x}Ca_xMn_{1-y}Fe_yO_3$ such influence is connected to additional microscopic disorder, induced by Fe [5, 6]. Another possible reason may be connected to suppression of the FM properties of a material by doping with Fe, because a direct replacement of Mn^{3+} by Fe^{3+} takes place in manganites and Fe^{3+} does not support the DE interactions in the Fe^{3+} – Mn^{4+} pairs [3, 4].

In this paper is investigated influence of Fe on the low-field magnetic properties of LSMFO.

2. Results and discussion

LSMFO samples with $x = 0.3$ and $y = 0.15$, 0.20 and 0.25 (marked below as # 15, # 20 and # 25) were obtained with the conventional solid-state reaction method [1]. According to the x-ray diffraction data all samples had a rhombohedrally distorted structure (space group R-3c) with lattice parameters a = 5.508(4), 5.513(2) and 5.513(4) Å and c = 13.365(6), 13.360(4) and 13.376(7) Å for # 15, # 20 and # 25, respectively. The size of the grains in ceramic LSMFO samples were of few micrometers, and the homogeneous and stoichiometric distribution of the elements over the volume of samples and in separate grains was observed with the microprobe and scanning-tunneling microscopy analyses. Magnetization M (T) was measured with an RF-SQUID magnetometer after cooling the sample from the room temperature down to 5 K in zero magnetic field (M_{ZFC} or zero-field cooled) or in fields of B = 10 G, 0.5 kG and 1 kG (M_{FC} or field-cooled). Temperature dependence of the thermoremanent magnetization (TRM) was measured after cooling the sample from 300 K down to 3 K in the field of 10 G and then reducing the field to zero.

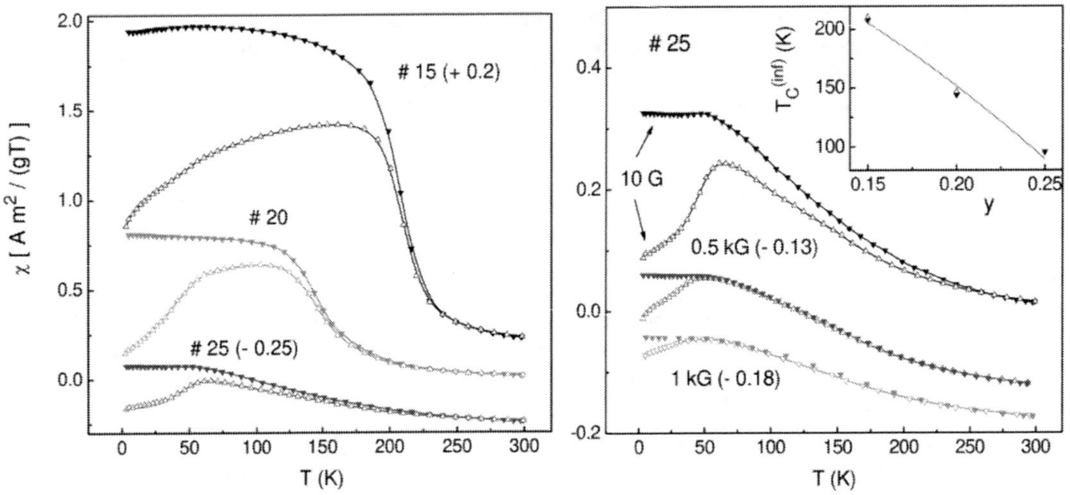

Figure 1. Temperature dependences of χ_{ZFC} (Δ) and χ_{FC} (▼) in LSMFO samples at B = 10 G (left panel) and in # 25 in various magnetic fields (right panel). Some of the plots are shifted along the vertical axis by the values given in parenthesis. Inset: The dependence of $T_C^{(inf)}$ on y, obtained in the ZFC (Δ) and FC (▼) regimes. The solid line is the fit with equation (1).

In the left panel of figure 1 one can see that both χ_{ZFC} (T) and χ_{FC} (T) (where $\chi \equiv M / B$) exhibit at B = 10 G a FM transition at a Curie temperature, T_C, defined initially by inflection of the corresponding curves ($T_C^{(inf)}$). In # 25 the transition is broaden, whereas the values of χ (T) are decreased substantially. It should be mentioned also a strong decay of $T_C^{(inf)}$ with y, as evident from the inset to the right panel of figure 1. In manganite perovskites analysis of T_C can be done with the Varma model [7], predicting the expression

$$kT_C \approx 0.05\ W c\ (1 - c), \qquad (1)$$

where W is the width of the electron band and c is the concentration of the holes or Mn^{4+} [7]. Because Fe^{3+} substitutes directly Mn^{3+} and do not support the DE interactions (see section 1), in LSMFO one can put $c \approx c_0 - y$, where $c_0 \sim x$ and deviations from x may be due to the cation vacancies [1]. The values of $W = 2.6 \pm 0.1$ eV and $c_0 = 0.31 \pm 0.01$ are obtained by fitting the dependence of $T_C^{(in)}(y)$ with equation (1) (the solid line in the inset to the right panel of figure 1). Hence, c_0 deviates from $x = 0.3$ only negligibly, whereas W coincides within the error with the value of $W \approx 2.5$ eV in $La_{1-x}Sr_xMnO_3$ [7]. Therefore, the decay of T_C with y in LSMFO is determined by breaking of the DE interaction by Fe^{3+}, whereas the disorder induced by doping with Fe plays a negligible role.

Another important feature of figure 1 is the magnetic irreversibility or deviation of $\chi_{ZFC}(T)$ from $\chi_{FC}(T)$. The magnetic irreversibility is damped strongly with increasing field in # 25 (right panel of figure 1) and disappears completely in # 15 and 20 already at $B = 1$ kG (not shown).

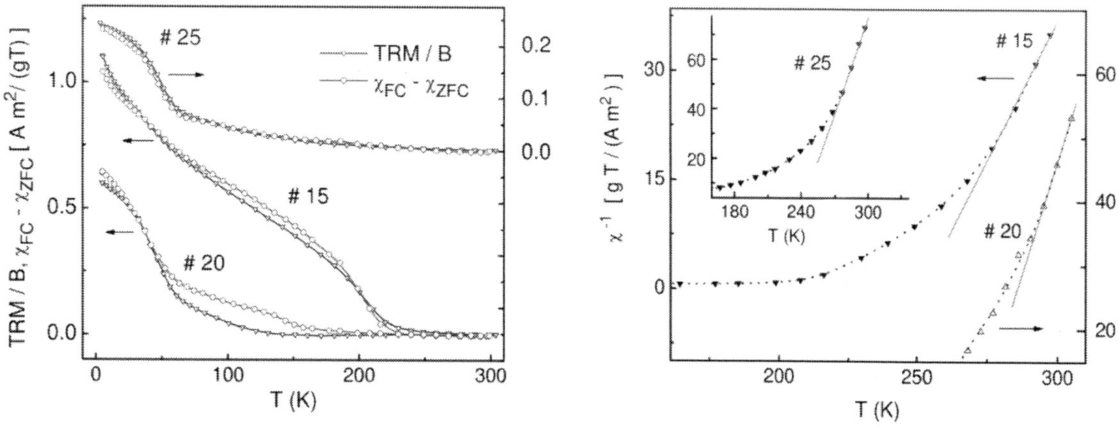

Figure 2. Temperature dependences of TRM / B (∇) and $\chi_{FC} - \chi_{ZFC}$ (O) (left panel) and the plots of χ^{-1} versus T in the ZFC (Δ) and FC (▼) regimes (right panel) for the LSMFO samples. The solid lines are linear fits and the dotted lines are to guide the eye.

Additional features of the magnetic irreversibility in LSMFO are displayed in the left panel of figure 2, where the plots of the TRM / B versus T are compared with the difference of $\chi_{FC}(T) - \chi_{ZFC}(T)$. One can see a reasonable coincidence of these plots for # 25 and clear divergence for # 15 and # 20. The irreversible magnetic behavior in figure 1 and in the left panel of figure 2 indicates a frustrated magnetic state of LSMFO. Such behavior is pertinent to spin-glass (SG) or cluster-glass (CG) phases, which set in below the onset of freezing of the magnetic moments in conditions of competing interactions between the moments [8]. In the SG phase the expression TRM $(T) = M_{FC}(T) - M_{ZFC}(T)$ reflects a symmetry of the energy distribution of potential barriers in the presence or absence of the external magnetic field [9]. In the CG phase this symmetry may be broken due to the anisotropy, associated with the shape and orientation of the magnetic clusters, leading to violation of the expression above [9].

The irreversibility disappears well above T_C and an asymptotic Curie-Weiss behavior is observed (the right panel of figure 2), given by the law $\chi(T) = C / (T - \theta)$, where $C = p_{eff}^2 \mu_B^2 N / (3k)$ is the Curie constant, p_{eff} is the effective Bohr magneton (μ_B) number per magnetic ion, N is the concentration of the magnetic ions and θ is the Weiss temperature. The values of $p_{eff}^2 \approx 240 - 280$, $280 - 300$ and $140 - 180$ for # 15, # 20 and # 25, respectively, are obtained with the linear fit of the plots of χ^{-1} versus T (the solid lines in the right panel of figure 2), if N is taken equal to the concentration $N_0 = 1.42\times10^{22}$ cm^{-3} of magnetic ions in LSMFO. Hence, p_{eff}^2 exceeds considerably the corresponding values of $p_{eff}{}^2$ = 24, 15 and 35 for Mn^{3+}, Mn^{4+} and Fe^{3+} single ions, respectively. This implies the phase separation with the onset already above the room temperature and well above T_C.

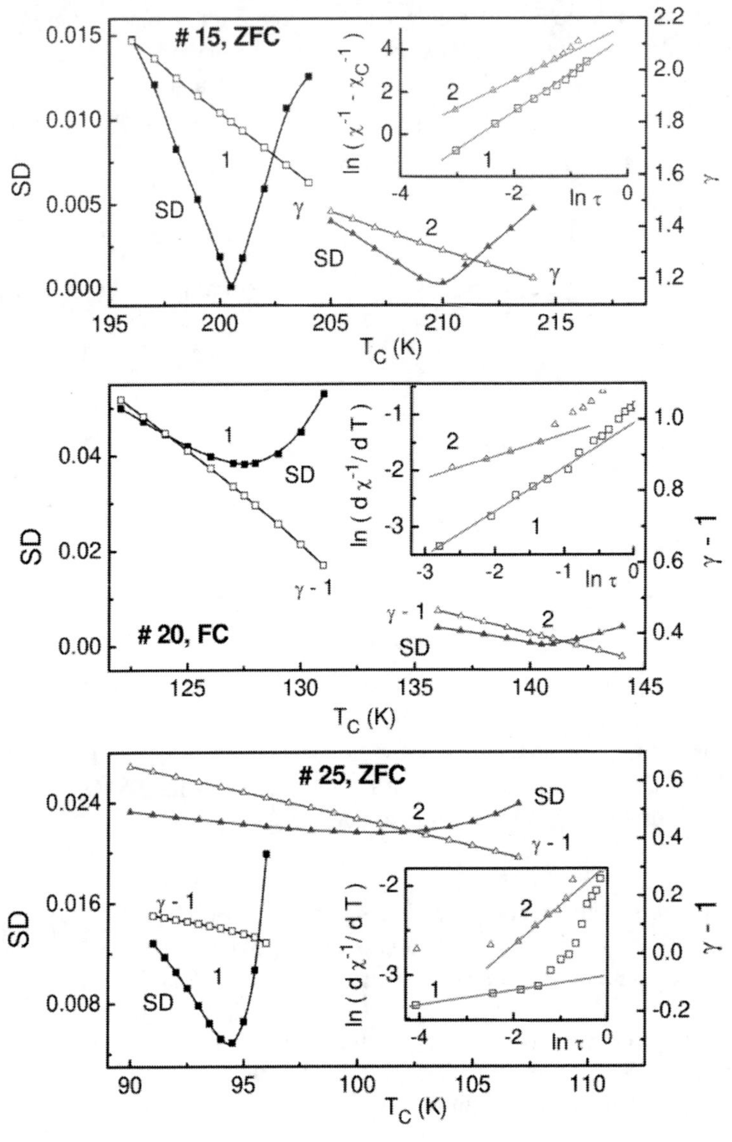

Figure 3. The dependences of SD and γ (top panel) or γ – 1 (middle and bottom panels) on T_C in the intervals ΔT_1 (1) and ΔT_2 (2) in the investigated LSMFO samples. Insets: the plots of ln $(\chi^{-1} - \chi_C^{-1})$ (in arbitrary units) vs. τ (top panel) and the plots of ln $(d\,\chi^{-1} / d\,T)$ (in arbitrary units) vs. τ (middle and bottom panels) in the intervals ΔT_1 (1) and ΔT_2 (2) for the LSMFO samples. The lines are linear fits.

The deviation from the Curie-Weiss dependence with lowering the temperature in the right panel of figure 2 is attributable to the critical behaviour of χ $(T) \sim (T / T_C - 1)^{-\gamma}$ at $T \to T_C$, where γ is the critical exponent, depending on the nature and dimensionality of a spin system [10]. The behavior of χ (T) is analyzed here using the critical law above, presented in the two following forms,

$$\chi^{-1} - \chi_C^{-1} \sim \tau^{\gamma} \text{ and } d\,\chi^{-1} / d\,T \sim \tau^{\gamma - 1}, \tag{2}$$

where $\tau = T / T_C - 1$ and $\chi_C^{-1}(T) \equiv \chi^{-1}(T_C)$. Such procedure is performed by interpolation of χ (T) and variation of T_C by a step of 0.5 – 1 K, to obtain the minimum standard deviation (SD) of the plots of ln $(\chi^{-1} - \chi_C^{-1})$ versus τ and ln $(d\,\chi^{-1} / d\,T)$ versus τ. This yields the pairs of T_C and γ or T_C and $\gamma - 1$, applying the first or the second of equations (2), respectively, and the optimum temperature interval, ΔT, corresponding to the minimum of the SD vs. T_C plots.

Some examples of the analysis described above are exhibited in figure 3. For all LSMFO samples in both ZFC and FC regimes at B = 10 G were found two different intervals, ΔT_1 and ΔT_2, where the formal dependence of SD (T_C) exhibited a distinct minimum, corresponding to $T_C^{(1)}$ and $T_C^{(2)}$ with $T_C^{(1)} < T_C^{(2)}$, and to different values of γ_1 and γ_2. Both $T_C^{(j)}$, j = 1 and 2, exhibit a decrease with y alike $T_C^{(\mathrm{inf})}$. The values of γ_1 and γ_2 in # 15 and # 20 are concentrated closely around the values of $\gamma_p = 1.80$ and $\gamma_H = 1.39$, respectively, which characterize a percolative spin system [11] and non-percolative (Heisenberg) spin system [10], respectively. In # 25 γ_2 is close to γ_H, as well, whereas γ_1 is found to lie near the mean-field value, $\gamma_{mf} = 1$ [10], implying the absence of the percolation behavior.

Therefore, in # 15 and # 20 the complex critical behavior of χ (T) reflects coexistence of two different spin systems. The first one is connected to generation of large and strongly correlated FM clusters, attributable to the phase separation and percolation by joining of nanosize FM particles into critical clusters when T is decreased [2]. The second system demonstrates the features, pertinent to an assembly of smaller and weakly correlated magnetic units and can be associated with the material, which do not enter the percolative critical clusters. Such units can exist because the volume fraction, η, of the second (FM) phase, has been estimated to be only ~ 0.29 [12] to achieve eventually the percolative FM transition at $T_C^{(1)} < T_C^{(2)}$.

In # 25 influence of the percolative spin system on the critical behavior of χ (T) is not observed. At this point it is important to note that in # 25 the FM properties are pronounced much weaker than in # 15 and # 20, but the irreversibility is enhanced (figure 1). In addition, the good coincidence of the plots of TRM/B versus T and $\chi_{FC}(T) - \chi_{ZFC}(T)$ is consistent with absence of large percolative clusters in # 25, whereas divergence of the corresponding plots for # 15 and # 20 is in line with presence of such clusters (the left panel of figure 2). Hence, it is possible to attribute ferromagnetism of LSMFO presumably to the first (percolative) spin system, whereas frustration and magnetic irreversibility to the second (non-percolative) system.

It is worth mentioning that influence of the percolative and non-percolative (Heisenberg) processes on the critical behavior of $\chi(T)$ has been observed in thin films of $La_{1-x}Ca_xMnO_3$ [13], as well as in bulk $La_{1-x}Ca_xMn_{1-y}Fe_yO_3$ [5] and $La_{1-x}Ba_xMnO_3$ [14], which reflects universality of the phase separation effect in manganite perovskites and related CMR compounds [2]. Similar to the aforementioned papers [5, 13, 14], one can estimate the following parameters, addressed to the onset of the percolation critical behavior in LSMFO: $\eta \approx 0.16$ and 0.11, the mean radius of FM particles $r \approx$ 3.4 nm and 3.8 nm and the mean magnetic moment $\mu/\mu_B \approx 0.9\times10^4$ and 1.1×10^4 in # 15 and # 20, respectively. All these data are quite typical of the nanosize hole-rich FM clusters in manganite perovskites and other CMR materials, obtained with macroscopic (magnetization) [5, 13 ,14], as well as with various microscopic [15 – 17] methods.

3. Conclusions

Investigations of the low-field magnetic properties of LSMFO give evidence for coexistence of the FM and irreversible magnetic properties. Strong decay of T_C with y is explained by damping of the DE interactions by doping with Fe. Asymptotic Curie-Weiss behavior well above T_C yields the values of p_{eff}, incompatible with those of single magnetic ions. The complex critical behavior of χ (T) demonstrates existence of the percolative and non-percolative or Heisenberg spin systems. The observed low-field magnetic properties of LSMFO are explained by the phase separation, pertinent to the manganite perovskites and leading to generation of nanosize FM clusters in the PM host matrix.

References

[1] Coey J M D, Viret M and von Molnar S 1999 *Adv. Phys.* **48** 167
[2] Dagotto E 2002 *Nanoscale Phase Separation and Colossal Magnetoresistance* (Berlin: Springer)
[3] Tiwari A and Rajeev K 1999 *J. Appl. Phys.* **86** 5175
[4] Chashka K B, Fisher B, Genossar J, Patlagan L and Reisner G M 2001 *Phys. Rev.* B **63** 064403.
[5] Laiho R, Lisunov K G, Lähderanta E, Salminen J and Zakhvalinskii V S 2002 *J. Magn. Magn. Mater.* **250** 267.
[6] Laiho R, Lisunov K G, Lähderanta E, Petrenko P A, Salminen J, Shakhov M A, Safontchik M O, Stamov V S, Shubnikov M V and Zakhvalinskii V S 2002 *J. Phys.: Condens. Matter* **14** 80439.
[7] Varma C M 1996 *Phys. Rev.* B **54** 7328.
[8] Chowdhury D 1986 *Spin Glasses and Other Frustrated Systems* (Singapore: World Scientific)
[9] Lähderanta E, Eftimova K, Laiho R, Kanani H Al and Booth J C 1994 *J. Magn. Magn. Mater.* **130** 23
[10] Stanley H E 1971 *Introduction to Phase Transitions and Critical Phenomena* (Oxford: Clarendon)
[11] Dunn A G, Essam J W and Ritchie D S 1975 *J. Phys.* C **8** 4219; Kirkpatrick S 1976 *Phys. Rev. Lett.* **36** 69.
[12] Shklovskii B I and Efros A L 1984 *Electronic Properties of Doped Semiconductors* (Berlin: Springer)
[13] Huhtinen H, Laiho R, Lähderanta E, Salminen J, Lisunov K G and Zakhvalinskii V S 2002 *J. Appl. Phys.* **91** 7944
[14] Laiho R, Lisunov K G, Lähderanta E, Zakhvalinskii V S, Kozhevnikov V L, Leonidov I A, Mitberg E B and Patrakeev M V 2005 *J. Magn. Magn. Mater.* **293** 892
[15] Hennion M, Moussa F, Biotteau G, Rodriguez-Carvajal J, Piusard L and Revcolevschi A 1998 *Phys. Rev. Lett.* **81** 1957.
[16] Chechersky V, Nath A, Isaac I, Franck J P, Ghosh K, Ju H and Greene R L 1999 *Phys. Rev.* B **59** 497
[17] Hennion M, Moussa F, Biotteau G, Rodriguez-Carvajal J, Piusard L and Revcolevschi A 2000 *Phys. Rev.* B **61** 9513

Andreev reflection and spin polarization of $SrRuO_3$ thin films on $SrTiO_3$ (111)

M. Foerster[1], S. Piano[2], F. Rigato[1], C.J. Mellor[2] and J. Fontcuberta[1]

[1] Institut de Ciència de Materials de Barcelona - CSIC, E-08193 Bellaterra, Spain
[2] School of Physics and Astronomy, University of Nottingham, NG7 2RD Nottingham, UK

mfoerster@icmab.es

Abstract. Epitaxial $SrRuO_3$ thin films with an extremely flat surface morphology suitable for the use in thin film heterostructures have been grown on $SrTiO_3$ (111) substrates. The transport spin polarization was measured by point contact spectroscopy for $SrRuO_3$ films on $SrTiO_3$ (111) and comparable samples on $SrTiO_3$ (001). The measured polarization for both types of samples is slightly smaller than previously measured by the same technique. Possible reasons for this difference are discussed.

1. Introduction

$SrRuO_3$ (SRO) is a metallic ($\rho \approx 200$ μΩcm at room temperature) and ferromagnetic (T_C =160K) oxide which has been attracting a lot of interest due to its potential use as electrode in oxide based electronic and spintronic devices. It is chemically extremely stable [1] and can be grown in heterostructures with other functional oxides, like superconductors [2,3,4], ferroelectrics [5], all oxide magnetic tunnel junctions [6] and spinel ferrites for spin filters [7,8]. SRO has an orthorhombic unit cell with a = 5.553 Å, b = 5.567 Å and c = 7.845 Å . The crystal structure derives from the cubic perovskite structure by distortion and a pseudocubic ($a_{pc} \approx 3.93$ Å) notation can be useful to describe SRO.

SRO can be grown in the pseudocubic $(001)_{pc}$ orientation on perovskite $SrTiO_3$ (001) (STO, a = 3.905 Å) and often vicinal cut substrates are used [9] to suppress the formation of a multidomain structure arising from the different growth modes that are possible due to the orthorhombic distortion of SRO (the pseudocubic $(100)_{pc}$, $(010)_{pc}$, $(001)_{pc}$ orientations are not equivalent) [10]. The properties of SRO grown on STO (001) have been extensively explored (see e.g. Ref. [9-11]).

However, for some applications SRO films in other orientations are more appropriate, e.g. for the use as bottom electrode in a spin filter structure with spinels as insulating ferrimagnetic tunnel barriers. Reason is that the spinel (001) surface is energetically unfavorable and therefore thin (001)-textured films are difficult to grow flat enough as needed for spin filters [12] on the most commonly used single crystalline perovskite (001) substrates [13]. Indeed, the growth of CFO on SRO buffered STO (001) was found to result in a very rough (rms roughness > 2nm) surface morphology [14]. In contrast, CFO (111) can be grown with flat surface (rms roughness < 0.3nm) on STO (111) with and without SRO buffer layers [7,8]. Using rf magnetron sputtering on STO (111) substrates, we have been able to obtain SRO films considerably more flat (rms roughness < 0.3nm) than previously reported [15]. These films can be used not only as bottom electrode for CFO based spin filters [7,8], but also in other situations, in which the pseudocubic $(001)_{pc}$ SRO surface is unfavorable.

For many of the possible applications of SRO thin films the transport spin polarization P is a key quantity. P is defined as:

$$P=\frac{I_{\uparrow}-I_{\downarrow}}{I_{\uparrow}+I_{\downarrow}}=\frac{N_{\uparrow}(E_F)v_{F\uparrow}^{n}-N_{\downarrow}(E_F)v_{F\downarrow}^{n}}{N_{\uparrow}(E_F)v_{F\uparrow}^{n}+N_{\downarrow}(E_F)v_{F\downarrow}^{n}} \qquad (1)$$

with n=1 for ballistic transport. $N_{\uparrow\downarrow}(E_F)$ is the density of states at the Fermi level and $v_{F\uparrow\downarrow}$ the Fermi velocity for the respective spin direction. In the case of SRO, it is believed that a negative transport polarization results mainly from differences in the Fermi velocities for spin up and spin down electrons rather than from the respective density of states, which is roughly equal [16,17].

The transport spin polarization of SRO films on STO (001) has been measured using planar tunnel junctions with magnetic [6] or superconducting counter electrode [1], yielding results of P ≈ -10%. In contrast, from point contact spectroscopy (PCS, explained in more detail below) a much higher polarization of $|P|$=50% is inferred [17,18] (note that PCS is not sensitive to the sign of P). It has been proposed [1,17] that this discrepancy may be due to different decay length of surface states in the tunnel barrier, underlining that electrode properties can appear drastically changed when measuring through a tunnel barrier (like e.g. for epitaxial MgO/Fe).

In principle, the transport spin polarization P as expressed in Equation 1 is dependent on the crystallographic direction through the Fermi velocity, which has to be taken into account. Therefore, for the new SRO films prepared on STO (111) an independent determination of P is needed.

In this work, we report on the growth of SRO thin films in (111) orientation with extremely flat surfaces as required for heterostructures (spin filters). Using PCS, a transport spin polarization of P ≈ 42% is determined by fitting experimental spectra with the extended Blonder-Tinkham-Klapwjik (BTK) model. Measurements on comparable SRO films grown on STO (001) yield the same P within the experimental accuracy. This result is in agreement with some available band structure calculations [16], while appearing to be in contrast with another [19]. However, it turns out that the measured P values of (111) and (001) SRO films are slightly smaller than previously measured by the same technique [17,18] and we discus the possible reasons for this difference.

2. Sample preparation and characterization

SRO//STO films were grown by radiofrequency magnetron sputtering from a stoichiometric target. Previous to deposition, the single crystal substrates, (111) oriented STO, were annealed for one hour in the sputtering chamber, in high vacuum at 700 °C. The coatings were produced in a mixed Ar/O_2 atmosphere (in ratio 3:2), with a total pressure of 100 mTorr. An RF power of 15 W was supplied to the target, and the distance between the target and the substrate was fixed at 6 cm, resulting in a growth rate of ≈ 0.54 nm/min. The typical deposition time was 50 min and the optimal substrate temperature during the growth was found to be 725 °C.

In Fig. 1a, X-ray diffraction (XRD) data taken on a 30nm thick SRO film on STO (111) is shown. SRO grows textured in pseudocubic $(111)_{pc}$ orientation on the STO (111) substrate. No impurity phases can be detected. Due to the orthorhombic distortion, the pseudocubic $(111)_{pc}$ can correspond to several, slightly different crystal directions, namely $(011)_o$ and$(101)_o$ of the orthorhombic structure. However, the differences between these orientations are so small that they cannot be resolved with standard XRD measurements. Importantly, the d(111)= 2.31 Å interplanar spacing deduced from the θ/2θ scans X-ray scans indicate that the out-of-plane parameter of SRO, as a result of epitaxial strain, is slightly expanded compared to the expected position for bulk STO (vertical lines in inset of Fig. 1a).

The surface morphology of SRO films deposited on STO (111), as observed by atomic force microscopy (AFM), is extremely flat (Fig. 1b). For example, the rms roughness of the sample in Fig. 1b (18nm thick) is only 0.15nm.

In the left inset of Fig. 2a, the temperature dependent resistance ρ(T) of a SRO film (30 nm thick) on STO (111) is shown. A kink at around T_C≈ 100 K signals the Curie temperature, which occurs at a

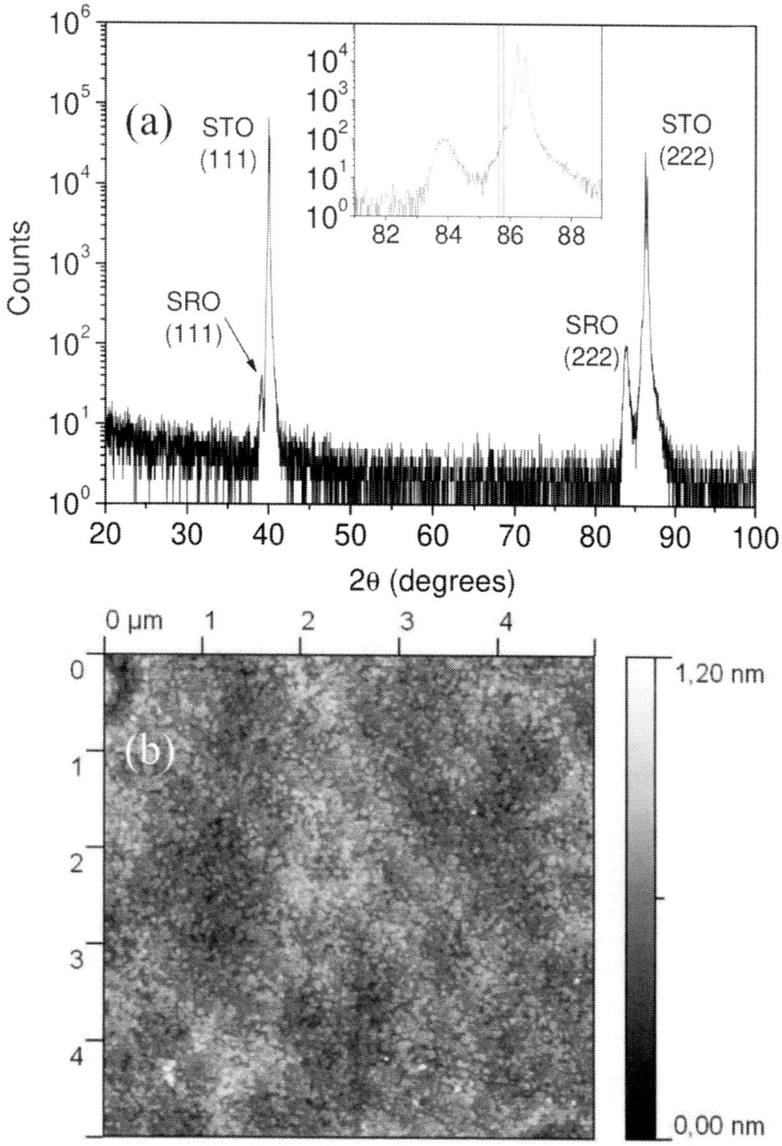

Fig. 1: (a) X-ray diffraction (θ/2θ scan) of a 30nm SRO film on STO (111). Intensity is observed around the position of the pseudocubic (hhh) reflections, indicating highly textured growth.
(b) Topography of a 18nm SRO film grown on STO (111) measured by AFM. No dependence of the typical roughness on the film thickness was found in the explored range up to 70nm.

temperature lower than that of bulk samples ($T_C \approx 160$ K). The reduction of T_C in SRO thin films has been attributed to strain [20] which indeed is observed in the XRD measurements. A certain degree of structural disorder is reflected in the relatively low residual resistance ratio RRR ≈ 1.8, which however is among the best reported for SRO grown on STO (111) [21], and the value of $\rho(5\text{ K}) \approx 200$ μΩcm which is larger than that measured in state-of-the-art (001)SRO thin films [9].

3. Point contact spectroscopy and Andreev reflection

The process of Andreev reflection (AR) can occur at superconductor(SC)-metal(M) interfaces for bias voltages V smaller than the superconducting energy gap Δ/e. Under this condition, incoming electrons from the metal cannot enter the superconductor, unless becoming part of a Cooper pair. This requires that a second electron from the opposite spin density of states is transferred from the M to the SC, leaving a hole that is reflected in the metallic electrode. Since two electrons are transferred in this process, ideally the conductance of such contact doubles within the superconducting energy gap.

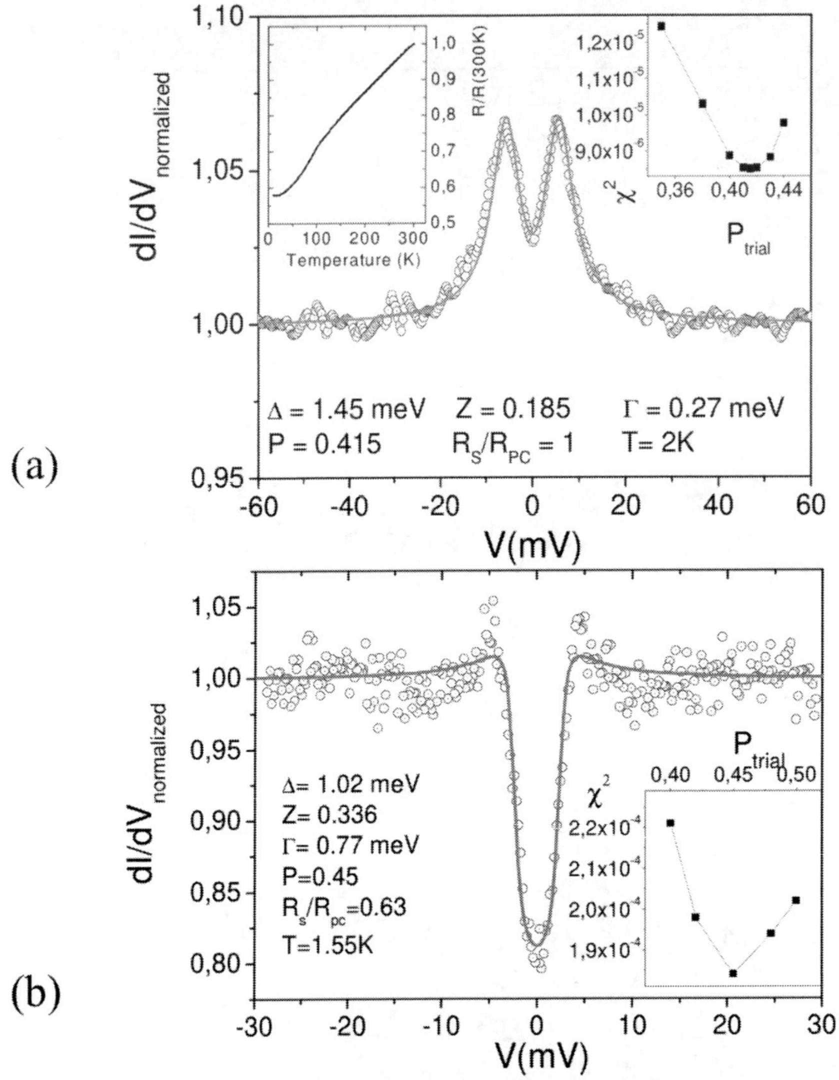

Fig. 2: Differential conductance dI/dV normalized to values at high bias voltages, measured for a) SRO in pseudocubic (111) orientation and b) SRO in pseudocubic (001) orientation. In the right insets the best fit χ^2 as function of P_{trial} is shown. Left inset in a) $\rho(T)$ of a 30nm thick SRO film on STO (111).

The AR process is partially suppressed when the metallic electrode is constituted of a magnetic material with a finite transport spin polarization P. This is most obvious in the case of a half metal where no electrons of the opposite spin direction are available and consequently no hole can be reflected. The degree by which the AR process is suppressed by P is used in point contact spectroscopy (PCS) to infer the transport spin polarization of ferromagnetic thin films [22]. PCS consists of measuring the dI/dV vs V characteristics of a small Metal(Ferromagnet)-SC contact, e.g. formed by pressing a chemically etched thin wire tip on a sample.

The dI/dV characteristics of point contacts can be described by the BTK (Blonder-Tinkham-Klapwjik) model and its extensions [23,24] for finite spin polarization P. Besides the transport spin polarization P, the model parameters are: the temperature T, the superconducting energy gap Δ, the broadening of the energy gap expressed by Γ, and a dimensionless interface barrier Z. A barrier of Z = 0 corresponds to a fully transparent M-SC contact while for larger values Z >1 the contact is in the M/I/SC tunneling regime. For materials with a non negligible resistance compared to the point contact, a serial resistance R_s is introduced to account for the voltage drop in the sample [24].

To determine the transport spin polarization of SRO films on STO (111), we have performed PCS measurements in a variable temperature (1.5–300 K) cryostat. The point contacts were formed by pushing a chemically etched Nb tip on the sample surface with the probe thermalized in ^{4}He gas. A piezo motor and scan tube were used to vary the distance and the position between tip and sample. The current-voltage (I-V) characteristics were measured by a conventional four probe method and a lock-in technique was used to measure the differential conductance dI/dV spectra as function of the applied voltage directly.

In Fig. 2a an experimental spectra recorded on SRO $(111)_{pc}$ is shown, normalized to the conductance value at high voltages. The best fit (line), obtained with the parameters given in the figure, yields a spin polarization of P ≈ 42%. Although the point contact is expected to be in the diffusive regime (estimated contact radius $r_c \approx 20$ nm using Wexler's formula [25] and ρ(5 K) ≈ 200 μΩcm), for simplicity ballistic formulas were used, as it has been shown that the error translates into Z instead of P [24]. As a possible interplay of the different parameters is not easy to control, it was proposed [26] to perform an optimization process in which for each fixed P_{trial} the best fit in (Δ, Γ and Z) is calculated. The resulting χ^2 values present a minimum at the correct spin polarization. In the right inset of Fig. 2a, the calculated χ^2 values for the spectra of Fig. 2a are shown. There is a clear minimum around P ≈ 42%, in agreement with the best fit of the data.

Although in a similar range, the value of P ≈ 42% for the SRO films grown on STO (111) is somewhat reduced compared to previously reported values P ≈ 53% [17,18] for high purity SRO samples grown on vicinal SRO (001). In principle, this difference could be either due to an anisotropy of the transport spin polarization P or a reduction of P in the present samples as an effect of structural disorder and/or strain. To check for these two possibilities, PCS was also performed on equivalent samples grown on SRO (100), with an RRR ≈ 2 (Fig. 2b). The best fit is obtained for P = 45% and again the value of P is confirmed by the χ^2 analysis in the inset. We thus conclude that our (111) and (001) SRO samples lead to similar spin polarization values: P(111) ≈ 42 % and P(001) ≈ 45 %.

Before discussing on the implications of this isotropic spin polarization, it is worth to comment on origin of the dissimilarity of the corresponding measured spectra (Figs. 2a and 2b). Some insight can be gained from an analysis of the other fit parameters. It is observed that the most remarkable changes refer to the value of the superconducting energy gap; the superconducting gap of the spectra in Fig. 2b (Δ ≈ 1.09 eV) is clearly reduced (in agreement with the experimental observation of a reduced T_c of the Nb tip) compared to that of Fig. 2a (Δ ≈ 1.45 eV) and that of bulk Nb, and also considerably more smeared out (Γ). Although the reason for this reduction (tip degradation, a proximity effect of the magnetic SRO, stronger in (001) direction) cannot be decided on the basis of the present data, the reduction of Δ and increase of Γ explains the different shapes of the spectra in Figs. 2a and 2b.

We turn now to the major question addressed in this manuscript, that close similarity of the experimental values of P(111) ≈ P(001). This finding gives support to available band structure calculations [16] predicting an isotropic spin-dependent Fermi velocity, while contradicting other theoretical calculations [19] indicating a substantial anisotropies in the Fermi surface and Fermi velocities. Our data do not support such scenario. However, since the residual resistivity ratio of our films is RRR ≈ 2 thus implying substantial scattering in the point contact area. To what extent these scattering events could average the spin-dependent Fermi velocity angular distribution and thus produce a more isotropic spin polarization of SRO cannot be addressed on the basis of present measurements. On the other hand, our data for P(111) ≈ P(001) ≈ 42-45 % are comparable but definitely smaller than that reported by Nadgorny et al [17] (P(001) ≈ 53 %); it can not be excluded that the above mentioned scattering mechanism could be at the origin of the observed relatively lower spin polarization.

4. Conclusions

In conclusion, flat epitaxial SRO thin films have been grown by rf magnetron sputtering on STO (111) oriented substrates. The transport spin polarization was measured by PCS and found to be P ≈ 42%. The finding of a very similar value of P ≈ 45% for SRO (001) films supports available band structure calculations predicting an isotropic transport spin polarization.

Acknowledgements:
Funding from the EU (Grant No. SemiSpinNano-237375) is acknowledged.

References

[1] Worledge D C and Geballe T H 2000 *Phys. Rev. Lett.* **85** 5182
[2] Eom C B, Cava R J, Fleming R M, Philips J M, van Dover R B, Mrashall J H, Hsu J W P, Krajewski J J and Peck W F 1992 *Science* **258** 1766
[3] Antognazza L, Char K, Geballe T H, King L L H and Sleight A W 1993 *Appl. Phys. Lett.* **63** 1005
[4] Aytug T, Wu J Z, Cantoni C, Verebelyi D T, Specht E D, Paranthaman M, Norton D P, Christen D K, Ericson R E and Thomas C L 2000 *Appl. Phys. Lett.* **76** 760
[5] Eom C B, Van Dover R B, Philips J M, Werder D J, Marshall J H, Chen C H, Cava R J, Fleming R M and Fork D K 1993 *Appl. Phys. Lett.* **63** 2570
[6] Takahashi K S, Sawa A, Ishii Y, Akoh H, Kawasaki and Tokura Y 2003 *Phys. Rev. B* **67** 094413
[7] Rigato F, Piano S, Foerster M, Giubileo F, Cucolo A M and Fontcuberta J 2010 *Phys. Rev. B* **81** 174415
[8] Foerster M, Rigato F, Bouzehouane K and Fontcuberta J 2010 *J. Phys. D: Appl. Phys.* **43** 295001
[9] Klein L, Dodge J S, Ahn C H, Reiner J W, Mieville L, Geballe G H, Beasley M R and Kapitulnik A 1996 *J. Phys.: Condens. Matter* **8** 10111
[10] Herranz G, Sanchez F, Fontcuberta J, Garcia-Cuenca, M V, Ferrater C, Varela M, Angelova T, Cros A and Cantarero A 2005 Phys. Rev. B **71** 174411
[11] Herranz G, Martínez B, Fontcuberta J, Sánchez F, Ferrater C, García-Cuenca M V and Varela M 2003 *Phys. Rev. B* **67** 174423
Sanchez F, Rigato F, Bachelet R and Fontcuberta J 2008 *Journal of Crystal Growth* **310,** 3348
[12] Moodera J, Santos T and Nagahma T 2007 *J. Phys.: Cond Matter* **19** 165202
Ramos A V, Guittet M and Moussy J 2007 *Appl. Phys. Lett.* **91** 122107
Takahashi Y, Kasai A, Furubayashi T, Mitani S, Inomata K and Hono K 2010 *Appl. Phys. Lett.* **96** 072512
[13] Lüders U, Sánchez F and Fontcuberta J 2004 *Phys. Rev. B* **70** 045403
[14] Rigato F 2010 *PhD thesis, Universidad Autonoma de Barcelona*
[15] Rubi D, Vlooswijk A H G and Noheda B 2009 *Thin solid films* **517** 1904
[16] Mazin I I and Singh D J 1997 *Phys. Rev. B* **56** 2556
Singh D J 1996 *J. Appl. Phys.* **79** 4818
[17] Nadgorny B, Osofsky M S, Singh D J, Woods G T, Soulen Jr. R J , Lee M K, Bu S D and Eom C B 2003 *Appl. Phys. Lett.* **82** 427
[18] Raychaudhuri P, Mackenzie A P, Reiner J W and Beasley M R 2003 *Phys. Rev. B* **67** 020411(R)
[19] Allen P B, Berger H, Chauvet O, Forro L, Jarlborg T, Junod A, Revaz B, and Santi G 1996 *Phys. Rev. B* **53** 4393
[20] Gan Q, Rao R A, Eom C B, Garrett and Lee M 1998 *Appl. Phys. Lett.* **72** 978
[21] Ito A, Masumoto H and Goto T 2007 Mater. *Trans.* **48** 227
[22] Soulen Jr. R J, Byers J M, Osofsky M S, Nadgorny B, Cheng S F, Ambrose T, Broussard P R, Tanaka C T, Nowak J, Moodera J S, Barry A and Coey J M D 1998 *Science* **282** 85
Upadhyay S K, Palanisami A, Louie R N and Buhrman R A 1998 *Phys. Rev. Lett.* **81** 3247
[23] Mazin I I, Golubov A A, and Nadgorny B 2001 *J. Appl. Phys.* **89** 7576
[24] Woods G T, Soulen Jr R J, Mazin I, Nadgorny B, Osofsky M S, Sanders J, Srikanth H, Egelhoff W F and Datla R 2004 *Phys. Rev. B* **70** 054416
[25] Nikolic B and Allen P B 1999 *Phys. Rev. B* **60** 3963
[26] Bugoslavsky Y, Miyoshi Y, Clowes S K, Branford W R, Lake M, Brown I, Caplin A D and Cohen L F 2005 *Phys. Rev. B* **71**, 104523

Joint European Magnetic Symposia – JEMS 2010 IOP Publishing
Journal of Physics: Conference Series **303** (2011) 012069 doi:10.1088/1742-6596/303/1/012069

Magnetic properties of the magnetoelectric compound Cu_2OSeO_3: Magnetization and ^{77}Se NMR study

M. Belesi[1], **T. Philippe**[1], **I. Rousochatzakis**[2], **H. C. Wu**[3], **H. Berger**[1], **S. Granville**[1], **I. V. Shvets**[3], **J. -Ph. Ansermet**[1]

[1] Institut de Physique de la Matière Condensée, Ecole Polytechnique Fédérale de Lausanne, Station 3, CH-1015 Lausanne-EPFL, Switzerland
[2] Max Planck Institut für Physik Komplexer Systeme, Nöthnitzer Str. 38, 01187 Dresden, Germany
[3] CRANN and School of Physics, Trinity College, Dublin 2, Ireland

E-mail: `mariaeleni.belesi@epfl.ch`

Abstract. We present magnetization and ^{77}Se Nuclear Magnetic Resonance (NMR) measurements in single crystals of the magneteoelectric compound Cu_2OSeO_3. The temperature and field dependence of the magnetization suggest a ferrimagnetic ordering at $T_c \simeq 60$ K in a 3up-1down configuration. The easy axis of the magnetization is along the [100] crystallographic direction. The ^{77}Se NMR line shape data collected at 14.09 T are consistent with the symmetries imposed by the cubic $P2_13$ space group in the temperature range 20–290 K and confirm that the magnetic transition is not accompanied by any lowering of the crystal symmetry as has recently been proposed by Bos *et al.* [*Phys. Rev.* B **78** 094416 (2008)].

1. Introduction

Materials that show cross coupling effects between magnetic and electric properties are currently attracting broad experimental and theoretical interest.[1, 2, 3, 4] The control of magnetic order by electric fields and vice versa is important for fundamental research but also for potential technological applications.[5] Even though the research in magnetoelectric materials goes back several decades,[6, 7] a revival of interest has taken place in recent years. This is mainly due to the developments in thin film technology (which have made possible the fabrication of thin films with enhanced magnetoelectric properties as compared to bulk materials[4]), but also due to the discovery of multiferroic compounds such as $TbMnO_3$[8] and $TbMn_2O_5$[9], where the ferroelectricity is driven directly by the spin order. Despite the significant progress over the last years, several problems remain still open in this field. For example, only a relatively small number of magnetoelectric materials exist in nature, or have been synthesized in the lab.[5] Furthermore, the coupling between spontaneous electric polarization P and magnetic moment M in multiferroic materials is weak and high electric (magnetic) fields are needed to switch the magnetization (electrical polarization).

Apart from the multiferroic compounds another interesting class of materials that exhibits magnetoelectric properties exists, such as a change in the dielectric constant at the magnetic ordering temperature or finite magnetocapacitance, but not a spontaneous polarization (in contrast to multiferroics). Materials exhibiting the so called magnetodielectric effect are, among others, $SeCuO_3$,[10] $TeCuO_3$,[10] $EuTiO_3$[11], $TmFeO_3$[12], and $Tb_3Fe_5O_12$[13]. Despite the fact that these materials are not ferroelectric, a direct coupling between the electric and magnetic ordering exists. The magnetodielectric effect is commonly accompanied by a distortion of the unit cell which suggests that the magnetoelectric coupling proceeds via lattice deformations. An exception to this behavior has recently been found in polycrystalline samples of Cu_2OSeO_3. [14] In Cu_2OSeO_3 the magnetic ordering is accompanied by a magnetocapacitance signal and an anomaly in the dielectric constant, but the lattice remains metrically cubic down to 10 K according to high resolution powder x-ray diffraction (XRD) data. [14] Motivated by the work of Bos *et al.*, [14] we have performed a magnetization and ^{77}Se NMR study on single crystals of Cu_2OSeO_3. Our results confirm the transition from the high temperature paramagnetic phase to a low-T ferrimagnetic state at $T_c \simeq 60$ K, whereby 3/4 of the Cu^{2+} ions are aligned parallel and 1/4 are aligned antiparallel to some axis. We further show that this magnetic configuration remains stable at least up to 14 T and that the preferential direction (easy axis) of the magnetization is along the [100] crystallographic direction. The transition to the low temperature ferrimagnetic state is probed by ^{77}Se NMR line shape measurements collected at 14.09 T, which show that the transition is not accompanied by any spontaneous lattice distortion in agreement with the high resolution XRD data on powder samples reported previously by Bos *et al.*[14]

2. Synthesis and characterization

We have grown single crystals of Cu_2OSeO_3 by the standard chemical vapor phase method. Mixtures of high purity CuO (Alfa-Aesar, 99.995%) and SeO_2 (Alfa-Aesar, 99.999%) powder in molar ratio 2:1 were sealed in quartz tubes with electronic grade HCl as the transport gas for the crystal growth. The ampoules were then placed horizontally into a tubular two-zone furnaces and heated very slowly by 50°C/h to 600°C. The optimum temperatures at the source and deposition zones for the growth of single crystals have been 610°C and 500°C, respectively. After six weeks, many dark green, almost black Cu_2OSeO_3 crystals with a maximum size of 8x6x3mm were obtained. X-ray powder diffraction analysis was conducted on a Rigaku X-ray diffractometer with Cu Kα radiation ($\lambda = 1.5418 \mathring{A}$) and the electron microprobe was used for chemical analysis of all solid samples. The diffraction data are similar to previously published data,[15, 14, 16] and show that the material crystallizes in the cubic $P2_13$ space group. The orientations of the crystal axes with respect to the crystal faces were determined by Laue X-ray back-scattering measurements. A detailed description of the crystal structure of this compound can be found in Refs. [15, 14, 16].

3. Experimental results

3.1. Magnetization measurements

For the magnetic measurements we have employed two Quantum Design Physical Properties Measurement Systems (at EPFL and Trinity College). We have measured the field dependence of the magnetization and have carried out zero-field cooled (ZFC) and field cooled (FC) magnetization measurements along various crystallographic directions in the temperature range 2-300 K and for fields of up to 14 T.

The field dependence of the magnetization was measured at 5 K (Fig. 1a) with the magnetic field applied along [111] (circles) and [100] (squares) crystallographic directions. The magnetization saturates at approximately 1 kOe when the magnetic field is applied along [100] axis, while a stronger field is needed (1.5 kOe) in order to saturate the magnetization along [111] direction. No coercive field was measured and the saturation value is 0.53 μ_B/Cu at 5 K, which is the same for both crystallographic directions. This saturation value is half of the full moment of all Cu^{2+} ions and suggests a simple collinear ferrimagnetic alignment where 3/4 of the copper

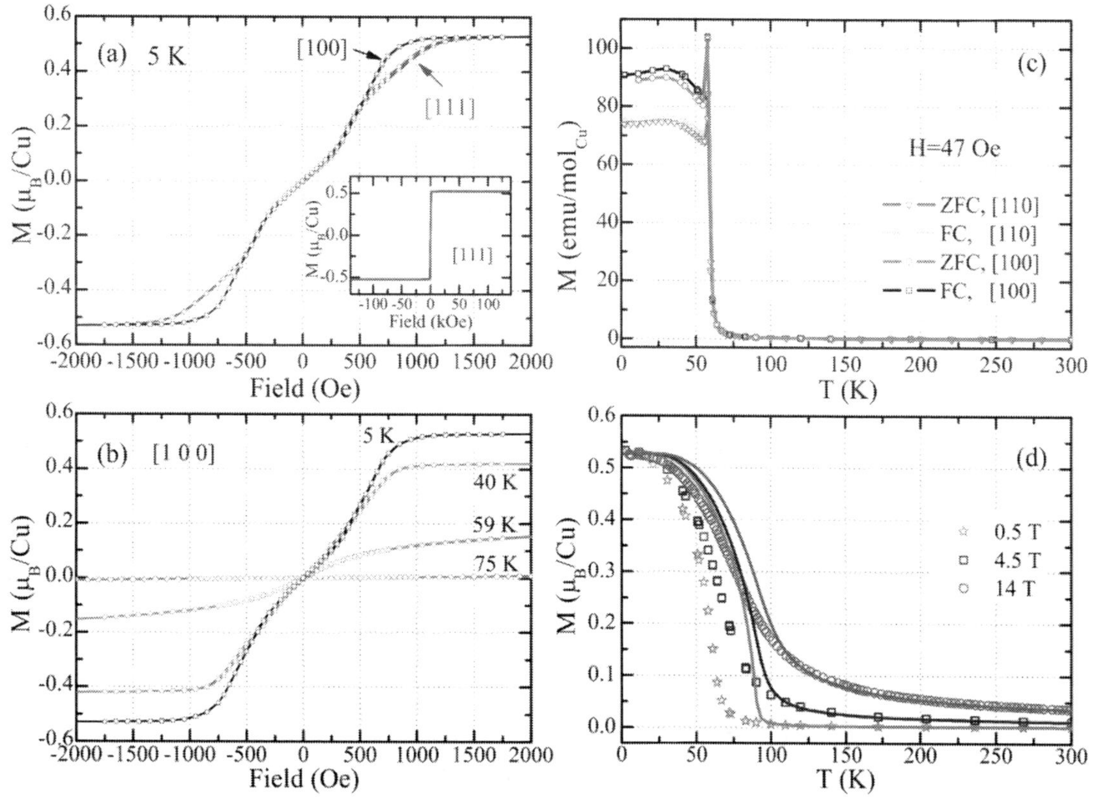

Figure 1. (a) Field dependence of the magnetization measured at 5 K with the magnetic field applied along [111] and [100] crystallographic directions. The inset shows the field dependence of the magnetization along [111] at 5 K where the magnetic field was swept up to 14 T. (b) The field dependence of the magnetization at various temperatures collected with the magnetic field applied along [100]. (c) The temperature dependence of the ZFC and FC magnetization measured in a magnetic field of 47 Oe applied along [110] and [100] direction. (d) The temperature dependence of the FC magnetization measured in applied fields of 0.5 T (stars), 4.5 T (squares) and 14 T (circles). The solid lines are high-T fits based on the mean field solution of a microscopic Heisenberg model introduced in Ref. [17].

ions are aligned parallel (Cu2 ions in 12b sites) and 1/4 are aligned antiparallel (Cu1 ions in 4a sites) to the applied magnetic field. This result is in agreement with the data reported by Bos *et al.*[14] in powder samples. We have extended our measurements to higher magnetic fields (Fig. 1a inset) and we have found that the 1/2 magnetization plateau remains stable at least up to 14 T. At low magnetic fields the magnetization grows linearly and at approximately 400 Oe a magnetization jump with a small amount of magnetic hysteresis is observed, in agreement with Ref. [14]. The field dependence of the magnetization with the magnetic field applied along the easy axis [100] was further measured at 5, 40, 59 and 75 K and the results are presented in Fig. 1b. At 75 K, i.e. in the paramagnetic phase, the typical linear field dependence is observed. In the ordered state we observe a decrease in the saturation value of the magnetization (and an increase in the saturation field) by increasing temperature due to thermal fluctuations.

The temperature dependence of the ZFC and FC magnetization collected at 47 Oe is shown in Fig. 1c. The measurements were performed for two different directions of the magnetic field, one along [110] and the other along the easy axis [100]. The magnetization jumps at 60 K and shows hysteresis between ZFC and FC data in agreement with published data in powder samples.[14] The magnetization measurements collected for magnetic fields higher than the metamagnetic transition are presented in Fig. 1d. The solid lines in Fig. 1d are theoretical fits based on a mean-field solution of a microscopic Heisenberg spin model introduced in Ref. [17]. This model consists of two Heisenberg exchange parameters between nearest-neighbor Cu^{2+} ions, one ferromagnetic $J_{\rm fm}$ between Cu2 sites and one antiferromagnetic $J_{\rm afm}$ between Cu1 and Cu2 sites. The optimal fit to the data with the self-consistent solution of the corresponding mean-field equations provides $g \simeq 2.11$,[16] $J_{\rm afm} \simeq 68$ K and $J_{\rm fm} \simeq -50$ K. The fits are quite accurate in a wide high temperature range above 100 K. At lower temperatures the mean-field theory is not accurate since it overestimates the critical temperature $T_c^{\rm MF} = \frac{1}{2}(\mid J_{fm} \mid + \sqrt{J_{fm}^2 + 3J_{afm}^2}) \simeq 89K$. In addition the mean field theory neglects the quantum fluctuations which reduce the length of the ordered moments.

3.2. Nuclear magnetic resonance

NMR is a powerful local probe which can provide information on both magnetic and structural properties. Here, we have applied pulsed NMR techniques on ^{77}Se nuclei with nuclear spin $I = 1/2$ and gyromagnetic ratio $\gamma_N/2\pi = 8.1179$ MHz/T. The NMR experiments were performed with a home made spectrometer at an Oxford 14.09 T magnet. The magnet is equipped with a variable temperature insert and the measurements were performed in the temperature range 20-295 K. The NMR spin-echo was produced by a standard Hahn echo pulse sequence with a typical $\pi/2$ pulse length of $6\mu s$. The separation between the two echo generating pulses was kept much shorter than the spin-spin relaxation time T_2. For narrow lines the NMR spectra were measured by a Fourier transform of the half spin-echo signal, while for broad lines they were obtained by plotting the area of the echo as a function of the irradiation frequency (frequency sweep).

The NMR line shapes were found to have strong orientational and temperature dependence. In Fig. 2(a) we show the NMR spectra at a representative high temperature value (220 K) for three different directions of the magnetic field with respect to the crystallographic axes. A comparison between the three directions shows that there are differences with respect to the number of peaks per direction, to the resonance frequency and also to their relative intensities. Two lines of the same integrated area, 4:4, are observed in [100] direction. These two lines split into four peaks when the field is applied along [110] or [111], while the ratio of the relative integrated areas along these directions is 2:2:2:2 and 3:1:3:1 respectively. An explanation of these findings can arise by considering the crystal symmetry but also the 3up-1down magnetic configuration. In Cu_2OSeO_3, there are two crystallographically inequivalent groups of selenium sites containing four selenium ions each.[14] The selenium ions occupy non-cubic sites thus in general more than one line per selenium group is expected. The dipolar contribution to the NMR line shape can be calculated[17] since the positions of the Cu ions are known.[15, 14, 16] It comes out that when the magnetic field is applied along [100] all the eight selenium ions experience the same dipolar field thus only one line should be observed, in contrast to our experimental results which show two lines. Hence apart from the dipolar interactions we must also invoke the transferred hyperfine interactions at the selenium sites. A detailed calculation of the different contributions in the ^{77}Se NMR lines of Cu_2OSeO_3 based on the local symmetry of selenium sites can be found in Ref. [17].

In order to probe the transition to the low temperature ferrimagnetic state we have studied the temperature dependence of the NMR spectra along the [111] axis. As can be seen in Fig. 2b by lowering the temperature the relative shift for each one of the four NMR peaks from the Larmor frequency (dotted line) increases. The temperature dependence of the NMR shifts is related to the spin density at the selenium sites, and the enhancement at low temperatures signals the transition to a magnetically ordered state. Two important observations can be made

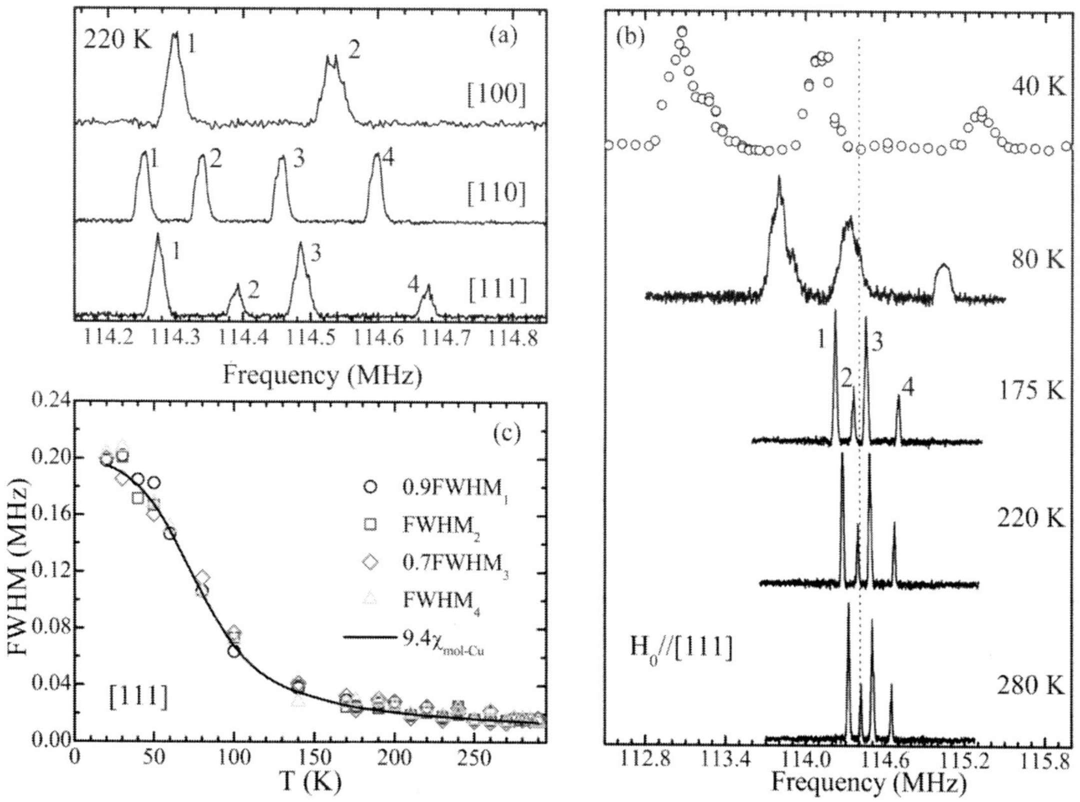

Figure 2. (a) ^{77}Se NMR line shape measurements collected at 220 K with the magnetic field, $H_0 = 14.09$T, applied parallel to three different crystallographic directions i.e., [100], [110] and [111]. (b) The ^{77}Se NMR spectra at representative temperatures obtained with the magnetic field applied along [111]. (c) The temperature dependence of the full width at half maximum (FWHM) for each one of the four ^{77}Se NMR lines shown in (b).

from the NMR spectra. First, no splitting of the NMR lines is observed and the ratio of the relative integrated areas for peaks 1-4 remains 3:1:3:1 down to 20 K, which is much lower than the transition temperature. This result is important since any symmetry lowering from the $P2_13$ crystal structure would result in a partial or full splitting of the 3:1:3:1 pattern observed. Second, the Full Width at Half Maximum (FWHM) for each one of the four spectral lines increases by decreasing temperature (Fig. 2c) and follows the temperature dependence of the bulk magnetization as measured by the magnetometer (solid line in Fig. 2c). This is the typical behavior of inhomogeneous broadening. Both of these results are important since they show that there is no modification of the crystal lattice to a state with symmetry lower than the original cubic $P2_13$ space group.

4. Conclusions

We have presented magnetization and high field ^{77}Se NMR data in single crystals of the magneteoelectric compound Cu_2OSeO_3. The temperature and the field dependence of the magnetization suggest that the high field magnetic state is a simple collinear ferrimagnetic arrangement with three majority and one minority Cu^{2+} magnetic moment. The ferrimagnetic ordering temperature is $T_c \simeq 60$ K and the preferential direction of the magnetization is along the [100] crystal axis. The magnetization data can be understood by a simple mean field model with two nearest neighbor magnetic interactions, one ferromagnetic between Cu2 ions and one antiferromagnetic between Cu1 and Cu2 sites. The values of the exchange parameters ($J_{\rm afm} \simeq 68$ K, $J_{\rm fm} \simeq -50$ K) were obtained by fitting a wide high temperature range of the magnetization data.

The ^{77}Se NMR spectra do not show any splitting or abrupt broadening through the magnetic transition thus there is no reduction in the crystal symmetry from the high temperature $P2_13$ space group. Our results are in agreement with previously published data in polycrystalline samples by Bos *et al.*,[14] and exclude the possibility that the magnetoelectric coupling in Cu_2OSeO_3 proceeds via a spontaneous lattice distortion below T_c. This is also confirmed by far-infrared[18] and Raman[19] studies in single crystals of Cu_2OSeO_3 prepared in the same way as our crystals.

Acknowledgments
We would like to thank F. Mila and K. Schenk for useful discussions. We also acknowledge experimental assistance from K. Schenk, and M. Zayed. One of the authors (H. B.) acknowledges financial support from the Swiss NSF and by the NCCR MaNEP.

References
[1] Eerenstein W, Mathur N D and Scott J F (2006) *Nature* **442** 759
[2] Spaldin N A, Fiebig M (2005) *Science* **309** 391
[3] Khomskii D (2009) *Physics* **2**, 20
[4] Wang K F, Liu J-M and Ren Z F (2009) *Advances in Physics* **58** 321
[5] Hill N A (2000) *J. Phys. Chem. B* **104** 6694
[6] Dzyaloshinskii I E (1959) *Sov. Phys. JETP* **10** 628
[7] Astrov D N (1609) *Sov. Phys. JETP* **11** 708
[8] Kimura T, Goto T, Shintani H, Ishizaka K, Arima T and Tokura Y (2003) *Nature* **426** 55
[9] Hur N, Park S, Sharma P A, Ahn J S, Guha S and Cheong S-W (2004) *Nature* **29** 392
[10] Lawes G, Ramirez A P, Varma C M and Subramanian M A (2003) *Phys. Rev. Lett.* **91** 257208
[11] Katsufuji T and Takagi H (2001) *Phys. Rev.* B **64** 054415
[12] Muralidharan R, Jang T-H, Yang C-H, Jeong Y H and Koo T Y (2007) *Appl. Phys. Lett.* **90** 012506
[13] Hur N, Park S, Guha S, Borissov A, Kiryukhin V and Cheong S-W (2005) *Appl. Phys. Lett.* **87** 042901
[14] Bos J-W G, Colin C V, and Palstra T T M (2008) *Phys. Rev.* B **78** 094416
[15] Effenberger H and Pertlik F (1986) *Monatsh. Chem.* **117** 887
[16] Larrañaga A, Mesa J L, Lezama L, Pizarro J L, Arriortua M I, and Rojo T (2009) *Materials Research Bulletin* **44** 1
[17] Belesi M, Rousochatzakis I, Wu H C , Berger H, Shvets I V, Mila F and Ansermet J P (2010) *Phys. Rev.* B **82** 094422
[18] Miller K H, Xu X S, Berger H, Knowles E S, Arenas D J, Meisel M W and Tanner D B, arXiv:1006.467v1 [cond-mat.mtrl-sci]
[19] Gnezdilov V P, Lamonova K V, Pashkevich Yu G, Lemmens P, Berger H, Bussy F and Gnatchenko S L (2010) *Fizika Nizkikh Temperatur* **36** 688

Magnetocaloric effect in polycrystalline La based manganites modified with Sr

A Kolano-Burian[a1], R Szymczak[b], R Kolano[a], H Szymczak[b], A Burian[c], Ł Hawełek[c], P Zackiewicz[a] and M Czepelak[a]
[a]*Institute of Non-Ferrous Metals, Gliwice, Poland*
[b]*Institute of Physics, Polish Academy of Sciences, Warszawa, Poland*
[c]*A. Chelkowski Institute of Physics, University of Silesia, Katowice, Poland*

E-mail: aleksandra.kolano@imn.gliwice.pl

Abstract. All In this paper, we present magnetocaloric properties of the polycrystalline $La_{0.7}Ca_{0.25}Sr_{0.05}MnO_3$ compound prepared by the ceramic method. X-ray diffraction indicated a single-phase compound with a perovskite structure. The value of magnetocaloric effect has been determined from the measurements of magnetization as a function of temperature and an external magnetic field. ΔS was found to reduce from 8 J/kgK to 4.5 J/kgK under $\Delta H = 2$ T, and the same time, the Curie temperature increase from 250 to 279 K for $x = 0$ and 0.05 respectively. From the functional point of view, the materials to be used for magnetic refrigeration should be characterized by the greatest possible entropy change near room temperature. The results from this work showed that $La_{0.7}Ca_{0.25}Sr_{0.05}MnO_3$ is interesting material and potential candidate for magnetic refrigerants.

Introduction

Since the discovery of giant magnetocaloric effect in $Gd_5Si_2Ge_2$ the search for materials exhibiting large magnetic entropy change has intensified. A large number of paper about the materials with magnetocaloric effect has been published, which demonstrates that many research centres are trying to develop new MCE materials which could be used in industrial production of refrigerators and heat pumps. The manganites such as $La_{0.7}Ca_{0.25}Sr_{0.05}MnO_3$ are characterised by chemical stability, increased resistivity and low production costs, which makes this compound an interesting material for magnetic refrigeration [1]. Our present study showed that $La_{0.7}Ca_{0.25}Sr_{0.05}MnO_3$ could be interesting material for the magnetic refrigeration around room temperature and good alternative for expensive gadolinium based refrigerators.

Experimental

The polycrystalline samples of composition $La_{0.7}Ca_{0.3}MnO_3$ and $La_{0.7}Ca_{0.25}Sr_{0.05}MnO_3$ were prepared by the ceramic method. The lanthanum oxide [La2O3] of 99.99% purity, manganese oxide [Mn2O3] of 99% purity, and analytically pure calcium carbonate [CaCO3] were used as starting materials. These materials were weighed directly after drying conducted at a temperature of 220°C. The mixtures were manually wet mixed for 15 min in an agate mortar with an addition of acetone and then dried and again ground to a powder form. The powders obtained by this method were pressed under the pressure of 150 MPa into pellets. The pellets were then subjected to sintering in small aluminum oxide crucibles at 1150°C in a chamber furnace under air atmosphere. In the next technological cycle, the fine powder was sintered in ceramic crucibles and the obtained sinters were again ground to the powder form. These operations were repeated during subsequent technological cycles, with the sintering temperature increased for each cycle. Before final sintering was conducted at 1350°C the fine powders were pressed under the pressure of 200 MPa into pellets with a diameter of 20 mm. After completion of the final sintering, the samples were slowly cooled down together with the furnace. After final sintering of the samples, their density was calculated based on the measurements of mass and dimensions. For the compositions under investigation, the values of density were similar and reached about 5.2 g/cm^3. Identification of the chase and structural analysis were carried out by X-ray diffraction technique using the Rigaku-Denki D/MAX RAPID II-R diffractometer with a rotating anode AgK_α tube (λ=0.5608 Å), an incident beam (002) graphite monochromator and an image plate in the Debye–Scherrer geometry as a detector. The pixel size was 100 μm x 100 μm. The beam width at the sample was 0.1 mm. Both samples placed inside the glass capillaries (1.0 mm in diameter) were mounted on the goniometer axis for measurements. During the measurements the samples were spinning with the speed of 10°/sec. The intensity of the empty capillary was then subtracted yielding the pure diffraction patterns of the samples. The two-dimensional diffraction patterns were converted into the one-dimensional intensity data using suitable software. The diffraction patterns were analyzed by using the Rietveld refinement program FULLPRO and the starting parameters were taken for the refinement procedure from. Magnetization versus temperature and magnetization versus magnetic field curves were measured by using Quantum Design MPMS-5 Superconducting Quantum Interference Device (SQUID) magnetometer. The magnetic entropy changes were calculated from the magnetization data, which is based on a Maxwell relation.

Result and discussion

Structural properties

Powder X-ray diffraction indicated a single-phase compound and X-ray diffraction reflections were successfully indexed in the orthorombic system (*Pnma* space group). The Rietveld refinement procedure converged with the discrepancy coefficients χ^2 of 1.26% for $La_{0.7}Ca_{0.25}Sr_{0.05}MnO_3$. The structural parameters of $La_{0.7}Ca_{0.25}Sr_{0.05}MnO_3$ are listed in Table 1. The Rietveld intensities for $La_{0.7}Ca_{0.25}Sr_{0.05}MnO_3$ are compared with the experimental data in Fig. 1.

Table 1. Refined structural parameters from X-ray powder diffraction data for $A_{1-x}A_x^{'}MnO_3$ compounds having Pnma (No. 62) space-group symmetry. Numbers in parentheses are statistical errors of the last significant digit. The site occupancies are kept at fixed values consistent with the nominal stoichiometry. The atomic positions are Mn: 4*b* $\bar{1}$ (0,0,0.5); *La/Ca/Sr* site: O(1): 4*c* .*m*. (x,0.25,z); O(2): 8*d* 1 (*x*,*y*,*z*).

		$La_{0.7}Ca_{0.25}Sr_{0.05}MnO_3$
a	(Å)	5.4752(4)
b	(Å)	7.7458(8)
c	(Å)	5.4890(5)
V	(Å^3)	232.79(3)
La/Ca/Sr site	x	0.0169(1)
	z	-0.0052(3)

O(1)	x	0.4933(8)
	z	0.0605(11)
O(2)	x	0.281(1)
	y	0.0289(8)
	z	0.722(1)
B	(Å^2)	0.999(7)
R_{WP}	(%)	9.86
χ^2	(%)	1.26

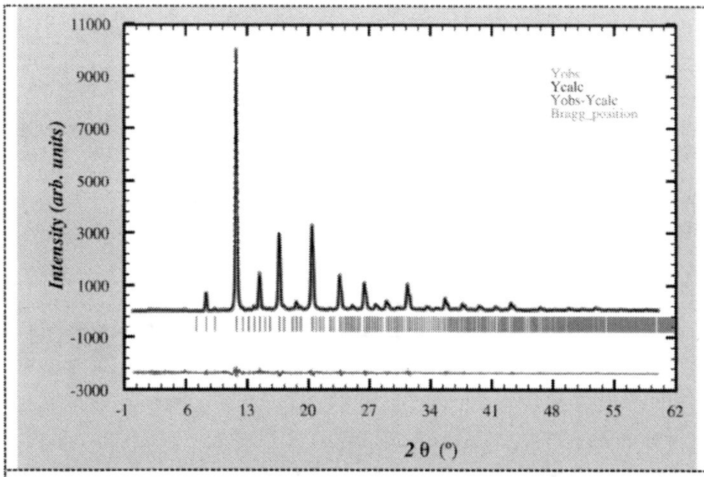

Figure 1. Rietveld plot of X-ray powder diffraction data at λ=0.5608 Å for $La_{0.7}Ca_{0.25}Sr_{0.05}MnO_3$ at room temperature.

The lattice parameters and the atomic coordinates of $La_{0.7}Ca_{0.25}Sr_{0.05}MnO_3$ are very close to those of reported in [2] for similar compositions and obtained using the neutron diffraction method. It indicates that the structure of the samples $La_{0.7}Ca_{0.25}Sr_{0.05}MnO_3$ investigated in the present work is practically the same as that of described in detail in [2] for $La_{0.7}Ca_{0.17}Sr_{0.13}MnO_3$.

Magnetic properties

Fig. 2 shows temperature dependence of magnetization for the $La_{0.7}Ca_{0.25}Sr_{0.05}MnO_3$ sample measured at different applied magnetic fields up to 50 kOe. From the dependence of magnetisation on temperature, measured at the magnetic field of 10 Oe, the Curie temperature T_c was determined by differentiation *($\delta M/\delta T$)*. It was found that the Curie temperature increase from 250 to 279 K for $La_{0.7}Ca_{0.3}MnO_3$ [3] and $La_{0.7}Ca_{0.25}Sr_{0.05}MnO_3$, respectively.

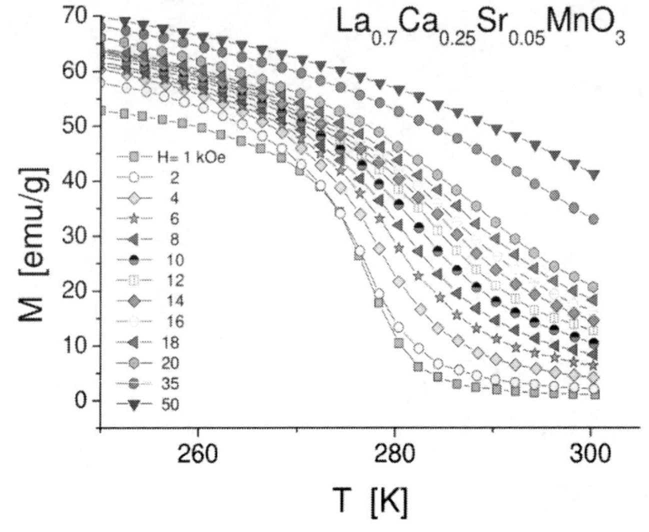

Figure 2. Temperature dependence of magnetization for the $La_{0.7}Ca_{0.25}Sr_{0.05}MnO_3$ sample.

As far as the magnetic transitions are concerned, one way of clarifying their nature is the determination of the respective critical exponents using a group of physical relations:

$$M(T) = M_0(-t)^\beta \text{ when } T < T_c \quad (1)$$

$$M(T) = DH^{1/\delta} \text{ when } T = T_c \quad (2)$$

$$H/M(T) = (h_0/M_0)\, t^\gamma \text{ when } T > T_c, H \to 0 \quad (3)$$

with M as magnetization in the applied magnetic field. The parameters β, δ and γ are the critical exponents and M_0, h_0 and D are the critical amplitudes. The parameter t is the reduced temperature, $(T-T_c)/T_c$. The critical exponents have been determined from the experimental data taken in a narrow temperature range around T_c from modified Arrott plots. This modified Arrott plot makes the $M^{1/\beta}$ versus $(H/M)^{1/\gamma}$ isotherms near the Curie temperature linear and parallel over a wide range of H/M values. Fig. 3 shows the modified Arrott plot for the $La_{0.7}Ca_{0.25}Sr_{0.05}MnO_3$ sample. The critical exponents β and γ were determined, which are equal to 0.3±0.02 and 1±0.02, respectively.

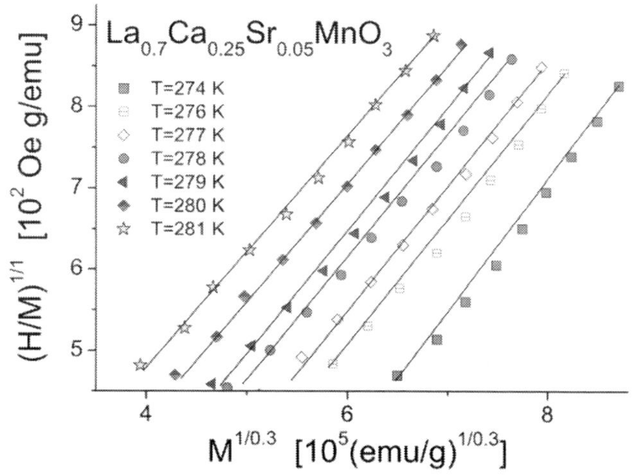

Figure 3. Arrott plot for the $La_{0.7}Ca_{0.25}Sr_{0.05}MnO_3$ sample.

It should be noted that the paramagnetic-ferromagnetic transition is very sharp, similar to that observed in case of first-order phase transitions.

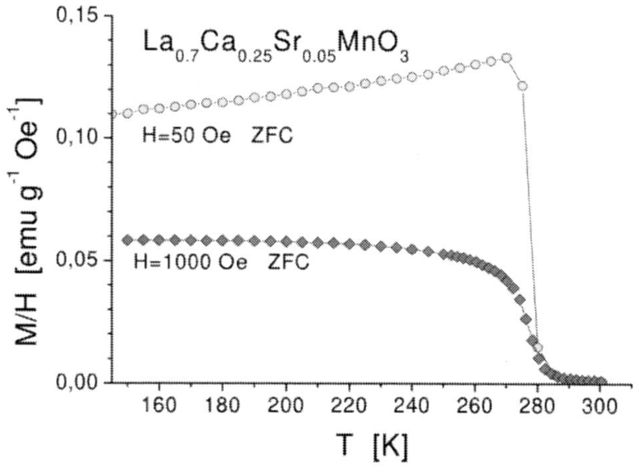

Figure 4. Dependence of M/H on T for the $La_{0.7}Ca_{0.25}Sr_{0.05}MnO_3$ sample.

Under this work, calculations of the magnetic entropy changes (ΔS_H) resulting from spin ordering were made for all alloys examined. For the isothermal processes, the total magnetic entropy change of the magnetic system due to the application of a magnetic field *H* is [4, 5]:

$$\Delta S_H(T,H) = \int_0^H \left(\frac{\partial M}{\partial T}\right)_H dH \qquad (4)$$

The maximum magnetic entropy change is obtained at the Curie temperature where the ferromagnetic – paramagnetic phase transition takes place. In order to evaluate ΔS_H, one usually resorts to numerical approximation to the integral in Equation (4) of the form:

$$\left|\Delta S_H(T,H)\right| = \sum_i \frac{M_i - M_{i+1}}{T_{i+1} - T_i} \Delta H_i \qquad (5)$$

where M_i is the experimental value of magnetization at the temperature T_i. Fig. 5 shows temperature dependence of the magnetic entropy changes at different applied magnetic fields for the $La_{0,7}Ca_{0,25}Sr_{0,05}MnO_3$ alloy. The obtained results, i.e. the entropy change of 4.5 J kg^{-1} K^{-1} occurring under the influence of magnetic field of 20 kOe near room temperature indicate that the magnetocaloric effect found in the polycrystalline La based manganites modified with 0.05% can be used for magnetic refrigeration. The entropy changes in the compounds without Sr (at the field of 20 kOe) was 8 J kg^{-1} K^{-1} [3], but the Curie temperature is much lower from a room temperature so that they cannot be used for magnetic refrigeration near room temperature.

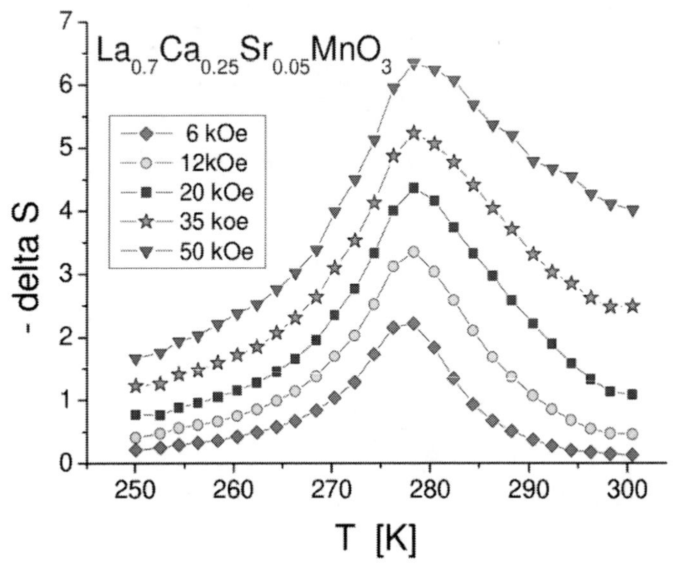

Figure 4. Temperature dependence of ΔS_M at different applied magnetic fields for the $La_{0,7}Ca_{0,25}Sr_{0,05}MnO_3$ alloy.

On the other hand, the magnetic cooling efficiency of magnetocaloric materials is evaluated by means of the relative cooling power (RCP) corresponding to the amount of heat transferred between the cold and the hot sinks in the ideal cycle [6]:

$$RCP(S) = -\Delta S_M^{\max}(T,H) \cdot \partial T_{FWHM} \qquad (6)$$

where ∂T_{FWHM} denotes the full width temperature span of ΔS_M versus T curie at its half maximum. The RCP value estimated for the $La_{0,7}Ca_{0,25}Sr_{0,05}MnO_3$ compound was 70 J kg^{-1} and 157 J kg^{-1} at the field of 20 kOe and 50 kOe, respectively.

Acknowledgement:

This work is supported by European Fund for Regional Development: POIG.01.03.01-00-058/08

References:

[1] M. Koubaa, W. Cheikh-Rouhou Koubaa, A. Cheikhrouhou, Journal of Alloys and Compounds, Volume 479, Issues 1-2, 24 June 2009, Pages 65-70,
[2] P.G. Radaelli, G. Iannone, M. Marezio, H.Y. Hwang, S-W. Cheong, J.D. Jorgensen and N. Argyriou, Phys. Rev. B 56 (1997) 8265.

[3] R. Szymczak, M. Czepelak, R. Kolano, A. Kolano-Burian, B. Krzymanska, H. SzymczakJ Mater Sci (2008) 43:1734–1739
[4] K.A Gschneider Jr., V.K. Pecharsky and A.O. Tsokol, *Rep. Prog. Phys.* 68 (2005) 1479-1483
[5] M.H. Phan and S.C. Yu, *Journal of Magnetism and Magnetic Materials* 308 (2007) 325-340
[6] K.A Gschneider Jr., V.K. Pecharsky Annu. Rev.Mater.Sci. 30 (2000) 387

Structural and magnetic characteristics of manganites in $La_{1-c+x}Sr_{c-x}Mn_{1-x}Ni_xO_3$ system

A M Smirnov[1], S Kh Estemirova[2], V K Karpasyuk[1]

[1]Astrakhan State University, 414056, Astrakhan, Russia

[2]Institute for Metallurgy UB RAS, 620016, Yekaterinburg, Russia

E-mail: vkarpasyuk@mail.ru

Abstract. Experimental data are shown for the influence of Ni and Sr concentration on phase composition, magnetic and electrical properties of La-Sr manganites with substitution of Mn by Ni. The system of chemical compositions $La_{1-c+x}Sr_{c-x}Mn_{1-x}Ni_xO_3$ is designed so that the sum of concentrations of divalent Sr and Ni remains constant (c). Under the condition that content of oxygen is stoichiometric, the values of c are equal to concentrations of Mn^{4+} (f.u.). These values are chosen near phase boundary "orthorhombic-rhombohedral structure" (c=0.15, 0.17, 0.19) on the classical phase diagram for $La_{1-c}Sr_cMnO_3$ system and at the region of rhombohedral metal phase ($0.20 \le c \le 0.35$). It was found that substituting Ni^{2+} ions shifted phase boundary to higher values of c. Of great interest is the phenomenon of phase conversion from rhombohedral to orthorhombic structure in the compositions with c=0.19 and 0.20 when x varies from 0.100 up to 0.125. Fragments of c-T and x-T phase diagrams "ferromagnetic-paramagnetic" have been plotted.

1. Introduction

Phase formation processes and phase properties in complex multicomponent manganites essentially depend on the concentration of heterovalent ions, their localization and radii, and the occurrence of cation and anion vacancies [1-3]. Magnetic and electric properties of manganites are determined by the probability of the transfer of e_g electrons between the positions of neighboring heterovalent ions Mn^{3+} and Mn^{4+}, the carriers concentration, the size mismatch of substituting cations, the number of unpaired electrons in their shells [4-5].

Divalent cation substitution (such as Mg^{2+}, Ni^{2+}, …) for Mn in manganites results in increasing of Mn^{4+} content and of Mn^{4+}/Mn^{3+} ratio. In order to study the doping effect at fixed ratio, a series of "double-doped" manganites $La_{2/3+4x/3}Sr_{1/3-4x/3}Mn_{1-x}Mg_xO_3$ was synthesized in [6]. Reported results are much different from that of the "single doped" manganites.

In the present work experimental data are shown for the influence of Ni concentration ($0.025 \le x \le 0.125$) on phase composition and magnetic properties of La-based manganites with substitution of Mn by Ni and, simultaneously, of La by Sr. In complex oxides, nickel ($3d^84s^2$) most commonly has an oxidation degree of +2. The system of chemical compositions $La_{1-c+x}Sr_{c-x}Mn_{1-x}Ni_xO_3$ is designed so that the sum of concentrations of divalent Sr and Ni remains constant (c). Under the condition that content of oxygen is stoichiometric, the values of c are equal to concentrations of Mn^{4+} (f.u.). These values are chosen near phase boundary "orthorhombic-rhombohedral structure"

Published under licence by IOP Publishing Ltd

(c=0.15; 0.17; 0.19) on the phase diagram for $La_{1-c}Sr_cMnO_3$ system [7] and at the region of rhombohedral metal phase (0.20≤c≤0.35).

2. Experimental

The experiments were performed on polycrystalline samples synthesized by traditional ceramic processing. The starting components (dried La_2O_3, $SrCO_3$, MnO_2, NiO powders) were mixed in stoichiometric proportions and ground in a ball mill with addition of alcohol. Pellets compacted of the obtained charge mixture were then preliminarily burned at 1000^0 C for 4 h. This operation was followed by grinding, introducing a binder (an aqueous solution of polyvinyl alcohol), pressing the samples, and burning out the binder. The final sintering step was performed at 1200^0 C for 10 h, and the samples were cooled together with the furnace. Then, in order to provide stoichiometric oxygen content, the samples were annealed at 950^0 C and partial pressure of oxygen $P_{O2}=10^{-1}$ Pa for 96 h. The choice of annealing conditions was based on the results of the works [8, 9].

Phase composition, and cell dimensions were determined by powder X-ray diffraction at room temperature (diffractometer Shimadzu XRD-7000, CuK_α radiation). The magnetization was determined by measuring the variation of magnetic flux that run through the measuring coil placed in the interpolar space of a constant magnet upon pulling a capsule with the powder sample out of it. The magnetic field strength in the interpolar space was 5600 Oe. Measurements of dc electrical characteristics and magnetoresistance were made using copper electrodes sputter-deposited onto opposite planes of pellets (thickness of about 4 mm). Magnetic field of 9200 Oe was parallel to the current direction. The temperature dependence of magnetic permeability (μ(T)) was measured by the induction method at a frequency of 98,6 kHz. The Curie point (T_c) was determined as the temperature corresponding to the maximum of $|d\mu/dT|$.

3. Results and discussion

Table 1 summarizes the structural parameters of initial (sintered) and annealed samples for the sets of composition. From the data obtained in [8, 9] it might be inferred that sintered manganites had an excess of oxygen content (γ>0). After annealing manganites became stoichiometric. Vacuum annealing results in decreasing of Mn^{4+} content and of cation vacancies concentration, that determines the changes of structural, magnetic, and electrical parameters.

Table 1. Phase composition and unit cell volume of synthesized manganites.

Sample				Phase composition	Unit cell volume/ formula unit, Å^3
Formula	c	x	State		
$La_{0.925}Sr_{0.075}Mn_{0.925}Ni_{0.075}O_{3+\gamma}$	0.15	0.075	Sintered	Orthorhombic	59.16
$La_{0.925}Sr_{0.075}Mn_{0.925}Ni_{0.075}O_3$			Annealed	Orthorhombic	59.03
$La_{0.905}Sr_{0.095}Mn_{0.925}Ni_{0.075}O_{3+\gamma}$	0.17	0.075	Sintered	Rhombohedral + Orthorhombic	-
$La_{0.905}Sr_{0.095}Mn_{0.925}Ni_{0.075}O_3$			Annealed	Orthorhombic	59.36
$La_{0.835}Sr_{0.165}Mn_{0.975}Ni_{0.025}O_{3+\gamma}$	0.19	0.025	Sintered	Rhombohedral	58.92
$La_{0.835}Sr_{0.165}Mn_{0.975}Ni_{0.025}O_3$			Annealed	Rhombohedral	58.79
$La_{0.860}Sr_{0.140}Mn_{0.950}Ni_{0.050}O_{3+\gamma}$	0.19	0.050	Sintered	Rhombohedral	58.79
$La_{0.860}Sr_{0.140}Mn_{0.950}Ni_{0.050}O_3$			Annealed	Rhombohedral	58.80

$La_{0.885}Sr_{0.115}Mn_{0.925}Ni_{0.075}O_{3+\gamma}$	0.19	0.075	Sintered	Rhombohedral	58.96
$La_{0.885}Sr_{0.115}Mn_{0.925}Ni_{0.075}O_3$			Annealed	Rhombohedral	
$La_{0.910}Sr_{0.090}Mn_{0.900}Ni_{0.100}O_{3+\gamma}$	0.19	0.100	Sintered	Rhombohedral	58.86
$La_{0.910}Sr_{0.090}Mn_{0.900}Ni_{0.100}O_3$			Annealed	Rhombohedral +Orthorhombic	58.82 58.93
$La_{0.935}Sr_{0.065}Mn_{0.875}Ni_{0.125}O_{3+\gamma}$	0.19	0.125	Sintered	Rhombohedral	58.90
$La_{0.935}Sr_{0.065}Mn_{0.875}Ni_{0.125}O_3$			Annealed	Orthorhombic	59.29
$La_{0.900}Sr_{0.100}Mn_{0.900}Ni_{0.100}O_{3+\gamma}$	0.20	0.100	Sintered	Rhombohedral	58.74
$La_{0.900}Sr_{0.100}Mn_{0.900}Ni_{0.100}O_3$			Annealed	Orthorhombic	59.06
$La_{0.840}Sr_{0.160}Mn_{0.900}Ni_{0.100}O_{3+\gamma}$	0.26	0.100	Sintered	Rhombohedral	58.66
$La_{0.840}Sr_{0.160}Mn_{0.900}Ni_{0.100}O_3$			Annealed	Rhombohedral	58.75
$La_{0.810}Sr_{0.190}Mn_{0.900}Ni_{0.100}O_{3+\gamma}$	0.29	0.100	Sintered	Rhombohedral	58.62
$La_{0.810}Sr_{0.190}Mn_{0.900}Ni_{0.100}O_3$			Annealed	Rhombohedral	58.59
$La_{0.780}Sr_{0.220}Mn_{0.900}Ni_{0.100}O_{3+\gamma}$	0.32	0.100	Sintered	Rhombohedral	58.49
$La_{0.780}Sr_{0.220}Mn_{0.900}Ni_{0.100}O_3$			Annealed	Rhombohedral	58.51
$La_{0.750}Sr_{0.250}Mn_{0.900}Ni_{0.100}O_{3+\gamma}$	0.35	0.100	Sintered	Rhombohedral	58.37
$La_{0.750}Sr_{0.250}Mn_{0.900}Ni_{0.100}O_3$			Annealed	Rhombohedral	58.33

Of great interest is the phenomenon of phase conversion from rhombohedral to orthorhombic structure in the compositions with c=0.19 when x varies from 0.100 up to 0.125, and in manganite $La_{0.900}Sr_{0.100}Mn_{0.900}Ni_{0.100}O_3$ (c=0.20) after annealing. These facts provide evidence for the shift of phase boundary "orthorhombic-rhombohedral structure" to higher values of c as compared with that established for $La_{1-c}Sr_cMnO_3$ system (c=0.175) [7].

The magnetization of samples with x = 0.1 slowly increases as a function of c (Figure 1 (a)), however compositions with c=0.35 exhibit decrease of magnetization when concentration of Ni increases (Figure 1 (b)).

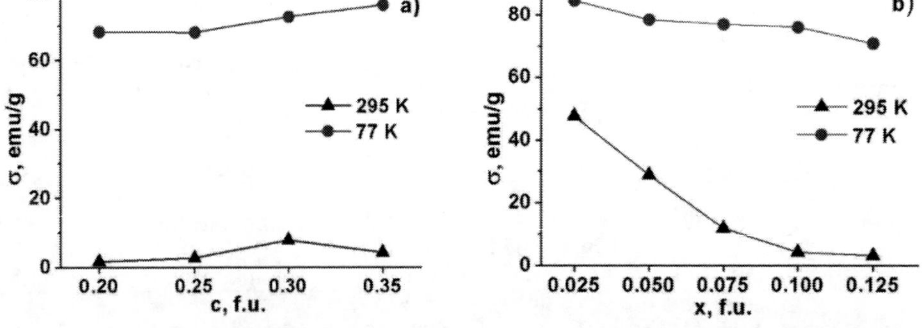

Figure 1. Dependences of magnetization on composition: a) x = 0.1; b) c= 0.35

Upon substitution, the Curie temperature of manganites increases as a function of c at fixed Ni concentration (Figure 2), and show a sharp decrease as a function of x (Figure 3).

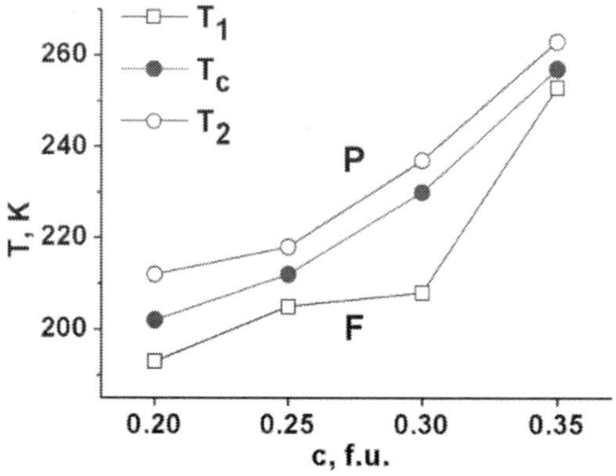

Figure 2. c-T phase diagram "ferromagnetic-paramagnetic" of manganites with x=0.1 (T_1, T_2 – temperature interval of phase transition)

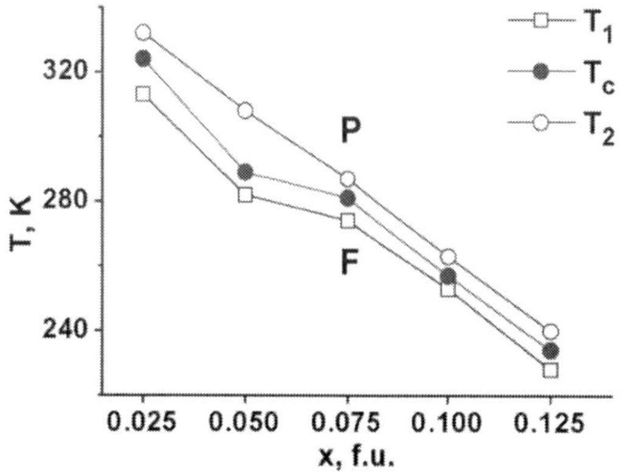

Figure 3. x-T phase diagram "ferromagnetic-paramagnetic" of manganites with c=0.35

The absolute value of magnetoresistance (MR=(R(H)-R(0))/R(0)) reaches 60% at 126 K in manganite $La_{0.85}Sr_{0.15}Mn_{0.9}Ni_{0.1}O_3$ (Figure 4), and oscillates in manganites with c=0.35 when Ni concentration rises (Figure 5). In latter case maximum value of 13 % takes place at 129 K in the sample $La_{0.7}Sr_{0.3}Mn_{0.95}Ni_{0.05}O_3$.

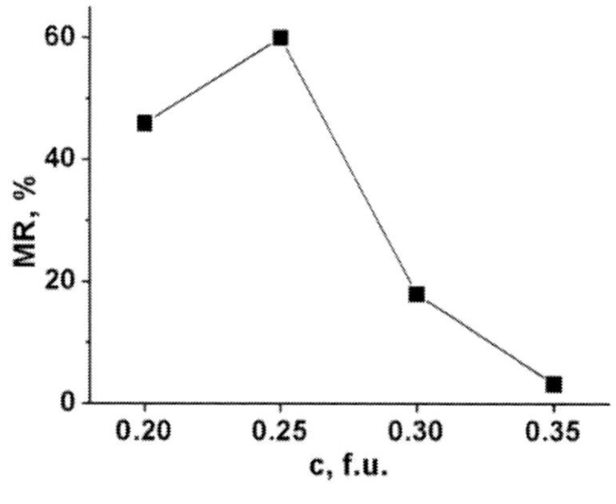

Figure 4. Magnetoresistance (maximum value) of manganites containing 0.1 f.u. of Ni depending on c

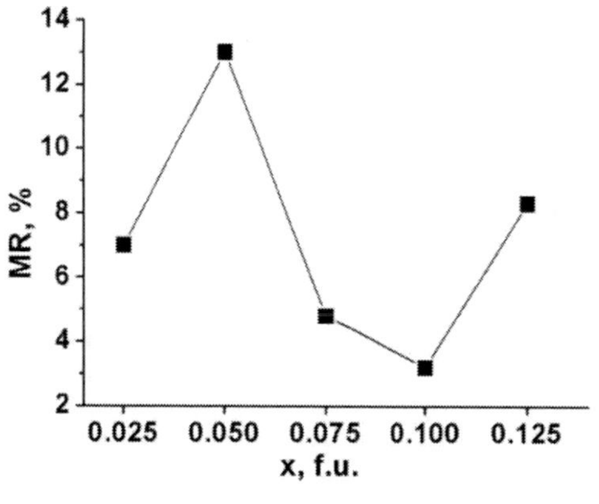

Figure 5. Magnetoresistance (maximum value) of manganites with c=0.35 depending on Ni concentration

4. Conclusion

The regularities of the influence of Sr^{2+} and Ni^{2+} concentrations on phase composition, saturation magnetization at different temperatures, Curie point, and magnetoresistance have been established for a series of "double-doped" manganites. It was found that substitution of Mn by Ni shifted the boundary of concentration phase transition "orthorhombic→ rhombohedral structure" in the system $La_{1-c+x}Sr_{c-x}Mn_{1-x}Ni_xO_3$ to higher values of c. Excess of oxygen content (γ>0) over stoichiometric one promotes the existence of rhombohedral phase at high values of x. Magnetization and T_c rise with c

increasing at fixed value of x, and decrease as a functions of x at fixed c. Absolute value of negative magnetoresistance in the system reaches 60% at low temperatures for magnetic field strength ~9 kOe.

5. Acknowledgments

The authors are grateful to Dr. A. G. Badelin for preparation of the samples and for assistance in the measuring of some electromagnetic parameters.

The research was supported by grant from the Russian Federal Agency of Education.

References

[1] Young S L, Chen Y C, Chen H Z, Horng L, Hsueh J F 2002 Effect of the substitution of Ni^{3+}, Co^{3+}, and Fe^{3+} for Mn^{3+} on the ferromagnetic states of the $La_{0,7}Pb_{0,3}MnO_3$ manganite *J. Appl. Phys.* **91** 8915

[2] Lobanov M V, Balagurov A M, Pomjakushin V.Ju et al. 2000 Structural and magnetic properties of the colossal magnitoresistance perovskite $La_{0.85}Ca_{0.15}MnO_3$ *Phys.Rev. B.* **61** 8941

[3] Salamon M B, Jaime M 2001 The physics of manganites: Structure and transport *Rev. Mod. Phys.* **73** 583

[4] Troyanchuk I O, Bushinsky M V, Szymchak H et al. 2002 Magnetic interaction in Mg, Ti, Nb doped manganites *Eur. Phys. J. B.* **28** 75

[5] Blundell S 2003 Magnetism in condensed Matter *Oxford: Oxford University Press* 238

[6] Zhe Qu, Li Pi, Fan Ji-Yu, Tan Shun, Zhang Bei, Zhang Meng and Zhang Yu-Heng 2007 Electric and magnetic behaviour in double doped $La_{2/3+4x/3}Sr_{1/3-4x/3}Mn_{1-x}Mg_xO_3$ *Chinese Phys.* **16** 258

[7] Urushibara A, Moritomo Y, Arima T, Asamitsu A, Kido G, Tokura Y 1995 Insulator-metal transition and giant magnetoresistance in $La_{1-x}Sr_xMnO_3$ *Phys. Rev. B.* **51** 14103

[8] Mizusaki J, Mori N, Takai H et al 2000 Oxygen nonstoichiometry and defect equilibrium in the perovskite-type oxides $La_{1-x}Sr_xMnO_{3+d}$ *Solid State Ionics* **129** 163

[9] Estemirova S Kh 2009 Thesis. *Yekaterinburg: Institute for Metallurgy UB RAS* 122

Magnetic and electrical properties of $La_{1-c}Sr_cMn_{1-x}Ga_xO_3$ manganites

A G Badelin[1], A A Shchepetkin[2], S Kh Estemirova[2], V V Sorokin[1], V K Karpasyuk[1]

[1]Astrakhan State University, 414056, Astrakhan, Russia

[2]Institute for Metallurgy UB RAS, 620016, Yekaterinburg, Russia

E-mail: vkarpasyuk@mail.ru

Abstract. The bulk manganites $La_{1-c}Sr_cMn_{1-x}Ga_xO_3$ (c=0.15; 0.17; 0.19; 0.025≤ x ≤0.125) were prepared by solid state reactions. The regularities of the influence of Ga^{3+} and Sr^{2+} concentrations on saturation magnetization at different temperatures, Curie point (T_c), semiconducting-metallic transition and magnetoresistance have been established. It was found that magnetization at 80 K and T_c rise with c increasing at each fixed value of x, and decrease as a functions of x at fixed c. Absolute value of negative magnetoresistance for some chemical compositions exceeds 80% at low temperatures for magnetic field strength ~9 kOe. Ga substitution for Mn shifts phase boundary "rhombohedral-orthorhombic structure" to lower value of c, moreover, excess of oxygen content over stoichiometric one promotes the existence of rhombohedral phase at c=0.15. Diamagnetic dilution of octahedral sublattice by Ga^{3+} ions and cation vacancies induce semiconducting behaviour in rhombohedral phase, even at c=0.19. Metallic type of temperature dependence of resistivity observes in appropriate low temperature range in sintered manganites with c=0.17; 0.19 only at small x.

1. Introduction

Numerous papers have been devoted to the study of Mn site doping effects in manganites exhibiting colossal magnetoresistance (CMR) properties. CMR is generally agreed to be a result of the competition between certain phases with different electronic, magnetic and structural orders, and is mainly connected with double exchange and Jahn-Teller effect of Mn^{3+} ions [1-4]. Phase formation and properties of manganites essentially depend on the concentration of heterovalent ions, their localization and radii, and the occurrence of vacancies in various sublattices [1,5-8]. Modifying the bonds between $Mn^{3+}$and Mn^{4+} ions, the doping by small amount of substituting cations and defects can lead to drastic changes in the main properties of manganites [1,7-10]. It is known that introduction of each Ga^{3+} ion in $LaMn_{1-x}Ga_xO_3$ increases the magnetic moment per Mn by ~ 16 μ_B [10]. This is a significant effect to occur by a diamagnetic ion, and therefore the $La_{1-c}Sr_cMn_{1-x}Ga_xO_3$ system generates considerable interest, but no systematic study of structural, magnetic and electrical properties of this system has been done. Ga substitution for Mn results in diamagnetic dilution, decreasing of Mn^{3+} content, and, hence, decreasing of Jahn-Teller distortions.

In the present work experimental data are shown for the influence of Ga concentration (0.025≤x≤0.125) on phase composition, magnetic and electrical properties of La-Sr manganites. Under the condition that content of oxygen is stoichiometric, the values of c in $La_{1-c}Sr_cMn_{1-x}Ga_xO_3$ are equal to concentrations of Mn^{4+} ions (f.u.). These values are chosen near phase boundary "ferromagnetic insulator-ferromagnetic metal" (c=0.15; 0.17; 0.19) on the phase diagram for $La_{1-c}Sr_cMnO_3$ system [11].

2. Experimental

The experiments were performed on polycrystalline samples synthesized by traditional ceramic processing. The starting components (dried La_2O_3, $SrCO_3$, MnO_2, Ga_2O_3 powders) were mixed in stoichiometric proportions and ground in a ball mill with addition of alcohol. Pellets compacted of the obtained charge mixture were then preliminarily burned at 1273 K for 4 h. This operation was followed by grinding, introducing a binder (an aqueous solution of polyvinyl alcohol), pressing the samples, and burning out the binder. The final sintering step was performed at 1473 K for 10 h, and the samples were cooled together with the furnace. Then, in order to provide stoichiometric oxygen content, the samples were annealed at 1223 K and partial pressure of oxygen $P_{O2}=10^{-1}$ Pa for 96 h. The choice of annealing conditions was based on the results of the works [6, 12] which demonstrate the achievement of stoichiometry at above-mentioned temperature and P_{O2} almost independently of composition in the systems $La_{1-c}Sr_cMnO_3$ and $La_{1-c}Ca_cMnO_3$.

Phase composition was determined by powder X-ray diffraction at room temperature (diffractometer Shimadzu XRD-7000, CuK_α radiation). The magnetization (per mass unit) was measured in magnetic field of 5,6 kOe. Measurements of dc electrical characteristics and magnetoresistance (MR=(R(H)-R(0))/R(0)) were made using copper electrodes sputter-deposited onto opposite planes of pellets (thickness of about 4 mm). Magnetic field of 9240 Oe was parallel to the current direction. The temperature dependence of magnetic permeability (μ(T)) was measured by the induction method at a frequency of 98,6 kHz. The Curie point was determined as the temperature corresponding to the maximum of $|d\mu/dT|$.

3. Results and discussion

It was found that all sintered samples in initial state were single phase and crystallized in the rhombohedral structure (Table 1). Dependence of unit cell volume on Ga concentration was nonmonotonic, because, on the one hand, ionic radius of Ga^{3+} is smaller than Mn^{3+}, and the dilution of Mn sublattice by Ga reduces the oxygen content (like to the effect of Al substitution for Mn in La-Ca manganite [13]), that results in reduction of cation vacancies concentration; but on the other hand, the number of Mn^{3+} ions increases simultaneously with the decrease of oxygen content. Rhombohedral structure of sintered manganites with c=0.15 may be due to the surplus of Mn^{4+} associated with excess of oxygen content (γ>0) over stoichiometric one. According to [6], the oxygen-excess nonstoichiometry can be observed in $La_{1-c}Sr_cMnO_{3+\gamma}$ manganite for c <0.4, and oxygen excess is attributed to the formation of metal vacancies.

After annealing in vacuum, the structure of manganites with small amount of Sr and Ga (c=0.15, x<0.125) became orthorhombic (Table 1). Samples with c=0.15, x=0.125 contained the mixture of rhombohedral and orthorhombic phases.

All manganites with $c \geq 0.17$ in initial state and after annealing were rhombohedral.

It seems that substituting Ga ions shifted phase boundary "rhombohedral-orthorhombic structure" to lower value of c.

It was found that magnetization at 80 K and T_c rise with c increasing at each fixed value of x, and decrease as a functions of x at fixed c (Tables 1, 2). These regularities are related to the rising of Mn^{4+} concentration when c increases, and to dilution and destabilization of Mn sublattice by diamagnetic ions.

After annealing, magnetization increased in most cases (Table 1) as a result of reduction of cation vacancies concentration.

Journal of Physics: Conference Series **303** (2011) 012072 doi:10.1088/1742-6596/303/1/012072

Table 1. Phase composition and magnetization (at 80 K) of synthesized manganites.

Sample				Phase composition	σ, emu/g
Formula	*c*	*x*	State		
$La_{0.850}Sr_{0.150}Mn_{0.975}Ga_{0.025}O_{3+\gamma}$	0.15	0.025	Initial	Rhombohedral	68.4
$La_{0.850}Sr_{0.150}Mn_{0.975}Ga_{0.025}O_3$			Annealed	Orthorhombic	71.0
$La_{0.850}Sr_{0.150}Mn_{0.950}Ga_{0.050}O_{3+\gamma}$	0.15	0.050	Initial	Rhombohedral	61.9
$La_{0.850}Sr_{0.150}Mn_{0.950}Ga_{0.050}O_3$			Annealed	Orthorhombic	68.1
$La_{0.850}Sr_{0.150}Mn_{0.925}Ga_{0.075}O_{3+\gamma}$	0.15	0.075	Initial	Rhombohedral	56.6
$La_{0.850}Sr_{0.150}Mn_{0.900}Ga_{0.100}O_{3+\gamma}$	0.15	0.100	Initial	Rhombohedral	53.2
$La_{0.850}Sr_{0.150}Mn_{0.900}Ga_{0.100}O_3$			Annealed	Orthorhombic	61.8
$La_{0.850}Sr_{0.150}Mn_{0.875}Ga_{0.125}O_{3+\gamma}$	0.15	0.125	Initial	Rhombohedral	48.9
$La_{0.850}Sr_{0.150}Mn_{0.875}Ga_{0.125}O_3$			Annealed	Rhombohedral + Orthorhombic	56.1
$La_{0.830}Sr_{0.170}Mn_{0.975}Ga_{0.025}O_{3+\gamma}$	0.17	0.025	Initial	Rhombohedral	75.7
$La_{0.830}Sr_{0.170}Mn_{0.975}Ga_{0.025}O_3$			Annealed	Rhombohedral	75.6
$La_{0.830}Sr_{0.170}Mn_{0.950}Ga_{0.050}O_{3+\gamma}$	0.17	0.050	Initial	Rhombohedral	67.8
$La_{0.830}Sr_{0.170}Mn_{0.950}Ga_{0.050}O_3$			Annealed	Rhombohedral	72.2
$La_{0.830}Sr_{0.170}Mn_{0.925}Ga_{0.075}O_{3+\gamma}$	0.17	0.075	Initial	Rhombohedral	62.3
$La_{0.830}Sr_{0.170}Mn_{0.900}Ga_{0.100}O_{3+\gamma}$	0.17	0.100	Initial	Rhombohedral	58.8
$La_{0.830}Sr_{0.170}Mn_{0.900}Ga_{0.100}O_3$			Annealed	Rhombohedral	62.2
$La_{0.830}Sr_{0.170}Mn_{0.875}Ga_{0.125}O_{3+\gamma}$	0.17	0.125	Initial	Rhombohedral	56.4
$La_{0.830}Sr_{0.170}Mn_{0.875}Ga_{0.125}O_3$			Annealed	Rhombohedral	58.0
$La_{0.810}Sr_{0.190}Mn_{0.975}Ga_{0.025}O_{3+\gamma}$	0.19	0.025	Initial	Rhombohedral	75.9
$La_{0.810}Sr_{0.190}Mn_{0.975}Ga_{0.025}O_3$			Annealed	Rhombohedral	79.5
$La_{0.810}Sr_{0.190}Mn_{0.950}Ga_{0.050}O_{3+\gamma}$	0.19	0.050	Initial	Rhombohedral	72.9
$La_{0.810}Sr_{0.190}Mn_{0.950}Ga_{0.050}O_3$			Annealed	Rhombohedral	72.8
$La_{0.810}Sr_{0.190}Mn_{0.925}Ga_{0.075}O_{3+\gamma}$	0.19	0.075	Initial	Rhombohedral	69.2
$La_{0.810}Sr_{0.190}Mn_{0.900}Ga_{0.100}O_{3+\gamma}$	0.19	0.100	Initial	Rhombohedral	64.6
$La_{0.810}Sr_{0.190}Mn_{0.900}Ga_{0.100}O_3$			Annealed	Rhombohedral	65.3
$La_{0.810}Sr_{0.190}Mn_{0.875}Ga_{0.125}O_{3+\gamma}$	0.19	0.125	Initial	Rhombohedral	58.5
$La_{0.810}Sr_{0.190}Mn_{0.875}Ga_{0.125}O_3$			Annealed	Rhombohedral	57.8

Table 2. Curie point, maximum value of magnetoresistance and corresponding temperature T_{max} of sintered manganites

Composition		T_c , K	$\lvert MR \rvert_{max}$, %	T_{max} , K
c	*x*			
0.15	0.025	202	58	118
0.15	0.050	178	65	126
0.15	0.075	167	60	142
0.15	0.100	160	59	100
0.15	0.125	158	68	132
0.17	0.025	225	21	119
0.17	0.050	203	34	115
0.17	0.075	184	67	130
0.17	0.100	172	33	112
0.17	0.125	167	19	150
0.19	0.025	242	16	100
0.19	0.050	225	16	110
0.19	0.075	210	88	132
0.19	0.100	192	65	80

All samples of manganites with c=0.15 exhibit semiconducting properties in the investigated temperature range (Fig. 1). Metallic type of temperature dependence of resistivity we observed in appropriate low temperature regions in sintered manganites with c=0.17; 0.19 only at x=0.025 (Figures 2, 3). After annealing, manganite $La_{0.830}Sr_{0.170}Mn_{0.975}Ga_{0.025}O_3$ revealed no metallic behaviour (Fig. 2), but $La_{0.810}Sr_{0.190}Mn_{0.975}Ga_{0.025}O_3$ remained metallic near 125 K (Fig. 3). Manganites with $x \geq 0.05$ have semiconducting type of temperature dependence of resistivity for all c.

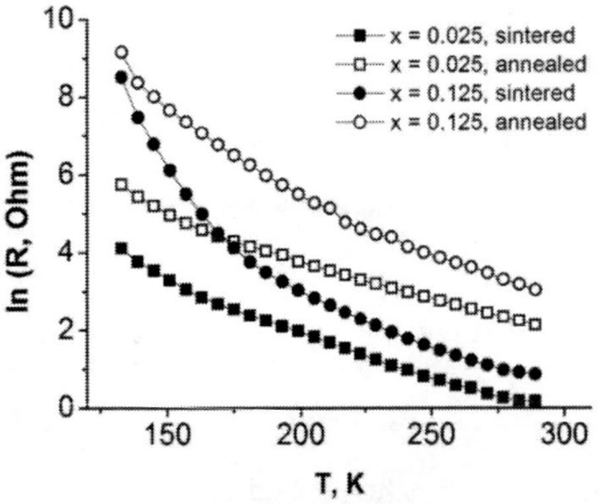

Figure 1. Temperature dependence of resistivity of manganites with c = 0.150

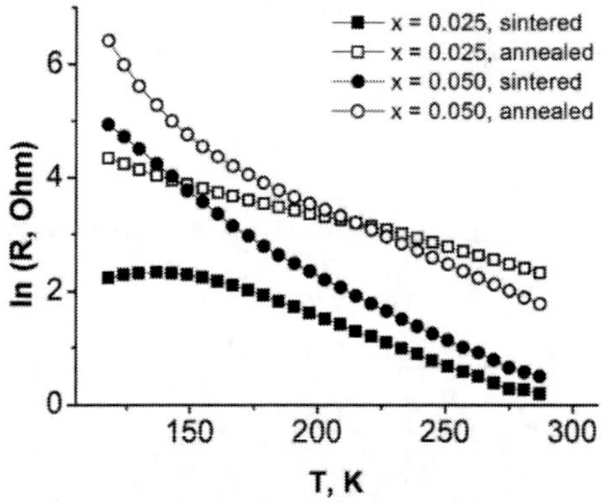

Figure 2. Temperature dependence of resistivity of manganites with c = 0.170

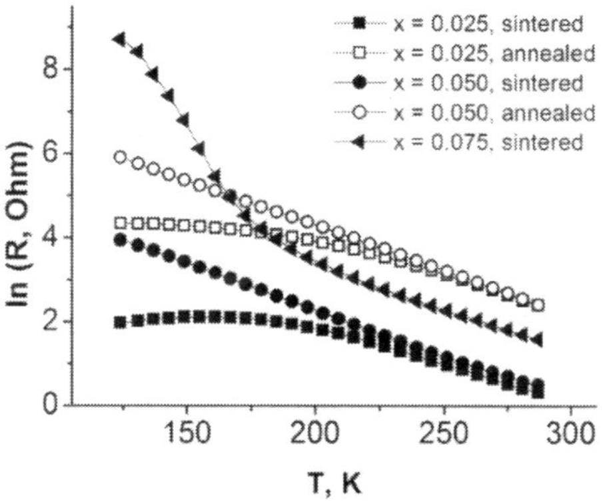

Figure 3. Temperature dependence of resistivity of manganites with $c = 0.190$

Characteristic temperature dependence of magnetoresistance is shown on Figure 4 for a certain sample. Maximum absolute value of negative magnetoresistance reached 88% at 132 K in manganite with $c = 0.19$, $x = 0.075$ (Table 2).

After annealing, the resistivity of all samples increased, and maximum absolute value of magnetoresistance decreased.

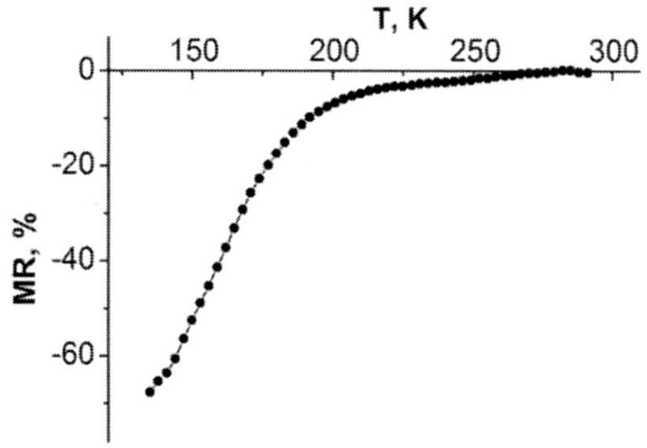

Figure 4. Temperature dependence of magnetoresistance of manganite with $c = 0.15$, $x = 0.125$

4. Conclusion

New data concerning the structure and properties of manganites $La_{1-c}Sr_cMn_{1-x}Ga_xO_{3+\gamma}$ (c=0.15; 0.17; 0.19; $0.025 \leq x \leq 0.125$; $\gamma \geq 0$) were obtained. Substituting Ga^{3+} ions shift phase boundary "rhombohedral-orthorhombic structure" in $La_{1-c}Sr_cMnO_3$ system to lower value of c, moreover, excess of oxygen content ($\gamma>0$) over stoichiometric one promotes the existence of rhombohedral phase at c=0.15.

Diamagnetic dilution of octahedral sublattice by Ga^{3+} ions and cation vacancies have significant influence on magnetic ordering and double-exchange interaction, inducing semiconducting behaviour in rhombohedral phase, even at c=0.19, when the charge carrier concentration is high enough. Metallic type of temperature dependence of resistivity observes in appropriate narrow low temperature regions in sintered manganites with c=0.17; 0.19 only at small Ga^{3+} content (x=0.025). After annealing, manganite $La_{0.830}Sr_{0.170}Mn_{0.975}Ga_{0.025}O_3$ revealed no metallic behaviour as a result of the reduction of Mn^{4+} ions concentration.

5. Acknowledgments

The authors are grateful to Dr. A.M.Smirnov for the assistance and fruitful discussions.

The research was supported by grant from the Russian Federal Agency of Education, project 1.2.09.

References

[1] Dagotto E, Hotta T, Moreo A 2001 Colossal magnetoresistant materials: the key role of phase separation *Physics reports* **344** 1

[2] Liu G.-L., Zhou J.-S., Goodenough J B 2001 Interplay between charge, orbital and magnetic ordering in $La_{1-x}Sr_xMnO_3$ *Phys.Rev. B.* **64** 144414 1

[3] Zhou J.-S., Yin H Q and Goodenough J B. 2001 Vibronic superexchange in single-crystal $LaMn_{1-x}Ga_xO_3$ *Phys. Rev. B.* **63** 184423

[4] Blundell S. 2003 Magnetism in condensed Matter *Oxford: Oxford University Press* 238

[5] Malavasi L, Ritter C, Mozzati M C, Tealdi C, Islam M S, Azzoni C B, Flor G. 2007 Effects of cation vacancy distribution in doped $LaMnO_{3+d}$ perovskites *Front for the arXiv. – Cond-mat* **0504334** 27 http://front.math.ucdavis.edu/0504.0334

[6] Mizusaki J, Mori N, Takai H et al. 2000 Oxygen nonstoichiometry and defect equilibrium in the perovskite-type oxides $La_{1-x}Sr_xMnO_{3+d}$ *Solid State Ionics* **129** 163

[7] De Leon-Guevara A M, Berthet P, Berthon J et al. 1997 Influence of controlled oxygen vacancies on the magnetotransport and magnetostructural phenomena in $La_{0.85}Sr_{0.15}MnO_{3-\delta}$ single crystals *Phys.Rev. B.* **56** 6031

[8] Wilson M L, Byers J M, Dorsey P C et al. 1997 Effects of defects on magnetoresistivity in $La_{0.7}Sr_{0.3}MnO_3$ *J.Appl. Phys* **81** 4971

[9] Orlova T S, Laval J Y, Monod Ph. et al. 2009 Influence of Mn-site doping on charge and orbital ordering in $La_{1/3}Ca_{2/3}Mn_{1-y}M_yO_3$ manganites (M=Ni, Ga) *Phys.Rev. B.* **79** 134407

[10] Farrell J, Gehring G A 2004 Interplay between magnetism and lattice distortions in $LaMn_{1-x}Ga_xO_3$ *New Journal of Physics* **6** 168

[11] Urushibara A, Moritomo Y, Arima T, Asamitsu A, Kido G, Tokura Y 1995 Insulator-metal transition and giant magnetoresistance in $La_{1-x}Sr_xMnO_3$ *Phys.Rev.B.* **51** 14103

[12] Estemirova S Kh 2009 Thesis. *Yekaterinburg: Institute for Metallurgy UB RAS* 122

[13] Blasco J, Garcia J, de Teresa J M et al. 1997 Structural, magnetic, and transport properties of the giant magnetoresistive perovskites $La_{2/3}Ca_{1/3}Mn_{1-x}Al_xO_{3-\delta}$ *Phys.Rev.B.* **55** 8905

New trends in the discussion of CMR-manganites and isostructural ferrites

K. Bärner[1], V. Morsakov[1] and I.V. Medvedeva[2]

[1]Department of Physics, University of Göttingen, F. Hund Platz 1, D-37077 Göttingen, Germany
[2] Institute of Metal Physics, S. Kovalevskaya 18, 620219 Ekaterinburg, Russia

E-mail: k.baerner@t-online.de

Abstract. Completing the Anderson and Hasegawa double exchange spin pair model by local spin quantization, ferrites as well as some Heusler alloys could be double exchange-superexchange mixed coupling systems. This concept also adds a third interpretation to the Kondo-like resistivity anomalies found recently in $FeSb_2$.

1. Introduction

Recently, by adding the local spin quantization to the Anderson & Hasegawa calculation of a double exchange pair state it has been found that the first excited state to the metallic ferromagnetic double exchange (DE) ground state is an antiferromagnetic charge ordered insulating (CO) state [1]. This has two main consequences: (1) one can now view the ferrites which are isostructural to the CMR-manganites as well as the spinel-type ferrites as insulating CO systems with only the ground and first excited state changing roles [2] and (2) one can view semimetals which could have originated from a spontaneous valence decomposition as all-metallic DE-systems [3]. The latter comes about as follows: an insulating CO state can be viewed as electrons (e) and holes (h) localized periodically along the band edges (Fig. 1).

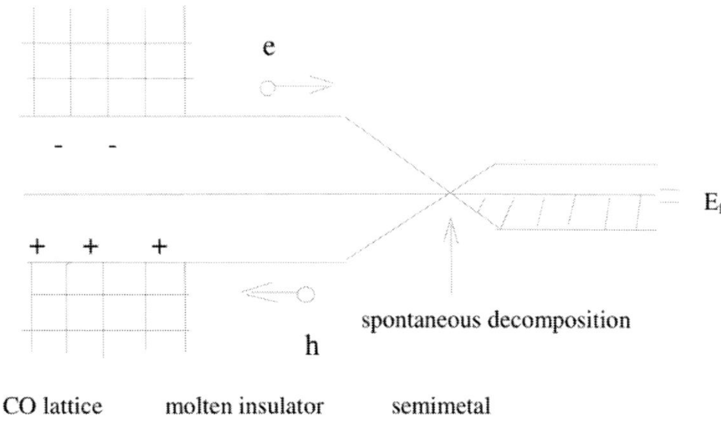

Figure 1. A charge ordered lattice is molten to yield a normal semiconductor and then - following a configurational coordinate- undergoes a spontaneous valence decomposition by letting the gap value go to zero; a semimetal with equal charge carrier concentration *n,p* occurs, which might develop an excitonic quasi-particle gap at the Fermi-level E_f - for more details see text, for an example see $CaFeO_3$ under pressure [4] .

Next, assuming some configurational coordinate change, the CO state could be molten, i.e. *e* and *h* now move into the near edge states, just like in ordinary semiconductors. In order to go over to a semimetal, one has to change the configuration again: for the occurrence of a spontaneous valence disproportion one has to let the electronic order parameter, i.e. the electronic gap, go through zero. The result is a semimetal with equal concentrations of electrons and holes *n,p*. All along the configurational changes, however, one can align the residual local spin moments either antiferromagnetically (AFM) or ferromagnetically (FM).

In this contribution we ask if this kind of DE-semimetals have unique excitations? While the doped CMR-manganites $R_{1-x}A_xMnO_3$ separate into electron and hole doped (quasiparticle-) regions, leaving the 1/2-1/2 compounds undecided, with a spontaneous valence decomposition, electrons and holes are excited simultaneously, opening the door for excitonic states, which, however, would have to be discrimated concerning their local spin array and which could also have a polaronic character because of the electron-spin-lattice subsystem coupling [5]. This yields in principle two kinds of (Frenkel-type, triangular) spin-excitons, one AFM (or spin frustrated) and the other spin parallel (Fig. 2)

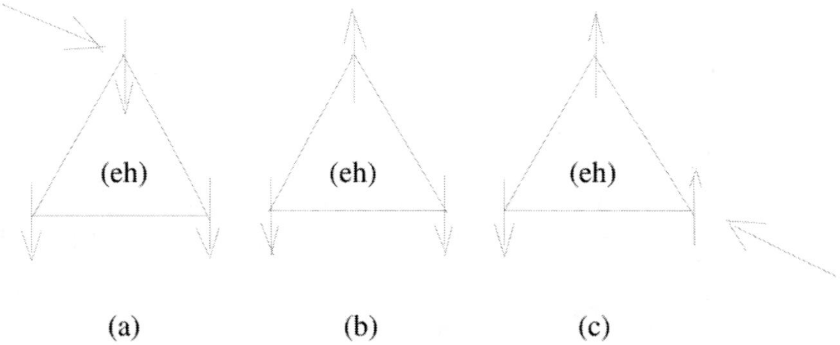

Figure 2. Triangular FM and AFM spin-excitons are designed and taken out of a FM (a) and AFM (b,c) layered spin lattice; while both collision initiated transitions a → b and c → b are possible, the latter only involves a momentum exchange while the former involves both a momentum and energy exchange; for more details see text.

The triangular units are taken out of a FM and AFM (layer) lattice array. As in periodic lattices one has exciton bands, this taking out assumes some kind of exciton confinement. In a layer antiferromagnet, the extended exciton would at least be confined to one layer, not so for the FM exciton. As the collision probability increases with mobility, the AFM exciton should have the longer lifetime. Introducing a strain potential well, say via a foreign atom, both excitons could be localized, but the AFM one would be easier to localize, as one of the linear independent directions is already closed. Moreover, the AFM exciton has a defence mechanism against electronic charge as well as against phonon collisions: by flipping the spin at one corner, the spin state and with it the energy of the scattering center would not change, while the FM exciton would have to take up energy, *i.e.* the difference between the AFM and FM exciton energies [3] - under a spin-flip. So, the AFM center can exchange momentum only while the FM one can change momentum only if it takes up energy. Exciton lifetimes in metals are generally thought to be in the order of 10^{-15} s to 10^{-13} s and so usually the excitons are thought to decompose fast. If, however, one or both spin exciton lifetimes could reach the phonon relaxation regime, 10^{-12} s, the (polaronic) excitons should show up in the resistivity versus temperature curves. There could even be a quasi-particle gap forming at the Fermi-level. In order to test the idea of a confinement enhanced lifetime, in what follows we have picked out three semimetallic compounds which also have random potentials, AFM or compensated spin order as well as resistivity anomalies, and $FeSb_2$.

2. Materials

If we now look for materials which might show such scattering effects, one should look for compounds which can be suspected to be valence decomposed, for example MnAs, 2x(+3) → (+2)+ (+4). In order to localize the exciton via (random) energy wells, we look for say, single crystals of $MnAs_{1-x}P_x$. Indeed, Fig. 3 shows unusual (Kondo-like) resistivity behaviour in the helical phase of the compound with $x = 0.11$, superimposed on spin disorder scattering.

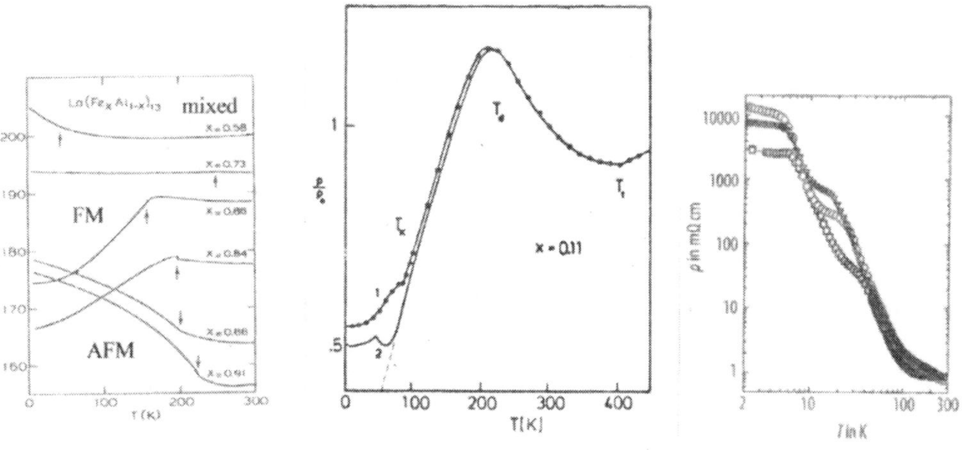

Figure 3. $La(Fe_{1-x}Al_x)_{13}$ [6], $MnAs_{1-x}P_x$, ($x = 0.11$), [7] mixed crystals showing spin disorder scattering in the FM (canted, $x = 0.11$, dashed line) state and a resistivity anomaly in the AFM (helical, $x = 0.11$, 1,2 $B = 0$, 0.8 T) magnetic phase, in comparison with the resistivity anomaly observed recently in $FeSb_2$ [8].

This compound also shows a helical-canted-helical-paramagnetic magnetic phase sequence and a high spin-low spin transition in the paramagnetic range, both of which are typical for DE-superexchange (SE) mixed coupled systems. Moreover, Hall experiments [9] show that the $n = p$ condition is very closely fulfilled, bearing in mind that electrons and holes have different mobilities. The next candidate are mixed crystals based on the compound $LaFe_{13}$, *i.e.* $La(Fe_{1-x}Al_x)_{13}$; here, via mixing Fe and Al, we will get a distribution of magnetic couplings and this should bring about the largest static (magnetic) potential fluctuations. Indeed, these compounds show the Kondo-like behaviour even more dramatically, but again only in the spin compensated, magnetically ordered regime. Complex magnetic phase sequences are visible here only under co-doping with cobalt [10]. Recently, an even larger Kondo-like effect has been observed in single crystal $FeSb_2$ which at first sight does not contain random potentials at all. $FeSb_2$ has been described as narrow bandgap semiconductor [12] or as a Kondo-lattice insulator [11]. That compound has been tentatively identified as weak ferromagnet (canting?) with a (local?) moment

Journal of Physics: Conference Series **303** (2011) 012073 doi:10.1088/1742-6596/303/1/012073

of $1\mu_B$ per Fe ion [11], so at least keeping within the possibilities of a double exchange band scenario. Note, however, that $FeSb_2$ shows a plateau between two drops, suggesting a two step process, just like the MnAs-derivate. So there are similarities between all three systems beyond the resistivity drop.

Viewing the three examples as the same or very similar phenomenon, we note that the Sommerfeld constant γ of the MnAs-derivate $x = 0.09$ is too low for a Kondo-lattice (11.6 mJ/K^2mol [9]) and that the resistivity of the mixed crystals is too low for a semiconductor - about 180 μΩcm for the $LaFe_{13}$ - derivates and ρ_0 of $x = 0.09$ is 33 μΩcm - see Fig. 3, while the negative MR found in the MnAs-derivate could be linked to a *H*-field induced transformation of some of the AFM excitons into FM ones. Thus, there is still room for alternative ideas. In what follows we try to connect the AFM (or frustrated) excitonic states with the scattering situation in the three selected semimetals. Specifically, we present a calculation of the resistivity versus temperature curve, ρ(T) to compare with experiment. Note, that while there might be a spin-flip scattering involved, we submit that the resistivity drop mainly results from the thermal/plasmonic destruction of the scattering centers on going to higher temperatures.

3. Calculation of the anomalous resistivity curve

In [2,13] a Matthiesen rule superposition of the anomalous, ρ_k, and a phonon (or spin disorder) scattering ρ_{ph} with $\rho_k >> \rho_{ph}$ in a *n,p* semimetal (or semiconductor) has been assumed. In addition, a coexistence of free carriers - of initial concentration n_o, p_o - and excitonic centers of number N_{ko} is assumed, with the possibility of the centers to thermally decompose and thus add to the free carrier concentration:

$$n(T) = n_0 + \Delta n,\ p = p_0 + \Delta p \quad (1)$$

The decomposition was assumed to follow a chemical reaction, in particular:

$$\Delta n \Delta p = N_k^2 K \quad (2)$$

with $K = K_0 \exp(-\Delta E/kT)$ the reaction constant and N_k the number of single excitonic scattering centers. Combining all this under the simplifications $n_0 = p_0$ and $\mu_{n0} = \mu_{p0}$, while setting aside the phonon or spin disorder contribution for a moment, yields:

$$\rho_0/\rho_k = (1/n_k + N_{k0}(K_0)^{1/2} \exp(-\Delta E/2kT)/n_0 \quad (3)$$

where $n_k = N_k/N_{k0}$ the center density with $0 < n_k < 1$. If initially, the free carrier concentration is low against the center density, $1/n_k$ can be approximated by a cut-off constant, *i.e*:

$$\rho_0/\rho_k = c + A\exp(-T_s/T) \text{ for } T < T_s \text{ with } kT_s = \Delta E/2 \quad (3a)$$

and $A = N_{k0}(K_0)^{1/2}/n_0 >> c$, while after the cut-off:

$$\rho_0/\rho_k = 1/n_k : T > T_s \quad (3b)$$

This unknown resistivity function (3b) we might obtain from the assumption that the single exciton centers surviving beyond T_s undergo a decomposition process $C \rightarrow e + h$ for $T > T_{s'}$, *i.e.*

$$\Delta n \Delta p = N_k K' \quad (4)$$

In [13], it has been argued that this reaction is diffusion controlled - in this present contribution, we assume a reaction control. Taking $kT_{s'} = \Delta E'$ and $\Delta n = \Delta p$ as before, for $T > T_{s'}$ but now assuming that most of the free carriers have already been created, so that. $\Delta n = (n - n_s) \approx (n^* - n_s)$, Nk' = Nk , n* high temperature carrier concentration, we obtain:

$$\rho_k/\rho_s = n_{k'} = 2(n^* - n_s)^2 \exp(\Delta E'/k(T-T_{s^*}))/N_{ks}K_{0'} \text{ for } T > T_{s'}. \quad (5)$$

Note, that this approach offers an interpretation of the plateau which is found experimentally to occur between the two stages of the falling resistivity curve of MnAs0.9P0.1 and that of FeSb2 (Fig. 3). For $T_s < T < T_{s'}$, $T_{s^*} = (T_s+T_{s'})/2$, we then would have a transitional regime between single and double excitons, where a biexciton-exciton condensation/melting and with it critical phenomena may well occur. Actually, this plateau-like regime is the one of major interest for applications, as one can calculate the highest figure of merit for TEP cycles exactly here. For an example of the calculations see Fig. 4, where the normalization to ρ_0 is introduced using $\rho_k/\rho_0 = (\rho_k/\rho_s)(\rho_s/\rho_0)$.

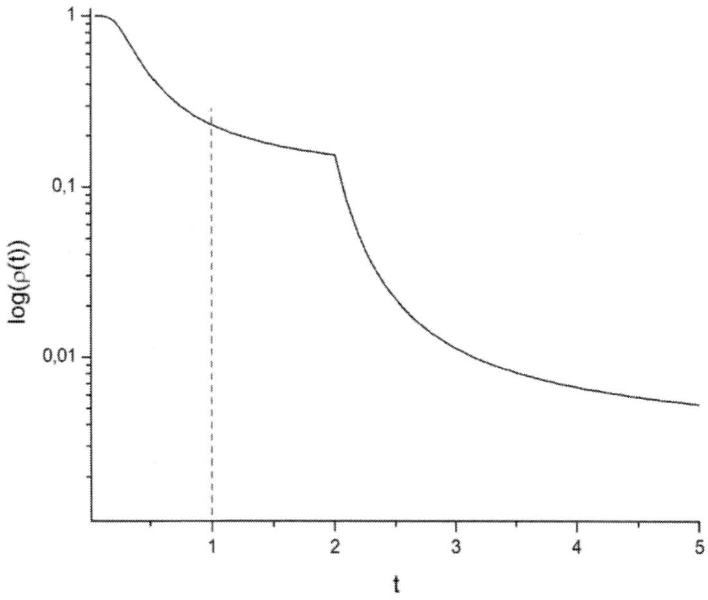

Figure 4. An example of the resistivity function: for $T<T_s = \Delta E/2k$, $1/(1+9\exp(-1/t)$ with $t = T/T_s$; for $T > T_{s'} = \Delta E'/k$, the function $0.003\exp(2/(t-1.5))$ takes over. For even higher temperatures the phonon (or spin disorder) resistivity takes over.

Unlike the mixed crystals, at first sight $FeSb_2$ gives no indication of random potentials. With the mixed crystals, the exciton-polarons could well be itinerant or hopping between particular sites, while they should be more (exciton-) band like in $FeSb_2$ - see also the differences along different crystal directions. As for the $LaFe_{13}$ derivates one expects a distribution of exchange interactions, here one could speak of spin-frustration in a band scenario, eventually helped by SE-frustration. The helically magnetically ordered MnAs-derivates, with As, P having only an ionic size difference, should be closer to $FeSb_2$.

Journal of Physics: Conference Series **303** (2011) 012073 doi:10.1088/1742-6596/303/1/012073

4. Conclusions

In a more comprehensive view, the anomalous Kondo-like resistivity curves obtained in several mixed and pure single crystal systems, could be connected with spin transfer magnetism. Specifically, electrons and holes could, at low temperatures, build up AFM polaron-exciton Kondo-like scattering centers. Then, the resistivity drop with increasing temperature would be rather due to their thermal decomposition as to an increase of the Kondo-mobility with temperature which could be connected to the spin-flip scattering on those centers.

References

[1] K. Bärner, JMMM 316 (2007) 400.
[2] K. Bärner, in: "New Trends in the Characterization of Ferrites and Related Materials, Ed. K. Bärner, Reserch Signpost, Trivandrum 695023, Kerala, India, 2010, Ch.5, ISBN: 978-81-308-0409-5.
[3] K. Bärner, in: "Double exchange in Heusler Alloys and related materials" Ch.II, Ed. K. Bärner, Reserch Signpost, Trivandrum-695023, Kerala, India, 2007, ISBN: 81-308-0152-3.
[4] T. Kawakami. S. Nasu, T.Sasaki, S. Moritomo, S.Endo, S. Kawasaki, M.Takano, J. Phys.Soc. Jpn 70 (2001) 1491.
[5] K. Bärner ""New trends in the Characterization of CMR-manganites and related Materials" Ed. K. Bärner, Reserch Signpost, Trivandrum 695023, Kerala, India, 2005, Ch.I, ISBN: 81-308-0043-8.
[6] T.T.M. Palstra, G.J. Nieuwenhyus, J.A. Mydosh, Phys. Rev. B 31 (1985) 4622.
[7] H. Berg, K. Bärner, JMMM 4 (1977) 69.
[8] A. Bentien, S. Johnsen, G.K.H. Madsen, B.B. Iversen and F. Steglich, Europhys. Letters 80 (2007) 17008 and 3990.
[9] H.J. Krokoszinski,"Transporteigenschsften der magnetischen Phasen in $MnAs_{1-x}P_x$ Misch-kristallen", thesis, Göttingen 1882; K. Bärner, H.J. Krokoszinski, U. Neitzel, VII. Int. Conf. on Transition Metal compounds, Grenoble 1982, IIB2.
[10] A.S. Eremulenko, E.V. Scherbakov, A.V. Andreev, N.V. Baranov, Physics of Metals and Metallography 65 (1988) 749.
[11] A.V. Lukoyanov, V.V. Mazurenko, V.I. Anisimov, M. Sigrist, T.M. Rice, Europhysics J. B 53 (2006) 205.
[12] P. Sun, N. Oeschler, S. Johnsen, B.B. Iversen, F. Steglich, J. Phys. Conf. Series 150 (2009) 012049.
[13] K. Bärner, M. Boshta, M.P. Annaorazov, BJP (2010) in progress.

Spontaneous generation of voltage in single-crystal $Sm_{0.55}Sr_{0.45}MnO_3$ during magneto-structural phase transition

D M Zashirinskii[1], L I Koroleva[1], A S Morozov[1], R Szymczak[2]

[1]M.V. Lomonosov Moscow State University, Vorobyevy Gory, 119992 Moscow, Russia

[2]Institute of Physics PAS, Lotnicov al. 32/46, 02668 Warsaw, Poland

E-mail: koroleva@phys.msu.ru

Abstract. In single-crystal $Sm_{0.55}Sr_{0.45}MnO_3$, grown by the floating zone method with the cooling in oxygen, the spontaneous generation of voltage (SGV) has been observed. Its maximum, reached 60 μV, occurs at temperature region where simultaneous destruction of the CE-type antiferromagnetic order and charge order take place in some clusters. SGV maximum decreases on ~ 40% by magnetic field 14.2 kOe. It is shown that SGV provokes of presence in sample of regions with different electrical charges.

1. Introduction

The spontaneous generation of voltage (SGV) was discovered in FeS [1], TiNi [2], FeNi [3], Ce [4], $Gd_5(Si_xGe_{1-x})_4$ [5,6] and $SmMn_2Ge_2$ [7]. In these materials the voltage observed between the sample ends, having the same temperature, was regarded as spontaneous since a current was not supplied on sample. SGV occurs near first-order magneto-structural phase transformation. Among them, the SGV is especially intriguing because it may result in the development of sensors, which can respond not only to changes in temperature, pressure, and/or magnetic field but most importantly to the rates of their changes without the need for a complicated analysis of signals. Furthermore, all of this can be done by a single sensor requiring no standby power.

Here we report on SGV in single-crystal $Sm_{0.55}Sr_{0.45}MnO_3$. It is known that compound $Sm_{1-x}Sr_xMnO_3$ system with $x = 0.5$ exhibits a charge-ordered state. Investigations of the neutron diffraction at 1.5 K ≤ T ≤ 300 K showed that the compound $^{154}Sm_{0.6}Sr_{0.4}MnO_3$ contains the clusters of three types: ferromagnetic (F), A-type antiferromagnetic (AF), and CE-type AF which are charge-order (CO) [8]. As this takes place an AF ordering of A-type disappears above T_N = 120 K while the AF order of CE-type clusters still exist at 150 K but are absent at 180 K. A spontaneous magnetic moment appears at Curie point T_C = 110 K; at the same temperature, an electrical resistivity ρ exhibits a maximum. The activation energy of the conductivity changes abruptly at 180 K; in compounds possessing a CO state, this is usually interpreted as a thermal breakdown of this order. The above values of T_C, T_N and the temperature of CO breakdown (T_{CO}) are close to those determined from the electron and neutron diffraction data and the temperature dependence of magnetization [9]. Runov *et al.* [8, 10,11] showed that the samples of $Sm_{1-x}Sr_xMnO_3$ compounds with x = 0.25 and 0.4 at low temperatures contain F clusters with 180 – 250 Å dimensions coexisting with large clusters up to several thousand angstroms in size. A the same time, no charge-ordered clusters were observed in the compound with $x = 0.25$ [8]. In work [12] the temperature dependence of the initial magnetic susceptibility in an alternating magnetic field with a frequency of 8 kHz $\chi(T)$ and paramagnetic susceptibility of the $Sm_{1-x}Sr_xMnO_3$ compounds with $x =$ 0.33, 0.4 and 0.45 were studied. The $\chi(T)$ curves exhibit a sharp drop at high temperature. The T_C-values of these compounds were determined as the temperatures of minima in the $(d\chi/dT)(T)$ curves and are equal to 79, 112 and 126 K respectively. These values are close to the published data. The paramagnetic susceptibility these compositions obeys the Curie-Weiss law at temperature above ~ $2T_C$ with the paramagnetic Curie points θ equal to 175, 194 and 250 K for $x =$ 0.33, 0.4 and 0.45 respectively. In the temperature interval $T_C < T < 2T_C$ the experimental curves deviate from the Curie-Weiss law.

Thus, the magnetic structure of $Sm_{1-x}Sr_xMnO_3$ compounds is more complicated as compared to that of the $La_{1-x}Sr_xMnO_3$ system. Compounds of the former system with x close to 0.5, in contrast to $La_{1-x}Sr_xMnO_3$ compounds, contain charge-ordered clusters. This order exhibits thermal breakdown at higher temperatures than does the magnetic order in the F and A-type AF phases of the sample. In the compounds with $x = 0.4$ and 0.45 at $T > T_C$ the magnetization isotherms exhibit a jump like increase in the interval of field strengths between H_{C1} and H_{C2} and than reach saturation. In the interval $H_{C1} < H < H_{C2}$ the magnetization exhibits a metastable behavior and conclusion was made that transformation of charge-ordered clusters with CE-type AF order into F state at the threshold field H_{C1} occurs. This conversion is analogous to a first-order phase transition. This magnetic transition coexists with structure transition since in CO clusters the crystal lattice is subjected to larger orthorhombic distortion than in F clusters.

2. Experimental methods

The $Sm_{0.55}Sr_{0.45}MnO_3$ single crystal was grown by the floating zone method with cooling at the oxygen by A.M. Balbashov. It is known that preparation of the manganite samples in the oxygen atmosphere improves their quality since the torn connections Mn-V-Mn (V is oxygen vacancy) are restored and the Curie point is raised [13]. That's why an anneal and cooling of single-crystal are produced in the oxygen atmosphere. Such preparation raises the T_C up to 134 K and paramagnetic Curie point θ up to 270 K that will be shown in section 3. Two parallelepiped-shaped samples were cut from single crystal by electric discharge machining. The longest side of samples was parallel to *c*-axis or *ab*-plane. The *dc* voltage across the sample was measured by standard two-probe method with use a B7-65 microvoltmeter. The sample was subjected to the temperature and magnetic field variations at rates 1.13 - 17.5 K/min for former and 170 kOe/min for second. The field was applied and the voltage were measured along the longest side of sample. The misorientation between the directions of the magnetic field vector and crystal axis were less than $\pm 5^0$. The magnetization was measured using SQUID-magnetometer and pm susceptibility was obtained using the Faraday method.

3. Results and discussion

Figure 1 shows the temperature dependence of magnetization M measured in weak magnetic field $H =$ 100 Oe of the $Sm_{0.55}Sr_{0.45}MnO_3$ single crystal. One can see from this Figure that the M-value abruptly decreases at temperature $T_C = 134$ K that is the Curie point of the F clusters. The figure 2*a* shows the SGV along *c*-axis and along *ab*-plane of this sample as function of temperature upon cooling at 17.5 K/min. The SGV anisotropy is essentially absent in *ab*-plane as our measurements are shown. The small difference of SGV measured along *c*-axis and *ab*-plane is presents as shown figure 2*a*. The maximum of SGV observes at T_m ~ 255 K. Figure 2*b* shows SGV along *c*-axis of the sample cooling with the rate 1.13 K/min and heating with the rate 1.38 K/min. From comparison figures *a* and *b* one can see that the SGV value near 255 K almost does not depend from the cooling rate, but the forms of their SGV(T) curves are distinguished. At rate 17.5 K/min the maximum is observed on its while at rate 1.13 K/min is plateau. This maximum and the plateau onset observe at

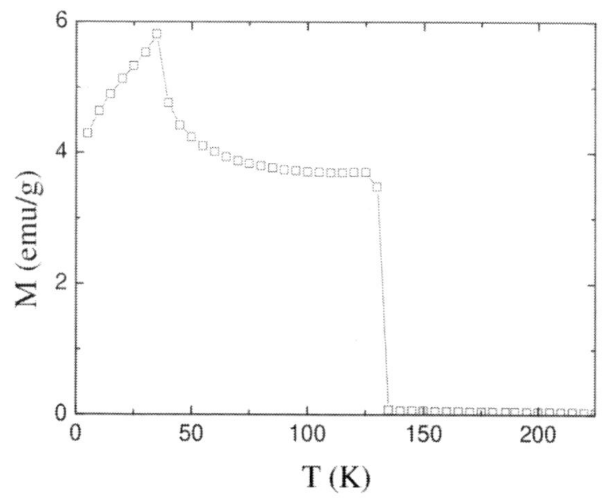

Figure 1. The temperature dependence of the magnetization M of the Sm0.55Sr0.45MnO3 single-crystal sample in weak magnetic field of 100 Oe.

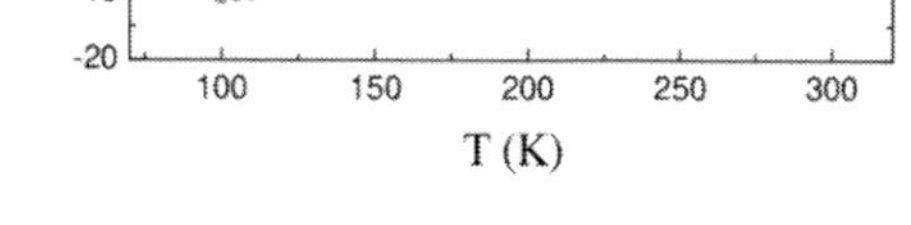

Figure 2a. $Sm_{0.55}Sr_{0.45}MnO_3$ single-crystal sample. The temperature dependence of spontaneous generation of voltage along the *c*-axis and along *ab*-plane upon cooling at the rate 17.5 K/min.

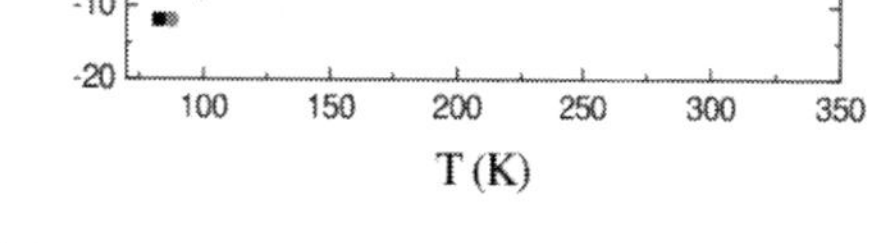

Figure 2b. $Sm_{0.55}Sr_{0.45}MnO_3$ single-crystal sample. The temperature dependence of spontaneous generation of voltage along the *c*-axis cooling with the rate1.13 K/min and heating with the rate 1.38 K/min.

T_m = 255 K which far exceeds the Curie point T_C = 134 K. Here the T_m-value is closer to paramagnetic Curie point θ = 270 K. (The θ -value was determined by extrapolation of the $1/\chi(T)$ curve up to intersection with *T*-axis as demonstrates Figure 3. Here χ is paramagnetic susceptibility).

As figure 3 shows, the $1/\chi(T)$ curve obeys the Curie-Weiss law at temperature above $\sim 2T_C$. Clearly in the temperature interval $T_C < T < 2T_C$ the clusters, obtained by magnetic order, are presented. It should be noted that the magnetocaloric effect maximum of this composition was observed at $T \sim 245$ K which is near to T_m.

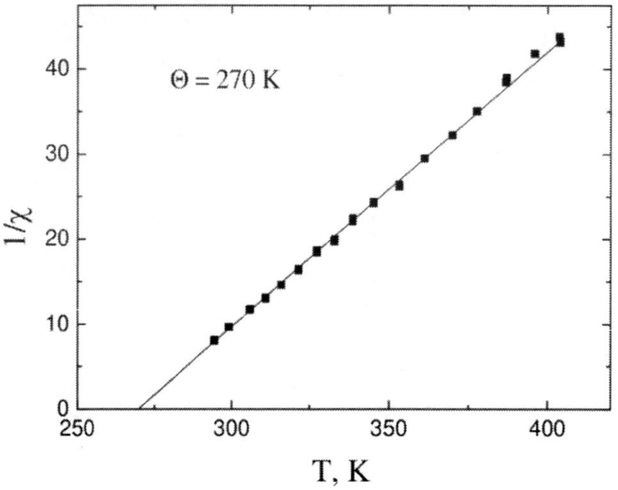

Figure 3. The temperature dependence of the paramagnetic susceptibility χ of the $Sm_{0.55}Sr_{0.45}MnO_3$ single-crystal sample.

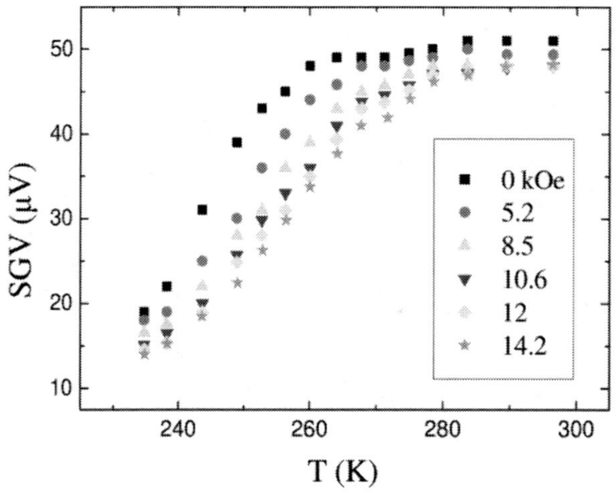

Figure 4. The temperature dependence of spontaneous generation of voltage along the *c*-axis of the $Sm_{0.55}Sr_{0.45}MnO_3$ single-crystal sample in some magnetic fields.

Figure 4 shows the magnetic field influence on SGV-value. From this Figure we notice that the maximum SGV-value decreases on ~ 45% at H = 14.2 kOe. We established that the maximum SGV value do not exchange for the time 24 hours. It is obvious that SGV is nonequilibrium process with big relaxation time. We emphasize that the maximum SGV-value is bigger than in another compounds mentioned above and can to be source to supply a nanoobjects with electricity.

The SGV behavior outlined above is connected with the presence in sample the clusters of three types: F, AF of A-type and AF of CE-type with charge-ordering. The maximum on SGV(T) curve can be caused by the simultaneous destruction of charge-ordering and AF order of CE-type since at $T > \sim$ 140 K the clusters of this type remain. Obviously, the generation of spontaneous voltage necessitates the presence in sample the regions with the opposite charges. In cluster with charge ordering Mn-ions, occupied by electrons (Mn^{3+}), alternate with Mn-ions without electron ($Mn^{4=}$). They are fixed in the cluster. So, there is voltage between cluster and paramagnetic host deprived of charge carriers. These clusters are symmetric disposed in sample from Coulombic repulsion if they are identical. In this case the electrical field from each cluster compensates the electrical field from neighboring clusters. As a whole the spontaneous voltage is absent. But the cluster sizes can distinguish and in this case the spontaneous generation of voltage can be. Obviously the distribution of cluster sizes wider in the temperature region of the phase transition from AF of CE-type phase to paramagnetic phase than in smaller temperature that explains the SGV maximum in $T = T_m$. It should be noted that SGV was observed at $T < T_m$ up to 78 K which is minimal temperature of measurement. Here the F clusters occur in which the electrical charges are localized from *s-d* exchange gain. As point out above, at low temperatures in these compounds F clusters were observed by Runov *et al.* [8,10] with 180 – 250 Å dimensions coexisting with large clusters up to several thousand angstroms in size that can be the case of SGV. It is known [14] that the external magnetic field near phase magnetic transition increases the clusters sizes and by doing so decreases SGV.

References

[1] Takahashi T and Yamada O 1976 *J. Phys. Chem. Solids* **17** 161
[2] Johnson A D and Katz P I 1977 *J. Appl. Phys.* **48** 73
[3] Koval Yu N and Molin A L 1980 *Fiz. Met. Metalloved.* **49** 1099
[4] Kokorin V V and Chernenko V A 1985 *Fiz. Met. Metalloved.* **60** 285
[5] Levin E M, Pecharsky V K and Gschneidner K A 2001 *Phys. Rev.* **B 63** 4110
[6] Zou M, Tang H, Schlagel D L, Lograsso T A, Gschneidner K A and Pecharsky V K 2006 *J. Appl. Phys.* **99** 304
[7] Zou M, Sampaio J A, Pecharsky V K and Gschneidner K A 2009 *Phys. Rev.* **B 80** 2403
[8] Runov V V, Chernyshov D Yu, Kurbakov A I, Runova M K, Trunov V A, Okorokov A I 2000 *J. Exp. and Theor. Phys.* **91** 1017
[9] Martin C, Maignan A, Hervieu M and Raveau B 1999 *Phys. Rev.* **B 60** 12191
[10] Runov V V, Glattli H, Kopitsa G V, Okorokov A I and Runova M K 1999 *JETP Lett.* **69** 353
[11] Runov V, Glattli H, Kopitsa G Okorokov and Runova M 2000 Physica B **276-278**, 795
[12] Abramovich A I, Koroleva L I and Michurin A V 2002 *J. Exp. and Theor. Phys.* **95** 917
[13] Koroleva L I 2003 *Magnetic Semiconductors*. (Moscow: Publishing of Phys. Dept. of MSU)
[14] Nagaev E L 2002 *Colossal Magnetoresistance and Phase Separation in Magnetic Semiconductors* (Singapore: Imperial College Press)

Spin glass state in the spinel families and stability conditions

P. Gusin[1], J. Krok-Kowalski[1], J. Warczewski[1], E. Malicka[2]

[1]Institute of Physics, University of Silesia, ul. Uniwersytecka 4, 40-007 Katowice, Poland
[2]Institute of Chemistry, University of Silesia, ul. Szkolna 9, 40-006 Katowice, Poland

E-mail: pgusin@us.edu.pl

Abstract. The conditions of stability of the spin glass state and its percolation limit in the spinel families have been analyzed on the base of the de Almeida-Thouless theory. The temperature range was determined as a function of the antimony concentration for which the spin glass state is stable.

1. Introduction

A spin glass is a frustrated and random system. Such system contains dilute magnetic spins scattered in the bulk of the sample, so that the individual spin-spin couplings J_{ij} vary in a quasi-random way. Frustration leads to a Hamiltonian of the system that has no global minimum and the partition function is not dominated by the lowest energy spin configuration. Instead there is a large number of local minima which are metastable and form an energy landscape. In the Sherrington-Kirkpatrick model where the mean-field theory approximation is valid the spin glass has a line of transitions in an external magnetic field, the de Almeida-Thouless (AT) line. This line separates a high-temperature high-field paramagnetic phase from a low temperature low-field phase where the energy landscape has valleys separated by barriers in the thermodynamic limit. In the two spinel families: $CuCr_{2-x}Sb_xS_4$ (x=0.3,0.4,0.5) and $Cu_{1+x}Cr_{1.5+y}Sb_{0.5+z}Se_{4+t}$ ($0.02\leq x\leq 0.01$, $0.03\leq y\leq 0.35$, $-0.2\leq z\leq -0.02$, $0.01\leq t\leq 0.08$) the spin glass states appear and it turned out that the concentrations of the antimony ions up to about 0.5 make the percolation limit of these spin glass states. Here we analyze the AT lines in these spinels and determine the temperature range as a function of the antimony concentration for which the spin glass states are stable (see also [1]).

The spin glass can be modelled by a statistical mechanical system whose degrees of freedom are N two-valued spins $s_i=\pm 1$ and the Hamiltonian H:

$$H=\sum_{i,j} J_{ij}s_is_j. \tag{1.1}$$

The partition function Z for the statistical system with the fixed set of couplings $\{J_{ij}\}$ is the sum over spins s_i:

$$Z[\beta,J_{ij}]=\sum_{s_i}\exp[-\beta H]. \tag{1.2}$$

Published under licence by IOP Publishing Ltd

Thus the Z is the function of $\beta=1/(kT)$ and the coupling constants J_{ij}. In the case of the spin glass the coupling constants are not fixed but are distributed randomly specified by giving a probability distribution $[DJ]$. In such system an expectation value of a variable X depending on s_i and J_{ij} is defined as:

$$\langle\langle X\rangle\rangle = \int \frac{DJ_{ij}}{Z[\beta, J_{ij}]}\sum_{i,j} X(s_i, J_{ij})\exp[-\beta H], \tag{1.3}$$

and is the function of β. As is well-known there exist four kind models of the spin glasses [e.g. 2]:

(1) the Sherrington–Kirkpatrick model,
(2) the Bethe lattice model,
(3) the long range Edwards–Anderson model and
(4) the Edwards–Anderson model.

In the Sherrington-Kirkpatrick (SK) model all J_{ij} are different from zero and the Gaussian probability distribution has variance $N^{-1/2}$. The coordination number z is N-1. In the infinity limit of N the mean field theory is valid. In the Bethe lattice (BL) model the spins live on a random lattice and only Nz/2 coupling constants J_{ij} are different from zero with the variance $z^{-1/2}$. In the large range Edwards-Anderson (LREA) model the spins live on a finite dimensional lattice of dimension D. Only nearest spins at a distance less than R interact and the variance of the J's is proportional to $D^{-R/2}$. The Edwards-Anderson (EA) model is obtained when R=1. These models in appropriate limits lead to the SK model:

$$\lim_{z\to\infty} BL = \lim_{R\to\infty} LREA = \lim_{D\to\infty} EA = SK. \tag{1.4}$$

Compared to a real spin glass, the main simplification of the SK model is that there is neglected the spatial distribution of spins, in effect taking the limit of spins in infinite spatial dimensions. In the SK model the partition function Z has two temperature regimes. At high temperature, the interactions are unimportant, and the spins are disordered. Conversely, at low temperature, there is ordering, but not detectable by a simple order parameter such as the average expectation value of the spins. This order is detected by the Edwards-Anderson order parameter q_{EA}:

$$q_{EA} = \frac{1}{N}\sum_i \langle s_i\rangle^2, \tag{1.5}$$

where $\langle s_i\rangle \equiv m_i$ means the local magnetization.

2. The AT line in the SK model

The order parameter q for a spin-glass state in the SK model is given by the formula [3]:

$$q = \frac{1}{\sqrt{2\pi}}\int_{-\infty}^{\infty} dz \exp(-z^2/2)\tanh^2\left(\frac{Jq^2 z + \mu H}{k_B T}\right), \tag{2.1}$$

where J is a measure of a coupling constant for the interaction resulting in the spin-glass state, H is an external magnetic field and μ is the magnetic moment of an ion. The integration variable z is related to the Gaussian probability distribution. The spin glass state is stable if the following condition holds [4, 5]:

$$\left(\frac{k_B T}{J}\right)^2 > \frac{1}{\sqrt{2\pi}}\int_{-\infty}^{\infty} dz \exp(-z^2/2)\operatorname{sec}h^4\left(\frac{Jq^2 z + \mu H}{k_B T}\right). \tag{2.2}$$

One can rewrite the above formula introducing a new function F:

$$F(J,H,T)=\frac{1}{\sqrt{2\pi}}\int_{-\infty}^{\infty}dz\exp(-z^2/2)\operatorname{sech}^4\left(\frac{Jq^2z+\mu H}{k_BT}\right). \tag{2.3}$$

Hence the condition (2.2) on stability of the spin glass state is: $(k_BT)^2-J^2F(J,H,T)>0$.

In the case when the above bound is saturated: $(k_BT)^2-J^2F(J,H,T)=0$ one can solve this equation (for fixed J) with respect to H so:

$$H=A(J,T). \tag{2.4}$$

The above relation is called the de Almeida-Thouless (AT) equation and define at the plane (T,H) a curve called the de Almeida-Thouless (AT) line. This line separates the region where the spin glass state is stable from the region where instabilities appear. Under assumption when coupling constant for a ferromagnetically ordered matrix is equal to J_0 and at high temperatures ($T\to T_G$) the AT equation can be solved and has the approximate form [5]:

$$H=\frac{k_BT_G}{\mu S}\left|1-\frac{J_0}{J}\right|\sqrt{C(S)\left(1-\left(\frac{k_BT}{J}\right)^3\right)}, \tag{2.5}$$

where the coefficient *C(S)* depends on spin S and has the following values: *C*(1/2)=5.333, *C*(1)=2.828 and T_G is the spin glass freezing temperature in zero field. The formula (3) is valid for the temperatures T which obey the relation $J/k_B \geq T$. In the considered compounds values of J_0/k_B are of the order of 100 K and the values of S are of the order of 1. The corresponding values J/k_B for these spinels are different. Thus for $CuCr_{1.6}Sb_{0.4}S_4$ the ratio J/k_B is J/k_B =31.55 K (see Table 5 in [6]) and the corresponding AT line is expressed by the following equation:

$$h=\left|1-\frac{100}{31.55}\right|\sqrt{2.828\left(1-\left(\frac{T}{31.55}\right)^3\right)}, \tag{2.6}$$

where $h=H\mu/(k_BT_G)$.

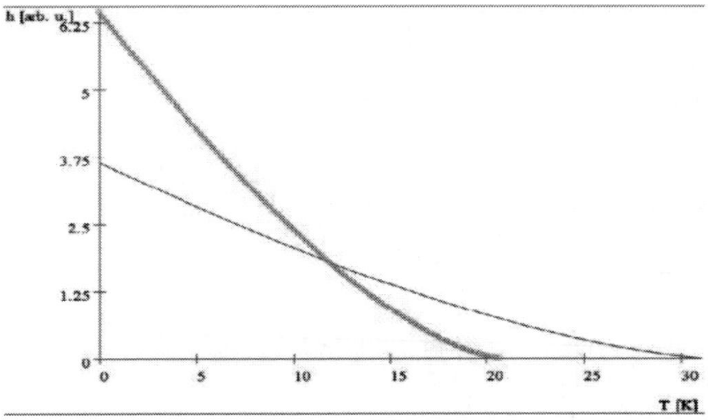

The Fig.1 on the left presents the AT lines for the spinel $CuCr_{1.6}Sb_{0.4}S_4$ with J/k_B=31.55 K (solid line) and for the spinel $CuCr_{1.7}Sb_{0.3}S_4$ with J/k_B=20.70 K (Table 5 in [6]) (dots line). On the other hand, for the spinel $CuCr_{1.5}Sb_{0.5}S_4$ the magnetic coupling constant J/k_B is equal to -1.85 K (see Table 5 in [6]). This negative value of the coupling constant corresponds to the experimental evidence of the collinear antiferromagnetism observed in this compound [7]. Thus in this case the AT line is described by the following equation:

$$h=\left|1+\frac{100}{1.85}\right|\sqrt{2.828\left(1+\left(\frac{T}{1.85}\right)^3\right)}. \tag{2.7}$$

The below Fig.2 presents the AT line for the spinel $CuCr_{1.5}Sb_{0.5}S_4$. Comparing the behaviour of the curves on Fig. 1 and Fig. 2 one can easily see that in Fig. 1 one deals with the temperature ranges of the stable spin glass states, whereas in Fig. 2 the AT line turns out to be divergent which indicates the absence of a spin glass state. This analysis is in a very good agreement with the experiment [6-9].

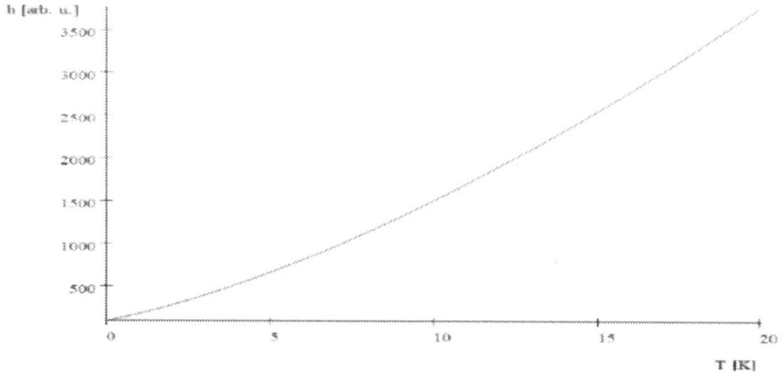

Taking into account that here J can be considered as a variable which depends on the Sb concentration let us consider the temperature T as a function of J and H (or h). In this case the temperature T can be interpreted as the freezing temperature of a spin glass state. This function has been obtained from Eq. (2.5) and has the form:

$$T(J,h) = J - \frac{J}{\sqrt[3]{C(S)}}\left(\frac{Jh}{J-J_0}\right)^{2/3}, \tag{2.8}$$

J and h have physical meaning only when T>0. If J>0, then the condition of the positive temperature T>0 is given by:

$$\left(C-h^2\right)\left(J-\frac{\sqrt{C}J_0}{\sqrt{C}-h}\right)\left(J-\frac{\sqrt{C}J_0}{\sqrt{C}+h}\right) > 0. \tag{2.9}$$

From the above inequality one gets two restrictions on J for fixed h. The first one is for $C\text{-}h^2>0$ $J > \frac{\sqrt{C}J_0}{\sqrt{C}-h} > 0$. And the second one is for $C\text{-}h^2<0$: $\frac{\sqrt{C}J_0}{\sqrt{C}+h} > J > 0$. For C(1)=2.828 and J_0/k_B=100 K and for the fixed values of h one obtains the corresponding plots temperature T as the function of J.

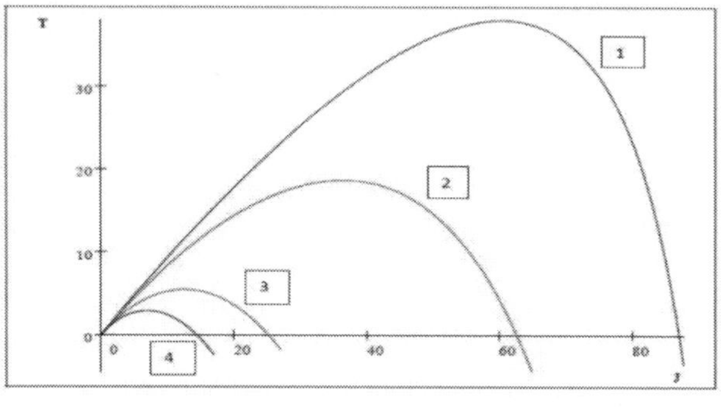

On the left Fig. 3 presents four T=T(J,h) dependencies for the fixed values of h. Curve 1 corresponds to h=0.25, curve 2 corresponds to h=1, curve 3 corresponds to h=5 and curve 4 corresponds to h=10. Note that on every curve presented there appears a maximum.

As one can see from the Fig. 3 the higher the value of the external magnetic field (h) the narrower both the T and J ranges in which the spin glass state can exist. The physical sense of this result is that the external magnetic field, leading to a magnetic ordering in the sample, liquidates the randomness of the orientation of the magnetic moments. In this way the spin glass state disappears in the sample. This result is in a good agreement with the experimental data (see above) for the compounds under consideration. In order to determine the maxima of T on all the curves $T=T(J)$ one has to solve the equation: $dT/dJ=0$ which assumes the form:

$$2j^5 - 5j^2 + 3\left(\frac{C}{h^2}\right)^{1/3} = 0, \qquad (2.10)$$

where $j=(J/(J-J_0))^{1/3}$. One can obtain from Eq. (2.10) the parameter h as a function of J (or of j) for the maxima mentioned above:

$$h = \frac{3\sqrt{3C}}{\sqrt{\left(5j^2 - 2j^5\right)^3}}. \qquad (2.11)$$

The normalized field h is real when: $5j^2-2j^5>0$. It means that: $-\infty<j<0$ or $0<j<(5/2)^{1/3}=1.36$. The plot h as a function of j is presented in Fig. 4 below

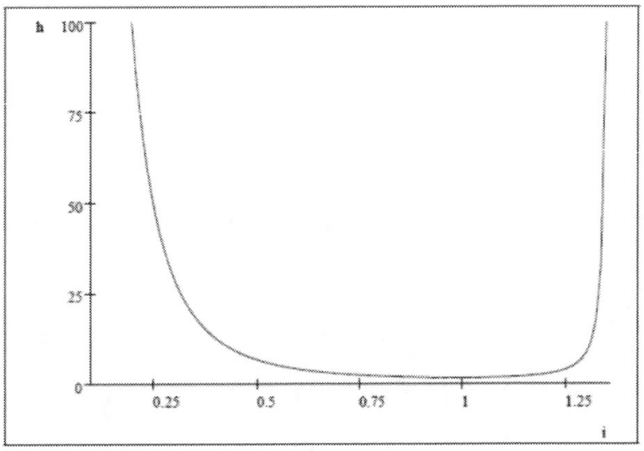

One can see that in this figure is an extended minimum. The minimum value of h - calculated from (2.11) - is equal to 1.68. For this value the minimum of the real field H assumes the value of $3.46 \cdot 10^{-23} T_G$ (tesla), if T is measured in Kelvin (K). Thus this is the minimum value of the external magnetic field H for the spin glass state to be stable. From Eqs. (2.8) and (2.10) one obtains the maximum freezing temperature T_{max} as a function of J in the form:

$$T_{max}(J) = J - J\left|\frac{J - J_0}{J - \frac{5}{3}J_0}\right|. \qquad (2.12)$$

Below on the Fig. 5 the temperature range of the existence of the spin glass state is presented. This temperature depends on J which is in turn dependent on Sb concentration. This range extends from 0 K to 100 K. On the other hand one can see in Fig. 5 that for the values of J/k_B greater than 130 K the spin glass state cannot exist. Such a situation is in a good agreement with the results obtained from the measurements of both the dc and ac magnetic susceptibility [6, 8].

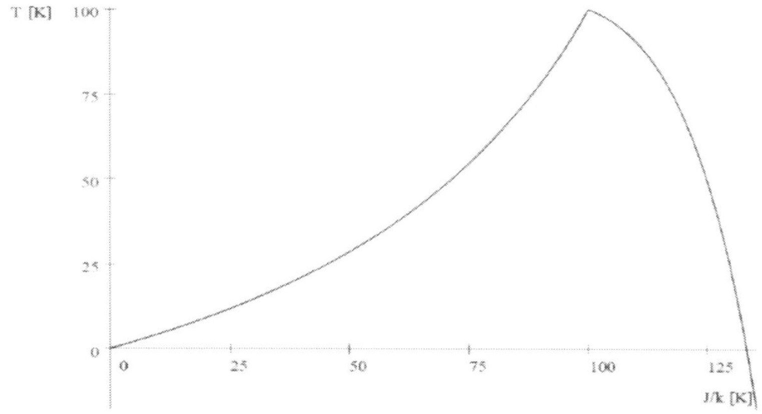

3. Conclusions

In the considered compounds the percolation limit of the spin glass state in terms of Sb concentration is about 0.5. The maximum freezing temperature (Eq. (2.12) and Fig. 5) turned out to be not greater than about 100 K. This temperature can be interpreted as the stability limit of the spin glass state in the compounds under study. The values of J/k_B can serve as another criterion of the existence of the spin glass state. For instance, for the existence of the spin glass states in these compounds the values of J/k_B should not be greater than 130 K. We also showed that a minimum value of the external magnetic field is necessary to stabilize the structure of the spin glass state. It seems that this stabilization consists in the partial but sufficient compensation of the thermal vibrations. The value of this field depends on the spin glass temperature T_G as follows:
$H_{min} = 3.46\cdot 10^{-23} T_G$ [tesla].

Acknowledgments

This work is partly founded from science Grant No. N N204 145938.

References

[1] J. Krok-Kowalski, J. Warczewski, P. Gusin, T. Śliwińska, T. Groń, G. Urban, H. Duda, E. Malicka, A. Pacyna, L.I. Koroleva, J. Phys.: Condens. Matter 21 (2009) 035402
[2] G. Parisi, Mean field theory of spin glasses: statistic and dynamics, [arXiv: 0706.0094]
[3] D. Sherrington, S. Kirkpatrick, Phys. Rev. Lett. 26, 1792 (1975)
[4] K. Binder, A. P. Young, Rev. Mod. Phys. 58, 801, (1986)
[5] P. Doussineau, A. Levelut, W. Schön, Solid State Comm. 70, 37 (1989)
[6] J. Krok-Kowalski, J. Warczewski, K. Krajewski, P. Gusin, T. Śliwińska, H. Duda, A. Pacyna, T. Mydlarz, S. Matyjasik, E. Malicka, A. Kita, Journal of Alloys and Compounds 430 (2007) 47-53
[7] J. Warczewski, J. Kusz, D. S. Filimonov, Ya. A. Kessler, L. I. Koroleva, M. G. Mikheyev, A. G. Odintsov, T. G. Aminov, E. V. Busheva, G. G. Shabunina, Journal of Magnetism and Magnetic Materials 175 (1997) 299-303
[8] J. Krok-Kowalski, J. Warczewski, P. Gusin, K. Krajewski, T. Śliwińska, H. Duda, A. Pacyna, T. Mydlarz, E. Malicka, A. Kita, Journal of Alloys and Compounds 430 (2007) 39-42
[9] J. Warczewski, J. Krok-Kowalski, L. I. Koroleva, T. Mydlarz, A. Gilewski, A. Pacyna, Journal of Alloys and Compounds 319 (2001) 7-13

Journal of Physics: Conference Series **303** (2011) 012076 doi:10.1088/1742-6596/303/1/012076

Ab-Initio Calculation of the Basic Magnetic Properties of (Mn, Cr, V)-doped $MgSiN_2$

J. Rufinus

Science Division, Widener University, Chester, PA 19013, USA

rufinus@cs.widener.edu

Abstract. The current interest in the emerging field of semiconductor spintronics is mostly focused on transition metal-doped binary materials, e.g. Mn-doped GaAs, GaN, etc. Recently, however, the explorations of transition metal-doped ternary semiconductors have intensified, due to some experimental confirmations of high Curie temperature in chalcopyrite compounds. A density functional theory within generalized gradient approximation study was performed on (Mn, Cr, V)-doped ternary material $MgSiN_2$. Our results show Mn-doped $MgSiN_2$ to be antiferromagnetic for Mn_{Mg} (Mn substitutes Mg site) and ferromagnetic for Mn_{Si} (Mn substitutes Si site). On the other hand, (Cr, V)-doped $MgSiN_2$ was found to be ferromagnetic, independent of the substitution sites.

1. Introduction

In the past several years, the study of room temperature Diluted Magnetic Semiconductors (DMS) has attracted great interest. DMS are semiconductors that are doped with metal atoms. It is believed that the interaction among the spins of these atoms leads to a ferromagnetic state. DMS received considerable attention when Mn-doped GaAs was found to have a Curie temperature above 100 K [1]. Since then the intensity of DMS research, both theoretical and experimental, has multiplied. The study of room temperature DMS would lead to the quest of building the "spintronic" devices.

Thus far, research on transition metal (TM)-doped diluted magnetic semiconductor (DMS) systems has been focused primarily on the binary compound. Some of the best-studied binary systems include InAs, GaAs, and GaN. [1-5] Mn-doped GaN DMS, for example, has for some time, been predicted to have transition temperature well above 300 K. [1]

TM-doped ternary compounds have attracted much attention due to the recent experimental confirmations of high transition temperature ferromagnetism in, e.g., $CdGeP_2$, $ZnGeP_2$, $ZnSnAs_2$, $CuGeSe_2$, $CdGeAs_2$, and $ZnSiAs_2$ compounds. [6-10] Ternary compounds are interesting systems due to the fact that it is possible to substitute TM atoms in one (or both) of the two cation sites. Based on which site was substituted by the metal atom, the ferromagnetic (FM) or antiferromagnetic (AFM) state could be produced. In the case of M-doped (M = Mn, Cr, or V) II-IV-V_2 ternary compound, M^{2+} will substitute a divalent (group II) cation site (M_{II}). In this case, the AFM spin arrangement is favored. On the other hand, if the M atom occupies the group IV cation site (M_{IV}), holes will be produced. Holes are expected to lead to ferromagnetism.

In this paper, we report the results of ab-initio calculations of the basic magnetic properties of M (M = Mn, Cr or V)-doped $MgSiN_2$. $MgSiN_2$ is a wide-band ternary semiconductor [11] with an interesting thermal property. Because of its high thermal conductivity, $MgSiN_2$ ceramic has recently been suggested as an alternative to replace the traditional heat sink material, Al_2O_3, in integrated circuits. [12] The crystal structure and lattice parameters of $MgSiN_2$ are closely related to those of AlN, whose fabrication technology is already known. To understand the magnetic properties of this material, we first study the effect of M substitutions on cation sites (Mg or Si). It is interesting to see whether substitutional M_{II} or M_{IV} of a ternary compound could lead to ferromagnetism or antiferromagnetism. Secondly, we also study the effect of M concentrations (3.125% and 6.25%) on the magnetic ordering of this structure. In this case, the role played by substitutional M_{II} and M_{IV} in ferromagnetism of $MgSiN_2$, with respect to metal concentration, is studied.

2. Computational Method

First-principles total energy calculations of M-doped $MgSiN_2$ were performed using the ABINIT code. [13-14] ABINIT is a plane-wave-based pseudopotential density functional theory (DFT) code. Specifically, the generalized gradient approximation (GGA) of DFT has been used in our calculations. The GGA is a widely used approximation for the exchange correlation functional in DFT and has proven to be very successful in predicting many material properties.

For our calculations, the GGA-PBE from Fritz-Haber-Institute (Troullier-Martins scheme) has been used for all elements. The automatic generation of k points in the Irreducible Brillouin Zone of 2 X 2 X 2 Monkhorst-Pack grids has been performed. The system was studied in a 2 X 2 X 1 supercell containing 64 atoms (accommodating 32 cations and 32 anions). The cutoff energy of the plane-wave basis set was fixed at 25 Ha.

The crystal structure of $MgSiN_2$ used in our calculations is orthorhombic. Its unit cell contains 16 atoms, compared to only 4 atoms in, e.g., wurtzite GaN. Its space group is $Pna2_1$ (#33). The following lattice constants have been used in our calculations: a = 5.2977 Å, b = 6.473 Å and c = 5.0139 Å. These parameters were obtained using a structural optimization calculation and in good agreement with the result found elsewhere. [15]

3. Results and Discussion

We substituted M atoms to replace either the group II (M_{II} or M_{Mg}) or group IV (M_{IV} or M_{Si}) atoms. For each substitutional case, the total energy for both the ferromagnetic (FM) and antiferromagnetic (AFM) spin alignments were calculated. To study the effect of metal concentration on $MgSiN_2$, we considered the following two cases: (i) 2 metal atoms in $MgSiN_2$ (corresponds to x = 3.125%); and (ii) 4 metal atoms in $MgSiN_2$ (corresponds to x = 6.25%). For each case of transition-metal concentration, we calculated the total energy of the systems for both the FM and AFM spin alignments.

Table 1. Magnetic configurations of (Mn, Cr, or V)-doped $MgSiN_2$. In this table, FM means ferromagnetic configuration and AFM means antiferromagnetic configuration. Positive ΔE means FM configuration is favourable.

Metal concentration	Metal type	Substitutional site: Mg		Substitutional site: Si	
		ΔE (meV/atom)	Magnetic structure	ΔE (meV/atom)	Magnetic structure
3.125%	Mn	-3.26	AFM	20.16	FM
	Cr	9.37	FM	7.27	FM
	V	4.53	FM	3.13	FM
6.25%	Mn	-0.83	AFM	9.44	FM
	Cr	2.16	FM	1.81	FM
	V	1.73	FM	3.46	FM

Table 1 shows the results of our calculations. The energy gain per magnetic ion, ΔE, can be expressed as ΔE = E(AFM) – E(FM), where E(AFM) denotes the total energy of the AFM state and E(FM) denotes the total energy of the FM state. The positive value of ΔE implies a FM state is preferred, while the negative value of ΔE implies an AFM state is preferred. The unit used for ΔE is meV/atom. Our results show different magnetic orderings, depending on which sites were substituted by the M atoms. Independent of the metal concentration, the substitutional Mn_{Mg} does lead to AFM interactions (negative ΔE), while the substitutional Mn_{Si} leads to FM interactions (positive ΔE). On the other hand, (Cr, V)-doped $MgSiN_2$ was found to prefer the FM state, independent of the substitution sites and concentration.

Table 2. The relationship between magnetic property and the number of 3d electrons

(Mg, M)SiN_2			V^{2+}	Cr^{2+}	Mn^{2+}
Mg(Si, M)N_2	V^{4+}	Cr^{4+}	Mn^{4+}		
d-electron	d^1	d^2	d^3	d^4	d^5
Magnetic property	FM	FM	FM	FM	AFM

Table 2 shows the relationship between calculated magnetic structures and the number of 3d electrons in each substitutional case. We have assumed that metal atoms are M^{2+} and M^{4+} when they substitute for Mg and Si, respectively. The AFM state is preferred in the case of (Mg, Mn)SiN_2 because the electron configuration is half-filled (d^5) and super-exchange interaction prevails. On the other hand, in the cases of d^1 to d^4, the FM state is preferred because there are plenty of donors available.

Since the case of $(Mn, Cr, V)_{Si}$ always leads to ferromagnetism, we show in Figure 1 the calculated values of ΔE with respect to (Mn, Cr, V) concentrations. Except for the case of V_{Si}, the value of ΔE decreases as we increase the concentration from 3.125% to 6.25%. Thus, adding more dopants into Si site could reduce the stabilization of the FM state in both Mn- and Cr-doped $MgSiN_2$. In a recent theoretical calculation, Mn-doped $MgSiN_2$ with high (12.5%) Mn concentration was found to have antiferromagnetism if Mn atom substitutes the group IV (Si) site. [16] Finally, independent of its concentration, ΔE(Mn) > ΔE(Cr or V). According to the mean-field approximation, the Curie temperature (T_c) is proportional to the value of ΔE. This, in turn, means Mn-doped $MgSiN_2$ would produce the highest T_c, compared to (Cr or V)-doped $MgSiN_2$.

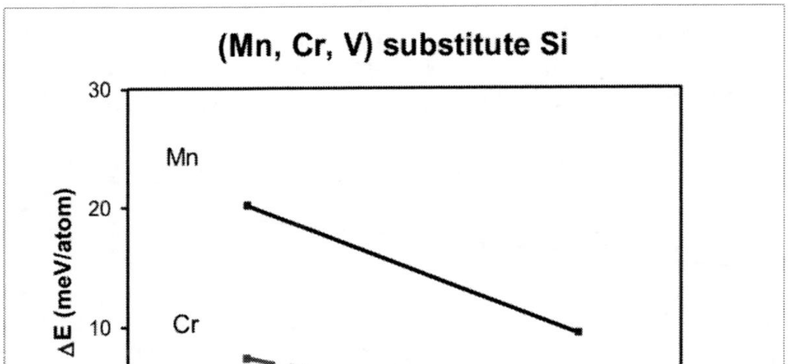

Figure 1. The calculated values of ΔE versus metal concentration.

4. Conclusion

Using the first principles DFT within the GGA, we have studied the basic magnetic properties of (Mn, Cr, V)-doped $MgSiN_2$. Our calculations show, Mn-doped $MgSiN_2$ to be antiferromagnetic for Mn_{Mg} (Mn substitutes Mg) and ferromagnetic for Mn_{Si} (Mn substitutes Si). Independent of the metal concentration, both $(Cr, V)_{Mg}$ and $(Cr, V)_{Si}$ lead to ferromagnetism. Increasing metal concentration into both sites, however, was predicted to lower the stabilization of the FM state, especially in Mn- and Cr- doped $MgSiN_2$.

Acknowledgment

This work was partially supported by the National Science Foundation (NSF) and Widener University grants. Computational time was provided by the National Science Foundation through Teragrid resources provided by the National Center for Supercomputing Applications and the Pittsburgh Supercomputing Center. The author also thanks the OU Supercomputing Center for Research and Education at the University of Oklahoma for additional supercomputer time.

References

[1] Ohno H, Shen A, Matsukura F, Oiwa A, Endo A, Katsumoto S and Iye Y 1996 *Appl. Phys. Lett.* **69** 363
[2] Dietl T, Ohno H, Cibert J and Ferrand D 2000 *Science* **287** 1019
[3] Dietl T 2003 *Nature Mater.* **2** 646
[4] Sonoda S, Shmizu S, Sasaki T, Yamamoto Y and Horia H 2002 *J. Cryst. Growth* **237** 1358
[5] Furdyna J1988 *J. Appl. Phys.* **64** R29
[6] Medvekin G A, Ishibashi T, Nishi T, Hayata K, Hasegawa Y and Sato K 2000 *Jpn. J. Appl. Phys.* Part 2 **39** L949
[7] Cho S, Choi S, Cha G –B, Hong S C, Kim Y, Zhao Y –J, Freeman A J, Ketterson J B, Kim B J, Kim Y C and Choi B –C 2002 *Phys. Rev. Lett.* **88** 257203
[8] Choi S, Cha G –B, Hong S C, Cho S, Kim Y, Ketterson J B, Jeong, S –Y and Yi G -C 2002 *Sol. State Commun.* **122** 165
[9] Demin R, Koroleva R, Marenkin S, Mikhailov S, Aminov T, Szymczak H, Szymczak R and Baran M 2005 *J. Magn. Magn. Mater.* **290** 1379
[10] Marenkin S F, Novotortsev V M, Fedorchenko I V, Varnavskiy S A, Koroleva L I, Zashchirinskii D M, Khapaeva T M, Szymczak R, Krzymanska B, Dobrowolski V and Kilanski L 2009 *J. Phys: Conf. Ser.* **153** 012058
[11] Petukhov A G, Lambrecht W R L and Segall B 1994 *Phys. Rev. B* **49** 4549
[12] Bruls R J, Kudyba-Jansen A A, Gerharts P, Hintzen H T and Metselaar R 2002 *J. Mat. Sci. Mat. In Electronics* **13** 63
[13] Gonze X *et al.* 2002 *Comput. Mater. Sci.* **25** 478
[14] Gonze X *et al.* 2005 *Z. Kristallogr.* **220** 558
[15] Fang C M, deGroot R A, Bruls R J, Hintzen H T and de With G 1999 *J. Phys.: Condens. Matter* **11** 4833
[16] Erwin S C and Zutic I 2004 *Nature Mater.* **3** 410

Spin-driven critical fields in a spinel series based on the matrix $ZnCr_2Se_4$

E. Malicka[1], T. Groń[2], A.W. Pacyna[3], A. Gągor[4] and T. Mydlarz[5]

[1]Institute of Chemistry, University of Silesia, ul. Szkolna 9, 40-006 Katowice, Poland
[2]Institute of Physics, University of Silesia, ul. Uniwersytecka 4, 40-007 Katowice, Poland
[3]The Henryk Niewodniczański Institute of Nuclear Physics, Polish Academy of Sciences, ul. Radzikowskiego 152, 31-342 Kraków, Poland
[4]Institute of Low Temperature and Structure Research, Polish Academy of Sciences, P. O. Box 1410, 50-950 Wrocław, Poland
[5]International Laboratory of High Magnetic Fields and Low Temperatures, ul. Gajowicka 95, 53-529 Wrocław, Poland

E-mail: Ewa.Malicka@us.edu.pl

Abstract. Influence of temperature and magnetic field on critical fields in the frustrated antiferromagnet and semiconductor $ZnCr_2Se_4$ doped with Ga, In and Ce is considered. The first critical field, H_{c1}, connected with metamagnetic transition remains unchanged with temperature while the second critical field, H_{c2}, corresponding to the break-down of the helical spin arrangement drops rapidly with temperature. The sharpness of the H_{c1} and H_{c2} peaks is reduced when the content of the non-magnetic Ga^{3+} and In^{3+} ions in the octahedral positions increases. This phenomenon seems to be connected with the elongation of the interatomic distances leading to the occurrence of the low spin state of the Cr^{3+} ions in the t_{2g} orbital.

1. Introduction

It is well known in literature that $ZnCr_2Se_4$ spinel is the matrix of various diluted systems where effects of site disorder, lattice frustration and random distribution of spin interactions create the novel potential applications in spin-based electronic technology. $ZnCr_2Se_4$ reveals the spectacular transition to the antiferromagnetic (AFM) state at a Néel temperature $T_N = 20$ K with the dominating ferromagnetic (FM) interactions evidenced by a large positive Curie-Weiss constant $\theta_{CW} = 115$ K [1,2] and accompanied by a structural transformation from cubic $Fd\bar{3}m$ to tetragonal $I4_1/amd$ symmetry with a small contraction along the c axis [3]. Although, the structural investigations using synchrotron radiation defined this latter symmetry as orthorhombic *Fddd* [4]. Previous neutron-diffraction investigations [1,2] of the $ZnCr_2Se_4$ spinel showed a spiral AFM spin structure below ordering temperature having a FM arrangement in the (001) planes with a turning angle of 42° between the spins in adjacent (001) planes. A detailed investigation of $ZnCr_2Se_4$ single crystal [5] showed the sharp first-order anomalies in the specific heat and negative thermal expansion (NTE) strongly suppressed by external magnetic fields at T_N suggesting a spin-driven origin of the structural transition. They also revealed a metamagnetic transition at the critical field H_{c1} of about 10 kOe for $T = 2$ K, the breakdown of the helical spin arrangement at the critical field H_{c2} of about 65 kOe for $T = 2$ K, and the full saturation magnetisation of about $3\mu_B$ per Cr ion above H_{c2} for the FM state. At a temperature smaller than

T_N the zero-field-cooling and field-cooling susceptibilities start diverging suggesting a spin frustration [5]. The magnetic isotherm at 4.3 K for a polycrystalline $ZnCr_2Se_4$ is shown in Fig. 1. The complex ac dynamic magnetic susceptibility used to study the influence of temperature on critical fields in a polycrystalline $ZnCr_2Se_4$ spinel [6] showed a parallel spin coupling in FM clusters in the range between T_N = 20.7 K and θ_{CW} = 55.1 K as well as the strong drop of the value of the second critical field H_{c2} with temperature suggesting a spin frustration of the re-entrant type (Fig. 2).

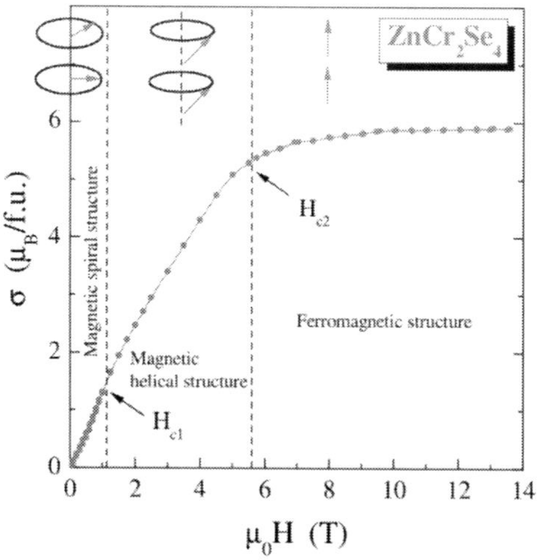

Figure 1. Magnetization σ vs. magnetic field $\mu_0 H$ at 4.3 K for a polycrystalline $ZnCr_2Se_4$. Critical fields H_{c1} and H_{c2} are indicated by arrows. Upper pictures show the evolution of magnetic structure in an external magnetic field.

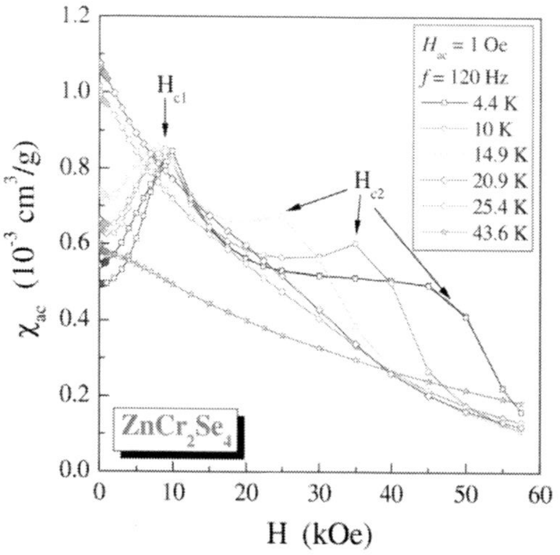

Figure 2. Ac magnetic susceptibility, χ_{ac}, vs. static magnetic field H at 4.3, 10, 14.9, 20.9, 25.4 and 43.6 K for a polycrystalline $ZnCr_2Se_4$. Critical fields H_{c1} and H_{c2} are indicated by arrows [6].

The magnetic and electrical studies carried out on the $Zn_{1-x}Ga_{2x/3}Cr_2Se_4$ poly- and single crystals in the compositional range $0.0 \leq x \leq 0.5$ showed an AFM order below the Néel temperature (22 K $\geq T_N \geq$ 12 K [7-9]) and the thermally activated *p*-type conduction [8-11]. With increasing Ga content a decrease of T_N [7,8], Curie-Weiss temperature (118 K $\geq \theta_{CW} \geq$ 62 [7-9]) and effective magnetic moment, (5.49 $\mu_B \geq \mu_{eff} \geq$ 5.27 μ_B [7-9]) as well as a disappearance of the critical fields [12] were observed.

The magnetic and electrical measurements carried out on the $ZnCr_{2-x}In_xSe_4$ single crystals in the compositional range $0.0 \leq x \leq 0.14$ revealed an AFM order with metamagnetic phase transitions below the Néel temperature (22 K $\geq T_N \geq$ 16 K [13,14]) and the thermally activated *p*-type conduction [15,16] for x $\leq$ 0.07. With increasing In content a decrease of T_N was observed while θ_{CW} (~49 K), μ_{eff} (~5.30 μ_B) and the critical fields remained almost constant [14].

The present contribution reports the magnetic properties for $ZnCr_2Se_4$ doped with the nonmagnetic Ga and In ions as well as the magnetic Ce ions for which the ion coordination changes from mixed for Ga and In to tetrahedral one for Ce. It seems natural to expect a correlation between the spin state and the critical fields in the disturbed magnetic system. For this purpose the magnetization and the ac dynamic magnetic susceptibility measurements were employed.

2. Experimental

Starting materials for single crystal growth of $Zn_xCr_yMe_zSe_4$ (Me = Ga, In, Ce) samples were the binary selenides ZnSe, Ga_2Se_3, In_2Se_3 and elementary cerium and selenium. Single crystals were obtained by chemical vapour transport reaction using anhydrous chromium chloride as transporting agent. The chemical composition of the $Zn_xCr_yMe_zSe_4$ single crystals were analyzed by energy-dispersive X-ray fluorescence spectrometry (EDXRF). The samples were excited using the air-cooled side-window and the Rh-target of the X-ray tube, which allowed a ca. 100 μm nominal focal spot size (XTF 5011/75, Oxford Instruments, USA). A single-crystal X-ray data collection was performed on a Kuma KM4CCD diffractometer with CCD data collector using Mo $K\alpha$ radiation. The CrysAlis software version 1.170.32 [17] was used for data processing. An empirical absorption correction was applied with the aid of spherical harmonics implemented using a scale 3 Abspack scaling algorithm. The structure was refined by the full-matrix least-squares method by means of SHELX-97 program package [18]. The chemical composition and some data of the crystal structure refinement of the single crystals under study are collected in Table 1. More details are described elsewhere [19].

Table 1. Chemical composition and crystal structure refinement of single crystals under study. The brackets () and [] mean tetrahedral and octahedral sites in the spinel structure, respectively. *a* is the lattice parameter, Cr/Ga-Se, Cr/In-Se and Cr-Se are the interatomic distances in the octahedral sites and Zn/Ga-Se, Zn/In-Se and Zn/Ce-Se are the interatomic distances in the tetrahedral sites [19].

Composition and refinement	a (pm)	Cr/Ga-Se Zn/Ga-Se (pm)	Cr/In-Se Zn/In-Se (pm)	Cr-Se Zn/Ce-Se (pm)
$(Zn_{0.98}Ga_{0.01})[Cr_{1.92}Ga_{0.06}]Se_4$	1049.71(2)	6x253.01(2)	-	-
		4x244.25(4)	-	-
$(Zn_{0.84}In_{0.1})[Cr_{1.78}In_{0.14}]Se_4$	1050.93(3)	-	6x253.22(7)	-
		-	4x244.67(4)	-
$(Zn_{0.86}Ce_{0.08})[Cr_2]Se_4$	1049.37(2)	-	-	6x252.83(2)
		-	-	4x244.34(4)

The magnetization, $\sigma(\mu_0 H)$, and the ac magnetic susceptibility, $\chi_{ac}(T)$, were measured using: 1) a vibrating sample magnetometer with a step motor [20] in the temperature range 4.2-80 K and in applied magnetic induction up to 14 T and 2) a Lake Shore 7225 ac susceptometer in the temperature range 4.2–22 K and in applied external magnetic fields up to 60 kOe, respectively.

3. Results and discussion

3.1. Crystal structure and cation distribution

$ZnCr_2Se_4$ crystallizes in a regular system in a spinel type structure and the space group symmetry $Fd\bar{3}m$ *(No. 227)*. In a normal spinel structure Zn^{2+} ions are located at 8*a* (1/8, 1/8, 1/8) (*A* site) with tetrahedral environment whereas the Cr^{3+} ions, that are known for a strong octahedral site preference [21], occupy octahedrally coordinated position 16*d*: (½, ½, ½) (*B* site). Se^{2-} ions are located at the 32*e* Wyckoff positions. The incorporation of Ga^{3+}, In^{3+} or Ce^{3+} into the $ZnCr_2Se_4$ does not change the crystal symmetry. Ga^{3+}, In^{3+} as well as Ce^{3+} replacement for Zn^{2+} in tetrahedral sites may introduce an increase of the overall positive charge in the crystal that has to be compensated. For first two dopants the compensation comes from cation vacancies that appear at both *A* and *B* sites. In $(Zn_{0.84}In_{0.1})[Cr_{1.78}In_{0.14}]Se_4$ there are ~4.6% cation vacant sites whereas in $(Zn_{0.98}Ga_{0.01})[Cr_{1.92}Ga_{0.06}]Se_4$ there are 1%. In $(Zn_{0.86}Ce_{0.08})[Cr_2]Se_4$ 2% of vacancies are observed only on *A* site [19]. The lattice parameters and interatomic distances for all compounds are given in Table 1.

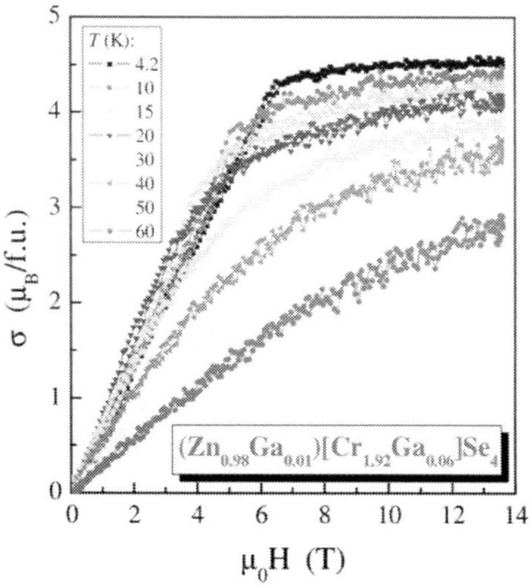

Figure 3. Magnetization σ vs. magnetic field μ_0H at 4.2, 10, 15, 20, 30, 40, 50 and 60 K for $(Zn_{0.98}Ga_{0.01})[Cr_{1.92}Ga_{0.06}]Se_4$.

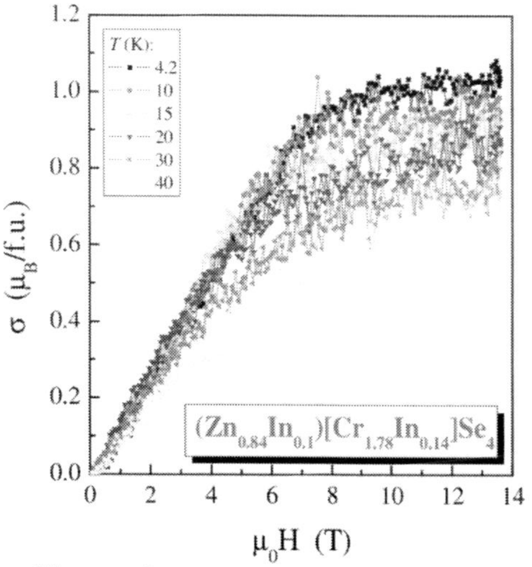

Figure 4. Magnetization σ vs. magnetic field μ_0H at 4.2, 10, 15, 20, 30 and 40 K for $(Zn_{0.84}In_{0.1})[Cr_{1.78}In_{0.14}]Se_4$.

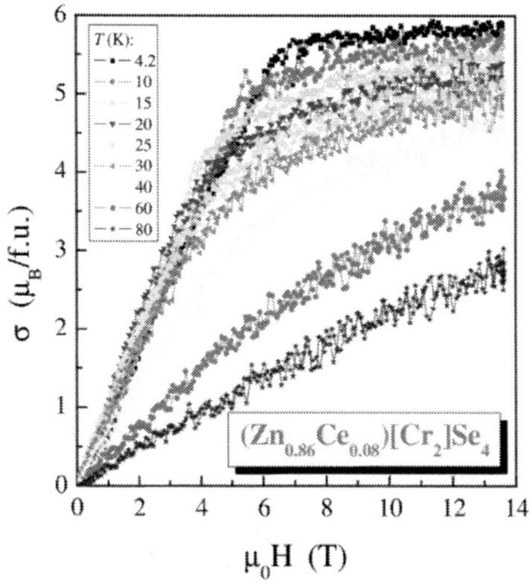

Figure 5. Magnetization σ vs. magnetic field μ_0H at 4.2, 10, 15, 20, 25, 30, 40, 60 and 80 K for $(Zn_{0.86}Ce_{0.08})[Cr_2]Se_4$.

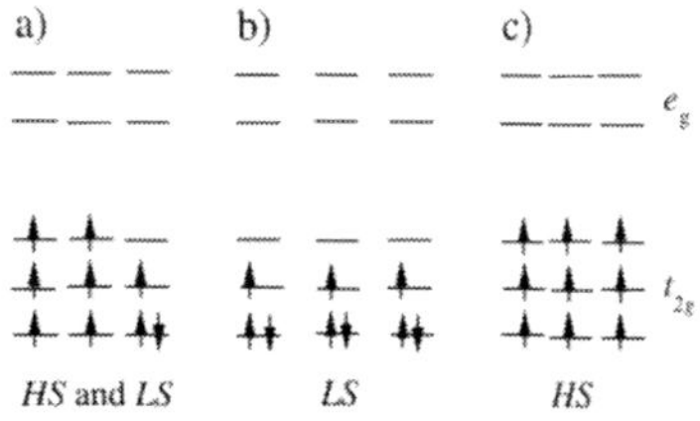

Figure 6. The proposed population of the HS and LS states on the example of three Cr^{3+} ions in the t_{2g} orbital for $(Zn_{0.98}Ga_{0.01})[Cr_{1.92}Ga_{0.06}]Se_4$ (a), $(Zn_{0.84}In_{0.1})[Cr_{1.78}In_{0.14}]Se_4$ (b) and $(Zn_{0.86}Ce_{0.08})[Cr_2]Se_4$ (c)

3.2. Magnetic properties

The result of measurements of magnetization, $\sigma(\mu_0H)$, and ac magnetic susceptibility, $\chi_{ac}(H)$, are presented in Figs. 3-5 and Figs. 7-9, respectively, in order to $(Zn_{0.98}Ga_{0.01})[Cr_{1.92}Ga_{0.06}]Se_4$, $(Zn_{0.84}In_{0.1})[Cr_{1.78}In_{0.14}]Se_4$ and $(Zn_{0.86}Ce_{0.08})[Cr_2]Se_4$. All the magnetic isotherms are similar (Figs. 3-5) and they are comparable to the matrix (Fig. 1) indicating a complex magnetic structure changing from spiral via helical to ferromagnetic one as magnetic field increases. With increasing temperature a saturation state disappears on going towards the paramagnetic (PM) limit. All the compounds differ in the value of the magnetic moment. When increasing the content of non-magnetic ions in octahedral position of the magnetic moment value is rapidly declining in comparison to the value of 6 μ_B per molecule for the $ZnCr_2Se_4$ matrix at 4.2 K. We have 4.56 μ_B/f.u. for $(Zn_{0.98}Ga_{0.01})[Cr_{1.92}Ga_{0.06}]Se_4$ and 1.06 μ_B/f.u. for $(Zn_{0.84}In_{0.1})[Cr_{1.78}In_{0.14}]Se_4$ in the saturation state. Only for $(Zn_{0.86}Ce_{0.08})[Cr_2]Se_4$ the saturation magnetic moment reaches the value of 5.89 μ_B/f.u. at 4.2 K which is close to the matrix one. It is known in literature that when only Cr ions occupy octahedral positions in the spinel structure, then they are in the high spin (HS) state with the total spin $S = 3/2$ in the t_{2g} orbital giving the full saturation of about 6 μ_B/f.u. [5].

In order to explain the dramatic undervaluation of the magnetic moment, especially for $(Zn_{0.84}In_{0.1})[Cr_{1.78}In_{0.14}]Se_4$, consider the low spin (LS) state of the Cr^{3+} ions in the t_{2g} orbital. This state seems be connected with a larger lattice parameter of the unit cell entailing the elongation of the interatomic distances observed for $(Zn_{0.98}Ga_{0.01})[Cr_{1.92}Ga_{0.06}]Se_4$ and $(Zn_{0.84}In_{0.1})[Cr_{1.78}In_{0.14}]Se_4$ both in the octahedral and tetrahedral positions (see Table 1). This is a consequence of the replacement of the Cr^{3+} ions by the non-magnetic Ga^{3+} and In^{3+} ions which have larger radii ($r_{Ga3+} = 62$ pm, $r_{In3+} = 80$ pm [22]) than Cr^{3+} ($r_{Cr3+} = 61.5$ pm [22]). In addition, the introduction of non-magnetic ions weakens the AFM superexchange interaction between the Cr ions. Schematically, the population of the HS and LS states for the spinels under study is depicted in Fig. 6. It depends on the content of the non-magnetic ion in the octahedral positions.

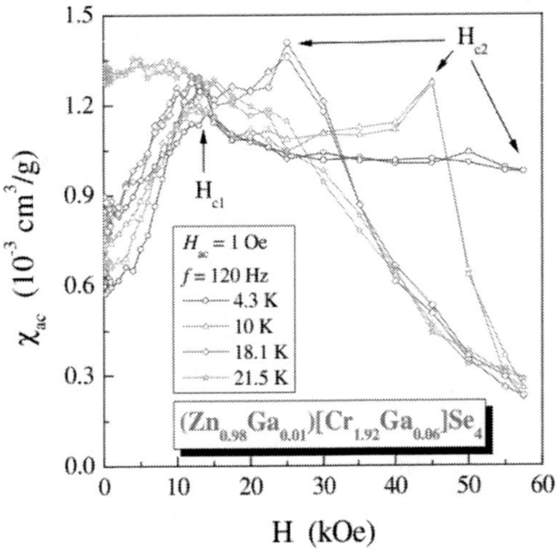

Figure 7. Ac magnetic susceptibility, χ_{ac}, vs. static magnetic field H at 4.3, 10, 18.1 and 21.5 K for $(Zn_{0.98}Ga_{0.01})[Cr_{1.92}Ga_{0.06}]Se_4$. Critical fields H_{c1} and H_{c2} are indicated by arrows.

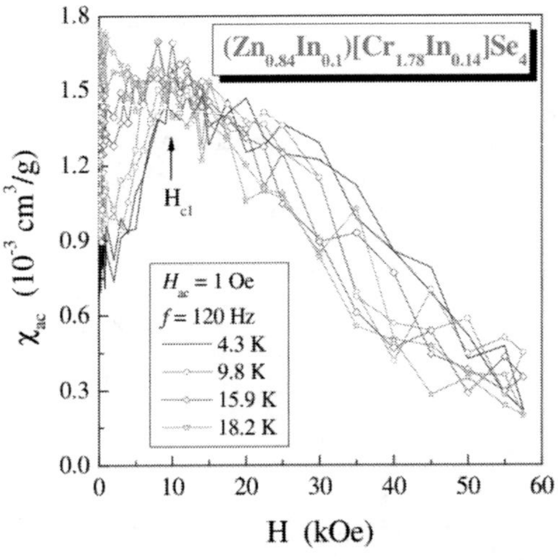

Figure 8. Ac magnetic susceptibility, χ_{ac}, vs. static magnetic field H at 4.3, 9.8, 15.9 and 18.2 K for $(Zn_{0.84}In_{0.1})[Cr_{1.78}In_{0.14}]Se_4$. Critical fields H_{c1} and H_{c2} are indicated by arrows.

Figures 7-9 show the critical fields for the spinels under study indicated by H_{c1} and H_{c2} on the curves of the ac magnetic susceptibility versus magnetic field, χ_{ac}(H), for different temperatures. The first critical field, H_{c1}, connected with the metamagnetic transition and the breakdown of the magnetic spiral structure [3,5], slightly decreases with temperature while the second critical field, H_{c2}, connected with the breakdown of the magnetic helical structure [5,6], drops strongly with temperature. Above T_N both critical fields disappear. Low sensitivity to temperature of H_{c1} well correlates with the AFM order in the orthorhombic structure. On the other hand, strong sensitivity to temperature of H_{c2} means that FM coupling is weaker than AFM one, although the FM short-range interactions exist far above T_N in the cubic (spinel) structure as the PM Curie-Weiss temperature (θ_{CW}) is positive and several times larger than T_N. The H_{c1} and H_{c2} peaks are not sharp for $(Zn_{0.98}Ga_{0.01})[Cr_{1.92}Ga_{0.06}]Se_4$ and the H_{c2} peak is totally fuzzy for $(Zn_{0.84}In_{0.1})[Cr_{1.78}In_{0.14}]Se_4$. Only for $(Zn_{0.86}Ce_{0.08})[Cr_2]Se_4$ the sharpness of the H_{c1} and H_{c2} peaks is comparable with the matrix ones. It means that the spin state strongly influences on the sharpness of the metamagnetic transitions connected with the critical fields. In our case the LS state reduces the sharpness of the magnetic phase transitions, and can even quench them, suggesting spin-driven critical fields in the doped $ZnCr_2Se_4$ spinels.

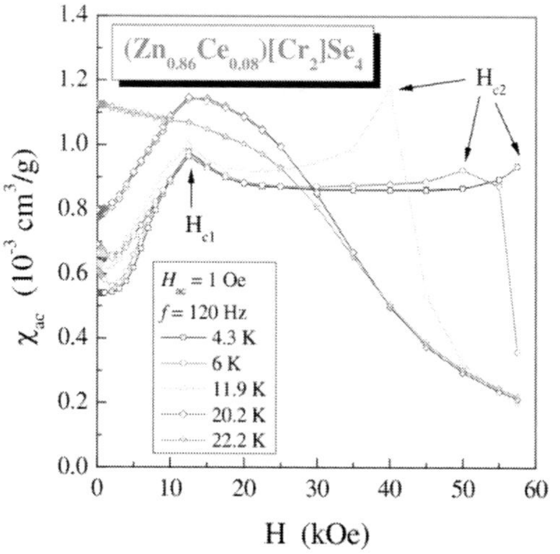

Figure 9. Ac magnetic susceptibility, χ_{ac}, vs. static magnetic field H at 4.3, 6, 11.9,20.2 and 22.2 K for $(Zn_{0.86}Ce_{0.08})[Cr_2]Se_4$. Critical fields H_{c1} and H_{c2} are indicated by arrows.

Acknowledgments
This work is partly founded from science Grant No. N N204 145938.

References

[1] F.K. Lotgering, in: Proceedings of the International Conference on Magnetism, Nottingham, 1964 (London, Institute of Physics, 1965).
[2] R. Plumier, J. Phys. (Paris) 27 (1966) 213-219.
[3] R. Kleinberger, R. de Kouchkovsky, C.R. Acad. Sci. Paris, Ser. B 262 (1966) 628.
[4] M. Hidaka, M. Yoshimura, N. Tokiwa, J. Akimitsu, Y.J. Park, J.H. Park, S.D. Ji, K.B. Lee, phys. status solidi (b) 236 (2003) 570-577.
[5] J. Hemberger, H.-A. Krug von Nidda, V. Tsurkan, A. Loidl, Phys. Rev. Lett. 98 (2007) 147203(4).
[6] T. Groń, E. Malicka, A.W. Pacyna, Physica B 404 (2009) 3554-3558.
[7] S. Juszczyk, J. Krok, I. Okońska-Kozłowska, T. Mydlarz, A. Gilewski, J. Magn. Magn. Mater. 46 (1984) 105-113.
[8] A.A. Zhukov, Ya.A. Kesler, V.F. Meshceriakov, A.V. Rozancev, Sov. Phys. Solid State (USA) 26 (1984) 1-9.
[9] I. Okońska-Kozłowska, H.D. Lutz, T. Groń, J. Krok, T. Mydlarz, Mater. Res. Bull. 19 (1984) 1-5.
[10] T. Groń, H. Duda, J. Warczewski, Phys. Rev. B 41 (1990) 12424-12431.
[11] T. Groń, H. Duda, J. Warczewski, J. Magn. Magn. Mater. 83 (1990) 487-489.

[12] S. Juszczyk, I. Okońska -Kozłowska, P. Byszewski, Phase Transitions 4 (1984) 291-296.
[13] J. Krok-Kowalski, J. Warczewski, T. Mydlarz, A. Pacyna, A. Bombik, J. Kopyczok, I. Okońska-Kozłowska, J. Magn. Magn. Mater. 111 (1992) 50-52.
[14] J. Krok-Kowalski, J. Warczewski, T. Mydlarz, A. Pacyna, A. Bombik, I. Okońska-Kozłowska, J. Magn. Magn. Mater. 166 (1997) 172-178.
[15] T. Watanabe, J. Phys. Soc. Jpn. 37 (1974) 140-144.
[16] T. Groń, J. Kopyczok, I. Okońska-Kozłowska, J. Warczewski, J. Magn. Magn. Mater. 111 (1992) 53-55.
[17] M. Mayer, CrysAlis Programs for Data Collection and Reduction, Version 1.170.32, Oxford Diffraction Ltd, 2007.
[18] G.M. Sheldrick, Acta Cryst. A 64 (2008) 112–122.
[19] E. Malicka, T. Groń, A. Ślebarski, A. Gągor, A.W. Pacyna, J. Goraus, R. Sitko, to be published.
[20] V.I. Nizhankovskii, L.B. Lugansky, Meas. Sci. Technol. 18 (2007) 1533-1537.
[21] A. Weiss, H. Witte, *Kristallstruktur und Chemische Bindung* (Verlag Chemie, Weinheim 1983).
[22] R. D. Shannon, Acta Cryst. A32 (1976) 751-767.

Magneto-Mechanical Energy Conversion in Magnetic Shape Memory Alloys

Herbert Schmidt
Eugen Seitz AG, Spitalstrasse 204, 8623 Wetzikon, Switzerland
E-mail: hschmidt@seitz.ch

Abstract. Magnetic Shape Memory (MSM) alloys present a novel means of transforming electric energy into mechanical work via the intermediate generation of magnetic field from electric current. They thereby follow a similar path of energy conversion as classical electromagnets. In this paper we will compare known designs of MSM actuators and electromagnets and discuss their characteristics based on properties of available MSM materials.

1. Introduction

Solid state actuators are materials that exhibit a noticable macroscopic shape-change upon application of a driving field, which e.g. could be electric, magnetic or thermal. Significant effects are seen and technically utilized in piezoelectrics (with electric field as driving field) and in (Giant) Magnetostrictive Materials (GMM, with magnetic field as driving field). In both cases strains on the order of 1,000...2,000 ppm are possible. On the other hand, there is a class of materials called (thermal) Shape Memory Alloys (SMA) that upon thermal cycling show shape-changes of several percent. In Fig. 1 the general idea is depicted. From a high-symmetric high-temperature phase the material transforms into a lower-symmetry low-temperature phase. There are different possible orientations to realize this reduced symmetry (in this two-dimensional sketch there are two), depicted as light and dark grey variants. Chosing the same variant over the entire sample would result in macroscopic shape-change, which normally will not be a preferred solution. Instead, macroscopic shape will be preserved by having different sections (twins) of the crystal transforming into different variants. These sections meet at twin boundaries that will be slightly off 45°, depending on the disparity between the tetragonal axes. These twin boundaries in principle are mobile, i.e., the sample may be significantly deformed without actually destroying the crystal structure anywhere — the sample simply accomodates such deformation by moving twin boundaries. Upon warming up into the high-temperature phase, however, this freedom in macroscopic shape disappears, since the cubic lattice does not allow for any of this. As a result, and with significant driving force, the crystal regains its "memorized" shape. So in this solid state active material high strains are combined with high stresses, however, since it depends on a phase transition, it is strongly dependent on environmental parameters (temperature). Also, it inherently is somewhat slow in the direction of cooling down.

Magnetic Shape Memory (MSM) alloys are named in some analogy to SMA, however, this resemblance in name is somewhat misleading. MSM work in the low-temperature, twinned phase only, so there really is no shape memory upon re-entering some high-symmetry high-temperature phase used in this effect (although these materials may also be used as thermal

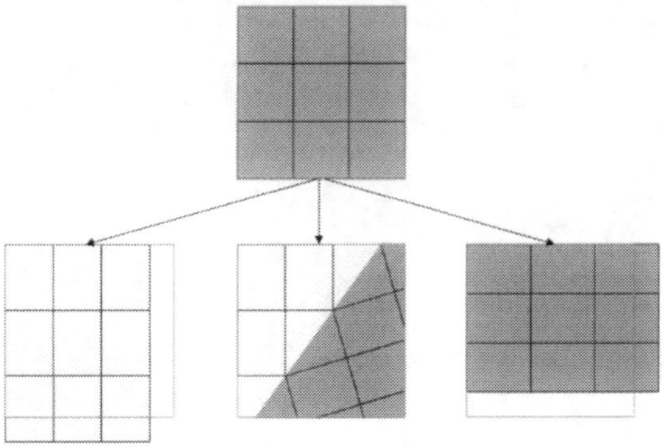

Figure 1. Schematic of thermal SMA. Top: cubic high-temperature phase. Bottom: tetragonal low-temperature phase, left to right: first twin; twin boundary; second twin.

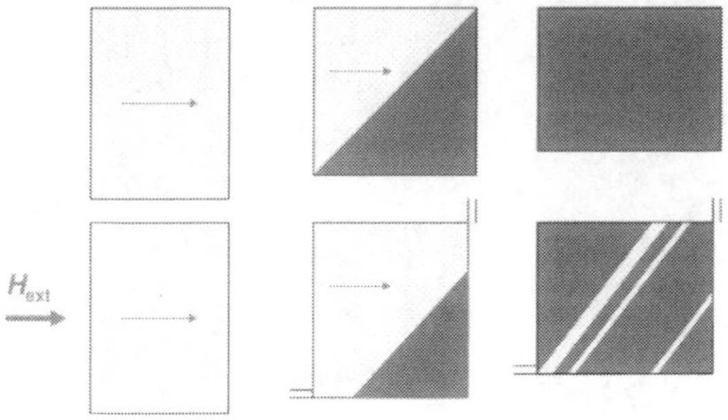

Figure 2. Schematic of MSM. Top: without external field. Bottom: with external field. Left to right: starting from easy axis variant; twinned state; hard axis variant.

SMAs). The driving force in this case is magnetic. In a ferromagnetic martensite with a pronounced disparity between magnetically hard and easy axes, magnetic field will provide a driving force favouring one variant over the other. In order for this to show some noticable effect, the twin boundaries need to be soft enough so the magnetically induced force will be sufficient to move them. This MSM effect is depicted in Fig. 2. Considering (top panel) the two variants by themselves, either will have a magnetically preferred axis (easy axis) and the moments will be oriented accordingly (arrows). When applying an external magnetic field (bottom panel), the effect will be as depicted. Starting from a sample that is oriented with the easy axis along the external field direction, nothing will happen (left). With a sample in a twinned state, the first reaction of the magnetic moments in the unfavourable twins will be to slightly rotate towards the direction of external magnetic field, which would constitute the "normal" magnetostrictive effect. In MSM alloys, however, there is an even more efficient reaction to external field, viz. twins of magnetically favourable orientation will grow (center), and at the same time the sample will macroscopically change its dimensions. With a sample starting from a single-variant state oriented along the hard axis, soft-axis twins will form and grow, as mentioned before (right). This effect is rather interesting, as it combines the strains of SMA (several percent) with the magnetic driving field of magnetostrictive actuators, thus it would seem it gives best of both worlds. There is one caveat, though, viz. that energy does not come for free, and of course an actuator having one or two orders of magnitude higher strain than GMMs will also have significantly less magnetic-field induced stress.

2. MSM Actuator Geometries

Still MSM alloys continue to receive much attention from physicists, material scientists and engineers alike, and the reason for this is that MSM alloys combine a number of properties that makes working with them challenging, but also promise interesting new applications provided its various properties can be used advantageously. Considering MSM as the basis for an actuator, first the different possible modes of operation need to be considered (see also [1]).

Twin boundary motion and thus macroscopic shape-change may be introduced either magnetically or mechanically as is depicted in Fig. 3. Assuming the magnetically preferred axis to be the short axis (as is the case in the most-studied class of alloys, Ni_2MnGa), elongation may be induced by either transverse field or transverse compressive stress, while contraction may be induced by either axial field or axial compressive stress. In order to build a useful actuator, at

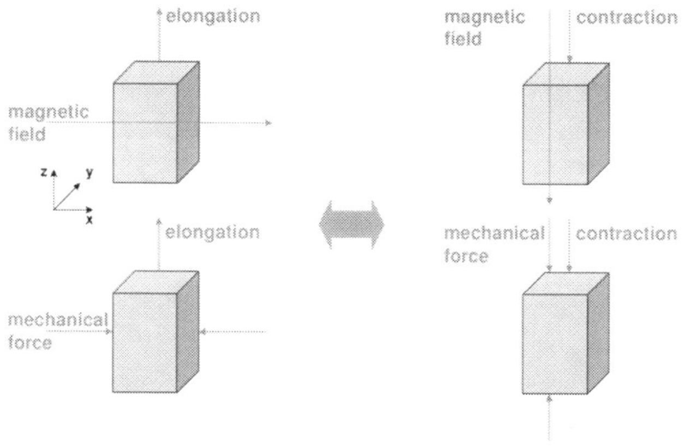

Figure 3. Schematic of operational modes in MSM actuators. Top: magnetic field induced shape change. Bottom: mechanically induced shape change. Left: elongation. Right: contraction.

least one of the two, elongation and compression, should be driven by an intelligent signal, which in most cases will mean that at least one shape-change should be induced by an electromagnet. The other may, e.g., be induced by yet another electromagnet, or by a mechanical return spring. The latter became a kind of standard solution, however, quite a number of different designs have been proposed to the present day (see, e.g., [2] for an overview).

Another design choice that needs to be made when using a single electromagnet and a return spring (which, again, is not necessarily the case, there are alternative options) is wheter to close the magnetic circuit such that induced mechanical deformation is in the plane of the magnetic circuit, or perpendicular to this plane. In Fig. 4 the option of induced deformation (vertical) perpendicular to the plane of the magentic circuit is shown. To the left, the most simple conceivable circuit is shown, while in practical applications typically a design closer to the right panel sketch of Fig. 4 will be chosen. In this design, very high field strength at the position of the MSM element can be achieved, while simultaneously stray-flux is efficiently minimized. In this design almost the full actuator height might be used for MSM active material, thus very high strains (in relation to actuator dimension) are possible. This geometry has become a kind of standard design described in various publications (see, e.g., [3]).

Alternatively, the direction of induced mechanical deformation can be within the plane of the magnetic circuit, i.e., the field lines do not close laterally around the MSM element, but under it. This geometry, and its evolution from the design previously described in Fig. 4, is shown in the left three panels of Fig. 5. In the right panel of Fig. 5 this design is rotated such that it results in a very compact design described in [4] (deviant from full rotational symmetry, an MSM circular ring obviously needs to be substituted by a polygon of MSM platelets). This design has a few shortcomings, e.g., having a number of MSM elements in a single magnetic

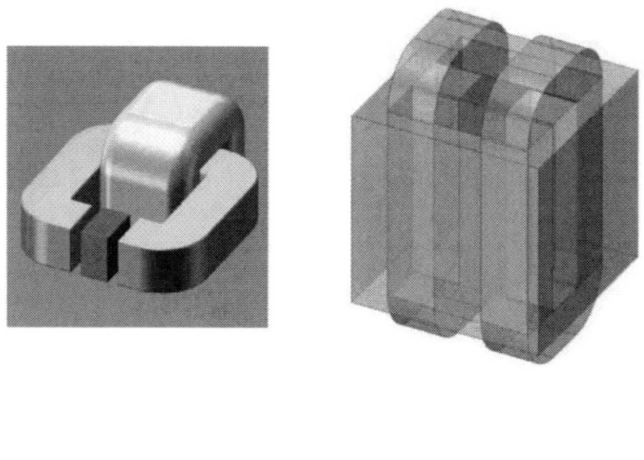

Figure 4. Schematic of a standard actuator design with magnetic field in xy-plane. MSM in blue, yoke in grey, coil in yellow.

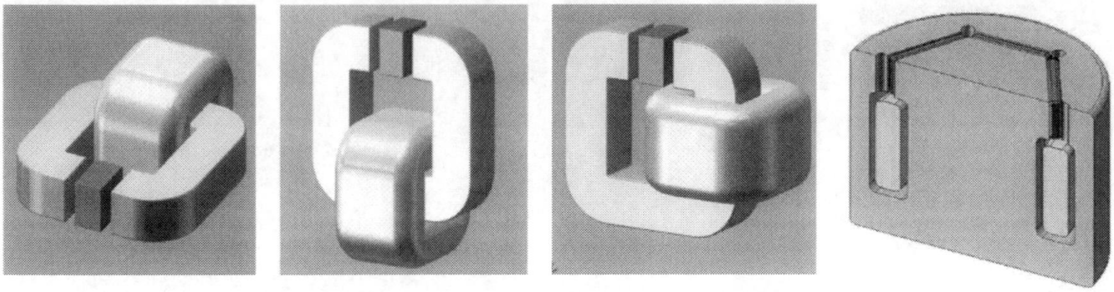

Figure 5. Schematic of an actuator design with magnetic field in xz-plane and its development into a rotational symmetric geometry.

circuit which will lead to problems if these elements are not rather similar in their properties. Furthermore, strain in relation to actuator height is significantly reduced, as only a fraction of the total actuator height is used for MSM elements. On the other hand, significangly higher forces (stresses in relation to the full actuator cross-section) theoretically are feasible with this design as compared to designs similar to the ones shown in Fig. 4.

3. Comparision of Actuator Principles

One fact has been hinted at a few times now, viz. that it is not sufficient to compare MSM (or any solid state actuator) to any another actuator principle by just comparing material properties. The reason for this is that solid state actuators in most cases are not operational by just taking a piece of active material. For one thing, they require their driving field, and if this, as with MSM, happens to be a magnetic field, this field cannot be supplied by sticking a wire into the MSM element; it needs to be generated at the location of the actuator, meaning there needs to be a yoke, a coil, connectors, fixtures, a return spring, housing etc., a lot of things that just take up space. A lot of space. The actual volume fraction of course varies depending on design, but it is realistic to think in terms of a few percent of actuator volume for actual MSM active material. This is not quite as bad as it sounds — it just means that by focusing on the material properties of the MSM material in order to compare it with other (solid state) actuators, one is prone to neglect what is actually the majority of volume and weight required in any prospective application.

Another fact that must not be forgotten is that the ultimate benchmark is neither GMM, piezoelectrics or SMA, but electromagnets. There are areas, where GMM and piezoelectrics, as well as SMA have their ideal applications, but the majority of small-stroke linear motion that is technically induced in this world is induced either electromagnetically, by pneumatics or by hydraulics. And as for the latter two, somewhere down the line there will be an electromagnet, e.g., in a pilot valve. Specifically with MSM, realistic strokes and forces (in actuator geometries realized so far) compare rather favourably with what is usually done by electromagnets. Note that this is different for, e.g., piezoelectrics — they provide a range of high stresses and low strains at very high precision that is hard to do in any other way. For MSM however, electromagnets are an obvious benchmark. An MSM actuator will need everything an electromagnet needs to generate its driving field, plus the MSM elements (introducing additional costs and complications). So there needs to be a significant advantage to render MSM an interesting alternative. In this paper we want to demonstrate, how an actuator comparison might look like, in order to give some indication, whether MSM as a solid state actuator compares at all favourably to an established, simple solution like an electromagnet.

4. A Direct Comparison between an MSM Actuator and an Electromagnet

It was pointed out above that a direct comparison of a solid state actuator and an electromagnet is not possible based on material properties of the active component alone. On the other hand, comparing these two actuator principles based on any arbitrary choice of design and technical requirements runs the risk of creating the impression of an absolute comparison of the two acutator principles. This it inherently cannot be. Keeping this in mind, we believe it is still instructive to go for such a direct comparison based on arbitrary, but meaningful assumptions of technical requirements.

In order to keep the geometries easily comparable, we will within a cylindrical volume consider the general actuator designs illustrated in Fig. 6. In the left panel, a closed magnetic circuit is shown (that obviously will not do any useful work). In the center panel, this same circuit is opened such that there is a working air gap that will exert a downwards force on the armature (electromagnet / pot magnet). The air gap volume indicated in blue is the nearest equivalent to the active element in a solid state actuator. The work done is limited by the electromagnetic energy stored in the opened air gap, which in turn is reasonably limited mostly by realistic working points of the yoke material. The air gap does not really impose any other limit. In the right panel, this same general magnetic circuit is opend such that MSM elements may be introduced. Note that the major reason to chose a design similar to that in [4] is that this allows for a close analogy between these circuits. The active volume here of course are the MSM elements. In contrast to free air, they do impose their own limitations. Their maximum work is limited by the difference between hard- and soft-axis magnetization, any further field applied will not result in useful work. As typical MSM saturation polarizations are significantly lower than those of typical yoke materials, the work per active volume is much lower than in the case of the electromagnet and its air gap. On the other hand, of course, its permeability is much higher, so one can assume to have the chance to introduce significantly more active volume. It is the major aim of this paper to illustrate how these different tendencies in principle interact.

This comparison aims at MSM and electromagnets as a means of producing a maximum of mechanical work from a given amount of electric energy in a given volume. This is a first important assumption — this needs not to be the case, there are applications with a different focus, and this comparison is not applicable for such applications.

The volume we will allow ourselves is a cylinder of one inch in diamter and one inch in height, and we will require ten mil in stroke (i.e. 1% of strain in relation to the overall external dimension of the actuator). This last assumption is not entirely random, obviously. A 5M martensite Ni_2MnGa will at most produce about 6% strain. Due to the assumed MSM actuator geometry, only a reasonable fraction of the height can be used for active material, so it appears most reasonable to require only about the above quoted stroke. Twice that might be conceivable without leaving this design concept, but probably not much more than that). It will be the aim of this comparison to meet this requirement and simultaneously provide as much force as possible at the beginning of the motion.

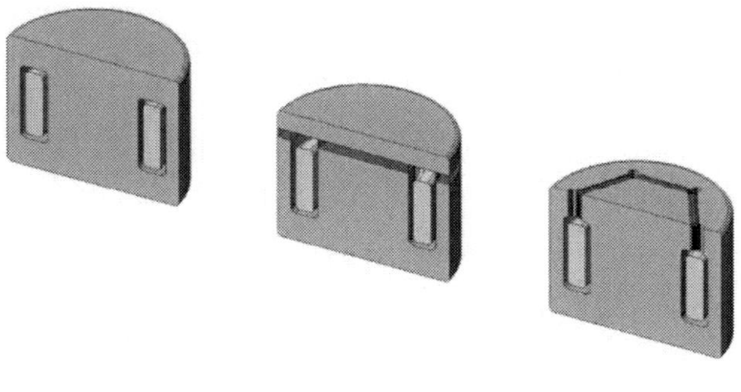

Figure 6. Schematic of a closed magnetic circuit (left) and its development into an electromagnet (center) and an MSM actuator (right) for direct comparison.

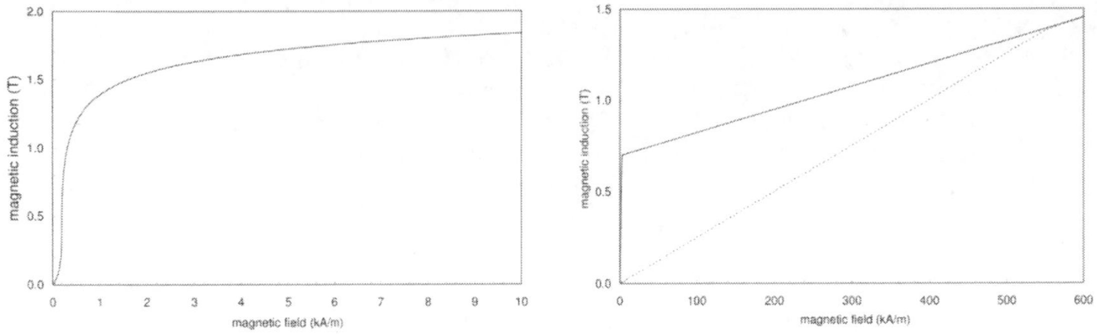

Figure 7. Induction curves for 11SMn30 (left) and for Ni_2MnGa (right, schematically, solid line: soft axis, dotted line: hard axis). Note the difference in scale in both, abscissae and ordinates.

The mechanical requirements thus being fixed, the amount of allowable input electric energy needs to be fixed — we choose 10 W of input power, P, and again this choice is not entirely random. It is adjusted such that it results in reasonable current densities in both actuators and in Joule losses that are conceivable to accept in an application for more than infrequent use at the given volume. This choice has significant influence on the outcome of the comparison, as it determines the advantageous volume fractions of copper and iron.

In first approximation, it is instructive to consider the magnetic circuit to be a single, closed loop without any stray-flux. This of course is a very significant simplification, but specifically with the rather short lengths of free space in the considered magnetic circuits and the moderate input power, one might argue that it is not expected that these circuits will suffer to much from stray-flux. This assumption very much simplifies analytical modelling and such facilitates an understanding of the circuit. This procedure is specifically instructive in the case of the MSM actuator, as in this simple model it is rather straight forward to keep track of the quantities that will determine the realistic output power. Under these assumptions, the magnetic behaviour can be derived from

$$\Theta = \Phi \sum_i R_{m,i}(\mu_r(B)) \tag{1}$$

where the Θ is the total magnetomotive force (amp turns) driving the system. Φ is the magnetic flux induced in the circuit and the $R_{m,i}$ are the reluctances for the various parts of the circuit that may be calculated based on knowledge of their permeabilities $\mu_r(B)$ in their respective working points (that in turn follow from geometry and Φ). This equation is easily solved self-consistendly.

For any such modelling, assumptions about the induction curves for the involved magnetic materials need to be made (i.e., about $\mu_r(B)$). For the yoke, we will consider a standard free cutting steel, 11SMn30, in annealed condition. This may be regarded as a good compromise between expense and magnetic properties. A representative curve $\mu_r(B)$ is shown in Fig. 7 (left panel). For MSM, we assume Ni_2MnGa as reportet in [4], somewhat simpliefied and reduced to a permeability for the hard axis of 2.0 up to a saturation polarization of 0.7 T, and a permeability fo the soft axis of 210 up to an identical saturation polarization of 0.7 T. This results in the curves $\mu_r(B)$ shown in Fig. 8 (left panel). Note that the circuit will have to be designed such that it will provide rather high fields in the MSM material. At such working points, the initial or maximum permeability at very low fields really is not of much consequence, since the relevant reluctances are defined by the permeabilities at the respective working points. So by simplifying the induction curve as described above, little harm is done.

Finally, a few assumptions about the coil need to be made. We will assume a Cu filling factor, η_{Cu}, of 60%, which is easily feasible, and we will further assume that there is a radial gap of twice 0.5 mm between iron and copper, and a total axial gap of 2.0 mm (used up by the

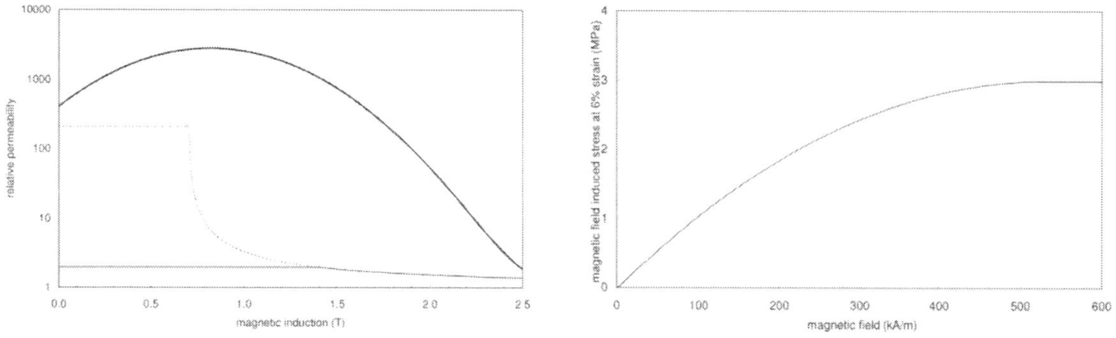

Figure 8. Left: relative permeability vs. magnetic induction for 11SMn30 (heavy line) and Ni_2MnGa (fine solid line: hard axis, fine dotted line: soft axis). Right: magnetic field induced stress vs. magnetic field in Ni_2MnGa.

bobbin of the coil, electrical connections etc.). We will consider the power consumption based on resistive losses alone at 50°C, i.e. at $\rho_{el} = 0.02\ \mu\Omega$m of resistivity. Total magnetomotive force then may easily be computed from

$$\Theta = \sqrt{P \frac{\eta_{Cu}}{\rho_{el}} \frac{h(r_o - r_i)}{\pi(r_o + r_i)}} \tag{2}$$

where r_i and r_o are the inner and outer radii of the coil, respectively, and h is its height. r_i and r_o are free parameters in the optimization, h is determined by the required stroke. It is assumed that 6% of stroke are actually available for external work, which for real materials is not doable. It still may give an impression of an upper limit for possible MSM actuator performance.

Assuming the MSM actuator geometry described above, in optimizing there are two additional parameters, viz. the radii m_i and m_o, which denote the inner and outer radius of the MSM element, respectively. Increasing m_i shifts the MSM elements outside, thus increasing the used circumference and active cross section and force at constant width $m_o - m_i$ of the MSM element, however, it also increases the volume that needs to be filled with sufficient magnetic field. So here it is rather useful to have this excessively simple model, as it allows to quickly check various options. This model of course does not directly result in an output force. The force is deduced from magnetic field in the MSM element(s) by

$$F = A \int_0^{H_0} (B_{soft}(H) - B_{hard}(H))dH \tag{3}$$

where A is the active cross section of the MSM element(s)

$$A = \pi(m_o^2 - m_i^2) \tag{4}$$

Obviously, this assumes an MSM crystal of rotational symmetry, which is impractical (though it would be neat in many respects). In reality, as was pointed out above, this ring would have to be approximated by a polygon of MSM crystals, as was described in [4]. There are specific challenges with this design, that were mentioned in the given reference. These technical problems, however, do not lessen the value of this design for the present consideration of MSM in direct comparison with an electromagnet. Here, it is rather a question of possible work per actuator volume; it is not primarily a question of optimum basic geometry.

It is further worth noting that the optimum of this design will not only depend on the curves $\mu_r(B)$ for all the involved materials in the reluctance circuit, but also on the assumed twinning stress. The reason for this is the non-linear shape of the curve $F(H)$ as depicted in Fig. 8 (right panel). The most force per field is realized at low working points, so for zero twinning stress the best design choice would be to keep the iron close to maximum permeability (so a minimum of reluctance is contributed by the yoke) and maximize MSM cross-section, i.e., use a low, but efficient working point and lots of active surface area (low stress and large area). With high twinning stress, however, it is crucial to have a large ratio of total magnetic stress and twinning stress, i.e., the working point needs to be higher, which usually will mean that the active surface area of MSM will be smaller (high stress and small area). Here, two examples will be considered. Zero twinning stress (to check which treasures may wait at the end of the rainbow) and 0.5 MPa of twinning stress, representing a realistic or even somewhat pessimistic value.

First we will consider zero twinning stress. The optimization will be done for soft-axis behaviour, i.e., for maximum force in the elongated state. As a result, one finds a suitable design with a force of 182 N at a working point of 91 kA/m. 62% of the total reluctance are then contributed by the MSM element(s), 22% by the air gap due to radial contraction of the MSM element(s) by 6% and just the remaining 16% are contributed by the yoke, i.e., the vast majority of magnetomotive force is indeed invested in the MSM effect. Note that even at this very low working point, the relative permeability of the assumed MSM material is a mere 7.1, compared to the assumed yoke material that in all elements that constitute the circuit is between 300 und 2,500. This disparity of course is the reason why it is possible to have such a large fraction of magnetomotive force useful for the MSM effect.

It is interesting to also consider the case of hard-axis behaviour for this geometry. In that case, the force increases to 294 N. It may come as a surprise that in hard-axis condition (contracted), with a relative permeability of only 2.0, the effective useful force is even higher than in soft-axis condition (elongated), where the relative permeability of the MSM element(s) was higher, viz. 7.1. The reason for this is twofold. First, the air-gap resulting from the shape-change is not present in this case, so the magnetomotive force expended on this in case of the elongated geometry is safed for better use. Second, it turns out that by increasing the reluctance of the MSM element(s), the yoke is forced into a more favorable working point (closer to maximum permeability), so effectively 98% of the total reluctance is contributed by the MSM element(s) in this condition. This second argument obviously is somewhat of a coincidence with the material properties chosen for the yoke in this example.

As advertized above, we want to consider a second example, viz. an MSM material of identical $\mu_r(B)$, yet with a finite twinning stress of 0.5 MPa. In the process of optimizing the assumed simplified circuit for these conditions, this just means that the effective force needs to be reduced by twice the twinning stress (once for overcoming the twinning stress in the first place and once for additionally overcoming the return spring, which at least needs to equate to the twinning stress). Also, it becomes desirable after optimizing the circuit under the assumption of soft-axis magnetic behaviour to check whether the field, assuming hard-axis magnetic behaviour, will be sufficient to induce twin-boundary motion in the first place. For the design discussed above this would not be the case. That actuator would not be able to do any useful work assuming a twinning stress of 0.5 MPa (effective externally available force is negative).

After re-optimizing the geometry for 0.5 MPa of twinning stress, the useful force for external work calculated in soft-axis conditions (elongated) is severely reduced to just 76 N, with a length of MSM in the circuit (=thickness of MSM platelets in polygon) of 1.7 mm only (less then half of the optimum at zero twinning stress, viz. 4.0 mm). The working point is increased to 226 kA/m, 67% of the magnetomotive force is used up in the MSM element(s) and 13% in the air gap resulting from the elongation. The effective relative permeability in the MSM element(s) in this case is 3.5 only. Again the useful force is increased in hard-axis condition as compared to soft-axis condition, in this case to 122 N. This confirms that there would not be a problem to induce motion in this actuator design.

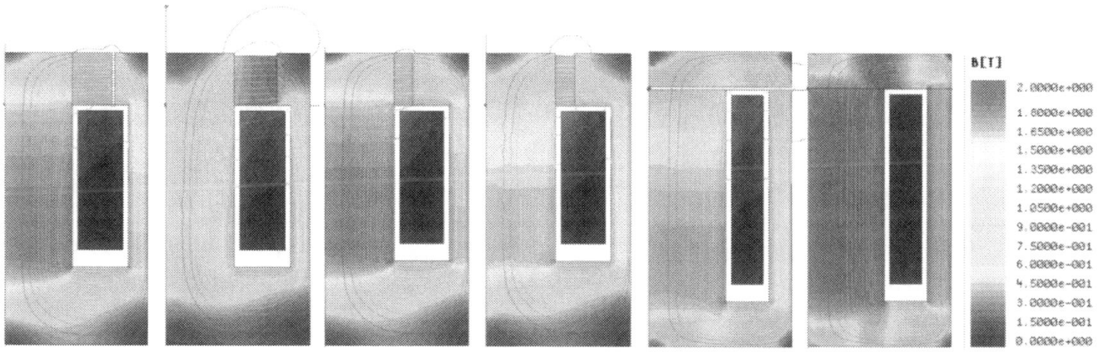

Figure 9. FEM analysis of the discussed actuator geometries. Left to right: Zero twinning stress MSM actuator, soft- and hard-axis condition; 0.5 MPa twinning stress MSM actuator, soft- and hard-axis condition; electromagnet, opened and closed armature condition.

For all four cases, zero twinning stress and 0.5 MPa twinning stress in soft- and hard-axis behaviour each, Finite Element Method (FEM) calculations were done to give some impression of the amount of stray-flux that is to be expected in these circuits (see Fig. 9, all calculations done using ANSYS/ANSOFT MAXWELL). A few remarks may be made. First, it is rather surprising that the bottom part of the magnetic circuit for the MSM actuators is as thick as it came out in the optimization. It certainly is rather over-dimensioned when comparing its working point to the the working points of the other parts of the yoke. So there would have been space to introduce more magnetomotive force by adding Cu into these circuits without any problem. Still, from the optimization this does not turn out to be desirable, since that would mean the working point would shift to higher induction, the permeability of the yoke would be decreased, a higher fraction of the total magnetomotive force would be invested in the yoke and effectively, *less* useful force would be generated in the MSM element(s). This again is due to this specific choice of yoke material and cannot be expected to be the case in every possible actuator design. Since in the case of MSM the force is driven by the field in the MSM element(s) (while its induction remains significantly lower than the yoke's anyway) it might, however, be expected to be a general tendency to best run the yoke close to its highest permeability so most reluctance is in the MSM. Furthermore, from these calculations it can be seen that with hard-axis behaviour (in the contracted state) there will be noticable stray-flux over the coil, so it is to be expected that the caculated increase in force in this state will not be realized in reality and would at least partly vanish using a slightly more elaborate model.

Finally getting to the comparison with the electromagnet, a similar optimization is done, however, in this case without the detour over the circuit model. This design is easily optimized in FEM directly, so it is not really necessary to look at a simplified circuit model. The results are shown along in Fig. 9. In this case, there is noticable stray-flux when the armature is opened, while there is none, when it is closed (as it should be). Note that since in closed condition the full circuit consists of 11SMn30, the working point becomes rather high, and accordingly the force increases signficiantle to 365 N. The appropriate comparison, however, is the opened-armature condition, and in this case, 132 N are possible.

In order to facilitate direct comparison, all these resulting forces over stroke are shown in a joint Fig. 10. For the electromagnet the full curve force vs. stroke was calculated (FEM results shown), for the MSM actuators only the forces at the initial and final state are reported (circuit model results shown). Filled symbols are zero twinning stress, open symbols are 0.5 MPa twinning stress. It is interesting to note that (even though the dashed line is really at most a guide to the eye) for zero twinning stress (i.e., ideal energy conversion, as is the case or the air

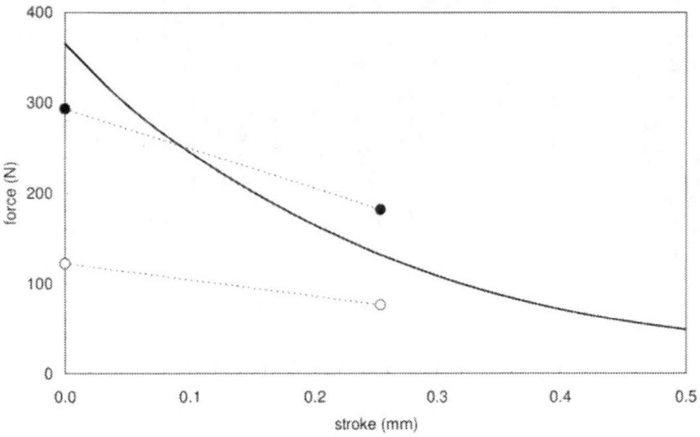

Figure 10. Force vs. stroke for electromagnet and MSM actuators. Solid line: electromagnet. Filled symbols: MSM actuator, zero twinng stress. Open symbols: MSM actuator, 0.5 MPa twinning stress.

gap of the electromagnet), the advantage that is with the MSM actuator at the inital state is with the elctromagnet at the final state.

At this point it should be mentioned that there is an important difference in that the MSM actuator by nature pushes away from the magnetic circuit, while the electromagnet pulls into the circuit, i.e., the forces both shown as positive here really are of opposite direction — this can easily be fixed by appropriate design measures and does not really influence this comparison. It means, however, that for the MSM actuator the initial position is at zero stroke, while for the electromagnet it is at 10 mil (0.254 mm). This means, the electromagnet starts at its lowest force and will show increasing force over stroke, while the MSM actuator will start at highest force and show decreasing force over stroke. Again, keep in mind that this increase of force in hard-axis conditions is somewhat of an artifact of the ommission of stray flux terms, still, even if the force would be constant and equal to the force in soft-axis conditions, this might in certain applications be regarded as advantageous.

5. Conclusion

In this paper we used an assumed set of technical requirements to directly compare MSM solid state actuators with conventional electromagnets. The assumed goal was producing maximum mechanical work at a given stroke, input power and actuator volume. All results need to be treated with applicable care — they are dependent on these assumptions. In the presently studied case it turns out that the feasible work is rather comparable assuming zero twinning stress for the MSM. However, it becomes notably reduced if 0.5 MPa twinning stress is assumed.

To our understanding this means that MSM actuators might be expected to be most competetive in areas that do not focus solely on their capability to transform magnetic field energy into work. It will be crucial to make use of all their properties, including finite twinning stress (and, e.g., concurrent damping properties). Even though MSM might come close to the properties of well-known, cheap electromagnets, or even surpass them in certain points, it needs to be kept in mind that there might be expected to be a tremendous reluctance to use such a novel actuator principle in a first technical application. This reluctance is best overcome by offering a qualitatively new behaviour. Considering the amount of interesting properties MSM have shown so far, it appears realistic to find such a qualitatively new and promising application.

References

[1] Holz B and Janocha H 2010, *Proc. Actuator* **2010** 307
[2] Schmidt H and Quandt E 2010, *Proc. Actuator* **2010** 289
[3] Tellinen J, Suorsa I, Jääskeläinen A, Aaltio I and Ullakko K 2002, *Proc. Actuator* **2002** 566
[4] Jansen S, Kessler M, Rammaier W, Schmidt H and Sugg B 2008, *Proc. Actuator* **2008** 279

Ni-Mn-Ga single crystals with very low twinning stress

L. Straka[1,2], H. Hänninen[1], A. Soroka[2], and A. Sozinov[2]

[1] Aalto University, Laboratory of Engineering Materials, PL 14200, FIN-00076 AALTO, Finland

[2] AdaptaMat Ltd., Yrityspiha 5, Helsinki, FIN-00390, Finland

E-mail: `ladislav.straka@tkk.fi`

Abstract. Twinning stress or mechanical hysteresis associated with the twin boundary motion is one of the most essential parameters which determine the actuating performance of magnetic shape memory alloys. Recent effort at AdaptaMat Ltd. to decrease the twinning stress resulted in a consistent production of Ni-Mn-Ga magnetic shape memory single crystals with the twinning stress of about 0.1 MPa, which is much lower than previously reported. In this work, the mechanical and magnetomechanical response of the developed crystals is discussed in detail and the importance of adjustment of the twin microstructure for obtaining an optimal actuating behavior is illustrated.

1. The MSM effect and importance of twinning stress

The magnetic-field-induced rearrangement of (ferro-)magnetic twinned martensite microstructure accompanied by a large macroscopic deformation, or magnetic shape memory (MSM) effect is a phenomenon which has attracted attention since its first demonstration on a Ni-Mn-Ga single crystal in 1996 [1]. Fast actuation seems to be one of the potential practical exploitations of the phenomenon as magnetic shape memory effect with large strains up to 10% was already demonstrated many times starting from 1999 [2–9] and the possibility of actuation in kHz range was also clearly shown [10]. Applications based on reverse mechanism were also suggested including sensing [11–13], energy harvesting [14,15], damping [16], etc. Ni-Mn-Ga alloys have been so far most studied and can be considered as a "prototype" magnetic shape memory material.

One way how to employ the MSM effect for linear reversible actuation is to apply a constant external compressive stress σ along one principal axis and variable magnetic field H perpendicularly along an other principal axis of a prismatic single crystalline Ni-Mn-Ga MSM element, Fig. 1 [13, 17]. The overview of the other possible arrangements can be found in [18]. The MSM element is typically cut along $\{100\}_P$ faces and contains a simple two-variant twin microstructure as this type of microstructure seems to exhibit the best performance [19]. One twin variant (yellow or light in Fig. 1) is favored energetically by the external stress while the other one (red or dark in Fig. 1) is favored by the magnetic field. When the magnetic field is increased to a certain magnitude, the variant favored by the magnetic field (red) starts to grow on the expense of the variant favored by the stress (yellow). When the magnetic field is decreased, the reverse process takes place during which the external stress induces the growth of twin variant favored by the stress (yellow) on the expense of the twin variant favored by

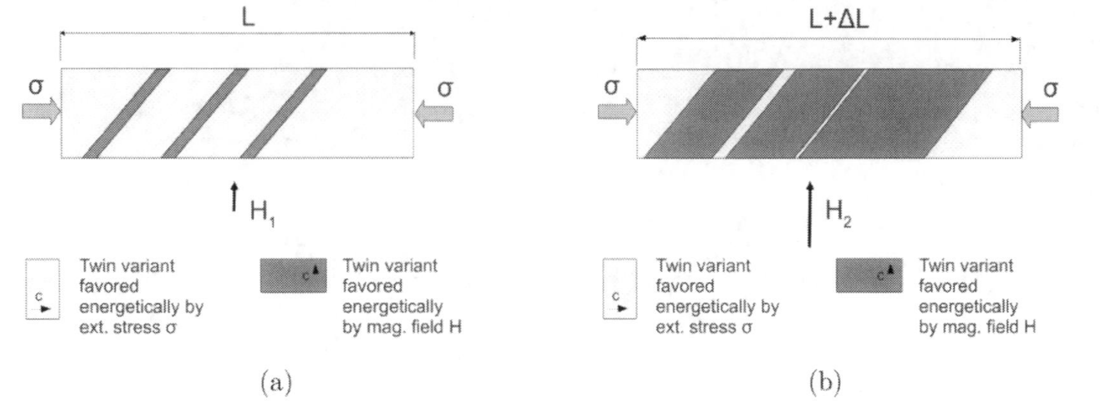

Figure 1. Reversible several percent actuation of a crystal showing magnetic shape memory effect can be achieved by the combined application of a magnetic field and a constant external compressive stress (orientation of short c-axis inside twin variants is indicated by arrows). a) In zero or small magnetic field H_1, the twin variant favored by the applied external stress σ (initial variant, yellow or light) occupies most of the crystal volume. b) With increasing magnetic field, the twin variant favored by magnetic field (red or dark) grows on the expense of the initial variant, which is accompanied by a several percent deformation ($\Delta L/L$) at final large field H_2. When the field is decreased back to zero or H_1, the twin microstructure rearranges to the initial state shown in (a) due to the effect of external stress which favors the initial (yellow or light) variant.

the magnetic field (red). The growth of the twin variants occurs mostly by twin boundary motion [4, 20].

The c-axis and a-axis of the two discussed twin variants point along the two different principal directions of the MSM element. Thus, by growth or consumption of one or other variant, the orientation of c-axis is changed from one principal direction to the other principal direction. This is the reason why magnetic shape memory effect is sometimes called *magnetic-field-induced reorientation*, or MIR. Exchange of c-axis and a-axis by growth or consumption of twin variants results in a large deformation as the axes differ considerably in length. The maximum possible macroscopic linear strain occurs when the whole volume of the element transforms from one variant to the other variant and it is given by the relative difference between the lattice parameters a and c. This is typically about 6% for 5M (five-layered, approximately tetragonal) Ni-Mn-Ga martensite studied and discussed in this article.

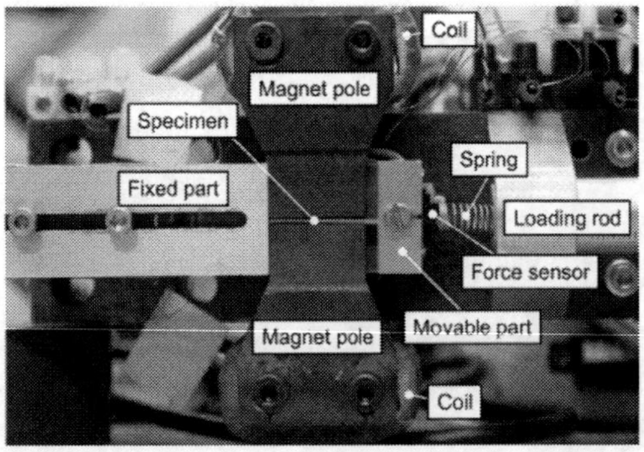

Figure 2. Apparatus for the investigation of MSM effect and of mechanical properties. Magnetic field and compressive stress were applied perpendicularly. Spring was not installed during compression tests but it was installed for measurement of magnetic-field-induced strain. Applied magnetic field was determined by a Hall probe without specimen installed.

Slow (≈1 Hz) and quasistatic actuation by the magnetic shape memory effect can be easily demonstrated and investigated using a simple magneto-mechanical apparatus as the one shown and described in Fig. 2. Using this apparatus with 20 mm long single crystals, MSM effect with strain close to 6% can be clearly visually observed by unaided eye. The same apparatus was used for determining the response of the studied crystals under compressive loading and in magnetic field. The twin microstructure, individual twin domains or variants, and twin boundaries were observed using polarized light optical microscopy. The microstructure was well observable by an optical microscope from certain faces of the specimen even without polarized light. X-ray diffraction was an additional tool which we used to confirm the observed or created twin configurations.

Twinning stress, σ_{TW}, as a measure of mobility of twin boundaries, is a crucial material parameter determining the existence of the MSM effect and the efficiency of the actuators based on the phenomenon [4, 21]. Twinning stress can be experimentally evaluated from the stress-strain curves [22] as the level of applied compressive (or tensile) uniaxial stress which causes the twin boundary motion. More specifically, it is often determined as the position of the detwinning plateau on the stress-strain curve, or it can be determined as the stress level at half transformation strain [19], or, two stress levels can be recorded – one near the beginning and the second near the end of the plateau on the stress-strain curve [4]. Such determinations, however, may not fully cover more general consideration of the twinning stress for which the stress level at which twinning initiates (i.e., twin variants nucleate) must be distinguished from the stress level at which twin variants grow [22, 23]. This is discussed more below.

There is a common agreement that the lower twinning stress results in a better actuating performance of the MSM material. From the model of MSM effect introduced by Likhachev and Ullakko [4] it follows that to obtain a reversible MSM effect, the twinning stress must be less than half of the magnetic stress [24], i.e. less than about 1.5 MPa as the maximum magnetic stress, σ_{MAG}, is about 3 MPa for 5M Ni-Mn-Ga martensite at room temperature (e.g. [24]). How low below the above determined limit 1.5 MPa is satisfactory for practical actuation can be answered when considering the coupling factor, or efficiency of the MSM effect determined as the ratio of obtained mechanical work and work spent on the magnetizing of the material during one magnetizing (actuating) cycle.

Table 1 summarizes the calculated efficiencies for selected magnitudes of twinning stress using $\eta = (\sigma_{\mathrm{MAG}} - 2\sigma_{\mathrm{TW}})/\sigma_{\mathrm{MAG}}$ [13, 21] and σ_{MAG} =3 MPa. As the generation of the magnetic field is typically costly and bulky, one would like to avoid unnecessary losses and keep the efficiency

Table 1. Efficiency of MSM effect $\eta = (\sigma_{\mathrm{MAG}} - 2\sigma_{\mathrm{TW}})/\sigma_{\mathrm{MAG}}$ determined using magnetic stress σ_{MAG} =3 MPa for selected levels of twinning stress σ_{TW}.

Twinning stress σ_{TW} MPa	Efficiency η %	Note
>3	—	No MSM effect
1.5–3	—	Irreversible (one way) MSM effect
1.4	7	Reversible MSM effect
1	33	
0.5	66	
0.2	87	
0.1	93	
0	100	Ideal MSM material

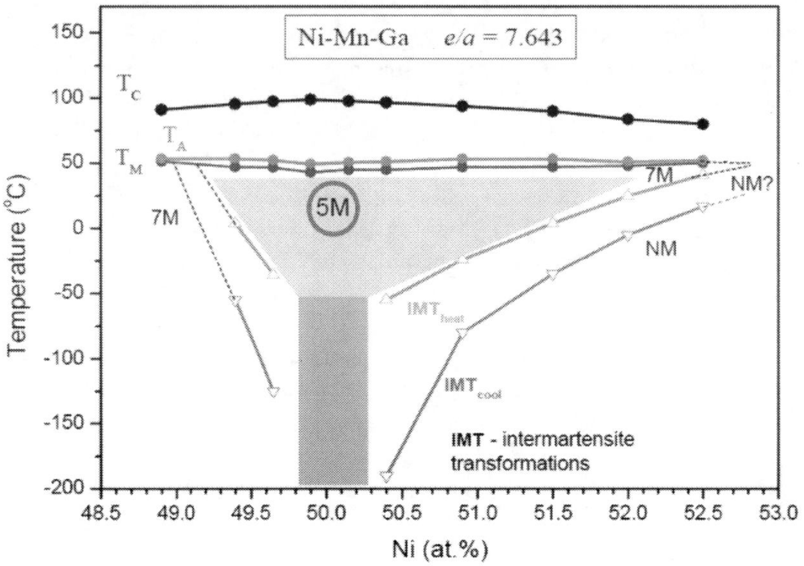

Figure 3. Illustration how nickel content of the studied alloys was selected. To avoid any problems with other martensitic phases, composition was selected to be in the region of 5M phase (green or dark) as far as possible (marked by circle) from intermartensitic transformations (IMT). This gives optimum nickel content of about 50% for the shown case with $e/a = 7.643$. 7M – seven-layered martensite, NM – tetragonal martensite with c/a>1, T_C – Curie temperature, T_M – martensitic transformation, T_A – reverse transformation to austenite.

not far from 100%. Using Table 1, it seems that this can be assured only if the twinning stress is not far from 0.1–0.2 MPa, where the efficiency is around 90%. It is noteworthy that 1 MPa twinning stress may seem low, when directly comparing to 3 MPa magnetic stress, but gives only 30% efficiency and, thus, may be too high for the practical usage of the MSM alloys.

2. Development of Ni-Mn-Ga single crystals

Our development of the crystals with a low twinning stress is based on the idea that the twin boundary mobility can be considerably increased by an efficient elimination of the obstacles for twin boundary motion from the lattice [25, 26]. As obstacles we consider chemical impurities from raw material and from the process resulting in inclusions and precipitates [25] and also inhomogeneities in composition, second phases, cell boundaries, etc. An example of the selection of nickel content for electron per atom ratio e/a =7.643 is demonstrated in Fig. 3. The nickel content was selected to be close to 50 at. % of Ni, as this composition is far from other phases and a possibility of occurrence of the other martensitic phases than 5M (five-layered modulated, approximately tetragonal, with c/a =0.94) in the crystal is minimized. The actual composition of the tested alloys was $Ni_{50.2}Mn_{27.8}Ga_{22.0}$ (±0.2 atomic percent).

3. Properties of the developed crystals and effect of twin microstructure

Additionally to the obstacles in the crystal, the twin microstructure also influences the twinning stress [27–29]. In other words, the twin boundaries themselves can present obstacles for the twin boundary motion [30]. Due to that the stress-strain response of the crystal depends on the twin microstructure, and, thus, attention must be paid to the twin microstructure when determining the twinning stress from the stress-strain curve. We created well defined twin

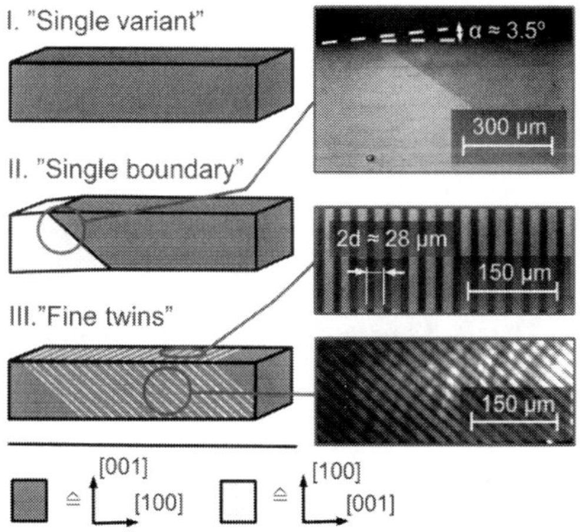

(a)

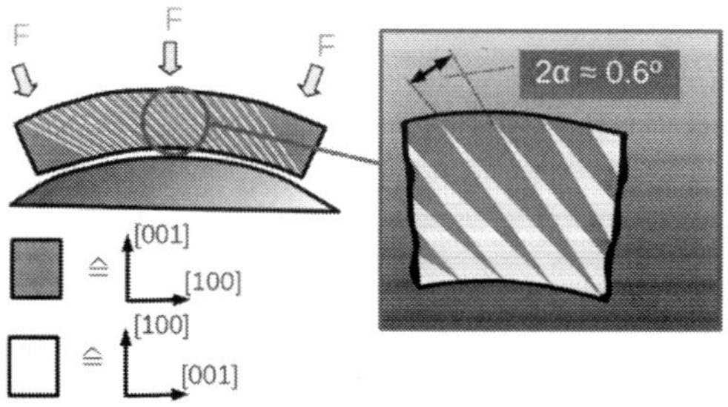

(b)

Figure 4. a) Twin microstructures investigated in this study (schematically on the left, polarized light microscopy observations on the right): i) single variant, ii) single boundary, and iii) fine twins. b) Illustration how the fine twins were created by bending of a single variant.

microstructures in the studied crystals using an external stress and magnetic field prior to the testing. The morphology of the created twin microstructures is demonstrated in Fig. 4a. How the microstructure with fine twins was created by bending of a single variant crystal is illustrated in Fig. 4b [31].

Mechanical testing using compressive stress and measurement in a quasistatic magnetic field under zero stress were performed for each microstructure presented in Fig. 4a using the apparatus demonstrated in Fig. 2. For a single twin boundary and for fine twins, magneto-mechanical testing under nonzero external stress was also performed. The results are shown and discussed below for each microstructure separately.

3.1. Single variant

The main feature observed on the stress-strain curves determined for the single variant, i.e., for an initially detwinned crystal, is large load drop at the beginning of loading by which the initial stress peak was formed on the stress-strain curve, Fig. 5. The load drop can be ascribed to the nucleation of the new twin variant [22,27,28]. The large height of the stress peak indicates that the nucleation barrier is rather high in Ni-Mn-Ga (detailed discussion on the topic was presented in [32]). High nucleation barrier had been confirmed also by a generally small amount of twin boundaries observed in Ni-Mn-Ga single crystals [20].

The height of the stress peak corresponds to the stress initiating twinning (nucleating new twin variants) in a single variant (detwinned crystal) and we will refer to the height of the stress peak as the *initial twinning stress*. The average height of the stress peak, or the *initial twinning stress*, in our experiments was about 2 MPa.

After the twinning was initiated and some twin boundaries were clearly observed in the crystal, the deformation of the crystal progressed by motion of twin boundaries (with only occasional nucleation of additional twin domains). Stress required for this motion was much smaller (0.2–1 MPa) than the *initial twinning stress* and we will refer to it here as the *conventional twinning stress*, or just the *twinning stress*.

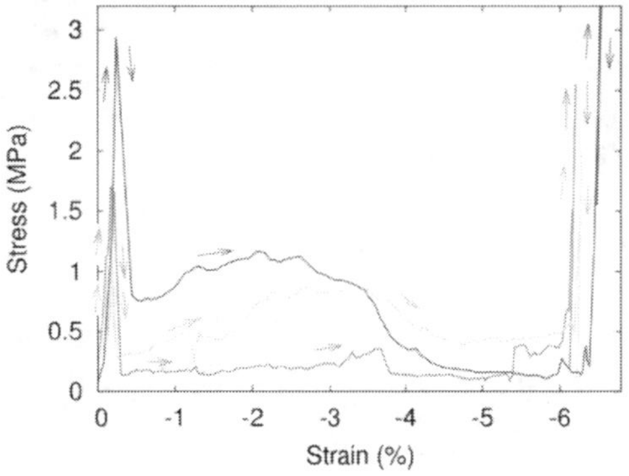

Figure 5. Response of single variant (detwinned crystal) to compressive stress. The three displayed measurements were performed on three different single variant crystals.

From the above observations we can conclude that to achieve actuation with a reasonable efficiency, some twin boundaries must be always present in the crystal. Otherwise, a high barrier (i.e., *initial twinning stress*) would have to be overcome by the magnetic stress and also by the external counterstress to achieve actuation. This was pointed out also in Ref. [32], where it was stated that the homogeneous magnetic field-induced nucleation of twin boundaries is not likely. It means that the twin boundary motion in the magnetic field may proceed only from the preexisting twin boundaries. In agreement with this statement, the observed response of a single variant in the magnetic field was unsteady and often no MSM effect was observed [33]. Note also that the initial twinning stress of 2 MPa is too high (>1.5 MPa) to allow a reversible MSM effect in the arrangement with the external counterstress, Fig. 1.

The stress-strain curves varied considerably between individual experiments made on single variants, Fig. 5. This can be ascribed to the different twin microstructure nucleated during each experiment with the different mobilities of twin boundaries. Nonetheless, as demonstrated also in Fig. 5, the stress necessary to move the twin boundaries often decreased to about 0.2 MPa. We concentrated on these cases, and after additional training of the crystal by repeated compression and extension we obtained crystals with a single highly mobile twin boundary described below.

3.2. Highly mobile twin boundary

The developed crystals showed tendency to nucleate only a pair or a few twin boundaries under compression and further deformation progressed by propagation of these nucleated twin boundaries [28]. The single twin boundary was formed from the nucleated pair when the pair was nucleated near the end of a crystal and one boundary vanished at the end of the crystal. The remnant boundary showed very low twinning stress either immediately or after a few compression-extension cycles performed manually while holding the crystal from its two ends.

Compression tests of the crystal with a single highly mobile twin boundary are demonstrated in Fig. 6a. The applied deformation (strain) induces motion of the highly mobile twin boundary present in the crystal and, typically no other twin boundary appeared in the specimen until the whole crystal had c-axis along the stress. At this point, the loading started to be elastic, which was indicated on the stress-strain curve by a sudden sharp increase of stress at the strain of about 6%. From the position of the detwinning plateau on the stress-strain curve we determined that the twinning stress of the crystal was 0.05–0.1 MPa. This is considerably less than previously reported (e.g. 0.5 MPa in [34]).

The magnetic shape memory effect in the crystal with a single highly mobile twin boundary in the quasistatic magnetic field is demonstrated in Fig. 6b. The field-induced straining occurred

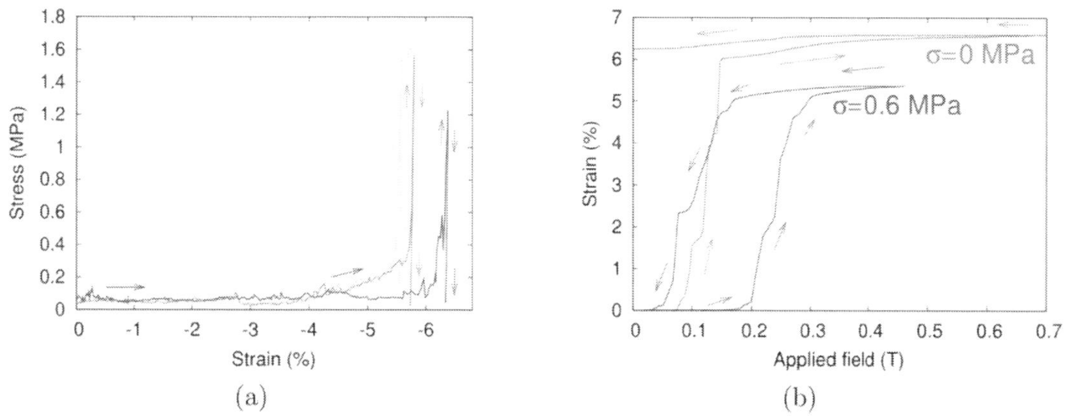

Figure 6. a) Response of crystal with single highly mobile twin boundary to compressive stress. b) Response of crystal with single highly mobile twin boundary to magnetic field for external stress 0 and 0.6 MPa (marked next to curves). All displayed measurements were performed on the same crystal.

at the applied fields of 0.12 T and 0.24 T (switching field) for the external stress of 0 MPa and 0.6 MPa, respectively. This difference in switching fields was present due to that for 0 MPa external stress only the twinning stress had to be overcome by the magnetic stress, but for 0.6 MPa external stress the twinning stress *and* external stress had to be overcome by the magnetic stress. This resulted in the higher magnetic field needed for field-induced straining of the crystal under load. In agreement with expectations, application of nonzero external stress made the effect reversible as the specimen contracted along the applied stress when the magnetic field was decreased.

Additional important factor influencing the switching field was the demagnetization factor of the crystal along the magnetizing direction. In the shown case, the field was applied perpendicularly to 2.5×20 mm^2 face of the $2.5 \times 20 \times 1$ mm^3 specimen, thus, the demagnetization factor was high. In spite of that, the observed switching field was rather low, which is in agreement with the observed very low twinning stress of the crystal. Magnetizing the crystal along direction with the low demagnetization revealed that the applied field as low as 0.03 T (300 Oe) was enough to achieve the MSM effect [35].

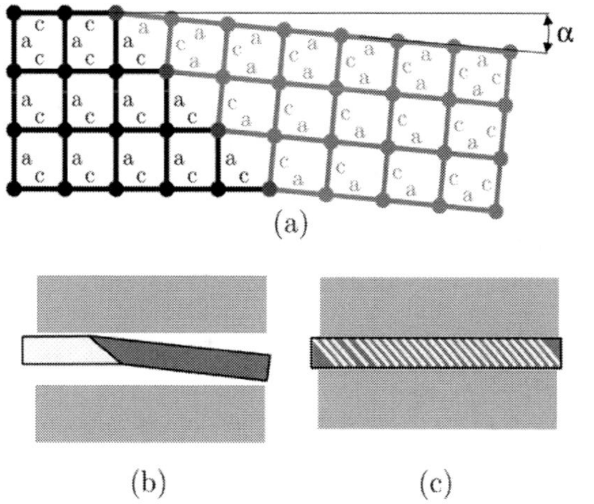

Figure 7. a) Due to twinning crystallography, a tilt of $\alpha = 90° - 2\arctan(c/a)$ is present at each twin boundary. b) This may cause problems when considering use of crystal with a single (or few) twin boundaries in an actuator with a narrow air gap. c) Creating many fine twins can eliminate the problem.

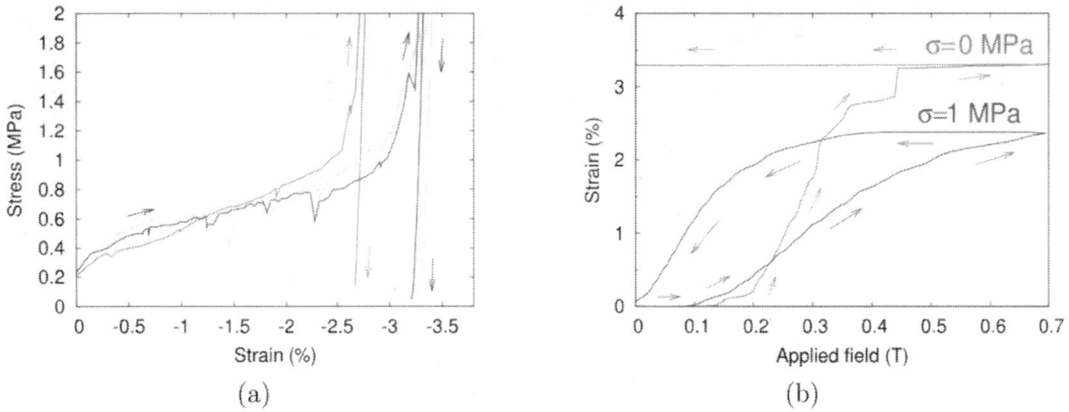

Figure 8. a) Response of fine twins to compressive stress (determined on the same crystal as in Fig. 6. b) Response of fine twins to applied magnetic field.

The demonstrated experiment and more detailed investigations (using tensile stress [28] and different temperatures [36]) show that a very low twinning stress can be achieved in Ni-Mn-Ga. However, the crystal with a single twin boundary does not seem to be the best candidate for practical usage in a magnetic actuator with a narrow air gap. One of the problems is demonstrated in Fig. 7. Due to 3.5° kink of the specimen on the twin boundary caused by the twinning crystallography, Fig. 7a, the specimen with only one twin boundary may be hard to operate in a "standard" magnetic actuator with a narrow air gap, Fig. 7b.

The stochastic nature of the single twin boundary motion [37] and the slower response of the crystals with small number of boundaries [32] may be listed as other reasons why to consider a twin structure with many twin boundaries or fine twins, Fig. 4, and Fig. 7c. Additionally, handling the large nontwinned volume of the crystal may not be trivial as in this volume twin boundaries with different orientations can appear. These can interfere with the existing highly mobile twin boundary and can hinder its motion.

The above mentioned reasons initiated the effort to create a fine twin structure in the material and to employ it in actuation. The effort, however, was only partly successful since the twinning stress of the crystal increased considerably when the fine twin structure was created. The response of the crystal with fine twins is described in more detail in the next chapter.

3.3. Fine twins

Fine twins with density about 80 twin boundaries/mm were created in the crystals by bending [31], Fig. 4b. When straight shape of the crystal was recovered, the twin boundaries persisted in the crystal and all were parallel, i.e., had the same twinning plane. The mechanical testing of the crystal with fine twins revealed that this type of microstructure exhibited about one order higher twinning stress, i.e. about 0.8 MPa, Fig. 8a, than the above described case with a single twin boundary (both twinning stresses were determined on the same crystal). Reason for this increase is under investigation. The response in the magnetic field under zero external stress, Fig. 8b, corresponded in its character to the observed stress-strain curves. Higher twinning stress resulted in the larger magnetic field necessary to induce the MSM effect and field-induced strain occured more gradually in comparison with the case of single boundary due to the considerable tilt of the stress-strain curve, compare Fig. 6a and Fig. 8a.

The fine twins annihilated completely from the crystal by the compression of several MPa stress or by the large magnetic field. Thus, achieving the reversible MSM effect in the crystal with fine twins was not straightforward. A method for stabilization of the fine twins based on a minor modification of the material is under development in AdaptaMat Ltd. An example response of a crystal with fine twins which do not annihilate under stress is shown in Fig. 8b. Due to the relatively high twinning stress and considerable tilt of the stress-strain curve, Fig. 8a, the straining achieved in the periodic actuation was less than 3% and a relatively large magnetic field was necessary to achieve straining of reasonable magnitude. Note that hand in hand with the high twinning stress goes necessity to apply at least the same high external compressive stress to maintain the reversibility. This further increase the field necessary for achieving actuation with reasonable strain.

4. Conclusions

We demonstrated that the very low twinning stress (0.1 MPa and lower) is possible to achieve in Ni-Mn-Ga 5M martensite. Thus, in principle, magnetic actuators with a high efficiency can be constructed based on Ni-Mn-Ga single crystals. Prior to that, however, several issues should be considered.

The measurements demonstrated that it is necessary to distinguish between the *initial twinning stress*, i.e., stress initiating twinning in material and the *conventional twinning stress* or just the *twinning stress*, i.e., the stress inducing motion of existing twin boundaries. As the former is much higher than the latter, the microstructure configuration without twin boundaries (i.e., single variant or detwinned crystal) should be avoided completely, when using the material for magnetic actuation.

Even the crystal with the pre-existing twin boundaries can show one order variation in the twinning stress, since the twinning stress and consequently the MSM effect depend strongly on the twin microstructure of the crystal. Our experiments demonstrate that a crystal with a single twin boundary which exhibits a very low twinning stress and an excellent actuating performance can be transformed easily, e.g. by bending, to a crystal with fine twins, moderate twining stress, and reduced actuation. Twin boundaries may also annihilate by the action of the external stress or magnetic field, which makes the crystal a single variant with an unsteady or zero actuating response.

Due to the conclusions presented above, we now focus our efforts towards the development of the suitable twin structures with a low twinning stress, which will be resistant to twin boundary annihilation. We expect that such modification will considerably increase the potential of the material for applications requiring fast and efficient actuation.

Acknowledgments

This work was supported by Tekes – the Finnish Funding Agency for Technology and Innovation and by the Academy of Finland.

References

[1] Ullakko K, Huang JK, Taunter C, O'Handley RC and Goran VV 1996 *Appl. Phys. Lett.* **69** 1966
[2] Tickle R and James RD 1999 *J. Magn. Magn. Mater.* **195** 627
[3] James RD, Tickle R and Wuttig M 1999 *Mat. Sci. Eng.* A **273–275** 320
[4] Likhachev AA and Ullakko K 2000 *Phys. Lett.* A **275** 142
[5] Murray SJ, Marioni M, Allen SM, O'Handley RC and T.A. Lograsso 2000 *Appl. Phys. Lett.* **77** 886
[6] Heczko O, Sozinov A and Ullakko K 2000 *IEEE Trans. Mag.* **36** 3266
[7] Sozinov A, Likhachev AA, Lanska N and Ullakko K 2002 *App. Phys. Lett.* **80** 1746
[8] Chmielus M, Zhang XX, Witherspoon C, Dunand DC and Müllner P 2009 *Nature Materials* **8** 863
[9] Chernenko VA, Chmielus M and Müllner P 2009 *Appl. Phys. Lett.* **95** 104103
[10] Marioni MA, O'Handley RC and Allen SM 2003 *Appl. Phys. Lett.* **83** 3966
[11] Sarawate N and Dapino M 2006 *Appl. Phys. Lett.* **88** 121923
[12] Suorsa I, Pagounis E and Ullakko K 2004 *Appl. Phys. Lett.* **84** 4658

[13] Suorsa I 2005 *Performance and Modeling of Magnetic Shape Memory Actuators and Sensors* (Doctoral Dissertation, Espoo 2005, ISBN 951-22-7644-5)
[14] Suorsa I, Tellinen J, Ullakko K and Pagounis E 2004 *J. App. Phys.* **95** 8054.
[15] Karaman I, Basaran B, Karaca HE, Karsilayan AI and Chumlyakov YI 2007 *Appl. Phys. Lett.* **90** 172505
[16] Aaltio I, Mohanchandra KP, Heczko O, Lahelin M, Ge Y, Carman GP, Söderberg O, Löfgren B, Seppälä J and Hannula S-P 2008 *Scripta Mat.* **59** 550
[17] Tellinen J, Suorsa I, Jääskeläinen A, Aaltio I and Ullakko K 2002 *Actuator 2002* 566 (ISBN 3-933339-04-9)
[18] Schmidt H and Quandt E 2010 *Actuator 2010* 289 (ISBN 978-3-933339-12-6)
[19] Straka L, Heczko O and Hänninen H, 2008 *Acta Mater.* **56** 5492
[20] Straka L, Novák V, Landa M and Heczko O 2004 *Mat. Sci. Eng.* A **374** 263
[21] Likhachev A, Sozinov A and Ullakko K 2006 *Mechan. Mater.* **38** 551
[22] Christian JW and Mahajan S 1995 *Prog. Mat. Sci.* **39** 1 (pp 79–83).
[23] Kibey S, Liu JB, Johnson DD and Sehitoglu H 2007 Acta Mat. **55** 6843
[24] Straka L, Heczko O and Hannula S-P. 2006 Scripta Mat. **54** 1497
[25] Richard ML 2005 *Systematic Analysis of the Crystal Structure, Chemical Ordering, and Microstructure of Ni-Mn-Ga Ferromagnetic Shape Memory Alloys* (Ph.D. thesis, M.I.T., Cambridge, MA http://hdl.handle.net/1721.1/33168)
[26] Rajasekhara S and Ferreira PJ 2005 *Scr. Mater.* **53** (2005) 817
[27] Aaltio I, Söderberg O, Ge Y and Hannula S-P 2010 *Scr. Mater.* **62** 9
[28] Straka L, Lanska N, Ullakko K and Sozinov A 2010 *Appl. Phys. Lett.* **96** 131903
[29] Chmielus M, Rolfs K, Wimpory R, Reimers W, Müllner P and Schneider R 2010 *Acta Mater.* **58** 3952
[30] Müllner P and King AH 2010 *Acta Mater.* **58** 5242
[31] Chopra HD, Bailly C and M. Wutig 1996 *Acta Mater.* **44** 747
[32] Marioni MA 2003 *Pulsed Magnetic Field-Induced Twin Boundary Motion in Ni-Mn-Ga* (Ph.D. thesis, M.I.T., Cambridge, MA http://hdl.handle.net/1721.1/7965)
[33] Straka L, Soroka A and Sozinov A 2010 *Actuator 2010* 727 (ISBN 978-3-933339-12-6)
[34] Rolfs K, Mecklenburg A, Guldbakke J-M, Wimpory RC, Raatz A, Hesselbach J and Schneider R 2009 *J. Magn. Magn. Mater.* **321** 1063
[35] Heczko O and Straka L, unpublished.
[36] Soroka A, Sozinov A, Lanska N and Straka L, AdaptaMat Ltd., unpublished.
[37] Marioni MA, Allen SM and O'Handley RC 2004 *Appl. Phys. Lett.* **84** 4071

Influence of grain size and training temperature on strain of polycrystalline $Ni_{50}Mn_{29}Ga_{21}$ samples

C. Hürrich[1], S. Roth[1], H. Wendrock[1], M. Pötschke[1], D. Y. Cong[1], B. Rellinghaus[1] and L. Schultz[1,2]

[1]IFW Dresden, Institute for Metallic Materials, P. O. Box 270116, 01069 Dresden, Germany, [2]Dresden University of Technology, Department of Mechanical Engineering, Institut for Materials Science, 01062 Dresden, Germany

E-mail: c.huerrich@ifw-dresden.de

Abstract. The alloy Ni-Mn-Ga aroused great interest for application as a magnetic shape memory (MSM) material. This effect is caused by reorientation of twin variants by an external magnetic field. So far, most of the experiments were concentrated on single crystals. But, the MSM effect can also be realised in polycrystals which can be prepared much more efficiently. Here, polycrystalline samples were prepared by directional solidification with a <100> fibre texture of the high temperature cubic austenitic phase parallel to the heat flow. Afterwards, a heat treatment was applied for chemical homogenisation and stress relaxation in the austenitic state. Then the samples were heated up to the austenitic state and cooled down under load. The microstructure was analysed by Electron Back Scatter Diffraction (EBSD) before and after that treatment. Mechanical training at room temperature and 40°C was tracked by recording stress-strain curves. By increasing the number of training cycles the strain also increases. The influence of different training temperatures was investigated on samples with different grain sizes.

1. Introduction

In recent years Ni-Mn-Ga magnetic shape memory alloys with a chemical composition close to the stoichiometric composition Ni_2MnGa have attracted increasing attention due to their large magnetic-field-induced strain (MFIS) of up to 10 %. This strain which is produced by twin boundary motion in the martensitic phase due to a magnetic field was demonstrated for the first time in 1996 by Ullakko et al. [1]. These materials can be used in sensors and actuators [2, 3].

Most studies reporting on MFIS in such alloys were performed on single crystals [4 - 7]. Polycrystalline samples are of great interest because they can be produced easier and for this reason they are cheaper. Usually polycrystalline samples have a low strain and a high twinning stress because of their differently oriented grains with varying microtwin orientation. The twinning stress is defined by the start of the movement of twin boundaries under load. Polycrystalline samples with a highly textured microstructure and coarse grains are close to single crystals. Such highly textured polycrystalline Ni-Mn-Ga samples were achieved by directional solidification in previous work [8]. Recently a residual strain of 2% was achieved after a maximal load of 40 MPa [9]. Polycrystalline Ni-Mn-Ga with addition of Si, In, Co or Fe was used to vary the transformation temperatures. But single crystals were prepared for investigation of the MFIS [10]. However, polycrystalline Ni-Fe-Ga samples

Published under licence by IOP Publishing Ltd

show a residual strain of more than 2.4% after a load of about 300 MPa. But no MFIS is reported for these alloys [11].
In order to find MFIS by twin boundary motion, the magnetically induced shear stress must be larger than the twinning stress [7]. The magnetically induced stress is given by $\sigma_{mag} = m \cdot K_u \cdot s^{-1}$, with the Schmid factor m, the magnetocrystalline anisotropy constant K_u and the twin shear $s = 0.5 \cdot \left(\frac{a}{c} - \frac{c}{a}\right)$, where a and c are the lattice parameters of the martensite [4, 12, 13]. Mechanical training is effective to increase the strain and to reduce the stress for twin boundary motion in polycrystals. The influence of mechanical training on the stress – strain relationship in martensitic polycrystalline $Ni_{50}Mn_{30}Ga_{20}$ has been shown by Gaitzsch et al. [14]. The influence of the magnetic field on the stress – strain behaviour was reported by Gaitzsch et al. [13].
The structure of $Ni_{50}Mn_{29}Ga_{21}$ alloys is pseudo tetragonal in the martensitic state (5M) at room temperature and cubic in the austenitic state above the A_f – temperature (~60°C) [15]. The Curie temperature is approximately 100°C.
This paper presents stress-strain curves from samples with different grain sizes (hot mould cast samples and Bridgman samples). Furthermore stress-strain curves with training at 20°C and 40°C will be shown. The microstructure of samples with different grain sizes will be evaluated for the initial state and after compression along the solidification direction. The orientation of the c-axes of the various twins will be investigated for the initial state and after compression.

2. Experimental

The polycrystalline shape memory alloy $Ni_{50}Mn_{29}Ga_{21}$ was melted inductively and cast into a hot cylindrical ceramic mould with a cold copper plate at the bottom. With this method it is possible to achieve a coarse-grained and a textured microstructure by directional solidification [8]. To achieve homogenisation the samples were annealed for 48 h at 1000°C under Ar/5 % H_2 atmosphere. Cubic samples with an edge length of 5 mm were cut by spark erosion. One edge of the cube is parallel to the direction of solidification which is preferably a [100] axis of the cubic high temperature phase (austenite). Afterwards, the samples were annealed for 14 h at 600°C to relieve stress. This procedure had been shown to be effective by Gaitzsch et al. [16]. The next step was grinding and polishing to achieve plane and parallel surfaces. Besides the production of these hot mould cast samples, we also produced polycrystalline samples using a modified Bridgman technique. The alloy was withdrawn from the heat zone with a growth rate of 100 mm/h. The drawing direction defines the growth direction. The ingot has a diameter of 10 mm. Samples with a size of 5x5x5 mm^3 were cut out erosively. The grains of the Bridgman samples and the hot mould cast samples have a diameter of 1 to 3 mm and 0.2 to 0.5 mm perpendicular to the solidification direction, respectively. The grains extend along the whole sample parallel to the direction of solidification. These polycrystalline samples were compressed along the 3 principal axes of the cube (3 compressions are 1 cycle). The properties of the samples with different grain sizes were investigated.
The microstructure was analysed by scanning electron microscopy (SEM) and the orientation of the grains was determined by electron backscatter diffraction (EBSD) with a step size of 18 μm.

3. Results and discussion

The microstructure of a hot mould cast and a Bridgman sample is shown in Fig. 1 for the initial state and after compression. The microstructure was characterised by EBSD after grinding and polishing. Fig. 1(a) shows the microstructure of a typical hot mould cast sample in the initial state. The orientation distribution of the c-axes of a hot mould cast sample after compression is shown in Fig. 1(c). Both figures show the same area of the sample. The grain diameter of the columnar grains is about 0.2 to 0.5 mm of the hot mould cast sample. For comparison the Bridgman sample has a grain diameter of 1 to 3 mm. The red lines in the EBSD image (colour online) indicate the grain boundaries of the parent austenitic phase. The grey scale symbolises the orientation of the c-axes in the grains and variants. The angles between the c-axis and the growth direction are indicated by the shade of grey from black (c-axis parallel to growth direction) to light grey (c-axis perpendicular to growth direction) as given in the legend. White areas are not indexed. The relative frequency versus deviation of the angle between the c-axis and the solidification direction is shown in Fig. 1(b). The amount of variants with c-axes within a cone of 30° around the solidification axis was calculated from the area under the histogram curve (marked red in Fig. 1(b) and (d)). For the hot mould cast sample the fraction of the c-axes in this cone is 4.9 % in the initial state. Afterwards, the sample was compressed with 20 MPa along the growth direction (Fig. 1(c)). This compression was realised by cooling under load (c.u.l.) which means heating up the sample to the austenitic state, compressing with 20 MPa and cooling down to room temperature and then releasing the stress. The view is along the growth direction which is also the compression direction. The relative frequency versus deviation (Fig. 1(d)) shows that after c.u.l. 46.5 % c-axes are orientated within this cone. During this procedure variants with c-axis close to the compression axis grew at the expense of others.

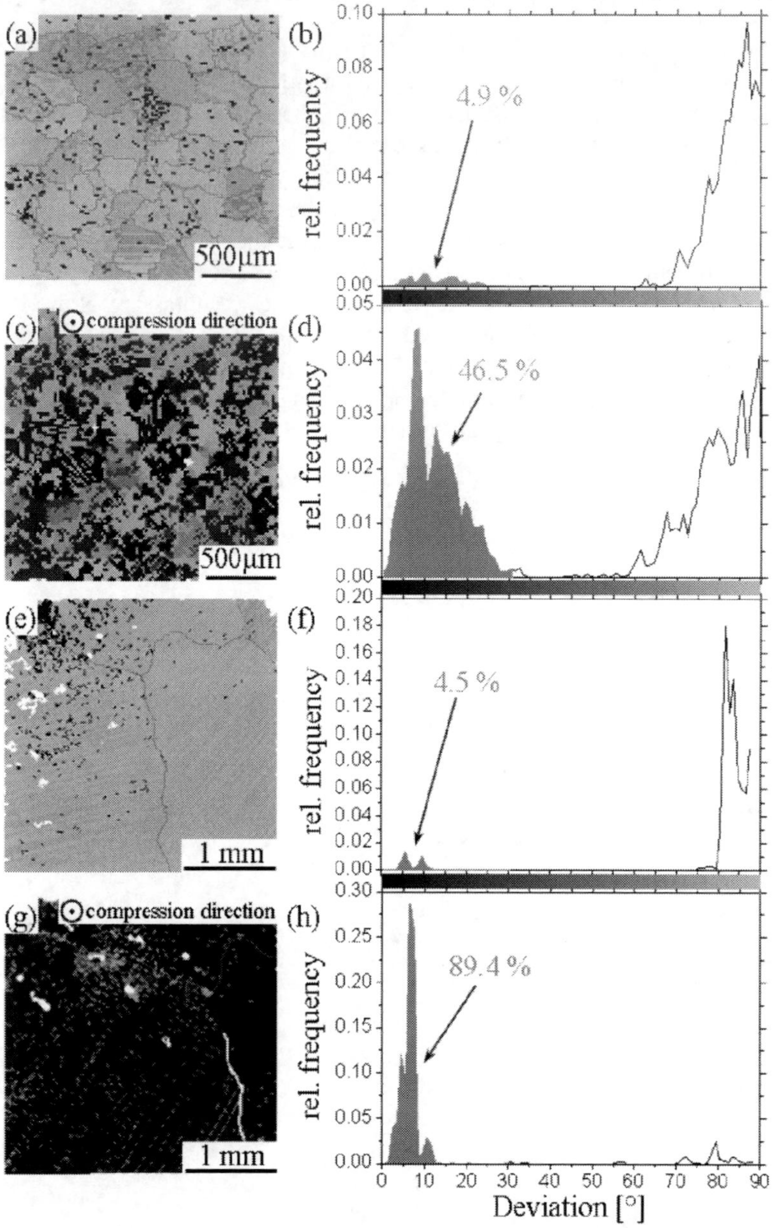

Fig. 1: hot mould cast sample (grain size 0.2 to 0.5 mm): a) Microstructure of the initial state c) Microstructure after cooling under load with 20 MPa load along solidification direction. e) Microstructure of the initial state, g) Microstructure after cooling under load with 20 MPa load along solidification direction. b), d), f) and h) Histogram of c-axes deviation from 0° to 90°. The amount of c-axes within a cone of 30° around the solidification (=compression) direction is marked red (colour online). The angle between the c-axes and the growth direction is indicated by the shade of grey from black (c-axis parallel to growth direction) to light grey (c-axis perpendicular to growth direction) as given in the legend. Bridgman samples (grain size 1 to 3 mm):

The microstructure of the Bridgman sample in the initial state with a grain size of 1-3 mm is shown in Fig. 1(e). The relative frequency as a function of the deviation (Fig. 1(f)) shows that only 4.5 % of the c-axes align within a cone of 30° in solidification direction. Thus, the c-axes fraction in solidification direction is equal and low in the initial state for the hot mould cast sample and the Bridgman sample. The sample from Fig. 1(e) was compressed by cooling under load (c.u.l.) with 20 MPa in solidification direction (Fig. 1(g)). Fig. 1(e) and g) show nearly the same area of the sample. Also in this case the view is along the growth direction which is also the compression direction. After compression, 89.4 % c-axes align in compression direction (0°-30° in Fig. 1(h)). The fraction of the c-axes in solidification direction increases significantly after compression in solidification direction.

Fig. 1(c) and Fig. 1(g) illustrate the effect of a compression which occurred with cooling under load at the same temperature and with the same stress of 20 MPa. The important difference between both samples is the grain diameter. With larger grains more c-axis can be aligned in compression direction.

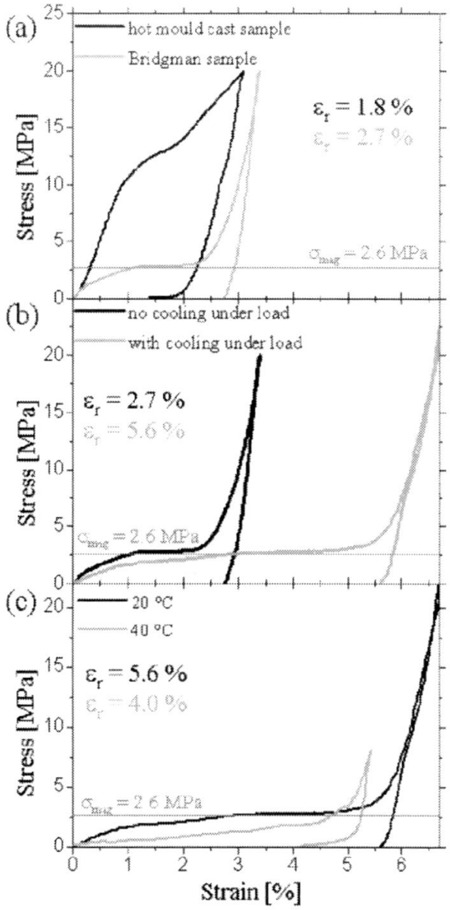

Fig. 2: Stress-strain curves of: a) hot mould cast and Bridgman samples at room temperature without cooling under load (c.u.l.). b) Bridgman samples at room temperature with and without c.u.l. c) Bridgman samples with c.u.l. at room temperature and at 40 °C. The maximal stress that can be achieved with a magnetic field is 2.6 MPa.

The comparison of the stress-strain curves of the hot mould cast and Bridgman samples, with and without cooling under load and at room temperature or at 40 °C is shown in Fig. 2. These stress-strain curves are measured after the 5th training cycle along the solidification direction. Here, successive compressions along the 3 principal axes of the cube are one cycle. Fig. 2(a) shows stress-strain curves of a hot mould cast (black) and a Bridgman sample (grey) at room temperature without cooling under load. The residual strain of the hot mould cast sample and the Bridgman sample was determined to be 1.8 % and 2.7 %, respectively. That means larger grains result in higher strain. Fig. 2(b) shows stress-strain curves of Bridgman samples at room temperature without cooling under load (black) and with c.u.l. (grey). Because of cooling under load the residual strain is 5.6 % compared to 2.7 % for without c.u.l. The use of cooling under load at the beginning of the training process results in a larger residual strain and a lower slope. This c.u.l. was applied only once at the beginning and only along solidification direction. The grey curve shows a pronounced plateau which is due to twin boundary motion. Fig. 2(c) shows stress-strain curves of Bridgman samples with cooling under load at room temperature (black) and at 40 °C (grey). During training at 40 °C a smaller residual strain of 4.0 % is observed than during training at 20 °C with a residual strain of 5.6 %. But the level of the plateau of the stress-strain curve is much lower at 40 °C than at 20 °C. The maximal stress that can be achieved with a magnetic field is 2.6 MPa, therefore a MFIS of ~4.5% could be expected in the Bridgman sample with c.u.l. and trained at 40°C. That means the use of c.u.l. at the beginning of the training process and training of the sample at 40 °C results in a lower stress level. This corresponds to the results of Heczko and Straka who found that the stress to move twin boundaries and the strain decrease with increasing temperature below the martensitic transformation in single crystals [17]. This effect is due to the temperature dependence of the c/a ratio which increases with increasing temperature [18, 19]. Larger grains cause higher strain because of easier twin boundary movement which results in preferred orientation of the c-axes in compression direction.

4. Conclusion

Polycrystalline $Ni_{50}Mn_{29}Ga_{21}$ samples with good MSM properties were achieved by directional solidification. The fraction of the c-axis in growth direction before and after compression were analysed by EBSD. The stress-strain curves give the following conclusions:

- Larger grains and cooling under load (i.e. heating up the sample to the austenitic state, compress the sample and cool it down to RT while compressed) result in a longer plateau of the stress-strain curve. Thus, higher MFIS can be expected for such samples.
- At higher temperatures a lower twinning stress is achieved, which is necessary to obtain high MFIS. Furthermore, training at higher temperatures is expected to reduce the risk of cracks.

Acknowledgment

The authors are grateful to A. Pöhl for technical assistance. This work is funded by DFG SPP 1239.

References

[1] Ullakko K, Huang J K, Kantner C, O´Handley R C and Kokorin V V 1996 *Appl. Phys. Lett.* **69** (13) 1966
[2] Müllner P, Chernenko V A and Kostorz G 2003 *Scripta Mater.* **49** 129
[3] Söderberg O, Aaltio I, Ge Y, Heczko O and Hannula S-P 2008 *Mater. Sci. Eng.* A **481–482** 80
[4] Müllner P, Chernenko V A, Wollgarten M and Kostorz G 2002 *J. Appl. Phys.* **92** (11) 6708
[5] O'Handley R C, Murray S J, Marioni M, Nembach H and Allen S M 2000 *J. Appl. Phys.* **87** (9) 47127
[6] Murray S J, Marioni M, Allen S M, O´Handley R C and Lograsso T A 2000 *Appl. Phys. Lett.* **77** (6) 886
[7] Müllner P, Chernenko V A and Kostorz G 2003 *J. Magn. Magn. Mater.* **267** 325
[8] Pötschke M, Gaitzsch U, Roth S, Rellinghaus B and Schultz L 2007 *J. Magn. Magn. Mat.* **316** 383
[9] Hürrich C, Wendrock H, Pötschke M, Gaitzsch U, Roth S, Rellinghaus B and Schultz L 2009 *Mater. Eng. Perf.* **18** (5) 554
[10] Glavatskyy I, Glavatska N, Söderberg O, Hannula S-P and Hoffmann J-U 2006 *Scripta Mater.* **54** 1891
[11] Masdeua F, Pons J, Santamarta R, Cesari E and Dutkiewicz J 2008 *Mater. Sci. Eng.* A **481–482** 101
[12] Sozinov A, Likhachev A A, Lanska N and Ullakko K 2002 *Appl. Phys. Lett.* **80** (10) 1746
[13] Gaitzsch U, Pötschke M, Roth S, Rellinghaus B and Schultz L *Acta Mater.* **57** 365

[14] Gaitzsch U, Pötschke M, Roth S, Rellinghaus B and Schultz L 2007 *Scripta Mater* **57** 493
[15] Roth S, Gaitzsch U, Pötschke M and Schultz L 2008 *Adv. Mater. Research* **52** 29
[16] Gaitzsch U, Pötschke M, Roth S, Mattern N, Rellinghaus B and Schultz L 2007 *J. All. Compds.* **443** 99
[17] Heczko O and Straka L 2003 *J. Appl. Phys.* **94** 7139
[18] Glavatska N, Mogilniy G, Glavatsky I, Danilkin S, Hohlwein D, Beskrovnij A, Söderberg O and Lindroos V K 2003 *J. Phys. IV France* **112** 963
[19] Okamoto N, Fukuda T, Kakeshita T, Takeuchi T and Kishio K 2004 *Science Techn. Adv. Mater.* **5** 29

Magnetic and magnetoelastic properties of Ni-Mn-Ga – Do they need a revision?

Oleg Heczko[1,3], Jaromír Kopeček[1], Dušan Majtás[1], and Michal Landa[2]

[1]Institute of Physics, Academy of Sciences of Czech Republic, Na Slovance 2, 182 02 Prague, Czech Republic

[2]Institute of Thermomechanics, Academy of Sciences of Czech Republic, Dolejškova 5, 18200 Prague 8, Czech Republic

[3]E-mail: heczko@fzu.cz

Abstract. The magnetic anisotropy, magnetostriction and behaviour of elastic constants in magnetic field are important for fundamental understanding of martensitic transformation and magnetic shape memory effects. Here we present the measurement of the properties of cubic austenite and tetragonal martensite as a function of temperature and compare it with previously published results. In austenite, while the magnetostriction broadly agrees with published data, measured anisotropy was about ten times smaller. Using single variant tetragonal martensite the magnetic anisotropy constants were revisited. The magnetic anisotropy of modulated (10M and 14M) and tetragonal martensites can be related using tetragonal distortion within the concept of adaptive phase consisting of nanotwinned tetragonal building blocks. The elastic constants measured by pulse echo methods vary only slightly with magnetic field with apparent exception of shear constant c`, which is field dependent. However, due to very low value of c`, the pulse echo method is not suitable and other method as resonant ultrasound spectroscopy has to be used.

Introduction

Ni-Mn-Ga alloys are key materials exhibiting magnetic shape memory effects [1-3]. Thus precise knowledge of the magnetic anisotropy, magnetostriction [4-6] and behavior of elastic constants [7-10] in magnetic field is vital for fundamental understanding of martensitic transformation in these materials and resulting magnetic shape memory effects. However, there are only few reports about these properties in cubic high temperature parent phase – austenite and some of them are controversial.

Ni-Mn-Ga austenite transforms to the lower symmetry phase called martensite. To accommodate large lattice deformation during transformation, the martensite is twinned. Depending on composition the martensite can be modulated 10M (called also 5M) and 14M (called 7M) or non-modulated martensite [11-13]. These modulated phases transform at lower temperature to non-modulated (NM) tetragonal martensite phase. The non-modulated phase is thermodynamically stable [14,15]. The modulated phases exhibit high mobility of twin boundaries. Large field-induced deformation due to twin boundary motion in magnetic field called magnetically induced reorientation (MIR) or magnetic shape memory (MSM) effect is observed in modulated 10M and 14M martensites [1-3, 11].

To explain the sequence of martensitic and intermartensitic transformations and existence of the modulated phases the concept of adaptive phase was developed [16, 17]. The concept asserts that the modulated phases are not thermodynamic phases in strict sense but they are adaptive phases to accommodate large difference between cubic austenite and tetragonal martensite during martensitic transformation. For Ni-Mn-Ga phases it was shown that modulated phases can be built from the unit cells of non-modulated tetragonal phase. The modulated phase consists of tetragonal building blocks which are twinned on the scale of five and two atomic layers in the case of 14M martensite and three and two layers in the case of 10M. As the matching is not perfect a large amount of stacking faults are introduced. These nano-twinned layers are then eliminated by coarsening resulting in macroscopic variant of NM martensite [17, 18]. This transition occurs at certain temperatures as intermartensitic transformations. The adaptive concept allows us to determine some properties of the modulated phases from the knowledge of the properties of non-modulated phase. On the other hand the agreement between calculated and measured values can be taken as a test of validity of the adaptive concept.

In the martensite the magnetic anisotropy strongly increased compared to cubic austenite [19-22]. The usual magnetostriction is masked by much larger magnetic shape memory effect and thus it is very difficult to determine it unambiguously [23]. The precise values of magnetic anisotropy constants of different phases can be determined only from single crystal. However, usually after martensitic transformation the single crystal is in twinned state with domains or variants with different crystal orientation connected by twin planes. Therefore the utmost care should be taken to prepare a single crystal with only single variant of martensite [24]. Here we compare some previous measurements of modulated and non-modulated phases of single variant martensite [19,22,25] and discuss the agreement with the adaptive concept.

During martensitic transformation large deformation is involved and thus the knowledge of elastic constant is crucial for understanding the transformation. Additionally as the transformation appears in ferromagnetic state magnetic and magnetoelastic properties and the effect of magnetic field on elastic constant can be important. However, there is a controversy of the elastic constants of austenite and particularly about their dependence on magnetic field and temperature. Here we re-measure magnetic properties of austenite, determine whole set of elastic constant of cubic austenite and compared them with previous measurements.

Experimental

Magnetic and elastic properties of austenite were measured using single crystals cut from same ingot with composition close to stoichiometric Ni_2MnGa. The samples were in the shape of discs of 13.5 mm diameter and thickness of 4.3 mm with two different nominal orientations 100 and 110 in axis of the discs. Precise orientation of the samples was determined by Laue and single crystal X-ray diffraction. It shows that there is a deviation from perfect orientation of about 2 deg for 100 and 9 deg for 110 orientation. The transformation temperature from austenite to martensite was about 221 K and Curie point at 385K.

Magnetic anisotropy of martensite was measured on various rectangular prismatic samples with approximate 100 orientation of the sides. The shape of the samples is not ideal but this shape is necessary to prepare samples in single variant state by mechanical compression or tension. We used the samples having NM martensite at room temperature. Magnetic anisotropy was determined from the magnetization curves.

Magnetization curves were measured by vibrating sample magnetometer (VSM) with electromagnet up to 2 T field. The same space of the electromagnet was used to measure magnetostriction and elastic properties as a function of the field. Magnetostriction was measured by semiconductor strain gages using four-wire resistivity of Keithley 2001 multimeter. The measurement of strain of the disc with 110 orientation in saturation field provides both anisotropy constants λ_{100} and λ_{111} [26].

The elastic constants were measured by pulse echo method using piezoelectric transducers. Details are described in ref. [10, 27]. By measuring phase velocity of the sound wave with different polarizations along [100] and [110] directions, we can determine all independent elastic constants of cubic material from the following equations. Errors in determination elastic constant are mainly due to imprecise planparalelity of the sample sides and the deviation of wave vector from precise crystal orientation.

For the wave propagation direction [100] measured velocity v_L of longitudinal wave and v_T of transversal wave gives elastic constants

$c_{11} = \rho v_L^2$, $c_{44} = \rho v_T^2$.

In the direction [110] by measuring velocity v_L of longitudinal wave we get elastic constant

$c_L = \rho v_L^2 = (c_{11} + c_{12} + 2c_{44})/2$.

In this case for transversal wave we must distinguish between two polarization modes along [100] and [110] (so called slow mode) and we get two elastic constants

$c_{44} = \rho v_{T1}^2$ and $c' = \rho v_{T2}^2 = (c_{11} - c_{12})/2$,

respectively. It is known that c' is very difficult to determine in shape memory alloys by pulse echo method due to very low velocity of the wave. Density ρ of Ni-Mn-Ga is 8.1 g/cm^3. From above equations we can also calculate $c' = c_{11} + c_{44} - c_L$, using directly measured constants c_{11}, c_{44}, and c_L.

Results and Discussion

Martensite

Magnetization curves of non-modulated martensite single crystal measured along principal axis [100] are in Fig. 1. The samples were prepared by compression and tensile stress along one of the axis. It was expected that this would result in single variant state of the martensite. The NM martensite has tetragonal structure with c > a = b and c/a ≅ 1.2, where a, b, and c are lattice parameters. Thus by compression a single variant sample cannot be created as there is ambiguous selection rule for variant orientation. The selection of long c axis along stress works only for the tensile stress. Thus observed difference of the magnetization curves of differently prepared samples can be ascribed to the fact that only by tensile stress the true single variant martensite can be formed. Both curves, however, indicates that NM martensite has easy plane anisotropy with hard axis along long c-axis. Small deviation of the easy curves can be due to the presence of residual variants with different orientation.

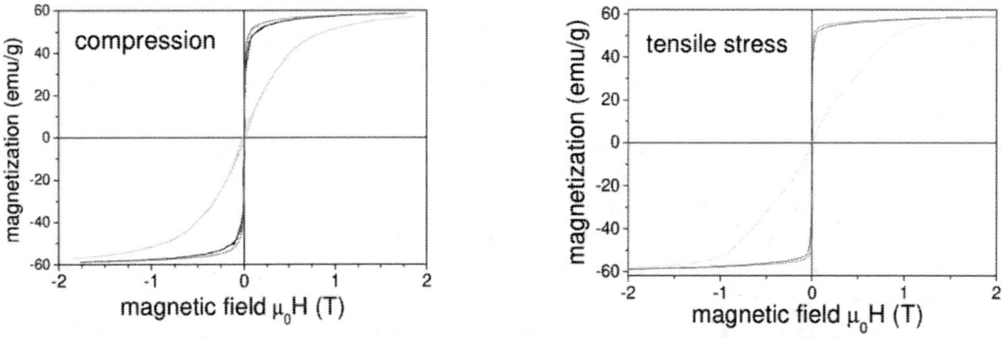

Figure 1. Magnetization curves of NM martensite measured along [100] directions at room temperature. Steep curves (black and red) are measured along a and b axis, tilted curve (green) along c-axis.

The curvature of the compression curve suggests that there are both anisotropy constants, K_1 and K_2. This was wrongly attributed before [19, 20]. Using sample prepared by tensile stress, K_2 became negligible. Small kink observed on hard axis curve in Fig. 2 is due to residual variants or variants re-created during preparation of the sample for measurement. Using correction for these residual twin variants one obtained a linear curve with $K_2 = 0$. The anisotropy constant K_1 is higher than that determined before. The absolute value of the anisotropy increases with decreasing temperature, from $2.6 \cdot 10^5$J /m^3 at room temperature up to $5 \cdot 10^5$J /m^3 at 10 K, which is about 1.5 times higher than for 10M martensite with c/a=0.94.

Using concept of adaptive phase for modulated martensites, 10M and 14M, we can calculate expected anisotropy constants in these phases from the knowledge of the anisotropy of non-modulated phase. Fig. 2 shows the comparison of calculated values from above described measurement and directly measured values taken from ref [22] for 14M martensite. The 14M martensite is considered as built from five and two layers of tetragonal unit cell connected by twinning. Thus we can calculate expected value just taking the proper fraction 5/7 and 2/7. The agreement between calculated and measured values is quite good in whole temperature range where 14M phase exists. The difference can be ascribed to different composition of compared martensites and to interaction between nanoscale twin variants of tetragonal martensite, which should lead to a decrease of high value constant and an increase of low value constant in 14M as observed.

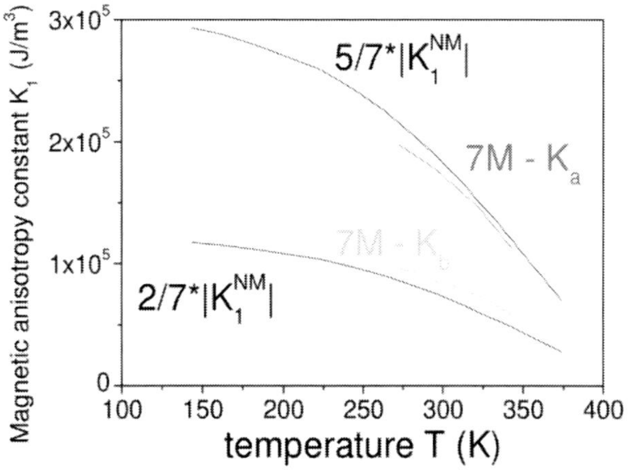

Figure 2. Calculated magnetic anisotropy constants for 14M martensite and comparison to measured ones (in color) marked in figure as 7M.

Similar comparison can be made for 10M and NM martensite. In this case 10M contains three and two layers of tetragonal martensite. The agreement cannot be expected to be very good as the tetragonal martensite as building block of this modulated phase should have a different c/a ratio (i.e. smaller) [17,18] than used here. The comparison is summarized in the table showing calculated and measured values for 10M [19]. The agreement is suspiciously good, which can be just accident.

Table 1. Calculated and measured values of magnetic anisotropy constant in 10M martensite

Temperature (K)	K_{NM} (J/m^3)	$-3/5K_{NM}$ (J/m^3)	K_{10M} (J/m^3)
10	$-5 \cdot 10^5$	$3 \cdot 10^5$	$3 \cdot 10^5$
300	$-2.6 \cdot 10^5$	$1.56 \cdot 10^5$	$1.65 \cdot 10^5$

One may also speculate about a second constant considering the arrangement of nano-twins in analogy with 14M case. The second constant should be given as $-2/5K_{NM}$. However, this was not observed. Reason for that can be that the difference between a and b are very small and it might not be possible to prepare unique single orientation i.e., a or b in large volume. Result is that the different orientations of modulation are mixed together and only some average is measured. In this case one can expect the constant close to $2.5/5K_{NM}$.

The calculation shows that we can predict anisotropy properties of modulated phases from non-modulated phase with reasonable precision. Viewing it from the other side, the presented calculation can be taken as another supporting evidence for the adaptive concept of modulated martensite. Suggested method can be also used for estimation of other properties of modulated phases which are not accessible in this phase due to MSM effect.

Austenite

Magnetic properties

Magnetization curves of the discs were measured in different directions. The curves were very similar. As example the magnetization curve measured in plane of the disc is shown in Fig. 3. The magnetization curve has very low hysteresis, less than 0.002 T and it is saturating fast. The magnetic anisotropy was evaluated as difference between the curves measured in [100], [110], [111] directions in the disc plane with 110 orientation (Fig. 3b). The difference between the curves are very small but it is clear that [100] direction is easy axis in agreement with Tickle and James [4]. However, the calculation of area between curves gives anisotropy $K_1 \cong 3 \cdot 10^2$ J/m^3, which is of one order magnitude smaller then their measurement, $K_1 = 2.7 \cdot 10^3$ J/m^3. Some of the difference might be ascribed to the composition difference and closeness of martensitic transformation in latter case, but in general the anisotropy is smaller than previously thought.

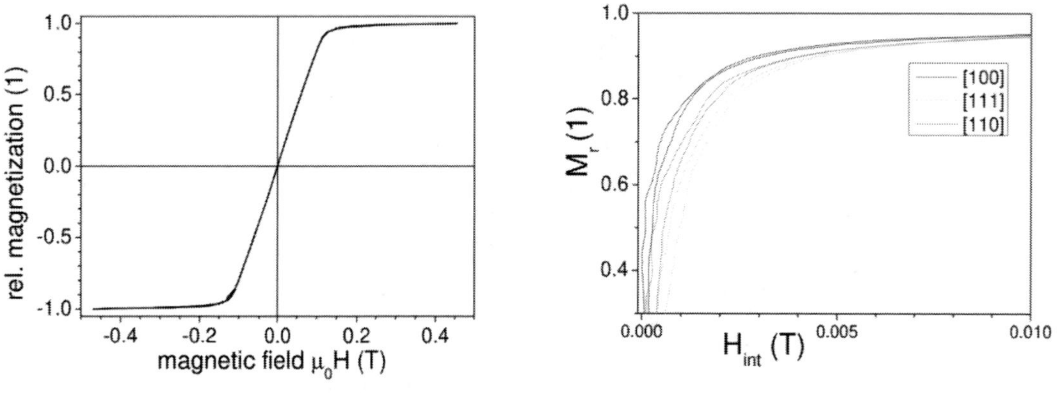

(a) (b)

Figure 3. a) Measured magnetization curve up to 0.5 T of austenite disc at room temperature. Saturation induction $J_s = 0.67$ T. b) Detail of magnetization curves measured at different directions as marked in the disc with 110 orientation. The curves are corrected for demagnetization to enhance the visibility of the difference.

The easy axis of magnetization of cubic austenite is along [100] directions. Magnetic domains are therefore separated by 90 deg domain walls. From the magnetic anisotropy constant one can determine domain wall width by minimization of exchange and anisotropy energies [28]

$$\delta_{dw90deg} = \pi/2\sqrt{A/K}.$$

Using approximate exchange constant $A \approx 1 \cdot 10^{-11}$ J/m the domain width is $\delta_{dw90deg} \approx 280$ nm. This very broad domain wall can explain very low hysteresis as in the single crystal the walls are not pinned or potential pinning sites are much smaller than the wall thickness.

Direct observation of magnetic domain is apparently difficult due to broad wall and in the bulk no observation has been published, yet. However, the observation in thin foil by TEM [29] shows quite random magnetic domain structure with large number of vortices. As authors pointed out, this most likely correlates with the low magnetocrystalline anisotropy of the austenitic phase. They also reported that magnetic domain walls are pinned on antiphase boundaries in the foil but apparently this pinning cannot be very strong in the bulk as measured magnetic hysteresis is negligible (Fig. 3).

Room temperature magnetostriction was determined from the angular dependence of strain at saturation field. The magnetostriction constants of cubic austenite are λ_{100} = 120 ppm and λ_{111} = 13 ppm. These values broadly agree with the literature for the temperature that is same far from martensite transformation [4, 6] and indicate relatively strong magnetoelastic coupling in this material.

Elastic properties

Fig. 4 shows the relative change of the elastic constants in magnetic field. It is apparent that the dependence is weak, below 1% and the constants always increase with increasing magnetic field. Additionally the constants are changing only in low field and then saturated. This is correlated with magnetization change (Fig. 3), i.e., the change of magnetic domain structure affects the value of the constants. Observed small stiffening is due to the disappearance of magnetic domain structure and magnetic saturation.

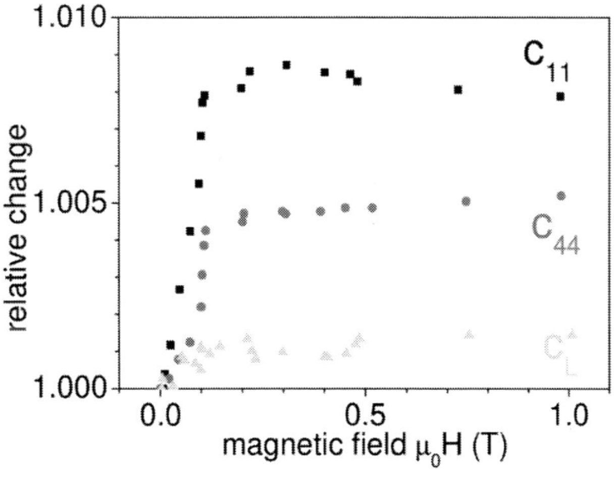

Figure 4. Measured relative change of the elastic constants of cubic austenite in magnetic field at room temperature. The constants were measured up to field of 2 T, for sake of clarity the saturation values are shown only to 1 T. The change of the constant above 0.2 T is within experimental error.

This very weak dependency on magnetic field is in contrast with previous report. It was suggested that constant c_{44} of ferromagnetic austenite is strongly dependent on magnetic field [8]. Our measurement indicated that only constant, which is discernibly field-dependent is c'. However, the value of the constant is so low and attenuation so high that reliable dependence cannot be given for this set-up. In fact the relatively strong magnetoelastic coupling as indicated by λ_{100} should result in field dependency of elastic constants and particularly c' which is related to shear deformation.

The measured constants at zero field and calculated c' are in Table 2 together with values from literature. Relative error of our pulse echo measurement is apparently even smaller than indicated in table as it can be inferred from the spread of the relative values of the elastic constants in saturation field which is less than 0.1% (Fig. 4). However, much larger error is introduced due to deviation of the crystal orientation from ideal one. Comparing the value of the constant c_{44} measured in both discs with different orientation, we can estimate that the error due to misorientation can easily exceed 10%. This strongly affects the calculated value of c' as it is obtained as difference from three large numbers and its value is only fraction of other constants. In this case the error exceeded easily the value of the constant. Thus better method is needed to determine c'. This will be published elsewhere [30].

Table 2. Elastic constants of cubic austenite at room temperature.

source	c_{11} (GPa)	c_{44} (GPa)	c_L (GPa)	c' (GPa)	c' (GPa) calculated*
this report	140 ± 3	104 ± 2	235 ± 4	$\sim 4 \pm 5$	9 ± 9
[9]	136 ± 3	102 ± 3	222 ± 9	22 ± 9	16 ± 15
[7]	-	103	250	4.4	

*$c' = (c_{11} - c_{12})/2 = c_{44} + c_{11} - c_L$

Measured values of elastic constants broadly agree with the literature as table indicates, with notable differences of c', which can be ascribed to difficulties of measuring pulse echo for this slow transversal mode. This is also apparent from the quoted error. Measuring this constant by resonance ultrasound spectroscopy (RUS) gives $c' = 3.6 \pm 0.03$ GPa. [10] On the other hand Zhao et al. [8] gives very low and field dependent value of constant c_{44} = 5-10 GPa, which might be due to cross-effect of c' on the measurement on this constant.

Conclusion

- Magnetic properties of NM martensite are revisited. It is important to consider microstructure i.e. proper preparation of single crystal single variant.
- Magnetic anisotropy of modulated martensites can be predicted from NM phase using adaptive phase concept.
- Magnetic anisotropy of austenite is even smaller than previously measured
- Elastic constants of austenite are different than in literature and do not depend strongly on magnetic field apart of c`.
- For proper determination of the c' the resonance ultrasound spectroscopy in magnetic field is better method, which on the other hand is not suitable for hard elastic constants as in the spectrum most of low lying resonances are given by shear constant c'.
- Presented measurements indicate that a revision of the magnetoelastic properties is timely.

Acknowledgements
Support from ASCR projects No. M100100913 and No. A2001006272 is acknowledged.

References

[1] Ullakko K, Huang J K, Kantner C, O'Handley R C, and Kokorin V V 1996 *Appl. Phys. Lett.* **69** 1966
[2] Sozinov A, Likhachev A A, Lanska N and Ullakko K 2002 *Appl. Phys. Lett.* **80** 1746
[3] Heczko O, Sozinov A, and Ullakko K 2000 *IEEE Trans. Magn.* **36** 3266
[4] Tickle R, James R.D 1999 *J. Magn. Magn. Mat.* **195** 627
[5] Albertini F, et al. 2001 *J. Appl. Phys.* **89** 5614
[6] Heczko O, 2010 Magnetoelastic Coupling in Ni-Mn-Ga Magnetic Shape Memory Alloy *Ferromagnetic Shape Memory Alloys II Book Series: Materials Science Forum* **635**, 125
[7] Worgull J, Petti E and Trivisonno J 1996 *Phys. Rev. B* **54**, 15695
[8] Zhao P, Dai L, Cullen J, and Wuttig M, Metallurgical and Material Transaction A, 38A, (2007), 745.
[9] Manosa L, Gonzalez-Comas A, Obrado E, Planes A, Chernenko V A, Kokorin V V and Cesari E, 1997 *Phys. Rev. B* **55** 11068
[10] Seiner H at al. 2009 *Mater. Sci. Eng. A.* **205-208**
[11] Chernenko VA, Segui C, Cesari E, Pons J, Kokorin VV 1998 *Phys. Rev. B* **57** 2659
[12] Chernenko VA et al. 2003 *Phys. Rev. B* **67** 064407
[13] Straka L, Heczko O and Lanska N 2002 *IEEE Trans. Magn.* **38** 2835
[14] Entel P, et al. 2006 *J. Phys. D: Appl. Phys.* **39** 865
[15] Ayuela A, Enkovaara J and Nieminen R M 2002 *J. Phys.: Condens. Matter.* **14** 5325
[16] Pons J, Chernenko V A, Santamarta R and Cesari E, 2000 *Acta Mater.* **48** 3027
[17] Kaufmann S, Rößler U K, Heczko O, Wuttig M, Buschbeck J, Schultz L and Fähler S, 2010 *Phys. Rev. Lett.* **104** 145702
[18] Fähler S et al. 2011 Playing Lego with martensite, to be published in *physica status solidi (RRL)*
[19] Straka L and Heczko O 2003 *J. Appl. Phys.* **93** 8636
[20] Sozinov A, Likhachev A A and Ullakko K, 2001 *Proc. SPIE* **4333** 189
[21] Albertini F et al. 2002 *Appl. Phys. Lett.* **81** 4032
[22] Straka L, Heczko O and Ullakko K, 2004 *J. Magn. Magn. Mater.* **272-276** 2049
[23] Heczko O 2005 *J. Magn. Magn. Mater.* **290** 846
[24] Soderberg O, Straka L, Novak V, Heczko O, Hannula SP and Lindroos V K, 2004 *Mater. Sci. Eng. A* **386** 27
[25] Heczko O, Straka L, Novak V and Fähler S 2010 *J. Appl. Phys.* **107** 09A914
[26] Zijlstra H, 1967 *Experimental Methods in Magnetism* 2, North-Holland, Amsterdam, 168.
[27] Landa M, Sedlak P, Sittner P, Seiner H, Heller L, 2008 *Mater. Sci. Eng. A* **481-482** 567
[28] Cullity D B, 1972 *Introduction to Magnetic Materials*, Addison-Wesley, New York
[29] Venkateswaran S P et al. 2007 *Acta Mater.* **55** 2621
[30] Seiner H. et al. 2011 Field and temperature dependence of elastic constants in Ni-Mn-Ga to be published.

Magnetic refrigeration at room temperature - from magnetocaloric materials to a prototype

L Theil Kuhn, N Pryds, C R H Bahl and A Smith

Fuel Cells and Solid State Chemistry Division, Risø DTU, Technical University of Denmark, DK-4000 Roskilde, Denmark

E-mail: `luku@risoe.dtu.dk`

Abstract. Based on the magnetocaloric effect, magnetic refrigeration at room temperature has for the past decade been a promising, environmentally friendly new energy technology predicted to have a significantly higher efficiency than the present conventional methods. However, so far only a few prototype refrigeration machines have been presented worldwide and there are still many scientific and technological challenges to be overcome. We report here on the MagCool project, which spans all the way from basic materials studies to the construction of a prototype. Emphasis has been on ceramic magnetocaloric materials, their shaping and graded composition for technological use. Modelling the performance of a permanent magnet with optimum use of the flux and relatively low weight, and designing and constructing a prototype continuous magnetic refrigeration device have also been major tasks in the project.

1. Introduction

In 1881 Warburg reported on the discovery of the magnetocaloric effect [1], and in the 1920-ies Debye and Giaugue suggested it for achieving low temperatures for laboratory use [2, 3]. The magnetocaloric effect is caused by the fundamental thermodynamic property of entropy conservation under adiabatic conditions [4, 5]: In a simple picture describing a second order magnetic phase transition in a magnetic material undergoing magnetization, the magnetic order is increased and correspondingly the magnetic entropy part is decreased. The adiabatic constraint assures entropy conservation and therefore the lattice entropy must increase, causing a raise in temperature of the magnetic material. The process is reversible in second order materials meaning that a cycle of magnetization and demagnetization of the material will cause a corresponding temperature raise and drop of equal magnitude. An expression for the magnetocaloric effect in terms of the field and temperature dependent magnetization can be derived using the basic thermodynamic expression for the change in entropy S of a given system as a function of the pressure p, the temperature T, and the applied magnetic field H,

$$dS(p,T,H) = \left(\frac{\partial S}{\partial p}\right)_{T,H} dp + \left(\frac{\partial S}{\partial T}\right)_{p,H} dT + \left(\frac{\partial S}{\partial H}\right)_{p,T} dH \tag{1}$$

together with the Maxwell relation

$$\left(\frac{\partial S}{\partial H}\right)_{p,T} = \left(\frac{\partial M}{\partial T}\right)_{p,H}. \tag{2}$$

In the simple case of a second order phase transition it can be assumed that the pressure is constant during magnetization, which leads to that the isothermal entropy change induced by a magnetic field change $H_1 \to H_2$ is

$$\Delta S_{H_1 \to H_2} = \int_{H_1}^{H_2} \left(\frac{\partial S}{\partial H} \right)_T dH. \tag{3}$$

Applying Maxwell's relations for an adiabatic magnetic field change, i.e. $dS = 0$, gives the easier measurable expression in terms of the magnetization M

$$\Delta S_{H_1 \to H_2} = \mu_0 \int_{H_1}^{H_2} \left(\frac{\partial M}{\partial T} \right)_H dH, \tag{4}$$

where μ_0 is the vacuum permeability. Applying the definition of the heat capacity at constant pressure C_p and the second law of thermodynamics, leads to the expression for the related adiabatic temperature change

$$\Delta T_{\text{ad},H_1 \to H_2} = -\mu_0 \int_{H_1}^{H_2} \frac{T}{C_p} \left(\frac{\partial M}{\partial T} \right)_H dH, \tag{5}$$

achieved by applying a magnetic field change $H_1 \to H_2$. Equations (4) and (5) describe well the characteristics of the magnetocaloric effect in terms of the directly measurable parameters M and T. It should be noted that both ΔS and ΔT_{ad} exhibits an extremum for the maximum of $\partial M / \partial T$, e.g. for a ferromagnetic material that is around the Curie temperature T_{C}. Further, the heat capacity C_p directly influences the adiabatic temperature change that can be achieved in a specific material.

The adiabatic temperature change that can be achieved in known magnetocaloric materials is only a few Kelvins [6]. However, in 1976 Brown demonstrated that by applying a regenerative cooling cycle a significant temperature gradient can be accumulated [7]. In Brown's so-called active magnetic regenerator (AMR) a temperature gradient of 46 K cooling from 319 K was generated with Gd in a magnetic field change of 7 T even though the adiabatic temperature change in Gd at 7 T is only $\Delta T_{\text{ad}} \sim 14$ K. It has been predicted that the cooling efficiency (COP) of magnetic refrigeration would result in at least 30% energy savings compared to conventional compressor based techniques working at room temperature [8]. Brown's pioneering work combined with the discovery of the so-called giant magnetocaloric effect [9] made magnetic refrigeration using environmentally friendly and non-hazardous gasses and materials a promising alternative cooling technique at room temperature to the conventional compressor based techniques. Further, other technologies like heat pumps and air conditioning that are also responsible for a significant part of the global energy consuming could in the future be replaced by systems using the magnetocaloric effect.

The challenges for converting the principles of magnetic refrigeration into a commercially relevant technology of the future are many and closely coupled. They can be organized in four major topics:

- Materials for the AMR. Designing a commercially relevant material that shows a large enough adiabatic temperature change (i.e. $\Delta T_{\text{ad}} \sim 1$ K) in the 250 K to 350 K region implying tunable magnetocaloric properties in this wide temperature span. The requirements to the materials are that they should be non-toxic, chemically stable, environmentally friendly, abundant, easy to fabricate and shape, recyclable, preferably not exhibit hysteresis effects that cause too significant losses, and last but not least cheap.
- Modelling of the thermodynamical and fluid dynamical properties of the AMR, so designs can be evaluated and optimal parameters predicted. This includes geometry of the AMR

with minimal pressure losses and optimum heat transfer between magnetocaloric material and heat transfer fluid. Also, modelling the temperature gradient and it's coupling to the magnetocaloric properties of the AMR materials and the magnetic properties of the AMR materials under varying temperature close to the magnetic phase transition is an essential problem to be addressed.

- Modelling of the permanent magnet supplying the magnetic field in the magnetic refrigeration system. As large as possible magnetic fields are required in a region of closed flux. However, to achieve magnetic flux densities above 1 T is both costly and requires large amounts of magnetic material. Modelling of permanent magnet designs with an optimized balance between flux density and volume and volume of magnet material.
- Demonstration of the feasibility of the principles by a prototype magnetic refrigeration device.

In the MagCool project, funded by the Danish Council for Strategic Research, we are emphasizing that all four topics needs to be addressed in close collaboration. Therefore we have investigated a selection of commercially relevant magnetocaloric materials (discussed in section 2), combined the materials' studies with modelling of both a permanent magnet and an AMR (briefly described in subsections 2.4 and 3.1) for finally designing a prototype magnetic refrigeration device (presented in section 3).

2. Commercially relevant magnetocaloric materials for room temperature magnetic refrigeration

Investigations of the magnetic properties of materials for magnetic refrigeration at room temperature have grown to an immense field of research and development. Materials candidates are found within both basic elements, alloys, and ceramics, among those are the well known Gd [7], $Fe_{49}Rh_{51}$ [10], the $Gd_5Si_2Ge_2$ and the related $Gd_5(Si_{4-x}Ge_x)$ series [9, 11], the $MnAs_{1-x}Sb_x$ [12] series, and the $LaFe_{13-x-y}Co_xSi_y$ series [13]. Gd has become the benchmark material for magnetic refrigeration at room temperature with an adiabatic temperature of 3.6 K at 1 T applied field change and a Curie temperature of 293 K. $Fe_{49}Rh_{51}$ has the so far largest known adiabatic temperature change of 5 K at 1 T. The latter materials are however, characterized by having only a relatively narrow working range of a few degrees due to the narrow magnetic phase transition. With the alloys of for instance $Gd_5(Si_{4-x}Ge_x)$ and $MnAs_{1-x}Sb_x$ a relatively large temperature change (ΔT_{ad} ~2 K at 1 T), the giant magnetocaloric effect, is found for a wider range of working temperatures as the Curie temperature can be tuned by the exact materials composition. These materials' magnetocaloric effect is related to several types of phase transitions inducing a 1st order magnetic phase transition, i.e. practical obstacles like hysteresis loses and volume changes are important. For commercialization both price, abundance, ease of production and shaping, and non-toxicity becomes highly relevant. By the newly developed industrial powder metallurgy methods [13] the $LaFe_{13-x-y}Co_xSi_y$ series presenting a ΔT_{ad} up to ~2.2 K at 1 T related to a second order phase transition could prove to be the magnetocaloric material of the future. Besides studying this materials series, we have during this project also focused on an alternative material candidate: the ceramic lanthanum manganite series (also known for the colossal magnetoresistance [14]), which also fulfill the requirements including a tunable T_{C} in a wide range around room temperature, negligible hysteresis effects and in addition also robustness of the magnetocaloric effect.

2.1. $La_{0.67}Ca_{0.33-x}Sr_xMn_{1.05}O_3$ - fine tuning T_{C} by industrial production methods

$La_{0.67}Sr_{0.33}MnO_3$ is a perovskite material with a ferromagnetic to paramagnetic second order phase transition at T_{C} = 354 K, and the corresponding $La_{0.67}Ca_{0.33}MnO_3$ has a first order magnetic (ferromagnetic to paramagnetic) and structural (orthorhombic to rhombohedral) phase transition at T_C = 260 K [15]. Mixing compositions of $La_{0.67}Ca_{0.33-x}Sr_xMnO_3$ (for $0 < x < 0.33$)(LCSM) results in an LCSM compound with tunable Curie temperature and a second order magnetic phase transition with no observable hysteresis and an adiabatic temperature change up to 1.6 K at 0.7 T applied field change in the temperature range 260 K to 354 K [16]. The ferromagnetism of LCSM is caused by double exchange between the Mn^{3+} and Mn^{4+} ions mediated via the O^{2-} ions. Therefore substituting Ca by Sr gradually increases the Mn-O-Mn bond angle and likewise increases the Curie temperature.

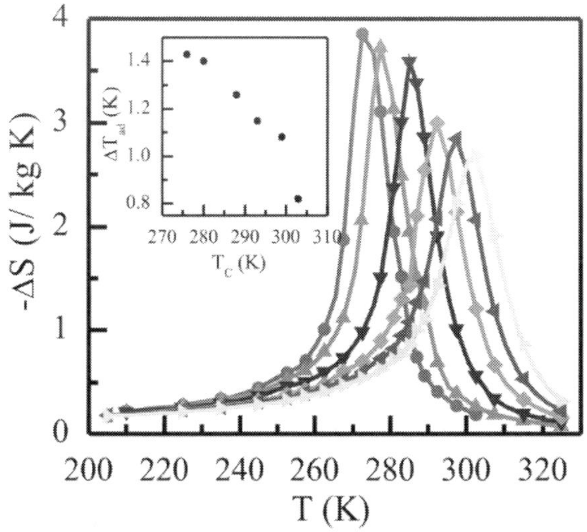

Figure 1. The magnetocaloric effect in $La_{0.67}Ca_{0.33-x}Sr_xMn_{1.05}O_3$ for increasing x (from left to right: $x = 0.0375$ to $x = 0.09$). The inset shows the corresponding adiabatic temperature change. ΔS has been determined indirectly by magnetization measurements and ΔT_{ad} has been measured directly, both are given for a 1 T applied field change.

LCSM can be prepared by for instance glycine-nitrate synthesis or solid state reaction. However, with industrial up-scaling in mind LCSM was prepared as a polycrystalline powder by spray pyrolysis (delivered by the company CerPoTech AS) and subsequently calcined and sintered (details of the materials fabrication will be published elsewhere [17]). Fig. 1 shows a predefined $La_{0.67}Ca_{0.33-x}Sr_xMn_{1.05}O_3$ series (the series was prepared with a slight overstoichiometry of Mn to avoid dilution of the magnetization caused by understoichiometry) prepared by gradually increasing x in the range $x = 0.0375$ to $x = 0.09$ to achieve a series of LCSM powders with T_C in the range 274 K to 303 K, which is relevant for magnetic refrigeration at room temperature. The magnetic entropy change ΔS was measured indirectly by magnetization measurements in a Lakeshore 7407 vibrating sample magnetometer. These measurements were supplemented by C_p measurements (not shown) by a custom built differential scanning calorimeter equipped with an in-house designed Halbach magnet [18] providing an adjustable magnetic field [19, 20]. Direct measurements of the adiabatic temperature change were also acquired by a custom built device using the Halbach magnet [20], these data are shown in the inset of Fig. 1.

The measured adiabatic temperature changes in a field change of 1 T are between 0.8 K and 1.5 K. The achieved ΔT_{ad} are qualitatively in good agreement with what has been determined indirectly using data for ΔS and C_p [16]. However, the direct method eliminates the use of assumptions about the magnetic field dependence of C_p and therefore gives a value for the adiabatic temperature change, which is closer to what would be achieved in the AMR.

The results in Fig. 1 demonstrates that it is indeed possible to fabricate large quantities (kilograms) of the LCSM material by industrial ceramic powder processing techniques with a precision good enough to tune the Curie temperature within a few degrees and achieving a practically usable ΔT_{ad}.

2.2. $La_{0.75}Ca_{0.25-x}Sr_xMnO_3$ - a broad temperature range of ΔS

As demonstrated above and by many other authors [5, 8, 11, 12, 13] tuning of the Curie temperature is a well known method to broaden the working range of the magnetocaloric materials, however, not the only possible one. For specific compositions of LCSM there will be co-existence of several phases, which may affect the magnetocaloric properties. The compound $La_{0.75}Ca_{0.25-x}Sr_xMnO_3$ ($x = 0.1$) lies close to the structural phase transition between the orthorhombic and rhombohedral regions [21] and may therefore exhibit such co-existence effects. We have prepared a series of $La_{0.75}Ca_{0.25-x}Sr_xMnO_3$ ($x = 0, 0.1, 0.25$) polycrystalline powders by solid state reaction and subsequently calcination and sintering. The powders were characterized structurally and the magnetocaloric properties were investigated as described above. Fig. 2 displays the entropy change for the $x = 0.1$ composition in magnetic field changes up to 1.6 T and a table-like magnetocaloric effect up to 1.2 J/kgK is observed in a broad temperature range between approximately 255 K and 315 K.

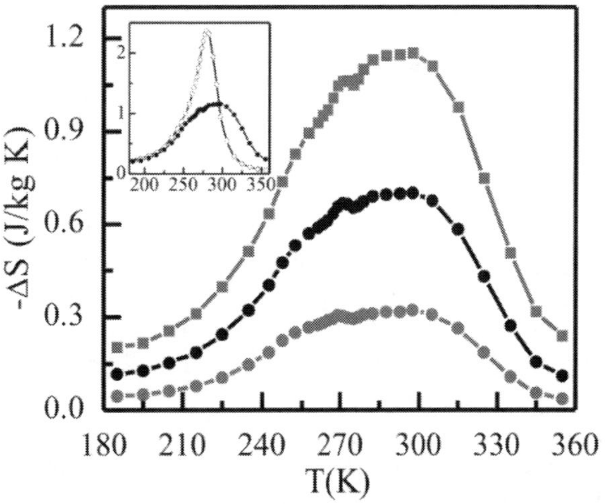

Figure 2. The broad table-like magnetocaloric effect in $La_{0.75}Ca_{0.25-x}Sr_xMnO_3$ ($x = 0.1$) for increasing applied magnetic field changes (from bottom to top: 0.5 T, 1.2 T, 1.6 T). The cusps on the curves indicate that several mechanisms are at play broadening the phase transition. The inset shows the comparison of the table-like magnetocaloric effect (filled symbols) to the pure entropy change related to the magnetic phase transition (open symbols) at 1.6 T, observed after annealing the material.

In the orthorhombic structure the dynamic Jahn-Teller effect plays a significant role in the magnetic properties. Due to the symmetry of the rhombohedral structure it is however not present in that structural phase. The competition between the structural and magnetic phases induces a so-called tri-critical point where an additional order parameter couples to the magnetic order parameter and a broad region of co-existence of several magnetic phases may occur. This is the origin of the table-like entropy change observed in Fig. 2.

The small cusps at the curves indicate the presence of competing phases. This co-existence of phases is induced by the structural phases and can be removed by annealing the $La_{0.75}Ca_{0.25-x}Sr_xMnO_3$ ($x = 0.1$) composition and the expected carrot-like entropy change is recovered (shown for comparison in the inset of Fig. 2). Though very complex, the nature of the table-like phase transition has shown to be of second order and not introducing hysteresis effects (detailed investigations of the mechanisms behind the table-like entropy is discussed in [22]).

Integration of the areas under the ΔS-curves show that the areas related to the table-like and the carrot-like entropy changes are comparable, i.e. the same cooling effect is achieved but spread in a much broader temperature range. A ΔT_{ad} ~0.6 K at a magnetic field change of 1 T is achieved for the LCSM composition exhibiting the table-like magnetocaloric effect. This study illustrates the perspectives for tailoring ΔS and thereby also ΔT_{ad} by manipulating the interplay between several magnetic and structural phases.

2.3. $(La_{1-x}A_x)_{0.67}Ba_{0.33}Mn_{1.05}O_3$ - robustness of ΔS against doping

For practical application of the magnetocaloric materials it is an advantage if the magnetocaloric effect is robust against impurities and that the tuning of T_C by doping does not significantly reduce ΔS and ΔT_{ad} as otherwise often observed [11, 13, 16]. We have studied the effect of substituting a part of the lanthanum in $La_{0.67}Ba_{0.33}Mn_{1.05}O_3$ [23], this composition shows a Curie temperature of $T_C = 347$ K, a magnetic entropy change of $\Delta S = 1.7$ J/kgK and a related adiabatic temperature change $\Delta T_{ad} = 0.65$ K at 1 T applied magnetic field change. A $(La_{1-x}A_x)_{0.67}Ba_{0.33}Mn_{1.05}O_3$ series was prepared by glycine-nitrate synthesis with A being a mixture of lanthanides containing 66% La, 22% Nd, 8% Pr and 4% Ce to mimic some of the naturally occurring lanthanide compositions. The structural and magnetocaloric properties of a series of compositions with $x = 0, 0.33, 0.67, 1$ was investigated, the entropy change in an applied magnetic field change of 1 T is shown in Fig. 3. In the $(La_{1-x}A_x)_{0.67}Ba_{0.33}Mn_{1.05}O_3$

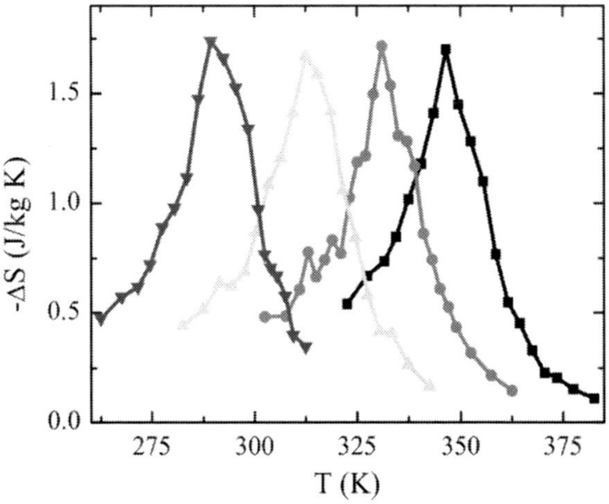

Figure 3. The magnetocaloric effect in $(La_{1-x}A_x)_{0.67}Ba_{0.33}Mn_{1.05}O_3$ ($x = 0, 0.33, 0.67, 1$ from right to left) acquired at 1 T field change. T_C is clearly reduced as x increases, whereas the corresponding ΔS is observed to be unaffected by the doping with lanthanides.

series the reduction of T_C with increasing doping on A is caused by the ionic size mismatch, but there is no change from a second to a first order magnetic phase transition accompanied by a structural phase transition as found in the $La_{0.67}Ca_{0.33-x}Sr_xMn_{1.05}O_3$ series. We suggest that the robust ΔS is caused by a counter acting mechanism that compensates the effect of ionic size mismatch. Such a mechanism could be the increased internal stresses induced by the smaller ionic sizes, in turn causing a higher degree of lattice disorder that may again relax the internal stresses and thereby actually preserve the achieved entropy change independently of the impurity composition. This is discussed in detail in [23].

2.4. Shaping of the AMR

Up to this point we have presented a family of material series that fulfill the requirements listed in the introduction of this section. However, for commercialization these polycrystalline ceramic powders also must be shaped into a form suited for an AMR by applying industrial techniques.

Shaping of the AMR is a very crucial point: an open structure that allows easy flow of the heat transfer fluid is required together with minimized pressure loss and an optimum contact area between the fluid and the magnetocaloric material. As a starting point we have chosen to work with a geometry of stacks of flat plates. These fulfill the criteria and can be modelled in great detail [24, 25]. We have applied tapecasting, sintering and subsequently laser cutting to fabricate thin and flat plates, an example is shown in Fig. 4. The thickness of the flat plates is ~ 0.3 mm. A characteristic feature of an AMR is the build-up of a temperature gradient, i.e. the optimum working temperature of the active magnetocaloric material changes gradually

Figure 4. A set of tape cast, sintered and laser cut LCSM plates ($25 \times 40 \times 0.3\,\mathrm{mm}^3$) prepared for the test machine. Inset: The extruded monolithic LCSM structure has approximately a mesh of channels with cross section $0.8 \times 0.8\,\mathrm{mm}^2$ and the walls have a thickness of 0.4 mm.

throughout the AMR in the direction of the heat transfer fluid flow. We have shown by modelling in [26] that tailoring the working temperature of the magnetocaloric material to the optimum working temperature in the AMR increases the achievable cooling power and COP by up to 30% when just grading the AMR with two in stead of one magnetocaloric material. For our prototype cooling machine modelling suggested that for optimal performance the LCSM plates should be graded with six materials with approximately 4 K difference between each T_C, an example of such a material set was shown in Fig. 1 and the principle of graded material is sketched in Fig. 5. For the tape casting slurries of each LCSM powder was prepared and a custom build multi compartment doctorblade was used for simultaneously tape casting all six (or up to ten) materials into one tape. The graded tape was subsequently dried and sintered and laser cut into narrow pieces prepared for stacking and mounting in the prototype [27]. An example of a graded plate with ten different LCSM compositions is shown in Fig. 5.

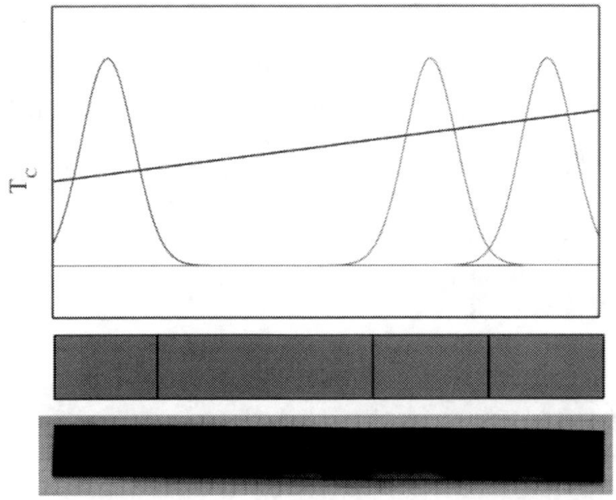

Figure 5. A plate of materials with graded T_C and therefore gradually changing working range for the magnetocaloric effect fitted to the hot and cold end of the prototype AMR. The flow direction of the heat transfer fluid is parallel to the plate. The lower panel shows a tape cast graded plate of 10 different compositions of LCSM ($250 \times 20\,\mathrm{mm}^2$ and thickness 0.3 mm).

Stacking of individual plates for a complete AMR can become a cumbersome activity and therefore we have also studied other ways of industrial shaping. One successful example is shown in the inset of Fig. 4, where we have applied extrusion to fabricate a monolithic AMR with channels of square cross section. The extruded monolithic structure was afterwards cut into a cylindrical AMR of length 50 mm and diameter of 34 mm to fit into our test machine.

In conclusion we can apply industrial ceramic processing and shaping techniques to tailor the magnetocaloric properties of the materials for the AMR including combined materials with gradual change of the magnetocaloric properties. The lanthanum manganite series have proven perfectly tunable to an optimized working temperature range and in the special case of LBMO also robustness towards doping and impurities without significant reduction of the magnetocaloric effect.

3. The MagCool prototype

A fast increasing number of prototype magnetic refrigeration devices is presented each year, so far at least 42 systems have been published. A selection of the so far most promising systems include those by Zimm et al. [28], Tura and Rowe [29], Okamura et al. [30] and Tusek et al. [31]. All of those are based on a rotating system to achieve continuous movement and maximize the use of the magnetic field in the AMR. In the design of the MagCool prototype we have applied a two-step approach: firstly, a simple test machine with a linear parallel plate AMR that could be used for experiments under very flexible conditions and where the measurements could be modelled numerically in a feasible way was constructed [24]. Secondly, an optimized rotary magnetic refrigeration device was constructed [27] based on the results acquired by the test machine. For both systems the numerical modelling included optimization of the permanent magnet configuration [32] and of the thermodynamical and fluid dynamical parameters for the AMR [25].

The basic idea of the prototype device is a rotating cylindrical AMR enclosed by two concentric multipole permanent magnets with alternating regions of high and low field. The heat transfer fluid flow is parallel to the axis of the cylinder. For minimal pressure loss in the AMR a geometry of stacked plates was chosen. This geometry also minimizes the magnetostatic demagnetization field from the plates [33]. The prototype consists of four symmetrically alternating high and low field regions. By rotation of the AMR cylinder equal amounts of magnetocaloric material are entering and leaving the high field regions. Therefore the magnetic forces are balanced and the only work that needs to be supplied to rotate the AMR is caused by the small differences in magnetization due to the temperature profile throughout the AMR, and of course, the mechanical work related to friction in the external parts. While rotating the AMR, the heat transfer fluid flows along the plates in alternating directions in phase with the magnetization and demagnetization of the plates to generate the temperature gradient between the hot and the cold end. For construction reasons the AMR is divided into 24 compartments of stacks of parallel plates. Fig. 6 shows the inner and the outer design of the prototype.

Figure 6. The inner (left) and outer (right) design of the rotary prototype. The inner part consists of 24 compartments of stacks of magnetocaloric plates and the heat transfer fluid flows parallel through the compartments in phase with the magnetization and demagnetization of the plates. The outer parts include a complicated manifold and a rotating seal to ensure correct flow with minimal losses.

The dimensions of the prototype where defined based on the numerical modelling with the target temperature difference between the hot and the cold end of 40 K and a continuous cooling power of 100 Watt using the benchmarking material Gd as magnetocaloric material. The prototype is currently in stage of final assembly. Future testing, also with the $La_{0.67}Ca_{0.33-x}Sr_xMnO_3$ and $LaFe_{13-x-y}Co_xSi_y$ series, will show how well these numbers can be reached in practice.

3.1. Permanent magnet configuration

For the test machine running with a linear parallel plate AMR, a permanent magnet with a Halbach configuration was constructed [18], but for the rotary prototype (see Fig. 6) a new permanent magnet had to be designed [34]. The performance of the magnet with regards to magnetic refrigeration can be evaluated using the Λ_{cool} parameter [35], which is defined as

$$\Lambda_{\text{cool}} = \left(\langle B_{\text{high}}^{0.7}\rangle - \langle B_{\text{low}}^{0.7}\rangle\right) \frac{V_{\text{field}}}{V_{\text{mag}}} P_{\text{field}}, \tag{6}$$

where B_{high} and B_{low} are the magnetic flux densities in the high and low field regions, respectively, and the $\langle\rangle$ denotes the average of the flux density. The flux density is evaluated to the power of 0.7 because this is the typical magnetic field dependence of the adiabatic temperature change for a second order magnetocaloric material at the Curie temperature [4]. The volume of magnet material V_{mag} should be minimized as this is the most costly of all components in the prototype, and the corresponding volume of magnetic field V_{field} should be maximized.

The prototype has a continuously rotating AMR with four alternating high and low magnetic field regions, and the magnetic field configuration was optimized to achieve the largest Λ_{cool} using 3D finite element magnetostatic modelling [32]. The magnet has been designed to accomodate a cylindrical AMR with stacks of magnetocaloric plates filling the gap of 250 mm length and of 30 mm width. The result of the modelling and the final constructed permanent magnet is shown in Fig. 7.

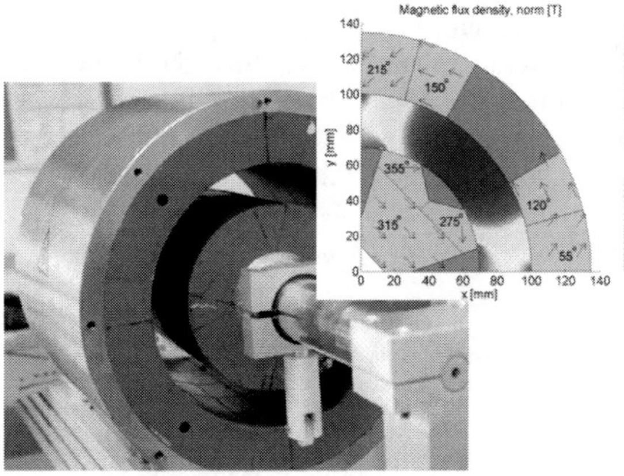

Figure 7. Modelling of the optimized magnetic field configuration for the continuously rotating prototype with 4 alternating regions of high (1.24 T) and low field (< 0.01 T). The two concentric cylindrical magnets measure 250 mm in length, the inner magnet has an inner diameter of 20 mm and an outer diameter of 140 mm, while the outer magnet has an inner diameter of 200 mm and an outer diameter of 270 mm.

Evaluating the Λ_{cool} parameter for the presented permanent magnet design shows a value of 0.21, which to the best knowledge of the authors matches the highest Λ_{cool} published so far [34].

4. Conclusion

In conclusion, we have fulfilled three of our four goals set up in the introduction: Industrial relevant material series with tunable Curie temperature and even robustness of the entropy change against doping has been demonstrated by the lanthanum manganites. Modelling of both a concentric permanent magnet with flux closure and a continuously rotating parallel plate AMR has resulted in a new improved design that is now being implemented as a prototype magnetic refrigeration machine. Only our last goal to demonstrate a running cooling machine is not yet achieved, though it will be in the near future.

Acknowledgments
The authors wish to acknowledge the work of the Risø DTU colleagues who during the MagCool project have contributed to the results presented here. In particular the invaluable work by J. Geyti and F. Saxild on the prototype, P. H. Nielsen and B. R. Sureddy on the materials synthesis, and K. B. Andersen for the tape casting of the magnetocaloric plates. Further, financial support from the Programme Commission on Energy and Environment EnMi Contract No. 2104-06-0032, which is part of the Danish Council for Strategic Research, is gratefully acknowledged.

References

[1] Warburg E 1881 *Ann. Phys.* **13** 141
[2] Debye P 1926 *Ann. Phys.* **81** 1154
[3] Giauque W F 1927 *J. Am. Chem. Soc.* **49** 1864
[4] Tishin A M, Spichkin Y I 2003 *The magnetocaloric effect and its applications* (Institute of Physics Series in Condensed Matter Phyics) Ed Coey J M D, Tilley D R, Vij D R (Bristol and Philidelphia: Institute of Physics) chapter 2 pp 4-68
[5] de Oliviera N A and von Ranke P J 2010 *Phys. Rep.* **489** 89
[6] Pecharsky V K, Geschneidner Jr K A 2006 *Int. J. Refrig.* **29** 1239
[7] Brown GV 1976 *J. Appl. Phys.* **47** 3673
[8] Geschneidner Jr K A, Pecharsky V K 2008 *Int. J. Refrig.* **31** 945
[9] Pecharsky V K, Geschneidner Jr K A 1997 *Phys. Rev. Lett.* **78** 4487
[10] Annaorazov M P, Nikitin S A, Tyurin A L, Asatryan K A, Dovletov A K 1996 *J. Appl. Phys.* **79** 1689
[11] Pecharsky V K, Geschneidner Jr K A 1997 *Appl. Phys. Lett.* **70** 3299
[12] Wada H, Tanabe Y 2001 *Appl. Phys. Lett.* **79** 3302
[13] Hansen B R, Theil Kuhn L, Bahl C R H, Lundberg M and Ancona-Torres C 2010 *J. Magn. Magn. Mat.* **322** 3447
[14] Jin S, Tiefel T H, McCormack M, Fastnacht R A, Ramesh R and Chen L H 1994 *Science* **264** 413
[15] Dagotto E, Hotta T, Moreo A 2001 *Phys. Rep.* **344** 1
[16] Dinesen A R, Linderoth S and Mørup S 2005 *J. Phys.: Cond. Mat.* **17** 6257
[17] Venkatesh R, Theil Kuhn L, Pryds N (to be submitted)
[18] Bjøerk R, Bahl C R H, Smith A, Pryds N 2008 *J. Appl. Phys.* **104** 013910
[19] Jeppesen S, Linderoth S, Pryds N, Theil Kuhn L, Buch Jensen J 2008 *Rev. Sci. Instr.* **79** 083901
[20] Bjørk R, Bahl C R H, Katter M 2010 *J. Magn. Magn. Mat.* **322** 3882
[21] Guo Z B, Yang W, Shen Y T and Du Y W 1998 *Solid State Commun.* **105** 89
[22] Venkatesh R, Theil Kuhn L, Pryds N and Bohr J *submitted*
[23] Ancona-Torres C, Pryds N, Kuhn L T, Bahl C R H and Linderoth S 2010 *J. Appl. Phys. (accepted)*
[24] Nielsen K K, Bahl C R H, Smith A, Bjørk R, Pryds N and Hattel J 2009 *Int. J. Refrig.* **32** 1478
[25] Nielsen K K, Bahl C R H, Smith A, Pryds N and Hattel J 2010 *Int. J. Refrig.* **33** 753
[26] Nielsen K K, Bahl C R H, Engelbrecht K, Smith A, Pryds N and Hattel J 2010 *Proc. of the 4th Int. Conf. Magnetic Refrigeration at Room Temperature (Baotou)* Ed Egolf P W (Paris: Int. Inst. Refrig.)
[27] Bahl C R H, Engelbrecht K, Bjørk R, Eriksen D, Smith A and Pryds N 2010 *Proc. of the 4th Int. Conf. Magnetic Refrigeration at Room Temperature (Baotou)* Ed Egolf P W (Paris: Int. Inst. Refrig.)
[28] Zimm C, Auringer J, Boeder A, Chell J, Russek S and Sternberg A 2007 *Proc. of the 2nd Int. Conf. Magnetic Refrigeration at Room Temperature (Portoroz)* Ed Egolf P W (Paris: Int. Inst. Refrig.) 341
[29] Tura A, Rowe A 2007 *Proc. of the 2nd Int. Conf. Magnetic Refrigeration at Room Temperature (Portoroz)* Ed Egolf P W (Paris: Int. Inst. Refrig.) 363
[30] Okamura T, Yamada K, Hirano N and Nagaya S 2006 *Int. J. Refrig.* **29** 1327
[31] Tusek J, Zupan S, Sarlah A, Prebil I and Poredos A 2009 *Proc. of the 3rd Int. Conf. Magnetic Refrigeration at Room Temperature (Des Moines)* Ed Egolf P W (Paris: Int. Inst. Refrig.) 409
[32] Bjørk R, Bahl C R H, Smith A, Christensen D V and Pryds N 2010 *J. Magn. Magn. Mat.* **322** 3324
[33] Smith A, Nielsen K K, Christensen D V, Bahl C r H, Bjørk R and Hattel J 2010 *J. Appl. Phys.* **107** 103910
[34] Bjørk R, Bahl C R H, Smith A and Pryds N 2010 *Int. J. Refrig.* **33** 437
[35] Bjørk R, Bahl C R H, Smith A and Pryds N 2008 *J. Appl. Phys.* **104** 13910

Domain model for the magnetic shape-memory effect in non-modulated tetragonal Ni-Mn-Ga

A.T. Onisan, A.N. Bogdanov, U. K. Rößler
IFW Dresden, Postfach 270116, D-01171 Dresden Germany

E-mail: u.roessler@ifw-dresden.de

Abstract. We employ a general phenomenological domain theory for ferromagnetic shape memory (FSM) materials to model equilibrium domain states in martensites with tetragonal twin variants having axes ratios $c/a > 1$ and easy-plane magnetic anisotropy. The model is evaluated for realistic material parameters of non-modulated (NM) Ni_2MnGa alloys. Phase diagrams under competing external stresses and magnetic fields are presented that show regions of fully magnetized single-variant states and various multiphase regions. The theoretical elastic stresses that block single-variant states are found to be much smaller in NM than in the 5M structure of Ni_2MnGa alloys.

1. Introduction

The magnetic shape-memory effect relies on the rearrangement of crystallographic twin variants in ferromagnetic martensitic microstructures [1]. It has been argued that modulated structures, as the 5M or 7M structure in the archetypical Ni-Mn-Ga FSM alloys, can be construed as adaptive martensite [2] and that these nano-twinned microstructures are crucial for the FSM effect to reach large field-driven strains up to 10%. However, experiments on single crystals of Ni-Mn-Ga Heusler alloys in the non-modulated (NM) state with tetragonal axis ratio $c/a = 1.18$ find sizeable field-driven strains up to 0.1%[3]. As these alloys own uniaxial magnetic anisotropy of the easy-plane type, their magnetic behavior and the FSM effects are different, as compared to the pseudo-tetragonal easy-axis behavior in the modulated Ni-Mn-Ga structures.

Based on a phenomenological domain model, we here report theoretical phase-diagrams depending on applied magnetic fields and stresses. These diagrams can be used to distinguish the FSM behavior in NM Ni-Mn-Ga from the behavior of the usual modulated easy-axis alloys. The model was presented previously [4] and was worked out for easy-axis systems and for the easy-plane case with relevant tetragonal anisotropy, yielding four-fold magnetic domains in each variants. It has been applied to calculate twin-variant and domain-state phase diagrams for modulated Ni-Mn-Ga with pseudo-tetragonal structure and axis ratio $c/a < 1$ and for Fe-Pd [5].

2. Domain model

The total energy E_t of the twinned ferromagnetic structure is composed of an elastic part E_{el} and a magnetic energy E_m:

$$E_t = E_{el}(\hat{u}) + E_m(\mathbf{M}) \tag{1}$$

depending on strain tensor $\hat{u}$ and magnetization $\mathbf{M}$. The domain state model for thermodynamic equilibrium states in such a ferromagnetic twinned microstructure assumes that the material can

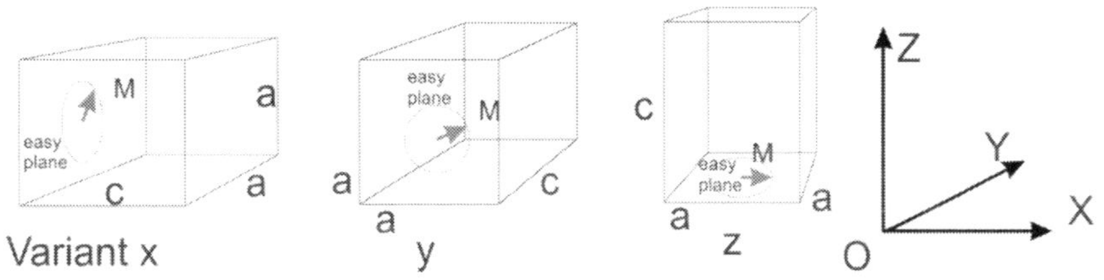

Figure 1. Variants for non-modulated Ni_2MnGa alloys: three tetragonal variants $c/a > 1$ with easy-plane uniaxial anisotropy.

be decomposed into homogeneous twin variants with uniform strain and matching boundary conditions at twin boundaries. These *crystallographic* domains may be further subdivided into different magnetic domains. For details, see Ref. [5]. The magnetic properties are described within a simplified micromagnetic approach appropriate for bulk systems. Thus, the magnetic energy contribution from one twin variant labelled l is given by

$$E_m^l = -\mu_o M_s \, \mathbf{m} \cdot \mathbf{H}^{(ext)} - \frac{1}{2}\mu_0 M_s \, \mathbf{m} \cdot \mathbf{H}^{(demag)} + w_a^l(\mathbf{m}) \tag{2}$$

with the external magnetic field $\mathbf{H}^{(ext)}$, the demagnetization field $\mathbf{H}^{(demag)}$, the intrinsic magnetic anisotropy energy w_a^l. Here, μ_0 is the magnetic vacuum permeability, M_s is the saturation magnetization, and $\mathbf{m} = \mathbf{M}/M_s$. For martensites with tetragonal crystal structure the three twin variants can be labelled $l = x, y, z$ by the orientation of the unique axis. The magnetic anisotropy of variant z is

$$\begin{aligned} w_a^z(\mathbf{m}) \;=\; & B_1 \, (\mathbf{m} \cdot \hat{\mathbf{z}})^2 + B_2 \, (\mathbf{m} \cdot \hat{\mathbf{z}})^4 \\ & + B_3 \, (\mathbf{m} \cdot \hat{\mathbf{x}})^2 \, (\mathbf{m} \cdot \hat{\mathbf{y}})^2 \, . \end{aligned} \tag{3}$$

In Ref. [5], magnetic phase diagrams have been analysed for modulated Ni-Mn-Ga with easy-axis anisotropy, $B_1 < 0$ and $B_2, B_3 \equiv 0$, and for easy-plane anisotropy with $B_1 > 0$, $B_1 > -B_2$ and strong 4-fold anisotropy $B_3 \neq 0$. Both systems have been modelled as pseudo-tetragonal crystals with axis ratio $c/a < 1$. In non-modulated (NM) Ni-Mn-Ga with the tetragonal structure $c/a > 1$ and easy plane anisotropy, the four-fold anisotropy is expected to be weak, and we set $B_3 \equiv 0$. As saturation of twins is achieved already in very small magnetic fields, we assume full saturation of the twin variants. This means that the model neglects magnetic domain structures within the single twins. Owing to the easy-plane character of its uniaxial anisotropy, the domain structure in NM is different from domain structures in easy-axes systems: the easy magnetization directions of two differently oriented twins intersect in a certain joint easy axis [6]. This is the reason why the magnetic anisotropy of modulated *adaptive* martensites, that are composed by NM variants with easy-plane anisotropy, acquires an effective easy-axis character [2]. Magnetic domains may encompass two different twins, if the magnetization oriented along this common easy direction, and twin boundaries are not necessarily associated with a magnetic domain wall. This type of domain structure has been observed in experiments on NM Ni_2MnGa[7].

The microstructure of a bulk sample is described by the volume fractions ξ_l of the three variants, $l = x, y, z$ and $\sum_l \xi_l = 1$, see Fig. 1. To solve the magnetostatic problem, the phase-theory approximation [8],[9] is used, as discussed in Ref. [5].

For the evaluation of phase diagrams, we have employed different sets of elastic constants and magnetic materials constants. Only one complete set of elastic constants from experiments has been reported for a Ni-Mn-Ga alloy with the pseudo-tetragonal 5M structure [10]. For stoichiometric NM Ni_2MnGa, theoretical values of the constants from density functional theory calculations (DFT) have been presented by Özdemir Kart et al. [11]. These constants are listed in Table 1 in our notation of [5]. Magnetic anisotropies for NM Ni-Mn-Ga have been reported earlier [12], but very recently a new measurement has been performed [13]. Corresponding magnetic constants and the deviatioric strain from Refs. [13] and [12] along with the demagnetization tensor components N are listed in Table 2. Our calculations have first been done using the earlier experimental data for elastic constants Table 1 (A) and magnetic anisotropy model Table 2 (I), as representative for the Ni-Mn-Ga system generally. This is model AI in an obvious notation. In order to quantify the dependence of the phase diagrams on the different materials constants, we have compared them also with the corresponding model BII, which may provide a better description of NM Ni-Mn-Ga. While there are quantitative differences, the qualitative behavior is identical for the two models.

Table 1. Elastic constants (in GPa) in Voigt notation and in our notation from Ref. [5]. (A) experimental data for 5M pseudo-tetragonal structure from Ref. [10] and (B) theoretical calculations for non-modulated Ni_2MnGa from Ref. [11].

	Material & method	c_{11} c_1	c_{22}, c_{33} c_2	c_{13} c_4	c_{12} c_3	c_{55}, c_{44} c_6	c_{66} c_5
(A)	$Ni_{0.50}Mn_{0.284}Ga_{0.216}$ 5M experiment	179	163	161.4	39.54	47.51	45.0
(B)	Ni_2MnGa NM by DFT	249	193	141	71	101	56

Table 2. Magnetic constants, deviatoric strain e_0 and demagnetization tensor used in the calculations. Experimental magnetic anisotropy constants for NM Ni-Mn-Ga are (I) from Ref. [12] and (II) new experiments, Ref. [13].

(I)		$B_1 =$	-0.24+0.06 MJ/m^3	$B_2 =$	0.06 MJ/m^3
(II)		$B_1 =$	-0.26 MJ/m^3		
$M_s =$	0.422 MJ/(Tm^3)	$e_0 =$	-0.21		
$N_x =$	0.2	$N_y =$	0.2	$N_z =$	0.4

3. Results

3.1. Phase diagrams

Phase diagrams have been analysed for bulk single crystals with ellipsoidal shape, having principal axes along the crystallographic directions of the 4 tetragonal variants, see Table 2. If such single crystals show good FSM behavior, the microstructure is expected to consist of macroscopic twins separated by few mobile twin boundaries. Moreover, usually only one twinning system is active, so that only two twin variants exist in a simple lamella structure. Nucleation of the third variant in such a structure creates large elastic stresses, as it is impossible to fulfill simultaneously compatibility conditions at the interfaces of all variants. Thus, nucleation of the third variant is effectively blocked. For the phase diagrams in Figs. 2, such a simple twinned structure is assumed to consist of y and z variants, while no x variant is permitted. The applied magnetic fields along z axis, left panel Fig. 2, or in the $x0z$-plane, right panel always favour the y variant. Thus, under tensile stresses $\sigma_{yy} > 0$ in y direction a pure y single-variant state exists. Compressive stresses $\sigma_{yy} < 0$ favour the z variant, but the magnetic field with finite components in z direction generates thermodynamically stable domain states.

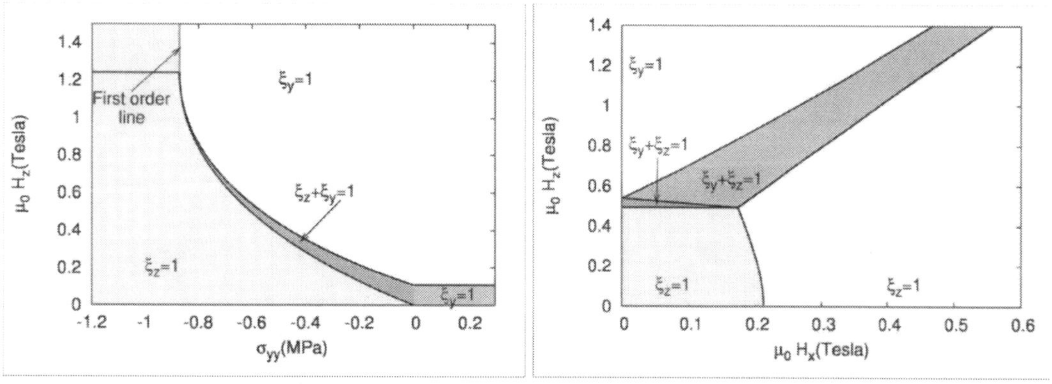

Figure 2. Phase diagram for non-modulated Ni_2MnGa alloys. Only the twinning system composed of x and y variants is active, and the nucleation of the x-variant is blocked. Left panel: magnetic field in z direction and stresses in y direction σ_{yy}. Right panel: Magnetic fields in xz-plane under strong compressive stresses $\sigma_{yy} = -0.6$ MPa< 0.

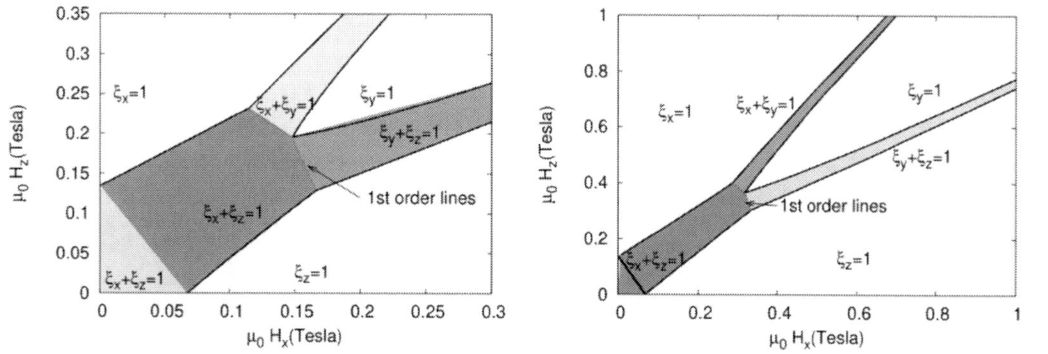

Figure 3. Phase diagram for non-modulated Ni_2MnGa alloys under applied fields in xz-plane under weaker compressive stresses allowing nucleation of the y-variant in oblique fields. Left panel: Model AI with $\sigma_{yy} = -0.1$ MPa< 0. Right panel: Model BII with $\sigma_{yy} = -0.25$ MPa< 0.

A rotating field in the $x0z$-plane under compressive stress $\sigma_{yy} < 0$ reveals a FSM behavior with a redistribution between the available y and z variants and a thermodynamically stable twinned structure for fields applied close to (101) directions. For large compressive stresses, the z single-variant state is blocked. Releasing the constraint of a y-z twinning system, the fields applied in $x0z$ plane will alternatingly favour z and x variants. In the simplest phase diagram under large compressive stresses $\sigma_{yy} < 0$, the occurrence of y variants is completely suppressed. However, for smaller stresses, the y variant can be nucleated for fields close to the (101) direction, Fig. 3. In a small rotating field, a succession of pure z and x single-variant states, separated by intermediate stable domain states, is found under these conditions. But for larger fields, a succession of pure z, y and x single-variant states takes place with intervening mixed states. In increasing fields under fixed angle close to (101) directions, transformations between intermediate states from an $x - z$ twinned structure towards $z - y$ or $y - x$ twinned structures are indicated as "1st-order lines" in Fig. 3. This domain process is hindered by the provision that the third y variant cannot be nucleated in the $z - x$ twinning structure. However, the 1st order process could take place via another intermediate domain state, where $x - z$ twinned plates co-exist either with $y - x$ or

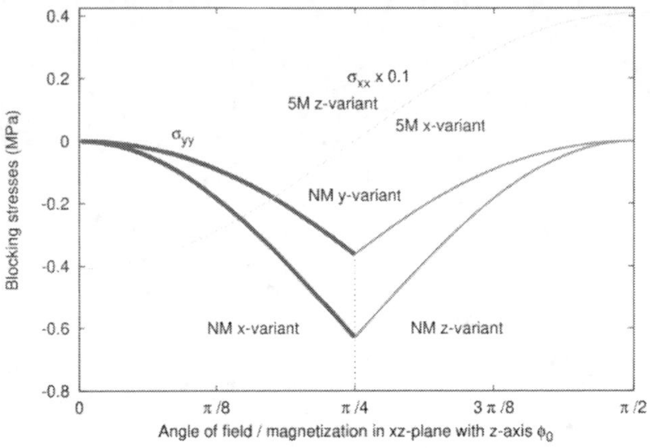

Figure 4. Blocking stress in dependence on angle of internal field ϕ_0. Thicker (blue and red) lines for non-modulated tetragonal Ni-Mn-Ga : σ_{yy} calculated from Eqs. (4,5) is the equilibrium blocking stress for the y-variant in magnetic saturation. The upper lines are for the model with materials parameters AI, the lower with parameters BII. Below the blue lines, only the x variant exists, below the red lines, only the z variant. At the dotted vertical line these two variants coexist. Above the lines, the y variant can be stabilized by a sufficiently strong applied magnetic field in the $x0z$-plane. Thinner (green) line gives the blocking stress for the pseudo-tetragonal 5M martensite with $c/a < 1$, according to Ref. [5]: large compressive stress $\sigma_x x < 0$ stabilizes the z variant, large tensile stresses $\sigma_x x > 0$ stabilize the z-variant. Note the scaling factor $\times 0.1$ for the blocking stress of 5M Ni-Mn-Ga.

$z - y$ twinned plates in a hierarchical structure.

3.2. Blocking stress

In Ref. [5], we have defined the *blocking stress* as the applied stress that enforces, in equilibrium, a single twin variant in its fully saturated state. It can be calculated by comparing the total energies of the different pure states. For the easy axis twins with $c/a > 1$, a sufficiently strong compressive stress $\sigma_{yy} < 0$ stabilizes the x or z and suppresses the y-variant. Conversely, an internal field applied in the $x0z$-plane favours the y variant. If the internal field subtends an angle $\phi_0 \in (0, \pi/4)$ with the $\hat{\mathbf{z}}$ axis, then the y variant will have lowest magnetic energy, followed by the x variant, and the least favoered variant is z. For $\phi_0 = \pi/4$ the x and z variants are equally favourable magnetically and can co-exist. At this point we will have the largest anisotropic magnetic energy that can be used to overcome applied stress. Excluding common terms the total energy can be written in the pure y case as:

$$E_t|_{\xi_y=1} = -\frac{c_2 e_0^2}{2} + \sigma_{yy}\left(e_0 - \frac{(c_1+c_3)\sigma_{yy}}{2c_2(c_1+c_3) - 4c_4^2}\right) - \mu_0 M_s h_0. \tag{4}$$

and for the x and z variants:

$$\begin{aligned} E_t|_{\xi_x+\xi_z=1} &= \frac{1}{2}\left(\frac{(-c_1c_2 + c_4^2)\sigma_{yy}^2}{(c_1-c_3)(c_2(c_1+c_3) - 2c_4^2)} - c_2 e_0^2\right) \\ &+ \sum_{l=x,z} \xi_l \left[-\mu_0 M_s h_0(\cos\phi_l \cos\phi_0 + \sin\phi_l \sin\phi_0) + B_1\,(\mathbf{m}_l\cdot\hat{\mathbf{l}})^2 + B_2\,(\mathbf{m}_l\cdot\hat{\mathbf{l}})^4\right]. \end{aligned} \tag{5}$$

where the magnetization vectors of the l-variants are written $\mathbf{m}_l = (\sin\phi_l, 0, \cos\phi_l)$ and h_0 is the magnitude of the internal field. For $\phi_0 \in [0, \pi/4)$ only the pure x variant exists $\xi_x = 1$, and for $\phi_0 \in (\pi/4, \pi/2]$ only the pure z variant. In both cases, the fully saturated states have $\phi_l = \phi_0$. Only at $\phi_0 \equiv \pi/4$ the x and z variant co-exist having the same magnetic energy. From symmetry, one sees that the elastic part of equation (5) also does not depend on the volume fraction of the x and z variants. At the equilibrium blocking stress, the energies (4) and (5) must become equal. The largest stress $|\sigma_{yy}|$ is reached for $\phi = \pi/4$. The dependence of the blocking stress for two different choices of materials constants on ϕ_0 is plotted in Fig. 4 and compared with blocking stress for 5M easy-axis Ni-Mn-Ga. In the isotropic case the expression of the equilibrium σ_{yy} simplifies to

$$\sigma_{yy} = \frac{B_1 \cos^2(\phi)}{e_0} \tag{6}$$

with a maximum value $B_1/(2e_0)$.

4. Conclusions

The theoretical model demonstrates important differences in the field- and stress-driven microstructure processes of non-modulated Ni-Mn-Ga. Most importantly, there are certain experimental settings combining stresses and fields as in Fig. 3, would allow to discern whether a FSM derives from an easy-axis or an easy-plane magnetic system. Moreover, the single-variant states in NM Ni-Mn-Ga are blocked by much smaller stresses than in the modulated 5M Ni-Mn-Ga. Therefore, applied stresses will tend to block variants rather than to drive twin boundary motion, unless twinning stresses are comparably small.

Acknowledgments

We acknowledge stimulating discussions with S. Fähler, O. Heczko, and S. Kaufmann. Supported by DFG SPP1239 (http:www.magneticshape.de) through project RO2238/8.

References

[1] Ullakko K, Huang J K, Kantner C, O' Handley R C and Kokorin V V 1996 *Appl. Phys. Lett.* **69** 1966
[2] Kaufmann S, Rößler U K, Heczko O, Wuttig M, Buschbeck J, Schultz L and Fähler S 2010 *Phys. Rev. Lett.* **104** 145702
[3] Chernenko V A, Chmielus M and Müllner P 2009 *Appl. Phys. Lett.* **95** 104103
[4] Bogdanov A N, Desimone A, Müller S and Rößler U K 2003 *J. Magn. Magn. Mater.* **261** 204–209
[5] Onisan A T, Bogdanov A N and Rößler U K 2010 *Acta Mater.* **58** 4378
[6] Lvov V A, Gomonaj E V and Chernenko V A 1998 *J. Phys.: Condens. Matter* **10** 4587
[7] Chernenko V A, Lvov V A, Besseghini S and Murakami Y 2006 *Scripta Mater.* **55** 307
[8] Néel L 1944 *J. de Physique Radium* **5** 241
[9] Hubert A and Schäfer R 1998 *Magnetic domains* (Berlin: Springer)
[10] Dai L, Cullen J and Wuttig M 2004 *J. Appl. Phys.* **95** 6957
[11] Özdemir Kart S, Uludoğn M, Karaman I and Çağin T 2008 *phys. stat. sol. (a)* **205** 1026
[12] Straka L and Heczko O 2003 *J. Appl. Phys.* **93** 8636
[13] Heczko O, Straka L, Novak V and Fähler S 2010 *J. Appl. Phys.* **107** 09A914

Monte Carlo modeling of exchange bias effect in $Ni_{50}Mn_{25+x}Sb_{25-x}$ Heusler alloys

V Buchelnikov[1], V Sokolovskiy[1], I Taranenko[1], S Taskaev[1], P Entel[2]
[1]Chelyabinsk State University, Chelyabinsk, 454001, Russia
[2]University of Duisburg-Essen, Duisburg, 47048, Germany

E-mail: buche@csu.ru

Abstract. A theoretical model for investigation of an exchange bias effect in Heusler $Ni_{50}Mn_{25+x}Sb_{25-x}$ alloys by Monte Carlo simulations is presented. It is shown that the exchange bias effect of Heusler $Ni_{50}Mn_{25+x}Sb_{25-x}$ alloys is observed in the composition range $7.5 < x < 15$ at the temperature $T = 13$ K. For case of $Ni_{50}Mn_{37.5}Sb_{12.5}$ alloy, we have obtained the exchange bias field $H_{EB} = 0.05$ T at 4.6 K and a blocking temperature $T_B = 46.7$ K. Theoretical results are in good qualitatively agreement with experimental data.

1. Introduction

The Heusler Ni-Mn-X (X = In, Sn, Sb) alloys have the unique properties, such as the shape memory effect, the giant inverse and direct magnetocaloric effect, the large magnetoresistence, the exchange bias effect and others interesting magnetic properties which are associated with the martensitic transition [1-7]. These properties are applicable in developing actuator materials, spintronic and magnetic cooling technology.

When a ferromagnet (FM) is in contact with an antiferromagnet (AFM) a shift of a hysteresis loop along a magnetic-field axis can occur which is called the exchange bias effect (EB). Usually, this shift of the hysteresis loop is observed after cooling an entire system in an external magnetic field below the Neel temperature T_N of the AFM. Though this effect has been well known for many years [8, 9] its microscopic origin is still discussed. Resent experiments have shown that in Heusler Ni-Mn-X alloys the EB is experimentally observed [5, 6]. An origin of the EB is next. When the temperature changes, a structural phase transition from a paramagnetic (PM) cubic (austenitic) phase to the FM tetragonal (martensitic) phase can occurs in Heusler Ni-Mn-X alloys [1-7]. It is well known, that in stoichiometric $Ni_{50}Mn_{25}X_{25}$ alloys all Mn_1 atoms on its regular sublattice sites interact FM (figure 1). However, resent experiments of non-stoichiometric $Ni_{50}Mn_{25+x}X_{25-x}$ alloys have revealed that in the martensite, the excess of Mn_2 atoms occupying the sites of X atoms, interact with Mn_1 atoms AFM (figure 2) [2, 3, 7]. Thus, in the martensitic state of $Ni_{50}Mn_{25+x}X_{25-x}$ alloys the FM and AFM interactions are coexisted. Opposite, there is the FM interaction between all Mn atoms only in the austenitic state. It should be note, the presence of AFM interactions in the martensitic phase are responsible for the EB.

2. The theoretical model

In the proposed model we use a simple three-dimensional cubic lattice with periodic boundary conditions and take into account the magnetic interactions only in two coordination spheres. Since the magnetic moment of the Ni atoms ($\approx 0.3\ \mu_B$) are small compared to the Mn magnetic moment (≈ 4.0

μ_B) and since the Sb atoms have not spin moment, we omit contributions to the magnetic interaction from the Ni and Sb atoms altogether. The Sb atoms are treated as nonmagnetic inclusions at lattice sites. Since the Ni concentration is constant in $Ni_{50}Mn_{25+x}Sb_{25-x}$ alloys we do not take into account the Ni lattice sites. For the case of stoihiometric $Ni_{50}Mn_{25}Sb_{25}$ alloy, the concentrations of Mn_1 and Sb atoms on the model lattice are taken as 50%. For the case of non-stoihiometric $Ni_{50}Mn_{25+x}Sb_{25-x}$ alloy the Mn_2 excess are located randomly at the Sb sites on the model lattice and the concentration of Mn_2 atoms is determined from the experimental compositions. So, in the proposed model, we consider FM and AFM Heisenberg's spin interactions with anisotropy term using *ab initio* exchange constants.If you don't wish to use the Word template provided, please set the margins of your Word document as follows.

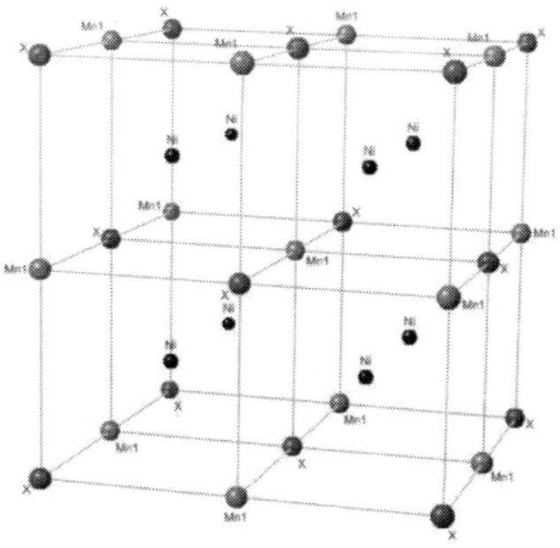

Figure 1. Stoichiometric $Ni_{50}Mn_{25}X_{25}$ alloys.

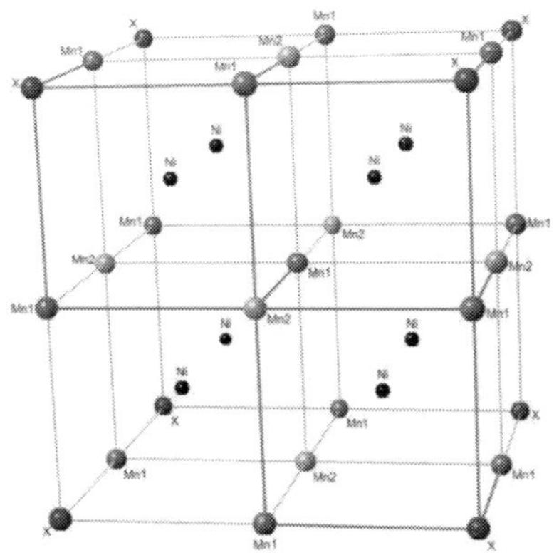

Figure 2. Non-stoichiometric $Ni_{50}Mn_{25+x}X_{25-x}$ alloys.

The generalized Hamiltonian (1) includes the FM part (2) and the AFM one (3) and the interaction between FM and AFM parts (4).

$$H = H_{\text{FM}} + H_{\text{AFM}} + H_{\text{FM-AFM}}, \tag{1}$$

$$H_{\text{FM}} = -J_{\text{FM}_1} \sum_{<i,j>\in \text{FM}_1} \mathbf{S}_i\mathbf{S}_j - J_{\text{FM}_2} \sum_{<i,j>\in \text{FM}_2} \mathbf{S}_i\mathbf{S}_j - \sum_{i\in \text{FM}} \left(K_{\text{FM}} S_{z_i}^2 + \mu_{\text{B}} g H S_{z_i} \right), \tag{2}$$

$$H_{\text{AFM}} = -J_{\text{AFM}_1} \sum_{<i,j>\in \text{AFM}_1} \mathbf{S}_i\mathbf{S}_j - J_{\text{AFM}_2} \sum_{<i,j>\in \text{AFM}_2} \mathbf{S}_i\mathbf{S}_j - \sum_{i\in \text{AFM}} \left(K_{\text{AFM}} S_{z_i}^2 + \mu_{\text{B}} g H S_{z_i} \right), \tag{3}$$

$$H_{\text{FM-AFM}} = -J_{int_1} \sum_{<i\in \text{AFM(FM)}, j\in \text{FM(AFM)}>} \mathbf{S}_i\mathbf{S}_j - J_{int_2} \sum_{<i\in \text{AFM(FM)}, j\in \text{FM(AFM)}>} \mathbf{S}_i\mathbf{S}_j. \tag{4}$$

Here, J_{FM1}, J_{FM2}, J_{AFM1} and J_{AFM2} are the FM and AFM exchange constants in the first and second coordination spheres; $\mathbf{S}_i$ is the three-dimensional vector of unit length; K_{FM} and K_{AFM} are the magnetic anisotropy terms for Mn_1 and Mn_2 atoms, respectively; H is the external magnetic field; g is the Lande factor; μ_B is the Bohr's magneton; J_{int1} and J_{int2} are the magnetic exchange constants between Mn_1 and Mn_2 atoms in the first and second coordination spheres. Sums are performed over all nearest neighbor and next nearest neighbor pairs.

In the proposed model we use renormalized parameters defined by ration over J_{FM1}: $A^* = A/J_{\text{FM1}}$, here $A = J_{\text{FM2}}, J_{\text{AFM1}}, J_{\text{AFM2}}, K_{\text{FM}}, K_{\text{AFM}}, J_{int1}, J_{int2}$. The reduced temperature defines as

$$T^* = \frac{k_B T}{J_{\mathrm{FM}_1}}, \tag{5}$$

where k_B is the Boltzmann constant.

The total normalized magnetization of the system can be presented as

$$m = m_{\mathrm{FM}} + m_{\mathrm{AFM}}, \quad {}^x m = \frac{1}{{}^x N}\sqrt{{}^x S_{i_x}^2 + {}^x S_{i_y}^2 + {}^x S_{i_z}^2}. \tag{6}$$

Here, m_{FM} and m_{AFM} are normalized magnetizations of Mn_1 and Mn_2 atoms respectively; x denotes Mn_1 and Mn_2 atoms.

The exchange bias field defines as

$$H_{\mathrm{EB}} = \frac{(H^+ + H^-)}{2}, \tag{7}$$

where H^+ and H^- are those fields of the hysteresis loop branches for increasing and decreasing field, where normalized magnetization becomes zero.

3. Model parameters

The corresponding simulations have been carried out using the standard Metropolis algorithm [10]. The changes in the spin states $\mathbf{S}_i$ (S_{ix}, S_{iy}, S_{iz}) are treated independently and accepted or rejected according to the single-site transition probability $W = \min\{1, \exp[-\Delta H^*/T^*]\}$. For simulations of the cubic lattice we used six near neighbors and twelve next nearest neighbors for the first and second coordination spheres, respectively. The number of the lattice sites was equal to $N = L^3$, where $L = 15$. The time unit is one Monte Carlo step, which consist of N attempts to change the $\mathbf{S}_i$ variables. For a given temperature number of the Monte Carlo steps on each site is 10^5. We start the Monte Carlo simulations from the ordered state at low temperature. For the case of Mn_1 atoms, the spin variables are $S_{ix} = 0$, $S_{iy} = 0$, $S_{iz} = 1$, and for the case of Mn_2 atoms the spin variables are $S_{ix} = 0$, $S_{iy} = 0$, $S_{iz} = -1$, respectively. The various quantities are averaged over 400 configurations taken every 100 Monte Carlo steps and discarding the first 10^4 Monte Carlo steps for equilibrium. The Lande factor was taken $g = 2$. In our simulations we used the fixed magnetic constants obtained from *ab initio* calculations for $Ni_{50}Mn_{37.5}Sb_{12.5}$ alloy (figure 3 and table 1) but magnetic anisotropy constants have varied.

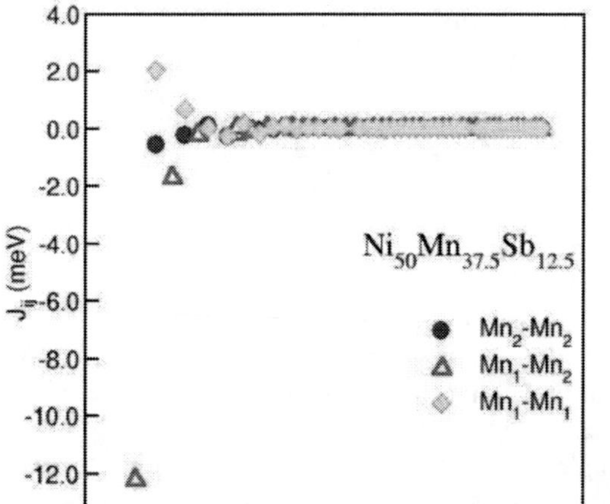

Figure 3. The magnetic exchange constants for $Ni_{50}Mn_{37.5}Sb_{12.5}$ alloy as a function of the distance *d/a* between pair atoms *i* and *j* obtained from *ab initio* calculations.

Table 1. Magnetic exchange constants for $Ni_{50}Mn_{37.5}Sb_{12.5}$ alloy.

Exchange constants (meV)	1st coord. sphere	2nd coord. sphere
J_{FM}	2	1
J_{AFM}	-1	-1
J_{int}	-2	-2

4. Numerical results

In this section we solve numerically our model by using Monte Carlo techniques. For all simulations we have used the exchange constants which list in Table 1 and the magnetic anisotropy terms K_{FM} = 0.05 meV and K_{AFM} = 0.015 meV.

At figure 4 the theoretical magnetic field dependences of the normalized magnetization m for ferromagnetic and antiferromagnetic components are shown. In the figure is clearly observable a shift of the hysteresis loops aside the positive field. Corresponding EB field for $Ni_{50}Mn_{37.5}Sb_{12.5}$ alloy is equal to H_{EB} = 0.05 T. Figure 5 presents the hysteresis loops of non-stoichiometric $Ni_{50}Mn_{25+x}X_{25-x}$ alloys (x = 5, 7.5, 7.75, 8, 9, 10, 12.5) at T = 13 K. As we can see, the hysteresis loops become broader and higher with increasing of concentration x. It is obvious because with bigger concentration we will have more excess Mn atoms and as a result bigger value of magnetization.

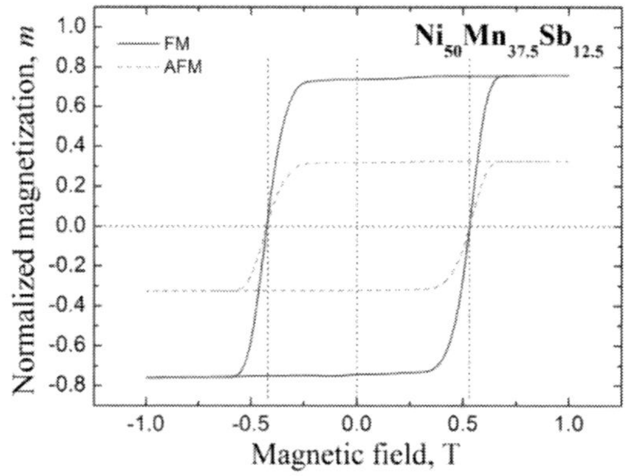

Figure 4. Theoretical ferromagnetic and antiferromagnetic hysteresis loops of $Ni_{50}Mn_{37.5}Sb_{12.5}$ alloy at T = 4.6 K.

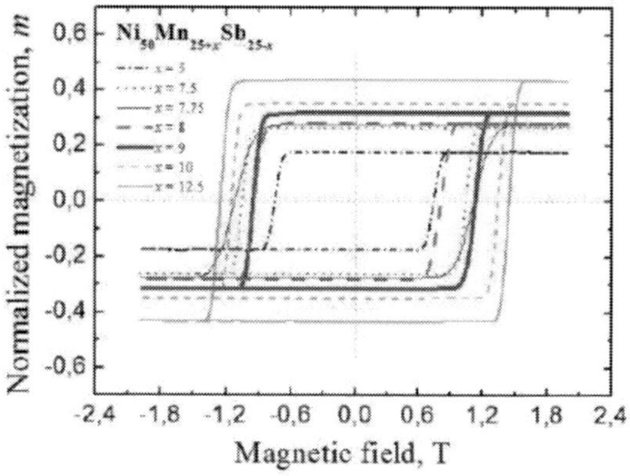

Figure 5. Theoretical hysteresis loops of $Ni_{50}Mn_{25+x}X_{25-x}$ alloys at T = 11.6 K.

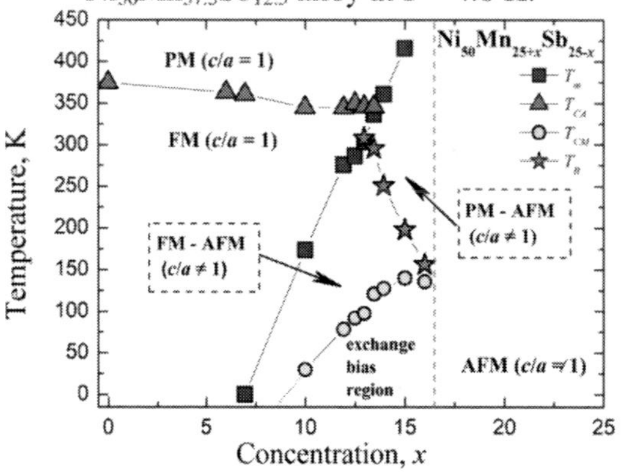

Figure 6. Experimental temperature–concentration (T-x) phase diagram of $Ni_{50}Mn_{25+x}X_{25-x}$ alloys [6].

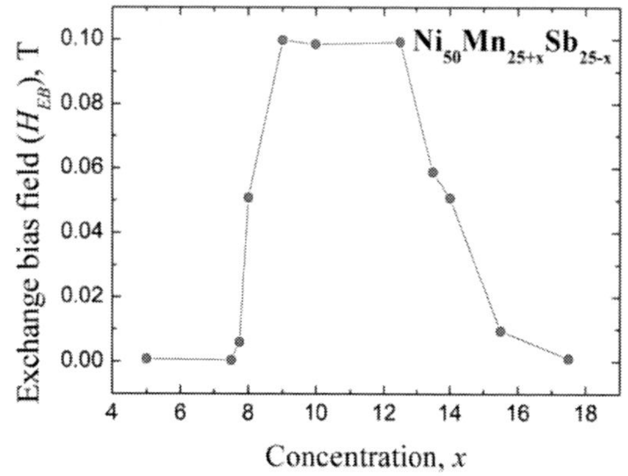

Figure 7. Theoretical concentration dependence of EB of $Ni_{50}Mn_{25+x}X_{25-x}$ alloys at T = 13 K.

As it is well known, the EB in Ni-Mn-Sb alloys exists only in the narrow range of excess Mn atoms ($7.5 < x < 15$). It can be seen from the experimental phase diagram (see figure 6) [6].

At figure 7 we present a theoretical concentration dependence of the EB for the $Ni_{50}Mn_{25+x}X_{25-x}$ alloys at $T = 13$ K. From this figure we can see that the EB exists in the concentration range from $x =$ 7.5 until $x = 15$. Some authors [11] explain this effect by the following way. At high concentration (amount of excess Mn_2 atoms) an exchange interaction between spins of FM and AFM atoms lost, and as a result lead to a decreasing of Cure temperature and corresponding disappearing of EB. On the other hand, because of the non-magnetic atoms involve in domain formation, the increasing of its amount (low concentration of excess Mn_2 atoms) lead to a decreasing of EB effect. The calculated behavior of the EB is in a good agreement with the experimental data.

Figure 8 presents the theoretical dependences of EB field and the coercive field H_C from the temperature for $Ni_{50}Mn_{37.5}Sb_{12.5}$ alloy. In agreement with experiments [5, 12, 13] (see Figure 9) the EB field and the coercive field H_C decrease with temperature almost linearly going to zero at the temperature T_B. T_B is called the blocking temperature. It is important to note that T_B is not equal to the Neel temperature of the AFM, which equal $T_N = 62.4$ K in our simulations. There are several reasons for this effect which have been clearly described in the paper [11].

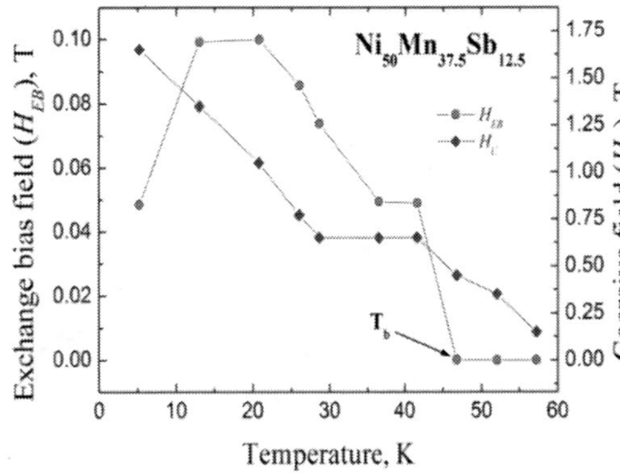

Figure 8. Theoretical temperature dependences of EB field (H_{EB}) and coercive field (H_C) of $Ni_{50}Mn_{37.5}Sb_{12.5}$.

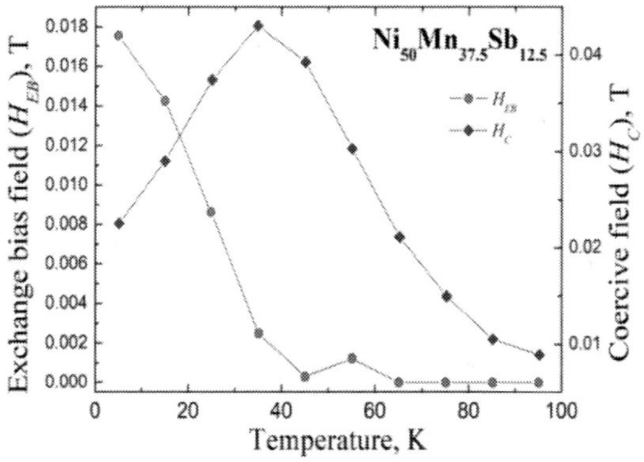

Figure 9. Experimental EB field (H_{EB}) and coercive field (H_C) as a function of temperature for $Ni_{50}Mn_{37.5}Sb_{12.5}$ [5].

5. Conclusions

In this work we have investigated exchange bias effect in the Heusler Ni-Mn-Sb alloys by classical Monte Carlo simulations. In the proposed model we have used a simple cubic lattice with periodic boundary conditions and take into account the magnetic interactions only in two coordination spheres. The whole lattice consists of magnetic atoms and non-magnetic impurities. We have considered the ferromagnetic and antiferromagnetic Heisenberg's spin interactions with anisotropy term using *ab initio* exchange constants. Configuration of antiferromagnetic atoms sets at non-magnetic sites of the lattice randomly.

Our simulations have shown that the exchange bias field depends on a concentration of the antiferromagnetic atoms, the temperature and number of hysteresis loops. Moreover the value of the bias field decreasing with increasing temperature and for $Ni_{50}Mn_{37.5}Sb_{12.5}$ alloy we have found a temperature of blocking of the exchange bias effect. Theoretical blocking temperature is close to experimental value.

6. Acknowledgments
This work was supported by grants RFBR Grant 10-02-92110, President of Russian Federation Grant MK-1891.2010.2 and Grant of support for young scientists OSPO 2010. We had further support by the SPP 1239 of the German Science Foundation (DFG).

7. References
[1] Entel P, Buchelnikov V D, Gruner M E, Hucht A, Khovailo V V, Nayak S K and Zayak A T 2008 *Mater. Sci. Forum* **21** 583
[2] Kainuma R, Imano Y, Ito W, Morito H, Sutou Y, Oikawa K, Fujita A, Ishida K, Okamoto S, Kitakami O and Kanomata T 2006 *Appl. Phys. Lett.* **88** 195513
[3] Sutou Y, Imano Y, Koeda N, Omori T, Kainuma R, Ishida K and Oikawa K 2004 *Appl. Phys. Lett.* **85** 4358
[4] Krenke T, Duman E, Acet M, Wassermann E, Moya X, Manosa L and Planes A 2005 *Nature Mater* **4** 450
[5] Khan M, Dubenko I, Stadler S and Ali N 2007 *Appl. Phys. Lett.* **91** 072510
[6] Khan M, Dubenko I, Stadler S and Ali N 2008 *J.Phys.: Condensed Matter* **20** 235204
[7] Planes A, Manosa L, Acet M 2009 *J.Phys.: Condensed Matter* **21** 233201
[8] Meiklejohn W H and Bean C P 1956 *Phys. Rev* **102** 1413
[9] Meiklejohn W H and Bean C P 1957 *Phys. Rev* **105** 904
[10] Landau D P and Binder K 2000 *A guide to Monte Carlo simulations in statistical Physics* (Cambridge: Cambridge University Press).
[11] Nowak U and Usadel K D 2002 *Phys. Rev.* B **66** 014430
[12] Miltenyi P, Gierlings M, Keller J, Beschoten B, Guntherodt G, Nowak U and Usadel K D 2000 *Phys. Rev. Lett.* **84** 4224
[13] Keller J, Miltenyi P, Beschoten B, Guntherodt G, Nowak U and Usadel K D 2002 *Phys. Rev.* B **66** 014431

Development of magnetically soft microwires with GMI effect

V. Zhukova, M. Ipatov and A. Zhukov
Dpto. de Fís. Mater., UPV/EHU San Sebastián 20009, Spain

E-mail: arkadi.joukov@ehu.es

Abstract. Thin amorphous magnetically soft microwires attract recently great attention because of their excellent soft magnetic properties and giant magneto-impedance (GMI) effect, thin dimensions and possibility for applications in magnetic micro-sensors. We overview research in the field of development of microwires with high GMI effect and improved features and report novel results on studies of GMI (diagonal and off-diagonal components) at high frequencies (between 10 MHz and 4 GHz) and its correlation with soft magnetic behaviour of thin amorphous microwires (Co-Fe-rich with nearly-zero magnetostriction constant) with metallic diameter between 3 and 20 μm and on optimization of GMI effect. We studied and analyzed low-field hysteresis of GMI effect and its dependence on circular magnetic field and discuss the nature the low-field hysteresis in terms of helical magnetic anisotropy and the effect of the bias field. Choosing samples composition, annealing conditions and geometry we were able to tailor their magnetoelastic anisotropy and respectively magnetic softness and GMI.

1. Introduction

Recently studies of soft magnetic properties and GMI effect of thin glass coated microwires attracted considerable interest [1-3]. Most reported results correspond to the microwires with the diameter about 20-30 μm.

As previously reported elsewhere [2,3] the hysteretic magnetic properties of glass-coated microwires are rather different from those of conventional amorphous wires mostly because of additional magnetoelastic energy related with strong internal stresses induced during the rapid solidification of the thin wire surrounded by the glass coating when using Taylor-Ulitovsky method [4]. Recently certain progress has been achieved in enhancement of magnetic softness and GMI effect of glass coated microwires, paying most attention on alloy composition and post-fabrication processing of microwires aiming improvement of the magnetic permeability and GMI effect through the diminishing of the magnetoelastic anisotropy related with strong internal stresses and induction of desirable magnetic anisotropy. Consequently, adequate choosing of the geometric parameters determined by the fabrication procedure and conditions of heat treatment can significantly enhance soft magnetic properties and GMI effect (up to 600%)[5,6].

In most of applications high linearity of signal and low hysteresis are desirable. Anti-symmetrical signal with linear region has been obtained in current pulsed scheme excitation of wires using detection of off-diagonal GMI component [7,8]. At the same time we recently showed, that linearity and shape of off-diagonal component in microwires can be tailored by thermal treatment [9]. At the same time, considerable GMI hysteresis has been observed and analyzed in microwires at low field region [10].

In this paper we review and analyze the results on GMI effect in thin microwires paying special attention to tailoring the GMI effect of thin glass-coated amorphous microwires (with the metallic nucleus diameter, d below 15 μm) by choosing the sample geometry (ratio, ρ, between metallic nucleus diameter, d and the total microwire diameter, D) and annealing conditions by the Joule heating with the aim to improve magnetic field and stress sensitivity of the GMI effect and diminish the GMI hysteresis.

2. Tailoring of magnetic properties and GMI effect in thin glass-coated microwires

2.1. Effect of microwire geometry

As mentioned above, the strength of the internal stresses, σ, is related with simultaneous solidification of metallic nucleus inside glass coating and most important contribution is induced by the difference in the thermal expansion coefficients of metallic nucleus and outer glass coating solidifying simultaneously [2-4, 11,12]. Chemical composition of the alloy has been fixed on the basis of previous knowledge of magnetically soft behavior of amorphous ribbons [13]. The effect of the ρ ratio on the hysteresis loops of nearly-zero magnetostrictive $Co_{67}Fe_{3.85}Ni_{1.45}B_{11.5}Si_{14.5}Mo_{1.7}$ microwires is shown in Fig.1. All studied samples exhibited inclined almost unhysteretic M(H) loops with extremely low coercivities (up to 4 A/m). Magnetic anisotropy field, H_k, is found to be determined by the ρ–ratio, increasing with ρ.

Such $H_k(\rho)$ dependence has been attributed to the magnetoelastic energy contribution given by

$$K_{me} \approx 3/2\ \lambda_s \sigma_i, \qquad (1)$$

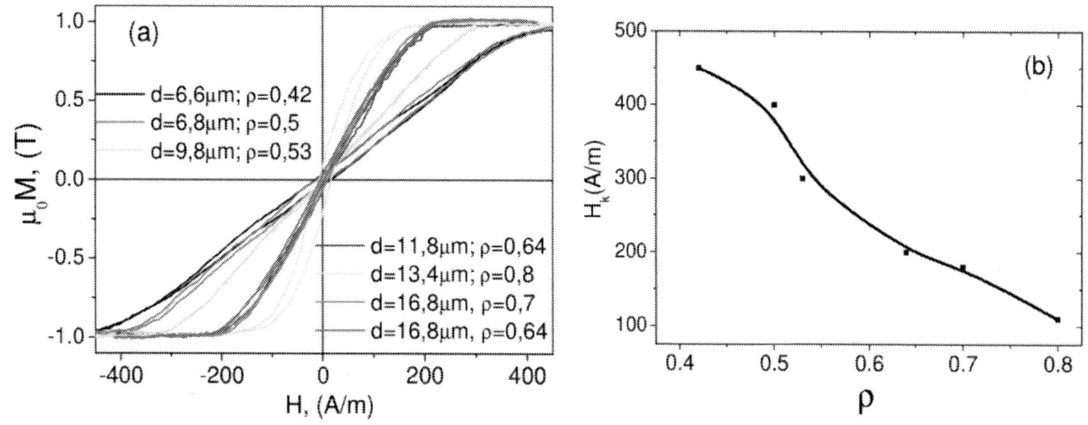

Fig.1. Hysteresis loops of $Co_{67.1}Fe_{3.8}Ni_{1.4}\ Si_{14.5}\ B_{11.5}\ Mo_{1.7}$ microwires with different geometry (a) and dependence of H_k on ρ–ratio (b).

where λ_s is the saturation magnetostriction and σ_i is the internal stress. The magnetostriction constant is mostly determined by the chemical composition and achieves almost nearly-zero values in amorphous alloys based on Fe-Co with Co/Fe ≈70/5 $\lambda_s \approx 0$ [2]. On the other hand, the estimated values of the internal stresses in these glass coated microwires arising from the difference in the thermal expansion coefficients of metallic nucleus and glass coating are of the order of 100-1000 MPa, depending strongly on the ratio between the glass coating thickness and metallic core diameter [2-4, 11,12], increasing with decreasing ρ–ratio. Considering elevated internal stresses stress dependence of magnetostriction constant should be taken into account as well [2-4]:

$$\lambda_s = (\mu_o M_s/3)(dH_k/d\sigma), \qquad (2)$$

where $\mu_0 M_s$ is the saturation magnetization.
Considering abovementioned we should assume that any method allowing change the internal stresses (by using thermal treatment, chemical etching, controlling microwire geometry) can change drastically magnetic anisotropy and consequently the hysteresis loops and the GMI behavior of glass coated microwires.

2.2. Tailoring of magnetic properties by heat treatment and induced magnetic anisotropy

Heat treatment resulting in structural and stress relaxation in amorphous alloys is a commonly used method for tailoring the magnetoelastic energy [2]. In the case if the alloy composition contains more than one transition metal the pair ordering mechanism of induced magnetic anisotropy especially if magnetic field and/or stress is applied during the heat treatment should be taken into account [2,4,14].

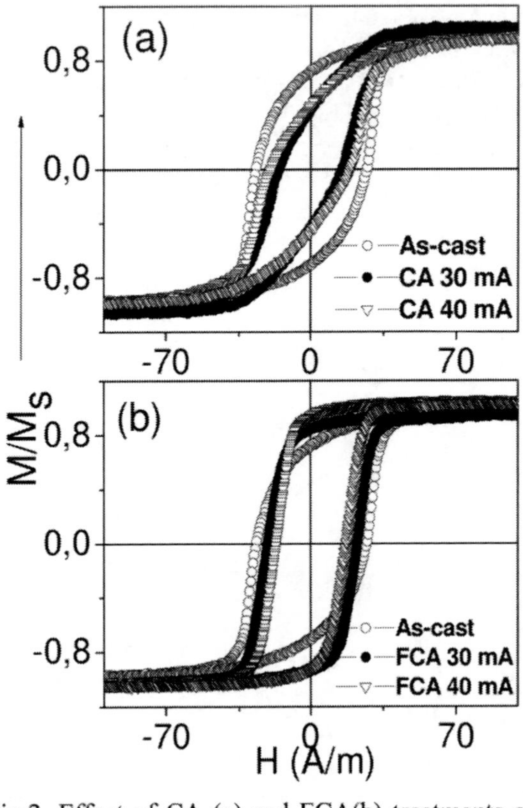

Fig.2. Effect of CA (a) and FCA(b) treatments on bulk hysteresis loops of $Co_{67}Fe_{3.85}Ni_{1.45}B_{11.5}Si_{14.5}Mo_{1.7}$ microwires (d=22.4 μm, D=22.8 μm).

Effect of current annealing (CA) and magnetic field current annealing (FCA) on hysteresis loop of $Co_{67}Fe_{3.85}Ni_{1.45}B_{11.5}Si_{14.5}Mo_{1.7}$ is shown in Fig.2. Axial magnetic field during FCA (about 8000A/m) has been much higher than circular magnetic field created by the current. As can be appreciated from comparison of Figs 2a and 2b, application of magnetic field during annealing results in induction of axial magnetic anisotropy.

Application of stress during stress annealing of $Fe_{74}B_{13}Si_{11}C_2$ microwires resulted in induction of considerable stress induced anisotropy [15]. In this case rectangular hysteresis loop associated with the strong axial magnetic anisotropy induced by residual stresses of mostly tensile origin after stress annealing converted into inclined hysteresis loop with large enough magnetic anisotropy field [15]. In fact the hysteresis loops can be tailored varying the time or temperature of stress annealing and consequently a variety of hysteresis loop with different magnetic anisotropy can be obtained in $Fe_{74}B_{13}Si_{11}C_2$ microwires, as can be observed in Fig.3a. Such behavior should be attributed to the compressive stresses induced by the SA compensating axial stresses due to the simultaneous solidification of the metal and glass.
The strength of such compressive stresses depends on both time and temperature of annealing.
Under applied stress, the hysteresis loop changes drastically, exhibiting enhanced stress sensitivity (Fig.3b).
Another interesting effect related with such stress sensitivity of hysteresis loop is stress-impedance, i.e. impedance change under applied stress observed in samples with induced stress anisotropy (see Fig.4)

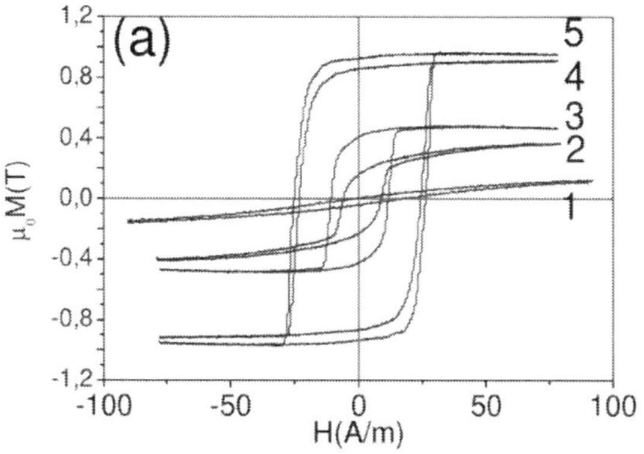

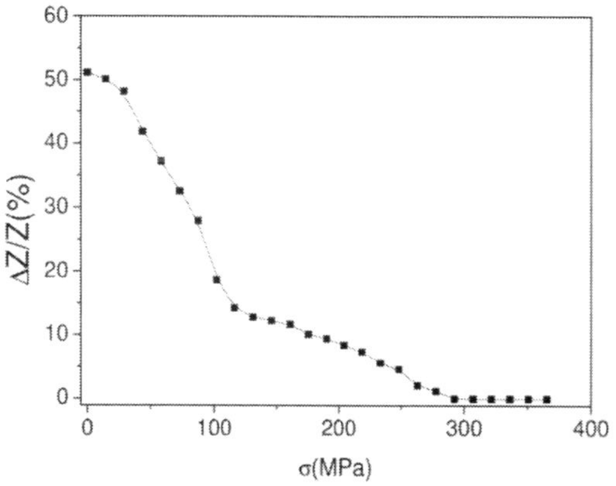

Fig.4 Stress impedance effect of stress annealed $Fe_{74}B_{13}Si_{11}C_2$ glass-coated microwire under stress (468 MPa) at 275°C for 0.5h measured at frequency, f=10 MHz for the driving current amplitude of 2 mA

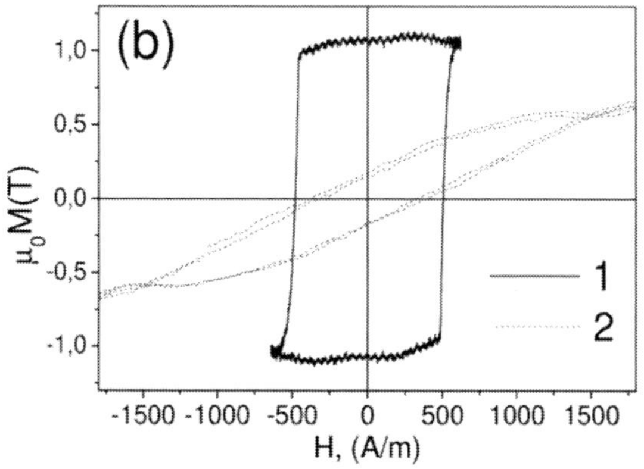

Fig.3 Hysteresis loops of $Fe_{74}B_{13}Si_{11}C_2$ microwire annealed under applied stress of 500 MPa (a) at (1) – 300 °C 3 hours, (2) – 280 °C 40 min, (3)- 265 °C 40 min, (4) 235 °C 40 min and (5)- 215 °C 40 min and (b) stress induced changes of hysteresis loops of the same microwires (1- measured under applied stress, 2- measured without stress).

2.3. Tailoring of the GMI effect (including off-diagonal component)

Like in the case of hysteresis loops the sample geometry strongly affects the GMI effect, i.e. magnetic field dependences of impedance (Fig.5). Here we present magnetic field, H, dependence of real part, Z_1 of the longitudinal wire impedance Z ($Z = Z_1+iZ_2$). Indeed we observe (see Fig.5) that the sample geometry strongly affects the DC axial magnetic field dependence of the real part of GMI, Z_1 of $Co_{67}Fe_{3.85}Ni_{1.45}B_{11.5}Si_{14.5}Mo_{1.7}$ microwires measured at different frequencies f.

In fact one should take into account that both magnetic permeability as well as magneto-impedance have tensor character. The off-diagonal components possess the asymmetrical dependence on magnetic field, that is a necessary condition for determination the magnetic field direction [7,8]. Figs. 6,7 show field dependence of the off-diagonal voltage response, V_{out} measured using pulsed scheme as described elsewhere [7-9] in $Co_{67.1}Fe_{3.8}Ni_{1.4}Si_{14.5}B_{11.5}Mo_{1.7}$ ($\lambda_s \approx 3 \cdot 10^{-7}$) microwire with different geometry: metallic nucleus diameter and total diameter with glass coating are 6.0/10.2, 7.0/11.0 and 8.2/13.7 μm.

Such considerations regarding dependence of the GMI response on the samples geometry should be taken into account for development of extremely thin microwires. It is important for microminiaturized sensor application since the demagnetizing factor starts to affect the domain structure and magnetic properties when geometric ratio of microwire diameter to its longitude becomes lower [10]. It should be assumed that the internal stresses relaxation after heat treatment should drastically change both the soft magnetic behavior and $\Delta Z/Z(H)$ dependence mostly because of the stress dependence of the magnetostriction described in ref [1]:

$\lambda_\sigma(\sigma)= \lambda_s(0)-A\sigma$ (3)

where λ_s (0) is the saturation magnetostriction constant without applied stresses and A is the positive coefficient of the order of 10^{-10} MPa.

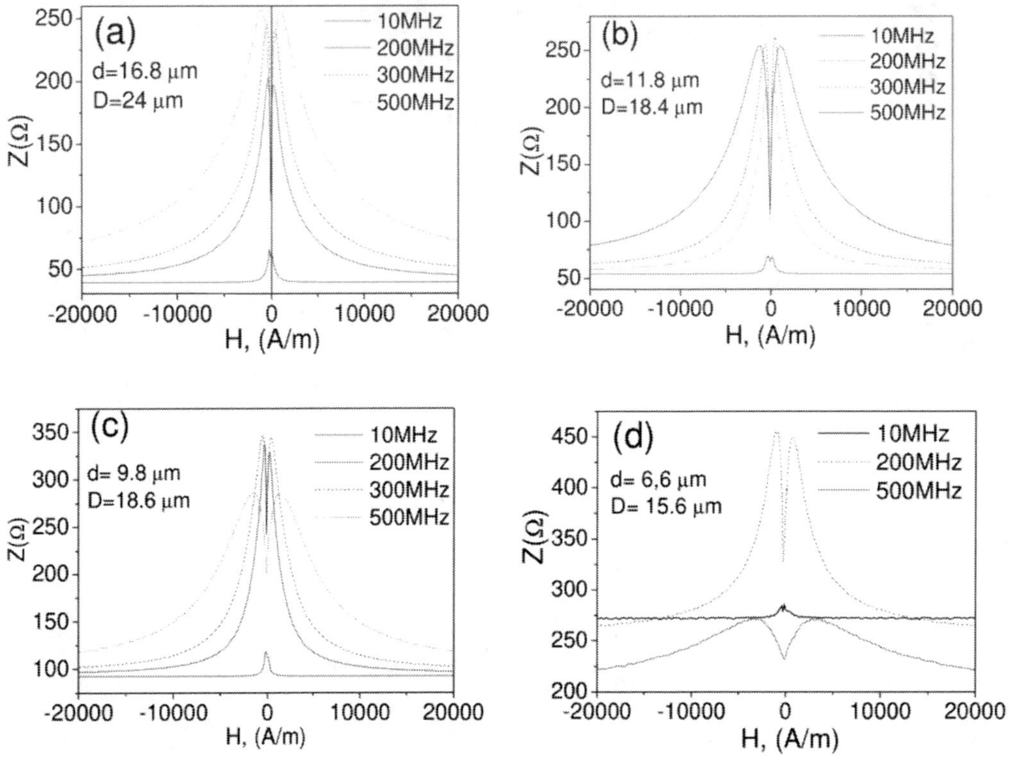

Fig. 5. Z (H) dependence $Co_{67}Fe_{3.85}Ni_{1.45}B_{11.5}$ $Si_{14.5}Mo_{1.7}$ microwires with different geometry measured at frequencies *f*=10, 200 and 500 MHz

It should be noted from Fig.6 that the $V_{out}(H)$ curves have asymmetrical shape exhibiting close to linear growth within the field range from $-H_m$ to H_m. The H_m limits the working range of MI sensor to 240 A/m and should be associated with the anisotropy field.

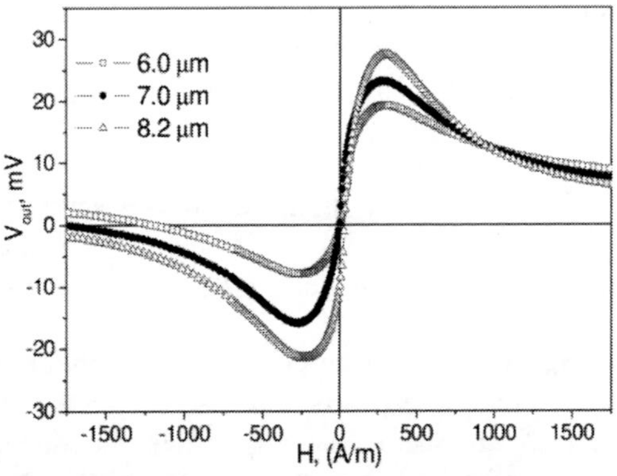

Fig.6. V_{out}(H) response of $Co_{67,1}Fe_{3,8}Ni_{1,4}$ $Si_{14,5}B_{11,5}Mo_{1,7}$ microwires with different d.

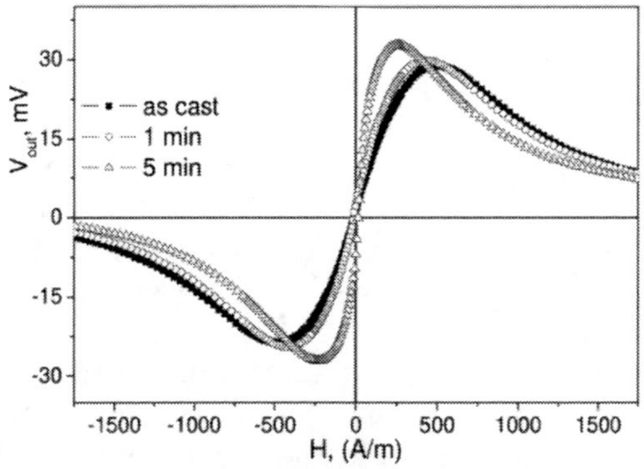

Fig.7. $V_{out}(H)$ of Joule-heated $Co_{67}Fe_{3.85}Ni_{1.45}$ $B_{11.5}Si_{14.5}Mo_{1.7}$ microwire annealed with 50 mA currents for different time.

The influence of Joule heating on off-diagonal field characteristic of nearly zero magnetostriction $Co_{67.1}Fe_{3.8}Ni_{1.4}Si_{14.5}B_{11.5}Mo_{1.7}$ microwire with diameters 9.4/17.0 μm is shown in Fig. 7. One can see that the thermal annealing with 50 mA DC current reduces the H_m from 480 A/m in as-cast state to 240 A/m after 5 min annealing.

Recent applications and reduced dimensionality of developed microwires require extending frequency range for GMI studies and applications. This particularly induced us to extend frequency range for GMI studies. Fig.8 presents results on magnetic field dependence of real part, Z_1 of the longitudinal wire impedance Z ($Z = Z_1+iZ_2$) till 4 GHz. General features of these dependences is that the magnetic field of maximum shifts to the higher field region increasing the f.

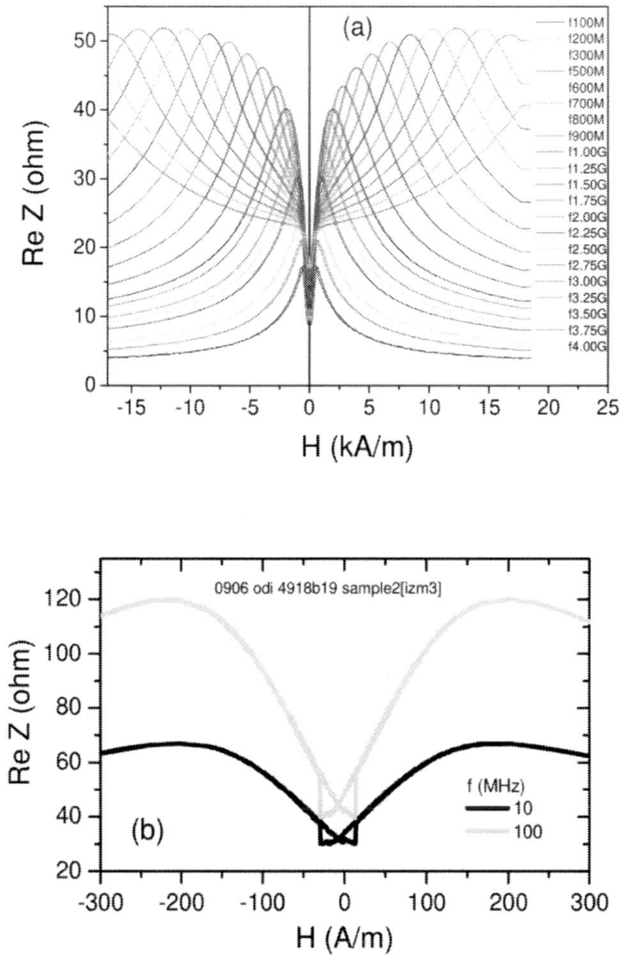

Fig.8. Z(H) dependence of $Co_{66}Cr_{3.5}Fe_{3.5}B_{16}Si_{11}$ microwires measured at different frequencies

2.4. GMI hysteresis

Another relevant feature is low field hysteresis observed even at high f (Fig.8b). Recently we developed a model for magnetization reversal and MI field dependence for nearly- zero-magnetostrictive amorphous microwires with consideration that low-field GMI hysteresis arises from deviation of the anisotropy easy axis from transversal direction and explaining why the circular bias magnetic field H_B produced by DC current I_B running through the wire affects the hysteresis and asymmetry of the MI dependence. The validity of the model has been confirmed by the experiments [10]. Thus it was demonstrated that the low-field hysteresis originates from deviation of easy axis from circular direction (helical magnetic anisotropy) and that the application of the bias field leads to suppressing of this hysteresis (see Fig.9, where effect of bias voltage on S_{21} parameter. proportional to off-diagonal GMI component is shown). Using the developed model, the main characteristics of the studied ferromagnetic microwire such as anisotropy field H_k, angle between the anisotropy easy axis and the transversal direction were obtained. In fact in pulsed exciting scheme when the sharp pulses with pulse edge time about 5 ns are produced by passing square wave multi-vibrator pulses through the differentiating circuit, overall pulsed current contains a DC component that produces bias circular magnetic field [6,7]. In this way low field hysteresis can be surpassed selecting adequate pulse amplitude.

In fact low field hysteresis can be also used for data storage in wire MI element. Thus, independently on the initial state, after applying of a constant current pulse I_B in one direction (right to left in Fig 10 a), the magnetic moments will orient along the easy axis in *up* direction (Fig. 10 b) which corresponds to the store logical '1' state. Then, submitting the wire to external magnetic field makes the magnetic moments rotate in *close-to-circumferential* direction which characterized by low impedance (Fig. 10c) that can be easily detected. To write logical '0' one need to pass current pulse *IB* in the opposite direction (left to right in d) that makes the magnetization orient along the easy axis in *down* direction (Fig 10e). In this case, the external magnetic field induces the magnetic moments rotatation in *close-to-longitudinal* direction characterized by high impedance as shown in Fig. 10f. Similarly, to rotate the magnetic moments from equi-impedance state (with the same impedance value), a small static current can be applied instead of external magnetic field H_E.

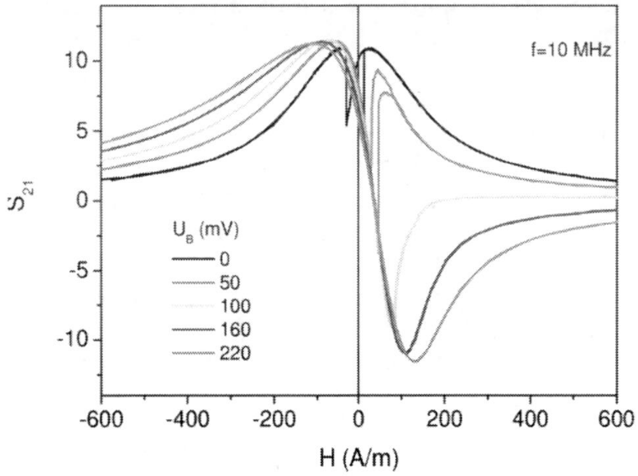

Fig.9. Effect of bias voltage U_B on magnetic field dependence of S_{21} parameter of $Co_{67}Fe_{3.85}Ni_{1.45}B_{11.5}Si_{14.5}Mo_{1.7}$ microwire.

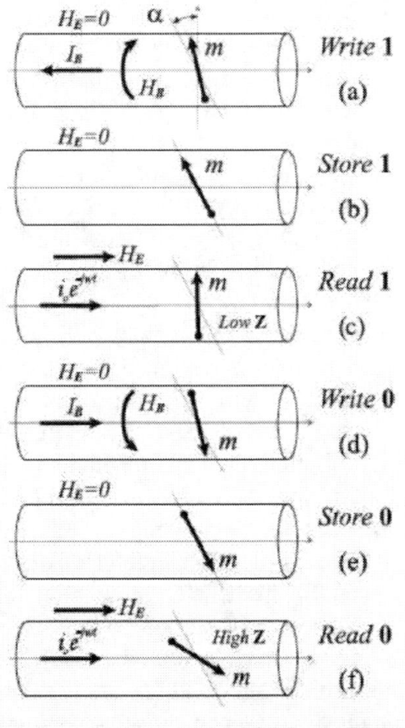

Fig.10 Applications of GMI hysteresis: Principle of data storage in wire element using MI hysteresis.

3. Conclusions

Thin amorphous wires with enhanced magnetic softness and GMI effect can be produced by the Taylor-Ulitovski technique. Magnetic properties and GMi effect of such microwires can be tailored by an appropriate selection of the metallic nucleus diameter, glass-coating thickness and chemical composition of the metallic nucleus and even by the heat treatment under magnetic field or without it. There are a number of interesting effects, such as induction of the transversal anisotropy in Fe-rich microwires allowing creating extremely stress sensitive elements. The investigation of diagonal and off-diagonal MI tensor component with pulse excitation in 6—12 μm amorphous glass-coated microwires has shown the grate potential of these materials for mi-crominiaturized magnetic field sensor application. Their main advantages are high sensitivity low-hysteresis field dependence, low power consumption, and very simple sensor scheme are very promising for technological application. By varying the alloys composition and applying post fabrication processing it is possible to control the sensor's operating range. The off-diagonal impedance component in microwires with nearly zero magnetostriction is anti-symmetrical with close to linear behavior within the working range. The Joule heating of such microwires tends to decreasing of magnetoelastic coupling and anisotropy field. Thermal treatment is an additional effective factor for tuning of MI dependence as well as the alloy composition and geometric parameters. Low field GMI hysteresis has been observed and explained in terms of helical magnetic anisotropy of microwires. Data storage in MI microwire element using low field hysteresis has been proposed.

ACKNOWLEDGMENTS

The work was supported by the EU ERA-NET program under Project DEVMAGMIWIRTEC "MANUNET-2007-Basque-3" and by the Basque Government under Saiotek 09 Mic Magn project. A. Zh. and V. Zh. wish to acknowledge the support of the Basque Government under Program of Mobility of the Investigating Personnel of the Department of Education, Universities and Investigation Grants MV-2009-2-21 and MV-2009-2-24.

[1] Jiles DC 2003 Acta Materialia **51** 5907
[2] Zhukova V, Ipatov M and Zhukov A 2009 *Sensors* **9** 9216
[3] Zhukov A and Zhukova V 2009 *Magnetic properties and applications of ferromagnetic microwires with amorphous and nanocrystalline structure*, (New York: Nova Science Publishers, ISBN: 978-1-60741-770-5)
[4] Antonov A S, Borisov V T, Borisov OV, Prokoshin A F and Usov N A 2000 *J. Phys. D: Appl. Phys.* **33** 1161
[5] Zhukov A, Gonzalez J, Blanco J M, Prieto M J, Pina E and Vazquez M 2000 *J. Appl. Phys.* **87** 1402
[6] Zhukova V, Chizhik A, Zhukov A, Torcunov A, Larin V and Gonzalez J 2002 *IEEE Trans Magn* **38** 3090
[7] Sandacci S I, Makhnovskiy D P, Panina L V, Mohri K and Honkura Y 2004 *IEEE Trans Magn.*, **35** 3505
[8] Mohri K and Honkura Y 2007 *Sensors letters,* **5** 267
[9] Zhukov A, Ipatov M, Gonzalez J, Blanco J M, Zhukova V 2009 *J. Magn. Magn. Mater.* **321** 822
[10] Ipatov M, Zhukova V, Zhukov A, Gonzalez J, and Zvezdin A 2010 *Phys. Rev. B* **81** 134421
[11] Velázquez J, Vazquez M and Zhukov A 1996 *J Mater Res.* **11** 2499.
[12] Chiriac H and Ovari TA 1996 *Progress in Material Science* **40** 333
[13] O'Handley R C, Hasegawa R, Ray R and Chou C-P 1976 *Appl. Phys. Lett.* **29** 330
[14] Morita H, Fujimori H and Obi Y, 1980 *J. Magn. Magn. Mater.* **15-18** 1359
[15] Zhukov A 2006 *Adv Func Mat* **16** 675.

The effect of compaction parameters and dielectric composition on properties of soft magnetic composites

Ling Xiao, Y H Sun[1], Lie Yu
Institute of Mechatronics and Information Systems, Key Laboratory of Ministry of Education for Strength and Vibration, Xi'an Jiaotong University, Xi'an, China

E-mail:sunyanhua@mail.xjtu.edu.cn

Abstract. This paper investigated the effect of compaction parameters and dielectric composition on mechanical, magnetic and electrical properties of iron-organosilicon epoxy resin soft magnetic composites. In this work, iron powders with high purity were covered by an organic material (organosilicon epoxy resin) and then by coupling agent (KH-550). The coated powders were then cold compacted at 600, 800 and 1000 MPa and cured under vacuum respectively. The results show that the saturation magnetic flux density and electrical resistivity are dependent on compaction pressure and resin content. Increase in the organic phase content leads to decrease of the saturation magnetic flux density, while increase of the electrical resistivity. Furthermore, the samples with 0.9 *wt*% resins + 0.1 *wt*% coupling agent at compaction pressure of 800 MPa shows better properties than the others.

1. Introduction

Soft magnetic composites (SMCs) can be described as pure iron powder particles coated with organic electrically insulating layers. SMCs have unique properties such as three-dimensional (3D) isotropy, low eddy current loss, high resistance (typically 70-1000 μΩ·m) and flexible shape design for low cost [1]. Although SMCs have many advantages, now it is seldom used in industrial production for its lower mechanical strength and permeability [2].

In order to be used in high speed rotors and meet the requirements of electrical motors and magnetic bearings, it is important that the mechanical strength of the SMCs should be increased without weakening the magnetic and electrical properties. For this reason, it is necessary to cover the iron particle with a very thin insulating layer to improve the densification and Vickers hardness. In this area, several researchers have investigated the effect of epoxy resin, phenol resin, polyepoxy on the properties of iron-epoxy SMCs [2,3,4]. In this paper, organosilicon epoxy resin was chosen for its good thermal stability and mechanical property. The object is to investigate the effects of compaction pressure and dielectric composition on mechanical, magnetic and electrical properties of SMCs.

2. Experimental

2.1 Raw materials

[1] School of ME, Xi'an Jiaotong Univ., Xianning West Road, Xi'an Shaanxi,710049

Published under licence by IOP Publishing Ltd

The iron powder was supplied by Fucheng chemistry factory. The purity of Fe was 98%, containing 0.1% sulphuric acid insoluble, 0.06% SO_4, 0.005% N, 0.005% Cu and 0.03% water soluble. The iron powders were coated by organosilicon epoxy resin (OER) and KH-550 coupling agent supplied by Xi'an resin factory.

2.2 Preparation of the coated powders
Firstly, the iron particles were degreased in acetone and dried in a vacuum furnace for 30 minutes. Then, the cleaned iron powders were mixed with the OER in acetone in a container by a spiral mixer for 20 minutes, and manual mixing for about 10 minutes. At last, the coated powders were obtained after the evaporation of the acetone, and then heated at 150℃ for 25 minutes.

2.3 Production
The powders coated with the OER were cold compacted at 600, 800 and 1000 MPa in a cylindrical die with a diameter of 10 mm and kept for 5 minutes respectively. The glycerol was used as the lubricant for the die wall when compacting. Finally, the samples were cured at 200 ℃ under vacuum for 35 minutes.

2.4 Characterization
The composition of coated powders was analyzed by a Fourier transform infrared spectrometer (FTIR) with the scan range from 400 to 4200 cm^{-1}. The microstructure and the powder morphology were examined by the scanning electron microscopy (JEOL JSM-7000F SEM/EDS). The average density was measured by the principle of Archimedes. The hardness was tested by a Vickers hardness tester. The value of the strength is three times more than that of Vickers hardness, which can reflect the strength to a certain extent, but can not be replaced. The saturation magnetic flux density was measured by vibrating sample magnetometer (VSM). The electrical resistivity was measured using four-point probe method. Then the density, Vickers hardness, saturation magnetic flux density and resistivity of the samples were measured and compared.

3. Results and discussion

3.1 Structure and morphology of coated powders and SMCs
In order to find the optimum proportion of resin and coupling agent, and to study its effect on mechanical, magnetic and electrical properties of SMCs, six different types of specimens were prepared as shown in Table 1. The samples are classified into two groups according to whether the coupling agent is used. The first three samples in group 1 have no coupling agent. The proportions of coupling agent in the last three samples in group 2 are 0.5, 0.3 and 0.1 *wt%* respectively. Furthermore, the corresponding samples in group 1 and group2 (sample 1 and 4, 2 and 5, 3 and 6) have the same percentage of iron powders.

Table 1 Powders prepared by different iron particles, resin and coupling agent content

Group	Series No.	Fe *wt%*	Resin *wt%*	*Coupling agent wt%*
1	1	95	5	-
	2	97	3	-
	3	99	1	-
2	4	95	4.5	0.5
	5	97	2.7	0.3
	6	99	0.9	0.1

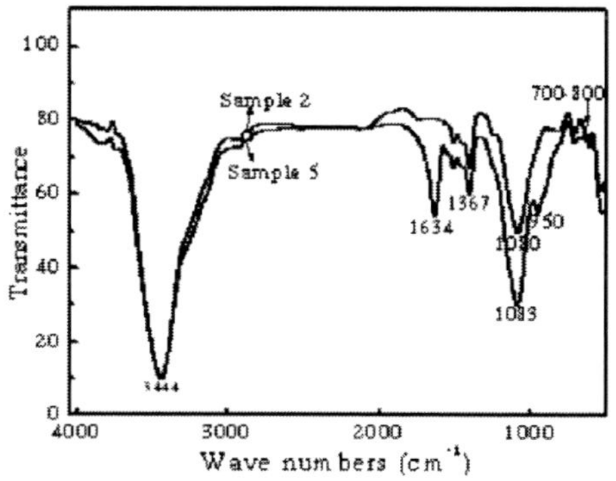

Figure 1. FTIR spectra of coated powders

The FTIR spectra of sample 2 and 5 are shown in figure 1. The main differences between sample 2 and 5 are the peaks at 700-800 cm^{-1}, 950 cm^{-1} and 1634 cm^{-1}. The peak at 950 cm^{-1} indicates the condensation reaction between silanol groups and iron surface hydroxyl group occurred in figure 2 [4,5] and the peaks 700-800 cm^{-1} in sample 5 is related to the stretching vibration of Si-C bond[6]. Additionally, the N–H bending vibration of primary amine is observed at 1634 cm^{-1} in sample 5. According to [6], the peaks at 1080 cm^{-1} may be considered as Si-O-C and cyclic ether in sample 2, at 1083 cm^{-1} in sample 5 is Si-O-Si. At around 1367 cm^{-1}, there is no distinct difference between two samples, may be caused by C-H bending vibration ($-CH_2-$, $C-CH_2-$) in organosilicon epoxy resin. The peak at 3444 cm^{-1} indicates the iron powder surface may be oxidized during degreasing process.

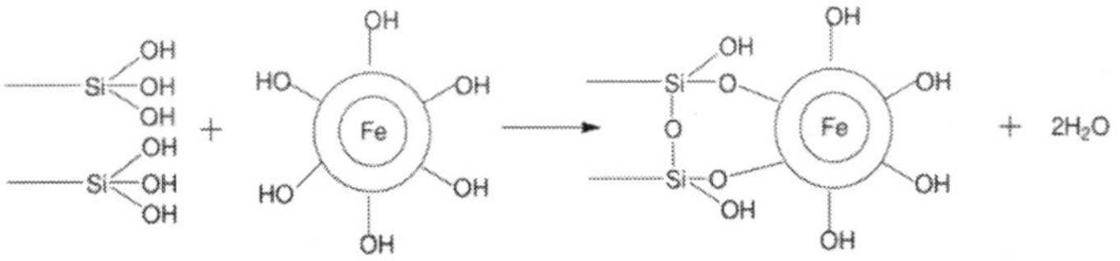

Figure 2. Condensation reaction between silanol groups and iron surface hydroxyl group

SEM micrographs of sample 6 on 800 MPa and coated powders are shown in figure 3. In figure 3a, it can be seen that porosities exist at the boundary of different iron particles and the morphology of iron powders in figure 3b is not regular. When the irregular powders were compacted together, the probability of the porosities is more at risk than that of regular powders. Thus, it can be considered that the densification of sample may be related to the morphology of iron powders.

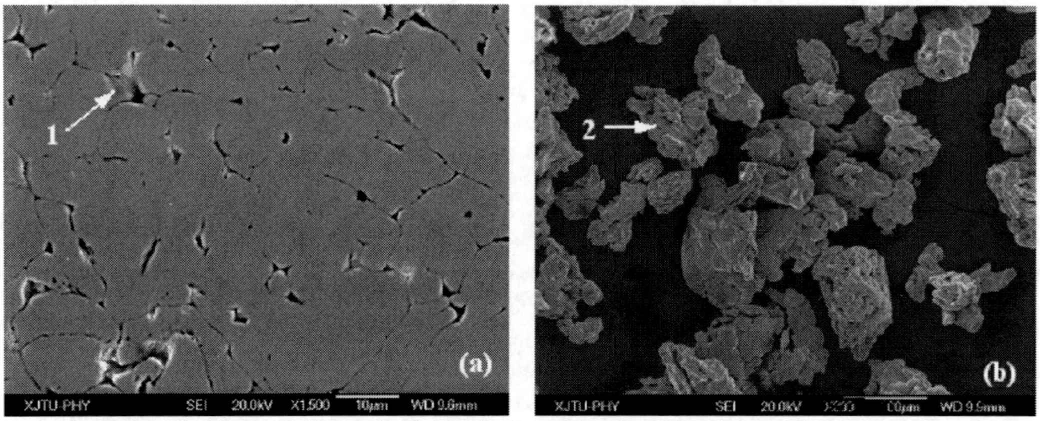

Figure 3. SEM micrograph of sample 6 on 800Mpa (a) and coated powders (b)

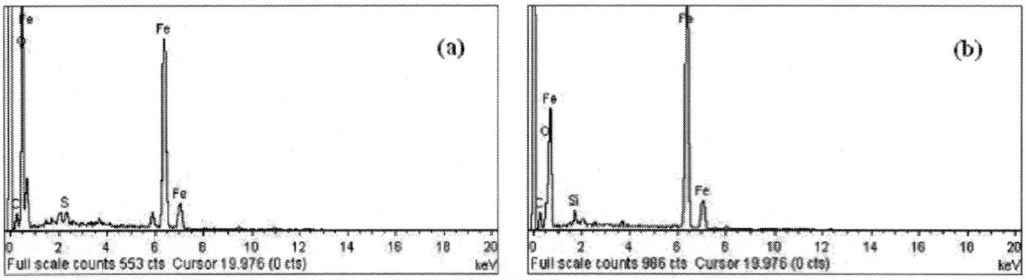

Figure 4. EDS spectrums in the scanning electron microscope (figure 3) of sample 6 at 800Mpa (a) and coated powders (b)

EDS spectrum (figure 4) shows that the composition of the porosity (figure 3a) and the surface layer (figure 3b) mainly consist of iron, oxygen and carbon elements. The sulfur in figure 4a is because of impurity of the iron powders. The silicon in figure 4b is from the resin element. Therefore, it confirms the existence of the insulating layer on the surface of the iron powders.

3.2 Mechanical properties

The densities of each sample at different pressures are shown in figure 5. With the increase of the iron content, the density of the sample increases. When the iron content is lower, the density increases with increase of the pressure from 600 to 800 MPa and as the pressure gets to 1000 MPa, the density starts to decrease. While the iron content is higher, the density has been increasing from 600 to 1000 MPa. Because iron particles are work hardened during compaction [7], heat treatment relieves the stresses and strains. Since the stresses between particles become larger when the pressure gets to 1000 MPa, the particles will spread out to reduce the stress during the heat treatment. This leads to the growing of the whole volume, and therefore the density will be decreased. The densities of sample 3 and 6 continue to increase for the total content of resin and coupling agent is less than other samples and heat expansion coefficient of the resin (6.0E-5/K) is greater than that of iron (about 1.2E-5/K). Thus, the volume decreasing during compaction (compared with 800 MPa) is more than the volume expansion in heat treatment process. Furthermore, by increasing the amount of epoxy resin, the density at the same pressure decreases, because the density of epoxy resin (about $1g/cm^3$) is much smaller than that of pure iron. Coupling agent will increase the density of the sample, which is related to the reactions of silane coupling agent with iron particle surface [4].

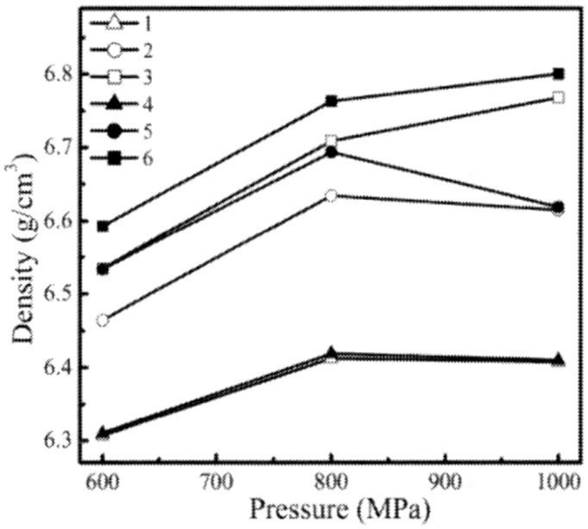

Figure.5 Density as a function of pressure

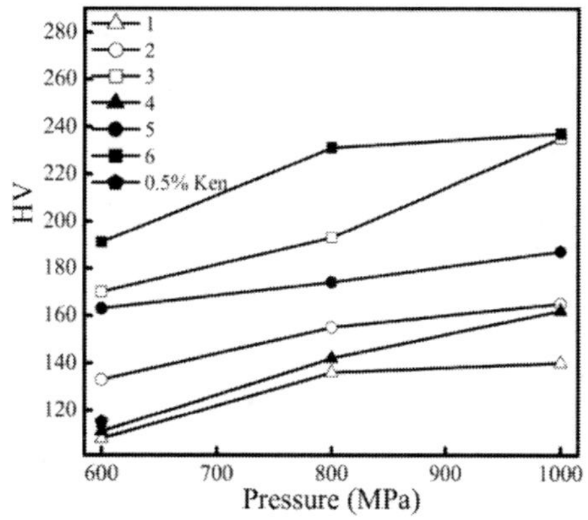

Figure.6 Vickers hardness as a function of pressure

The Vickers hardness (HV) of the SMCs was measured using four point method with a load 20 g for 10s on the polished samples and the results are shown in figure 6. With the increase of the pressure and iron content, the HV increases. This is according to the increase of density, work hardening in the grains and decrease of porosities. Furthermore, the HV of the same iron content sample with coupling agent is higher than that without it. This is confirmed by the cross linking reactions among iron particles, resin and coupling agent. To samples without coupling agent, there is only mechanical combination of iron particles and resin, no reaction with each other. Comparing with Somaloy 500 + 0.5 *wt*% Ken [7], the HV of samples with 97 *wt*% and 99 *wt*% Fe at 600 MPa are higher.

3.3 Magnetic properties

Figure 7 shows the saturation magnetic flux density increases with the increase of the pressure from 600 to 1000 MPa. The saturation magnetic flux densities of the SMCs are lower than that of pure iron sample (about 2T). The soft magnetic properties are related to microstructure change during compaction and heat treatment process. On the one hand, the internal stress is generated during compaction process. This stress, having a significant effect on metal magnetization, results in giant magneto impedance effects [8]. When eliminating the stress below the Curie temperature, it promotes the magnetization and improves the magnetic properties. On the other hand, there are also other microstructure changes such as a decrease in the defects from 600 to 800 MPa, which also improve the soft magnetic properties. The coupling agent has no obvious effect on the saturation magnetic flux density.

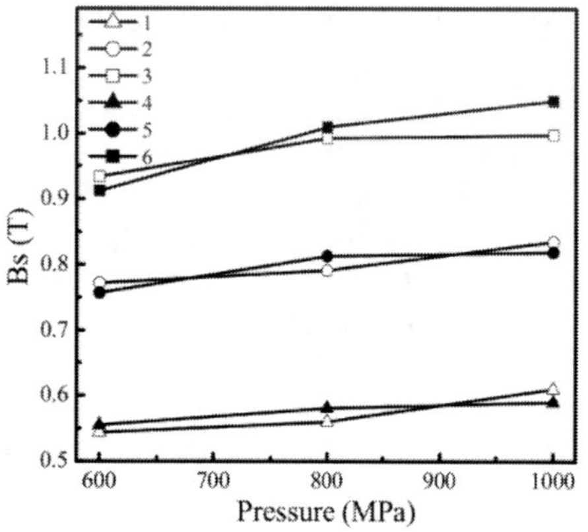

Figure 7. Saturation magnetic flux density as a function of pressure for different resin contents

3.4 Electrical properties

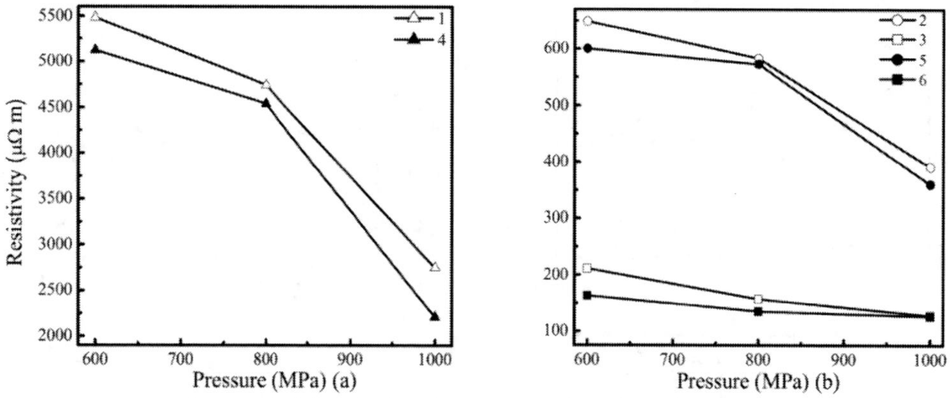

Figure 8. Electrical resistivity as a function of pressure

Figure 8 depicts the electrical resistivity of the samples on different pressures. Firstly, in both groups, the resistivity decreases as the content of iron powder increases at the same pressure, and the resistivity of sample with coupling agent is smaller than that without coupling agent. Since the resistivity of resin (1.0e17 Ω·m) is much greater than that of pure iron (1.0e-7 Ω·m), the resistivity increases as the proportion of resin increase because that better insulating layer between iron particles may be obtained. Furthermore, the resistivity of SMCs strongly depends on the pressure. With the increase of the pressure, the resistivity decreases. Because under very high pressure, the distance between atoms becomes smaller, and the morphology, electronic structure, Fermi surface, energy band structure and electron scattering mechanism in inner defect will be changed, which might finally affect the resistivity of the SMCs. The relationship between resistivity and pressure can be written as [9]:

$$\rho = \rho_0(1 + \psi p) \tag{1}$$

where ρ_0 is the resistivity in vacuum, ψ is the pressure coefficient (negative), p is the pressure (Pa). It can be seen from this formula that the relationship between the resistivity and pressure is linear. However, since only three sets of data were obtained in the experiment, it is not enough to confirm the linear relationship.

4. Conclusion

In this paper, the effect of compaction parameters and dielectric composition on mechanical, magnetic, and electrical properties of SMCs is investigated. The results show that the insulating layer was prepared successfully and the optimum compaction pressure is about 800 MPa and proportion of content is 0.9 wt % resins + 0.1 wt % coupling agent. The saturation magnetic flux densities of the SMCs fabricated from iron and resin are lower than 1.1 T for lower permeability of resin and defects in SMCs. In particular, the coupling agent improves the density and Vickers hardness of the SMCs as the proportion of resin is decreased.

5. Acknowledgements

The work described in this paper was supported by the National Natural Science Foundation of China (Grant No. 50635060) and the National High Technology Research and Development Program of China (863 program No. 2007AA050501).

References

[1] Information on *http://www.hoganas.com/*
[2] Shokrollahi H, Janghorban K, Mazaleyratc F, LoBuec M, Ji V and Tcharkhtchi*et* A 2009 *Mater. Chem. Phys.* **114** 588
[3] Shokrollahi H and Janghorban K 2007 *J. Magn. Magn. Mater.* **317** 61
[4] Taghvaei A H, Shokrollahi H, Ebrahimi A and Janghorban K 2009 *Mater. Chem. Phys.* **116** 247
[5] Ukaji E, Furusawa T, Sato M and Suzuki N 2007 *Appl. Surf. Sci.* **254** 563
[6] Liu J X 2008 *Applied IR spectrum analysis*(Beijin, Science press)
[7] Gilbert I, Bull S and Evans T 2004 *J Mater Sci.* **39** 457
[8] Yu R H, Basu S, Ren L, Zhang Y, Parvizi-Majidi A, Unruh K M and Xiao J Q 2000 *IEEE Trans. Magn.* **36** 5
[9] Qiu C J, Wang Y H and Wang Y J 2003 *Material Physical Properties*(Harbin, Harbin Institute of Technology press)

Magnetic and thermomagnetic studies of the formation of the Rhometal powders by high energy mechanical milling

I Chicinaş[1,a], O Isnard[2], H Chiriac[3], F Popa[1], V Pop[4], C V Prică[1], B V Neamţu[1], T F Marinca[1]

[1]Materials Sciences and Technology Dept., Technical University of Cluj-Napoca, Cluj-Napoca, Romania,
[2]Institut Néel, CNRS/ Université Joseph Fourier, Grenoble, France,
[3]National Institute of R@D for Technical Physics, Iaşi, Romania,
[4]Babes-Bolyai University, Cluj-Napoca, Romania

[a]Corresponding author: Ionel.Chicinas@stm.utcluj.ro

Abstract. Nanocrystalline Rhometal (36Ni64Fe, wt. %) powders have been obtained by mechanical alloying under argon atmosphere. The initial mixture was milled up to 20 h. In order to eliminate internal stresses and to improve the solid state reaction annealing at 350 °C was performed for 4h. The alloy formation is obtained after 8 h of milling. A mean crystallite size of 10 ± 4 nm is obtained after 20 hours of milling. The magnetization values for the as-milled samples are between that of the starting sample and that of the as cast one. The evolution of the magnetization versus the milling time is discussed. The thermomagnetic analysis shows the Curie temperatures confirming the alloy formation by milling. As a result of the solid state reaction by milling means, a continuous decrease of the magnetization with increasing the temperature is recorded, similarly to the behaviour of the classical cast alloy.

1. Introduction

The nanocrystalline materials may exhibit superior properties compared with materials having polycrystalline structure. From the properties that could be amplified by the presence of the interfaces and by the large surface/volume ratio can be noted: superior hardness, amplified diffusion, superior ductility, reduced elastic module, higher electrical resistivity, higher specific heat, higher thermal dilatation coefficient, smaller thermal conductivity and higher soft magnetic properties [1].

In principle, the obtaining of the nanocrystalline materials can be performed using several techniques [1, 2, 4]. For the production of nanocrystalline/nanostructured materials in larges quantities only two techniques seem to be appropriates: the mechanical alloying and the re-crystallization of the amorphous ribbons.

Mechanical alloying (MA) is a technique for alloy obtaining in powder form, consisting from plastic deformation, mixing and plastic deformation repeatedly of the elemental or prealloyed powder [1, 2]. Many studies have been carried out in order to understand the process [4]. The final product of the MA technique depends on the energy involved. It was shown that for particular milling conditions a final product in nanocrystalline or amorphous state can be obtained. In order to predict the state of final milled product simulations have been made [5 - 7]. Also the effect of milled on the phase transition have been studied [8].

Soft magnetic materials from the Ni-Fe system were obtained by mechanical alloying in the whole composition range [9 - 16]. Our researches are focused on the obtaining of the commercial Ni-Fe soft magnetic alloys by mechanical alloying. In our previous studies we successfully obtained the Permalloy alloy (75Ni25Fe, wt. %) [13-16]. In this study, we focus our attention on an interesting alloy called Rhometal (36Ni64Fe, wt. %) which is known to present the highest resistivity among Ni-Fe alloys. In opposition with Permalloy, the Rhometal alloy is a biphasic alloy and this offers an interesting perspective on the structural and magnetic properties of the alloy obtained by mechanical alloying. In this study we discuss the alloy formation by mechanical alloying and some magnetic properties of the milled samples versus milling time.

2. Experimental

The Rhometal (36Ni64Fe, wt. %) alloy was obtained in a planetary ball mill Pulverisette 4 Fritch starting from a mixture of elemental iron and nickel powder weighed in the proper ratio. The powders used were commercial iron powder NC 100.24 and nickel powder 123 – carbonyl type. The mixture was first homogenized for 15 min in a Turbula type apparatus, and then milled in the planetary mill for time ranging from 1 to 20 hours. The starting powders mixture, unmilled, is called in papers "ss". Argon gas was used as milling atmosphere. The powder/ball mass ratio was 8:1 at a filling factor of 40%. Several samples were sealed in silica tubes and annealed for 4h at 350 °C in order to improve the solid state reaction and to remove the internal stresses.

The alloy formation was investigated by X-ray diffraction in the angular range $2\theta = 32 - 103$ °. The X-ray patterns were recorded with a Bruker D8 powder diffractometer, operating with Cu – Kα radiation (λ = 1.54183 Å). The mean crystallite size was calculated using the Williamson & Hall method [17]. In order to confirm the alloy formation a mixture of iron and nickel powders having the same nominal composition was melted and used as reference (called in paper "cast ss" or "cast sample") for the obtained phase by milling.

The magnetization curves were recorded at room temperature by extraction method technique in a continuous field of up to 8 T. The thermomagnetic analysis was performed by means of a LakeShore 7410 vibrating sample magnetometer in a field of 1 T.

3. Results and discussion

Rhometal alloy formation was studied by X-ray diffraction. Figure 1 presents the recorded patterns for the as-milled samples and for the starting sample mixture ("ss" refers to 0 h of milling) and for a cast alloy obtained from the starting mixture.

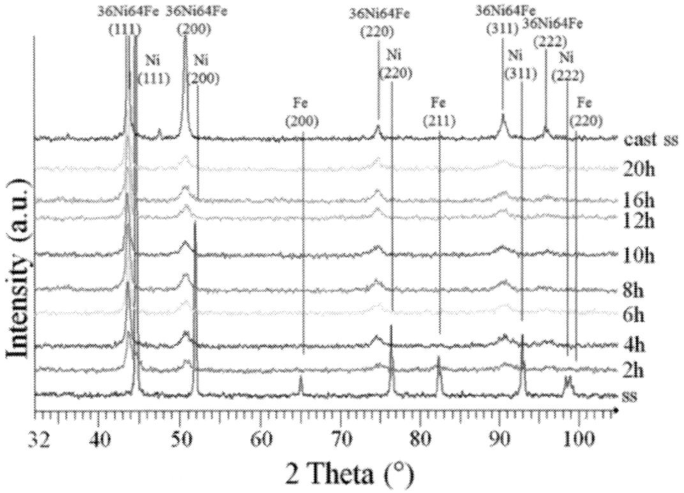

Figure 1. X-ray diffraction patterns for the as-milled samples. For clarity the patterns are vertically shifted and for comparison the starting sample (ss) and the cast sample are also showed.

The alloy formation is observed by disappearance of the Bragg peaks of Ni and Fe elements clearly observed in the 0 h milled sample and the shift of the remaining peaks toward the alloy position as compared with the cast sample. A close analysis of the diffraction patterns (figure 2) reveals the existence of an asymmetry of the Bragg peaks on the high angle side until 10 h of milling.

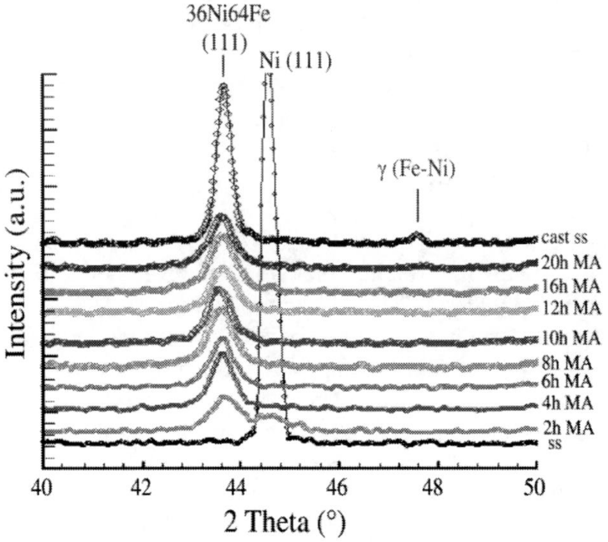

Figure 2. Detail of the diffraction patterns representing the (111) peak for the as-milled samples

The alloy formation by milling is accompanied by internal stress introduction by the milling process. First order internal stresses induced by milling leads to peak shifts toward lower angles. This evolution suggests the formation mechanism of the alloy: by high energy milling, stresses are induced at lattice level in the materials leading to enhanced atoms diffusion and favouring solid state reaction to form the alloy.

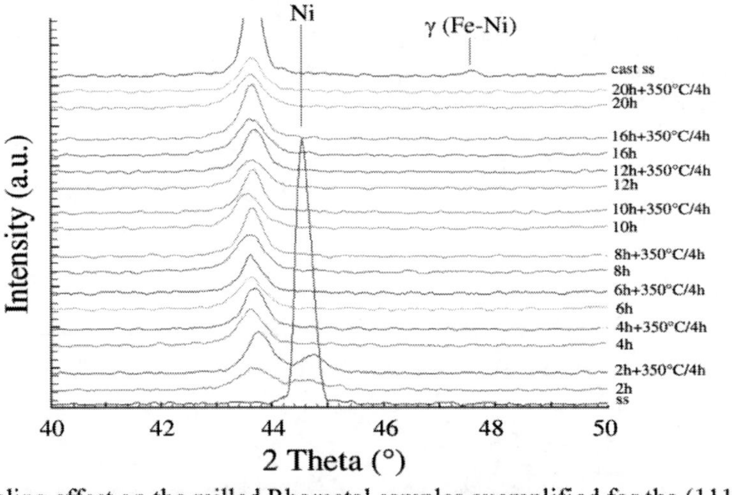

Figure 3. Annealing effect on the milled Rhometal samples exemplified for the (111) Bragg peak.

In order to eliminate the internal stresses induced by the milling process and to enhance the alloy formation, an annealing was performed at low temperature (for preventing the powders recrystallization) at 350 °C for 4 h. The effect of annealing is to release the internal stresses and to improve the solid state reaction. The improvement of the solid state reaction rate can be observed in figure 3 by the peak asymmetry disappearance for samples milled more than 8h. After the X-ray studies can be concluded that for high milling times only one FCC phase is obtained and the Rhometal alloy is formed after 8h of milling followed by annealing at 350 °C for 4h.

A close analysis of the cast sample diffraction patterns reveals the existence of two phases: a fcc (Ni-Fe) phase and γ (Fe-Ni) phase. The existence of both phases is found also in practical Ni-Fe phase diagram [18]. Figure 2 presents the observed mixture of phases. For the 2h milled sample a very broad peak can be seen at a position corresponding to γ (Fe-Ni) phase. This peak cannot be seen for longer milling times, except for the 20h milled sample.

The mean crystallite size was calculated from the X-ray diffraction patterns using the Williamson – Hall plots. The obtained values are presented in figure 4.

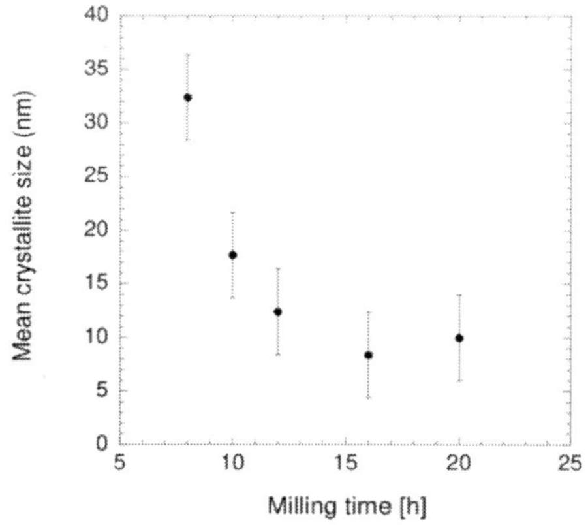

Figure 4. Mean crystallite size versus the milling time for the Rhometal alloy.

The evolution of the mean crystallite size versus the milling time, illustrates a strong decrease for long milling times. After the phase obtaining in the whole sample volume, 8 h of milling, the mean crystallite size decrease toward a value of 10 ± 4 nm for 20h milled samples. The crystalline size reduction is seen by the broadening of the Bragg peaks for the milled sample. It is worth to note, that the broadening of the peaks is not entirely due to the crystallite size refinement but also to the introduction of the second order internal stress. Due to the Williamson Hall method procedure, the effects of stresses are eliminated from the crystallite size calculation.

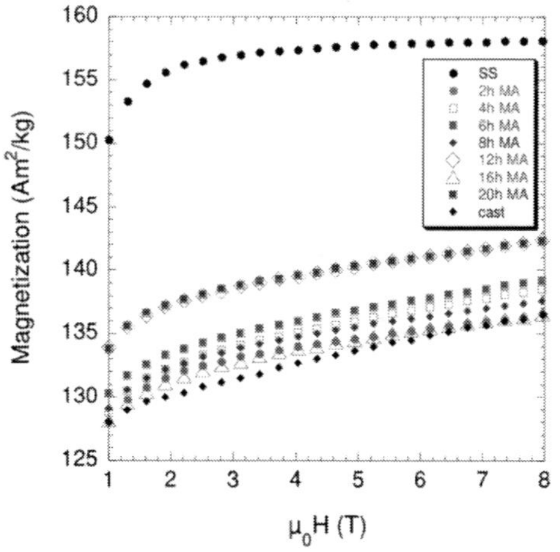

Figure 5. Room temperature magnetization curves recorded for the as-milled samples. For comparison the cast and starting sample are also presented.

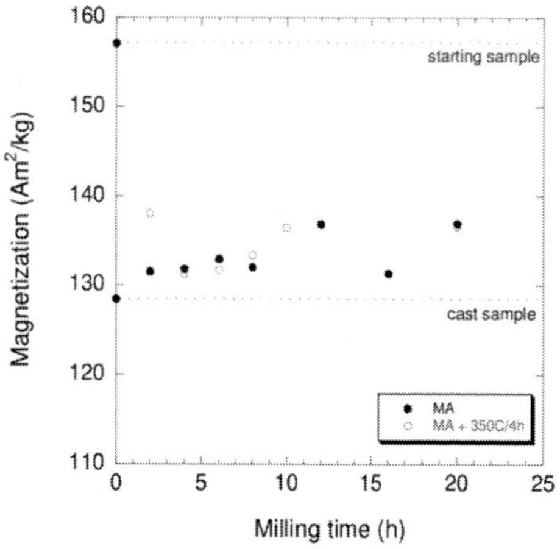

Figure 6. Spontaneous magnetization versus milling time. For comparison the magnetizations of the starting sample and of the cast sample are presented.

The obtained powders were magnetically characterized in fields up to 8 T. The measured curves are plotted in figures 5. The spontaneous magnetisation values have been derived from an extrapolation to zero field of the magnetisation obtained in magnetic fields between 4 and 8 T. The spontaneous magnetization versus milling time is shown in figure 6.

From figure 5 and 6, it can be concluded that the magnetization of the milled samples is ranging between the magnetization of the starting mixture (the „ss" sample) and that of the alloy obtained by melting (the as cast sample). For low milling times, interesting magnetic behaviour is observed. Due to the γ (Fe-Ni) phase presence in the samples, after 2 h of milling, the magnetization decreases drastically to the value of the cast sample. Saturation magnetization increases versus milling time (for milling times longer than 2h) and is higher than the magnetization of the cast alloy, due to the elimination of the γ (Fe-Ni) phase, which is antiferromagnetic, and to formation of the face centred cubic Rhometal alloy. The effect of the annealing is more pronounced for the short milling times (2h, where the phase γ (Fe-Ni) is present and has limited effect at long milling times. Investigations of the structure effect on the magnetic behaviour of the milled powder are now in progress.

The thermomagnetic measurements are used to characterize the formation and magnetic behaviour of the Rhometal powders. The recorded curves are presented in figures 7 and 8.

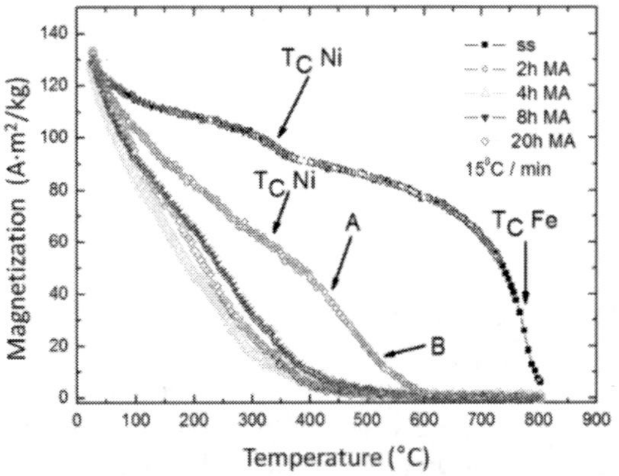

Figure 7. Thermomagnetic curves of the as-milled samples. For comparison the curves of the starting sample is also presented.

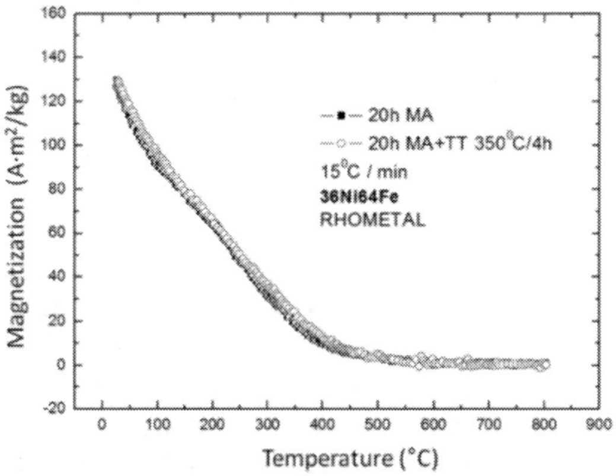

Figure 8. Comparison between the thermomagnetic curves of as-milled and annealed samples milled for 20h.

The thermal evolution of the magnetization for the starting sample presents two transitions corresponding to the elemental Ni and Fe. The beginning of milling (2h of milling), induces the alloy formation, thus leading to reduction of the first fall of magnetisation corresponding to the Curie temperature of the elemental Ni phase, followed by a change of slope (point A in figure 7). The slope change corresponds to the obtained Rhometal (36Ni64Fe). At higher temperature a large transition is observed, corresponding to the progressive formation of the alloy during heating (point B). The large transition expresses the existence of a large field of compositions. For higher milling times, a single Curie temperature is recorded (at 420 °C), the Curie temperature of Rhometal single phase. Similar thermomagnetic behaviour was found for the Ni-Fe-Mo alloys [19].A supplementary analysis of the annealed samples was performed and the obtained curves for the as-milled and annealed samples are presented in figure 8. It is seen that in the curves are barely identical for the samples behaviour with or without annealing, suggesting that for this milling duration the Rhometal alloy is formed in the whole sample volume.

4. Summary

The Rhometal (36Ni64Fe, wt. %) powders were obtained by mechanical alloying and annealing. The X-ray and magnetic measurements indicate the complete alloy formation after 8 – 10 h of milling in the whole sample volume. The obtained powders are nanocrystalline, with a mean crystallite size of 10 ± 4 nm after 20 h of milling.

The spontaneous magnetization of the milled samples is between the magnetization of the Ni-Fe starting sample and Rhometal alloy obtained by melting. After 2 h of milling, the spontaneous magnetization decreases toward a value close to the magnetization of the as cast sample, due to the presence of the γ (Fe-Ni) in the structure. Magnetization increases for further milling times higher than 2h, as a results of the face centred cubic Rhometal structure formation by milling. The thermomagnetic measurements show the continuous formation of the alloy by heating for low milling times. The Curie temperature of the alloy obtained by mechanical alloying is 420 °C.

Acknowledgements
The authors would like to thank the Romanian Ministry of Education and Research for the Grant PNCD II 71-015/2007 supporting the researches leading to these results.

References
[1] Suryanarayana C 1995 International Mater. Reviews **40** 41,
[2] Tjiong S C, Chen H 2004 Mater. Sci. Eng. R **45** 1,
[3] Gleiter H 1989 Progress Mater. Sci. **33** 223,
[4] Iwasaki T, Kim J H, Satoh M 2006 Chem. Eng. Sci. **61** 1061,
[5] Gaffet E 1991 Mater. Sci. Eng. A **132** 181,
[6] Gaffet E, Abdellaoui M, Malhouroux-Gaffet N 1995 Materials transactions **36** 198,
[7] Lund A C, Schuh C A 2004 Acta Mater. **52** 2123,
[8] Kang P C, Yin Z D, Celestine O 2005 Mater. Sci. Eng. A **395** 167,
[9] Chinasamy C N, Narayasamy A, Chattopadyay K, Ponpandian N 1999 NanoStruct. Mater. **12** 951,
[10] Djekoun A, Bouzabata B, Otmani A, Greneche J M 2004 Catalysis Today **89** 319,
[11] Hamzaoui R, Elkedim O, Greneche J M, Gaffet E 2005 J. Magn. Magn. Mater. **294** 145,
[12] Hays V, Marchand R, Saindrenan G, Gaffet E 1996 NanoStruct. Mater. **7** 411,
[13] Chicinaş I, Pop V, Isnard O, Le Breton J. M, Juraszek J 2003 J. All. Comp. **352** 34,
[14] Pop V, Isnard O, Chicinas I 2003 J. All. Comp. **361** 144,
[15] Neamţu BV, Chicinaş I, Isnard O., Popa F, Pop V 2011 Intermetallics **19** 19,
[16] Neamţu BV, Isnard O, Chicinaş I, Vagner C, Jumate N, Plaindoux P 2011 Mater. Chem. Phys. **125** 364,
[17] Langford J.I. 1992 Proc. Internat. Conf. Accuracy in Powder Diffraction II, Gaithersburg, MD, Maz **26-29** 110,
[18] *A hundred years after the discovery of invar... The iron – nickel alloys* 1996 Ed. by G. Beranger, F. Duffaut, J. Morlet, J. F. Tiers (Intercept Limited Andover, UK),
[19] Popa F, Isnard O, Chicinaş I, Pop V 2010 J. Magn. Magn. Mater. **322** 1548.

Efficient dynamic magnetization measurement in a Py/Ru/Py trilayer

A. Bonneau-Brault [1, a, b], **S. Dubourg** [a], **V. Dubuget** [a, b], **D. Plessis** [a], **F. Duverger** [a]
[a] CEA, DAM, Le Ripault, F-37260 Monts, France
[b] Univ. Tours, CNRS, CEA, LEMA, UMR 6157, Parc Grandmont, F-37041 Tours 1, France

E-mail: aurelien.bonneau-brault@cea.fr

Abstract. It has previously been shown that the quantity $M_\mu = \frac{1}{4\pi\bar{\gamma}}\sqrt{\frac{2}{\pi}\int_0^\infty \mu''(f)fdf}$, called the "efficient dynamic magnetization", is a powerful tool for characterizing magnetic layers dedicated to microwave applications. It correlates the imaginary part of permeability and the magnetization distribution. This paper presents microwave permeability spectra of a Py/Ru/Py trilayer, which have been measured in the layer plane under an external magnetic static field (H_{DC}) using a single coil cell. The efficient dynamic magnetization versus H_{DC} was found to exhibit a significant decrease for $0<H_{DC}<5$ Oe, and it is demonstrated that this decrease was induced by an optic mode excitation.

1. Introduction

Microwave applications such as magnetic recording write heads, broadband skin antennas, microwave filters, noise suppressors and radar-absorbing materials require the development of soft magnetic materials for high broadband permeability levels at elevated frequencies. However, it has been shown [1] that there exists a trade-off between permeability levels and the frequency according to

$$\int_0^\infty \mu''(f)fdf \approx k_A \frac{\pi}{2}(\bar{\gamma}4\pi M_s)^2 \qquad (1)$$

Where k_A is a dimensionless factor associated with the distribution of the magnetization orientation of the sample. For anisotropic uniform soft thin films, $k_A = 1$ and for isotropic materials, such as bulk sintered ferrites, $k_A = 1/3$.

[1] To whom any correspondence should be addressed.

For a thin layer, it has also been established that the above relation can be refined by taking into account experimental factors [2]. Once the sample presents no angular magnetic dispersion, the measured imaginary part of permeability along the hard axis is bounded according to

$$\int_0^F \mu''(f)fdf \approx \frac{\pi}{2} N_Z \left(\bar{\gamma} 4\pi M_s\right)^2 \left[1 - t - s \pm e\right] \qquad (2)$$

Here, $N_Z \approx 1$ is the out-of-plane demagnetizing factor and t, s and e are the correcting factors for respectively the finite truncation, skin effect and measurement uncertainty. F is the maximum frequency measurement.

The quantities M_μ^x, M_μ^y and M_μ, known as the "efficient dynamic magnetizations" [2] have been considered along the x, y directions and in the sample plane:

$$M_\mu^x = \frac{1}{4\pi\bar{\gamma}} \sqrt{\frac{2}{\pi} \int_0^\infty \mu_x''(f)fdf} \qquad M_\mu^y = \frac{1}{4\pi\bar{\gamma}} \sqrt{\frac{2}{\pi} \int_0^\infty \mu_y''(f)fdf} \qquad M_\mu = \sqrt{M_\mu^{x^2} + M_\mu^{y^2}} \qquad (3)$$

According to equation (1), for a single layer with a well-defined easy axis, M_μ equals the static magnetization of the layer material. This assumption maintains its validity when a static field is applied for the condition $H_{int} \approx H_k + H_{DC} << 4.\pi.M_s$ [2]. It remains available in the whole field range applied to the sample in the present experiments.

Figure 1 displays a plot of M_μ^x, M_μ^y and M_μ for a 0.6-μm CoNbZr layer under a static field. As can be seen, M_μ remained equal (within experimental errors) to the static layer magnetization deduced from Vibrating Sample Magnetometry, VSM.

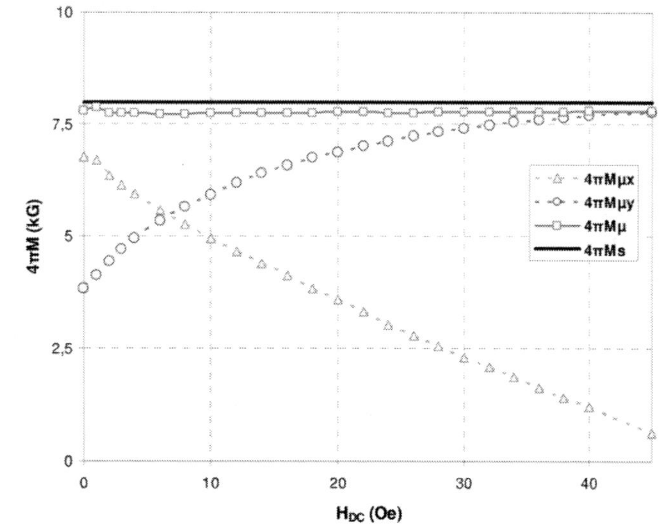

Figure 1. The efficient dynamic magnetization and saturation magnetization of a 0.6-μm CoNbZr layer.

A similar experiment was carried out with an artificial antiferromagnet made of a two Permalloy (Py) layers separated by a Ru spacer. The two Py layers were antiferromagnetically exchange-coupled by a RKKY interaction. Recently, it has been shown that such a system presents both optic and acoustic modes when a static in-plane field is applied [5-9]. The in-plane permeability under a static field generally exhibits a reduced M_μ value at low field when the optic mode is excited. Moreover, when the static field is strong enough, the optic mode vanishes while M_μ reaches M_s.

2. Experimental procedure

After first depositing a 5-nm Ta sub-layer to induce a Py texture with good softness properties, a Py(60 nm)/Ru(0.6 nm)/Py(60 nm) trilayer was sputter-deposited on 9 mm x 9 mm flat glass substrates.. This was followed by addition of another 5-nm Ta layer on the top as a capping layer. The in-plane edge directions were respectively designated X and Y.

Static magnetic properties were investigated using a Vibrating Sample Magnetometer (VSM). The microwave permeability measurement setup was based on the impedance measurement of a shorted microstrip line with the ferromagnetic sample between the ground plane and the conducting strip [3]. The microwave permeability spectra were measured in the range [10 MHz ..6 GHz] along the X and Y directions of the sample, as shown in Figure 2. A Helmholtz coil was used to induce an external magnetic static field, H_{DC}, parallel to the X sample direction. H_{DC} was swept from 0 to 100 Gauss.

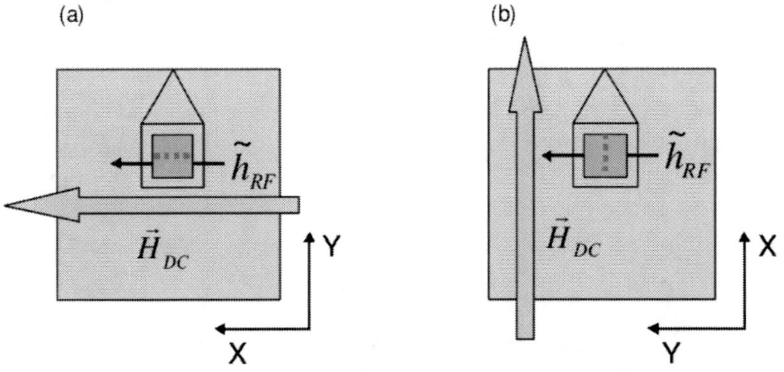

Figure 2. The setup for the permeability measurement: a single coil cell under a static field. M_μ^x and M_μ^y were respectively measured in configurations (a) and (b).

3. Results and discussion

Experimental and simulated hysteresis loops of the artificial antiferromagnet Py/Ru/Py have been plotted in Figure 3.a. As can be seen, the hysteresis loop exhibited a singular point and the saturation magnetization was equal to 9560 G. The fitting line in Figure 3.a was obtained with a macro-spin model [8] for which the magnetization was considered uniform in a same layer. The total energy (equation 4) - which includes the Zeeman energy, anisotropic energy and exchange-coupling energy - is minimized with respect to the angle of the magnetization of the top and bottom layers.

$$E_T = -M_s H[t_1 \cos(\theta_1 - \theta_H) + t_2 \cos(\theta_2 - \theta_H)] + K_u [t_1 \sin^2(\theta_1 - \theta_{K1}) + t_2 \sin^2(\theta_2 - \theta_{K2})] - J_1 \cos(\theta_1 - \theta_2) - J_2 \cos^2(\theta_1 - \theta_2)$$
(4)

Here, E_T represents the total energy to minimize; t_1 and t_2 are the thicknesses of the two layers; θ_H, θ_1 and θ_2 are the angles between H_{DC}, the first- or second-layer magnetization and the X axis of the sample. θ_{K1} and θ_{K2} are respectively the angles between the easy axes of the first and second layer, and the sample's X axis. K_u is the anisotropy constant, and M_s the magnetization saturation. Finally, J_1 and J_2 are the bilinear and biquadratic coupling parameters of the RKKY interaction through the Ru layer.

A good agreement between experiment and simulation was obtained with $\theta_{K1} = \theta_{K2} = 15°$, K_u = 760 erg/cm^3, J_1= 0.06 erg/cm² and J_2 = 0.018 erg/cm². The J_1 and J_2 exchange-coupling values were consistent with data from the literature [5, 6, 7, 9].
The magnetization orientation of each layer is presented in Figure 3.b. At zero field, the two magnetizations were in anti-parallel states. In the range [-5 Oe ..+5 Oe], there was a spin-flop transition where the two magnetizations were forced to adopt a canted configuration. Then, the two magnetizations slowly tended toward the magnetic field direction, to finally reach a saturated state at 40 Oe.

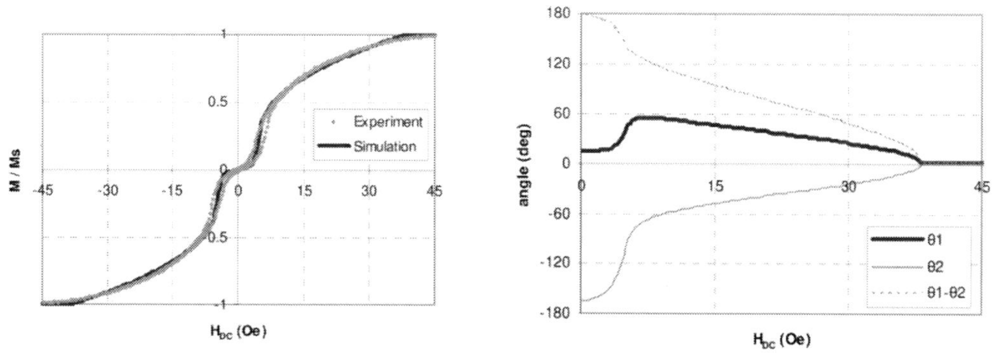

Figure 3. (a) Experimental and simulated VSM hysteresis loops. (b) The variation of the magnetization orientation.

Figure 4 illustrates the efficient dynamic magnetization M_μ versus the static field H_{DC}. As can be seen, the curve exhibited a drop at low field. Moreover, there was a discrepancy between the static M_s value and M_μ; this gap was observed when the Py magnetizations were in canted configuration (H_{DC} < 5 Oe). For higher values of external static field, the quantity $4.\pi.M_{\mu x}$ decreases because the magnetization turns up to be aligned to the external static field at saturation. So, the dynamic response along x axis is almost zero rather than it's maximum along y axis. One can moreover see that $4\pi M_\mu$ = 8700 G when H_{DC} = 0 Oe and reached 9700 G when H_{DC} > 5 Oe.

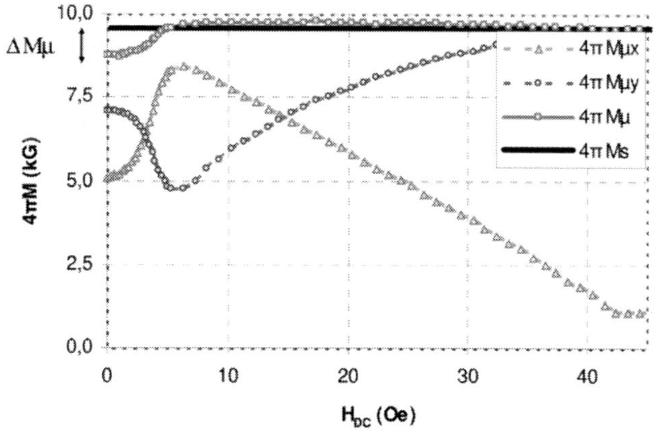

Figure 4. The efficient dynamic magnetization and saturation magnetization of the Py/Ru/Py trilayer.

Figures 5.a and 5.b present the imaginary part of permeability spectra and the deduced resonance frequencies. One can observe three behaviors. At low field (< 3 Oe), the resonance frequency decreased from 1.4 GHz to 1.26 GHz. At high field (> 6 Oe), a single peak with a resonance frequency increasing with the static field could be evidenced, in accord with an $Fr^2 \sim H_{DC}$ law. In between, two resonance peaks were observed.

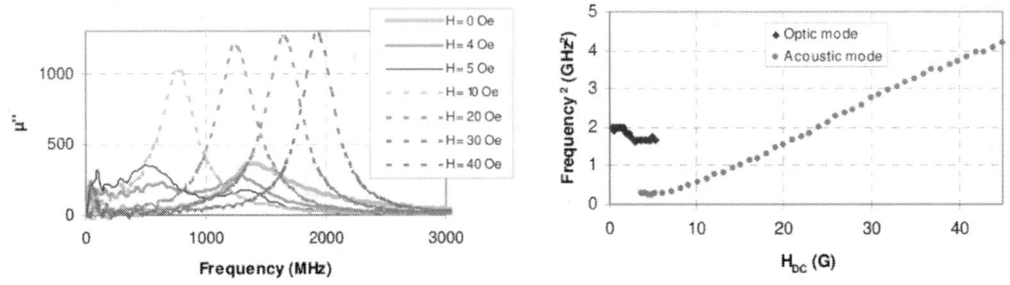

Figure 5. (a) Imaginary part of permeability spectra. (b) Deduced optic and acoustic resonance mode frequencies under a static field.

This feature has previously been evidenced in artificial antiferromagnetic trilayers [5-9], and is typical of the coexistence of an optic and an acoustic mode. Such optic and acoustic modes have also been observed in simple systems, such as Co-Py bilayers [4].

As shown in figure 5.b, the optic mode became excited at low field while its acoustic counterpart was excited at high field. The pure optic mode was obtained when the two ferromagnetic layers oscillated in phase in the out-of-plane direction and in opposite phases within the sample plane. One should notice that the observed resonance could not be a pure optic mode since the resulting permeability measured in the sample plane was not perfectly compensated. However, it generated the observed drop in efficient dynamic magnetization at low field (see figure 4). This discrepancy can be corrected by taking into account the out-of-plane efficient dynamic magnetization component M_μ^z (see equation 5). This component is usually neglected and can be enhanced by the optic mode.

$$M_s = \sqrt{{M_\mu^x}^2 + {M_\mu^y}^2 + {M_\mu^z}^2} \qquad (5)$$

4. Conclusions

The determination of the efficient dynamic magnetization is a powerful tool for characterizing magnetic samples dedicated to microwave applications. It has been successfully carried out with an artificial antiferromagnetic trilayer where optic and acoustic modes were excited through a static field. A decrease of the dynamic magnetization was evidenced when measured in the sample plane at low field induced by the optic mode.

Acknowledgement

The authors would be like to thank B. Rodmacq for providing the sample, and A. Thiaville for fruitful discussions.

References

[1] Acher O and Adenot A-L 2000 *Phys. Rev.* B **62** 11324
[2] Acher O and Dubourg S 2008 *Phys. Rev.* B **77** 104440
[3] Pain D et al 1999 *J. Appl. Phys.* **85** 5151
[4] Crew D.C 2005 *J. Appl. Phys.* **97** 10A707
[5] Belmeguenai et al. 2007 *Phys. Rev.* B **76** 104414
[6] Strijkers G.J et al. 2000 *Phys. Rev.* B **62** 13896
[7] Kuanr B et al. 2002 *J. Appl. Phys.* **91** 7209
[8] Layadi A 2002 *Phys. Rev.* B **65** 104422
[9] Zhao Z 2007 *IEEE Trans. Magn.* **43** 4056

Magnetic properties of $Co_{1-x}Ni_{x/2}Sr_{x/2}Fe_2O_4$ (x = 0–1) ferrite nanoparticles synthesized by self-assembling media of reverse micelles

A Ghasemi[1,*], A Kaewrawang[2], X Liu[2] and A Morisako[2]
[1]Materials Engineering Department, Malek Ashtar university of Technology, Shahin Shahr, Iran
[2]Spin Device Technology Center, Faculty of Engineering, Shinshu University, 4-17-1 Wakasato, Nagano 380-8553, Japan

*E-mail: ali13912001@yahoo.com

Abstract. $Co_{1-x}Ni_{x/2}Sr_{x/2}Fe_2O$ ferrite nanoparticles were prepared using reverse micelle process. Mean particle size of pure ferrite nanoparticles is approximately 9 nm as determined by transmission electron microscopy (TEM). The heating thermograms of samples confirmed that there are no hydroxides in the ferrites fabricated via reverse micelles. Vibrating sample magnetometer (VSM) was employed to probe magnetic properties of nanosized. It was found that with increasing the amount of Ni and Sr in nanosized ferrites, the saturation magnetization decreases. Magnetic dynamics of the samples was studied by the measurement of AC magnetic susceptibility versus temperature at different frequencies. The phenomenological Néel – Brown and Vogel - Fulcher models were employed to distinguish between interacting or non-interacting system. Results show that there is strong interaction between fine particles.

Introduction

Soft ferrite nanoparticles have been fabricated by different processing including sol-gel process, mechanical alloying, thermolysis, shock waves synthesis, electro-deposition, severe plastic deformation consolidation, hydrothermal processing, hydrolysis method, sonochemical decomposition and co-precipitation [1-10]. These methods are able to produce fine particles; however control of size distribution is arbitrary. In most of the mentioned technique, the particle size is varied by changing post-annealing temperature. It is well known that although the size of particles changes according to the sintering temperature, however the distribution of cations in the crystallographic sites (octahedral and tetrahedral) is also varied by temperature. Therefore, when sintering process is used to control the particle size, the correlation between particle size and magnetic characteristics is not possible, due to the re-distribution of cations which is another important variable. In order to correlate size of particle with magnetic properties, it is essential to employ a synthesis technique that allows for preparation of nanoparticles at relatively low temperature. Precipitation in reverse micelles is a very promising technique to prepare nanoparticle in the narrow distribution size and at low temperature. In this process self-assembling media of reverse micelles leads to synthesis nanoparticles of ferrite. Briefly, reverse micelles are water-in-oil microemulsions in which molar ratio of constitutive components and pH control the size of water pools and resultant particles. There are several literatures about the synthesizing of Ni-Zn and Mn-Zn ferrite with reverse micelle method [11-15]. In the recent years, we have focused our studies on the preparation and magnetic characteristics of ferrites [16-20]. In the present research, $Co_{1-x}Ni_{x/2}Sr_{x/2}Fe_2O_4$ferrite has cubic spinel structure and is widely used in many electronic devices because of high electrical resistivity, hard mechanical properties, high Curie temperature and environmental stability. To the best of our knowledge no studies on the preparation of this ferrite via reverse micelle are report. With this view in mind, current interest is to make $Co_{1-x}Ni_{x/2}Sr_{x/2}Fe_2O_4$ferrite nanoparticles using reverse micelle technique at low temperature. The structural and magnetic characteristics were evaluated by employing XRD, TEM, TGA, VSM, and AC susceptometer. We experimentally analyze magnetic susceptibility behavior of prepared samples and interpret our experimental results using Néel – Brown and Vogel – Fulcher laws.

Experiments

$Co_{1-x}Ni_{x/2}Sr_{x/2}Fe_2O_4$ ferrite nanoparticles were synthesized by reverse micelle process at 100 °C. X-ray powder diffraction was performed on the nanoparticles using a Phillips x-ray diffractometer employing Cu *Kα* radiation from a sealed tube (50 kV, 30 mA) source. A JEOL 2010 model TEM was employed to characterize the morphologies and particle size distribution of the prepared samples. The nanoparticles were suspended in methanol and agitated in an ultrasonic bath. The TEM samples were prepared by placing a few drops of this suspension on a copper grid.
TGA and DTA measurements were performed in air up to 500 °C, with heating and cooling rate of 5 °C/min. The vibrating sample magnetometer was employed to investigate the variation of magnetization with magnetic field at room temperature. The temperature dependence of the magnetic susceptibility of nanoparticles was studied under an applied field of 115 Oe in the temperature range of 100 – 300 K by magnetic susceptometer.

Results and Discussion

Structural and magnetic properties

The XRD patterns of samples in Fig. 1 revealed that in all samples, by substituting Ni and Sr with Co, no extra peak corresponding to any secondary phases were identified. Based on the XRD results, it seems that substituted cations could be rearranged in the spinel structure of cobalt ferrite. It is reasonable to conclude from the patterns that the spinel phases can be formed in the all specimens without existence of any secondary phases at relatively low temperature.
The heating thermograms of sample x=1 prepared by reverse micelle is presented in Fig. 2. In the TGA graph of sample fabricated by reverse micelle 3 % of samples evaporated while heating to 500 ^{o}C. This reduction in mass is related to the evaporation of water and partly due to the oxidized remaining surfactant. Normally the heating analysis of ferrite nanoparticles exhibit one or more exothermic peaks in the temperature range of 200-350 °C. The mentioned peaks are attributed to the amorphous-to-crystalline phase transition and oxidation of remaining surfactant molecules. It is worthy to note that the absence of exothermic peaks in the graphs clarifying that the sample prepared at 100 °C has no secondary hydroxide phases, while in another synthesizing method the existence of sharp weight-loss peak in the TGA curve demonstrate that the intermediate hydroxides are present in the sample.

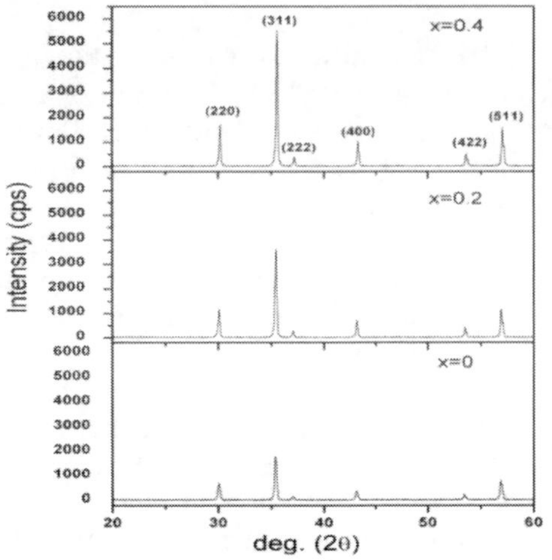

Figure 1. XRD patterns of nanoparticles fabricated via reverse micelle.

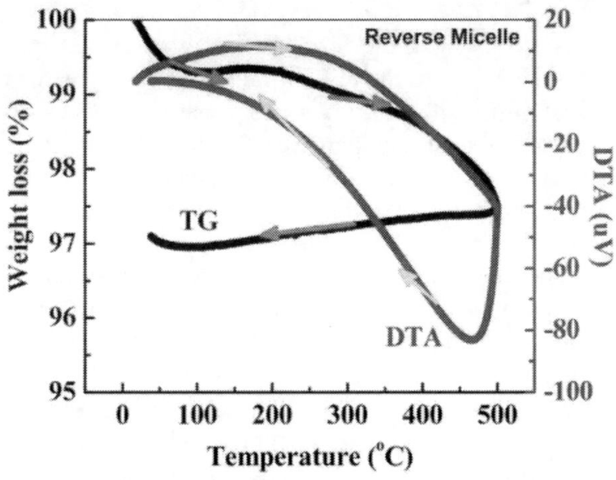

Figure 2. Thermal analysis of nanosize ferrites.

It was found from the VSM loops (not shown here) that the coercivity and remanence are almost zero for all samples. The samples were not saturated even at magnetic field of 20 kOe, which reflects the characteristic of superparamagnetism. The value of saturation magnetization obtained by fitting the high field data to the function $M=M_s(1-\alpha/H)$, where M and M_s are magnetization and saturation magnetization respectively, α is the fitting parameter and H is the applied field, appear in Table 1

.

It was found that with an increase in Ni and Sr content from x=0 to x=1, the values of saturation of magnetization decline from 40 to 23 emu/g. The variation of magnetization of synthesized ferrite nanoparticles prepared by reverse micelle indicated that up to a specific magnetic field, the magnetic moments of core align with applied field. Beyond that field, any increase in the magnetic field has an effect on the shell of the particles and therefore the slope in the increasing of magnetization slows down. The lower values of saturation magnetization of nanocrystalline ferrites prepared by reverse micelle compared to the bulk ferrites is attributed to the surface effects which leads to the non-colinearity of the magnetic moments on their surface and can be explained in terms of core-shell model of nanoparticles in which the core contains ferrimagnetically aligned spin and the surface or interface with a certain degree of spin canting. The disordering in the shell is caused due to local chemical disorder, broken exchange bonds and different local symmetry for those atoms near the surface [21]. Some researchers report that this partially disordered layer gives rise to a spin canting due to the imbalance of ferromagnetic sublattice [22, 23]. However, other authors claimed that the spin canting involves the whole particle and is not restricted to the surface layer [24]. The vacancies or modified cation distribution weakens the super-exchange paths and therefore causes the spin canting in the core of ferrite nanoparticles. It must be noted that if the surfactant molecules from reverse micelle method are absorbed on the surface of nanoparticles and the electrons can no longer participate in the super-exchange interaction, then the pinning of spin occurs [25].

Table 1. Saturation of magnetization of synthesized samples.

x	0	0.2	0.4	0.6	0.8	1
M_s *(emu/g) (nanoparticles)*	40	36	30	27	25	23
M_s *(emu/g) (bulk)*	72	65	59	51	46	40

Interacting or non-interacting nanoparticles

To study the magnetic dynamic behavior of the nanoparticles, we measured the AC magnetic susceptibility of prepared ferrite nanoparticle versus temperature at different frequencies. Figure 3 reflects the temperature dependence of real and imaginary parts of AC magnetic susceptibility of sample x=1 at different frequencies in the range of 111-1000 Hz and at an AC magnetic field of 115 Oe. In Fig. 3 the real parts of effective magnetic susceptibility has a peak value. This characteristic maximum is the signature of blocking/freezing process of the superparamagnetic/spin glass systems [26-30]. The blocking temperature is the threshold point of thermal activation. Above blocking temperature magnetocrystalline anisotropy is overcome by thermal activation and the magnetization direction of each nanoparticle simply follows the applied field direction. Consequently, the nanoparticles show paramagnetic properties. Below the blocking temperature, thermal activation is no longer able to overcome the magnetocrystalline anisotropy of the nanoparticles. As a consequence, the magnetization direction of each nanoparticle rotates from the field direction back to its own easy axis without any movement of the nanoparticle. Since the nanoparticles and consequently their easy axes are randomly orientated, overall susceptibility is reduced with decreasing temperature as Fig. 3 shows. It was found that the blocking temperature increases almost linearly with an increase in Ni and Sr concentration in synthesized samples. The main reason is attributed to the size of nanoparticles. It is well known that blocking temperature has linear proportional to the volume of particles and anisotropy constant. It was found that with an increase in Ni and Sr content the size of particles increases. Consequently, it is reasonable to have the mentioned trend in blocking temperature versus substitutions content.

The interacting or non-interacting behavior of the fine powders was evaluated by Neél- Brown (non-interacting model) and Vogel-Fulcher (interacting law). According to Neél- Brown model, the blocking temperature measured at a given working frequency, is related to the considered frequency as

$$\ln 1/2\pi\nu = \ln \tau_0 + KV/ kT_B \quad (1)$$

T_B can be obtained by assuming that at the maximum of the out-of phase component of the susceptibility, the relaxation time matches the working frequency. Attempt time is calculated from the curves and it is found that it spreads between $\tau_0 = 10^{-25}$ to 10^{-23} s. These values, however, are unphysical, being several orders of magnitude lower than the typical value found in the literature for a non-interacting assembly of superparamagnetic nanoparticles (10^{-8}-10^{-11} s) [26]. This suggests that the Neél- Brown model is not appropriate to describe the dynamic behavior of our system. In this case, deviations from the model can be ascribed to the presence of strong interactions. The interaction between nanoparticles affected the blocking/freezing temperature by modifying the potential barrier.

By increasing the strength of interaction, T_B shifts to higher temperatures. For interacting magnetic nanoparticles, the frequency dependence of T_B is given by the Vogel-Fulcher law [26]

$$\ln 1/2\pi\nu=\ln\tau_0+E_a/k(T-T_0) \quad (2)$$

here E_a is the enegy of barrier, T_0 is an effective temperature which reveals the existence of the interaction between nanoparticles and T is the characteristic temperature indicating the onset of the blocking process. We tried to fit the experimental data of susceptibility for our samples, using Eq. (2). It was found that the values of attempt time were spread between 4.2×10^{-9} to 5.1×10^{-10} s. The good agreement between the experimental data and Vogel-Fulcher model confirms the existence of strong interactions between nanoparticles.

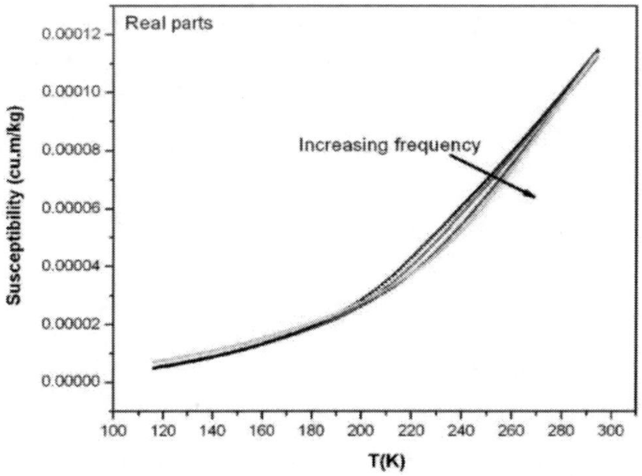

Figure 3. Thermal variation of real parts of effective magnetic susceptibility of synthesized samples.

Conclusions

Fine particles single phase of spinel phase are prepared by reverse micelle method. The value of calculated saturation magnetization for fine particles is much lower than that of large particles. This behavior can be explained by core-shell model. The value of blocking temperature has straightforward relation with particle size. Temperature dependence of real and imaginary parts of effective magnetic susceptibility was measured. Fitting the experimental data of susceptibility with Neél-Brown model gives unphysical high values for relaxation time and indicates that there are strong interactions between nanoparticles of ferrite. The good agreement between the experimental data and Vogel-Fulcher model confirms the existence of strong interactions between nanoparticles.

References

[1] Hyeon T, Chung Y, Park J, Lee S S, Kim Y W, Park B H 2002 *J. Phys. Chem. B* **106** 6831.
[2] Kagawa M, Kikuchi M, Ohno R, Nagae T 1983 *J. Am. Ceram. Soc.* **66** 751.
[3] Blesa M A, Matijevic C 1989 *Adv. Colloid Interface Sci.* **29** 173.
[4] Xiong G, Mai Z, Xu M, Cui S, Ni Y, Zhao Z, Wang Z, Lu L 2001 *Chem. Mater.* **13** 1943.
[5] Cote L J, Teja A S, Wilkinson A P, Zhang Z J 2002 *J. Mater. Res.* **17** 2410.

[6] Chen Q, Rondinone A J, Chakoumakos B C, Zhang Z J 1999 *J. Magn. Magn. Mater.* **194** 1.
[7] Zhang Z J, Wang Z L, Chakoumakos B C, Yin J S 1998 *J. Am. Chem. Soc.* **120** 1800.
[8] Cabanas A, Poliakoff M 2001 *J. Mater. Chem.* **11** 1408.
[9] Bao N Z, Shen L M, Wang Y H, Padhan P, Gupta A 2007 *J. Am. Chem. Soc.* **129** 12374.
[10] Liu J J, He H L, Jin X G, Hao Z P, Hu Z Q 2001 *Mater. Res. Bull.* **36** 2357.
[11] Thakur S, Latyal S C, Singh M 2009 *J. Magn. Magn. Mater.* **321** 1.
[12] Uskokovic V, Drofenik M, Ban I 2004 *J. Magn. Magn. Mater.* **284** 294.
[13] Shultz M D, Allsbrook M J, Carpenter E E 2007 *J. Appl. Phys.* **101** 09M518.
[14] Morrison S A, Cahill C L, Carpenrer E E, Calvin S, Swaminathan R, McHenry M E, Harris V G 2004 *J. Appl. Phys.* **95** 6392.
[15] Morrison S A, Cahill C L, Carpenrer E E, Calvin S, Harris V G 2003 *J. Appl. Phys.* **93** 7489.
[16] Ghasemi A, Šepelák V, Liu X X, Morisako A 2010 *J. Appl. Phys.* 107 09A734.
[17] Ghasemi A, Šepelák V, Liu X X, Morisako A 2010 *J. Appl. Phys.107* 09A743.
[18] Ghasemi A, Šepelák V, Liu X X, Morisako A 2009 *IEEE Trans. Magn.* **45** 2456.
[19] Ghasemi A, Liu X X, Morisako A 2009 *IEEE Trans. Magn.* **45** 4420.
[20] Ghasemi A, Morisako A 2008 *J. Magn. Magn. Mater.* **320** 1167.
[21] Tronc E, Fiorani D, Noguès M, Testa A M, Lucari F, D'Orazio F, Grenèche J M, Wernsdorfer W, Galvez N, Chanéac C, Mailly D, and Jolivet J P 2003 *J. Magn. Magn. Mater.* **262**, 6.
[22] Tronc E, Ezzir A, Cherkaoui R, Chanéac C, Noguès M, Kachkachi H, Fiorani D, Testa A M, Grenèche J M, and Jolivet J P 2000 *J. Magn. Magn. Mater.* **221** 63.
[23] Jiang J Z, Goya G F, and Rechenberg H R 1999 *J. Phys.: Condens. Matter* **11** 4063.
[24] Šepelák V 2002 *Am. Chim. Sci. Mater.* **27** 61.
[25] Linderoth S, Pedersen M S 2002 *J. Appl. Phys.* **35** 3035.
[26] Dormann J L, Fiorani D, Tronc D 1999 *J. Magn. Magn. Mater.* **202** 251.
[27] P. E. Jönsson P E 2004 *Adv. Chem. Phys.* **128** 191.
[28] M. Suzuki M, Fullem S I, Suzuki I S 2009 *Phys. Rev. B* **79** 024418.
[29] Parker D ,Dupuis V , Ladieu F, Bouchaud J P, Dubois E, Perzynski R, Vincent E, Phys. Rev. B **77,** 104428 (2008).
[30] Vasseur S, Duguet E, Portier J, Goglio G, Mornet S, Hadová E, Knížek K, Maryško M, Veverka P, Pollert E, J. Magn. Magn. Mater. **3**
02, 315 (2006).

α-Fe nanopowder synthesised in low-pressure microwave plasma and studied by Mössbauer spectroscopy

B David[1], N Pizúrová[1], O Schneeweiss[1], E Šantavá[2], O Jašek[3] and V Kudrle[3]

[1] Institute of Physics of Materials, Academy of Sciences of the Czech Republic, v.v.i., Žižkova 22, CZ-61662 Brno, Czech Republic
[2] Institute of Physics, Academy of Sciences of the Czech Republic, v.v.i., Na Slovance 2, CZ-18221 Praha 8, Czech Republic
[3] Institute of Physical Electronics, Faculty of Science, Masaryk University, Kotlářská 2, CZ-61137 Brno, Czech Republic

E-mail: david@ipm.cz

Abstract. An iron-based nanocomposite has been prepared by the microwave plasma method: $Fe(CO)_5$ vapour was introduced into an argon discharge at ~1 kPa. The synthesised nanopowder was passivated in situ with a mixture of Ar and air. The as-prepared nanopowder was characterised by XRD, TEM, Raman and Mössbauer spectroscopies. In the XRD pattern of the nanopowder the α-Fe (d_{XRD} = 14 nm, 76 wt.%) and Fe_3O_4 (d_{XRD} = 3 nm, 24 wt.%) phases were identified only. α-Fe cores covered with oxide shells were observed under TEM. A huge increase of Mössbauer absorption was observed after the sample was cooled down to 5 K. The results of magnetic and thermal properties studies at low and high temperature are presented. The heat capacity of the sample exhibited an unexpectedly high value of 599 J/kg/K.

1. Introduction

Magnetic nanoparticles (MNPs) attract the attention of the materials science community because of their unique magnetic and electronic properties [1,2]. Among MNPs iron-based nanoparticles gained high attention [3], especially for their prospective medical applications. α-Fe, γ-Fe and Fe_3C nanoparticles were studied predominantly as the catalysts for growing carbon nanotubes (CNTs). Recently, zero-valent iron nanoparticles are considered for environmental applications [4].

Microwave-induced plasma operated at low pressure [5] offers high power density, moderate temperature, easy scalability and various applications in materials processing. Experimental devices employ freely expanding microwave-induced plasma (MIP), in which case, in the most of used configurations, a quartz tube intersects a rectangular wave guide with a surfaguide wave launcher. MIPs were used to synthesise Fe nanopowders [6-8], Co and Fe-Co nanoparticles [9].

In the present article, we describe the synthesis of α-Fe nanoparticles in a microwave induced plasma. Iron pentacarbonyl $Fe(CO)_5$ has been used as a precursor.

2. Experimental

Our experimental set up was straightforward and similar to those previously used (Fig. 1a in Ref. [10]; Fig. 1 in Ref. [11]). The Ar discharge was generated at low pressure in a cylindrical 45 mm outer diameter fused silica reactor passing through a special launcher, surfaguide (SURFG 439 by Sairem,

France) [11], powered by a microwave (2.45 GHz) generator with regulated power up to 2 kW maximum. $Fe(CO)_5$ vapour was transported from a reservoir to the discharge by argon. Particulate products were trapped at a filter before a rotary pump.

The operating conditions for the LP16 sample were as follows: gas pressure 1.36 kPa, microwave power 500 W, Ar flow 280 sccm and $Fe(CO)_5$ flow 1.75 g/min. The $Fe(CO)_5$ injection point was situated 33 cm downstream from the launcher. Deposition time was 4 minutes. After the deposition, the powder was passivated under 280 sccm of Ar for 2 hours and then the reactor was filled with Ar to 10^5 Pa and left so for 72 hrs before collecting the powder. The as-synthesised amount was 0.5 g.

The composition of the nanopowder was examined by X-ray diffraction (XRD) on a PANalytical X'Pert Pro MPD device. The XRD pattern fitting was done with a commercial software and database and it yielded weight fraction F and mean crystallite size d_{XRD} for a given phase [12,13]. Transmission electron microscopy (TEM) observations were carried out on a Philips CM12 microscope with a wolfram cathode. Raman micro-spectroscopy measurements were performed on a Jobin Yvon microscope Labram HR-800. The spectra were excited with Kr^+ laser radiation (the wavelength of 532.2 nm). In order to avoid thermally induced phase transformation during the collecting of the Raman spectrum, the power of 0.3 mW was used. Mössbauer spectra were obtained in standard transmission geometry with ^{57}Co in Rh matrix. A CCS-800 Mössbauer closed cycle refrigerator system from Janis was used for low-temperature measurements. The spectra were fitted with the CONFIT program using a transmission integral and we obtained the values of relative spectrum area A for a given phase and spectral component parameters: hyperfine magnetic induction B_{HF}, quadrupole shift ε_Q, quadrupole splitting ΔE_Q and isomer shift δ (against α-Fe) [14]. A physical properties measurement system PPMS 9 from Quantum Design was used for low temperature magnetic (ACMS option) and heat capacity measurements (HC option). A vibrating sample magnetometer EG&G PARC was employed for high-temperature magnetic measurements.

3. Results and discussion

3.1. X-ray diffraction

The XRD pattern of the LP16 sample (Fig. 1a) was fitted with cubic α-Fe (ICSD #53451, space group Im-3m, a = 0.2867 nm) and cubic magnetite Fe_3O_4 (ICSD #75627, space group Fd-3mZ, a = 0.8397 nm). The Rietveld refinement yielded for α-Fe values d_{XRD} = 14 nm, a = 0.2867 nm, F = 76 wt.%, and for Fe_3O_4 values d_{XRD} = 3 nm, a = 0.8418 nm, F = 24 wt.%. Fe_3O_4 has its structure (and the XRD pattern) similar to γ-Fe_2O_3 (maghemite), but both tetragonal maghemite Q (ICSD #87121, space group $P4_32_12$, a = 0.8346 nm) and cubic maghemite C (ICSD #87119, space group $P4_332$, a = 0.8346 nm) have lower values of the lattice parameter a. Nevertheless, it is necessary to employ other suitable analytical methods for distinguishing Fe_3O_4 from γ-Fe_2O_3.

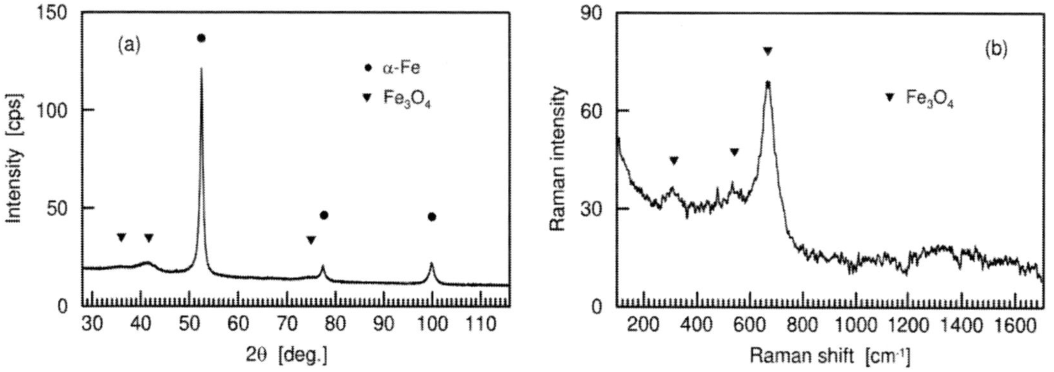

Figure 1. (a) XRD pattern and (b) Raman spectrum for the as-synthesised LP16 nanopowder.

3.2. Raman spectroscopy

In the Raman spectrum of the LP16 sample (Fig. 1b) only Fe_3O_4 characteristic bands at 308, 537 and 664 cm^{-1} were found. It is known that iron oxide magnetite Fe_3O_4 has a spinel structure with space group O_h^7 (above 119 K), giving rise to five Raman bands: three T_{2g}, one E_g and one A_{1g}. Nevertheless it follows from the experiments that the strongest Raman band of magnetite is positioned at 670 cm^{-1} and other significant but much weaker bands are at 300 and 534 cm^{-1} [15]. The characteristic γ-Fe_2O_3 bands (broad structures at 350, 500 and 700 cm^{-1}, a peak at 210 cm^{-1} and very broad bands at 1370 and 1590 cm^{-1} [15]) were not found in the Raman spectrum. Hematite, α-Fe_2O_3, if it were present, would have characteristic bands at 225, 293, 412 and 613 cm^{-1} [15].

3.3. Transmission electron microscopy

The obtained electron diffractograms for the as-synthesised nanopowder confirmed the phase composition found: only the characteristic α-Fe and Fe_3O_4 diffraction circles were observed. Morphology and agglomeration of nanoparticles (Fig. 2a) were similar to nanopowders produced by combustion methods. The large amount of very small Fe_3O_4 crystallites observed in TEM pictures corresponded to the broad Fe_3O_4 diffraction peaks found in the XRD pattern. Smaller α-Fe crystallites

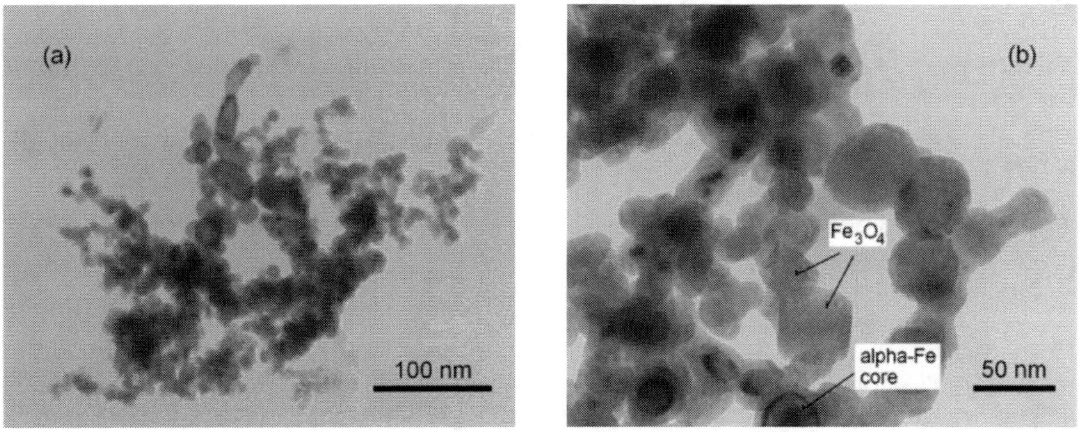

Figure 2. TEM pictures for the as-synthesised LP16 nanopowder.

were imbedded in iron oxide Fe_3O_4 and the larger ones exhibited iron oxide shells (Fig. 2b). These shells are supposed to have defective coordination configuration [16]. Particle size, morphology and connections among particles affected the measured Mössbauer spectra of this nanopowder.

3.4. Mössbauer spectroscopy

The Mössbauer spectrum (MS) of the LP16 sample measured at 293 K (Fig. 3a) exhibits a sunken shape (valley), which is the consequence of the superparamagnetic effect: heating a sample, a sextet belonging to a magnetically ordered phase gradually collapses into a doublet when approaching the blocking temperature $T_B = (KV)/(25k_B)$ (V particle volume, K magnetocrystalline anisotropy constant, k_B Boltzmann constant) [17]. Superparamagnetic state is described by relaxation time $\tau_{SP} = \tau_0 \cdot \exp[(KV)/(k_B T)]$ of flipping of particle magnetic moment orientation ($\tau_0 \approx 10^{-11}$–10^{-9} s) and blocking temperature T_B under which the probability of flipping is very low. Hence, a distribution of particle sizes induces a distribution of blocking temperatures, i. e. a distribution of hardly assignable sextets. The only clearly identifiable components in the MS measured at 293 K were: <u>α-Fe</u> sextet (B_{HF} = 33.1 T, ε_Q = 0.01 mm/s, δ = 0.02 mm/s, A = 0.43; light red filler) [18], <u>doublet</u> (δ = 0.16 mm/s, ΔE_Q = 0.38 mm/s, A = 0.04; dashed line) and <u>singlet</u> (δ = 0.36 mm/s, A = 0.02; full line). The doublet and the singlet belong to Fe impurities in the Al foil in which the nanopowder was wrapped [19].

The superparamagnetic effect was much weaker at 5 K (Fig. 1a). The MS measured at 5 K (with broader lines due to the vibrations caused by the closed cycle cryostat) could be better fitted with: α-Fe sextet (B_{HF} = 34.1 T, ε_Q = 0.00 mm/s, δ = 0.14 mm/s, A = 0.55; light red filler), the superposition of five sextets (#1: 52.2 T, 0.28 mm/s 0.49 mm/s, 0.04; #2: 50.4 T, 0.03 mm/s, 0.61 mm/s, 0.10; #3: 50.1 T, -0.11 mm/s, 0.24 mm/s, 0.09; #4: 48.2 T, -0.17 mm/s, 1.03 mm/s, 0.05; #5: 36.8 T, 0.45 m/s, 1.04 mm/s, 0.04; values given in the format B_{HF}, ε_Q, δ, A) belonging to the low-temperature Fe_3O_4 phase [20], sextet S1 (B_{HF} = 38.5 T, ε_Q = 0.01 mm/s, δ = 0.10 mm/s, A = 0.04; black line), sextet S2 (B_{HF} = 25.9 T, ε_Q = 0.07 mm/s, δ = 0.28 mm/s, A = 0.06; violet line), sextet S3 (B_{HF} = 11.6 T, ε_Q = 0.00 mm/s, δ = 0.13 mm/s, A = 0.02; yellow line), and singlet L1 (δ = 0.24 mm/s, A = 0.01; black line). The components S1, S2, S3 and L1 belong to Fe^{3+} atoms probably at surfaces or interfaces.

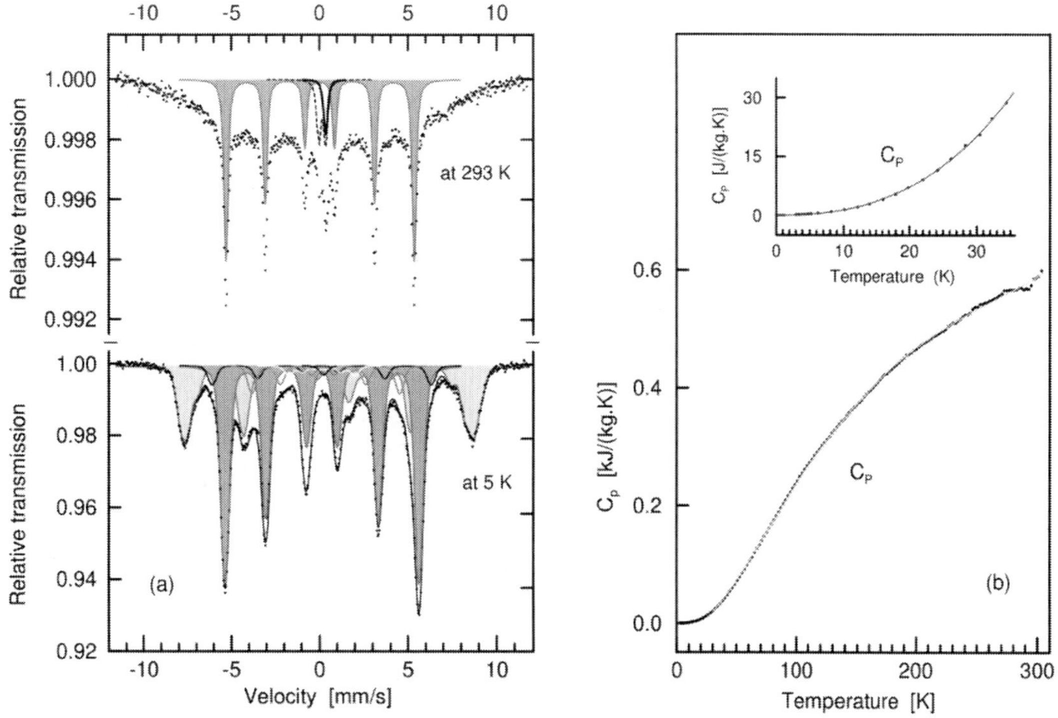

Figure 3. (a) Mössbauer spectra (MS) for the as-synthesised LP16 nanopowder. Sextets with light red filler correspond to α-Fe. (b) Heat capacity C_P for the compacted as-synthesised nanopowder. The low-temperature region is enlarged in the inset.

It can be seen that the absorption in the MS measured at 5 K is seven times stronger than in the MS measured at 293 K, which is a large effect rarely observed. The anomalous decrease of the absorption at 293 K can be caused by diffusive tilting motions of nanoparticles as it was already suggested [21]. Similar decrease was observed for a nanopowder synthesised by laser induced pyrolysis [22].

3.5. Thermal measurement

The as-synthesised powder was compacted and the heat capacity of the pellet (m = 5.51 mg) was measured at constant pressure of $1.2 \cdot 10^{-7}$ bar in the temperature range 3–303 K (Fig. 3b). The literature values C_P(bulk α-Fe|300 K) = 449.5 J/kg/K [23] and C_P(bulk Fe_3O_4|300 K) = 619.5 J/kg/K [24] allow to calculate, based on the XRD result given in the subsection 3.1, the corresponding heat capacity for the bulk sample (0.76·449.5 + 0.24·619.5) J/kg/K = 490.3 J/kg/K, which is significantly lower than the experimentally obtained value of 599.0 J/kg/K for the pellet at 300 K.

The C_P curve exhibited no peak at the Verwey temperature of Fe_3O_4 [24], which is the consequence of small particle size [25]. A simple mathematical fit in the range 30–305 K was $C_P = -6.02 \cdot 10^{-3}T + 1.64 \cdot 10^{-3}T^{3/2} - 8.63 \cdot 10^{-5}T^2 + 6.08 \cdot 10^{-8}T^3 + 6.00 \cdot T^{-2}$ [kJ/kg/K] (green line). In the range 3–35 K C_P behaves basically as T^3 and either $aT^{3/2}$ term or bT^2 term had to be used to obtain a satisfactory fit. The best fit was $C_P = 10.48 \cdot 10^{-3}T^{3/2} + 6.12 \cdot 10^{-3}T^2 + 4.82 \cdot 10^{-4}T^3$ [J/kg/K] (blue line). The bT^2 term probably arises due to the vibrations of the atoms at surfaces [26] and the $aT^{3/2}$ term belongs to spin waves. In the framework of the Debye model [27], the low-temperature approximation $C_V = (12/5)k_B\pi^4N(T/\theta_D)^3$ holds, where k_B is the Boltzmann constant, N is the total number of atoms, and θ_D is the Debye temperature. Using this equation, however, does not seem to provide correct θ_D for either α-Fe particles or Fe_3O_4 particles due to their particle size distributions and morphology. We note only that it holds: θ_D(bulk α-Fe) = 462 K (valid below 20 K) [28] and θ_D(bulk Fe_3O_4) = 550 K (valid below 100 K) [24].

3.6. Magnetic measurements

A loose nanopowder has been pressed into pellets for magnetic measurements. The measurement of the hysteresis loop (HL) at 293 K provided the values: H_C = 31.1 kA/m, σ_R = 45.2 Am²/kg, σ_S = 123.4 Am²/kg (at 1227 kA/m). The measured zero field cooled (ZFC) and field cooled (FC) curves (Fig. 4a) show the split between σ_{ZF} and σ_{FC} curves over the whole measurement range. σ_{ZFC} exhibits no maximum and σ_{FC} remains nearly constant which indicates a broad range of blocking temperatures T_B reaching above room temperature and the presence of single- and multi-domain nonsuper-paramagnetic particles in the sample. The HL measured at 4 K provided the values: H_C = 57.1 kA/m, σ_R = 14.2 Am²/kg, σ_S = 129.0 Am²/kg (at 1227 kA/m).

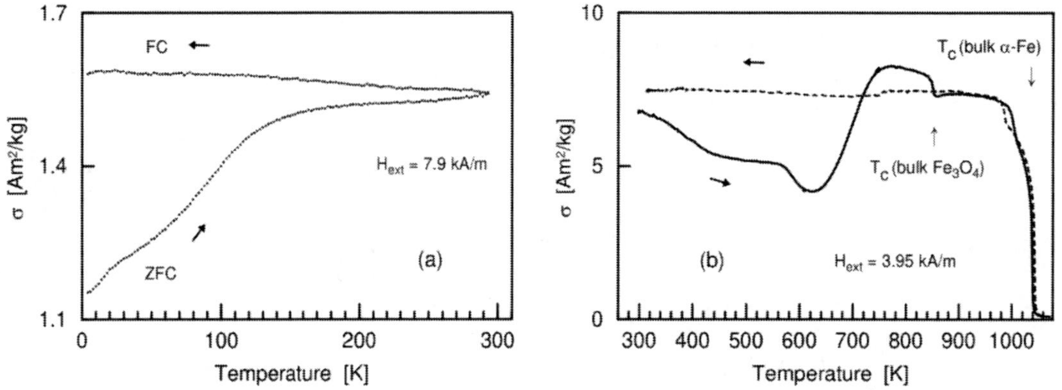

Figure 4. (a) ZFC/FC curves and (b) TM curve for the as-synthesised nanopowder.

By the thermomagnetic curve measurement (TM curve) a pellet was held in the evacuated cell (~1 Pa), heated up to 1073 K (hold time 30 minutes) at the rate of 4 K/min in the external field of 3.95 kA/m and then cooled down to room temperature. The Curie temperatures T_C(bulk Fe_3O_4) = 858 K and T_C(bulk α-Fe) = 1043 K [27] can be recognised on the curve measured with increasing temperature. After starting to heat the sample, σ decreased (minimum at 622 K) and then grew (maximum at 740 K). The increase of σ was caused by the recrystallisation/growth of Fe_3O_4 and α-Fe.

After the whole TM measurement cycle, a HL was measured at 293 K and it exhibited the values: H_C = 3.2 kA/m, σ_R = 5.6 Am²/kg, σ_S = 142.9 Am²/kg (at 1227 kA/m). Subsequently, the phase composition was determined with XRD: α-Fe (d_{XRD} = 128 nm, a = 0.2866 nm, F = 62 wt.%), Fe_3O_4 (d_{XRD} = 41 nm, a = 0.8401 nm, F = 27 wt.%), and FeO (d_{XRD} = 42 nm, a = 0.4330 nm, F = 11 wt.%). Wustite FeO (ICSD #82223, space group Fm-3m, a = 0.4326 nm) appeared as the result of the described heating procedure performed in vacuum.

4. Conclusions

We have synthesised α-Fe nanoparticles in a low-pressure microwave discharge and described its structural, morphological and magnetic properties. The thermal properties of the as-synthesised nanopowder are significantly different from those of bulk samples and ought to be studied on more samples prepared by this method. The reason for the anomalously low absorption in the transmission Mössbauer spectrum measured at 293 K (compared to the spectrum measured at 5 K) was seen in the diffusive tilting motions of nanoparticles and should be further investigated.

Acknowledgement We thank to Dr. V. Machovič (Institute of Chemical Technology, Prague) for the measurement of the Raman spectrum. This work was supported by the GA CR (202/08/0178), the AS CR (AV0Z20410507), and the MEYS CR (MSM0021622411, 1M6198959201).

References

[1] Kodama R H 1999 *J. Magn. Magn. Mater.* **200** 359
[2] Lu A, Salabas E L and Schüth F 2007 *Angew. Chem.- Int. Ed.* **48** 1222
[3] Huber D L 2005 *Small* **1** 482
[4] Antony J, Qiang Y, Baer D R and Wang C 2006 *J. Nanosci. Nanotechnol.* **6** 568
[5] Vollath D and Szabo D V 2006 *J. Nanopart. Res.* **8** 417
[6] Kalyanaraman R, Yoo S, Krupashnkara M S, Sudarshan T S and Dowding R J 1998 *Nanostruct. Mater.* **10** 1379
[7] Poddar P, Wilson J L, Srikanth H, Ravi B G, Wachsmuth J and Sudarshan T S 2004 *Mat. Sci. Eng.* B **106** 95
[8] David B, Pizúrová N, Schneeweiss O, Hoder T, Kudrle V and Janča J 2007 *Deffect and Diffusion Forum* **263** 147
[9] Chou C H and Phillips J 1992 *J. Mater. Res.* **7** 2107
[10] Anthony R, Thimsen E, Johnson J, Campbell S and Kortshagen U 2006 *Mater. Res. Soc. Symp. Proc.* **892** 0892-FF11-05.1
[11] Moisan M and Zakrzewski Z 1991 *J. Phys. D-Appl. Phys.* **24** 1025
[12] X'Pert High Score Plus 2.0a, PANalytical BV, Lelyweg 1, Almelo, the Netherlands
[13] ICSD Inorganic Crystal Structure Database, release 2009/1, FIZ Karlsruhe, Germany
[14] Žák T and Jirásková Y 2006 *Surf. Interface Anal.* **38** 710
[15] de Faria D L A, Silva S V and de Oliveira M T 1997 *J. Raman Spectrosc.* **28** 873
[16] Wang C, Baer D R, Amonette J E, Engelhard M H, Antony J and Qiang Y 2009 *J. Am. Chem. Soc.* **131** 8824
[17] Dorman J L, Fiorani D and Tronc E 1997 *Adv. Chem. Phys.* **98** 283
[18] Ron M 1980 Iron-Carbon and Iron-Nitrongen Systems *Applications of Mössbauer Spectroscopy* ed Cohen R L (New York: Academic Press) pp 329-392
[19] Nasu S, Gonser U, Shingu P H and Murakami Y 1974 *J. Phys. F (Metal Phys)* **4** L 24
[20] Berry J F, Skinner S and Thomas M F 1998 *J. Phys.: Condens. Matter* **10** 215
[21] von Eynatte, Ritter T, Bömmel H E and Dransfeld K 1987 *Z. Phys. B Cond. Matter* **65** 341
[22] David B, Pizúrová N, Schneeweiss O, Klementová M, Šantavá E, Dumitrache F, Alexandrescu R and Morjan I 2007 *J. Phys. Chem. Solids* **68** 1152
[23] *CRC Handbook of Chemistry and Physics, 80th Edition* 1999 ed D R Lide (New York: CRC Press), p 12-193
[24] Westrum E F and Grønvold F 1969 *J. Chem. Thermodyn.* **1** 543
[25] Goya F, Berquó T S and Fonseca F C 2003 *J. Appl. Phys.* **94** 3520
[26] Zhang H and Banfield J B 1998 *NanoStructured Mater.* **10** 185
[27] Kittel C 2005 *Introduction to solid state physics* 2005 (New York: Wiley) p 328
[28] Keesom W H and Kurrelmeyer B 1939 *Physica* **6** 633

The influence of assist gas on magnetic properties of electrotechnical steel sheets cut with laser

Dominika Gaworska-Koniarek[1], Bronisław Szubzda[1], Wiesław Wilczyński[2], Jerzy Drosik[3], Kazimierz Karaś[3]

[1]Electrotechnical Institute, Division of Electrotechnology and Materials Science, Wrocław, Poland, [2]Electrotechnical Institute, Warsaw, Poland, [3]ALSTOM Power Sp. z o.o. Wroclaw Branch, Wrocław, Poland

gaworska@iel.wroc.pl

Abstract. The paper presents the influence of assist gas (air and nitrogen) during laser cutting on magnetization, magnetic permeability and loss characteristics of non-oriented electrical steels. The research was made on an non-oriented M330-50A grade electrical steels by means of single sheet tester. In order to enhance the effect of cutting and the same degradation zone on magnetic properties, strips with different width were achieved. Measurements results indicate that application of air as assist gas has more destructive effect on magnetic properties of electrical steels than nitrogen one.

1. Introduction

Electrotechnical steels – in relation to the rate of production – constitute the most important magnetically soft materials which are used in the production of magnetic cores of electric machines. In order to follow the market needs, the manufacturers of electric machines still look for the cheapest and quickest technologies of magnetic cores production. Increasingly, manufacturers resign from the mass production of machines and they produce them only to a special order as well as they adjust sizes and parameters to a specific order. Small and short series of production as well as a continuous improvement of machine constructions force manufacturers of machines to change production lines quickly and economically.

Nowadays, in the majority of electrotechnical industry factories which produce the magnetic circuits elements of electric machines, punching of sheets by means of punching dies on the fast-cutting presses have been prevailing for decades invariably (over 300 hits a minute). The process of punching is quick, while the time which is needed for its preparation is very long. A considerable cost and long time of making punching dies, the necessity of sharpening as well as a limited time of their existence have become, among others, the reason for searching other methods of cutting out shapes, for example, by means of laser. In spite of the fact that the process of cutting out by means of laser is more time-consuming than the mechanical punching, it has a number of advantages which increasingly decide about its application. Cutting by means of laser opens up possibilities of making changes in forms and sizes of shapes without bearing additional costs. This requires only the change in the computer program which controls the laser work. Moreover, considering a non-contact cutting effect, in this process the burr edge effect is avoided, which, in turn, can lead to the short-circuit between the laminated sheets and in this way can become a reason of the increase of losses from eddy currents. Nevertheless, the performance of the impulse laser beam with a big power density in a short time, which causes melting and partial vaporization of the material, can change a microstructure of the electrotechnical steel sheet along the cutting line. Laser cutting out of steel is commonly used in the industry but its influence on magnetic properties of electrotechnical steel sheets is still not recognized well in spite of many years' studies.

The results presented in the literature are often discrepant and to a large extent they depend upon the employed type of eletrotechnical steel, sample sizes, laser beam power, laser type, etc. For instance, in [1], it was shown that laser cutting of non-oriented electrotechnical steels leads to the increase in total energy loss in the range 3 to 17 %. However, this increase is lower than what we observe after mechanical punching. The research by Lanott [2] conducted on 3% Si grain-oriented electrotechnical steels also confirmed the advantage of laser cutting over mechanical cutting as regards the total energy loss. On the other hand, Belhadj in [3] proved that laser cutting in the assisting atmosphere of oxygen deteriorates the magnetic properties of electrotechnical steels to a larger extent than punching. It particularly has an impact on the increase of losses as well as the decrease of magnetic permeability. In the case of the coercion value, depending on the selected cutting parameters and the type of an electrotechnical steels, the result was the coercion decrease or increase in relation to the punched samples. Dupre in [4] showed that laser cutting causes the greatest increase of loss from hysteresis in electrotechnical steels in comparison to punching.

All of the above mentioned publications investigate the influence of different parameters of the laser cutting process on the magnetic properties of electrotechnical steels, however, none of them takes into account the influence of the associated gas which is employed during this process and whose type and pressure can influence the final magnetic properties of the shape. The most commonly encountered results in the literature of the subject are those for laser cutting in the associated atmosphere of nitrogen, and very rarely of oxygen [3]. However, there are no publications known to the authors, which take into account the influence of the associated gas employed during this process whose composition and type can affect the final magnetic properties of the shape. This paper, in which we present the influence of the two most commonly used associated atmospheres – air and nitrogen on the magnetic properties of electrotechnical laser cut steels, partly fills in this gap.

2. Magnetic measurements

The aim of the magnetic investigations was to determine the influence of cutting and the type of the employed associated gas on the magnetic properties of electrotechnical steels – the magnetization curve, magnetic permeability and specific total energy loss. The measurements were executed by means of the computerized measuring position MAG-RJJ-3.0 with the usage of the Single Sheet Tester (SST), according to the norm PN-EN 10280:2003 [5] on two sets of M330-50A grade electrotechnical steel samples cut out by means of laser in two different associated atmospheres. These were: 1) laser cut out samples in the atmosphere of the air, 2) laser cut out samples in the protective atmosphere of nitrogen. The samples were magnetized perpendicular to the direction of rolling. The set consisted of samples with the following dimensions:

- 500x500 mm sheet – two cutting edges,
- 500x500 mm sheet cut into two equal parts perpendicular to the direction of rolling (dimension 250x500 mm) – four cutting edges,
- 500x500 mm sheet cut into four equal parts perpendicular to the direction of rolling (dimension 125x500 mm) – eight cutting edges,
- 500x500 mm sheet cut into eight equal parts perpendicular to the direction of rolling (dimension circa 62.5x500 mm) – 16 cutting edges,
- 500x500 mm sheet cut into 16 equal parts perpendicular to the direction of rolling (dimension circa 31.25x500 mm) – 32 cutting edges,

500x500 mm sheets constituted the so called reference sample. Along with the sheet dimension getting smaller, there appears the increase of the influence of the deformed and, if necessary, oxidized cutting zone on the magnetic properties of the material. In the case of 500x500 mm sheets, on account of a big distance of the cutting edges and at the same time an insignificant part of the deformed zone in the total mass and volume of the sample, it can be assumed that they are deprived of stresses connected with the sample preparation. The appearance of lateral surfaces of M330-50A grade steel cut by means of the punching die and laser is presented in Fig.1.The magnetization curve, magnetic permeability and specific total energy loss were measured. The chosen measurements results are presented in table 1. The course of magnetization curves of the examined samples was shown in Fig. 2. Relative changes of maximum magnetic permeability (mmax) and total energy loss (Pc) are presented in Figs. 3a and 3b. For comparison purposes, the diagrams also present the results obtained for 16 punched strips of M330-50A grade steel size 31.25x500 mm.

Table 1. Chosen magnetic properties of laser cut out sheets in the atmosphere of nitrogen and air

Sample	Width of strip [mm]	B_m [T] for 100 A/m	500 A/m	1000 A/m	μ_{max} [-]	$H_{\mu max}$ [A/m]	P_c [W/kg] 1.0T	1.5 T
Reference	500	0.82	1.40	1.46	6548	100	1.36	3.19
Laser cut–nitrogen atmosphere	250	0.79	1.40	1.46	6318	100	1.38	3.21
	125	0.74	1.40	1.46	5886	100	1.42	3.30
	62.5	0.62	1.39	1.45	5108	150	1.50	3.42
	31.25	0.40	1.37	1.45	3713	150	1.73	3.72
Laser cut – air atmosphere	250	0.79	1.40	1.46	6242	100	1.38	3.21
	125	0.70	1.39	1.46	5572	100	1.44	3.33
	62.5	0.54	1.38	1.46	4683	150	1.57	3.59
	31.25	0.33	1.35	1.44	3258	200	1.83	3.83
Punched	31.25	0.70	1.35	1.44	5569	100	1.48	3.44

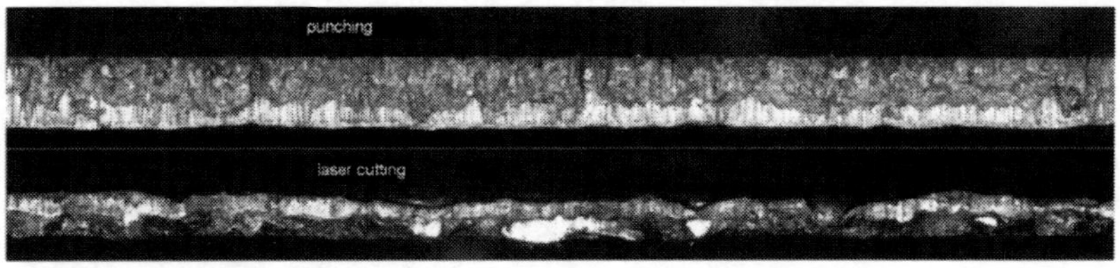

Fig.1. Sheet edges after cutting by means of the punching die and laser

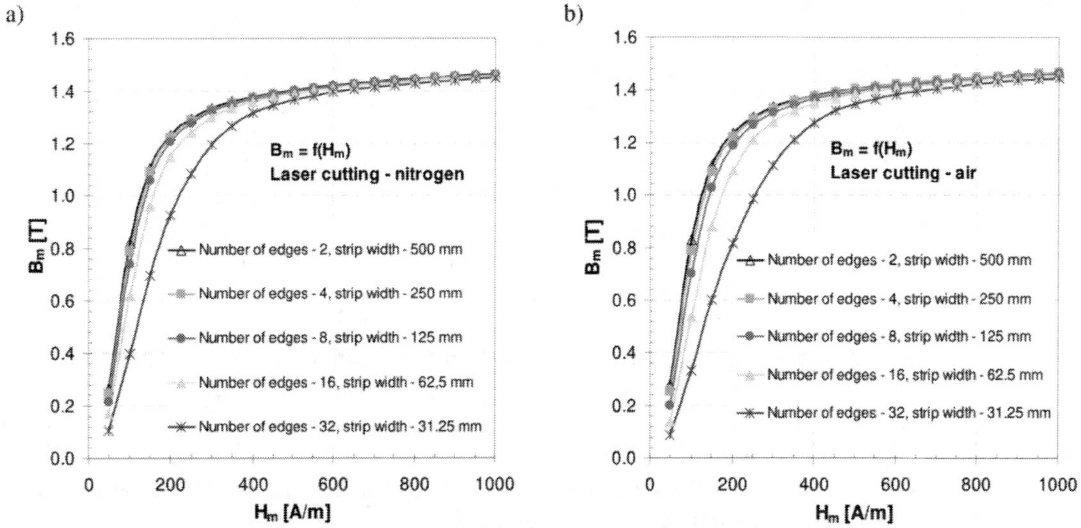

Fig. 2. The magnetization curves for different strip widths cut out with laser in the nitrogen atmosphere, with laser in the air atmosphere

Thermally degraded areas of lateral surfaces of laser cut out sheets prove the suspicion as to a possible influence of the chemical composition of the protective gas atmosphere in which this process takes place on the level of degradation, which can be observed in Fig. 1. Particularly, by the application of the oxidized atmosphere – air, we can expect foreign layers on surfaces which appear as a result of high energy laser cutting.

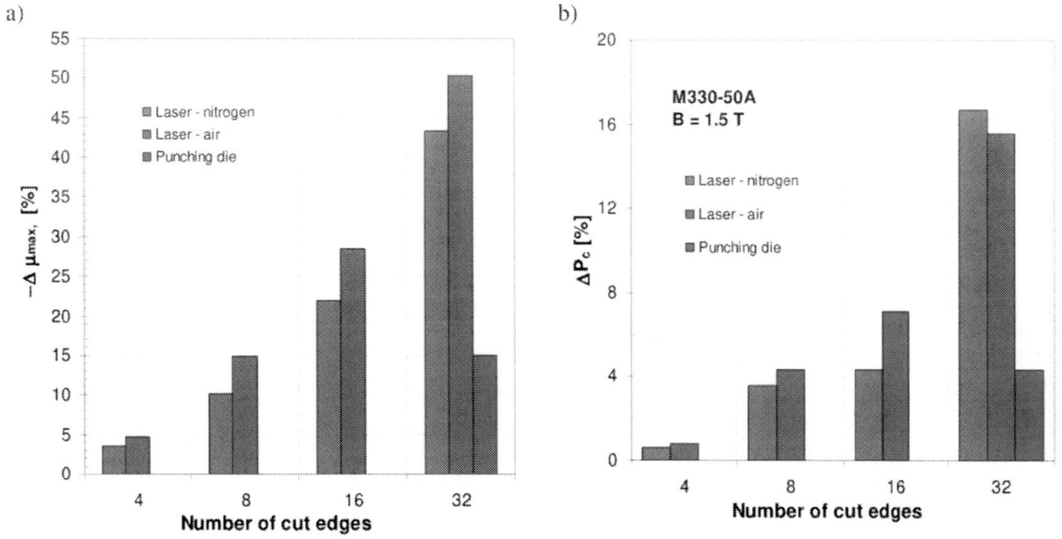

Fig. 3. A relative change of maximum magnetic permeability (a) and total energy loss ΔP_c for B_m=1.5 T (b), for different strip widths cut out with laser in the nitrogen atmosphere, with laser in the air atmosphere and with a punching die

3. X-ray examinations

X-ray measurements were carried out on DRON-2 prom. Co/Fe diffractometer on samples of laminated sheets stuck together and placed in a frontal position. X-ray diffraction patterns of samples are presented cumulatively in Fig. 4.

Barium sulfate $BaSO_4$ which was identified in diffraction patterns, comes from the superficial insulating layer of electrotechnical steels. Additionally, in the sample, which was cut in the protective atmosphere of the air, the presence of iron oxides was noticed. As far as magnetic properties are concerned, a particularly unfavourable phenomenon is the formation of magnetite Fe_3O_4, which shows hard magnetic properties (coercion H_c of the order of some kA/m).

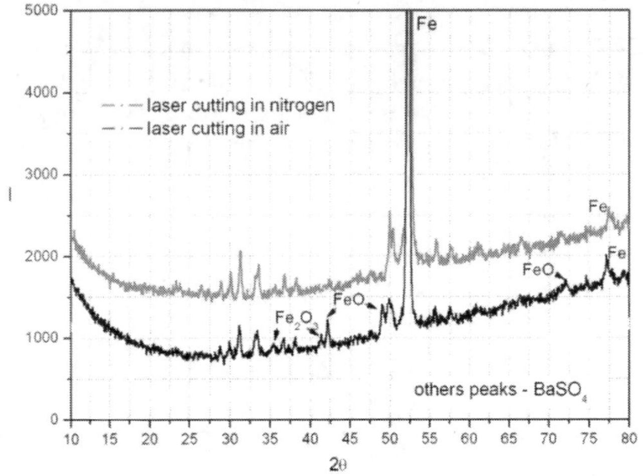

Fig. 7. Cutting surface diffractometers of the examined electrotechnical sheets

Samples cut with laser in the air atmosphere also show a smaller size of crystallites (~ 52 nm) than steels cut in the nitrogen atmosphere (~ 55 nm). This can happen as a result of greater structure defecting. Dimensions of the crystalline cell were not changed.

4. Research results analysis

The research results show that the strips widths, which are part of the examined sample, i.e. the increase of the degraded zone share as a result of cutting/punching, have an influence on the electrical steels magnetic properties. The increase of the degraded zone share causes the flattening of magnetization characteristics of the examined samples – Fig. 2 and at the same time, a decrease of magnetic permeability – Fig. 3a. This effect is particularly seen for the narrowest samples.
The width of component strips and at the same time, the cutting zone share in the sample volume has also an influence on the total specific loss of the examined electrotechnical sheets – Fig. 3b. Total energy losses after laser cutting in the air atmosphere increased by 35% for 1.0 T induction and by circa 20% for 1.5 T induction. The application of the atmosphere of nitrogen at cutting limited the increase of total energy losses up to 27% for 1.0 T induction and 17% for 1.5 T induction. When using punching the increase of losses is much smaller and is circa 9% for 1.0 T induction and circa 5% for 1.5 T induction.

The reasons for deterioration of magnetic properties are compressive stresses which appear during cutting/punching and which result in changing the magnetic structure of the material and the increase of 90° domain volume at the cost of 180° domains [6]. Therefore, the process of magnetization changing requires the use of greater external magnetizing field in order to change the domain structure in the stress area along the lines of punching. This is the reason of the increase of hysteresis losses and anomalous losses [6]. Movements of domain wall take place at the lower inductions (a steep part of the magnetic curve) and in this range they are mainly responsible for the hysteresis loss. During the magnetization process when inductions are higher, the processes which dominate are connected with domain magnetization vector rotations and their influence on hysteresis losses is smaller. Thus, a greater percentage-wise increase of a total specific loss can be seen for the lower values of induction. A big increase of hysteresis loss after laser cutting in comparison to the increase of loss after punching shows that during this process bigger stresses and/or in a bigger sample volume than in the case of punched samples are introduced into the material.

After punching, as a result of plastic deformations the zone of the deformed material is created (the zone width measured from the edge is from 0,1 mm for sharp blades, up to 0.23 mm for the blunt ones [7]) as well as the stressed zone in which there are no visible grain strains. On the other hand, after laser cutting the shrinkage of the material appears as a result of rapid heating at first and then equally fast cooling by means of the protective gas or air flux. An impulse laser beam which acts within a short time with a great power density produces a very high unsteady gradient temperature field along the cutting line. A very quick heating makes the material boil, which is immediately cooled with a high speed by means of the associated gas which blows out a liquid metal and its vapours from the gap. Within this layer at such a high temperature dislocations of atoms take place. In the grains along the cutting line, displacement of atoms may cause edge dislocations to climb; whereas, on the grain boundaries it may cause the slipping of one grain after the other. It may even lead to the creation of micro-cracks. These phenomena can become the reason of a significant increase of dislocations and stresses at the edge of the material, decrease of crystallites, as a result of which the magnetic properties may deteriorate [3]. The total width of deformed and stressed zones in the case of punching as well as in the case of cutting out is much wider and makes the so called slightly magnetic zone. For the non-oriented sheet it is circa 0,6 mm after punching [8,9] and up to 6.0 mm after laser cutting [10].

In the case of laser cut electrotechnical steels in the air atmosphere, iron oxide (Fe_3O_4) – magnetite which is created as a result of air cooling additionally hinders the process of magnetization; the presence of magnetite is proved by the conducted X-ray examinations. As it is a material magnetically hard, in order to change its domain structure we need a greater energy of the external magnetic field the consequence of which is the flattening of magnetization characteristics, decrease of magnetic permeability and increase of hysteresis and anomalous losses. Moreover, foreign inclusions such as oxides appearing on the cutting edges can act as an pinning centres which inhibit the movement of domain walls and additionally increase the coercion of material and hysteresis loss.

5. Summary

The executed research explicitly indicate that the participation of the zone degraded as a result of cutting of the examined electrotechnical steels measured with the number of cutting edges, has a significant influence on magnetic properties of electrotechnical steelss. Decreasing the widths of strips included in the examined sample and at the same time increasing the participation of degraded zone causes flattening of the magnetization curves as well as the decrease of maximum magnetic permeability. We can also observe the increase of hysteresis and additional losses.

A negative influence of cutting/punching on magnetic properties of electrotechnical steels is more visible in the case of laser cut samples than in the case of punched samples. Moreover, we can see the influence of the applied assist atmosphere during laser cutting on the level of deterioration of magnetic properties of electrotechnical steels – the employment of the air atmosphere has a more unfavourable impact on the steel magnetic properties than the nitrogen atmosphere. Most probably, this results from the creation of iron oxides on the cutting edges, which can act as an anchor centres inhibiting the movement of domain walls. Moreover, iron oxide Fe_3O_4 shows hard magnetic properties, which additionally increases the material coercion and hysteresis losses.

References

[1] Dickmann K., Influence of laser cutting process on the magnetic properties of electrical sheets, Anales de Fisica B, 86, 82, 1990.

[2] Lanotte L., Luponio C., De Iorio I., Tagliaferii V., Tammaro G, Effect of laser cutting on magnetic properties of grain oriented Fe-Si ribbons, Mater. Sci. Technol., 8, 252-256, 1998.

[3] Belhaldj A., Baudouin P., Breaban F., Deffontaine A., Dewulf M., Houbaert Y., Effect of laser cutting on microstructure and on magnetic properties of grain non-oriented electrical steels, J. Magn. Magn. Mater. 256, 20 -31, 2003.

[4] Dupre L.R., Van Kerr R., Melkebeek J.A.A, A study of the influence of laser cutting and punching on the electromagnetic behaviour of electrical steel sheets using a combined finite element-dynamic Preisach model., Fourth International Workshop on Electric and Magnetic Fields, Marseille, France, 12-15 May, 195 - , 1998.

[5] PN-EN 10280:2003 Magnetic materials – Measurement methods of sheets and electrotechnical

tapes magnetic properties by means of the device for single sheet examination.
[6] LoBue M., Sasso C., Basso V., Fiorillo F., Bertotti G., Power losses and magnetization process in Fe -Si non-oriented steels under tensile and compressive stress, J. Magn. Magn. Mater. 215 - 216, 124-126, 2000.
[7] Matheisel Z., Cold-rolled electrotechnical sheets, WNT Warszawa, 1973.
[8] Wilczyński W.: The influence of technology on magnetic properties of electric machine cores, Prace Instytutu Elektrotechniki, zeszyt 215/2003
[9] Wilczyński W., Szubzda B., Talik S., Lipiec W., Aspects of the punching and laser cutting effect on the power losses and flux density distribution in electrical steel, Proceedings of SMM16, 499, Düsseldorf 2003.
[10] Loisos G., Moses A.J., Effect of mechanical and Nd:YAG laser cutting on magnetic flux distribution near the cut edge of non-oriented steels, Journal of materials processing technology, 161, 151-155, 2003.

Subtractively Prepared Permalloy Nanowires for Spin-Torque Experiments

Gesche Nahrwold, Sedat Dogan, Lars Bocklage, Toru Matsuyama, Ulrich Merkt, and Guido Meier
Institut für Angewandte Physik und Zentrum für Mikrostrukturforschung, Universität Hamburg, Jungiusstrasse 11, 20355 Hamburg, Germany

Kouta Kondou, Gen Yamada, and Teruo Ono
Institute for Chemical Research, Kyoto University, Kyoto 611-0011, Japan

E-mail: gesche.nahrwold@physik.uni-hamburg.de

Abstract. Physical properties of Permalloy ($Ni_{80}Fe_{20}$) can be considerably enhanced by sputtering on substrates heated to temperatures around 300 °C. To enable the use of sputtered Permalloy in micro- and nanostructures for spintronic experiments a subtractive preparation process has to be established, as common lift-off processing is incompatible with high temperatures. Two subtractive fabrication processes for curved Permalloy nanowires are executed: Ion-milling and rf-sputter-etching. The results of current-induced domain-wall depinning experiments with these nanowires are compared.

1. Introduction

Permalloy ($Ni_{80}Fe_{20}$) is used in many experiments on domain-wall dynamics, spin-torque effects [1, 2, 3], spin valves [4, 5, 6, 7, 8, 9], and dynamics of magnetic vortices and antivortices [10, 11, 12, 13, 14, 15]. We demonstrate that sputtering Permalloy on substrates heated to 300 °C leads to advantageous properties for such experiments [3, 16], e.g. low specific resistances and reasonable depinning fields. Most commonly lift-off processing is employed to prepare micro- and nanostructures. To make use of the advantageous properties of Permalloy sputtered on heated substrates micro- and nanostructures have to be prepared subtractively. Two promising ways on this path are either to ion-mill or by rf-sputter etching Permalloy films sputtered on heated substrates. Both approaches have been studied and will be described and compared.

2. Sample Preparation

Films of Permalloy with thicknesses of 20 or 30 nm are sputtered on Si substrates covered with a 300 nm thick SiO_2 layer. During the sputter process the substrates are heated to 280 °C, 300 °C, or 320 °C. The sputter system is a three-target magnetron-sputtering system with a transfer chamber and a base pressure of around 1×10^{-9} mbar. Argon pressure and DC power have been optimized for controlling deposition rates and film roughnesses [17]. The Permalloy films studied in this work were sputtered at a rate of 0.11 nm/s, with a DC power of 50 W and an argon pressure of 1×10^{-2} mbar.

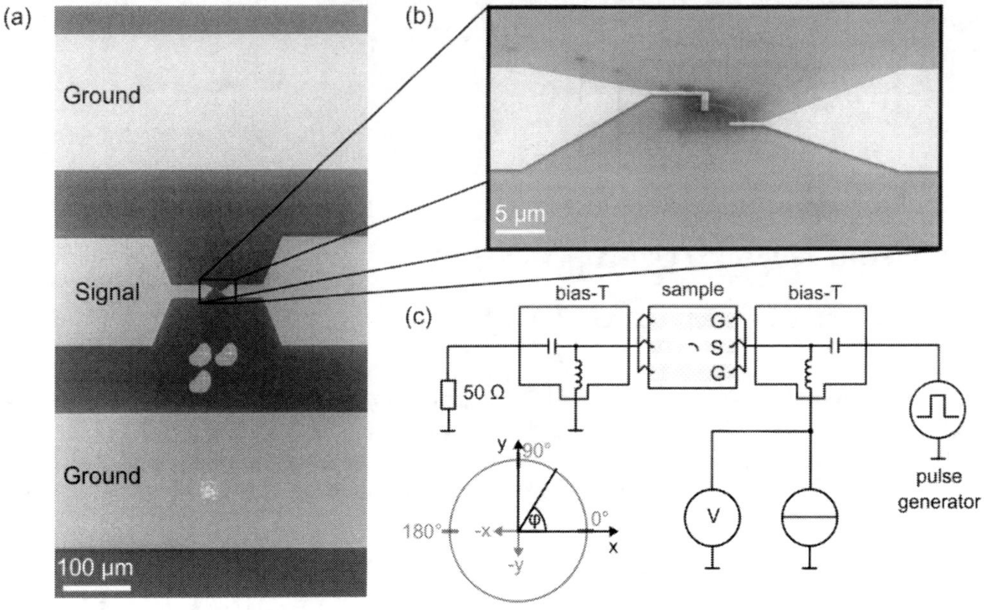

Figure 1. (a) Scanning electron micrograph of the sample, including the curved Permalloy nanowire in the middle and the coplanar waveguide made of gold. (b) Blowup of the inner part of (a) with the curved Permalloy nanowire in the middle, contacted by the signal leads. (c) Schematic of the measurement setup.

To subtractively prepare curved Permalloy nanowires for the investigation of domain-wall motion from these films the structures are defined by electron-beam lithography with negative resists. Resists used here are ma-N 2403 by micro-resist in case of the ion-milled wires and ARN 7520 by allresist in case of the rf-sputter-etched wires. After development the desired structure on the sample remains covered by resist while the leftover parts of the metal film that are meant to be removed are uncovered. The uncovered parts of the samples are then removed by ion-milling and rf-sputter-etching, respectively. Both processes are based on physical etching by argon-ion bombardement. In case of ion-milling a power of 18 W at an argon gas flow of 8.9 sccm yield an etching rate of nanometers per second (here 0.22 nm/s). For rf-sputter etching a power of 10 W at an argon-pressure of 1×10^{-2} mbar results in an etch rate of nanometers per minute (here 0.02 nm/s, i.e. by a factor of 11 slower than ion-milling).

3. Principle of Measurement

The samples consist of a 30 nm thick (20 nm in case of the sputter etched wires), 200 nm wide curved Permalloy wire with a radius of curvature of 3 μm. It is contacted by two gold leads in coplanar waveguide geometry that are used as voltage probes to measure the resistance as well as to inject the current pulse as visible in Fig. 1. Between the contacts the Permalloy wires show resistances of several hundred Ω depending on the actual geometry. The Permalloy wires and their contacts are prepared on a silicon substrate covered with a thin SiO_2 layer to prevent bypass conductivity. Samples are produced in a two-step process. First the curved Permalloy wire is prepared by subtractive etching from a film that has been sputtered onto the heated substrate; second the larger electrode structures are lithographically defined and then

Figure 2. Scanning electron micrograph of a 200 nm wide and 20 nm thick Permalloy wire subtractively prepared by rf-sputter etching from a film sputtered onto a substrate heated to 300 °C. The white circles at the edge of the wire with diameters of 5, 10, and 15 nm illustrate the roughness of the wire.

deposited by thermal evaporation of 100 nm gold. A schematic of the measurement setup is shown in Fig. 1(c). The resistance of the wire is measured by a dc current of I = 0.1 mA. To induce a head-to-head domain wall in the curved region of the wire a magnetic field of 140 mT is applied under an angle of φ=45°, as shown in Fig. 1. By decreasing the magnetic field the magnetization relaxes along the wire and a domain wall is created in the curved region of the wire [18, 3, 19]. The domain wall is sensed via its contribution to the resistance by the anisotropic magnetoresistance [3]. Subsequently current-induced motion of a domain wall is caused by single current pulses with variable length while an additional magnetic field is applied under an angle of φ=180°. Then the wire resistance is measured again. If the domain wall is depinned by spin-torque transfer and moved out of the region between the taps, the difference in resistance before and after the pulse ΔR is increased by about 0.05 Ω - 0.3 Ω.

4. Experimental Results

Permalloy nanowires can be subtractively prepared by ion-milling as well as by rf-sputter etching. A wire prepared by the latter technique is shown in the scanning electron micrograph in Fig. 2. The edge roughness of the wires of 5-15 nm for both fabrication processes are around the limit of accuracy of the lithography system.

Before starting current-induced domain-wall depinning experiments each wire is characterized by applying increasing magnetic fields under various angles between the negative x- and the positive y-direction (see Fig. 1, φ=180°-90° in steps of 5° or 10°). By measuring the resistance during the increase of the magnetic field a jump in the measured resistance indicates the field value which is strong enough to depin the domain wall. These measurements are repeated several times to determine the angle with the lowest and most stable depinning fields. For both subtractive fabrication processes low depinning field values, often below 1 mT, are observed. In comparison to that, wires fabricated by thermal evaporation and lift-off processing show depinning fields of 5-10 mT [3, 20].

Measurements on the ion-milled wires with current pulses with a maximum current density of $j = 1.37 \times 10^{12}$ A/m^2 revealed several resistance difference levels ΔR as shown in Fig. 3. The wire exhibits a resistance of 438 Ω, the magnetic background field was varied between 0 and 3 mT in steps of 0.01 mT, and the current pulse length was varied between 0.2 and 10.2 ns in steps of 0.2 ns. Each combination of magnetic background field and pulse length has been repeated five times, summing up to more than 150,000 measurements of the resistance difference ΔR with and without a domain wall. A histogram of all measured ΔR is shown in Fig. 3(a).

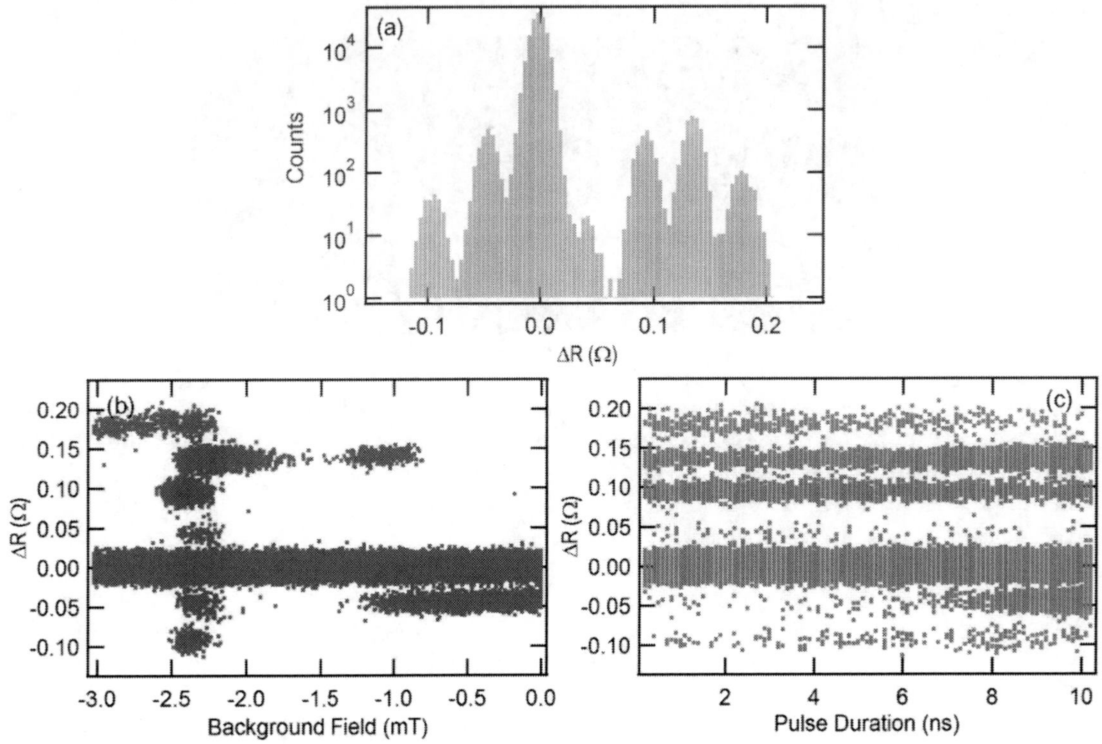

Figure 3. Experimental results of current-induced domain-wall depinning in an ion-milled Permalloy wire: (a) Histogram of measured resistance differences ΔR with and without a domain wall, (b) ΔR vs. background field, and (c) ΔR vs. pulse duration.

The highest peak at $\Delta R = 0$ corresponds to no change in resistance before and after sending the current pulse through the wire. Thus in these cases the domain wall has not been moved by the current pulse. The four countable positive peaks in the resistance difference can easily be understood as different types of domain walls being depinned by the current pulse. The negative values of ΔR are more difficult to understand and most probably correspond to changes in the magnetic configuration of the domain wall rather than real depinning events. The negative ΔR values correspond to a larger resistance of the wire after a current pulse has been sent through the wire, meaning that a larger amount of magnetization is aligned perpendicular to the current flow. This can be interpreted as a change in the domain-wall structure [21] due to the current pulse. This has been observed with imaging techniques like magnetic force microscopy [22] and x-ray microscopy [2]. However, statistical interpretation with a large number of repetitions is not possible with imaging techniques. We deduce that the observed behavior is caused by a specific unknown potential landscape that influences the motion of the domain wall when moving in the negative x-direction. Measurements with a current pulse pushing the domain wall along the wire in the negative y-direction with a background magnetic field assisting the motion of the domain wall in the same direction (not shown) indicate that these domain-wall transformations are reproducible and thus contradict the above speculation. Furthermore similar observations in ion-milled wires have been reported by Hayashi et al. [21] as well as by Malinowski et al. [23]. Figures 3(b) and (c) show the measured values of the resistance change ΔR versus the magnetic background field and the pulse duration, respectively.

The corresponding data of an experiment on current-induced domain-wall depinning in a

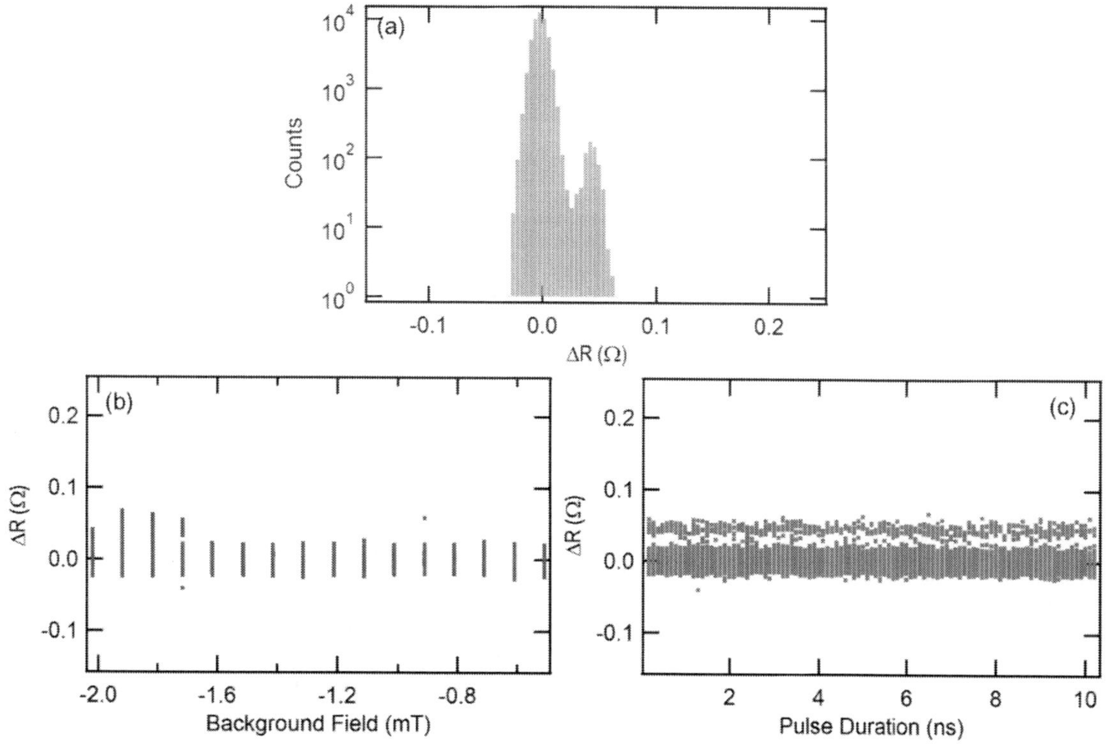

Figure 4. Experimental results of current-induced domain-wall depinning in a rf-sputter etched Permalloy wire: (a) Histogram of measured resistance differences ΔR, (b) ΔR vs. background field, and (c) ΔR vs. pulse duration.

rf-sputter etched Permalloy nanowire is displayed in Fig. 4. This wire exhibits a resistance of 405 Ω, the magnetic background field was varied between 0.5 and 2.1 mT in steps of 0.1 mT and the current pulse length was varied between 0.2 and 10.2 ns in steps of 0.2 ns and had a maximum current density of $j = 1.32 \times 10^{12}$ A/m^2. Each combination of magnetic background field and current pulse length has been repeated 30 times, summing up to more than 48,000 measurements of ΔR. In this rf-sputter etched wire only two resistance levels have been observed. The two resistance levels of either zero or 0.05 Ω shown in Fig. 4(a) indicate three possibilities. Either there was a wall in the wire and it still is there after the current pulse, or the wall got moved out of the wire by the background magnetic field before the current pulse arrived, or there was a wall inside the wire that has been moved out of the wire by the current pulse. The latter corresponds to the resistance difference of ΔR = 0.05 Ω. However, no dependence of the domain wall depinning probability on current pulse length [3] or current pulse shape [19] could be observed in the rf-sputter etched wires. Note, that the value ΔR ≈ 0.05 Ω is comparable to the lowest measured positive ΔR in the ion-milled wire described above. This could imply that the same domain-wall type has been observed in both cases.

5. Summary and Discussion

In summary we are able to subtractively prepare nanostructures from Permalloy films sputtered on heated substrates. This process enables using the advantageous properties of this material which is prevented by additive preparation processes due to two reasons: First, additive sputtering may alter the cross-sectional geometry of the wires and second, lift-off of resists

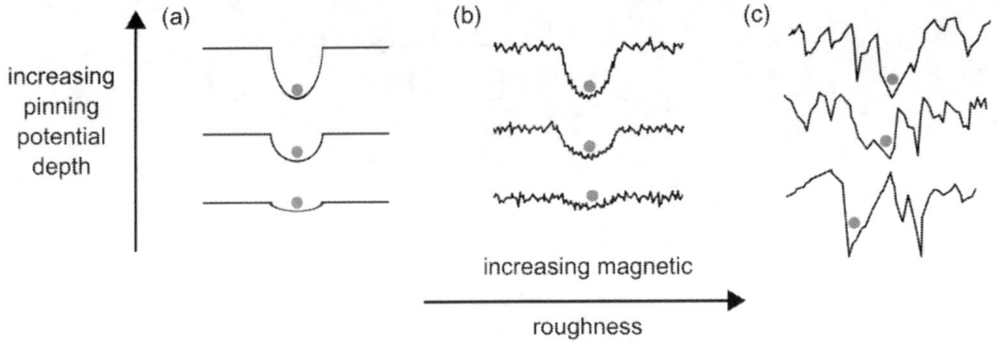

Figure 5. Sketch of pinning potential landscapes. (a) In the ideal case with no magnetic roughness, (b) with moderate magnetic roughness, and (c) with strong magnetic roughness.

heated to 300 °C is impossible. The subtractively prepared Permalloy nanowires feature small edge roughnesses and low depinning fields, independent of the etching process. No influence of the substrate temperature between 280 °C and 320 °C has been observed. The differences in the experimental results on current-induced domain-wall depinning between the ion-milled and the rf sputter-etched wires imply that rf-sputter etching with PMMA resist masks is preferable versus ion-milling with PMMA resist masks as subtractive preparation process for these experiments. This is most probably due to the higher energy of the argon ions penetrating the resist mask and bombarding the Permalloy surface during preparation of the wires. An interesting catchword in this context is magnetic roughness, brought up by Hayashi et al. [18]. Interpreting magnetic roughness as some sort of noise of whatever reason in the pinning potential landscape of the wire it is possible to imagine why it gets more difficult to observe the dependence of the domain wall depinning probability on current pulse length [3] or current pulse shape [19] with decreasing depinning fields. Figure 5 shows a sketch of pinning potential landscapes of different strengths, in the ideal case with no magnetic roughness, with moderate magnetic roughness, and with strong magnetic roughness. The lower the initial pinning potential the more important magnetic roughness becomes. The domain wall in the pinning potential in an ideally magnetic flat landscape as depicted in Fig. 5(a) will exhibit the same dependence on current pulse length as well as on current pulse shape independent of the depth of the pinning potential. Contrary to that in situations as depicted in Fig. 5(b) and (c), when the magnetic roughness and the depinning energy are of the same order of magnitude, this will not be the case. These speculations suggest to further refine the subtractive preparation of nanostructures, which could be obtained by hard masks that protect the magnetic film reasonably during the etching process.

Acknowledgments

We thank Kim Martens for sputter deposition of several thin films and Michael Volkmann for outstanding technical support. Financial support by the Deutsche Forschungsgemeinschaft via the Graduiertenkolleg 1286 "Functional Metal-Semiconductor Hybrid Systems", the Sonderforschungsbereich 668 "Magnetism - from the single atom to the nanostructure", by the Forschungs- und Wissenschaftsstiftung Hamburg via the Cluster of Excellence Nanospintronic, and by the Japan Society for the Promotion of Science via the "JSPS Summer Program 2009" is gratefully acknowledged.

References

[1] Parkin S S P, Hayashi M and Thomas L 2008 *Science* **320** 190

[2] Meier G, Bolte M, Eiselt R, Krüger B, Kim D H and Fischer P 2007 *Phys. Rev. Lett.* **98** 187202
[3] Nahrwold G, Bocklage L, Matsuyama T, Krüger B, Scholtyssek J M, Merkt U and Meier G 2009 *J. Appl. Phys.* **105** 07D511
[4] van Staa A, Wulfhorst J, Vogel A, Merkt U and Meier G 2008 *Phys. Rev. B* **77** 214416
[5] Vogel A, Wulfhorst J and Meier G 2009 *Appl. Phys. Lett.* **94** 122510
[6] Bocklage L, Scholtyssek J M, Merkt U and Meier G 2007 *J. Appl. Phys.* **101** 09J512
[7] Jedema F J, Filip A T and van Wees B J 2001 *Nature* **410** 345
[8] Miura K, Kasai S, Kobayashi K and Ono T 2006 *Jpn. J. Appl. Phys.* **45** 2888
[9] Yang T, Kimura T and Otani Y 2008 *Nature Physics* **4** 851
[10] Van Waeyenberge B, Puzic A, Stoll H, Chou K W, Tyliszczak T, Hertel R, Fähnle M, Brückl H, Rott K, Reiss G, Neudecker I, Weiss D, Back C H and Schütz G 2006 *Nature* **444** 461
[11] Bolte M, Meier G, Krüger B, Drews A, Eiselt R, Bocklage L, Bohlens S, Tyliszczak T, Vansteenkiste A, van Waeyenberge B, Chou K W, Puzic A and Stoll H 2008 *Phys. Rev. Lett.* **100** 176601
[12] Kamionka T, Martens M, Chou K W, Curcic M, Drews A, Schütz G, Tyliszczak T, Stoll H, Van Waeyenberge B and Meier G 2010 *Phys. Rev. Lett.* **105** 137204
[13] Yamada K, Kasai S, Nakatani Y, Kobayashi K, Kohno H, Thiaville A and Ono T 2007 *Nature Materials* **6** 269
[14] Shigeto K, Okuno T, Mibu K, Shinjo T and Ono T 2002 *Appl. Phys. Lett.* **80** 4190
[15] Shinjo T, Okuno T, Hassdorf R, Shigeto K and Ono T 2000 *Science* **289** 930
[16] Nahrwold G, Scholtyssek J M, Motl-Ziegler S, Albrecht O, Merkt U and Meier G 2010 *J. Appl. Phys.* **108** 013907
[17] Pels C, Barthelmess M, Bolte M, Thieme A and Meier G 2005 *J. Magn. Magn. Mat.* **293** 885
[18] Hayashi M, Thomas L, Rettner C, Moriya R, Bazaliy Y B and Parkin S S P 2007 *Phys. Rev. Lett* **98** 037204
[19] Bocklage L, Krüger B, Matsuyama T, Bolte M, Merkt U, Pfannkuche D and Meier G 2009 *Phys. Rev. Lett.* **103** 197204
[20] Langner H, Bocklage L, Krüger B, Matsuyama T and Meier G 2010 *Appl. Phys. Lett.* **97** in press
[21] Hayashi M, Thomas L, Rettner C, Moriya R, Jiang X and Parkin S S P 2006 *Phys. Rev. Lett* **97** 207205
[22] Yamaguchi A, Ono T, Nasu S, Miyake K, Mibu K and Shinjo T 2004 *Phys. Rev. Lett.* **92** 077205
[23] Malinowski G, Lörincz A, Krzyk S, Möhrke P, Bedau D, Boulle O, Rhensius J, Heyderman L J, Cho Y J, Seo S and Kläui M 2010 *J. Phys. D: Appl. Phys.* **43** 045003

Microstructure and Magnetic Properties of Fe and Fe-alloy Thin Films Epitaxially Grown on MgO(100) Substrates

Katsuki Matsubara[1], **Mitsuru Ohtake**[1], **Fumiyoshi Kirino**[2] **and Masaaki Futamoto**[1]

1 Faculty of Science and Engineering, Chuo University, 1-13-27 Kasuga, Bunkyo-ku, Tokyo 112-8551, Japan
2 Graduate School of Fine Arts, Tokyo National University of Fine Arts and Music, 12-8 Ueno-koen, Taito-ku, Tokyo 110-8714, Japan

E-mail: matsubara@futamoto.elect.chuo-u.ac.jp

Abstract. Fe, $Fe_{50}Co_{50}$, and $Fe_{80}Ni_{20}$ (at. %) single-crystal films with the $(100)_{bcc}$ plane parallel to the substrate surface were prepared on MgO(100) single-crystals heated at 300 °C by ultra high vacuum molecular beam epitaxy. The film growth mechanism, the film structure, and the magnetic properties were investigated. *In-situ* reflection high energy electron diffraction and X-ray diffraction analyses indicate that the strains in the films are very small though there are fairly large mismatches of −3.7~−4.3% at the film/substrate interface. Cross-sectional high-resolution transmission electron microscopy shows that misfit dislocations are introduced in the film at the interface. Dislocations are also observed in the film up to around 10~20 nm distance from the interface. The presence of such dislocation relieves the strain caused by the lattice mismatch. The in-plane magnetization properties of these films reflect the magnetocrystalline anisotropies of respective bulk Fe, $Fe_{50}Co_{50}$, and $Fe_{80}Ni_{20}$ crystals.

1. Introduction

Fe, FeCo, and FeNi are 3*d* ferromagnetic transition metals which crystallize with bcc structure in Fe-rich compositional regions in the bulk phase diagrams. Various bcc 3*d* metal multilayer films combined with thin MgO layers are widely investigated for magnetic tunneling junction (MTJ) applications like tunneling magnetoresistance (TMR) heads and magnetic random access memory (MRAM) devices [1,2]. For such applications, high-quality single-crystal films are the strong candidates, since the film uniformity and the magnetic anisotropy are well controlled. Theoretical calculations show that the TMR ratio of MTJ prepared using single-crystal films is one order of magnitude greater than that obtained with amorphous films [3,4]. The metal/MgO interface structure gives strong influence on the TMR effect.

The magnetocrystalline anisotropy of bulk bcc 3*d* metal varies depending on the kind and the composition of metal elements. The easy magnetization direction of bulk Fe, Fe_xCo_{100-x} (x>59), and Fe_yNi_{100-y} (y>75) crystals is along the $<100>_{bcc}$ direction, whereas that of bulk Fe_xCo_{100-x} (59>x>41) crystals is along the $<111>_{bcc}$ direction. Furthermore, the magnetocrystalline anisotropy of bcc 3*d* metal film is delicately influenced by the strain caused by film growth conditions [5,6]. On the contrary, the bcc 3*d* metal crystals have similar lattice constants ranging between 0.283 and 0.287 nm [7], and similar epitaxial film growth is thus expected on MgO layers. Fe [8,9] and FeCo [10,11] single-crystal films with the $(100)_{bcc}$ plane parallel to the substrates surface have been prepared on MgO(100) substrates. The purpose of the present study is to investigate the film growth mechanism, the microstructure, and the magnetic properties of Fe, FeCo, and FeNi single-crystal films grown on MgO(100) substrates.

2. Experimental procedure

Thin films were prepared on polished MgO(100) substrates at a substrate temperature of 300 °C by using a molecular beam epitaxy chamber equipped with a reflection high energy electron diffraction (RHEED) facility. The base pressures were lower than 3×10^{-8} Pa. Substrates were heated at 500 °C for 1 hour in the ultra high vacuum chamber to obtain clean surfaces. The surface structure was checked by RHEED. As shown in figure 1(d), the RHEED pattern observed from the substrate exhibited a *Kikuchi* pattern, indicating that the surface was clean and smooth. The roughness R_a examined by *ex-situ* atomic force microscopy (AFM) was 0.06 nm. Fe, $Fe_{50}Co_{50}$, and $Fe_{80}Ni_{20}$ (at. %) sources were employed in *e*–beam evaporation. The Fe-alloy film compositions were confirmed by energy dispersive X-ray spectroscopy and the errors were less than ±3 at. %. The film thickness was fixed at 40 nm. The deposition rate was kept constant at 0.01 nm/s.

The surface structure during film deposition was studied by *in-situ* RHEED. The structural property was investigated by out-of-plane ($2\theta/\omega$-scan) and in-plane ($2\theta\chi/\phi$-scan) X-ray diffraction (XRD) with Cu–Kα radiation (λ=0.15418 nm). The cross-sectional microstructure was observed by high-resolution transmission electron microscopy (HR-TEM). The sample was first thinned mechanically and then ion-milled to be transparent for the TEM electron beam accelerated at 300 kV. The in-plane magnetization curves were measured by using a vibration sample magnetometer.

3. Results and discussion

Fe, FeCo and FeNi epitaxial thin films were obtained on MgO(100) substrates. Figures 1(a)–(c) show the RHEED patterns observed for 40-nm-thick films. The patterns correspond to $(100)_{bcc}$ texture for all samples, as shown in the RHEED spot map of figure 1(i). Fe, FeCo, and FeNi single-crystal films with the $(100)_{bcc}$ plane parallel to the substrate surface are obtained. The epitaxial orientation relationships of $Fe(100)[011]_{bcc}$ || MgO(100)[001], $FeCo(100)[011]_{bcc}$ || MgO(100)[001], $FeNi(100)[011]_{bcc}$ || MgO(100)[001] are determined from RHEED observation. The orientation of $(100)_{bcc}$ is rotated around the film normal by 45° with respect to that of MgO(100). In this configuration, the lattice mismatches of $Fe(100)_{bcc}$, $FeCo(100)_{bcc}$, and $FeNi(100)_{bcc}$ with respect to MgO(100) are respectively –3.8%, –4.3%, and –3.7%. The values are calculated from the lattice constants of the bulk Fe, $Fe_{50}Co_{50}$, and $Fe_{80}Ni_{20}$ crystals.

Figures 1(d)–(f) show the RHEED patterns observed in an early stage of $Fe_{50}Co_{50}$ film growth on an MgO(100) substrate. A pattern corresponding to $(100)_{bcc}$ reflection is observed after 0.3-nm-thick

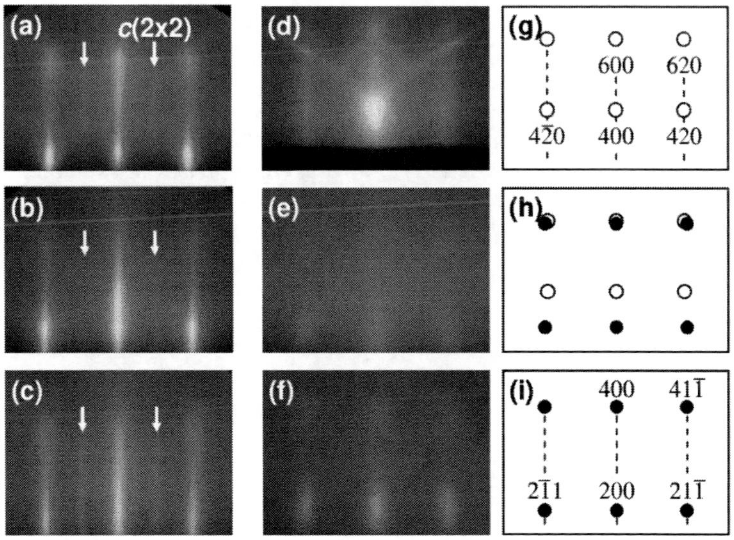

Figure 1. RHEED patterns observed for 40-nm-thick (a) Fe, (b) $Fe_{50}Co_{50}$, and (c) $Fe_{80}Ni_{20}$ thin films grown on MgO(100) substrates. RHEED patterns observed for (d) an MgO(100) substrate and (e), (f) an $Fe_{50}Co_{50}$ film during growth process on the substrate and (g)–(i) the RHEED maps. The film thicknesses are (e) 0.3 and (f) 1.0 nm. The incident electron beam is parallel to the MgO[001] direction. The open and the filled circles, respectively, correspond to the reflections from MgO substrate and $Fe_{50}Co_{50}$ crystal.

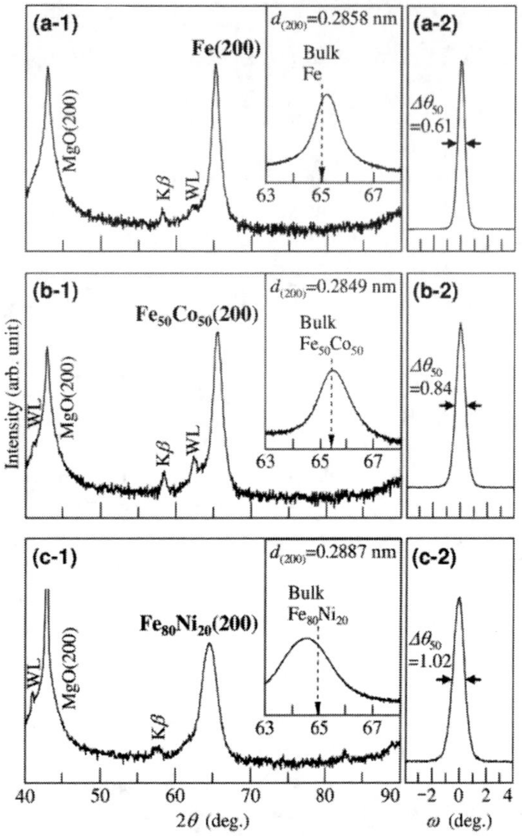

Figure 2. Out-of-plane XRD spectra of (a-1) Fe, (b-1) $Fe_{50}Co_{50}$, and (c-1) $Fe_{80}Ni_{20}$ films grown on MgO(100) substrates. The rocking curves observed for the $(200)_{bcc}$ XRD peaks of (a-2) Fe, (b-2) $Fe_{50}Co_{50}$, and (c-2) $Fe_{80}Ni_{20}$. The scattering vector of the out-of-plane XRD is parallel to the MgO[200] direction. The intensity is shown in (a-1)–(c-1) a logarithmic scale or (a-2)–(c-2) a linear scale.

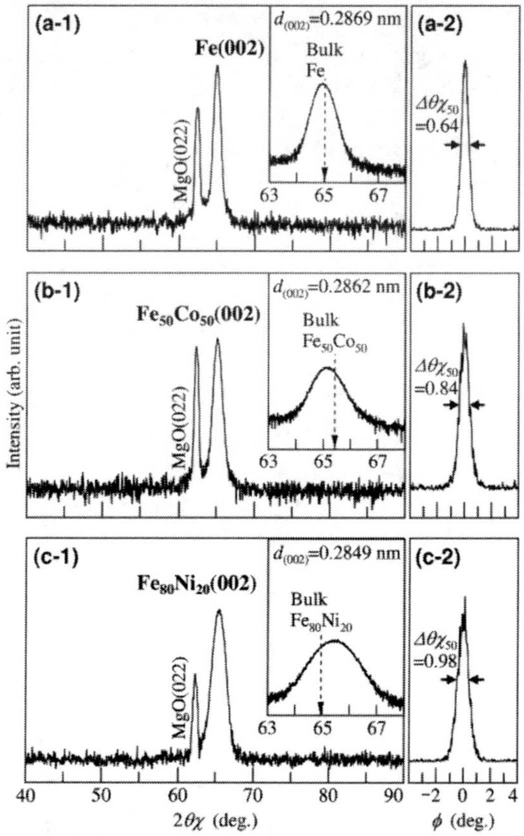

Figure 3. In-plane XRD spectra of (a-1) Fe, (b-1) $Fe_{50}Co_{50}$, and (c-1) $Fe_{80}Ni_{20}$ films grown on MgO(100) substrates. The rocking curves observed for the $(002)_{bcc}$ XRD peaks of (a-2) Fe, (b-2) $Fe_{50}Co_{50}$, and (c-2) $Fe_{80}Ni_{20}$. The scattering vector of the in-plane XRD is parallel to the MgO[022] direction. The intensity is shown in (a-1)–(c-1) a logarithmic scale or (a-2)–(c-2) a linear scale.

FeCo deposition, overlapping with that from the substrate, as shown in figure 1(e). The RHEED data indicate that the two materials, FeCo and MgO, are exposed to the surface at this stage. Isolated FeCo island nucleation is taking place on the MgO substrate, which corresponds to the film growth mode of Volmer-Weber (island growth) type [12]. With increasing the film thickness, the reflection intensity from the film becomes stronger, while that from the substrate weakens. For the thickness exceeding 1 nm, a sharp pattern from only the film starts to be observed, as shown in figure 1(f). Furthermore, the in-plane lattice spacing estimated from the RHEED reflection is close to that of the bulk $Fe_{50}Co_{50}$ crystal and stays almost unchanged till the end of 40-nm-thick deposition. Similar film growth processes are observed for the cases of Fe and $Fe_{80}Ni_{20}$ deposition. These RHEED results indicate that the film strains are small, suggesting that fairly large mismatches around –4% existing at the film/substrate interfaces are relieved.

Figures 2 and 3 show the out-of-plane and the in-plane XRD spectra, respectively. The out-of-plane XRD spectra show the $(200)_{bcc}$ reflections in addition to the MgO(200) reflections. In the in-plane XRD spectra measured by making the scattering vector parallel to the MgO[022] direction, the $(002)_{bcc}$ reflections are recognized. The XRD results also indicate that single-crystal films are formed on the MgO(100) substrates. The out-of-plane and the in-plane lattice spacings estimated from the reflection peaks are in agreement with the values of respective bulk Fe, $Fe_{50}Co_{50}$, and $Fe_{80}Ni_{20}$ crystals

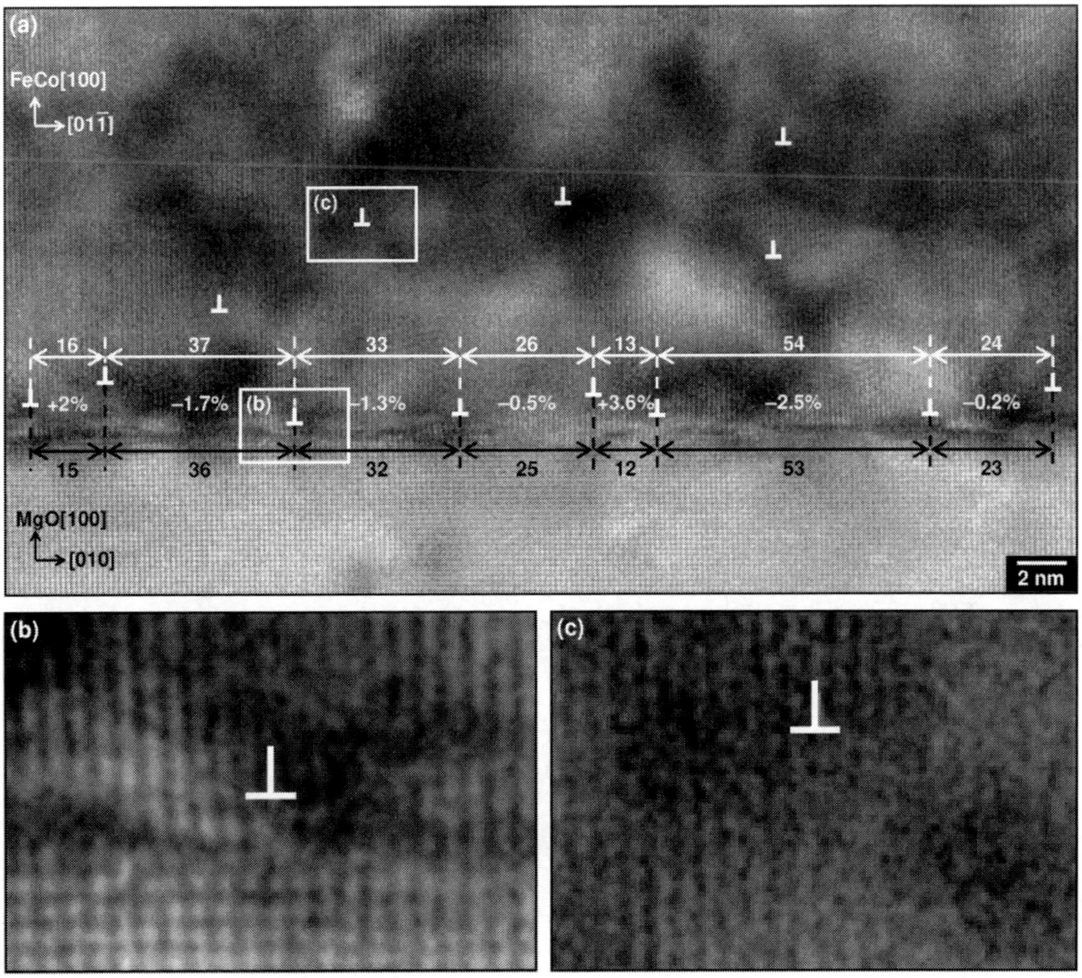

Figure 4. (a) High-resolution cross-sectional TEM image of $Fe_{50}Co_{50}(100)_{bcc}$/MgO(100) interface observed along the MgO[001] direction. (b), (c) Enlarged views of the area surrounded by white square lines in (a). ⊥ marks show the positions of misfit dislocation.

with very small errors less than ±0.8%. Figures 2(a-2)–(c-2) and 3(a-2)–(c-2) show the rocking curves observed for the $(200)_{bcc}$ out-of-plane and the $(002)_{bcc}$ in-plane XRD peaks, respectively. The values of full width at half maximum of rocking curves, $\Delta\theta_{50}$ and $\Delta\theta\chi_{50}$, are less than 1.1 degrees. The XRD results also indicate that the film strains are small.

It is known that misfit dislocations are easily introduced in a film at the interface to reduce the mismatch between immiscible elements when a weak binding force is working between the deposited atoms and the substrate atoms. This type of epitaxial growth where misfit dislocations are introduced at the interface is reported for the bcc–Cr/MgO [13] and the fcc–Co/MgO [14] systems. The epitaxial growth of Fe, FeCo, and FeNi crystals on MgO(100) is also considered to follow the growth mode.

Figure 4 shows the cross-sectional HR-TEM image of an $FeCo(100)_{bcc}$/MgO(100) specimen observed along the MgO[001] direction. An atomically sharp boundary is recognizable between the film and the substrate. Misfit dislocations are preferentially introduced in the film at the interface, as shown by the ⊥ marks in figure 4. Variations of lattice matching between $FeCo(01\bar{1})_{bcc}$ and MgO(020) are observed at the interface like 16:15, 37:36, 33:32, 26:25, 13:12, 54:53, 24:23 etc. When the $FeCo(01\bar{1})_{bcc}$ matches with the MgO(020) in a ratio of 24:23 at the interface, the lattice mismatch decreases from –4.3% to –0.2%. Dislocations are also observed in the film up to around 10~20 nm

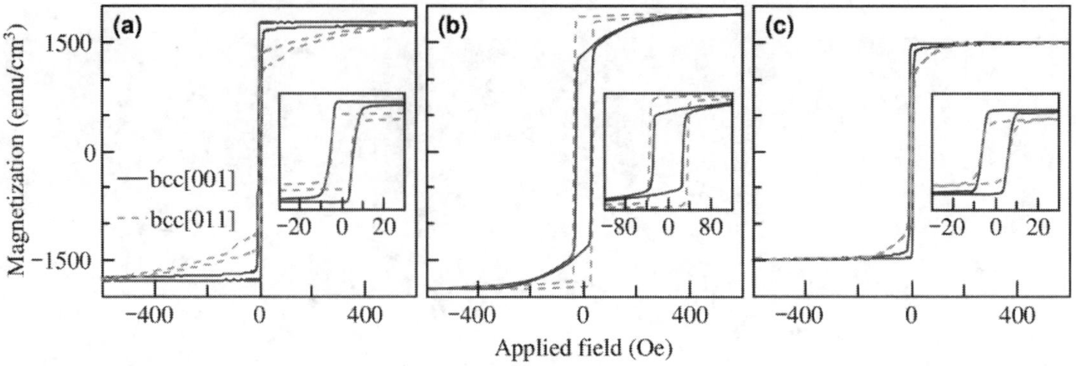

Figure 5. Magnetization curves of (a) Fe, (b) $Fe_{50}Co_{50}$, and (c) $Fe_{80}Ni_{20}$ films grown on MgO(100) substrates. The inset figures show the enlarged views around coercivities.

distance from the interface, further reducing the strain. Similar behavior of dislocation introduction is taking place for the Fe and the $Fe_{80}Ni_{20}$ films. The small strains in these films are apparently realized by introduction of such dislocations.

Figure 5 shows the magnetization curves. The saturation magnetization (M_s) values of Fe, $Fe_{50}Co_{50}$, and $Fe_{80}Ni_{20}$ films are similar to those of respective bulk Fe, $Fe_{50}Co_{50}$, and $Fe_{80}Ni_{20}$ crystals. The Fe and the FeNi films are easily magnetized when the magnetic field is applied along the $[001]_{bcc}$ direction, whereas the magnetization curves along the $[011]_{bcc}$ direction saturate at higher applied fields. On the contrary, the FeCo film is easily magnetized when the magnetic filed is applied along the $[011]_{bcc}$ direction. The anisotropy field (H_k) values of Fe, FeCo, and FeNi films are 668, 294, and 151 Oe, respectively. Here the H_k value is measured as the magnetic field when magnetization measured along the Fe[011], the FeCo[001], or the FeNi[011] direction saturates. The decreasing order of H_k is corresponding to the magnetocrystalline anisotropy constants (K_1) of the bulk Fe crystal [15], the bulk $Fe_{50}Co_{50}$ crystal [15], and the $Fe_{80}Ni_{20}$ film [16] of $+4.8\times10^5$, -1.6×10^5, and $+5.2\times10^4$ erg/cm^3, respectively. The coercivities measured by applying magnetic field in $[001]_{bcc}$ and $[011]_{bcc}$ directions are nearly equal for these films as shown in the inset figures. The magnetization properties of Fe, $Fe_{50}Co_{50}$, and $Fe_{80}Ni_{20}$ single-crystal films grown on MgO(100) substrates are reflecting the magnetocrystalline anisotropies of respective bulk Fe, $Fe_{50}Co_{50}$, and $Fe_{80}Ni_{20}$ crystals.

4. Conclusion

The film growth mechanism, the film structure, and the magnetic properties of Fe, $Fe_{50}Co_{50}$, and $Fe_{80}Ni_{20}$ (at. %) single-crystal films on MgO(100) substrates were investigated. The film growth follows the Volmer-Weber mode. The out-of-plane and the in-plane lattice spacings are in agreement with the values of respective bulk Fe, $Fe_{50}Co_{50}$, and $Fe_{80}Ni_{20}$ crystals with small errors less than ±0.8%. The film strains are very small, which is associated with introduction of dislocations. The in-plane magnetization properties are reflecting the magnetocrystalline anisotropies of bulk Fe, FeCo, and FeNi crystals.

Acknowledgements

This work was supported by NEDO–Japan, ISIJ (Iron and Steel Institute of Japan), STFJ (Science and Technology Foundation of Japan), and Chuo University Grant for special Research. M. Ohtake is a Research Fellow of the Japan Society for the Promotion of Science.

References

[1] Wulfhekel W, Klaua M, Ullmann D, Zavaliche F and Kirschner J 2001 *Appl. Phys. Lett.* **78** 509
[2] Yuasa S, Nagahama T, Fukushima A, Suzuki Y and Ando K 2004 *Nature Mater.* **3** 868
[3] Butler W H, Zhang X –G, Schulthness T C and MacLaren J M 2001 *Phys. Rev.* B **63** 054416
[4] Mathon J and Umerski A 2001 *Phys. Rev.* B **63** 220403
[5] Krebs J J, Jonker B T and Prinz G A 1987 *J. Appl. Phys.* **61** 3744

[6] Dumm M, Uhl B, Zölfl M, Kipferl W and Bayreuther G 2002 *J. Appl. Phys.* **91** 8763
[7] Bozorth R M 1993 *Ferromagnetism* (New York: Wiley–IEEE press)
[8] Urano T and Kanaji T 1988 *J. Phys. Soc. Jpn.* **57** 3403
[9] Shikada K, Tabuchi K, Ohtake M, Kirino F and Futamoto M 2008 *J. Magn. Soc. Jpn.* **32** 296
[10] Mühge T, Zeidler T, Wang Q, Morawe C, Metoki N and Zabel H 1995 *J. Appl. Phys.* **77** 1055
[11] Shikada K, Ohtake M, Kirino F and M Futamoto 2009 *J. Magn. Soc. Jpn.* **33** 85
[12] Venable J A 2000 *Introduction to Surface and Thin Film Process* (New York: Cambridge Univ Press) pp 145–147
[13] Nakamura A, Koguchi M and Futamoto M 1995 *J. Jpn. Appl. Phys.* **34** 2307
[14] Ohtake M, Nukaga Y, Sato Y, Kirino F and Futamoto M 2009 *J. Appl. Phys.* **106** 123921
[15] Hall R C 1959 *J. Appl. Phys.* **30** 816
[16] Yin L F, Wei D H, Lei N, Zhou L H, Tian C S, Dong G S, Jin X F, Guo L P, Jia Q J, Wu R Q 2006 *Phys. Rev. Lett.* **97** 067203

Magnetic properties of FeCoC thin films prepared by various sputtering methods

V. Edon[1], S. Dubourg[1,a], J. Vernières[2], J.-F. Bobo[2]
[1] CEA, DAM, LE RIPAULT, F-37260 Monts, France
[2] LNMH-CEMES-CNRS-ONERA, F-31055 Toulouse, France

a) E-mail: sebastien.dubourg@cea.fr

Abstract. In order to grow nanocrystallized soft magnetic thin films, FeCoC alloys were first deposited by reactive sputtering in Ar/C_2H_2 plasma. This deposition process rendered it possible to incorporate a carbon content between 0 and 30 at.% into the FeCo samples. The films were then compared to FeCoC samples obtained from a $Fe_{65}Co_{35}/C$ composite target, with an adjustable amount of C slots. Layers with soft magnetic properties were achieved when increasing the C_2H_2 rate during the reactive deposition, whereas films deposited by sputtering of FeCo and C on the same target demonstrated a very high coercivity. Permeability spectra measurements (and published elsewhere) demonstrated that FeCoC prepared with acetylene is a very promising material for high-frequency magnetic devices.

1. Introduction

$Fe_{100-x}Co_x$ ($30 < x < 40$) magnetic alloys are well-known for their high saturation magnetization ($4\pi Ms \geq 24$ kG) [1]. Among possible additives (such as B, Si, N…) to FeCo-based thin films, carbon is one of the most promising elements for obtaining soft magnetic properties [2]. In fact, both the magnetocrystalline anisotropy and coercivity have been reported to become decreased - by a reduction of the crystallite size - when a third element such as carbon is added to the FeCo alloys [3,4]. Nanocrystallized structures combining soft magnetic properties, a high resistivity and a high Ms have already been grown, but the best way to add carbon into FeCo layers has yet to be clearly demonstrated. Due to its high ferromagnetic resonance, the FeCoC material has been extensively investigated for a wide range of applications [5]. In this study, the atomic composition and the microstructure were related to the static and dynamic magnetic properties of $Fe_{65}Co_{35}(C)$ thin films formed by sputtering in a reactive Ar/C_2H_2 plasma [6,7]. A careful adjustment of the acetylene content rendered it possible to grow soft magnetic materials with a relevant magnetic response at GHz frequencies. Moreover, $Fe_{65}Co_{35}(C)$ samples prepared from a FeCo/C composite target were also investigated, and these layers were clearly not as soft as those obtained by reactive sputtering.

2. Experimental section

$(Fe_{65}Co_{35})_{100-x}C_x$ alloys with a nominal thickness of 100 nm have been obtained according to one of two processes: i) by reactive sputtering in argon/acetylene plasma with a C_2H_2/Ar ratio comprised between 0 and 1.5%, and ii) by using a $Fe_{65}Co_{35}/C$ composite target with an adjustable amount of C slots. Fe and Co contents were measured by Rutherford backscattering spectrometry (RBS) using a 2.4 MeV $^4He^+$ particle beam, and the C content was determined by nuclear reaction analysis (NRA) based on the $^{12}C(d,p)^{13}C$ nuclear reaction [8]. X-ray diffraction (XRD) analysis was performed to evaluate the microstructure of the films. The resistivity of the layers was measured by a four-probe device, and static magnetic characteristics were determined using a vibrating sample magnetometer (VSM). The saturation magnetostriction coefficient was deduced from the permeability spectra, measured in the frequency range of 100 MHz – 6 GHz under a stress up to 100 MPa.

3. Results and discussion

Figure 1a displays the atomic concentration of Fe, Co and C elements as a function of the C_2H_2 percentage in the discharge. Before introduction of the reactive gas, the atomic concentrations of Fe and Co were respectively 0.6 and 0.4, i.e., close to the target stoichiometry. Both atomic ratios decreased significantly when introducing up to 0.8% of acetylene, and more slowly for higher gas concentrations. The estimated stoichiometry of the films deposited with 0.8% of C_2H_2 was $Fe_{49\pm2}Co_{29\pm1}C_{22\pm1}$. Figure 1b presents the atomic concentrations of the same three elements based on the FeCo/C composite target. As we may expect, the increase of the carbon ratio seemed more continuous than in the former case, the cobalt content was nearly constant whereas the iron ratio decreased somewhat more slowly. The final amount of C was also lower than in the case of the reactive deposition. The resulting stoichiometry obtained for a sample prepared with 6% of carbon slots on the target was estimated to $Fe_{61\pm2}Co_{27\pm1}C_{12\pm1}$.

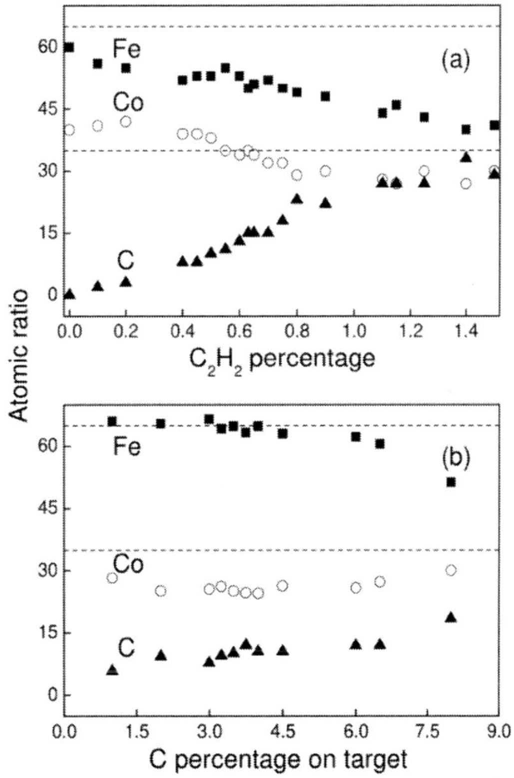

Figure 1. Atomic ratios of Fe (square), Co (circle) and C (triangle) determined by RBS/NRA (a) vs the C_2H_2 rate in the discharge for $Fe_{65}Co_{35}$ films deposited by reactive sputtering, and (b) vs the carbon percentage for $Fe_{65}Co_{35}$ films deposited by using the FeCo/C composite target.

Figure 2a shows the XRD spectra obtained as a function of the acetylene percentage. The initial microstructure of the FeCo(C) films corresponded to a bcc (200) lattice, then changed into a (110) diffraction line with an increase in the acetylene rate. Films prepared with at least 0.70% of C_2H_2 did not show the same crystallographic texture, and a reasonable explanation could be the decrease of the grain sizes with the carbon addition. Previous transmission electron microscopy observations have shown the formation of nanocrystallites with the addition of carbon to FeCo samples [6,7]. Since the (110) diffraction line disappeared with a further increase of the acetylene rate, one may suppose that the nanocrystalline structure was formed of very small grains., The X-ray diffraction spectra in Figure 2b reveal that the bcc (110) diffraction line disappeared when introducing 8% of carbon into the FeCoC film. A diffraction line corresponding to an fcc (200) lattice was also evidenced for only 5% of carbon. The distinct preferential crystalline orientations induced by the two processes may lead to very different magnetic properties [2].

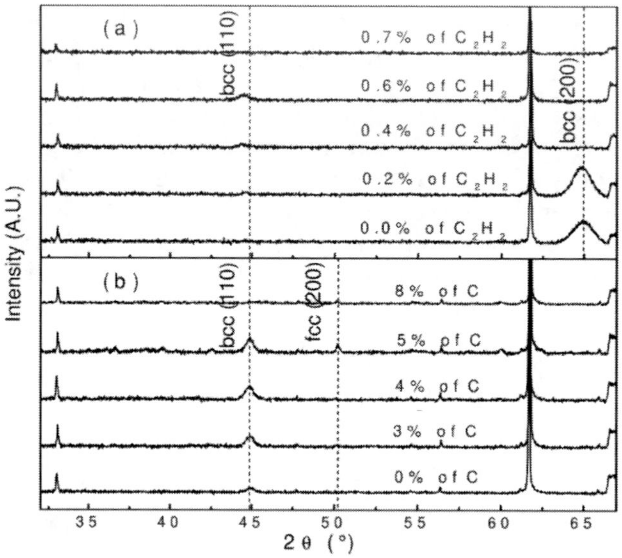

Figure 2. XRD spectra (a) vs the C_2H_2 rate in the discharge for $Fe_{65}Co_{35}$ films deposited by reactive sputtering and (b) vs the carbon percentage for $Fe_{65}Co_{35}$ films deposited by means of the FeCo/C composite target.

Figure 3a exhibits the evolution of the resistivity and the saturation magnetization when increasing the C_2H_2 concentration from 0 to 1.5%. The resistivity was found to be 25 μΩ.cm and rapidly increased up to 150 μΩ.cm for an acetylene rate higher than 0.6%. Simultaneously, the saturation magnetization decreased from 24.8 kG down to 10 kG for a high reactive ratio. The behaviors of the saturation magnetization and resistivity when increasing the C_2H_2 pressure corresponded to a transition zone between a crystalline and nanocrystalline/amorphous structure in the films [9]. This transition was well-defined, with a distinct location, for a C_2H_2/Ar ratio in the range 0.5-0.8%. Beyond 0.75% of C_2H_2, the formation of the nanocrystallites induced highly resistive and weakly magnetized films. The same transition is shown in Figure 3b for films prepared with the FeCo/C composite target. Here, the saturation magnetization was initially close to 25.0 kG and decreased down to 5.2 kG when the number of C slots increased. At the same time, the resistivity increased from 80 to 283 μΩ.cm. As compared to the reactive process, the non-reactive sputtering deposition led to highly resistive and weakly magnetized FeCoC layers. As reported elsewhere, the presence of different C layers among the FeCo ones may also imply some magnetic decoupling and therefore magnetic hardening [6]. Moreover a partial oxidation during the deposition of the FeCoC samples may explain these two behaviors at high carbon content. Further investigations are still in progress to confirm this assumption.

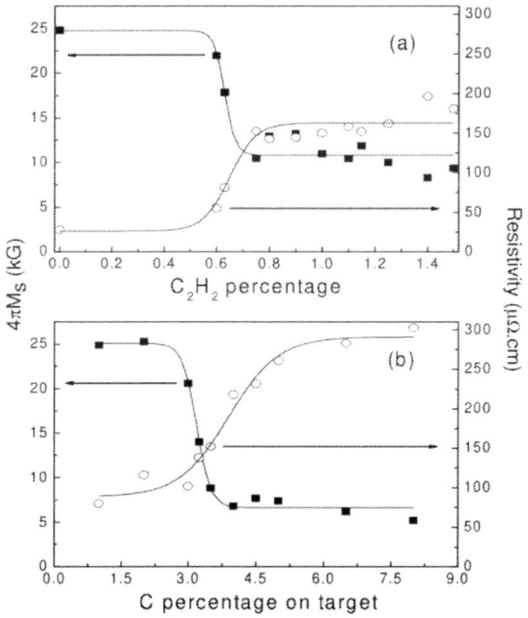

Figure 3. The saturation magnetization (square) and resistivity (circle) (a) vs the C_2H_2 rate in the discharge for $Fe_{65}Co_{35}$ films deposited by reactive sputtering and (b) vs the carbon percentage for $Fe_{65}Co_{35}$ films deposited by means of the FeCo/C composite target.

Figure 4 compares the hysteresis loops obtained from a FeCoC alloy with enough acetylene to be located in the structural transition domain, with a FeCoC film prepared by means of the $Fe_{65}Co_{35}$/C composite target. The FeCoC layer with 0.75% of C_2H_2 exhibited well-defined hysteresis loops along the easy and hard axes. Measurements in the easy direction led to a low coercivity, Hc, of 0.7 Oe, whereas in the hard direction the anisotropy field Ha was found to be 9.8 Oe. The Hc value was comparable to those previously reported for FeCoC films obtained by various techniques [6,10]. On the contrary, the sample prepared by sputtering of FeCo and C exhibited a coercivity larger than 80 Oe in both the easy and hard axis directions. As previously assumed, the oxidation of the FeCoC samples may be specifically induced by the non-reactive process, which may also be one of the reasons for the larger coercivity measured on these films. Since no diffraction line was detected by XRD analysis, it may reasonably be supposed that the oxidized fraction in FeCoC samples is amorphous.
Figure 5a shows the dependence of the acetylene concentration on Hc, as deduced from VSM measurements along the easy axis and obtained for FeCoC samples deposited by reactive sputtering. The evolution of Hc followed a biphasic sigmoïdal law, with strong variations when the C_2H_2 content was in the range 0.5-0.8%. When the C_2H_2 rate was higher than 0.75%, Hc stabilized around 1 Oe. The soft magnetic properties induced by the addition of carbon were enhanced once the transition zone was exceeded, as predicted by the random anisotropy model proposed by Herzer [3]. Figure 5a also gives the evolution of the saturation magnetostriction coefficient (λs) as a function of the acetylene rate. Due to the highly-dispersive magnetocrystalline anisotropy of the films deposited with less than 0.5% of C_2H_2, it was not possible to describe the dynamic properties of the samples by a gyromagnetic model. Consequently, we were unsuccessful in determining the saturation magnetostriction coefficient from the permeability spectra for this range of acetylene content. Moreover, it is interesting to note that λs follows the behavior of Hc but is slightly shifted to higher values of the acetylene rate. The detailed mechanism for this shift remains unclear. Nevertheless, one can observe that the coefficient changed from negative to positive values once 0.8% of the acetylene had been incorporated into the FeCoC layers. Finally, the measured coefficient close to 1.10^{-6} after carbon addition was lower than the values reported for a pure $Fe_{65}Co_{35}$ composition [11], or for FeCoC samples prepared by various techniques [10,12]. A comparison to previous $Fe_{9.5}Co_{90.5}$(C) samples - theoretically weakly

magnetostrictive - and also deposited in reactive plasma (data not shown) provided similar values of λs. The samples prepared with acetylene gas showed proper dynamic properties in the GHz range, as previously published [7]. In addition, Figure 5b shows the evolution of the coercivity with the number of C slots. Close to the region described by the dotted line, a large increase followed by a sudden decrease in Hc may be related to the transition from a crystalline into the nanocrystalline structure depicted by the XRD measurements, similarly to the behavior observed for the films deposited with acetylene. Nonetheless, after this transition, coercivity values remained very much higher than those measured with the reactive deposition, perhaps due to a partial oxidation of the FeCoC layers or a magnetic decorrelation of the FeCo clusters by carbon.

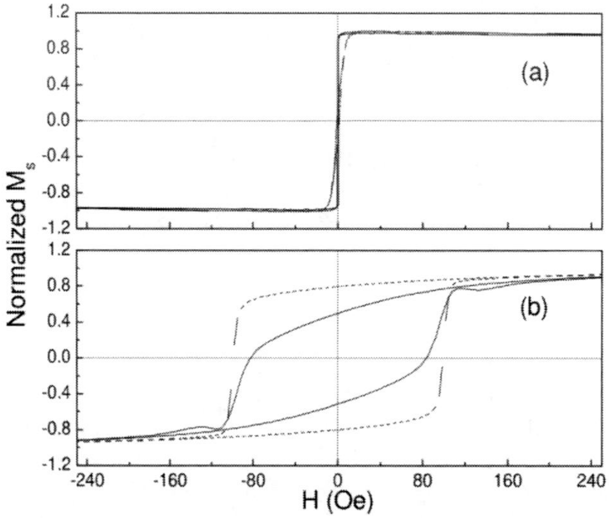

Figure 4. Hysteresis loops along the easy (solid line) and the hard (dashed line) axes vs the magnetic field, for (a) a $Fe_{65}Co_{35}$ film deposited by reactive sputtering with 0.75% of C_2H_2 and (b) a $Fe_{65}Co_{35}$ film deposited by means of the FeCo/C composite target with 8% of carbon.

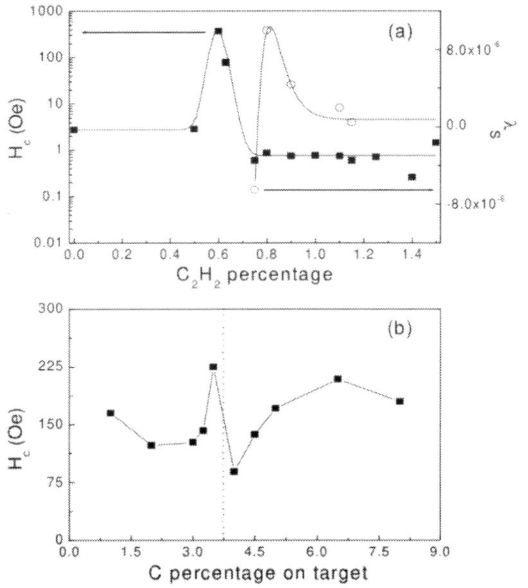

Figure 5. The coercivity (square) and saturation magnetostriction coefficient (circle) (a) vs the C_2H_2 rate in the discharge for $Fe_{65}Co_{35}$ films deposited by reactive sputtering and (b) vs the carbon percentage for $Fe_{65}Co_{35}$ films deposited by means of the FeCo/C composite target.

4. Summary

The effects adding carbon to $Fe_{65}Co_{35}$ thin films, when deposited by reactive sputtering in Ar/C_2H_2 gas or by sputtering from a FeCo/C composite target, were investigated. Carbon contents up to 22% were incorporated into the samples prepared with only 0.8% of acetylene. A change of the diffraction spectra, i.e., starting from a bcc (200) lattice and going to a (110) microstructure with the increase of the reactive gas rate, corresponded to the transition from a crystalline into a nanocrystalline structure. This effect was also confirmed by the large decrease in saturation magnetization and by an improvement of the resistivity. Films deposited through the use of a conventional FeCo/C target also exhibited a structural change when the carbon content was increased, quite similar to the observations on the reactive process. A comparison between the samples prepared with acetylene and those obtained by means of the composite target demonstrated very different static magnetic properties, with an Hc of less than 1 Oe in the former case, and above 80 Oe in the latter case. The relatively low saturation magnetostriction coefficient measured for the samples obtained with reactive gas rendered this sputtering method suitable for preparing soft magnetic films with relevant dynamic properties in the GHz range.

References

[1] Bozorth R.M 1993 *Ferromagnetism* (Willey, New York)
[2] Dong X.L, Zhang Z.D, Chuang Y.C, Jin S.R 1999 *Phys. Rev.* B **60** 3017
[3] Herzer G 1990 *IEEE Trans. Mag.* **26** 1397
[4] Desvaux C, Lecante P, Respaud M, Chaudret B 2010 *J. Mat. Chem.* **20** 103
[5] Liu X.G, Li J, Geng D.Y, Zhang Z.D 2009 *Appl. Phys. Lett.* **95** 023114
[6] Liu X, Morisako A 2008 *IEEE Trans. Mag.* **44** 3910
[7] Edon V, Dubourg S, Vernières J, Warot-Fonrose B, Bobo J-F 2010 *J. Appl. Phys.* **107** 09A321
[8] Amsel G, Battistig G 2005 *Nucl. Inst. Meth. B* **240** 1
[9] Wang W, Chen Y, Yue G.H, Sumiyama K, Hihara T, Peng D.L 2009 *J. Appl. Phys.* **106** 013912
[10] Fu Y, Miyao T, Cao J.W, Yang Z, Matsumoto M, Liu X.X, Morisako A. 2007 *J. Mag. Mag. Mat.* **308** 165

[11] Wohlfarth E.P. 1980 *Ferromagnetic Materials* (North-Holland Publishing Co)
[12] Favieres C, Vergara J, Madurga V 2007 *J. Phys. D* **40** 4101

Monte Carlo Study of Temperature and Bias dependence of Spin Transport in GaAs

Sheetal Kapoor[1], Gionni Marchetti[1], Fausto Rossi[2], Irene D'Amico[1]

[1]Department of Physics, University of York, Heslington, York, YO10 5DD, UK
[2]Department of Physics, Politecnico di Torino, Torino, Italy

E-mail: irene.damico@york.ac.uk

Abstract. An ensemble Monte Carlo approach has been used to simulate spin relaxation in GaAs devices. Effect of varying temperature and applied bias on spin relaxation time has been studied in n-type bulk GaAs. Our results show that the spin relaxation times are longer at low temperatures and that increasing the applied bias yields enhancement of the spin polarization in the system.

1. Introduction

Semiconductor spintronic devices based on control and manipulation of electron spins are considered to be the future extension of present electronics. Spin lifetime of more than 100 ns has been experimentally observed in n-type semiconductors [1, 2]. Clearly, the longer the spin lifetime, the better and more reliable will be the spintronic devices. Study of spin transport in semiconductors has become extremely important for the development of the novel spin based devices, and GaAs is one of the widely studied materials both experimentally and theoretically for spin related phenomena [1]-[7]. In this paper we present a study of a three-dimensional, n-doped GaAs sample in which a spin-polarized current is electrically injected via a ferromagnetic contact. This study is conducted using Monte Carlo methods and we analyse the effect of varying temperature and applied bias on the decay of spin polarization within the GaAs sample.

2. Monte Carlo Method

The Ensemble Monte Carlo method has been used extensively in modeling charge transport in semiconductor materials and devices. It can be used to simulate specific devices and to include various scattering mechanisms, material properties and different boundary conditions, which makes it a highly flexible technique [8, 9]. This Monte Carlo method is a semiclassical approach in which the basic transport simulation in any semiconductor device begins with random generation of a free flight time for each particle which ends with a scattering event changing the energy and momentum of the particle. The process is then repeated for the next set of free flights. The free flight time is given by $\tau = -ln(r)/\Gamma$ where r is any random number between 0 and 1 and Γ is the total scattering rate (including self-scattering) calculated at the beginning of the simulation. In between the scattering events the charge carriers are considered to propagate along a classical trajectory and are influenced by external forces due to applied electric and magnetic fields. The equations of motion for the free flight make use of the electric field obtained by solving the Poisson equation consistent with the device boundary conditions. At the end of each free flight the type of scattering responsible for terminating it is chosen by using a random number between 0 and Γ, and then the new energy and momentum of the particle are calculated. The scattering mechanisms considered in our simulation are scattering between electrons and phonons and scattering of electrons with ionized impurities. Phonon scattering involves both acoustic and optical modes as well as absorption and emission of phonons. The corresponding scattering rates are calculated using Fermi's Golden rule. Data sampling is done at regular time intervals Δt to allow for estimation of dynamic carrier parameters like position, velocity, energy as well as to update potentials and fields. The choice of this time interval Δt depends on the stability criteria [9].

The next step is to integrate the spin dynamics in the Monte Carlo simulation. At the beginning of the simulation when positions and velocities are assigned to carriers, a single electron spin density matrix ρ_i is defined for each particle, $\rho_i(t) = \begin{bmatrix} \rho_i \uparrow\uparrow(t) & \rho_i \uparrow\downarrow(t) \\ \rho_i \downarrow\uparrow(t) & \rho_i \downarrow\downarrow(t) \end{bmatrix}$.

During the free flight sequence the spin of each particle evolves coherently which can be represented as

$$\rho_i(t+\tau) = e^{-i(H_D)\tau/\hbar}\rho_i(t)e^{i(H_D)\tau/\hbar}. \tag{1}$$

Here, H_D is the spin dependent Hamiltonian to incorporate the spin-orbit mechanism responsible for the relaxation of spin polarization. In the present calculations we have included only Dyakonov-Perel (DP) mechanism for spin relaxation as, in n-type GaAs, Elliot-Yafet(EY) mechanism is less important [1, 4, 7], at least in the range of temperatures and densities we consider. In DP mechanism, between the scattering events, electron spins precess with an effective, momentum dependent, Larmor frequency $\mathbf{\Omega}$. The direction of the momentum $\mathbf{p}$ changes due to electron scattering events, which results in spin reorientation. In our simulation

$$H_D = \hbar\mathbf{\Omega}\cdot\boldsymbol{\sigma}, \tag{2}$$

where $\mathbf{\Omega} = \frac{\beta}{\hbar^4}\{p_x(p_y^2 - p_z^2), p_y(p_z^2 - p_x^2), p_z(p_x^2 - p_y^2)\}$ [10], $\boldsymbol{\sigma}$ are the Pauli spin matrices and β is the spin-orbit coupling constant.

Thus, the evolution operator in equation (1) is given by

$$e^{-i(H_D)\tau/\hbar} = \begin{bmatrix} cos(A\tau) - i\frac{C}{D}sin(A\tau) & -i\frac{B}{D}sin(A\tau) \\ -i\frac{B^*}{D}sin(A\tau) & cos(A\tau) + i\frac{C}{D}sin(A\tau) \end{bmatrix}. \tag{3}$$

In the above equation,

$$A = \frac{\beta}{\hbar^3}\{[p_x(p_y^2 - p_z^2)]^2 + [p_y(p_z^2 - p_x^2)]^2 + [p_z(p_x^2 - p_y^2)]^2\}^{1/2},$$
$$B = [p_x(p_y^2 - p_z^2)] - i[p_y(p_z^2 - p_x^2)],$$
$$C = [p_z(p_x^2 - p_y^2)], \text{ and}$$
$$D = \{[p_x(p_y^2 - p_z^2)]^2 + [p_y(p_z^2 - p_x^2)]^2 + [p_z(p_x^2 - p_y^2)]^2\}^{1/2}.$$

The normalized spin polarization density can be written as

$$P_\alpha = \Sigma_i Tr(\sigma_\alpha \rho_i)/\Sigma_i Tr(\rho_i). \tag{4}$$

After each scattering event the particle momentum is updated and hence its Larmor frequency $\mathbf{\Omega}$. The spreading of these frequencies as the simulation proceeds ultimately causes the spin polarization to decay. In our simulation we consider the lowest conduction band in the effective mass approximation.

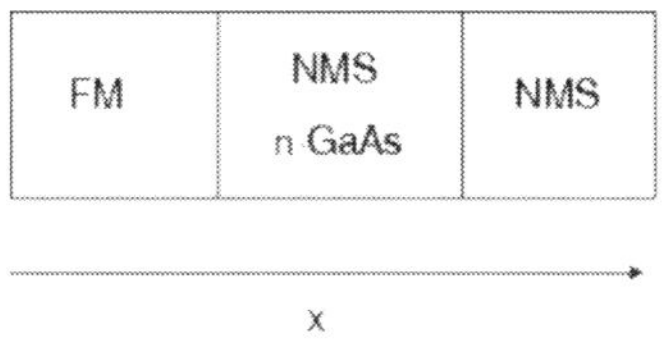

Figure 1. Sketch of the studied system

3. System

We have simulated a simple spin-transport device consisting of a 3-dimensional n-type GaAs sample sandwiched between a ferromagnet(FM) and a non-magnetic material (NM) as shown in Fig. (1). It is a layered structure with electrical spin injection from the ferromagnet into the semiconductor layer. The presence of the FM and NM layers is simulated by implementing a fully polarized $\left(\rho_i(0) = \begin{bmatrix} 1 & 0 \\ 0 & 0 \end{bmatrix}\right)$ and an unpolarized $\left(\rho_i(0) = \begin{bmatrix} 0.5 & 0 \\ 0 & 0.5 \end{bmatrix}\right)$ spin reinjection from the left and right-end side respectively, into the GaAs layer. A bias is applied to the GaAs layer. Charge neutrality condition is implemented i.e. when an electron exits the GaAs layer, a new particle is reinjected at the other end of the device with a thermal velocity. The electric- field profile along the x direction is computed self-consistently at each time step via a coupled Poisson-Monte Carlo solution. Spin dynamics is calculated via the evolution of the spin polarization vector according to Eqs.(1-4).

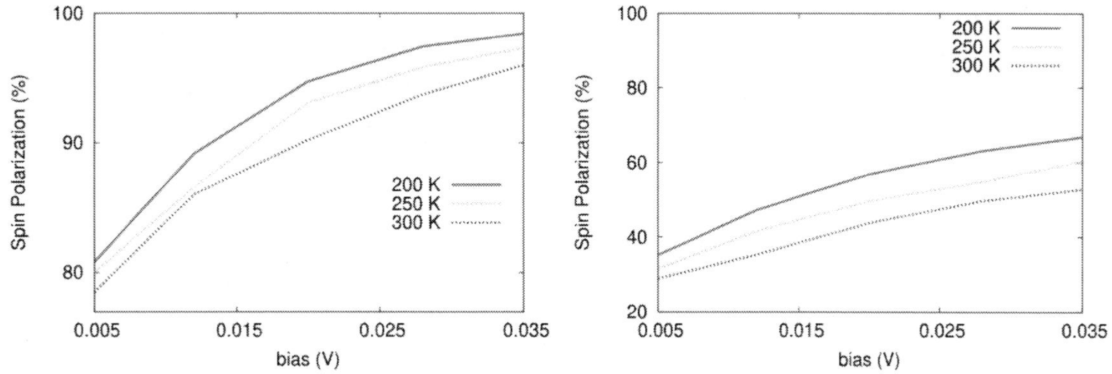

Figure 2. Variation of spin polarization with bias (a) at the start and (b) at the end of the simulated n-type GaAs layer at different temperatures after 20ps.

4. Results

Values of the spin-orbit coupling parameter β, reported in the literature show a wide variation. Our studies on the temperature dependence of spin relaxation time in n-type *bulk* GaAs yields similar results to the ones obtained recently by Jiang et al. [7] using a kinetic spin Bloch equations approach when the same values of β are used. Similar to [7], we also find a good agreement with experimental results in [1] for temperatures greater than 20 K with $\beta = 8.2\ eV\AA^3$. Following [7] have used $\beta = 23.9\ eV\AA^3$ for the range of temperatures and densities considered in the simulation results presented here.

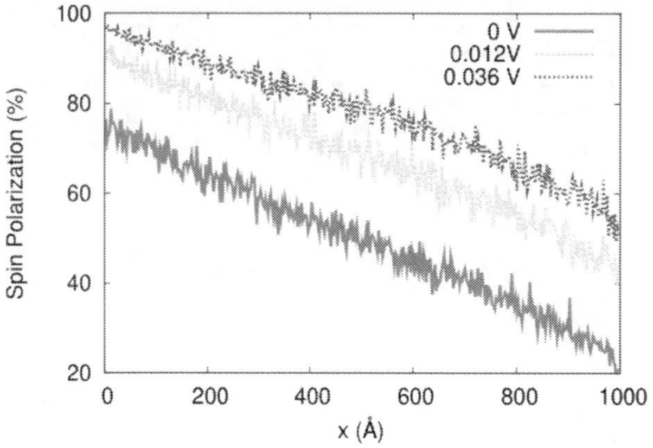

Figure 3. Spin polarization over the device length after 20 ps at 300 K and for varying applied bias.

In this section we analyze the effect of varying temperature and applied bias on the spin relaxation time in an n-type GaAs sample. The GaAs layer is uniformly doped with a carrier density of $1 \times 10^{17}\ cm^{-3}$ and its length is taken as $0.1\ \mu m$. The total number of simulated particles is 200000 and the sampling time step is $\Delta t = 0.1fs$. Each simulation is run for 200000 time steps, i.e. for 20 ps. Figure (2) shows the temperature and bias dependence of spin relaxation for the n-doped GaAs layer. The spin polarization decreases with the increasing temperature. For a particular temperature an increase in applied field enhances the spin polarization in the system. Figure (2a) is plotted for the spin polarization (as defined by Eq (4)) at the left end of the GaAs layer [11] whereas Fig. (2b) shows the residual spin polarization at the right end of the GaAs layer [11] after a simulation time of 20 ps. Our results agree with the experimental [1, 6] and theoretical [5, 7] results which also predict that spin relaxation time decreases with increase in the temperature of the system.

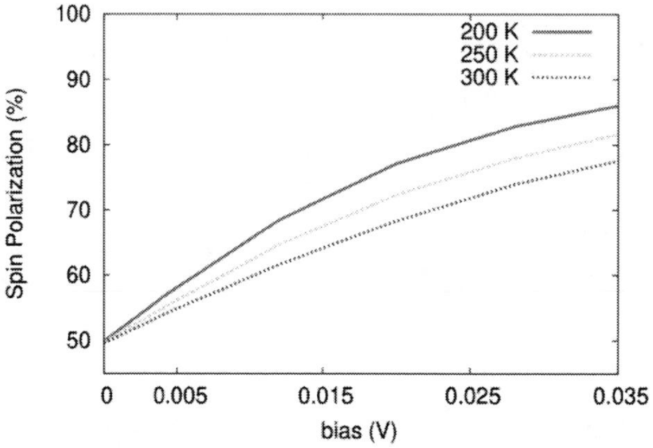

Figure 4. Spin polarization averaged over the device vs bias after 20 ps and for three different temperatures.

In the high temperature regime, the effect of temperature on DP mechanism of spin relaxation is mainly determined by inhomogeneous broadening which is proportional to T^3 [7].

The spin dephasing is considerably affected by the applied fields. We note that even though the injection condition from the FM interface is 100% polarization, the spin polarization close to the FM interface is decreased to less than 80% for no applied fields due to particle diffusion. This value decreases to about 20% at the end of the device. This can be seen in Fig. (3) in which we have plotted the spin polarization over the whole length of device after 20 ps, for a temperature of 300 K and for varying applied bias. At high applied fields the spin polarization is enhanced and at the end of the simulation more spin-coherence is retained in the system (approx. 50% at the right end of the GaAs sample for a bias of 0.036 V). This seems in agreement with the high-field effect predicted in [12]. We also have calculated the average spin polarization in the system at the end of the simulation time and this is shown in Fig. (4). It is seen that for zero applied bias the system has an average of 50% spin polarization and this increases to nearly 80% for high bias (0.036 V) at 300 K. A similar trend is seen for lower temperatures but the system retains more spin-coherence in these cases, up to $\sim$ 85% at 200 K.

5. Conclusions

Spin relaxation in GaAs has been analyzed using the ensemble Monte Carlo technique incorporating various scattering mechanisms. We have analyzed the effects of temperature and applied bias on electrically injected spin polarization. Our results show that increasing temperature reduces the spin relaxation times whereas high applied bias corresponds to a longer spin memory in the system.

6. Acknowledgments

Authors would like to thank Marco Saraniti, Jeremy Coe and Phil Hasnip for helpful discussions. The use of White Rose Grid Computational Resources at the University of York is gratefully acknowledged. This work is supported by EPSRC through grant number EP/F016719/1. Sheetal Kapoor, the principal author, sadly died shortly after this manuscript was completed: the other authors wish to dedicate it to her memory.

7. References

[1] Kikkawa J M and Awschalom D D 1998 *Phys. Rev. Lett.* **80**, 4313
[2] Dzhioev R I, Kavokin K V, Korenev V L, Lazarev M V, Meltser B Ya and Stepnova M N 2002 *Phys. Rev.* B **80**, 245204
[3] Kurebayashi H, Trypiniotis T, Lee K, Easton A, Ionescu A, Farrer I, Ritchie D A, Bland J A C and Barnes C H W 2010 *Appl Phys. Lett.* **96**, 022505
[4] Oertel S, Hubner J and Oestreich M 2008 *Appl Phys. Lett.* **93**, 132112
[5] Krishnamurthy S, Schilfgaarde M van and Newman N 2003 *Appl Phys. Lett.* **83**, 1761
[6] Miah M I, (2008) *J. Phys. D: Appl.Phys.* **41**, 185503
[7] Jiang J H and Wu M W 2009 *Phys. Rev.* B **79**, 125206
[8] Saikin S, Pershin Y V and Privman V 2005 *Proc.-Circuits Devices Syst.* **142**, 366
[9] Jacoboni C and Lugli P 1989 *The Monte Carlo Method for Semiconductor Device Simulation* (Newyork, Springer-Verlag)
[10] Dyakonov M I 2008 *Basics of Semiconductors and Spin Physics* (Spin Physics in Semiconductors) ed M I Dyakonov (Berlin, Springer)
[11] Data have been averaged over 25 Å to improve statistics
[12] Yu Z G and Flatté M 2002 *Phys. Rev.* B **66**, 235302

Journal of Physics: Conference Series **303** (2011) 012096 doi:10.1088/1742-6596/303/1/012096

Electric detection of the spin-Seebeck e ect in magnetic insulator in the presence of interface barrier

K Uchida[1], T Ota[1], Y Kajiwara[1], H Umezawa[2], H Kawai[2] and E Saitoh[1,3,4]

[1]Institute for Materials Research, Tohoku University, Sendai 980-8577, Japan
[2]FDK Corporation, Shizuoka 431-0495, Japan
[3]PRESTO, Japan Science and Technology Agency, Sanbancho, Tokyo 102-0075, Japan
[4]Advanced Science Research Center, Japan Atomic Energy Agency, Tokai 319-1195, Japan

E-mail: `kuchida@imr.tohoku.ac.jp`

Abstract. The spin-Seebeck e ect (SSE), the spin-voltage generation as a result of a temperature gradient, has recently been observed in ferrimagnetic insulator $LaY_2Fe_5O_{12}$ lms by means of the inverse spin-Hall e ect in Pt lms. Here we investigate the SSE using $LaY_2Fe_5O_{12}/SiO_2(Cu)/Pt$ systems, where the $LaY_2Fe_5O_{12}$ and Pt layers are separated by SiO_2 (Cu) thin- lm barriers. The experimental results show that the SSE signal disappears in the $LaY_2Fe_5O_{12}/SiO_2/Pt$ system, but the nite signal appears in the $LaY_2Fe_5O_{12}/Cu/Pt$ system, indicating that the direct contacts between the $LaY_2Fe_5O_{12}$ and normal metals is necessary for generating the SSE signal.

In the elds of spintronics [1, 2] and spin caloritronics [3], spin-voltage generation from heat is highly desired for driving spin-based devices and for enhancing their e ciency. Here, spin voltage refers to the potential for electron's spins, which drives a spin current, i.e., a ow of spin angular momentum [4-6]. The spin-Seebeck e ect (SSE) is one enabling the generation, which converts a heat current into the spin voltage in ferromagnetic metals [7-9], semiconductors [10], and insulators [11]. The SSE was recently observed in a garnet-type ferrimagnetic insulator $LaY_2Fe_5O_{12}$ lm by means of the inverse spin-Hall e ect (ISHE) [12-19] in Pt lms attached on the $LaY_2Fe_5O_{12}$ [11]. In this paper, to buttress the previous results on the SSE in the $LaY_2Fe_5O_{12}/Pt$ systems, we report supplementary experiments using $LaY_2Fe_5O_{12}/SiO_2/Pt$ and $LaY_2Fe_5O_{12}/Cu/Pt$ systems. The experiments show that the direct contacts between $LaY_2Fe_5O_{12}$ and normal metals is necessary for the SSE experiments.

Figure 1 schematically shows the experimental setup for measuring the SSE-induced ISHE. The device consists of a rectangular-shaped ferromagnet (F) with two (or more) normal-metal (N) wires attached on the top of the F layer, where F and N correspond to $LaY_2Fe_5O_{12}$ and Pt in the previous experiments, respectively [11]. Here, an in-plane temperature gradient ∇T is applied to the F layer along the x direction (see gure 1). Since the localized magnetic moments in F and conduction electrons in N are coupled by the interface spin-exchange interaction, or spin-mixing conductance [20, 21], the thermally generated spin voltage injects a spin current into the N wire with a spatial direction $\mathbf{J}_s$ (z direction) and a spin-polarization vector $\boldsymbol{\sigma}$ parallel to the magnetization $\mathbf{M}$ direction of the F layer (x direction). In the N layer, this spin current is converted into an electric eld $\mathbf{E}_{ISHE}$ due to the ISHE. When $\mathbf{M}$ is along the ∇T direction, $\mathbf{E}_{ISHE}$ is generated along the N layer (y direction) because of the relation $\mathbf{E}_{ISHE} \propto \mathbf{J}_s \quad \boldsymbol{\sigma}$

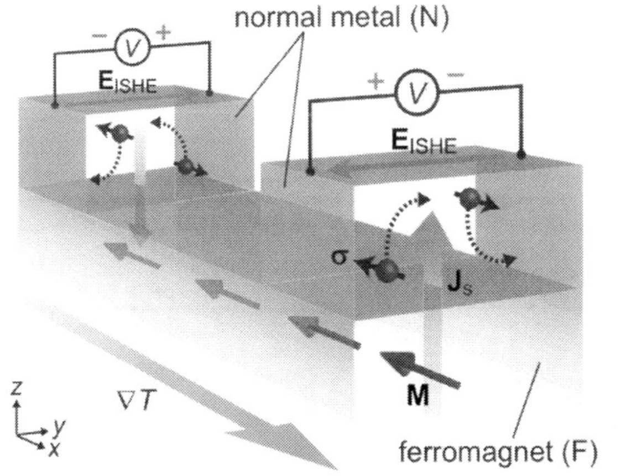

Figure 1. A schematic illustration of the spin-Seebeck e ect (SSE) in a ferromagnet (F) and the inverse spin-Hall e ect (ISHE) in normal metals (Ns). Here, **M**, $\mathbf{J}_s$, and $\mathbf{E}_{\rm ISHE}$ denote the magnetization vector of the F layer, the spatial direction of the spin current, and the electric eld generated by the ISHE in the N layer, respectively. In Ref. [11], $LaY_2Fe_5O_{12}$ and Pt lms were used as the F and N layers, respectively. While $LaY_2Fe_5O_{12}$ is a ferrimagnet, it behaves like a ferromagnet under the present experimental conditions [11].

[12]. Therefore, by measuring $\mathbf{E}_{\rm ISHE}$, we can detect the SSE in the F layer electrically. Since the sign of the spin voltage generated by the SSE is reversed between the higher- and lower-temperature ends of the F layer, the sign of $\mathbf{E}_{\rm ISHE}$ is also reversed between the ends of the F/N device (see gure 1) [11, 21]. In this setup, collinear orientation of the temperature gradient and magnetization suppresses experimental artifacts with the same symmetry as the Nernst-Ettingshausen e ect [22]. The following supplementary experiments further support the above interpretation on the SSE measurements.

In the present study, in addition to the conventional $LaY_2Fe_5O_{12}$/Pt sample, we fabricated $LaY_2Fe_5O_{12}/SiO_2$/Pt and $LaY_2Fe_5O_{12}$/Cu/Pt systems, where the $LaY_2Fe_5O_{12}$ and Pt layers are separated by thin SiO_2 and Cu lms, respectively. First, the single-crystal $LaY_2Fe_5O_{12}$ (111) lm with the 8 4 mm^2 rectangular shape and 3.9 m thickness was grown on a $Gd_3Ga_5O_{12}$ (111) substrate by liquid phase epitaxy. Then, the two 10-nm-thick SiO_2 (12-nm-thick Cu) wires were fabricated on the $LaY_2Fe_5O_{12}$ layer by a sputtering (an electron-beam evaporation) method. Finally, the Pt lms were sputtered on the SiO_2 (Cu) wires. The thickness of the Pt wires is 15 nm (10 nm) for the $LaY_2Fe_5O_{12}/SiO_2$(Cu)/Pt sample. The length and width of the Pt, SiO_2, and Cu wires are 4 mm and 0.1 mm, respectively. An in-plane external magnetic eld H and a uniform ∇T were applied along the x direction (see gure 2). Here, the temperatures of both the ends of the sample were stabilized to 300 K and 300 K+ T, where the temperature di erence T was xed at 20 K. We measured electric voltage di erence V between the ends of the Pt layer of each sample using a micro probing system.

Figure 2(a) shows the measured voltage V as a function of H in the $LaY_2Fe_5O_{12}$/Pt sample at $T = 20$ K, measured when the probe needles are attached to the ends of the Pt layer placed at the higher-temperature end of the sample. By reversing H, the nite voltage step appears; the sign of V is reversed in response to the magnetization reversal of the $LaY_2Fe_5O_{12}$ layer. This situation is consistent with the feature of the ISHE induced by the SSE. In Ref. [11], we con rmed that the results of the T dependent and Pt-wire-position dependent measurements are also consistent with the SSE behavior.

Figure 2(b) shows the H dependence of V at $T = 20$ K in the $LaY_2Fe_5O_{12}/SiO_2$/Pt sample. In this sample, the V signal completely disappears (see also the inset in gure 2(b)). In contrast, in the $LaY_2Fe_5O_{12}$/Cu/Pt sample, the small but nite V signal appears although the Cu layers are thicker than the SiO_2 layers (see the inset in gure 2(c)), indicating that the V signals in gures 2(a) and 2(c) are attributed to the direct contact and thermally activated interface spin exchange between the $LaY_2Fe_5O_{12}$ and normal metals [11, 21]. The results in gures 2(b) and 2(c) also become evidence that the Pt layers in the samples are not magnetized by the proximity

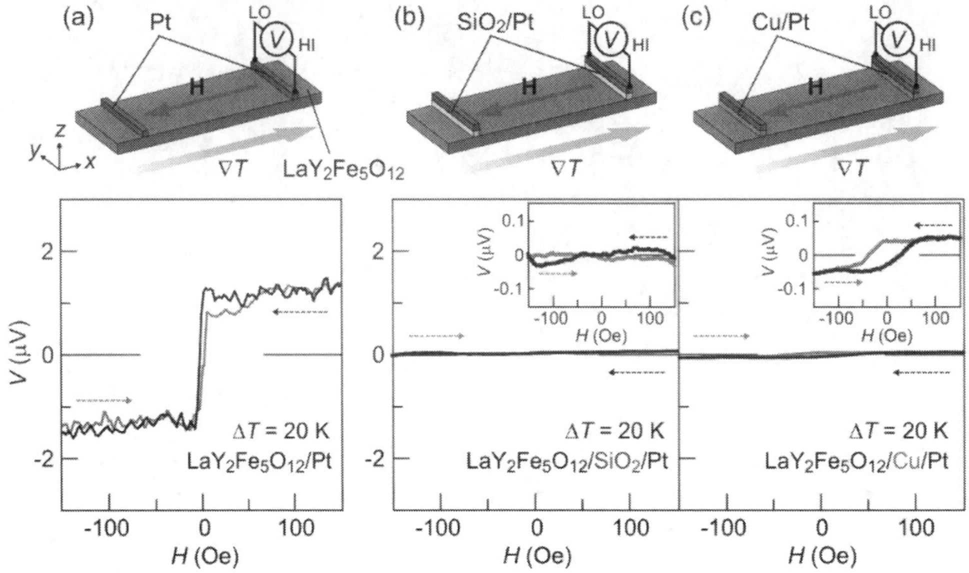

Figure 2. H dependences of V in the $LaY_2Fe_5O_{12}$(3.9 m)/Pt(15 nm) sample (a), the $LaY_2Fe_5O_{12}$(3.9 m)/SiO_2(10 nm)/Pt(15 nm) sample (b), and the $LaY_2Fe_5O_{12}$(3.9 m)/Cu(12 nm)/Pt(10 nm) sample (c) at $T = 20$ K. Here, a parenthetic number represents the thickness of each layer. The insets in (b) and (c) show the H dependences of V in the $LaY_2Fe_5O_{12}/SiO_2$/Pt and $LaY_2Fe_5O_{12}$/Cu/Pt samples at $T = 20$ K in which the vertical axes are magni ed. All the results were measured in the Pt wires attached to the higher-temperature ends of the sample. The di erences of the noise levels and coercive forces between the data in (a) and (c) are due to those of the smoothing parameters. The V-H curve in (a) re ects the actual magnetization curve of the $LaY_2Fe_5O_{12}$ layer (see Ref. [11]).

e ect from the $LaY_2Fe_5O_{12}$ interface. Therefore, we can conclude that the V signals are due entirely to the SSE in the magnetic insulator $LaY_2Fe_5O_{12}$.

In summary, to buttress our previous experiments [11], we have investigated the spin-Seebeck e ect (SSE) in the ferrimagnetic insulator $LaY_2Fe_5O_{12}$ lms with the SiO_2/Pt and Cu/Pt wires. Although the inverse spin-Hall voltage induced by the SSE disappears in the $LaY_2Fe_5O_{12}/SiO_2$/Pt sample, the nite signal appears in the $LaY_2Fe_5O_{12}$/Cu/Pt system. These results con rm that the observed V signals are attributed to the ISHE induced from the thermally generated spin voltage in the $LaY_2Fe_5O_{12}$ layer.

Acknowledgments
The authors thank S. Maekawa, G. E. W. Bauer, J. Xiao, H. Adachi, J. Ohe, S. Takahashi, J. Ieda, B. J. van Wees, and J. Sinova for valuable discussions. This work was supported by a Grant-in-Aid for Scienti c Research in Priority Area 'Creation and control of spin current' (19048028), a Grant-in-Aid for Scienti c Research A (21244058), the global COE for the 'Materials integration international center of education and research' all from MEXT, Japan, a Grant for Industrial Technology Research from NEDO, Japan, and Fundamental Research Grants from PRESTO-JST, TRF, and TUIAREO, Japan.

References

[1] Wolf S A, Awschalom D D, Buhrman R A, Daughton J M, von Molnar S, Roukes M L and Chtchelkanova A Y 2001 *Science* **294** 1488-95
[2] Zutic I, Fabian J and Sarma S D 2004 *Rev. Mod. Phys.* **76** 323-410
[3] Hatami M, Bauer G E W, Zhang Q-F and Kelly P J 2007 *Phys. Rev. Lett.* **99** 066603
[4] Slonczewski J C 1996 *J. Magn. Magn. Mater.* **159** L1-7
[5] Takahashi S and Maekawa S 2008 *J. Phys. Soc. Jpn.* **77** 031009
[6] Kajiwara Y, Harii K, Takahashi S, Ohe J, Uchida K, Mizuguchi M, Umezawa H, Kawai H, Ando K, Takanashi K, Maekawa S and Saitoh E 2010 *Nature* **464** 262-6
[7] Uchida K, Takahashi S, Harii K, Ieda J, Koshibae W, Ando K, Maekawa S and Saitoh E 2008 *Nature* **455** 778-81
[8] Uchida K, Ota T, Harii K, Takahashi S, Maekawa S, Fujikawa Y and Saitoh E 2010 *Solid State Commun.* **150** 524-8
[9] Slachter A, Bakker F L, Adam J-P and van Wees B J 2010 arXiv:1004.1566
[10] Jaworski C M, Yang J, Mack S, Awschalom D D, Heremans J P and Myers R C 2010 arXiv:1007.1364
[11] Uchida K, Xiao J, Adachi H, Ohe J, Takahashi S, Ieda J, Ota T, Kajiwara Y, Umezawa H, Kawai H, Bauer G E W, Maekawa S and Saitoh E 2010 *Nature Mater.* (accepted)
[12] Saitoh E, Ueda M, Miyajima H and Tatara G 2006 *Appl. Phys. Lett.* **88** 182509
[13] Valenzuela S O and Tinkham M 2006 *Nature* **442** 176-9
[14] Kimura T, Otani Y, Sato T, Takahashi S and Maekawa S 2007 *Phys. Rev. Lett.* **98** 156601
[15] Seki T, Hasegawa Y, Mitani S, Takahashi S, Imamura H, Maekawa S, Nitta J and Takanashi K 2008 *Nature Mater.* **7** 125-9
[16] Harii K, Ando K, Inoue H Y, Sasage K and Saitoh E 2008 *J. Appl. Phys.* **103** 07F311
[17] Ando K, Kajiwara Y, Takahashi S, Maekawa S, Takemoto K, Takatsu M and Saitoh E 2008 *Phys. Rev.* B **78** 014413
[18] Ando K, Ieda J, Sasage K, Takahashi S, Maekawa S and Saitoh E 2009 *Appl. Phys. Lett.* **94** 262505
[19] Mosendz O, Pearson J E, Fradin F Y, Bauer G E W, Bader S D and Ho mann A 2010 *Phys. Rev. Lett.* **104** 046601
[20] Tserkovnyak Y, Brataas A, Bauer G E W and Halperin B I 2005 *Rev. Mod. Phys.* **77** 1375-421
[21] Xiao J, Bauer G E W, Uchida K, Saitoh E and Maekawa S 2010 *Phys. Rev. B* **81** 214418
[22] Callen H B 1948 *Phys. Rev.* **73** 1349-58

Magnetoresistance in CuPc Based Organic Magnetic Tunnel Junctions

H Tokuc, K Oguz, F Burke and J M D Coey
CRANN and School of Physics, Trinity College, Dublin 2, Ireland

E-mail: htokuc@tcd.ie

Abstract. We investigate the magnetoresistance of exchange biased spin valves with hybrid barrier layers, MgO(2 nm)/CuPc (t = 0, 1, 2, nm) where CuPc is the organic semiconductor copper phthalocyanine, which are patterned into micron size tunnel junctions by UV lithography. The stacks are produced by sputtering in a Shamrock tool and the organic layer is subsequently deposited by thermal evaporation. Devices in the first set showed 22 %, 10 % and 5% MR at 300 K for t = 0, 1 and 2 nm, respectively. These values increased up to 33 %, 28 % and 23% for the same junctions at 50 K. From these data we infer that the spin polarization is reduced by CuPc in the barrier and the spin scattering by Cu^{2+} is strongly temperature-dependent. By comparison, if an Alq_3 layer is used, where there are no paramagnetic scattering centres, the magnetoresistance is independent of the thickness of the organic layer, at least up to 8 nm.

1. Introduction

Spin injection, transport and dynamics in organic semiconductors are of growing interest as the new field of organic spin electronics begins to take shape. Long spin life times due to the low spin-orbit coupling and weak hyperfine interactions should allow the spin polarized carriers to travel over significant distances in the organic semiconductors despite their small mobility [1]. From the application point of view, cheap and flexible organic electronic devices have been developed for commercial displays OLED and there are prototype organic field effect transistors (OFETs) and organic solar panels. The challenge is to add spin functionality to these organic devices.

A common approach is to look for magnetoresistance in organic based spin valve structures [3-5]. A typical spin valve device consists of two ferromagnetic layers (FM) separated by a nonmagnetic (NM) spacer layer. The operating principle of these devices depends on the relative alignment of magnetizations of the two FM layers in a pseudo spin valve, the two layers have different coercivity so a symmetric, butterfly-like plot of resistance versus applied field is obtained. In an exchange-biased spin valve one of the FM layers is pinned while the other one is free to rotate under external magnetic field. An asymmetric magnetoresistance plot is obtained with switches close to zero field when the magnetization directions of the magnetic layers are antiparallel, the system normally shows high resistance when the two FM layers are made of the same material. When they are aligned parallel, the resistance reaches its minimum value.

The possibility of spin injection into an organic layer (Alq_3) was suggested by Dediu *et al.* [2], but the first successful spin valve structure was reported by Xiong *et al.* [3]. They reported 40% negative magnetoresistance at 11 K by using $La_{0.7}Sr_{0.3}MnO_3$ (LSMO) and Co FM electrodes with a 120 nm Alq_3 organic spacer. The magnetoresistance disappeared when the temperature was increased up to

Published under licence by IOP Publishing Ltd

200 K. Another negative magnetoresistance effect was seen by Dediu and his group on the LSMO/Alq_3/AlO_x/Co/Al structure [4]. They observed nearly 11% negative MR at 20 K but the MR fell to -0.15% at room temperature. Use of an oxide layer between organic layer and a ferromagnetic electrode was suggested by Santos *et al.* [5]. The oxide layer was intended to overcome the conductivity mismatch problem between the semiconductor and metal layers and assist the electron injection into the semiconductor [6,7]. They used Co and NiFe as FM electrodes with an AlO_x (1.5nm)/Alq_3 (1-4 nm) hybrid barrier. In this structure 6% positive MR was observed at room temperature. Another group was unable to detect any magnetoresistance with 50-100 nm thick Alq_3 barrier [8]. Later, Szulczewski *et al.* made lithographically patterned junctions using an exchange biased spin valves structure with MgO(2nm)/Alq_3(0-8nm) hybrid barrier [9]. In this experiment organic Alq_3 layers (0-8nm) were deposited on crystalline MgO barrier and the bottom stack was made with exchange biased CoFeB. They observed up to 16 % MR at room temperature. The remarkable observation was that the MR was independent of Alq_3 barrier thickness in a range where the transport mechanism changed from tunnelling (2nm) to diffusion (8nm). Three quarter of spins preserved their orientations on passing through the 8 nm organic layer.

The crossover from tunnelling to diffusive (or hopping) transport has also been studied by Yoo *et al* with rubrene [10]. Schoonus et al. studied spin valves with on AlO_x(1.5nm)/Alq_3(0-4nm) hybrid barrier [11]. In the tunnelling regime (t = 1nm) they observed 12 % room temperature MR but there was a significant reduction when $t \geq 2$ nm which they identify as a two-step tunneling where the largest contribution to the current comes from tunnelling via a site in the AlO_x/Alq_3 interface. Another study was done by Drew et al. on the FeCo/TPD/Alq_3/LiF/NiFe vertical stack, where the Alq_3 is 200 nm thick, by low-energy muon spin rotation technique [12]. They observed that spin diffusion length at low temperature is nearly 35 nm, but MR is less than 0.2 %.

After these promising experiments on Alq_3, investiganions have spread to other organic semiconductors. Another widely used material is copper phthalocyanine (CuPc), a prototypical p-type organic semiconductor, also used in OLEDs. The possibility of spin injection into CuPc was first demonstrated by Cinchetti et al. [13]. Their approach was using spin-resolved two-photon photoemission spectra to inject the spins into CuPc from Co/CuPc interface. By using this technique they successfully injected the spins into a CuPc layer and spins retained their polarization up to 16 monolayers (~ 5nm) of CuPc. Another experiment was done by Liu et al. to show the possibility of injection of spin polarized carriers into the CuPc [14]. In their experiment they used Fe and Co as FM electrodes and 100 nm CuPc as a spacer layer. Their result showed 6.4% MR at 40 K but it disappeared at room temperature.

Here we focus on the CuPc based magnetic tunnel junctions with a hybrid MgO/CuPc barrier. We prepared lithographically patterned junctions of various sizes using a bottom-pinned exchange biased stack.

2. Experimental methods and materials

The CuPc layers were deposited in a bell jar system with a base pressure of ~$5x10^{-6}$ mbar. CuPc was supplied from Sigma Aldrich in a powder form and it was purified by heating at 300°C under an Ar flow. The CuPc was evaporated from an alumina crucible which is heated by tungsten filament. Firstly, we made a series of singe CuPc layers on SiO_2 substrate to study the surface morphology by atomic force microscopy (AFM). One of the key elements in organic spin valves is the roughness of ferromagnetic/organic interfaces which is directly related to the spin injection from FM layer so we have also studied the structural and magnetic characterization of FM/organic interfaces by TEM and SQUID, respectively.

For the magneto transport measurements, the following multilayer stack Si/SiO_2(500)/Ta(5) /Ru(30) /Ta(5) /NiFe(5) /IrMn(10) CoFe(2.5)/Ru(0.9)/CoFeB(3) /MgO(2) was prepared by sputtering in the Shamrock tool which has a base pressure of $2x10^{-8}$ torr (the numbers in parenthesis are in nanometer). Then, vacuum was broken and samples were transferred to an annealing furnace. Here they were annealed at 350°C in a field of 800 mT during 1 hour to set the exchange bias and to get crystallize the

CoFeB layer. After that, samples were transferred to the bell jar organic evaporation system, where the base pressure is ~5x10^{-6} mbar, to deposit the CuPc organic layers (0-2 nm). After the organic depositions, the samples were transferred to Shamrock tool to deposit top FM electrode (CoFe 3nm) and capping layers (Ta 5nm/Ru5nm). These completed stacks were patterned by a three step UV lithography technique with junction sizes ranging from 4x4 μm^2 to 50x150 μm^2. Positive photoresist (S1813) was spun on the samples and exposed to define a dog bone structure. Then, the samples were Ar ion milled down to substrate and covered with 90 nm SiO_2 to protect the organic layer from any solvents. Then, SiO_2 layer was removed from the dog bone structure by lifting off. Due to pure yield after lift off we did not succeed to make devices with an organic layer thicker than 2 nm. The junction areas were defined by the second lithography process. The positive photoresist was spun again on the samples and exposed to UV with a metal mask defining the mesas. Later, they were Ar ion milled down to MgO barrier and 50 nm SiO_2, which was removed from mesas by lifting off, deposited to isolate the bottom stacks. A final exposure and development step creates strips for the top and bottom contacts. Here, Ta(5nm)/Cu(50nm) bilayer was sputtered to form the top contacts and the lithography was successfully finished by the last lift off step. We made two sets of samples. The thickness of the organic layers is 0, 1, 2 nm and 0, 2 nm for the first and second set, respectively.

3. Results and discussion

Figure 1 shows the AFM image of SiO_2/CuPc(2nm) single layer and the RMS roughness of CuPc layers in a range of 2-8 nm. The RMS roughness of the 2 nm CuPc layer is found to be 0.39 nm and it is nearly independent of thickness. This low roughness in the CuPc layer allows us to make very smooth organic barrier in our devices.

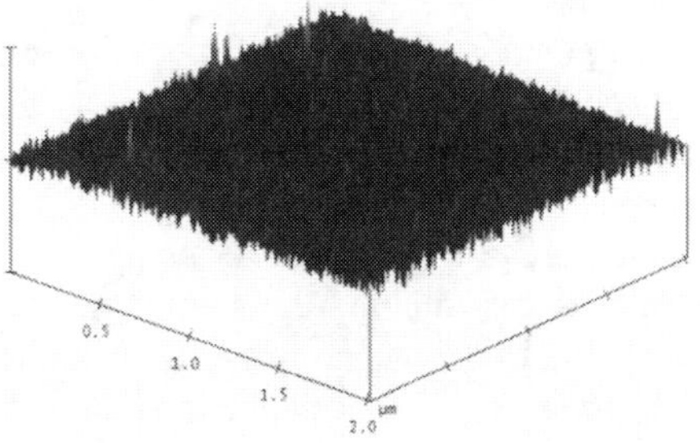

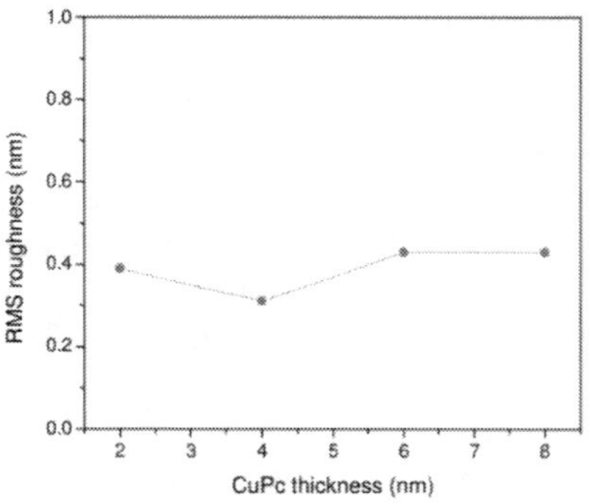

Figure 1. (a) AFM image of SiO_2/CuPc(2nm) single layer and (b) RMS roughness of the different CuPc single layers with thickness ranging from 2 to 8 nm.

Another key element of the spin injection into an organic is the chemical nature of the organic /FM interface. The chemical reaction between the Co or Fe and CuPc has been reported by Aristov *et al.* [15]. Figure 2 shows the magnetic moment measurements of the bilayers with and without CuPc. It is very clear that inserting CuPc between SiO_2 and Co results in a significant reduction of the magnetic moment of Co layer, by nearly 40%. The reason for this significant reduction might be the formation of magnetic dead layer due to chemical reaction at the organic/FM interface, which can influence carrier injection into the organic. The figure 3 shows the TEM image of one unpatterned stack. The CoFeB and MgO layers seem to have a good crystalline structure and a sharp interface but the CuPc/CoFe interface seems to be rougher which might lead to spin scattering at the interface. This rough CuPc/CoFe interface is a critical issue for the spin injection into the CuPc layer.

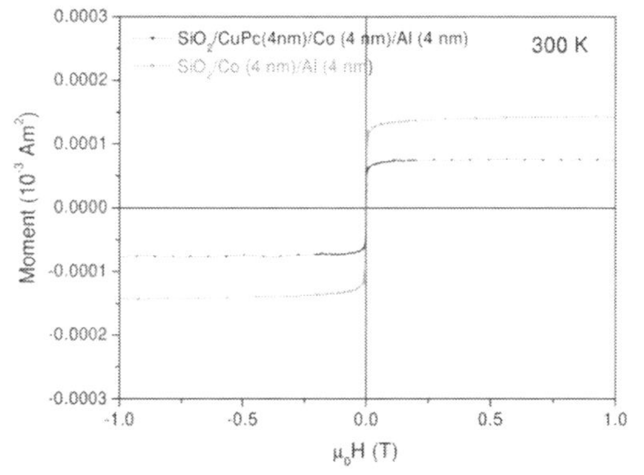

Figure 2. Magnetic moment measurements of the with and without CuPc bilayers.

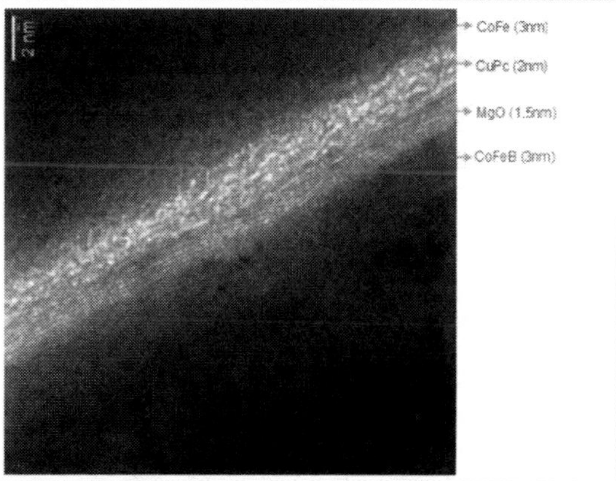

Figure 3. TEM image of an unpatterned stack.

The figure 4a and 4b show the TMR characteristics of the first set of CuPc based MTJs at 300 K and 50 K, respectively. Here, the junction sizes are 50x150 μm^2. In the case it is obvious that TMR strongly depends on the thickness of the CuPc layer as well as temperature. At room temperature, inserting 1 nm CuPc layer between MgO and CoFe layer destroyed nearly 10% MR in the tunnelling regime. The further increase of the thickness of CuPc layer (t = 2 nm) caused to lose another 7% MR. We infer that this is due to the spin scattering from the paramagnetic Cu^{2+} ions in CuPc.

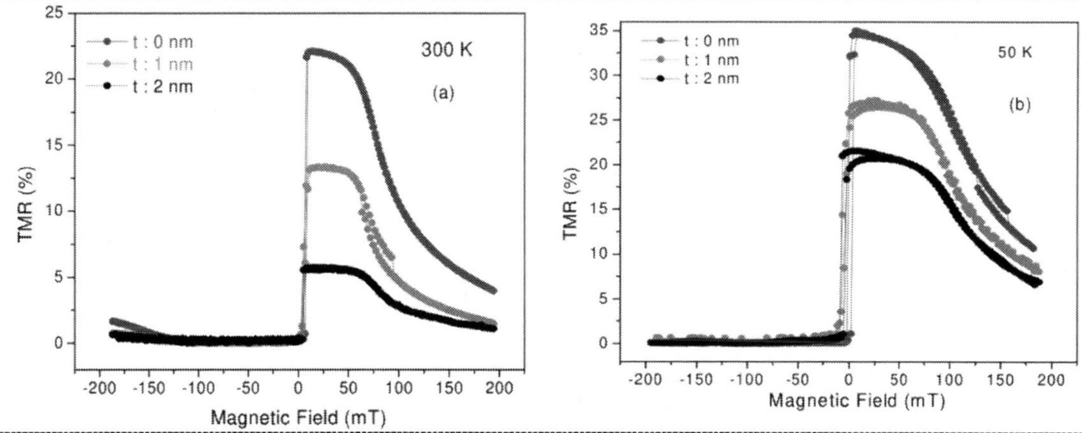

Figure 4. (a) MR characteristics of the first set of samples at 300 K and (b) at 50 K

Figure 5a shows the temperature-dependent resistance behaviour of the devices. When t is 1 nm, the resistance is nearly independent of temperature as expected for the tunnelling. However, when t is 2 nm, resistance changes nearly by a factor of 2 over the temperature range suggesting the onset of a parallel conduction channel due to diffusion through the thicker CuPc layer. Figure 5b shows a plot of resistance area product (RA) versus the square root of area of the MgO(2nm)CuPc(0nm) samples. RA changes with area, unlike the second set where RA was nearly constant. We attribute this difference to a non-uniform barrier in the first set of samples.

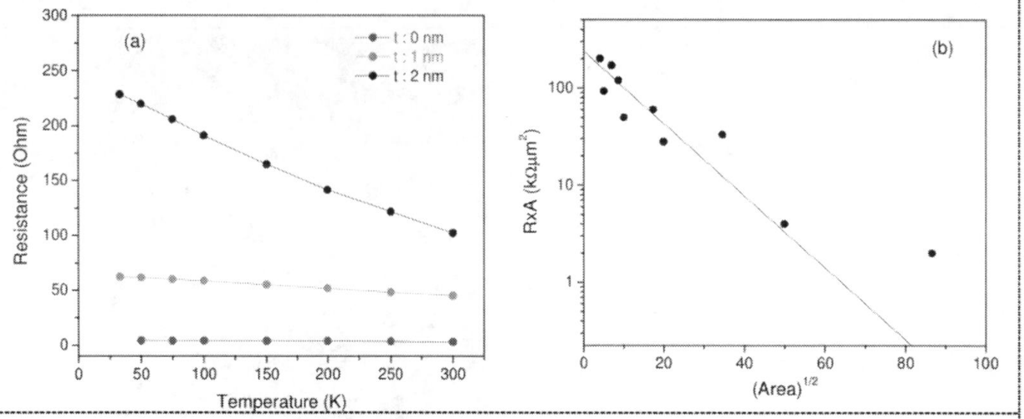

Figure 5. (a) Temperature dependent resistance behaviour of the first st of samples and (b) RA product versus square root of area plot.

Figure 6a and 6b show MR characteristics of second set of CuPc based junctions. Here, the junction sizes are 10x10 μm^2 and 50x50 μm^2 for the MgO(2nm)/CuPc(2nm) and MgO(2nm)/CuPc(0nm) samples, respectively. In this set, the MR values are bigger than for the first set and the RA is approximately 1 MΩμm^2 for t = 0 nm, regardless of junction area. There is still a significant reduction of the MR nearly 10% at 300 K, when 2 nm CuPc is inserted between MgO and CoFe.

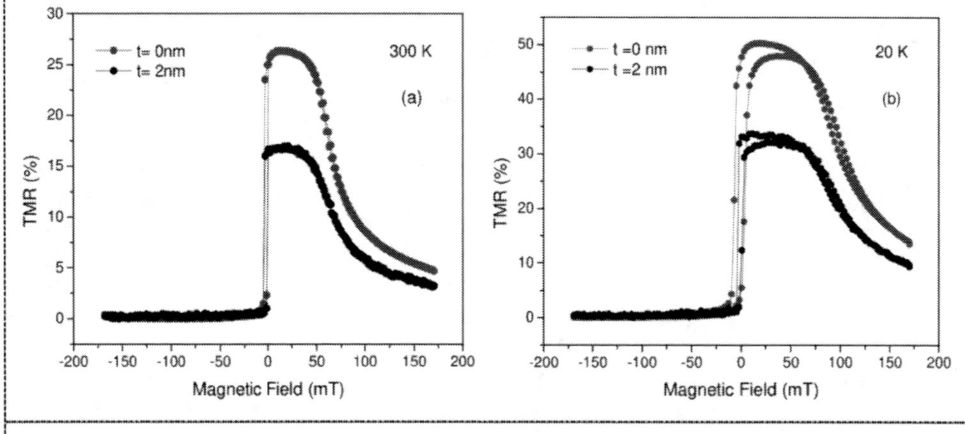

Figure 6. (a) MR characteristics of the second set of samples at 300 K and (b) at 20 K.

From these results we can deduce that the spin diffusion length in CuPc may be 3-4 nm at room temperature and that depends strongly on two factors which are the paramagnetic nature of CuPc and the quality of CuPc/CoFe interface. One can compare these results with Alq_3 based MTJs. Spin diffusion length is more than 8 nm for Alq_3 at room temperature while it is less in CuPc.

4. Summary

Lithographically patterned and exchange-biased CuPc based vertical junctions show useful room-temperature magnetoresistance .The room temperature spin scattering length in the CuPc inferred to be ~ 3-4 nm and it is depended on the paramagnetic scattering centres in the CuPc and the CuPc/CoFe interface. We believe that spin diffusion length can be improved by using a fabrication process that does not break vacuum. Our results suggest that it is possible to inject and recover spin polarized electrons to the organic layer.

Acknowledgements

This work was supported by Science Foundation Ireland as part of the MANSE project 05/IN/1850. This work conducted under the framework of the INSPIRE programme, funded by the Irish Government's Programme for Research in Third Level Institutions, Cycle 4, National Development Plan 2007-2013. We are also grateful to Dr. Colm Faulkner for preparation of TEM sample.

References:

[1] Szulczewski G, Sanvito S and Coey M 2009 *Nature Materials* **8** 693
[2] Dediu V, Murgia M, Matacotta F C, Taliani C and Barbanera S 2002 *Solid State Community* **122** 181
[3] Xiong Z H, Wu D, Vardeny Z V and Shi J 2004 *Nature* **427** 821
[4] Dediu V, Hueso L E, Bergenti I, Riminucci A, Borgatti F, Graziosi P, Newby C, Casoli F, De Joung M P, Taliani C and Zhan Y 2008 *Phys. Rev. B* **78** 115203
[5] Santos T S, Lee J S, Migdal P, Lekshmi I C, Satpati B and Moodera J S 2007 *Phys. Rev. Lett.* **98** 016601
[6] Rashba E I 2000 *Phys. Rev. B* **62** 16267
[7] Schmidt G, Ferrand D, Molenkamp L W, Filip A T and van Wees B J 2000 *Phys. Rev. B* **62** 4790
[8] Jiang S J, Pearson E J and Bader S D 2008 *Phys. Rev. B* **77** 035303
[9] Szulczewski G, Tokuc H, Oguz K and Coey J M D 2009 *Appl. Phys. Lett.* **95** 202506
[10] Yoo J W, Jang H W, Prigodin V N, Kao C, Eom C B and Epstein A J 2010 *Synthetic Metals* **160** 216
[11] Schoonus J J H M, Lumens P G E, Wagemans W, Kohlhepp J T, Bobbert P A, Swagten H J M and Koopmans B 2009 *Phys. Rev. Lett.* **103** 146601
[12] Drew A J, Hoppler J, Schulz L, Pratt F L, Desai P, Shakya P, Kreouzis T, Gillin W P, Suter A, Morley N A, Malik V K, Dubroka A, Kim K W, Bouyanfif H, Bourqui F, Bernhard C, Scheuermann R, Nieuwenhuys G J, Prokscha T and Morenzoni E 2008 *Nature Materials* **8** 109
[13] Cinchetti M, Heimer K, Wüstenberg J P, Andreyev O, Bauer M, Lach S, Ziegler C, Gao Y and Aeschlimann M 2009 *Nature Materials* **8** 115
[14] Liu Y, Lee T, Katz H E and Reich D H 2009 *J. Appl. Phys.* **105** 07C708
[15] Aristov V Yu, Molodtsova O V, Ossipyan Yu A, Doyle B P, Nannarone S and Knupfer M 2009 *Organic Electronics* **10** 8

Annealing effect on low frequency noise in MgO-based magnetic tunnel junctions

J. F. Feng, Z. Diao, and J. M. D. Coey
CRANN and School of Physics, Trinity College, Dublin 2, Ireland

E-mail: jfeng@tcd.ie

Abstract. The annealing temperature (T_a) dependence of the low frequency noise in exchange-biased magnetic tunnel junctions (MTJs) with an MgO barrier and CoFeB electrodes has been investigated for 200 < T_a < 425 °C. A *1/f* noise spectrum is observed for the parallel and antiparallel states, where the Hooge parameter (α) for the parallel state remains unchanged with T_a after the CoFeB electrodes is fully crystallized, while α for the antiparallel state increases over the range of T_a from 275 to 425 °C. The *1/f* noise for the antiparallel state is largely of magnetic origin, and it can be attributed to magnetization fluctuations of ferromagnetic electrodes due to the loss of the exchange bias, whereas there is little magnetic noise in the parallel state. An annealing temperature of 325 °C yields an optimized signal-to-noise ratio, as the CoFeB electrodes become fully crystallized. The bias dependence of α is discussed based on the Glazman-Matveev model, which considers both elastic and inelastic tunneling.

1. Introduction

Magnetic tunnel junctions (MTJs) were originally developed using amorphous aluminum oxide tunnel barriers [1-4]. These AlO_x-based MTJs with CoFeB magnetic electrodes have attained a tunneling magnetoresistance (TMR) value up to 80% [$(R_{AP}-R_P)/R_{AP}$] at room temperature and low field [4], where R_P and R_{AP} are the resistances of a MTJ when the magnetizations of the two electrodes are in parallel and antiparallel, respectively. More recently, MTJs with single crystalline MgO tunnel barriers have been found to exhibit much higher TMR values; more than 600% at room temperature and 1100% at low temperature, have been reported in poseudo spin valve structures [5]. This TMR ratio is close to the theoretical maximum [6,7]. MTJ devices have already found important applications in magnetic memory and as sensors [8, 9]. MgO-based MTJs present a seven-fold increase in their output signal compared to those with AlO_x barriers at room temperature. However, sensor applications require not only a large output signal but also a good signal-to-noise ratio. Therefore, minimizing the noise in an MTJ becomes an important consideration.

Electrical noise in MTJs has various mechanisms, such as *thermal noise*, *shot noise* at nonzero current bias, and *1/f noise* [10]. The thermal noise and shot noise are both frequency-independent, and together they define the ultimate noise floor. *1/f* noise is often dominant in the low frequency range, and it is a performance-limiting factor for magnetic field sensors [11, 12]. The *1/f* noise is quantified by the Hooge parameter

$$\alpha = AfS_V/V^2, \qquad (1)$$

A is the junction area, f the frequency, S_V the noise power spectrum, and V the voltage applied to the MTJ. The origin of $1/f$ noise in MTJs can be either magnetic or nonmagnetic. The magnetic contribution to the $1/f$ noise is attributed to thermally-activated magnetization fluctuations of the

Published under licence by IOP Publishing Ltd

ferromagnetic electrodes, which can be understood in terms of the fluctuation-dissipation theorem [12]. The nonmagnetic $1/f$ noise manifests itself as spin-independent resistance fluctuation [13] due, for example, to localized charge traps in the tunnel barrier or at the electrode/barrier interface [14,15]. Thermal annealing, which leads to a crystallized CoFeB layer, is widely used to improve the TMR ratio of MTJs [16]. However, the effect of thermal annealing on electrical noise in MTJs is more controversial. In the early report, Scola *et al.* [14] found no marked dependence of $1/f$ noise in MTJs on the annealing temperature. Liou *et al.* [17] and Stearrett *et al.* [18], however, found a significant drop of low frequency noise in MTJs after annealing.

In this work, we investigate the low frequency noise of MgO-based MTJs in a large T_a range, from 200 to 425 °C in the parallel and antiparallel states. Noise in the perpendicular configuration, required for sensors, is expected to the intermediate between these two extremes. Micro-scale MTJs having high TMR ratio, with MgO tunnel barrier and CoFeB ferromagnetic electrodes, have been fabricated. The annealing temperature dependence of the magnetotransport properties is similar to that reported by Dimopoulos *et al.* [16]. Under a moderate bias of 10 - 30 mV, a $1/f$ noise spectrum is observed at low frequency in both R_P and R_{AP} states. It is well known that the annealing temperature T_a can influence magnetotransport properties of exchange-biased MgO-based MTJs, i.e., TMR, resistance-area (RA) products, interface quality, and Mn diffusion. We further study the relation between these properties and the Hooge parameter (α) of our MTJs under different annealing conditions.

2. Experimental detail

The typical layer structure of our bottom-pinned synthetic antiferromagnetic MTJ samples is Ta 5 /Ru 20 /Ta 5 / $Ni_{81}Fe_{19}$ 5 /$Ir_{22}Mn_{78}$ 10 /$Co_{90}Fe_{10}$ 2 /Ru 0.85 / $Co_{40}Fe_{40}B_{20}$ (CoFeB) 3 /MgO 2.5 /CoFeB 3 /Ta 5 /Ru 10 (from the bottom electrode to the top with thickness given in nanometers). All metallic layers were deposited by dc-magnetron sputtering in a Shamrock sputtering tool. The MgO barrier was deposited from a target-facing-target source using rf sputtering. The MTJ stack was patterned into square-shaped pillars, with an area of 24 × 24 and 24 × 48 μm^2, using UV lithography. High vacuum annealing of MTJ samples in the temperature range of 200 - 425 °C was carried out in a magnetic field of 800 mT in the sample plane for an hour. All magnetotransport and electrical noise measurements were performed by a four-probe method at room temperature. Further details on the sample fabrication and measurements can be found in Refs. [19, 20].

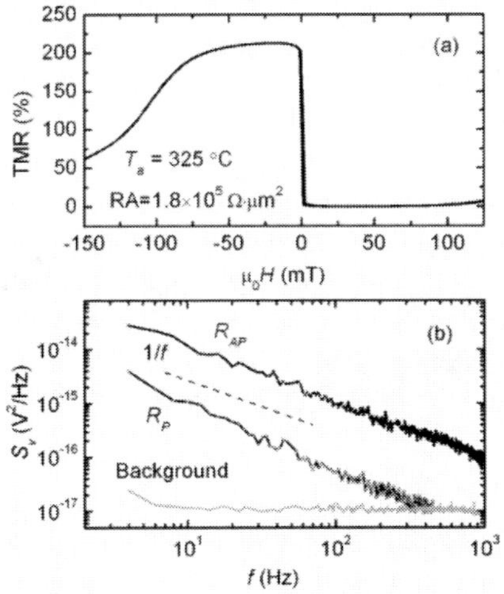

Figure 1. (a) The TMR - H curve for an MTJ at T_a = 325 °C. (b) The noise power spectral density (S_V) of the same device in the R_P and R_{AP} states.

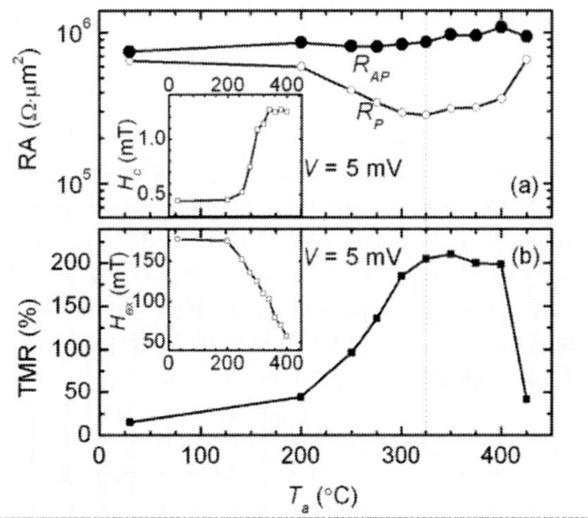

Figure 2. The annealing temperature T_a dependence of the RA value (a) and TMR (b) for an MTJ measured at V = 5 mV. The insets in (a) and (b) show the H_c and H_{ex}.

3. Results and discussion

Figure 1 (a) shows the TMR - H curve for an MTJ with RA = 1.8×10^5 $\Omega\cdot\mu m^2$, annealed at T_a = 325 °C. A TMR of 210% at 5 mV is obtained at room temperature, suggesting that the MgO barrier is of high quality. The well-separated switching of the pinned and free CoFeB electrodes creates well-defined parallel (R_P) and antiparallel (R_{AP}) states. Fig. 1 (b) shows the noise power spectral density (S_V) of the same MTJ below 1 kHz. Magnetic fields of 50 mT and -20 mT are applied to set the R_P and R_{AP} states, respectively. The amplifier noise and thermal noise have been subtracted from the measured S_V. Similar $1/f$ noise is always observed for both states of MTJs studied in this work. In the following, we focus on the influence of annealing temperature on the noise power of our MTJs, and correlate it with the magnetotransport and magnetic properties during annealing.

The annealing temperature dependence of RA and TMR for an MTJ with RA = 2.9×10^5 $\Omega\cdot\mu m^2$ is plotted in Fig. 2 (a) and (b). The RA values remain almost unchanged in the low T_a range until the CoFeB begins to crystallize on the MgO; TMR changes from 15% at the as-grown state to 44% at T_a = 200 °C. The TMR value is 96% at T_a = 250 °C, and then it increases monotonically to reach 210% at T_a = 325 °C due to the improvement of the MgO/CoFeB interface. The crystallization process of the amorphous CoFeB electrodes with the increase of T_a has been observed at the electrode/barrier interfaces [16, 21]. Crystallization of CoFeB is accompanied by a change in coercivity (H_c) of the free CoFeB layer, as shown in the inset of Fig. 2 (a). It is found that the crystallization process starts at $T_a \approx$ 250 °C. Full crystallization of the free CoFeB is found at an annealing temperature of around 325 °C [16], which corresponds to the highest H_c. The H_c value triples from 0.4 mT to 1.2 mT in this annealing temperature range. Together with the high TMR ratio at T_a = 325 °C, it suggests that the large TMR in MgO-based MTJs resides in the fact that the CoFeB layer at the electrode/barrier interface is of crystalline nature [21]. After that, the TMR ratio remains almost unchanged, though Mn has been found at the bottom CoFeB/MgO interface at around T_a = 350 °C [16]. However, a lower TMR ratio is obtained at T_a = 425 °C due to the enhanced Mn diffusion.

Figure 3 (a) shows the Hooge parameter (α) for the R_P and R_{AP} states as a function of the annealing temperature. The lines are guided to the eye. For each T_a, at least three individual MTJs with similar RA were measured, and they all show a similar α value. The α values remain unchanged when T_a is below 250 °C, where RA of the junctions remains steady as described above. Furthermore, α for the R_P state is very close to that for the R_{AP} state. This suggests that the electrical noise of these MTJs at low T_a mainly comes from the amorphous CoFeB/MgO interfaces and the MgO barrier. When the interfaces improve with T_a, α for the R_P state falls and reaches a roughly constant value of 2.0×10^{-9} μm^2, at T_a = 300 °C. This can be explained by the fact that defects formed at the CoFeB/MgO interfaces during the sputtering deposition are released from the system when the CoFeB layer crystallizes. Hence, thermal annealing reduces the electrical noise from these defects in the MTJs. The α values observed in this work are comparable to those reported in Refs. [22], but are higher than those in fully epitaxial Fe/MgO/Fe MTJs [23]. Above 300 °C, α for the R_P state remains unchanged, despite Mn diffusion. However, the α value for the R_{AP} state behaves differently. Firstly, α in the R_{AP} state is larger than that in the corresponding R_P state (6.5×10^{-9} μm^2 at T_a = 300 °C, more than 3 times higher). Similar observations have been reported for junctions with an MgO or AlO_x barrier [10, 14, 15, 20, 22-26]. The difference of α between two states increases to more than one order of magnitude at higher T_a. Above 350 °C, α increases significantly with T_a for the R_{AP} state. This may be attributed to the fact that Mn diffusion at high T_a leads to the degradation of IrMn as the antiferromagnetic pinning layer. Hence, magnetization in the pinned layer becomes more vulnerable to thermal excitations. The decrease of TMR at T_a = 425 °C is not accompanied by a change in α for either R_P or R_{AP} state.

A high signal-to-noise ratio is desirable for sensor applications, which requires a high TMR ratio and low noise in the devices. The optimum annealing temperature of 325 °C for these MgO-based MTJs, corresponds to a full crystallization of the CoFeB ferromagnetic electrodes, and lower diffusion of Mn.

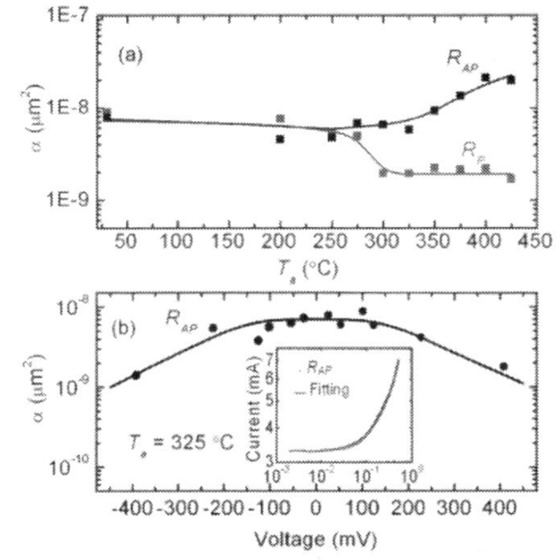

Figure 3. The Hooge parameter (α) as a function of the annealing temperature (a) and bias (b). The inset in (b) shows the *I-V* curve at the R_{AP} state, and the blue line is a fit using Eq. (2).

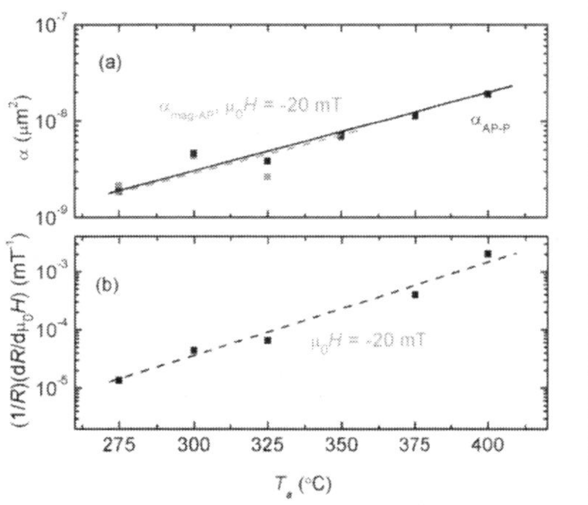

Figure 4. The annealing temperature dependence of α_{mag} (a) and $(1/R)(dR/d\mu_0 H)$ (b) at R_{AP} state. Fig.4 (a) also shows the difference of α between R_{AP} and R_P states above 250 °C. The lines are guided to the eye.

The bias dependence of α for the R_{AP} state is shown in Fig. 3 (b) for another MTJ at T_a = 325 °C. The α value is practically constant in the voltage range of ±100 mV [14], but a three-fold decrease in α is observed under higher voltage bias. It may be related to the nonlinearity of the *I-V* curve of MTJs (as shown in the inset of Fig. 3 (b)), where a clear nonlinear behavior starts to appear beyond also about 100 mV [22-24, 27]. The Glazman-Matveev model [28] has previously been used to describe the nonlinear *I-V* curves in MTJs [22, 27],

$$I = G_1 V + G_2 V^{7/3} + G_3 V^{7/2} + \cdots, \tag{2}$$

where G_1 is the elastic contribution of direct and resonant tunneling via a single localized state, and G_2, G_3, ··· gives the inelastic contribution to the total current from tunneling involving 2, 3, ··· impurity states [27, 28]. The direct tunneling and resonant tunneling are independent of bias, while the inelastic tunneling contribution is bias dependent. The fit in the inset of Fig. 3(b) shows that the nonlinear *I-V* curves of our MTJs can also be well described within the framework of this model. If we consider fluctuations both in the elastic and the inelastic tunneling conductance, Eq. (2) gives that

$$\delta V \approx V\frac{\delta G_1}{\tilde{G}} + V^{7/3}\frac{\delta G_2}{\tilde{G}} + V^{7/2}\frac{\delta G_3}{\tilde{G}}, \tag{3}$$

where the total tunneling conductance is $\tilde{G} = G_1 + 7/3 G_2 V^{4/3} + 7/2 G_3 V^{5/2}$ [27]. It can be shown from a simple analysis of Eq. (3) that a decrease in the normalized noise power under high voltage bias can be attributed to a mixing between the different tunneling channels. The bias dependence of α suggests that the voltage bias should always be specified when the α parameters in different junctions are compared.

A related point is that, under low voltage bias, the inhomogenity of the tunneling barrier plays an important role in defining the noise of the device, since fewer localized fluctuators may contribute. While under high voltage bias, different localized fluctuators are active, which can lead to a decrease of the α value. Nowak *et al.* [29] have previously measured the correlation of low frequency noise signatures at different biases in MTJs. Their results suggest that there exists a current redistribution when the bias across the MTJ changes. This leads to the appearance of different fluctuators under different biases. This effect is closely related to the strong sensitivity of bias current to the defect distribution in MTJs [23].

$1/f$ noise in magnetic devices has two origins -- magnetic and nonmagnetic. In the high field range (500 mT), only residual electrical noise is assumed to remain, and there is no noise of magnetic origin. We subtract the high field residual electrical noise from the measured power spectral density S_V to obtain the magnetic contribution S_V'. Defining the magnetic Hooge parameter

$$\alpha_{mag} = Af S_V'/V^2, \quad (4)$$

this parameter allows to compare the noise level in different magnetic states [12,20]. In Fig. 4 (a), we plot $\alpha_{mag\text{-}AP}$ for the R_{AP} state as a function of T_a in the range of 275 - 400 °C. Compared to the values of the conventional Hooge parameter α (Eq. (1)) at T_a = 300 °C, $\alpha_{mag\text{-}AP}$ at the same annealing temperature is 4.2×10^{-9} μm^2, 3 - 4 times higher than the nonmagnetic contribution to the noise in these MTJs. This shows that during the magnetization switching, the noise source in a MTJ in its R_{AP} state is mainly of magnetic origin, whereas non-magnetic noise plays the major role when the MTJ is in its parallel state. The R_P state is much quieter magnetically and there is little hint of instability with T_a (see Fig. 3 (a)) [26]. It is found that the α_{mag} value for the R_P state ($\alpha_{mag\text{-}P}$), typically under a 50 mT magnetic field, is much lower (around 8×10^{-10} μm^2).

Figure 4 (a) also shows the difference of the conventional Hooge parameter α between two magnetic states ($\alpha_{AP\text{-}P}$). It is found that $\alpha_{AP\text{-}P}$ also increases with T_a. This further supports our conclusion that magnetic fluctuations contribute very little to the noise of a MTJ in its R_P state. To understand the origin of the evolution of $\alpha_{mag\text{-}AP}$ with T_a, we use the definition of the sensitivity of an MTJ device, $(1/R)(dR/d\mu_0 H)$, but extend it to the part of the MR curve where the pinned layer switches. This can be treated as a characterizatic of the pinning quality of the antiferromagnetic layer. As shown in Fig. 4 (b), an increase of $(1/R)(dR/d\mu_0 H)$ at $\mu_0 H$ = -20 mT is found with the increase of T_a. From the inset of Fig. 2 (b), it is also found that the exchange bias (H_{ex}) decreases with T_a. Hence, both facts suggest that the loss of the exchange bias, resulting in a loose pinning effect of IrMn, can increase the instability of the magnetic configuration of the pinned layer when the MTJ is in the R_{AP} state. This effect leads to more magnetization fluctuations in the pinned layer which increases $\alpha_{mag\text{-}AP}$ with T_a. Since an increase in T_a can change the noise behavior of MTJs at the R_{AP} state, the annealing conditions prior to noise measurements should be mentioned in order to allow the correct comparison of low-frequency noise.

4. Summary

The annealing temperature dependence of low frequency noise in MgO-based MTJs with high TMR provides some insight into the origin of the noise. Annealing at a moderate temperature helps reduce the defects in the MgO tunnel barrier and at the MgO/CoFeB interfaces, which leads to a decrease of the nonmagnetic noise level in these MTJs. A high annealing temperature is shown to significantly increase the noise of the MTJ device in its antiparallel state due to the degradation of the exchange-bias pinning, whereas it has little effect on its noise level in the parallel state. Our results suggest that magnetic stability of the pinned layer is a crucial factor in reducing the low frequency noise in MTJs. A carefully selected annealing temperature and an antiferromagnet with good thermal stability will improve the properties of MTJ sensors. Further improvement in the signal-to-noise ratio in MgO-based MTJs will depend on the ability to reduce the nonmagnetic noise level of the barrier which is mainly related to the parallel state.

Acknowledgements

This work was supported by Science Foundation Ireland as part of the MANSE project No.05/IN/1850. Some support was also provided by Enterprise Ireland, as part of the 'Spincurrents' FoNE network. The authors would like to thank Prof. E. R. Nowak for advice on the measurement setup.

References

[1] Miyazaki T and Tezuka N 1995 *J. Magn. Magn. Mater.* **139** L231
[2] Moodera J S, Kinder L R, Wong T M, and Meservey R 1995 *Phys. Rev. Lett.* **74** 3273

[3] Wang D, Nordman C, Daughton J M, Qian Z, and Fink J 2004 *IEEE Tran. Magn.* **40** 2269
[4] Wei H X, Qin Q H, Ma M, Sharif R, and Han X F 2007 *J. Appl. Phys.* **101** 09B501
[5] Ikeda S, Hayakawa J, Ashizawa Y, Lee Y M, Miura K, Hasegawa H, Tsunoda M, Matsukura F, and Ohno H 2008 *Appl. Phys. Lett.* **93** 082508
[6] Butler W H, Zhang X-G, Schulthess T C, and MacLaren J M 2001 *Phys. Rev. B* **63** 054416
[7] Mathon J and Umersky A 2001 *Phys. Rev. B* **63** 220403
[8] Parkin S S, Xin J, Kaiser C, Panchula A, Roche K, and Samant M 2003 *Proc. IEEE* **9**, 661
[9] Tondra M, Daughton J M, Nordman C, Wang D X, and Taylor J A 2000 *J. Appl. Phys.* **87** 4679-81
[10] Weissman M B 1988 *Rev. Mod. Phys.* **60** 537
[11] Stutzke N A, Russek S E, Pappas D P, and Tondra M 2005 *J. Appl. Phys.* **97** 10Q107
[12] Ozbay A, Gokce A, Flanagan T, Stearrett R A, Nowak E R, and Nordman C 2009 *Appl. Phys. Lett.* **94** 202506
[13] Reed D, Nordman C, and Daughtoon J 2001 *IEEE Tran. Magn.* **37** 2028
[14] Scola J, Polovy H, Fermon C, and Pannetier-Lecoeur M, Feng G, Fahy K, and Coey J M D 2007 *Appl. Phys. Lett.* **90** 252501
[15] Nowak E R, Weissman M B, and Parkin S S P 1999 *Appl. Phys. Lett.* **74** 600
[16] Dimopoulos T, Gieres G, Wecker J, Wiese N, and Sacher M. D 2004 *J. Appl. Phys.* **96** 6382
[17] Liou S H, Zhang R, Russek S E, Yuan L, Halloran S T, and Pappas D P 2008 *J. Appl. Phys.* **103** 07E920
[18] Stearrett R, Wang W G, Shah L R, Gokce A, Xiao J Q, and Nowak E R, 2010 *J. Appl. Phys.* **107** 064502
[19] Feng J F, Feng G, Coey J M D, Han X F, and Zhan W S 2007 *Appl. Phys. Lett.* **91** 102505
[20] Feng J F, Diao Z, Feng G, Nowak E R, and Coey J M D 2010 *Appl. Phys. Lett.* **96** 052504
[21] Hayakawa J, Ikeda S, Mastsukura F, Takahashi H, and Ohno H 2005 *Jpn. J. Appl. Phys.* Part 2 **44** L587
[22] Gokce A, Nowak E R, Yang S H, and Parkin S S P 2006 *J. Appl. Phys.* **99** 08A906
[23] Aliev F G, Guerrero R, Herranz D, Villar R, Greullet F, Tiusan C, and Hehn M 2007 *Appl. Phys. Lett.* **91** 232504
[24] Almeida J M, Wisniowski P, and Freitas P P 2008 *IEEE Tran. Magn.* **44** 2569
[25] Jiang L, Nowak E R, Scott P E, Johnson J, Slaughter J M, Sun J J, and Dave R W 2004 *Phys. Rev. B* **69** 054407
[26] Ingvarsson S, Xiao G, Parkin S S P, Gallagher W J, Grinstein G, and Koch R H 2000 *Phys. Rev. Lett.* **85** 3289
[27] Philipp J B, Alff L, Marx A, and R. Gross 2002 *Phys. Rev. B* **66** 224417
[28] Glazman L I, and Matveev K A 1988 *Sov. Phys. JETP* **67** 1267
[29] Nowak E R, Merithew R D, Weissman M B, Bloom I, and Parkin S S P 1998 *J. Appl. Phys.* **84** 6195

Annealing effect on perpendicular [CoFe /Pd]$_n$ multilayers

J. F. Feng, and J. M. D. Coey
CRANN and School of Physics, Trinity College, Dublin 2, Ireland

E-mail: jfeng@tcd.ie

Abstract. Perpendicular magnetic anisotropy is investigated in [CoFe/Pd]$_n$ multilayers grown on Si/SiO_2 substrates. Coercivity is measured for [CoFe(t_1)/Pd(t_2)]$_n$ multilayers with t_1 = 0.4 or 0.5 nm when t_2 varies from 0.4 to 2.7 nm. Furthermore, there is an enhancement of the perpendicular anisotropy on increasing the thickness of the Pd layer and the annealing temperature. We also compare the [CoFe/Pd]$_n$ multilayers to those made with a thin CoFeB layer in contact with Pd inserted at the top or bottom of the [CoFe/Pd]$_n$ stacks. The perpendicular magnetic anisotropy is maintained.

1. Introduction

Magnetic multilayers with perpendicular anisotropy have attracted much attention due to their easily tunable magnetic properties and potential applications in high density read heads, magnetoresistive random access memory (MRAM), and magnetic logic [1-8]. There are several candidate material systems which show the perpendicular magnetic anisotropy (PMA), but [Co(Fe)/Pt]$_n$ or [Co(Fe)/Pd]$_n$ multilayers are most often used [1-4, 6,7]. The thickness of the magnetic layer is small, usually in the range of 0.2 to 0.6 nm, since the PMA is a surface/interface effect. Besides, the PMA is also influenced by the repetition number (n) and the relative thickness of Co(Fe), Pt or Pd. Recently, Fowley *et al.* [9] and Jung *et al.* [10] reported PMA for a single CoFeB film. Multilayer systems with PMA have been successfully used as the soft or hard magnetic layers in perpendicular magnetic tunnel junctions (p-MTJs) and spin valves [6, 7, 11, 12]. A high tunneling magnetoresistance (TMR) ratio of around 90% has been achieved after annealing a p-MTJ [7], and a TMR up to 124% has been reported for p-MTJs with thin single CoFeB layer [13]. Besides, they offer lower critical current density for spin-transfer-torque switching and better thermal stability than in-plane free layer.

In this work, we investigate the PMA of [CoFe/Pd]$_n$ and [CoFe/Pd]$_n$/CoFeB multilayers by changing the thickness of Pd, CoFe and CoFeB, and the repetition number. Strong PMA observed in the as-grown state is much enhanced after thermal annealing. Using the optimized [CoFe/Pd]$_n$/CoFeB multilayers as the soft and hard magnetic layers, MgO-based p-MTJs have been fabricated and a TMR ratio of around 17% has been achieved without annealing.

2. Experimental detail

[CoFe(t_1)/Pd(t_2)]$_n$ multilayers were grown under high vacuum ($P_{base} \leq 5 \times 10^{-8}$ Torr) in a Shamrock sputtering tool [3, 14]. The thickness (t_1) of CoFe is 0.4 or 0.5 nm, while Pd (t_2) varies from 0.4 to 2.7 nm. We also compare the [CoFe/Pd]$_n$ multilayers to those made by inserting a thin CoFeB layer at the top or bottom of the [CoFe/Pd]$_n$ stacks, i.e., [CoFe(t_1)/Pd(t_2)]$_{n-1}$ /CoFeB (t_3). Here the thickness (t_3) of the inserted CoFeB is in the range of 0.4 - 5 nm. All metallic layers are grown at room temperature. High vacuum annealing in the temperature range of 150 < T_a < 350 °C was carried out in an applied magnetic field of 800 mT for half an hour. Extraordinary Hall Effect (EHE) was used to measure the out-of-plane magnetic hysteresis loops with a Van der Pauw geometry [3].

3. Results and discussion

Figure 1 (a) shows the out-of-plane EHE hysteresis loops for the perpendicular $[CoFe(t_1)/Pd(t_2)]_5$ multilayers with t_1 = 0.4 nm and t_2 = 0.8, 1.2 and 1.8 nm, respectively. The three loops are square, suggesting that $[CoFe(t_1)/Pd(t_2)]_5$ multilayers have almost 100% remnant magnetization. The X-ray reflectivity measurements show that these multilayers have (111)-oriented texture, as observed in Ref. [3]. With the increase of t_2, the coercivity (H_c) of the multilayers increases then becomes stable, as shown in Fig. 1 (b). It is found that the H_c with t_1 = 0.4 nm is almost double compared to that with t_1 = 0.5 nm, which may be related to the interlayer exchange coupling between two ajdacent CoFe layers in addition to the interface effect of PMA. The coercivity difference between two perpendicular multilayers may make them easy to use as the soft and hard layers in perpendicular MTJs.

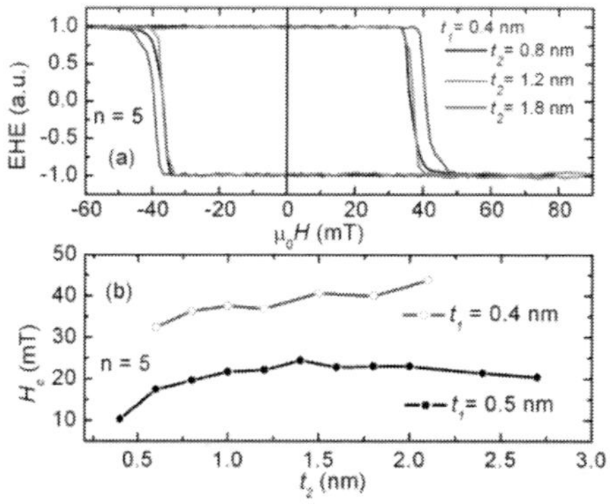

Figure 1. (a) The out-of-plane EHE loops with different t_2 for $[CoFe(0.4)/Pd(t_2)]_5$ multilayers, and (b) the coercivity as a function of the thickness of Pd.

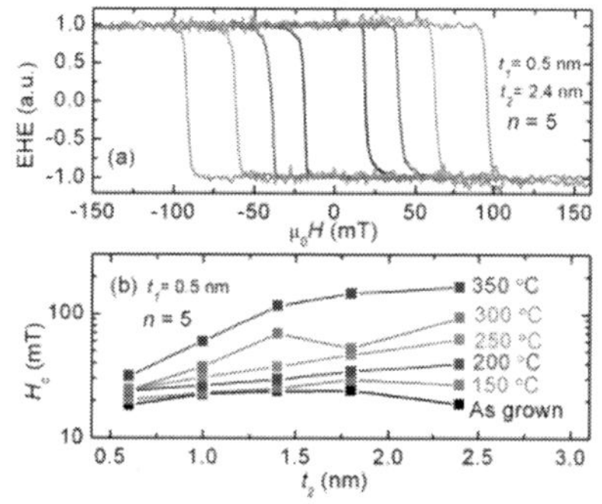

Figure 2. The out-of-plane EHE loops with t_2= 2.4 nm (a) and the coercivity varying with t_2 (b) as a function of the annealing temperature

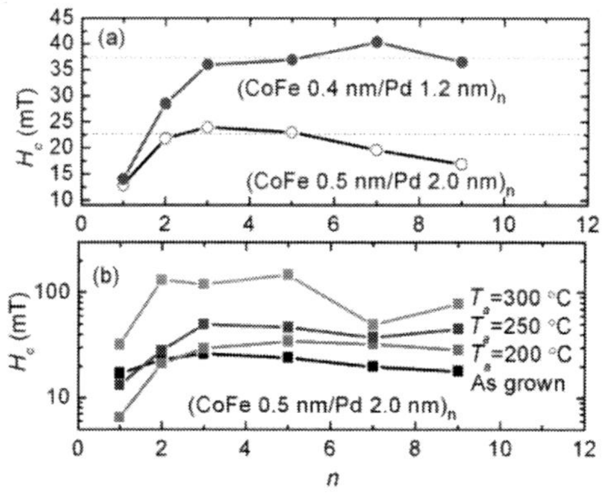

Figure 3. (a) The coercivity as a function of the n and (b) the coercivity variation with T_a for $[CoFe(0.5)/Pd(2.0)]_n$ stacks.

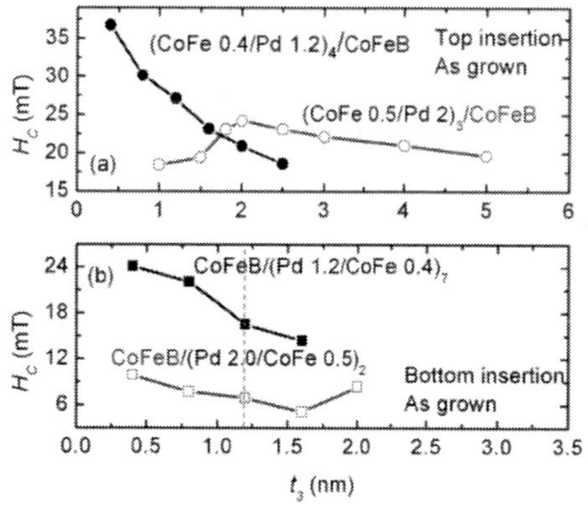

Figure 4. The coercivity variation with the top (a) and bottom (b) CoFeB insertion for $[CoFe(0.4 \text{ or } 0.5)/Pd(t_2)]_n$ multilayers

Figure 2 presents the annealing effect on the $[CoFe(0.5)/Pd(t_2)]_5$ multilayers with t_2 = 0.6 - 2.4 nm. The PMA remains up to 350 °C for all multilayers. With increasing the annealing temperature (T_a), the

PMA is enhanced, especially for the thicker Pd layer; see Fig. 2 (a). Moreover, H_c increases significantly with T_a. For example, the coercivity is 18.8 mT in the as-grown state and reaches 164.4 mT at T_a = 350 °C for t_1 = 0.5 nm and t_2 = 2.4 nm, almost nine-fold increase. Similar results with annealing have been reported for the single CoFeB layer which shows a PMA [10,13], but no coercivity was observed in the as-grown state.

To further optimize the [CoFe(t_1)/Pd(t_2)]$_n$ multilayers, we change the repetition number as shown in Fig. 3. We use t_1 = 0.4 nm and t_2 = 1.2 nm, t_1 = 0.5 nm and t_2 = 2.0 nm, respectively, see Fig. 3 (a). The dashes lines denote the optimized coercivity. The H_c becomes stable after n = 3 for t_1 = 0.4 nm and t_2 = 1.2 nm, while it is unchanged when n is 2 - 5 for t_1 = 0.5 nm and t_2 = 2.0 nm. However, H_c decreases when n > 6 for t_1 = 0.5 nm and t_2 = 2.0 nm. This can be attributed to an increase in the number of structural defects with increasing bilayer number affecting the Co/Pd interface quality and thus eventually resulting in an overall reduction of the magnetic anisotropy [15]. Besides, H_c is a little lower at the beginning and increases with T_a for n > 1 in [CoFe(0.5)/Pd(2)]$_n$, which may indicate that the magnetic properties of thin CoFe are sensitive to the annealing effect. For n > 2, H_c continuously increases with T_a for [CoFe(0.5)/Pd(2)]$_n$ multilayers.

To achieve a high TMR ratio in MgO-based perpendicular MTJs, the amorphous CoFeB should crystallize the (001)-orientation after annealing. Here we also try to insert a CoFeB layer at the top or bottom of the [CoFe/Pd]$_n$ stacks. For the top insertion case as shown in Fig. 4 (a), the thickness (t_3) of CoFeB varies from 0.4 to 2.5 nm for [CoFe(0.4) /Pd(1.2)]$_4$ /CoFeB. The PMA remains, despite the decease of H_c with t_3. However, for [CoFe(0.5) /Pd(2.0)]$_3$ /CoFeB multilayers, H_c increases by increasing t_3 and then decreases slightly. Interestingly, the PMA remains good for t_3 up to 5 nm for [CoFe(0.5) /Pd(2.0)]$_3$ /CoFeB. Actually, a thicker CoFeB insertion layer has been used in the in-plane MTJs [6, 14] to obtain a high TMR effect. Fig. 4 (b) shows the t_3 dependence of H_c for the bottom insertion case. The H_c deceases with t_3 for both CoFeB/[CoFe(0.4) /Pd(1.2)]$_7$ and CoFeB/[CoFe(0.5) /Pd(2.0)]$_2$ multilayers due to the amorphous nature of the inserted CoFeB.

Using the optimized results as mentioned above, micron-sized p-MTJs with structures [CoFe(0.4)/Pd(1.2)]$_4$ /CoFeB (1.0 or 1.5)/MgO 2.5/CoFeB (1.0 or 1.5)/ [Pd(2) /CoFe(0.5)]$_2$ have been fabricated. More details about MTJ fabrication can be found in Ref. [14]. Magnetotransport properties were measured by a four-probe system with a perpendicularly applied magnetic field. A TMR ratio of 16.3% and 17.2% in the as-grown state has been obtained in these MgO-based p-MTJs when the thickness of the inserted CoFeB is 1.0 and 1.5 nm, respectively. This value is comparable to that in Ref. [16] in the as-grown state. Unfortunately, a decrease in TMR is found on annealing. This might be due to the diffusion of boron of the inserted CoFeB layers.

4. Summary

Strong PMA is observed in [CoFe/Pd]$_n$ multilayers. The annealing effect on these [CoFe/Pd]$_n$ multilayers is to enhance the PMA on increasing the annealing temperature up to 350 °C. By inserting a CoFeB layer at the top or bottom of the [CoFe/Pd]$_n$ stacks, it is found that the [CoFe/Pd]$_n$ /CoFeB multilayers are suitable for p-MTJs as the soft and bottom layers. However, further work is necessary to improve the TMR value. These results may be helpful for understanding the realization of high tunneling magnetoresistance in perpendicular MTJ devices.

Acknowledgements

This work was supported by Science Foundation Ireland as part of the MANSE project No.05/IN/1850.

References

[1] Carcia P F, Meinhaldt A D, and Suna A 1985 *Appl. Phys. Lett.* **47** 178
[2] Hellwig O, Maat S, Kortright J B, and Fullerton E E 2002 *Phys. Rev. B* **65** 144418
[3] Rodmacq B, Baltz V, and Dieny B 2006 *Phys. Rev. B* **73** 092405
[4] Chen J Y, Feng J F, Diao Z, Feng G, Coey J M D, and Han X F 2010 *IEEE Trans. Mag.* **46** 1401

[5] Nishimura N, Hirai T, Koganei A, Ikeda T, Okano K, Sekiguchi Y, and Osada Y *J.* 2002 *Appl. Phys. Lett.* **91** 5246
[6] Mizunuma K, Ikeda S, Park J H, Yamamoto H, Gan H, Miura K, Hasegawa H, Hayakawa J, Matsukura F, and Ohno H 2009 *Appl. Phys. Lett.* **95** 232516
[7] Yakushiji K, Noma K, Saruya T, Kubota H, Fukushima A, Nagahama T, Yuasa S, and Ando K 2010 *Appl. Phys. Exp.* **3** 053003
[8] Reiss G and Meyners D, 2007 *J. Phys.:Condens. Matter* **19** 165220
[9] Fowley C, Decorde N, Oguz K, Rode K, Kurt H, and Coey J M D, 2010 *IEEE Trans. Mag.* **46** 2116
[10]Jung J H, Lim S H, and Lee S R, 2010 *Appl. Phys. Lett.* **96** 042503
[11] Mangin S, Ravelosona D, Katine J A, Carey M J, Terris B D and Fullerton P F, 2006 *Nature Mater.* **5** 210
[12] Law R, Sbiaa R, Liew T, Chong T C, 2007 *Appl. Phys. Lett.* **91** 242504
[13] Ikeda S, Miura K, Yamamoto H, Mizunuma K, Gan H D, Endo M, Kanai S, Hayakawa J, Matsukura F, and Ohno H 2010 *Nature Mater.* doi:10.1038/nmat2804 Letter
[14] Diao Z, Feng J F, Kurt H, Feng G, and Coey J M D 2010 *Applied Physics letters* **96** 202506
[15] Suzuki T 1995 *Scr. Mater.* **33** 1609
[16] Mizunuma K, Ikeda S, Yamamoto H, Gan H D, Miura K, Hasegawa H, Hayakawa J, Ito K, Matsukura F, and Ohno H 2009 *Jpn. J. Appl. Phys.* **49** 04DM04

Magnetic tunnel junctions using Co/Pt multilayered free layers with perpendicular magnetic anisotropy

K Machida[1], K Furukawa[2], T Nakayama[2], N Funabashi[1], K Aoshima[1], K Kuga[1], H Kikuchi[1], T Ishibashi[3] and N Shimidzu[1]

[1]*Science & Technology Research Labs., Japan Broadcasting Corp. (NHK),*
1-10-11 Kinuta, Setagaya-ku, Tokyo 157-8510, Japan
[2]*Department of Electrical Engineering, Tokyo Denki University,*
2-2 Kandanishikicho, Chiyoda-ku, Tokyo 101-8457, Japan
[3]*Department of Materials Science and Technology, Nagaoka University of Technology,*
1603-1 Kamitomioka, Nagaoka, Niigata 940-2188, Japan

E-mail: machida.k-ge@nhk.or.jp

Abstract. Co/Pt multilayered films with perpendicular magnetic anisotropy have a large magneto-optical Kerr effect. To use the films with a submicron magneto-optical light modulator driven by spin transfer switching, we fabricated two types of magnetic tunnel junctions (MTJs) with Co/Pt multilayered films for the free layers. One is an fcc-based MTJ, another is a bcc-based MTJ with CoFeB/MgO/CoFeB junction. The fcc-based MTJ with a Ag buffer layer on the bottom electrode showed a large coercive force of the pinned layer, a large Kerr rotation angle of 0.3 degree in the free layer and a tunnel magnetoresistance (TMR) ratio of 3.8%. In the CoFeB/MgO/CoFeB junction, an X-ray diffraction pattern of an MgO layer showed a large MgO(002)-orientation. However, the TMR ratio was less than 3 %. An MTJ with a Ta buffer layer between the CoFeB layer and the Co/Pt multilayered films in the free layer was prepared. The Ta buffer was used to alleviate a lattice mismatch between bcc-CoFeB/MgO/CoFeB and fcc-Co/Pt multilayer. The peak intensity of the MgO(002)-orientation was increased up to 2 times. This result suggests that the crystalline texture of the bcc-CoFeB/MgO/CoFeB junction is strongly influenced by the fcc-Co/Pt multilayered films.

1. Introduction

Submicron spatial light modulators (SLMs) are promising for use in high-density holographic recording and wide-viewing-angle holographic 3D displays that are not requiring special glasses. We have advanced a magneto-optical light modulation device driven by spin transfer switching (STS MO-LM)[1]-[3], as shown in figure 1. The free layer of a current perpendicular to plane giant magnetoresistive (CPP-GMR) pillar or a tunnel magnetoresistive (TMR) pillar works as a light modulation layer. The top electrode of the device is formed with a transparent electrode such as Indium Tin Oxide (ITO) or Indium Zinc Oxide (IZO). Incident polarized light entered through the transparent top electrode is subject to diffraction at the free layer. The polarization plane of the diffracted light is modulated by the magnetization direction of the free layer, reversed by STS. The diffraction angle (φ) is defined by the following equation; $\varphi = 2\sin^{-1}(\lambda/2p)$, where λ and p are a wavelength of the incident light and a pitch length between pixels[4]. In conventional SLMs, the value of p is larger than 5 μm, resulting in $\varphi < 6°$. On the other hand, the STS MO-LM is a high-resolution device with high-speed modulation of the polarization plane. The value of p in the STS MO-LM can be set at less than 1μm, which leads to a sufficient wide-viewing-angle of $\varphi > 30°$. However, a large magneto-optical effect is required for practical use of this device. It is well known that Co/Pt and Co/Pd multilayered films show a large Kerr rotation angle and a strong perpendicular magnetic anisotropy[5]. Magnetoresistance (MR) properties of CPP-GMR and TMR devices with

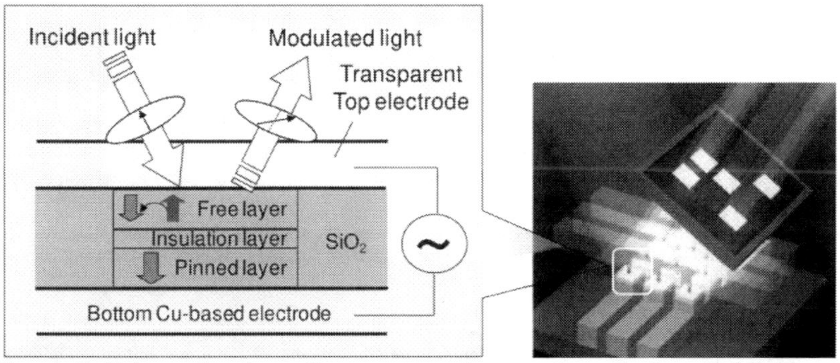

Figure 1. Schematic view of our proposed magneto-optical spatial light modulator and cross-section view of the light modulation device driven by spin transfer switching in the left inset.

ferromagnetic/noble metal multilayered films with perpendicular magnetic anisotropy such as CoFe/Pt and CoFe/Pd have been investigated for use in high-density magnetic random access memories[6],[7].

In this paper, we reports on two types of magnetic tunnel junctions (MTJs) with Co/Pt multilayered films for free layers. One is an fcc-based MTJ, another is a bcc-based MTJ with CoFeB/MgO/CoFeB junction. The magneto-optical Kerr effects, TMR properties, and X-ray diffraction patterns of MgO films were investigated.

2. A fcc-based MTJ

In this section, magnetic properties, X-ray diffraction (XRD) patterns, and MR characteristics of an fcc-based MTJ with perpendicular magnetic anisotropy are discussed.

2.1. Experimental details

The film configuration of the MTJ stack is as follows; a Cu-based bottom electrode layer of $[Ta(3)/Cu(15)]_{10}/Ta(3)/Ru(3)$, a Ag(10) buffer layer, a pinned layer of $[Co(0.2)/Pd(1.2)]_{11}/Co(0.2)$, an MgO(0.8) barrier, a free layer of $[Co(0.3)/Pt(1.0)]_3$, a capping layer of Ta(3), and a metal top electrode layer of Cu(350)/Ru(10) for a pillar. The numbers in parentheses are thicknesses in nanometers. The bottom electrode layer was deposited onto a thermally oxidized silicon wafer by dc ion beam sputtering. The buffer layer and the pinned layer were deposited by dc magnetron sputtering. The MgO barrier was formed by an rf ion beam sputtering method. The free layer and the other layers were

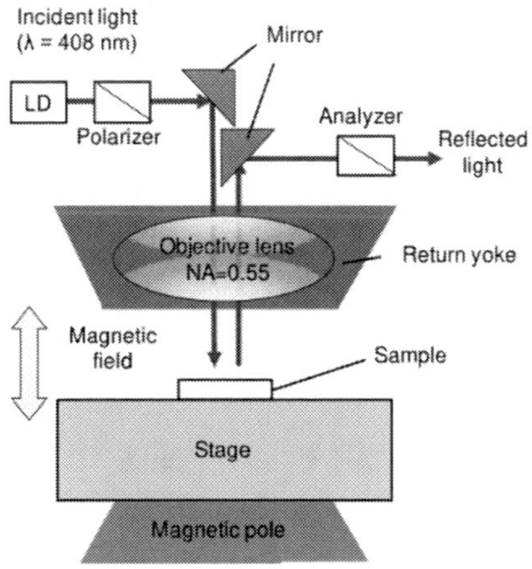

Figure 2. Schematic illustration of micro-magneto-optical Kerr effect microscopy with polar Kerr mode.

deposited using dc ion beam sputtering. The MTJ stack was post-annealed at 350°C for 1 hour in vacuum of 3×10^{-5} Pa. Rectangular pillar of the stack was fabricated by electron beam lithography using a two-step liftoff process [8].

Kerr hysteresis loops were measured using micro-magneto-optical Kerr effect microscopy with a polar Kerr mode as shown in figure 2. An incident light emitted by a laser diode (LD) with a wavelength of 408 nm is linearly polarized and enters the sample vertically, focused by an objective lens. The spot size of the incident light with Gaussian distribution is approximately 1 μm. The microscopy utilizes polarization modulation with a photoelastic modulator and has high sensitivity of 1×10^{-4} degree for detecting the magneto-optical Kerr effect.

2.2. Results and discussions

The full loop of polar Kerr hysteresis at 408-nm wavelength after annealing at 350°C for 1 hour is shown in figure 3, compared to that without the Ag buffer layer. The coercive force (H_c) of the pinned layer with the Ag layer as deposited was 0.72 kOe, while H_c of the pinned layer without the Ag layer as deposited was 0.93 kOe (not shown). After annealing at 350°C, the H_c with the Ag layer increased up to 2.8 kOe, in contrast with the H_c of 1 kOe without the Ag layer as shown in figure 3. The Kerr rotation angles of the free layers with the Ag buffer layer and without the layer were 0.3 degree and 0.24 degree, respectively. Factors in the increase of the Kerr rotation angle may be multiple reflection effect by the high reflectance of the Ag layer or a surface plasmon excitation around the Ag layer. Further detailed investigations are required.

XRD measurements were done for the MTJ stacks. The XRD pattern of the sample with Ag(10)/[Co(0.2)/Pd(1.2)]$_{11}$/Co(0.2)/MgO(90) deposited on the Cu-based bottom electrode is shown in figure 4, compared to that without the Ag layer. The source of the X-rays is CoKα. The Ag layer sample had the oriented textures of fcc-based Pd(111), Ag(111), Cu(111), and MgO(111). No peaks related to Co were detectable. The Ru(3nm) layer as the top of the Cu-based bottom electrode shows a peak of hcp-(0002). An origin of perpendicular anisotropy on ferromagnetic/noble metal multilayered films is considered to be a surface magnetic anisotropy, which is caused by a strain of a lattice misfit. A lattice misfit for a Ag/Co interface is approximately 2 times larger than that of a Ru/Co interface. The strong perpendicular anisotropy of the Ag buffer layer stack is attributed to the large lattice misfit of the Ag/Co interface. On the other hand, MgO(002) peaks associated with a high TMR ratio were never seen in either stack.

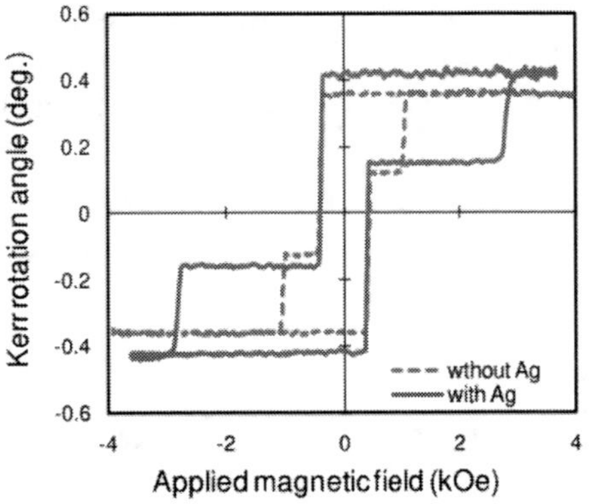

Figure 3. Polar Kerr hysteresis loops of the stacks without (dotted blue line) and with (solid red line) Ag buffer layers, after 350℃-annealing, at 408-nm wavelength.

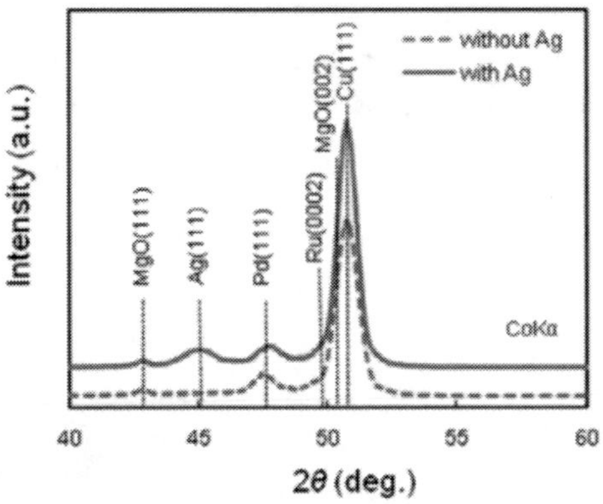

Figure 4. X-ray diffraction patterns for the samples of [Co(0.2)/Pd(1.2)]$_{11}$/Co(0.2)/MgO(90) on the Cu-based bottom electrode without (dotted blue line) and with (solid red line) Ag buffer layers, after 350℃-annealing.

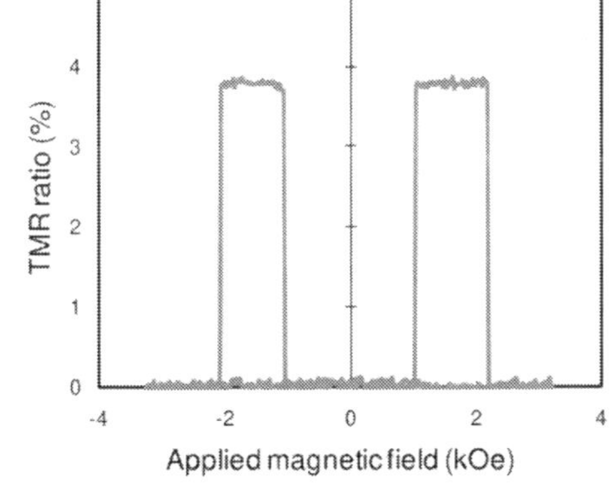

Figure 5. MR curve of the pillar with Ag buffer layer after 350°C-annealing, 100 nm × 100 nm.

The MR curve of the rectangular pillar with the Ag buffer layer after annealing at 350°C is shown in figure 5. The size of the pillar is 100 nm × 100 nm. The applied voltage was 2 mV. The TMR ratio and resistance-area product (*RA*) were 3.8 % and 10.5 $\Omega\mu m^2$, respectively. The relatively low TMR ratio in contrast with MgO(100)-based MTJs was caused by the undesirable MgO(111)-oriented texture.

3. A CoFeB-based MTJ

For a high TMR ratio, we fabricated a CoFeB-based MTJ using Co/Pt multilayer films for free layers with perpendicular magnetic anisotropy.

3.1. Experimental details

The film configuration of the MTJ stack is as follows; a bottom electrode layer of Ta(50)/Ru(3), a pinned layer of $Tb_{28}Fe_{60}Co_{12}$(30)/$Co_{40}Fe_{40}B_{20}$(1.2), an MgO(1) barrier, a free layer of $Co_{40}Fe_{40}B_{20}$(0.6)/$[Pt(1.5)/Co(0.3)]_3$, a capping layer of Ta(1.8)/Pt(2), and a metal top electrode layer of Cu(350)/Ru(10) for a pillar. The numbers in parentheses are thicknesses in nanometers. The MgO barrier was formed by an rf ion beam sputtering method. The layers except for the MgO barrier were deposited using dc ion beam sputtering. The MTJ stack was post-annealed at 260°C or 300°C for 1 hour in vacuum of 3×10^{-5} Pa. Rectangular pillar of the stack was fabricated by electron beam lithography using a two-step liftoff process [8].

3.2. Results and discussions

The minor loop of Kerr hysteresis at 408-nm wavelength after annealing at 300°C for 1 hour is shown in figure 6. The loop indicates magnetization reversal of the free layer and shows that the free layer has a binary condition at zero magnetic fields. The H_c of the pinned layer was more than 22 kOe even after annealing, which is the value of the measurement limit for our magneto-optical Kerr effect measuring system. The Kerr rotation angle was 0.23 degree and the H_c of the free layer was no more than 0.1 kOe because of the small H_c of CoFeB in the free layer.

The MR curve of the pillar, 100 nm × 180 nm, is shown in figure 7. The TMR ratio and the RA as deposited were 2.0 % and 17.2 $\Omega\mu m^2$. After annealing at 260°C, the TMR ratio increased up to 2.6 %. However that was smaller than the fcc-based MTJ pillar discussed in the previous section.

We measured the XRD patterns of the stacks with CoFeB-based MTJs as shown in figure 8. In these measurements, the source of the X-rays is CuKα. The film configuration of the stack (stack I) is Ta(50)/Ru(3)/$Tb_{28}Fe_{60}Co_{12}$(30)/$Co_{40}Fe_{40}B_{20}$(1.2)/MgO(30)/$Co_{40}Fe_{40}B_{20}$(0.6)/$[Pt(1.5)/Co(0.3)]_3$/capping. The as-deposited MgO layer showed a large (002) crystalline orientation. By contrast, the MgO(111)-

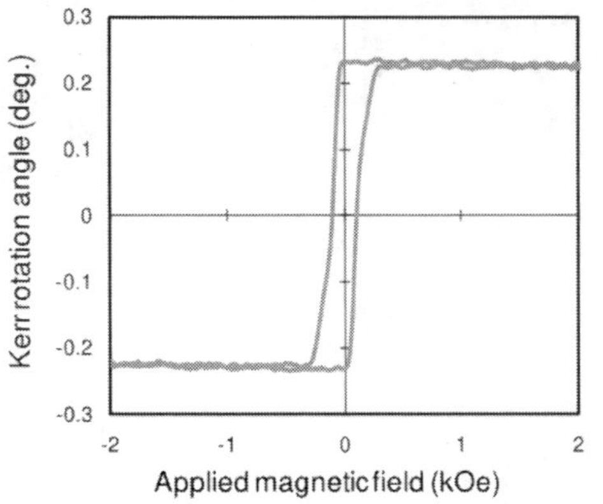

Figure 6. Polar Kerr hysteresis loop at 408 nm-wavelength after annealing at 300℃ for 1 hour.

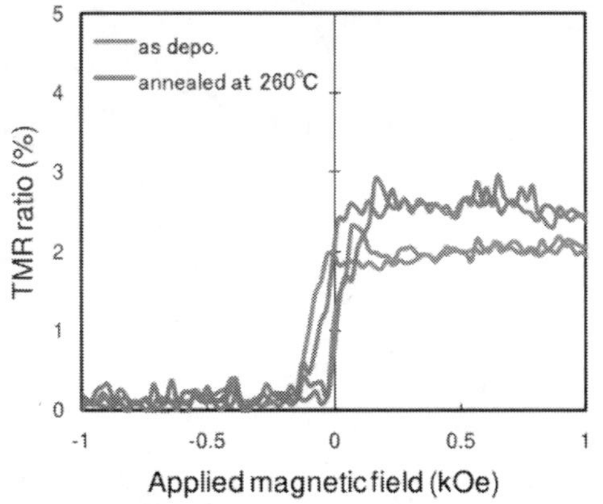

Figure 7. MR curves of as-deposited (dotted blue line) and annealed (solid red line) pillar, 100nm x 180nm.

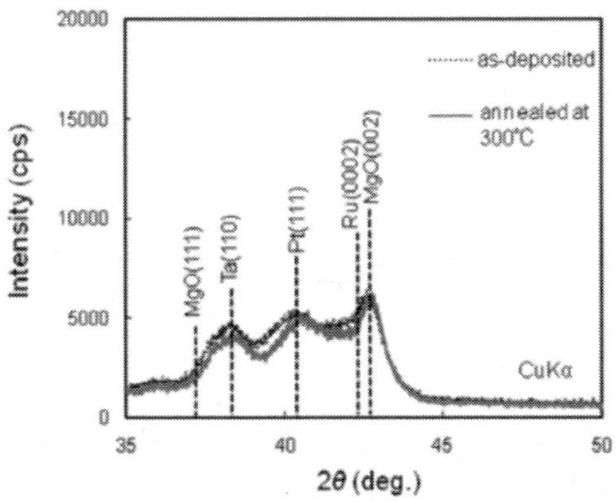

Figure 8. X-ray diffraction (XRD) patterns for as-deposited (dotted blue line) and annealed (solid red line) samples of stack I.

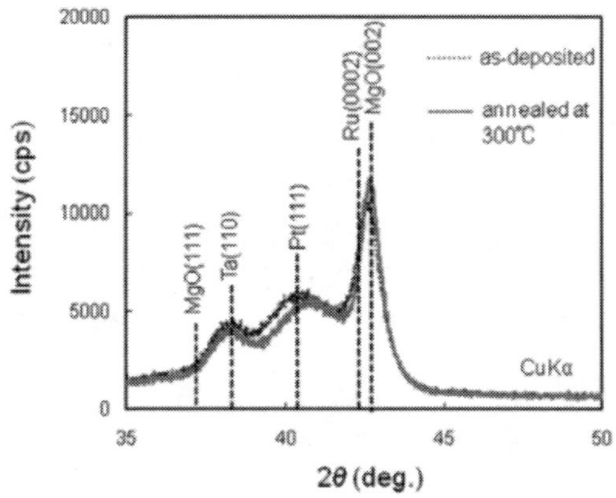

Figure 9. XRD patterns for as-deposited (dotted blue line) and annealed (solid red line) samples of stack II.

orientation was never seen. After annealing at 300°C, there was no significant change in the peak intensity of the MgO(002). The Co/Pt multilayered films and the junction of CoFeB/MgO/CoFeB have the fcc and bcc structures, respectively. A current research on MTJs with fcc-based Co/Pd multilayered films and bcc-based CoFeB/MgO/CoFeB junction indicates that a low TMR ratio may be caused by the lattice mismatch between the fcc and bcc phases[9]. The free layer of our stack was modified as follows; $Co_{40}Fe_{40}B_{20}$ (0.6)/Ta(1)/$[Pt(0.3)/Co(0.3)]_2$/$[Pt(1.5)/Co(0.3)]_2$(stack II). The Ta buffer layer was used to alleviate the lattice mismatch. Figure 9 shows the XRD patterns of the stack II as deposited and annealed at 300°C for 1 hour. The peak intensity of the MgO(002)-orientation was increased up to 2 times. After annealing, the intensity increased by 15%, unlike in the case of the stack I. These results suggest that the crystalline texture of the CoFeB/MgO/CoFeB junction is strongly influenced by the fcc-based Co/Pt multilayered films. According to figure 9, the Ta buffer layer may appear to work properly as a transition layer from the bcc to fcc phases. However, the perpendicular magnetic anisotropy of the stack II for the case of MgO(1 nm) barrier was inadequate, in addition an in-plane magnetization process showed a slight hysteresis around zero magnetic field (not shown). These results may be attributed to an insufficient magnetic connection between the CoFeB layer and the Co/Pt multilayer in the free layer. The perpendicular magnetic anisotropy of the stack II should be improved by reducing the thickness of the Ta buffer layer and substituting another material such as Ru for the Ta buffer.

4. Conclusions

To use the Co/Pt multilayered films with a submicron magneto-optical light modulator driven by spin transfer switching, we fabricated two types of MTJs. One is an fcc-based MTJ, another is a bcc-based MTJ with CoFeB/MgO/CoFeB junction. The magneto-optical Kerr effects, TMR properties, and X-ray diffraction patterns of MgO layer were investigated. The fcc-based MTJ with a Ag buffer layer on the bottom electrode showed a large coercive force of the pinned layer, a large Kerr rotation angle of 0.3 degree in the free layer and a tunnel magnetoresistance (TMR) ratio of 3.8%. In the CoFeB/MgO/CoFeB junction, the TMR ratio was less than 3 %, though an X-ray diffraction pattern of an MgO (30 nm) layer showed a large MgO(002)-orientation. An MTJ with a Ta buffer layer inserted between the CoFeB layer and the Co/Pt multilayered films in the free layer was prepared. The peak intensity of the MgO(002)-orientation was increased up to 2 times. The Ta buffer layer may appear to work properly as a transition layer from the bcc to fcc phases. These results suggest that the crystalline texture of the bcc-CoFeB/MgO/CoFeB junction is strongly influenced by the fcc-Co/Pt multilayered films.

5. References

[1] Aoshima K, Funabashi N, Machida K, Miyamoto Y, Kawamura N, Kuga K, Shimidzu N, Kimura T, Otani Y and Sato F 2007 *Appl. Phys. Lett.* **91** 052507
[2] Machida K, Furukawa K, Funabashi N, Aoshima K, Kuga, K, Ishibashi T and Shimidzu N 2010 *IEEE Trans. Magn.* **46** 2171
[3] Funabashi N, Aoshima K, Machida K, Kuga, K, Ishibashi T and Shimidzu N 2010 *IEEE Trans. Magn.* **46** 1998
[4] Mishina T, Okano F and Yuyama I 1999 *Appl. Opt.* **38** 3703
[5] Carcia P F 1988 *J. Appl. Phys.* **63** 5066
[6] Meng H and Wang J P 2006 *Appl. Phys. Lett.* **88** 172506 .
[7] Park J -H Ikeda S, Yamamoto H, Gan H, Mizunuma K, Miura K, Hasegawa H, Hayakawa J, Ito K, Matsukura F and Ohno H 2009 *IEEE trans. Magn.* **45** 3476
[8] Machida K, Funabashi N, Aoshima K, Miyamoto Y, Kawamura N, Kuga K and Shimidzu N 2008 *J. Appl. Phys.* **103** 07A713
[9] Tadisina Z R, Natarajarathinam A, Clark B D, Highsmith A L, Mewes T, Gupta S, Chen E and Wang S 2010 *J. Appl. Phys.* **107** 09C703

Acknowledgments

This research has been supported by the National Institute of Information and Communications Technology in Japan, (program period; 2009–2011).

Journal of Physics: Conference Series **303** (2011) 012101 doi:10.1088/1742-6596/303/1/012101

Simulation of tunneling magneto-resistance in Fe/MgO/Fe junction

K Shiiki
Department of Applied Physics and Physico-informatics,
Keio University, Yokohama 223-8522, Japan

E-mail: shiiki@appi.keio.ac.jp

Abstract. The conductivity asymmetry of junctions with insulating MgO barriers embedded between ferromagnetic Fe electrodes has been investigated by the first principles calculation program of Atomistix ToolKit 2008. The conductivity in up-spin electron is larger than that in down-spin electron for the ideal structure. It leads to the large TMR ratio. The defect of O or Fe atoms hardly varies the conductivity in both up and down-spin electrons. The defect of Mg atoms increases the conductivity in down-spin electron, although the up-spin conductivity does not change very much. So the defect of Mg decreases the TMR ratio. The defect of Mg near the interface especially shows this degradation. The derivative conductivity dI/dV vs. V characteristics are asymmetric for the polarity of the bias voltage V according to the location of the atom defect. The quality of the junction in devices can be evaluated by the asymmetry.

1. Introduction

Tunneling magneto-resistance (TMR) junctions have been used for high sensitive read devices in high-density magnetic recording systems. It was predicted by calculating spin dependent conductance using first principles modeling that Fe/MgO/Fe sandwiches have a large magneto-resistance ratio [1]. The conductivity in up-spin electron is larger than that in down-spin electron for the ideal structure. It leads to the large TMR ratio. Yuasa fabricated fully epitaxial Fe/MgO/Fe magnetic tunnel junction and observed a large TMR ratio at room temperature [2]. However, the sensitivity varies in practical devices, although it is thought to be dependent on the interface structure of the junction [3]. It is difficult very much to evaluate the interface structure in practical devices by non-destructive testing. In this paper, relationship between the non-ideal structure and the asymmetric behavior in the current-voltage characteristics of the junction is investigated.

2. Calculation method

TMR ratio of junctions with insulating MgO barriers embedded between ferromagnetic Fe electrodes has been investigated by the first principles calculation program of Atomistix ToolKit 2008, which gives essentially the same result as Butler [4]. The calculation method is reliable because the calculated results are reasonable. Large magneto-resistance ratio of Fe-MgO-Fe junction can be understood by the difference in transmission coefficients corresponding to the up and down electrons. Figure 1 shows the geometry of NaCl-type MgO layers between two bcc Fe electrodes. In the x-y plane, the unit cell is repeated periodically. In the z direction, the system is divided into a central region containing the MgO layers surrounded by three Fe layers on each side, a left electrode and a right electrode with three Fe layers each in the parallel or anti-parallel spin configuration. The central region is solved as a scattering region and the electrode is treated as an isolated bulk. Electronic structure is calculated by using the generalized gradient approximation based on the local spin density approximation. Perdew-Zunger parameterization is used for the exchange-correlation potential. Non-

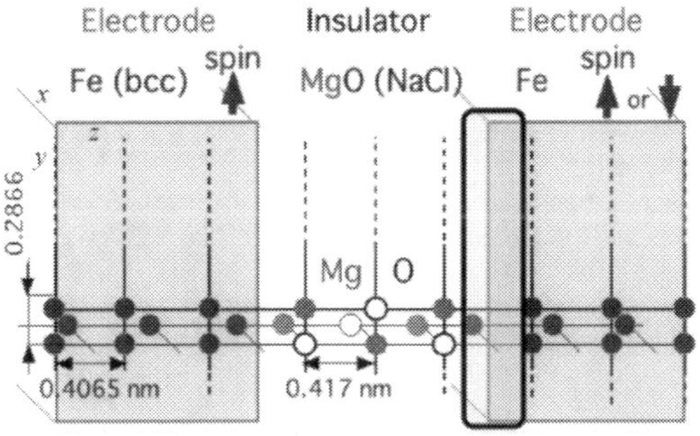

Figure 1. The schematic atom geometry of the Fe-MgO-Fe junction.

conserving pseudo-potential approximation is used for core electrons. The number of k points is 900 and cut off energy is 200 eV. Initial distances between atoms shown in the figure were determined to minimize the total energy. The lattice constant of Fe is 0.2866 nm. Effects on the lattice defect and strain on the right side of the interface between MgO and Fe layers are investigated. The polarity of the current is defined as positive, when the current flows from left to right side of the junction.

3. Results and discussion

Figure 2 shows the dependence of the tunneling current i at bias voltage of 0.1 V for various MgO layer thicknesses d in the parallel spin configuration. The current can be expressed as $i \propto \exp[-5.42d]$ which agrees to the experimental results and Simmons' equation [5] for a simple tunneling junction with barrier potential of 0.28 eV. MgO thickness dependence of the TMR ratio in the Fe-MgO-Fe junction is shown in Figure 3. TMR ratio increases and saturates after a small vibration as the thickness increases. It is predicted by the theory of the square well potential [6]. Figure 4 shows tunneling conductivity for up and down spins of the Fe-insulator-Fe junctions in the parallel spin configuration. The conductivity difference, which leads to the large TMR ratio, is very large and the insulator shows the spin filter characteristics especially in the case of MgO and BeO.

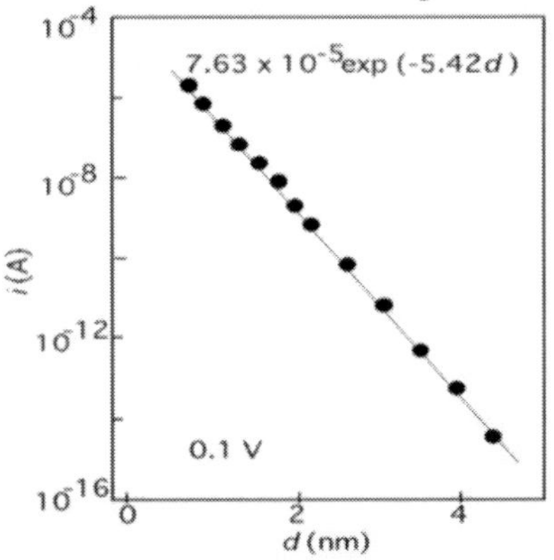

Figure 2. MgO thickness dependence of the tunneling current at 0.1 V.

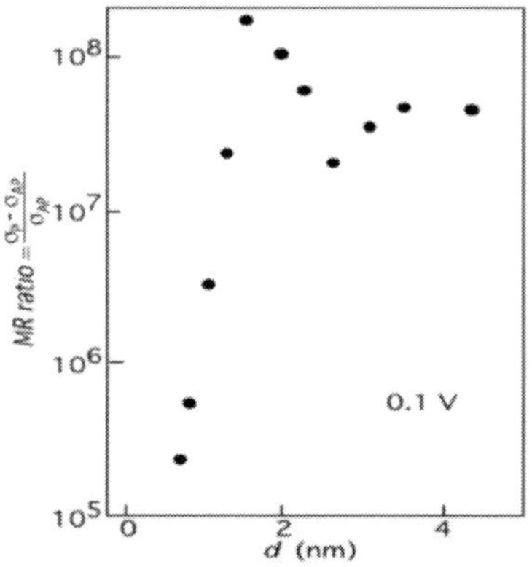

Figure 3. MgO thickness dependence of the TMR ratio in the Fe-MgO-Fe junction.

Figure 5 shows the effect of lattice defect on the conductivity for up and down spin in the parallel spin configuration. In the figure, I and II denote the atom site on the side of the Fe electrode and the MgO insulator, respectively. The defect of O or Fe atoms hardly varies the conductivity in both up and down-spin electrons. However, the defect of Mg atoms increases the conductivity in down-spin

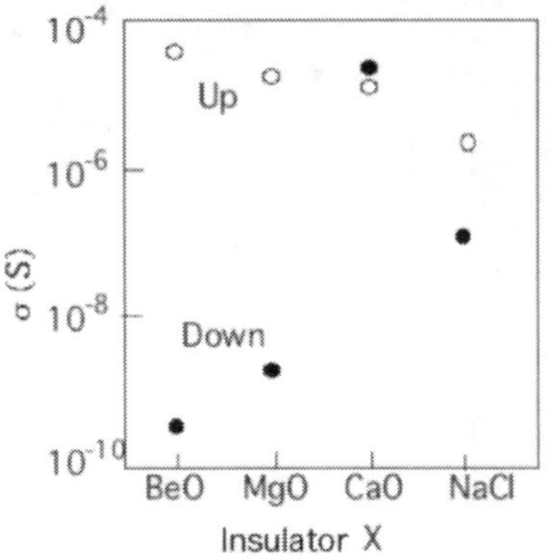

Figure 4. Tunneling conductivity σ of the Fe-X-Fe junctions.

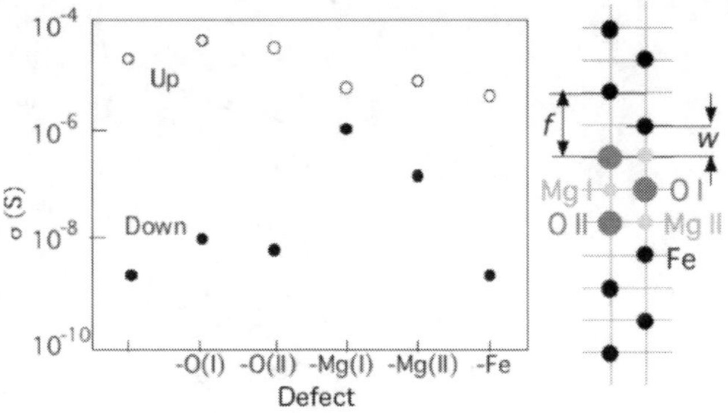

Figure 5. Effect of the defect on the tunneling conductivity σ of the Fe-MgO-Fe junctions.

electron, although the up-spin conductivity does not change very much. So the defect of Mg decreases the TMR ratio. The defect of Mg near the interface (site I) especially shows this degradation.

The derivative conductivity dI/dV vs. voltage V characteristics are asymmetric for the polarity of the bias voltage V according to the location of the atom defect. Figures 6 and 7 show the asymmetry for Fe and O defects. The conductivity increases on the positive bias by introducing the defect. Especially, it increases very much for Fe defects and goes up as the bias voltage in the case of the Fe defect of site 1.

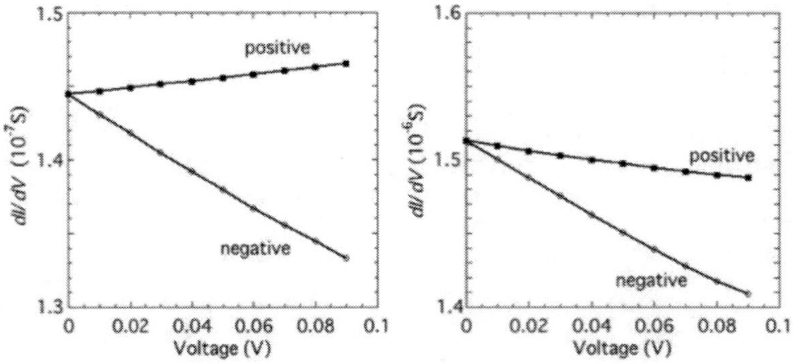

(a) MgO side (b) Fe side

Figure 6. Conductivity asymmetry of the bias voltage dependence due to Fe defect.

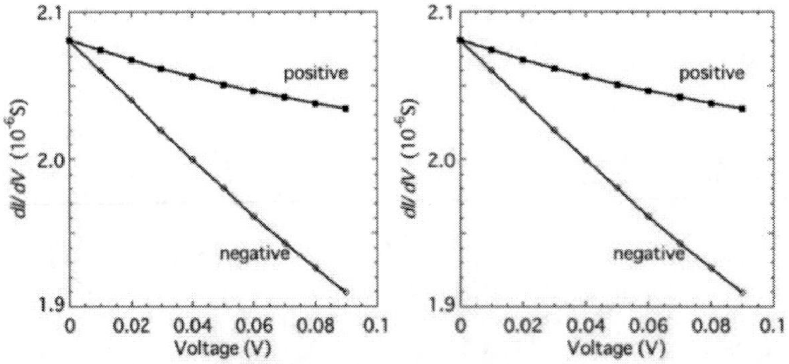

(a) MgO side (b) Fe side

Figure 7. Conductivity asymmetry of the bias voltage dependence due to O defect.

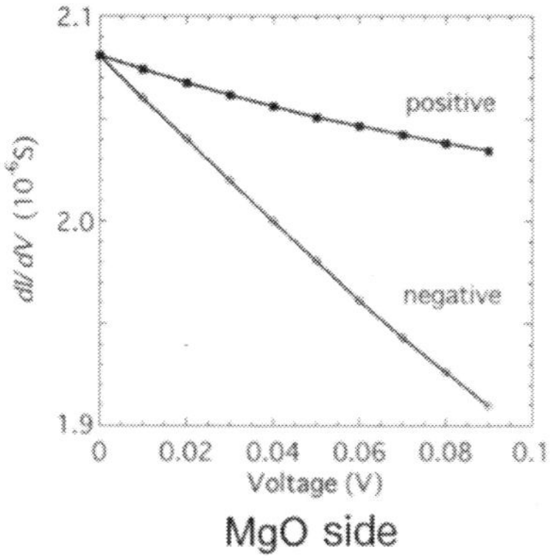

Figure 8. Conductivity asymmetry of the bias voltage dependence due to Mg defect.

The Mg defect makes almost the same effect as O defect. Figure 8 shows the asymmetry for Mg defect of site 1. However, the calculation does not converge for Mg defect of site 2. The conductivity increases on the positive bias by introducing the defect, although the conductivity does not change very much on the negative bias. Especially, it increases very much for Fe defects and goes up as the bias voltage in the case of the Fe defect of site 1.

Figures 9 (a) and (b) show the effect of lattice strain on the conductivity for up and down spin. The distances *w* and *f* denoted in Figure 5. change from the equilibrium value of 0.417 nm and 0.287 nm, respectively. The strain hardly varies the conductivity in both up and down-spin electrons. So the strain does not decrease the TMR ratio very much. And the effect of the change in distance *w* on the asymmetry is little as shown in Figures 10 (a) and (b). The quality of the junction can be evaluated by an asymmetric behavior of the current-voltage characteristics.

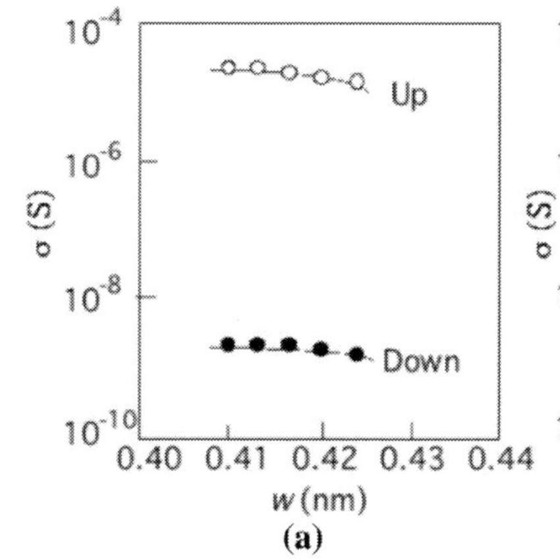

Figure 9. Effect of the lattice strain on the tunneling conductivity σ of the Fe-MgO-Fe junctions.

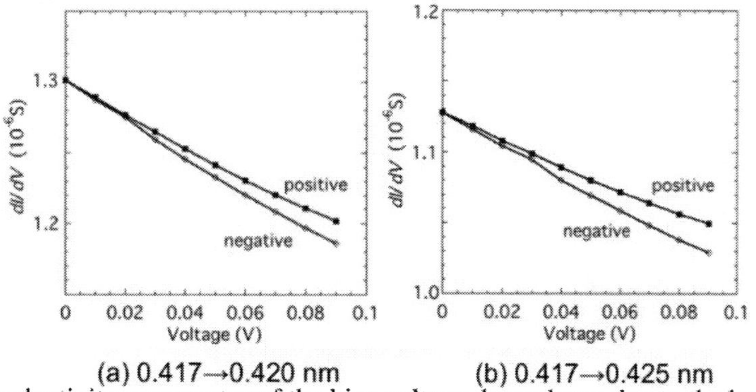

Figure 10. Conductivity asymmetry of the bias voltage dependence due to the lattice strain.

4. Conclusion

The defect of O or Fe atoms hardly varies the conductivity in both up and down-spin electrons. The defect of Mg atoms increases the conductivity in down-spin electron, although the up-spin conductivity does not change very much. So the defect of Mg decreases the TMR ratio. It is found that the asymmetric behavior of the conductivity is caused from the atom defect around the junction, which leads to the degradation of the TMR ratio.

The author would like to thank Mr. Yoshino for his support and valuable discussions.

References

[1] Butler W H, Zhang X G and Schulthess T C 2001 *Phys. Rev.* **B63** 054416
[2] Yuasa S, Fukushima A, Nagahama T, Ando K and Suzuki Y 2004 *Jpn. J. of Appl. Phys.* **43** L588
[3] Zhang X G, Butler W H and Bandyopadhyay Amrit 2003 *Phys. Rev.* **B68** 092402
[4] Stokbro K 2008 *J. Phys.: Condens. Matter* **20** 064216
[5] Simmons J G 1963 *J. Appl. Phys.* **34** 1793
[6] Ito H and Inoue J 2003 *J. Magn. Soc. Japan* **27** 1013 (in Japanese)

Change in Co/Al-oxide/Co tunneling junction under constant voltage stress

K Shiiki, K Takeda and H Horikiri
Department of Applied Physics and Physico-informatics,
Keio University, Yokohama 223-8522, Japan

E-mail: shiiki@appi.keio.ac.jp

Abstract. The tunneling junctions Co(10nm)/Al-oxide/Co(50nm) were fabricated on a glass substrate using a metal mask by ion-beam sputtering. On the bottom Co electrode, a 2.5 nm thick Al film was deposited and was oxidized in pure oxygen gas for 12 - 216 h at 200 C. The electrical properties of the junction were measured by the four-probe method under a constant voltage and the current-voltage characteristics at low voltages were measured every half hour. The barrier parameters of Al-oxide layers are evaluated from a fit to the calculated current-voltage characteristics for the junction by using the Particle Swarm Optimization method with high accuracy. The oxide layer was assumed to comprise two different barriers. In case of the barrier with low potential height, oxidation proceeds by applying the voltage and the resistivity increases. In the case of hard breakdown, the barrier retains uniform at the breakdown with sudden change of the resistivity. In other case of the soft breakdown, the barrier shows segregation just before the breakdown.

1. Introduction

Magnetic tunneling junctions (MTJs), which consist of two ferromagnetic electrodes separated by an insulator, are expected for future memories such as magneto-resistive random memory (MRAM), because of a large magneto-resistance ratio at room temperature [1]. However, MTJs suffer from electrical breakdown by induced voltage stress because of the very thin insulator. It is an important reliability issue for commercial device applications. Various investigations on electrical breakdown of MTJs have been performed [2, 3]. However, the mechanism of the breakdown was not clarified. In this article, the change in the insulator has been studied under constant voltage stress.

2. Experimental and calculation method

The tunneling junctions Co(10nm)/Al-oxide/Co(50nm) were fabricated on a glass substrate using a metal mask by ion-beam sputtering. On the bottom Co electrode, a 2.5 nm thick Al film was deposited and was oxidized in pure oxygen gas for 12 - 216 h at 200 C. The electrical properties of the junction were measured by the four-probe method under a constant voltage more than 0.5 V. And the current-voltage characteristics at low voltages from -0.3 V to 0.3 V were measured every half hour.

Two potential model is used for the evaluation of the barrier parameters. Figure 1 shows the schematic illustration of the model. The insulator film is assumed to comprise two different barriers. Here, the ϕ_a and ϕ_b are the barrier heights, d is the thickness, and β is the ratio of potential barrier thickness. In this model, tunneling probability is numerically calculated from Schrödinger equation. The barrier parameters of Al-oxide layers are evaluated by fitting the current-voltage characteristics

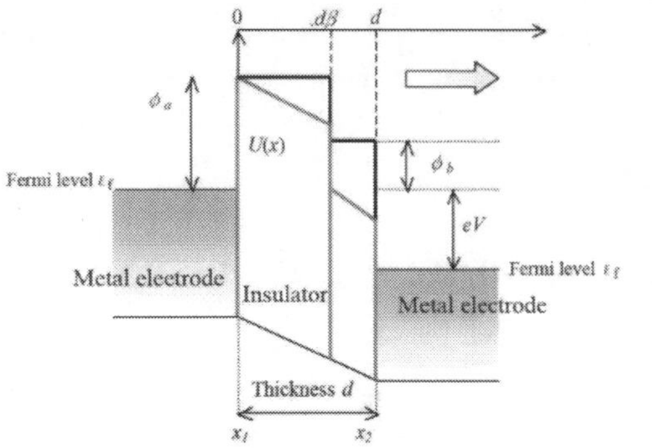

Figure 1. Schematic illustration of the two potential model.

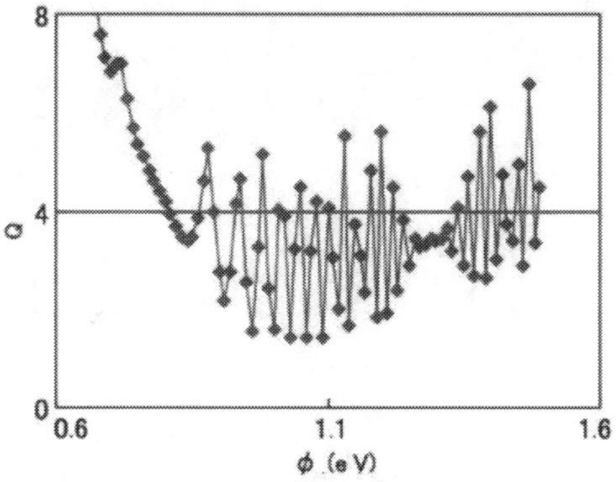

Figure 2. Typical result of the barrier-height dependence of the fitting error, Q.

for the junction to the calculated result obtained from the two-potential model [4].

Figure 2 shows a typical result of the barrier-height dependence of the fitting error, Q. The fitting error is defined as sum of the absolute differences between the observed value and calculated value. In this case, we set parameters $\phi_a = \phi_b$ and $d = 1.7$ nm, which are obtained as the suitable values by the steepest descend method of a preliminary search. The fitting error oscillates extremely. So the conventional steepest descend method cannot find the real minimum value due to an interruption by local minimum values. In order to solve this problem, the Particle Swarm Optimization method [5] is applied. Table 1 shows that this method can determine the barrier parameters with high accuracy.

Table 1. Barrier parameters determined by the PSO method with different convergence parameters.

	d (nm)	ϕ (eV)	Q	endstep
Steepest descent method	1.80	0.90	2.78	1200
PSO method (1)	1.67	1.07	1.41	617
(2)	1.66	1.09	1.41	728
(3)	1.69	1.04	1.42	482

3. Results and discussion

3.1. Progress of oxidation in the Al-oxide layer

Figure 3 shows the change in tunnel resistance under constant voltage stress at 0.5 V. The change in thickness and barrier heights evaluated are shown in Figures. 4(a) and (b).

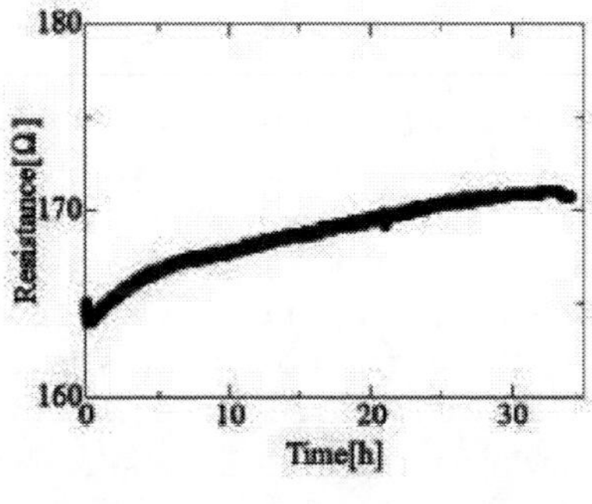

Figure 3. Change in tunnel resistance under constant voltage stress at 0.5 V.

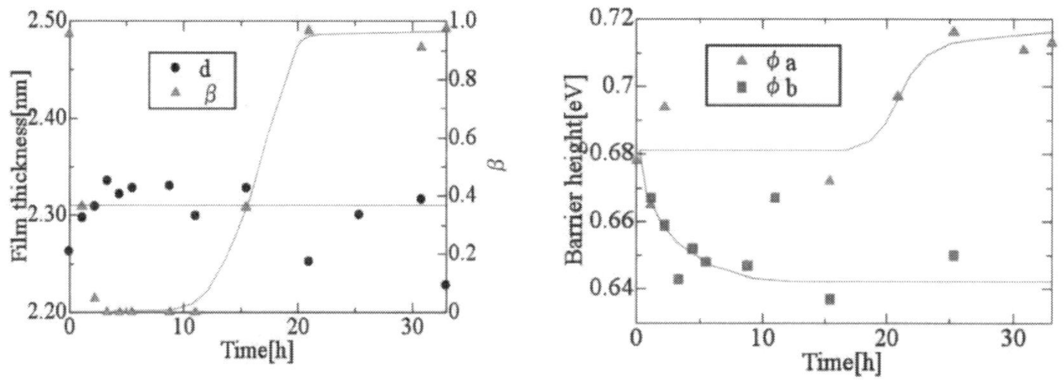

(a) (b)

Figure 4. Change in thickness and barrier heights evaluated under constant voltage stress at 0.5 V.

At first, the oxide layer is almost the homogeneous insulator ($\beta \sim 0$) with low barrier height of about 0.64 eV, which is estimated to be AlO_2 by XPS analysis. Then oxidation proceeds by applying the voltage and the resistivity increases. The phase changes to Al_2O_3 with high barrier height of about 0.72 eV at 15 h. The total thickness of about 2.3 nm does not change during the oxidation process, so the density of the oxide layer increases. Finally, the oxide layer becomes a homogeneous ($\beta \sim 1$) and dense Al_2O_3 phase with high barrier height.

Figure 5 shows the change in tunnel resistance at voltage of 0.6 V. The change in thickness and barrier heights evaluated are shown in Figures. 6(a) and (b). The parameters hardly change at 0.6 V. The Al-Ox phase is almost the homogeneous, because $\phi_a \sim \phi_b$.

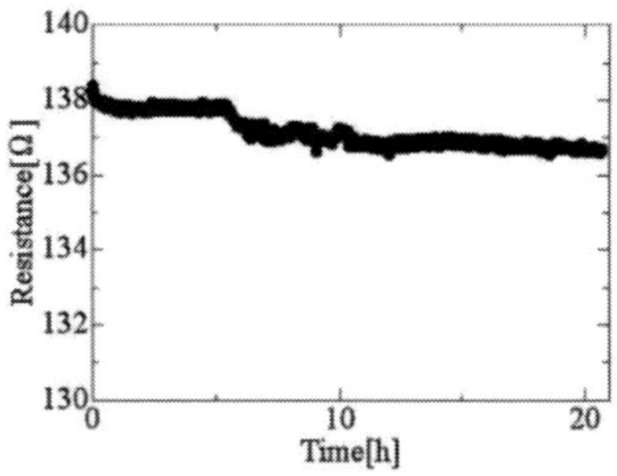

Figure 5. Change in tunnel resistance at voltage of 0.6 V.

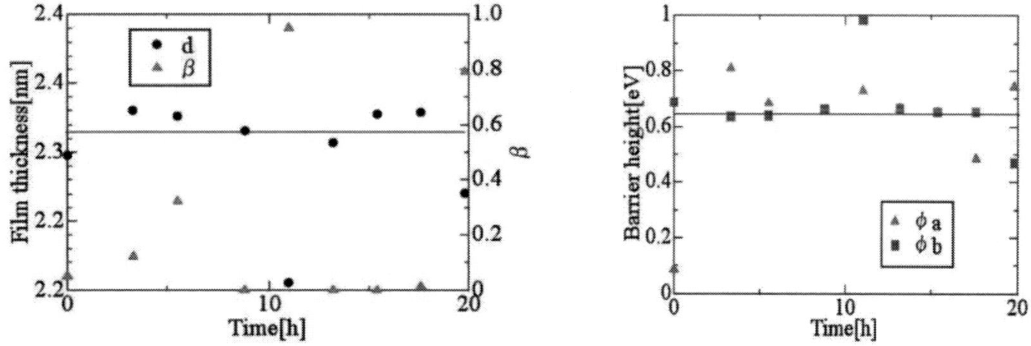

(a) (b)

Figure 6. Change in thickness and barrier heights evaluated at voltage of 0.6 V.

3.2. Hard breakdown of the Al-oxide layer

As shown in Figure 7, changes in tunnel resistance, evaluated thickness and barrier heights at voltage of 0.65 V are almost the same as changes at 0.6 V. However, a sudden decrease in the tunnel resistivity, which means an electrical breakdown, occurs at about 39 h. The barrier parameters do not change before the breakdown. It is thought that the breakdown is probabilistic.

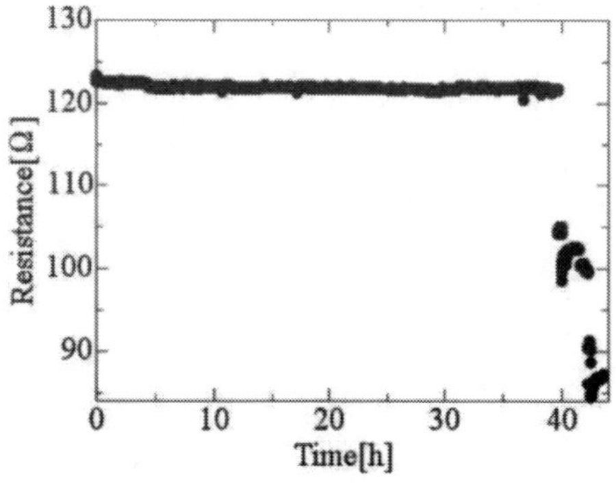

Figure 7. Change in tunnel resistance at voltage of 0.65 V.

3.3. Soft breakdown of the Al-oxide layer

Figure 8 shows the change in tunnel resistance at voltage of 0.6 V for another specimen. The change in thickness and barrier heights evaluated are shown in Figures 9(a) and (b). There are sudden changes in tunnel resistance and barrier parameters. These changes, called as soft breakdown, seem to be caused by a reaction such as phase change in the insulator.

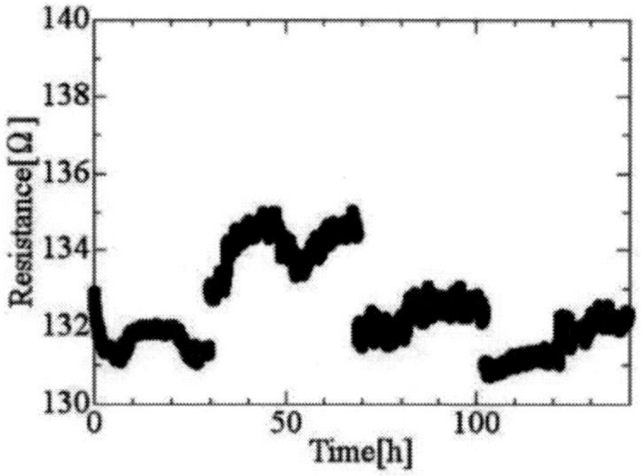

Figure 8. Change in tunnel resistance at voltage of 0.6 V for another specimen.

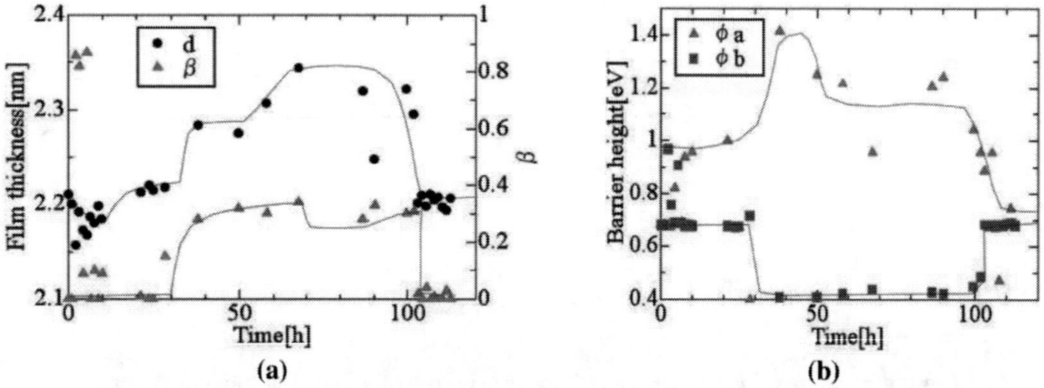

Figure 9. Change in thickness and barrier heights evaluated at voltage of 0.6 V.

4. Conclusions

There are several different cases of the change under constant voltage stress. In case of the barrier with low potential barrier, oxidation proceeds by applying the voltage and the resistivity increases. The barrier retains uniform at the breakdown with abrupt change of the resistivity. However, the barrier sometimes shows segregation, which leads to a soft breakdown.

References

[1] Miyazaki T and Tezuka N 1995 *J. Magn. Magn. Mater.* **139** L231
[2] Kim K S and Cho B K 2005 *Appl. Phys. Lett.* **86** 142106-1
[3] Horikiri K and Shiiki K 2007 *phys. stat. sol.* (c) **4** 4532
[4] Arakawa N, Otaka Y and Shiiki K 2006 *Thin Solid Films* **505** 67
[5] Kennedy J and Eberhart R C 1995 *Proc of IEEE ICNN* **4** 1942

Temperature dependence of exchange bias in Co/FeMn-structure induced by heating and cooling in magnetic field

I O Dzhun[1], **S A Dushenko**[1,2], **N G Chechenin**[1,2,3] **and E A Konstantinova**[2]

[1]Lomonosov Moscow State University Skobeltsyn Institute of Nuclear Physics (MSU SINP), 1(2), Leninskie gory, GSP-1, Moscow 119991, Russian Federation

[2]Faculty of Physics Lomonosov Moscow State University Leninskie Gory, Moscow 119991, Russian Federation

[3]E-mail: chechenin@sinp.msu.ru

Abstract. Using the method of angular dependence of ferromagnetic resonance field the magnetic properties of Si/SiO_2/Cu/Co/FeMn/Cu and Si/SiO_2/Cu/Co/Cu/FeMn/Cu structures were investigated. The layer deposition was carried out by magnetron sputtering in absence of an external magnetic field. It was established that thermal annealing with further cooling down in presence of a magnetic field can generate an exchange bias at anneal temperature significantly below the bulk antiferromagnetic Néel temperature. It was also shown that a thin interlayer between ferromagnetic and antiferromagnetic layers reduces the exchange bias effect at low anneal temperatures, however, makes this effect more stable at high annealing temperatures.

1. Introduction

A common method to generate the exchange bias in ferromagnetic (F)/antiferromagnetic (AF) thin film structures is to anneal the sample above the Néel temperature of the antiferromagnet and below the ferromagnetic Curie point with subsequent cooling down in presence of a magnetic field high enough to saturate the F-layer and applied in the plane of the sample. With such a treatment the magnetic anisotropy direction can be induced both in F and AF layers.

Recently, a number of exchange bias investigations were carried out using $Ir_{20}Mn_{80}$ as an antiferromagnetic layer (see for instance [1]. This alloy in contact with F layer demonstrates high exchange bias and blocking temperature values. Another prospective material for spinvalve devices is FeMn alloy with approximately 50% elements composition (see review [2]). FeMn is much cheaper material than IrMn, but characterized by lower exchange bias, blocking temperature. Néel temperature, and melting temperature values [3] and more susceptible to thermal effects, serving a convenient sample to investigate the fundamental aspects of temperature dependence of F/AF exchange biasing in a fairly narrow range near the room temperature.

In this paper the results of investigation of exchange bias generation by thermal annealing-cooling in magnetic field are reported. It was founded that in case of 7 nm AF layer thickness the exchange bias does not arise. At the thickness of 15 nm the exchange bias appears at the annealing temperature as low as T_{ann}=50°C. This temperature value is much lower than the Néel temperature for the bulk

Published under licence by IOP Publishing Ltd

$Fe_{50}Mn_{50}$ (T_N = 490 K [2]). The annealing temperature effects on the exchange bias field and intrinsic FMR field is reported as well as influence of a thin interlayer between F and AF layers on these parameters.

2. Experimental techniques

Experimental samples were deposited by magnetron sputtering using the magnetron ATC ORION-5 produced by AJA INTERNATIONAL. The chamber was vacuumed to the base pressure of 10^{-7} Torr and the DC and RF depositions of the layer structures Si/Cu(50nm)/Co(5nm)/FeMn(t_{AF})/Cu(50nm) were made in the argon atmosphere at the pressure of $3*10^{-3}$ Torr on Si (100) substrate with the AF-layer thicknesses t_{AF} of 7 nm and 15 nm. There were also deposited the Si/Cu50nm/Co5nm/Cu0,5nm/FeMn20nm/Cu100nm structures with thin Cu interlayer. The deposition rates for each layer were evaluated by measuring the thickness of the calibrating layers using the Rutherford backscattering. Prior the deposition the Si substrates were cleaned in ethanol in an ultrasonic bath. The deposited samples were thermally annealed at T_{ann} = 50, 100, 150, 200 and 250 °C for 0.5 hour and cooled down to the room temperature all in presence of a magnetic field of the order about 1 kOe applied in the plane of the sample.

Magnetic properties of the samples were studied by measuring the angular dependence of the ferromagnetic resonance (FMR) field. The FMR DC magnetic field for a ferromagnetic sample, H_R, is determined by the FMR frequency $\omega = 2\pi f$ of the RF field of the spectrometer, as well as by internal magnetic properties of the sample, such as the saturation magnetization, I_S, magnetic anisotropy field, H_K, and sample orientation relative to the DC and RF fields directions. For the sample aligned with the easy axis (EA) and hard axis (HA) along the DC field the Kittel relations applied [4]:

$$\omega^2 = \gamma^2 (H_R^{EA} + H_k)(H_R^{EA} + H_k + 4\pi I_s) \quad (1a)$$

$$\omega^2 = \gamma^2 (H_R^{HA} - H_k)(H_R^{HA} - H_k + 4\pi I_s) \quad (1b)$$

where, $\gamma = ge/(2mc) \approx g \cdot 8.79 \cdot 10^6 (\mathrm{G\ s})^{-1}$ is the gyromagnetic constant, H_R^{EA} and H_R^{HA} are the resonance fields for EA and HA sample alignment, respectively. The Kittel relations can be generalized to include the unidirectional exchange bias field [5, 6, 7, 8]. Assuming that θ is the angle between the FMR DC magnetic field and the unidirectional exchange bias field, which is parallel to the EA of the F-layer, and for $4\pi I_s >> H_r$, the resonance field can be written approximately as [Błąd! Nie zdefiniowano zakładki.5, Błąd! Nie zdefiniowano zakładki.6]:

$$H_r = \frac{\omega^2 / \gamma^2}{4\pi I_s} - H_{EB} \cos\theta - H_K \cos 2\theta \quad (2)$$

Though FMR is rather unordinary method to investigate the exchange bias phenomenon it has a high sensitivity to low magnetic moments of the FM, which can be obtained only on the most advanced VSM devices, and has an advantage in investigation the magnetic properties of structures with F layer covered by insulating, diamagnetic or antiferromagnetic layers, which have their restrictions in using other methods like MOKE or SQUID. In addition, the FMR can give the value of the sample saturation magnetization regardless to its porosity, in contrast to the VSM. In this report, from the angular dependence of the resonance field the unidirectional and uniaxial anisotropies were determined using equation (2). The measurements were done at room temperature, using BRUKER ELEXSYS e500 spectrometer with the frequency of the RF field of 9,65 GHz and the rectangular resonator of E_{102} mode.

3. Experimental results and discussion

The typical differential FMR absorption spectra are shown in figure 1 for a structure with non-zero exchange bias at several sample orientation relative to the FMR DC field which was set in the sample plane.

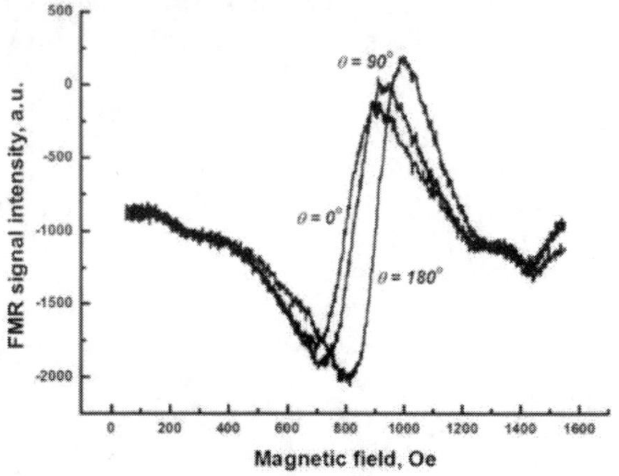

Figure 1. FMR absorption spectra at different sample orientations.

From the resonance field, H_r, using equation (2) one can determine H_{r0}, H_K, H_{EB} values. For this it is enough to measure the FMR spectra for three orientations, for example for $\theta = 0$, 90 or 270 and 180°.

The investigation of the structures with 7 nm AF thickness did not reveal any exchange bias at the annealing temperature in the range used. The exchange bias dependence on the annealing temperature for structures with $t_{AF} = 15$ nm is shown in figure 2. The exchange bias field showed up at $T_{ann} = 50$°C. In other words, starting from this annealing temperature FeMn layer begins to arrange a certain predominating antiferromagnetic order at the AF/F interface and pins the F-layer. We note that this temperature is much lower that the Néel temperature [2], the issue to be discussed later on. At higher annealing temperatures H_{EB} increases to reach the maximum value of 180 Oe at $T_{ann} = 150$°C then falls down and zeroes at $T_{ann} = 250$°C.

The reduction of the H_{EB} at the T_{ann} above 150°C is a consequence of weakening of the F and AF layers interaction. It is, probably, due to intensification of interdiffusion and layer intermixing processes on the interface at higher temperatures. An occurrence of these effects is also indicated by an FMR signal weakening and increasing of signal-to-noise ratio at these annealing temperatures. Possibly, the Mn atoms became more mobile and penetrate into Co layer causing a weakening of the ferromagnetic properties of Co layer by reducing the volume fraction of ferromagnetic Co. The FMR signal weakening was observed in the structures with 7 nm AF thickness where the interdiffusion effect can be more essential because of smaller thickness. It can also explain the absence of the exchange bias in the structures with AF layer of 7 nm reported above.

From the other hand the absence of exchange bias in these structures might result from AF small thickness. Following the reference [9], the reduction of the Néel temperature can be expressed as

$$\frac{T_N(\infty) - T_N(t_{AF})}{T_N(\infty)} = (\frac{\xi_0}{t_{AF}})^{\lambda} \quad , \tag{3}$$

where $T_N(\infty)$ and $T_N(t_{AF})$ are the Néel temperatures for the bulk material and for a film with the thickness t_{AF}, respectively. Assuming $\xi_0 = 3.04$ nm, $\lambda = 1.5$ as in [3] one could obtain $\Delta T_N = (T_N(\infty)- T_N(t_{AF})) \approx 140^0$, or $T_N(t_{AF}) \approx 350$ K for the film of 7 nm. It is a significant reduction of the Néel temperature, but could be even more, because of uncertainties in the parameters used. If the Néel point lowers down to or below the room temperature than the AF layer stays in paramagnetic state and did not pin the F-layer, which is, possibly, the case with 7 nm AF thickness.

The influence of the thermal effects on the intrinsic resonance field, H_{r0}, is illustrated on figure 3. Puzzling enough, the annealing temperature dependence of H_{r0} has an opposite tendency compared with that obtained earlier with IrMn AF layer. In [10] the intrinsic resonance field increased at increasing T_{ann} and always remained above the H_{r0} for the free (i.e. not pinned) Co layer which is about 600 Oe (see accompanying paper [11]). In case of FeMn the intrinsic resonance field decreases with the temperature and tends to the value for the free Co layer. Keeping in mind the inverse dependence of H_{r0} on I_s, one can conclude that up to 150^{0}C the magnetization of Co layer does not change significantly being lower than that in Co free layer. At a higher T_{ann} the H_{r0} starts to decrease more steeply indicating an increase of I_s still keeping it at a lower value compared to that in Co unpinned layer. Evidently, the origin of such a behavior of the intrinsic resonance field is complicated and not completely clear, however, the trend to the intrinsic value for unpinned Co layer is an indication of a weakening of the antiferromagnetic properties of FeMn layer with an increase of T_{ann}, while the magnetization of Co layer does not drastically change or even improve.

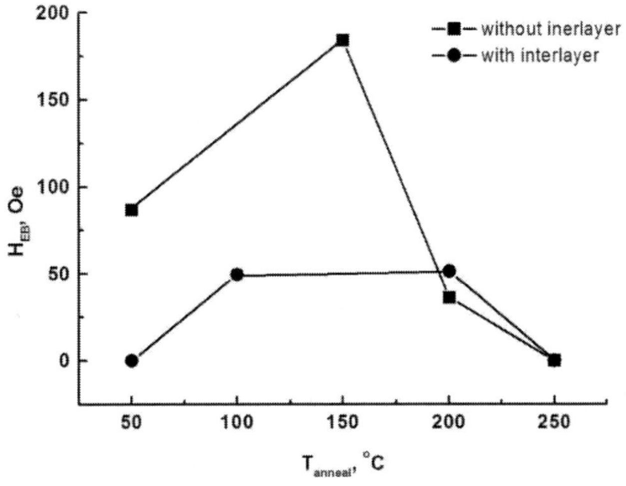

Figure 2. The dependence of exchange bias on the anneal temperature for Co/FeMn and Co/Cu/FeMn structures

In order to prevent the thermal interdiffusion and intermixing, a set of trilayer structures with thin Cu interlayer between F and AF layers was made. As can be seen from figure 2 the presence of the interlayer leads to the increase of the annealing temperature necessary to generate the exchange bias. At T_{ann} = 50°C the exchange bias field is missed and appears only at T_{ann} = 100°C. An exponential decay of the interaction between F and AF layers and, consequently, the exchange bias with the spacer thickness was observed before [12]. The positive influence of the spacer is that it provides a certain thermal stability of the magnetic properties of the structure. As illustrated in figure 2, the exchange bias for the structure with interlayer is even higher than that one without interlayer at 200°C annealing temperature.

The intrinsic resonance field decreases in the F/Cu/AF structures with the annealing temperature increase as it does in case of the F/AF structures, see figure 3. At T_{ann} =100°C the H_{r0}, for the trilayer structure is lower than that for bilayer structure that means that the magnetization of F layer in the trilayer is higher. At the subsequent increasing of the anneal temperature the intrinsic resonance field value decreases more smoothly in the trilayer system that demonstrates a higher thermal stability than that of the bilayer structure.

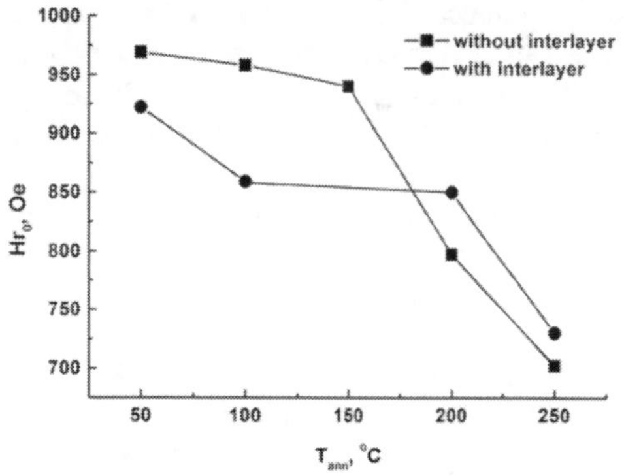

Figure 3. The intrinsic resonance field dependence on the annealing temperature for the Co/FeMn and Co/Cu/FeMn structures

4. Conclusions

The results of this paper show that the exchange bias at the room measuring temperature in the bilayer system Co/FeMn either can not be generated by heating-cooling in magnetic field at any temperature for thinner AF layer thickness (7 nm) or can be induced at the annealing temperature significantly lower than the bulk AF Néel temperature for larger AF layer thickness (15 nm). We conclude that this effect originates from a combination of two major sources. The first one is the low-dimensional reduction of the Néel temperature due to small film thickness, in accordance with [9]. The second reason is the interdiffusion of the layers in contacts which changes the composition with reduction of the Néel and blocking temperatures of the AF layer and lowering the saturation magnetization of the F layers, making the intrinsic FMR DC field higher. Both, the interdiffusion and the low-dimensional effects bring stronger consequences for the thinner AF films.

In accordance with other publications, see for instance [12], we observe that the nonmagnetic interlayer deposited between F and AF layers reduces the exchange bias at low anneal temperatures, however, makes this effect more stable at high anneal temperatures. This spacer weakens the interaction strength between F and AF layers. At low annealing temperature the F/AF interaction is too weak and only a small fraction of the FeMn grains with a relatively small size is set in the antiferromagnetic order. This fraction has too small unidirectional anisotropy energy to withstand the thermal fluctuations to keep the antiferromagnetic order in the AF layer and to keep the F-layer pinned at room temperature of the FMR measurements, showing no exchange bias at T_{ann} = 50 °C in figure 2. However, a weak interdiffusion in the trilayer structure keeps unchanged H_{r0} and I_s of the Co-layer up to 200^{0}C and making I_s larger compared with the bilayer structure.

Acknowledgments

This work is supported by Federal Agency of Science and Innovations (contract 02.740.11.0242, contract 02.740.11.0389). The FMR study was performed at User Facilities Center of M.V. Lomonosov Moscow State University.

References

[1] O'Grady K, Fernandez-Outon L E and Vallejo-Fernandez G 2010 *J. Mag. Mag. Mat.* **322** 883
[2] Nogués J, and Schuller I K 1999 *J. Mag. Mag. Mat.* **192** 203
[3] Devasahayam A J, Sides P J, and Kryder M.H. *J. Appl.Phys.* **83** 7216
[4] Kittel Ch 1948 *Phys. Rev.* **73** 155
[5] Scott J C 1985 *J. Appl. Phys.* **57**, 3681

[6] Layadi A A, Cain W C, Lee J-W, and Artman J O 1987 *IEEE Trans. Magn.* MAG-**23**, 2993.
[7] Khomenko E V, Chechenin N G, Goikhman A Yu, and Zenkevich A V 2008 *JETP Letters* **88**, 602
[8] McMichael R D, Stiles M D, Cheng P J and Egelhoff W F, Jr 1998 *Phys. Rev.* B **58**, 8605.
[9] Ambrose T, Chien C L 1996 *J. Appl. Phys.* **79**, 5920
[10] Khomenko E V, Chechenin N G, Dzhun I O, Perov N S, Samsonova V V, Goikhman A Yu, and Zenkevich A V 2010 *Phys. Sol. State*, **52** 1701
[11] Dzhun I O, Dushenko S A, Chechenin N G and Konstantinova E A Acompanying paper in this issue
[12] Wang J, Wang W N, Chen X, Zhao H W, Jhao J G and Zhan W Sh 2002 *J. Appl. Phys.* **91** 7236

The dependence of magnetic properties of Co/FeMn bilayer structure on the magnitude of magnetic field applied during the layer deposition

I O Dzhun[1], S A Dushenko[1,2], N G Chechenin[1,2,3] and E A Konstantinova[2]

[1]Lomonosov Moscow State University, Skobeltsyn Institute of Nuclear Physics 1(2), Leninskie Gory, GSP-1, Moscow 119991, Russian Federation

[2]Faculty of Physics, Lomonosov Moscow State University, Leninskie Gory, Moscow 119991, Russian Federation

[3]E-mail: chechenin@sinp.msu.ru

Abstract. By measuring the angular dependence of ferromagnetic resonance field at room and low temperatures, it is demonstrated that the magnitude of magnetic field applied during magnetron deposition of Ta/Co/FeMn/Ta structures influences their magnetic properties such as uniaxial and unidirectional anisotropy, magnetization and the exchange bias blocking temperature. The deposition field effects on the bilayer structure are compared with the effects on a similar structure, but without antiferromagnetic layer. The exchange bias blocking temperature of investigated structures is found to be significantly lower than the Néel temperature of a bulk antiferromagnet. The origin of the observed effects is shortly discussed.

1. Introduction

In spite of more than 50 years history since the discovery and a high importance for development of magnetic information storage media, the exchange bias effect in a bilayer ferromagnet(F)/antiferromagnet (AF) structure still has a number of open issues. The exchange bias is normally detected by a shift of hysteresis loop. It can be induced in F/AF structures by layer deposition in magnetic field, first F-layer in the magnetically saturated state and then, on top of it, an AF-layer. The exchange bias also can be induced by thermal annealing at a temperature which is higher than the Néel point of the antiferromagnet transition into the paramagnetic state, but lower than the Curie point of the ferromagnetic layer, followed by cooling in magnetic field. In both cases the external magnetic field induces magnetic anisotropy and saturates the F-layer, which in its turn, induces the anisotropy in the AF-layer. In such a simplified picture, the magnetic moment of the F-layer plays a major role in the magnetic ordering and initiating the anisotropy of the AF-layer, while the magnitude of the external magnetic field is not of importance as soon as it is higher than the value for F-layer saturation. However, the experimental findings are not definite to support this picture. In fact, there is a number of evidences showing the field dependence of the magnitude and even the sign of the exchange bias (see a summary of the observations and references in the reviews [1, 2]). Shortly, the nature of the observed effects is unclear and experimental data are insufficient to make an

Published under licence by IOP Publishing Ltd

unambiguous conclusion about the role of the magnitude of the external field in appearance and strength of the exchange bias in F/AF bilayers.

This paper reports on the results of our study of the exchange bias generation in Co/FeMn bilayer structures during the layer deposition in presence of the magnetic field. The influence of the applied magnetic field magnitude on the magnetic properties of this bilayer structures is demonstrated and compared with that of a free Co layer.

2. Experimental techniques

The specimens were deposited by DC magnetron sputtering, using the magnetron sputtering system ATC ORION-5 produced by AJA INTERNATIONAL. The base pressure prior the deposition was of the order of 10^{-7} Torr and the pressure of argon atmosphere during the deposition was about 2,5 mTorr. Two permanent magnet plates were fixed on the specimen holder with the substrates being placed in the magnet poles gap. The magnitude of the magnetic field was varied by changing the gap width. To clean the surface and remove a native oxide, Si (100) substrates were dipped for 30 seconds in hydrofluoric acid before the sputtering. The multilayer structures Si/Ta/Co7nm/FeMn15nm/Ta were deposited onto the substrates in presence of 40, 420 and 1000 Oe magnetic field applied in plane of the substrate at ambient temperature. For comparison, also the structures Si/Ta/Co7nm/Ta, where F-layer did not contact with AF-layer (free Co-layer), in 40 and 420 Oe magnetic fields were deposited.

The magnetic properties of the structures were studied using the angular dependence of the ferromagnetic resonance (FMR) field. This method is characterized by a high sensibility, simplicity and possibility to investigate ferromagnetic layers covered by insulating, diamagnetic or antiferromagnetic layers that is hard to realize in MOKE and SQUID methods. The FMR is used for investigation of high-frequency properties of ferromagnetic materials [3], their anisotropy and saturation magnetization [4].The FMR method can be also effectively used for exchange bias phenomenon investigation [5, 6, 7].

In this method the absorbed energy of microwave field, applied perpendicular to the direction of the external DC magnetic field is measured. The saturation magnetization Is and uniaxial anisotropy field H_K can be obtained scanning the FMR DC field aligned along the easy (EA) and hard (HA) axes and using the Kittel equations [8]:

$$\omega^2 = \gamma^2 (H_R^{EA} + H_k)(H_R^{EA} + H_k + 4\pi I_s) \qquad (1a)$$

$$\omega^2 = \gamma^2 (H_R^{HA} - H_k)(H_R^{HA} - H_k + 4\pi I_s) \qquad (1b)$$

Here $\omega = 2\pi f$ - the frequency of RF field applied, $\gamma = ge/(2mc) \approx g \cdot 8.79 \cdot 10^6 (\text{G s})^{-1}$ - gyromagnetic constant, H_R^{EA} and H_R^{HA} - the value of resonance fields along the easy and hard axis, respectively. These equations could be also generalized for the case of an arbitrary orientation of the FMR DC field with respect to the vector of the unidirectional anisotropy field [4, 5, 9].

For a sample with a large saturation magnetization, $4\pi I_s >> H_r$, the resonance peak position H_r is governed by intrinsic resonance field $H_{r0} = (\omega/\gamma)^2/(4\pi I_s)$ of the Co-layer, bidirectional uniaxial crystallographic anisotropy field H_K and unidirectional anisotropy field H_{EB}. If the sample is aligned with an angle θ between the unidirectional exchange bias field, which is parallel to EA of the F-layer, and the FMR DC magnetic field direction, then [4, 5]:

$$H_r = \frac{\omega^2 / \gamma^2}{4\pi I_s} - H_{EB} \cos\theta - H_K \cos 2\theta \qquad (2)$$

Since the EA directions in our samples are set by the direction of the field during the deposition, in our experiment θ is the angle between the FMR DC magnetic field direction and the direction of deposition field.

The FMR investigations were carried out, using BRUKER ELEXSYS e500 setup with RF frequency of 9,65 GHz. A rectangle type resonator E_{102} was used. The temperature dependence of the specimens' magnetic properties in the range of 115K to 300K was investigated by setting a cryostat into the resonator and cooling the cryostat by a liquid nitrogen vapor blowing.

3. Experimental results

3.1. Exchange bias

The exchange bias of 14 Oe was detected at room temperature only in the Co/FeMn sample deposited in presence of 1000 Oe magnetic field. The samples deposited in magnetic field of 40 Oe and 420 Oe did not exhibit exchange bias at room temperature. The temperature dependence of exchange bias for the sample deposited in 420 Oe field is shown in figure 1. The magnitude of exchange bias decreases with increasing of the measuring temperature. The blocking temperature at which the exchange bias vanished was about 250K, which is much lower than the Néel temperature for the bulk $Fe_{50}Mn_{50}$ (T_N = 490 K [1]).

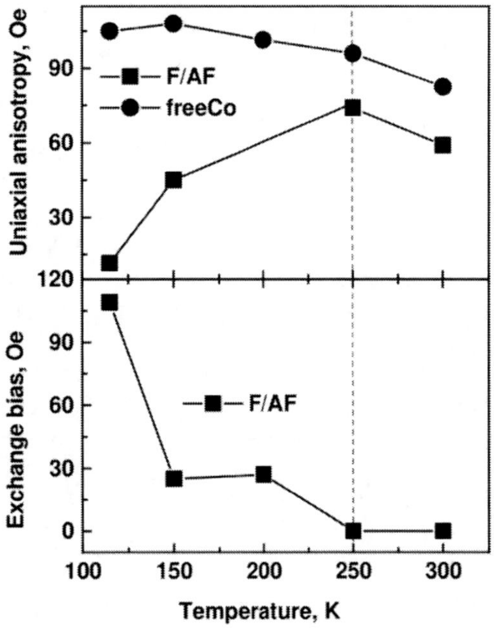

Figure 1. The temperature dependence of the uniaxial anisotropy (at the top) and of the exchange bias (at the bottom) for F/AF bilayer structure (squares) and free Co structure (circles) deposited in applied field of 420 Oe.

3.2. Uniaxial anisotropy

The temperature dependences of uniaxial anisotropy for F/AF structure and free Co layer deposited in the magnetic field of 420 Oe are given in figure 1. While the value of H_K for the free Co layer almost monotonically decreases when the temperature rises, the dependence of H_K for F/AF system shows a peak shape with a maximum at the temperature where the exchange bias vanishes for this structure, i.e. at $T = T_B$. The higher the temperature the closer H_K for the F/AF system to that for the free F-layer samples, however, these values do not become equal even above the blocking temperature. This comparison signalizes that with increasing the temperature the exchange interaction between F and AF layers weakens but does exist even above the blocking temperature.

The magnitude of the uniaxial magnetic anisotropy varied with alteration of the deposition field. The dependence shown in figure 2 reveals that the uniaxial anisotropy H_K increases approximately linearly with increase of the deposition field. The field dependence of H_K for the free Co layer, also

Journal of Physics: Conference Series **303** (2011) 012104 doi:10.1088/1742-6596/303/1/012104

shown in figure 2, behaves in a similar way. With increasing of the deposition field the difference between H_K for F/AF bilayer structure and for the free F-layer also increases.

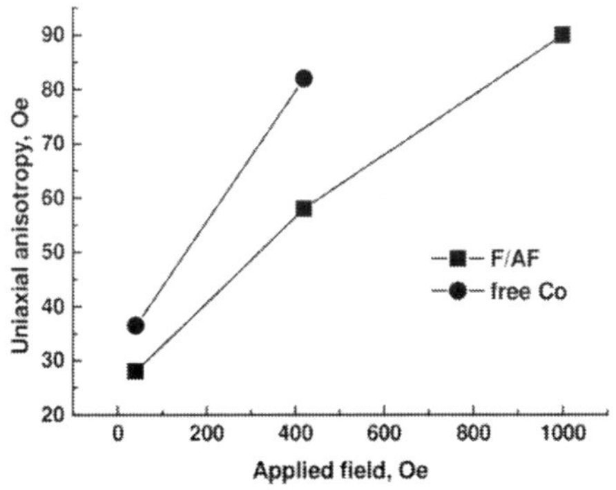

Figure 2. The dependence of the uniaxial anisotropy on the magnetic field applied during the deposition for F/AF bilayer (squares) and free Co (circles) structures

3.3. Intrinsic resonance field

The intrinsic resonance field H_{r0} characterizes only ferromagnetic layer and is determined by its saturation magnetization via relation $H_{r0} = (\omega/\gamma)^2/(4\pi I_s)$. The magnitude of H_{r0} also depends on the magnitude of the magnetic field applied during the structures deposition. The field dependence of H_{r0} (figure 3) is more complicated than field dependence of H_K. With increasing of the deposition field from 40 to 420 Oe the intrinsic resonance field slightly increases, then with the further increase of deposition field H_{r0} sharply decreases. At the same time, the intrinsic resonance field of the free Co layer remains constant and is very close to that one of the Co layer deposited in absence of magnetic field, thus no field dependence was observed. This H_{r0} value is also equal to the H_{r0} value for the F/AF system deposited in absence of the magnetic field. This observation evidences the absence of exchange interaction between the F and AF layers in this case because of antiferromagnetic disorder on the interface.

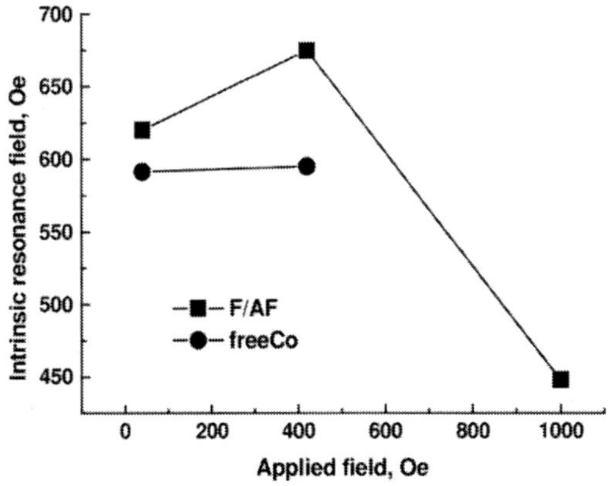

Figure 3. The intrinsic resonance field as a function of the deposition field for F/AF bilayer structures (squares) and free Co layer (circles).

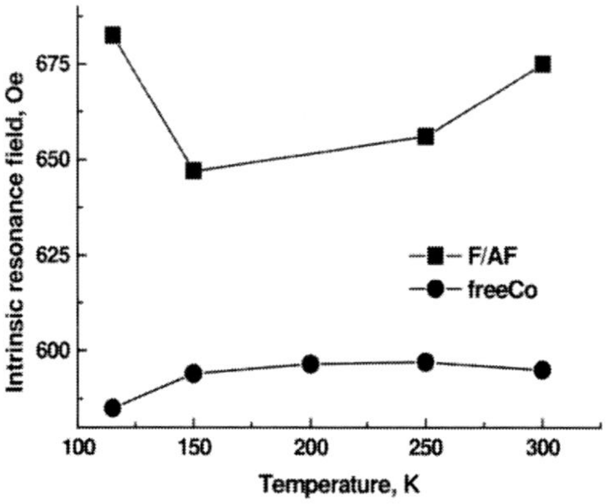

Figure 4. The temperature dependence of the intrinsic resonance field for F/AF bilayer structure (squares) and free Co layer (circles) deposited in magnetic field of 420 Oe.

In figure 4 the temperature dependences of H_{r0} for the deposited in 420 Oe field F/AF and free-F structures are represented. One can see that such dependence for bilayer structure is not monotonic, while the intrinsic resonance field for the free-Co layer almost does not vary. One can also notice that the intrinsic resonance field of the bilayer structure remains at all temperatures higher than that for the free-F sample. Keeping in mind the inverse dependence of H_{r0} on I_s, the saturation magnetization of the free-F layer remains higher that the saturation magnetization of the pinned F-layer in the F/AF structure.

4. Discussion

The fact that the exchange bias blocking temperature is below the Néel temperature of a bulk antiferromagnet is explained in a number of papers as a result of low-scaling effect [10]. According to this model the Néel temperature of AF nanofilms is lower than the Néel temperature of bulk AF and this temperature becomes lower while the AF film becomes thinner. But even taking into account this effect the Néel temperature still is not as low as blocking temperature we have obtained. It is essential to bear in mind that the Néel critical point marks the transition to paramagnetic state, while the blocking temperature marks the point of vanishing exchange bias, meaning an averaging out to zero balance of the magnetic moments of the AF layer at the interface in close contact with the F-interface layer. In other words, exchange interaction may exist even at temperatures above the blocking point, but not in average. Our results reported above give a certain evidence for this suggestion.

It is also important to notice that the blocking temperature normally has a certain local variation. As far as an ideal interface is an approximation and the interface roughness always reflects the layer thickness variation from point to point that is followed by widening of the blocking temperature local variation. It also seems to be true that as the roughness increases the average blocking temperature decreases. Our analysis did not reveal any correlation between the deposition field and the surface roughness of the samples. The largest roughness was observed for the sample deposited at 420 Oe. The sample deposited at 40 Oe was characterized by almost smooth surface, but the blocking temperature of this sample still remained below the room temperature. As for the structure deposited at 1000 Oe the surface was also almost smooth but the blocking point in this case was above the room temperature. From this we conclude that roughness is not the most important factor affecting the blocking temperature. In our case the blocking temperature variation is probably caused by variation of the grain size distribution in AF layer which may be somewhat different for different deposition field.

The comparison of the temperature dependences of uniaxial anisotropies of Co in F/AF system and of free Co structure shows that the interaction between the F and AF layers still remains even at temperatures above the blocking point. A persistent exchange interactions above the blocking temperature was also observed in comparison of coercivity of Co layer in contact with IrMn and of free Co layer [11].

The comparison of the field dependencies of the uniaxial anisotropies for the F/AF and free Co layer structures proves that the interaction between the F and AF layers becomes stronger for stronger deposition field. In the cases of layer deposition in magnetic field of 40 and 420 Oe the interaction between F and AF layers was too weak to withstand the thermal oscillation energy.

The fact that the intrinsic resonance field of the free Co layer is close to that of the bilayer structure deposited in absence of magnetic field proves that there is no exchange interaction between F and AF layers in this case. The magnetic properties of the structures deposited at 40 Oe are close to those of the structures deposited without magnetic field. With the increase of the deposition field the difference in H_K and H_{r0} also increases. These observations evidence that the increase of the deposition field leads to the enhancement of the exchange interaction between F and AF layers at the interface.

Acknowledgments

This work is supported by Federal Agency of Science and Innovations (contract 02.740.11.0242, contract 02.740.11.0389). The FMR study was performed at User Facilities Center of M.V. Lomonosov Moscow State University.

References

[1] Nogués J, Schuller I K 1999 *J. Mag. Mag. Mat.* **192** 203
[2] O'Grady K, Fernandez-Outon L E and Vallejo-Fernandez G 2010 *J. Mag. Mag. Mat.* **322** 8
[3] Chechenin N G 2004 *Phys. of the Sol. State* **46** 466
[4] Chechenin N G, Khomenko E V, Vainchtein D I, and De Hosson J Th M 2008 *J. Appl. Phys.* **103** 07E738
[5] Scott J C 1985 *J. Appl. Phys.* **57** 3681
[6] Layadi A A, Cain W C, Lee J - W, Artman J O 1987 *IEEE Trans. Magn.* MAG-**23** 2993
[7] Khomenko E V, Chechenin N G, Goikhman A Yu, Zenkevich A V 2008 *JETP Lett.* **88** 693
[3] Kittel Ch 1948 *Phys. Rev.* **73** 155
[9] McMichael R D, Stiles M D, Cheng P J, Egelhoff W F 1998 *J. Phys. Rev.* B **58** 8605
[10] Ambrose T, Chien C L 1996 *J. Appl. Phys.* **79** 5920
[11] Ali M, Marrows C H, Al-Jawad M, Hickey B J, Misra A, Nowak U and Usadel K D 2003 *Phys. Rev.* B **68** 214420

Chiral Skyrmionic matter in non-centrosymmetric magnets

Ulrich K. Rößler, Andrei A. Leonov, Alexei N. Bogdanov
IFW Dresden, Postfach 270116, D-01171 Dresden, Germany
E-mail: u.roessler@ifw-dresden.de

Abstract. Axisymmetric magnetic strings with a fixed sense of rotation and nanometer sizes (chiral magnetic *vortices* or *Skyrmions*) have been predicted to exist in a large group of non-centrosymmetric crystals more than two decades ago. Recently these extraordinary magnetic states have been directly observed in thin layers of cubic helimagnet (Fe,Co)Si. In this report we apply our earlier theoretical findings to review main properties of chiral Skyrmions, to elucidate their physical nature, and to analyse these recent experimental results on magnetic-field-driven evolution of Skyrmions and helicoids in chiral helimagnets.

1. Introduction

In non-centrosymmetric magnetic systems, antisymmetric *Dzyaloshinskii-Moriya* (DM) exchange causes particular magnetic couplings and inhomogeneous states [1]. These *chiral* interactions stabilize two- and three- dimensional localized structures [2, 3, 4]. Phenomenologically the inhomogeneous Dzyaloshinskii energy contributions [1, 5] are described by Lifshitz invariants, antisymmetric differential forms linear in first spatial derivatives of the magnetization $\mathbf{M}$

$$\Lambda_{ij}^{(k)} = M_i \frac{\partial M_j}{\partial x_k} - M_j \frac{\partial M_i}{\partial x_k} \tag{1}$$

where x_k are Cartesian components of the spatial variable $\mathbf{r}$. Solutions for chiral magnetic *Skyrmions* as static states localized in two dimensions have been derived first in 1989 [2]. It was shown that these topological solitonic field configurations exist in magnetic systems for all non-centrosymmetric crystallographic classes that allow Lifshitz invariants in their magnetic free energy [2]. The unique character of these textures stems from the general instability of multidimensional solitonic states in field theories [3, 6]. Nonlinear continuum models for condensed matter systems do not contain solutions for static and smooth multidimensional localized states. Such states appear only as dynamic excitations, while static configurations are generally unstable and collapse spontaneously into topological singularities.

For a long time investigations of chiral Skyrmions have been restricted to theoretical studies [3, 7, 8, 9, 10]. The solutions for two-dimensional localized and bound states (isolated Skyrmions and Skyrmion lattices) have been studied in non-centrosymmetric ferromagnets [7, 8] and antiferromagnets [5], in cubic helimagnets [3, 8, 9], and in confined centrosymmetric magnetic systems with surface/interface-induced chiral interactions (e.g. nanolayers of magnetic metals) [10]. In [3], we formulated the idea of Skyrmionic matter in non-centrosymmetric crystals predicting the existence of mesophases, composed of Skyrmions as 'molecular units', similar to vortex matter in type-II superconductors [11]. Various effects observed in MnSi and other cubic helimagnets with B20 structure [12] indicate multidimensionally modulated magnetic states conforming with our theoretical predictions of Skyrmions and their properties [3, 7]. The theoretical ideas of Skyrmions in chiral magnets have triggered various experimental efforts to find evidence for these twisted textures [13, 14]. These experiments collected an impressive range of data that suggest complex magnetic order phenomena. But, mainly using diffraction or indirect evidence by transport measurements, these experimental results remained essentially inconclusive and have been contested, see, e.g., [15]. Moreover, the interpretation of the experimental data has been based on approximate solutions [14, 16] to the Dzyaloshinskii model by using variational approaches in terms of a *mode instability*. The corresponding results do not describe the properties of Skyrmions and the phase transition behavior of chiral magnets, which is governed by the *nucleation* of a localized mesoscale entity [1, 7, 17]. Our theoretical developments [8, 9] show that multiply modulated chiral states in non-centrosymmetric magnets are composed of localized solitonic states with particle-like behavior. The thermodynamic stability of condensed Skyrmionic phases has been shown [3, 8, 9] for the standard and modified Dzyaloshinskii models devised for cubic helimagnets [1, 18]. Meanwhile, experimental efforts culminated in the direct microscopic observations of chiral Skyrmions in thin layers of (Co,Fe)Si [19]. This break-through is the first clear experimental evidence for the existence of Skyrmions as *axisymmetric chiral localized states* that are stabilized by a complex interplay of nonlinear and chiral effects, as predicted earlier [2, 3].

Here, we address the problem of multidimensional solitonic states in nonlinear systems lacking inversion symmetry. This emerging field of nonlinear physics is based on solutions of nonlinear partial differential equations [3, 7, 9], and intricate mathematical methods of micromagnetics [20], and physics of solitons [21]. Concentrating on the physical side of the problem rather than on mathematical details we give an elementary introduction into the properties of chiral *Skyrmions* in magnetism.

2. Dzyaloshinskii theory for cubic helimagnets

2.1. Phenomenological energy and equations

Within the phenomenological theory introduced by Dzyaloshinskii [1] the magnetic energy density of a cubic non-centrosymmetric ferromagnet with the magnetization **M** can be written as [1, 18]

$$w = \underbrace{A\,(\mathrm{grad}\mathbf{M})^2 - D\,\mathbf{M}\cdot\mathrm{rot}\mathbf{M} - \mathbf{M}\cdot\mathbf{H}}_{w_0} - \sum_{i=1}^{3}\left[B\,(\partial M_i/\partial x_i)^2 + K_c M_i^4\right] - f(M) \qquad (2)$$

including the exchange stiffness with constant A. The inhomogeneous DM exchange with constant D is a combination of Lifshitz invariants (1), $w_D = D\,(\Lambda_{yx}^{(z)} + \Lambda_{xz}^{(y)} + \Lambda_{zy}^{(x)}) = D\,\mathbf{M}\cdot\mathrm{rot}\mathbf{M}$. These isotropic interactions together with the Zeeman energy are the *essential* couplings to stabilize chiral modulations (w_0). The sum in Eq. (2) includes exchange (K) and cubic (K_c) magnetic anisotropy contributions; $f(M)$ comprises magnetic interactions imposed by variation of the magnetization modulus $M \equiv |\mathbf{M}|$. In a broad temperature range the magnetization vector practically does not change its length, and nonuniform magnetic states display only rotation of **M**. The low temperature properties, thus, can be described by the model (2) with a fixed magnetization modulus $M = \mathrm{const}$.

2.2. One-dimensional chiral modulations: cones and helicoids

In non-centrosymmetric magnets, chiral couplings of type (1) favour rotation of the magnetization vector. They destabilize the homogeneous magnetic structure and induce long-

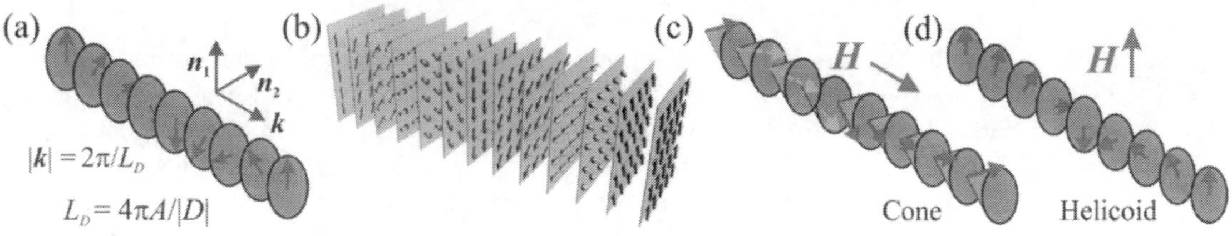

Figure 1. One-dimensional chiral modulations in cubic helimagnets. In a helical "array" (a) the magnetization rotates in the plane spanned by the orthogonal unity vectors $\mathbf{n}_1$ and $\mathbf{n}_2$ and the rotation sense is determined by the sign of Dzyaloshinskii constant D. Chiral *helices* (b) are composed of arrays (a) with the same propagation direction. Under the influence of the applied field the helix (b) transforms into longitudinal distorted *cones* (c) or into transversally distorted *helicoids*(d).

range modulations of the magnetization [1]. At zero field helical modulations

$$\mathbf{M} = M\left[\mathbf{n}_1 \cos\left(\mathbf{k}\cdot\mathbf{r}\right) + \mathbf{n}_2 \sin\left(\mathbf{k}\cdot\mathbf{r}\right)\right], \quad |\mathbf{k}| = 2\pi/L_D, \quad L_D = 4\pi A/|D| \tag{3}$$

with period L_D and wave vector $\mathbf{k}$ correspond to the absolute minimum of the isotropic functional w_0 (Fig. 1 a, b). In Eq. (3) $\mathbf{n}_1$, $\mathbf{n}_2$ are orthogonal unit vectors in the plane of the magnetization rotation (Fig. 1). The modulations (3) have a fixed rotation sense determined by the sign of constant D and are continuously degenerate with respect to propagation directions in the space. An applied magnetic field lifts the degeneracy of the helices (3) and stabilizes a *cone* solution with the propagation direction along the magnetic field

$$\cos\theta = |\mathbf{H}|/H_D, \quad \psi = 2\pi z/L_D, \quad H_D = D^2 M/(2A), \tag{4}$$

where $\mathbf{M} = M(\sin\theta\cos\psi; \sin\theta\sin\psi; \cos\theta)$ is written in spherical coordinates. In such a helix the magnetization component along the applied field has a fixed value $M_\perp = M\cos\theta = M(H/H_D)$, and the magnetization vector $\mathbf{M}$ rotates within a cone surface. The critical value H_D marks the saturation field of the cone phase. The helix periods (L_D) and critical fields (H_D) for some non-centrosymmetric cubic ferromagnets (*helimagnets*) are presented in Table 1.

Table 1. Néel temperatures (T_N), helix periods (L_D), and saturation fields (H_D) for some cubic helimagnets, data from Refs. [12].

Compound	MnSi	FeGe	$Fe_{0.3}Co_{0.7}Si$	$Fe_{0.5}Co_{0.5}Si$	$Fe_{0.8}Co_{0.2}Si$
T_N [K]	29.5	278.7	8.8	43.5	32.2
L_D [nm]	18.0	68.3- 70.0	230	90.0	29.5
H_D [T]	0.62	0.2	$(6.0 \pm 1.5)\cdot 10^{-3}$	$(4.0 \pm 0.5)\cdot 10^{-2}$	0.18

In real non-centrosymmetric magnets anisotropic forces fix the propagation of chiral modulations along certain *easy axes* directions imposed by the crystal symmetry. A magnetic field perpendicular to the propagation direction violates the uniform rotation of the magnetization ($\psi \propto z$, Eq. (3)). In a sufficiently high magnetic field $H = H_H$ distorted *helicoids* infinitely expand their period into a system of isolated 2π domain walls separating domains with the magnetization along the applied field (Figs. 3 (d), 5 (a)) [1, 7].

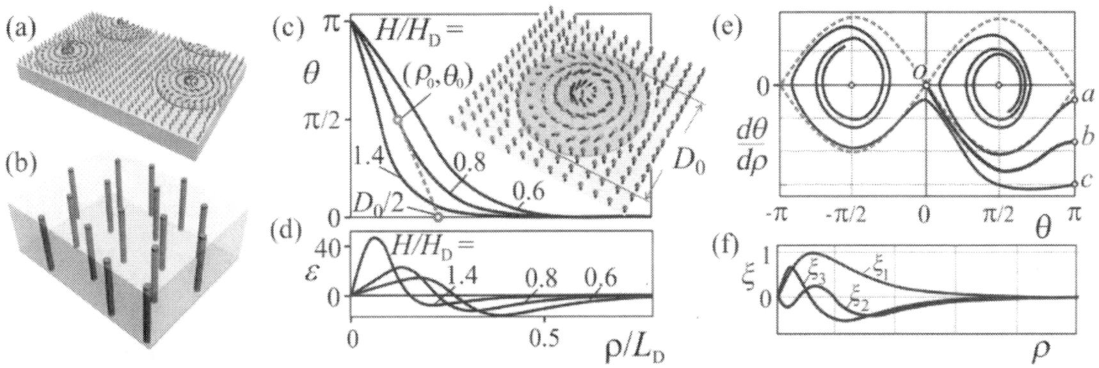

Figure 2. Isolated Skyrmions in a nanolayer (a) and a bulk sample (b) of a cubic helimagnet. The magnetization profiles $\theta(\rho)$ (c) and corresponding energy densities $\varepsilon(\rho)$ (d) for typical solutions of Eq. (5). A cross-section through an isolated Skyrmion shows axisymmetric distribution of the magnetization (shaded area indicates the core with the diameter D_0 ((a) and Inset in (c)). Phase trajectories of Skyrmions correspond to separatrix curves in the phase space (curve $b-o$)(e). Typical eigen functions for radial excitations (f).

3. Chiral flux-lines: the building blocks for Skyrmionic textures

3.1. Chiral Skyrmions in cubic helimagnets

The equations minimizing energy (2) include axisymmetric localized solutions $\psi = \varphi + \pi/2$, $\theta = \theta(\rho)$, $M =$ const (Fig. 2, (a), (b)) where the spatial variable $\boldsymbol{r} = (\rho\cos\varphi; \rho\sin\varphi; z)$ is written in cylindrical coordinates and the magnetization in spherical coordinates (4). The Euler equation for $\theta(\rho)$ to determine the equilibrium structure of isolated Skyrmions [2, 7] is

$$A\left(\frac{d^2\theta}{d\rho^2} + \frac{1}{\rho}\frac{d\theta}{d\rho} - \frac{1}{\rho^2}\sin\theta\cos\theta\right) - \frac{H}{2M}\sin\theta - \frac{D}{\rho}\sin^2\theta = 0 \tag{5}$$

with boundary conditions $\theta(0) = \pi$, $\theta(\infty) = 0$. In *non-centrosymmetric uniaxial magnets* functionals w_D have anisotropic forms depending on crystallographic class and may include several terms [2]. The solutions $\psi(\varphi)$ have different form depending on the crystal symmetry, however, the Euler equation for $\theta(\rho)$ has the same funcional form for all non-centrosymmetric uniaxial ferromagnets [2].

3.2. Solutions for Skyrmions

Typical solutions of Eq. (5), $\theta(\rho)$, consist of arrow-like cores ($\pi - \theta \propto \rho$ for $\rho \leq L_D$) and exponential "tails", $\theta \propto \exp(-\rho)$ for $\rho \gg L_D$ [2, 7] (Fig. 2). In *phase space* $(\theta, d\theta/d\rho)$ these localized solutions correspond to *separatrix* trajectories (e.g. curve $b-o$ in Fig. 2(e)). A Skyrmion core diameter D_0 can be defined as two times the value of R_0, which is the coordinate of the point where the tangent at the inflection point (ρ_0, θ_0) intersects the ρ-axis (Fig. 2(d)), in analogy to definitions for domain wall width [20]. Skyrmion energy per unit length and diameter are

$$E = 2\pi\int_0^\infty \varepsilon(\theta, \rho)d\rho, \qquad D_0 = 2L_0[\rho_0 - \theta_0(d\theta/d\rho)^{-1}_{\rho=\rho_0}] \tag{6}$$

where $\varepsilon(\theta, \rho)$ is the "linear" energy density [7]. The energy density distributions $\varepsilon(\rho)$ (Fig. 2(f)) reveal two distinct regions: positive energy "bags" are concentrated in the Skyrmion center and are surrounded by extended areas with negative energy density, where the DM exchange

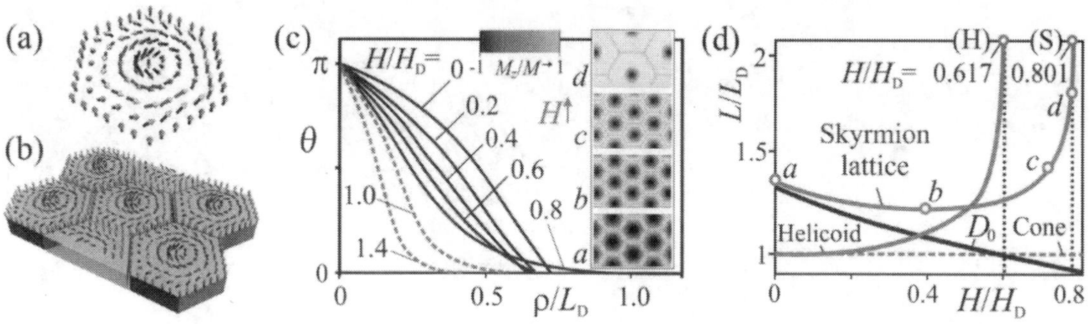

Figure 3. Skyrmion lattices: (a) unit cells have axisymmetric magnetization structure near the center; (b) a thin layer with a hexagonal lattice. The magnetization profiles $\theta(\rho/L_D)$ of a Skyrmion cell (solid blue lines) and of an isolated Skyrmion (dashed red lines) for different values of applied magnetic field (c). Equilibrium sizes of the cell core (D_0) and lattice period compared to helicoid periods (d). Inset shows the transformation of the hexagonal lattice into a set of isolated Skyrmions in increasing magnetic field. Snapshots correspond to the field values indicated in the diagram for period (L/L_D) vs. applied field (H/H_D) by hollow circles (d).

dominates. The radial stability of Skyrmion solutions has been proved by solving the corresponding spectral problem [7], which also yields excitation modes of Skyrmion cores, as in Fig. 2(f),

4. Skyrmion lattices: where does "double-twist" become beneficial?

4.1. Solutions for bound Skyrmions

The equilibrium parameters of a Skyrmion lattice are derived from a system of differential equations for $\theta(x, y)$, $\psi(x, y)$ minimizing the system energy. For Skyrmions at low temperatures the *circular cell* approximation provides a very elegant and reliable approximation [7]. In this method the lattice cell is replaced by a circle of equal area. The equilibrium parameters of Skyrmion lattices are derived by integration of Eq. (5) with boundary conditions $\theta(0) = \pi$, $\theta(R) = 0$ and a subsequent minimization of the lattice energy density $W_S = (2/R^2) \int_0^R w(\rho)\rho d\rho$ with respect to the cell radius R [7]. The solutions for magnetization profiles $\theta(\rho)$ and the lattice periods L are plotted in Fig. 3. The reduced perpendicular magnetization of a Skyrmion lattice averaged over the Skyrmion cell $m_S = (2/R^2) \int_0^R \cos(\theta)\rho d\rho$ is plotted together with the magnetization of the helicoid, m_H (Fig. 5). Contrary to the helicoid the Skyrmion lattice has a finite magnetization at zero field, $m_S(0) = 0.124$.

Table 2. Critical fields and characteristic parameters of the hexagonal Skyrmion lattice: H_1 transition field between the helicoid and Skyrmion lattice; H_S saturation field of the Skyrmion lattice; last column for isolated Skyrmions as excitations of the saturated state.

Reduced magnetic field, H/H_D	0	H_1 0.216	H_S 0.801	1.4
Lattice cell period, L/L_D	1.376	1.270	∞	-
Core diameter, D_0/L_D	1.362	1.226	0.920	0.461
Averaged magnetization, m_S	0.124	0.278	1	1

4.2. Skyrmions compete with helicoids. Transition field H_1

In the Skyrmion lattices rotation of the magnetization in two directions leads to a larger reduction in the Dzyaloshinskii-Moriya energy than single-direction rotation in helical phases. On the other hand, such *double-twist* modulations increase the exchange energy. The equilibrium energy of the Skyrmion cell at zero field $\widetilde{w}_S(\zeta) = (2/\zeta^2)\int_0^\zeta w(\rho)\rho d\rho$ plotted as a function of the distance from the center ζ (Fig. 4 (a)) shows that an energy excess near the border outweighs the energy gain at the Skyrmion center (for details see [3]). At higher magnetic fields, however, the Skyrmion lattice has lower energy than the helicoid. The first order transition between these two modulated states occurs at $H_1 = 0.2168 H_D$ [7].

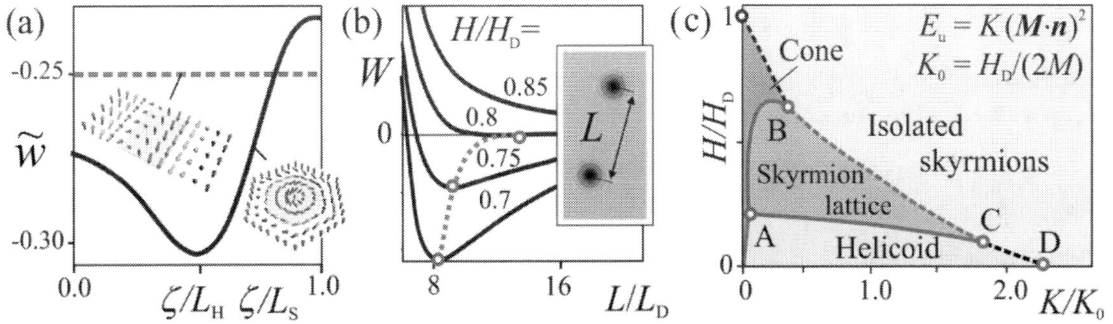

Figure 4. Local energies $\widetilde{w}(\zeta)$ of the Skyrmion lattice and helicoid at zero field (a). Below the critical field H_S the negative internal energies of Skyrmions overcome the repulsive Skyrmion-Skyrmion interaction and a dense packed (hexagonal) lattice is created (b). Under the influence of induced uniaxial anisotropies (E_u) the Skyrmion lattice becomes globally stable in a broad range of the applied fields, adopted from [8], K_0 is the critical value for uniaxial distortions suppressing the cones in zero field (c).

4.3. "Receding" Skyrmions. Critical field H_S

Properties of the Skyrmion lattice solutions are collected in Figs. 3, 5 and in Table 2. With increasing magnetic field, a gradual localization of the Skyrmion core D_0 is accompanied by the expansion of the lattice period. The lattice transforms into the homogeneous state by infinite expansion of the period at the critical field $H_S = 0.80132 H_D$. Remarkably, the Skyrmion core retains a finite size, $D_0(H_S) = 0.920 L_D$ and the lattice releases a set of *repulsive* isolated Skyrmions at the transition field H_S, owing to their topological stability. These free Skyrmions can exist far above H_S. On decreasing the field again below H_S, they can re-condense into a Skyrmion lattice (Fig. 4 (b)). A similar type of sublimation and resublimation of particle-like textures occurs in helicoids at the critical field $H = H_H/H_D = \pi^2/8 = 0.6168$: the period infinitely expands and the helicoid splits into a set of isolated 2π domain walls or kinks [1, 7].

This peculiar transformation of chiral modulations into homogeneous states constitutes a *nucleation* type of continuous phase transition, according to a classification introduced by De Gennes [17]. In contrast to the common *instability* type of continuous (2nd order) transitions that is described by the amplitude of a sole fundamental mode acting as the order parameter, the nucleation transitions retain a full spectrum of lattice modes at the transition. Magnetic-field-driven transitions of multidomain states into the homogeneous phase also belong to the nucleation type of phase transformations [20].

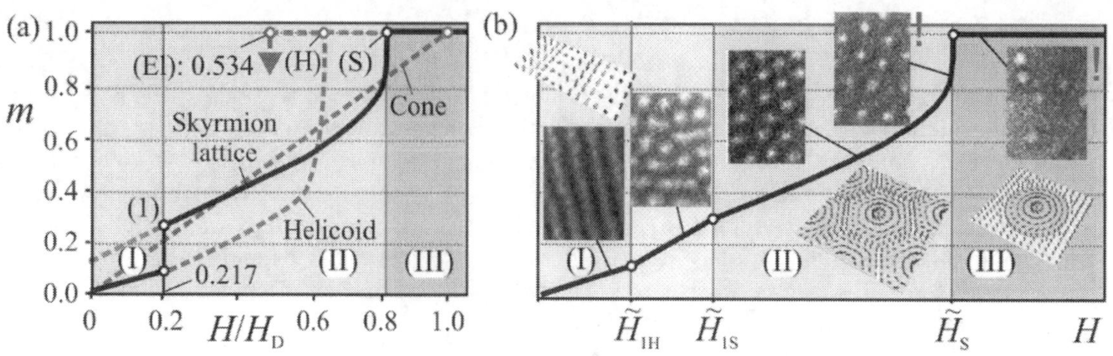

Figure 5. The ideal magnetization curves for a bulk sample (based on results of [7]) (a) and for a thin layer (b) of a cubic helimagnet with suppressed cone phases. Solid lines indicate the thermodynamically stable states; dashed lines in Fig. (a), metastable configurations. The 1st order transition line at H_1 is expanded into a region of multidomain states in the thin layer. Fragments of experimentally observed images [19] demonstrate excellent agreements with theoretically calculated magnetization curves. The patterns marked by "(!)" display *isolated chiral Skyrmions.*

4.4. Stabilization effects of magnetic anisotropy

For isotropic model w_0 (2) the cone phase (4) is the global minimum in the whole range of the applied fields where the modulated states exist ($0 < H < H_D$). The helicoids and Skyrmion lattices can exist only as metastable states. Under the influence of anisotropies (2), however, the Skyrmion lattices may gain thermodynamic stability in a certain interval of applied fields, for details see [8, 9]. E.g., uniaxial anisotropies can be induced in bulk helimagnets by uniaxial stresses or they arise due to surface effects in thin layers. These anisotropies are described by $E_u = K(\mathbf{M} \cdot \mathbf{n})^2$, where K is a constant of uniaxial anisotropy, $\mathbf{n}$ is a unity vector along the distortion axis. Such uniaxial distortions suppress the cone phase in cubic helimagnets and establish global stability of the Skyrmion lattice in broad ranges of the thermodynamic parameters (Fig. 4 (c)) [8].

5. On the observation of Skyrmionic and helical textures in $Fe_{0.5}Co_{0.5}Si$ nanolayers

Real-space images of Skyrmion states in a thin layer of cubic helimagnet $Fe_{0.5}Co_{0.5}Si$ have recently been obtained by using Lorentz transmission electron microscopy [19]. This is the first clear experimental manifestation of *chiral Skyrmion states.* The first-order transition of a helicoid into a Skyrmion lattice and its subsequent transformation into a system of isolated Skyrmions observed in bias magnetic fields (Figs. 1, d-f, 2, 3 (a-d) in [19]) are in excellent agreement with the theoretical predictions on the behavior of Skyrmions and the field-driven transitions into densely packed Skyrmion lattices according to the magnetic phase diagrams calculated earlier [3, 7] (Fig. 5).

In the experiments, the thickness 20 nm of the magnetic layer $Fe_{0.5}$ $Co_{0.5}Si$ is much smaller than the helix period $L_D = 90$ nm [19]. But, even for such a small thickness, the conical state propagating only for a fraction of a period perpendicularly through the layer has lower energy than a Skyrmion lattice, absent additional effects that stabilize it in applied fields. Usually in magnetic nanolayers strong perpendicular uniaxial anisotropy arises, either as a result of surface effects [22] or of lattice strains. Thus, induced anisotropies give a possible explanation for the experimental observation of the Skyrmions in these $Fe_{0.5}Co_{0.5}Si$ layers, in accordance with the phase diagram for cubic helimagnets with uniaxial distortions (Fig. 4 (c)) [8]. Fig. 5

presents the magnetization curve for a bulk isotropic helimagnet (a) (based on results of [7], Fig. 12) and the corresponding magnetization curve for a thin layer involving demagnetization effects [20] (b). Compared to theoretically calculated values in a bulk material (H_S, H_H) the corresponding critical fields in a thin layer are shifted, and their values can be estimated as $\widetilde{H}_{S(H)} = H_{S(H)} + 4\pi M$. Due to demagnetization effects multidomain states can be stabilized in the vicinity of the transition field H_1 [7]. The boundaries of these regions with coexisting phases can be estimated as $\widetilde{H}_{1H} = H_1 + 4\pi M m_H(H_1)$, $\widetilde{H}_{1S} = H_1 + 4\pi M m_H(H_1)$. The magnetizations of the competing phases at the transition field equal $m_H(H_1) = 0.111$ and $m_S(H_1) = 0.278$. The jump of the magnetization during the transition equals $\Delta M = [m_S(H_1) - m_H(H_1)]M = 0.167M$, i.e., it reaches about 17 % of the saturation value.

The magnetization curves in Fig. 5 are constructed for ideally soft magnetic material under the condition that only the equilibrium states are realized in the magnetic sample. In real materials the formation of the equilibrium states is often hindered (especially during the phase transitions), and evolution of metastable states and hysteresis effects play an important role in the magnetization processes. Particularly, the formation of the Skyrmion lattice below H_S can be suppressed. Then isolated Skyrmions exist below this critical field. At a critical field H_{El} the Skyrmions become unstable with respect to elliptical deformations and "strip-out" into isolated 2π domain walls. In a bulk material $H_{El} = 0.534\, H_D$ (indicated in Fig. 5 with a red arrow). In a thin layer, one estimates $\widetilde{H}_{El} = H_{El} + 4\pi M$. As discussed earlier [2, 3, 7] the evolution of chiral Skyrmions in magnetic fields has many features in common with that of bubble domains in perpendicular magnetized films [20], and with Abrikosov vortices in superconductors [11].

The fragments of images from Ref. [19] (Fig.5 (b)) reflect in details the theoretically predicted evolution of the chiral modulations in the applied magnetic field: the helicoid phase is realized at low fields (region (I)); at higher field this transforms into the Skyrmion lattice (region (II)) via an intermediate state ($\widetilde{H}_{1H} < H < \widetilde{H}_{1S}$); finally the Skyrmion lattice by extension of the period transforms into the homogeneous phase where isolated Skyrmions still exist as topologically stable 2D solitons.

Two patterns indicated in Fig. 5 (b) with exclamation mark manifest the main result of Ref. [19]: the first images of *static two-dimensional localized states* aka *chiral Skyrmions*! In Ref. [19] this result has been overlooked and misinterpreted as a coexisting ferromagnetic and Skyrmion lattice phases. As it was expounded in the previous section, the transition of the Skyrmion lattice into the homogeneous state is a *continuous* transition, but of the particular *nucleation* type. Such transitions exclude the formation of coexisting states.

The condensed Skyrmion phases in the micrograph of Ref. [19] also appear as heavily distorted densely packed two-dimensional lattice configurations. This is expected for Skyrmionic matter. As these mesophases are composed from elastically coupled radial strings, dense Skyrmion configurations generally do not form ideal crystalline lattices but various kinds of partially ordered states, e.g. hexatic ordering implying only orientational order of bonds without positional long-range order, or other glassy arrangements following standard arguments put forth for the similar vortex matter in type-II superconductors [11]. The observation derives from the particle-like (or string-like) nature of Skyrmions and suggests that Skyrmionic mesophases may display rich phase diagrams.

6. Chiral modulations near the ordering temperature

6.1. Solutions for high-temperature Skyrmions. Confinement temperature T_p

Near the ordering temperature the magnetization amplitude varies under the influence of the applied field and temperature. Basic properties of chiral modulations in this region can be derived by minimization of the isotropic model $w = w_0 - f(M)$ derived from Eq. (2) with respect to *all* components of the magnetization vector (M, θ, ψ). Within this model axisymmetric isolated structures are described by equations $\psi = \varphi + \pi/2$ and the solutions of the Euler

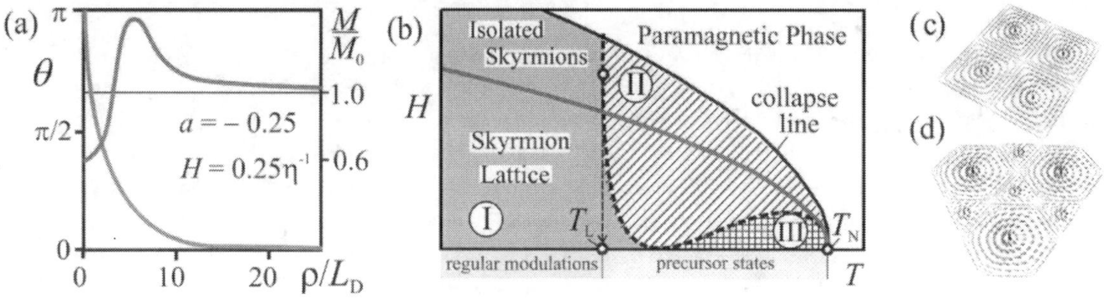

Figure 6. Typical solutions for magnetization profiles $\theta(\rho)$, $M(\rho)$ (Eqs. 8) (a). Magnetic phase diagram of an *isotropic* helimagnet near the ordering temperature includes areas with *repulsive* (I), *attractive* (II) Skyrmions, and *confined* Skyrmion states (III) (b). *Confinement temperature* T_L separates the main part of phase diagram with regular chiral modulations ($T < T_L$) from the region of "precursor state" ($T_L < T < T_N$). The confined Skyrmion textures include square (*half*-Skyrmion) (c) and hexagonal (d) lattices.

equations for $\theta(\rho)$, $M(\rho)$ [3, 9]

$$A\left(\frac{d^2\theta}{d\rho^2}+\frac{1}{\rho}\frac{d\theta}{d\rho}-\frac{\sin\theta\cos\theta}{\rho^2}\right)-D\frac{2}{\rho}\sin^2\theta-H\sin\theta+A\frac{2}{M}\frac{dM}{d\rho}\left(\frac{d\theta}{d\rho}-1\right)=0, \quad (7)$$

$$A\left[\frac{1}{M}\left(\frac{d^2M}{d\rho^2}+\frac{1}{\rho}\frac{dM}{d\rho}\right)+\left(\frac{d\theta}{d\rho}\right)^2+\frac{\sin^2\theta}{\rho^2}\right]+D\left[\frac{d\theta}{d\rho}+\frac{\sin\theta\cos\theta}{\rho}\right]-\frac{H}{M}\cos\theta+\frac{1}{M}\frac{\partial f}{\partial M}=0.$$

With $f(M) = a_1M^2 + a_2M^4$ ($a_1 = J(T_c - T)$, $a_2 > 0$) which is commonly used to describe magnetization processes in cubic helimagnets at high temperatures the solutions of Eqs. (8) have been investigated in [3, 9]. With boundary conditions $\theta(0) = \pi$, $\theta(\infty) = 0$, $M(\infty) = M_0$, Eqs. (8) describe the structure of isolated Skyrmions. The magnetization of the homogeneous phase M_0 is derived from equation $2a_1M + 4a_2M^3 - H = 0$. It is convenient to introduce new parameters, an effective temperature $a = (T_c - T)/T_D$ and $H_0 = (D^2/2A)^{3/2}/\sqrt{a_2}$ where "chiral shift", $T_D = D^2/(8AJ)$, is a characteristic temperature equal to the difference between the ordering temperature of the helimagnet ($T_N = T_c + T_D$) and the ferromagnet with $D = 0$ (T_c).

Typical solutions $\theta(\rho)$, $M(\rho)$ of Eqs. (8) demonstrate an inhomogeneity of M associated with the Skyrmion cores, Fig. 6(a). Detailed analysis of the solutions for isolated Skyrmions has revealed a number of remarkable phenomena imposed by the interplay between the angular (θ) and modulus (M) order parameters [9].

(i) *Collapse of Skyrmions at high fields.* The solutions of Eqs. (8) exist only below a critical *collapse* line (Fig. 6). As the applied field approaches this line the magnetization in the Skyrmion center $M(0)$ gradually shrinks, and as $M(0)$ becomes zero the Skyrmion collapses. This is in contrast with low-temperature Skyrmions, which exist without collapse even at very large magnetic fields [7] because the stiffness of the magnetization modulus maintains topological stability of Skyrmions. At high temperatures the softening of the magnetization allows the order parameter to pass through zero and to unwind the Skyrmion core.

(ii) *Crossover of Skyrmion-Skyrmion interactions.* The coupling of Skyrmions is repulsive in a broad temperature range, but it starts to oscillate at high temperatures, Fig. 6 (b). The equation to describe this crossover in the inter-Skyrmion coupling is $\eta H^* = \sqrt{2 \pm \sqrt{3+4a}}\,(a + 1 \pm \sqrt{3+4a}/2)$. For $a < -0.5$ this equation yields a crossover line in the phase diagram, Fig. 6(b).

This is ended in turning point (-0.5, 0). Another turning point (-0.75, $\sqrt{2}\eta/4$) indicates the lowest temperature, i.e., a *confinement temperature* $T_L = T_N - 4T_D$, where Skyrmions begin to attract each other at certain distances, owing to the oscillatory character of their interactions. This results in energetic confinement of Skyrmions, as they can form clusters to lower their energy.

(iii) *Confinement.* For $-0.5 < a < 0.25$ the crossover line delimits a small pocket (III) in the vicinity of the ordering temperature. Within this region Skyrmions can exist only as bound states. In the *confinement* region Skyrmionic states drastically differ from those in the main part of the phase diagram. Confined Skyrmion textures (hexagonal and square half-Skyrmion lattices, Fig. 6 (c)) arise from the disordered state through a rare case of an *instability-type nucleation transition*, and their field-driven transformation evolves by distortions of the modulus profiles $M(\rho)$ while the equilibrium periods of the lattices do not change strongly with increasing applied field (for details see [9]).

6.2. Confinement phenomenon and precursor states

The magnetic phase diagram of isotropic helimagnet (Fig. 6 (b)) includes two distinct temperature intervals: (i) in the main part ($0 < T < T_L$) the rotation of the local magnetization vector determines the chiral modulation, while the magnetization amplitude remains constant. (ii) At high temperatures ($T_L < T < T_N$) spatial variation of the magnetization modulus becomes a sizeable effect, and strong interplay between M and angular variables is the main factor in the formation and peculiar behaviour of chiral modulations in this region. The *confinement temperature* T_L [9] provides the scale that delineates the border between these two regimes in the phase diagram. The characteristic temperature T_L is of fundamental importance for chiral magnets. It is of the same order of magnitude as the temperature interval $(T_N - T_c) \propto D^2/A$, where chiral couplings cause inhomogeneous precursor states around the magnetic order temperature. Due to the relativistic origin and corresponding weakness of the DM exchange the shift $\Delta T = T_N - T_L$ is small (for MnSi this is estimated to be about 2 K [9]).

Since discovery of chiral modulations in MnSi family of B20 compounds numerous magnetic anomalies have been observed near the ordering temperature of these helimagnets [12]. However, the nature of the chiral spin textures in this region of the phase diagram is largely unresolved in experiment. During last years these *precursor* effects have become a subject of intensive investigations [13, 14] motivated by the expectation to identify Skyrmionic states [3].

The rigorous solutions for helical and Skyrmionic modulations recently derived within the basic model $w = w_0 + f(M)$ (2) and the theoretical description of novel phenomena attributed to this region [9] give a first explanation of these anomalies. The stabilization of Skyrmionic textures and other chiral modulations near magnetic ordering involve the confinement of localized state, but also a strong influence of minor energy contributions. E.g., the phase diagram of cubic helimagnets contains defected half-Skyrmion lattices at small applied fields and densely packed full-Skyrmion lattices. However, the conformation of these mesophases and their relative stability will depend on various small additional effects, such as dipolar couplings, thermal fluctuations, quenched defects etc. It would be naive to expect these phase diagrams to be simple and to be determined by one dominating mechanism able to stabilize Skyrmions over helicoids. Owing to the hierarchy of interactions, the properties and stability of the Skyrmion cores will always be provided by direct exchange and Dzyaloshinskii-Moriya couplings via the double-twist mechanism. But the mesophase formation will be ruled by much weaker couplings, owing to the localized and frustrated nonlinear character of these solitonic entities.

7. Topogical solitons, vortices, Skyrmions...

7.1. How the chiral Skyrmion got its name

Localized solutions of Eq. (5) have been initially introduced under the name *magnetic vortices* [2] because, to a certain extent, they are similar to 2D topogical defects investigated in magnetism and known as "two-dimensional topological solitons" or "vortices" (e.g. well-known Belavin-Polyakov solutions for magnetic vortices [23]). On the other hand, the term *Skyrmion* has been conceived in a field rather distant from condensed-matter physics and initially was related to the localized solutions derived by Skyrme within his model for low-energy dynamics of mesons and baryons [24]. In fact, the Skyrme model [24] comprises three spatial dimensions, and the name "baby Skyrmion" was used by some field theorists to distinguish two-dimensional localized states from "mature" three-dimensional solutions in the original Skyrme model [24], both types of them being topological static solitons. During the last decades the "Skyrmion" has progressively won currency in general physics to designate *any* non-singular localized and topologically stable field configuration. Complying with this trend, in 2002 we have renamed "chiral magnetic vortices" into "chiral Skyrmions" [5]. Ironically, the fate of the localized states near ordering transition, as they can decay by longitudinal magnetization processes, betrays that these "chiral Skyrmions" are not topologically stable, at high enough temperature.

7.2. 'What exactly is a chiral Skyrmion'?

Thus, the term *Skyrmion* is an umbrella title for smooth localized structures to distinguish them from singular localized states, e.g., disclinations in liquid crystal textures [25]. This convention provides only a formal *label* for a large variety of very different solitonic states from many fields of physics [26]. With respect to the subject of this paper a "Skyrmion" designates well-defined solutions of Eq. (5) which are **(i) localized**, **(ii) axisymmetric**, and have **(iii) fixed rotation sense**. Examples of chiral Skyrmions have been presented in Fig. 2. The axisymmetric structure of the Skyrmion core and its localized character are retained in bound states as Skyrmion lattices [3, 4]. This reflects the particle-like character of chiral Skyrmions and the most general features of their energetics (Fig. 4 (a), and Refs. [3, 4]). Skyrmions as countable entities can be arranged in various ways to create dense magnetic textures. This is the essence of Skyrmionic matter and entails the possibility to form a variety of mesophases with crystalline, but also with liquid-like or glassy configurations in chiral magnetism.

Alternative approaches construct Skyrmionic textures from crossing plane waves (helices) as fundamental modes of so-called spin-spiral crystals [14, 16]. In [14] a Skyrmion lattice is composed of three helices superimposed under an angle of 120 degree. According to [14] such a "triple-Q anti-Skyrmion lattice" ansatz even reaches the global free energy minimum in the A phase of MnSi. This construction is inconsistent with the properties of Skyrmions. Chiral Skyrmionic textures incorporate isolated or embedded axisymmetric lines. The notions of a "spin-spiral crystal" [16] or "multi-Q Skyrmions" [14] are misconceptions because they blend mutually exclusive ideas of particle-like localized Skyrmions from one side and delocalized plane waves from the other. The radial structure of the chiral Skyrmion cannot be reduced to a superposition of harmonic helical waves. Calculations based on such multi-Q ansaetze predict incorrect phase diagrams. E.g., the hexagonal Skyrmion lattice is stable in remanent state at $H = 0$ [7], but in [14] an instability at a finite field is presented for the variational hexagonal spin-spiral solution. The idea of a spin-spiral crystal ruled by one harmonic mode, where contributions of higher-order modes are small in some sense [14, 16], also is at variance with the nucleation type of transition actually observed in the chiral Skyrmion textures. This idea wrongly places the Dzyaloshinskii theory in a class of models for modulated states with instability type of transitions.

8. Conclusions: What makes chiral Skyrmions interesting?

In non-centrosymmetric magnets chiral magnetic Skyrmions arise as a result of the specific stabilization mechanism imposed by the handedness of the underlying crystal structure [1, 2, 3]. In centrosymmetric magnets such solutions are radially unstable and collapse spontaneously under the influence of the applied magnetic field or intrinsic short-range interactions. In nonlinear field models Skyrmion states can be stabilized by higher order spatial derivatives (often this is referred to as *Skyrme mechanism*). In condensed-matter systems there are no physical interactions providing such energy contributions. *Chiral interactions* present a unique mechanism to stabilize Skyrmion states in ordered condensed-matter systems. This singles out chiral systems (including non-centrosymmetric magnets, multiferroics, liquid crystals, and metallic nanostructures with induced chiral interactions) into a particular class of materials with Skyrmionic states.

Acknowledgment. We thank S. Blügel, S.V. Grigoriev, G. Güntherodt, R. Schäfer, W. Selke, and H. Wilhelm, for discussions. Support by DFG project RO 2238/9-1 is gratefully acknowledged.

References

[1] Dzyaloshinskii I E 1964 *Sov. Phys. JETP* **19** 960
[2] Bogdanov A N D. and Yablonsky A 1989 *Sov. Phys. JETP* **68** 101; Bogdanov A N et al. 1989 *Sov. Phys. Solid State* **31** 1707
[3] Rößler U K *et al.* 2006 *Nature* **442** 797; Bogdanov A 1995 *JETP Lett.* **62** 247; Bogdanov A N *et al.* 2005 *Physica B* **359** 1162
[4] Rößler U K *et al.* 2010 *J. Phys.: Conf. Series* **200** 022029
[5] Bogdanov A N *et al.* 2002 *Phys. Rev. B* **66** 214410
[6] Derrick G H 1964 *J. Math. Phys.* **5** 1252
[7] Bogdanov A and Hubert A 1994 *J. Magn. Magn. Mater.* **138** 255; 1999 **195** 182; 1994 *phys. stat. sol. (b)* **186**
[8] Butenko A B *et al.* 2010 *Phys. Rev. B* **82** 052403
[9] Leonov A A *et al.* 2010 arXiv:1001.1992v2
[10] Bogdanov A N and Rößler U K 2001 *Phys. Rev. Lett.* **87** 037203
[11] Blatter G *et al.* 1994 *Rev. Mod. Phys.* **66** 1125
[12] Beille J *et al.* 1983 *Sol. State Comm.* v. 47, pp. 399-402; Sato T *et al.* 1987 *J. Magn. Magn. Mater.* **70** 411; Lebech B *et al.* 1989 *J. Phys.: Condens. Matter* **1** 6105; Gregory C I *et al.* 1992 *J. Magn. Magn. Mater.* **104-107** 689; Lebech B *et al.* 1995 *J. Magn. Magn. Mater.* **140** 119
[13] Lamago D *et al.* 2006 *Physica B* **385-386** 385; Pedrazzini P *et al.* 2007 *Phys. Rev. Lett.* **98** 047204; Pappas C *et al.* 2009 *Phys. Rev. Lett.* **102** 197202; Lee M *et al.* 2009 *Phys. Rev. Lett.* **102** 186601
[14] Mühlbauer S *et al.* 2009 *Science* **323** 915
[15] Grigoriev S V *et al.* 2010 *Phys. Rev. B* **29** 144413
[16] Binz B *et al.* 2006 *Phys. Rev. Lett.* **96** 207202
[17] De Gennes P G 1975 *Fluctuations, Instabilities, and Phase transitions* (NATO ASI Ser. B vol 2) ed T Riste (Plenum, New York)
[18] Bak P and Jensen M H 1980 *J. Phys.C: Solid State Phys.* **13** L881
[19] Yu X Z *et al.* 2010 *Nature* **465** 901
[20] Hubert A and Schäfer R 1998 *Magnetic Domains* (Berlin: Springer)
[21] Novikov S P *et al.* 1984 *Theory of Solitons* (Berlin: Springer-Verlag)
[22] Johnson M T *et al.* 1996 *Rep. Prog. Phys.* **59** 1409
[23] Belavin A A and Polyakov A M 1975 *JETP Lett.* **22** 245
[24] Skyrme T H R 1961 *Proc. Roy. Soc. Lon.* **260** 127
[25] De Gennes P G and J. Prost 1993 *The Physics of Liquid Crystals* (Oxford: Oxford University Press)
[26] Brown G E and Rho M (Eds.) 2010 *The multifaceted skyrmion*(New Jersey: World Scientific)

Easy axis switching in magnetite

Z. Kąkol[1], G. Król[1], W. Tabiś[1], T. Kołodziej[1], A. Wiśniewski[2], H. Stepankova[3], V. Chlan[3], J. Kusz[4], Z. Tarnawski[1], A. Kozłowski[1], J. M. Honig[5]

[1]Solid State Physics Department, AGH UST, al. Mickiewicza 30, 30—59 Kraków, Poland
[2]Institute of Physics, Polish Academy of Sciences, al. Lotników 32/46, 02-668 Warszawa, Poland
[3]Faculty of Mathematics and Physics, Charles University, V Holesovickach 2, 180 00 Prague 8, Czech Republic
[4]Institute of Physics, University of Silesia, ul. Uniwersytecka 4, 40-007 Katowice,Poland
[5]Department of Chemistry, Purdue University, West Lafayette, Indiana 47906, USA

e-mail: kakol@agh.edu.pl

Abstract. The easy magnetic axis switching in magnetite is investigated. Magnetization data confirmed activation character of the process with activation energy of the same order as that of the Verwey transition, suggesting common origin. On the other hand this activation energy rises with pressure (up to 1.2GPa), unlike T_V. The axis switching is clearly reflected in field dependence of resistivity and the direct structural data showed that it is simultaneous with the reorganization of structure. Thus, control of the structure can be possible with the application of magnetic field, as in shape memory materials. Finally, NMR showed that all, possibly decoupled entities: lattice distortion and charge and orbital orderings, change simultaneously while the axis switching occurs.

1. Introduction

The Verwey transition -the first order phase transformation at T_V=124K – is mainly known due to the drop of resistance by two orders of magnitude while heating above T_V. The Verwey temperature T_V diminishes with cation substitution, nonstoichiometry (usually caused by iron atoms deficiency), and pressure; in case of Ti, Zn and nonstoichiometry the T_V vs. lattice defects relation breaks into two parts, separating first and second order Verwey transition [1].

Following the first papers [2, 3], the transition was usually believed to be related to Fe^{+3} and Fe^{+2} ions on octahedral positions in the spinel lattice, Fig. 1a, lacking its individuality at high temperature (electron is equally spread across all octahedral Fe positions above T_V), but maintaining it below T_V (electron is frozen on particular positions at low T forming well specified Fe^{+2} and Fe^{+3} cations). Thus, magnetite undergoes order-disorder electronic transition, as symbolically presented on Fig. 1b.

However, neither the exact iron of +2 and +3 valance, nor the mixed valence above T_V were supported by experiments and this strict view of ionic charge order is not accepted now [4, 5]. Instead, the more relaxed charge ordering, $Fe^{+2.4}$ and $Fe^{+2.6}$, on octahedral iron positions have been found [6] together with the ordered Fe octahedral orbitals [7, 8, 9]. Moreover, Resonant X Ray Diffraction (RXD) technique allowed to find that both these types of orderings are decoupled [10, 11]. Finally, it became clear that the transition is due to intertwined electron-electron strong correlations as well as strong electron-lattice confinement [12].

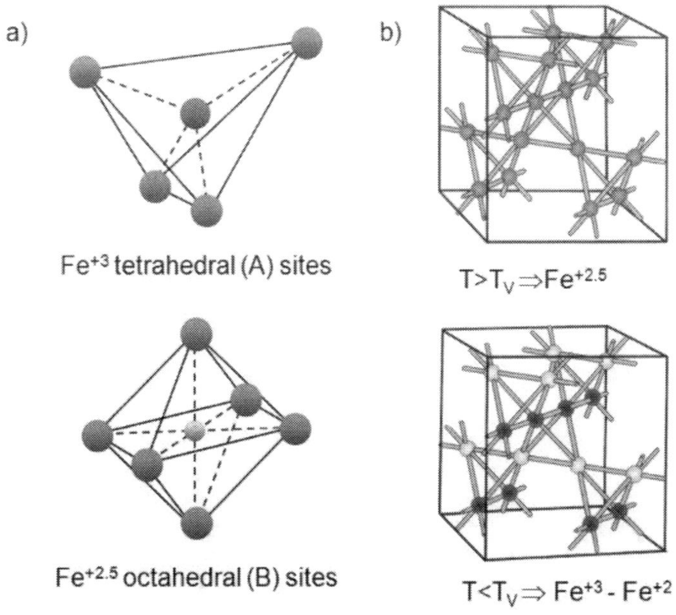

Figure 1. a) Tetrahedral and octahedral sites in the structure of magnetite and their formal valances at room temperature, b) Tetrahedra formed by octahedrally coordinated Fe with formal valence of $Fe^{+2.5}$ above T_V and with +2 and +3 valences ordering suggested originally by Verwey (and not realized in reality)..

The Verwey transition is reflected in virtually all physical characteristics including the abrupt change of lattice symmetry from high T cubic $Fd\bar{3}m$ to low T monoclinic [13, 6], as shown schematically on Fig. 2. There is also the accompanying change of magnetic anisotropy primary axes: while <111>, <110> and <100> are easy, intermediate and hard direction above 130K, monoclinic c axis becomes the easy axis below T_V [14, 15], Fig. 3.

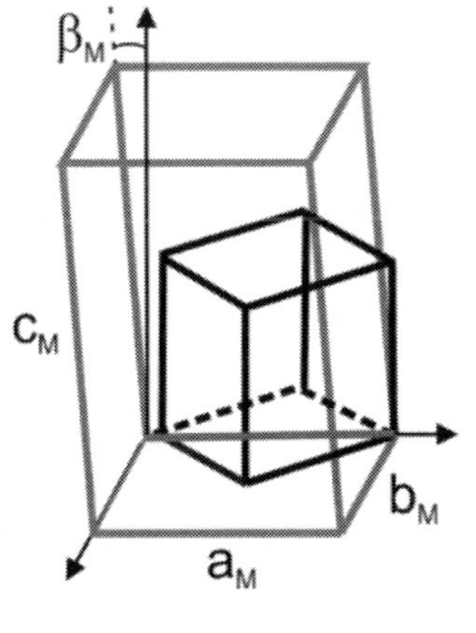

Figure 2. Schematic view of the changes of magnetite unit cell from cubic (black) to monoclinic *Cc* (red) once the sample is cooled below T_V.

Since magnetite structure changes at T_V from high temperature cubic $Fd\bar{3}m$ to monoclinic *Cc*, each of the high temperature cubic <100> directions may become the low temperature monoclinic *c*-axis. As a result, the material breaks into several structural domains unless the external magnetic field B>0.2T along particular <001> is applied while cooling through the transition. This particular direction will then become both the unique *c*-axis and also the magnetically easy axis (Fig.3).

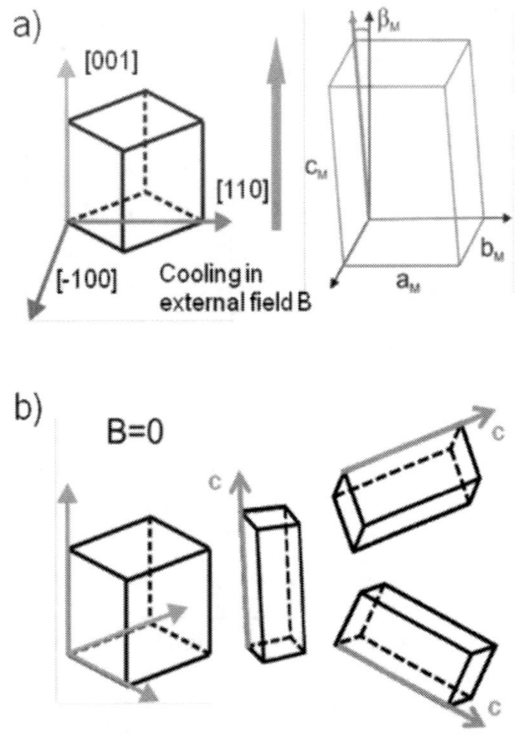

Figure 3. a) Magnetic easy (green), intermediate (blue) and hard (red) directions of cubic magnetite phase. Once magnetic field is applied along, e.g. [001], while cooling below T_V, the cell is doubled in this direction which also becomes the easy magnetic axis in low *T* phase, b) Zero-field cooling results in structural domains with *c* doubled along three original <001> directions.

If now the sample is magnetized along other <100> direction below T_V (and higher than 50K), as in Fig. 4 (e.g. at T=70K), a reorientation of magnetic moments, i.e. axis switching (AS), may take place and this <100> direction becomes a new easy axis [16, 17]. The phenomenon was first noticed in magnetization measurements by Calhoun and his analysis showed its activation character with the characteristic activation energy of 363K, i.e. of the same order of magnitude as the energy relevant to the Verwey transition.

The immediate question then arises, if AS concerns only magnetic axes switching or magnetic axes are moved simultaneously with crystal axes, especially with monoclinic c axis. If this were the case, one might expect that the magnetic field can trigger some electronic processes similar to those that occur at the Verwey transition. We would then have the unique chance to manipulate with Verwey transition-like processes, but at lower temperatures and with magnetic field. The purpose of the present paper is to elaborate on the axis switching phenomenon taking into account that it might be also linked to the mechanism of Verwey transition.

Magnetite is a very special material, where all important interactions (electron- electron, electron-phonon), vital in solid state physics, in particular, in oxides, are represented and the structure is relatively simple (just two kinds of atoms); for these reasons magnetite started to be considered as a test ground for experimental and theoretical techniques. It is, thus, important to realize that although the axis switching is a particular phenomenon, possibly linked to the phase transition in a particular material, the problem is universal for condense matter science, specifically to the physics of oxides. Since very small doping or nonstoichiometry levels greatly affect the Verwey transition it is obvious that proper sample preparation procedures are crucial for the reliability of the experimental results. All the data presented below were obtained on single crystalline samples grown at Purdue University by the skull melter, crucibleless technique [18]. This technique allows for the control of the oxygen partial pressure during growth, thereby ensuring that the melt remains within the stability range of the material. After preparation the crystals were subjected to subsolidus annealing under CO/CO2 gas mixtures to establish the appropriate metal/oxygen ratio [19].

In the paper, magnetic studies of AS are presented first, also under pressure. In the next section the other global characteristic, electrical resistivity, will be shown to reflect the changing axis. We will then prove by the direct experiment that, indeed, changing of magnetic easy axis is similar to the change of c axis. Finally, the introductory data concerning microscopic observation of AS (by NMR) will be presented.

2. Magnetic observation of AS

As already mentioned above, when magnetite single crystal is field cooled below T_V with a magnetic field along one of cubic a axes (as [001] in Fig. 3), this particular axis becomes the c axis of monoclinic low temperature Cc structure and, simultaneously, magnetic easy axis. This can easily be verified by magnetic measurements in this direction (Fig. 4a, hollow triangles): the initial slope is highest possible and the saturation is reached easily [20]. The measurements along perpendicular direction, an unspecified magnetic axis for low T phase ([100] cubic direction), at T=4K and 55K on Fig. 4a result in $M(B)$ curve that does not show saturation and the slope is much lower. At T close to T_V (e.g. bulk triangles, T=70K, on Fig. 4) application of field causes axis switching, easily seen as the sudden change of a slope, after which the signal finally reaches the saturation, while the field lowering results in $M(B)$ curve identical to that in an easy direction.

A sudden change of slope may be used for switching field B_{sw} estimation (in fact, this is the external switching field B_{sw_ext}, where demagnetizing field is included; note that in all figures and relations B_{sw}, after correcting for demagnetization field is presented), although the process is time-dependant and B_{sw} estimated this way is not precise (in this type of activation processes external parameter that finally triggers the process to occur is not a well defined parameter at all). Nevertheless, we have measured several samples and at different temperatures, and it was found that with raising temperature the field at which the axis switches gets lower, as shown on Fig.4b, in accord with Calhoun's description of the process. The probability of the magnetic-field -induced electronic transfer to change the axis reads [16]:

$$P(B_i) = \exp(-U/kT)(1-\exp(-\mu_B B_i/kT)) \cong (\mu_B B_i/kT)\exp(-U/kT) \tag{1}$$

where U is the activation energy of the electron to be able to jump to the other position without magnetic field and $\mu_B B_i$ is the energy excess of unfavorable magnetic axis position in the internal switching field B_i (μ_B stands for Bohr magneton). To start the cascade of experimentally observable electronic jumps (this parameter is very ill-defined and this is a reason why B_{sw} is difficult to measure and use), the probability (1) has to achieve the critical value p_c

$$(\mu_B B_i/kT)\exp(-U/kT) = p_c \tag{2}$$

Thus, the magnetic field required to switch the axis fulfils the relation $B_{sw} = B_i = p_c T e^{\frac{U}{kT}}$, shown, in other parameterization on Fig.4b. Note that activation energy for this particular sample is ca 350K; for other samples, we found U in the range of 290-300K.

Apparently, the observation of the axis switching by the magnetization method is very easy and efficient. This was the reason why this method was applied to observe the pressure dependence on AS.

Pressure affects the Verwey transition in an complicated way, forcing T_V, at ca 8GPa down to zero, i.e. exhibiting quantum phase transition [21]. It was also reported that $T_V(p)$ resembles T_V vs. doping relation [22], in particular the transition changes its character from discontinuous to continuous.

Finally, the conjecture was made [23, 24] that the crossover from the normal to inverse spinel takes place in magnetite under pressure; this was, however, refuted based both on Mossbauer [25], and structural [26] studies, and recent XMCD measurements [27]. In any case, the effect of pressure is important and should be measured for the phenomenon like axis switching.

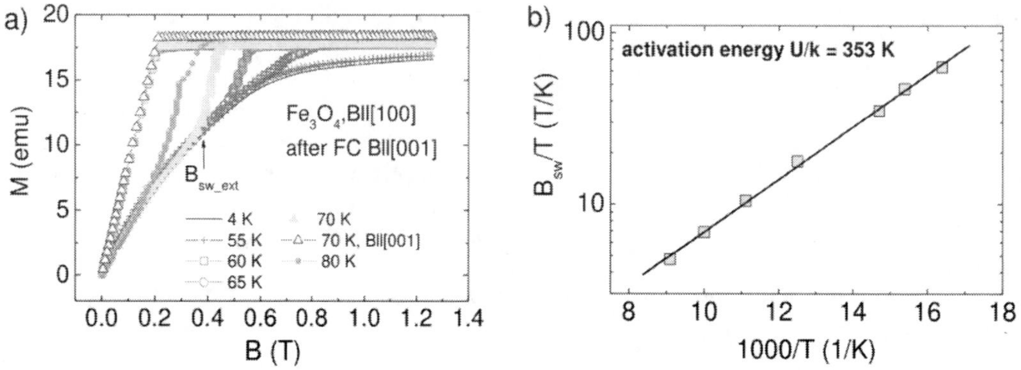

Figure 4. a) In $M(B)$ isotherms along [100] measured after field cooling in B = 1T along [001] across T_r (easy axis establishing). As long as the axis switching does not occur, $M(B)$ is reversible and typical for an unspecified magnetic direction (T=4, 55K). The arrow shows the onset of axis switching and defines B_{sw}; for still higher field and with field lowering, the data reproduce an easy direction (black triangles), proving the permanent character of AS, b) Temperature dependence of switching field, supporting the relation from (2), i.e. Calhoun's [16] prediction.

We have measured p dependence of axis switching the same way as above, by performing magnetization measurements under pressure below 1.2GPa in the pressure cell described elsewhere [28]. Verwey temperature was measured in magnetic moment vs. temperature ramp under remanence field of electromagnet (the transition is then marked as a sudden change of signal) and is presented on Fig.5 on top of other literature results.

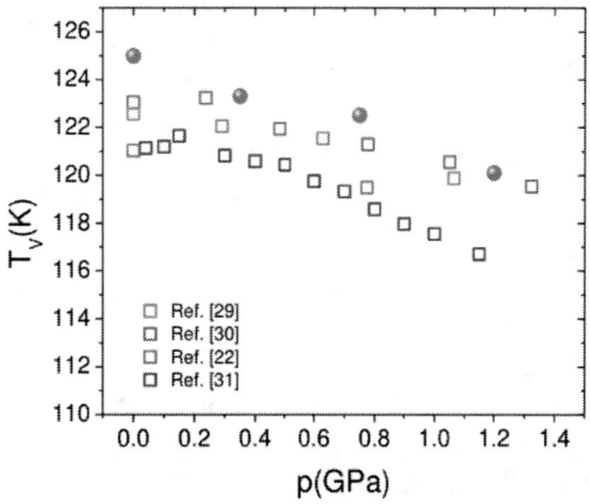

Figure 5. Pressure dependence of the Verwey transition temperature in a low p region studied here.

The representative $M(B_{eff})$ curve (for p=0.35GPa) is shown on Fig.6, together with switching fields, for this pressure value, at different temperatures. Although the determination of B_{sw} is not precise, as was already mentioned above, we have estimated the B_{sw} for four values of p, at different temperatures and the result, for p=0.37GPa, is presented on Fig.6b.

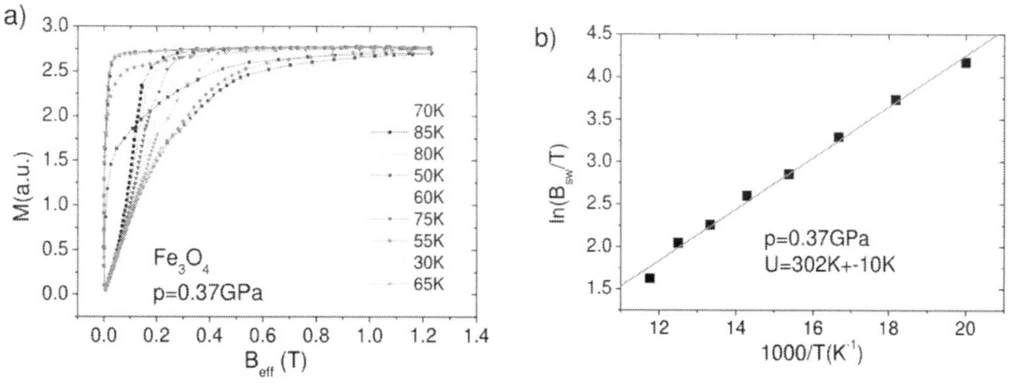

Figure 6. a) Results of magnetization measurements of stoichiometric magnetite under hydrostatic pressure of 0.37GPa; the axis switching is seen as a sudden change of M vs. B_{eff} relation, b) typical switching field B_{sw} vs. inverse temperature relation.

Pressure dependence of switching field at 60K and activation energy calculated from linear $\log(B_{sw}/T)$ vs $1000/T$ relations are stored in Fig.7. We have found that B_{sw} lowers with p, while the activation energy increases. It is thus easier to change the axis at higher p, but the sample is less susceptive to temperature.

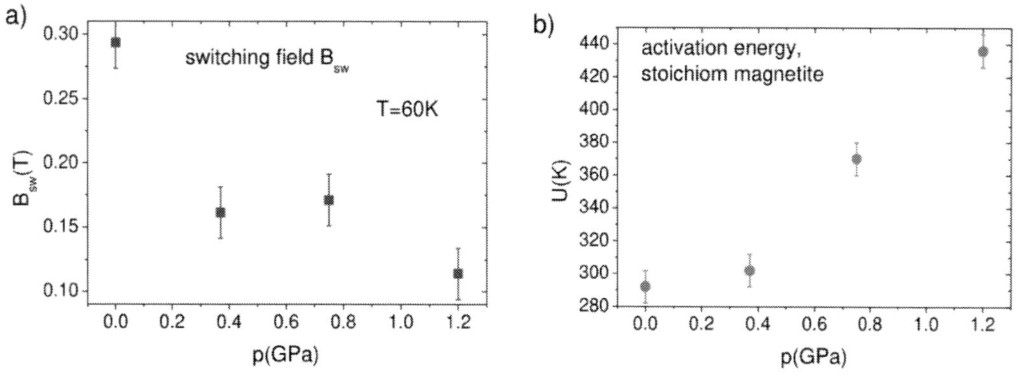

Figure 7. Results from the switching field B_{sw} at 60K: a) and activation energy U, b) estimation for all studied pressure values of p=0, 0.37, 0.75 and 1.2 GPa.

Also, the effect of hydrostatic pressure is different than that of chemical pressure, as for nonstoichiometric magnetite [32] (Fig.8): there, both activation energy and the switching fields increase with nonstoichiometry parameter that mimics external pressure.

Since also Zn substitution (that expands the lattice) does exactly the same, we may say that both substitution and doping creates much more complicated disturbance to the lattice than common relation of T_V vs. pressure and disturbance would suggest.

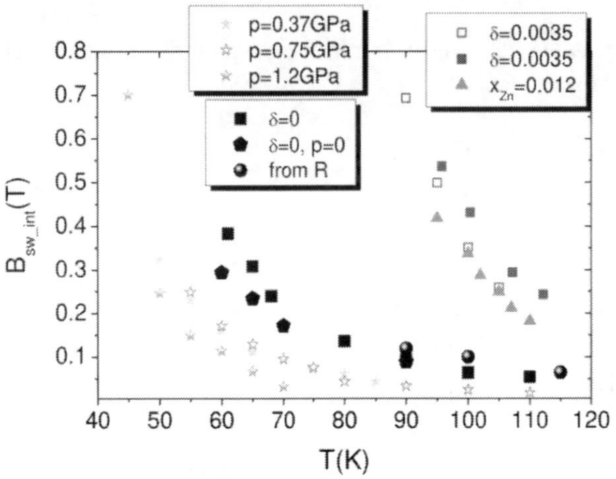

Figure 8. Temperature dependence of switching field B_{sw} in Zn-doped, stoichiometric and nonstoichiometric magnetite under ambient and increased pressure. Note that the effect of chemical substitution (or lattice disturbance, as with nonstoichiometry) is much higher, and different, than that caused by the pressure.

3. Impact of axis switching on electrical resistivity

One of the most spectacular manifestations of the Verwey transition is the 2 orders of magnitude jump of the resistance. Some change of transport properties should thus be observed during the axis switching, if AS were linked to the Verwey transition, as suggested before. To check it, magnetic field dependence of electrical resistance while the AS occurs were measured on PPMS system, equipped with horizontal rotator option, with the experimental layout presented on Fig.9. Two sets of contacts, along [010], and along [100], were attached to the plate-shaped sample. The sample was first field-cooled to the specified temperature below T_V in an external magnetic field B_{ext}=1T set parallel to [100]. The subsequent measurements with B_{meas} (=0→4T→ -4T→0) along cooling direction reflect magnetic field dependence of resistivity without axis switching. The axis should be switched in the next experiment, with B_{meas} perpendicular to the cooling field B_{ext}. Similar experiments were also performed after field cooling along [010]. In any case, the magnetic domain structure is affected, so the experiment should also show magnetic domain structure impact on resistivity.

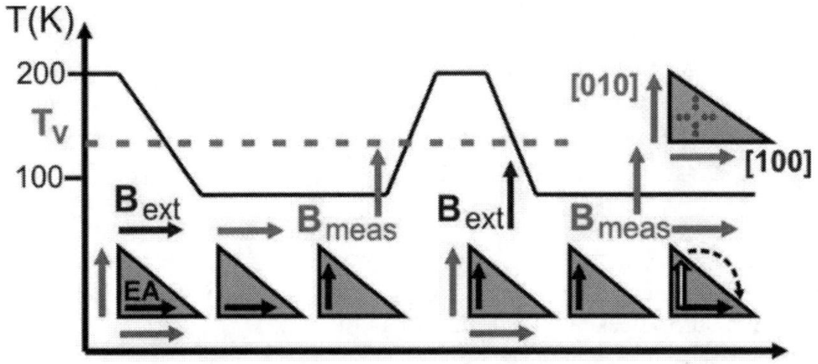

Figure 9. Experimental layout for checking AS impact on electrical transport. The sample (triangle shaped with two pars of contacts upper right corner) is field cooled (in external field B_{ext}=1T, along [100]) below T_V and measured in field B_{meas} aligned in the same direction (no AS). Then, the resistance is measured with field along [010], forcing AS. The sample is then heated, field cooled again, but with B_{ext} pointing along [010] and the whole measuring procedure is repeated. Black arrow shows magnetic easy axis direction.

Fig.10a presents the results of measurements after the sample was FC along [100] direction and the field was subsequently applied along this line. No AS should occur in this geometry and the curve should reflect magnetoresistance and a possible magnetic domains influence on resistance. Although the resistance rises initially with field, it results from still unset temperature (and due to the fact that resistance of magnetite is extremely sensitive to temperature change below T_V). The linear R vs. B decrease will be addressed later on.

The representative result of the experiment when the sample was magnetized in the direction perpendicular to field cooling axis, causing AS to occur, is shown on Fig.10b. Here, the rapid changes (shown in the expanded form on Fig. 10c) of much higher amplitude than any other effects observed previously, mark the switching field; note that all figures are presented in the same scale. Similar effects, in the form of sharp spikes, were observed in the symmetrical experiment mentioned on Fig. 9 and also after zero field cooling when the field was applied in any of cubic <100> directions. However, neither these spikes are identical for both directions, nor in any other resistivity feature any clear systematics can be observed. For example, the rapid maximum in R along [100] could be found with simultaneous drop of R along [010] (Fig. 10c) and this results could be completely different in subsequent measurements even for the same conditions.

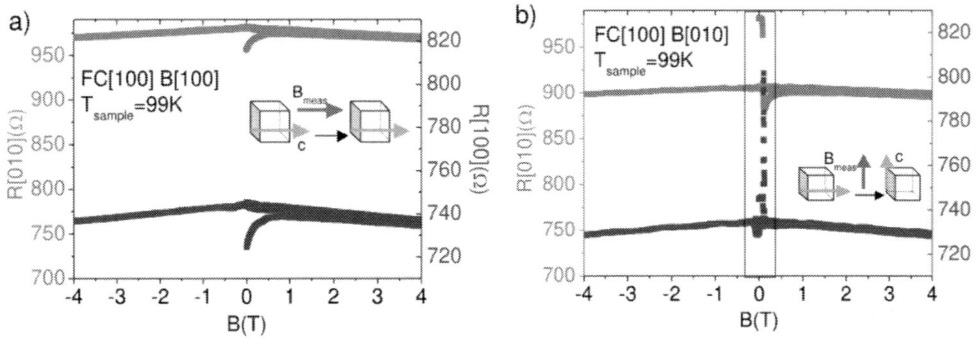

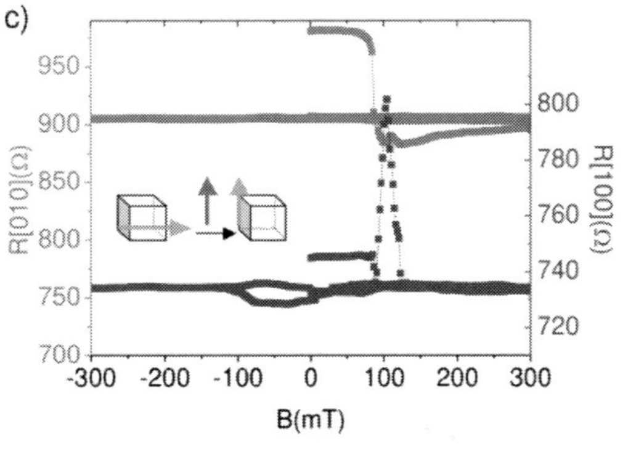

Figure 10. Resistance dependence on magnetic field after the sample was field cooled with B_{ext}=1T set along [100] (FC[100]). The sample was then measured either in measuring field B_{meas} in the same direction as cooling field B[100] (a), or with B perpendicular to it (B[010], (b). In (c) the region marked in (b) is expanded, showing sudden resistance changes while AS occurs, as schematically suggested. Note that a full hysteresis loop in the field range 0T→4T→ -4T→4T→0 was measured.

As already mentioned, for all the experiments, the negative magnetoresistance can be observed. To see if this is a real effect, we have estimated the resistance changes in case the changing sample temperature (which was monitored for each field) was taken into consideration. As we have reported separately [33], the effect can entirely be explained by the T flow with the magnetic field: magnetite does not exhibit any substantial magnetoresistance at $T<T_V$, contrary to literature reports [34, 35, 36].

Summarizing electrical resistivity measurements, the axis switching phenomenon can be observed in electrical transport. On the other hand, no clear systematics, whatsoever, as to R dependence on magnetic field along, or perpendicular to the current flow direction, or the magnetic domain structure could be found. It means, most probably, that all parameters we control, i.e. *c*-axis order and magnetic order, are not the most important parameters that affect the electrical resistance.

4. Structural Studies

In all the experiments presented above we have observed the switching of magnetic axis and confirmed its activation character; however, no insight in the mechanism of this phenomenon was proposed. In particular, it was not discerned if AS is solely a switching of magnetic easy axis, without ion displacement, or by changing the axis we actually change the charge arrangement, i.e. redirect monoclinic *c* axis. Below, we are showing that this last situation occurs: the application of magnetic field changes the ionic order established on cooling the sample below T_V, i.e. some new superstructure reflections appear due to the switched monoclinic *c* axis.

We have mounted stoichiometric magnetite single crystal on the four-circle KM4 diffractometer with graphite-monochromatized Mo Kα radiation, observing (4 4 1/2) and (7/2 0 4)-type reflections (in cubic notation), caused by doubling of cubic lattice constant in monoclinic *c* direction. When the sample was zero-field cooled below T_V (we used Oxford Cryosystems nitrogen gas cooling), the domains with three *c* axes were formed, and all six superstructure reflections were observed, as on Fig. 11a. Application of a magnetic field (supplied by specially prepared NdFeB magnet system) along [001] resulted in only one peak for each type (Fig.11b), proving doubling of lattice parameter in this particular direction. Thus, external magnetic field not only changes magnetic axes, but also affects crystal structure by changing the direction of monoclinic *c* axis.

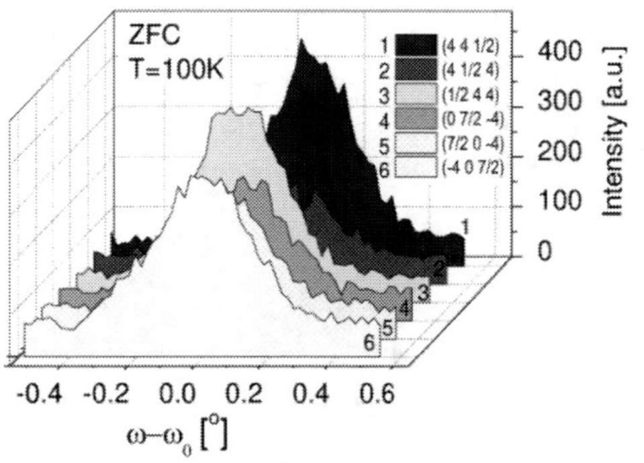

Figure 11a. The ω scan of (4 4 1/2) and (-4 0 7/2) type reflections after ZFC the sample to 100K. Note similar intensity within two groups of reflections, proving similar abundance of structural twins

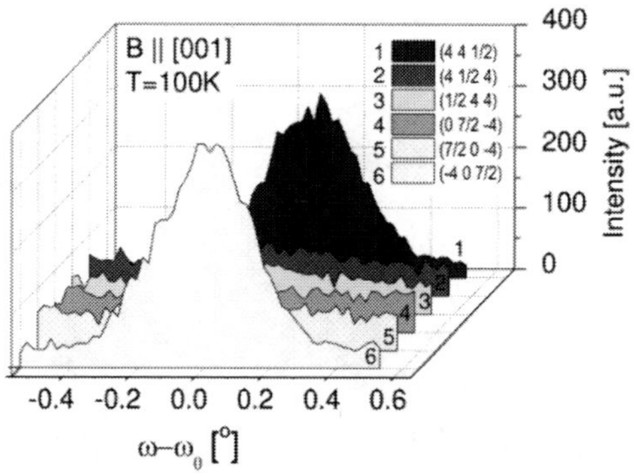

Figure 11b. With the application of ca. 0.2T along [001] twins with c along [100] and [010] greatly diminish, leaving only [001] twin identified by (4 4 1/2) and (-4 0 7/2) superstructure peaks.

Qualitatively the same results were obtained when B was applied along [010] (the third direction was not available do to goniometer vs. magnet geometry). We have thus proved by direct method that magnetic and structural monoclinic axes are switched simultaneously. In other words, we could, to some extend, manipulate the crystal structure with a magnetic field.

Similar conclusion that structure can be modified by magnetic field was presented in [37]. However, the structural effects described there originated from the subtle interplay between frustrated FM and AFM exchange interactions, unbalanced by an applied magnetic field. These processes are not present in our situation due to the large difference between T_N=858K and T_V.

5. NMR observation

After we have shown that the magnetic axis switching is simultaneous with the change of monoclinic c axis, it is interesting to check if this change of *c* axis is linked to the switching of charge and orbital ordering; as mentioned above, three Verwey transitions are present in magnetite, with lattice partially decoupled from orbital and charge orderings. The Resonant X ray scattering studies of this effect, allowing the observation of each type of ordering separately, will be published elsewhere; here, we will focus on microscopic observation of the process by Nuclear Magnetic Resonance.

The axis switching was not studied by NMR method before; the former NMR measurements in external magnetic field were performed at temperatures too low for this phenomenon to occur. On the other hand, NMR spectroscopy can easily see all different iron positions (both tetrahedral and octahedral) in monoclinic *Cc* structure [38, 39], allowing, potentially, a precise observation of all different ordering entities. The results of extended studies of AS by this technique are described in the separate paper [40]; here, only the most important results are summarized.

The sample was the sphere of stoichiometric magnetite oriented with [001] axis vertical, parallel to the RF coil axis. External magnetic field, up to 1T, could be set horizontally and the sample could be rotated to allow all the directions in [001] zone to be checked (see Fig.12). Only tetrahedral lines were checked what amounts to A iron nucleus being a probe for its nearest neighbors (B atoms) arrangement.

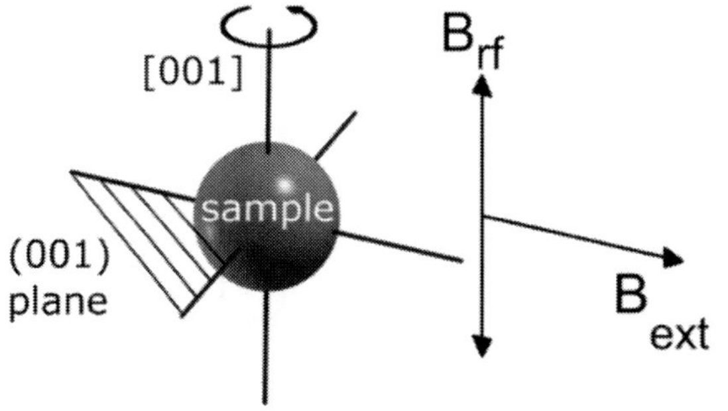

Figure 12. Experimental arrangement for NMR measurements: rotational axis [001], directions of external magnetic field and radiofrequency field are designated.

We have first field cooled the sample to 20K, which sets the c axis in the field direction (say [100]), establishing some atomic arrangement. Then the field of 0.3T was applied along [100] and the corresponding NMR signal for all A lines is shown on Fig. 13a. The field was then rotated such that it points in other cubic direction (Fig.13b); since the field was small enough not to force AS, the NMR pattern reflects the different hyperfine field but with the same atomic arrangement as before. Axis switching was subsequently forced (with the field B=1.3T) after heating the sample to T=80K just below T_V and the sample was measured again in B=0.3T and at T=20K, revealing the NMR pattern (Fig. 13c) almost identical to the original one. It thus looks as if all octahedral surroundings of the A atom we observe are, in relation to the field B direction, the same as before. This suggests that not only lattice distortion, but also orbital and charge orders, were changed simultaneously after the application of external magnetic field.

Finally, when the magnetic field axis was reverted to its initial direction and the whole procedure was repeated, the pattern almost came back to the one found before (Fig 13d-f).

Cooling the sample in 1.3 T through the Verwey transition down to 57 K (the field was along [100] axis) and applying 0.5T field along [010] direction enabled the observation of axis switching dynamics (only lines A_3-A_7 were measured in order to obtain reasonable signal/noise ratio). It is apparent (see Fig. 14) that the electronic system evolved from that on Fig. 13b (the external field was slightly higher) to the one on Fig. 13c within approximately 20 hours.

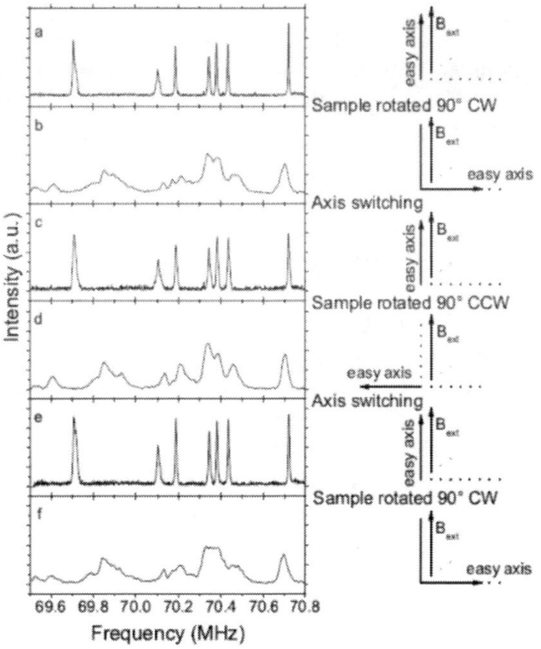

Figure 13. Selected tetrahedral sites spectra revealing axis switching. Figures 13a, 13c and 13e correspond to external magnetic field (0.3T, T=20K) set along the easy axis (however c and d show the pattern after axis switching) while in 13b, 13d and 13f the external field is perpendicular to the easy axis. Between b and c, and d and e, the sample was heated up to 80K and the field of 1.3T was applied to switch the easy axis. Scheme on right side describes the steps of the experiment.

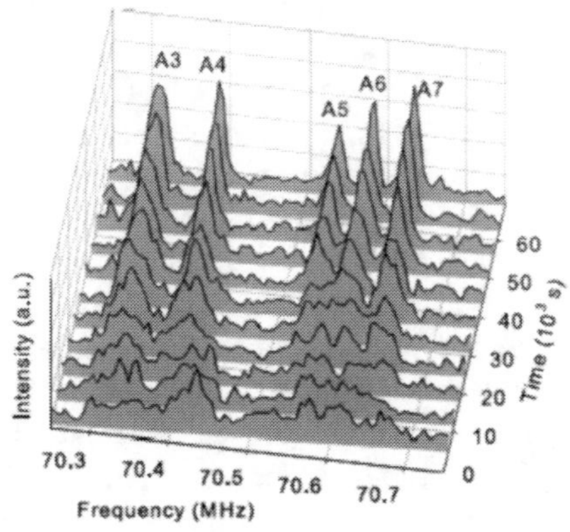

Figure 14. Axis switching dynamics observed by NMR (T=57K, B=0.5T). See text for details.

We hope that similar procedure, when analyzed together with the spectra for different field direction in (001) plane may result in finding those octahedral positions that had to change most to reorganize charge, orbital and lattice distortion arrangement, as observed experimentally.

6. Conclusions

We have studied the axis switching process with various techniques, both bulk and microscopic. By the magnetization measurements, we have confirmed the activation character of this process. Since the activation energy was similar to Verwey temperature itself, it appeared that this phenomenon may be closely linked to the electronic processes active also in the Verwey transition. However, both magnetization studies in nonstoichiometric and Zn doped magnetite, as well as magnetization measurements under hydrostatic pressure have shown that activation energy increases rather than lowers with T_V, rendering the simple proof of a close link of AS with the transition questionable. On the other hand, AS is very well reflected by electric transport, where the strong oscillations are found at the magnetic field that switches the axis. Direct measurements of the superstructure reflections due to lattice constant doubling below T_V proved that AS is simultaneous with the partial lattice reorganization, especially the change of c monoclinic lattice axis. Finally, NMR proved that this reorganization is complete, i.e. tetrahedral iron surroundings with magnetic field along easy axis is identical to that after the axis has been switched.

Acknowledgements
This work is financially supported by the Polish Ministry of Science and Higher Education and by the Faculty of Physics and Applied Computer Science of AGH UST

References

[1] Aragón R, Rasmussen R J, Shepherd J P, Koenitzer, and J W,Honig J M 1986 *J. Magn. Magn. Mat.* **54-57** 1335
Shepherd J P, Koenitzer J W, Aragón R, Spałek J, and Honig J M 1991 *Phys. Rev.* **B43** 8461
[2] Verwey E J 1939 *Nature* (London) **144** 327
[3] Anderson P W 1956 *Phys. Rev.* **102** 1008
[4] Garcia J, Subias G, Proietti M G, Blasco J, Renevier H, Hodeau J L, and Joly Y 2001 *Phys. Rev.* **B63**, 54110
[5] Garcia J and Subias G 2004 *J. Phys. Condens. Matter* **16**, R145
[6] Wright J P, Attfield J P, and Radaelli P G 2001 *Phys. Rev. Lett.* **87** 266401
Wright J P, Attfield J P, and Radaelli P G 2002 *Phys. Rev.* **B66** 214422
[7] Huang D J, Lin D H-J, Okamoto J, Chao K S, Jeng H-T, Guo G Y, Hsu C-H, Huang C-M, Ling, D C , Wu W B, Yang C S, and Chen C T 2006 *Phys. Rev. Lett.* **96** 096401
[8] Schlappa J, Schußler-Langeheine C, Chang C F, Ott H, Tanaka A, Hu Z, Haverkort M W, Schierle E, Weschke E, Kaindl G, and Tjeng L H 2008 *Phys. Rev. Lett.* **100** 026406
[9] Nazarenko E, Lorenzo J E, Joly Y, Hodeau J L, Mannix D, and Marin C 2006 *Phys. Rev. Lett.* **97** 056403.
[10] Lorenzo J E,Mazzoli C, Jaouen N, Detlefs C, Mannix D, Grenier S, Joly Y, and Marin C 2008 *Phys. Rev. Lett.* **101** 226401
[11] Joly Y, Lorenzo J E,. Nazarenko E, Hodeau J-L, Mannix D, and Marin C *Phys. Rev.* **B 78** 134110
[12] Piekarz P, Parlinski K, and Oleś A M 2006 *Phys. Rev. Lett.* **97** 156402
Piekarz P, Parlinski K, and Oleś A M 2007 *Phys. Rev.* **B 76** 165124
[13] Iizumi M, Koetzle T F, Shirane G, Chikazumi C, Matsui M, and Todo S 1982 *Acta Cryst.* B **38** 2121
[14] Aragón R 1992 *Phys. Rev.* **B46** 5334
Aragón R 1992 *Phys. Rev.* **B46** 5328
[15] Kąkol Z and Honig J M 1989 *Phys. Rev.* **B40** 9090
[16] Calhoun B A 1954 *Phys. Rev.* **94** 1577
[17] Vittoratos E, Baranov I, and Meincke P P M 1971 *J. Appl. Phys.* **42** 1633
[18] Aragón R and McCallister R H 1982 *Phys. Chem. Minerals* **8** 112
[19] Aragón R, Harrison H R, McCallister R H, and Sandberg C J 1983 *Cryst. Growth.* **61**, 221
[20] Król G, Kusz J, Tarnawski Z, Kąkol Z, Tabiś W, and Kozłowski A 2006 *Acta Phys. Pol.* **A109** 601
Król G, Kusz J, Tarnawski Z, Kakol Z, Kozlowski A, and Honig, J M 2007 *J. All. Comp.* **83** 442
[21] Spałek J, Kozłowski A, Tarnawski Z, Kąkol Z, Fukami Y, Ono F, Zach R, Spalek L J, and Honig J M 2008 *Phys. Rev.* **B 78** 100401R
Spałek J, Kozłowski A, Tarnawski Z, Kąkol Z, Fukami Y, Ono F, Zach R, Spalek L J and, Honig J M 2009 *Physica* **B 404** 2894
[22] Rozenberg G, Hearne G R, Pasternak M P, Metcalf P A, and Honig J M 1996 *Phys. Rev.* **B53**, 6482
[23] Pasternak M P, Xu W M, Rozenberg G, Taylor R D, and Jeanloz R 2003 *J. Magn. Magn. Mat.* **265** L10
Pasternak M P, Xu W M, Rozenberg G, Taylor R D, and Jeanloz R 2004 *J. Phys. Chem. Sol.* **65** 1531
[24] Rozenberg G Kh, Pasternak M P, Xu W M, Amiel Y, Hanfland M, Amboage M, Taylor R D,

and Jeanloz R 2006 *Phys. Rev. Lett.* **96** 045705
Rozenberg G Kh, Pasternak M P, Xu W M, Amiel Y, Hanfland M, Amboage M, Taylor R D, and Jeanloz R 2007 *Phys. Rev.* **B 75** 020102
[25] Żukrowski J, Wiecheć A, Zach R, Tabiś W, Tarnawski Z, Król G, Kim-Ngan N-T H, Kąkol Z, and Kozłowski A 2007 *J. All. Comp.* **442** 219
[26] Klotz S, Rousse G, Strässle Th, Bull C L, and Guthrie M 2006 *Phys. Rev.* **B 74** 012410
Klotz S, Rousse G, Strässle Th, Bull C L, and Guthrie M 2008 *Phys. Rev.* **B 77**, 012411
[27] Baudelet F, Pascarelli S, Mathon O, Itié J-P, Polian A, and Chervin J-C 2010 *Phys. Rev.* **B 82** 140412R
[28] Baran M, Dyakonov V, Gladczuk L, Levchenko G, Piechota S, and Szymczak H 1995 *Physica* C **241** 383
[29] Kakudate Y, Mori N, and Kino Y 1979 *J. Magn. Magn. Mat.* **12** 22
[30] Tamura S 1990 *J. Phys.Soc. Jap.* **59** 4462
[31] Wiecheć A, Zach R, Kąkol Z, Tarnawski Z, Kozłowski A, and Honig J M 2005 *Physica* B **359-361** 1342
[32] Król G, Fabiś W, and Kołodziej T unpublished results
[33] Król G, Tabiś W , Przewoźnik J, Kołodziej T, Kakol Z , Kozłowski A, and Tarnawski Z 2009 *J. All. Comp.* **480** 128
[34] M Ziese and H J Blythe 2000 *J. Phys.: Condens. Matter* **12** 13
[35] Mertens D C, Montfrooij W, McQueeney R J, Yethiraj M, and Honig J M 2005 *J. Appl. Phys.* **97** 10A303
[36] Coey J M D, Berkowitz A E, Balcells L, Putris F F, and Parker F T 1998 *Appl. Phys. Lett.* **72** 734
[37] Yokaichiya F, Krimmel A,Tsurkan V, Margiolaki I, Thompson P, Bordallo H N, Buchsteiner A, Stüßer N, Argyriou D N, and Loidl A 2009 *Phys. Rev.* **B79** 064423
[38] Novak P, Stepankova H, English J, Kohout J, and Brabers V A M 2000 *Phys. Rev.* **B61** 1256
[39] Mizoguchi M 2001 *J. Phys. Soc. Jpn.* **70** 2333
[40] Chlan V, Kouřil K, Štěpánková H, Řezníček R, Štěpánek J, Tabiś W, Król G, Tarnawski Z, Kąkol Z, and Kozłowski A 2010 *J. Appl. Phys.* **108** 083914

Non-collinear ferromagnetic Luttinger liquids

N. Sedlmayr[1], S. Eggert[1,2], J. Sirker[1,2]
[1]Fachbereich Physik and [2]Optimas Research Centre, Technische Universität Kaiserslautern, Erwin-Schrödinger-Strasse Gebäude 46, 67663 Kaiserslautern, Germany
E-mail: `sedlmayr@physik.uni-kl.de`

Abstract. The presence of electron-electron interactions in one dimension profoundly changes the properties of a system. The separation of charge and spin degrees of freedom is just one example. We consider what happens when a system consisting of a ferromagnetic region of non-collinearity, *i.e.* a domain wall, is coupled to interacting electrons in one-dimension (more specifically a Luttinger liquid). The ferromagnetism breaks spin charge separation and the presence of the domain wall introduces a spin dependent scatterer into the problem. The absence of spin charge separation and the effects of the electron correlations results in very different behaviour for the excitations in the system and for spin-transfer-torque effects in this model.

1. Introduction

The behaviour of the electronic and magnetic degrees of freedom in quasi one-dimensional wires and films has already received considerable interest and study, see for example the review of Marrows[1]. So far most works focus on how the transport properties of free electrons behave in a ferromagnetic wire with a domain wall, and how these spin polarized currents set the domain wall itself into motion. One obvious question to ask is whether interactions are important in such cases. A straightforward yes was answered by Dugaev et al.[2] who explicitly demonstrated that one must include at least mean field interaction corrections to correctly describe the charge build up around the domain wall. This of course should not be surprising but it is a timely reminder that correlation effects are often important even for nominally "free" electron systems. In this case failure to include these correlation effects results in an un-physical charge excess. In addition to the mean field interaction work of Dugaev et al. there also exists in the literature some consideration of Hartree-Fock corrections[3, 4]. Here we will consider a more strongly correlated system: the Tomonaga-Luttinger liquid.

Experimentally[5] the construction of chains of single magnetic atoms is already possible and one dimensional magnetic systems are experimentally available. Several obvious questions must be answered at this point. Traditionally it is thought that long range ferromagnetic order does not exist in one dimension. However, there are several things which mitigate this bald statement for these systems. The long range ferromagnetic order is in fact broken by the presence of domain walls, precisely the situation we are interested in. We also note that effects from the substrate mean that one can not really consider the chain as an isolated spin chain to be solved by a Heisenberg or Ising type model. Experimentally ferromagnetically ordered chains of atoms are indeed observed[5, 6, 7, 8]. Furthermore they exhibit non-collinear ferromagnetic order[9]. One should of course bear in mind that the condition for defining the system as one-dimensional may be different for the magnetic degrees of freedom and for the electronic degrees of freedom. It is only the latter that we require to be one-dimensional to be in the Luttinger regime.

These considerations lead us to our model of a ferromagnetic Tomonaga-Luttinger model[10, 11, 12, 13]. Unfortunately the chains of atoms do not easily display Luttinger physics due to the interference of the substrate[14], though they remain an intriguing possible future application of the model in a more complicated scenario. Nonetheless there is still a class of possible candidates left for our model: dilute magnetic semiconductors[15]. Systems where the magnetic and electronic degrees of freedom belong to different layers would also be a possible realization.

We will consider an "s-d" like model in which the bulk magnetization and the conduction electrons are treated as separate (though of course still interacting). As such we introduce two timescales into the problem, a fast electronic one and a slow magnetic one. This allows us to first answer the question of how the presence of the domain wall affects the Luttinger liquid, forgetting for the moment the effect the *moving* domain wall will have on the conduction electrons.

The question of a non-collinear ferromagnetic Luttinger liquid has not yet been fully addressed[13], we set up the model in section 2. Firstly we would like to analyze the low energy effective field theories and their low temperature properties, the results of this are presented in section 3. Secondly we wish to understand the behaviour and nature of the appropriate excitations of the model. Finally we want to consider how this affects the magnetization dynamics of the domain wall. Due to the radically different nature of the excitations of a Luttinger, as opposed to Fermi, liquid we expect these dynamics to be very different. We offer some thoughts on this based on our preliminary results in section 4 later in this article.

2. The model

The magnetization of the wire is described by $\vec{M}(x) = M_s \cos[\Theta(x)]\hat{\mathbf{z}} + M_s \sin[\Theta(x)]\hat{\mathbf{y}}$ where M_s is the saturation magnetization and, using $\cos[\Theta(x)] = -\tanh[x/\lambda]$, we have a description of a domain wall of length λ situated at $x = 0$. This magnetization is coupled to the conduction electrons with a strength given by the exchange coupling J. Naturally we consider a screened, and hence short range, interaction $V(x-x')$. This allows us to start from the following standard "s-d" Hamiltonian[16]:

$$\tilde{H} = \sum_\sigma \int \mathrm{d}x \tilde{\psi}^\dagger_\sigma(x)\Big[-\frac{1}{2m}\partial_x^2 - \mu\Big]\tilde{\psi}_\sigma(x) - J\sum_{\sigma\sigma'}\int \mathrm{d}x \tilde{\psi}^\dagger_\sigma(x)\tilde{\psi}_{\sigma'}(x)\vec{\sigma}_{\sigma\sigma'}.\vec{M}(x)$$
$$\underbrace{+\frac{1}{2}\sum_{\sigma\sigma'}\int \mathrm{d}x\,\mathrm{d}x'\tilde{\psi}^\dagger_\sigma(x)\tilde{\psi}^\dagger_{\sigma'}(x')V(x-x')\tilde{\psi}_{\sigma'}(x')\tilde{\psi}_\sigma(x)}_{=H_I}. \qquad (1)$$

μ is the chemical potential.

In order to be able to linearize the system our first step must be to remove the spatially dependent, and in principle perhaps very large, magnetization. To this end we rotate the spin direction to get a collinear ferromagnet via the following gauge transformation[17, 18]: $\mathbf{H} = \mathbf{U}^\dagger(x)\tilde{\mathbf{H}}\mathbf{U}(x)$, $\psi_\sigma(x) = U^\dagger_{\sigma\sigma'}(x)\tilde{\psi}_{\sigma'}(x)$, and $\mathbf{U}(x) = e^{\frac{i}{2}\theta(x)\sigma^x}$. The interaction is left unaffected, the magnetization is locally rotated to a Zeeman term, and the kinetic energy operator introduces a new potential when it acts on the local rotation. Thus we have

$$H = \sum_\sigma \int \mathrm{d}x \psi^\dagger_\sigma(x)\Big[-\frac{1}{2m}\partial_x^2 - JM_s\sigma^z_{\sigma\sigma} - \mu\Big]\psi_\sigma(x) + H_I + H_w \text{ and} \qquad (2)$$

$$H_w = -\frac{1}{2m}\sum_{\sigma\sigma'}\int \mathrm{d}x\psi^\dagger_\sigma(x)\big[i\theta'(x)\sigma^x_{\sigma\sigma'}\partial_x + \frac{i\theta''(x)}{2}\sigma^x_{\sigma\sigma'} - \frac{1}{4}[\theta'(x)]^2\delta_{\sigma\sigma'}\big]\psi_{\sigma'}(x). \qquad (3)$$

This new potential can be approximated if we assume that the Fermi wavelength $\lambda_F \ll \lambda$. In this case only the first term of equation (3) is relevant, the next two terms are of order $(\lambda_F/\lambda)^2$.

The next step is linearization via the ansatz $\psi_\sigma(x) = e^{ik_{F\sigma}x}\psi_{R\sigma}(x) + e^{-ik_{F\sigma}x}\psi_{L\sigma}(x)$, where $k_{F\sigma} = \sqrt{2m(\mu + JM_s\sigma^z_{\sigma\sigma})}$. The R and L indices denote right and left moving electrons respectively. Note that if the Zeeman term is large we must linearize around the spin split Fermi points. This leads to the physically relevant phenomena of the breakdown of spin-charge separation[19]. Umklapp processes scattering two left movers into right movers and vice versa are always neglected here due to the non-commensurate nature of the Fermi wavevectors.

Finally we have our model to be bosonized[10]. Naturally the bosonic fields satisfy $[\phi_\sigma(x), \Pi_\sigma(x')] = i\delta(x - x')$ where $\Pi_\sigma(x) = \partial_x\theta_\sigma(x)$. We first rotate to the spin and charge degrees of freedom. We can then diagonalize the quadratic part of the Hamiltonian with $\phi_{c/s}(x) = [\phi_1(x) \pm \phi_2(x)]/\sqrt{2}$ (and similar for the $\theta(x)$ fields) giving us for the quadratic part of the bosonic Hamiltonian

$$H_0 = \int \mathrm{d}x \Bigg[[\partial_x\theta_c(x)]^2\frac{v_cK_c}{2} + [\partial_x\theta_s(x)]^2\frac{v_sK_s}{2} + [\partial_x\theta_c(x)\partial_x\theta_s(x)]v_a + [\partial_x\phi_c(x)]^2\frac{v_c}{2K_c} + [\partial_x\phi_s(x)]^2\frac{v_s}{2K_s} + [\partial_x\phi_c(x)\partial_x\phi_s(x)]v_b\Bigg]. \quad (4)$$

K_s and K_c are the spin and charge Luttinger parameters. v_a and v_b describe the coupling between the spin and charge sectors. These parameters are functions of the interaction strengths and Fermi velocities.

The diagonalization of the quadratic terms introduces new velocities, u_i, for H_0 and a set of $\{T_i^{\theta,\phi}\}$ parameters. These are all known in terms of the previously mentioned Luttinger parameters. The final Hamiltonian is $H = H_0 + H_1 + H_w^f + H_w^b$ where

$$H_0 = \int \mathrm{d}x \sum_{i=1,2} \frac{u_i}{2}[(\partial_x\tilde{\phi}_i(x))^2 + (\tilde{\Pi}_i(x))^2] \quad (5)$$

$$H_w^f = \frac{1}{2\pi\alpha}\int \mathrm{d}x[v_{F\uparrow} + v_{F\downarrow}]\frac{\mathrm{sech}[x/\lambda]}{\lambda}\sin[\sqrt{2\pi}(T_1^\theta\tilde{\theta}_1(x) + T_2^\theta\tilde{\theta}_2(x))] \sin[x(k_{F\uparrow} - k_{F\downarrow}) + \sqrt{2\pi}(T_1^\phi\tilde{\phi}_1(x) + T_2^\phi\tilde{\phi}_2(x)] \quad (6)$$

$$H_w^b = \frac{1}{2\pi\alpha}\int \mathrm{d}x[v_{F\uparrow} - v_{F\downarrow}]\frac{\mathrm{sech}[x/\lambda]}{\lambda}\sin[\sqrt{2\pi}(T_1^\theta\tilde{\theta}_1(x) + T_2^\theta\tilde{\theta}_2(x))] \sin[x(k_{F\uparrow} + k_{F\downarrow}) - \sqrt{2\pi}(\tilde{T}_1^\phi\tilde{\phi}_1(x) + \tilde{T}_2^\phi\tilde{\phi}_2(x))] \quad (7)$$

$$H_1 = \frac{2}{(2\pi\alpha)^2}\int \mathrm{d}x\cos[2\sqrt{2\pi}(T_1^\phi\tilde{\phi}_1(x) + T_2^\phi\tilde{\phi}_2(x))] \big[[g_{1\perp\uparrow} + g_{1\perp\downarrow}]\cos[2x(k_{F\uparrow} - k_{F\downarrow})] - i[g_{1\perp\uparrow} - g_{1\perp\downarrow}]\sin[2x(k_{F\uparrow} - k_{F\downarrow})]\big]. \quad (8)$$

We have both forward and backward scattering terms, H_w^f and H_w^b respectively. We also find an oscillating sine-Gordon interaction term H_1, describing "slow" Umklapp processes which are not averaged out, with some strength $g_{1\perp\sigma}$. These all couple our otherwise diagonal bosonic degrees of freedom. The appropriate excitations of such an $SU(2)$ asymmetric model have no obvious physical interpretation. This summarizes the effective bosonic field theory which is the foundation of our mathematical analysis of the problem.

3. Low energy physics

We start by writing a functional integral partition function[20]:

$$\mathcal{Z} = \int D\tilde{\phi}D\tilde{\Pi}e^{-\int_0^\beta \mathrm{d}\tau\left[\int \mathrm{d}x[-i\tilde{\Pi}_i(x)\partial_\tau\tilde{\phi}_i(x)] + H[\tilde{\Pi}(x),\tilde{\phi}(x)]\right]} \quad (9)$$

with periodic boundary conditions in imaginary time τ. Following the standard procedure we split the fields into fast, $\phi^>$, and slow, $\phi^<$, fields. Our fast fields are defined for $\Lambda' < |k|, |\omega|/u_i < \Lambda$, and the slow for $|k|, |\omega|/u_i < \Lambda'$. Expanding the exponent in terms of H_1 and H_w and performing the averaging over the fast modes we then re-exponentiate the expression to find the appropriate scaling equations. Parameterizing $\Lambda^{(l)} = \Lambda_0 e^{-l(-dl)}$ we find for the first order H_1 term

$$\frac{dg_{1\perp\sigma}}{dl} = g_{1\perp\sigma}\gamma^1. \tag{10}$$

γ^1, and also the $\gamma^{f,b}$ used below, are known functions of the rotation terms $\{T_i^{\theta,\phi}\}$. This term is an irrelevant perturbation for any situation we wish to look at. In the limit of weak magnetization we can simplify the expressions to find $\gamma^1 \to 2 - 2K_s$. In this limit $K_s > 1$ and it becomes clear that the term is irrelevant.

The second order equation for our model is more difficult than for the sine-Gordon model. A diagonal equation in the $\tilde{\phi}_i$'s is not recovered and to perform any further analysis we would have to rediagonalize the problem and then renormalize the model once again. This is perhaps not totally unexpected as the scattering term we are dealing with explicitly couples these terms. We leave the more involved second order renormalization group analysis to a future work and stick here to the first order equations.

The same analysis is performed on the scattering terms, we define the scattering coupling constants $g^{f/b}(0) \equiv v_{F\uparrow} \pm v_{F\downarrow}$. For back scattering we find:

$$\frac{dg^b}{dl} = g^b\gamma^b. \tag{11}$$

This term is a relevant perturbation. In the limit of weak magnetization this simplifies to $\gamma^b \to 2 - [(K_s)^{-1} + K_c]/2$. Similarly the forward scattering equation stands as

$$\frac{dg^f}{dl} = g^f\gamma^f. \tag{12}$$

This is also a relevant term. In the limit of weak magnetization this becomes $\gamma^f \to 2 - [(K_s)^{-1} + K_s]/2$.

In contrast to the scaling equations of Pereira and Miranda[13, 21] we find that we can not neglect forward scattering. They were interested in the limit of very sharp walls and the small magnetization limit. The model they solve leads to the consideration of a magnetic impurity in an otherwise collinear ferromagnetic system. In such a case it is perhaps not surprising that this model leads to the magnetic analog of the Kane-Fisher problem[22, 23]. Their final bosonic Hamiltonian is appropriate only for very sharp walls where the domain wall profile will be different to the longer walls which we wish to consider, which leads to the different scaling equations and hence a different effective low temperature model. It should also be pointed out that depending on the length of the domain wall forward scattering can become the *dominant* form of scattering. Of course in the physical model "forward scattering" refers to an electron which passes through the wall *without* changing its spin.

We have three natural length scales present in the problem: $\lambda_+ = 2\pi(k_{F\uparrow} + k_{F\downarrow})^{-1}$, $\lambda_- = 2\pi(k_{F\uparrow} - k_{F\downarrow})^{-1}$, and λ. In the limit of weak magnetization $\lambda_- \to \infty$ and $2\lambda_+ \to \lambda_F$ and the limit of large magnetization, when one spin channel becomes frozen out gives $\lambda_+ \approx \lambda_- \approx \lambda_F$. We are generally interested in long domain walls when $\lambda \geq \lambda_F$ and we are in the physical regime where both forward and backward scattering are important. As one might expect in the large magnetization limit the scattering terms can be neglected as only one spin channel is available. For very sharp delta function like walls one finds only back scattering relevant[13], a case we do not consider further. For long domain walls we are back in the adiabatic limit where scattering can be completely neglected. Interestingly there is a regime in between, $\lambda_+ < \lambda < \lambda_-$, where forward scattering is the dominant mode of scattering. This regime can also be reached by tuning the magnetization to be suitably weak.

4. Outlook

We are principally interested in the spin density and the spin and electronic currents. The spin density and current will also necessary for understanding the magnetization dynamics of the domain wall. The spin density[24, 25] is defined as $\vec{S}(x) = \psi^\dagger(x)\boldsymbol{U}^\dagger\vec{\boldsymbol{\sigma}}\boldsymbol{U}\psi(x)$. It can be calculated perturbatively in the domain wall profile using the bosonic model derived above. Preliminary results show an important development. The corrections to the zeroth order density caused by scattering are concentrated around the edges of the domain wall. In a normal quasi one-dimensional wire these corrections are found around the centre of the domain wall[2] and contribute to a spin torque which causes precessional motion of the domain wall, in analogy with a moment in a magnetic field. In contrast we expect the domain wall to be distorted rather than set into precessional motion by these terms in the Luttinger liquid. The temperature dependence of the spin density corrections can be extracted from these corrections for the small temperature limit. One finds power law contributions in the forward and backward scattering exponents: $T^{-\gamma^{f,b}}$. These become sharper for larger values of JM.

The magnetization dynamics are described by the Landau-Lifschitz-Gilbert equation, or some suitable generalization thereof[26, 27, 28]. There are two different aspects to this. One is the straightforward point that the dynamics, *over the length of the wall*, will be affected by the different spin density of the Luttinger liquid compared to the Fermi liquid or non-interacting case. The second, more interesting point, is whether the derivation of the non-adiabatic terms in the LLG equation are valid for a Luttinger liquid.

Following Zhang and Li[28] one can derive contributions to the magnetization dynamics which allow for the fact that the electrons do not instantaneously follow the magnetization profile. One first writes a continuity equation for the spins, assuming part of it to be always parallel to the bulk magnetization and allowing a small deviation from this. In order to derive the current dependent (so called β-) terms, those which drive the domain wall along the wire, one assumes that $j_s^{x,y,z}(x,t) = -\mu_B P j_e^{x,y,z}(x,t) M^{x,y,z}(x,t)/eM_s$. $\vec{j}_e(x,t)$ is the charge current and P is the magnitude of the polarization, whilst $\vec{j}_s(x,t)$ is the spin current. A quite reasonable assumption in a Fermi liquid, this of course starts to look more dubious in the case of a Luttinger liquid. When $SU(2)$ symmetry is present, not our case one is reminded, spin and charge are of course uncorrelated and possess different velocities. Thus this assumption would completely fail. For us the situation is not so simple, nonetheless what is obvious from the model is that spin and charge are not fully correlated. One is forced to work with the spin current and not the electric current and, as we have already seen, the spin degrees of freedom can behave rather differently for this model.

Other than the magnetic behaviour we also wish to analyze the transport through the wire.

5. Conclusion

We have considered a non-collinear ferromagnetic Luttinger liquid. We assessed its low energy properties and the relevant lengthscales, presenting the results of renormalization group calculations. It is possible to tune the relevant types of scattering operators by considering different domain wall lengths and magnetization magnitudes. Here 'magnetization' refers to both its magnitude and the magnitude of the exchange coupling. The regime of most interest to us which we go on to assess is the intermediate regime where we have both forward and backward scattering terms and an appreciable magnetization. We discuss some preliminary results for the spin density correlation function and the effect that this will have on the subsequent domain wall dynamics. We propose that in our Luttinger liquid system the magnetization dynamics will be altered by the very different nature of the Luttinger liquid excitations.

Acknowledgments
The authors wish to thank J. Berakdar and R. Pereira for useful and stimulating discussions.

References

[1] Marrows C H 2005 *Advances in Physics* **54** 585
[2] Dugaev V K, Barnaś J, Łusakowski A and Turski L A 2002 *Phys. Rev. B* **65** 224419
[3] Araújo M A N, Dugaev V K, Vieira V R, Berakdar J and Barnaś J 2006 *Phys. Rev. B* **74** 224429
[4] Araújo M A N, Berakdar J, Dugaev V K and Vieira V R 2007 *Phys. Rev. B* **76** 205107
[5] Gambardella1 P, Dallmeyer A, Maiti K, Malagoli M C, Eberhardt W, Kern K and Carbone C 2002 *Nature* **416** 301
[6] Shen J, Skomski R, Klaua M, Jenniches H, Manoharan S S and Kirschner J 1997 *Phys. Rev. B* **56** 2340–2343
[7] Elmers H J, Hauschild J, Höche H, Gradmann U, Bethge H, Heuer D and Köhler U 1994 *Phys. Rev. Lett.* **73** 898–901
[8] Hauschild J, Elmers H J and Gradmann U 1998 *Phys. Rev. B* **57** R677–R680
[9] Wiesendanger R 2009 *Rev. Mod. Phys.* **81** 1495–1550
[10] Giamarchi T 2003 *Quantum Physics in One-Dimension* (Oxford)
[11] Tomonaga S 1950 *Progress of Theoretical Physics* **5** 544–569
[12] Luttinger J M 1963 *Journal of Mathematical Physics* **4** 1154–1162
[13] Pereira R G and Miranda E 2004 *Phys. Rev. B* **69** 140402
[14] Dash L K and Fisher A J 2001 *Journal of Physics: Condensed Matter* **13** 5035
[15] Jungwirth T, Sinova J, Mašek J, Kučera J and MacDonald A H 2006 *Rev. Mod. Phys.* **78** 809–864
[16] Blundell S 2009 *Magnetism in Condensed Matter* (Oxford)
[17] Tatara G and Fukuyama H 1997 *Phys. Rev. Lett.* **78** 3773–3776
[18] Korenman V, Murray J L and Prange R E 1977 *Phys. Rev. B* **16** 4032–4047
[19] Pereira R G and Sela E 2009 Coulomb drag from spin-charge coupling at zero magnetic field arXiv:0911.1391
[20] Negele J W and Orland H 1998 *Quantum Many-Particle Systems* (Westview Press)
[21] Pereira R G 2004 *Transporte em Sistemas Unidimensionais Ordenados* Master thesis Universidade Estadual de Campinas Instituto de Fsica Gleb Wataghin
[22] Eggert S and Affleck I 1992 *Phys. Rev. B* **46** 10866–10883
[23] Kane C L and Fisher M P A 1992 *Phys. Rev. B* **46** 15233–15262
[24] Eggert S and Affleck I 1995 *Phys. Rev. Lett.* **75** 934–937
[25] Rommer S and Eggert S 2000 *Phys. Rev. B* **62** 4370–4382
[26] Lifschitz E M and Pitaevskii L P 2002 *Statistical Physics Part 2: Theory of the Condensed State* (Butterworth-Heinemann)
[27] Gilbert T 2004 *IEEE Transactions on Magnetics* **40** 3443
[28] Zhang S and Li Z 2004 *Phys. Rev. Lett.* **93** 127204

Joint European Magnetic Symposia – JEMS 2010 IOP Publishing
Journal of Physics: Conference Series **303** (2011) 012108 doi:10.1088/1742-6596/303/1/012108

Exchange interaction as the source of pairing in correlated electron systems: a brief overview

Jozef Spałek[1,2]

[1] Marian Smoluchowski Institute of Physics, Jagiellonian University, Reymonta 4, PL-30-059 Kraków, Poland

[2] Faculty of Physics and Applied Computer Science, AGH University of Science and Technology, Reymonta 19, PL-30-059 Kraków Poland

E-mail: `ufspalek@if.uj.edu.pl`

Abstract. We address the question of real space pairing induced by exchange interaction, customarily associated with the presence of the broken-symmetry magnetism. For such mechanism of pairing, the appearance of superconductivity (or superfluidity) without magnetism presents itself still as a highly nontrivial problem and it seems, it can be practically realized only when the kinetic energy of individual-particle motion is comparable or even smaller than the magnitude of exchange interaction that takes place among *strongly correlated fermions*, where the kinetic exchange and/or the Hund's rule coupling can become relatively strong in the vicinity of the onset of the Mott insulating state. In that situation, the pairing arises from a coherent correlated pair motion rather than from a free motion of a tightly bound pair. We elaborate on this question in general terms, as well as provide selected results in the mean-field approximation in concrete situations.

1. Introduction

The original Heisenberg-Dirac exchange interaction concept [1] reflects the transposition symmetry of the many-particle systems and has been applied to atomic [2], solid state [3], and nuclear [4] systems of fermions. In the case of atomic physics, it allows to define the high- and low-spin multiplets of 3d and 4f shells according to the Hund's rule. In the condensed matter physics it is customarily used to describe symmetry-broken magnetic states (ferromagnetism, antiferromagnetism, spin density or spin-glass states) [5], whereas in nuclear physics it serves to classify high angular-momentum states, as well as to introduce the pairing corrections to the nuclei bound energy in the filled-shell systems [6]. In brief, the exchange interaction is instrumental in explaining the magnetism of magnetic insulators and other systems with localized magnetic moments. In real systems, it may be augmented with anisotropy, spin-orbit, crystal field, and other interactions [5]. It is also crucial as one of the principal factor in classifying the high-spin states of finite Fermi systems such as a quantum dots or nuclei.

A related but highly nontrivial question concerns the role of exchange interactions in itinerant-fermion (metallic) systems, particularly when describing ferromagnetic metals, heavy-fermion and high-temperature superconductors, where magnetism either coexists or is intimately connected with the onset of superconductivity, and involves the same set of electrons responsible for the appearance of both phenomena [7]. The natural question is then whether the spin correlations in real space and induced by the short-range exchange interaction, can also become the mechanism of pairing resulting in either superfluid or superconducting state of fermion systems. This is an involved problem, since the magnetic ordering is described by the (staggered) magnetization and the superconducting order is described by the off-diagonal correlation function (in the simplest Hartree-Fock limit by the binding of electron and hole of opposite momentum and spin). So, in principle, those two types of ordering are not directly related, as they are described by different type of correlations. In particular, it must by noted that while the magnetism of localized spins reflects the static long-range spin-spin correlations, in the superfluid state the spin-spin correlations contain a coherent translational motion of thus mutually correlated pairs of particles with spin. Therefore, in the latter situation the strength of the exchange interaction must represent a sizable portion of the kinetic (band) energy of the particles, so the coherent (or bound) moving pair states can be sustained, not necessarily in the milieu of already ordered spins. This is the reason why this type of pairing can be effective only in the strongly correlated systems, where the kinetic energy of individual particles is drastically reduced by the correlations resulting from strong repulsive Coulomb interaction as is the case in e.g. t-J model [8] (in the extreme case of the Mott-Hubbard insulators the kinetic energy is reduced to zero by the correlations).

The magnetic mechanism of pairing interesting us here and induced by exchange interaction in real space differs from the superconducting or superfluid states induced by paramagnons (spin fluctuations), which represent collective boson (paramagnon) virtual excitations in either spin disordered or weakly ordered systems [9]. In the latter situation, we deal with pairing analogous to that induced by phonons [10], except that in the case of paramagnons spin-triplet (high-spin) pairing may be preferred, as is in the case of superfluid ^{3}He [11], neutron stars [6], and possibly, in UGe_2 [12] and Sr_2RuO_4 [13]. In the case of exchange-interaction induced pairing the spin-correlations in real space are instrumental in the pairing, i.e. there is no "*boson glue*" to produce a coherent superfluid state of pairs [14]. Nonetheless, as we shall see below, any mechanism of pairing, must eventually be brought to the "effective" BCS form, since the condensed state is either metallic (for superconductors) or liquid (for superfluids), which is described by paired (quasi)particles, with well defined momentum (in nuclear physics, the angular momentum).

The purpose of this paper is to overview briefly the role of the real-space magnetic pairing as the driving force of superconductivity. Such mechanism must be first considered for strongly correlated systems: heavy fermions, high temperature superconductors, and selected magnetic systems, particularly close to the Mott-Hubbard localization of relevant particles, i.e. when their kinetic energy is drastically reduced and the purely magnetic states are on the verge of (in) stability. In Section 2 we characterize briefly the strongly correlated system with their kinetic exchange interaction leading to spin-singlet-pairing, whereas in Section 3 we characterize briefly the mean-field solution for them. The case of spin-triplet pairing will not be discussed here. One of the conclusions (cf. Sec. 4) is that the topic announced in the title is by no means a finished story. We provide some reasons for that opinion, as well as point out to a possibility of further development.

2. Kinetic exchange and spin-singlet pairing: narrow band case

The role of the so-called kinetic [15] exchange interaction in spin-singlet-pairing case was raised for the first time in the context of high temperature superconductivity [16] and a bit later, in the context of heavy-fermion superconductivity [17]. In the former case, the superconducting state evolves from antiferromagnetic Mott insulator and in the latter, it can also coexist with a very weak antiferromagnetic state of the spin-density-wave type. Under such circumstances, a natural question to be posed is whether the source of the magnetism can also become simultaneously the mechanism of pairing. Models of these two types should be considered first in the absence of evidence for any other pairing mechanism such as the electron-phonon interaction. This last presumption seems to be legitimate, since the superconducting state in those systems is highly unconventional in many respects not to be discussed here [18]. We introduce first the idea of kinetic exchange for strongly correlated metals [8] and then discuss the paired states.

2.1. Kinetic exchange for strongly correlated metals

The relevant for our purposes here history of antiferromagnetic exchange interactions in strongly correlated and insulating magnetic oxides bears its origin from the pioneering work of Anderson [14] who introduced the idea of kinetic exchange between the neighboring atomic 3d states of transition - metal cations, say Ni^{2+} each of which has an admixture of 2p states of surrounding O^{2-} anions. The Anderson's superexchange interaction via the nonmagnetic anions is effectively expressed by that between the magnetic 3d cations dressed with an admixture of the 2p oxygen states. The Anderson's *kinetic exchange*, applicable first of all to the Mott insulators, has been applied also to the charge-transfer insulators [19].

The idea of kinetic exchange was extended subsequently to strongly correlated metals [17]. Its essence can be summarized as follows. In the simplest situation, we start from the Hubbard model representing a single narrow band and expressed in the Wannier representation, which has the form

$$H = \sum_{ij\sigma} t_{ij}\, a^{\dagger}_{i\sigma}\, a_{j\sigma} + U \sum_{i} n_{i\uparrow}\, n_{i\downarrow}, \tag{1}$$

where the summation means that we take only the sites $i \neq j$ in the hopping part with the hopping integral t_{ij} between them. The hopping range is determined by the number of coordination spheres of 3d ions, for which $t_{ij} \neq 0$. The second term represents the intraatomic part of the Coulomb interaction only between the two particles with opposite spins ($\sigma =\uparrow$ and $\sigma =\downarrow$) located on the same orbital. Here i denotes symbolically the Wannier state $|w_i(\underline{r})\rangle = |w(\underline{r} - \underline{R}_i)\rangle$ centered on site i. In the strong-correlation limit, we have $|t_{ij}| \ll U$ and hence the kinetic (bare band) energy, as represented by the first term, is much smaller than U and thus can be regarded as a perturbation. In effect, the Hubbard Hamiltonian may be transformed in the (second) order in t_{ij}/U to the following form [17, 20] including exchange interaction in an explicit form:

$$\mathcal{H}_{tJ} = \sum_{ij\sigma} t_{ij}\, b^{\dagger}_{i\sigma}\, b_{j\sigma} + \sum_{ij} J_{ij} \left(\mathbf{S}_i \cdot \mathbf{S}_j - \frac{1}{4} c_1 \nu_i \nu_j \right) + c_2 \sum_{ijk} \frac{t_{ij} t_{jk}}{U} \left(b^{\dagger}_{i\sigma}\, S^{\bar{\sigma}}_j\, b_{k\bar{\sigma}} - b^{\dagger}_{i\sigma}\, \nu_{j\bar{\sigma}}\, b_{k\sigma} \right), \tag{2}$$

where normally $c_1 = c_2 = 1$ for their role see below. The processes representing the dynamical processes are sketched in Fig. 1. In the above expression the projected fermions operators are: $b^{\dagger}_{i\sigma} \equiv a_{i\sigma}(1 - n_{i\bar{\sigma}}), \nu_{i\sigma} \equiv n_{i\sigma}(1 - n_{i\bar{\sigma}}), \nu_i \equiv \sum_\sigma \nu_{i\sigma}$, and $S^{\sigma}_i \equiv b^{\dagger}_{i\sigma}\, b_{i\bar{\sigma}}$, with $\bar{\sigma} \equiv -\sigma$. The kinetic - exchange integral is $J_{ij} = 2t^2_{ij}/U$. In this Hamiltonian, we project out the local (site) double occupancies, as well as include both the two-site (virtual-exchange) and the three-site hopping terms, the second and the third terms, respectively. In the Mott-Hubbard insulating limit we have the conservation of particle number *at each site*, i.e. $n_{i\sigma} + n_{i\bar{\sigma}} = 1$, and hence the effective Hamiltonian reduces to that of Anderson [15] in this single-band situation (with spin $S = 1/2$). One has also to note that the intersite Coulomb interaction $\sim \sum_{ij} K_{ij} n_i n_j$ has been neglected in (1). This term can be included in the analysis [21]. In general, the first term describes a nontrivial form of single-particle hopping, with the particles avoiding each other (no double occupancy), while the third term represents the nontrivial three-site hopping $j \to i$, with and without spin flip on the intermediate site j, respectively. In effect, one can say that in the metallic state (with the band filling $n = \langle n_i \rangle \neq 1$), we have nN moving spins (or equivalently, of $N - Nn = N(1 - n)$ holes among the spins of the N-site system) that are correlated via the antiferromagnetic exchange interaction, with its range equal to the number of coordination spheres we take, for which $t_{ij} \neq 0$. The state describing such liquid of moving spins and holes in between them is customarily called the *resonating valence bond* (RVB) state [16] although,

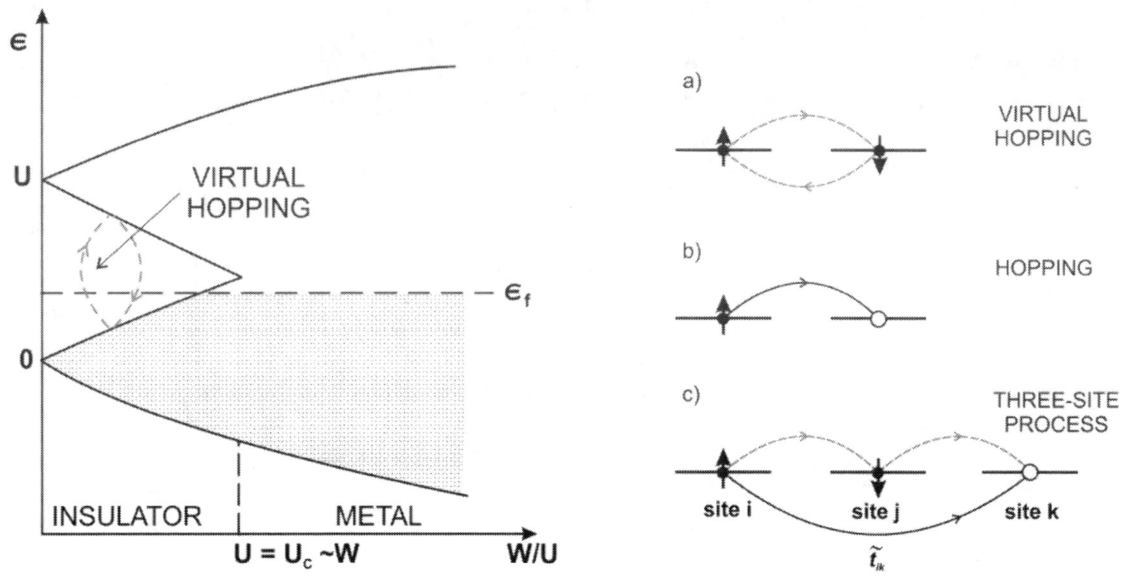

Figure 1. On the left: schematic representation of the Hubbard subbands. The virtual hopping processes correspond to the virtual transitions lower → upper → lower Hubbard subbands (process a)). On the right: three possible intersite hopping processes (in real space) determining the dynamics of Hamiltonian (2) in the lower Hubbard subband (i.e. for the band filling $n \leq 1$): a) virtual hopping between single occupied neighboring sites that leads to their antiferromagnetic (kinetic) exchange interaction; b) two-site single-particle hopping between the single occupied and the empty sites; and c) three-site hopping between the singly occupied and the empty site via on intermediate single occupied site with opposite spin.

strictly speaking, in the present situation the various pair spin-singlet configurations, taken into account in the RVB theory of Mott insulator [6], are intermixed with the hopping processes, due to the presence of holes. So, strictly speaking we should speak about a quantum spin liquid (no double occupancies!) with antiferromagnetic correlations.

2.2. Kinetic exchange as real space pairing

The effective Hamiltonian (2) contains essentially the intertwinned hopping (kinetic motion) and the explicit spin correlations (kinetic exchange and 3-site hopping), as any pair of the terms do not commute with each other. So quite few nontrivial types the behavior may appear, since minimally the band and the frozen spin-ordering effects compete *viciously*, as they become comparable near the Mott insulating limit [16]. Therefore, it was totally surprising at first (at least for the author), that the Hamiltonian (2) contains also an essentially new physics connected with the real space pairing [6], which effectively can be reduced to some form of the BCS theory. A systematic approach to the concept of this new type of pairing starts with definition of the local pairing operators [19] in the form:

$$\begin{cases} B_{ij}^{\dagger} \equiv \frac{1}{2}\left(b_{i\uparrow}^{\dagger}b_{j\downarrow}^{\dagger} - b_{i\downarrow}^{\dagger}b_{j\uparrow}^{\dagger}\right), \\ B_{ij} \equiv \frac{1}{2}\left(b_{i\downarrow}b_{j\uparrow} - b_{i\uparrow}b_{j\downarrow}\right) = \left(B_{ij}^{\dagger}\right)^{\dagger}. \end{cases} \quad (3)$$

In effect, one can rewrite the kinetic-exchange energy part in the completely equivalent forms as follows:

$$\mathbf{S}_i \cdot \mathbf{S}_j - \frac{1}{4}\nu_i\nu_j \equiv -B^\dagger_{ij}B_{ij}. \tag{4}$$

What is even more surprising, the Hamiltonian (2), even with inclusion of the three-site terms, can be rewritten in the following closed form [19] for $c_1 = c_2 = 1$, namely:

$$\mathcal{H}_{tJ} = \sum_{ij\sigma} t_{ij}\, b^\dagger_{i\sigma}\, b_{j\sigma} - \sum_{ijk} \frac{4t_{ij}t_{jk}}{U} B^\dagger_{ij}B_{kj}\ . \tag{5}$$

This is the full form of the so-called *t-J model*. In this form we see that the effective Hamiltonian contains in a direct manner an individual spin hopping (the first term) and the local pair binding (the $i = k$ part), as well as the 3-site hopping (and spin-flip) processes for $i \neq k$. The beauty (and the complication) of the problem is that the two terms composing the above Hamiltonian can become comparable right at the lower (lower critical) concentration for the transition from an insulating antiferromagnetic to the superconducting state in representing high temperature superconductors [17]. This is the reason why the superconductivity and antiferromagnetism can become intimately connected in the t-J model if only B_{ij} are the forerunners of a condensed spin-singlet superconducting state. Parenthetically, because of the exclusion of the double-occupied site configurations we have $B^\dagger_{ii} = B_{ii} \equiv 0$, and the pairing in the superconducting will have a nontrivial momentum $\hbar\mathbf{k}$ dependence (d-wave) of the superconducting gap. The superconducting state emerging out of this representation of the kinetic-exchange part is discussed in the next Section. We should mention that the ideas of real-space pairing have been extended also to the case with orbital degeneracy in [22], and to the case of hybridized systems as we discuss next.

2.3. Kondo-interaction in hybridized-band case

The Anderson-lattice model represents another example and the simplest two-band situation, in which (originally) atomic and strongly correlated states (of 3d or 4f type, for example) are intermixed with the conduction (4s or d5-6s type, respectively), which form correlated and hybridized bands. This is easy to understand, since the valence states as more extended collectivize first. The specific feature of the Anderson-lattice model is that the outermost valence states loose their atomic parity due to their band character, so it is possible to have a nonzero hybridization (e.g. the interatomic mixing between the atomic (4f) and the Wannier 5d-6s states). This last circumstance constitutes the physical origin of the model. Its formal structure from the first-quantization point of view is not so simple. However, in the second - quantization language it is easy to formulate the microscopic Hamiltonian, as the qualitative argument provided above is contained in the microscopic parameters. In effect, the starting Hamiltonian in the Wannier representation has the form

$$\mathcal{H} = \sum_{mn\sigma} t_{mn}c^\dagger_{n\sigma}c_{n\sigma} + \epsilon_f \sum_{i\sigma} N_{i\uparrow} + U\sum_{i\downarrow} n_{i\uparrow}n_{i\downarrow} + \sum_{im\sigma} V_{im}\left(f^\dagger_{i\sigma}c_{m\sigma} + c^\dagger_{m\sigma}f_{i\sigma}\right)$$
$$-\mu\left(\sum_{i\sigma} N_{i\sigma} + \sum_{m\sigma} n_{m\sigma}\right), \tag{6}$$

where i labels correlated atomic (f=3d or 4f) states, (m,n) label delocalized c states, $N_{i\sigma} \equiv f^\dagger_{i\sigma}f_{i\sigma}$, and V_{im} represents the hybridization magnitude. In the limit when $|V_{im}| \ll |\epsilon_f|$, U, the above periodic Anderson Hamiltonian transforms into the Kondo Hamiltonian with the antiferromagnetic (kinetic-exchange) Kondo coupling between itinerant and localized electrons [23]. In this limit, the numbers of correlated and of the itinerant electrons are conserved

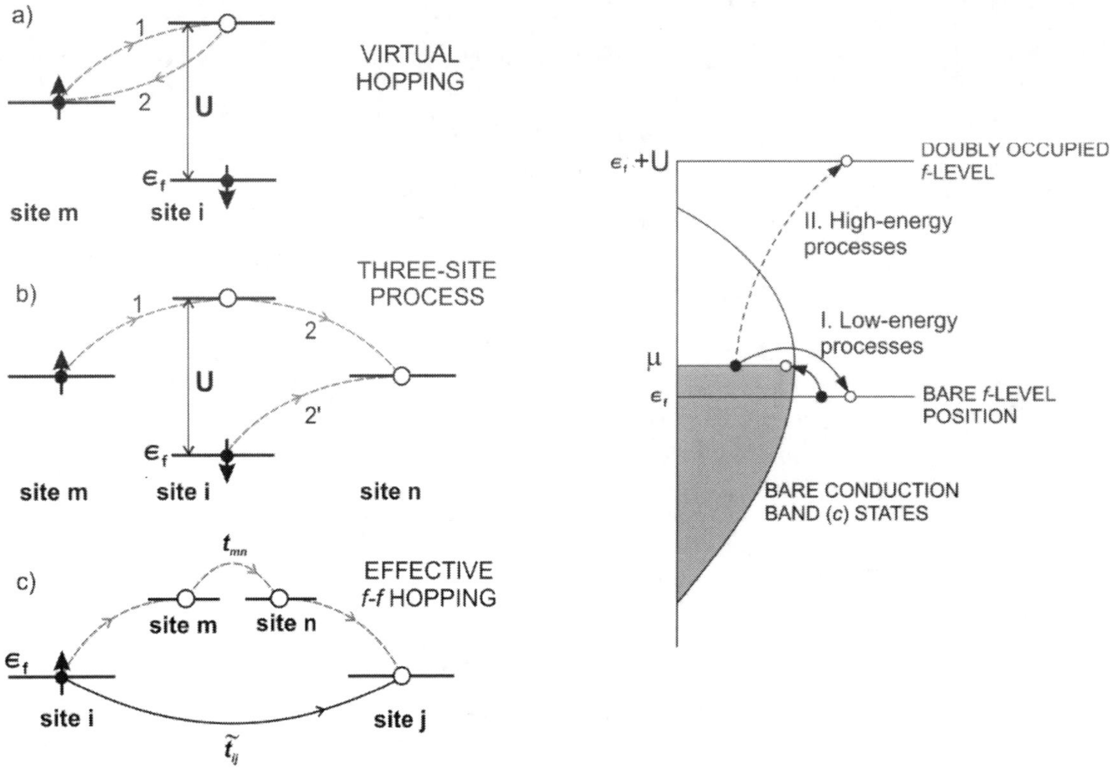

Figure 2. On the left: the possible interband hopping process in real space involving virtual hopping process to the double occupied f-level a) and the three-state (site) hopping process (b) and c), 2 and 2' alternative steps). The arrow (labels) by $\widetilde{t}_{ij}$ represents the resulting intersite $f-f$ hopping process. On the right: schematic representation of hybridization-induced process as f-level occupation dependent hopping process and its division into the low- and the high-energy processes. The former (I) corresponds to the presence of the residual hybridization term in (8); the other (II) leads to the Kondo-type coupling which in turn is expressed as a real-space hybrid pairing in (11) in the second order in V/ϵ_f.

separately. This limit of the Anderson-lattice Hamiltonian, the so-called *Kondo-lattice limit*, does not concern us here, even though it has been also discussed in the context of pairing.

The situation of interest to us here is provided by the condition that $|V_{in}| \ll U$, but $|V_{in}| \lesssim |\epsilon_f|$, as represented schematically in Fig 2.

This means that we transform out only the high-energy terms of the order $\sim V_{im}/U$ and replace them with the effective kinetic-exchange interactions of the Kondo type, but leave at the same time the residual hybridization term $\sim V_{im}\,(1-N_{\bar{\sigma}})\,f^{\dagger}_{im}c_m + H.c.$, untouched. In result, the effective Hamiltonian in the second order acquires the following form [24]

$$\begin{aligned} \mathcal{H} &= \sum_{mn\sigma} t_{mn} c^{\dagger}_{n\sigma} c_{n\sigma} + \epsilon_f \sum_{i\sigma} \nu_{i\sigma} + \sum_{im\sigma} V_{in}\,(1-N_{i\bar{\sigma}}) \left(f^{\dagger}_{i\sigma} c_{m\sigma} + c^{\dagger}_{m\sigma} f_{i\sigma} \right) \\ &+ \sum_{im} \frac{2V_{im}^2}{\epsilon_f + U} \left(\mathbf{S}_i \cdot \mathbf{s}_m - \frac{1}{4}\nu_i n_m \right) + \frac{1}{\epsilon_f + U} \sum_{imn\sigma} V_{mi} V_{in} S_i^{\sigma} c^{\dagger}_{m\sigma} c_{n\bar{\sigma}}, \end{aligned} \tag{7}$$

where now $t_{mn} \equiv t_{mn} - \mu\delta_{mn}$, and $\epsilon_f \equiv \epsilon_f - \mu$. As in the single-narrow-band case, we have projected out completely the double occupancies on f-site, as well as have defined the quantities $\nu_{i\sigma} \equiv N_{i\sigma}(1 - N_{i\bar{\sigma}})$, $\nu_i = \sum_\sigma \nu_{i\sigma}$, $\mathbf{S}_i \equiv (S_i^\sigma, S_i^z) \equiv \left[f_{i\sigma}^\dagger f_{i\bar{\sigma}}, \frac{1}{2}(\nu_{i\uparrow} - \nu_{i\downarrow})\right]$, and $\mathbf{s}_m = (s_m^\sigma, s_m^z) = \left[c_{m\sigma}^\dagger c_{m\bar{\sigma}}, (1/2)(n_{m\uparrow} - n_{m\downarrow})\right]$. The third and the fourth terms represent the Kondo coupling the truncated a-c mixing respectively. The Kondo antiferromagnetic exchange integral here is $J_{im}^{(K)} = 2V_{im}^2/(\epsilon_f + U)$. The dynamical process leading to the above Hamiltonian have been sketched in Fig. 2. In this situation, the antiferromagnetic Kondo coupling corresponds to the kinetic exchange in the narrow-band case. Note that in this form a second-order renormalization of the hopping part has been neglected.

2.4. Kondo interaction as hybrid pairing

In analogy to the Section 2.2 we introduce here the so-called hybrid pairing operators in real space

$$\begin{cases} A_{im}^\dagger = & \frac{1}{2}\left[f_{i\uparrow}^\dagger c_{m\downarrow}^\dagger (1 - N_{i\uparrow}) - f_{i\downarrow}^\dagger c_{m\uparrow}^\dagger (1 - N_{i\downarrow})\right] \\ A_{im} = & \left(A_{im}^\dagger\right)^\dagger . \end{cases} \tag{8}$$

Note that the f-electron operators contain the same type of projections as those in Subsection 2.1. In that situation, the hole affective Hamiltonian can be cast in a bit more compact form as follows

$$\mathcal{H} = \sum_{mn\sigma} t_{mn} c_{m\sigma}^\dagger c_{n\sigma} + \sum_{im\sigma}\left[V_{in}(1 - N_{i\bar{\sigma}}) f_{i\sigma}^\dagger c_{m\sigma} + H.c.\right] - \sum_{imn} \frac{2V_{im}V_{in}^*}{\epsilon_f + U} A_{im}^\dagger A_{in} . \tag{9}$$

In this form we have taken into account the fact, that the hybridization matrix element V_{in} can be complex. In principle, we can also include the higher order terms. In the fourth order the pairing part between f-electrons takes the form (5). The basic question when the Hamiltonian for e.g. magnetic oxides can be taken in the simpler single-band form (5). The question has not answered definitely as yet, although there exist a subdivision into Mott- and charge-transfer insulators [19].

2.5. Specific features of the magnetic real-space pairing

The pairing operators $B_{ij}^\dagger B_{ij}$ and $A_{im}^\dagger A_{im}$ have not a direct relation to the diagonal particle-number operator $a_{i\sigma}^\dagger a_{j\bar{\sigma}}^\dagger a_{j\bar{\sigma}} a_{i\sigma} = n_{i\sigma} n_{j\bar{\sigma}}$, as the latter represents intersite (for $i \neq j$) particle-particle, density-density correlation function, whereas the former represent the pair spin-singlet correlations. However, the pairing part of the total energy diminishes if the average $\langle B_{ij}^\dagger B_{ij}\rangle$ is nonzero. This average has a nontrivial structure, namely $B_{ii}^\dagger B_{ii} \equiv 0$. The last property follows from the circumstance that the double occupancies are strictly projected out on each site. This circumstance leads in turn, to a nontrivial structure of the superconducting gap, as we shall see below. Both of the above properties follow from fact that the pairing operators are composed of the projected fermion operators $b_{i\sigma}^\dagger \equiv a_{i\sigma}(1 - n_{i\bar{\sigma}})$and $b_{i\sigma} \equiv a_{i\sigma}(1 - n_{i\bar{\sigma}})$, and similarly for f operators. Obviously, there is a long way to go from forms (3) and (9), representing the real-space correlations to the condensed paired state.

The second characteristic feature of the whole approach is that the projected Fermi operators have non-fermionic anticommutation relations. For example,

$$\left\{b_{i\sigma}, b_{j\sigma}^\dagger\right\} = \left[(1 - n_{i\bar{\sigma}})\delta_{\sigma\sigma'} - S_i^{\bar{\sigma}}(1 - \delta_{\sigma\sigma'})\right]\delta_{ij}. \tag{10}$$

This introduces a completely new feature of resulting quantum-liquid states, although this property has not been explored in full. Additionally, the hopping and the exchange-interaction terms do not commute with each others, so the motion of electrons and the spin-interaction (e.g. the spin-flip processes) are entangled with each other and result in a *strongly correlated* (non-Fermi liquid) metallic state. What is equally important, the kinetic-energy ($\sim t_{ij}$) and the exchange $\left(\sim t_{ij}^2/U\right)$ can easily become comparable for filing close to the half filling (n) and we can have a transformation form such metal to the localized antiferromagnetic state (or vice versa) or to the phase separation into antiferromagnetic islands immersed in the sea of $(1-n)N$ holes with ferromagnetic ordering of the spin inside the sea. The second state is rather unstable as the Coulomb repulsion introduced by the hole condensation in part of the system will charge the system and destabilize such phase-separated state. Nonetheless, an itinerant antiferromagnetic (or spin density wave) state may coexist with the paired state. Such possibility arises from the identity which for the single narrow-band case reads

$$\mathbf{S}_i \cdot \mathbf{S}_j - \frac{1}{4}\nu_i\nu_j \equiv -B_{ij}^{\dagger}B_{ij}, \tag{11}$$

and hence, the nonzero averages of the type $\langle B_{ij}\rangle$ may imply at the same time nonzero average of magnetic moment $\langle S_i^z\rangle$ and vice versa. Likewise, from the analogous identity for the $f-c$ system, i.e.

$$\mathbf{S}_i \cdot \mathbf{s}_m - \frac{1}{4}\nu_i n_m \equiv -A_{im}^{\dagger}A_{im}, \tag{12}$$

it follows, that the nontrivial structure of magnetic correlations expressed via $\langle \mathbf{S}_i \cdot \mathbf{s}_m\rangle$ may implicate nontrivial structure of the averages $\langle A_{im}^{\dagger}\rangle$ (and vice versa) in some sort of mean field approximation. Obviously, there is a problem, to what extent the magnetic and the paired states can be regarded as different states in view of the identities (11) and (12). This question can be answered in a way empirically only when studying relative stability of different states obtained from an approximate treatment, as the exact solution of any of these models is not available.

3. Renormalized mean-field solution: pure paired states

We concentrate here next on the analysis of the so-called *renormalized mean-field* solution [25-30] for both t-J model (5) and the Anderson-lattice with the Kondo-type pairing (9) in the narrow-band limit. The standard method of approach is either the Gutzwiller approximation (GA) or the slave-boson approach, the latter constructed in such a manner that it reduces to the Gutzwiller approximation for the cases of the simplest normal and paramagnetic states. However, in the case of magnetic and/or superconducting state, the SB approach additionally introduces specific effective fields which also renormalize the chemical potential. Recently, we have proposed a new method of approach [28] which combines the feature of the GA and SB methods in the mean-field approximation. We called it the *statistically-consistent Gutzwiller approximation* (SGA). We discuss this method on two examples of the models introduced above leaving the detailed elaboration of the method to separate publications.

3.1. General features of mean-field approximation for strongly correlated electrons in a narrow band

Hamiltonian (5) (or its equivalent form (2)), contains principal features two principal features. First of them is the restriction on motion (hopping) of the single particles due to the exclusion of double occupied sites (it is called the projection). The second is a kind of real-space correlations, which are composed of the pairing part (for $k=i$) and the pair-hopping processes ($k \neq i$). Both the Gutzwiller projection and the real-space pairing must be treated jointly. The treat the first of them a certain type of approximation has to be used which starts back to the original Gutzwiller considerations for the Hubbard model (1) [20], and in the present context, applied to t-J and related models was developed later [25, 27]. Its essence relies in replacing the operator part of

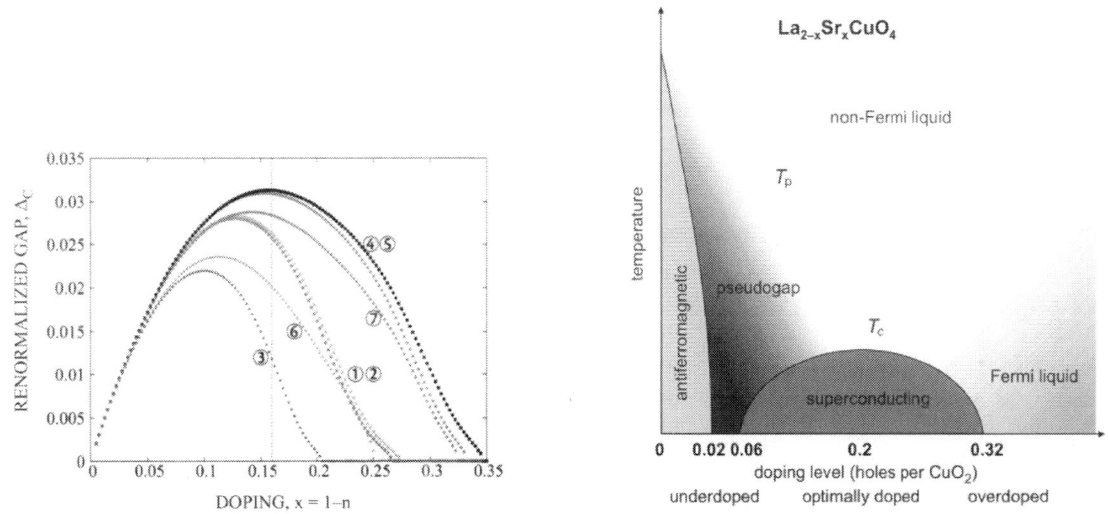

Figure 3. On the left: Doping dependence of the renormalized superconducting order parameter Δ_c. The curves 1-7 are explained in detail in Ref. [11] and correspond to various methods of solving Hamiltonian (2). The most important feature of the solution is the appearance of the upper critical concentration for the d-wave superconductivity disappearance. The vertical line defines roughly the optimal doping. On the right: overall phase diagram for $La_{2-x}Sr_xCuo_4$ on the temperature T - doping x plane.

the hopping part: $b^\dagger_{i\sigma}b_{i\sigma} \cong g^t_{ij}a^\dagger_{ij}a^\dagger_{i\sigma}a_{j\sigma}$ and the paired operators $B^\dagger_{ij} \cong g^J_{ij}\left(a^\dagger_{i\uparrow}a^\dagger_{j\downarrow} - a^\dagger_{i\downarrow}a^\dagger_{j\uparrow}\right)/2$, where g^t_{ij} and g^J_{ij} are called renormalization factors and take into account on average scale the hopping restriction. In the simplest (Gutzwiller) version $g^t_{ij} = (1-n)/(1-n/2))$ and $g^J_{ij} = 1$, this approximation leads to good description of antiferromagnetism, but not of the superconducting state, as it provides then a maximal pairing strength in the limit Mott insulating limit (i.e. when the band filling $n \to 1$).

3.2. Results

In general, with the help of the above procedure we get the renormalized hopping probability and the pairing gap magnitude. Namely, if we define $\chi_{ij} \equiv \langle a^\dagger_{i\sigma}a_{j\sigma}\rangle$ and $\Delta_{ij} \equiv \langle a_{i\bar{\sigma}}a_{j\sigma}\rangle$, then the corresponding physical quantities have the renormalized form: $g^t_{ij}\chi_{ij}$ and $g^J_{ij}\Delta_{ij}$.

The second factor is to make some sort of mean-field (Hartree-Fock) approximation on the pairing part. In the simplest version, it is decomposed in the following manner

$$B^\dagger_{ij}B_{ij} \cong \langle B^\dagger_{ij}\rangle + B^\dagger_{ij}\langle B_{ij}\rangle - \langle B^\dagger_{ij}\rangle\langle B_{ij}\rangle \ . \tag{13}$$

This type of the decoupling $\lambda\sigma$ a real space version of the Bogoliubov-Valatin approximation commonly utilized in BCS theory. In practice, the H-F approximation scheme is much more involved, as it comprises both using a more sophisticated than Gutzwiller method [22], as well as including in the Hartree-Fock scheme all possible decoupled terms [23] and the so-called statistical consistency conditions [27] ensuring agreement between the self-consistently and variationally obtained results for mean fields and other parameters. There is no need for repeating this quite involved analysis here. Therefore, we only illustrate the final results and discuss them briefly in physical terms.

In Fig. 3a we plot the renormalized gap value $\Delta_c = g^J \Delta_{\langle ij \rangle}$ for nearest neighbors $\langle ij \rangle$ in the square lattice as a function of hole number (doping) $x \equiv 1 - n$. This dependence reflects the overall trend of the experimental data for $la_{2-x}Sr_xCuO_4$ displayed in Fig.4. Before explaining various curves (i.e. specifying the microscopic parameters and different versions of the model leading to the curves 1-7) let make the following note. The presence of the upper critical concentration is in our view a crucial factor to be reproduced by any approach pretending to describe correctly the phenomenon of high temperature superconductivity. This is because can appear only in the situation with the real space pairing. Explicitly, as the kinetic energy of particle is roughly $\sim tx$ and the exchange energy $\sim t(1-x)^2$ and for parameters chosen (see below) they are comparable for $x \sim 0.1$. For $x \geq 0.3$ the kinetic energy is already substantially larger than the pairing part, so the pairing correlations between neighboring spin is radically reduced. On the other hand, the hopping is reduced to zero in the complementary limit $x \to 0$ and the metallic paired state cases to exist (the Mott insulating state becomes stable).

What concerns detailed information concerning the calculation of the curves 1-7 the situation i as follows. The ratio $T/|t| = 0.3$, which in the Hubbard model terminology means $U/|t| = 12$. Taking into account the fact that the experimental value of $T = 1500$K, we obtained that $|t| \simeq 400$ meV $\simeq 5000$K. The last value is important since the magnitude Δ_c of the calculated renormalized super-conducting gaps (of the d-wave $\Delta_k = \Delta_c (\cos k_x - \cos k_y)$); this means that its maximum value at the optimal doping (the dotted vertical line) is $\Delta_c \simeq 150$K. The group labels 1-3 and 4-6 correspond respectively to the two situation: the second-neighbor hopping $t'/|t| = 0$ and 0.25. Additionally, the curves within 1-3 ($t' = 0$) and 4-6 ($t' = 0.25|t|$) represent three additional conditions imposed, namely 1(4): $c_1 = c_2 = 0$; 2(5): $c_1 = 1, c_2 = 0$; and 3(6): $c_1 = c_2 = 1$. We can now thus explain the meaning of the extra constants c_1 and c_2. The full t-J model is that with $c_1 = c_2 = 1$. The situation with $c_1 = c_2 = 0$ [25] and $c_1 = 1, c_2 = 0$ [26] are also used. Hence with the help of those extra parameters we can single out the particular role of particular terms of (2) in the pairing. One can see clearly that the presence of the three-site terms suppresses both the concentration range in which the superconducting solution exists (i.e. reduces the upper critical concentration), as well as reduces the superconducting gap. The role of term $\sim c_2$ is minor even though it express an attractive density-density interaction between the neighbors, the reason being it does not differentiate between the singlet and the triplet of diagonal correlations $\langle c^\dagger_{i\sigma} c^\dagger_{j\sigma'} \rangle$. Note also that the curve labeled as 7 comprises the fit closest to experiment and corresponds to slightly changed values of parameters. The detailed discussion of quasiparticle energies, their Fermi velocity, and of the upper critical concentration is provided elsewhere [27].

The summarize this Section, we can say that from the day of the complete t-J model derivation for strongly correlated metals [20] to its reinterpretation in terms of real-space pairing operators for cuprate superconductors [17], we have reaching the (semi) quantitative stage of description the concrete single plane high-T_c systems [27] in the overdoped regime. However, we still require the detailed interpretation of the underdoped regime $x \leq 0.15$ and in particular the coexistence of magnetic density order with superconductivity the understanding evaluation of the Fermi-surface topology with the doping [32], and the quantitative explanation of the pseudogap appearance [33]. We should be able to see a progress along those lines soon.

3.3. Kondo-interaction induced pairing: spin-dependent effective masses and Fulde-Ferrell-Larkin-Ovchinnikov (FFLO) phase

In this Section we concentrate on a novel state discovered in strongly correlated system, namely on the Fulde-Ferrell-Larkin-Ovchinnikov state proposed theoretically many years ago [34].The motivation for us for carrying out the overviewed below work was the discovery that this state has been observed in the heavy-fermion system $CeCoIns_5$ [35]. The interesting thing is that in the same system the spin-dependent masses have been observed earlier [36]. This last discovery was predicted earlier theoretically [37] and leads to rather strong Zeeman spin splitting of distortion induced by correlations of the band states. Therefore, it seems natural to combine the two effects appearing simultaneously when the electronic correlations are considered. However, one has to include then also the appearance of the effective field driven by the correlations and the real space pairing introduced in Section 2.2 when the correlations are strong . All of these factors have been included the recent series of papers for both electron-gas [38, 39] and two-dimensional-band systems [40].

Spin-dependent masses
For a quasiparticle gas of correlated particles, their energy in the applied field $h = \frac{1}{2}g\mu_B H_a$, when the Landau quantization is neglected, is expressed as [38, 39]

$$\xi_{\mathbf{k}\sigma} = \frac{\hbar^2 k^2}{2m\sigma} - \sigma h - \mu - \sigma h_{corr}, \tag{14}$$

where h_{corr} is the field induced by the correlations and spin-dependent mass enhancement m_σ / m_B is of the form in the limit of $U \to \infty$ for a single narrow band [37]

$$\frac{m_\sigma}{m_B} = \frac{1 - n_\sigma}{1 - n} = \frac{1 - n/2}{1 - n} - \sigma \frac{\bar{m}}{2(1 - n)} \equiv \frac{1}{m_B}\left(m_{av} - \frac{\sigma}{2}\Delta m\right), \tag{15}$$

where m_B is the bare (band) mass, $\bar{m} = n_\uparrow - n_\downarrow$ is the system spin polarization, and n is the band filling ($n = n_\uparrow + n_\downarrow$). Also, $\Delta m = m_\downarrow - m_\uparrow$ is the mass difference and $m_{av} = (m_\uparrow + m_\downarrow)/2$ is the average quasiparticle mass. It is interesting to note that in the magnetic-saturation limit we recover the band limit for the spin-majority subband, i.e. $m_\uparrow / m_B = 1$, whereas the heavy quasiparticles in the spin-minority band (with $m_\downarrow / m_B = 1/(1 - n)$) disappear. Two features are important here: (i) the masses are high in the almost-localized limit $(1 - n) \ll 1$ and (ii) the mass $m_\downarrow$ in the spin-minority band is the heaviest, since due to the spin imbalance, those quasiparticles scatter very strongly (due to the presence of the large-magnitude Hubbard term $\sim U \sum_i n_{i\uparrow} n_{i\downarrow}$). Additional features follows from the circumstance that we can "switch-off" completely the Hubbard interaction by applying the magnetic field and saturating magnetically the system. This is possible only (and is the case) because the field induced by the correlations enhances strongly the effect of applied magnetic field.

The brief analysis provided above delineates the principal message about what we mean by nonstandard quasiparticles in a (strongly) correlated system. First, they can become quite heavy, i.e. $m_{av}/m_B \gg 1$. Second, they depend on spin direction, what makes them distinguishable in the quantum-mechanical sense, since the mass in nonrelativistic quantum mechanics is an external (input) characteristic in the problem at hand. Third, the effective field driven by correlations can become very strong i.e. much stronger than any molecular field appearing in traditional magnetism. All these microscopic properties must be determined self-consistently. This feature of nonstandard quasiparticles distinguishes the way they are introduced, as compared to that they are defined within the original phenomenological Landau- Fermi-liquid theory, where the enhancement of the effective mass and of the magnetic susceptibility is expressed in terms of interaction parameters, which are not determined within the scheme. Additionally, in the Landau scheme the enhancement factors are determined by including the interaction only among the quasiparticles at and/or in close vicinity of the Fermi surface. Here, *all the particles* mutually influence each other, what is expressed via an integration over all occupied states when solving appropriate self-consistent equations.

Superconducting state: Fulde-Ferrell-Larkin-Ovchinnikov state
We will describe next the Fulde-Ferrell-Larkin-Ovchinnikov (FFLO) state for the model heavy-fermion system starting from Hamiltonian (9) with the hybrid pairing. This Hamiltonian in the

simplest Gutzwiller approximation yields the following effective BCS-type Hamiltonian

$$\mathcal{H} = \sum_{\mathbf{k}\sigma} \left[\epsilon_{\mathbf{k}\sigma} c^{\dagger}_{\mathbf{k}\sigma} c_{\mathbf{k}\sigma} + \tilde{\epsilon}_{f\sigma} f^{\dagger}_{\mathbf{k}\sigma} f_{\mathbf{k}\sigma} + \tilde{V}_{\mathbf{k}\sigma} f^{\dagger}_{\mathbf{k}\sigma} c_{\mathbf{k}\sigma} + \tilde{V}^{*}_{\mathbf{k}\sigma} c^{\dagger}_{\mathbf{k}\sigma} f_{\mathbf{k}\sigma} \right] - \frac{2}{\epsilon_f + U} \frac{1}{N} \sum_{\mathbf{k}\mathbf{k}'Q} V_{\mathbf{k}} V^{*}_{\mathbf{k}'} A^{\dagger}_{\mathbf{k}\mathbf{Q}} A_{\mathbf{k}'\mathbf{Q}}, \tag{16}$$

with

$$A_{\mathbf{k}\mathbf{Q}} = \frac{1}{\sqrt{2}} \left[f^{\dagger}_{\mathbf{k}+\mathbf{Q}/2\uparrow} c^{\dagger}_{-\mathbf{k}+\mathbf{Q}/2\downarrow} - f^{\dagger}_{\mathbf{k}+\mathbf{Q}/2\downarrow} c^{\dagger}_{-\mathbf{k}+\mathbf{Q}/2\uparrow} \right]. \tag{17}$$

The single-particle part can be easily diagonalized by moving to the hybridized basis, whereas the pairing part is represented by a separable pairing potential. We proceed with the transformation to the hybridized basis first, which yields the following transformed pairing part among the quasiparticles in the lower hybridized band (appropriate for $\langle f^{\dagger}_{i\sigma} f_{i\sigma} \rangle + \langle c^{\dagger}_{i\sigma} c_{i\sigma} \rangle \leq 1$)

$$\mathcal{H} = \sum_{\mathbf{k}\sigma} E_{\mathbf{k}\sigma} \alpha^{\dagger}_{\mathbf{k}\sigma} \alpha_{\mathbf{k}\sigma} - \frac{4}{\epsilon_f + U} \sum_{\mathbf{k}\mathbf{k}'\mathbf{Q}} \frac{|V_{\mathbf{k}} V_{\mathbf{k}'}|^2 (q_{\sigma} q_{\bar{\sigma}})^{1/2}}{[(\epsilon_{\mathbf{k}\sigma} - \tilde{\epsilon}_{f\sigma})^2 + |\tilde{V}_{\mathbf{k}\sigma}|^2]^{1/2} [(\epsilon_{\mathbf{k}'\bar{\sigma}} - \tilde{\epsilon}_{f\bar{\sigma}})^2 + |\tilde{V}_{\mathbf{k}'\bar{\sigma}}|^2]^{1/2}} \times \alpha^{\dagger}_{\mathbf{k}+\mathbf{Q}/2\uparrow} \alpha^{\dagger}_{-\mathbf{k}+\mathbf{Q}/2\downarrow} \alpha_{-\mathbf{k}'+\mathbf{Q}/2\downarrow} \alpha_{\mathbf{k}'+\mathbf{Q}/2\uparrow}, \tag{18}$$

where the quasiparticle energy in the normal state is given by

$$E_{\mathbf{k}\sigma} \equiv \frac{\epsilon_{\mathbf{k}\sigma} + \tilde{\epsilon}_{f\sigma}}{2} - \left[\left(\frac{\epsilon_{\mathbf{k}\sigma} + \tilde{\epsilon}_{f\sigma}}{2} \right)^2 + |\tilde{V}_{\mathbf{k}\sigma}|^2 \right]^{1/2}. \tag{19}$$

In all these equations $\epsilon_{\mathbf{k}\sigma} = \epsilon_{\mathbf{k}} - \sigma h - \mu$, $\epsilon_{\mathbf{k}\sigma} = \epsilon_f - \sigma h - \mu$, $\tilde{V}_{\mathbf{k}\sigma} = V_{\mathbf{k}} q_{\sigma}^{1/2}$, where the normalization factor is takes in the Gutzwiller approximation [31]. For the constant (intraatomic) form of the hybridization $V_{\mathbf{k}} = V = -|V|$, we obtain effectively single-band model discussed in detail in [36], with heavy electrons of primary f type, i.e $\alpha_{\mathbf{k}\sigma} \simeq f_{\mathbf{k}\sigma}$ if $1 - n \ll 1$. Below we present only the principal features of the phase diagram leaving the details for the full papers [36, 37, 38]. Note that in general we have here the center-of-mass momentum of the Cooper pair $\mathbf{Q} \neq 0$ to include a possibility of the FFLO phase appearance (strictly speaking, we consider only the Fulde-Ferrell case).

In the standard Hartree-Fock approximation with anomalous averages, Hamiltonian (18) in the narrow band limit reduces to the form

$$\mathcal{H} = \sum_{\mathbf{k}\sigma} (\epsilon_{\mathbf{k}\sigma} - \mu) f^{\dagger}_{\mathbf{k}\sigma} f_{\mathbf{k}\sigma} - g\mu_B H \sum_{\mathbf{k}} \left(f^{\dagger}_{\mathbf{k}\uparrow} f_{\mathbf{k}\uparrow} - f^{\dagger}_{\mathbf{k}\downarrow} f_{\mathbf{k}\downarrow} \right) + \sum_{\mathbf{k}} \left(\Delta^{*}_{\mathbf{k}\mathbf{Q}} f_{\mathbf{k}\uparrow} f_{-\mathbf{k}+\mathbf{Q}\downarrow} + H.c. \right) + N \frac{|\Delta_{\mathbf{k}\mathbf{Q}}|^2}{V_0}, \tag{20}$$

with the single-particle energy parameterized in the hight-binding approximation, which in for the case of square lattice takes the form

$$\epsilon_{\mathbf{k}\sigma} \equiv q_{\sigma}[-2t(\cos k_x + \cos k_y) + 4t' \cos k_x \cos k_y], \tag{21}$$

and q_{σ} is, as before, the Gutzwiller band narrowing factor (the spin-dependent mass enhancement) and the superconducting gap is determined from the self-consistent equation

$$\Delta_{\mathbf{k}\mathbf{Q}} = -\frac{V_0}{N} \sum_{\mathbf{k}'} \gamma_{\mathbf{k}} \gamma_{\mathbf{k}'} \langle f_{-\mathbf{k}'+\mathbf{Q}\downarrow} f_{\mathbf{k}'\uparrow} \rangle. \tag{22}$$

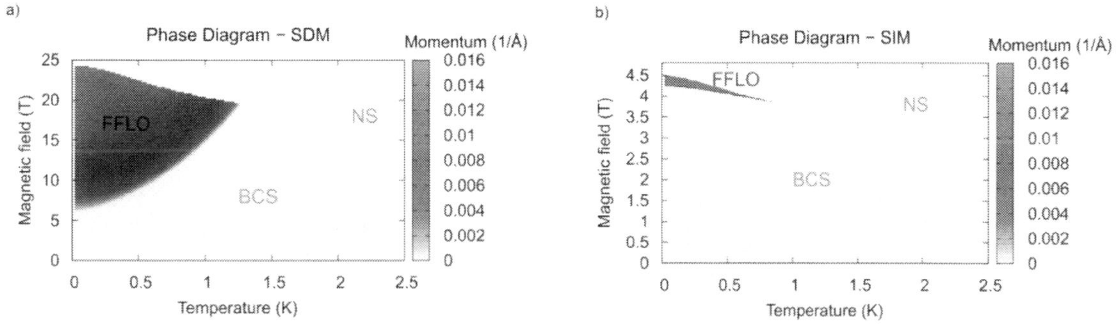

Figure 4. Phase diagram for the cases with the spin-dependent (a) and the spin-independent masses (b). Yellow region corresponds to $\mathbf{Q} = 0$ (BCS phase), the darker one to $\mathbf{Q} \neq 0$ (FFLO phase) and the white to normal state. Note that with increasing temperature, the transition from BCS to FFLO state occurs at higher fields, in qualitative agreement with experimental results [28]. The FFLO phase is stable in an extended H_a-T regime only in the spin-dependent-masses (SDM) case.

Additionally, V_0 is the pairing magnitude which for the constant hybridization has the form

$$V_0 = -\frac{4V^2 \left(q_\sigma q_{\bar{\sigma}}\right)^{1/2}}{\epsilon_f + U}, \tag{23}$$

and the factors $\gamma_{\mathbf{k}}$ and $\gamma_{\mathbf{k}'}$ correspond to the separable $\mathbf{k}$-dependent factors in (18) divided by V^2.

Next, we carry out the approximate form of the Bogoliubov transformation which in the meantime acquired the name of the Bogoliubov-Valatin-de Gennes-Nambu transformation! For that purpose we represent (20) in the matrix form [38-40]

$$\mathcal{H} = \sum_{\mathbf{k}} (f^\dagger_{\mathbf{k}\uparrow}, f_{-\mathbf{k}+\mathbf{Q}\downarrow}) \begin{pmatrix} \epsilon_{\mathbf{k}\uparrow} - g\mu_B H - \mu & \Delta_{\mathbf{k},\mathbf{Q}} \\ \Delta^*_{\mathbf{k},\mathbf{Q}} & -\epsilon_{-\mathbf{k}+\mathbf{Q}\downarrow} - g\mu_B H + \mu \end{pmatrix} \begin{pmatrix} f_{\mathbf{k}\uparrow} \\ f^\dagger_{-\mathbf{k}+\mathbf{Q}\downarrow} \end{pmatrix} + \sum_{\mathbf{k}} (\epsilon_{\mathbf{k}\downarrow} + g\mu_B H - \mu) + N\frac{\Delta^2_{\mathbf{Q}}}{V_0}. \tag{24}$$

The transformation to the quasiparticle representation has the usual form

$$\begin{pmatrix} \tilde{\alpha}_{\mathbf{k}} \\ \tilde{\beta}^\dagger_{\mathbf{k}} \end{pmatrix} = \begin{pmatrix} u_{\mathbf{k}} & v_{\mathbf{k}} \\ -v_{\mathbf{k}} & u_{\mathbf{k}} \end{pmatrix} \begin{pmatrix} f_{\mathbf{k}\uparrow} \\ f^\dagger_{-\mathbf{k}+\mathbf{Q}\downarrow} \end{pmatrix}, \tag{25}$$

with the Bogoliubov coherence factors given now by the relations

$$u_{\mathbf{k}} = \left[\frac{1}{2}\left(1 + \frac{\epsilon_{\mathbf{k}\uparrow} + \epsilon_{-\mathbf{k}+\mathbf{Q}\downarrow} - 2\mu}{\sqrt{(\epsilon_{\mathbf{k}\uparrow} + \epsilon_{-\mathbf{k}+\mathbf{Q}\downarrow} - 2\mu)^2 + 4\Delta^2_{\mathbf{kQ}}}}\right)\right]^{1/2}, \tag{26}$$

$$v_{\mathbf{k}} = \left[\frac{1}{2}\left(1 - \frac{\epsilon_{\mathbf{k}\uparrow} + \epsilon_{-\mathbf{k}+\mathbf{Q}\downarrow} - 2\mu}{\sqrt{(\epsilon_{\mathbf{k}\uparrow} + \epsilon_{-\mathbf{k}+\mathbf{Q}\downarrow} - 2\mu)^2 + 4\Delta^2_{\mathbf{kQ}}}}\right)\right]^{1/2}. \tag{27}$$

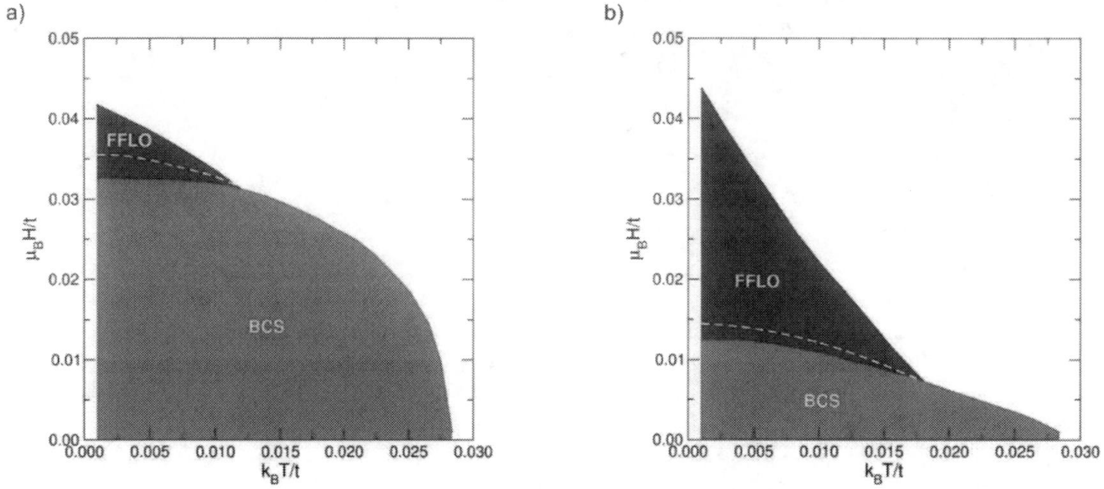

Figure 5. Phase boundaries for a two-dimensional d-wave superconductor with both the spin-independent masses (SIM) (a) and with the spin-dependent masses (SDM) (b). The FFLO-BCS transition line is of discontinuous nature. The dashed line marks the stability limit of the BCS state as determined by the value of the second critical field H_{c2} for the BCS state. The values of parameters are $n = 0.97$ and $V_0 = 12.5\,K$. For these values of the parameters, the superconducting transition temperature is $T = 2.5\,K$ and the uppermost critical field for the FFLO phase is above $6\,T$. Note that the FFLO state is robust in the situation with SDM and this result is one of the principal features of the present discussion.

The quasiparticle energies in the phase with $\mathbf{Q} \neq 0$ are

$$E_{\mathbf{kQ}\alpha} = \frac{1}{2}\left(\epsilon_{\mathbf{k}\uparrow} - \epsilon_{-\mathbf{k}+\mathbf{Q}\downarrow}\right) - g\mu_B H + \alpha\frac{1}{2}\left[\left(\epsilon_{\mathbf{k}\uparrow} + \epsilon_{-\mathbf{k}+\mathbf{Q}\downarrow} - 2\mu\right)^2 4|\Delta_{\mathbf{kQ}}|^2\right]^{1/2}, \tag{28}$$

where sign factor $\alpha = \pm$ corresponds to the electron or hole excitations, respectively.

Having discussed the explicit expression for the fermionic quasiparticle excitations, we can construct the free energy functional, as well as determine the system of self-consistent equations for $\Delta_{\mathbf{kQ}}$, μ, and $\bar{m}$; we also optimize the energy with respect to the magnitude of the wave vector $\mathbf{Q}$. One should also mention that for the electron-gas situation we have included explicitly in the effective Hamiltonian the correlation field h_{corr}, which we optimize, whereas for the two-dimensional band structure (21) we have been able so far to carry out only the whole analysis in the Gutzwiller approximation for the Hamiltonian (24), before the pairing is included. In Fig. 4 and 5 we provide the exemplary phase diagrams on the plane temperature T - applied magnetic field H for the three-dimensional gas [38] and square-lattice cases [40], respectively.

The most important feature coming out of these Figures are: (i) the BCS state (i.e. that with $\mathbf{Q} = 0$) is quite robust in the lower fields and higher temperatures, (ii) the inclusion of the effective-mass spin dependence leads to a remarkable extension of regime of the FFLO stability, (iii) in the FFLO phase the upper critical field of the transition to the normal phase is much higher than that for the BCS superconducting state, and (iv) the first-order BCS→FFLO transition can be accompanied by a weak metamagnetic transition. One should also not that the detailed analysis of the FFLO state is carried out separately. Also, the full SGA analysis of the superconducting states within the full statistically-consistent Gutzwiller approach (SGA) for the present model is still to be finalized.

4. Concluding remarks

In this brief review we have tried to demonstrate the usefulness of the concept of an *unconventional Fermi liquid* for such diverse phenomena as high temperature superconductivity (as represented by t-J model) and heavy-fermion superconductivity (represented by the Kondo-type pairing) and particularly, the Fulde-Ferrell-Larkin-Ovchinnikov-state stability the latter (on the example of narrow-band limit with the Kondo-interaction induced pairing). The statistically-consistent (SGA) method proposed by us [28] is probably the most general Fermi-liquid approach possible for almost localized correlated systems, as we renormalize all relevant quantities: introduce the spin-dependent mass renormalization, as well as the effective field and the chemical-potential shift, all induced by the interparticle correlations. In addition to that, specific purely electronic mechanism of pairing in real space, leads to the BCS-type of Hamiltonian *for all electrons* and with a nontrivial (d-wave) form of the superconducting gap. Inclusion of quantum fluctuations in the mean-field SGA solution may be possible [27], but first one has to implement the SGA method completely by incorporating in it also the possibility of coexistence of superconductivity with antiferromagnetism. Also, the role of the intersite Coulomb (for the t-J model) and of the interorbital Coulomb (for the model with Kondo pairing) has to clarified, as the form works against real-space pairing driven by the electronic correlations. We should be able to see a progress along these lines in the near future.

Acknowledgments

The wok was supported by Grant No. N N202 128736 from our Ministry of Science and Higher Education. The results reviewed in this paper are based largely on the detailed work of my Ph.D. students Jakub Jędrak and Jan Kaczmarczyk. I am grateful to my students for many discussions. Numerous discussions with Profs. Maciej Maśka and Marcin Mierzejewski are appreciated. Technical help of Dr. Danuta Goc-Jagło is also acknowledged. The reviewed work represents part of the project TEAM supported by the Foundation for Polish Science (FNP).

References

[1] Dirac P A M *The Principles of Quantum Mechanics*, (Clarendon Press, Oxford, 1958) chapter V, §58
[2] Landau L D and Lifshitz E M *Quantum Mechanics: Non-Relativistic Theory* (Pergamon Press 1980) chapter IX, §62
[3] Mattis D C *The Theory of Magnetism I: Statics and Dynamics* (Springer Verlag, Berin 1981) chapters 2, 4, and 6
[4] cf. Ref. [2], chapter XVI, §116 - §117
[5] Vonsovsky S V *Magnetism* (Wiley, New York, 1974); Morrish A H *Physical Principles of Magnetism* (IEEE Press, New York 2001)
[6] Brink D M and Broglia R A *Nuclear Superfluidity*; cf. also articles in Pairing in Fermionic System (World Scientific Publications, Singapore, 2006) edited by Sedrakian A, Clark J W and Alford M; Fernandez M. A. and Egido J L 2005, Eur. Phys. J. B 48, 305 and References therein
[7] Pugh E, Saxena S and Lonzarich G in *Handbook of High-Temperature Superconductivity*, edited by Schrieffer J R and Brooks J, Springer Verlag, 2007, chapter 16 p. 597ff; see also Tranguada, chapter 6, p. 251ff
[8] For a didactical account of the model see: Spałek J 2007 Acta Phys. Polon. A **111** 409 [cf. arXiv: 0706.4236]
[9] Monthoux P, Pines D and Lonzarich G G 2007 *Nature* (London) **450** 1177 and References therein
[10] De Gennes P G *Superconductivity of Metals And Alloys* (Addison-Wesley Publ. Co, Reading MA, 1989)
[11] Leggett A J *Quantum Liquids* (Oxford University, 2006), chapter 6; Superconductivity and Superconductivity (Cambridge University Press, 1998) chapter 6
[12] Huxley A, Sheiken I, Ressouche E, Kernavanois N, Braithwaite D, Calemczuk R and Flouquet J 2001 *Phys. Rev.* B **63** 144519
[13] Annett J F, Györffy B I, Litak G and Wysokiński K I, 2003 Eur. Phys. J. B **36** 301
[14] Anderson P W 2007 *Science* **317** 1705

[15] Anderson P W 1959 *Phys. Rev.* **115** 2
[16] Anderson P W in *Frontiers and Borderlines in Many Particle Physics* edited by Broglia R A and Schrieffer R J (North Holland, Amsterdam, 1988), p.1ff; Spałek J 1988 *Phys. Rev.* B **37** 533; Ruckenstein A, Hirschfeld P and Appel J 1987 *Phys. Rev.* B **36**, 857
[17] Spałek J 1988 *Phys. Rev.* B **37** 533; 2007 Acta Phys. Polon. A **111** 409
[18] Cf. *The Physics of Superconductors* edited by Bennemann K H and Ketterson J B (Springer Verlag, Berlin 2003) vol. I, chapter 6, 9, and vol II, chapter 3, 5, and 7
[19] Zaanen J, Sawatzky G A and Allen J W, 1986 J. Magn. Magn. Mater. **54-57** 607; Zaanen J and Sawatzky G A 1990 J. Solid St. Chem. **88** 8
[20] Spałek J and Oleś A M 1977 *Physica B* **86-88** 375; Chao K A, Spałek J and Oleś A M 1977 *J. Phys. C* **10** L271
[21] Spałek J 2008 Condens. Matter Phys. **11** 455
[22] Klejnberg A and Spałek J 2000 Phys. Rev. B **61** 15542
[23] Shrieffer J R and Wolff P A 1966 Phys. Rev. **149** 491
[24] Spałek J 1988 Phys. Rev. B **38** 208
[25] Zhang F C, Gros C, Rice T M and Shiba H 1988 *Supercond. Sci. Technol.* **1** 36; Sigrist M, Rice T M and Zhang F C 1994 *Phys. Rev.* B **49** 12058; Ogata M and Himeda A 2003 *J. Phys. Soc. Jpn.* **72** 374; Raczkowski M 2008 *Acta Phys. Pol.* A **114** 107 and References therein
[26] Fukushima N 2008 *Phys. Rev.* B **78** 11510
[27] Jędrak J and Spałek J 2010 *Phys. Rev.* B **81** 073108; Jędrak J and Spałek J 2011 *Phys. Rev.* B **83** 104512
[28] Jędrak J, Kaczmarczyk J and Spałek J 2010, arXiv:1008.0021 (unpublished)
[29] Ogata M and Himeda A 2003 *J. Phys. Soc. Jpn.* **72** 374; For review see Ogata M and Fukuyama H 2008 *Rep. Prog. Phys.* **71** 036501
[30] Edegger B, Muthukumar V N, Gros C and Anderson P W (2006) *Phys Rev Lett* **96** 207002
[31] Gutzwilerr M C 1963 *Phys. Rev. Lett.* **10** 159; Gutzwilerr M C 1965 *Phys. Rev.* **137** A1726
[32] Norman M R 2010 *Pysics* **3** 86; Chubukov A V 2010 *Physics* **3** 54 and References therein
[33] Laughlin R B 1997 *Phys. Rev. Lett.* **79** 1726; Lee P A, Nagaosa N, and Wen X-G, 2006 *Rev. Mod. Phys.* **78** 17
[34] Fulde P and Ferrell R A 1964 *Phys. Rev.* **135** A550; Larkin A I and Ovchinnikov Yu N 1964 *Zh. Exp. Teor. Fiz.* **47** 1136; 1965 *Sov. Phys. JETP* **20** 762
[35] Kumagai K, Saitoh M, Oyaizu T, Furukawa Y, Takashima S, Nohara M, Takagi H and Matsuda Y 2006 *Phys. Rev. Lett.* **97** 227002; Miclea C F, Nicklas M, Parker D, Maki K, Sarrao J L, Thompson J D, Sparn G and Steglich F 2006 *Phys. Rev. Lett.* **96** 117001; Kakuyanagi K, Saitoh M, Kumagai K, Takashima S, Nohara M, Takagi H and Matsuda Y 2005 *Phys. Rev. Lett.* **94** 047602
[36] McCollam A,Julian S R, Rourke P M C, Aoki D and Flouquet J 2005 *Phys. Rev. Lett.* **94** 186401
[37] Spałek J and Gopalan P 1990 *Phys. Rev. Lett.* **64** 2823; Korbel P, Spałek J, Wójcik W and Acquarone M 1995 *Phys. Rev. B* **52** R2213; Spałek J 2006 *Phys. Stat. Sol.* (b) **243** 78 - Editor's Choice
[38] Kaczmarczyk J and Spałek J 2009 *Phys. Rev. B* **79** 214519
[39] Kaczmarczyk J and Spałek J 2010 *J. Phys.: Condens. Matter* **22** 355702
[40] Maśka M, Mierzejewski M, Kaczmarczyk J and Spałek J 2010 *Phys. Rev. B* **82** 054509

Grüneisen ratio in Kondo-lattice compound CeRhSn

Andrzej Ślebarski
Institute of Physics, University of Silesia, Uniwersytecka 4, 40-007 Katowice, Poland

E-mail: andrzej.slebarski@us.edu.pl

Abstract. In this contribution we present the Grüneisen-ratio analysis for polycrystalline sample of CeRhSn. We find that the critical Grüneisen ratio $\Gamma^{cr} \sim \beta^{cr}/C^{cr}$ is divergent below 8 K, which is characteristic for a system with quantum critical point.

1. Introduction

An increasing number of Ce based heavy fermion (HF) metals were shown to display a pronounced deviation from the properties of conventional Landau-Fermi liquids (LFL). In a number of cases this non-Fermi-liquid (NFL) behaviour can be attributed to the proximity to a quantum-critical transition at T = 0 between magnetic and nonmagnetic ground states which can be tuned by varying a control parameter such as pressure (p), magnetic fields (H), or doping. Such zero-temperature critical points (QCP) can determine an unusual behaviour in thermodynamics, as well as in transport over a wide range of temperatures. The essential question is how the heavy quasiparticles evolve if these materials are tuned from the paramagnetic to a long-range magnetically ordered state. The spin-density-wave theory [1-3] describes well a quantum critical behaviour.

Very recently, it has been shown that the thermal expansion, $\beta = V^{-1}(dV/dT)$, compared to the specific heat C(T)/T behaviour, is much more singular as the system approaches to the QCP. Scaling analysis [4] showed that the Grüneisen ratio $\Gamma \propto \beta/C$ can be used as a sensitive probe of quantum criticality, which is measured by the ratio of thermal volume expansion β and molar specific heat C. It has been shown theoretically [4] and experimentally [5-7] that Γ has to diverge at any QCP. This divergence is entirely determined by the scaling dimension of the control parameter r, where, e.g., $r = (p - p_c)/p_c$ or $r = (H - H_c)/H_c$. As shown in [4] the critical exponent ϵ in the divergence $\Gamma \propto 1/T^{\epsilon}$, $\epsilon = 1/\nu z$, where ν is the critical exponent for the correlation length and z is the dynamical critical exponent in the divergence of the correlation time. For the three-dimensional (3D) antiferromagnetic (AF) QCP, $\nu = ½$ and z = 2, yielding $\epsilon = 1$.

The first observation of a Grüneisen ratio divergence in the Ce-based HF system has been made on $CeNi_2Ge_2$ with AF QPC [5]. In this report we discuss an effective Grüneisen ratio in CeRhSn. CeRhSn is known [8,9] to be a NFL compound which is located near QCP. In the lowest temperatures the electrical resistivity $\rho \propto T$ and $\chi \propto C/T \propto T^{-n}$ (n < 1), which suggest the presence of the Griffiths phases. However, the thermal expansion coefficient α divided by temperature T, $\alpha(T)/T \propto \ln T$, is not divergent at the low temperatures T < 10 K [10]. One reason of this abnormal T-behavior could be the onset of weak AF order or presence of magnetic correlations. We found, however, that the critical Grüneisen ratio $\Gamma^{cr} \propto \beta^{cr}/C^{cr}$ shows the $1/T^{\epsilon}$ behaviour below ~8 K with exponent $\epsilon \approx 1$ characteristic

Journal of Physics: Conference Series **303** (2011) 012109 doi:10.1088/1742-6596/303/1/012109

of the 3D antiferromagnetic QCP, where β^{cr} and C^{cr} are the thermal volume expansion and specific heat with the background contributions subtracted.

2. Grüneisen parameter of CeRhSn; results, discussion and analysis

Polycrystalline samples CeRhSn and LaRhSn were prepared by arc-melting stoichiometric amounts of the elemental metals in an ultra-high-purity argon atmosphere and annealing for 8 days at 800°C. The specific heat C was measured in the temperature range 1.8-300 K using a PPMS Heat Capacity Measurement System. The linear thermal expansion coefficient α was measured using a high-resolution capacitance dilatometer capable of measuring length changes down to ~0.1 Å [10].
The low-temperature specific heat data for CeRhSn and LaRhSn are displayed as C/T and ΔC/T vs. T in figure 1, where ΔC = C(CeRhSn) – C(LaRhSn). The ΔC/T increase has either been described by $\Delta C/T \propto -\log T$ or $\Delta C/T \propto \gamma_0 - cT^{1/2}$ (the best fit with $\gamma_0 = 0.229\ \mathrm{Jmol^{-1}K^{-2}}$ and $c = 0.211\ \mathrm{Jmol^{-1}K^{-5/2}}$ is represented by the dotted line), and is characteristic for NFL. As has been reported previously [8,9] the magnetic susceptibility also shows a power law behaviour and $\chi \propto \Delta C/T \propto T^{-n}$, that was interpreted in terms of the Griffiths phase model [8, 11,12].

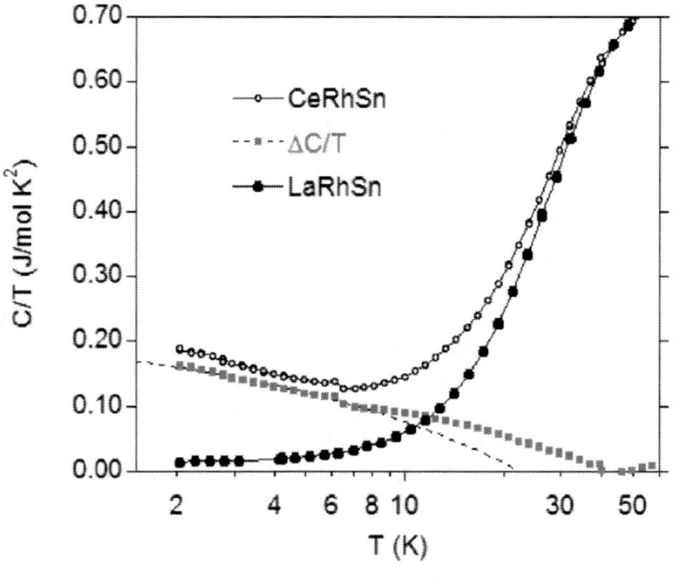

Figure 1. Specific heat as C/T vs. T for CeRhSn and LaRhSn. The ΔC(T)/T, where ΔC specific heat increment defined as ΔC = C(CeRhSn)-C(LaRhSn) is well approximated by the expression: $\Delta C = \gamma_0 T - cT^{3/2}$ below 8 K (dotted line).

The thermal expansion data for CeRhSn and LaRhSn are displayed as α/T and Δα/T vs. T in figure 2. The magnetic contribution to the volume thermal expansion is isolated by subtracting the volume thermal expansion of the non-magnetic LaRhSn. The thermal expansion shows a broad maximum at ~8 K and it decreases below T_{max}. This behaviour suggests a short range magnetic ordering resulting, e.g., from atomic disorder (a very similar α/T T-dependence was observed in $CeRhIn_{5-x}Sn_x$ for concentration x: 0.4 < x < 0.48, with the onset of AF ordering [13]).
The variation of the Grüneisen ratio $\Gamma^{eff}(T) = (V_m/\kappa)\beta(T)/C(T)$ for CeRhSn (in figure 3) is very similar to that found in α/T, here $V_m = 3.44\ 10^{-5}\ \mathrm{m^3/mol}$ [14] and β = 3α, whereas compressibility has not

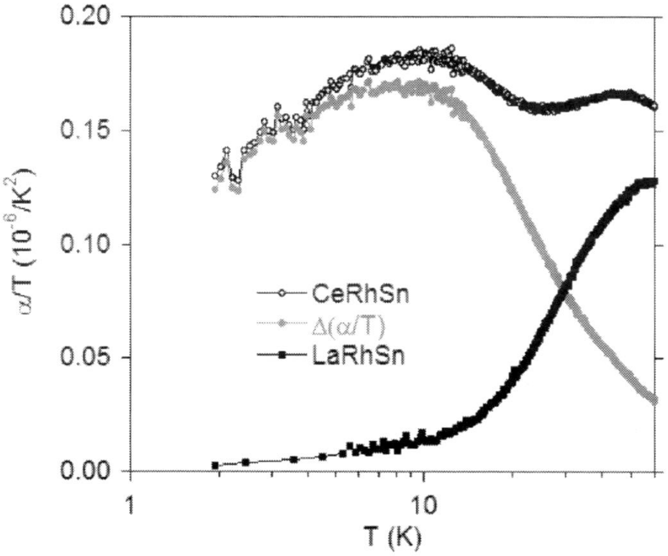

Figure 2. Thermal expansion of CeRhSn and LaRhSn as α/T vs. T. Also is shown the magnetic contribution $\Delta\alpha/T$ vs. T. $\Delta\beta(T) = 3\Delta\alpha(T)$ is well fitted by the expression $b_0T + b_1T^{1/2}$ below 9 K with the parameters for the best fit: $b_0 = 0.615\ 10^{-6}\ K^{-2}$ and $b_1 = -\ 0.30\ 10^{-6}\ K^{-3/2}$.

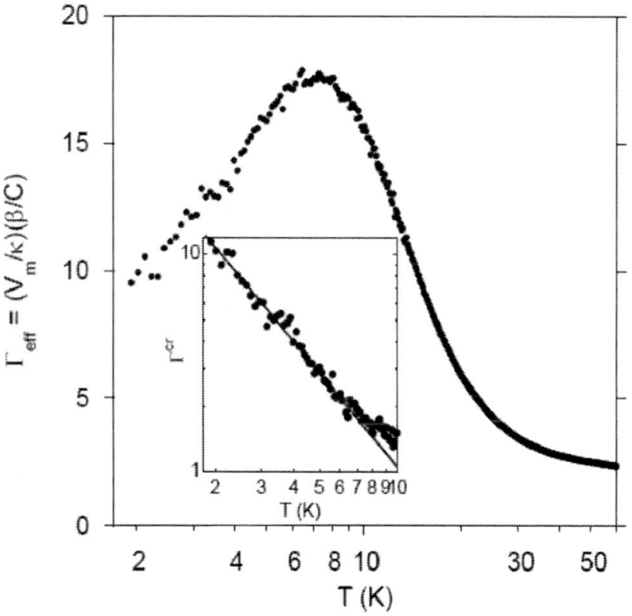

Figure 3. Temperature dependence of the effective Grüneisen ratio $\Gamma^{eff} = (V_m/\kappa)\ \beta/C$ vs. log T. The inset displays the log-log plot of Γ^{cr}.

been obtained experimentally yet, therefore we roughly estimated $\kappa \sim 1\ 10^{-11}$ m^2/N from the elastic moduli [14] obtained for CeRhSn. The value of Γ^{eff} is rather large, much larger than the corresponding one for the linear electronic term in normal metals, indicating a strong volume dependence of the quasiparticle bands. The low temperature Grüneisen ratio is consistent with moderate heavy fermion behaviour in CeRhSn, but is an order of magnitude smaller than those observed for the *classic* heavy fermion compounds [15].

In order to compare the Grüneisen ratio $\Gamma(T)$ of CeRhSn, which is located near to the QCP [8], we calculated the *critical* Grüneisen ratio $\Gamma^{cr}(T) \propto \beta^{cr}/\Delta C^{cr}$ of critical contributions to thermal expansion and specific heat. Within the itinerant theory for 3D AF, the critical contribution to the specific heat [4] is $C^{cr}(T) = C(T) - \gamma_0 T$, with $C^{cr}/T \rightarrow 0$ for $T \rightarrow 0$, and $\gamma_0 = C/T\,|_{T=0} \approx 0.229$ $Jmol^{-1}K^{-2}$. The γ_0 value was obtained from the fit of equation $C/T = \gamma_0 - cT^{1/2}$ to the experimental data $\Delta C/T$ in figure 1. For thermal expansion coefficient $\beta^{cr}(T) = \Delta\beta(T) - b_0 T$, with $b_0 = 0.615\ 10^{-6}\ K^{-2}$ as determined from the best fit below 10 K in figure 2. We found $\Gamma^{cr} \propto T^{-\epsilon}$, with an exponent $\epsilon = 1.2 \pm 0.1$ (the inset to figure 3), which is close to $\epsilon = 1$ predicted by the itinerant theory for 3D critical spin fluctuations. The divergent $\Gamma^{cr} \propto T^{-\epsilon}$ behaviour also suggests that CeRhSn is near the QCP.

Acknowledgments

The work was supported by the Ministry of Science and Higher Education within the research project No. N N202 032137.

References:

[1] Hertz J A 1976 *Phys. Rev.* B **14** 1165
[2] Millis A J 1993 *Phys. Rev.* B **48** 7183
[3] Moriya T and Takimoto T 1995 J. Phys. Soc. Jpn. **64** 960
[4] Zhu L, Garst M, Rosch A, and Si Q 2003 Phys. Rev. Lett. **91** 066404
[5] Küchler R, Oeschler N, Gegenwart P, Cichorek T, Neumaier K, Tegus O, Geibel C, Mydosh J A, Steglich F, Zhu L, and Si Q 2003 Phys. Rev. Lett. **91** 066405
[6] Küchler R, Gegenwart P, Heuser K, Scheidt E –W, Stewart G R, and Steglich F 2004 Phys. Rev. Lett. **93** 096402
[7] Küchler R, Gegenwart P, Custers J, Stockert O, Caroca-Canales N, Geibel C, Sereni J G, and Steglich F 2006 Phys. Rev. Lett. **96** 256403
[8] Ślebarski A, Maple M B, Freeman E J, Sirvent C, Radłowska M, Jezierski A, Granado E, Huang Q, and Lynn J W 2002 Philos. Mag. B **82** 943
[9] Kim M S, Echizen Y, Umeo K, Kobayashi S, Sera M, Salamakha P S, Sologub O L, Takabatake T, Chen X, Jung M H, and Maple M B 2003 Phys. Rev. B **68** 054416
[10] Ślebarski A, Grube K, Lortz R, Meingast C, and , vLöhneysen H 2004 J. Magn. Magn. Mater. **272-276** 234
[11] Griffiths R B 1969 Phys. Rev. Lett. **23** 17
[12] Castro Neto 1998 Phys. Rev. Lett. **81** 3531
[13] Donath J G, Steglich F, Bauer E D, Ronning F, Sarrao J L, Gegenwart P 2009 Eur. Phys. Lett. **87** 57011
[14] Higaki H, Ishii I, Hirata D, Kim Muo-Sung, Takabatake T, and Suzuki T 2006 J. Phys. Soc.Jpn **75** 024709
[15] de Visser A, Franse J J M, and Floquet J 1989 Physica B **161** 324

Joint European Magnetic Symposia – JEMS 2010 IOP Publishing
Journal of Physics: Conference Series **303** (2011) 012110 doi:10.1088/1742-6596/303/1/012110

Entanglement and position-space information entropy: Hubbard model as an approximation to nanostructure systems.

J. P. Coe[1], V. V. França[2], I. D'Amico[1]

[1] Department of Physics, University of York, York YO10 5DD, UK

[2] Physikalisches Institut, Albert-Ludwigs-Universität, Hermann-Herder-Straße 3, D-79104 Freiburg, Germany

E-mail: jpc503@york.ac.uk, vivian.franca@physik.uni-freiburg.de, ida500@york.ac.uk

Abstract.
We consider the position-space information and linear entropies as proxy measures to the average single-site entanglement—quantified using the von Neumann entropy—of the one-dimensional Hubbard model and of a one-dimensional nanostructure system comprised of an array of quantum-dots. Spatial entanglement in the quantum-dot system is also investigated via the three entropies. We appraise the use of the possible proxy measures in the Hubbard model as an approximation to their use for the nanostructure system.

1. Introduction

Systems based on quantum-dot (QD) nanostructures could be used as components for quantum information processing devices. Since entanglement is considered a key ingredient for quantum information processes, ascertaining the availability of entanglement in these nanostructures would appear to be useful when designing them. However it is often practically impossible to exactly model many-electron systems, such as complex nanostructures, and therefore precisely calculate their entanglement degree. Approximation methods such as the Hubbard model allow systems of many strongly-correlated electrons to be modelled efficiently and often capture much of the important physics. We therefore consider the one-dimensional Hubbard model (HM) as an approximation to the entanglement of a one-dimensional nanostructure model consisting of two electrons interacting via a contact interaction and confined in an array of QDs.

We first assess the position-space information and linear entropies' ability to indicate the average *single-site entanglement*, as quantified by the von Neumann entropy, in the two systems: if more computationally/experimentally amenable quantities were shown to reproduce the relevant behaviour of the entanglement in these systems then they could be used for the investigation of more complex nanostructure systems. It has previously been shown [1, 2] that the position-space information entropy behaves similarly in some ways to the spatial entanglement, as measured by the von Neumann entropy, for certain parameter regimes in one- and three-dimensional systems, but does not capture all of the features of the spatial entanglement.

The average single-site entanglement of the HM [3] was found to be a good approximation to that of the QD system [4] when quantified by the von Neumann entropy. Therefore here we investigate if this holds true for other possible entanglement indicators. By doing this we can hope to better understand if the HM is an appropriate approximation to the entanglement in a QD system with a contact interaction when other entropies, perhaps simpler than the von Neumann, are used.

2. The systems

The Hamiltonian for the QD system is

$$H = \sum_{i=1,2} \left(-\frac{1}{2}\frac{d^2}{dx_i^2} + v(x_i) \right) + C_U \delta(x_1 - x_2). \tag{1}$$

Here the potential $v(x_i)$ is used to model an array of regularly spaced, identical square wells, symmetric about the origin, and defined by four quantities: M the number of wells, d the barrier width between two consecutive wells, and w and v_0 the width and depth of each well. Here we use $v_0 = -10$ Hartrees, $w = 2\ a_0$ and $d = 0$ or $d = 2\ a_0$, a_0 the (effective) Bohr radius. C_U controls the strength of the interaction and allows us to determine the corresponding interaction strength of the HM.

The HM Hamiltonian is

$$H = -t \sum_{j=1}^{M} \sum_{\sigma=\uparrow,\downarrow} \left(\hat{c}^{\dagger}_{j,\sigma}\hat{c}_{j+1,\sigma} + \hat{c}^{\dagger}_{j+1,\sigma}\hat{c}_{j,\sigma} \right) + \tilde{U} \sum_{j=1}^{M} \hat{n}_{j,\uparrow}\hat{n}_{j,\downarrow},$$

which we solve for two electrons and M sites, with average particle density $n = 2/M$. We use the hopping parameter t to rescale the interaction $\tilde{U}$ in the dimensionless interaction strength $U = \tilde{U}/t$. We may then estimate the appropriate U for the QD system as $U \approx \frac{\tilde{U}_w}{t_w}$ where $\tilde{U}_w = \frac{C_U}{2} \int \phi_w^4(x)dx$, and $t_w = \langle \phi_L(x) | \left(-\frac{1}{2}\frac{d^2}{dx^2} + v(x) \right) | \phi_R(x) \rangle$. Here ϕ_w is the non-interacting, single-particle ground-state wavefunction in any one of the square wells, while ϕ_L and ϕ_R are two such wavefunctions situated in adjacent wells (see [4]).

3. Comparison of possible entanglement measures

We use the von Neumann entropy of the reduced density matrix, $S = -Tr[\rho_{\text{red}} \log_2 \rho_{\text{red}}]$, to calculate the bipartite entanglement of a pure state. The reduced density matrix is obtained by tracing out one of the subsystems $\rho_{\text{red}} = Tr_B |\psi\rangle\langle\psi|$. We compare the often computationally simpler linear entropy, $L = 1 - Tr[\rho^2_{\text{red}}]$, with S.

We consider the single-site entanglement, which quantifies how much information about the quantum state of the other sites can be obtained by measurement at a single-site i. For the HM the ρ_{red} at site i is diagonal in the basis of occupation [5, 6], $\rho_{\text{red,i}} = \text{diag}\,[P_i(\uparrow\downarrow), P_i(\uparrow), P_i(\downarrow), P_i(0)]$. For the QD system, wells and their surrounding barrier regions are assigned to sites so that all of space is accounted for. This also results in a diagonal $\rho_{\text{red,i}}$ (see Ref.[4]). For a system of M sites or wells, the average single-site entanglement S_{site} is then defined by the value of S_i averaged over all sites. In addition, we calculate the spatial entanglement [1] of the QD system, S_{spatial}. This entanglement can be thought of as quantifying how much extra information is gained upon measuring one particle's position about the position of the remaining particle. The corresponding entropies using L instead of S are also calculated.

We also analyse the spatial and single-site position-space information entropies defined as $I_{\text{spatial}} = -\int n(x) \log_2 n(x) dx$ and $I_{\text{site}} = -\frac{1}{M}\sum_i^M n_i \log_2 n_i$. These entropies have an appealing form as they depend only on the particle density. In addition, I_{spatial} has been used to study entanglement in the Moshinsky atom [7], and can be considered as an approximation to S_{spatial} where only the diagonal terms of the reduced density matrix are considered [1].

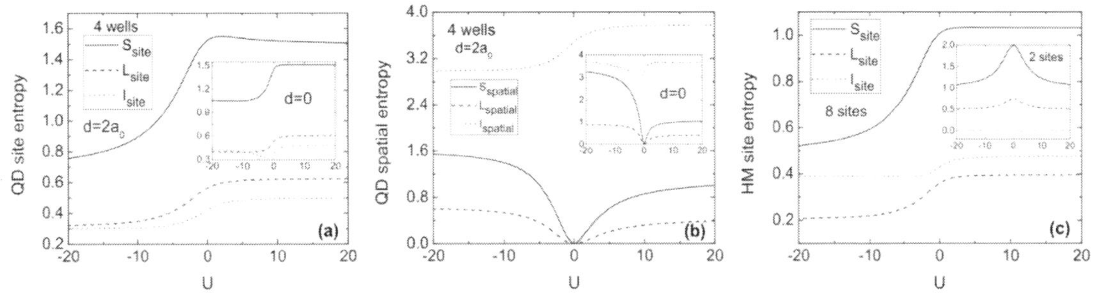

Figure 1. Linear (L), von Neumann (S), and position-space information (I) entropies versus interaction strength U: (a) single-site entropies and (b) spatial entropies for the QD system with 4 wells $d = 2\ a_0$ and $d = 0$ (insets); (c) single-site entropy in the HM with 8 sites and 2 sites (inset).

We see in Fig. 1(a) that for 4 wells with wide barriers ($d = 2\ a_0$), L_{site} is a fairly good proxy for S_{site}: the general trend is captured, but one small discrepancy is the absence of the maximum for small U in L_{site}. Interestingly, I_{site} behaves like L_{site} so is also a fairly good approximation to S_{site}. For smaller barrier width, we consider the limiting case of $d = 0$ (inset of Fig. 1(a)). Here L_{site} seems to be an even better approximation to S_{site}, and there is no maximum in both S_{site} and L_{site} for small U. I_{site} is similar enough to L_{site} for positive U and hence S_{site}.

For the spatial entanglement in the QD nanostructure, Fig. 1(b), we see that the linear entropy behaves similarly to S_{spatial} for any value of d, while the position-space information entropy fails to describe the behaviour of S_{spatial} for $d = 2\ a_0$, for which it is more similar to the site entanglement, S_{site}.

For the HM with 8 sites, Fig. 1(c), L_{site} and I_{site} are also good approximations to the behaviour of S_{site}, except I_{site} increases slightly at strongly negative U. For the HM with 2 sites (inset), L_{site} and S_{site} can be used interchangeably, while I_{site} is null for any U, as the site density is frozen at $n_i = 1$. We note that for the QD system with two wells, I_{site} and I_{spatial} do exhibit very small changes, but the change would be indiscernible on the scale of these graphs, and we can not confidently attribute this to a physical effect rather than a numerical phenomenon.

4. Hubbard model as an approximation to the quantum-dot system

Here we compare the average single-site linear and position-space information entropies of the HM to those of a nanostructure composed of 8 wells/sites. From Fig. 2(a) we see that, for the QD system with $d = 2\ a_0$, the HM L_{site} and I_{site} are almost identical, respectively, to L_{site} and I_{site} of the QD.

For $d = 0$ (Fig. 2(b)), the HM L_{site} is a little less accurate than for $d = 2\ a_0$, but still a very good approximation to the QD L_{site}, while I_{site} for each model behaves in a very different way. Here the QD I_{site} for $d = 0$ is more similar in behaviour to the spatial entanglement. This strong change in the QD I_{site} when changing d was also observed in the QD I_{spatial} (Fig. 1(b)): by passing from $d = 0$ to $d = 2\ a_0$ the QD I_{spatial} becomes similar in behaviour to the single-site entanglement. This leads to a very interesting conclusion: the barrier width d in the QD nanostructure seems to define for which kind of entanglement the position-space information entropy can be a good entanglement indicator; spatial ($d = 0$) or site entanglement ($d = 2\ a_0$).

In Fig. 2(c) we also see that the behaviour of I is the same whether the spatial or site form is considered. This is interesting as I_{spatial} depends on the shape of the density not just the amount of density at each site, which suggests that the behaviour of I is relatively insensitive

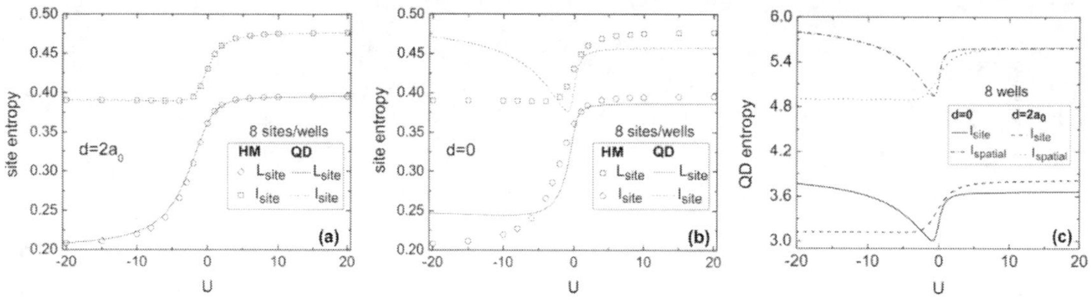

Figure 2. Linear (L) and information (I) entropies versus interaction strength U: single-site entropy in the HM and in the QD with 8 wells, for (a) $d = 2\ a_0$ and (b) $d = 0$; (c) spatial and single-site information entropy in the QD with 8 wells for $d = 2a_0$ and $d = 0$. In (c) the I_{site} was rescaled ($I_{\text{site}} \equiv MI_{\text{site}}$).

to the fine details of the density profile.

5. Summary

We verified that the linear entropy, in both the QD and HM systems, can reproduce qualitatively the von Neumann entropy behaviour and hence the entanglement. We found that the position-space information entropy can indicate different types of entanglement in the QD system, depending on the barrier width d: for $d = 2\ a_0$ it is a good indicator of the average single-site entanglement, while for $d = 0$ its trend is similar to the spatial entanglement. For the HM the single-site position-space information entropy indicated the general behaviour of entanglement as a function of U except for the two site case.

We verified that both the single-site linear and position-space entropies in the QD nanostructure with $d = 2\ a_0$ can be quantitatively described by the equivalent entropy of the HM, with excellent precision. This corroborates our previous findings [4] for the von Neumann entropy. For the limit of vanishing barrier width, the linear entropy of the HM is also a good approximation to describe the entanglement in the QD system. But the information entropy in the QD for this limit seems to behave as an indicator of spatial entanglement rather than single-site entanglement.

Acknowledgments

JPC and IDA gratefully acknowledge funding from EPSRC grant EP/F016719/1. VVF is supported by Brazilian funding from CAPES.

References

[1] Coe J P, Sudbery A and D'Amico I 2008 *Phys. Rev. B* **77** 205122
[2] Coe J P, Abdullah S and D'Amico I 2010 *J. Appl. Phys.* **107** 09E110
[3] França V V and Capelle K 2006 *Phys. Rev. A* **74** 042325
[4] Coe J P, França V V and D'Amico I 2010 *Phys. Rev. A* **81** 052321
[5] Zanardi P 2002 *Phys. Rev. A* **65** 042101
[6] Larsson D and Johannesson H 2006 *Phys. Rev. A* **73** 042320
[7] Amovilli C and March N H 2004 *Phys. Rev. A* **69** 054302

Interorbital pairs in clean and impure superconductors

Anna Ciechan and Karol I Wysokiński

Institute of Physics and Nanotechnology Centre, M. Curie-Skłodowska University, ul. Radziszewskiego 10, Pl 20-031 Lublin, Poland

E-mail: ankcie@kft.umcs.lublin.pl (AC), karol@tytan.umcs.lublin.pl (KIW)

Abstract. We consider two orbital superconductor with spin singlet inter-orbital pairs only. The calculations of the local and global properties of such superconductor have been presented with special attention to the features which distinguish such system from the typical BCS one. In particular, in clean system we have found the existence of critical interaction U necessary to induce superconducting transition and the asymmetry (with respect to the chemical potential) in the partial density of states. In the disordered system, in which the characteristics change from place to place, we have calculated local densities of states $N(\vec{r}, E)$ and local values of the order parameters $\Delta(\vec{r})$. Both $N(\vec{r}, E)$ and $\Delta(\vec{r})$ can be measured with help of scanning tunneling microscope.

1. Introduction

Materials with many bands near the Fermi energy offer a number of possible scenarios for the exotic quantum superconducting state. Despite the typical symmetries related to spin degrees of freedom we have new ones related to orbital quantum numbers. The discoveries of superconducting cuprates [1] and even more strontium ruthenate [2], magnesium diboride [3] and iron pnictides [4] have all greatly contributed to the development of the many band theory of superconductivity. Theoretical studies of superconductors with "overlapping bands" have been started by Suhl, Matthias and Walker and independently by Moskalenko in fifties [5] and were later continued by many others [6]. The understanding of many interesting properties of the above systems require new superconducting states. A good example are iron pnictides, which seem to be dominated by the inter band pairing scenario leading to the so called $s_{\pm}$ symmetry state characterized by constant gaps Δ_1, Δ_2 on two sheets of the Fermi surface, which for identical bands differ in sign ($\Delta_1 = -\Delta_2$) only [7, 8].

Of particular interest are the scenarios of superconductivity which rely on the unconventional spin or orbital symmetries. As an example let us mention the inter band superconductors. They are defined by the only (many particle) pairing interaction having, in the band picture, the form [9]

$$H_{int} = U^{12} \sum_{\vec{k},\vec{k}'} \left[c^{\dagger}_{1,\vec{k},\uparrow} c^{\dagger}_{1,-\vec{k},\downarrow} c_{2,-\vec{k}',\downarrow} c_{2,\vec{k}',\uparrow} + c^{\dagger}_{2,\vec{k},\uparrow} c^{\dagger}_{2,-\vec{k},\downarrow} c_{1,-\vec{k}',\downarrow} c_{1,\vec{k}',\uparrow} \right]. \quad (1)$$

The Hamiltonian (1) describes the scattering of intra-band Cooper pairs and does not allow for inter-band pairs. In this work we are interested in the opposite situation and study the properties of two orbital superconductor with inter-orbital pairs. It means we consider the Cooper pairs

Published under licence by IOP Publishing Ltd

in which each member belongs to different orbital. We shall call them superconductors with inter-orbital pairs. The properties of such systems markedly differ from the properties of inter-orbital superconductors or standard two-orbital superconductors with intra-orbital pairs. Similar models have earlier been considered in the context of photoinduced superconductivity in semiconductors [10], high temperature superconductors [11], cold atoms or quantum chromodynamic superfluids [12]. Recently superconductors with inter band pairs have been studied in the context of iron pnictides [13] and heavy fermions superconductors [14].

2. The model for two orbital superconductors

We consider the metallic system described by the two-orbital tight-binding model. The orbital description is especially important as we are also interested in the real space properties of the model. We consider the square lattice with two orbitals in each site. We assume the spin-independent effective pairing interaction between fermions in different orbital states. The randomness in the system is easily incorporated *via* site dependence of parameters. The Hamiltonian reads

$$\begin{aligned} H &= \sum_{ij,\lambda\lambda',\sigma} \left[(e_\lambda - \mu)\delta_{ij}\delta_{\lambda\lambda'} - t_{ij}^{\lambda\lambda'} + V_{imp}^{\lambda\lambda'}(\vec{r_j})\delta_{ij} \right] c_{i\lambda\sigma}^{+} c_{j\lambda'\sigma} \\ &- \sum_i U(\vec{r_i})(c_{i1\uparrow}^{+} c_{i2\downarrow}^{+} + c_{i2\uparrow}^{+} c_{i1\downarrow}^{+})(c_{i1\downarrow} c_{i2\uparrow} + c_{i2\downarrow} c_{i1\uparrow}), \end{aligned} \quad (2)$$

where $c_{i\lambda\sigma}^{+}$, $c_{i\lambda\sigma}$ are the creation and annihilation operators of electrons with spin $\sigma =\uparrow,\downarrow$ at the lattice site $\vec{r_i} = i$ in the orbital λ. e_λ is the electron on - site energy and μ the chemical potential. $t_{ij}^{\lambda\lambda'}$ are the hopping integrals between the same or different orbitals (if $\lambda \neq \lambda'$). $U(\vec{r_i})$ denotes the attractive interactions responsible for inter-orbital pairing. They are allowed to vary from site to site in impure system. For such systems we also introduce local potential $V_{imp}^{\lambda\lambda'}$ at impurity sites $\vec{r_i} = i$ scattering electrons between two orbitals for $\lambda \neq \lambda'$. We limit our discussion to inter - orbital impurity scattering as it is expected to more severely affect the superconducting state with inter-orbital pairs in comparison to intra - orbital scatterers $V_{imp}^{\lambda\lambda}$.

We use the standard mean-field decoupling valid for a spin singlet superconductor and get the effective Hamiltonian

$$\begin{aligned} H^{MFA} &= \sum_{ij,\lambda\lambda',\sigma} \left[(e_\lambda - \mu)\delta_{ij}\delta_{\lambda\lambda'} - t_{ij}^{\lambda\lambda'} + V_{imp}^{\lambda\lambda'}(\vec{r_j})\delta_{ij} \right] c_{i\lambda\sigma}^{+} c_{j\lambda'\sigma} \\ &+ \sum_{i,\lambda\neq\lambda'} \left[\Delta(\vec{r_i}) c_{i\lambda\uparrow}^{+} c_{i\lambda'\downarrow}^{+} + h.c. \right], \end{aligned} \quad (3)$$

where the order parameters $\Delta(\vec{r_i})$ are related to the pairing correlation functions through

$$\Delta(\vec{r_i}) = U(\vec{r_i})(< c_{i1\downarrow} c_{i2\uparrow} > + < c_{i2\downarrow} c_{i1\uparrow} >). \quad (4)$$

Here we ignore Hartree terms $V_{\lambda\sigma}(\vec{r_i}) = U(\vec{r_i}) n_{\lambda'-\sigma}(\vec{r_i})(1 - \delta_{\lambda\lambda'})$, which depend on the number of particles at a given site $n_{\lambda\sigma}(\vec{r_i}) =< c_{i\lambda\sigma}^{+} c_{i\lambda\sigma} >$.

The Hamiltonian (3) is diagonalized by means of the Bogolubov - Valatin transformation [15]

$$c_{i\lambda\uparrow} = \sum_\nu \left[u_{\lambda\nu}(\vec{r_i})\gamma_{\nu\uparrow} - v_{\lambda\nu}^{*}(\vec{r_i})\gamma_{\nu\downarrow}^{+} \right], \quad (5)$$

$$c_{i\lambda\downarrow} = \sum_\nu \left[u_{\lambda\nu}(\vec{r_i})\gamma_{\nu\downarrow} + v_{\lambda\nu}^{*}(\vec{r_i})\gamma_{\nu\uparrow}^{+} \right] \quad (6)$$

leading to the Bogolubov- de Gennes (BdG) equations for the amplitudes $u_{\lambda\nu}(\vec{r_i})$, $v_{\lambda\nu}(\vec{r_i})$ and the eigenenergies E_ν

$$\sum_{\lambda'}\left[\sum_j K_{ij}^{\lambda\lambda'} u_{\lambda'\nu}(\vec{r_j}) + \Delta(\vec{r_i})v_{\lambda'\nu}(\vec{r_i})(1-\delta_{\lambda\lambda'})\right] = E_\nu u_{\lambda\nu}(\vec{r_i}), \tag{7}$$

$$\sum_{\lambda'}\left[-\sum_j K_{ij}^{\lambda\lambda'} v_{\lambda'\nu}(\vec{r_j}) + \Delta^*(\vec{r_i})u_{\lambda'\nu}(\vec{r_i})(1-\delta_{\lambda\lambda'})\right] = E_\nu v_{\lambda\nu}(\vec{r_i}), \tag{8}$$

where the operator $K_{ij}^{\lambda\lambda'}$ reads

$$K_{ij}^{\lambda\lambda'} = (\varepsilon_{\lambda'} - \mu)\,\delta_{ij}\delta_{\lambda\lambda'} + V_{imp}^{\lambda\lambda'}(\vec{r_i})\delta_{ij} - t_{ij}^{\lambda\lambda'}. \tag{9}$$

The pairing parameters $\Delta(\vec{r_i})$ are, in turn, expressed in terms of the eigenfunctions $u_{\lambda\nu}(\vec{r_i})$, $v_{\lambda\nu}(\vec{r_i})$ and eigenenergies E_ν as [16]

$$\Delta(\vec{r_i}) = U\sum_\nu (u_{1\nu}(\vec{r_i})v_{2\nu}^*(\vec{r_i}) + u_{2\nu}(\vec{r_i})v_{1\nu}^*(\vec{r_i}))(1-2f_\nu). \tag{10}$$

The mean number of particles is given by $n = \sum_{i\lambda} n_\lambda(\vec{r_i})/L$, where L is a number of sites in the cluster and

$$n_\lambda(\vec{r_i}) = \sum_\nu \left[|u_{\lambda\nu}(\vec{r_i})|^2 f_\nu + |v_{\lambda\nu}(\vec{r_i})|^2(1-f_\nu)\right]. \tag{11}$$

In the above formulae $f_\nu = \left(e^{E_\nu/k_BT}+1\right)^{-1}$ denotes the Fermi-Dirac distribution function of quasi-particles.

The local density of states (LDOS) $N(\vec{r_i}, E)$ is directly accessible in scanning tunneling microscope (STM) measurements and is proportional to the local conductance $dI(\vec{r_i}, V)/dV$. In the two - orbital system it is a sum over orbitals of the local densities of states $N(\lambda, \vec{r_i}, E)$ projected onto respective orbitals

$$N(\lambda, \vec{r_i}, E) = \sum_\nu [|u_{\nu\lambda}(\vec{r_i})|^2\delta(E-E_\nu) + |v_{\nu\lambda}(\vec{r_i})|^2\delta(E+E_\nu)]. \tag{12}$$

Obviously, we have in each site $N(\vec{r_i}, E) = N(1, \vec{r_i}, E) + N(2, \vec{r_i}, E)$. For a clean system equations (7,8) can be Fourier transformed and written in reciprocal space. For the impure systems with broken translational symmetry, the Bogolubov - de Gennes equations (7 - 10) are solved self-consistently in the real space for a small n×m cluster with periodic boundary conditions.

3. Clean system

For numerical calculation we include only hopping elements to the nearest-neighbor sites on a square lattice and no hybridization term $t_{ij}^{\lambda\lambda'} = \delta_{\lambda\lambda'}\delta_{j,i+\delta}t_\lambda$ ($\delta = \pm x, \pm y$). The resulting two bands are described by energies $\varepsilon_\lambda(k_x, k_y) = e_\lambda - 2t_\lambda(cos(k_x) + cos(k_y)) - \mu$, and we use the following set of parameters: $t_1 = t$, $t_2 = 2t$ and $e_2 - e_1 = 2t$, the interaction strength $U = 2.5t$ and the number of particles in the system $n = 1.2$. t is our energy unit. With these parameters the bands cross each other along $\Gamma - X$ and $\Gamma - M$ lines of the square lattice Brillouin zone for $\mu = -2t$. It means that for that value of the chemical potential both bands have the same value of the Fermi wave vector $k_{F1} = k_{F2}$. For the assumed number of electrons in the system $n = 1.2$, the narrower band which we denote '1' has larger Fermi wave vector $k_{F1} > k_{F2}$.

We consider here only the pairs with zero center of mass momentum. The gap parameter in this case is $\Delta = U\sum_k(< c_{1,\vec{k},\uparrow}c_{2,-\vec{k},\downarrow} > + < c_{2,\vec{k},\uparrow}c_{1,-\vec{k},\downarrow} >)$ and due to the local character of pairing interaction it includes all the $\vec{k}$ states in the Brillouin zone. The dependence of Δ on the inter-orbital interaction U for fixed values of the chemical potential μ is shown in the figure 1. Note the existence of the critical value of pairing attraction $U_C(\mu)$ for all positions of the chemical potential except the case where two bands cross each other at the Fermi energy and the system effectively behaves as a typical BCS one. For bands crossing at the Fermi level (homothetic bands) the pairing take place for arbitrarily small attractive interaction. Below U_C the normal metal has lower free energy than the superconductor.

In a clean two-orbital system one defines partial (projected onto respective orbitals) and total densities of states. They are shown in the figure 2. The partial densities of states are not centered at the chemical potential $\mu = 0$. Their sum is centered at the chemical potential and features a number of peaks of different height at positive and negative energies, which are coherence peaks observed in partial densities of states. To understand this let us note, that the eigenvalues of the Bogoliubov-de Gennes equations have the form $E_\nu = \pm\frac{\varepsilon_1-\varepsilon_2}{2} \pm \sqrt{(\frac{\varepsilon_1+\varepsilon_2}{2})^2 + \Delta^2}$, where we have suppressed the $\vec{k}$ dependence of $\varepsilon_{1//2}$. It follows that the gaps in the partial densities of states are centered at the energies $(\varepsilon_1(k_{F1}) - \varepsilon_2(k_{F1}))/2$ and $(\varepsilon_1(k_{F2}) - \varepsilon_2(k_{F2}))/2$ instead at the Fermi level.

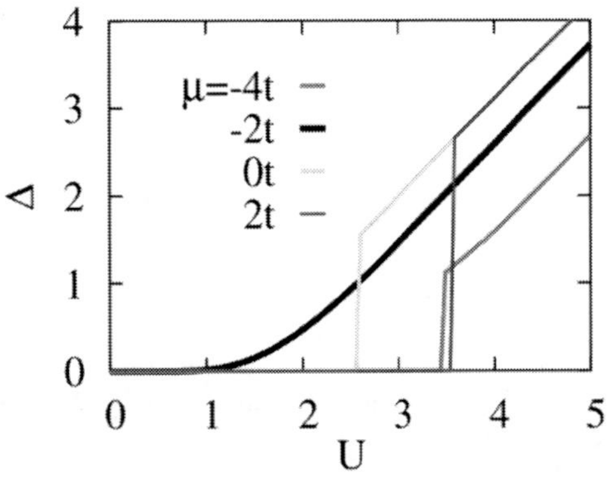

Figure 1. The dependence of $\Delta(U)$ for a few values of the chemical potential μ. For $\mu = -2t$ two bands cross each other at the Fermi energy.

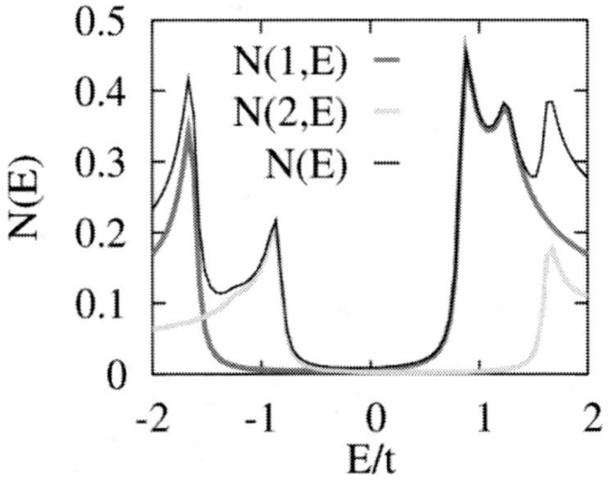

Figure 2. Partial $N(\lambda, E)$ and total $N(E) = \sum_\lambda N(\lambda, E)$ densities of states for interaction $U = 2.5t$ and the number of particle in the system $n = 1.2$.

4. Impure system

To study the properties of disordered systems we consider a cluster of 23×17 sites on a square lattice with unit lattice constant with 15% inter-orbital impurities introduced in random sites. Impurities are extended $V^{12}_{imp}(\vec{r_i}) = V^{12}f_{id}$ with f_{id} being a number from the Gaussian distribution at sites distance $id = 0, 1, \sqrt{2}, 2$ from the impurity center. Other parameters have been assumed to take the same values as in clean superconductor. For a given distribution of impurities, which is shown in the figure 3, and their strength V^{12} we have solved Bogolubov-de Gennes equations and calculated local values of the order parameter and density of states. The typical map of the order parameter $\Delta(\vec{r})$ in the impure system normalized to its clean system value Δ_0 is shown in the figure 4. Comparison of both maps clearly shows that inter-orbital impurities locally suppress the order parameter.

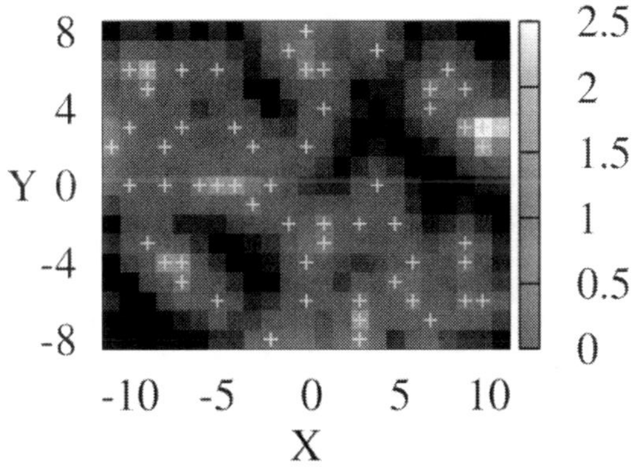

Figure 3. Distribution of the impurity potential $V^{12}_{imp}(\vec{r_i})/V^{12}$ in the system of 23×17 sites. Symbols (+) indicate impurity sites.

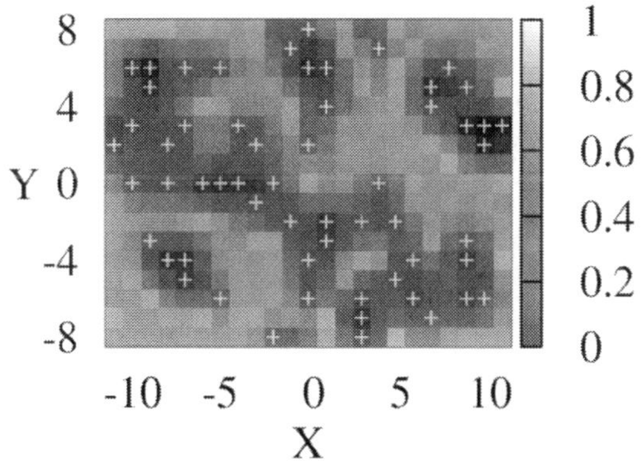

Figure 4. The gap maps $\Delta(\vec{r})/\Delta_0$ for the system with inter-orbital impurities $V^{12} = 4t$ and inter-orbital interaction $U = 2.5t$.

More detailed information can be obtained from the local density of states. In the actual calculations of this function we have replaced the Dirac distribution entering formula (12) by the Lorentzian of width $\gamma = 0.05t$. The local densities of states $N(\vec{r}, E)$ for all sites along the line $y = 0$ are shown in the figure 5. The red curves show the density of states at the sites where the impurities are centered. Note the smaller gap values and less distinctive coherence features at the impurity sites. This is in contrast to the typical BCS superconductors for which the coherence peaks are sharper in the impurity sites [17]. The local values of the gap fluctuate from site to site and are visibly smaller in impurity sites.

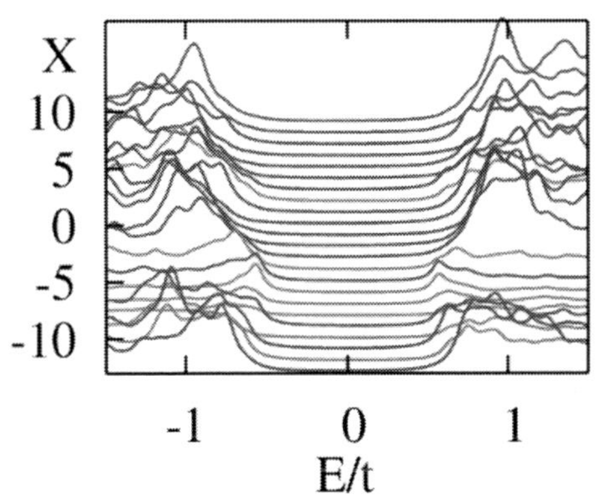

Figure 5. The local densities of states $N(\vec{r}, E)$ along the line $y = 0$ for the considered system. Red lines indicate position of the impurity sites.

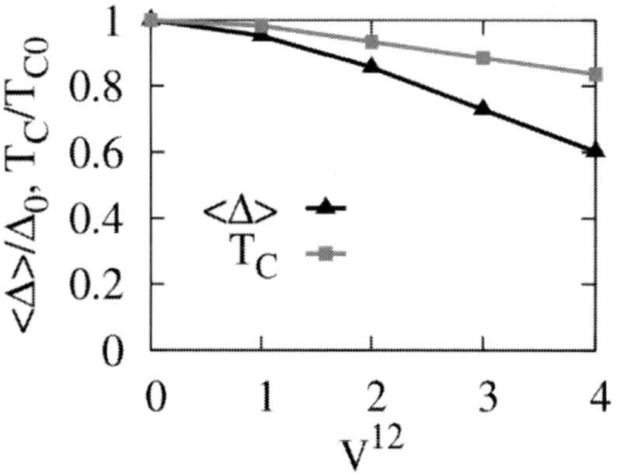

Figure 6. The dependence of the mean order parameter $< \Delta > /\Delta_0$ and the critical temperature T_C/T_C^0 on V^{12}.

The dependence of the average value of the order parameter normalized to its value in the system without impurities and the superconducting transition temperature on the strength of impurity scattering is shown in the figure 6. The slightly different reaction of $< \Delta >$ and T_c is related to inhomogeneous structure of the local gaps as visible in the figure 4. Δ_0 and T_{C0} refer to the clean system.

5. Summary and conclusions

We have studied superconducting properties of the system with inter-orbital pairs. Such system differs from classic one band or two band systems. In particular we have found that:

- there exist critical inter-orbital interaction U_C necessary to stabilize the superconducting state. The value of U_C strongly depends on relative position of both bands and increases with minimal distance between bands at the Fermi level.
- Inter-orbital impurities V^{12} act as a local hybridization. They rather weakly suppress a superconducting order parameter Δ with respect to the clean case.
- The local density of states is characterized by suppressed coherence features at the impurity sites.

As both, the local density of states $N(\vec{r}, E)$ and the gap values $\Delta(\vec{r})$ can be measured with help of scanning tunneling microscope, the results of this work are subject to experimental verification provided there exist systems in which the present scenario is realized. More detailed analysis of the model in the context of superconducting iron pnictides will be presented elsewhere.

Acknowledgments

This work has been partially supported by the Polish Ministry of Science and Education under the grant No. N N202 2631 38.

References

[1] Bednorz J G and Müller K A 1986 *Z. Phys.* B **64** 189
Müller K A and Bednorz J G 1987 *Science* **237** 1133
[2] Maeno Y, Hashimoto H, Yoshida K, Nishizaki S, Fujita T, Bednorz J G and Lichtenberg F 1994 *Nature* **372** 532
[3] Nagamatsu J, Nakagawa N, Muranaka T, Zenitani Y and Akimitsu J 2001 *Nature* **410** 63
[4] Kamihara Y, Hiramatsu H, Hirano M, Kawamura R, Yanagi H, Kamiya T and Hosono H 2006 *J. Am. Chem. Soc.* **128** 10012
Kamihara Y, Watanabe T, Hirano M and Hosono H 2008 *J. Am. Chem. Soc.* **130** 3296
[5] Suhl H, Matthias B T and Walker L R 1959 *Phys. Rev. Lett.* **3** 552
Moskalenko V A 1959 *Fiz. Met. Metalloved.* **8** 503
[6] Suffczynski M 1962 *Phys. Rev.* **128** 1538
Kondo J 1963 *Prog. Theor. Phys.* **29** 1
Leggett A J 1966 *Prog. Theor. Phys.* **36** 901
[7] Golubov A A and Mazin I I 1997 *Phys. Rev.* B **55** 15146
Bang Y and Choi H Y 2008 *Phys. Rev.* B **78** 134523
Seo K, Bernevig B A and Hu J 2008 *Phys. Rev. Lett.* **101** 206404
Mazin I I and Schmalian J 2009 *Physica C: Superconductivity* **469** 9
[8] Cvetkovic V and Tesanovic Z 2009 *Europhysics Lett.* **85**, 37002.
[9] Dolgov O V, Mazin I I, Parker D and Golubov A A 2009 *Phys. Rev.* B **79** 060502(R)
[10] Kumar N and Sinha K P 1968 *Phys. Rev* **174** 482
[11] Tahir-Kheli J 1998 *Phys. Rev.* B **58** 12307
[12] Liu W V and Wilczek F 2003 *Phys. Rev. Lett.* **90** 047002
Forbes M M, Gubankova E, Liu W V and Wilczek F 2005 *Phys.Rev.Lett.* **94** 017001
Gubankova E, Mishchenko E and Wilczek F 2006 *Phys.Rev.* B **74** 184516
[13] Moreo A, Daghofer M, Nicholson A and Dagotto E 2009 *Phys. Rev.* B **80** 104507
Gao Y, Su W-P and Zhu J-X 2010 *Phys. Rev.* B **81** 104504
[14] Neto F D, Continentino M A and Lacroix C 2010 *J. Phys.: Condens. Matter* **22** 075701
[15] Bogoliubov N N, Tolmachev V V and Shirkov D V 1958 *Fortschr. Phys.* **6** 605
Valatin J G 1958 *Nuovo Cimento* **7** 843
[16] Ketterson J B and Song S N 1999 *Superconductivity* (Cambridge: Cambridge University Press)
[17] Ciechan A, Krzyszczak J and Wysokiński K I 2009 *Journal of Physics: Conference Series* **150** 052283
Ciechan A and Wysokiński K I 2009 *Phys. Rev.* B **80** 224523

Local properties of an inhomogeneous two-component correlated superconductor

Jaromir Krzyszczak and Karol Izydor Wysokiński
Institute of Physics and Nanotechnology Centre, M. Curie-Skłodowska University, Radziszewskiego 10, Pl 20-031 Lublin

E-mail: jkrzyszczak@kft.umcs.lublin.pl (JK); karol@tytan.umcs.lublin.pl (KIW)

Abstract. The effect of random point-like impurities on correlated superconductor has been studied. We use the two component (also called boson-fermion (BF)) model formulated in a real space supplemented with strong on-site electron-electron repulsion U and solve it by means of the real space Bogoliubov-de Gennes approach. It has been shown that due to electron correlations the gap is larger near the impurity sites, contrary to the results for the same model but without U. This naturally leads to positive correlations between the gap magnitude and the position of impurities, as was observed in various scanning tunneling measurements performed on d-wave symmetry BSCCO-2122 superconductors. Strong correlations are responsible for the protection of the low energy quasiparticles against disorder.

1. Introduction

Strong correlations play a crucial role in high-temperature superconductors (HTS). They are commonly believed to be responsible for both the Mott insulating and the d-wave superconducting state. Disorder is introduced into the system in the process of doping and is unavoidable in these materials. Indeed, the recent scanning tunneling microscopy (STM) measurements have revealed nanoscale inhomogeneities in high temperature superconducting cuprates [1, 2]. It has also been observed in pnictide [3] superconductors. Detailed analysis of the results obtained in these experiments unveiled the unexpected correlation between positions of dopant oxygen atoms in $Bi_2Sr_2CaCuO_{8+\delta}$ systems and the large value of the gap. These positive correlations have been successfully analyzed in terms of $t - J$ [4] and boson-fermion [5] models of superconductivity.

Another important experimental fact about the superconducting state, at least in some HTSs is its apparent insensitivity to disorder [6], in marked contrast to conventional theories for d-wave pairing, which predict just the opposite [7]. Recently it has been shown that strong correlations may be responsible for this behavior [8]. In other words strong correlations make the d-wave superconductor described by the $t-J$ model robust against disorder. It is an interesting question if the same reasoning can be applied to Boson-Fermion model, *i.e.* if the addition of strong on-site repulsion between fermions in the boson - fermion mixture will also lead to renormalization of disorder effects and in particular to smearing of the low energy parts of the local densities of states as observed in experiment. To answer this question we use two component model [9] in its real space version [10] and study the local properties of materials with d-wave symmetry of the order parameter in the presence of 'atomic' disorder. This so called boson-fermion (BF) model has earlier been proposed by many groups [12] to describe HTS.

2. Model and approach

The boson-fermion model describes two kinds of quantum objects coexisting in the system - local electron pairs (hard-core bosons) and itinerant fermions - coupled with each other via mechanism of charge transfer. Hamiltonian of this two component model (in standard notation) has the form [5, 11]

$$\begin{aligned}\hat{H}^{\rm BF} &= \sum_{i,j,\sigma} t_{ij}\hat{c}^{\dagger}_{i,\sigma}\hat{c}_{j,\sigma} + \sum_{i,\sigma}\left(V_i^{\rm imp}-\mu\right)\hat{c}^{\dagger}_{i,\sigma}\hat{c}_{i,\sigma} + \sum_{i,\sigma} U\hat{c}^{\dagger}_{i,\sigma}\hat{c}_{i,\sigma}\hat{c}^{\dagger}_{i,-\sigma}\hat{c}_{i,-\sigma} + \\ &+ \sum_{i}\left(E_i^B-2\mu\right)\hat{b}^{\dagger}_i\hat{b}_i + \sum_{i,j}\frac{g_{ij}}{2}\left(\hat{b}^{\dagger}_i\left(\hat{c}_{i,\downarrow}\hat{c}_{j,\uparrow}-\hat{c}_{i,\uparrow}\hat{c}_{j,\downarrow}\right)+h.c.\right). \qquad (1)\end{aligned}$$

Third term in (1), with $U \geq 0$ describes electron-electron interaction. The interaction between fermions and charged bosons is provided by the coupling g_{ij}, which for a d-wave symmetry takes form g_{ij}=g_0 for nearest neighbour sites $\langle i,j\rangle$ only, positive along X-axis and negative along Y-axis.

If $U \ll zt$ we shall use a mean field approximation. This approximation merely changes the local impurity potential in the above Hamiltonian $V_i^{\rm imp} \to \tilde{V}_i^{\rm imp} = V_i^{\rm imp} + 0.5Un_i^f$. Due to the site dependence of both $V_i^{\rm imp}$ and n_i^f one expects some interplay of disorder and correlation even at this level of approximation. For $U \gg zt$ the Gutzwiller approximation has been used. This approximation, which projects out doubly occupied sites leaves the on-site parameters unchanged but renormalizes hoping t_{ij} and electron-boson interaction g_{ij} as $t_{ij} \to \tilde{t}_{ij} = t_{ij}r^t_{ij}$ and $g_{ij} \to \tilde{g}_{ij} = g_{ij}r^g_{ij}$. The factors r^t_{ij} and r^g_{ij} are known as Gutzwiller renormalization factors [13] and for the present model they read

$$r^t_{ij} = r^g_{ij} = \sqrt{\frac{2\left(1-n_i^f\right)}{2-n_i^f}}\sqrt{\frac{2\left(1-n_j^f\right)}{2-n_j^f}}. \qquad (2)$$

The boson-fermion Hamiltonian with correlations treated in mean field, respectively Gutzwiller, approximation can be written in a concise form as

$$\begin{aligned}\hat{H}^{\rm BF} &= \sum_{i,j,\sigma}\tilde{t}_{ij}\hat{c}^{\dagger}_{i,\sigma}\hat{c}_{j,\sigma} + \sum_{i,\sigma}\left(\tilde{V}_i^{\rm imp}-\mu\right)\hat{c}^{\dagger}_{i,\sigma}\hat{c}_{i,\sigma} + \sum_{i}\left(E_i^B-2\mu\right)\hat{b}^{\dagger}_i\hat{b}_i + \\ &+ \sum_{i,j}\frac{\tilde{g}_{ij}}{2}\left(\hat{b}^{\dagger}_i\left(\hat{c}_{i,\downarrow}\hat{c}_{j,\uparrow}-\hat{c}_{i,\uparrow}\hat{c}_{j,\downarrow}\right)+h.c.\right). \qquad (3)\end{aligned}$$

It still contains many body part describing charge exchange between both subsystems. The relevant parameters, however, have been changed as a result of strong electron-electron interaction. To decouple fermionic and bosonic degrees of freedom we use Hartree-Fock-Bogoliubov approximation $\hat{b}^{\dagger}_i\left(\hat{c}_{i,\downarrow}\hat{c}_{j,\uparrow}-\hat{c}_{i,\uparrow}\hat{c}_{j,\downarrow}\right) \simeq \langle\hat{b}^{\dagger}_i\rangle\left(\hat{c}_{i,\downarrow}\hat{c}_{j,\uparrow}-\hat{c}_{i,\uparrow}\hat{c}_{j,\downarrow}\right) + \hat{b}^{\dagger}_i\langle(\hat{c}_{i,\downarrow}\hat{c}_{j,\uparrow}-\hat{c}_{i,\uparrow}\hat{c}_{j,\downarrow})\rangle$. This leads to the decomposition of (3) onto the separate parts $\hat{H}^{\rm BF} = \hat{H}^{\rm B} + \hat{H}^{\rm F}$ with

$$\hat{H}^{\rm B} = \sum_{i}\left[\left(E_i^B-2\mu\right)\hat{b}^{\dagger}_i\hat{b}_i + \chi_i\hat{b}^{\dagger}_i + \chi_i^*\hat{b}_i\right], \qquad (4)$$

$$\hat{H}^{\rm F} = \sum_{i,j,\sigma}\left[\tilde{t}_{ij}+\delta_{ij}\left(\tilde{V}_i^{\rm imp}-\mu\right)\right]\hat{c}^{\dagger}_{i,\sigma}\hat{c}_{j,\sigma} + \sum_{i,j}\left(\Delta_{ij}\left[\hat{c}^{\dagger}_{j,\uparrow}\hat{c}^{\dagger}_{i,\downarrow}-\hat{c}^{\dagger}_{j,\downarrow}\hat{c}^{\dagger}_{i,\uparrow}\right]+h.c.\right), \qquad (5)$$

where $\chi_i = \sum_{\langle j\rangle}(\tilde{g}_{ij}/2)\langle\hat{c}_{i,\downarrow}\hat{c}_{j,\uparrow}-\hat{c}_{i,\uparrow}\hat{c}_{j,\downarrow}\rangle$ and $\Delta_{ij} = (\tilde{g}_{ji}/2)\langle\hat{b}_j\rangle$.

Boson part of this problem is exactly solvable [5, 10, 11] by the means of standard statistical approach and we get

$$\langle \hat{b}_i \rangle = -\frac{\chi_i}{2\epsilon_i} \tanh\left(\frac{\epsilon_i}{k_B T}\right), \tag{6}$$

$$\langle \hat{b}_i^\dagger \hat{b}_i \rangle = \frac{1}{2} - \frac{(E_i^B - 2\mu)}{4\epsilon_i} \tanh\left(\frac{\epsilon_i}{k_B T}\right), \tag{7}$$

with $\epsilon_i = \frac{1}{2}\sqrt{(E_i^B - 2\mu)^2 + 4|\chi_i|^2}$.

The fermion part $\hat{H}^F$ has the standard BCS structure and could be diagonalized by the Bogoliubov-Valatin transformation. This results in the following Bogoliubov-de Gennes equations, with $\tilde{\Delta}_{ij} = \Delta_{ij} + \Delta_{ji}$

$$\sum_j \begin{pmatrix} \tilde{t}_{ij} + \left(\tilde{V}_i^{\rm imp} - \mu\right)\delta_{ij} \ ; \ \tilde{\Delta}_{ij} \\ \tilde{\Delta}_{ij} \ ; \ -\tilde{t}_{ij} - \left(\tilde{V}_i^{\rm imp} - \mu\right)\delta_{ij} \end{pmatrix} \begin{pmatrix} u_j^l \\ v_j^l \end{pmatrix} = E^l \begin{pmatrix} u_i^l \\ v_i^l \end{pmatrix}, \tag{8}$$

solution of which enable us calculation of all parameters of interest. In particular the local density of states (LDOS) $N_i(\omega)$ and local fermion number density n_i^f is found to be

$$N_i(\omega) = \sum_l \left[|u_i^l|^2 \delta(\omega - E^l) + |v_i^l|^2 \delta(\omega + E^l)\right], \tag{9}$$

$$n_i^f = \sum_l \left[|u_i^l|^2 f(E^l) + |v_i^l|^2 (1 - f(E^l))\right], \tag{10}$$

where $f(E^l) = 1/[\exp(E^l/k_B T) + 1]$ is the Fermi-Dirac distribution function.

The above equations are solved self-consistently on a finite cluster of size 31×47. All energies are measured in units of nearest neighbour hoping t of homogeneous system. The next nearest neighbour hoping $t' = -0.25t$. The position of the bosonic level $E_i^B - 2\mu$=0.15t, the interaction g_0=0.50t and total concentration of carriers n^{tot}=0.9. We take point like impurities with the strength $V^{imp} = 1t$ and varying concentration n_d.

3. Results

To understand how correlations and disorder acting together affect the superconducting properties of the boson-fermion model we compare the results of the model without correlations to those with U. Figure 1 shows the evolution of spatially averaged density of states (DOS), defined as $\frac{1}{L}\sum_{i=1}^{L} N_i(\omega)$ (L is a total number of sites) with the concentration of 'atomic' point-like impurities V_i^{imp}. As one can notice, with increasing concentration of impurities in uncorrelated d-wave boson fermion model or if mean field approximation is used for weak correlations ($U = 1.0t$) the gap in DOS is losing its characteristic d-wave shape. The coherence features are smeared by impurities and their maxima only slightly move away of the chemical potential located at $\omega = 0$. With higher concentration of impurities there appear low energy (resonance) states which 'fill' the gap. As mentioned above the mean field approach to electron-electron interaction results in site dependent modification of the 'effective' potential $\tilde{V}_i^{imp}$. For a given parameters the change is of the order of 10% and thus hardly seen in the figures. The low energy resonances are almost absent in the strong correlation limit $U \to \infty$ (lower panels of Fig. (1)). In this case the gapless nodal quasiparticles are more robust against disorder, leading to the low-energy density of states that is quite insensitive to impurities. The structures seen

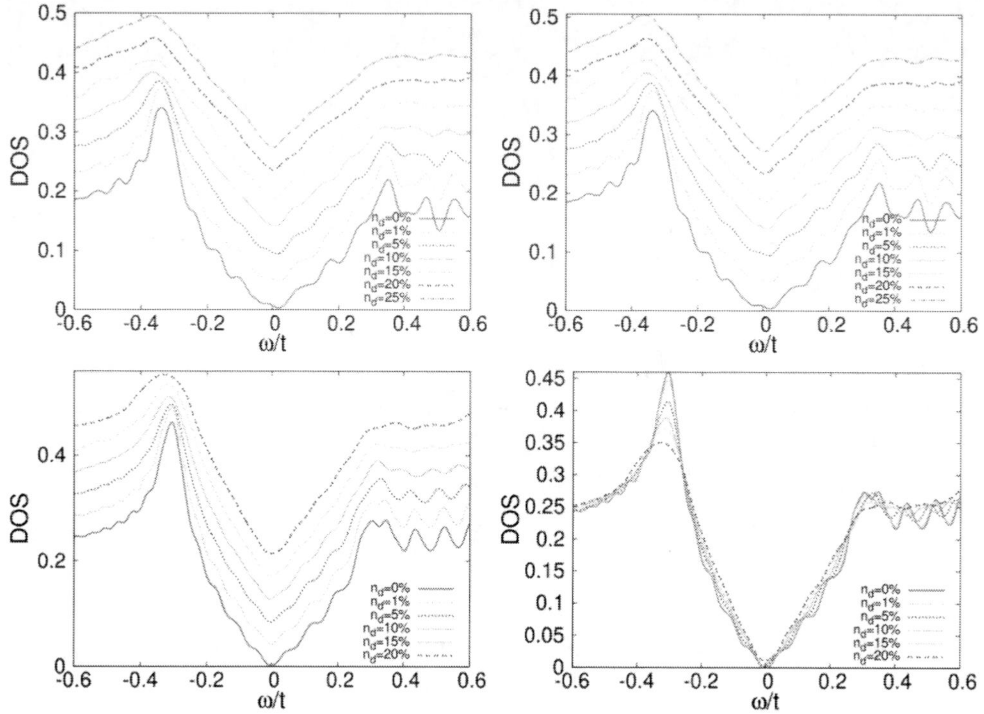

Figure 1. (Colour online) Averaged over the system density of states for an uncorrelated model (upper left panel) is compared with that for correlated model. The correlations are treated in the mean field approximation (upper right panel) and in the Gutzwiller approximation - lower panels which offer two different views of the same data. Various curves on each panel correspond to various concentration of point like potential impurity scatterers.

in the energy dependence of the density of states in all panels is related to the finite size of our system and its discrete energy spectrum. The asymmetry of the spectrum is related to the Van Hove singularity in the normal state density of states, which in the model with the present set of parameters lies below the chemical potential.

We also look how strong correlations affect local pairing amplitudes in a disordered superconductor. Figure (2) shows that in the absence of strong correlations (uncorrelated d-wave boson fermion model or a mean field treatment of Coulomb repulsion, left and middle panels) impurities reduce the value of the gap at the dopant sites and in a surrounding area. When a Gutzwiller approximation is applied (right panel) the value of the gap at the impurity sites is larger than the average value. It is obvious, that this gives simple explanation to the discovered positive correlation between the value of the spectral gap in the local density of states (LDOS) and the position of dopant atoms [2]. Uncorrelated boson-fermion model offers the explanation of this puzzling behavior under the assumption [5] that dopant impurities provide not only potential scattering but also change the positions of the bosonic levels. The strongly correlated impure boson-fermion model, in a close analogy with impure $t-J$ model [4], indicates the important role of impurity scattering in establishing the positive correlation between position of dopants and the value of the gap. It is thus a matter of future experiments to decide between various scenarios. Closer look at Fig.(2) shows that in uncorrelated system the response to the impurity spreads over a distance of few lattice constant, whereas in the correlated one it is highly localized in the close vicinity of the impurity site.

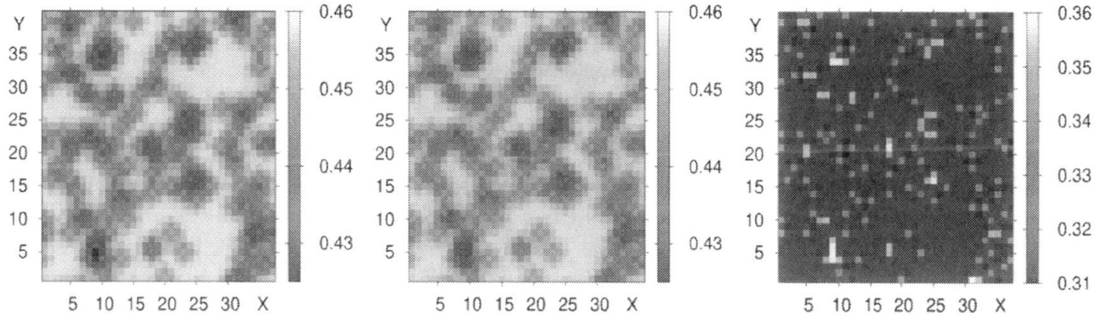

Figure 2. (Colour online) Maps of order parameter $\Delta_i = \sum_{<j>} |r_{ij}^{\Delta}\Delta_{ij}|$ in a system with $n_d = 10\%$ of potential impurities. Positions of dopant atoms are marked by green points. Left panel stands for uncorrelated d-wave boson fermion model. Middle panel present the results of the mean field approximation while right one the Gutzwiller approximation.

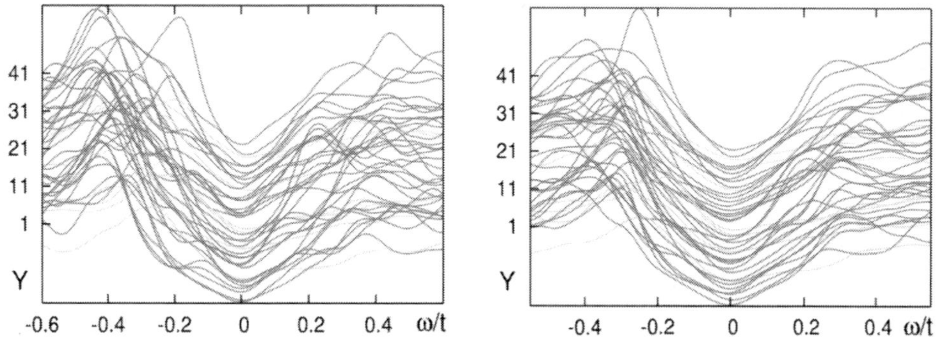

Figure 3. (Colour online) The local densities of states (LDOS) for sites along a cut through the lattice for $X = 5$. Left panel refers to uncorrelated d-wave boson fermion model and right one shows LDOS obtained within Gutzwiller approximation for a system with $n_d = 10\%$ of impurities. Energy dependence of LDOSs at impurity sites are plotted in green.

In the short coherence length superconductors the response to impurities obtained within Bogoliubov-de Gennes (BdG) approach differs considerably from the expectations of the standard Abrikosov-Gorkov theory [14, 15]. The BdG wave function distorts itself near the impurity site. In a weak scattering limit this distortion is similar for s- and d-wave symmetry of the order parameter. The strong correlations additionally protect the condensate against the disorder, at least at low energies. This leads to almost complete absence of impurity induced low energy excitations, in agreement with the results of scanning tunneling microscopy (STM) measurements [1, 2], where very low-energy spectra do not vary from one region of the sample to another.

In summary, our results elucidate the interplay between disorder and strong correlations. Even though we concentrate on the boson-fermion model some of the findings may have more general application. We have treated strong correlations by means of Gutzwiller approximation [13]. In disordered system it leads to random (but well defined) modifications of the hoping parameters and interactions. In the real space BdG approach the disorder is treated exactly. The protection of the low energy excitations against destructive impurity scattering we observe here is similar to that earlier observed [8] for correlated $t - J$ model. The obtained results provide an explanation of the observed positive correlations between the value of the spectral gaps in the local density of states (LDOS) and the positions of dopant atoms [2]. The detailed analysis will be presented elsewhere.

Acknowledgments

This work has been partially supported by the Ministry of Science and Education under the grant No. N N202 2631 38.

References

[1] Cren T, Roditchev D, Sacks W, Klein J, Moussy J-B, Deville-Cavellin C and Lagues M 2000 *Phys. Rev. Lett.* **84** 147
Pan S H, O'Neal J P, Badzey R L, Chamon C, Ding H, Engelbrecht J R, Wang Z, Eisaki H, Uchida S, Gupta A K, Ng K-W, Hudson E W, Lang K M and Davis J C 2001 *Nature (London)* **413** 282
Howald C, Fournier P and Kapitulnik A 2001 *Phys. Rev.* B **64** 100504(R)
Lang K M, Madhavan V, Hoffman J E, Hudson E W, Eisaki H, Uchida S and Davis J C 2002 *Nature (London)* **415** 412
[2] McElroy K, Lee J, Slezak J A, Lee D H, Eisaki H, Uchida S and Davis J C 2005 *Science* **309** 1048
[3] Massee F, Huang Y, Huisman R, de Jong S, Goedkoop J B and Golden M S 2009 *Phys. Rev.* B **79** 220517(R)
[4] Maśka M M, Śledź Ż, Czajka K and Mierzejewski M 2007 *Phys. Rev. Lett.* **99** 147006
[5] Krzyszczak J, Domański T, Wysokiński K I, Micnas R and Robaszkiewicz S 2010 *J. Phys.: Condens. Matter* **22** 255702
[6] Xiao G, Cieplak M, Xiao J Q and Chien C L 1990 *Phys. Rev.* B **42** 8752
Tolpygo S K, Lin J-Y, Gurvitch M, Hou S Y and Phillips J M 1996 *Phys. Rev.* B **53** 12454
Alloul H, Bobroff J, Gabay M and Hirschfeld P J 2009 *Rev. Mod. Phys.* **81** 45-108
[7] Abrikosov A A and Gorkov L P 1961 *Sov. Phys. JETP* **12** 1243
Larkin A I 1965 *JETP Lett.* **2**, 130
Martin A M, Litak G, Gyorffy B L, Annett J F and Wysokiński K I 1999 *Phys. Rev.* B **60** 7523
Balatsky A V, Vekhter I and Zhu J-X 2006 *Rev. Mod. Phys.* **78** 373-433
[8] Garg A, Randeria M and Trivedi N 2008 *Nature Physics* **4** 762-765
[9] Ranninger J and Robaszkiewicz S 1985 *Physica* B **135** 468
Robaszkiewicz S, Micnas R and Ranninger J 1987 *Phys. Rev.* B **36** 180
Micnas R, Ranninger J and Robaszkiewicz S 1990 *Rev. Mod. Phys* **62** 113
Domański T and Ranninger J 2004 *Phys. Rev.* B **70** 184503
[10] Krzyszczak J, Domański T and Wysokiński K I 2008 *Acta Phys. Polon.* A **114** 165
Ciechan A, Krzyszczak J and Wysokiński K I 2009 *J. Phys. Conf. Series* **150** 052283
[11] Robaszkiewicz S and Pawłowski G 2001 *Mol. Phys. Rep.* **34** 76
Domański T and Wysokiński K I 2002 *Phys. Rev.* B **66** 064517
Domański T, Ranninger J and Wysokiński K I 2003 *Acta Phys. Polon.* B **34** 493.
[12] Friedberg R, Lee T D and Ren H C 1992 *Phys. Rev.* B **45** 10732
Geshkenbein V B, Ioffe L B and Larkin A I (1997) *Phys. Rev.* B **55** 3173
Castro Neto A H 2001 *Phys. Rev.* B **64** 104509
[13] Edegger B, Muthukumar V M and Gros C 2007 *Advances in Physics* **56** 927
[14] Wysokiński K I, Ciechan A and Krzyszczak J 2010 *Acta Phys. Pol.* A **118** 232
[15] Atkinson W A, Hirschfeld P J and MacDonald A H 2000 *Phys. Rev. Lett.* **85** 39223925
Ghosal A, Randeria M and Trivedi N 2000 *Phys. Rev.* B **63** 020505

Quasiclassical Eilenberger approach to the vortex state in pnictide superconductors

P. Belova[1,2], **M. Safonchik**[1,3], **K. B. Traito**[1], **E. Lähderanta**[1]
[1] Lappeenranta University of Technology, P.O.Box 20, FI-53851, Lappeenranta, Finland
[2] Petrozavodsk State University, Lenin str. 33, RU-185640, Petrozavodsk, Russia
[3] A. F. Ioffe Physico-Technical Institute, St. Petersburg, 194021, Russia

E-mail: Polina.Belova@lut.fi

Abstract. Quasiclassical Eilenberger equations are solved for $s^{\pm}$-wave superconductors in the mixed state. This symmetry has been proposed for multiband superconductors as pnictides. This mechanism can be realized because of Umklapp scattering between the electron and the hole Fermi surface pockets resulting in opposite sign of pairing gap in these pockets. The applicability of the phenomenological Hao-Clem theory is investigated. Magnetic, temperature and impurity scattering rate dependencies of vortex core size are calculated. It is found that the accuracy of the effective London model gets better with the presence of the impurity scattering and even near the second critical field it is below 6%. The model with the parameters of intraband and interband impurity scattering, describing well superfluid density in $BaFe_2As_2$, is also considered.

The discoveries of the superconductivity in the layered FeP [1] and $FeAs$ [2] systems have ignited tremendous research activities for understanding of the superconductivity in Fe pnictides. The superconducting transition temperature T_c of electron-doped $LaFeAsO_{1-x}F_x$ ("1111") reaches 43 K under pressure [3], and $NdFeAsO_{1-x}F_x$ and $SmFeAsO_{1-x}F_x$ show T_c higher than 40 K at ambient pressure [4]. $BaFe_2As_2$ ("122") also shows superconductivity by hole doping with the highest T_c of 38 K in $(Ba,K)Fe_2As_2$ [5] and by electron doping with the highest T_c of 25 K in $Ba(Fe,Co)_2As_2$ [6]. Stoichiometric iron pnictides such as $LiFeAs$ or $NaFeAs$ ("111") are also interesting because they become superconducting without doping [7]. These Fe pnictides commonly have the $FeAs$ layers, where the Fe ions form a square lattice and each Fe ion is tetrahedrally coordinated by four As ions.

One key feature for understanding the origin of the high critical temperature and the pairing mechanism in superconductors is the symmetry of the order parameter. For iron pnictides, one of the attractive idea is that antiferromagnetic spin fluctuations mediate interband pairing. The sign of the Umklapp scattering between the electron and the hole Fermi surface, which is responsible for the superconducting pairing, is positive and the order parameter is fully gapped but changes sign between different Fermi sheets [8]. This situation is referred to as the extended $s^{\pm}$ symmetry. Using an $s^{\pm}$ model for the superconducting gap, a good explanation of experimental results on penetration depth in electron-doped and hole-doped $BaFe_2As_2$ pnictides has been obtained [9, 10]. In this model the Fermi surface is approximated by two cylindrical pockets centered at Γ (hole) and M (electron) points of the Fermi surface, i.e. two dimensional limit of five-band model is proposed. Recently, strong evidence of the $s^{\pm}$ symmetry in iron pnictides has been found by scanning tunneling microscopic measurements [11]. In spite of success of $s^{\pm}$ model, there are some indications, e.g. weak sensitivity of T_c to the impurity concentration [12], that a conventional s-wave state without sign reversal (so called s_{++}-wave state) is also possible candidate for iron pnictides. It has been proposed that the moderate electron-phonon interaction due to Fe-ion oscillation can induce the critical orbital fluctuation, without being prohibited by the Coulomb interaction. These fluctuations give rise to the strong pairing interactions for the s_{++}-wave superconductivity [13].

The aim of our paper is to apply quasiclassical Eilenberger approach to the vortex state of stoichiometric and nonstoichiometric iron pnictides considering $s^{\pm}$ and s_{++} pairing symmetries as presumable states [9] and to calculate magnetic coherence length ξ_h [14, 15]. As described in Ref. [16], ξ_h is important for the description of the muon spin rotation (μSR) experiments and can be directly measured. On theoretical ground, the magnetic coherence length can be found from the fitting of the calculated magnetic field distribution $h_E(\mathbf{r})$ to the Eilenberger - Ginzburg-Landau field distribution $h_{EGL}(\mathbf{r})$ [14, 15]

$$h_{EGL}(\mathbf{r}) = \frac{\phi_0}{S} \sum_{\mathbf{G}} \frac{F(G) e^{i\mathbf{Gr}}}{1 + \lambda^2 G^2}, \tag{1}$$

where $F(\mathbf{G}) = u K_1(u)$, $K_1(u)$ is the modified Bessel function, $u = \xi_h G$, $\mathbf{G}$ is a reciprocal lattice vector and S is the area of the vortex lattice unit cell. It is important to note that ξ_h in Eq. (1) is obtained from solving the Eilenberger equations and doesn't coincide with the variational parameter ξ_v (analytical Ginzburg-Landau (AGL) model), "improved" analytical GL solution [17] or numerical GL solution [18]. We will call the obtained field distribution as an Eilenberger - Ginzburg-Landau field distribution $h_{EGL}(\mathbf{r})$. Using the GL type of the field distribution does not mean direct connection to the GL theory and it is taken as a reasonable starting point of the investigation similar to the empirical approach to the problem [16, 19]. In Eq. (1) $\lambda(T)$ is calculated in Ref. [9]

$$\frac{\lambda_{L0}^2}{\lambda^2(T)} = 2\pi T \sum_{\omega_n > 0} \frac{\bar{\Delta}_n^2}{\eta_n (\bar{\Delta}_n^2 + \omega_n^2)^{3/2}}, \quad \eta_n = 1 + 2\pi \frac{\Gamma_0 + \Gamma_\pi}{\sqrt{\bar{\Delta}_n^2 + \omega_n^2}}, \tag{2}$$

where λ_{L0} is the London penetration depth at $T = 0$ K in the absence of the impurities. Here, $\bar{\Delta}_n = \Delta(T) - 4\pi\Gamma_\pi \bar{\Delta}_n / \sqrt{\bar{\Delta}_n^2 + \omega_n^2}$ for the $s^{\pm}$ pairing and $\bar{\Delta}_n = \Delta(T)$ for the s_{++} pairing symmetry. The order parameter $\Delta(T)$ is determined by the selfconsistent equation

$$\Delta(T) = 2\pi T \sum_{0 < \omega_n < \omega_c} \frac{V^{SC} \bar{\Delta}_n}{\sqrt{\bar{\Delta}_n^2 + \omega_n^2}}. \tag{3}$$

With the Riccati transformation of the Eilenberger equations quasiclassical Green functions f and g can be parameterized via functions a and b [20]

$$f = \frac{2a}{1 + ab}, \quad f^\dagger = \frac{2b}{1 + ab}, \quad g = \frac{1 - ab}{1 + ab}, \tag{4}$$

satisfying the nonlinear Riccati equations. In Born approximation for impurity scattering we have

$$\mathbf{u} \cdot \nabla a = -a[2(\omega_n + G) + i\mathbf{u} \cdot \mathbf{A}] + (\Delta + F) - a^2(\Delta^* + F^*), \tag{5}$$

$$\mathbf{u} \cdot \nabla b = b[2(\omega_n + G) + i\mathbf{u} \cdot \mathbf{A}] - (\Delta^* + F^*) + b^2(\Delta + F), \tag{6}$$

where $\omega_n = \pi T(2n+1)$, $G = 2\pi \langle g \rangle(\Gamma_0 + \Gamma_\pi) \equiv 2\pi \langle g \rangle \Gamma^*$, $F = 2\pi \langle f \rangle(\Gamma_0 - \Gamma_\pi)$ for $s^\pm$ pairing symmetry and $F = 2\pi \langle f \rangle \Gamma^*$ for the s_{++} pairing symmetry. Here, $\Gamma_0 = \pi n_i N_F |u_0|^2$ and $\Gamma_\pi = \pi n_i N_F |u_\pi|^2$ are the intra- and interband impurity scattering rates, respectively ($u_{0,\pi}$ are impurity scattering amplitudes with correspondingly small, or close to $\pi = (\pi, \pi)$, momentum transfer) and $\mathbf{u}$ is a unit vector of the Fermi velocity. The FLL create the anisotropy of the electron spectrum. Therefore the impurity renormalization correction in Eq. (5) and (6), averaged over Fermi surface, can be reduced to averages over the polar angle θ, i.e. $\langle \ldots \rangle = (1/2\pi) \int \ldots d\theta$. To take into account the influence of screening the vector potential $\mathbf{A}(\mathbf{r})$ in Eqs. (5) and (6) is obtained from the equation $\nabla \times \nabla \times \mathbf{A_E} = \frac{4}{\kappa^2}\mathbf{J}$, where the supercurrent $\mathbf{J}(\mathbf{r})$ is given in terms of $g(\omega_n, \theta, \mathbf{r})$ by

$$\mathbf{J}(\mathbf{r}) = 2\pi T \sum_{\omega_n > 0} \int_0^{2\pi} \frac{d\theta}{2\pi} \frac{\hat{\mathbf{k}}}{i} g(\omega_n, \theta, \mathbf{r}). \tag{7}$$

Here $\mathbf{A}$ and $\mathbf{J}$ are measured in units of $\phi_0/2\pi\xi_0$ and $2ev_F N_0 T_c$, respectively. The self-consistent condition for the pairing potential $\Delta(\mathbf{r})$ is given by

$$\Delta(\mathbf{r}) = V^{SC} 2\pi T \sum_{\omega_n > 0} \int_0^{2\pi} \frac{d\theta}{2\pi} f(\omega_n, \theta, \mathbf{r}), \tag{8}$$

where V^{SC} is the coupling constant and ω_c is the ultraviolet cutoff determining T_{c0} [15].

All over this paper the energy, the temperature and the length are measured in units of T_{c0} and the coherence length $\xi_0 = v_F/T_{c0}$, where v_F is the Fermi velocity. The magnetic field $\mathbf{h}$ is given in units of $\phi_0/2\pi\xi_0^2$. The impurity scattering rates are in units of $2\pi T_{c0}$. In calculations the ratio $\kappa = \lambda_{L0}/\xi_0 = 10$ is used. It corresponds to $\kappa_{GL} = 43.3$ [20]. To obtain the quasiclassical Green function, the Riccati equations [Eq. (5, 6)] are solved by the Fast Fourier Transform (FFT) method for triangular FLL [15]. This method is reasonable for dense FLL discussed in this paper. In high field the pinning effects are weak and they are not considered in our paper. To study high field regime we should calculate upper critical field $B_{c2}(T)$. It can be found from the similarity of the considered model to the model of spin-flip superconductors [21].

Solid lines in Fig. 1 demonstrate magnetic field dependence $\xi_h(B)$ in reduced units for superconductors with impurity scattering at $T/T_{c0} = 0.5$ with $\Gamma_0 = \Gamma_\pi = 0$; $\Gamma_0 = 3$, $\Gamma_\pi = 0.03$ and $\Gamma_0 = 0.5$, $\Gamma_\pi = 0.01$. Dash line demonstrates the result of the AGL theory for ξ_v[22]

$$\xi_v = \xi_{c2}(\sqrt{2} - \frac{0.75}{\kappa_{GL}})(1 + b^4)^{1/2}[1 - 2b(1-b)^2]^{1/2}, \tag{9}$$

where ξ_{c2} is determined from the relation $B_{c2} = \Phi_0/2\pi\xi_{c2}^2$ (in our units $\xi_{c2} = 1/\sqrt{B_{c2}}$). This dependence with ξ_{c2} as a fitting parameter is used often for the description of the μSR experimental results [16, 19]. As can be seen from Fig. 1 (a) the magnetic field dependence of ξ_h/ξ_{c2} is nonuniversal because it depends not only from B/B_{c2} (as in the AGL theory, dash line in Fig. 1 (a)), but also on interband and intraband impurity scattering parameters. In the case when, $\Gamma_0 = \Gamma_\pi = 0$, the results are the same for $s^\pm$ and s_{++} pairing symmetries. We mark this curve as "clean" one. In this figure is considered the case $\Gamma_0 \gg \Gamma_\pi$ and the value of ξ_h is reduced considerably in comparison with clean case. One can compare the observed behavior with that in s_{++} pairing model. In s_{++} pairing symmetry the intraband and interband scattering rates act in similar way and ξ_h/ξ_{c2} decreases always with impurity scattering. In contrast, in $s^\pm$ model $\xi_h/\xi_{c2}(B/B_{c2})$ dependences show different behavior with Γ_π: ξ_h/ξ_{c2} increases with Γ_π at $B/B_{c2} < 0.8$ and decreases at higher fields, i.e. the curves are getting more flat. A crossing point appears in the dependences $\xi_h/\xi_{c2}(B/B_{c2})$ for $s^\pm$ and s_{++} pairing. This can be explained by the fact that in superconductors without interband pair breaking the increasing in high field is

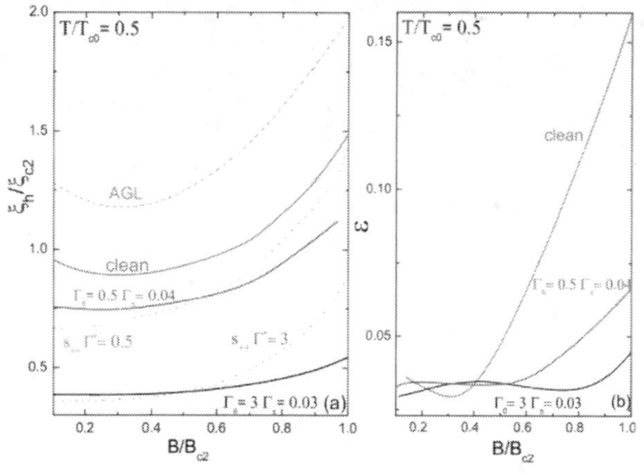

Figure 1. (a) The magnetic field dependence of ξ_h/ξ_{c2} for superconductors with impurity scattering. Dashed line demonstrates the result of the AGL theory for ξ_v from Eq. 9. Solid lines represent our solution of Eilenberger equations at $T/T_{c0} = 0.5$ for the "clean" case ($\Gamma_0 = \Gamma_\pi = 0$) and $s^\pm$ model ($\Gamma_0 = 0.5$, $\Gamma_\pi = 0.04$ and $\Gamma_0 = 3$, $\Gamma_\pi = 0.03$). Dotted lines show result for s_{++} model ($\Gamma^* = 0.5$ and $\Gamma^* = 3$). (b) Magnetic field dependence of mean square deviation of the h_{EGL} distribution from the Eilenberger distribution normalized by the variance of the Eilenberger distribution, ε, for $T/T_{c0} = 0.5$ for the "clean" case, $\Gamma_0 = 3$, $\Gamma_\pi = 0.03$ and $\Gamma_0 = 0.5$, $\Gamma_\pi = 0.04$ for the $s^\pm$ model.

connected with field-dependent pair breaking under approaching to the upper critical field. We also calculate magnetic field dependence of mean square deviation of the h_{EGL} distribution of the magnetic field from the Eilenberger distribution normalized by the variance of the Eilenberger distribution $\varepsilon = \sqrt{\overline{(h_E - h_{EGL})^2}/\overline{(h_E - B)^2}}$, where $\overline{\cdots}$ is average over unit vortex cell. Fig. 1 (b) demonstrates $\varepsilon(B)$ dependence for $T/T_{c0} = 0.5$ at $\Gamma_0 = 0$, $\Gamma_\pi = 0$; $\Gamma_0 = 3$, $\Gamma_\pi = 0.03$ and $\Gamma_0 = 0.5$, $\Gamma_\pi = 0.04$. It can be seen from this figure that accuracy of effective London model getting worse with increasing magnetic field, but in superconductors with impurity scattering the accuracy is below 6% even near the second critical field (Fig. 1 (b)).

The superfluidity density in pnictides shows often a power law dependence with exponent approximately equal to two at low temperatures [9, 10]. This law was explained by $s^\pm$ model with parameters $\Gamma_0 = 3$ and $\Gamma_\pi = 0.04 - 0.06$. Fig. 2 (a) shows $\xi_h/\xi_{c2}(B/B_{c2})$ dependence with $\Gamma_0 = 3$ and $\Gamma_\pi = 0.06$ at different temperatures. All curves demonstrate growing behavior with values much less that one in whole field range, i.e. they are under the AGL curve of ξ_v. This shows strong effect of interband scattering. The inset to Fig. 2 (a) presents $\xi_h/\xi_{c2}(B/B_{c2})$ results for $\Gamma_0 = 3$, $\Gamma_\pi = 0.06$ ($s^\pm$ pairing) and $\Gamma^* = 3$ (s_{++} pairing) at $T/T_{c0} = 0.15$. This type of the behavior is cut off by the impurity pair breaking and introducing characteristic field B^* in the field dependence by the substitution $B/B_{c2} \to (B + B^*(\Gamma_\pi))/B_{c2}(\Gamma_\pi)$. There is additional low-field crossing point between $s^\pm$ and s_{++} curves in this low-temperature case comparing with $T/T_{c0} = 0.5$ (Fig. 1 (a)). It can be explained by the restoration of the Usadel dirty limit behavior (where $\Gamma \gg 1$ and monotonously decreasing $\xi_h(B)$ is expected [20, 23]) which is not realized for $s^\pm$ symmetry due to the pair breaking there. Opposite, slowly increasing $\xi_h/\xi_{c2}(B/B_{c2})$ function is obtained for $s^\pm$ case in low-field range (Fig. 2 (a) main plot). It can be explained by the field-dependent splitting of the low-energy spectrum of bound state in the vortex core similar to the case of the surface bound states in d-wave superconductors [24]. The same effect is realized for extended state in high-field (for $B/B_{c2} > 0.5$ in Fig. 2 (a)).

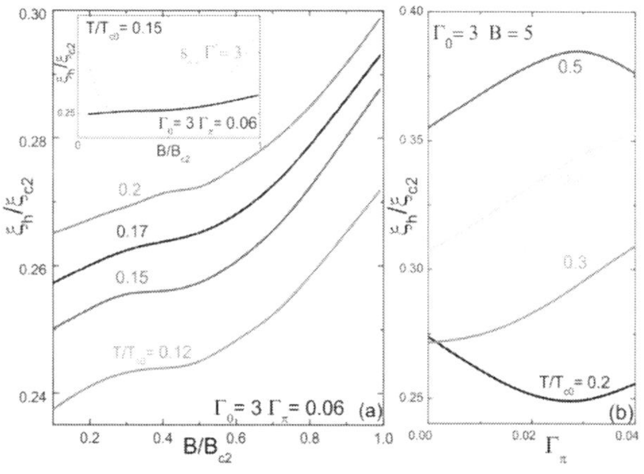

Figure 2. (a) The magnetic field dependence of coherence length ξ_h with different temperatures T_c/T_{c0} for $\Gamma_0 = 3$, $\Gamma_\pi = 0.06$. The inset shows the magnetic field dependence of ξ_h/ξ_{c2} for $s^{\pm}$ model ($\Gamma_0 = 3$, $\Gamma_\pi = 0.06$) and s_{++} model ($\Gamma^* = 3$, dotted line) at $T/T_{c0} = 0.15$. (b) The interband scattering Γ_π dependence of ξ_h/ξ_{c2} at different temperatures T/T_{c0} (intraband scattering $\Gamma_0 = 3$ and $B = 5$) for the $s^{\pm}$ pairing.

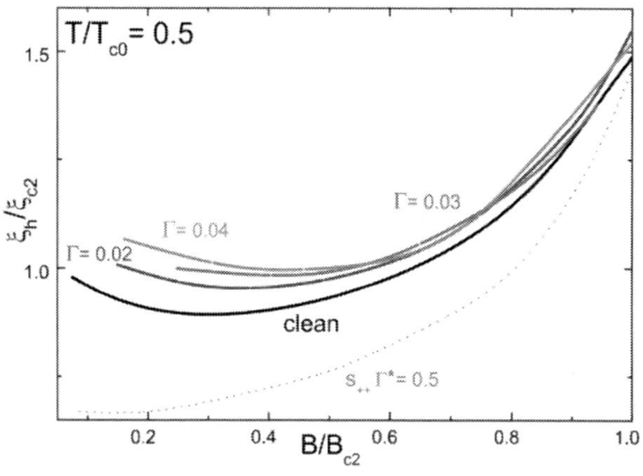

Figure 3. The magnetic field dependence of coherence length at $T/T_{c0} = 0.5$ with the similar values of intraband Γ_0 and interband Γ_π scattering Γ ($\Gamma = 0$ for "clean" case and $\Gamma = 0.02, 0.03, 0.04$ for the $s^{\pm}$ pairing). Dotted line shows result for s_{++} model ($\Gamma^* = 0.5$).

Study of the field dependence $\xi_h/\xi_{c2}(B/B_{c2})$ can clarify the both branches [25] of the energetic spectrum of the mixed state which can not be done in the phenomenological GL theory. Thus, the field-dependent suppression of ξ_h/ξ_{c2} is expected in $s^{\pm}$ model in comparison to s_{++} one in nonstoichiometric iron pnictides with high Γ_π (like doped 122 compounds). Also nonmonotonous $\xi_h/\xi_{c2}(\Gamma_\pi)$ dependence is possible in general as is shown in Fig. 2 (b).

We also study the case of weak intraband scattering. This case can be realized in stoichiometric pnictides such as $LiFeAs$. Fig. 3 presents the ξ_h/ξ_{c2} magnetic field dependence with scattering parameters $\Gamma_0 = \Gamma_\pi = \Gamma$ equal to 0, 0.02, 0.03 and 0.04. Dotted line shows the result for s_{++} model ($\Gamma^* = 0.5$). The $\xi_h(B)$ dependence shifts upward from the "clean" curve and have higher values in $s^{\pm}$ model. In contrast, ξ_h/ξ_{c2} curve shifts downward with impurity scattering in s_{++} model. The high values of ξ_h observed in μSR measurements in $LiFeAs$ [7] supports the $s^{\pm}$ pairing.

To conclude, Eilenberger equations have been solved in the mixed state for superconductors with $s^{\pm}$ pairing symmetry. This symmetry is proposed to realize in iron pnictide superconductors. Effects of interband (Γ_0) and intraband (Γ_π) impurity scattering on coherence length ξ_h are investigated. It is found that $\xi_h/\xi_{c2}(B/B_{c2})$ dependence is nonuniversal and different from the GL theory prediction. In the case of intraband scattering ξ_h/ξ_{c2} decreases with Γ_0. The effects of interband scattering on ξ_h depends on Γ_0. The predictions for $\xi_h/\xi_{c2}(B/B_{c2})$ for doped 122 compounds (nonstoichiometric iron pnictides), where $\Gamma_0 \gg \Gamma_\pi$, are done. These dependencies demonstrate growing behavior defined by Γ_0 with values much less that one in whole field range, i.e. they are under the "clean" ($\Gamma_0 = \Gamma_\pi = 0$) curve. In the case of weak impurity scattering, $\Gamma_0 = \Gamma_\pi \ll 1$, the $\xi_h/\xi_{c2}(B/B_{c2})$ dependence shifts upward from the "clean" curve. This case can be realized in stoichiometric 111 compounds. A comparison with s_{++} pairing model is done. Opposite tendencies with interband scattering for $\xi_h/\xi_{c2}(B/B_{c2})$ dependences are found for $s^{\pm}$ and s_{++} models for stoichiometric and nonstoichiometric iron pnictides. The predictions can be tested by μSR experiments.

This work was supported by the Finnish Cultural Foundation.

References

[1] Kamihara Y, Hiramats H, Hirano M, Kawamura R, Yanagi H, Kamiya T and Hosono H 2006 *J. Am. Chem. Soc.* **128** 10012
[2] Kamihara Y, Watanabe T, Hirano M and Hosono H 2008 *J. Am. Chem. Soc.* **130** 3296
[3] Takahashi H, Igawa K, Arii K, Kamihara Y, Hirano M and Hosono H 2008 *Nature (London)* **453** 376
[4] Chen G F, Li Z, DWu, GLi, ZHu W, Dong J, Zheng P, Luo J L and Wang N L 2008 *Phys. Rev. Lett.* **100** 247002
[5] Rotter M, Tegel M and Johrendt D 2008 *Phys. Rev. Lett.* **101** 107006
[6] Sefat A S, Jin R, McGuire M A, Sales B C, Singh D J and Mandrus D 2008 *Phys. Rev. Lett.* **101** 117004
[7] Inosov D S, White J S, Evtushinsky D V, Morozov I V, Cameron A, Stockert U, Zabolotnyy V B, Kim T K, Kordyuk A A, Borisenko S V, Forgan E M, Klingeler R, Park J T, Wurmehl S, Vasiliev A N, Behr G, Dewhurst C D and Hinkov V 2010 *Phys. Rev. Lett.* **104** 187001
[8] Mazin I I, Singh D J, Johannes M D and Du M H 2008 *Phys. Rev. Lett.* **101** 057003
[9] Vorontsov A B, Vavilov M G and Chubukov A V 2009 *Phys. Rev. B* **79** 140507(R)
[10] Martin C, Gordon R T, Tanatar M A, Kim H, Ni N, Bud'ko S L, Canfield P C, Luo H, Wen H H, Wang Z, Vorontsov A B, Kogan V G and Prozorov R 2009 *Phys. Rev. B* **80** 020501(R)
[11] Hanaguri T, Niitaka S, Kuroki K and Takagi H 2010 *Science* **328** 474
[12] Tanatar M A, Reid J P, Shakeripour H, Luo X G, Doiron-Leyraud N, Ni N, Bud'ko S L, Canfield P C, Prozorov R and Taillefer L 2010 *Phys. Rev. Lett.* **104** 067002
[13] Kontani H and Onari S 2010 *Phys. Rev. Lett.* **104** 157001
[14] Laiho R, Safonchik M and Traito K B 2007 *Phys. Rev. B* **76** 140501(R)
[15] Laiho R, Safonchik M and Traito K B 2008 *Phys. Rev. B* **78** 064521
[16] Sonier J E 2007 *Rep. Prog. Phys.* **70** 1717
[17] Pogosov W V, Kugel K I, Rakhmanov A L and Brandt E H 2001 *Phys. Rev. B* **64** 064517
[18] de Oliveira I G and Thompson A M 1998 *Phys. Rev. B* **57** 7477
[19] Sonier J E 2004 *J. Phys.: Condens. Matter* **16** S4499
[20] Miranović P, Ichioka M and Machida K 2004 *Phys. Rev. B* **70** 104510
[21] Ovchinnikov Y N and Kresin V Z 1995 *Phys. Rev. B* **52** 3075
[22] Hao Z, Clem J R, McElfresh M W, Civale L, Malozemoff A P and Holtzberg F 1991 *Phys. Rev. B* **43** 2844
[23] Sonier J E, Kiefl R F, Brewer J H, Chakhalian J, Dunsiger S R, MacFarlane W A, Miller R I, Wong A, Luke G M and Brill J W 1997 *Phys. Rev. Lett.* **79** 1742
[24] Zare A, Markowsky A, Dahm T and Schopohl N 2008 *Phys. Rev. B* **78** 104524
[25] Gygi F and Schlüter M 1991 *Phys. Rev. B* **43** 7609

Joint European Magnetic Symposia – JEMS 2010 IOP Publishing
Journal of Physics: Conference Series **303** (2011) 012114 doi:10.1088/1742-6596/303/1/012114

Coherence length of magnetic field in the mixed state of type-II superconductors

P. Belova[1,2], M. Safonchik[1,3], K. B. Traito[1], E. Lähderanta[1]

[1] Lappeenranta University of Technology, P.O.Box 20, FI-53851, Lappeenranta, Finland
[2] Petrozavodsk State University, Lenin str. 33, RU-185640, Petrozavodsk, Russia
[3] A. F. Ioffe Physico-Technical Institute, St. Petersburg, 194021, Russia

E-mail: `Polina.Belova@lut.fi`

Abstract. Influence of impurities on coherence length ξ_h in the mixed state of s-wave superconductors is investigated in framework of quasiclassical Eilenberger theory. The increasing of impurity scattering rate results in decreasing of ξ_h. The obtained field dependence of ξ_h for clean superconductors has a minimum and it is similar to that in Hao-Clem and Miranović-Ichioka-Machida theories for order parameter of coherence length. It is found that growing behavior of ξ_h with magnetic field in dirty superconductors is different from order parameter coherence length determining by pairing potential near with vortex core. The magnetic field dependence of coherence length in normalized units, $\xi_h/\xi_{c2}(B/B_{c2})$, is nonuniversal and depends on impurity scattering potential.

Last ten years much attention has been paid to the investigation of the field distribution in high-κ superconductors [1–3]. On the theoretical level, there are four widely used methods: solving of the Bogoliubov-de Geenes (BdG) equations [4], the quasiclassical nonlocal Eilenberger theory [5–7] (this is the quasiclassical limit of the BdG theory for $k_F\xi_{BCS} \gg 1$), solving of the Usadel theory [8] (this is the dirty local limit of the Eilenberger equations with the strong impurity scattering rates ($\Gamma/T_c \gg 1$)) and the phemenological Ginzburg-Landau (GL) theory [9–13] which is valid near T_c. Because BdG method is very time consuming for the self-consistent numerical calculation [4], the Eilenberger or Usadel theories are used in the microscopical consideration. In analysis of the experimental data, the analytical GL model (AGL) with penetration depth λ and cutoff parameter ξ_h as a fitting parameters is used very often. The cutoff parameter ξ_h (in the notation of the AGL ξ_v) is connected with GL coherence length ξ_{c2}, determined by the relation $B_{c2} = \Phi_0/2\pi\xi_{c2}^2$. From theoretical reasons (see discussion in Ref. [14]), λ can not be taken arbitrarily and should be taken as a differential operator L_{ij} [15], for the description of the nonlocal effects giving the additional field dependence in the mixed state, or its local limit obtained from microscopical consideration of the Meissner state [16] independent on the magnetic field.

There is no consensus about the meaning of ξ_h, the problem was discussed originally by de Geenes group [17]. There are several proposes for the value of ξ_h: it can be taken as a coherence length ξ_{c2} with some numerical coefficient [18], or as the order parameter coherence length ξ_1, or as a proportional to the superconducting current coherence length ξ_2. Characteristic length ξ_1 is determined as $1/\xi_1 = (\partial|\Delta(r)|/\partial r)_{r=0}/|\Delta_{NN}|$, where $|\Delta_{NN}|$ is the maximum value of the order parameter along the nearest-neighbor direction which is the direction of taking the derivative [19] and ξ_2 is determined by maximum of screening current around the vortex [20].

But connections between ξ_h, ξ_1 and ξ_2 is not investigated in detail yet. The microscopical model allowing to obtain analytical solution for $\xi_h(B)$ has been suggested in Ref. [21] (the KZ model). In this model linearized Eilenberger equation has been solved and uniform magnetic field has been suggested. It means than Kramer-Pesch effect is not included in the consideration. The exact form equation of $\xi_h(B)$ for the zero-T clean case for both Fermi sphere and cylinder has been obtained. The result can be represented as $\xi_{KZ}(B)/\xi_{c2}(B_{c2}) = U(B/B_{c2})$ with U being an universal function. The most important features of the KZ model and $\xi(B)$ dependence are as follows: (i) this dependence is weakened by scattering and disappears in the dirty limit; (ii) the B dependence of ξ vanishes as $T \to T_c$; (iii) in reduced variables, the dimensionless coherence length $\xi^* = \xi/\xi_{c2}$ should be nearly universal function of the reduced field $b = B/B_{c2}$ for clean materials in high fields and low temperatures; and (iv) for materials on the clean side ($\Gamma < 1$) the low-T slope $d\xi^*/db^{(-1/2)}$ is nearly universal in high fields ($b \to 1$). It is found that the microscopical calculations of ξ_1 [5] do not agree with the KZ theory. In particular, these calculations don't confirm the KZ assertion about weakening of the field dependence of the core size with the increasing scattering. As noted in Ref. [14] the question still remains: which of these two theoretical approaches, Ref. [5] or Ref. [21], describes better various data on $\xi_h(B)$? It is important to note also that the GL theory predictions is not reproduced by the KZ theory.

Recently, an effective London model with the magnetic coherence length $\xi_h(B)$ as a fitting parameter has been obtained for clean [22] and dirty [23] superconductors, using self-consistent solution of quasiclassical nonlocal Eilenberger equations. Such theory looks appropriate for the description of the vortex core where strong nonlinear and nonlocal effects are expected. In this approach the coherence length obtained from the Ginzburg-Landau model is extended over the whole field and temperature range. The Fourier components of magnetic field in this model are described by London equation with GL type cutoff function

$$h_{EGL}(\mathbf{r}) = \frac{\phi_0}{S}\sum_{\mathbf{G}} \frac{F(G)e^{i\mathbf{Gr}}}{1+\lambda^2G^2}, \quad (1)$$

where $F(\mathbf{G}) = uK_1(u)$, $u = \xi_h G$. It is important to note that ξ_h in Eq. (1) is obtained by solving the Eilenberger equations and ξ_h doesn't coincide with the variational parameter of the AGL model. We will call obtained field distribution as an Eilenberger - Ginzburg-Landau field distribution $h_{EGL}(\mathbf{r})$. In Eq. (1) $\lambda(T)$ is calculated from microscopical theory for the Meissner state and renormalized by impurity scattering [24]. In dirty superconductors the value of λ increases considerable and gives the main effect of impurities in the field distribution (Eq. (1)) suppressing deviation of the field from the mean value B. Thus, in this model there is only one fitting parameter for the description of the vortex state, ξ_h, similar to Ref. [14].

The aim of our paper is to calculate $\xi_h(B)$ in the framework of the Eilenberger theory and to study the applicability of the above mentioned theories in wide temperature range and at different impurity scattering rates. In particular, we are interested in looking for possible predicted universal behavior. With the Riccati transformation of the Eilenberger equations, quasiclassical Green functions f and g can be parameterized via functions a and b [5]

$$\bar{f} = \frac{2a}{1+ab}, \quad f^\dagger = \frac{2b}{1+ab}, \quad g = \frac{1-ab}{1+ab}, \quad (2)$$

satisfying the nonlinear Riccati equations. In Born approximation for impurity scattering we have

$$\mathbf{u}\cdot\nabla a = -a[2(\omega_n + G) + i\mathbf{u}\cdot\mathbf{A}] + (\Delta + F) - a^2(\Delta^* + F^*), \quad (3)$$

$$\mathbf{u}\cdot\nabla b = b[2(\omega_n + G) + i\mathbf{u}\cdot\mathbf{A}] - (\Delta^* + F^*) + b^2(\Delta + F), \quad (4)$$

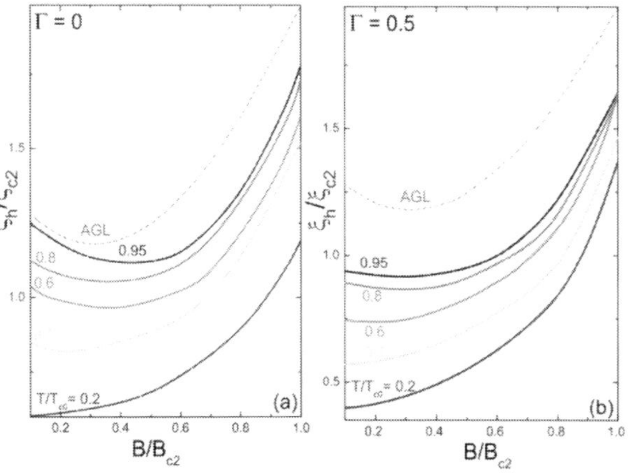

Figure 1. The magnetic field dependence of ξ_h/ξ_{c2} at different temperatures (a) with impurity scattering $\Gamma = 0$ (b) with impurity scattering $\Gamma = 0.5$. Dashed lines dshow the result of the AGL theory for ξ_v from Ref. [9].

where $\omega_n = \pi T(2n+1)$, $F = 2\pi \langle f \rangle \cdot \Gamma$ and $G = 2\pi \langle g \rangle \cdot \Gamma$. Here, $\Gamma = \pi n_i N_F |u|^2$ is the impurity scattering rate, u is impurity scattering amplitude and $\mathbf{u}$ is a unit vector of the Fermi velocity. The FLL create the anisotropy of the electron spectrum [19]. Therefore the impurity renormalization correction in Eq. (3) and (4) are averaged over Fermi surface and can be reduced to averages over the polar angle θ, i.e. $\langle \ldots \rangle = (1/2\pi)\int \ldots d\theta$.

To take into account the influence of screening the vector potential $\mathbf{A}(\mathbf{r})$ in Eqs. (3) and (4) is obtained from the equation $\nabla \times \nabla \times \mathbf{A_E} = \frac{4}{\kappa^2}\mathbf{J}$, where the supercurrent $\mathbf{J}(\mathbf{r})$ is given in terms of $g(\omega_n, \theta, \mathbf{r})$ by

$$\mathbf{J}(\mathbf{r}) = 2\pi T \sum_{\omega_n > 0} \int_0^{2\pi} \frac{d\theta}{2\pi} \frac{\hat{\mathbf{k}}}{i} g(\omega_n, \theta, \mathbf{r}). \tag{5}$$

Here $\mathbf{A}$ and $\mathbf{J}$ are measured in units of $\phi_0/2\pi\xi_0$ and $2ev_F N_0 T_c$, respectively. The self-consistent condition for the pairing potential $\Delta(\mathbf{r})$ is given by

$$\Delta(\mathbf{r}) = V^{SC} 2\pi T \sum_{\omega_n > 0} \int_0^{2\pi} \frac{d\theta}{2\pi} f(\omega_n, \theta, \mathbf{r}), \tag{6}$$

where V^{SC} is the superconducting coupling constant and ω_c is the ultraviolet cutoff determining T_{c0}[23]. All over this paper the energy, the temperature, and the length are measured in units of T_{c0} and the coherence length $\xi_0 = v_F/T_{c0} = \xi_{BCS}\pi\Delta_0/T_{c0}$. Here $\xi_{BCS} = v_F/\pi\Delta_0$, where v_F is the Fermi velocity and Δ_0 is temperature dependent uniform gap. The magnetic field $\mathbf{h}$ is given in units of $\phi_0/2\pi\xi_0^2$. The impurity scattering rates are in units of $2\pi T_{c0}$. In calculations the ratio $\kappa = \lambda_{L0}/\xi_0 = 10$ is used. It corresponds to $\kappa_{GL} = 43.3$ [5]. The Riccati equations [Eq. (3 and 4)] are solved by the Fast Fourier Transform (FFT) method [23]. This method is reasonable for dense FLL discussed in this paper. In high field the pinning effects are weak and they are not considered in our paper. After solving the Eilenberger equations the obtained magnetic field distribution $h_E(\mathbf{r})$ is fitted with the London field distribution $h_{EGL}(\mathbf{r})$ (Eq. (1)). To study high field regime we should calculate upper critical field $B_{c2}(T)$ [25].

Our calculations show that in clean superconductors $\xi_h(B)$ dependence has minimum which disappears at low temperatures. The absolute values of ξ_h are smaller than those of the AGL theory predictions, with increasing temperature ξ_h dependences move to higher values.

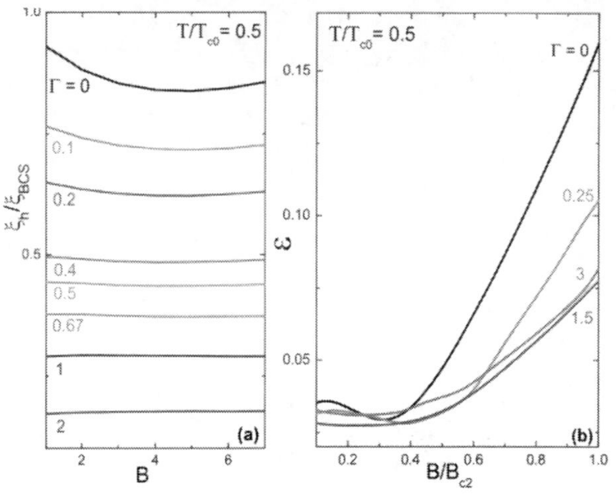

Figure 2. (a) The magnetic field dependence of ξ_h/ξ_{BCS} at $T/T_{c0} = 0.5$ with different impurity scattering Γ. (b) Magnetic field dependence of mean square deviation of the London distribution from the Eilenberger distribution normalized by the variance of the Eilenberger distribution, ε, for $T/T_{c0} = 0.5$ with different impurity scattering Γ.

These effects can be seen in Fig. 1 (a), where $\xi_h(B/B_{c2})$ are presented with $\Gamma = 0$ at $T/T_{c0} = 0.2, 0.4, 0.6, 0.8, 0.95$. The same tendency is also visible in the presence of impurity scattering, but shifting of ξ_h to direction of AGL curve is slower. Fig. 1 (b) presents $\xi_h(B/B_{c2})$ dependence with $\Gamma = 0.5$ at $T/T_{c0} = 0.2, 0.4, 0.6, 0.8, 0.95$. Strong decreasing of ξ_h with decreasing of temperature can be explained by Kramer-Pesch effect [7]. The change of the shape of $\xi_h(B)$ curve in low fields with increasing of scattering rate Γ is shown in details at Fig. 2 (a) at $T/T_{c0} = 0.5$. At high scattering rate, a flat dependence is clearly visible.

We also calculate magnetic field dependence of mean square deviation of h_{EGL} distribution of the magnetic field from the Eilenberger distribution normalized by the variance of the Eilenberger distribution $\varepsilon = \sqrt{\overline{(h_E - h_{EGL})^2}/\overline{(h_E - B)^2}}$, where $\overline{\cdots}$ is average over unit vortex cell. Fig. 2 (b) shows $\varepsilon(B)$ dependence for $T = 0.5$ with different impurity scattering Γ. It can be seen from this picture that accuracy of EGL model is getting better with increasing impurity scattering and saturates at the $\Gamma \approx 1.5$.

We should note that the the similarity of our results to the AGL theory (which is suppose to be quantitavely incorrect [11]) can be considered only as a coincidence. First, in our methods there is only one fitting parameter ξ_h, while in the AGL theory there are two of them, ξ_v and f_∞ ($f_\infty \equiv 1$ in our case). Because of the boundary condition of the field distribution ($h(r) \to B$ at $B \to B_{c2}$) ξ_h is growing function of B at the high fields resulting in the appearance of the minimum. In the AGL theory (and "improved" analytical GL theory [11]) the boundary condition is satisfied by the limit $f_\infty \to 0$ at $B \to B_{c2}$, so the behavior of $\xi_v(B)$ dependence is not predetermined. For example, in the "improved" analytical GL theory there is no minimum in $\xi_v(B)$ [11]. Absence of the minimum in $\xi_h(B)$ results also from local Usadel theory for $\xi_1(B)$ [26] and $\xi_2(B)$ [27] dependences. Second, there is clear impurity dependence of the ξ_h/ξ_{c2} value even at high temperatures (compare Fig. 1 (a) and Fig. 1 (b)), which can not be explained by the local Usadel or "improved" analytical GL theories, where scaling $\xi_h/\xi_{c2} = Const$ (independent on Γ) is expected.

In Fig. 1 the normalization constant ξ_{c2} is dependent on impurity scattering rate Γ. It is well known that at high Γ $\xi_{c2} \sim \sqrt{1/\Gamma} \sim \sqrt{l}$, where l is the mean-free path. Therefore, the

Journal of Physics: Conference Series **303** (2011) 012114 doi:10.1088/1742-6596/303/1/012114

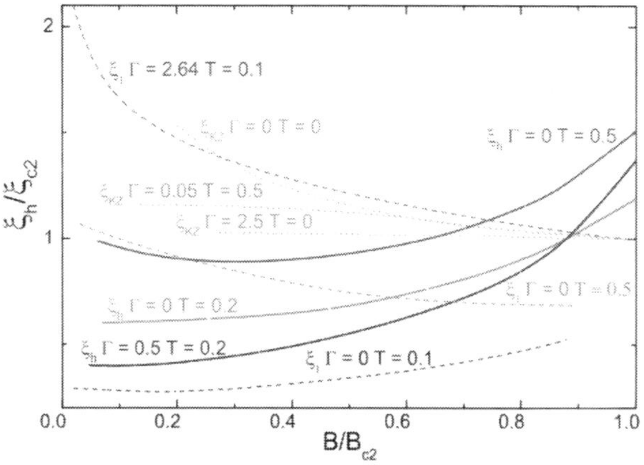

Figure 3. Comparison between field dependences of ξ_h/ξ_{c2} (at $T = 0.2$, $\Gamma = 0; 0.5$ and $T = 0.5$, $\Gamma = 0$, solid lines), ξ_1/ξ_{c2} (at $T = 0.1$, $\Gamma = 0$ and $T = 0.1$, $\Gamma = 2.64$ from Ref. [5] and $T = 0.5$, $\Gamma = 0$ from Ref. [19], dashed lines) and ξ_{KZ}/ξ_{c2} (at $T = 0$, $\Gamma = 0$; $T = 0.5$, $\Gamma = 0.05$ and $T = 0$, $\Gamma = 2.5$ from Ref. [21], dotted lines).

decreasing of the ratio ξ_h/ξ_{c2} with Γ implies the most strong dependence ξ_h on l. It is found that ξ_h in dirty superconductors can be scaled with relaxation time τ

$$\xi_h(B,T,\tau) = \frac{\xi_{pure}(B,T)}{1 + \frac{\tau_0(B,T)}{\tau}} \tag{7}$$

where $\xi_{pure}(B,T)$ is the effective coherence length in clean superconductors [22] and τ_0 is a characteristic relaxation time. It results in $\xi_h \sim l$ dependence at high Γ similar to the behavior of nonlocality radius [18] resulting in decreasing of ξ_h/ξ_{c2} at high Γ. Such fast decreasing of ξ_h can be compared with the behavior of the another characteristic length ξ_1. It has been found that at low temperatures impurity scattering suppresses Kramer-Pesch effect in $\xi_1(T)$ dependence resulting in nonmonotonous behavior of $\xi_1(\Gamma)$. On the another hand, it is clearly visible from Fig. 2 (a) that ξ_h monotonously decreases with Γ, where normalization constant ξ_{BCS} is used (ξ_{BCS} is not dependent on Γ).

The data in Figs. 1 and 2 demonstrate nearly universal behavior near T_c and small scattering rates: (i) the nonmonotonous field dependence with a minimum and (ii) the similar slope $d(\xi_h/\xi_{c2})/db$ at $b = 1$ which is weakly dependent on temperature and scattering rate. But the results are very different from the prediction of the KZ theory because of neglection of the Kramer-Pesch effect there. This can be seen from Fig. 3, where predictions of the various theories for clean superconductors and different temperatures are presented. Strong suppression of ξ_h (the ξ_h curves) and ξ_1 (the ξ_1 curves) with temperature lowering is visible in contrast to the increasing of ξ_{KZ} (the KZ curves). But impurity induced behavior is similar for ξ_h/ξ_{c2} and ξ_{KZ}/ξ_{c2}: both decreases with increasing impurity rates.

To conclude, the magnetic coherence length ξ_h (cutoff parameter) in the mixed state of high-κ s-wave superconductors is investigated in framework of quasiclassical Eilenberger theory. Nearly universal field dependence with a minimum is found near critical temperature in clean superconductors. A similar slope $d(\xi_h/\xi_{c2})/d(B/B_{c2})$ at $B/B_{c2} = 1$ weakly dependent on temperature and scattering rate is discovered. Quasiparticle scattering by impurities and lowering of the temperature reduce the value of ξ_h shifting it considerably downward from the the AGL curve and at low temperatures strong influence of the Kramer-Pesch effect is found. It can explain muon spin rotation experimental results in some low temperature superconductors, where the ratio $\xi_h/\xi_{c2} \ll 1$ [3] is observed in intermediate fields. A comparison with the behavior of the order parameter coherence length ξ_1 and another theories is done. It is found that impurities influence by different way on ξ_h and ξ_1.

This work was supported by the Finnish Cultural Foundation.

References

[1] Sonier J, Brewer J and Kiefl R 1977 *Rev. Mod. Phys.* **72** 769
[2] Sonier J E 2004 *J. Phys.: Condens. Matter* **16** S4499
[3] Sonier J E 2007 *Rep. Prog. Phys.* **70** 1717
[4] Atkinson W A and Sonier J E 2008 *Phys. Rev. B* **77** 024514
[5] Miranović P, Ichioka M and Machida K 2004 *Phys. Rev. B* **70** 104510
[6] Ichioka M, Hasegawa A and Machida K 1999 *Phys. Rev. B* **59** 8902
[7] Nakai N, Miranović P, Ichioka M and Machida K 2006 *Phys. Rev. B* **73** 172501
[8] Usadel K 1970 *Phys. Rev. Lett.* **25** 507
[9] Hao Z, Clem J R, McElfresh M W, Civale L, Malozemoff A P and Holtzberg F 1991 *Phys. Rev. B* **43** 2844
[10] Yaouanc A, de Reotier P D and Brandt E H 1997 *Phys. Rev. B* **55** 11107
[11] Pogosov W V, Kugel K I, Rakhmanov A L and Brandt E H 2001 *Phys. Rev. B* **64** 064517
[12] de Oliveira I G and Thompson A M 1998 *Phys. Rev. B* **57** 7477
[13] Brandt E H 2003 *Phys. Rev. B* **68** 054506
[14] Kogan V G, Prozorov R, Bud'ko S L, Canfield P C, Thompson J R, Karpinski J, Zhigadlo N D and Miranović P 2006 *Phys. Rev. B* **74** 184521
[15] Laiho R, Safonchik M and Traito K B 2007 *Phys. Rev. B* **75** 174524
[16] Lemberger T R, Ginsberg D M and Rickayzen G 1978 *Phys. Rev. B* **18** 6057–6065
[17] de Gennes P 1989 *Superconductivity of Metals and Alloys* Addison–Wesley, New York
[18] Kogan V G, Gurevich A, Cho J H, Johnston D C, Xu M, Thompson J R and Martynovich A 1996 *Phys. Rev. B* **54** 12386
[19] Ichioka M, Hasegawa A and Machida K 1999 *Phys. Rev. B* **59** 184
[20] Kadono R, Higemoto W, Koda A, Larkin M I, Luke G M, Savici A T, Uemura Y J, Kojima K M, Okamoto T, Kakeshita T, Uchida S, Ito T, Oka K, Takigawa M, Ichioka M and Machida K 2004 *Phys. Rev. B* **69** 104523
[21] Kogan V G and Zhelezina N V 2005 *Phys. Rev. B* **71** 134505
[22] Laiho R, Safonchik M and Traito K B 2007 *Phys. Rev. B* **76** 140501(R)
[23] Laiho R, Safonchik M and Traito K B 2008 *Phys. Rev. B* **78** 064521
[24] Lemberger T, Ginsberg D and Rickayzen G 1978 *Phys. Rev. B* **18** 6057
[25] Ovchinnikov Y N and Kresin V Z 1995 *Phys. Rev. B* **52** 3075
[26] Golubov A A and Hartmann U 1994 *Phys. Rev. Lett.* **72** 3602
[27] Sonier J E, Kiefl R F, Brewer J H, Chakhalian J, Dunsiger S R, MacFarlane W A, Miller R I, Wong A, Luke G M and Brill J W 1997 *Phys. Rev. Lett.* **79** 1742

Frustrated magnetic structure of Y-substituted CePdAl studied by powder neutron diffraction

Petr Čermák[1], Tommy Hofmann[2] and Pavel Javorský[1]

[1] Charles University, Faculty of Mathematics and Physics, Department of Condensed Matter Physics, Ke Karlovu 5, 121 16 Prague 2, The Czech Republic

[2] Helmholtz-Zentrum Berlin, Lise-Meitner Campus, Glienicker Str. 100, 14109 Berlin, Germany

E-mail: cermak@mag.mff.cuni.cz

Abstract. CePdAl is a heavy-fermion antiferromagnet with T_N = 2.7 K, crystallizing in the ZrNiAl-type structure. The magnetic structure is described by the propagation vector k = (1/2, 0, τ), τ = 0.35, with the cerium magnetic moments aligned along the c-axis. One third of magnetic moments remains disordered due to the geometrical frustration. Specific heat measurements on substituted $Ce_{1-x}Y_xPdAl$ compounds revealed strong reduction of T_N with Y substitution and the antiferromagnetic order vanishes around $x = 0.2$. To investigate the microscopic details of the changes in the magnetic structure evoked by nonmagnetic ion substitution, we have performed an experiment on the powder neutron diffractometer E6 at HZB on the samples with $x = 0.02, 0.06$ and 0.1. Measurements showed the magnitude reduction of the ordered cerium moments with Y substitution while the propagation vector and other magnetic structure characteristics remain unchanged.

1. Introduction

Rare-earth intermetallic compounds containing Ce are known to exhibit often unusual magnetic properties at low temperatures. CePdAl crystallizes in the hexagonal ZrNiAl-type structure (space group P$\bar{6}$2m) [1], orders antiferromagnetically below T_N = 2.7 K [2]. This order is characterized by an incommensurate propagation vector $k = (1/2, 0, \tau)$ with $\tau \cong 0.35$, and a longitudinal sine-wave modulated spin arrangement oriented along the hexagonal c-direction [3]. One third of the Ce magnetic moments is strongly frustrated in a close relation to the Kagome-like triangular arrangement of Ce atoms within the basal planes. The neutron diffraction reveals that the frustrated moments do not order down to 180 mK at least [4]. The compound exhibits several metamagnetic phase transitions to the ferromagnetic state in fields around 4 T applied along c-axis [5]. These transitions could be connected with the magnetic moment appearance on the frustrated 1/3 of the Ce sites and subsequent appearance of a ferromagnetic component on the remaining Ce sites. By contrast the magnetic field applied perpendicular to the c-axis does not break the antiferromagnetic order up to 7.5 T [6].

The substitution of Ce ions by nonmagnetic Y ions leads to a rapid suppression of the magnetic order and it vanishes around 20 % of yttrium concentration. The persistency of antiferromagnetic order in magnetic field applied perpendicularly to the c-axis is broken above 6 % of yttrium concentration. It is speculated that it is an indication of reducing the strong magnetocrystalline anisotropy in CePdAl [7].

In this paper, we present the results of our experiment on the powder neutron diffractometer E6 at Helmholtz Zentrum Berlin (HZB) on samples with 2, 6 and 10 % of yttrium atoms. The goal of the experiment was to clarify the effect of Y substitution on the magnetic structure of CePdAl, namely possible change of propagation vector and appearance of frustrated Ce moments.

2. Experimental

Samples of $Ce_{1-x}Y_xPdAl$ with $x = (0.02; 0.06$ and $0.1)$ were prepared by arc-melting stoichiometric mixtures of pure elements (3N for Ce and Y, 3N5 for Pd and 5N for Al) in a mono-arc furnace under protection of argon atmosphere. The samples were turned and re-melted several times to achieve better homogeneity. These samples were tested with X-ray diffraction for the phase purity. Powder neutron diffraction studies were performed at the Helmholtz Zentrum Berlin (HZB) in Germany on the multi detector powder diffractometer E6 with neutron focusing optics. An oscillating fan collimator in front of the detector array was utilized for background reduction. Prior to the experiments wavelength ($\lambda = 2.438$ Å) and resolution were determined by measuring a standard YIG sample. Measured data were first analyzed with a software package called BEAN. Bean merges data from different detectors in the detector array and integrates data along Debye-Scherrer rings to obtain the intensity vs. 2θ dependence. The Rietveld method was utilized to refine simultaneously magnetic and nuclear structure by the Fullprof program [8]. Samples with $x = 0.02$ and 0.06 were measured in a standard orange cryostate at two temperatures, that are 1.3 K to probe the magnetically ordered region and 8 K to elucidate the paramagnetic state. An ordering temperature below 1.3 K (limit of the orange cryostate) was expected for the sample with $x = 0.1$ [7]. Therefore this sample was probed in ^{3}He-^{4}He dilution refrigerator at temperatures of 0.4 K, 1.3 K (for comparison with the other two compounds) and 8 K.

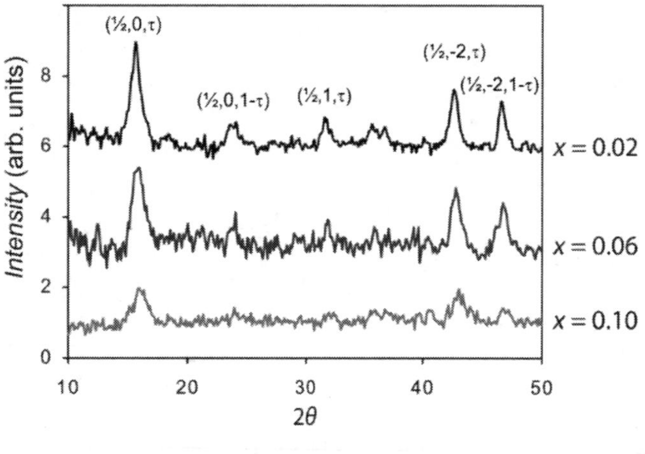

Figure 1. Differences between patterns recorded below the magnetic ordering temperature (1.3 K for x=0.02 and 0.06; 0.4 K for x=0.10) and in the paramagnetic region (8 K) for all measured $Ce_{1-x}Y_xPdAl$ compounds.

3. Results

Diffraction pattern analysis at 8 K confirms the ZrNiAl-type crystal structure of all compounds. Diffraction patterns recorded at 1.3 K contain additional Bragg reflections arising from scattering on the ordered magnetic moments of Ce atoms. We have determined from the positions of these magnetic peaks, that the propagation vector keeps unchanged for all measured concentrations and is in agreement with a measurement done by Dönni at al. on the pure CePdAl [3]. Standard overall magnetic structure refinement was rather difficult, because magnetic reflections are very weak - especially in the sample with $x = 0.1$. Therefore we decided to reveal possible changes in a magnetic structure following the two strongest magnetic peaks $(1/2, 0, \tau)$ and $(1/2, -2, \tau)$ - see Fig. 1. The ratio of the integrated intensities of these two peaks remains almost unchanged in the series and also compared to the pure CePdAl. This indicates that the magnetic structure of CePdAl is not affected by the yttrium substitution. However if we would remove frustration on 1/3 of Ce atoms, ratio of the area of the examined peaks does not change radically and possible removing of the frustration with increasing yttrium concentration cannot be excluded.

The refinement of $Ce_{0.98}Y_{0.02}PdAl$ is presented in Fig. 2. Reflections on $2\theta = 55$ and around $2\theta = 66$ are caused by sample can. Determined magnetic structure parameters were the propagation vector $k = (0.5, 0, \tau)$, with $\tau = 0.36(1)$, and the amplitude of the ordered magnetic Ce moments $\mu = 1.6(2)\ \mu_B$. These values are in exact agreement with the values obtained for the pure CePdAl [3]. Refinement of the sample with 6 % of yttrium atoms gives us $\tau = 0.35(2)$ and $\mu = 1.6(4)\ \mu_B$. The last sample with 10 % of yttrium showed only small magnetic reflections and therefore refinement of its magnetic structure parameters is not possible.

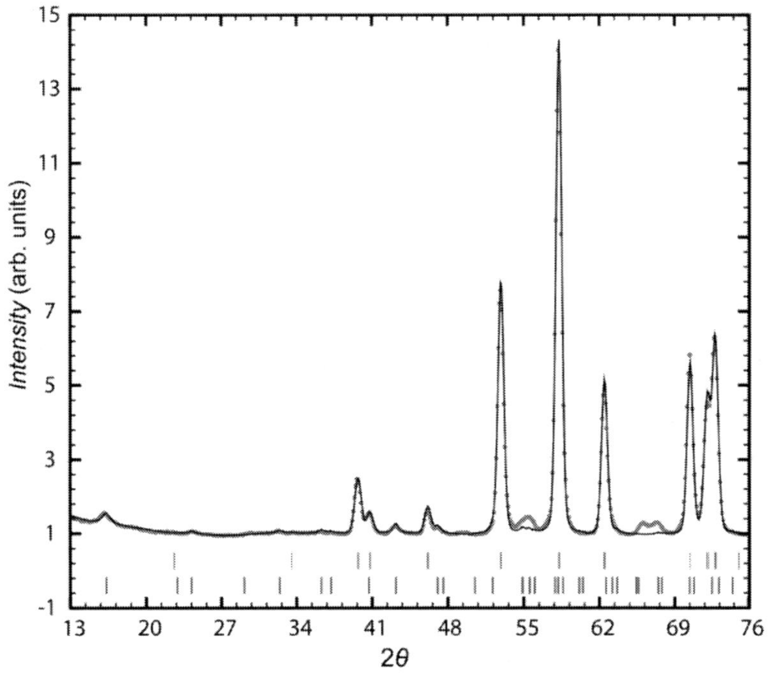

Figure 2. FullProf refinement of the crystal and magnetic structures of $Ce_{0.98}Y_{0.02}PdAl$ at 1.3 K. Red dots represent measured data while the black solid line show fitted dependency. Small bars under the plot represents Bragg peak positions - the upper one are for nuclear peaks and the lower one are for magnetic peaks.

4. Conclusion

Measurements reveal persistence of propagation vector of the pure CePdAl $k = (0.5, 0, \tau)$ for all samples. Its incommensurate component $\tau = 0.35$ remains unchanged. Also magnetic structure does not change significantly with yttrium substitution, although this can not be surely confirmed because of very weak magnetic reflections.

Acknowledgments

This research project has been supported by the European Commission under the 7th Framework Programme through the Key Action: Strengthening the European Research Area, Research Infrastructures. Contract n: 226507 (NMI3). The work was supported by the Grant Agency of the Czech Republic under the Grant No. 202/09/1027 and is a part of the research program MSM 0021620834 financed by the Ministry of Education of the Czech Republic. The work of P. Čermák was also supported by the grant SVV-2010-261303.

References

[1] Hulliger F 1993 *J. Alloys and Compounds* **196** 225
[2] Kitazawa H, Matsuhita A, Matsumoto T and Suzuki T 1994 *Physica* B **199-200** 28
[3] Dönni A, Ehlers G, Maletta H, Fischer P, Kitazawa H and Zolliker M 1996 *J. Phys.: Condens. Matter* **8** 11213
[4] Keller L, Dönni A, Kitazawa H and van den Brandt B 2002 *Appl. phys.* A **74** S686
[5] Prokeš K, Manuel P, Adroja D T, Kitazawa H, Goto T and Isikawa Y 2006 *Physica* B **385-386** 359
[6] Isikawa Y, Mizushima T, Fukushima N, Kuwai T, Sakurai J and Kitazawa H 1996 *J. Phys. Soc. Jpn.* **65** Suppl. B 117
[7] Čermák P, Kitazawa H, Prchal J and Javorský P 2010 *J. Phys.: Condens. Matter* **22** 126002
[8] Rodriguez-Carvajal J 1993 *Physica* B **192** 55

Calculation of the electronic structure of the Mott insulator NiO, $LaMnO_3$ and of GaN:Mn

R. J. Radwanski[1,2], **Z. Ropka**[1]

[1] Center of Solid State Physics, S^{nt} Filip 5, 31-150 Krakow, Poland
[2] Institute of Physics, Pedagogical University, 30-084 Krakow, Poland

E-mail: `sfradwan@cyf-kr.edu.pl; www.actaphysica.eu.`

Abstract. We have shown that energy values of the $d-d$ splittings of 2.06 eV in $LaMnO_3$ and 1.41 eV in Mn-doped GaN can be consistently explained within the ordinary crystal-field theory. We claim that the $d-d$ splitting is of the (almost) purely electrostatic origin and it has been calculated from first principles. Finally, contrary to the band-theory, we claim the existence of the discrete electronic structure, in a meV scale, in $3d$-atom containing compounds.

1. Introduction

Theoretical description of the magnetism and the electronic structure of compounds containing $3d$ transtition-metal atoms is still a challenging problem in the modern solid-state physics. In last years we put our contribution to the theoretical research starting from the simplest compounds. We have managed to consistently describe properties of NiO, reconciling its insulating ground state and a strong (antiferro-)magnetism with T_N of 525 K [1, 2]. We exactly calculated electronic structure associated with eight $3d$ electrons at each Ni site in the incomplete $3d$ shell. We solved a problem of origin of electron correlations in NiO. We calculated value of the magnetic moment and its temperature dependence. We derived the orbital moment to be substantial of 0.54 μ_B. We calculated the low-energy electronic structure of NiO from first principles [2, 3]. We take it as a remarkable result because NiO has been considered by more than 60 years as a prominent Mott insulator exhibiting very strong electron correlations. In our calculations we include spin-orbit interactions - thanks it our description goes beyond commonly made text-book spin-only/quenched-orbital-moment approximation [4], p.71.

In this contribution the same procedure is applied to the Mn doped GaN compounds. The obtained electronic structure is compared to the one found in $LaMnO_3$. We have found that the electronic structure in these two different compounds with manganesse atoms can be consistently theoretically described. Good description of NiO, $LaMnO_3$ and Mn-doped GaN allows better understanding of interplay of electron correlations, magnetism, localized states and the detailed crystal structure.

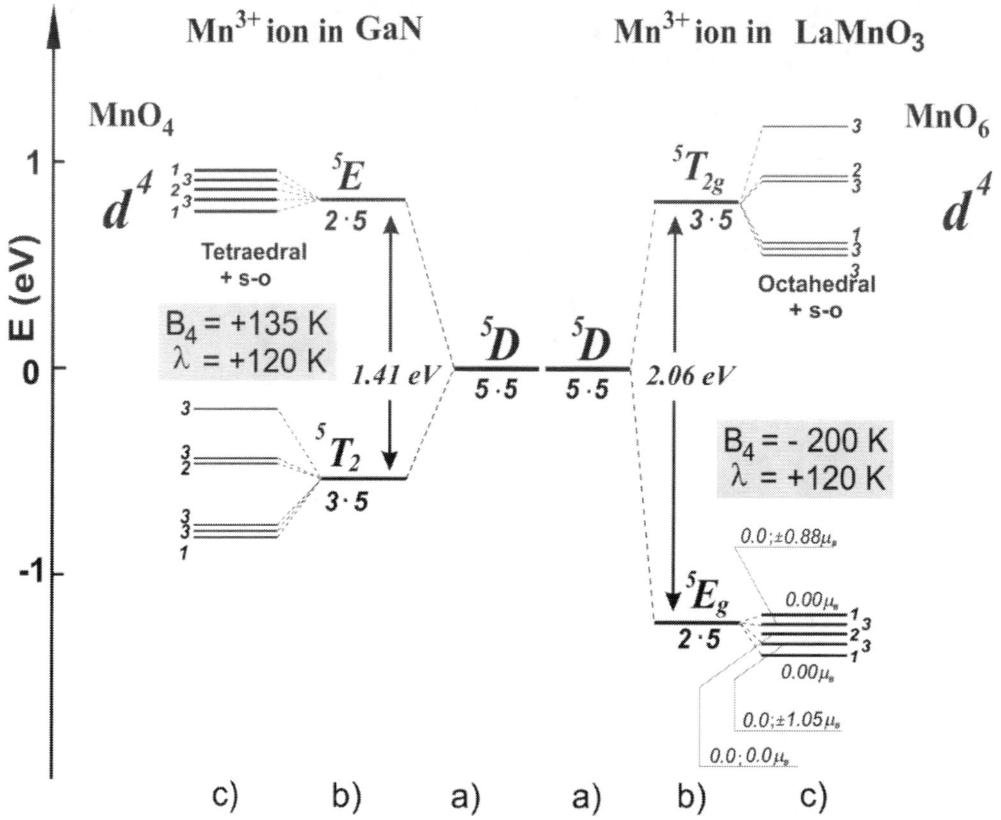

Figure 1. Electronic structure, associated with the Hund rules ^5D subterm (a), of the Mn^{3+} ion in the octahedral and tetrahedral (b) sites realized, respectively, in $LaMnO_3$ and in the Mn-doped GaN with taking into account the intra-atomic spin-orbit interactions (c).

2. Theoretical outline and discussion

Two compounds, $LaMnO_3$ and Mn-doped gallium nitride GaMnN, have been chosen owing to their completely different crystallographhic structure, different character of bonds and different formal valences but in both compounds the involved $3d$-ion (Mn) is the same and practically the same is the distance to the nearest-neighbour (n-n) anions; 200 pm in $LaMnO_3$ (the mean value of the Mn-O bonds) and 196 pm (Mn-N distance) in the doped GaN. We work in the frame of the Quantum Atomistic Solid-State Theory (QUASST) [5, 6] assuming that the involved $3d$ atoms (becoming ions in the solid) preserve their atomic-like integrity, and consequently the respective atomic-like electronic structure, becoming the full part of a solid. In other words, we assume that a better physical start for description of electronic and magnetic properties of any compound containing $3d/4f/5f$ atoms is single-ion phenomena than solid-state effects.

$LaMnO_3$ crystallizes in the perovskite-based structure (regular) and belongs to the class of compounds known as Mott insulators. The scientific interest to $LaMnO_3$-based compounds has increased considerably about 1995 after revealing its colossal magnetoresistance properties [7, 8]. Its magnetism and the electronic structure is quite well understood within the crystal-field approach with the $10Dq$ of 2.1 eV, if completed with the spin-orbit coupling and the detailed description of off-octahedral distortions [9, 10]. This value of $10Dq$ is in perfect agreement with the reality taking an experimentally-observed excitation at 2.06 eV as related with the t_{2g}-e_g excitation [11]. Mn-doped gallium nitride GaMnN is a wide-band semiconductor with the wurzite hexagonal structure but with a sharp energy structure peaked at 1.4116 eV. Recently this sharp line was identified with the intra-atomic $d-d$ transition, $^5T_2$ - ^5E, of the d^4 configuration [12, 13].

In both these completely different compounds manganese atoms occur as the trivalent ions Mn^{3+} [9, 10, 11, 12, 13] with four outer electrons. In the many-electron CEF approach these four d electrons form a strongly-correlated intra-atomic $3d^4$ electron system. The strong correlations among the $3d$ electrons are accounted for, in a zero-order approximation, by two Hund's rules. These two Hund's rules yield for the d^4 system the ground term ^5D with S=2 and L=2. Its 25-fold degeneracy is removed by the crystal field and spin-orbit interactions as is shown in Fig. 1b and 1c. Under the action of the dominant cubic crystal field, the ^5D term splits into the orbital triplet $^5T_{2g}$ and the orbital doublet ^5E_g with the energy separation Δ =$10Dq$ of about 1-3 eV.

From the crystal-field theory [14] is well known that the crystal-field coefficient A_4, originating from the charge surroundings in the lattice, for the tetrahedral site (in GaMnN) is 4/9 of that for the octahedral site, and of the opposite sign. Taking into account the trivalency of the N ions with respect to the divalent oxygen ions, one comes to a conclusion that for the same nearest-neighbour Mn-anion distance the value of the CEF splitting, $10Dq$, in case of the tetrahedral site should be 2/3 of that for the octahedral site.

Thus one expects on basis of 2.06 eV in $La^{3+}Mn^{3+}O_3^{2-}$ a value of 1.373 eV for the Mn^{3+} ion in $Ga^{3+}N^{3-}$ - it is in very good agreement with the observed value of 1.41 eV [12, 13]. In fact, one comes to a perfect agreement taking into account a slightly smaller Mn-N distance in GaN. Very good reproduction of the ratio of the octupolar crystal-field interactions with the up-side-down electronic structure of the Mn-doped GaN compared to $LaMnO_3$ gives very strong argument for **the purely electrostatic origin of the crystal-field splittings** in these both compounds.

The strength of $10Dq$ of 2.06 eV in $LaMnO_3$ and the ^5E_g ground subterm can be calculated from the ordinary point-charge model with the O^{2-} ions forming the octahedron and taking into account the mean value $\langle r_d^4 \rangle$ as 17.4 a_B^4 [15]. For the same value of $\langle r_d^4 \rangle$ and the local N^{3-} tetrahedron one gets 1.41 eV for the $^5T_2$-^5E splitting in the Mn-doped GaN with the $^5T_2$ ground subterm, i. e. exactly as is observed. A necessary removal of large degeneracies of the ground subterm (10 and 15-fold) goes via the spin-orbit coupling and the lattice distortions (Jahn-Teller effect) as it was discussed in Refs [9, 10]. Also in these papers we have discussed formation of the (anti-)ferromagnetic state in $LaMnO_3$ below T_N of 140 K.

3. Conclusions

We claim that the energy values of the $d-d$ splittings, of 2.06 eV in $LaMnO_3$ and 1.41 eV in Mn-doped GaN, with their up-side-down ground subterms can be consistently explained within the ordinary crystal-field theory. We claim that **the $d-d$ splitting is predominantly of the purely electrostatic origin**, contrary to a textbook knowledge of a significant contribution due to the $3d$(TM) - $2p$(O) hybridization [4]. Finally, **contrary to the band-theory we claim the existence of the discrete electronic structure, in a meV scale, in $3d$-atom containing compounds.** A discrete electronic structure with such large accuracy has been calculated by us for a quite great number of $3d$ compounds, not mentioning $4f$ and $5f$ compounds, like $FeBr_2$ [16], $LaCoO_3$ [17], CoO, FeO [3]. At present the Resonant Inelastic X-ray Scattering (RIXS) experiments allow to reveal such $d-d$ excitations [18] - a theoretical explanation of the observed

1-2 eV excitations used by authors of Ref. [18] is quite similar to our approach. In such a situation we can express a large satisfaction developing a theoretical approach which becomes step by step confirmed in novel experiments.

References

[1] R. J. Radwanski and Z. Ropka, Acta Phys. Pol. A **97**, 963 (2000).
[2] R. J. Radwanski and Z. Ropka, Acta Physica **1**, 26 (2006).
[3] R. J. Radwanski and Z. Ropka, Physica B **403**, 1453 (2008).
[4] J. M. D. Coey - this Conference; *http://confer.uj.edu.pl/jems2010/JEMS_tutorial_Coey.pdf.*
[5] R. J. Radwanski and Z. Ropka, Acta Phys. Pol. B **31**, 3079 (2000).
[6] R. J. Radwanski and Z. Ropka, in: *New development of in Field Theory*, Ed. O. Kovras, Nova Science Publ., New York 2006, p. 93 arXiv/cond-mat/0404713 (2004).
[7] Y. Tokura and N. Nagaosa, Science **288**, 462 (2000).
[8] D. D. Sarma, N. Shanthi, S. R. Barman, N. Hamada, H. Sawada, and K. Terakura, Phys. Rev. Lett. **75**, 1126 (1995).
[9] R. J. Radwanski and Z. Ropka, Acta Physica **9-10**, 1 (2007).
[10] R. J. Radwanski and Z. Ropka, Acta Physica **9-10**, 39 (2007).
[11] M. W. Kim, P. Murugavel, S. Parashar, J. S. Lee, and T. W. Noh, New J. Phys. **6**, 156 (2004).
[12] A. Wolos, A. Wysmolek, M. Kaminska, A. Twardowski, M. Bockowski, I. Grzegory, S. Porowski, and M. Potemski, Phys. Rev. B **70**, 245202 (2004).
[13] S. Marcet, D. Ferrand, D. Halley, S. Kuroda, H. Mariette, E. Gheeraert, F. J. Teran, M. L. Sadowski, R. M. Galera, and J. Cibert, Phys. Rev. B **74**, 125201 (2006).
[14] A. Abragam and B. Bleaney, *Electron Paramagnetic Resonance of Transition Ions* (Clarendon, Oxford) 1970.
[15] R. J. Radwanski and Z. Ropka, Acta Physica **25**, 34 (2008).
[16] Z. Ropka, R. Michalski, and R. J. Radwanski, Phys. Rev. B **63**, 172404 (2001).
[17] Z. Ropka and R. J. Radwanski, Phys. Rev. B **67**, 172401 (2003).
[18] M. Moretti Sala, V. Bisogni, C. Aruta, G. Balestrino, H. Berger, N. B. Brookes, G. M. de Luca, D. Di. Castro, M. Grioni, M. Guarise, P. G. Medaglia, F. Miletto Granozio, M. Minola, P. Perna, M. Radovic, M. Salluzzo, T. Schmitt, K. J. Zhou, L. Braicovich, G. Ghiringhelli, arXiv:1009.4882v4.

AUTHOR INDEX

Adamska, A M 88
Affronte, M 213
Agnus, G 385, 391
Akimov, I A 261
Akjouj, A 120
Albu, M 293
Aleshkevych, P 411
Algarabel, P A 345
Al-Wahsh, H 120
Amoretti, G 213
Andreev, A V 83
Ansermet, J -Ph 439
Aoshima, K 288, 638
Apertet, Y 405
Atoneche, F 309
Auban-Senzier, P 405
Avdeev, M 175
Badelin, A G 457
Bahl, C R H 526
Baiotto, R 191
Bajpai, A 367
Balanda, M 76, 223, 241, 255
Balogh, A G 88
Bärner, K 463
Basso, V 303
Bayer, M 261
Beavillain, P 270
Belesi, M 439
Bellini, V 213
Belotelov, V I 261
Belova, P 733, 739
Berakdar, Jamal 395
Berger, H 439
Bessais, L 185, 303
Blinc, Robert 416
Bobo, J -F 606
Bocklage, Lars 593
Bogacz, B F 179
Bogdanov, A N 339, 536, 666
Bonneau-Brault, A 568
Breedlove, B K 231
Buchelnikov, V 542
Budziak, A 76
Burian, A 445
Burke, F 622
Burzo, E 147
Cabrera, L 293
Cadogan, J M 153, 175
Camarero, J 379
Campbell, S J 153
Campos, M A 141
Candini, A 213
Cannies, G 356, 361
Carretta, S 213
Cermák, Petr 745
Chechenin, N G 654, 660
Chicinas, I 562
Chiriac, H 70, 562
Chistyakov, O 163
Chlan, V 678
Chmaissem, O 373
Chuang, T -H 39
Chumak, A V 59, 130
Ciechan, Anna 721
Coe, J P 717
Coey, J M D 622, 628, 634
Cong, D Y 511
Costa, L F T 191
Crivello, J-C 185
Cwik, J 159, 163, 169
Czapla, M 223
Czepelak, M 445
Da Silva, F S 197
Dabek, J 361
Dabkowska, H A 411
Dabkowski, A 411
Dabrowski, B 373
Dagens, B 270
D'Amico, I 613, 717
Danis, S 83, 88, 203
Das, P 367
David, B 580
De Campos, M F 141, 191, 197, 327
Diao, Z 628
Djafari-Rouhani, B 120
Dobrzynski, L 120
Dogan, Sedat 593
Dou, S X 153
Drath, O 255
Drosik, Jerzy 586
Dubourg, S 568, 606
Dubuget, V 568
Dugaev, V K 333
Dushenko, S A 654, 660
Duverger, F 568
Dyvorne, H 361
Dzhun, I O 654, 660
Edon, V 606
Eggert, S 691
Eloirdi, R 88
Eniu, D 147
Entel, P 542
Estemirova, S Kh 452, 457
Esteve, D 391
Feng, J F 628, 634
Fermon, C 65, 356, 361, 385, 405
Figiel, H 76
Fink-Finowicki, J 411
Foerster, M 433
Fontcuberta, J 96, 433

AUTHOR INDEX

Fördös, T ... 282
França, V V ... 717
Franco, F A ... 197
Funabashi, N ... 288, 638
Furukawa, K ... 288, 638
Futamoto, Masaaki ... 102, 108, 114, 315, 351, 600
Gagor, A ... 484
Gallego, Silvia ... 1
García-García, A ... 345
Gaworska-Koniarek, Dominika ... 586
Gerhardt, G ... 191
Ghasemi, A ... 299, 574
Ghirri, A ... 213
Girotto, F ... 191
Gliga, Sebastian ... 46
Golub, V ... 345
Gopal, A V ... 261
Gouder, T ... 88
Granitzer, P ... 293, 321
Granville, S ... 439
Gregg, J F ... 59, 130
Grigoras, M ... 70
Gron, T ... 484
Guerrero, R ... 385, 405
Gusin, P ... 474
Gutfleisch, O ... 169
Haase, Wolfgang ... 241
Halagacka, L ... 270
Hänninen, H ... 501
Havela, L ... 83, 88
Hawelek, L ... 445
Heczko, Oleg ... 518
Hellwig, O ... 339
Herreros, Q ... 361
Hertel, Riccardo ... 46
Higuchi, Jumpei ... 351
Hillebrands, B ... 59, 130
Hilscher, G ... 321
Hofmann, Tommy ... 745
Hole, Janez ... 416
Honig, J M ... 678
Honma, T ... 288
Horikiri, H ... 649
Horley, Paul P ... 395
Hrabovský, D ... 282
Huber, F ... 88
Hürrich, C ... 511
Huth, M ... 367
Ibarra, M R ... 345
Inglot, M ... 333
Ipatov, M ... 548
Ishibashi, T ... 288, 638
Isnard, O ... 562
Jaglicic, Zvonko ... 416
Jagodic, Marko ... 416
Jašek, O ... 580
Javorský, P ... 203, 745
Jia, Chenglong ... 395
Jiménez, E ... 379
Joch, V ... 53
Kaczorowski, D ... 179
Kaewrawang, A ... 299, 574
Kajiwara, T ... 231
Kajiwara, Y ... 618
Kakol, Z ... 678
Kalish, A N ... 261
Kapoor, Sheetal ... 613
Karas, Kazimierz ... 586
Karenowska, A D ... 59, 130
Karpasyuk, V K ... 452, 457
Karpenkov, A ... 163, 209
Karpenkov, D ... 163, 209, 159
Kasture, S ... 261
Katoh, K ... 231
Kawai, H ... 618
Kikuchi, H ... 638
Kim-Ngan, Nhu-T H ... 88
Kirilyuk, A ... 309
Kirino, Fumiyoshi ... 108, 600
Kirschner, J ... 39
Kiselev, N S ... 339
Klementová, M ... 282
Klicpera, M ... 203
Kolano, R ... 445
Kolano-Burian, A ... 445
Kolesnik, S ... 373
Kolodziej, T ... 678
Kolwicz-Chodak, L ... 76
Komatsu, T ... 288
Kondou, Kouta ... 593
Konstantinidis, Nikolaos P ... 21
Konstantinova, E A ... 654, 660
Kopecek, Jaromír ... 518
Koroleva, L I ... 401, 469
Korzeniak, T ... 223
Kosec, Marija ... 416
Kothapalli, K ... 83
Kotov, V A ... 261
Kozlowski, A ... 678
Krawiec, K ... 179
Krok-Kowalski, J ... 474
Król, G ... 678
Krzyszczak, Jaromir ... 727
Kudrle, V ... 580
Kuga, K ... 288, 638
Kuhn, L Theil ... 526
Kumar, P ... 175, 276
Küpferling, M ... 303
Kusz, J ... 678
Lähderanta, E ... 421, 427, 733, 739

AUTHOR INDEX

Laiho, R 421, 427
Lancok, J 282
Landa, Michal 518
Landgraf, F J G....141, 327
Lashkul, A V....421, 427
Lassalle-Balier, Rémy 65
Le Goff, G 361
Lecoeur, Ph....385, 391, 405
Lemberger, L....405
Leonov, Andrei A 666
Lins, J F C 197
Lisunov, K G 421, 427
Liu, X 299, 574
Liu, Yaowen....46
Lobue, M....303
Lorusso, G 213
Loyau, V 303
Luke, G M....411
Lupu, N 70
Lyubina, J....169
Machens, Anna....21
Machida, K 288, 638
Macl, J....88
Magdenko, L 270
Magén, C....345
Mais, J....373
Majcher, A....255
Majtás, Dušan....518
Malicka, E....474, 484
Malik, D 309
Marchetti, Gionni....613
Marinca, T F....562
Maroutian, T 391
Martí, X 96
Maskova, S....83
Matsubara, Katsuki 108, 600
Matsuyama, Toru....593
Mazaleyrat, F 303
Méchin, L....379
Medvedeva, I V 463
Meier, Guido 593
Mellor, C J....433
Merkt, Ulrich....593
Miglierini, M 282
Mikuszeit, N....379
Miranda, R 379
Missell, F P 191, 327
Molho, P....53
Morales, M P 293
Morisako, A 299, 574
Morozov, A S....401, 469
Morsakov, V 463
Müller, J....367
Muñoz, M Carmen....1
Mydlarz, T 484
Nagano, Katsumasa.... 102, 315
Naganuma, M 288
Nahrwold, Gesche.... 593
Nakayama, T.... 638
Nasreen, F 83
Neamtu, B V.... 562
Nehrkorn, Joscha 21
Nekrasova, Yu S 421, 427
Nenkov, K 169
Nikitin, S 159, 163, 209
Nitek, W 255
Nizhankovskii, N I.... 209
Nukaga, Yuri 114
Oguz, K 622
Ohno, H 367
Ohno, Y 367
Ohtake, Mitsuru 102, 108, 114, 351, 600
Onisan, A T.... 536
Ono, Teruo.... 593
Ota, T 618
Pacyna, A W 484
Padovese, L R 141, 197
Pajic, Damir.... 416
Palewski, T....159, 163, 169, 209
Pannetier-Lecoeur, M 356, 361, 385, 405
Pardo, J A 345
Parkkonen, L 356
Pasko, A.... 303
Pasquier, C.... 405
Pastushenkov, Yu G 159
Pedziwiatr, A T.... 179
Pelka, R 223, 255
Perna, P.... 379
Pešicka, J.... 88
Petrenko, P A.... 421, 427
Philippe, T.... 439
Piano, S.... 433
Pillard, V 391
Pinkowicz, D 223, 255
Pištora, J 282
Pizúrová, N 580
Plessis, D.... 568
Podgajny, R.... 223
Poelt, P.... 293, 321
Pohl, M 261
Politova, G.... 163
Polovy, H.... 356, 361
Pop, V 562
Popa, F 562
Porrati, F.... 367
Postava, K 270, 282
Pötschke, M.... 511
Prabhakar, A 276
Pratt, F L.... 223
Prica, C V.... 562

AUTHOR INDEX

Prokop, J ... 39
Pryds, N ... 526
Przewoznik, J ... 179
Radwanski, R J ... 749
Rams, M ... 241, 255
Ranno, L ... 53
Rasing, Th ... 309
Rellinghaus, B ... 511
Ribeiro, S B ... 197
Rigato, F ... 96, 433
Robler, U K ... 339, 536, 666
Rodrigo, C ... 379
Romero, S A ... 327
Ropka, Z ... 749
Rossi, Fausto ... 613
Roth, S ... 511
Rousochatzakis, I ... 439
Rufinus, J ... 480
Rumpf, K ... 293, 321
Russier, V ... 185
Ryan, D H ... 175
Safonchik, M ... 733, 739
Saitoh, E ... 618
Salyuk, O ... 345
Sanchez, Nadiezhda ... 1
Šantavá, E ... 580
Santini, P ... 213
Santos, R ... 197
Sasso, C P ... 303
Sato, Yoichi ... 351
Schlottmann, P ... 11
Schmidt, Herbert ... 491
Schneeweiss, O ... 580
Schultz, L ... 511
Sedlmayr, N ... 691
Serga, A A ... 59, 130
Sergeeva-Chollet, N ... 356, 361
Shchepetkin, A A ... 457
Shiiki, K ... 644, 649
Shimidzu, N ... 288, 638
Shvets, I V ... 439
Sieklucka, B ... 223, 255
Sirker, J ... 691
Skokov, K P ... 135, 209
Slebarski, Andrzej ... 713
Smirnov, A M ... 452
Smith, A ... 526
Sokolovskiy, V ... 542
Solignac, A ... 385, 405
Soroka, A ... 501
Sorokin, V V ... 457
Sozinov, A ... 501
Spalek, Jozef ... 697
Stankiewicz, Jolanta ... 135
Stepankova, H ... 678
Straka, L ... 501
Štrichovanec, P ... 345
Studer, A J ... 153
Sukhov, Alexander ... 395
Sun, Y H ... 556
Sundt, Alexander ... 21
Suresh, K ... 175
Szubzda, Bronislaw ... 586
Szymczak, H ... 411, 445
Szymczak, R ... 411, 445, 469
Tabis, W ... 678
Takeda, K ... 649
Takeya, J ... 231
Tanaka, Takahiro ... 108
Tang, S ... 288
Tang, W X ... 39
Taranenko, I ... 542
Tarnawski, Z ... 76, 678
Taskaev, S ... 542
Teran, F J ... 379
Tereshina, E ... 163
Tereshina, I ... 159, 163
Timco, G ... 213
Tobari, Kousuke ... 102, 315
Tokuc, H ... 622
Tomkowicz, Zbigniew ... 241
Toonen, A J ... 309
Traito, K B ... 733, 739
Troiani, F ... 213
Trontelj, Zvonko ... 416
Uchida, K ... 618
Umezawa, H ... 618
Urse, M ... 70
Van Etteger, A F ... 309
Vanwolleghem, M ... 270
Vengurlekar, A S ... 261
Venkatesan, M ... 293
Vernières, J ... 606
Vlaic, P ... 147
Vovk, A ... 345
Waldmann, Oliver ... 21
Wang, J L ... 153
Warczewski, J ... 474
Wasiutynski, T ... 223
Wendrock, H ... 511
Wernsdorfer, W ... 213
Wilczynski, Wieslaw ... 586
Winpenny, R E P ... 213
Wirth, S ... 367
Wisniewski, A ... 678
Wróbel, S ... 179
Wu, H C ... 439
Wysokinski, Karol I ... 721, 727
Xiao, Ling ... 556
Yabuhara, Osamu ... 114

AUTHOR INDEX

Yakovlev, D 261
Yamada, Gen...593
Yamamoto, K 231
Yamashita, M 231
Younsi, K 185
Yu, Lie...556
Zackiewicz, P...445
Zakeri, Kh 39
Zakhvalinskii, V S.......................................421, 427
Zashichirinskii, D M 401
Zashirinskii, D M 469
Zeng, R...153
Zhang, Y 39
Zhukov, A 548
Zhukova, V 548
Zielinski, P M 223
Životský, O 282
Zvezdin, A K...261